Free Student Aid.

Log on.

Tune in.

Succeed.

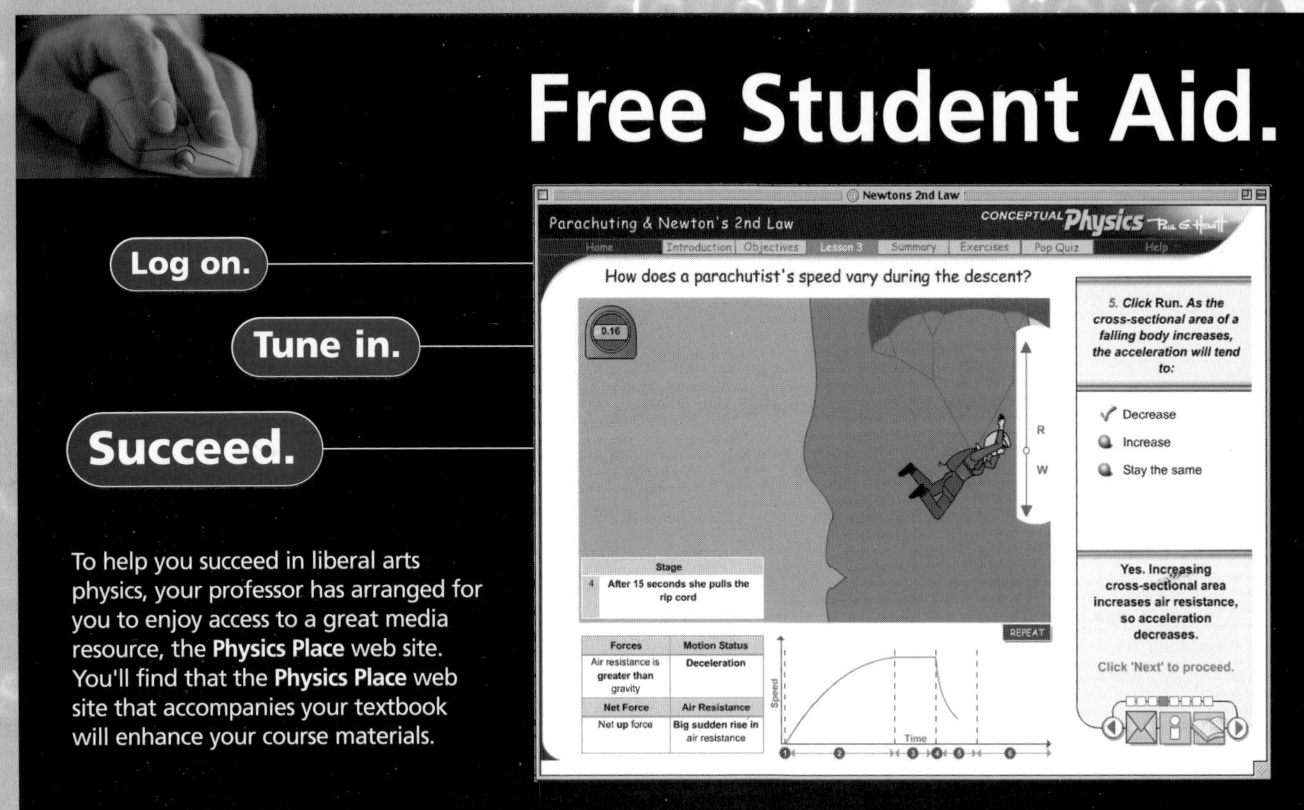

To help you succeed in liberal arts physics, your professor has arranged for you to enjoy access to a great media resource, the **Physics Place** web site. You'll find that the **Physics Place** web site that accompanies your textbook will enhance your course materials.

What your system needs to use these media resources:

WINDOWS
250 MHz
Windows 98, NT, 2000, XP
32 MB RAM installed, 64 preferred
800 X 600 screen resolution
Thousands of colors
56K modem or better
Browser: Internet Explorer 5.0 or Netscape Communicator 4.7, 7.0
Plug-Ins: Flash Player 5, QuickTime 4, Shockwave Player 8, Adobe Acrobat Reader
NOTE: Use of Netscape 6.0 and 6.1 are not recommended due to a known compatibility issue between Netscape 6.0 and 6.1 and the Flash and Shockwave plug-ins.

MACINTOSH
233 MHz PowerPC
OS 9.2 or higher
32 MB RAM minimum
800 x 600 screen resolution
Thousands of colors
56K modem or better
Browser: Internet Explorer 5.0 or Netscape Communicator 4.7, 7.0
Plug-Ins: Flash Player 5, QuickTime 4, Shockwave Player 8, Adobe Acrobat Reader
NOTE: Use of Netscape 6.0 and 6.1 are not recommended due to a known compatibility issue between Netscape 6.0 and 6.1 and the Flash and Shockwave plug-ins.

Got technical questions?

For technical support, please visit www.aw.com/techsupport and complete the appropriate online form. Technical support is available Monday-Friday, 9 a.m. to 6 p.m. Eastern Time (US and Canada)

Here's your personal ticket to success:

How to log on to www.physicsplace.com:

1. Go to www.physicsplace.com.
2. Click *Conceptual Physical Science - Explorations.*
3. Click **Register**.
4. Scratch off the silver foil coating below to reveal your pre-assigned access code.
5. Enter your pre-assigned access code exactly as it appears below.
6. Complete the online registration form to create your own personal user Login Name and Password.
7. Once your personal Login Name and Password are confirmed by email, go back to www.physicsplace.com, type in your new Login Name and Password, and click **Log In**.

Your Access Code is:

If there is no silver foil covering the access code above, the code may no longer be valid. In that case, you need to either:

- Purchase a new student access kit at your campus bookstore.
- Purchase access online using a major credit card. Go to www.physicsplace.com, click *Conceptual Physical Science - Explorations*, and click **Buy Now**.

Conceptual Physical Science—Explorations

Conceptual Physical Science—Explorations

Paul G. Hewitt

City College of San Francisco

John Suchocki

Leeward Community College

Leslie A. Hewitt

Westlake School

with special editorial contributions
from Suzanne Lyons

San Francisco Boston New York
Capetown Hong Kong London Madrid Mexico City
Montreal Munich Paris Singapore Sydney Tokyo Toronto

Editorial Director: Frank Ruggirello
Acquisitions Editor: Adam Black, Ph.D.
Assistant Editor: Liana Allday
Marketing Manager: Christy Lawrence
Media Producer: Claire Masson
Managing Editor: Joan Marsh
Manufacturing Buyer: Vivian McDougal
Cover Designer: Blakeley Kim
Composition, Prepress Services: The GTS Companies

Cover Photo: Courtesy of Michael Stewart, Lava Images

Library of Congress Cataloging-in-Publication Data

Hewitt, Paul G.
Exploring conceptual physical science / Paul G. Hewitt, John Suchocki, Leslie A. Hewitt.— 1st ed.
p. cm.
Includes bibliographical references and index.
ISBN 0-321-10663-6 (case)
1. Physical sciences. I. Suchocki, John. II. Hewitt, Leslie A. III. Title.

Q161.2 .H48 2002
500.2—dc21 2001055996

ISBN 0-321-10663-6

08 09 10 11 12 13 14 15–VHP–07 06

www.aw.com/physics

To

Millie Hewitt

wife aunt mother

Brief Table of Contents

1 About Science 2

Part 1 Mechanics 15

2 Newton's First Law of Motion—The Law of Inertia 16

3 Newton's Second Law of Motion—Force and Acceleration 34

4 Newton's Third Law of Motion—Action and Reaction 53

5 Momentum 69

6 Work and Energy 83

7 Gravity 102

8 Projectile and Satellite Motion 117

Part 2 Heat 133

9 Thermal Energy 134

10 Heat Transfer and Change of Phase 151

Part 3 Electricity and Magnetism 169

11 Electricity 170

12 Magnetism 191

Part 4 Waves—Sound and Light 209

13 Waves and Sound 210

14 Light and Color 234

15 Reflection and Refraction 253

16 Properties of Light 273

Part 5 The Atom 289

17 Atoms and the Periodic Table 290

18 Atomic Models 307

19 Radioactivity 322

20 Nuclear Fission and Fusion 338

Part 6 Chemistry

21 Elements of Chemistry

22 Mixtures

23 Chemical Bonding

24 Molecular Mixing

25 Acids and B

26 Oxida

27

28 Th

29 Plastics

Part 7 Earth Science 535

30 Minerals and Their Formation 536

31 Rocks 552

32 The Architecture of the Earth 571

33 Our Restless Planet 584

34 Water on Our World 600

35 Our Natural Landscape 620

36 A Brief History of the Earth 638

37 The Atmosphere, the Oceans, and Their Interactions 659

38 Weather 680

Part 8 Astronomy 699

39 The Solar System 700

40 The Stars 721

Appendix A Systems of Measurement 738

Appendix B Linear Motion 742

Appendix C Vectors 747

Appendix D Physics of Fluids 752

Appendix E Exponential Growth and Doubling Time 761

Appendix F Safety 766

Detailed Table of Contents

The Conceptual Physical Science—Explorations Photo Album xviii

To the Student xix

To the Teacher xx

Acknowledgments xxiv

1 About Science 2

1.1 A Brief History of Advances in Science 3

1.2 Mathematics and Conceptual Physical Science 4

1.3 The Scientific Method—A Classic Tool 4

1.4 Scientific Hypotheses 6

1.5 A Scientific Attitude Underlies Good Science 7

1.6 Science Has Limitations 7

Pseudoscience 8

1.7 The Search for Order—Science, Art, and Religion 8

1.8 Technology—Practical Use of the Findings of Science 10

1.9 The Physical Sciences: Physics, Chemistry, Geology, and Astronomy 10

Risk Assessment 11

1.10 In Perspective 12

Part 1 Mechanics 15

2 Newton's First Law of Motion—The Law of Inertia 16

Aristotle (384–322 BC) 17

2.1 Aristotle's Classification of Motion 17

2.2 Galileo's Concept of Inertia 17

Galileo Galilei (1564–1642) 19

2.3 Galileo Formulated the Concepts of Speed and Velocity 19

Speed 19

Velocity 20

2.4 Motion Is Relative 21

Isaac Newton (1642–1727) 22

2.5 Newton's First Law of Motion—The Law of Inertia 23

2.6 Net Force—The Combination of All Forces That Act on an Object 24

Paul Hewitt Personal Essay 25

2.7 Equilibrium for Objects at Rest 26

2.8 The Support Force—Why We Don't Fall Through the Floor 27

2.9 Equilibrium for Moving Objects 28

2.10 The Earth Moves Around the Sun 29

3 Newton's Second Law of Motion—Force and Acceleration 34

3.1 Galileo Developed the Concept of Acceleration 35

3.2 Force Causes Acceleration 37

3.3 Mass Is a Measure of Inertia 38

Mass Is Not Volume 38

Mass Is Not Weight 38

One Kilogram Weighs 9.8 Newtons 40

3.4 Mass Resists Acceleration 40

3.5 Newton's Second Law Links Force, Acceleration, and Mass 41

Calculation Corner 42

3.6 Friction Is a Force That Affects Motion 43

Practicing Physics 43

3.7 Objects in Free Fall Have Equal Acceleration 44

3.8 Newton's Second Law Explains Why Objects in Free Fall Have Equal Acceleration 45

Practicing Physical Science 46

3.9 Acceleration of Fall Is Less When Air Drag Acts 46

Hands-On Exploration 48

4 Newton's Third Law of Motion—Action and Reaction 53

4.1 A Force Is Part of an Interaction 54
4.2 Newton's Third Law—Action and Reaction 55
Hands-On Exploration 56
4.3 A Simple Rule Helps Identify Action and Reaction 56
Calculation Corner 57
4.4 Action and Reaction on Objects of Different Masses 58
4.5 Action and Reaction Forces Act on Different Objects 60
4.6 The Classic Horse-Cart Problem—A Mind Stumper 61
4.7 Action Equals Reaction 62
Hands-and-Feet-On Exploration 63
4.8 Summary of Newton's Three Laws 65

5 Momentum 69

5.1 Momentum Is Inertia in Motion 70
5.2 Impulse Changes Momentum 70
Case 1: Increasing Momentum—Increase Force, Time, or Both 71
Case 2: Decreasing Momentum in a Long Time Means Less Force 72
Case 3: Decreasing Momentum Over a Short Time Means More Force 73
5.3 Momentum Change Is Greater When Bouncing Occurs 74
5.4 When No External Force Acts, Momentum Doesn't Change—It Is Conserved 74
Hands-On Exploration: Skateboard Throw 76
5.5 Momentum Is Conserved in Collisions 77
Elastic Collisions 77
Inelastic Collisions 77
Calculation Corner 79

6 Work and Energy 83

6.1 Work—Force x Distance 84
6.2 Power—How Quickly Work Gets Done 85
6.3 Mechanical Energy 86
6.4 Potential Energy Is Stored Energy 87
6.5 Kinetic Energy Is Energy of Motion 88
6.6 Work-Energy Theorem 89
6.7 Conservation of Energy 91
Calculation Corner 92
6.8 Machines—Devices to Multiply Forces 92
Hands-On Exploration 94
6.9 Efficiency—A Measure of Work Done for Energy Spent 95
6.10 Sources of Energy 96
6.11 Energy Is Needed for Life 97

7 Gravity 102

7.1 The Legend of the Falling Apple 103
7.2 The Fact of the Falling Moon 103
7.3 Newton's Grandest Discovery—The Law of Universal Gravitation 104
7.4 Gravity and Distance: The Inverse-Square Law 105
7.5 The Universal Gravitational Constant, *G* 107
7.6 The Mass of the Earth Is Measured 108
7.7 Ocean Tides Are the Result of Differences in Gravitational Pulls 109
7.8 Gravitation Is Universal 112

8 Projectile and Satellite Motion 117

8.1 Projectile Motion 118
Hands-On Dangling Beads Exploration 121
8.2 Projectile Altitude and Range 122
8.3 The Effect of Air Drag on Projectiles 122
Calculation Corner 123
8.4 Fast-Moving Projectiles—Satellites 123
8.5 Earth Satellites 124
Pretend Corner: A Cannon Shoot 125
Hands-On Exploration: The Water Bucket Swing 126
8.6 Elliptical Orbits 126
Hands-On Exploration 127
8.7 Escape Speed 127

Part 2 Heat 133

9 Thermal Energy 134

9.1 Thermal Energy—The Total Energy in a Substance 135

9.2 Temperature—Average Kinetic Energy Per Molecule in a Substance 135
9.3 Absolute Zero—Nature's Lowest Possible Temperature 136
9.4 Heat Is the Movement of Thermal Energy 138
We Know What Heat Is—What Is Cold? 139
9.5 Heat Units Are Energy Units 139
9.6 The Laws of Thermodynamics 140
9.7 Specific Heat Capacity—A Measure of Thermal Inertia 141
9.8 Thermal Expansion 144
Expansion of Water 146

10 Heat Transfer and Change of Phase 151
10.1 Conduction—Heat Transfer via Particle Collision 152
10.2 Convection—Heat Transfer via Movements of Fluid 153
Breath-On Exploration 154
10.3 Radiation—Heat Transfer via Radiant Energy 155
Emission of Radiant Energy 156
Absorption of Radiant Energy 157
Reflection of Radiant Energy 158
10.4 Heat Transfer Occurs Whenever Matter Changes Phase 158
10.5 Evaporation—A Change of Phase from Liquid to Gas 158
10.6 Condensation—A Change of Phase from Gas to Liquid 160
10.7 Boiling Is Evaporation Within a Liquid 162
10.8 Melting and Freezing—Opposite Directions in Phase Change 163
10.9 Energy Is Needed for Changes of Phase 164
Link to Entomology: Life at the Extremes 165

Part 3 Electricity and Magnetism 169

11 Electricity 170
11.1 Electric Charge Is a Basic Characteristic of Matter 171
Link to Electronics Technology: Electrostatic Charge 172
11.2 Coulomb's Law—The Force Between Charged Particles 172
11.3 Charge Polarization 174
11.4 Electric Current—The Flow of Electric Charge 174
11.5 An Electric Current Is Produced by Electrical Pressure—Voltage 175
11.6 Electrical Resistance 176
11.7 Ohm's Law—The Relationship Between Current, Voltage, and Resistance 177
Link to Technology: Superconductors 178
11.8 Electric Shock 178
Link to Safety: Electric Shock 180
11.9 Direct Current and Alternating Current 181
11.10 Electric Power—The Rate of Doing Work 181
Link to History of Technology: 110 Volts 181
11.11 Electric Circuits—Series and Parallel 182
Series Circuits 182
Parallel Circuits 184
Electrical Energy and Technology 186

12 Magnetism 191
12.1 Magnetic Poles—Attraction and Repulsion 192
12.2 Magnetic Fields—Regions of Magnetic Influence 193
Hands-On Exploration 194
12.3 Magnetic Domains—Clusters of Aligned Atoms 194
12.4 The Interaction Between Electric Currents and Magnetic Fields 195
Electromagnets 196
Superconducting Electromagnets 196
12.5 Magnetic Forces Are Exerted on Moving Charges 197
Magnetic Force on Current-Carrying Wires 198
Electric Meters 198
Electric Motors 199
12.6 Electromagnetic Induction—How Voltage Is Created 200
Faraday's Law 201
12.7 Generators and Alternating Current 202
12.8 Power Production—A Technological Extension of Electromagnetic Induction 203

12.9 The Induction of Fields—Both Electric and Magnetic 204

Magnetic Therapy 205

Part 4 Waves—Sound and Light 209

13 Waves and Sound 210

13.1 Special Wiggles—Vibrations and Waves 211

13.2 Wave Motion—Transporting Energy 212

Wave Speed 213

13.3 Two Types of Waves—Transverse and Longitudinal 214

13.4 Sound Travels in Longitudinal Waves 214

Link to Technology: Loudspeakers 216

Speed of Sound 216

13.5 Sound Can Be Reflected 217

13.6 Sound Can Be Refracted 217

Link to Zoology: Dolphins and Acoustical Imaging 219

13.7 Forced Vibrations and Natural Frequency 219

13.8 Resonance and Sympathetic Vibrations 219

13.9 Interference—The Addition and Subtraction of Waves 221

Sound-Off Exploration 223

Beats—An Effect of Sound Interference 223

Standing Waves—The Effect of Waves Passing Through Each Other 223

13.10 The Doppler Effect—Changes in Frequency Due to Motion 225

13.11 Wave Barriers and Bow Waves 226

13.12 Shock Waves and the Sonic Boom 228

14 Light and Color 234

14.1 The Electromagnetic Spectrum—A Tiny Bit of Which Is Light 235

14.2 Why Materials Are Either Transparent or Opaque 236

14.3 Color Science 240

Selective Reflection 240

Selective Transmission 241

14.4 Mixing Colored Lights 242

Additive Primary Colors—Red, Green, and Blue 243

Complementary Colors 244

14.5 Mixing Colored Pigments 245

14.6 Why the Sky Is Blue 246

14.7 Why Sunsets Are Red 248

Shine-On Exploration 249

14.8 Why Clouds Are White 249

15 Reflection and Refraction 253

15.1 Reflection of Light 254

Law of Reflection 254

Diffuse Reflection 256

15.2 Refraction—The Bending of Light Due to Changing Speed 257

Hands-On Explorations: Playing with Mirrors 257

15.3 Illusions and Mirages Are Caused by Atmospheric Refraction 259

15.4 Light Dispersion and Rainbows 260

Snap-Shot Explorations: Making a Pinhole Camera 262

15.5 Lenses Are a Practical Application of Refraction 263

Hands-On Explorations: Playing Around with Lenses 264

Link to Physiology: Your Eye 265

15.6 Image Formation by a Lens 266

Lens Defects 268

Link to Optometry 269

16 Properties of Light 273

16.1 Diffraction—The Spreading of Light 274

16.2 Interference—Constructive and Destructive 276

Link to Optometry: Seeing Star-Shaped Stars 278

16.3 Interference Colors by Reflection from Thin Films 278

Hands-On Exploration: Swirling Colors 279

16.4 Polarization—Evidence for the Transverse Wave Nature of Light 280

Hands-On Exploration: Interference Colors with Polaroids 281

16.5 Wave-Particle Duality—Light Travels as a Wave, and Strikes Like a Particle 283

Part 5 The Atom 289

17 Atoms and the Periodic Table 290

17.1 Elements Contain a Single Kind of Atom 291

Link to Physiology: A Breath of Air 293

17.2 Atoms Are Mostly Empty Space 294

17.3 The Atomic Nucleus Is Made of Protons and Neutrons 294

17.4 Protons and Neutrons Determine Mass Number and Atomic Mass 296

17.5 Elements Are Organized in the Periodic Table by Their Properties 297

17.6 A Period Is a Horizontal Row, a Group Is a Vertical Column 299

18 Atomic Models 307

18.1 Models Help Us Visualize the Invisible World of Atoms 308

18.2 Atoms Can Be Identified by the Light They Emit 310

Specs-On Exploration: Spectral Patterns 313

18.3 Niels Bohr Used the Quantum Hypothesis to Explain Atomic Spectra 313

Hands-On Exploration: Quantized Whistle 316

18.4 A Shell Is a Region of Space Where an Electron Can Be Found 317

19 Radioactivity 322

19.1 Alpha, Beta, and Gamma Radiation Result from Radioactivity 323

19.2 Radioactivity Is a Natural Phenomenon 325

19.3 Radioactivity Results from an Imbalance of Forces in the Nucleus 327

19.4 A Radioactive Element Can Transmute to a Different Element 330

19.5 The Shorter the Half-Life, the Greater the Radioactivity 331

19.6 Isotopic Dating Measures the Age of a Material 333

20 Nuclear Fission and Fusion 338

20.1 Nuclear Fission Is the Splitting of the Atomic Nucleus 339

20.2 Nuclear Reactors Convert Nuclear Energy to Electrical Energy 341

The Breeder Reactor Breeds Its Own Fuel 342

20.3 Nuclear Energy Comes from Nuclear Mass, and Vice Versa 343

20.4 Nuclear Fusion Is the Combining of Atomic Nuclei 346

20.5 An Important Goal of Nuclear Research Is Controlled Fusion 347

Part 6 Chemistry 351

21 Elements of Chemistry 352

21.1 Chemistry Is a Central Science Useful to Our Lives 353

21.2 The Submicroscopic World Is Made of Atoms and Molecules 354

21.3 Matter Has Physical and Chemical Properties 356

Hands-On Exploration: Fire Water 360

21.4 An Element Is Made of a Collection of Atoms 361

21.5 Elements Can Combine to Form Compounds 362

Hands-On Exploration: Oxygen Bubble Bursts 363

21.6 Chemical Reactions Are Represented by Chemical Equations 364

You Can Balance Unbalanced Equations 365

22 Mixtures 372

22.1 Most Materials Are Mixtures 373

22.2 Mixtures Can Be Separated by Physical Means 375

Hands-On Exploration: Bottoms Up and Bubbles Out 376

22.3 Chemists Classify Matter as Pure or Impure 377

22.4 A Solution Is a Single-Phase Homogeneous Mixture 379

Hands-On Exploration: Overflowing Sweetness 384

23 Chemical Bonding 388

23.1 An Atomic Model Is Needed to Understand How Atoms Bond 389

23.2 Atoms Can Lose or Gain Valence Electrons to Become Ions 390

23.3 Ionic Bonds Result from a Transfer of Valence Electrons 393

Hands-On Exploration: Up Close with Crystals 397

23.4 Covalent Bonds Result from a Sharing of Valence Electrons 397

23.5 Polar Covalent Bonds Result from an Uneven Sharing of Electrons 401

23.6 Molecular Polarity Results from an Uneven Distribution of Electrons 403

24 Molecular Mixing 410

24.1 Submicroscopic Particles Electrically Attract One Another 411

Polar Molecules Attract Other Polar Molecules 412

24.2 Polar Molecules Can Induce Dipoles in Nonpolar Molecules 412

Hands-On Exploration: Circular Rainbows 414

24.3 Solubility Is a Measure of How Well a Solute Dissolves 414

24.4 Solubility Changes with Temperature and Pressure 417

24.5 Soap Works by Being Both Polar and Nonpolar 419

Detergents Are Synthetic Soaps 421

Hands-On Exploration: Crystal Crazy 422

25 Acids and Bases 425

25.1 Acids Donate Protons, Bases Accept Them 426

A Salt Is the Ionic Product of an Acid–Base Reaction 430

25.2 Some Acids and Bases Are Stronger Than Others 431

25.3 Solutions Can Be Acidic, Basic, or Neutral 434

The pH Scale Is Used to Describe Acidity 435

25.4 Rainwater Is Acidic and Ocean Water Is Basic 436

Hands-On Exploration: Rainbow Cabbage 437

26 Oxidation and Reduction 444

26.1 Oxidation Is the Loss of Electrons and Reduction Is the Gain of Electrons 445

Hand-On Exploration: The Silver Lining 447

26.2 The Energy of Flowing Electrons Can Be Harnessed 447

26.3 The Electricity of a Battery Comes from Oxidation–Reduction Reactions 450

26.4 Fuel Cells Are Highly Efficient Sources of Electrical Energy 454

26.5 Electrical Energy Can Produce Chemical Change 456

26.6 Oxygen Is Responsible for Corrosion and Combustion 458

Hands-On Exploration: Splitting Water 458

27 Organic Compounds 465

27.1 Organic Chemistry Is the Study of Carbon Compounds 466

27.2 Hydrocarbons Contain Only Carbon and Hydrogen 467

27.3 Unsaturated Hydrocarbons Contain Multiple Bonds 471

Hands-On Exploration: Twisting Jellybeans 473

27.4 Organic Molecules Are Classified by Functional Group 474

Alcohols Contain the Hydroxyl Group 475

The Oxygen of an Ether Group Is Bonded to Two Carbon Atoms 477

Amines Form Alkaline Solutions 478

Ketones, Aldehydes, Amides, Carboxylic Acids, and Esters All Contain a Carbonyl Group 480

28 The Chemistry of Drugs 488

28.1 There Are Several Ways to Classify Drugs 489

28.2 The Lock-and-Key Model Guides Chemists in Creating New Drugs 491

28.3 Chemotherapy Cures the Host by Killing the Disease 494

28.4 The Nervous System Is a Network of Neurons 496

Neurotransmitters Include Norepinephrine, Acetylcholine, Dopamine, Serotonin, and GABA 498

28.5 Psychoactive Drugs Alter the Mind or Behavior 499

Stimulants Activate the Stress Neurons 499

Depressants Inhibit the Ability of Neurons to Conduct Impulses 504

28.6 Pain Relievers Inhibit the Transmission or Perception of Pain 506

Detailed Table of Contents xv

29 Plastics 513

29.1 Organic Molecules Can Link to Form Polymers 514

29.2 Addition Polymers Result from the Joining Together of Monomers 514

29.3 Condensation Polymers Form with the Loss of Small Molecules 519

Hands-On Exploration: Racing Water Drops 521

29.4 The Development of Plastics Involved Experiments and Discovery 523

Collodion and Celluloid Begin with Nitrocellulose 524

Bakelite Was the First Widely Used Plastic 525

The First Plastic Wrap Was Cellophane 526

29.5 Polymers Win in World War II 527

29.6 Attitudes About Plastics Have Changed 530

Part 7 Earth Science 535

30 Minerals and Their Formation 536

30.1 Minerals Can Be Identified by Their Properties 537

Crystal Form Expresses the Arrangement of Atoms in a Mineral 537

Crystal Power 538

Hardness Is the Resistance of a Mineral to Scratching 539

Cleavage and Fracture Are Ways in Which Minerals Break 540

Hands-On Exploration: Salt Crystals 540

Luster Is the Appearance of a Mineral's Surface in Reflected Light 541

A Mineral's Color May Vary, but Its Streak Is Always the Same 541

Specific Gravity Is a Ratio of Densities 542

Asbestos: Friend and Foe 543

Chemical Properties—The Taste Test and the Acid Test 544

30.2 Minerals That Form Rock Fall into Five Main Groups 544

Silicates Make Up Nearly 90% of the Earth's Crust 545

Oxides Are Important Ore Minerals 546

Carbonate Minerals Make Limestone 546

Sulfides and Sulfates Are Also Important Ore Minerals 546

30.3 On the Way to Rocks 547

Minerals and Rock Formed from Magma 547

Minerals and Rock Formed from a Water Solution 549

Minerals—The Link to Rocks 549

31 Rocks 552

31.1 Rocks Are Divided into Three Main Groups 553

31.2 Igneous Rocks Form When Magma Cools 553

Some Igneous Rocks Form at the Earth's Surface 554

Fissure Eruptions Occur Under Water and on Land 554

Link to Mythology 555

Volcanoes Come in a Variety of Shapes and Sizes 555

Some Igneous Rocks Form Beneath the Earth's Surface 557

31.3 Sedimentary Rocks Blanket Most of the Earth's Surface 559

From Weathering to Sedimentation—Making Sedimentary Rock 559

Clastic Sediments Are Classified by Particle Size 560

Limestones and Evaporites Are Sedimentary Rocks 562

Fossils, Clues to Life in the Past 562

Fossil Fuels 563

31.4 Metamorphic Rocks Are Changed Rocks 563

Two Kinds of Metamorphism: Contact and Regional 564

Foliated Metamorphic Rocks Have a Layered Appearance 565

Nonfoliated Metamorphic Rocks Have a Smoother Appearance 566

31.5 The Rock Cycle, a Descriptive Key 566

32 The Architecture of the Earth 571

32.1 Earthquakes Make Seismic Waves 572

32.2 Seismic Waves Show the Earth Has Layers 573

The Core Has Two Parts: A Solid Inner Core, and a Liquid Outer Core *575*

The Mantle Is Dynamic *575*

The Crustal Surface Makes Up the Ocean Floor and Continental Land *576*

32.3 Folds, Faults, and Earthquakes 577

Rock Folds, an Expression of Compressive Force *577*

Faults Are Made by the Forces of Compression or Tension *578*

Earthquake Measurements—Mercalli and Richter Scales *581*

33 Our Restless Planet 584

33.1 The Theory of Continental Drift 585

A Scientific Revolution *587*

33.2 The Theory of Plate Tectonics 589

33.3 There Are Three Types of Plate Boundaries 590

Hot Spots and Lasers—A Measurement of Tectonic Plate Motion *590*

Divergent Plate Boundaries *591*

Convergent Plate Boundaries *592*

Transform-Fault Plate Boundaries *595*

33.4 The Theory That Explains Much 596

Calculation Corner *596*

34 Water on Our World 600

34.1 The Hydrologic Cycle 601

34.2 Water Below the Surface 602

The Water Table *603*

Aquifers and Springs *604*

Groundwater Movement *604*

34.3 Streams Come in Different Shapes and Sizes 605

Stream Speed Changes as a Stream Moves *605*

Drainage Networks Are Made Up of Many Streams *607*

34.4 Glaciers Are Flowing Ice 608

Glacier Formation and Movement *608*

Glacial Mass Balance *609*

34.5 Most of Earth's Water Is in the Oceans 611

Ocean Waves *612*

Wave Refraction Occurs When Waves Encounter Obstacles *613*

34.6 Can We Drink the Water? 614

Water-Supply Contamination *615*

How Can We Conserve Water? *616*

35 Our Natural Landscape 620

35.1 The Work of Air 621

35.2 The Work of Groundwater 622

Pumping Can Cause Land Subsidence *622*

Some Rocks Are Dissolved by Groundwater *623*

35.3 The Work of Surface Water 624

Erosion, Transport, and Deposition of Sediment *625*

Stream Valleys and Floodplains *626*

Deltas Are the End of the Line for a River *629*

Even Dry Places Are Affected by Surface Water *629*

35.4 The Work of Glaciers 630

Glacial Erosion and Erosional Landforms *630*

Glacial Sedimentation and Depositional Landforms *632*

35.5 The Work of Oceans 633

36 A Brief History of the Earth 638

36.1 The Geologic Clock 639

36.2 Relative Dating—The Placement of Rocks in Order 639

36.3 Radiometric Dating Reveals the Actual Time of Rock Formation 642

36.4 The Precambrian Era, the Time of Hidden Life 644

Precambrian Tectonics *645*

36.5 The Paleozoic Era, a Time of Life Diversification 646

The Cambrian Period, an Explosion of Life Forms *646*

The Ordovician Period, the Explosion of Life Continues *646*

The Silurian Period, Life Begins to Emerge on Land *646*

The Devonian Period, the Age of the Fishes *647*

The Carboniferous Period, a Time of Great Swampy Forests *647*

The Permian Period, the Beginning of the Age of Reptiles *648*

Detailed Table of Contents xvii

Paleozoic Tectonics 649
36.6 The Mesozoic Era, When Dinosaurs Ruled the Earth 650
Mesozoic Tectonics 651
36.7 The Cenozoic Era, the Time of the Mammal 651
Link to Global Thermodynamics: Is It Cold Outside? 652
Cenozoic Tectonics 653
Human Geologic Force 654

37 The Atmosphere, the Oceans, and Their Interactions 659
37.1 Earth's Atmosphere and Oceans 660
Earth's Oceans Moderate Land Temperatures 660
Evolution of the Earth's Atmosphere and Oceans 662
37.2 Components of the Earth's Atmosphere 663
Vertical Structure of the Atmosphere 664
Calculation Corner 664
37.3 Solar Energy 666
The Seasons 666
Terrestrial Radiation 667
The Greenhouse Effect and Global Warming 668
37.4 Driving Forces of Air Motion 669
37.5 Global Circulation Patterns 672
Upper Atmospheric Circulation 673
Oceanic Circulation 674

38 Weather 680
38.1 Water in the Atmosphere 681
38.2 Air Masses—Movement and Temperature Changes 682
Hands-On Exploration: Atmospheric Can-Crusher 683
Adiabatic Processes in Air 683
Atmospheric Stability 684
Hands-On Exploration: Adiabatic Expansion 685
38.3 There Are Many Different Clouds 686
High Clouds 686
Middle Clouds 686
Low Clouds 688
Clouds That Have Vertical Development 688
38.4 Air Masses, Fronts, and Storms 689
Atmospheric Lifting Creates Clouds 690
38.5 Weather Can Be Violent 692
Thunderstorms 692
Tornadoes 693
Hurricanes 694
38.6 The Weather—Number One Topic of Conversation 695
Weather Maps 696

Part 8 Astronomy 699

39 The Solar System 700
39.1 The Moon—Our Closest Celestial Neighbor 701
39.2 Phases of the Moon—Why Appearance Changes Nightly 702
39.3 Eclipses—The Shadows of the Earth and the Moon 703
Appearance of the Moon During a Lunar Eclipse 704
39.4 Why One Side of the Moon Always Faces Us 705
39.5 The Sun—Our Nearest and Most-Loved Star 706
39.6 How Did the Solar System Form? 707
39.7 Planets of the Solar System 708
39.8 The Inner Planets—Mercury, Venus, Earth, and Mars 710
39.9 The Outer Planets—Jupiter, Saturn, Uranus, and Neptune 712
39.10 Asteroids, Meteoroids, and Comets 716

40 The Stars 721
40.1 The Constellations 722
The Big Dipper and the North Star 724
40.2 Birth of Stars 725
40.3 Life and Evolution of Stars 726
40.4 Death of Stars 727
Astrology—A Famous Pseudoscience 728

40.5 The Bigger They Are, the Harder They Fall—Supernova 729
40.6 Black Holes—The Fate of the Supergiants 730
40.7 Galaxies 732
40.8 The Big Bang 735

Appendix A Systems of Measurement 738

Appendix B Linear Motion 742

Appendix C Vectors 747

Appendix D Physics of Fluids 752

Appendix E Exponential Growth and Doubling Time 761

Appendix F Safety 766

Glossary 769
Photo Credits 781
Index 783

Conceptual Physical Science–Explorations Photo Album

This is a very personal book, a family undertaking shown in many photographs throughout. Millie Hewitt, to whom this book is dedicated, is shown on page 53 with granddaughter Emily Kate Abrams, illustrating the principle that you can't touch without being touched.

The touching principle is also illustrated on pages 64 and 294 by John's wife and son, Tracy and Ian Suchocki (pronounced Su-hock-ee, with a silent c). A smaller photo of the same principle on page 67 is Paul's brother Steve with daughter Gretchen at their coffee farm in Costa Rica. Paul's other brother, Dave, is shown on page 758 with wife Barbara.

The late Charlie Spiegel, to whom our first physical science book was dedicated, is shown with great granddaughter Sarah Stafford on page 1, and again on page 239. His optimistic flavor remains in this book.

All 8 part opener photos are of family and friends. Part 1 opens on page 15 with the children of San Francisco friends Herman and Hideko Limogan, daughters Debbie and Natalie, with Genichiro Nakada in between. Part 2, page 133, is of niece Corine Jone's son, Terrence. Part 3 on page 169 shows Megan Hewitt Abrams, daughter of Leslie. Megan's cousin, Alexander Hewitt, opens Part 4 on page 209. On page 289, John's daughter Maitreya Rose opens Part 5, her brother Ian opens Part 6 on page 351, and brother Evan opens Part 7 on page 535. Lastly, Part 8 is opened on page 699 by Alexander's sister, Grace.

Author Leslie is shown on page 306 in a colored rendition of a black and white photo at the age of 16. This photo has been in all her dad's books since then. Leslie is shown more recently with her dad on page 634. Husband Bob Abrams is shown on pages 542, 608 and 631. Leslie's and Bob's children Emily and Megan colorfully open Chapter 14 on page 234. Emily also opens Chapter 31 on page 552. Leslie's brother Paul is shown demonstrating thermodynamics on pages 141 and 154, and his lovely wife Ludmila is shown with crossed Polaroids on page 283, and again with son Alexander on page 245. Even Alexander's dog, Hanz, is shown on page 159. Paul and Leslie's younger brother James, who was killed in an auto accident in 1988, is shown on page 268. He left a son, Manuel, shown on page 233. Manuel's cousin, another James Hewitt, pulls the tablecloth on page 16 (grandson of author Paul's brother Dave), and cousin Lisa Hewitt blows bubbles on page 279.

Author John dramatically walks barefoot on hot coals on page 151. John's mom, Marjorie Hewitt Suchocki (a theologian and author of several books), shows the niceties of reflection on page 255. John's brother-in-law, Peter Elias, who helped produce the chemistry videos that accompany this book, is shown on page 525. John's nephew Graham Orr is shown enjoying one of this planet's most valued resources on page 373. John's sister Joan rides horseback on page 20. Friends of the Suchocki family include Rinchen Trashi on page 311, Jill Rabinov and daughter Michaela on page 361, Maya Stevens on page 465, and Alex Feldman on page 513.

Author Paul's personal friends include, foremost, Lillian Lee, pages 153, 215, and with her pet conure on page 246. Lillian helped with all production stages of this book and its ancillaries (the ninth edition of Conceptual Physics is gratefully dedicated to her). The fellow most influential in instilling Paul's love of science is Burl Grey, page 26. Paul's physics mentor, Ken Ford, glider enthusiast, is shown with his noise-canceling earphones in his airplane, page 222. (Two previous books by Paul have been dedicated to Ken Ford.) Close friend Marshall Ellenstein, one of Chicago's finest physics teachers, is shown on page 126. Marshall is the producer of the physics videos and DVDs that accompany this book. Howie Brand, dear friend from college days, is seen on page 74. Paul's buddy Tim Gardner is seen on page 721 and again on page 759. That's teacher friend Pablo Robinson on pages 83 and 100, bravely sandwiched between two beds of nails. Ouch! And Suzanne Lyons, good friend and also the editor of this book, is shown with children Tristan and Simone on page 251. Honolulu friend Tin Hoy Hu is shown on page 195, and his cousin Andrea on page 101.

City College of San Francisco friends include Will Maynez on pages 78 and 168. Tenny Lim, former teaching assistant to Paul, is seen on page 87. Tenny is now a space engineer at Jet Propulsion Labs. Another of Paul's former teaching assistants who turned to science is Helen Yan, shown on page 158. Helen presently monitors satellite launches for Lockheed Martin. Paul's former student Cassy Cosme nicely severs bricks with her bare hand on page 73.

The inclusion of these people who are so dear to the authors makes this book all the more our labor of love.

To the Student

Rules, rules, rules—who needs them? Quite frankly, we all do. When we were children we were taught the rules of right and wrong. These rules guided us toward good experiences and away from harm. As a teenager, you know that before you can enjoy the privilege of driving, you have to demonstrate your knowledge of the rules of the road—not only to protect yourself, but others. Likewise, before you can fully appreciate any game or sporting event, you need to know its rules. Rules are important—without them we are lost.

And now it's time to learn the rules of the physical world. Like knowing the rules of any game, you'll better appreciate nature if you're familiar with its rules. The exploration of nature's rules is what this book is about.

We enjoy science, and you will too—for you'll understand it. Welcome to the exploration of *Conceptual Physical Science!*

To the Teacher

Conceptual Physical Science—Explorations melds physics, chemistry, earth science, and astronomy, in a manner that captivates student interest. We have taken care to match the reading level with the average high school student. We have done this without watering down the content, so what your students learn here is serious science. Only the most central concepts are treated. (We're all mindful of courses that dwell on difficult topics that, once understood, are hardly worth the effort. Not here.) This is solid stuff, in a very readable and student-friendly format. More than enough material is included for a one-year course, which allows for a variety of course designs to fit your taste.

We begin with physics, the most basic of the physical sciences, because it reaches up to chemistry, the central science, which in turn reaches to the Earth sciences, and finally to astronomy. This sets the foundation for serious study of the life sciences (not in this book).

To say a science course is conceptual is not to say it is non-mathematical. The mathematical foundation of physics, particularly, is quite evident in the many equations throughout the book. The equations are guides to thinking. They show the connections between concepts, rather than being used as recipes for plugging and chugging. Our emphasis is on qualitative analysis, helping students to get a gut feel for the science they're studying. That's why qualitative exercises greatly outnumber math-based problems. The challenge to the student is understanding concepts. Students appreciate and differentiate among major scientific ideas—rather than reduce them to algebraic problem solving. That can be done in another course—not this one. This course is too valuable for that!

Our presentation of chemistry also focuses on concepts and their interconnections. We emphasize models your class can visualize. Electron configurations are treated via the easy-to-visualize shell model, while chemical bonding is treated in terms of overlapping of these shells and Coulomb's law. Throughout, chemistry relates to the world familiar to the student—the fluoride in their toothpaste, the Teflon on their frying pans, and how medicines work. There is also an emphasis on the many environmental aspects of chemistry.

Earth science encompasses the sciences of geology and meteorology. Our Earth science chapters will guide your students in a study of our home planet, including its rocks, minerals, and the dynamics that makes it a continually changing place. Emphasis is on processes, with the theme that geological and atmospheric changes are ongoing—the present is the key to the past.

Applications of physics, chemistry, and geology to other massive bodies in the universe bring us to astronomy. Astronomy deals with what happens "out there," which is fascinating to just about everyone.

Chapter Review and Practice

Each chapter in this book concludes with a list of Key Terms and Matching Definitions, Review Questions and Exercises. Some chapters also include added Explorations, a reading and Internet resource list, and a few mathematical problems.

Key Terms are listed alphabetically and followed by **Matching Definitions** listed in order of appearance in the chapter. To review terms, students do more than merely read definitions—they choose the term that matches each definition. To get the most out of this format, students match as many definitions as they can before checking back in the chapter. Students gain extra practice with vocabulary terms, making it easier for them to apply the concepts.

The **Review Questions** help students fix ideas firmly in their minds and catch the essentials of the chapter material. Like the Key Terms and Matching Definitions, they are meant to familiarize—not to challenge. The answers to the Review Questions can be found in the chapter. Studying only the Key Terms and Review Questions is a good and essential beginning. But students will want to learn more and go further.

The **Explorations** occur in most chapters. They are simple activities that can be done at home, extending the in-class Hands-On Explorations found inside the chapters.

In contrast to the Review Questions, the **Exercises** are designed to challenge understanding of the chapter material. They emphasize thinking rather than mere recall. They are meant as mental push-ups to be attempted only after students are well acquainted with the chapter and have gone through the Matching Definitions and Review Questions. In many cases, Exercises apply concepts to familiar situations. Answers should be in complete sentences, with an explanation or sketch when applicable. And since Exercises often make for great exam material, they also make for great test preparation.

Problems feature concepts that are more clearly understood with numerical values and straightforward calculations. Students should include units in answers. The Problems are relatively few in number to avoid an emphasis on problem-solving that could obscure the primary goal of *Conceptual Physical Science—Explorations.* That goal is teaching concepts and how they relate to everyday living.

Supplementary Materials

The **Teacher's Guide to Text and Laboratory Manual** is very different from most teaching guides. Every section of every chapter has discussion pointers for conceptual teaching. There are suggested demonstrations, suggested Check-Your-Neighbor questions [with answers in brackets], and topics different from but related to the topics in the textbook. It has answers to all the *Review Questions*, solutions to all the *Exercises*, and step-by-step solutions to the *Problems*.

The **Test Bank** is available both in booklet form and on Test Gen software CD-ROM (PC and Macintosh compatible).

There is also a smaller test booklet, **Assessment Masters: Pre Tests and Semester Tests.** These are blackline master tests for all 40 chapters, which contain true/false, multiple choice, and short answer questions. Included also are multiple-choice tests for all 8 Parts of the textbook.

The Minds-On Hands-On Activities Book goes beyond the exploration activities in the textbook and provides teachers a variety of pedagogical approaches for shaping science skills in diverse settings. Teaching strategies include groupwork (cooperative learning), concept mapping, student-designed investigations, quick hands-on activities, skill building (construction of data tables), activity-based assessment, and oral presentations. This collection is written by physics educator, Suzanne Lyons, who also edited the textbook.

The **Laboratory Manual** is written by the authors, and by Dean Baird, new to the writing team. Dean is a prize-winning physics teacher at Rio Americano High School in Sacramento. The Laboratory Manual has both *activities* and *experiments* for most chapters.

Your students will enjoy the **Explorations Practice Book**, which guides them to a somewhat computational way of developing concepts. Written by the authors (who consider it the most creative of their ancillaries) the Practice Book's user-friendly tone makes wide use of analogies and intriguing situations.

A book of **Next-Time Questions** is available. It features intriguing full-page questions appropriate for posting after a concept has been covered in class. The answers on the reverse side of each question make a nice jumping off place for following concepts. Each is illustrated with cartoons drawn by the authors. Post selected questions in a glass case, wait a week or so, then post the answer.

Also available are 100 **Transparency Acetates** that feature select figures from the textbook. Discussions with questions and answers for each transparency are in the guide booklet, **Overhead Transparencies Teaching Guide.**

Last but not least, *Conceptual Physical Science—Explorations* has many strong multimedia components. This includes the resources posted on the *Conceptual Physical Science—Explorations* Web site, which may be found through www.PhysicsPlace.com. Videos, DVDs, and CD-ROMs of classroom demonstrations and lectures in physics and chemistry performed by the authors are also available through your Addison Wesley sales representative—or from Conceptual Productions (www.CPro.cc).

Go to it! Your conceptual physical science course can be one of the most interesting, informative, and worthwhile classes your students will ever enjoy.

Acknowledgments

For suggestions and ideas for the physics chapters we are grateful to Dean Baird, Marshall Ellenstein, Ken Ford, Bob Friedhoffer, John Hubisz, Jules Layugan, Lillian Lee, Suzanne Lyons, and Chuck Stone.

For suggestions and ideas for the chemistry chapters we are thankful to Ted Brattstrom, Hilair Chism, Peter Elias, Sharon Hopwood, Steven Jacquier, Michael Reese, and Tracy Suchocki.

For suggestions and ideas for the Earth science chapters we are very grateful to Bob Abrams. Bob's reorganization and restructuring of the chapters truly helped make them more conceptual. We are also grateful to Trayle Kulshan for her careful content editing.

For contributions to the astronomy chapters we are grateful to Richard Crowe and Lynda Williams.

For first-rate service and composition we're very grateful to Claudia McCowan and her team at GTS Companies.

Wow, Great Grandpa Charlie! Before this chickie exhausted its inner space resources and poked out of its shell, it must have thought it was at its last moments. But what seemed like its end was a new beginning. Are we like chickies, ready to poke through to a new environment and a new understanding of our place in the universe?

Chapter 1: About Science

Science has given us much. Our modern world is built on it. Nearly all forms of technology—from medicine to space travel—are applications of science. But what exactly is this amazing thing called *science*? How should science be used? Where did science come from? And what would the world be like without it?

Science is an organized body of knowledge about nature. It is the product of observations, common sense, rational thinking, and (sometimes) brilliant insights. Science has grown out of group efforts as well as individuals' discoveries. It has been built up over thousands of years and gathered from places all around the Earth. It is a huge gift to us today from the thinkers and experimenters of the past.

Yet, science is not just a body of knowledge. It is also a method, a way of exploring nature and discovering the order within it. Importantly, science is also a tool for solving problems.

Science began back before recorded history, when people first discovered repeating patterns in nature such as star patterns in the night sky, weather patterns, and patterns in animal migration. From these patterns, people learned to make predictions that gave them some control over their surroundings.

Science is a way of knowing about the world and making sense of it.

1.1 A Brief History of Advances in Science

When a sore throat keeps you home from school, you probably ask yourself, "How did I catch this cold?" You then speculate about how you could have been exposed to germs, and how you might avoid exposure next time. When a light goes out in your room, you ask, "How did that happen?" You might check to see if the lamp is plugged in, check the bulb, or even look at your neighbors' houses to see if there has been a power outage. When you think like this, you are searching for *cause and effect* relationships—trying to find out what events cause what results. This type of thinking is *rational thinking.* Rational thinking is basic to science.

Today we use rational thinking so much that it's hard to imagine other ways of interpreting our experiences. But it wasn't always this way. At times, people have relied more on superstition and magic to interpret the world around them—or have simply failed to ask, "Why?"

Rational thought became very popular in Greece in the 3rd and 4th centuries BC. From there it spread throughout Rome and other parts of the Mediterranean world. When the Roman Empire fell in the 5th century AD, advancements in science came to a halt in Europe. Barbarians destroyed much in their paths as they conquered Europe and brought in the Dark Ages. But during this time science continued to advance in other parts of the world. The Chinese and Polynesians were charting the stars and the planets. Arab nations developed mathematics and learned to make glass, paper, metals, and certain chemicals. Finally the Greek philosophy of rational thinking was brought back into Europe by Islamic people who entered Spain during the 10th, 11th, and 12th centuries. Then universities emerged. When the printing press was invented in the 15th century, science made a great leap forward. This invention did much to advance scientific thought (just as computers and the Internet are doing today).

Up into the 16th century most people thought the Earth was the center of the universe. They thought the sun circled the stationary Earth. This thinking was challenged when the Polish astronomer Nicolaus Copernicus quietly published a book proposing that the sun is stationary and that the Earth revolves around it. These ideas conflicted with Church teachings and were banned for 200 years.

Modern science began in the 16th century when the Italian physicist Galileo Galilei revived the Copernican view. Galileo used experiments, rather than speculation, to study nature's behavior (we'll say more about Galileo in chapters to follow). Galileo was arrested for popularizing the Copernican theory and for his other contributions to scientific thought. Yet a century later his ideas and those of Copernicus were accepted by most thinking people.

Scientific discoveries are often opposed, especially if they conflict with what people want to believe. Every age has its intellectual rebels who are persecuted, condemned, or suppressed at that time. But later they seem harmless and are often essential to the elevation of human conditions. "At every crossway on the road that leads to the future, each progressive spirit is opposed by a thousand men appointed to guard the past."*

1.2 Mathematics and Conceptual Physical Science

Pure mathematics is different from science. Math is a study of relationships among numbers. When used as a tool of science, the results are fantastic. Measurements and calculations are essential parts of the powerful science we practice today. For example, it would not be possible to send missions to Mars if we couldn't measure the positions of spacecraft or calculate their trajectories.

You will use some math in this course, especially when you make measurements in lab. In this book, we don't make a big deal about math. Our focus is on understanding concepts in everyday language. We use equations as guides to thinking rather than as recipes for "plug-and-chug" math work. We believe that focusing on math too early, especially math-based problem solving, is a poor substitute for learning the concepts. That's why the emphasis in this book is on building concepts. Only then, does solving problems make sense.

You'll see many more conceptual exercises than problems at the ends of the chapters that follow. *Conceptual Physical Science—Explorations* puts comprehension comfortably before computation.

1.3 The Scientific Method—A Classic Tool

In the 16th century the Italian physicist Galileo and the English philosopher Francis Bacon developed a formal method for doing science—the **scientific method.** Based on rational thinking and experimentation, this method works as follows:

1. Recognize a question or a problem.
2. Make an educated guess—a **hypothesis**—to answer the question.
3. Predict consequences that can be observed if the hypothesis is correct. The consequences should be *absent* if the hypothesis is not correct.
4. Do experiments to see if predicted consequences are present.
5. Formulate the simplest general rule that organizes the three ingredients—hypothesis, predicted effects, and experimental findings.

---

* From Count Maurice Maeterlinck's "Our Social Duty."

A scientific hypothesis is an educated guess. When a hypothesis has been tested over and over again and has not been contradicted, it may become known as a **law** or *principle.* A scientific **fact,** on the other hand, is something that competent observers can observe and agree to be true. For example, it is a fact that an amputated limb of a salamander can grow back. Anyone can watch it happen. It is not a fact—yet—that a severed limb of a human can grow back.

Scientists use the word *theory* in a way that differs from everyday speech. In everyday speech a theory is the same as a hypothesis—a statement that hasn't been tested. But scientifically speaking, a **theory** is a synthesis of facts and well-tested hypotheses. Physicists, as we will learn, speak of the quark theory of the atomic nucleus. Chemists have the theory of metallic bonding. Geologists use the theory of plate tectonics, and astronomers speak of the theory of the Big Bang.

Theories are a foundation of science. They are not fixed, but evolve. They pass through stages of refinement. For example, since the theory of the atom was proposed 200 years ago it has been refined many times in light of new evidence. Some people argue that scientific theories can't be taken seriously because they change. Those who understand science, however, see that theories grow stronger as they evolve to include new information.

(a) (b)

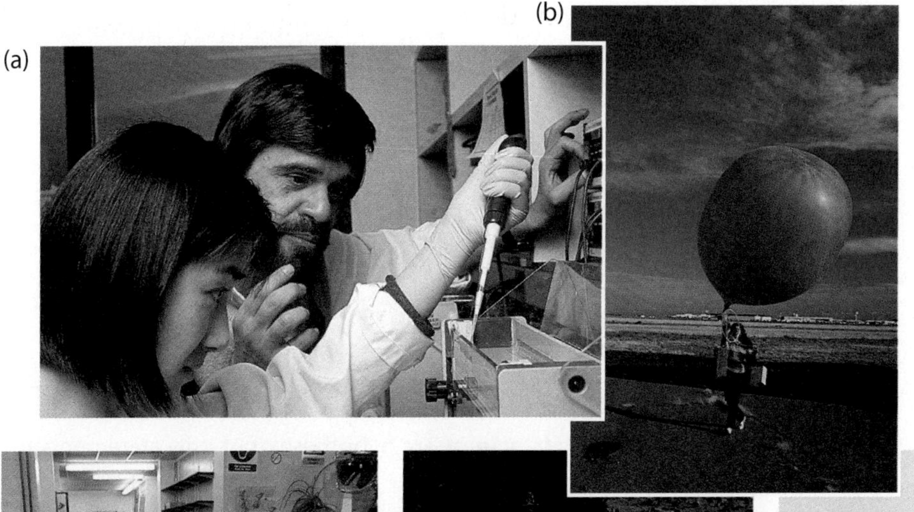

(c) (d) (e)

Figure 1.1
Physical science covers many disciplines. (a) Biochemists preparing samples. (b) Meteorologist releasing a weather balloon to study the composition of the upper atmosphere. (c) Technicians conducting DNA research. (d) Paleontologist preparing fossilized dinosaur bones for carbon dating. (e) Physicist simulating lightning strokes.

1.4 Scientific Hypotheses

In order for a hypothesis to be scientific, it must be testable. A test or series of tests can determine whether a hypothesis is valid. In scientific work, most hypotheses turn out to be wrong. Scientists have to be patient though, and keep testing their ideas for accuracy.

A well-known scientific hypothesis that was incorrect was that of the greatly respected Greek philosopher Aristotle (384–322 BC), who claimed that heavy objects naturally fall faster than light objects. This hypothesis was considered true for nearly 2000 years—mainly because of respect for this great man. Also, air resistance was not recognized as an influence on how quickly objects fall. We've all seen that stones fall faster than leaves fluttering in the air. Without investigating further it is easy to accept false ideas.

Galileo very carefully examined Aristotle's hypothesis. Then he did something that caught on and changed science forever. He *experimented.* Galileo showed the falseness of Aristotle's claim with a single experiment—dropping heavy and light objects from the Leaning Tower of Pisa. Legend tells us that they fell at equal speeds. In the scientific spirit, one experiment that can be reproduced outweighs any authority, regardless of reputation or the number of advocates. Albert Einstein put it well when he stated, "No number of experiments can prove me right; a single experiment can prove me wrong." In science, the test of knowledge is experiment.

Which statements are *scientific* hypotheses?

a. Better stock market decisions are made when the planets Venus, Earth, and Mars are aligned.
b. Atoms are the smallest particles of matter that exist.
c. Albert Einstein was the greatest physicist of the 20th century.

Check Your Answer All statements are hypotheses, but only statements *a* and *b* are scientific hypotheses—because they are testable. Statement *a* can be tested (and proven wrong) by going to the library and researching the performance of the stock market during times when these planets were aligned. Statement *b* not only can be tested, but has been tested. Although it is untrue (many particles smaller than atoms have been discovered), the statement is nevertheless a scientific one. Lastly, statement *c* is an assertion that has no test. What possible test, beyond collective opinion, could prove Einstein was the greatest physicist? How could we know? Greatness is a quality that cannot be measured in an objective way.

Because the name Einstein is held in high esteem, it is a favorite of quacks (see the box on page 8). Take notice when the name of Einstein, Jesus, and other highly respected sources are cited, often by quacks who wish to bring respect to themselves and their points of view. In all fields we should be skeptical of people who wish to credit themselves by calling upon the authority of others.

While the scientific method is powerful, good science is often done differently, in a less systematic way. Many scientific advances involve trial and error, experimenting without guessing, or just plain accidental discovery. Trained observation, however, is essential for noticing questions in the first place and for making sense of evidence. But more than a particular method, the success of science has to do with an attitude common to scientists. This attitude is one of inquiry, experimentation, and humility before the facts.

1.5 A Scientific Attitude Underlies Good Science

Scientists must accept their experimental findings even when they wish they were different. They must strive to distinguish between the results they see and those they wish to see. This is not easy. Scientists, like most people, are very capable of fooling themselves. People have always tended to adopt general rules, beliefs, creeds, and ideas without thoroughly questioning their validity. And sometimes we retain these ideas long after they have been shown to be meaningless, false, or at least questionable. The most widespread assumptions are often the least questioned. Too often, when an idea is adopted, great attention is given to the instances that support it. Contrary evidence is often distorted, belittled, or ignored.

None of us has the time or resources to test every idea. So most of the time we take somebody's word. How do we know whose word to accept? To reduce the likelihood of error, scientists listen to people whose findings are testable—if not in practice, then at least in principle. Ideas that cannot be tested are regarded as "unscientific."

The fact that scientific statements will be thoroughly tested helps keep science honest. Sooner or later, mistakes (or deception) are found out. A scientist exposed for cheating doesn't get a second chance in the community of scientists. Honesty, so important to the progress of science, thus becomes a matter of self-interest. There is relatively little bluffing in a game where all bets are called.

1.6 Science Has Limitations

Science deals only with hypotheses that are testable. Its domain is therefore restricted to the observable natural world. While scientific methods can be used to debunk various paranormal claims, they have no way of accounting for testimonies involving the supernatural. The term supernatural literally means "above nature." Science works within nature, not above it. Likewise, science is unable to answer philosophical questions such as "What is the purpose of life?" Though these questions are valid and important ones, they lie outside the realm of science.

Pseudoscience

Some belief systems are not scientific but pretend to be. For example, people in the early United States believed in the "science" of phrenology. Phrenology was the study of the surface bumps on a person's head. Phrenologists claimed to predict all sorts of things about a person's health and personality based on the bumps on their head. But phrenology was fake science—**pseudoscience.** No experimental findings backed up its claims. Can you tell real science apart from pseudoscience?

Consider astrology. Astrology tells us that human affairs are influenced by the positions and movements of planets and other celestial bodies. Yet there is no solid body of experimental evidence to back up this claim. This non-scientific view can be quite appealing. No matter how insignificant we may feel at times, astrologers assure us that we are deeply connected to the workings of the cosmos, which has been created just for us humans. Astrology as ancient magic or entertainment is one thing, but astrology disguised as science is another. When it poses as a science related to astronomy, astrology is full-fledged pseudoscience.

Pseudoscience, like science, makes predictions. The predictions of a dowser, who locates underground water supplies with a dowsing rod, have a very high rate of success—nearly 100%. Whenever the dowser goes through his or her ritual and points to a spot on the ground, a well digger is sure to find water. Dowsing works. Of course, the dowser can hardly miss, because there is ground water within 100 meters of the surface at nearly every spot on Earth. (The real test of a dowser would be finding a place where water wouldn't be found!). Dowsing is another example of pseudoscience.

We humans have learned much since the onset of science four centuries ago. Only by enormous effort did people along the way gain this knowledge and overthrow superstition. We have come far in comprehending nature and freeing ourselves from ignorance. We should rejoice in what we've learned. We no longer have to die whenever an infectious disease strikes. We no longer live in fear of demons. We no longer torture women accused of witchery, as was done for nearly three centuries during medieval times. Today we have no need to pretend that superstition is anything but superstition, or that junk notions are anything but junk notions—whether voiced by street-corner quacks, or by loose thinkers who write promise-heavy health books.

Yet there is reason to fear that what people of one time fight for, a following generation surrenders. The grip that belief in magic and superstition had on people took centuries to overcome. Yet today the same magic and superstition are enchanting a growing number of people. James Randi reports in his book *Flim-Flam!* that more than twenty thousand practicing astrologers in the United States service millions of credulous believers. Science writer Martin Gardner reports that a greater percentage of Americans today believe in astrology and occult phenomena than did citizens of medieval Europe. Few newspapers carry a daily science column, but nearly all provide daily horoscopes. And then there are the flourishing television psychics who gain adherents daily.

Some people believe that the human condition is slipping backward because of growing technology. More likely, however, it is slipping backward because science and technology will bow to the irrationality of the past. Watch for the spokespeople of pseudoscience. It is a huge and lucrative business.

In your education it's not enough to be aware that other people may try to fool you. More important is being aware of your own tendency to fool yourself.

1.7 The Search for Order— Science, Art, and Religion

The search for order and meaning in the world has taken different forms: One is science, another is art, and another is religion. The domains of science, art, and religion are different, although they often overlap. Science is mostly engaged with discovering and recording natural phenomena. The arts are concerned with personal interpretation and creative expression. And religion addresses the source, purpose, and meaning of it all.

Science and the arts have certain things in common. In the art of literature, we find out about what is possible in human experience. We can learn about emotions from rage to love, even if we haven't yet experienced them. The arts do not necessarily give us those experiences, but they describe them and suggest what may be possible for us. A knowledge of science similarly tells us what is possible in nature. Scientific knowledge helps us predict possibilities in nature even before these possibilities have been experienced. It provides us with a way of connecting things, of seeing relationships between and among them, and of making sense of the great variety of natural events around us. Science broadens our perspective of nature. A knowledge of both the arts and the sciences makes up a wholeness that affects the way we view the world. It aids the decisions we make about it and ourselves. A truly educated person is knowledgeable in both the arts and the sciences.

Science and religion have similarities also, but they are basically different. Science is concerned with physical things, while religion is concerned with spiritual matters. Simply put, science asks *how;* religion asks *why.* The practices of science and religion are also different. Whereas scientists experiment to find nature's secrets, many religious practitioners worship God and work to build human community. In these respects, science and religion are as different as apples and oranges and do not contradict each other. Science and religion are two different yet complementary fields of human activity.

When we study the nature of light later in this book, we will treat light first as a wave and then as a particle. To the person who knows a little bit about science, waves and particles are contradictory. Light can be only one or the other, and we have to choose between them. But to the enlightened person, waves and particles complement each other and provide a deeper understanding of light. In a similar way, it is mainly people who are either uninformed or misinformed about the deeper natures of both science and religion who feel that they must choose between believing in religion and believing in science. Unless one has a shallow understanding of either or both, there is no contradiction in being religious and being scientific in one's thinking.*

Many people are troubled about not knowing the answers to religious and philosophical questions. Some avoid uncertainty by eagerly accepting any comforting answer. An important message from science, however, is that uncertainty is acceptable. For example, in Chapter 16 you'll learn that it is not possible to know with certainty both the momentum and position of an electron in an atom. The more you know about one, the less you can know about the other. Uncertainty is a part of the scientific process. It's okay not to know the answers to fundamental questions. Why are apples gravitationally attracted to the Earth? Why do electrons repel one another? Why

* Of course this doesn't apply to certain extremists, Christian, Moslem, or otherwise, who steadfastly assert that one cannot embrace both their brand of religion and science.

does energy have mass? At the deepest level, scientists don't know the answers to these questions—at least not yet. Scientists in general are comfortable about not knowing. We know a lot about where we are, but nothing really about *why* we are. Perhaps we can apply a lesson from science to our religious questions. Maybe it's okay not to know the answers to religious questions—especially if we keep exploring with an open mind and heart.

1.8 Technology—Practical Use of the Findings of Science

Science and technology are also different from each other. Science is concerned with gathering knowledge and organizing it. Technology lets humans use that knowledge for practical purposes, and it provides the instruments scientists need to conduct their investigations.

Technology is a double-edged sword. It can be both helpful and harmful. We have the technology, for example, to extract fossil fuels from the ground and then burn the fossil fuels to produce energy. Energy production from fossil fuels has benefited society in countless ways. On the flip side, the burning of fossil fuels damages the environment. It is tempting to blame technology itself for problems such as pollution, resource depletion, and even overpopulation. These problems, however, are not the fault of technology any more than a stabbing is the fault of the knife. It is humans who use the technology, and humans who are responsible for how it is used.

Remarkably, we already possess the technology to solve many environmental problems. This 21st century will likely see a switch from fossil fuels to more sustainable energy sources. We recycle waste products in new and better ways. In some parts of the world, progress is being made toward limiting the human population explosion, a serious threat that worsens almost every problem faced by humans today. Difficulty solving today's problems results more from social inertia than failing technology. Technology is our tool. What we do with this tool is up to us. The promise of technology is a cleaner and healthier world. Wise applications of it *can* improve conditions on planet Earth.

If it crawls, it's biology.
If it smells, it's chemistry.
If it doesn't work, it's physics!

1.9 The Physical Sciences: Physics, Chemistry, Geology, and Astronomy

Science is the present-day equivalent of what used to be called *natural philosophy.* Natural philosophy was the study of unanswered questions about nature. As the answers were found, they became part of what is now called *science.* The study of science today branches into the study of living things and nonliving things: the life sciences

Risk Assessment

Technology comes with risks as well as benefits. When the benefits are seen to outweigh risks, a technology can be accepted and applied. X-rays, for example, continue to be used to diagnose disease despite their potential risk for causing cancer. The benefits outweigh the risks. Of course, when the risks of a technology outweigh its benefits, it should be used sparingly or not at all.

A hard ethical problem arises when a technology benefits one group of people but poses risk to a different group of people. For example, aspirin is useful for adults, but it can cause a potentially fatal condition in children, called *Reye's Syndrome.* Are the benefits of aspirin to adults worth the risk that children will die by getting Reye's Syndrome? Dumping raw sewage into the local river may pose little risk for a town located upstream, but for towns downstream the untreated sewage is a health hazard. Technologies involving different risks for different people raise questions that are often hotly debated. Which medications should be sold to the general public over-the-counter and how should they be labeled? Should food be irradiated to eliminate the food poisoning that kills more than 5000 Americans each year? Or are the unknown potential hazards of food irradiation sufficient cause for banning food irradiation?

People seem to have a hard time accepting the fact that zero risk is impossible. Airplanes cannot be made perfectly safe. Processed foods cannot be completely free of toxicity, for all foods are toxic to some degree. You cannot go to the beach without risking skin cancer no matter how much sunscreen you apply. You cannot avoid radioactivity, for it's in the air you breathe and the foods you eat, and has been that way before humans first walked the Earth. Even the cleanest rain contains radioactive carbon-14; our bodies do as well. Between each heartbeat in the human body, there have always been about 10,000 naturally-occurring radioactive decays. You might hide yourself in the hills, eat the most natural foods, practice obsessive hygiene and still die from cancer caused by radioactivity. The probability of eventual death is 100%. We have to accept that. Nobody is exempt.

Science helps to determine the most probable results. As the tools of science improve, risks can be evaluated more and more accurately. Acceptance of risk, on the other hand, is more of a social issue than a scientific one. If society were to demand zero risk from its technology, this goal would not only be impractical, but selfish. Any society striving toward a policy of zero risk would consume its present and future economic resources. A society that accepts no risks receives no benefits.

and the physical sciences. The life sciences branch into such areas as biology, zoology, and botany. The *physical sciences* branch into such areas as physics, chemistry, geology, meteorology, and astronomy—the areas addressed in this book.

Physics is the study of basic concepts such as motion, force, energy, matter, heat, sound, light, and the components of atoms. Chemistry builds on physics and tells us how matter is put together, how atoms combine to form molecules, and how the molecules combine to make the materials around us. Physics and chemistry applied to the Earth and its processes makes up Earth science—geology. When we apply physics, chemistry, and geology to other planets and to the stars, we are speaking about astronomy.

Biology is more complex than physical science, for it involves matter that is alive. Underneath biology is chemistry, and underneath chemistry is physics. So physics is basic to both physical science and life science. That is why we begin with physics, then follow with chemistry and geology, and conclude with astronomy. All are treated conceptually, with the twin goals of enjoyment and understanding.

Concept Check ✓

Which of the following activities involves the utmost human expression of passion, talent, and intelligence?

a. painting and sculpture
b. literature
c. music
d. religion
e. science

Check Your Answer All of them! In this book we focus on science, which is an enchanting human activity shared by a wide variety of people. With present-day tools and know-how, science types are reaching further and finding out more about themselves and their environment than people in the past were ever able to do. The more you know about science, the more passionate you feel toward your surroundings. There is physical science in everything you see, hear, smell, taste, and touch!

1.10 In Perspective

Only a few centuries ago, the most talented and skilled artists, architects, and artisans of the world directed their genius to the construction of the great cathedrals, synagogues, temples, and mosques. Some of these architectural structures took centuries to build. This meant that nobody witnessed both the beginning and the end of construction. The architects and early builders who lived to a ripe old age never saw the finished results of their labors. Entire lifetimes were spent in the shadows of construction that must have seemed without beginning or end. This enormous focus of human energy was inspired by a vision that went beyond worldly concerns—a vision of the cosmos. To the people of that time, the structures they built were their "spaceships of faith," firmly anchored but pointing to the cosmos.

Today the efforts of many skilled scientists, engineers, artists, and artisans are directed to building the spaceships that already orbit the Earth and others that will voyage beyond. The time required to build these spaceships is extremely brief compared to the time spent building the stone and marble structures of the past. Many people working on today's spaceships were alive before the first jetliner aircraft carried passengers. Where will younger lives lead in a comparable time?

We seem to be at the dawn of a major change in human growth. For as little Sarah suggests in the photo at the beginning of this book, we may be like the hatching chicken who has exhausted the resources of its inner-egg environment and is about to break through to a whole new range of possibilities. The Earth is our cradle and has served us well. But cradles, however comfortable, are one day outgrown. So with the inspiration that in many ways is similar to the inspiration of those who built the early cathedrals, synagogues, temples, and mosques, we aim for the cosmos.

We live in an exciting time!

Chapter Review

Key Terms and Matching Definitions

_____ fact
_____ law
_____ hypothesis
_____ pseudoscience
_____ science
_____ scientific method
_____ technology
_____ theory

1. Organized common sense. Also the collective findings of humans about nature, and a process of gathering and organizing knowledge about nature.
2. An orderly method for gaining, organizing, and applying new knowledge.
3. An educated guess; a reasonable explanation that is not fully accepted as factual until tested over and over again by experiment.
4. A phenomenon about which competent observers can agree.
5. A general hypothesis or statement about the relationship of natural quantities that has been tested over and over again and has not been contradicted. Also known as a *principle.*
6. A synthesis of a large body of information that encompasses well-tested hypotheses about certain aspects of the natural world.
7. Fake science that has no tests for its validity.
8. Method and means of solving practical problems by applying the findings of science.

Review Questions

1. What is science? (Name the two major aspects of science that are discussed in the first paragraphs of this chapter.)

A Brief History of Advances in Science

2. What discovery in the 15th century greatly advanced progress in science?
3. Throughout the ages, has acceptance or resistance usually been the general reaction to new ideas about established "truths"?

Mathematics and Conceptual Physical Science

4. When was the mathematical structure of science discovered?
5. Why is mathematical problem solving not a major feature of this book?

The Scientific Method—A Classic Tool

6. Outline the steps of the scientific method.
7. Distinguish among a scientific fact, a hypothesis, a law, and a theory.

Scientific Hypotheses

8. What is the hallmark of a scientific hypothesis?
9. How many experiments are necessary to invalidate a scientific hypothesis?

A Scientific Attitude Underlies Good Science

10. In science, what kind of ideas are generally accepted?
11. Why is honesty a matter of self-interest to a scientist?

Science Has Limitations

12. What is meant by the term supernatural, and why does science not deal with it?

The Search for Order—Science, Art, and Religion

13. How are science and the arts similar?
14. Why are students of the arts encouraged to learn about science and science students encouraged to learn about the arts?
15. Why do many people believe they must choose between science and religion?
16. How do scientists regard "not knowing" in general?

Technology—Practical Use of the Findings of Science

17. Clearly distinguish between science and technology.

The Physical Sciences: Physics, Chemistry, Geology, and Astronomy

18. Cite at least two examples of the physical sciences, and two from the life sciences.

19. Of physics, chemistry, and biology, which science is the least complex? The most complex? (At your school, which of these is the "easiest" and which is the "hardest" as a science course?)

In Perspective

20. How does the material in this section relate to the opening photo on page 1 of this book?

Exercises

1. In daily life, people are often praised for maintaining some particular point of view, for the "courage of their convictions." A change of mind is seen as a sign of weakness. How is this different in science?

2. In daily life, we see many cases of people who are caught misrepresenting things and who soon thereafter are excused and accepted by their contemporaries. How is this different in science?

3. Which of the following are scientific hypotheses?
  a. Chlorophyll makes grass green.
  b. The Earth rotates about its axis because living things need an alternation of light and darkness.
  c. Tides are caused by the moon.

4. In answer to the question, "When a plant grows, where does the material come from?" Aristotle hypothesized by logic that all material came from the soil. Do you consider his hypothesis to be correct, incorrect, or partially correct? What experiments do you propose to support your choice?

5. What is probably being misunderstood by a person who says, "But that's only a scientific theory"?

6. a. Make an argument for bringing to a halt the advances of technology.
  b. Make an argument that advances in technology should continue.
  c. Contrast your two arguments.

Suggested Reading and Web Sites

Feynman, Richard P. *Surely You're Joking, Mr. Feynman.* New York: Norton, 1986.

Park, Robert. *Voodoo Science—The Road from Foolishness to Fraud.* Oxford, 2000.

Sagan, Carl. *The Demon-Haunted World.* New York: Random House, 1995.

Shermer, Michael. *The Borderlands of Science—Where Sense Meets Nonsense.* Oxford, 2001.

www.howstuffworks.com
An intriguing Web site about technological devices.

www.phschool.com
This is the Web site for the Prentice Hall version of Hewitt's *Conceptual Physics* text for high schools. Much should be useful for this course.

www.aw.com/physics
This is the Web site for the 9th edition of Hewitt's textbook for college, *Conceptual Physics.* Much on it is useful at the high-school level.

Part 1 Mechanics

You cannot touch without being touched — that's Newton's third law!

Chapter 2: Newton's First Law of Motion—The Law of Inertia

If you want to move something, you apply a force to it—you either push it or pull it. To move dishes on a tablecloth, you can move them by pulling on the tablecloth. Pull slowly, and they'll move slowly. But what if you pull the cloth very quickly—quick enough so the force of friction between the cloth and dishes is very small and very brief? Will the dishes move? It takes a force to get things moving, but once moving, is a force needed to keep them moving? How about satellites in orbit? They are in a friction-free environment. What keeps them moving? What rules of nature guide the answers to these questions?

Aristotle (384–322 BC)

Aristotle was the most famous philosopher, scientist, and educator, in ancient Greece. He was the son of a physician who personally served the king of Macedonia. At 17 he entered the Academy of Plato, where he worked and studied for 20 years until Plato's death. He then became the tutor of young Alexander the Great. Eight years later he formed his own school. Aristotle's aim was to arrange existing knowledge in a system, just as Euclid had earlier done with geometry. Aristotle made careful observations, collected specimens, and gathered together and classified almost all existing knowledge of the physical world. His systematic approach became the method from which Western science later arose. After his death, his voluminous notebooks were preserved in caves near his home and were later sold to the library at Alexandria. Scholarly activity came to a stop in most of Europe through the Dark Ages, and the works of Aristotle were forgotten and lost. Various texts were reintroduced to Europe during the eleventh and twelfth centuries and translated into Latin. The Church, the dominant political and cultural force in Western Europe, at first prohibited the works of Aristotle. But soon thereafter the Church accepted them and incorporated them into Christian doctrine.

2.1 Aristotle's Classification of Motion

The idea that motion requires a force (a push or a pull) goes back to the 4th century BC when the Greeks were developing scientific ideas. Aristotle, the most famous Greek scientist, classified motion into two kinds: *natural motion* and *unnatural motion.* Aristotle taught that **natural motion** on the Earth was directed either up or down. For example, he said it was natural for a boulder to fall down toward the Earth and a puff of smoke to rise in air. He believed that objects have resting places that they naturally seek. He went on to say that it was natural for heavy objects to fall faster than light objects.

Natural motion, Aristotle believed, occurred without force. For example, motions of the sun, moon, and other objects in the sky were considered natural. These celestial objects moved continuously without the need for any force to push or pull them.

Unnatural motion, on the other hand, required forces such as those imposed by people or animals. For example, the only way to get a cart moving across the ground was to push or pull on it. Unnatural motion required forces.

2.2 Galileo's Concept of Inertia

Aristotle's ideas were taken as fact for nearly 2000 years. But in the early 1500s the Italian scientist Galileo demolished Aristotle's belief that heavy things fall faster than light things. As mentioned in Chapter 1, legend tells us that Galileo dropped a heavy object and a light

Figure 2.1
Galileo's famous demonstration.

object from the Leaning Tower of Pisa. He showed that except for the effects of air friction, objects of different weights fell to the ground at the same time.

Galileo made another huge discovery. He showed that Aristotle was wrong about forces being necessary to keep objects moving. Galileo said that a force is required to start an object moving, but once moving, no force is required to keep it moving—except for the force needed to overcome friction. (We'll study friction, a force that opposes motion, in the next chapter). When friction is absent, a moving object does not need a force to keep it moving.

Galileo found a simple and powerful way to test his revolutionary idea. He rolled balls along flat plane surfaces tilted at different angles. He noted that a ball rolling down an inclined plane picks up speed. This is shown in Figure 2.3. Gravity increases the ball's speed. He also noticed the ball slows down when rolling up an inclined plane. Then gravity decreases the ball's speed. What about a ball rolling on a level surface? While rolling level, the ball does not roll with nor against gravity. Galileo saw that a ball rolling on a smooth horizontal plane doesn't speed up and doesn't slow down. It has a constant speed. Galileo reasoned that a ball moving horizontally would move forever if friction were entirely absent. A ball would move of itself.

All objects show the same property of motion as the balls rolling on Galileo's planes. The tendency of things is to remain as they are. If moving, they tend to remain moving. If at rest, they tend to remain at rest. This property of objects is called **inertia.**

Figure 2.2
Motion of balls on various planes.

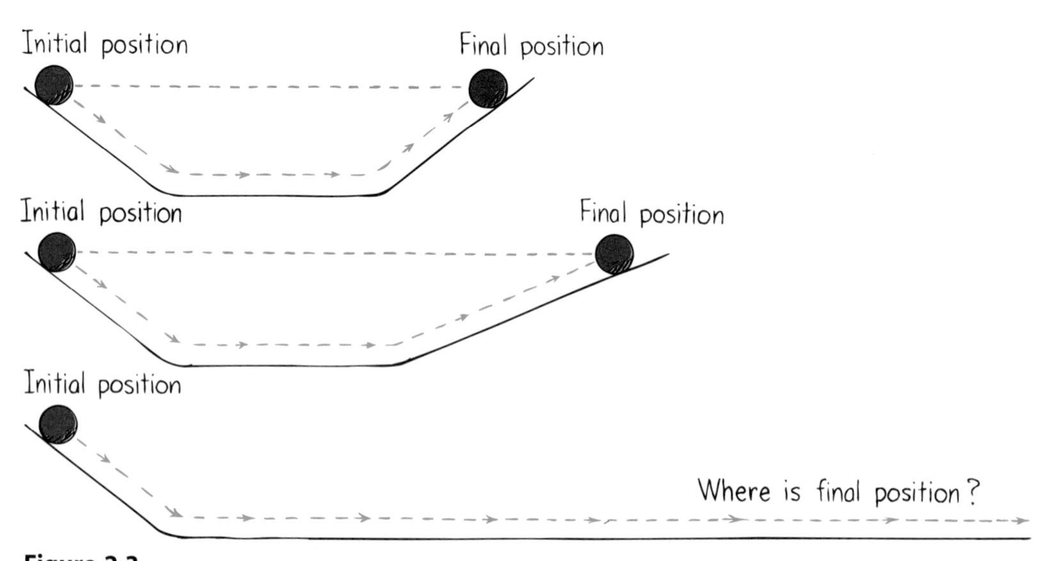

Figure 2.3
A ball rolling down an incline on the left tends to roll up to its initial height on the right. The ball must roll a greater distance as the angle of incline on the right is reduced.

Galileo Galilei (1564–1642)

Galileo was born in Pisa, Italy, in the same year Shakespeare was born and Michelangelo died. He studied medicine at the University of Pisa and then changed to mathematics. He developed an early interest in motion and was soon at odds with others around him, who held to Aristotelian ideas on falling bodies. He left Pisa to teach at the University of Padua and became an advocate of the new theory of the solar system advanced by the Polish astronomer Copernicus. Galileo was one of the first to build a telescope, and he was the first to direct it to the nighttime sky. He discovered mountains on the moon and the moons of Jupiter. Because Galileo published his findings in Italian instead of the Latin typically used by scholars, and because of the recent invention of the printing press, Galileo's ideas reached many people. He soon ran into disagreements with the Church and was warned not to teach and not to hold to Copernican views. He restrained himself publicly for nearly 15 years. Then Galileo defiantly published his observations and conclusions, which didn't agree with Church doctrine. The outcome was a trial in which he was found guilty, and he was forced to renounce his discoveries. By then an old man broken in health and spirit, he was sentenced to perpetual house arrest. Nevertheless, Galileo completed his studies on motion and his writings were smuggled from Italy and published in Holland. Earlier he damaged his eyes looking at the sun through a telescope, which led to blindness at the age of 74. He died 4 years later.

Concept Check ✓

A ball rolling on a pool table slowly comes to a stop. How would Aristotle explain this behavior? How would Galileo explain it? How would you explain it?

Check Your Answers *Did you think about the questions and arrive at your own answers before reading this? Please do so, and you'll find yourself learning more. Much more!*

Aristotle would probably say that the ball stops because it seeks its natural state of rest. Galileo would probably say that friction overcomes the ball's natural tendency to continue rolling. Friction overcomes the ball's *inertia*—and brings it to a stop. Only you can answer the last question!

2.3 Galileo Formulated the Concepts of Speed and Velocity

Speed

Before the time of Galileo, people described moving things as simply "slow" or "fast". Such descriptions were vague. Galileo was the first to measure speed by considering the distance covered and the time it takes. He defined **speed** as the distance covered per unit of time. The word *per* means "divided by."

$$\text{Speed} = \frac{\text{distance}}{\text{time}}$$

Figure 2.4
The greater the distance traveled each second, the faster the horse gallops.

For example if a horse covers 20 kilometers in a time of 1 hour, its speed is 20 km/h. Of if you run 6 meters in 1 second, your speed is 6 m/s. These values are *average speeds,* for the speed at any instant, *instantaneous speed,* may be different.

Any combination of distance and time units can be used for speed—kilometers per hour (km/h), centimeters per day (the speed of a sick snail), or whatever is useful and convenient. The slash symbol (/) is read as "per." In science the preferred unit of speed is meters per second (m/s). Table 2.1 shows some comparative speeds in different units.

Table 2.1

Approximate Speeds in Different Units

mi/h		km/h		m/s
12 mi/h	=	20 km/h	=	6 m/s (bowling ball)
25 mi/h	=	40 km/h	=	11 m/s (very good sprinter)
37 mi/h	=	60 km/h	=	17 m/s (sprinting rabbit)
50 mi/h	=	80 km/h	=	22 m/s (tsunami)
62 mi/h	=	100 km/h	=	28 m/s (sprinting cheetah)
75 mi/h	=	120 km/h	=	33 m/s (batted softball)
100 mi/h	=	160 km/h	=	44 m/s (batted baseball)

Figure 2.5
A speedometer gives readings in both miles per hour and kilometers per hour.

Velocity

When we know both the speed and direction of an object, we know its **velocity.** For example, if a car travels at 60 km/h, we know its speed. But if we say it moves at 60 km/h to the north, we specify its *velocity.* Speed is a description of how fast; velocity is how fast *and* in what direction. A quantity such as velocity that specifies direction as well as magnitude is called a **vector quantity.** Velocity is a vector quantity, and is developed nicely in the *Conceptual Physical Science Explorations Practice Book.* Vectors are further discussed in Appendix C.

Constant speed means steady speed. Something with constant speed doesn't speed up or slow down. Constant velocity, on the other hand, means both constant speed *and* constant direction. Constant direction is a straight line—the object's path doesn't curve. So constant velocity means motion in a straight line at constant speed.

In the next chapter we will consider motion that is not constant—*acceleration.*

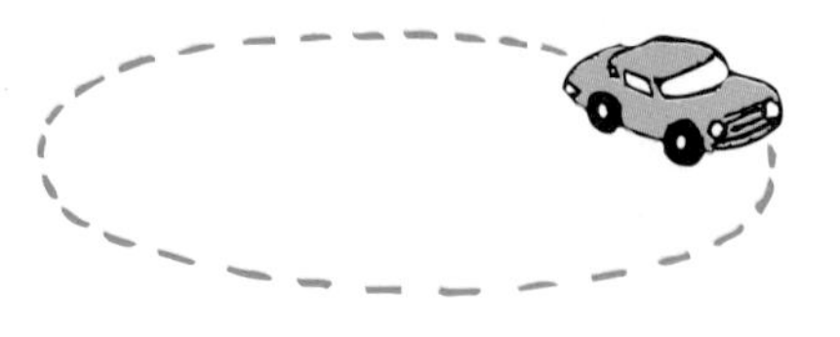

Figure 2.6
The car on the circular track may have a constant speed, but its velocity is changing every instant. Why?

Concept Check ✓

1. What is the average speed of a cheetah that sprints 100 m in 4 s? How about if it sprints 50 m in 2 s?
2. The speedometer on a bicycle moving east reads 50 km/h. It passes another bicycle moving west at 50 km/h. Do both bikes have the same speed? Do they have the same velocity?
3. "She moves at a constant speed in a constant direction." Say the same sentence in fewer words.

Check Your Answers

1. In both cases the answer is 25 m/s:

$$\text{Average speed} = \frac{\text{distance covered}}{\text{time interval}} = \frac{100 \text{ meters}}{4 \text{ seconds}}$$
$$= \frac{50 \text{ meters}}{2 \text{ seconds}} = 25 \text{ m/s}$$

2. Both bicycles have the same speed, but they have opposite velocities because they move in opposite directions.
3. "She moves at constant velocity."

2.4 Motion Is Relative

Everything is always moving. Even when you think you're standing still, you're actually speeding through space. You're moving relative to the sun and stars—though you are at rest relative to the Earth. Right now your speed relative to the sun is about 100,000 kilometers per hour. And you're moving even faster relative to the center of our galaxy.

When discussing motion, we mean motion relative to something else. When we say a space shuttle moves at 30,000 kilometers per hour, we mean relative to the Earth below. When we say a racing car reaches a speed of 300 kilometers per hour, we mean relative to the track. Unless stated otherwise, all speeds discussed in this book are relative to the surface of the Earth. Motion is relative.

Aristotle used logic to establish his ideas of motion. Galileo used experiment. Galileo showed that experiments are better than logic in testing knowledge. Galileo was concerned with *how* things move rather than *why* they move. The path was clear for Isaac Newton (1642–1727) to make further connections of concepts of motion.

Figure 2.7
When you sit on a chair, your speed is zero relative to the Earth but 30 km/s relative to the sun.

Isaac Newton (1642–1727)

Isaac Newton was born prematurely and barely survived on Christmas Day, 1642, the same year that Galileo died. Newton's birthplace was his mother's farmhouse in Woolsthorpe, England. His father died several months before his birth, and he grew up under the care of his mother and grandmother. As a child he showed no particular signs of brightness, and at the age of 14 1/2 he was taken out of school to work on his mother's farm. As a farmer he was a failure, preferring to read books he borrowed from a neighboring druggist. An uncle sensed the scholarly potential in young Isaac and prompted him to study at the University of Cambridge, which he did for 5 years, graduating without particular distinction.

A plague swept through London, and Newton retreated to his mother's farm—this time to continue his studies. At the farm, at age 23, he laid the foundations for the work that was to make him immortal. Seeing an apple fall to the ground led him to consider the force of gravity extending to the moon and beyond. He formulated the law of universal gravitation (which he later proved). He invented the calculus, a very important mathematical tool in science. Newton extended Galileo's work and formulated the three fundamental laws of motion. He also formulated a theory of the nature of light and showed with prisms that white light is composed of all colors of the rainbow. It was Newton's experiments with prisms that first made him famous.

When the plague subsided, Newton returned to Cambridge and soon established a reputation for himself as a first-rate mathematician. His mathematics teacher resigned in his favor and Newton was appointed the Lucasian professor of mathematics. He held this post for 28 years. In 1672 he was elected to the Royal Society, where he exhibited the world's first reflector telescope. It can still be seen, preserved at the library of the Royal Society in London with the inscription: "The first reflecting telescope, invented by Sir Isaac Newton, and made with his own hands."

It wasn't until Newton was 42 that he began to write what is generally acknowledged as the greatest scientific book ever written, the *Principia Mathematica Philosophiae Naturalis.* He wrote the work in Latin and completed it in 18 months. It appeared in print in 1687 and wasn't printed in English until 1729, 2 years after his death. When asked how he was able to make so many discoveries, Newton replied that he solved his problems by continually thinking very long and hard about them—and not by sudden insight.

At the age of 46 he was elected a member of Parliament. He attended the sessions in Parliament for 2 years and never gave a speech. One day he rose and the House fell silent to hear the great man. Newton's "speech" was very brief; he simply requested that a window be closed because of a draft.

A further turn from his work in science was his appointment as warden and then as master of the mint. Newton resigned his professorship and directed his efforts toward greatly improving the workings of the mint, to the dismay of counterfeiters who flourished at that time. He maintained his membership in the Royal Society and was elected president, then was re-elected each year for the rest of his life. At the age of 62, he wrote *Opticks,* which summarized his work on light. Nine years later he wrote a second edition to his *Principia.*

Although Newton's hair turned gray at 30, it remained full, long, and wavy all his life. Unlike others in his time, he did not wear a wig. He was a modest man, very sensitive to criticism, and never married. He remained healthy in body and mind into old age. At 80, he still had all his teeth, his eyesight and hearing were sharp, and his mind was alert. In his lifetime he was regarded by his countrymen as the greatest scientist who ever lived. In 1705 he was knighted by Queen Anne. Newton died at the age of 85 and was buried in Westminister Abbey along with England's kings and heroes.

Newton showed that the universe ran according to natural laws—a knowledge that provided hope and inspiration to people of all walks of life and that ushered in the Age of Reason. The ideas and insights of Isaac Newton truly changed the world and elevated the human condition.

2.5 Newton's First Law of Motion—The Law of Inertia

Isaac Newton was born on Christmas Day in the year Galileo died. At the age of 24, Newton extended Galileo's concept of inertia and gave it the status of a fundamental law that underlies all motion. He developed two other laws of motion as well (which we'll study in the following chapters). Newton's laws were built on Galileo's findings. Once and for all, they put to rest mistaken Aristotelian ideas that dominated the thinking of the best minds during the previous 2000 years.

Newton's first law, usually called the **law of inertia,** is a restatement of Galileo's idea.

> **Every object continues in a state of rest, or in a state of motion in a straight line at constant speed, unless it is compelled to change that state by forces exerted upon it.**

Figure 2.8
Inertia in action.

This says that things tend to keep on doing what they're already doing. Objects at rest tend to remain at rest—a force is needed to set them in motion. For example, dishes on a table are in a state of rest. The dishes tend to remain at rest even if you snap a tablecloth from beneath them.*

The law also says that when an object is moving, its tendency is to remain moving, along a straight-line path. For example, if you slide a hockey puck on the surface of slippery ice, it moves a long way until ice and air friction finally stops it. If you slide it along the surface of a city street, it is quickly brought to rest by the force of friction. It tends to keep sliding but friction acts against it. If you toss an object where there is no friction, such as in the vacuum of outer space, it will move forever in a straight-line path. An object moves by its own inertia.

While the ancients thought continual forces were needed to maintain motion, the law of inertia provides a completely different way of thinking about motion. We now know that objects continue to move by themselves. If an object is at rest you'll have to apply force to get it moving, but once in motion no force is needed (except that to overcome any friction). The object moves in a straight line indefinitely. In the next chapter we'll see that forces are needed to change the speeds and directions of objects, but not to maintain motion if there is no friction.

Why will the coin drop into the glass when a force accelerates the card?

Why does the downward motion and sudden stop of the hammer tighten the hammerhead?

Why is it that a slow continuous increase in the downward force breaks the string above the massive ball, but a sudden increase breaks the lower string?

Figure 2.9
Examples of inertia.

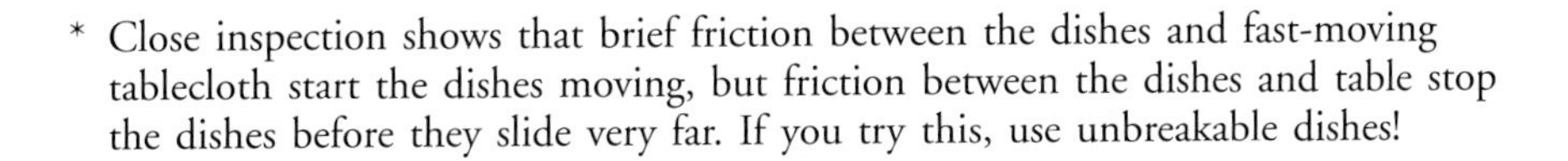

* Close inspection shows that brief friction between the dishes and fast-moving tablecloth start the dishes moving, but friction between the dishes and table stop the dishes before they slide very far. If you try this, use unbreakable dishes!

Concept Check ✓

When the space shuttle moves in a nearly circular orbit around the Earth, is a force needed to maintain its high speed? If suddenly the force of gravity were cut off, what type of path would the shuttle follow?

Check Your Answers No force in the direction of the shuttle's motion exists. The shuttle coasts by its own inertia. The only force acting on it is the force of gravity, which acts at right angles to its motion (toward the Earth's center). We'll see later that this right-angled force holds the shuttle in a circular path. If it were cut off, the shuttle would move in a straight line at constant speed (constant velocity).

2.6 Net Force—The Combination of All Forces That Act on an Object

So we see that without force, objects don't speed up, slow down, or change direction. When we say "force," we mean the total force, or *net* force, acting on an object. Often more than one force acts. For example, when you throw a basketball, the force of gravity, air friction, and the pushing force you apply with your muscles all act on the ball. The **net force** on the ball is the combination of all these forces. It is the net force that changes an object's state of motion.

For example, suppose you pull on a box with a force of 5 pounds. If your friend also pulls with 5 pounds in the same direction, the net force on the box is 10 pounds. If your friend pulls on the box with the same force as you but in the opposite direction, the net force on it is zero. Now if you increase your pull to 10 pounds and your friend pulls oppositely with 5 pounds, the net force is 5 pounds in the direction of your pull. We see this in Figure 2.10, where instead of pounds, the scientific unit of force is used—the **newton,** abbreviated N.

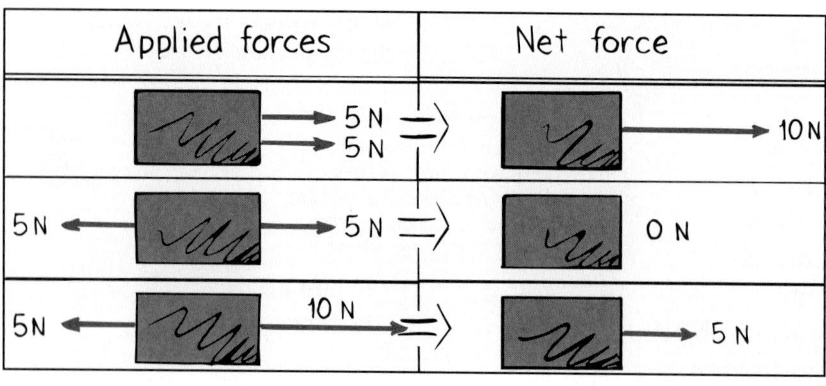

Figure 2.10
Net force.

In Figure 2.10, forces are shown by arrows. Arrows are used because forces are vector quantities. As mentioned earlier, a vector quantity has both magnitude (how much) and direction (which way). When an arrow represents a vector quantity, the arrow's length represents magnitude and its direction shows the direction of the quantity. Such an arrow is called a *vector.* (Again, more on vectors in Appendix C.)

Paul Hewitt Personal Essay

When I was in high school my counselor advised me not to take science and math classes. Instead I was asked to focus on what seemed to be my gift for art. I took this advice. I was then interested in drawing comic strips and in boxing, but neither of these earned me much success. After a stint in the army I tried my luck at sign painting, and the cold Boston winters drove me south to warmer Miami, Florida. There, at age 26, I got a job painting billboards and met my intellectual mentor, Burl Grey. Like me, Burl had never studied physics in high school. But he was passionate about science in general. He shared his passion by discussing many fascinating questions as we painted together.

I remember Burl asked me about the tensions (stretching forces) in the ropes that held up the staging we were standing on. The staging was simply a heavy horizontal plank suspended by a pair of ropes. Burl twanged the rope nearest his end of the staging. He asked me to do the same with mine. He was comparing the tensions in both ropes to see which was greater. Burl was heavier than I was and he guessed the tension in his rope was greater. Like a more tightly stretched guitar string, the rope with greater tension twangs at a higher pitch. The finding that Burl's rope had a higher pitch seemed reasonable because his rope supported more of the load.

When I walked toward Burl to borrow one of his brushes he asked if tensions in the ropes changed. Does tension in his rope increase as I get closer? We agreed that it should because even more of the load was supported by Burl's rope. How about my rope? Would rope tension there decrease? We agreed that it would, for it would be supporting less of the total load. I was unaware that I was discussing physics.

Burl and I used exaggeration to bolster our reasoning (just as physicists do). If we both stood at an extreme end of the staging and leaned outward, it was easy to imagine the opposite end of the staging rising like the end of a seesaw. And then the opposite rope would go limp. There would be no tension in that rope. We then reasoned the tension in my rope would gradually decrease as I walked toward Burl. It was fun posing such questions and seeing if we could answer them.

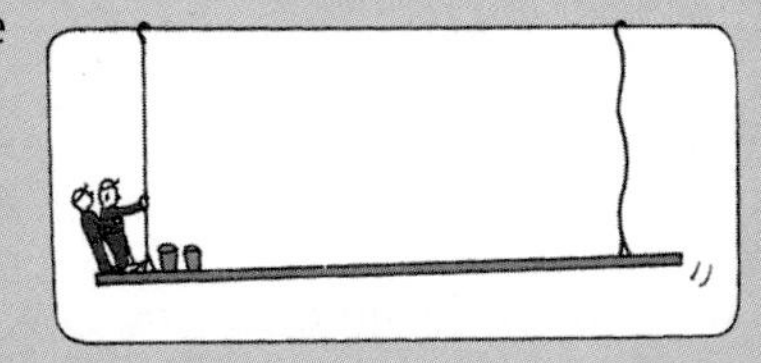

A question that we couldn't answer was whether or not a decrease of tension in one rope would be *exactly* compensated by an increase of tension in the other. For example, if my rope underwent a decrease of 50 newtons, would Burl's rope gain 50 newtons? (We talked pounds back then, but here we use the scientific unit of force, the *newton*—abbreviated N.) Would the gain be *exactly* 50 N? And if so, would this be a grand coincidence? I didn't know the answer until more than a year later. That was when Burl's stimulating questions prompted me to leave full-time painting and go to college to study science.*

In my science classes I learned that any object at rest, such as the sign-painting staging I worked on with Burl, experiences no net force. It is said to be in *equilibrium*. In other words, the individual forces on the object add up to a net force with zero magnitude. So the sum of the upward forces supplied by the supporting ropes indeed do add up to the downward forces of our weights plus the weight of the staging. A 50 N loss in one rope would have to be accompanied by a 50 N gain in the other. Only then do the upward and downward forces cancel.

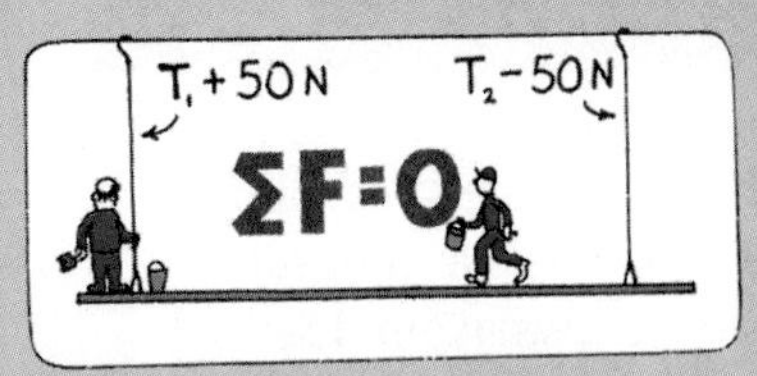

I tell this true story to make the point that one's thinking is very different when there is a rule to guide it. Now, when I look at any motionless object I know right away that all the forces acting on it cancel out. We see nature differently when we know its rules. It makes nature seem simpler and easier to understand. Without the rules of physics, we tend to be superstitious and see magic where there is none. Quite wonderfully, everything is connected to everything else by a surprisingly small number of rules, and in a beautifully simple way. The rules of nature are what the study of physics is about.

* I am forever indebted to Burl Grey for the stimulation he provided, for when I continued with formal education, it was with enthusiasm. I lost touch with Burl for 40 years. A student in my class, Jayson Wechter, doing some detective work, located him in 1998 and put us in contact. Friendship renewed, we once again continue in spirited conversations.

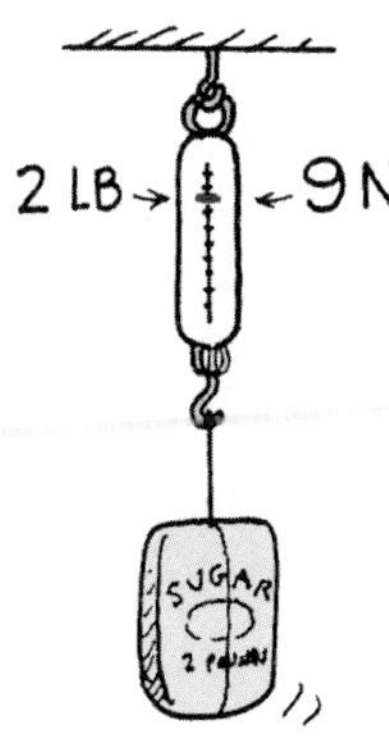

Figure 2.11
The upward tension in the string has the same magnitude as the weight of the bag, so the net force on the bag is zero.

2.7 Equilibrium for Objects at Rest

Suppose you tie a string around a 2-pound bag of sugar and hang it on a weighing scale (Figure 2.11) similar to the scales found in grocery stores. A spring in the scale stretches until the scale reads 2 pounds. The stretched spring experiences a "stretching force" called *tension.* The same scale in a science lab is likely in units of newtons. This scale will show the weight of the bag of sugar as 9 newtons rather than 2 pounds. Both pounds and newtons are units of weight. Units of weight in turn are units of force. The bag of sugar is attracted to the Earth with a gravitational force of 2 pounds—or equivalently, 9 newtons. Hang twice as much sugar from the scale and the reading will be 18 newtons.

Note there are two forces acting on the bag of sugar—tension force acting upward and weight acting downward. The two forces on the bag are equal and opposite, and cancel to zero. Hence the bag remains at rest.

When the net force on something is zero, we say that something is in *mechanical equilibrium.* In mathematical notation, this **equilibrium rule** is

$$\Sigma \mathbf{F} = 0.$$

The symbol Σ stands for "the vector sum of" and F stands for "forces." The rule says that the forces acting upward on something at rest must be balanced by other forces acting downward—to make the vector sum equal zero. Vector quantities take direction into account, so if upward forces are +, downward ones are −. Then when added, (+) + (−), they actually subtract.

In Figure 2.13 we see the forces involved for Burl and Paul on their sign-painting staging. The sum of the upward tensions is equal to the sum of their weights plus the weight of the staging. Note how the magnitudes of the two upward vectors equal the magnitude of the three downward vectors. Net force on the staging is zero, so we say it is in mechanical equilibrium.

Figure 2.12
Burl Grey, who first introduced Paul Hewitt to the concept of tension, shows a 2-lb bag producing a tension of 9 N (actually slightly more than 2 lb and 9 N).

Figure 2.13
The sum of the upward vectors equals the sum of the downward vectors. $\Sigma F = 0$ and the staging is in equilibrium.

Concept Check ✓

Consider the gymnast hanging from the rings.

1. If she hangs with her weight evenly divided between the two rings, how would scale readings in both supporting ropes compare with her weight?
2. Suppose she hangs with slightly more of her weight supported by the left ring. How would a scale on the right read?

Check Your Answers

*(Again, are you reading this before you have thought about and formulated **your** reasoned answers? If so, do you also exercise your body by looking at others do push-ups? Exercise your thinking: When you encounter the many Concept Checks as above in this book, think before you look at these answers!)*

1. The reading on each scale will be half her weight. The sum of the readings on both scales then equals her weight.
2. When more of her weight is supported by the left ring, the reading on the right is less than half her weight. No matter how she hangs, the sum of the scale readings equals her weight. For example, if one scale reads two-thirds her weight, the other scale will read one-third her weight. Get it?

2.8 The Support Force—Why We Don't Fall Through the Floor

A book lies at rest on a desk. The book is in equilibrium—the net force on it is zero. What forces act on the book to make up this zero net force? One is the force of gravity—the *weight* of the book. Since the book is in equilibrium we know there must be another force acting on it that cancels gravity. The force must be equal to gravity but pointing upward.

Where is the upward force coming from? It is coming from the desk that is supporting the book. We call this upward force the **support force,** or the *normal force.** The support force must equal the weight of the book. We say the upward support force is positive and the downward weight is negative. Then the support force plus gravity add mathematically to become zero. So the net force on the book is zero. Another way to say the same thing is

$$\Sigma \mathbf{F} = 0.$$

To better understand that the desk pushes up on the book, think about a spring being compressed (Figure 2.14). Push the spring down and you can feel the spring pushing up on your hand. Similarly, the book lying on the desk compresses atoms in the desk, which behave

Figure 2.14
(Left) The desk pushes up on the book with as much force as the downward force of gravity on the book. (Right) The spring pushes up on your hand with as much force as you exert to push down on the spring.

* This force acts at right angles to the surface. Mathematically, "normal to" means "at right angles to." Hence the name normal force.

Figure 2.15
The upward support is as much as your weight.

like tiny springs. The weight of the book squeezes downward on the atoms, and the atoms squeeze upward on the book. In this way the compressed atoms produce the support force.

When you step on a bathroom scale, two forces act on the scale. One is the downward pull of gravity, your weight, and the other is the upward support force of the floor. These forces compress a spring that is calibrated to show your weight (Figure 2.15). In effect, the scale shows the support force. When you weigh yourself on a bathroom scale at rest, the support force and your weight have the same magnitude.

Concept Check ✓

1. What is the net force on a bathroom scale when a 110-pound person stands on it?
2. Suppose you stand on two bathroom scales with your weight evenly divided between the two scales. What will each scale read? How about if you lean with more of your weight on one scale than the other?

Check Your Answers

1. Zero, for the scale remains at rest. The scale reads *support force*, which has the same magnitude as weight—not the net force.
2. The reading on each scale is half your weight. If you lean more on one scale than the other, more than half your weight will be read on that scale but less on the other. In this way they add up to your weight. Like the example of the gymnast hanging by the rings, if one scale reads two-thirds her weight, the other scale will read one-third her weight.

2.9 Equilibrium for Moving Objects

When an object isn't moving, it's in equilibrium. The forces on it add up to zero. But the state of rest is only one form of equilibrium. An object moving at constant speed in a straight-line path is also in equilibrium. We say the same thing when we say an object moving at constant velocity is in equilibrium. The forces on this object are also zero (in accord with Newton's first law).

Equilibrium is a state of no change. A bowling ball rolling at constant velocity is in equilibrium—until it hits the pins. Whether at rest or steadily rolling in a straight-line path, the sum of the forces on the bowling ball is zero:

$$\Sigma \mathbf{F} = 0.$$

It follows from Newton's first law that an object under the influence of only one force cannot be in equilibrium. Net force couldn't be zero. Only when two or more forces act on an object can it be in equilibrium. We can test whether or not something is in equilibrium by noting whether or not it undergoes changes in motion.

Consider a crate being pushed horizontally across a factory floor. If it moves at constant velocity, it is in equilibrium. This tells us that more than one force acts on the crate—likely the force of friction between the crate and the floor. The fact that the net force on the crate equals zero means that the force of friction must be equal and opposite to our pushing force.

We say objects at rest are in *static* equilibrium, and objects moving at constant velocity are in *dynamic* equilibrium. These are examples of mechanical equilibrium. There are other types of equilibrium. For example, when we study heat, we'll talk about thermal equilibrium, where temperature doesn't change.

Figure 2.16
When the push on the crate is as great as the force of friction between the crate and the floor, the net force on the crate is zero and it slides at an unchanging speed.

Concept Check ✓

An airplane flies at constant velocity. In other words, it is in equilibrium. Two horizontal forces act on the plane. One is the thrust of the propeller that pushes it forward. The other is the force of air resistance that acts in the opposite direction. Which force is bigger?

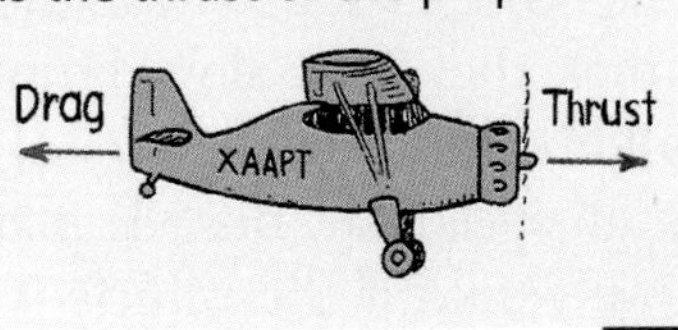

Check Your Answer Both forces have the same magnitude. Call the forward force exerted by the propeller positive. Then the air resistance is negative. Since the plane is in equilibrium, can you see that the two forces combine to equal zero?

2.10 The Earth Moves Around the Sun

Before the 16th century, it was believed that the Earth was the center of the universe. People then believed that the sun circles the Earth. As mentioned in Chapter 1, Galileo advanced the idea that the Earth moves around the sun, instead of the other way around. There was much arguing and debate about this idea. People thought like Aristotle, and the existence of a force big enough to keep the Earth moving was beyond their imagination. One of the arguments against a moving Earth was the following.

Consider a bird sitting at rest at the top of a tall tree. On the ground below is a fat, juicy worm. The bird sees the worm and drops vertically below and catches it. It was argued that this would be impossible if the Earth were moving. A moving Earth would have to travel at an enormous speed to circle the sun in one year. While

Figure 2.17
Can the bird drop down and catch the worm if the Earth moves at 30 km/s?

the bird is in the air descending from its branch to the ground below, the worm is swept far away along with the moving Earth. It seemed that catching a worm on a moving Earth would be an impossible task. The fact that birds do catch worms from high tree branches seemed to be clear evidence that the Earth must be at rest.

Can you see the mistake in this argument? You can if you use the concept of inertia. You see, not only is the Earth moving at a great speed, but so are the tree, the branch of the tree, the bird that sits on it, the worm below, and even the air in between. Things in motion remain in motion if no unbalanced forces are acting. So when the bird drops from the branch, its initial sideways motion remains unchanged. It catches the worm quite unaffected by the motion of its total environment.

We live on a moving Earth. Stand next to a wall. Jump up so that your feet are no longer in contact with the floor. Does the moving wall slam into you? Why not? Because you are also traveling at the same speed, before, during, and after your jump. The speed of the Earth relative to the sun is not the speed of the wall relative to you.

Figure 2.18
When you flip a coin in a high-speed airplane, it behaves as if the airplane were at rest. The coin keeps up with you—inertia in action!

Four hundred years ago, people had difficulty with ideas like these. One reason is that they didn't yet travel in high-speed vehicles. Rather, they took slow, bumpy rides in horse-drawn carts. People couldn't notice the effects of inertia as much. Today we flip a coin in a high-speed car, bus, or plane and catch the vertically moving coin as we would if the vehicle were at rest. We see evidence for the law of inertia when the horizontal motion of the coin before, during, and after the catch is the same. The coin keeps up with us.

Our ideas of motion today are very different from those of our distant ancestors. Aristotle did not recognize the idea of inertia. He imagined different rules for motion in the heavens and for motion on the Earth. He saw horizontal motion as "unnatural," and requiring a steady force. Galileo and Newton, on the other hand, saw that all moving things follow the same rules. To them, moving things require *no* force to keep moving if friction is absent. We can only wonder how differently science might have progressed if Aristotle had recognized the unity of all kinds of motion.

Chapter Review

Key Terms and Matching Definitions

_____ equilibrium rule
_____ force
_____ inertia
_____ net force
_____ Newton
_____ Newton's first law of motion—the law of inertia
_____ speed
_____ support force
_____ vector quantity
_____ velocity

1. The property of things to remain at rest if at rest, and in motion if in motion.
2. The distance traveled per time.
3. The speed of an object and specification of its direction of motion.
4. A quantity that specifies direction as well as magnitude.
5. Every object continues in a state of rest, or in a state of motion in a straight line at constant speed, unless it is compelled to change that state by forces exerted upon it.
6. A push or a pull.
7. The combination of all forces that act on an object.
8. The scientific unit of force.
9. $\Sigma F = 0$.
10. The force that supports an object against gravity.

Review Questions

Aristotle's Classification of Motion

1. According to Aristotle, what tendency of moving objects governed their motions?

2. According to Aristotle, what kinds of motion required no forces?

Galileo's Concept of Inertia

3. What two main ideas of Aristotle did Galileo discredit?

4. What is the name of the property of objects to maintain their states of motion?

Galileo Formulated the Concepts of Speed and Velocity

5. Distinguish between speed and velocity.

6. Why do we say velocity is a vector quantity and speed is not?

Motion is Relative

7. How can you be both at rest and also moving at 100,000 km/h at the same time?

8. Between Aristotle and Galileo, who relied on experiments?

Newton's First Law of Motion—The Law of Inertia

9. Who was the first to discover the concept of inertia, Galileo or Newton?

10. What is the tendency of a moving object when no forces act on it?

Net Force—The Combination of All Forces That Act on an Object

11. When only a pair of equal and opposite forces act on an object, what is the net force acting on it?

12. We've learned that velocity is a vector quantity. Is force also a vector quantity? Why or why not?

Equilibrium for Objects at Rest

13. What is the name given to the force that occurs in a rope when both ends are pulled in opposite directions?

14. How much tension is there in a rope that holds a 20-N bag of apples at rest?

15. What does $\Sigma F = 0$ mean?

The Support Force—Why We Don't Fall Through the Floor

16. Why is the support force on an object often called the *normal* force?

17. When you weigh yourself, are you actually reading the support force acting on you, or are you really reading your weight?

Equilibrium for Moving Objects

18. Give an example of something moving when a net force of zero acts on it.

19. If we push a crate at constant velocity, how do we know how much friction acts on the crate compared to our pushing force?

The Earth Moves Around the Sun

20. If you're in a smooth-riding bus that is going at 50 km/h and you flip a coin vertically, what is the horizontal velocity of the coin in midair?

Explorations

1. By any method you choose, determine your average walking speed. How do your results compare with those of your classmates?

2. Ask a friend to drive a small nail into a piece of wood placed on top of a pile of books on your head. Why doesn't this hurt you? (Be careful! Wear a helmet in case your partner misses, use a very small nail, safety glasses, and a wooden mallet instead of a hammer.)

Exercises

1. Galileo found that a ball rolling down one incline will pick up enough speed to roll up another. How high will it roll compared to its initial height?

2. Correct your friend who says, "The race-car driver rounded the curve at a constant velocity of 100 km/h."

3. If the speedometer of a car reads a constant speed of 50 km/h, can you say that the car has a constant velocity? Why or why not?

4. If a huge bear were chasing you, its enormous mass would be very threatening. But if you ran in a zigzag pattern, the bear's mass would be to your advantage. Why?

5. A space probe may be carried by a rocket into outer space. What keeps the probe going after the rocket no longer pushes it?

6. Consider a ball at rest in the middle of a toy wagon. When the wagon is pulled forward, the ball rolls against the back of the wagon. Interpret this observation in terms of Newton's first law.

7. Why do you lurch forward in a bus that suddenly slows? Why do you lurch backward when it picks up speed? What law applies here?

8. Push a shopping cart and it moves. When you stop pushing, it comes to rest. Does this violate Newton's law of inertia? Defend your answer.

9. When a car moves along the highway at constant velocity, the net force on it is zero. Why, then, do you continue running your engine?

10. Roll a bowling ball down a lane and you'll find it moves slightly slower with time. Does this violate Newton's law of inertia? Defend your answer.

11. Consider a pair of forces, one having a magnitude of 20 N, and the other 12 N. What maximum net force is possible for these two forces? What is the minimum net force possible?

12. Can an object be in mechanical equilibrium when only a single force acts on it? Explain.

13. The sketch shows a painting staging in mechanical equilibrium. The person in the middle weighs 250 N, and the tensions in each rope are 200 N. What is the weight of the staging?

14. A different staging that weighs 300 N supports two painters, one 250 N and the other 300 N. The reading in the left scale is 400 N. What is the reading in the right hand scale?

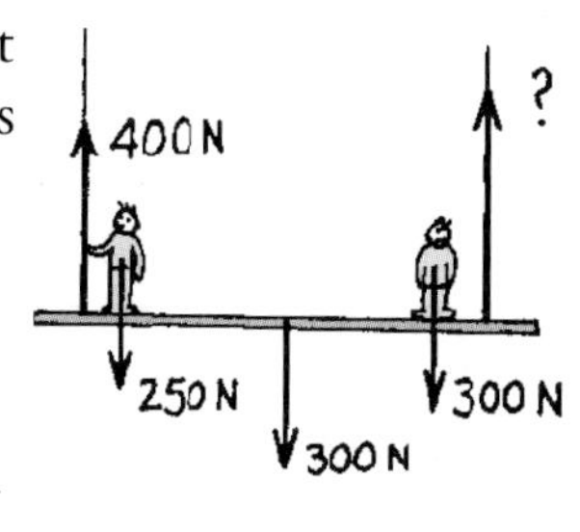

15. Nellie Newton hangs at rest from the ends of the rope as shown. How does the reading on the scale compare to her weight?

16. Harry the painter swings year after year from his bosun's chair. His weight is 500 N and the rope, unknown to him, has a breaking point of 300 N. Why doesn't the rope break when he is supported as shown at the left below? One day Harry is painting near a flagpole, and, for a change, he ties the free end of the rope to the flagpole instead of to his chair as shown at the right. Why did Harry end up taking his vacation early?

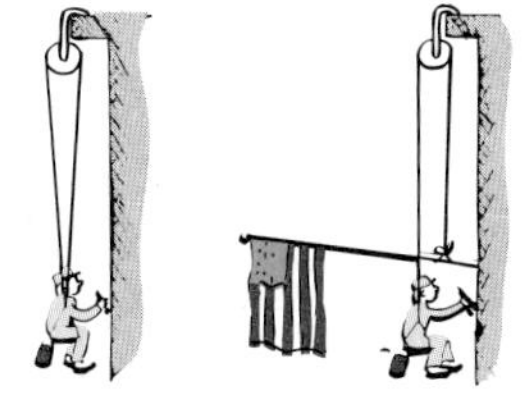

17. As you stand at rest on a floor, does the floor exert an upward force against your feet? If so, what exactly is this force?

18. A child learns in school that the Earth is traveling faster than 100,000 kilometers per hour around the sun, and in a frightened tone asks why we aren't swept off. What is your explanation?

19. If you toss a coin straight upward while riding in a train, where does the coin land when the motion of the train is uniform along a straight-line track? When the train slows while the coin is in the air? When the train is turning?

20. As the Earth rotates about its axis, it takes 3 hours for the United States to pass beneath a point above the Earth that is stationary relative to the sun. What is wrong with the following scheme? To travel from Washington D.C. to San Francisco and use very little fuel, simply ascend in a helicopter high over Washington D.C. and wait three hours until San Francisco passes below.

Problems

1. What is your average speed if you run 50 meters in 10 seconds?

2. A tennis ball travels the full length of the court, 24 meters, in 0.5 second. What is its average speed?

3. Find the net force produced by a 30-N and 20-N force in each of the following cases:
a. Both forces act in the same direction.
b. Both forces act in opposite directions.

4. A horizontal force of 100 N pushes a box across a floor at a constant speed.
a. What is the net force acting on the box?
b. What is the force of friction on the box?

5. Phil Physicer weighs 600 N (132 lb) and stands on two bathroom scales. He stands so one scale reads twice as much as the other. What are the scale readings?

Suggested Reading and Web Site

Asimov, Isaac. *Understanding Physics.* New York: Barnes and Noble, 1996.
Excellent and accurate introductory physics.

www.merlot.org
A rich smorgasbord of teaching and learning materials, mostly at the college level, but also with links to good material in science (and other fields) at the high-school and middle-school levels. Well organized and searchable.

Chapter 3: Newton's Second Law of Motion—Force and Acceleration

Kick a soccer ball and it moves. Does the ball's motion depend on its mass? Does its motion depend on how hard it's kicked? Or does how far the ball moves depend both on its mass and on how hard you kick it? What about air resistance? Does that also affect how far the soccer ball goes?

Consider some more examples. Suppose that you drop two basketballs from a cliff, one ball with air inside and the other filled with heavy sand. Which hits the ground first? Does air resistance make a difference? Suppose you and a much-heavier friend sky dive from a high-flying plane. You open your same-size parachutes at the same time. Who reaches the ground first? Does air resistance make a difference in this case? What rules guide the answers to these questions?

3.1 Galileo Developed the Concept of Acceleration

In addition to speed and velocity (previous chapter), Galileo developed the concept of *acceleration* in his experiments with inclined planes. He found that balls rolling down inclines rolled faster and faster. Their speed changed as they rolled. Further, the balls gained the same amount of velocity in equal time intervals. In other words, their speed increased by a given amount each second.

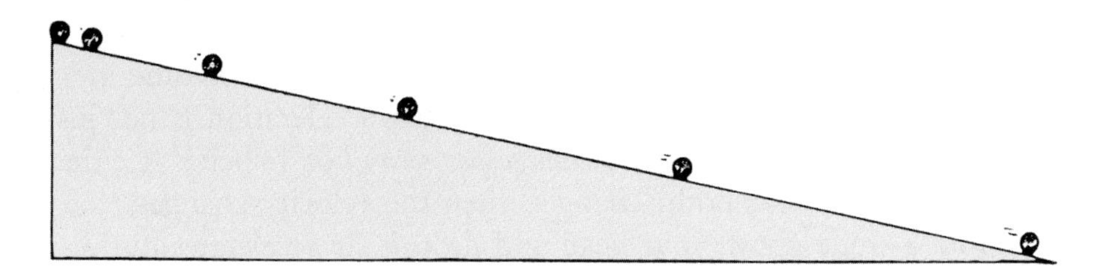

Figure 3.1
A ball gains the same amount of speed each second as it rolls down an incline.

Galileo defined the rate of change of velocity as **acceleration:***

$$\textbf{Acceleration} = \frac{\textbf{change of velocity}}{\textbf{time interval}}$$

You observe acceleration every time you ride in a car. When the driver steps on the gas pedal the automobile gains speed. We say it accelerates. We see why the gas pedal is called the "accelerator"!

When the brakes are applied, the car slows. This is also acceleration because the velocity of the car is changing. When something slows down, we often call this *deceleration,* or *negative acceleration.*

When a car makes a turn, even if its speed does not change, it is accelerating. Can you see why? Acceleration occurs because the car's direction is changing. Acceleration is a change in velocity. So acceleration is a change in speed, a change in direction, or a change in both speed *and* direction. Figure 3.2 illustrates this.

Figure 3.3
Rapid deceleration is sensed by the driver who lurches forward (in accord with Newton's first law).

Figure 3.2
We say that a body undergoes acceleration when there is a *change* in its state of motion.

* The Greek letter Δ (delta) is often used as a symbol for "change in" or "difference in." In "delta" notation, $a = \frac{\Delta v}{\Delta t}$, where Δv is the change in velocity, and Δt is the change in time (the time interval).

Suppose we are driving. In 1 second, we steadily increase our velocity from 30 kilometers per hour to 35 kilometers per hour. In the next second, we go from 35 kilometers per hour to 40 kilometers per hour, and so on. We change our velocity by 5 kilometers per hour each second. We see that:

$$\textbf{Acceleration} = \frac{\textbf{change of velocity}}{\textbf{time interval}} = \frac{\textbf{5 km/h}}{\textbf{1 s}} = \textbf{5 km/h·s}$$

In this example the acceleration is 5 kilometers per hour-second (abbreviated as 5 km/h·s). Note that a unit for time enters twice: Once for the unit of velocity and again for the interval of time in which the velocity is changing. Also note that acceleration is not just the change in velocity; it is the *change per second* of velocity. If either speed or direction (or both) changes, then the velocity changes.

Hold a stone above your head and drop it. It accelerates during its fall. When air resistance doesn't affect the motion of a falling object, we say the object is in **free fall.** Interestingly, the amount of acceleration is the same for all freely falling objects in the same vicinity. We find that a freely falling object gains speed at the rate of 10 m/s each second.

$$\textbf{Acceleration} = \frac{\textbf{change in speed}}{\textbf{time interval}} = \frac{\textbf{10 m/s}}{\textbf{1 s}} = \textbf{10 m/s}^2$$

We read the acceleration of free fall as 10 meters per second squared. This is the same as saying that acceleration is 10 meters per second per second. Note again that the unit of time, the second, appears twice. It appears once for the unit of velocity, and again for the time during which velocity changes.

In Figure 3.4, we imagine a freely-falling boulder with a speedometer attached. As the boulder falls, the speedometer shows that the boulder goes 10 m/s faster each second. This 10 m/s gain each second is the boulder's acceleration.

The acceleration of free fall is further developed in Appendix B and in the Practice Book.

Downward-falling objects gain speed because of the force of gravity. How about an object thrown straight upward? Once it leaves your hand, it continues moving upward for a while and then comes back down. While going up, it moves against gravity and loses speed. Guess how much speed it loses each second while going upward? And guess how much speed it gains each second while coming down? That's right: The change in speed per second is 10 m/s—whether moving upward or downward. At the highest point, when it changes direction from upward to downward, its instantaneous speed is zero. Then it starts downward *just as if it had been dropped from rest at that height.* It will return to its starting point with the same speed it had when thrown.

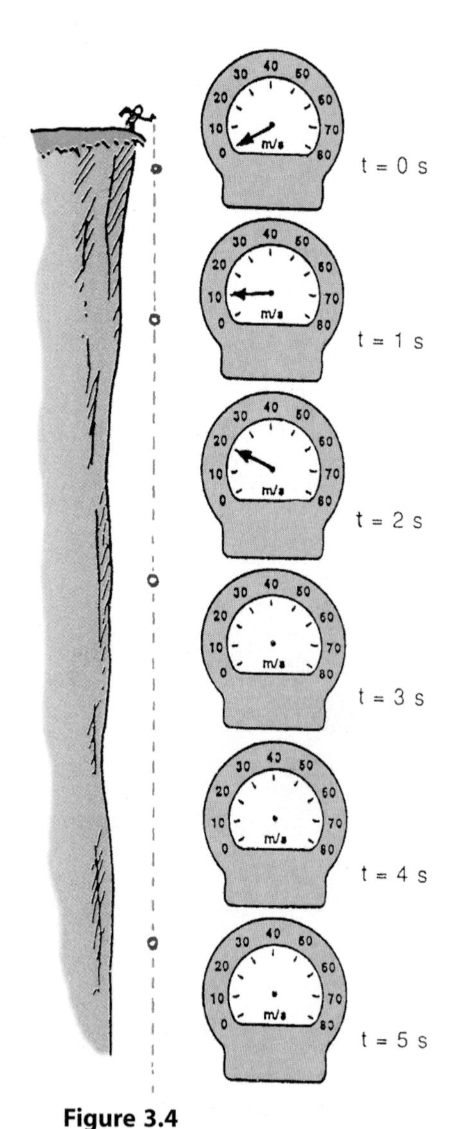

Figure 3.4
Pretend that a falling boulder is equipped with a speedometer. In each succeeding second of fall, you'd find the boulder's speed increasing by the same amount; 10 m/s. Sketch in the missing speedometer needle at t = 3 s, 4 s, and 5 s.

3.2 Force Causes Acceleration

Any object that accelerates is acted on by a push or a pull—a force of some kind. It may be a sudden push, like hitting a punching bag, or the steady pull of gravity. Acceleration is caused by applying force.

Most often, more than one force acts on an object. Recall from the previous chapter that the combination of forces that act on an object is the *net force.* Acceleration depends on the *net force.* For example, if you push with 25 N on an object, and somebody else pushes in the opposite direction with 15 N, the net force applied to the object is 10 N. The object will accelerate as if a single 10-N force acts on it.

Figure 3.5
Kick the ball and it accelerates.

Suppose you pull a wagon with a net force of 20 N. The wagon accelerates. Now double the force and pull with 40 N. How much more will the wagon accelerate now? There is a general rule here. If the net force is doubled, the acceleration also doubles. Three times the net force produces three times the acceleration. We say that the acceleration produced is directly proportional to the net force. We write:

$$\textbf{Acceleration} \sim \textbf{net force}$$

The symbol ~ stands for "is directly proportional to." That means any change in one quantity matches the same amount of change in the other.

The direction of acceleration is always in the direction of the net force. When a force is applied in the direction of the object's motion, the speed increases. When a force is applied in the opposite direction, the speed decreases. When a force acts at right angles, it will deflect the object.

Force of hand accelerates the brick

Twice as much force produces twice as much acceleration

Twice the force on twice the mass gives the same acceleration

Figure 3.6
Acceleration is directly proportional to force.

Concept Check ✓

1. If you push on a shopping cart it will accelerate. If you apply four times the net force, how much greater will the acceleration be?
2. If the net force acting on a sports car is increased by five, how much greater will the acceleration be?

Check Your Answers

1. It will have four times as much acceleration.
2. It will have five times as much acceleration.

So we see that force produces acceleration. How much acceleration, however, also depends on something else. It depends on the mass of the object being pushed or pulled.

When one thing is **directly proportional** to another, then as one gets bigger the other gets bigger too.

3.3 Mass Is a Measure of Inertia

What happens when you kick a tin can? It accelerates—changes its state of motion. Now kick the same can filled with rocks. What happens? It doesn't accelerate as much as when it was empty. If the can is full of something really heavy like lead, say, it will hardly move. Ouch!

The more massive full can has more inertia than the empty can. In other words, it is more resistant to a change in motion. This suggests that the **mass** of an object corresponds to its inertia. The greater an object's mass, the greater its inertia. This is why powerful engines are required in tractor trailers that pull massive loads—and why they have powerful brakes for stopping. Heavy loads have lots of inertia.

Figure 3.7
The greater the mass, the greater the force needed for a given acceleration.

Mass is more than an indication of an object's inertia. Mass is also a measure of how much material an object contains. Mass depends on the number and kinds of atoms making up the object. A dense material such as lead is made up of many tightly packed atomic particles. So it has a lot of mass.

Mass Is Not Volume

Do not confuse mass and volume. Volume is a measure of space. It is measured in units such as cubic centimeters, cubic meters, or liters. Mass is measured in **kilograms.** If an object has a large mass, it may or may not have a large volume. For example, equal size bags of cotton and rocks may have equal volumes, but very unequal masses. How many kilograms of matter an object contains and how much space the object occupies are two different things. Mass is different from volume.

Mass Is Not Weight

Do not confuse mass and weight. They are different from each other. Mass is a measure of the amount of matter in an object. As already mentioned, mass depends on the number and kinds of atoms in the object. Weight, however, depends on gravity. You would weigh less on the moon, for example, than you do on Earth. Why? The moon's gravity is weaker than Earth's so you'd be pulled to the moon's surface with less force than on Earth. On the other hand, the numbers and kinds of atoms in your body are the same on the moon as on Earth. There is just as much material in your body no matter where you are. So, unlike weight, your mass doesn't change if gravity varies.

A pillow is bigger than an auto battery, but which has more matter—more *inertia*—more *mass*?

You can sense how much mass is in an object by feeling its inertia. When you shake an object back and forth you can feel its inertia. If it has a lot of mass, it's hard to change the object's direction. If it has a small mass, shaking the object is easier. To-and-fro shaking requires the same force even in regions where gravity is different—on the moon, for example. The rock's inertia, or mass, is a property of the object itself and not its location.

Mass is different than weight. We can define mass and weight as follows:

> **Mass is the amount of matter in an object. Also, a measure of the inertia, or "laziness," that an object shows when you try to change its state of motion.**
>
> **Weight is the force due to gravity that acts on an object's mass.**

Although mass and weight are different from each other, they are directly *proportional* to each other. Objects with large mass have large weight; objects with little mass have little weight. In the same location, twice the mass weighs twice as much. That's what we mean when we say that mass and weight are proportional to each other. Remember that mass has to do with the amount of matter in the object and its inertia, while weight has to do with how strongly that matter is attracted by gravity.

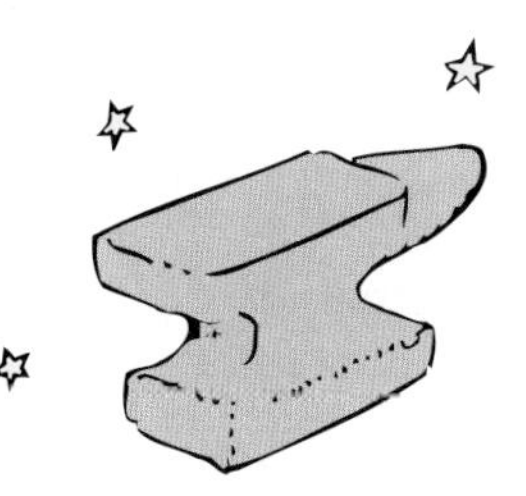

Figure 3.8
An anvil in outer space, between the Earth and moon for example, may be weightless, but it is not massless.

Figure 3.9
The astronaut in space finds it is just as difficult to shake the "weightless" anvil as it would be on Earth. If the anvil is more massive than the astronaut, which shakes more—the anvil or the astronaut?

Concept Check ✓

1. Does a 2-kilogram iron block have twice as much *inertia* as a 1-kilogram iron block? Twice as much *mass*? Twice as much *volume?* Twice as much *weight* when weighed in the same location?
2. Does a 2-kilogram iron block have twice as much *inertia* as a 1-kilogram bunch of bananas? Twice as much *mass*? Twice as much *volume*? Twice as much *weight* when weighed in the same location?
3. How does the mass of a bar of gold vary with location?

Check Your Answers

1. The answer is yes to all questions. A 2-kilogram block of iron has twice as many iron atoms, and therefore twice the amount of matter, mass, and weight. The blocks are made of the same material, so the 2-kilogram block also has twice the volume.
2. Two kilograms of *anything* has twice the inertia and twice the mass of one kilogram of anything else. Since mass and weight are proportional in the same location, two kilograms of anything will weigh twice as much as one kilogram of anything. Except for volume, the answer to all the questions is yes. Volume and mass are proportional only when the materials are the same—when they have the same *density.* Iron is much more dense than bananas, so two kilograms of iron must occupy less volume than one kilogram of bananas.
3. Not at all! It is made of the same number of atoms no matter what the location. Although its weight may vary with location, it has the same mass everywhere. This is why mass is preferred to weight in scientific studies.

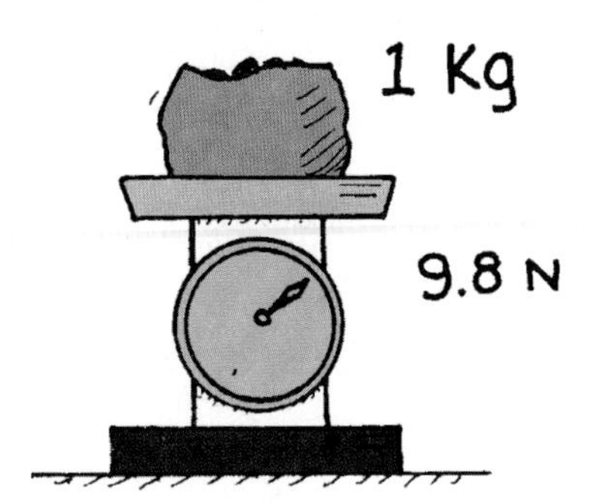

Figure 3.10
One kilogram of nails weighs 9.8 newtons, which is equal to 2.2 pounds.

One Kilogram Weighs 9.8 Newtons

The standard unit of mass is the kilogram, abbreviated kg. The standard unit of force is the newton as discussed in Chapter 2. The standard symbol for the newton is N. The abbreviation is written with a capital letter because the unit is named after a person. A 1-kg bag of any material has a weight of 9.8 N in standard units. Away from the Earth's surface where the force of gravity is less, the bag would weigh less.

Except in cases where precision is needed, we will round off 9.8 and call it 10. So 1 kilogram of something weighs about 10 newtons. If you know the mass in kilograms and want weight in newtons, multiply the number of kilograms by 10. Or, if you know the weight in newtons, divide by 10 and you'll have the mass in kilograms. Weight and mass are proportional to each other.

The relationship between kilograms and pounds is that 1 kg weighs 2.2 lb at the Earth's surface. (That means 1 lb is the same as 4.45 N.)

Concept Check ✓

Why is it okay to say a 1-kg bag of sand weighs 10 N, but a 1-kg bag of gold weighs 9.8 N? Don't they weigh the same?

Check Your Answers Both 1-kg bags have the same weight, and both weigh 10 N. However, since gold is more valuable, saying 9.8 N rather than the rounded off 10 N is usually a good idea. But we won't make a big deal in this book about whether or not rounding off is okay. It's more important to learn the main idea.

Force of hand accelerates the brick

The same force accelerates 2 bricks 1/2 as much

3 bricks, 1/3 as much acceleration

Figure 3.11
Acceleration is inversely proportional to mass.

3.4 Mass Resists Acceleration

More massive objects are more difficult to accelerate. Experiments show that for the same force, twice as much mass results in half as much acceleration; three times the mass results in one third the acceleration, and so forth. In other words, for a given force the acceleration produced is *inversely* proportional to the mass. We write:

When one thing is **inversely proportional** to another, then as one gets bigger the other gets smaller.

$$\textbf{Acceleration} \sim \frac{\textbf{1}}{\textbf{mass}}$$

By **inversely** we mean that the two values change in opposite ways. When one gets larger, the other gets smaller. Or when one gets smaller, the other gets larger. More mass means less acceleration, because more mass means more resistance to changes in motion. (Mathematically we see that as the denominator increases, the whole quantity decreases. For example, the quantity 1/100 is less than the quantity 1/10.)

Concept Check ✓

1. Suppose you're offered either 1/4 of an apple pie or 1/8 of the pie. Which piece is larger?
2. Suppose you apply the same amount of force to two carts, one cart with a mass of 4 kg, and the other with a mass of 8 kg.
   a. Which cart will accelerate more?
   b. How much greater will the acceleration be?

Check Your Answers

1. The larger piece is 1/4, one-quarter of the pie. If you choose the 1/8 piece, you'll have half as much pie as 1/4.
2. a. The 4-kg cart will have more acceleration.
   b. The 4-kg cart will have *twice* the acceleration because it has half as much mass—which means half as much resistance to changes in motion!

3.5 Newton's Second Law Links Force, Acceleration, and Mass

Isaac Newton was the first to realize the connection between force and mass in producing acceleration. He discovered one of the most important rules of nature ever proposed—his *second law of motion.* **Newton's second law** states:

> **The acceleration produced by a net force on an object is directly proportional to the net force, is in the same direction as the net force, and is inversely proportional to the mass of the object.**

Or in shorter notation,

$$\textbf{Acceleration} \sim \frac{\textbf{net force}}{\textbf{mass}}$$

By using consistent units such as newtons (N) for force, kilograms (kg) for mass, and meters per second squared (m/s^2) for acceleration, we get the exact equation.

$$\textbf{Acceleration} = \frac{\textbf{net force}}{\textbf{mass}}$$

Figure 3.12
Acceleration depends on the mass being pushed.

In briefest form, where a is acceleration, F is net force, and m is mass:

$$a = \frac{F}{m}$$

Acceleration equals the net force divided by the mass. If the net force acting on an object is doubled, the object's acceleration will be doubled. Suppose instead that the mass is doubled. Then the acceleration will be halved. If both the net force and the mass are doubled, then the acceleration will be unchanged.

Learn your concepts now! Problem solving later will be much more meaningful.

Concept Check ✓

If you push on a shopping cart, it will accelerate.

1. If you push five times harder, what happens to the acceleration?
2. If you push the same, but the cart is loaded so it has five times as much mass, what happens to the acceleration?
3. If you push five times harder when it is loaded with five times as much mass, what happens to the acceleration?

Check Your Answers

1. Acceleration will be five times greater.
2. Acceleration will be less—only one-fifth as much.
3. It will have the same acceleration as it had to begin with.

Physics problems are often more complicated than these examples. We don't focus on solving complicated problems in this book, but instead emphasize equations as guides to thinking about the connections of basic concepts. That's why there are more exercises than problems at the end of the chapters. Mastering the techniques of problem solving may have a higher priority in a follow-up course.

Calculation Corner

Consider a 1000-kg car pulled by a cable with 2000 N of force. What will be the acceleration of the car? Using Newton's second law we find

$$a = \frac{F}{m} = \frac{2000 \text{ N}}{1000 \text{ kg}} = \frac{2000 \text{ kg} \cdot \text{m/s}^2}{1000 \text{ kg}} = 2 \text{ m/s}^2$$

Here we see that the units of force, the N, is the same as the units kg·m/s^2. From now on we'll take a shortcut and simply say that the ratio N/kg equals m/s^2.

Suppose that the force were 4000 N. What would be the acceleration?

$$a = \frac{F}{m} = \frac{4000 \text{ N}}{1000 \text{ kg}} = 4 \text{m/s}^2$$

Doubling the force on the same mass simply doubles the acceleration.

3.6 Friction Is a Force That Affects Motion

Friction occurs when one object rubs against something else. Friction occurs for solids, liquids, and gases. An important rule of friction is that it always acts in a direction to oppose motion. If you drag a solid block along a floor to the left, the force of friction on the block will be to the right. A boat propelled to the east by its motor experiences water friction to the west. When an object falls downward through the air, the force of friction (**air drag**) acts upward. Friction always acts in a direction to oppose motion.

The amount of friction between two surfaces depends on the kinds of material and how much they are pressed together. Friction between a crate and a rough wooden floor is greater than between the same crate and a polished linoleum floor. And if the surface is inclined, friction is less because the crate doesn't press as much on the inclined surface.

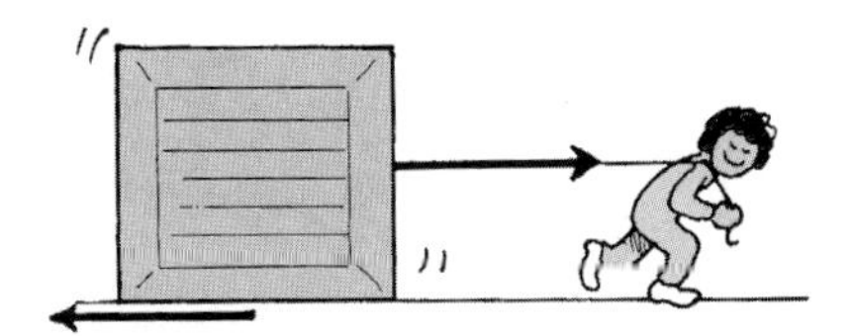

Figure 3.13
Applied force just overcomes friction so the crate slides at constant velocity.

Practicing Physical Science

1. A crate filled with delicious peaches rests on a horizontal floor. The only forces acting on the crate are gravity and the support force of the floor. We show these forces with vectors. $\mathbf{F_w}$ is the weight vector, and $\mathbf{F_N}$ is the support force. We use the subscript $_N$ to indicate the force is **normal** to the floor. Normal means at a right angle with respect to the floor. Support forces, interestingly, are always normal to the surfaces of support.

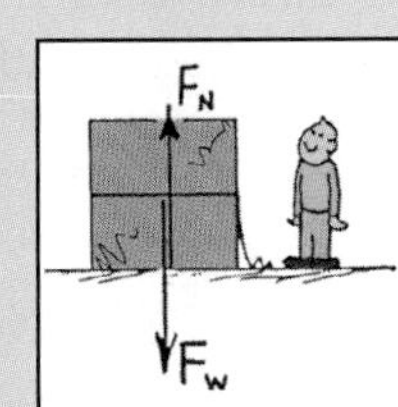

Choose the correct words.

a. The net force on the crate is (zero) (greater than zero).

b. Evidence for this is that there is no (velocity) (acceleration).

2. A small pull **P** is exerted on the crate, but not enough to move it.

a. A force of friction acts, which is (less than **P**) (equal to **P**) (greater than **P**).

b. Net force on the crate is (zero) (greater than zero).

3. Pull **P** is increased until the crate begins to slide. It is pulled so that it moves at constant velocity across the floor.

a. Friction force $\mathbf{F_f}$ is (less than **P**) (equal to **P**) (greater than **P**).

b. Constant velocity means acceleration is (zero) (greater than zero).

c. Net force on the crate is (less than zero) (zero) (greater than zero).

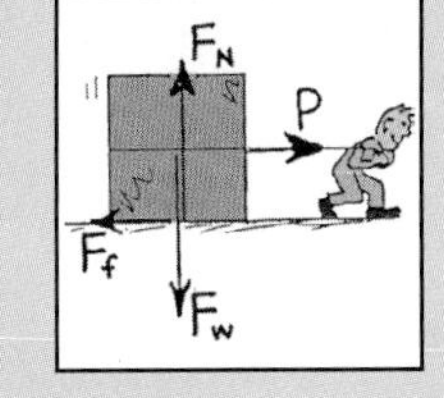

4. Pull **P** is further increased and is now greater than friction $\mathbf{F_f}$.

a. Net force on the crate is (zero) (greater than zero).

b. The net force acts to the right, so acceleration acts toward the (left) (right).

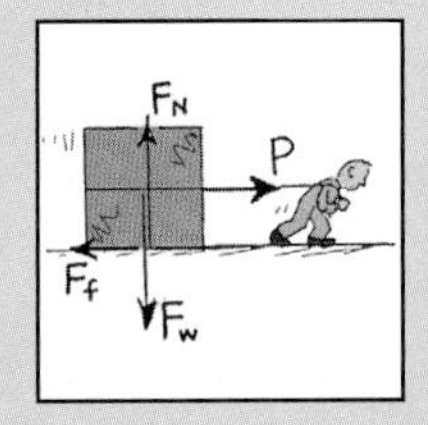

We see here that the friction force varies with different pulling strengths. When the pull is small enough not enough to move the crate, friction equals the pull, and net force is zero. As the pull is increased, the friction increases the same, and net force remains zero. When the pulling force is greater than the friction, the net force is greater than zero and the crate accelerates. Once moving, the pull can be reduced until it matches the friction and the crate moves steadily. Then acceleration is zero and the net force is zero.

Discuss this with your classmates until you understand it. When you do, you're understanding good physics!

When you pull horizontally on a crate and it slides across a factory floor, both your force and the opposite force of friction affect the motion. When you pull hard enough to match friction, the net force on the crate is zero and it slides at constant velocity. Notice that we are talking about what we learned in the previous chapter—no change in motion occurs when $\Sigma F = 0$.

Concept Check ✓

1. Two forces act on a bowl resting on a table: the bowl's weight and the support force from the table. Does a force of friction also act on the bowl?
2. Suppose a high-flying jumbo jet flies at constant velocity when the thrust of its engines is a constant 80,000 N. What is the *acceleration* of the jet? What is the force of air drag acting on the jet?

Check Your Answers

1. No, not unless the bowl tends to slide or does slide across the table. For example, if it is pushed toward the left by another force, then friction between the bowl and table will act toward the right. Friction forces occur only when an object tends to slide or is sliding.
2. The acceleration is zero because the velocity is constant (not changing). Since the acceleration is zero, it follows from $a = F/m$ that the net force is zero. This says that the force of air drag must be equal to the thrusting force of 80,000 N and act in the opposite direction. So the air drag is 80,000 N.

3.7 Objects in Free Fall Have Equal Acceleration

Galileo developed the concept of acceleration on inclined planes. He was interested in falling objects, but because he lacked suitable timing devices he used inclined planes to effectively slow down acceleration. He found that as the planes were tipped at greater angles, the acceleration of the balls was greater. When tipped all the way vertical, acceleration was that of free fall. We define free fall as falling only under the influence of gravity, where other forces such as air drag can be neglected.

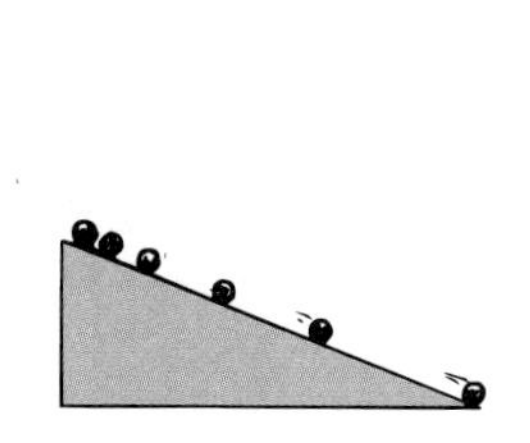
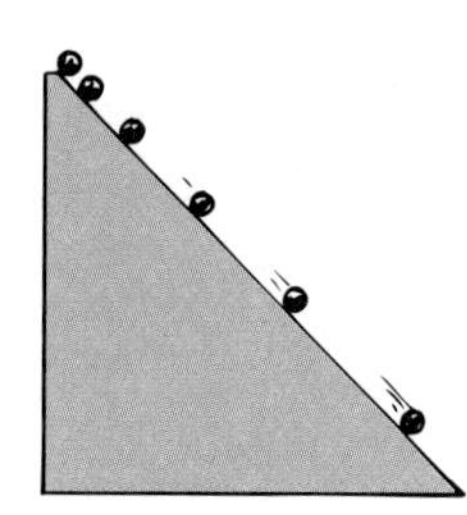
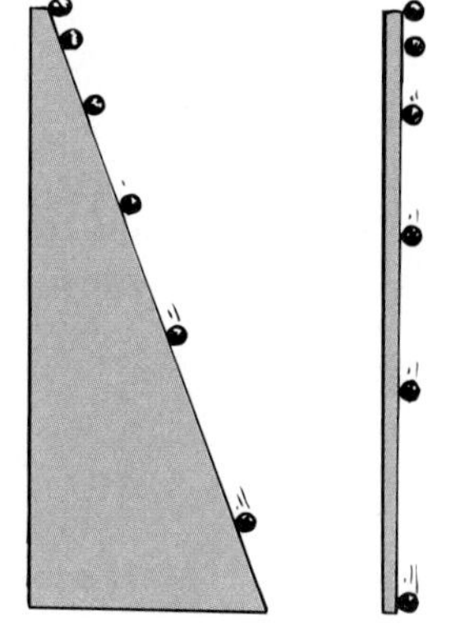

Figure 3.14
The greater the slope of the incline, the greater the acceleration of the ball. What is the acceleration when the incline is vertical?

Galileo further discovered that the acceleration didn't depend on mass. On any incline all balls have the same acceleration. Likewise, the acceleration of free fall doesn't depend on mass.

A 10-kilogram boulder and a 1-kilogram stone dropped from an elevated position at the same time will fall together and strike the ground at practically the same time. This experiment, said to be done by Galileo from the Leaning Tower of Pisa, destroyed the Aristotelian idea that an object that weighs ten times as much as another should fall ten times faster. Galileo's experiment and many others demonstrated the same result. But Galileo couldn't say *why* the accelerations were equal. The explanation comes from Newton's second law.

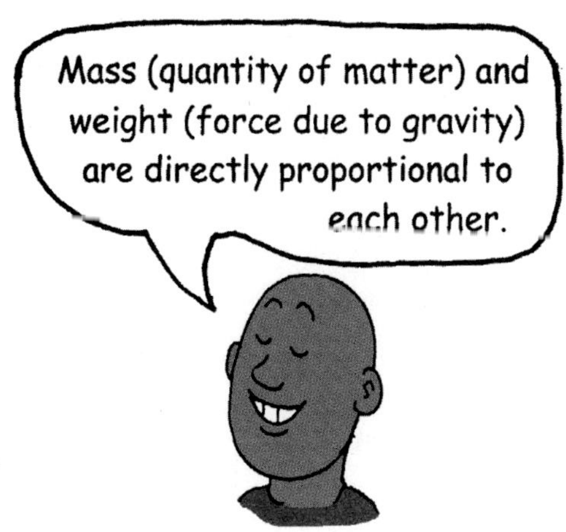

3.8 Newton's Second Law Explains Why Objects in Free Fall Have Equal Acceleration

A falling 10-kg boulder "feels" 10 times the force of gravity (weight) as a 1-kg stone. Followers of Aristotle believed the boulder should therefore accelerate ten times as much as the stone—because they considered only the greater weight. But Newton's second law tells us to also consider the mass. Can you see that ten times as much force acting on ten times as much mass produces the same acceleration as the smaller force acting on the smaller mass? In symbolic notation,

$$\frac{F}{m} = \frac{\boldsymbol{F}}{\boldsymbol{m}}$$

where F stands for the force (weight) acting on the boulder, and m stands for its correspondingly large mass. The small F and m stand for the smaller weight and mass of the stone. We see that the *ratio* of weight to mass is the same for these or any objects. All freely falling objects have the same force/mass ratio and undergo the same acceleration at the same location. This acceleration, due to gravity, is represented by the symbol g.

We can show the same result with numerical values. The weight of a 1-kg stone (or 1 kg of *anything*) is 10 N at the Earth's surface. The weight of 10 kg of matter, such as the boulder, is 100 N. The force acting on a falling object is its weight. The acceleration of the stone is

$$a = \frac{F}{m} = \frac{weight}{m} = \frac{10\ \text{N}}{1\ \text{kg}} = 10\ \text{m/s}^2 = g$$

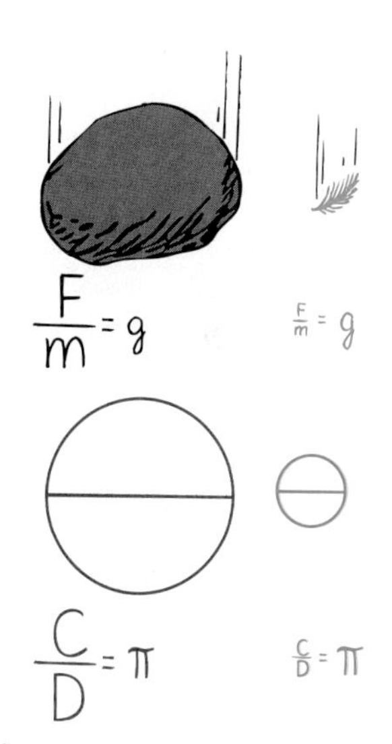

Figure 3.15
The ratio of weight (F) to mass (m) is the same for the large rock and the small feather; similarly, the ratio of circumference (C) to diameter (D) is the same for the large and the small circle.

and for the boulder,

$$a = \frac{F}{m} = \frac{weight}{m} = \frac{100\text{N}}{10\text{kg}} = 10 \text{ m/s}^2 = g$$

We all know that a feather drops more slowly than a coin when both are dropped in air. Air drag more greatly affects the feather. But in a vacuum where air drag isn't present, a feather and coin dropped together will fall side by side. This is shown in the demonstration of Figure 3.16, where a vacuum pump removes air from the glass tube. We can see why acceleration is the same for both. With no air drag the force/mass ratio is the same for both.

Figure 3.16
A feather and a coin fall at equal accelerations in a vacuum.

It's of historical interest to note that although Galileo spoke about force, and was the first to propose the concepts of acceleration and inertia, he did not make the connection between these three concepts. It took the genius of Isaac Newton to show the connection—namely, $a = \frac{F}{m}$. This rule of mechanics is one of the most profound in physics. With it, scientists and engineers have been able to put people on the moon.

Practicing Physical Science

Fill in the blanks:

1. A 5-kg bag of sand has a weight of 50 N. When dropped its acceleration is

   $a = \frac{50 \text{ N}}{5 \text{ kg}} =$ _____ m/s^2.

2. A 10-kg bag of sand has a weight of 100 N. When dropped its acceleration is

   $a = \frac{\_\_\_\_}{10 \text{ kg}} =$ _____ m/s^2.

3. Calculate the free-fall acceleration of a 20-kg bag of sand.

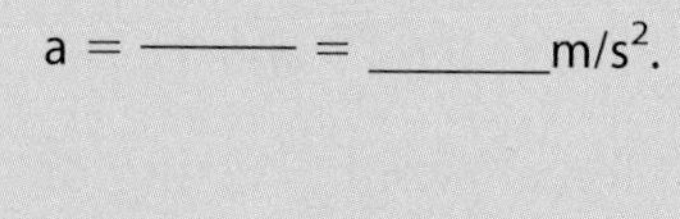

   a = ——— = _______ m/s^2.

3.9 Acceleration of Fall Is Less When Air Drag Acts

Most often, air drag is not negligible for falling objects. Then acceleration of fall is less. Air drag depends on two things: speed and surface area. When a sky diver steps from a high-flying plane, air drag builds up as speed increases. The result is reduced acceleration. More reduction can occur by increasing surface area. A diver does this by orienting the body so more air is encountered—by spreading out like a flying squirrel. So air drag depends on speed and the surface area encountered by the air.

For free fall the downward net force is weight. Only weight! But when air is present, the downward net force = weight − air drag. Can you see that the presence of air drag reduces net force? And that less net force means less acceleration? So as a diver falls faster and faster the acceleration of fall gets less and less. What happens to the net force if air drag builds up to equal weight? The answer is, net force becomes zero. Here we see $\Sigma F = 0$ again! Then acceleration becomes zero. Does this mean the diver comes to a stop? No! What it means is the diver no longer picks up speed. Acceleration terminates—it no longer occurs. We say the diver has reached **terminal speed.** If we are concerned with direction, down for falling objects, we say the diver has reached **terminal velocity.**

Terminal speed for a human skydiver varies from about 150 to 200 km/h, depending on weight and orientation of the body. A heavier person falls faster for air drag to balance weight. The greater weight is more effective in "plowing through" air. This means more terminal speed for a heavier person. Increasing surface area reduces terminal speed. That's where a parachute comes in. A parachute greatly increases air drag, and terminal speed can be reduced to a safe 15 to 25 km/h.

When we previously discussed the interesting demonstration of the falling coin and feather in the glass tube, we found that the feather falls more slowly because of air drag. The feather's weight is very small so it reaches terminal speed very quickly. The feather doesn't have to fall very far or fast before air drag builds up to equal its small weight. The coin, on the other hand, doesn't have a chance to fall fast enough for air drag to build up to equal its weight. Interestingly, if you drop a coin from a very high location, like off a tall building, terminal speed will be reached when the speed of the coin is about 200 km/h. This is a much, much higher terminal speed than that for a feather!

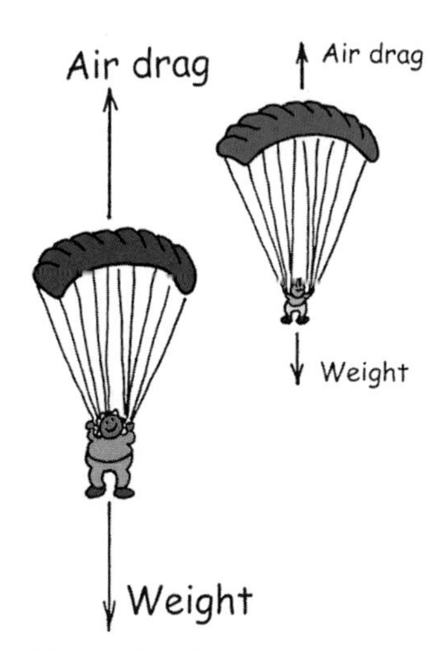

Figure 3.17
The heavier parachutist must fall faster than the lighter parachutist for air resistance to cancel her greater weight.

Concept Check ✓

Consider two parachutists, a heavy person and a light person, who jump from the same altitude with parachutes of the same size.

1. Which person reaches terminal speed first?
2. Which person has the greatest terminal speed?
3. Which person gets to the ground first?
4. If there were no air drag, like on the moon, how would your answers to these questions differ?

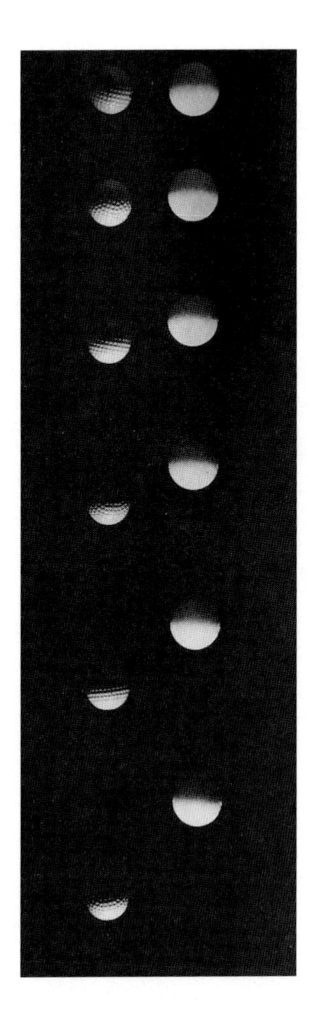

Figure 3.18
A stroboscopic study of a golf ball (left) and a Styrofoam ball (right) falling in air. The air resistance is negligible for the heavier golf ball, and its acceleration is nearly equal to g. Air resistance is not negligible for the lighter Styrofoam ball, which reaches its terminal velocity sooner.

Check Your Answers

To answer these questions think of a coin and feather falling in air.

1. Just as a feather reaches terminal speed very quickly, the lighter person reaches terminal speed first.
2. Just as a coin falls faster than a feather through air, the heavy person falls faster and reaches a higher terminal speed.
3. Just like the race between a falling coin and feather, the heavier person falls faster and will reach the ground first.
4. If there were no air drag, there would be no terminal speed at all. Both would be in free fall and hit the ground at the same time.

When Galileo tried to explain why all objects fall with equal accelerations, wouldn't he have loved to know the rule: $a = \frac{F}{m}$?

When Galileo reportedly dropped objects of different weights from the Leaning Tower of Pisa, they didn't actually hit at the same time. They almost did, but because of air drag, the heavier one hit a split second before the other. But this contradicted the much longer time difference expected by the followers of Aristotle. The behavior of falling objects was never really understood until Newton announced his second law of motion. Isaac Newton truly changed our way of seeing the world.

Hands-On Exploration

If you drop a sheet of paper and a book side-by-side, the book will fall faster than the paper. Why? The book falls faster because of its greater weight compared to the air drag it encounters. If you place the paper against the lower surface of the raised book and again drop them at the same time, it will be no surprise that they hit the surface below at the same time. The book simply pushes the paper with it as it falls. Now, repeat this, only with the paper on *top* of the book, not sticking over its edge. How will the accelerations of the book and paper compare? Will they separate and fall differently? Will they have the same acceleration? Try it and see! Then see if you can explain what happens.

Chapter Review

Key Terms and Matching Definitions

_____ acceleration
_____ air drag
_____ free fall
_____ friction
_____ inertia
_____ inversely
_____ kilogram
_____ Newton's second law
_____ mass
_____ terminal speed
_____ terminal velocity
_____ volume
_____ weight

1. The rate at which velocity changes with time; the change may be in magnitude or direction or both.
2. The property of things to resist changes in motion.
3. The quantity of matter in an object. More specifically, it is the measure of the inertia or sluggishness that an object exhibits in response to any effort made to start it, stop it, deflect it, or change in any way its state of motion.
4. When two values change in opposite directions, so that if one increases and the other decreases by the same amount, they are said to be inversely proportional to each other.
5. The quantity of space an object occupies.
6. The force due to gravity on an object.
7. The fundamental SI unit of mass. One kilogram (symbol kg) is the mass of 1 liter (l) of water at 4°C.
8. The acceleration produced by a net force on an object is directly proportional to the net force, is in the same direction as the net force, and is inversely proportional to the mass of the object.
9. The resistive force that opposes the motion or attempted motion of an object past another with which it is in contact, or through a fluid.
10. Motion under the influence of gravitational pull only.
11. Frictional resistance due to motion through air.
12. The speed at which the acceleration of a falling object terminates because air resistance balances its weight.
13. Terminal speed with direction of motion (down for falling objects).

Review Questions

Galileo Developed the Concept of Acceleration

1. Distinguish between velocity and acceleration.
2. When are you most aware of motion in a moving vehicle—when it is moving steadily in a straight line or when it is accelerating?
3. What is the acceleration of free fall?

Force Causes Acceleration

4. Is acceleration proportional to net force or does acceleration equal net force?

Mass Is a Measure of Inertia

5. What relationship does mass have with inertia?
6. What relationship does mass have with weight?
7. Fill in the blanks: Shake something to and fro and you're measuring its _____. Lift it against gravity and you're measuring its _____.

8. What is the weight of a 1-kilogram brick?

Mass Resists Acceleration

9. Is acceleration *directly* proportional to mass, or is it *inversely* proportional to mass? Give an example.

Newton's Second Law Links Force, Acceleration, and Mass

10. If the net force acting on a sliding block is somehow tripled, by how much does the acceleration increase?

11. If the mass of a sliding block is somehow tripled at the same time the net force on it is tripled, how does the resulting acceleration compare to the original acceleration?

Friction Is a Force That Affects Motion

12. Suppose you exert a horizontal push on a crate that rests on a level floor, and it doesn't move. How much friction acts compared with your push?

13. As you increase your push, will friction on the crate increase also?

14. Once the crate is sliding, how hard do you push to keep it moving at constant velocity?

Objects in Free Fall Have Equal Acceleration

15. What is meant by *free fall*?

Newton's Second Law Explains Why Objects in Free Fall Have Equal Acceleration

16. Why doesn't a heavy object accelerate more than a light object when both are freely falling?

17. The ratio of circumference/diameter for all circles is π. What is the ratio of force/mass for freely-falling bodies?

Acceleration of Fall Is Less When Air Drag Acts

18. What two principal factors affect the force of air resistance on a falling object?

19. What is the acceleration of a falling object that has reached its terminal velocity?

20. If two objects of the same size fall through air at different speeds, which encounters the greater air resistance?

Explorations

1. Drop a sheet of paper and a coin at the same time. Which reaches the ground first? Why? Now crumple the paper into a small, tight wad and again drop it with the coin. Explain the difference observed. Will they fall together if dropped from a second-, third-, or fourth-story window? Try it and explain your observations.

2. Drop two balls of different weight from the same height, and at small speeds they practically fall together. Will they roll together down the same inclined plane? If each is suspended from an equal length of string, making a pair of pendulums, and displaced through the same angle, will they swing back and forth in unison? Try it and see; then explain using Newton's laws.

3. The net force acting on an object and the resulting acceleration are always in the same direction. You can demonstrate this with a spool. If the spool is pulled horizontally to the right, in which direction will it roll?

Exercises

1. What is the net force on a bright red Mercedes convertible traveling along a straight road at a steady speed of 100 km/h?

2. On a long alley a bowling ball slows down as it rolls. Is any horizontal force acting on the ball? How do you know?

3. In the orbiting space shuttle you are handed two identical boxes, one filled with sand and the other filled with feathers. How can you tell which is which without opening the boxes?

4. Your empty hand is not hurt when it bangs lightly against a wall. Why is it hurt if it does so while carrying a heavy load? Which of Newton's laws is most applicable here?

5. What happens to your weight when your mass increases?

6. When a junked car is crushed into a compact cube, does its mass change? Its weight? Its volume? Explain.

7. What is the net force on a 1-N apple when you hold it at rest above your head? What is the net force on it after you release it?

8. Does a stick of dynamite contain force?

9. If it takes 1 N to push horizontally on your book to make it slide at constant velocity, how much force of friction acts on the book?

10. A bear that weighs 4000 N grasps a vertical tree and slides down at constant velocity. What is the friction force that acts on the bear?

11. A crate remains at rest on a factory floor while you push on it with a horizontal force *F*. How big is the friction force exerted on the crate by the floor? Explain.

12. Aristotle claimed the speed of a falling object depends on its weight. We now know that objects in free fall, whatever their weights, undergo the same gain in speed. Why does weight not affect acceleration?

13. Two basketballs are dropped from a high building through the air. One ball is hollow and the other filled with rocks. Which accelerates more? Defend your answer.

14. A parachutist, after opening the chute, finds herself gently floating downward, no longer gaining speed. She feels the upward pull of the harness, while gravity pulls her down. Which of these two forces is greater? Or are they equal in magnitude?

15. Why will a sheet of paper fall slower than one that is wadded into a ball?

16. Upon which will air resistance be greater; a sheet of falling paper or the same paper wadded into a ball that falls at a faster terminal speed? (Careful!)

17. How does the force of gravity on a raindrop compare with the air drag it encounters when it falls at constant velocity?

18. How does the terminal speed of a parachutist before opening a parachute compare with terminal speed after? Why is there a difference?

19. How does the gravitational force on a falling body compare with the air resistance it encounters before it reaches terminal velocity? After?

20. Why is it that a cat that accidentally falls from the top of a 50-story building hits the ground no faster than if it falls from the 20th story?

Problems

1. One pound is the same as 4.45 newtons. What is the weight in pounds of 1 newton?
2. What is your own mass in kilograms? Your weight in newtons?
3. What is the acceleration of a 40-kg block of cement when pulled sideways with a net force of 200 N?
4. If a mass of 1 kg is accelerated 1 m/s^2 by a force of 1 N, what would be the acceleration of 2 kg acted on by a force of 2 N?
5. How much acceleration does a 747 jumbo jet of mass 30,000 kg experience in takeoff when the thrust for each of four engines is 30,000 N?
6. Gravity on the surface of the moon is only 1/6 as strong as gravity on the Earth. What is the weight in newtons of a 10-kg object on the moon and on the Earth? What is its mass on each?

Remember, review questions provide you with a self check of whether or not you grasp the central ideas of the chapter. The exercises and problems are extra "pushups" for you to try after you have at least a fair understanding of the chapter and can handle the review questions.

Suggested Reading

Asimov, Isaac. *Understanding Physics.* New York: Barnes and Noble, 1996.

Excellent and accurate introductory physics.

Chapter 4: Newton's Third Law of Motion—Action and Reaction

Here we see Millie, to whom this book is dedicated, touching her grandchild, Emily. And Emily is touching Millie. Can Millie touch Emily without Emily also touching Millie? Newton's third law says no—that you can't touch without being touched! Consider a heavy truck hitting a small car in a head-on collision. Can the heavy truck hit the car without the car also hitting the truck? And does it hit the car with more force, or the same amount of force, with which the car hits back? Can the heavyweight champion of the world punch a piece of paper in midair any harder than the paper hits back? In this chapter, we'll see how Newton's third law of motion guides our answers to these and other intriguing questions.

4.1 A Force Is Part of an Interaction

Figure 4.1
In the interaction between the car and the truck, is the force of impact the same on each? Is the damage the same?

In the previous chapters we've looked at force as a push or a pull. Looking closer, Newton realized that a force is more than just a single push or pull. A force is part of a mutual action—an **interaction**—between one thing and another. When a truck crashes into a car there is an interaction between the truck and the car. Part of the interaction is the truck exerting a force on the car. The other part is the car exerting a force on the truck. The forces are equal in strength and opposite in direction and they occur at exactly the same time.

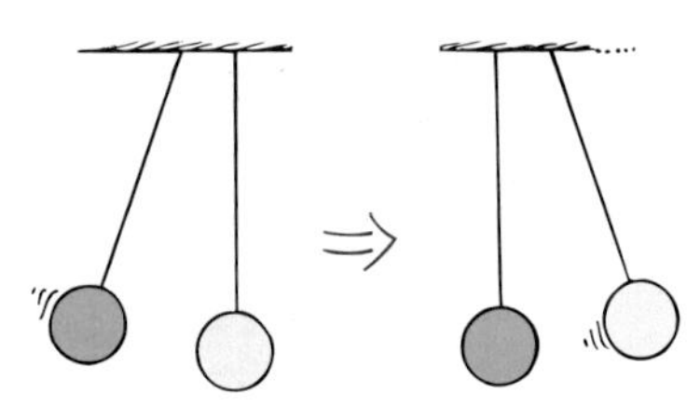

Figure 4.2
The impact forces between the blue and yellow balls move the yellow ball and stop the blue ball.

In every interaction, forces always occur in pairs. For example, you interact with the floor when you walk on it—you push backward against the floor, and the floor simultaneously pushes forward on you. Likewise, the tires of a car interact with the road—the tires push against the road, and the road pushes back on the tires. In swimming you interact with the water—you push the water backward, and the water pushes you forward. There is a pair of forces acting in each interaction. The interactions in these examples depend on friction. But a person or car on slippery ice, by contrast, may not be able to exert a force against the ice to produce the needed opposite force. Then the ice cannot push back and allow the person or car to move along.

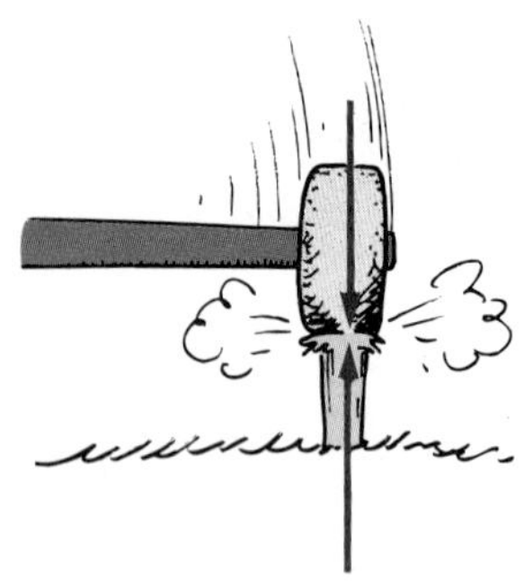

Figure 4.3
In the interaction between the hammer and the stake, each exerts the same amount of force on the other.

Can things like floors, car tires, and water exert forces? Your friends (not taking this course) may think that only living things like people and animals can exert forces. For example, when you push on a wall, how can the wall push back? It's not alive. It doesn't have muscles. But look at your fingers as you push on a wall. They're bent a little. Something must have pushed on them. The wall has pushed back on your fingers as hard as your fingers have pushed on the wall.

So in this chapter we expand our thinking about forces. We see that non-living things can exert them. A force is more than a push or pull. It is part of a mutual interaction between objects.

Does a speeding baseball have force? The answer is no. Force is not something an object possesses, like mass. A speeding baseball *exerts* a force when it hits something. How much force it exerts

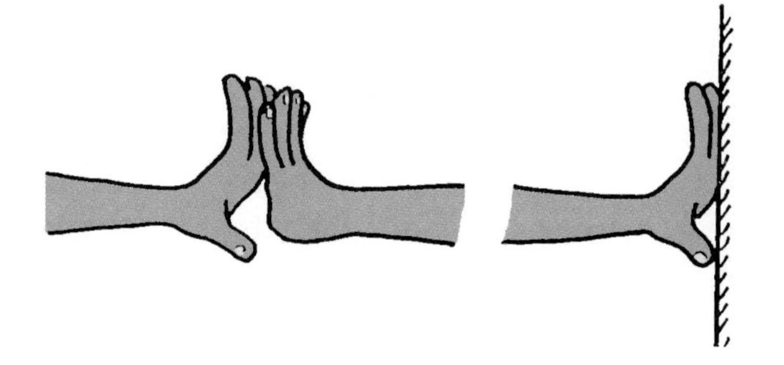

Figure 4.4
You can feel your fingers being pushed by your friend's fingers. You also feel the same amount of force when you push on a wall and it pushes back on you. As a point of fact, you can't push on the wall *unless* it pushes back on you!

depends on how quickly the ball decelerates. Objects don't possess force as a thing in itself. As we will see in the following chapters, a speeding object possesses *momentum* and *kinetic energy*—but not force. A force is an interaction between one object and another.

Concept Check ✓

A car accelerates along a horizontal road. Strictly speaking, exactly what is it that pushes the car?

Check Your Answer It is the road that pushes the car along. Really! Except for some road friction and air drag, only the road provides a horizontal force on the car. How? The rotating tires push back on the road (action). The road simultaneously pushes forward on the tires (reaction). The next time you see a car moving along a road, tell your friends that the road pushes the car along. If at first they don't believe you, convince them that there is more to the physical world than meets the eye of the casual observer. Turn them on to some physical science.

4.2 Newton's Third Law—Action and Reaction

In his investigation of many interactions, Newton discovered an underlying principle, called **Newton's third law:**

> **Whenever one object exerts a force on a second object, the second object exerts an equal and opposite force on the first.**

We can call one force the *action force,* and the other the *reaction force.* Then we can express Newton's third law in the form:

> **To every action there is always an opposed equal reaction.**

It doesn't matter which force we call *reaction.* The important thing is that they are co-parts of a single interaction and that neither force exists without the other. Action and reaction forces are equal in strength and opposite in direction.

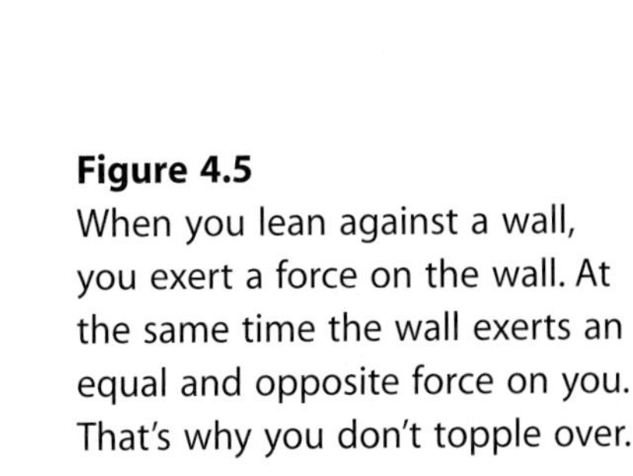

Figure 4.5
When you lean against a wall, you exert a force on the wall. At the same time the wall exerts an equal and opposite force on you. That's why you don't topple over.

Hands-On Exploration

Playing with magnets is fun. Applying Newton's third law to magnets is also fun. Hold a toy magnet near another magnet. Notice that when one magnet moves another, it is also moved by the other. For equal mass magnets the effect is most noticeable. That's because the changes in motion (acceleration) are the same for each. For different size magnets, the smaller magnets move more. Can you see how this ties into Newton's second law? (Newton's second law tells us that the acceleration of the magnet depends not only on force, but on mass.)

Concept Check ✓

1. Which exerts more force, the Earth pulling on the moon, or the moon pulling on the Earth?
2. When a heavy football player and a light one run into each other, does the light player *really* exert as much force on the heavy player as the heavy player exerts on the light one?
3. Is the damage to the heavy player the same as the damage to the light one?

Check Your Answers

1. This is like asking which is greater, the distance between New York and San Francisco, or the distance between San Francisco and New York. Both distances are the same, but in opposite directions. Likewise for the pulls between the Earth and the moon.
2. Yes. In the interaction between the two players, the forces each exert on the other have equal strengths.
3. No. Although the forces are the same on each, the *effects* of these equal forces are quite unequal! The low-mass player may be knocked unconscious while the heavier one may be completely unharmed. There is a difference between the *force* and the *effect* of the force.

4.3 A Simple Rule Helps Identify Action and Reaction

Here's a simple rule for identifying action and reaction forces. First, identify the interaction: One thing, say object A, interacts with another, say object B. Then action and reaction forces can be stated in the form:

Action: Object A exerts a force on object B.
Reaction: Object B exerts a force on object A.

This is easy to remember. If the action is A on B, the reaction is B on A. We see that A and B are simply switched around. Consider the case of your hand pushing on the wall. The interaction is between your hand and the wall. We'll say the action is your hand (object A) exerting a force on the wall (object B). Then the reaction is the wall exerting a force on the your hand.

Action: tire pushes on road Reaction: road pushes on tire

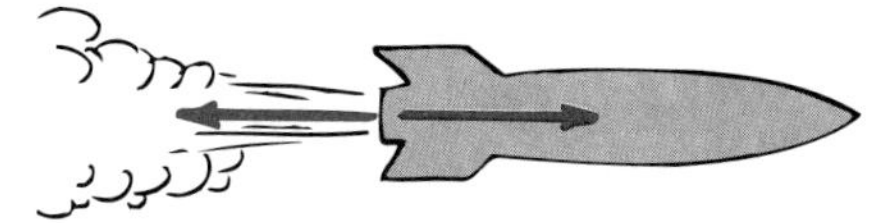

Action: rocket pushes on gas Reaction: gas pushes on rocket

Action: man pulls on spring Reaction: spring pulls on man

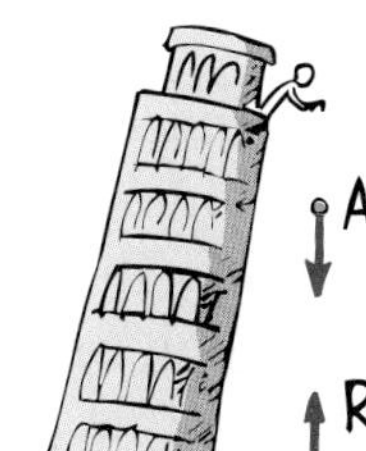

Action: earth pulls on ball

Reaction: ball pulls on earth

Figure 4.6
Action and reaction forces. Note that when action is "A exerts force on B," the reaction is simply "B exerts force on A."

Calculation Corner

Below we see two vectors on the sketch of the hand pushing the wall. The wall also pushes back on the hand. Note the others show only the action force. Draw appropriate vectors showing the reaction forces. Can you specify the action-reaction pairs in each case?

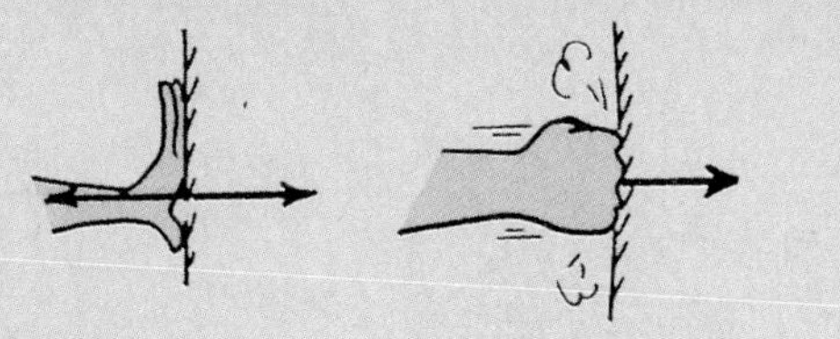

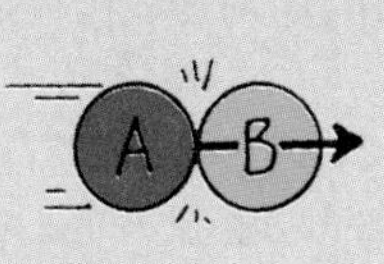

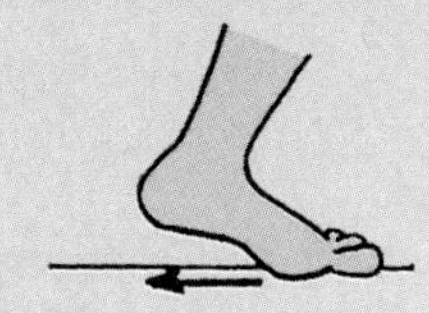

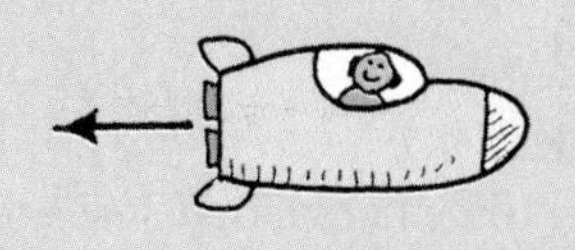

Concept Check ✓

Can you identify the action and reaction forces in the case of an object that is falling in a vacuum? (A vacuum is a region of space that is completely empty—no air.)

Check Your Answer To identify a pair of action-reaction forces in any situation, first identify the pair of interacting objects. In this case the Earth is interacting with the falling object through the force of gravity. So the Earth pulls the falling object downward (call it *action*). Then *reaction* is the falling object pulling the Earth upward. (Hmm. . . you say, the falling object pulls the Earth upward? Yes, this may be hard to imagine at first but it is true. You don't notice the Earth being pulled upward by a falling object because of the Earth's large mass. More on this in Section 4.4.)

4.4 Action and Reaction on Objects of Different Masses

When a cannon is fired, there is an interaction between the cannon and the cannonball. The sudden force that the cannon exerts on the cannonball is exactly equal and opposite to the force the cannonball exerts on the cannon. This is why the cannon recoils (kicks). But the effects of these equal forces are very different. This is because the forces act on different masses. Recall Newton's second law:

$$a = \frac{F}{m}$$

Let F represent both the action and reaction forces, ${\large m}$ the mass of the cannon, and m the mass of the cannonball. Different sized symbols are used to indicate the differences in relative masses and resulting accelerations. Then the acceleration of the cannonball and cannon are:

$$\text{Cannonball: } \frac{F}{m} = {\Large a}$$

$$\text{Cannon: } \frac{F}{\Large m} = a$$

Do you see why the change in velocity of the cannonball is so large compared to the change in velocity of the cannon? A given force exerted on a small mass produces a large acceleration, while the same force exerted on a large mass produces a small acceleration.

Figure 4.7
The force exerted against the recoiling cannon is just as great as the force that drives the cannonball along the barrel. Why, then, does the cannonball undergo more acceleration than the cannon?

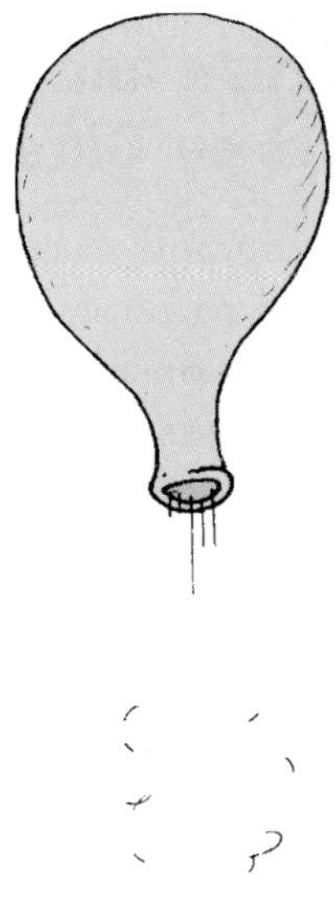

Figure 4.8
The balloon recoils from the escaping air and climbs upward.

We can extend the idea of a cannon recoiling from the ball it fires, to understanding rocket propulsion. Consider an inflated balloon recoiling when air is expelled. If the air is expelled downward, the balloon accelerates upward. A rocket accelerates the same way. It continually "recoils" from the ejected exhaust gas. Each molecule of exhaust gas is like a tiny cannonball shot from the rocket (Figure 4.9).

A common misconception is that a rocket is propelled by the impact of exhaust gases against the atmosphere. In fact, before the advent of rockets, it was commonly thought that sending a rocket to the moon was impossible. Why? Because there is no air above the Earth's atmosphere for the rocket to push against. But this is like saying a cannon wouldn't recoil unless the cannonball had air to push against. Not true! Both the rocket and recoiling cannon accelerate because of the reaction forces by the material they fire—not because of any pushes on the air. In fact, a rocket works better above the atmosphere where there is no air drag.

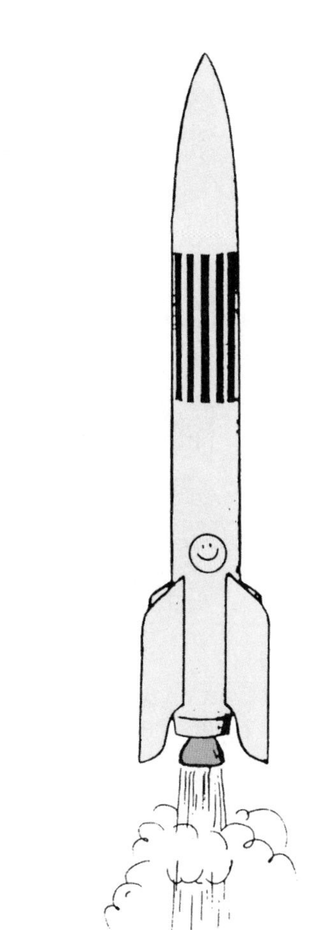

Figure 4.9
The rocket recoils and rises from the "molecular cannonballs" it fires.

Concept Check ✓

A high-speed bus and an innocent bug have a head-on collision. The force of the bus on the bug splatters the poor bug all over the windshield. Is the corresponding force of the bug on the bus greater, less, or the same? Is the resulting deceleration of the bus greater than, less than, or the same as that of the bug?

Check Your Answers The magnitudes of both forces are the same, for they constitute an action-reaction **force pair** that makes up the interaction between the bus and the bug. The accelerations, however, are very different because the masses involved are different! The bug undergoes an enormous and lethal deceleration, while the bus undergoes a very tiny deceleration—so tiny that the very slight slowing of the bus is unnoticed by its passengers. But if the bug were more massive, as massive as another bus, for example, the slowing down would be quite evident.

Figure 4.10
A acts on B, and B accelerates.

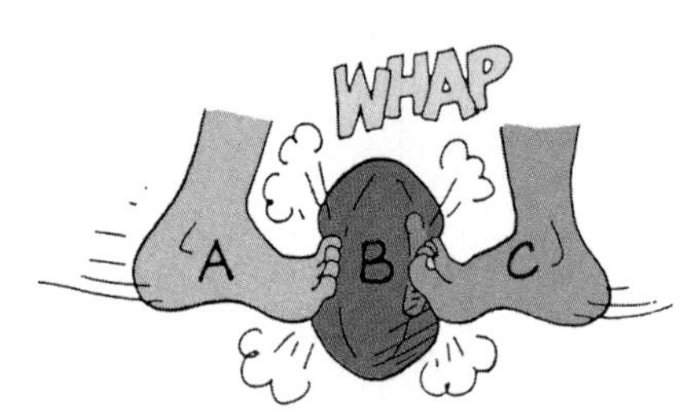

Figure 4.11
Both A and C act on B. They can cancel each other so B does not accelerate.

4.5 Action and Reaction Forces Act on Different Objects

Since action and reaction forces are equal and opposite, why don't they cancel to zero? They don't cancel out because they act on different bodies. Consider kicking a football (Figure 4.10). Call the force your foot exerts action. That's the only horizontal force on the football, so the football accelerates. Reaction is the football exerting a force on your foot, which tends to slow your foot down a bit. You can't cancel the force on the football with a force on your foot. Forces cancel only when they act on the *same* body. Now, what would happen if two players kicked the same football with opposite and equal forces at the same time, as shown in Figure 4.11? In this case, two interactions occur. Two different forces act on the football and these forces together cancel to zero. Is there a reaction force to each of these two forces that act on the football? The answer is yes, in accordance with Newton's third law.

If this is confusing, it may be well to point out that Newton himself had difficulties with the third law.

Concept Check ✓

Why does a flower pot sitting on a shelf never accelerate "spontaneously" in response to the trillions of inter-atomic forces acting within it?

Check Your Answer Every one of these inter-atomic forces is part of an action-reaction pair within the flower pot. These forces add up to zero, no matter how many of them there are. This is what makes Newton's *first* law apply to the pot. It has zero acceleration unless an *external* force acts on it.

4.6 The Classic Horse-Cart Problem—A Mind Stumper

A situation similar to the kicked football is shown in the comic strip "Horse Sense" (facing page). Here we think of the horse as believing its pull on the cart will be canceled by the opposite and equal pull by the cart on the horse, making acceleration impossible. This is the classic horse-cart problem that is a stumper for many students at the university level. By thinking carefully, you can understand it here.

The horse-cart problem can be looked at from different points of view. One is the farmer's point of view where his only concern is getting his cart (the cart system) to market. Then, there is the point of view of the horse (the horse system). Finally, there is the point of view of the horse and cart together (the horse-cart system).

First look at the farmer's point of view—the cart system. The net force on the cart, divided by the mass of the cart, will produce an acceleration. The farmer doesn't care about the reaction on the horse.

Now look at the horse's point of view—the horse system. It's true that the opposite reaction force by the cart on the horse restrains the horse. This force tends to hold the horse back. Without this force the horse could freely gallop to the market. So how does the horse move forward? By interacting with the ground. At the same time the horse pushes backward against the ground, the ground pushes forward on the horse. If the horse pushes the ground with a greater force than its pull on the cart, then there will be a net force on the horse. Acceleration occurs. When the cart is up to speed, the horse needs only to push against the ground with enough force to offset the friction between the cart's wheels and the ground.

Finally, look at the horse-cart system as a whole. From this viewpoint, the pull of the horse on the cart and the reaction of the cart on the horse are internal forces—forces that act and react within the system. They contribute nothing to the acceleration of the horse-cart system. The forces cancel and can be neglected. To move across the ground, there must be an interaction between the horse-cart system and the ground. This is similar to pushing a car while you're sitting in it, as discussed earlier. To get a car moving you must get outside and make the ground push you and the car. The horse-cart system is similar. It is the outside reaction by the ground that pushes the system.

Figure 4.12
All the pairs of forces that act on the horse and cart are shown: (1) the pull *P* of the horse and the cart on each other; (2) the push *F* of the horse and the ground on each other; and (3) the friction *f* between the cart wheels and the ground. Notice that there are two forces applied to the cart and to the horse. Can you see that the acceleration of the horse-cart system is due to the net force $F - f$?

Concept Check ✓

1. What is the net force that acts on the cart in Figure 4.12? On the horse? On the ground?
2. Once the horse gets the cart moving at the desired speed, must the horse continue to exert a force on the cart?

Check Your Answers

1. The net force on the cart is *P-f;* on the horse, *F-P;* on the ground, *F-f.*
2. Yes, but only enough to counteract wheel friction and air resistance. Interestingly enough, air resistance would be absent if there were a wind blowing in the same direction and just as fast as the horse and cart. If the wind blows fast enough to provide a force to counteract friction, the horse could wear rollerblades and simply coast with the cart all the way to the market.

4.7 Action Equals Reaction

When you tie a rope to a wall and pull on it, you produce a tension in the rope. Your pull on the rope and the pull by the supporting wall are equal and opposite. Otherwise there would be a net force on the rope and it would accelerate. The same is true if a friend holds one end of the rope and you have a tug-of-war. Rope tension when pulled at opposite ends is the same as the force provided by each end. Both pulls are the same in magnitude. This leads to a fascinating discovery for people who play tug-of-war. The team to win is not the team to exert the greatest force on the rope, but the greatest force *against the ground*! In this way a greater net force acts on the winning team.

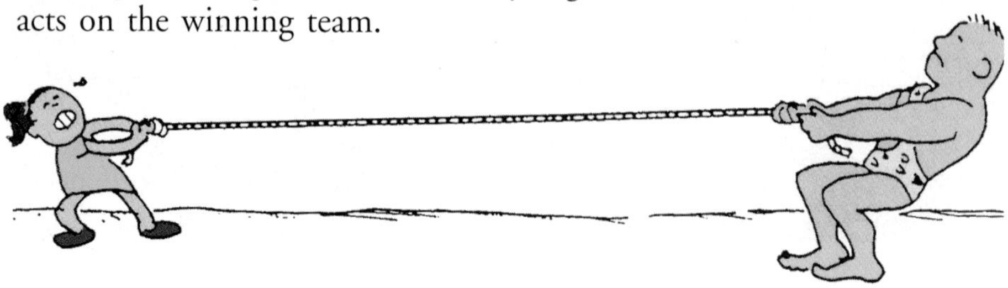

Figure 4.13
Arnold and Suzie pull on opposite ends of the rope. Can Arnold pull any harder on the rope than Suzie pulls on it? If Suzie lets go, could Arnold provide tension in the rope?

Hands-and-Feet-On Exploration

Perform a tug-of-war between boys and girls. Do it on a polished floor (that's somewhat slippery). Have the boys wear socks and the girls wear rubber-soled shoes. Who will surely win, and why?

Concept Check ✓

1. We said earlier that a car accelerates along a road because the road pushes it. Can we say that a team wins in a tug-of-war when the ground pushes harder on them than on the other team?
2. Does the scale read 100 N, 200 N, or zero?

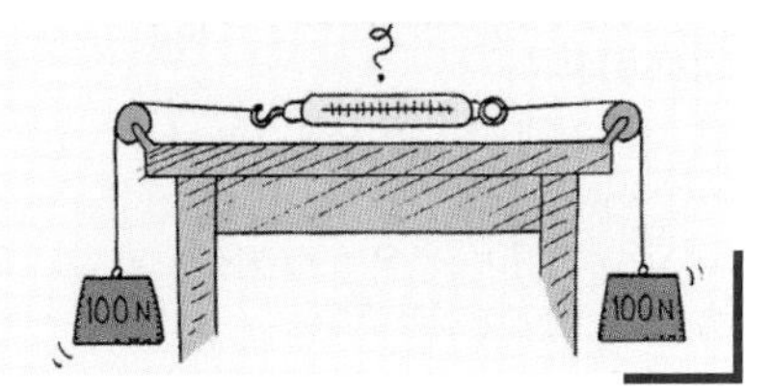

Check Your Answers

1. Yes!
2. Although the net force on the system is zero (as evidenced by no acceleration), the scale reading is 100 N, the tension in the string. Note that the string tension is 100 N in all the positions shown.

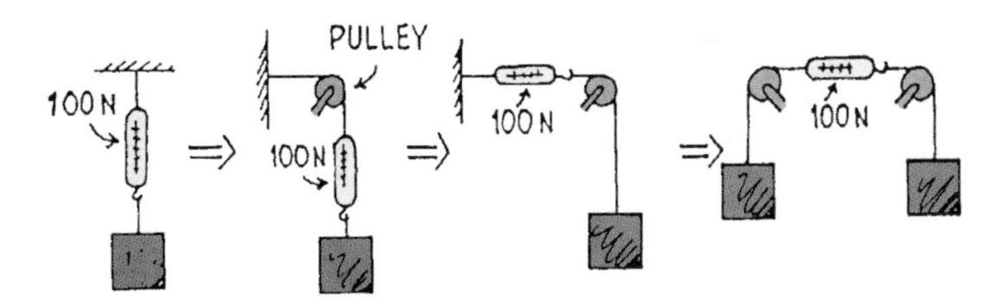

Newton's third law tells us how a helicopter gets its lifting force. The whirling blades are shaped to force air particles down (action), and the air forces the blades up (reaction). This upward reaction force is called *lift*. When lift equals the weight of the craft, the helicopter hovers in midair. When lift is greater, the helicopter rises.

The same is true for airplanes and birds. Birds fly by pushing air downward. The air simultaneously pushes the bird upward. Interestingly, when some birds deflect the air downward, the moving air meets air below and is swirled upward, mostly near the edges of the wings. Off to the side another bird can position itself to get added lift from this updraft. This bird, in turn, creates an updraft for a following bird and so on. This is the physics of why geese and ducks fly in a V formation!

Figure 4.14
The pair of vectors represents the force each wing exerts on the air. What forces act on the bird?

Figure 4.15
When a duck pushes air downward with its wings, air at the tips of its wings swirls upward, creating an updraft that is strongest off to its side. A trailing bird gets added lift by positioning itself in this updraft. This bird pushes air down and creates an updraft for the next bird, and so on. The result is a flock flying in a V formation.

Figure 4.16
He can hit the massive bag with considerable force. But with the same punch he can exert only a tiny force on the tissue paper in midair.

Have you ever heard the expression that someone "can't fight their way out of a paper bag?" There's some interesting physics beneath this statement. According to Newton, you can't hit a piece of paper any harder than the paper can hit you back. Hold a sheet of paper in midair and tell your friends that nobody can hit the paper with a force of 20 N (4.5 lb). You're correct even if the heavyweight boxing champion of the world hits the paper. The reason is that a 20-N interaction between the champ's fist and the sheet of paper in midair isn't possible—the paper is not capable of exerting a reaction force of 20 N. You cannot have an action force without its reaction force. Now, if you hold the paper against the wall, that is a different story. The wall will easily assist the paper in providing 20 N of reaction force, and more if needed!

You can't exert a force on anything unless it also exerts a force on you!

For every interaction between things, there is always a pair of oppositely-directed forces that are equal in strength. If you push hard on the world, for example, the world pushes hard on you. If you touch the world gently, the world touches you gently in return. The way you touch others is the way others touch you.

Figure 4.17
You cannot touch without being touched—Newton's third law.

4.8 Summary of Newton's Three Laws

Newton's first law, the law of inertia: An object at rest tends to remain at rest; an object in motion tends to remain in motion at constant speed along a straight-line path. This property of objects to resist change in motion is called *inertia.* Mass is a measure of inertia. Objects will undergo changes in motion only in the presence of a net force.

Newton's second law, the law of acceleration: When a net force acts on an object, the object will accelerate. The acceleration is directly proportional to the net force and inversely proportional to the mass. Symbolically, $a \sim F/m$. Acceleration is always in the direction of the net force. When objects fall in a vacuum, the net force is simply the weight, and the acceleration is g (the symbol g denotes that acceleration is due to gravity alone.) When objects fall in air, the net force is equal to the weight minus the force of air drag, and the acceleration is less than g. If and when the force of air drag equals the weight of a falling object, acceleration terminates, and the object falls at constant speed (called *terminal speed*).

Newton's third law, the law of action-reaction: Whenever one object exerts a force on a second object, the second object exerts an equal and opposite force on the first. Forces come in pairs, one action and the other reaction, both of which comprise the interaction between one object and the other. Action and reaction always act on different objects. Neither force exists without the other.

There has been a lot of new and exciting physics since the time of Isaac Newton. Nevertheless, and quite interestingly, it was primarily Newton's laws that got us to the moon.

Chapter Review

Key Terms and Matching Definitions

_____ force pair
_____ interaction
_____ Newton's third law

1. Mutual action between objects where each object exerts an equal and opposite force on the other.
2. The action and reaction pair of forces that occur in an interaction.
3. Whenever one object exerts a force on a second object, the second object exerts an equal and opposite force on the first. Or put another way, "To every action there is always an opposed equal reaction."

Review Questions

A Force Is Part of an Interaction

1. In the simplest sense, a force is a push or a pull. In a deeper sense, what is a force?
2. How many forces are required for an interaction?
3. When you push against a wall with your fingers, they bend because they experience a force. Identify this force.
4. Why do we say a speeding object doesn't have force?

Newton's Third Law—Action and Reaction

5. State Newton's third law of motion.
6. Consider hitting a baseball with a bat. If we call the force on the bat against the ball the *action* force, identify the *reaction* force.
7. If a bat hits a ball with 1000 N of force, how much force does the ball hit back on the bat?

A Simple Rule Helps Identify Action and Reaction

8. If the world pulls you downward, what is the reaction force?

Action and Reaction on Objects of Different Masses

9. If the forces that act on a cannonball and the recoiling cannon from which it is fired are equal in magnitude, why do the cannonball and cannon have very different accelerations?
10. Identify the force that propels a rocket.

Action and Reaction Forces Act on Different Objects

11. How can the net force on the ball be zero when you kick it?
12. Why does a push on the dashboard of a stalled car not accelerate the car?

The Classic Horse-Cart Problem—A Mind Stumper

13. Referring to Figure 4.12, how many forces are exerted on the cart? What is the horizontal net force on the cart?
14. How many forces are exerted on the horse? What is the net force on the horse?
15. How many forces are exerted on the horse-cart system? What is the net force on the horse-cart system?

Action Equals Reaction

16. Which is most important in winning in a tug-of-war; pulling harder on the rope, or pushing harder on the floor?
17. How does a helicopter get its lifting force?
18. A boxer can hit a heavy bag with great force. Why can't he hit a sheet of newspaper in midair with the same amount of force?
19. Can you physically touch another person without that person touching you with the same magnitude of force?

Summary of Newton's Three Laws

20. Fill in the blanks: Newton's first law is often called the law of ________; Newton's second law highlights the concept of ________; and Newton's third law is the law of ________ and ________.

Exploration

Hold your hand like a flat wing outside the window of a moving automobile. Then slightly tilt the front edge upward and notice the lifting effect. Can you see Newton's laws at work here?

Exercises

1. The photo shows Steve Hewitt and daughter Gretchen. Is Gretchen touching her dad, or is dad touching her? Explain.

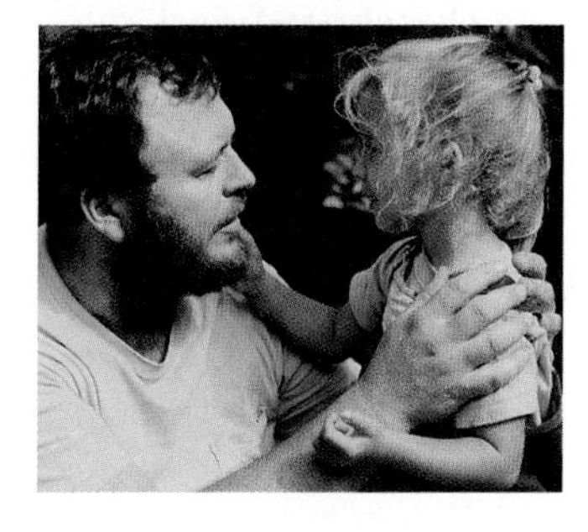

2. For each of the following interactions, identify action and reaction forces. (a) A hammer hits a nail. (b) Earth gravity pulls down on you. (c) A helicopter blade pushes air downward.

3. You hold an apple over your head. (a) Identify all the forces acting on the apple and their reaction forces. (b) When you drop the apple, identify all the forces acting on it as it falls and the corresponding reaction forces. Neglect air drag.

4. Identify the action-reaction pairs of forces for the following situations: (a) You step off a curb. (b) You pat your tutor on the back. (c) A wave hits a rocky shore.

5. Consider a tennis player hitting a ball. (a) Identify the action-reaction pairs when the ball is being hit, and (b) while the ball is in flight.

6. When you drop a rubber ball on the floor it bounces almost to its original height. What causes the ball to bounce?

7. Within a book on a desk there are billions of forces pushing and pulling on all the molecules. Why is it that these forces never by chance add up to a net force in one direction, causing the book to accelerate "spontaneously" across the desk?

8. You push a heavy car by hand. The car, in turn, pushes back with an opposite but equal force on you. Doesn't this mean the forces cancel one another, making acceleration impossible? Why or why not?

9. A farmer urges his horse to pull a wagon. The horse refuses, saying to try would be futile for it would flout Newton's third law. The horse concludes that she can't exert a greater force on the wagon than the wagon exerts on her, and therefore won't be able to accelerate the wagon. What is your explanation to convince the horse to pull?

10. Suppose two carts, one twice as massive as the other, fly apart when the compressed spring that joins them is released. How fast does the heavier cart roll compared with the lighter cart?

11. If you exert a horizontal force of 200 N to slide a crate across a factory floor at constant velocity, how much friction does the floor exert on the crate? Is the force of friction equal and oppositely directed to your 200-N push? If the force of friction isn't the reaction force to your push, what is?

12. If a massive truck and small sports car have a head-on collision, upon which vehicle is the impact force greater? Which vehicle experiences the greater acceleration? Explain your answers.

13. Ken and Joanne are astronauts floating some distance apart in space. They are joined by a safety cord whose ends are tied around their waists. If Ken starts pulling on the cord, will he pull Joanne toward him, or will he pull himself toward Joanne, or will both astronauts move? Explain.

14. Which team wins in a tug-of-war: the team that pulls harder on the rope, or the team that pushes harder against the ground? Explain.

15. In a tug of war between two physics types, each pulls on the rope with a force of 250 N. What is the tension in the rope? If both remain motionless, what horizontal force does each exert against the ground?

16. A stone is shown at rest on the ground. (a) The vector shows the weight of the stone. Complete the vector diagram showing another vector that results in zero net force on the stone. (b) What is the conventional name of the vector you have drawn?

17. Here a stone is suspended at rest by a string. (a) Draw force vectors for all the forces that act on the stone. (b) Should your vectors have a zero resultant? (c) Why, or why not?

18. Here the same stone is being accelerated vertically upward. (a) Draw force vectors to some suitable scale showing relative forces acting on the stone. (b) Which is the longer vector, and why?

19. Suppose the string in the preceding exercise breaks and the stone slows in its upward motion. Draw a force vector diagram of the stone when it reaches the top of its path.

20. What is the acceleration of the stone in Exercise 19 at the top of its path?

Problems

1. If you apply a net force of 5 N on a cart with a mass 5 kg, what is the acceleration?

2. If you increase the speed of a 2.0-kg air puck by 3.0 m/s in 4.0 s, what force do you exert on it?

3. A boxer punches a sheet of paper in midair, and brings it from rest up to a speed of 25 m/s in 0.05 s. If the mass of the paper is 0.003 kg, what force does the boxer exert on it?

4. If you stand next to a wall on a frictionless skateboard and push the wall with a force of 30 N, how hard does the wall push on you? If your mass is 60 kg, what's your acceleration?

Chapter 5: Momentum

How does the speed of the athlete affect the clearing of hurdles in his path? How can a karate expert break a stack of cement bricks with the blow of her bare hand? And why is her blow stronger if her hand bounces off of the bricks? But doesn't bouncing reduce the impact on a trapeze artist when he falls into a circus net? And why is an extended swing "follow through" important in golf, tennis, and boxing? A related question involves the game of pool. In playing pool, why does a cue ball stop short when it hits another one at rest head on? And why does that struck ball continue with the speed of the first? The answers to these questions are related. They all involve more than the concepts of inertia, force, etc., covered in previous chapters. Now we concern ourselves with a new concept—*momentum*.

5.1 Momentum Is Inertia in Motion

We know that a massive truck is harder to stop than a small car moving at the same speed. We say the truck has more momentum than the car. By **momentum** we mean "inertia in motion."

Figure 5.1
Why are the engines of a supertanker normally cut off 25 km from port?

We define momentum as:

$$\text{momentum} = \text{mass} \times \text{velocity}$$

or, in shorthand notation,

$$\text{momentum} = mv$$

When direction is not an important factor, we can say:

$$\text{momentum} = \text{mass} \times \text{speed},$$

which we still abbreviate mv.

From the definition we see that a moving object can have a large momentum if its mass is large, its speed is large, or if both its mass and speed are large. A huge ship moving at a low speed has a large momentum because its mass is large, and a fast-moving small bullet has a large momentum because of its high speed. A massive truck with no brakes rolling down a steep hill has a large momentum, whereas the same truck at rest has no momentum at all—because the v part of mv is zero.

Concept Check ✓

Can you think of a case where a car and a truck with twice the car's mass would have the same momentum?

Check Your Answer They'd have the same momentum if the car were traveling twice as fast as the truck. Then $(m \times 2v)$ for the car would equal $(2m \times v)$ for the truck. Get it? And if they were both at rest, they'd certainly have the same momentum—zero.

5.2 Impulse Changes Momentum

When the momentum of an object changes, either its mass changes, its velocity changes, or both change. If its mass doesn't change, as is most often the case, then the object's velocity changes. So the object accelerates. And what produces an acceleration? The answer is a *force.*

The greater the force acting on an object, the greater will be the change in velocity, and hence, the change in momentum.

But something else is important also: *Time*—how long the force acts. When you briefly push a stalled automobile, you'll change its momentum only a little. If you exert the same force over a longer period of time, a greater change in momentum results. A long sustained force produces more change in momentum than the same force applied briefly. So when changing the momentum of an object, both force and time are important.

Figure 5.2
The boulder, unfortunately, has more momentum than the runner.

The quantity "force × time interval" is called **impulse.** The greater the impulse exerted on something, the greater will be the change in momentum of that something. This is known as the **impulse-momentum relationship.** Mathematically, the exact relationship is:

Impulse = change in momentum

or

$$Ft = \textbf{change in } mv$$

which reads, "force multiplied by the time-during-which-it-acts equals change in momentum."*

The impulse-momentum relationship helps us analyze a variety of examples of changing momentum. Let's consider familiar examples of impulse for the cases of increasing momentum, and decreasing momentum.

Case 1: Increasing Momentum—Increase Force, Time, or Both

If you want to get maximum increase in momentum of something, you not only apply the greatest force you can, you also extend the time of application as much as possible. Hence the different results in pushing briefly on a stalled automobile and giving it a sustained push.

Long-range cannons have long barrels. The longer the barrel, the greater the velocity of the emerging cannonball or shell. Why? The force of exploding gunpowder in a long barrel acts on the cannonball for a longer time. This increased impulse produces a greater momentum. Of course the force that acts on the cannonball is not steady—it is strong at first and weaker as the gases expand. Most often the forces involved in impulses vary over time. The force that acts on the golf ball in Figure 5.3, for example, increases rapidly as the ball is distorted and then gets less as the ball comes up to speed and returns to its original shape. When we speak of forces in this chapter, we mean the *average* force.

Figure 5.3
Impact force against a golf ball.

* Note that this comes from a rearrangement of Newton's second law, $a = F/m$. If we equate the cause of acceleration ($a = F/m$) to what acceleration is ($a =$ change in v/t), $F/m =$ (change in $v)/t$ simple rearrangement gives $Ft =$ change in (mv).

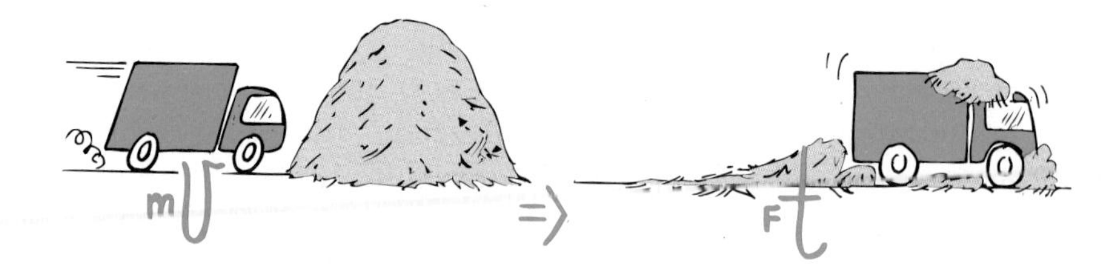

Figure 5.4
A large change in momentum in a long time requires a small force.

Case 2: Decreasing Momentum in a Long Time Means Less Force

Imagine you are in a fast-moving truck and the brakes fail, and you have a choice of stopping by either hitting a concrete wall or a haystack. You don't have to be a rocket scientist to make a correct decision. Of course you'd hit the haystack, because it would be a "softer" hit. Let's examine *why* hitting something softly is entirely different from hitting something hard. Follow carefully the logic of the following paragraph.

Whether you hit the wall or the haystack, your momentum decreases the same—for you come to rest. According to the impulse-momentum relationship, the same change in momentum means the same impulse. The same impulse doesn't mean the same force, or the same time—it means the same *product* of force and time. By hitting the haystack instead of the wall, you extend the contact time—*the time during which your momentum is brought to zero.* A longer time reduces the force and decreases the resulting deceleration. For example, if the time of contact is extended 10 times, the force of contact is reduced 10 times. Whenever you wish the force of contact to be small, extend the time of contact.

Figure 5.5
A large change in momentum in a short time requires a large force.

A wrestler thrown to the floor tries to extend his time of arrival on the floor by relaxing his muscles and spreading the crash into a series of impacts as foot, knee, hip, ribs, and shoulder fold onto the floor in turn. The increased time of contact reduces the force. Of course, falling on a mat is better than falling on a solid floor, because the force experienced is further decreased by increased time of contact.

When you jump from an elevated position to a floor below, you bend your knees when you make contact. This extends the time during which your momentum is being reduced. Suppose that you extend your time of landing by 10 times that of a stiff-legged, abrupt landing. Then you'll reduce the forces experienced by your bones by 10 times.

Figure 5.6
A large change in momentum over a long time requires a safely small average force.

Bungee jumping puts the impulse-momentum relationship to a thrilling test (Figure 5.6). The momentum gained during fall must be decreased to zero by an impulse of the same amount. The long stretching time of the cord results in a small average force to bring the jumper to a safe halt before hitting the ground. Bungee cords typically stretch to about twice their original length during the fall.

Ballet dancers prefer a wooden floor with "give" to a hard cement floor with little or no "give." The wooden floor allows a longer time of contact whenever the dancer lands, reducing the force of contact and possible injury. A safety net used by acrobats is an obvious example of small contact force over a long time.

If you're about to catch a fast baseball with your bare hand, you extend your hand forward so you'll have plenty of room to let your hand move backward after you make contact with the ball. You extend the time of contact and thereby reduce the force. Similarly, a boxer rides or rolls with the punch to reduce the force the boxer encounters (Figure 5.7).

Figure 5.7
In both cases the boxer's jaw provides an impulse that reduces the momentum of the punch.
(a) The boxer is moving away when the glove hits, thereby extending the time of contact. This means the force is less than if the boxer had not moved.
(b) The boxer is moving into the glove, thereby lessening the time of contact. This means that the force is greater than if the boxer had not moved.

Case 3: Decreasing Momentum Over a Short Time Means More Force

When boxing, if you move into a punch instead of away, you're in trouble. This is similar to catching a high-speed baseball while your hand moves toward the ball instead of away upon contact. For short contact times, the forces are large.

The idea of short contact time explains how a karate expert can sever a stack of bricks with the blow of her bare hand (Figure 5.8). She brings her arm and hand swiftly against the bricks with a lot of momentum. This momentum is quickly reduced when she hits the bricks. The impulse is the force of her hand against the bricks multiplied by the time her hand makes contact with the bricks. By swift execution she makes the time of contact very short, which makes the force huge. If her hand bounces upon contact, the force is even greater.

Figure 5.8
Cassy imparts a large impulse to the bricks in a short time and produces a considerable force.

Concept Check ✓

1. If the boxer in Figure 5.7 makes the time of contact 3 times as long by riding with the punch, by how much is the force reduced?
2. If the boxer instead moves into the punch and shortens the contact time by half, by how much is the force increased?
3. A boxer being hit with a punch tries to extend time for best results, whereas a karate expert delivers a force in a short time for best results. Isn't there a contradiction here?

Check Your Answers

1. The force will be three times less than if he didn't pull back.
2. The force will be two times greater than if he held his head still. Forces of this kind account for many knockouts.
3. There is no contradiction because the best results for each are quite different. The best result for the boxer is reduced force, accomplished by maximizing time. The best result for the karate expert is increased force delivered in minimum time.

5.3 Momentum Change Is Greater When Bouncing Occurs

You know that if a flower pot falls from a shelf onto your head, you may be in trouble. If it bounces from your head, you may be in even more trouble. Why? Because impulses are greater when an object bounces. The impulse required to bring an object to a stop and then to "throw it back again" is greater than the impulse required merely to bring it to a stop. Suppose, for example, that you catch the falling pot with your hands. You provide an impulse to reduce its momentum to zero. If you throw the pot upward again, you have to provide additional impulse. It takes more impulse to catch it and throw it back up than merely to catch it. This increased amount of impulse is supplied by your head if the pot bounces from it.

An interesting application of the greater impulse that occurs for bouncing was employed with great success in California during the gold rush days. The water wheels used in gold-mining operations were not very effective. A man named Lester A. Pelton saw that the problem had to do with their flat paddles. He designed curved-shape paddles that would cause the incident water to make a U-turn—to "bounce." In this way the impulse exerted on the water wheels was increased. Pelton patented his idea and made more money from his invention, the Pelton wheel, than most gold miners made from gold.

Figure 5.9
Teacher Howie Brand shows that the block topples when the swinging dart bounces from it. When he changes the head of the dart so it doesn't bounce when it hits the block, no tipping occurs.

Concept Check ✓

1. Refer to Figure 5.8. How does the force that Cassy exerts on the bricks compare with the force exerted on her hand?
2. How will the impulse differ if her hand bounces back when striking the bricks?

Check Your Answers

1. In accord with Newton's third law, the forces will be equal. Only the resilience of the human hand and the training she has undergone to toughen her hand allow her to do this without breaking bones.
2. The impulse will be greater if her hand bounces from the bricks. If the time of contact is not increased, a greater force is then exerted on the bricks (and her hand!).

5.4 When No External Force Acts, Momentum Doesn't Change—It Is Conserved

Newton's second law tells us that to accelerate an object, you apply a force. We say much the same thing, but in different language. To change the momentum of an object, exert an impulse on it.

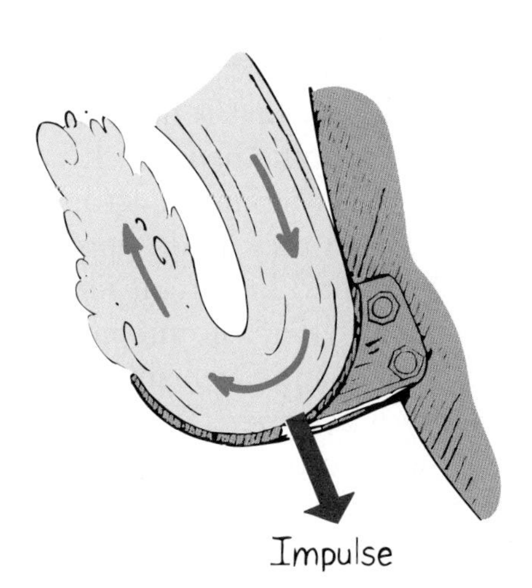

Figure 5.10
The Pelton wheel. The curved blades cause water to bounce and make a U-turn which produces a greater impulse to turn the wheel.

In either case, the force or impulse must be exerted on the object by something outside the object. Internal forces won't work. For example, the molecular forces within a baseball have no effect upon the momentum of the baseball, just as your push against the dashboard of a car you're sitting in does not change the momentum of the car. Molecular forces within the baseball and a push on the dashboard are internal forces. They come in balanced pairs that cancel within the object. To change the momentum of the baseball or car, an outside force is required. Without an outside force, no change in momentum is possible.

When a cannon fires a cannonball, the explosive forces are internal forces. That means the total momentum of the cannon-cannonball system doesn't change (Figure 5.11). Can you see that the impulses must be the same, only in opposite directions? Think of Newton's third law of action and reaction. Then we see that the force on the cannonball is equal and opposite to the force on the cannon. Since these forces act for the same time, equal but oppositely directed impulses are produced. That means equal and oppositely directed momenta (the plural form of *momentum*). The recoiling cannon has just as much momentum as the speeding cannonball.* Together as one system, there is no net momentum change. In summary, when only internal forces act on a system, no change in momentum occurs. No momentum is gained and no momentum is lost.

Figure 5.11
The momentum before firing is zero. After firing, the net momentum is still zero, because the momentum of the cannon is equal and opposite to the momentum of the cannonball.

* Here we are neglecting the momentum of the ejected gases from the exploding gunpowder. Firing a gun with blanks at close range is a definite no-no because of the momentum of ejected gases. People have been killed by firing of close-range blanks. Although no slug emerges from the gun, exhaust gases do—enough to be lethal.

Two important ideas are to be learned from the cannon-and-cannonball example. The first is that momentum, like velocity, is a vector quantity that is described by both magnitude and direction; we measure both "how much" and "which direction." Therefore, like velocity, when momenta act in the same direction, they are simply added. When they act in opposite directions, they are subtracted.

The second important idea is *conservation.* For the cannon-cannonball system, no momentum was gained; none was lost. When a physical quantity remains unchanged during a process, we say that quantity is *conserved.* We say momentum is **conserved.**

The concept that momentum is conserved when no external force acts is so important it is considered a law of mechanics. It is called the **law of conservation of momentum**:

> **In the absence of an external force, the momentum of a system remains unchanged.**

A system can undergo changes in which all forces are internal. This happens, for example, when tennis rackets hit tennis balls, cannons fire cannonballs, cars collide, or stars explode. In such cases, the net momentum of the system before and after the event remains the same.

Concept Check ✓

A high-speed bus and an innocent bug have a head-on collision. The sudden change of momentum for the bug spatters it all over the windshield. Is the change in momentum of the bus greater, less, or the same as the change in momentum of the unfortunate bug?

Check Your Answer

The momentum of both bug and bus change by the same amount because both the amount of force and the time, and therefore the amount of impulse, is the same on each. Momentum is conserved. Speed is another story. Because of the huge mass of the bus, its reduction of speed is very tiny—too small for the passengers to notice.

Hands-On Exploration: Skateboard Throw

Stand at rest on a skateboard and throw a massive object to the front or the rear. Note that you recoil in the opposite direction. This is understandable if you understand momentum. The net momentum before the throw was zero. The net momentum just after is also zero—because your recoil momentum is equal and opposite to the momentum of the tossed object. You'll see that momentum is conserved. Now repeat, but don't let go when you "throw" the object. Do you still recoil when you go through the motions of throwing, but don't really release the object? Explain.

5.5 Momentum Is Conserved in Collisions

Momentum is conserved in collisions because the forces that act are internal forces—acting and reacting within the system itself. There is only a redistribution or sharing of whatever momentum exists before the collision.

In any collision, we can say

Net momentum before collision = net momentum after collision.

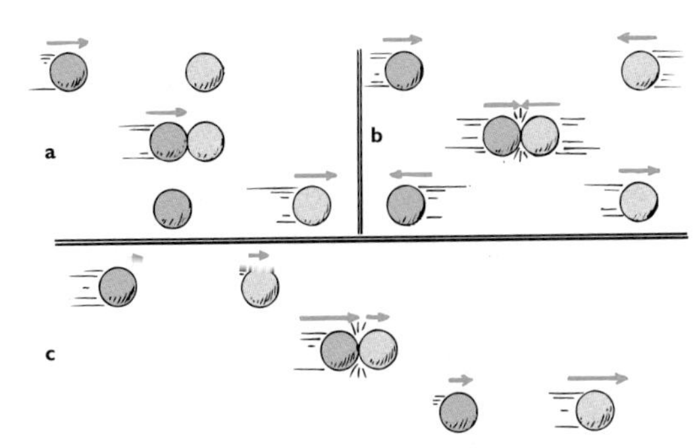

Figure 5.12
Elastic collisions of equally massive balls. (a) A green ball strikes a yellow ball at rest. (b) A head-on collision. (c) A collision of balls moving in the same direction. In each case, momentum is transferred from one ball to the other.

Elastic Collisions

When a moving billiard ball hits another ball at rest head on, the first ball comes to rest and the second ball moves away with a velocity equal to the initial velocity of the first ball. We see that momentum is transferred from the first ball to the second ball. When objects collide without being permanently deformed and without generating heat, we say the collision is an **elastic collision**. Colliding objects bounce perfectly in perfectly elastic collisions (Figure 5.12).

Inelastic Collisions

Momentum is conserved even when the colliding objects become distorted and generate heat during the collision. Whenever colliding objects become tangled or coupled together, we have an **inelastic collision.** In a perfectly inelastic collision, both objects stick together. Consider, for example, the case of a freight car moving along a track and colliding with another freight car at rest (Figure 5.13). If the freight cars are of equal mass and are coupled by the collision, can we predict the velocity of the coupled cars after impact?

Figure 5.13
Inelastic collision. The momentum of the freight car on the left is shared with the freight car on the right after collision.

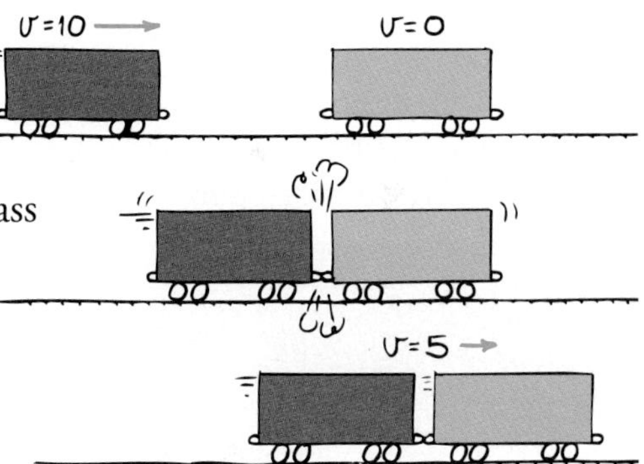

Suppose the single car is moving at 10 meters per second, and we consider the mass of each car to be m. Then, from the conservation of momentum,

$$(\text{net } mv)_{\text{before}} = (\text{net } mv)_{\text{after}}$$

$$(m \times 10)_{\text{before}} = (2m \times v)_{\text{after}}$$

By simple algebra, $v = 5$ m/s. This makes sense, for twice as much mass moves after the collision, with half as much as the velocity as before collision. Then both sides of the equation are equal.

Note the inelastic collisions in Figure 5.14. Can you see the net momentum after collisions is the same?

Figure 5.14
Inelastic collisions. The net momentum of the trucks before and after collision is the same.

Concept Check ✓

Refer to the gliders on the air track in Figure 5.15 to answer these questions.

1. Suppose both gliders have the same mass. They move toward each other at the same speed and experience an elastic collision. Describe their motion after the collision.
2. Suppose both gliders have the same mass and move toward each other at equal speed. This time they stick together when they collide. Describe their motion after the collision.
3. Suppose one of the gliders is at rest and is loaded so that it has twice the mass of the moving glider. Again, the gliders stick together when they collide. Describe their motion after the collision.

Check Your Answers

1. Since the collision is elastic, the gliders reverse directions upon colliding and move away from each other at the same speed as before.
2. Before the collision, the gliders had equal and opposite momenta, since their equal masses were moving in opposite directions at the same speed. The net momentum of both was zero. Since momentum is conserved, their net momentum after they stick must also be zero. They slam to a dead halt.
3. Before collision, the net momentum equals the momentum of the unloaded, moving glider. After the collision, the net momentum is the same as before, but now the gliders are stuck together and moving as a single unit. The mass of the stuck-together gliders is 3 times that of the unloaded glider. Thus, the velocity is 1/3 of the unloaded glider's velocity before collision and in the same direction as before, since the direction as well as the amount momentum is conserved.

Figure 5.15
An air track. Blasts of air from tiny holes provide a friction-free surface for the carts to glide upon.

If your teacher has an air track similar to that shown in Figure 5.15, you may be treated to fascinating demonstrations of momentum conservation. The air that spurts from the tiny holes in the track let you see an almost friction-free performance. In the everyday world, friction usually shows itself. Ideally, the net momentum of a couple of freight cars that collide is the same before and just after collision. But as the combined cars move along the track, friction provides an impulse to decrease momentum.

Another thing: Perfectly elastic collisions are not common in the everyday world. Usually some heat is generated in collisions. Drop a ball, and after it bounces from the floor, both the ball and the floor are a bit warmer. Even a dropped "superball" will not bounce to its initial height. At the microscopic level, however, perfectly elastic collisions are commonplace. For example, gas molecules bounce off one another without generating heat; they don't even touch in the classic sense of the word. (As later chapters show, the notion of touching at the atomic level is different from touching at the everyday level.)

Conservation of momentum and, as the next chapter will discuss, conservation of energy are the two most powerful tools of mechanics. These laws let us understand the details of interactions among subatomic particles or entire galaxies.

Calculation Corner

For a numerical example of momentum conservation, consider a pair of carts, A and B, on the air track shown in Figure 5.15. Suppose that cart B is at rest and that A slides against it and they stick together. Let cart A have a mass of 5 kg and move 1 m/s toward cart B, which has a mass of 1 kg. What will be the velocity of both carts when they link together?

$$\text{Net momentum before collision} = \text{net momentum after collision.}$$

$$[m_A v_A + m_B v_B]_{\text{before}} = [(m_A + m_B)v]_{\text{after}}$$

$$(5\text{ kg})(1\text{ m/s}) + (1\text{ kg})(0\text{ m/s}) = (5\text{ kg} + 1\text{ kg})v$$

$$5\text{ kg m/s} = (6\text{ kg})v$$

$$v = 5/6\text{ m/s}$$

Here we see that the small cart has no momentum before collision because its velocity is zero. After collision, the combined mass of both carts moves at velocity *v*, which by simple algebra is seen to be 5/6 m/s. This velocity is in the same direction as that of cart A.

Now suppose cart B is not at rest, but moves toward the left at a velocity of 4 m/s. The carts are moving toward each other ready for a head-on collision. Let cart A's direction be +, and cart B – (because it moves in a direction negative relative to cart A). Then we see that:

$$\text{Net momentum before collision} = \text{net momentum after collision.}$$

$$[m_A v_A + m_B v_B]_{\text{before}} = [(m_A + m_B)v]_{\text{after}}$$

$$(5\text{ kg})(1\text{ m/s}) + (1\text{ kg})(-4\text{ m/s}) = (5\text{ kg} + 1\text{ kg})v$$

$$(5\text{ kg m/s}) + (-4\text{ kg m/s}) = (6\text{ kg})v$$

$$1\text{ kg m/s} = 6\text{ kg } v$$

$$v = 1/6\text{ m/s}$$

Note that the negative momentum of cart B before collision has more effect in slowing cart A after collision. If cart B were moving twice as fast, then

$$\text{Net momentum before collision} = \text{net momentum after collision.}$$

$$[m_A v_A + m_B v_B]_{\text{before}} = [(m_A + m_B)v]_{\text{after}}$$

$$(5\text{ kg})(1\text{ m/s}) + (1\text{ kg})(-8\text{ m/s}) = (5\text{ kg} + 1\text{ kg})v$$

$$(5\text{ kg m/s}) + (-8\text{ kg m/s}) = (6\text{ kg})v$$

$$-3\text{ kg m/s} = 6\text{ kg } v$$

$$v = -1/2\text{ m/s}$$

Here we see the final velocity is −1/2 m/s. What is the significance of the minus sign? It means that the final velocity of the two-cart system is *opposite* to the initial velocity of cart A. After collision the two-cart system moves toward the left.

Here we have discussed carts on an air track. Our example could well have been football players or even swimming fish. We leave as a chapter-end problem finding the initial velocity of a small fish halting the motion of a larger fish. Try the same type of calculations!

Chapter Review

Key Terms and Matching Definitions

_____ elastic collision
_____ impulse
_____ impulse-momentum relationship
_____ inelastic collision
_____ law of conservation of momentum
_____ momentum

1. The product of the mass of an object and its velocity.
2. The product of the force acting on an object and the time during which it acts. In an interaction, impulses are equal and opposite.
3. Impulse is equal to the change in the momentum of the object that the impulse acts on. In symbol notation, $Ft = \Delta mv$
4. When no external net force acts on an object or a system of objects, no change of momentum takes place. Hence, the momentum before an event involving only internal forces is equal to the momentum after the event: $mv_{\text{(before event)}} = mv_{\text{(after event)}}$
5. A collision in which colliding objects rebound without lasting deformation or the generation of heat.
6. A collision in which the colliding objects become distorted, generate heat, and possibly join together.

Review Questions

Momentum Is Inertia in Motion

1. Which has a greater momentum, a heavy truck at rest or a moving automobile?

2. How can a supertanker have a huge momentum when it moves relatively slowly?

Impulse Changes Momentum

3. Why is it incorrect to say that impulse equals momentum?

4. To impart the greatest momentum to an object, what should you do in addition to exerting the largest force possible?

5. For the same force, which cannon imparts the greater speed to a cannonball—a long cannon or a short one? Explain.

6. If you're in a car with faulty brakes and you have to hit something to stop, the momentum will change to zero whether you hit a brick wall or a haystack. So why is hitting a haystack a safer bet?

7. Why is it less damaging if you fall on a mat than if you fall on a solid floor?

8. Why is it a good idea to extend your hand forward when catching a fast-moving baseball with your bare hand?

9. In boxing, why is it advantageous to roll with the punch?

10. In karate why is a short time of the applied force advantageous?

Momentum Change Is Greater When Bouncing Occurs

11. Which is the greater change in momentum, a stop of something dead in its tracks, or a stop and then a reversal of direction?

12. Which requires the greater impulse, stopping something dead in its tracks, or stopping it and then reversing its direction?

When No External Force Acts, Momentum Doesn't Change—It Is Conserved

13. When can the momentum of two moving objects be cancelled?

14. What does it mean to say that momentum (or any quantity) is *conserved*?

15. When a cannonball is fired, its momentum does change! Is momentum conserved for the cannonball?

16. When a cannonball is fired, the cannon recoils. Is momentum conserved for the cannon?

17. When a cannonball is fired, is momentum conserved for the cannon-cannonball system as a whole? (Why is your answer different than in the previous two questions?)

Momentum Is Conserved in Collisions

18. Distinguish between an *elastic* collision and an *inelastic* collision. For which type of collision is momentum conserved?

19. Railroad car A rolls at a certain speed and makes a perfectly elastic collision with car B of the same mass. After the collision, car A is observed to be at rest. How does the speed of car B compare with the initial speed of car A?

20. If the equally massive cars of the previous question stick together after colliding inelastically, how does their speed after the collision compare with the initial speed of car A?

Exercises

1. When rollerblading, why is a fall less harmful on a wooden floor than on a concrete floor that has less give? Explain in terms of impulse and momentum.

2. In terms of impulse and momentum, why do air bags in cars reduce the chances of injury in accidents?

3. In terms of impulse and momentum, why are nylon ropes (which stretch considerably under tension) favored by mountain climbers?

4. If you throw an egg against a wall, the egg will break. But if you throw it at the same speed into a sagging sheet, it may not break. Why?

5. A lunar vehicle is tested on Earth at a speed of 10 km/h. When it travels as fast on the moon, is its momentum more, less, or the same?

6. Which has the greater momentum when they move at the same speed—an automobile or a skateboard? Which requires the greatest stopping force?

7. In answering the preceding exercise, perhaps you stated that the automobile requires more stopping force. Make an argument that the skateboard could require more stopping force, depending on how quickly you want to stop it.

8. Why is a punch more forceful with a bare fist than with a boxing glove?

9. Why do 6-ounce boxing gloves hit harder than 16-ounce gloves?

10. Which undergoes the greatest change in momentum: (1) a baseball that is caught, (2) a baseball that is thrown, or (3) a baseball that is caught and then thrown back, if the baseball has the same speed just before being caught and just after being thrown?

11. In the preceding question, in which case is the greatest impulse required?

12. An apple gains momentum as it falls from a tree. Does this violate the conservation of momentum? Defend your answer.

13. In the preceding exercise, would momentum be conserved in the larger apple-Earth system? In this system, can it be said that the momentum of the falling apple is equal and opposite to the momentum of the Earth as it "races" up to meet the apple? What does this say about the upward speed of the Earth?

14. If only an external force can change the velocity of a body, how can the internal force of the brakes bring a car to rest?

15. You are at the front of a floating canoe near a dock. You jump, expecting to land easily on the dock. Instead you land in the water. Explain.

16. A fully dressed person is at rest in the middle of a pond on perfectly frictionless ice and must get to shore. How can this be accomplished?

17. Two football players have a head-on collision and both stop short in their paths. If one player is twice as heavy as the other, how does his speed compare to the smaller player?

18. In the previous chapter, rocket propulsion was explained in terms of Newton's third law. That is, the force that propels a rocket is from the exhaust gases pushing against the rocket—the reaction to the force the rocket exerts on the exhaust gases. Explain rocket propulsion in terms of momentum conservation.

19. When you are traveling in your car at highway speed, the momentum of a bug is suddenly changed as it splatters onto your windshield. Compared to the change in momentum of the bug, by how much does the momentum of your car change?

20. If an 18-wheeler tractor-trailer and a sports car have a head-on collision, which vehicle will experience the greater force of impact? The greater impulse? The greater change in momentum? The greater acceleration?

Problems

1. What is the impulse needed to stop a 10-kg bowling ball moving at 6 m/s?

2. A car with a mass of 1000 kg moves at 20 m/s. What braking force is needed to bring the car to a halt in 10 s?

3. A car crashes into a wall at 25 m/s and is brought to rest in 0.1 s. Calculate the average force exerted on a 75-kg test dummy by the seat belt.

4. Lillian (mass 40.0 kg), standing on slippery ice, catches her leaping dog (mass 15 kg) moving horizontally at 3.0 m/s. What is the speed of Lillian and her dog after the catch?

5. A 2-kg ball of putty moving to the right has a head-on inelastic collision with a 1-kg putty ball moving to the left. If the combined blob doesn't move just after the collision, what can you conclude about the relative speeds of the balls before they collided?

6. A railroad diesel engine weighs four times as much as a freight car. If the diesel engine coasts at 5 km/h into a freight car that is initially at rest, how fast do the two coast after they couple together?

7. A 5-kg fish swimming 1 m/s swallows an absent minded 1-kg fish swimming toward it at a velocity that brings both fish to a halt immediately after lunch. What is the velocity v of the smaller fish before lunch?

8. Can you run fast enough to have the same momentum as an automobile rolling at 1 mi/h? Make up reasonable figures to justify your answer.

Chapter 6: Work and Energy

The author wields a sledge-hammer blow to fellow physics teacher Pablo Robinson, bravely sandwiched between beds of nails. Why is Pablo unharmed by this dramatic classroom demonstration? At an amusement park, why is the first summit on a roller coaster always the highest one? And why are following summits lower in height? In ancient times, people had great difficulty lifting huge loads, yet today a child can lift much heavier loads with a simple pulley system. How does a pulley so greatly increase output force? The answers to these questions involve the concept of *energy*—what this chapter is about.

Energy is the most central concept in physical science. Energy comes from the sun in the form of sunlight. Energy is in the food you eat and it sustains your life. Our study of energy begins with a related concept: *work*.

6.1 Work—Force × Distance

It takes energy to push something and make it move. How much energy depends on the force exerted and how long it is exerted. If "how long" means time, then we're talking about impulse, as in the previous chapter. Recall that impulse equals a change in momentum. In this chapter, by "how long," we mean distance. The quantity "force × distance" is equal to the change in energy. We call this quantity **work.**

Work = force × distance

We do work when we lift a load against Earth's gravity. The heavier the load or the higher we lift it, the more work we do. The amount of work done on an object depends on (1) how much force is applied and (2) how far the force causes the object to move.*

Figure 6.1
When a load is lifted two stories high, twice the work is done because the *distance* is twice as much.

When a weight lifter raises a heavy barbell, he does work on the barbell. He gives energy to the barbell. Interestingly, when a weight lifter simply holds a barbell overhead, he does no work on it. He may get tired holding the barbell still, but if the barbell is not moved by the force he exerts, he does no work *on the barbell.* Work may be done on his muscles as they stretch and contract, which is force × distance on a biological scale. But this work is not done *on the barbell. Lifting* the barbell is different than *holding* the barbell.

The unit of work combines the unit of force (N) with the unit of distance (m), the newton-meter (N·m). We call a newton-meter the *joule* (J) (rhymes with *cool*). One joule of work is done when a force of 1 newton is exerted over a distance of 1 meter, as in lifting an apple over your head. For larger values we speak of kilojoules (kJ), thousands of joules, or megajoules (MJ), millions of joules. The weight lifter in Figure 6.3 does work in kilojoules. The work done to vertically raise a heavily loaded truck can be in megajoules.

Figure 6.3
Work is done in lifting the barbell. Lifting it twice as high requires twice as much work.

Figure 6.2
When two loads are lifted to the same height, twice as much work is done because the *force* needed to lift them is twice as much.

* Force and distance must be in the same direction. When force is not along the direction of motion, then work equals the *component* of force in the direction of motion × distance moved.

Figure 6.4
He may expend energy when he pushes on the wall, but if it doesn't move, no work is done on the wall.

Concept Check ✓

1. How much work is needed to lift an object that weighs 500 N to a height of 4 m?
2. How much work is needed to lift it twice as high?
3. How much work is needed to lift a 1000 N to a height of 8 m?

Check Your Answers

1. $W = F \times d = 500 \text{ N} \times 4 \text{ m} = 2000 \text{ J}$.
2. Twice the height requires twice the work. That is, $W = F \times d = 500 \text{ N} \times 8 \text{ m} = 4000 \text{ J}$.
3. Lifting twice the load twice as high requires four times the work. That is, $F \times d = 1000 \text{ N} \times 8 \text{ m} = 8000 \text{ J}$.

6.2 Power—How Quickly Work Gets Done

Lifting a load quickly is more difficult than lifting the same load slowly. If equal loads are lifted to the same height, the forces and distances are equal, so the *work* is the same. What's different is the *power.* **Power** is the rate at which energy is changed from one form to another. Also, power is the rate at which work is done. It equals the amount of work done divided by the time interval during which the work occurs.

$$\textbf{power} = \frac{\textbf{work done}}{\textbf{time interval}}$$

A high-power auto engine does work rapidly. An engine that delivers twice the power of another, however, does not necessarily go twice as fast. Twice the power means the engine can do twice the work in the same amount of time—or it can do the same amount of work in half the time. A powerful engine can speed up a car more quickly.

The unit of power is the joule per second, called the **watt.** This is in honor of James Watt, the eighteenth-century developer of the steam engine. One watt (W) of power is used when one joule of work is done in one second. One kilowatt (kW) equals 1000 watts. One megawatt (MW) equals one million watts.

Figure 6.5
The space shuttle can develop 33,000 MW of power when fuel is burned at the enormous rate of 3400 kg/s. This is like emptying an average-size swimming pool in 20 seconds!

Which of these does a speeding baseball not **possess**? Force, momentum, energy. (*Hint:* The answer begins with an *F*)

Concept Check ✓

1. You do work when you do push-ups. If you do the same number of push-ups in half the time, how does your power output compare?
2. How many watts of power are needed when a force of 1 N moves a book 2 m in a time of 1 s?

Check Your Answers

1. Your power output is twice as much.
2. The power expended is 2 watts: $P = \frac{W}{t} = \frac{F \times d}{t} = \frac{1\text{N} \times 2\text{m}}{1\text{ s}} = 2\text{ W}$.

6.3 Mechanical Energy

Work is done in lifting the heavy ram of a pile driver. When raised, the ram then has the ability to do work on a piling beneath it when it falls. When work is done by an archer in drawing a bow, the bent bow has the ability to do work on the arrow. When work is done to wind a spring mechanism, the spring then has the ability to do work on various gears to run a clock, ring a bell, or sound an alarm.

In each case, the ability to do work has been acquired by an object. This ability to do work is **energy.** Like work, energy is measured in joules.

Energy appears in many forms, such as heat, light, sound, electricity, and radioactivity. It even takes the form of mass, as celebrated in Einstein's famous $E = mc^2$ equation. In this chapter we focus on potential energy and kinetic energy. **Potential energy** is energy that arises because of an object's position. **Kinetic energy** is energy of motion possessed by moving objects. Potential and kinetic energy are both considered to be kinds of mechanical energy. So mechanical energy may be in the form of either potential energy, kinetic energy, or both.

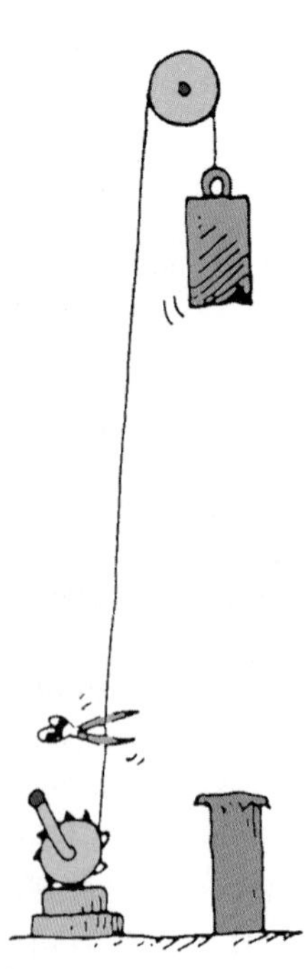

Figure 6.6
Work is required to lift the massive ram of the pile driver.

Concept Check ✓

1. She pushes the block five times farther up the incline than the man lifts it to the same height. How much more force does the man exert when he lifts the ice?
2. Who does more work on the ice?
3. If both jobs are done in the same time, who expends more power?

Check Your Answers

1. The man exerts 5 times as much force as the woman exerts.
2. Although he exerts more *force,* both do the same amount of work on the ice.
3. They both do the same amount of work in the same time, so both expend the same power.

6.4 Potential Energy Is Stored Energy

An object can store energy because of its position. This energy is stored and held in readiness. Therefore, it is called *potential energy* (PE). In the stored state, energy has the potential to do work. For example, when an archer draws an arrow with a bow, energy is stored in the bow. When released, energy is transferred to the arrow.

There are various kinds of potential energy. The potential energy that is easiest to visualize is stored in the object when work is done on the objects to elevate them against Earth's gravity. The potential energy due to elevated position is called *gravitational potential energy.* The elevated ram of a pile driver and of water in an elevated reservoir both have gravitational potential energy.

The amount of gravitational potential energy possessed by an elevated object is equal to the work done against gravity in lifting it—the force required to move it upward multiplied by the vertical distance moved ($W = F \times d$). Once upward motion begins, the upward force to keep an object moving at constant speed equals its weight. So the work done in lifting an object is its weight × height. We say:

Gravitational potential energy = weight × height

An object's weight is its mass m multiplied by the acceleration of gravity g. We write weight as mg. So the work done in lifting mg through a height h is equal to its gain in gravitational potential energy (PE):

$$\mathbf{PE} = mgh$$

Note that the height h is the distance above some base level, such as the ground or the floor of a building. The potential energy is relative to that level and depends only on weight and height h. You can see in Figure 6.8 that the potential energy of the boulder at the top of the structure depends on height only, and not on the path taken to get it there.

Figure 6.7
The potential energy of Tenny's drawn bow equals the work (average force × distance) she did in drawing the arrow into position.

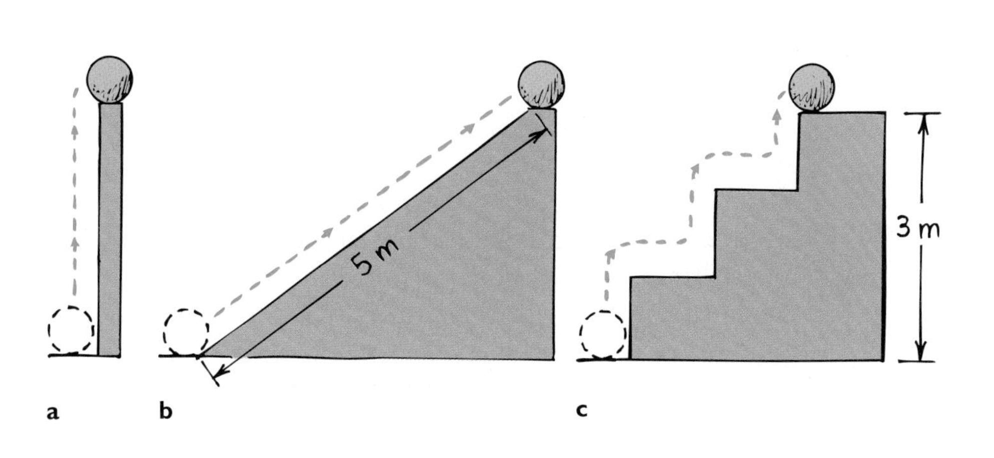

Figure 6.8
The PE of the 10-N ball is the same (30 J) in all three cases. That's because the work done in elevating it 3 m is the same whether it is (a) lifted with 10 N of force, (b) pushed with 6 N of force up the 5-m incline, or (c) lifted with 10 N up each 1-m stair. No work is done in moving it horizontally (neglecting friction).

Figure 6.9
The man raises a block of ice by lifting it vertically. The girl pushes an identical block of ice up the ramp. When both blocks are raised to the same height, both have the same potential energy.

Concept Check ✓

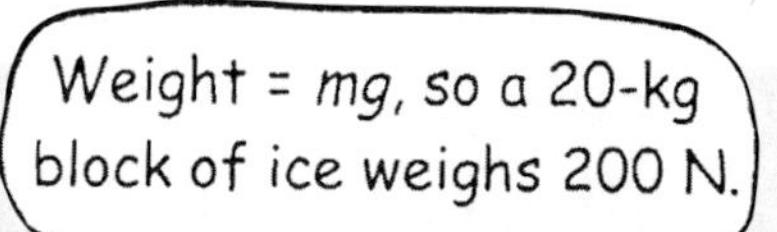

1. How much work is done in lifting the 200-N block of ice shown in Figure 6.9 a vertical distance of 2.5 m?
2. How much work is done in pushing the same block of ice up the 5-m long ramp? The force needed is only 100 N (which is why inclines are used).
3. What is the increase in the block's potential energy in each case?

Check Your Answers

1. 500 J. (We get this either by *Fd* or *mgh*.)
2. 500 J. (She pushes with half the force over twice the distance.)
3. Either way increases the block's potential energy by 500 J. The ramp simply makes this work easier to perform.

Figure 6.10
The downhill "fall" of the roller coaster results in its roaring speed in the dip, and this kinetic energy sends it up the steep track to the next summit.

6.5 Kinetic Energy Is Energy of Motion

When you push on an object, you can make it move. Then a moving object becomes capable of doing work. It has energy of motion, or **kinetic energy** (KE). The kinetic energy of an object depends on its mass and speed. Specifically, kinetic energy is equal to one half the product of the mass and the square of the speed:

$$\textbf{Kinetic energy} = \textbf{1/2 mass} \times \textbf{speed}^2$$

$$\textbf{KE} = 1/2\, mv^2$$

Since kinetic energy depends on mass, heavy objects have more kinetic energy than light ones moving at the same speed. For example, a car moving along the road has a certain amount of kinetic energy. A twice-as-massive car moving at the same speed has twice as much kinetic energy.

Kinetic energy also depends on speed. In fact, kinetic energy depends on speed more than it depends on mass. Why? Look at the equation. Kinetic energy depends on speed multiplied by speed, or speed squared. So if a car moving along the road has a certain amount of kinetic energy, a twice-as-fast car (with the same mass) has 2^2 or *four* times as much kinetic energy! The same car moving with three times the speed has 3^2 or nine times as much kinetic energy. So we see that small changes in speed produce large changes in kinetic energy.

Concept Check ✓

1. A car travels at 30 km/h and has kinetic energy of 1 MJ. If it travels twice as fast, 60 km/h, how much kinetic energy will it have?
2. If it travels three times as fast, at 90 km/h, what will be its kinetic energy?
3. If it travels four times as fast, at 120 km/h, what will be its kinetic energy?

Check Your Answers

1. Twice as fast means (2^2) four times the kinetic energy, or 4 MJ.
2. Three times as fast means (3^2) nine times the kinetic energy, or 9 MJ.
3. Four times as fast means (4^2) sixteen times the kinetic energy, or 16 MJ.

6.6 Work-Energy Theorem

To increase the kinetic energy of an object, work must be done on it. Or, if an object is moving, work is required to bring it to rest. In either case, the change in kinetic energy is equal to the work done. This important relationship is called the **work-energy theorem.** We abbreviate "change in" with the delta symbol, Δ, and say

$$\textbf{Work} = \Delta\textbf{KE}$$

Work equals change in kinetic energy. The work in this equation is the *net* work—that is, the work based on the net force.

The work-energy theorem emphasizes the role of *change.* If there is no change in an object's energy, then we know no work was done on it. This theorem applies to changes in potential energy also. Recall our previous example of the weight lifter raising the barbell. When work was being done on the barbell, its potential energy was being changed. But when it was held stationary, no further work was being done *on the barbell*—as evidenced by no further change in its energy.

Similarly, push against a box on a floor. If it doesn't slide, then you are not doing work on the box. Put the box on a very slippery floor and push again. If it slides, then you're doing work on it. When the amount of work done to overcome friction is small, the amount of work done on the box is practically matched by its gain in kinetic energy.

The work-energy theorem applies to decreasing speed as well. The more kinetic energy something has, the more work is required to stop it. Twice as much kinetic energy means twice as much work. When we apply the brakes to slow a car, we do work on it. This work is the friction force supplied by the brakes, multiplied by the distance over which the friction force acts.

Figure 6.11
When the car goes twice as fast, it has four times the kinetic energy (and will need four times the stopping distance when braking).

Interestingly, the friction supplied by the brakes is the same whether the car moves slowly or quickly. Friction doesn't depend on speed. The variable is the *distance* of braking. This means that a car moving at twice the speed as another takes four times ($2^2 = 4$) as much work to stop. Therefore it takes four times as much distance to stop. Accident investigators are well aware that an automobile going 100 kilometers per hour has four times the kinetic energy as it would have at 50 kilometers per hour. So a car going 100 kilometers per hour will skid four times as far when its brakes are applied as it would going 50 kilometers per hour. Kinetic energy depends on speed *squared.*

Concept Check ✓

1. When the brakes of a car are locked, the car skids to a stop. How much farther will the car skid if it's moving 3 times as fast?
2. Can an object have energy?
3. Can an object have work?

Check Your Answers

1. Nine times farther. The car has nine times as much energy when it travels three times as fast: $1/2\ m(3v)^2 = 1/2\ m9v^2 = 9(1/2\ mv^2)$. The friction force will ordinarily be the same in either case. Therefore, to do nine times the work requires nine times as much sliding distance.
2. Yes, but only in a relative sense. For example, an elevated object may possess PE relative to the ground, but none relative to a point at the same elevation. Similarly, the kinetic energy of an object is relative to a frame of reference, usually taken to be the Earth's surface.
3. No, unlike energy, work is not something an object *has*. Work is something an object *does* to some other object. An object can *do* work only if it has energy.

Kinetic energy often appears hidden in different forms of energy, such as heat, sound, light, and electricity. Random molecular motion is sensed as heat: When fast-moving molecules bump into others in the surface of your skin, they transfer kinetic energy to your molecules similar to the way colliding billiard balls transfer energy. Sound consists of molecules vibrating in rhythmic patterns. When a vibrating object pushes nearby molecules, those molecules are pushed into action. In turn, they disturb neighboring molecules that disturb others, preserving the rhythm of the vibration throughout the region. When the moving molecules hit your ears, you hear sound. Even light energy comes from the motion of electrons within atoms. Electrons in motion make electric currents. We see that kinetic energy has many applications in our lives, and is far-reaching.

6.7 Conservation of Energy

By studying how energy changes from one form to another, scientists have developed one of the greatest generalizations in physical science—the law of **conservation of energy**:

> **Energy cannot be created or destroyed; it may be transformed from one form into another or transferred from one object to another, but the total amount of energy never changes.**

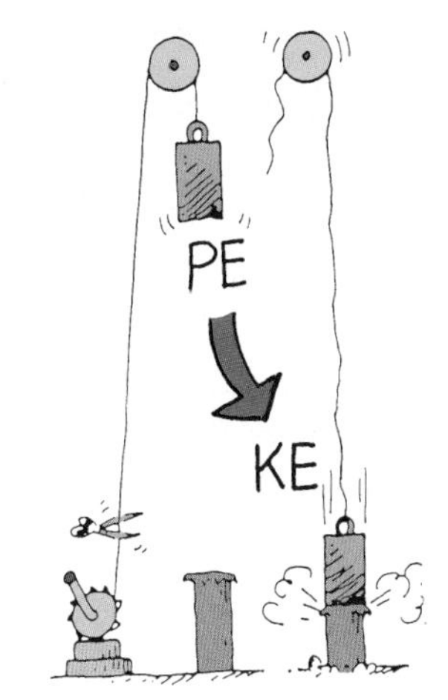

Figure 6.12
The potential energy of the elevated ram is converted to kinetic energy when released.

For any system, whether as simple as a swinging pendulum or as complex as an exploding star, energy remains the same. Energy may change form or may be transferred from one place to another, but the total energy score stays the same.

This energy score takes into account the fact that atoms that make up matter are themselves concentrated bundles of energy. When the nuclei of atoms rearrange themselves, enormous amounts of energy can be released. We will learn in Chapter 18 that enormous gravitational forces in the deep, hot interior of the sun push hydrogen nuclei together to form helium. This welding together of atomic cores is called *thermonuclear fusion.* This process releases radiant energy, some of which reaches the Earth as sunshine.

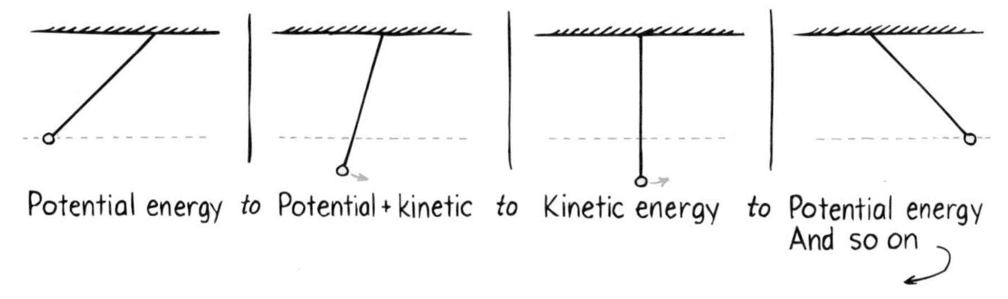

Figure 6.13
Energy transitions in a pendulum. PE is relative to the lowest point of the pendulum when it is vertical.

Part of the energy of sunshine falls on plants, and part of this in turn later becomes coal. So the energy in coal began in the sun. Animal life is sustained by plant life, and eventually becomes oil. So the energy in oil began in the sun. Part of the sun's energy goes into evaporating water from the ocean, and part of this returns to the Earth as rain that may be trapped behind a dam. The potential energy of the dammed water may be used to power a generating plant below, where it will be transformed to electric energy. So the energy generated at dams began in the sun. And this

Figure 6.14
The pendulum bob will swing to its original height whether or not the peg is present.

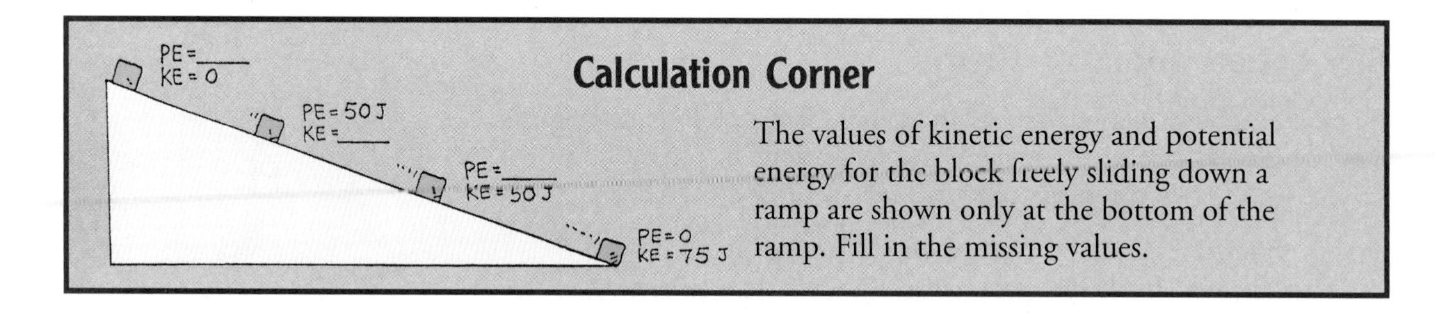

Calculation Corner

The values of kinetic energy and potential energy for the block freely sliding down a ramp are shown only at the bottom of the ramp. Fill in the missing values.

energy travels through wires to homes, where it is used for lighting, heating, cooking, and operating electric gadgets. How wonderful that energy changes from one form to another!

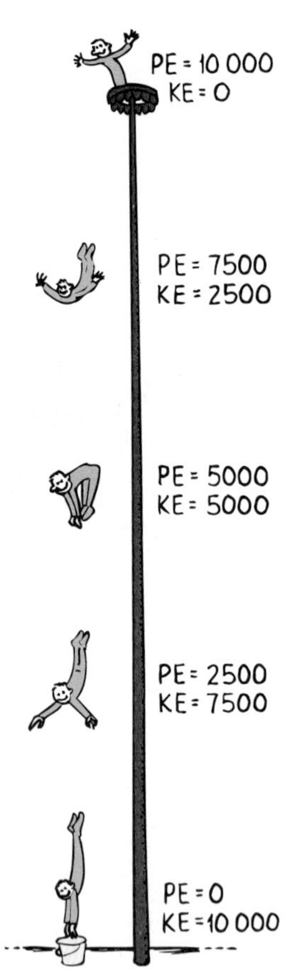

Figure 6.15
A circus diver at the top of a pole has a potential energy of 10,000 J. As he dives, his potential energy converts to kinetic energy. Note that at successive positions one-fourth, one-half, three-fourths, and all the way down, the total energy is constant. (Adapted from K. F. Kuhn and J. S. Faughn, *Physics in Your World,* Philadelphia: Saunders, 1980.)

Concept Check ✓

1. Does an automobile consume more fuel when its air conditioner is turned on? When its lights are on? When its radio is on while the auto is sitting in the parking lot?
2. Rows of wind-powered generators are used in various windy locations to generate electric power. Does the power generated affect the speed of the wind? Would locations behind the "windmills" be windier if they weren't there?

Check Your Answers

1. The answer to all three questions is yes, for the energy consumed ultimately comes from the fuel. Even energy from the battery must be given back to the battery by the alternator, which is turned by the engine, which runs from the energy of the fuel. All energy that is used has to come from some source. There's no free lunch!
2. Windmills generate power by taking kinetic energy from the wind, so the wind is slowed by interaction with the windmill blades. So yes, it would be windier behind the windmills if they weren't there.

6.8 Machines—Devices to Multiply Forces

A **machine** is a device for multiplying forces or simply changing the direction of forces. All machines employ the conservation of energy. Consider one of the simplest machines, the *lever.* A lever is shown in Figure 6.16. When you do work by pushing one end of the lever down, work is done at the other end. The work done on the output side of the lever raises a load. If heat from friction forces is small enough to neglect, the work input is equal to the work output:

Work input = work output.

Since work equals force multiplied by distance, we can say that input force × input distance = output force × output distance.

$$(\text{Force} \times \text{distance})_{\text{input}} = (\text{force} \times \text{distance})_{\text{output}}$$

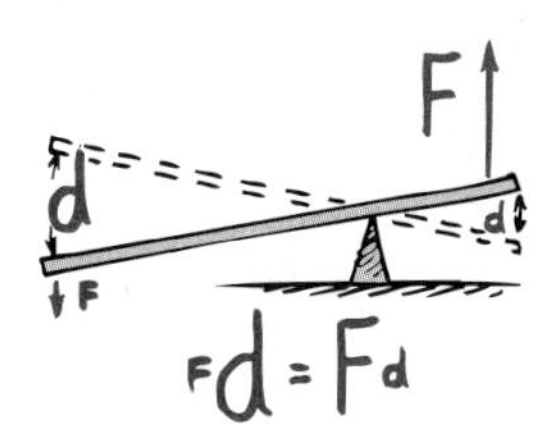

Figure 6.16
A simple lever.

The support or point of support on which a lever rotates is called a *fulcrum.* When the fulcrum is close to the load, a large output force is produced by a small input force. This is because the input force is exerted through a large distance and the load is moved over a short distance. In this way, a lever can multiply force. But no lever or machine has been found that can multiply work or energy. Our understanding of energy suggests that none ever will be found. We are so confident of this that we say energy is never created or destroyed.

Figure 6.17
Work done on one end equals the work done on a load at the other end.

Figure 6.18
Force is multiplied. Note that a small input force × large distance = large output force × small distance.

The principle of the lever was understood by the Greek scientist Archimedes in the 3rd century BC. He said that he could move the whole world if he had a long enough lever and a place to put the fulcrum. Some good science has been around for a long time!

Concept Check ✓

If a lever is arranged so that input distance is twice output distance, can we predict that energy output will be doubled?

Check Your Answer No, no, a thousand times no! We can predict output *force* will be doubled, but never *energy*. Work and energy stay the same, which means force × distance stays the same. Shorter distance means greater force, and vice versa. Be careful to distinguish between the concepts of *force* and *energy!*

Another simple machine is a pulley. Can you see that it is a lever "in disguise"? When used as in Figure 6.19, it changes only the direction of the force. But when used as in Figure 6.20, the output force is doubled. Force is increased and distance moved is decreased.

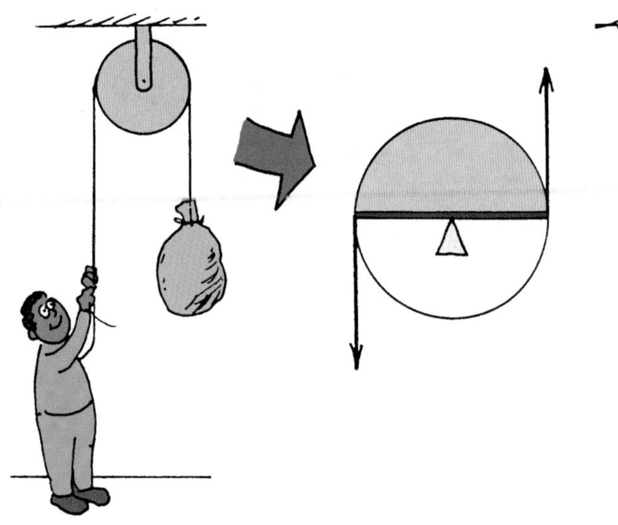

Figure 6.19
This pulley acts like a lever. It changes only the direction of the input force.

Figure 6.20
In this arrangement, a load can be lifted with half the input force.

Figure 6.21
Input force × input distance = output force × output distance. Note the load is supported by 7 strands of rope. Each strand supports 1/7 the load. The tension in the rope pulled by the man is likewise 1/7 the load.

Forces can be nicely multiplied with a system of pulleys. Such pulley arrangements are common wherever heavy loads are lifted, like in automobile service centers or machine shops. An ideal pulley system is shown in Figure 6.21. The man pulls 7 meters of rope with a force of 50 newtons and lifts 500 newtons through a vertical distance of 0.7 meter. The work the man does when pulling the rope is numerically equal to the increased potential energy of the 500-N block.

Any machine that multiplies force does so at the expense of distance. Likewise, any machine that multiplies distance does so at the expense of force. No machine or device can put out more energy than is put into it. No machine can create energy; it can only transfer it or transform it from one form to another.

Hands-On Exploration

Rub your hands briskly together. The friction between them multiplied by the distance of rubbing produces work that becomes heat. Note how quickly your palms are warmed.

6.9 Efficiency—A Measure of Work Done for Energy Spent

Given the same energy input, some machines can do more work than others. The machines that can do more work are said to be more efficient.

Efficiency can be expressed by the ratio:

$$\textbf{Efficiency} = \frac{\textbf{work done}}{\textbf{energy used}}$$

Even a lever converts a small fraction of input energy into heat when it rotates about its fulcrum. We may do 100 joules of work but get out 98 joules. The lever is then 98 percent efficient, and we waste 2 joules of work input on heat. In a pulley system, a larger fraction of input energy goes into heat. If we do 100 joules of work, the forces of friction acting through the distances through which the pulleys turn and rub about their axles may dissipate 60 joules of energy as heat. So the work output is only 40 joules, and the pulley system has an efficiency of 40 percent. The lower the efficiency of a machine, the greater the amount of energy wasted as heat.

Concept Check ✓

Consider an imaginary miracle car that has a 100 percent efficient engine and burns fuel that has an energy content of 40 megajoules per liter. If the air drag plus frictional forces on the car traveling at highway speed is 500 N, what is the maximum distance the car can go on one liter of fuel?

Check Your Answer From the definition work = force × distance, simple rearrangement gives distance = work/force. If all 40 million J of energy in 1 liter is used to do the work of overcoming the air drag and frictional forces, the distance covered is:

$$\text{distance} = \frac{\text{work}}{\text{force}} = \frac{40{,}000{,}000\ \text{J}}{500\ \text{N}} = 80{,}000\ \text{m} = 80\ \text{km}$$

The important point here is that even with a perfect engine, there is an upper limit of fuel economy dictated by the conservation of energy.

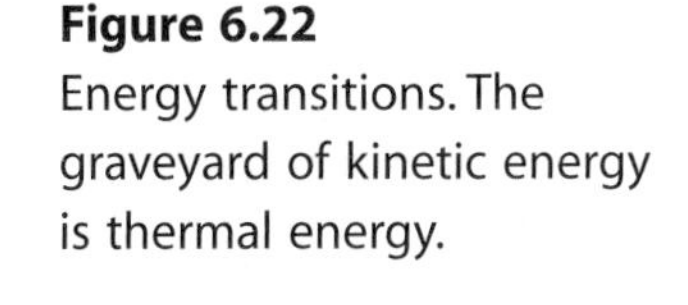

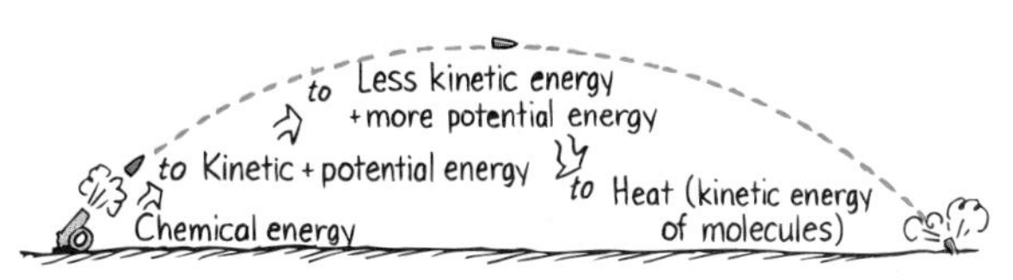

Figure 6.22
Energy transitions. The graveyard of kinetic energy is thermal energy.

An automobile engine is a machine that transforms chemical energy stored in gasoline into mechanical energy. But only a fraction of the energy in the gas is used by the car to move forward. Some of the fuel energy in the gas goes out in the hot exhaust gases and is wasted. Also, nearly half of the energy stored in the gas is wasted in the friction of the moving engine parts. In addition to these inefficiencies, some of the gas doesn't even burn completely. So the energy in the unburned gasoline also goes unused.

6.10 Sources of Energy

Except for nuclear power, the source of practically all our energy is the sun. Even the energy we obtain from petroleum, coal, natural gas, and wood comes from the sun. That's because these fuels are created by photosynthesis, the process by which plants trap solar energy and store it as plant tissue.

Figure 6.23
Except for nuclear power, all the Earth's energy comes from the sun.

Sunlight is also directly transformed into electricity by photovoltaic cells, like those found in solar-powered calculators. We use the energy in sunlight to generate electricity indirectly as well. Sunlight evaporates water, which later falls as rain; rainwater flows into rivers and turns water wheels, or it flows into modern generator turbines as it returns to the sea.

Wind, caused by unequal warming of the Earth's surface, is another form of solar power. The energy of wind can be used to turn generator turbines within specially equipped windmills. Because wind power is not reliable, concentrated energy sources such as fossil and nuclear fuels are the choice contenders for large-scale power production.

The most concentrated form of usable energy is stored in uranium and plutonium, which are nuclear fuels. Public fear about nuclear processes prevents the growth of nuclear power. But it is interesting to note that the Earth's interior is kept hot because of nuclear power, which has been with us since time zero.

A by-product of nuclear power in the Earth's interior is geothermal energy. Geothermal energy is held in underground reservoirs of hot water. Geothermal energy is predominantly limited to areas of volcanic activity, such as Iceland, New Zealand, Japan, and Hawaii. In these places, heated water near the Earth's surface is tapped to provide steam for running turbo-generators.

In locations where heat from volcanic activity is near the ground surface and ground water is absent, another method holds promise for producing electricity. That's dry-rock geothermal power (Figure 6.24). With this method, water is put into the cavities in deep, dry, hot rock. When the water turns to steam it is piped to a turbine at the surface. After turning the turbine, it is returned to the cavity for re-use. In this way electricity is produced cheaply and cleanly.

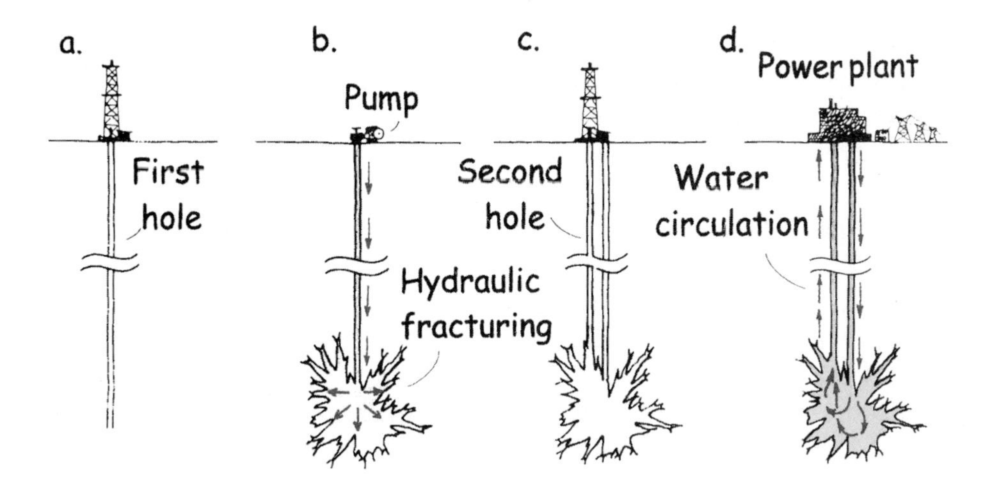

Figure 6.24
Dry-rock geothermal power.
(a) A hole is sunk several kilometers into dry granite.
(b) Water is pumped into the hole at high pressure and fractures surrounding rock to form a cavity with increased surface area.
(c) A second hole is sunk to intercept the cavity.
(d) Water is circulated down one hole and through the cavity where it is superheated before rising through the second hole. After driving a turbine, it is recirculated into the hot cavity again, making a closed cycle.

Except for geothermal power, methods for obtaining energy have serious environmental consequences. Although nuclear power doesn't pollute the atmosphere, it is controversial because of the nuclear wastes generated. The combustion of fossil fuels, on the other hand, leads to increased atmospheric concentrations of carbon dioxide, sulfur dioxide, and other pollutants. Methods of using solar energy are limited in that they require proper atmospheric conditions.

As the world population increases, so does our need for energy. Common sense dictates that as new sources are being developed, we should continue to optimize present sources and use what we consume efficiently and wisely.

6.11 Energy Is Needed for Life

Your body is a machine—a wonderful machine. It is made up of smaller machines, the living cells. Like any machine, a living cell needs a source of energy. Most living organisms on this planet feed on molecules called hydrocarbons. Hydrocarbon compounds release energy when they react with oxygen. Like gasoline burned in an auto engine, there is more potential energy in the hydrocarbons than there is in the molecules that result after the hydrocarbons are "burned" for food. The energy difference is what sustains life.

Can you get something for nothing? Not when it comes to energy.

Combustion, or burning, takes place when certain molecules combine with oxygen in air and release carbon dioxide molecules plus large amounts of energy. Combustion occurs when food is processed, or metabolized, in your body's digestive system. The combustion in your body is similar to the combustion that occurs in an engine when it burns gasoline. The main difference is the rate of combustion. During metabolism, the reaction rate is much slower and energy is released as needed by the body. Once the reaction starts, it is self-sustaining, similar to burning fossil fuels.

The reverse process is more difficult. Only green plants and certain one-celled organisms can make carbon dioxide combine with water to produce hydrocarbons like sugar. This process is *photosynthesis.* It requires an energy input, which normally comes from sunlight. Sugar is the simplest food. All other foods, such as carbohydrates, proteins, and fats, are made up of combinations of carbon, hydrogen, oxygen, and other elements. So green plants use the energy of sunlight to make food that gives us and all other organisms energy.

When you look at a tall tree, you might wonder where the atoms came from that compose it. Interestingly, the atoms that make up a tree come mostly from the air. Carbon dioxide molecules in the air make their way into tiny pores in the leaves. By photosynthesis, the carbon is taken from the molecules and incorporated into the tree. What happens to the oxygen? It's expelled by the tree. And that's nice for us, for we need oxygen to live!

Chapter Review

Key Terms and Matching Definitions

_____ conservation of energy
_____ efficiency
_____ energy
_____ kinetic energy
_____ machine
_____ potential energy
_____ power
_____ watt
_____ work
_____ work-energy theorem

1. The product of the force and the distance through which the force moves: $W = Fd$
2. The time rate of doing work: Power = work/time.
3. The unit of power, the joule per second.
4. The property of a system that enables it to do work.
5. The stored energy that a body possesses because of its position.
6. Energy of motion, described by the relationship: Kinetic energy = $1/2\ mv^2$
7. The work done on an object is equal to the energy gained by the object. Work = ΔE
8. Energy cannot be created or destroyed; it may be transformed from one form into another, but the total amount of energy never changes. In an ideal machine, where no energy is transformed into heat, $\text{work}_{\text{input}} = \text{work}_{\text{output}}$ and $(Fd)_{\text{input}} = (Fd)_{\text{output}}$.
9. A device such as a lever or pulley that increases (or decreases) a force or simply changes the direction of a force.
10. The percent of the work put into a machine that is converted into useful work output.

Review Questions

Work—Force × Distance

1. A force sets an object in motion. When the force is multiplied by the time of its application, we call the quantity *impulse,* which changes the *momentum* of that object. What do we call the quantity *force* × *distance*?

2. Cite an example where a force is exerted on an object without doing work on the object.

3. Which requires more work—lifting a 50-kg sack a vertical distance of 2 m or lifting a 25-kg sack a vertical distance of 4 m?

Power—How Quickly Work Gets Done

4. If both sacks in the preceding question are lifted their respective distances in the same time, how does the power required for each compare? How about for the case where the lighter sack is moved its distance in half the time?

5. What are the two main forms of mechanical energy?

Mechanical Energy

6. Exactly what is it that a body having energy is capable of doing?

Potential Energy Is Stored Energy

7. A car is lifted a certain distance in a service station and therefore has potential energy relative to the floor. If it were lifted twice as high, how much potential energy would it have?

8. Two cars are lifted to the same elevation in a service station. If one car is twice as massive as the other, how do their potential energies compare?

9. How many joules of potential energy does a 1-N book gain when it is elevated 4 m? When it is elevated 8 m?

Kinetic Energy Is Energy of Motion

10. A moving car has kinetic energy. If it speeds up until it is going four times as fast, how much kinetic energy does it have in comparison?

Work-Energy Theorem

11. Compared to some original speed, how much work must the brakes of a car supply to stop a car moving four times as fast? How will the stopping distance compare?

Conservation of Energy

12. What will be the kinetic energy of pile driver ram when it undergoes a 10 kJ decrease in potential energy? (Assume no energy goes to heat.)

Machines—Devices to Multiply Forces

13. Can a machine multiply input force? Input distance? Input energy? (If your three answers are the same, seek help, for the last question is especially important.)

14. If a machine multiplies force by a factor of four, what other quantity is diminished, and how much?

15. If the man in Figure 6.21 pulls 1 m of rope downward with a force of 100 N, and the load rises 1/7 as high, what is the maximum load that can be lifted?

Efficiency—A Measure of Work Done for Energy Spent

16. What is the efficiency of a machine that miraculously converts all the input energy to useful output energy?

17. Is a machine physically possible that has an efficiency greater than 100%? Discuss.

Sources of Energy

18. What is the ultimate source of energies of fossil fuels, dams, and windmills?

19. What is the source of geothermal energy?

Energy Is Needed for Life

20. The energy we require for existence comes from the chemically stored potential energy in food, which is transformed into other forms when it is metabolized. What happens to a person whose work output is less than the energy he or she consumes? Whose work output is greater than the energy he or she consumes? Can an undernourished person perform extra work without extra food? Briefly discuss.

Explorations

1. Fill two mixing bowls with water from the cold tap and take their temperatures. Then run an electric or hand beater in the first bowl for a few minutes. Compare the temperatures of the water in the two bowls.

2. Pour some dry sand into a tin can with a cover. Compare the temperature of the sand before and after vigorously shaking the can for a couple of minutes.

Exercises

1. When the mass of a moving object is doubled with no change in speed, by what factor is its momentum changed? Its kinetic energy?
2. When the velocity of an object is doubled, by what factor is its momentum changed? Its kinetic energy?
3. Consider a ball thrown straight up in the air. At what position is its kinetic energy a maximum? Where is its gravitational potential energy a maximum?
4. At what point in its motion is the KE of a pendulum bob a maximum? At what point is its PE a maximum? When its KE is half its maximum value, how much PE does it have?
5. A physical science teacher demonstrates energy conservation by releasing a heavy pendulum bob, as shown in the sketch, allowing it to swing to-and-fro. What would happen if in his exuberance he gave the bob a slight shove as it left his nose? Explain.
6. Discuss the design of the roller coaster shown in the sketch in terms of the conservation of energy.
7. Suppose that you and two classmates are discussing the design of a roller coaster. One classmate says that each summit must be lower than the previous one. Your other classmate says this is nonsense, for as long as the first one is the highest, it doesn't matter what height the others are. What do you say?
8. Consider molecules of hydrogen (tiny ones) and oxygen (bigger ones) in a gas mixture. If they have the same average kinetic energy (they will at the same temperature), which molecules have the greatest average *speed*?
9. On a slide a child has potential energy that decreases by 1000 J while her kinetic energy increases by 900 J. What other form of energy is involved, and how much?
10. According to the work-energy theorem, in the absence of friction, if you do 100 J of work on a cart, how much will you increase its kinetic energy?
11. Does speed affect the friction between a road and a skidding tire?
12. The photo shows Paul Hewitt delivering a blow to a cement block that rests on a bed of nails. Sandwiched bravely between beds of nails is San Mateo High School physics teacher Pablo Robinson. Since the blow is shared by many nails on Robinson's body, the force per nail won't puncture his skin. Discuss what Robinson's fate might be if the block were less massive and unbreakable, and the beds contained fewer nails.

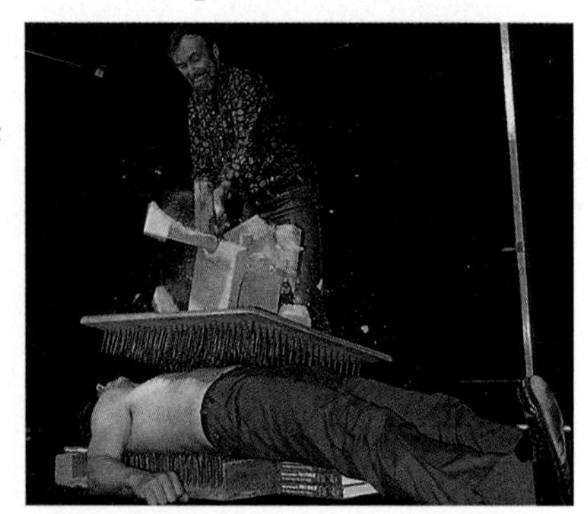

13. Consider the identical balls released from rest on Tracks A and B as shown. When they reach the right ends of the tracks, which will have the greater speed? (Hint: Will their KEs be the same at the end?) Which will get to the end in the shortest time? (Hint: Considering the extra speed in the lower part of track B, which ball has the greatest average speed on the ramps?)

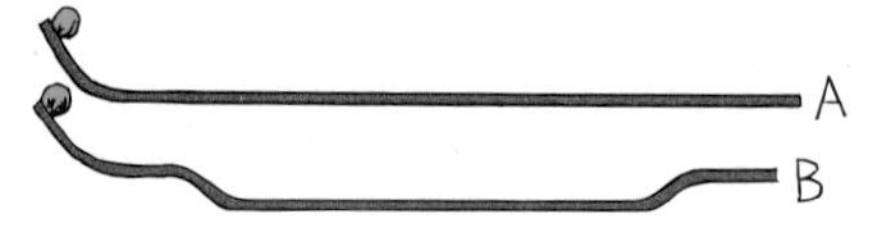

14. You tell your friend that no machine can possibly put out more energy than is put into it, and your friend states that a nuclear reactor puts out more energy than is put into it. What do you say?

15. Two lumps of clay with equal and opposite momenta have a head-on collision and come to rest. Is momentum conserved? Is kinetic energy conserved? Why are your answers the same or different?

16. Scissors for cutting paper have long blades and short handles, whereas metal-cutting shears have long handles and short blades. Bolt cutters have very long handles and very short blades. Why is this so?

17. Consider the swinging-balls apparatus. If two balls are lifted and released, momentum is conserved as two balls pop out the other side with the same speed as the released balls at impact. But momentum would also be conserved if one ball popped out at twice the speed. Can you explain why this never happens? (Hint: if the collision is perfectly elastic, what beside momentum would have to be conserved? Can you see why this exercise is here rather than in the previous chapter on momentum?)

18. Does a high-efficiency machine degrade a relatively high or relatively low percentage of energy to thermal energy?

19. If an automobile had a 100% efficient engine, transferring all of the fuel's energy to work, would the engine be warm to your touch? Would its exhaust heat the surrounding air? Would it make any noise? Would it vibrate? Would any of its fuel go unused?

20. A friend says the energy of oil and coal is actually a form of solar energy. Is your friend correct, or mistaken?

Problems

1. How many joules of work are done when a force of 1 N moves a book 2 m?

2. (a) How much work is done when you push a crate horizontally with 100 N across a 10-m factory floor? (b) If the force of friction on the crate is a steady 70 N, how much KE is gained by the crate? (c) How much of the work you do converts to heat?

3. This question is typical on some driver's license exams: A car moving at 50 km/h skids 15 m with locked brakes. How far will the car skid with locked brakes at 150 km/h?

4. A force of 50 N is applied to the end of a lever, which is moved a certain distance. If the other end of the lever moves one-third as far, how much force can it exert?

5. Consider an ideal pulley system. If you pull one end of the rope downward with 50 N a distance of 1 meter, how high will you lift a 200-N load?

6. In the hydraulic machine shown, when the small piston is pushed down 10 cm, the large piston is raised 1 cm. If the small piston is pushed down with a force of 100 N, what is the most force that the large piston could exert?

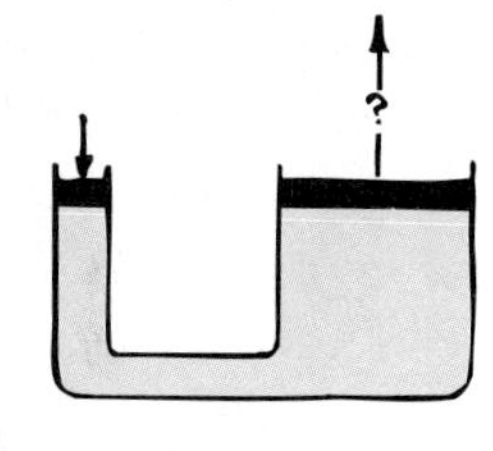

7. How many watts of power are expended when a force of 1 N moves a book 2 m in a time interval of 1 s?

8. Which produces the greater change in kinetic energy: exerting a 10-N force for a distance of 5 m, or exerting a 20-N force over a distance of 2 m? (Assume that all of the work goes into KE.)

9. Consider the inelastic collision between the two freight cars in chapter 5 (Figure 5.13). The momentum before and after the collision is the same. The KE, however, is less after the collision. How much less, and what becomes of this energy?

Chapter 7: Gravity

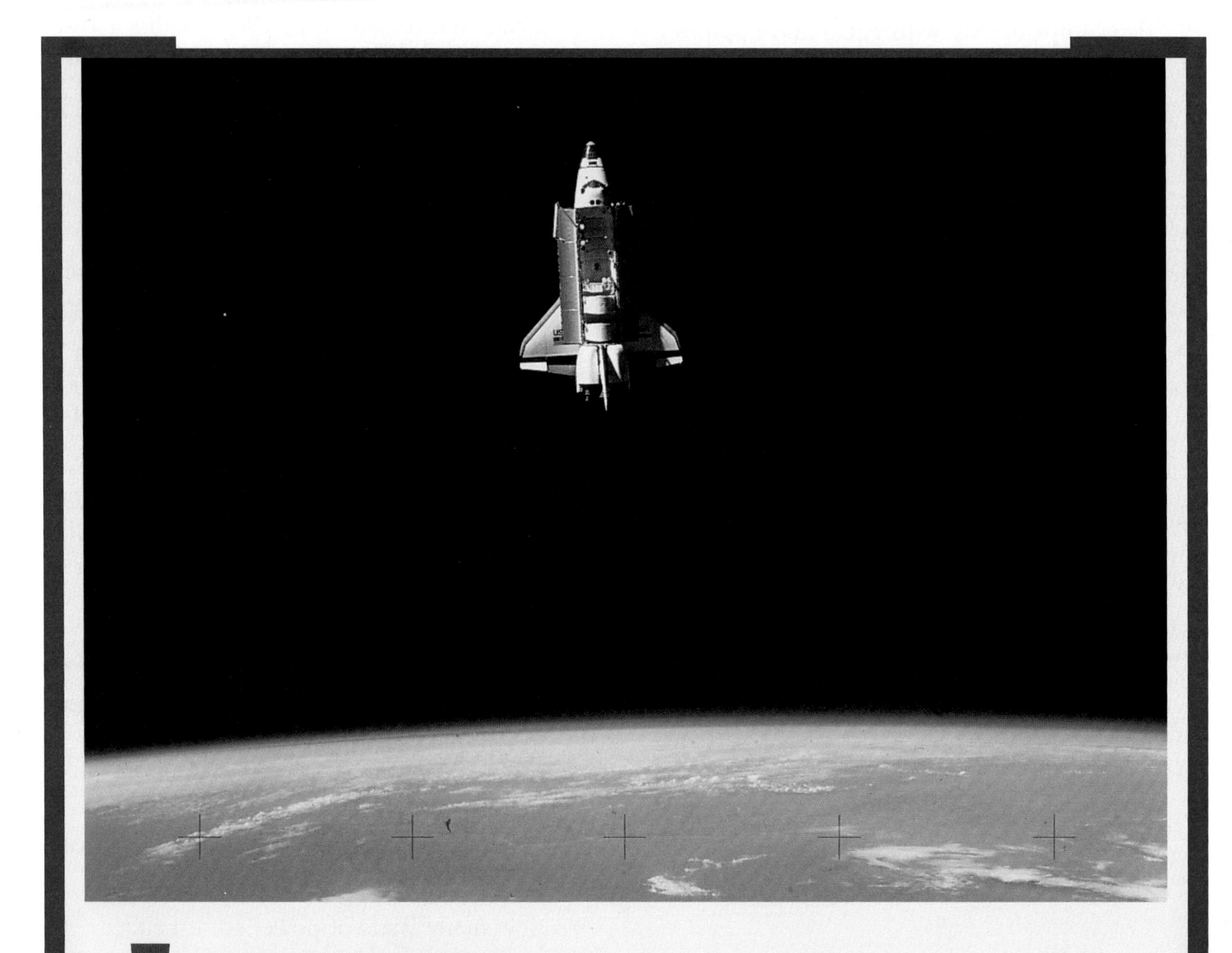

Is the space shuttle in the grips of Earth gravity, or beyond it? Was gravity discovered by Isaac Newton? Or was gravity discovered by earlier people who fell from trees or from their caves? If Newton didn't discover gravity, what did he discover about gravity? Does gravity reach to the moon? Does it reach to the planets? To the stars? How far does gravity reach? Why are there ocean tides? Are there also tides in the solid earth and in its atmosphere? We'll learn the answers to these questions in this chapter.

7.1 The Legend of the Falling Apple

Legend tells us that when Newton was a young man sitting under an apple tree, he made a connection that changed the way we see the world. He saw an apple fall. Perhaps he looked up through the tree branches toward the origin of the falling apple and noticed the moon. In any event, Newton had the insight to realize that the force pulling on a falling apple is the same force that pulls on the moon. Newton realized that Earth gravity reaches to the moon.

Figure 7.1
Newton realizes that Earth's gravity affects both the apple *and* the moon.

7.2 The Fact of the Falling Moon

Why doesn't the moon fall toward the Earth, like an apple from a tree falls? If the apple or anything else drops from rest, it falls in a vertical straight-line path. To get a better idea of this, consider a tree in the back of a truck (Figure 7.2). If the truck is at rest when the apple falls, we see that its path is vertical. But if the truck is moving when the apple begins its fall, the apple follows a curved path. Can you see that the faster the truck moves, the wider the curved path of the falling apple? In the next chapter we'll see that if the apple or anything else moves fast enough so that its curved path matches the Earth's curvature, it becomes a satellite.

Figure 7.2
If an apple falls from a tree at rest, it falls straight downward. But if it falls from a moving tree, it falls in a curved path.

As the moon traces out its orbit around the Earth, it maintains a **tangential velocity**—a velocity parallel to the Earth's surface. Newton realized that the moon's tangential velocity keeps it continually falling *around* the Earth instead of directly into it. Newton further realized that the moon's path around the Earth is similar to the paths of the planets around the sun.

These ideas have very much changed the way people think.

Figure 7.3
The tangential velocity of the moon allows it to fall around the Earth rather than directly into it.

Concept Check ✓

In Figure 7.3, we see that the moon falls around the Earth rather than straight into it. If the tangential velocity were zero, how would the moon move?

Check Your Answer If the moon's tangential velocity were zero, it would fall straight down and crash into the Earth! (Compare this idea with Figure 7.2.)

7.3 Newton's Grandest Discovery— The Law of Universal Gravitation

Newton further realized that everything pulls on everything else. He discovered that a force of gravity acts between all things in a beautifully simple way—involving only mass and distance. According to Newton, every mass attracts every other mass with a force that is directly proportional to the product of the two interacting masses. This statement is known as the **law of universal gravitation.** The force is inversely proportional to the square of the distance separating them.

$$\text{Force} \sim \frac{\text{mass}_1 \times \text{mass}_2}{\text{distance}^2}$$

Expressed in symbol shorthand,

$$F \sim \frac{m_1 m_2}{d^2}$$

where m_1 and m_2 are the masses, and d is the distance between their centers. Thus, the greater the masses m_1 and m_2, the greater the force of attraction between them. The greater the distance of separation d, the weaker is the force of attraction—weaker as the inverse square of the distance between their centers.

Concept Check ✓

1. According to the equation for gravity, what happens to the force between two bodies if the mass of one body is doubled?
2. What happens if instead the mass of the other body is doubled?
3. What happens if the masses of both bodies are doubled?
4. What happens if the mass of one body is doubled, and the other tripled?

Check Your Answers

1. When one mass is doubled, the force between them doubles.
2. The force is still doubled, for it doesn't make any difference *which* mass doubles. ($2 \times 1 = 1 \times 2$; same product either way!)
3. The force is four times as much.
4. Double × triple = six. So the force is six times as much. (If you don't see why, discuss this with a friend before going further.)

7.4 Gravity and Distance: The Inverse-Square Law

Gravity gets weaker with distance the same way a light gets dimmer as you move farther away from it. Consider the candle flame in Figure 7.4. Light from the flame travels in all directions in straight lines. A patch is shown 1 meter from the flame. Notice that at a distance of 2 meters away, the light rays that fall on the patch spread to fill a patch twice as tall and twice as wide. The same light falls on a patch with 4 times the area. The same light 3 meters away spreads to fill a patch 3 times as tall and 3 times as wide. The light would fill a patch with 9 times the area.

Figure 7.4
Light from the flame spreads in all directions. At twice the distance, the same light is spread over 4 times the area; at 3 times the distance it is spread over 9 times the area.

As the light spreads out, its brightness decreases. Can you see that when you're twice as far away, it appears 1/4 as bright? And can you see that when you're 3 times as far away, it appears 1/9 as bright? There is a rule here: The intensity of the light gets less as the inverse square of the distance. This is the **inverse-square law.**

The inverse-square law also applies to a paint sprayer. Pretend you hold a paint gun at the center of a sphere with a radius of 1 meter (Figure 7.5). Suppose that a burst of paint produces a square patch of paint 1 millimeter thick. How thick would the patch be if the experiment were done in a sphere with twice the radius—that is, with the spray gun twice as far away? The answer is not half as thick, because the paint would spread to a patch twice as tall *and* twice as wide. It would spread over an area *four times* as big, and its thickness would be only 1/4 millimeter. Can you see that for a sphere of radius 3 meters the thickness of the paint patch would be only 1/9 millimeter? Do you see that the thickness of paint decreases as the *square* of the distance? The inverse-square law holds for light, for paint spray, and for gravity. It holds for all phenomena where something from a localized source spreads uniformly throughout the surrounding space. We'll see this to be true of the electric field about an electron, light from a match, radiation from a piece of uranium, and sound from a cricket.

Saying that *F* is inversely proportional to the **square** of *d* means, for example, that if *d* gets bigger by 5, *F* gets *smaller* by 25. Get it?

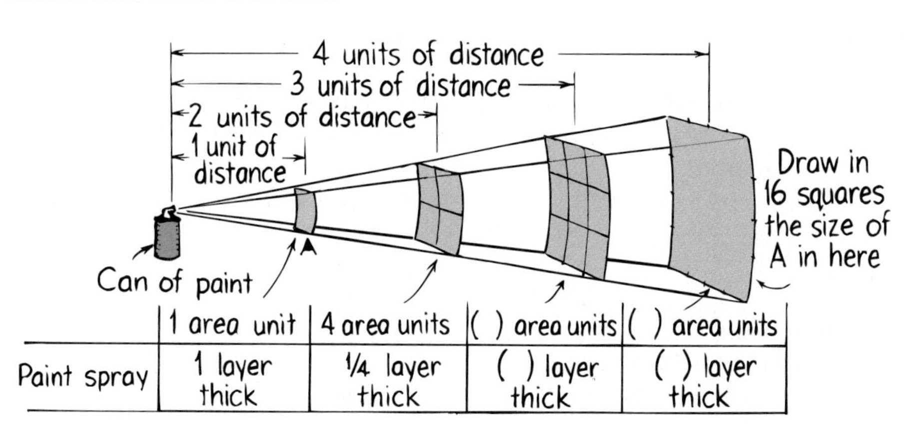

Figure 7.5
The inverse-square law. Paint spray travels in straight lines away from the nozzle of the can. Like gravity, the "strength" of the spray obeys the inverse-square law. Fill in the blanks.

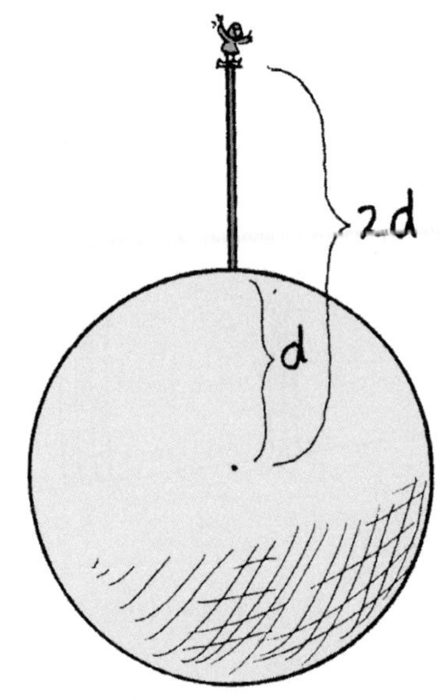

Figure 7.6
At the top of the ladder she is twice as far from the Earth's center, and weighs only 1/4 as much as at the bottom of the ladder.

The greater the distance from the Earth's center, the less the gravitational force on an object. In using Newton's equation for gravity, the distance term d is the distance between the *centers* of the masses of objects attracted to each other. Note in Figure 7.6 that the girl at the top of the ladder weighs only 1/4 as much as she weighs at the Earth's surface. That's because she is twice the distance from the Earth's *center*.

Concept Check ✓

1. How much does the force of gravity change between the Earth and a receding rocket when the distance between them is doubled? Tripled? Ten times as much?
2. Consider an apple at the top of a tree. The apple is pulled by Earth's gravity with a force of 1 N. If the tree were twice as tall, would the force of gravity be only 1/4 as strong? Defend your answer.

Check Your Answers

1. When the distance is doubled, the force is 1/4 as much. When tripled, 1/9 as much. When 10 times, 1/100 as much.
2. No, because the twice-as-tall apple tree is not twice as far from the Earth's center. The taller tree would have to be 6,370 km tall (the Earth's radius) for the apple's weight to reduce to 1/4 N. For a decrease in weight by 1 percent, an object must be raised 32 km—nearly four times the height of Mt. Everest. So as a practical matter we disregard the effects of everyday changes in elevation for gravity. The apple has practically the same weight at the top of the tree as at the bottom.

So gravity gets weaker with increasing distance. But no matter how far away, the Earth's gravitational force approaches, but never reaches, zero. Even if you traveled to the far reaches of the universe, the gravitational influence of home would still be with you. It may be overwhelmed by the gravitational influences of nearer and/or more massive bodies, but it is there. The gravitational influence of every material object, however small or far, is exerted through all of space.

Figure 7.7
As the rocket gets farther from the Earth, gravitation between the rocket and the Earth gets less.

Concept Check ✓

1. Light from the sun, like gravity, obeys the inverse-square law. If you were on a planet twice as far from the sun, how bright would the sun look?
2. How bright would the sun look if you were on a planet twice as close to the sun?

Check Your Answers

1. One-quarter as bright.
2. Four times brighter.

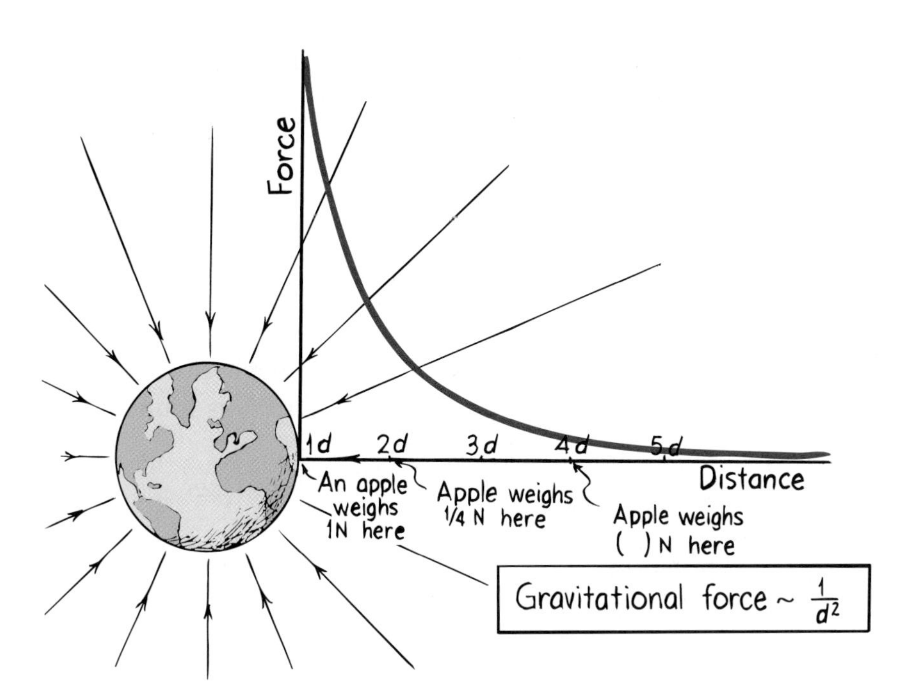

Figure 7.8
If an apple weighs 1 N at the Earth's surface, it weighs only 1/4 N twice as far from the Earth's center. At three times the distance, it weighs only 1/9 N. What would it weigh at four times the distance? Five times?

7.5 The Universal Gravitational Constant, *G*

The universal law of gravitation can be written as an exact equation when the universal constant of gravitation, *G*, is used. Then we have

$$F = G\frac{m_1 m_2}{d^2}$$

The units of *G* make the force come out in newtons. The magnitude of G is the same as the gravitational force between two 1-kilogram masses that are 1 meter apart: 0.0000000000667 newton.

$$G = 6.67 \times 10^{-11}\ \text{N}\cdot\text{m}^2/\text{kg}^2$$

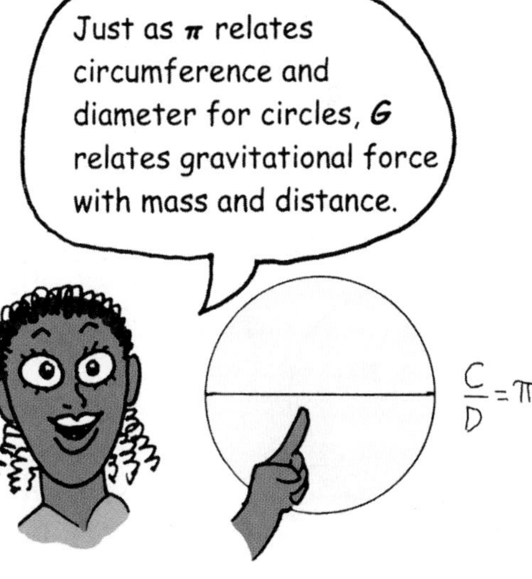

This is an extremely small number. It shows that gravity is a very weak force compared with electrical forces. The large net gravitational force we feel as weight is because of the enormity of atoms in planet Earth that are pulling on us.

To better understand the constant *G*, consider the analogous case of the geometry constant, π, in the equation for the circumference of a circle,

$$C = \pi D.$$

The equation tells you that the circumference of a circle, *C*, is equal to π multiplied by the diameter *D*. If we didn't know π, we could say

$$C \sim D$$

This expression tells us only that the circumference of a circle is *proportional* to its diameter. This means a small-circumference circle will have a small diameter, and that a large-circumference circle will have a large diameter. How much smaller or larger requires that we know π. We find this by dividing C by D. That is,

$$\frac{C}{D} = 3.14... = \pi.$$

Similarly with the proportion form of Newton's gravitational law on the previous page. The constant of proportionality G is found by

$$G = \frac{F}{m_1 m_2 / d^2} = 6.67 \times 10^{-11}\ \text{N}\cdot\text{m}^2/\text{kg}^2$$

Regardless of the masses and distance between them, the gravitational force will have a value that results in the same value for G.

7.6 The Mass of the Earth Is Measured

The value of G wasn't measured until a century after the publication of Newton's theory of universal gravitation. One method of measuring it, though not the first, is shown in Figure 7.9. Once we know the value of G, we have enough information to calculate the mass of the Earth! Here's how: The force that the Earth exerts on a 1-kilogram mass at its surface is 9.8 newtons. The distance between the 1-kilogram mass and the center of the Earth is the Earth's radius, 6.4×10^6 meters. Using

$$F = G\frac{m_1 m_2}{d^2},$$

where F is 9.8 N, m_1 is the mass of the 1-kilogram mass, and m_2 is the mass of the Earth,

$$9.8\ \text{N} = 6.67 \times 10^{-11}\ \frac{\text{N}\cdot\text{m}^2}{\text{kg}^2} \times \frac{1\ \text{kg} \times \text{m}_2}{(6.4 \times 10^6\text{m})^2}$$

The only unknown quantity is m_2, the mass of the Earth. Solving, we find $m_2 = 6 \times 10^{24}$ kilograms.

When G was first measured in the eighteenth century, people all over the world were excited about it. That's because newspapers everywhere announced the discovery as one that measured the mass of planet Earth. How exciting that Newton's formula gives the mass of the entire world, with all its oceans, mountains, and inner parts yet to be discovered. G and the mass of the Earth were measured when much of the Earth's surface was still undiscovered.

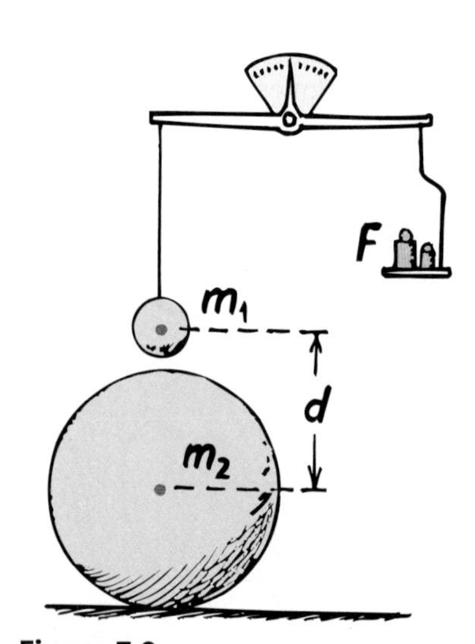

Figure 7.9
Philipp von Jolly's method of measuring G. A small ball of known mass is attracted to a 6-ton sphere rolled beneath it. The force of attraction can be measured by the weights needed to restore balance.

Concept Check ✓

1. What value will result if you let your mass be m_1, the mass of the Earth m_2, and d the Earth's radius, in the equation for gravity?
2. If your mass increases, does your weight increase also?

Check Your Answers

1. Your weight.
2. Yes, in direct proportion. That is, if you double your mass, your weight also doubles.

7.7 Ocean Tides Are the Result of Differences in Gravitational Pulls

Sailors have always known there is a connection between ocean tides and the moon. Newton was the first to show that tides are caused by *differences* in the gravitational pull by the moon on the Earth's opposite sides. Since gravitational force gets weaker with distance, the gravitational force between the Earth and moon is stronger on the side of the Earth nearer to the moon than on the opposite side of the Earth.

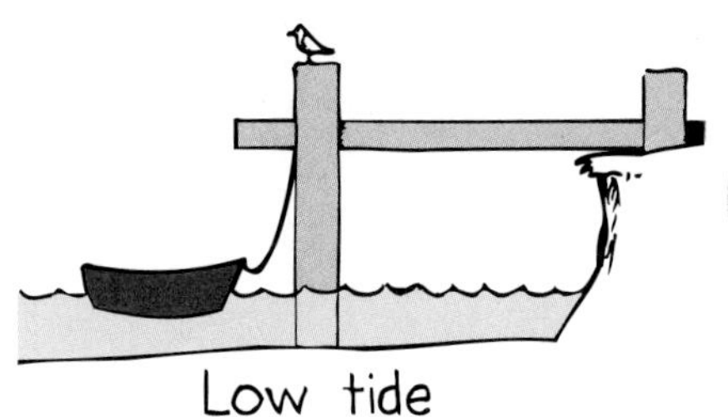

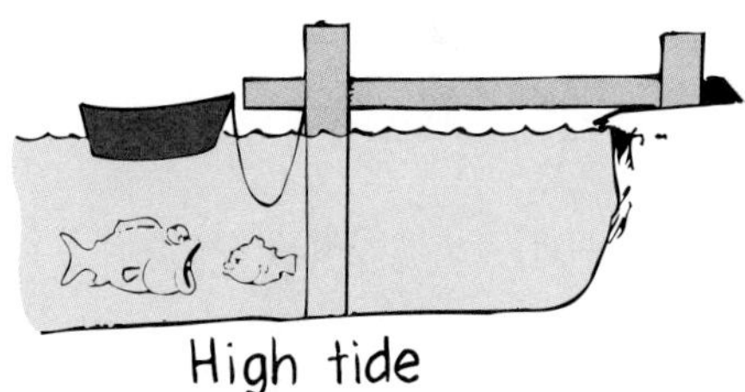

Figure 7.10
Low tide; high tide.

To understand why these different pulls produce tides, let's look at a spherical ball of Jell-O (Figure 7.11). If you exerted the same force on every part of the ball, it would remain spherical as it accelerated. But if you pull harder on one side than the other, the different pulls would stretch the ball. That's what's happening to this big ball on which we live. Different pulls of the moon stretch the Earth, most notably in its oceans. The stretch produces an average ocean bulge of nearly 1 meter on each side of the Earth. That's why the oceans on opposite sides of the Earth bulge about 1 meter above the ocean's average surface level. The Earth rotates once per day, so a fixed point on Earth passes beneath both of these bulges each day. This produces two sets of ocean tides per day—two high tides and two low tides.

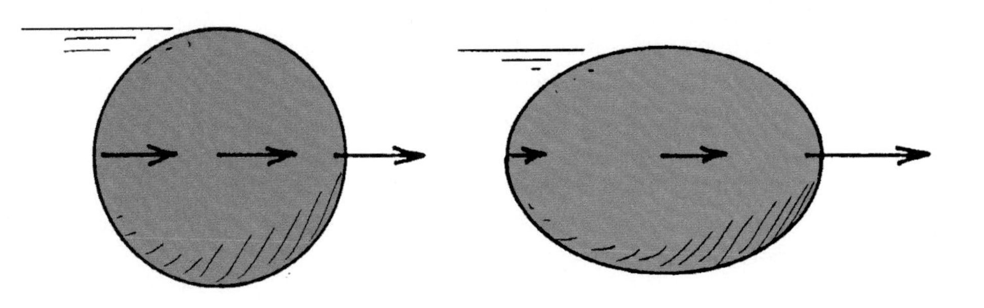

Figure 7.11
A ball of Jell-O stays spherical when all parts are pulled equally in the same direction. When one side is pulled more than the other, its shape is distorted.

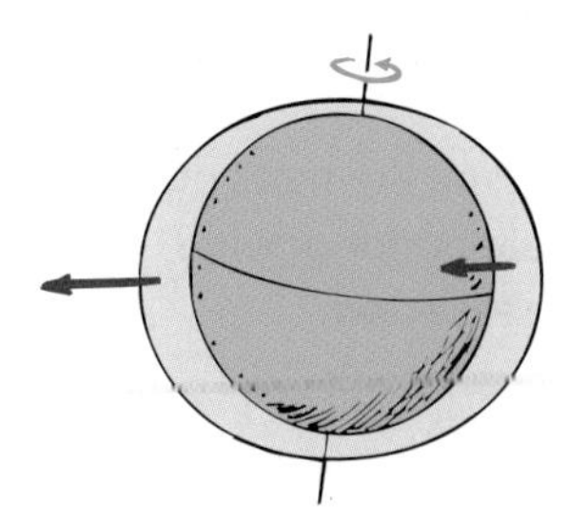

Figure 7.12
Two tidal bulges are produced by differences in gravitational pulls by the moon.

While the Earth rotates, the moon also moves in its orbit. The moon appears at the same position in our sky every 24 hours and 50 minutes, so the two-high-tide cycle is actually at 24-hour-and-50-minute intervals. This means that tides do not occur at the same time every day.

The sun also contributes to ocean tides, but it's about half as effective as the moon. Interestingly, the sun pulls 180 times harder on the Earth than the moon. Why aren't tides due to the sun 180 times greater than tides due to the moon? Because of the sun's great distance, the *difference* in gravitational pulls on opposite sides of the Earth is very small. In other words, the sun pulls almost as hard on the far side of the Earth as it does on the near side. The difference is only about 0.02 percent, compared with a difference of about 7 percent for the moon.

Figure 7.13
Tidal bulges due only to the sun are small because the *differences* in pulls by the sun are small.

Concept Check ✓

1. If you pull a blob of Jell-O equally on all parts, it will keep its shape as it moves. But if you pull harder on one end than the other, it will stretch. How does this relate to tides?
2. If the moon didn't exist, would the Earth still have ocean tides? If so, how often?
3. We know that both the moon and the sun produce our ocean tides. And we know the moon plays the greater role because it is closer. Does its closeness mean it pulls the oceans with more gravitational force than the sun?

Check Your Answers

1. Just as differences in pulls on the Jell-O will distort it, differences in pulls on the oceans distort the ocean and produce tides.
2. Yes, the Earth's tides would be due only to the sun. They'd occur twice per day (every 12 hours instead of every 12.4 hours) due to the Earth's daily rotation.
3. No, the sun's pull is much stronger. But tides are not caused by gravitational pulls. Tides are caused by *differences* in pulls across a body. Differences in pulls, not pulling strength, is the key to tides. (Amazingly, if the moon were much closer to Earth, increased tides on both the Earth and the moon could tear the moon into pieces. Astronomers believe that the planetary rings of Saturn and other planets formed this way.)

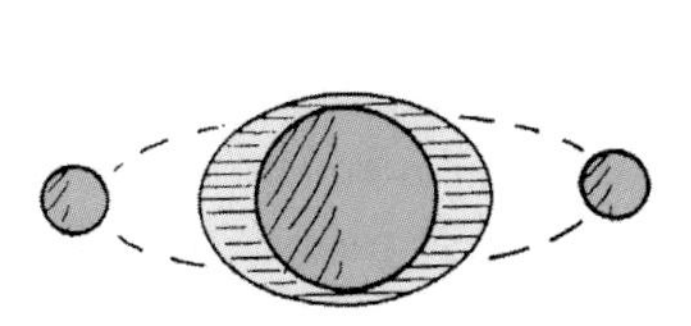

Figure 7.14
When the pulls by the sun and moon are lined up, spring tides occur.

When the sun, Earth, and moon are all lined up, the tides due to the sun and the moon overlap, and we have higher-than-average high tides and lower-than-average low tides. These are called **spring tides.** (Spring tides have nothing to do with the spring season.) Spring tides occur at the times of a new or full moon.

When the moon's phase is half way between a new moon and a full moon, in either direction, tides due to the sun and moon partly cancel each other. Then high tides are lower than average and the low tides are not as low as average low tides. These are called **neap tides.**

Because of the Earth's tilt, the two tides per day are normally unequal at a given location. Figure 7.16 shows how a person in the Northern Hemisphere may find the tide nearest the moon much lower (or higher) than the tide half a day later. The heights of tides vary with the positions of the moon and the sun.

Ocean tides occur because the Earth's oceans are thousands of kilometers apart and are connected. There are large differences in pulls across such great distances. Why don't we see tides in lakes? We don't because no part of a lake is significantly closer to the moon than any other part—there is no significant *difference* in moon pulls on the lake. So lake water doesn't "pile up" on one side and empty out of the other side. Similarly with the fluids in your body. Although much of your body is composed of water and other fluids, any tides caused by the moon in your body are negligible. You're not tall enough for tides. A 1-kilogram melon held one meter above your head produces more microtides in your body than the moon over your head! Tell this to anyone who claims the gravitational attraction of the moon has an influence on humans!

Now I have an idea of why some of my friends are weird at the time of a full moon!

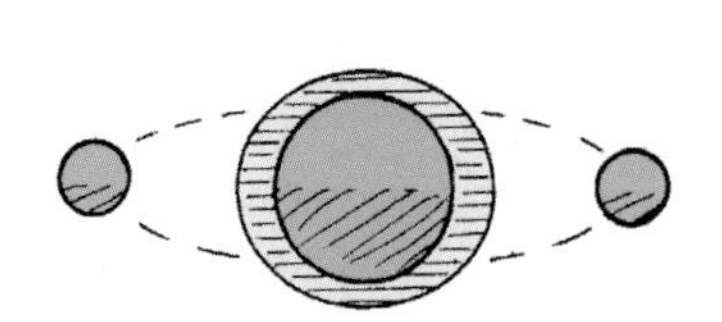

Figure 7.15
When the pulls of the sun and moon are at right angles to each other, neap tides occur.

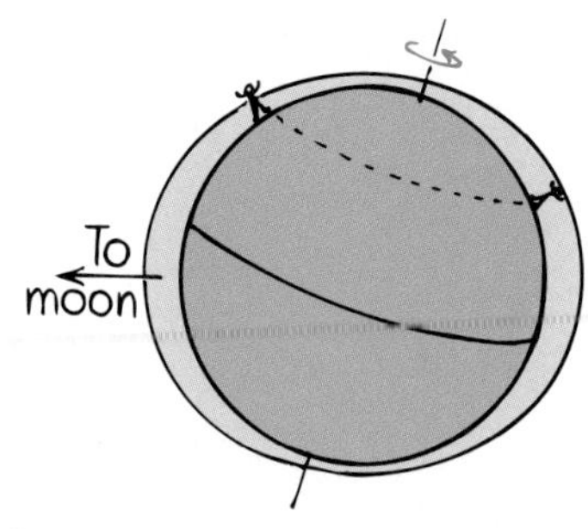

Figure 7.16
Unequal tides are due to the Earth's tilt.

The part of the Earth beneath the crust is molten—fluid. Because of this we have Earth tides—actual rises and falls in the Earth's crust. Earth tides, however, are much smaller than ocean tides. There are also atmospheric tides. These regulate the cosmic rays that reach the Earth's surface. The greatest fluctuation of tides between high and low (ocean, Earth, or atmospheric) occurs during the alignments that make a new and full moon. The fluctuations in atmospheric tides produce changes in the intensity of cosmic rays reaching the Earth's surface—which in turn affects some life forms.

Our brief treatment of tides is quite simplified, for the tilt of the Earth's axis, interfering landmasses, friction with the ocean bottom, and other factors complicate tidal motions. Tides are fascinating!

Figure 7.17
An overhead moon and an overhead melon both exert gravitational forces on you. A 1-kilogram mass 1 meter over your head produces a greater difference in gravitational force between your head and your feet than does the moon.

7.8 Gravitation Is Universal

We all know that the Earth is round. But why is it round? It is round because everything attracts everything else, and all parts of the Earth have attracted themselves together as much as they can! Any "corners" of the Earth have been pulled in, making it a sphere. Likewise, the sun, moon and stars are spherical. Rotational effects make them bulge slightly at their equators.

The shapes of distant galaxies show that the law of gravitation applies to large distances. Gravity underlies the fate of the entire universe. Current scientific speculation is that the universe originated in the explosion of a primordial fireball some 8 to 15 billion years ago. This is the **Big Bang** theory of the origin of the universe. The explosion was space itself, with all the matter of the universe hurled outward. Space is still stretching out, carrying the galaxies with it. This expansion may go on indefinitely, or it may eventually be overcome by the combined gravitation of all the galaxies and come to a halt. Like a stone thrown upward, whose speed comes to a stop when it reaches the top of its trajectory and then begins falling to the place of its origin, the universe may contract and fall back into a single unity. This would be the *Big Crunch.* After that, we can only speculate that the universe might re-explode to produce a new universe. The same course of action might repeat itself, and the process may well occur in cycles. If this speculation is true, we live in an oscillating universe.

We do not know whether the expansion is indefinite because we are uncertain about whether enough mass exists to halt the expansion. Recent evidence suggests the presence of a **dark matter** throughout the universe. This is mysterious matter, unlike the matter we know. We can't see it, but its gravitational presence is felt by stars and galaxies. If the expansion of the universe halts and is followed by contraction, the time from Big Bang to Big Crunch is estimated to be somewhat less than 100 billion years. Our universe is still young. But humankind is younger by far.

Chapter Review

Key Terms and Matching Definitions

_____ Big Bang
_____ dark matter
_____ law of universal gravitation
_____ inverse-square law
_____ neap tide
_____ spring tide
_____ tangential velocity
_____ universal constant of gravitation, G

1. Velocity that is parallel (tangent) to a curved path.
2. Every body in the universe attracts every other body with a mutually attracting force. For two bodies, this force is directly proportional to the product of their masses and inversely proportional to the square of the distance separating them:

$$F = G\frac{m_1 m_2}{d^2}$$

3. A law relating the intensity of an effect to the inverse square of the distance from the cause:

$$\text{Intensity} \sim \frac{1}{\text{distance}^2}$$

4. The proportionality constant in Newton's law of universal gravitation.
5. A high or low tide that occurs when the sun, Earth, and moon are all lined up so that the tides due to the sun and moon coincide, making the high tides higher than average and the low tides lower than average.
6. A tide that occurs when the moon is midway between new and full, in either direction. Tides due to the sun and moon partly cancel, making the high tides lower than average and the low tides higher than average.
7. The primordial explosion that is thought to have resulted in the expanding universe.
8. Mysterious matter different than known matter that can't be seen, but with gravitational effects.

Review Questions

The Legend of the Falling Apple

1. What connection did Newton make between a falling apple and the moon?

The Fact of the Falling Moon

2. What does it mean to say something moving in a curve has a tangential velocity?
3. In what sense does the moon "fall?"

Newton's Grandest Discovery—The Law of Universal Gravitation

4. State Newton's law of universal gravitation in words. Then do the same with one equation.

Gravity and Distance: The Inverse-Square Law

5. How does the force of gravity between two bodies change when the distance between them is doubled?
6. How does the thickness of paint sprayed on a surface change when the sprayer is held twice as far away?
7. How does the brightness of light change when a point source of light is brought twice as far away?
8. At what distance from Earth is the gravitational force on an object zero?

The Universal Gravitational Constant, *G*

9. What is the magnitude of gravitational force between two 1-kilogram bodies that are 1 meter apart?
10. What is the magnitude of the gravitational force between the Earth and a 1-kilogram body?
11. What do we call the gravitational force between the Earth and your body?

The Mass of the Earth Is Measured

12. When G was first measured, the experiment was called the "weighing the Earth experiment." Why?

Ocean Tides Are the Result of Differences in Gravitational Pulls

13. Do tides depend more on the strength of gravitational pull or on the *difference* in strengths? Explain.

14. Why do both the sun and the moon exert a greater gravitational force on one side of the Earth than the other?

15. Which pulls with greater force on the Earth's oceans, the sun or the moon? Which is more effective in raising tides? Why are your answers different?

16. Distinguish between *spring tides* and *neap tides.*

17. Are all tides greatest at the time of a full moon or new moon? Why?

18. Do tides occur in the molten interior of the Earth for the same reason that tides occur in the oceans?

Gravitation Is Universal

19. What makes the Earth round?

20. Distinguish between the *Big Bang* and the *Big Crunch.*

Explorations

1. Hold your hands outstretched with one hand twice as far from your eyes as the other. Make a casual judgment about which hand looks bigger. Most people see them to be about the same size, while many see the nearer hand as slightly bigger. Very few people see the nearer hand as four times as big. But by the inverse-square law, the nearer hand should appear twice as tall and twice as wide. Twice times twice means four times as big! That's four times as much of your visual field as the farther hand. Your belief that your hands are the same size is so strong that you likely overrule this information. Try it again, only this time overlap your hands slightly and view them with one eye closed. Aha! Do you now more clearly see that the nearer hand is bigger? This raises an interesting question: What other illusions do you have that are not so easily checked?

2. Repeat the eyeballing experiment, only this time use two dollar bills—one regular, and the other folded in half length-wise, and again width-wise, so it has 1/4 the area. Now hold the two in front of your eyes. Where do you hold the folded bill so that it looks the same size as the unfolded one? Share this with your friends!

Exercises

1. Comment on whether or not this label on a consumer product should be cause for concern. *CAUTION: The mass of this product pulls on every other mass in the universe, with an attracting force that is proportional to the product of the masses and inversely proportional to the square of the distance between them.*

2. Gravitational force acts on all bodies in proportion to their masses. Why, then, doesn't a heavy body fall faster than a light body?

3. What would be the path of the moon if somehow all gravitational forces on it vanished to zero?

4. Is the force of gravity stronger on a piece of iron than a piece of wood if both have the same mass? Defend your answer.

5. Is the force of gravity stronger on a piece of paper when it is crumpled? Defend your answer.

6. What is the magnitude and direction of the gravitational force that acts on a teacher who weighs 1000 N at the surface of the Earth?

7. The Earth and the moon are attracted to each other by gravitational force. Does the more massive Earth attract the less massive moon with a force that is greater, smaller, or the same as the force with which the moon attracts the Earth?

8. What do you say to a friend who says that if gravity follows the inverse-square law, that when you are on the 20th floor of a building gravity on you should be one-fourth as much as if you're on the 10th floor?

9. Most people today know that the ocean tides are caused principally by the gravitational influence of the moon. They therefore think that the gravitational pull of the moon on the Earth is greater than the gravitational pull of the sun on the Earth. What do you think?

10. If somebody tugged on your shirt sleeve, it would likely tear. But if all parts of your shirt were pulled equally, no tearing would occur. How does this relate to tidal forces?

11. Would ocean tides exist if the gravitational pull of the moon (and sun) were somehow equal on all parts of the world? Explain.

12. Why aren't high ocean tides exactly 12 hours apart?

13. With respect to spring and neap ocean tides, when do the lowest tides occur? That is, when is it best for digging clams?

14. Whenever the ocean tide is unusually high, will the following low tide be unusually low? Defend your answer in terms of "conservation of water." (If you slosh water in a tub so it is extra deep at one end, will the other end be extra shallow?)

15. The human body is composed of mostly water. Why does the moon overhead cause appreciably less biological tides in the fluid compartment of the body than a 1-kg melon held over your head?

16. If the moon didn't exist, would the Earth still have ocean tides? If so, how often?

17. What would be the effect on the Earth's tides if the diameter of the Earth were a lot larger than it is? If the Earth were as it presently is, but the moon a lot larger in size with the same mass?

18. Does the strongest tidal force on our bodies come from the Earth, moon, or sun?

19. Some people dismiss the validity of scientific theories by saying they are "only" theories. The law of universal gravitation is a theory. Does this mean that scientists still doubt its validity? Explain.

20. Ultimately, the universe may expand without limit, or it may coast to a stop, or it may turn around and collapse to a "Big Crunch." What is the single most important quantity that will determine which of these fates is in store for the universe?

Problems

1. If you stood atop a ladder that was so tall that you were three times as far from the Earth's center, how would your weight compare with its present value?

2. Find the change in the force of gravity between two planets when the masses of both planets are doubled, but the distance between them stays the same.

3. Find the change in the force of gravity between two planets when masses remain the same, but the distance between them is increased by ten times.

4. Find the change in the force of gravity between two planets when the distance between them is *decreased* by ten times.

5. Find the change in the force of gravity between two planets when the masses of the planets don't change, but the distance between them is decreased by five times.

6. By what factor would your weight change if the Earth's diameter were doubled and its mass were also doubled?

7. Find the change in the force of gravity between two objects when both masses are doubled and the distance between them is also doubled.

8. Consider a bright point light source located 1 m from a square opening of area one-square meter. Light passing through the opening illuminates an area of 4 m^2 on a wall 2 m from the opening. (a) Find the area illuminated if the wall is moved to a distance of 3 m, 5 m, or 10 m. (b) How can the same amount of light illuminate more area as the wall is moved farther away?

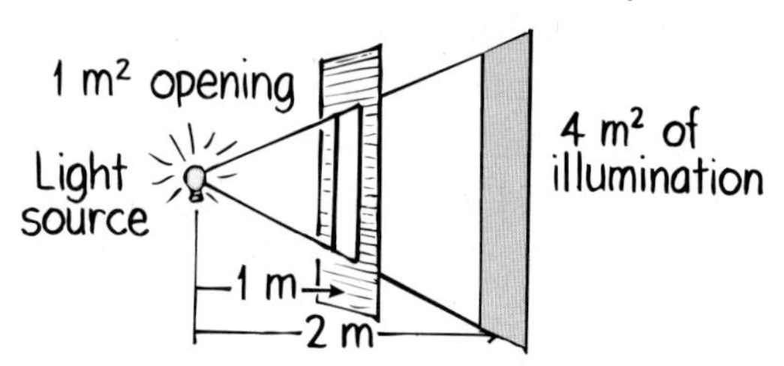

9. Calculate the force of gravity between the Earth (6×10^{24} kg) and the sun (2×10^{30} kg). The average distance between the two is 1.5×10^{11} m.

10. A 3-kg newborn baby at the Earth's surface is gravitationally attracted to Earth with a force of about 30 N. (a) Calculate the force of gravity with which the baby on Earth is attracted to the planet Mars when Mars is closest to Earth. (The mass of Mars is 6.4×10^{23} kg and its closest distance is 5.6×10^{10} m). (b) Calculate the force of gravity between the baby and the physician who delivers it. Assume the physician has a mass of 100 kg and is 0.5 m from the baby. (c) How do the forces compare?

Chapter 8: Projectile and Satellite Motion

How did the astronaut get so high above the Earth's surface? When you throw a ball straight upward, how does the time going up compare with the time coming down? Why does the path of a ball curve when it's thrown at an angle? If you wanted to see how far you could throw something, at what angle should you throw it to get the longest horizontal distance? Is it true that if a ball or any object is thrown fast enough, and above the atmosphere, it becomes a satellite? Why doesn't Earth's gravitational pull make a satellite crash to Earth? What holds a satellite up? To answer these questions we need to learn the physics of *projectile motion.*

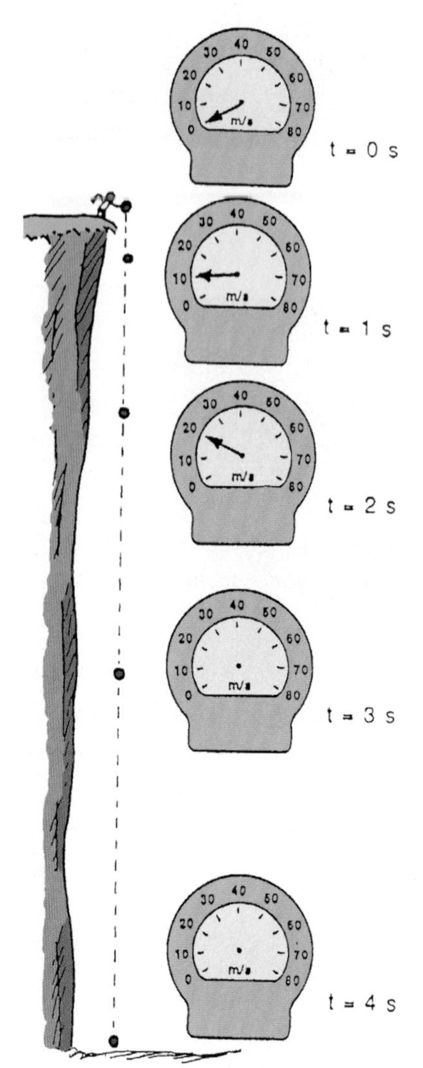

Figure 8.1
The falling stone gains a speed of 10 m/s each second. Fill in the speedometer readings for the times 3 and 4 seconds.

8.1 Projectile Motion

A tossed stone, a cannonball, or any object projected by any means that continues in motion is called a **projectile.** A very simple projectile is a falling stone, Figure 8.1. This is a version of Figure 3.4, which we studied in Chapter 3. The stone gains speed as it falls straight downward, as indicated by a speedometer. Remember that a freely falling object gains 10 meters/second during each second of fall. This is the acceleration due to gravity, 10 m/s^2. If it begins its fall from rest, 0 m/s, then at the end of the first second of fall its speed is 10 m/s. At the end of 2 seconds, its speed is 20 m/s, and at the end of 3 seconds, it is 30 m/s—and so on. It keeps gaining 10 m/s each second it falls.

Although the change in speed is the same each second, the *distance* of fall keeps increasing. That's because the average speed of fall increases each second. Let's apply this to a new situation—throwing the stone horizontally from a high cliff.

First, imagine that gravity doesn't act on the stone (of course it *does* act, but we're just pretending for now). In Figure 8.2 we see the positions that a horizontally thrown stone would have with *no gravity.* Note that the positions each second are the same distance apart. That's because there is no force acting on it. The motion of the stone is like the motion of a bowling ball rolling along a bowling lane. Horizontal motion is constant because no horizontal force acts. Both the stone and the ball move without accelerating. They move at constant velocities, covering equal distances in equal times.

In the real world there is gravity, and the thrown stone falls beneath the straight line it would follow with no gravity (Figure 8.4). The stone curves as it falls. Interestingly, this familiar curve is the result of *two* kinds of

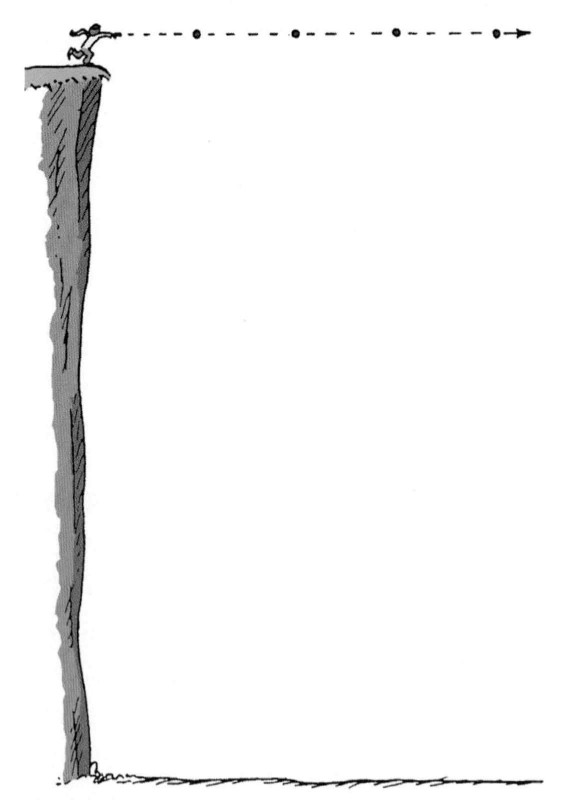

Figure 8.2
If there were no gravity, a stone thrown horizontally would move in a straight-line path and cover equal distances in equal time intervals.

Figure 8.3
A bowling ball rolling along a lane similarly covers equal distances in equal times. It rolls at constant velocity.

motion occurring at the same time. One kind is the straight-down vertical motion shown in Figure 8.1. The other is the horizontal motion of constant velocity, as imagined in Figure 8.2. Both occur simultaneously. As the stone moves horizontally, it also falls straight downward—beneath the place it would be if there were no gravity. This is indicated in Figure 8.4.

The curved path of a projectile is the result of constant motion horizontally and accelerated motion vertically under the influence of gravity. This curve is a **parabola.**

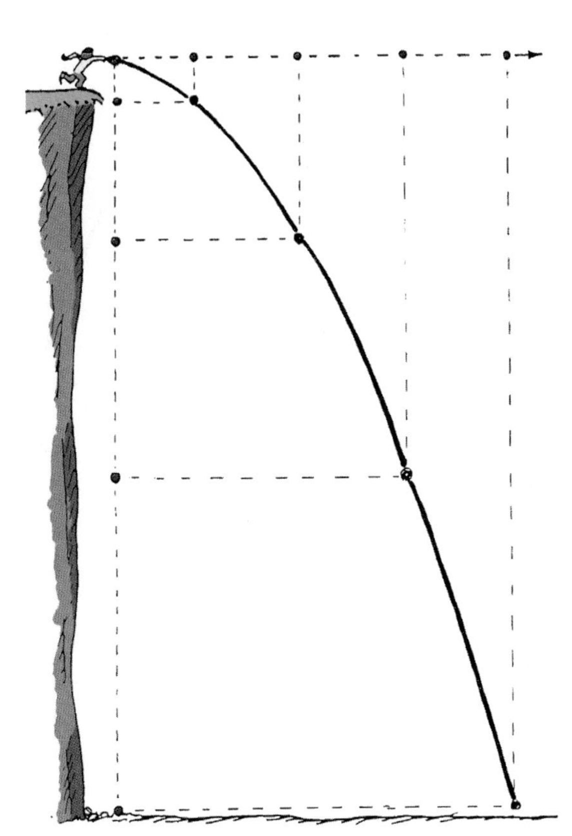

Figure 8.4
The vertical path (dashed line) is for a stone dropped at rest. The horizontal path (dashed line) would occur with no gravity. The solid line shows the path that results from both the vertical and horizontal motions.

Concept Check ✓

At the instant a horizontal cannon fires a cannonball from atop a high cliff, another cannonball is simply dropped from the same height. Which hits the ground below first, the one fired downrange, or the one that drops straight down?

Check Your Answer Both cannonballs hit the ground at the same time, for both fall *the same vertical distance.* Can you see that the physics is the same as the physics of the figures above? We can reason this another way by asking which one would hit the ground first if the cannon were pointed at an *upward* angle. Then the dropped cannonball would hit first, while the fired ball is still in the air. Now consider the cannon pointing *downward.* In this case the fired ball hits first. So projected upward, the dropped one hits first; downward, the fired one hits first. Is there some angle at which there is a dead heat—where both hit at the same time? Can you see that this occurs when the cannon is horizontal?

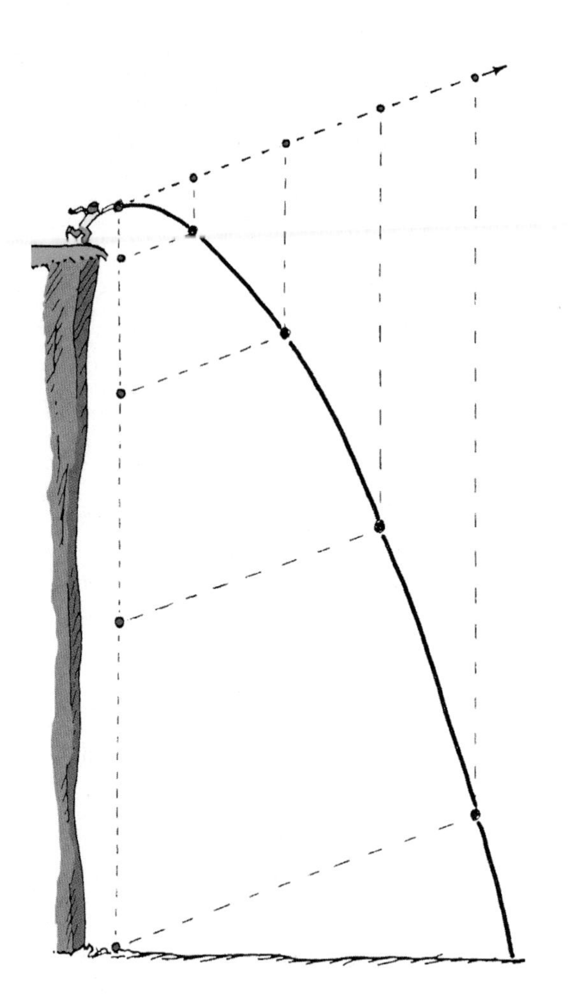

Figure 8.5
A stone thrown at an upward angle would follow the dashed line in the absence of gravity. Because of gravity, it falls beneath this line and describes the parabola shown by the solid curve.

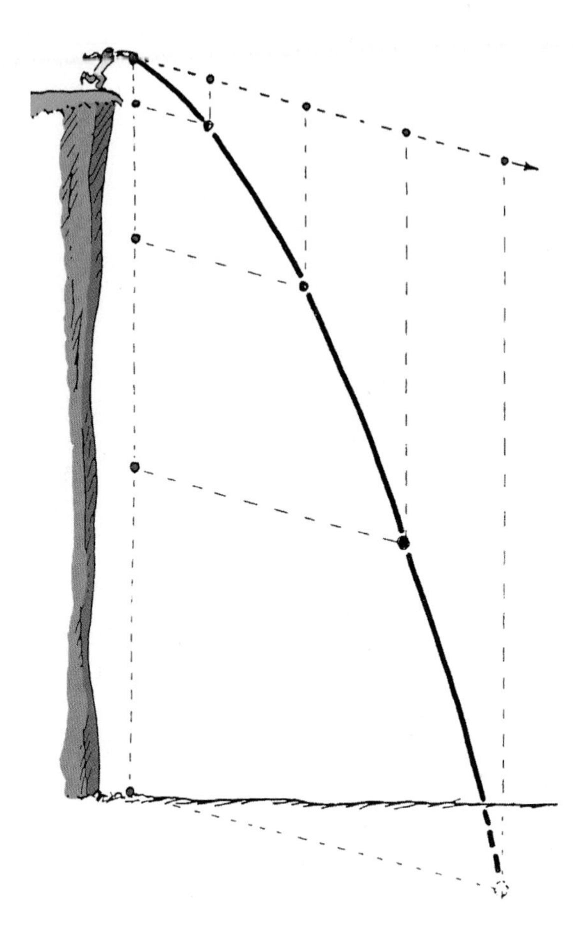

Figure 8.6
A stone thrown at a downward angle follows a somewhat different parabola.

In Figure 8.5, we consider a stone thrown upward at an angle. If there were no gravity, the path of the stone would be along the dashed line with the arrow. Positions of the stone at 1-second intervals along the line are shown by light dots. Because of gravity, the actual positions (dark dots) are below these points. How far below? The answer is, the same distance an object would fall if dropped from the light-dot positions. When we connect the dark dots to plot the path, we get a different parabola.

In Figure 8.6, we consider a stone thrown at a downward angle. The physics is the same. If there were no gravity, it would follow the dashed line with the arrow. Because of gravity, it falls beneath this line, just as in the previous cases. The path is a somewhat different parabola.

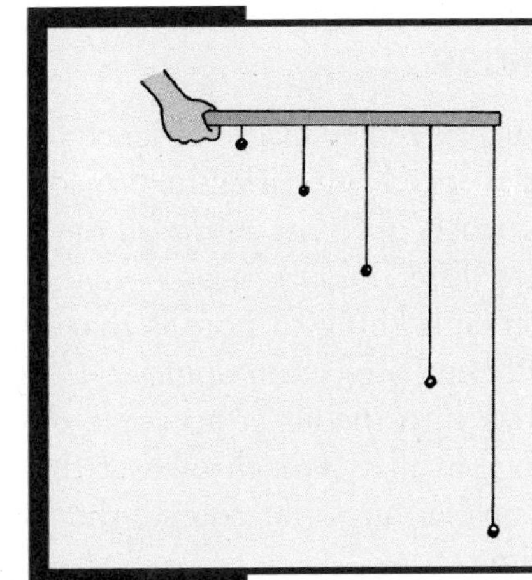

Hands-On Dangling Beads Exploration

Make your own model of projectile paths. On a ruler or a stick, at position 1, hang a bead from a string 1 cm long as shown. At position 2 hang a bead from a string 4-cm long. At position 3, do the same with a 9-cm length of string. At position 4, use 16 cm of string, and for position 5, 25 cm of string. Hold the stick horizontally and you have a version of Figure 8.4. Hold it at a slight upward angle to show a version of Figure 8.5. Held at an angle downward, you have Figure 8.6.

The curved path of a projectile is a combination of horizontal and vertical motions. Consider the girl throwing the stone in Figure 8.7. The velocity she gives the stone is shown by the light blue vector. Notice that this vector has horizontal and vertical *components.* These components, interestingly, are completely independent of each other. The horizontal component is completely independent of the vertical component. They act as if the other didn't exist. Their combined effects produce the curved paths of projectiles.

A typical projectile path in Figure 8.8 shows velocity vectors and their components. Notice that the horizontal component remains the same at all points. That's because no horizontal force exists to change this component of velocity (assuming negligible air drag). The vertical component, however, changes because of the vertical influence of gravity.

Figure 8.7
The velocity of the ball (light blue vector) has vertical and horizontal components. The vertical component relates to how high the ball will go. The horizontal component relates to the horizontal range of the ball.

Concept Check ✓

1. At what part of its trajectory does a projectile have minimum speed?
2. (Challenge Question) A tossed ball changes speed along its parabolic path. When the sun is directly overhead, does the shadow of the ball across the field also change speed?

Check Your Answers

1. The speed of a projectile is a minimum at the top of its path. If it is launched vertically, its speed at the top is zero. If it is projected at an angle, the vertical component of speed is zero at the top, leaving only the horizontal component. So the speed at the top is equal to the horizontal component of the projectile's velocity at any point.
2. No, for the shadow moves at constant velocity across the field, showing exactly the motion due to the horizontal component of the ball's velocity.

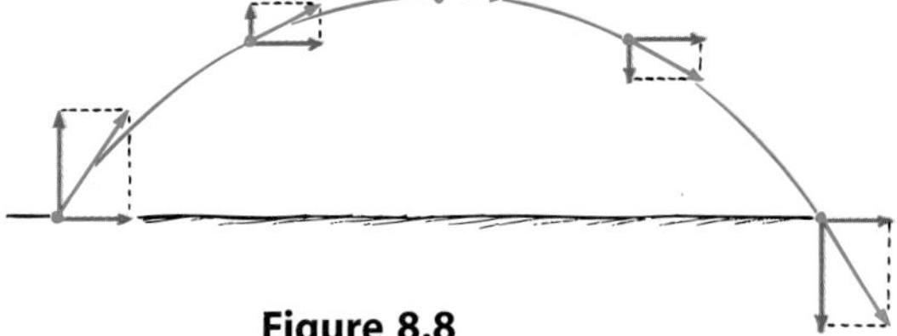

Figure 8.8
The velocity of a projectile at various points. Note that the vertical component changes while the horizontal component is the same everywhere.

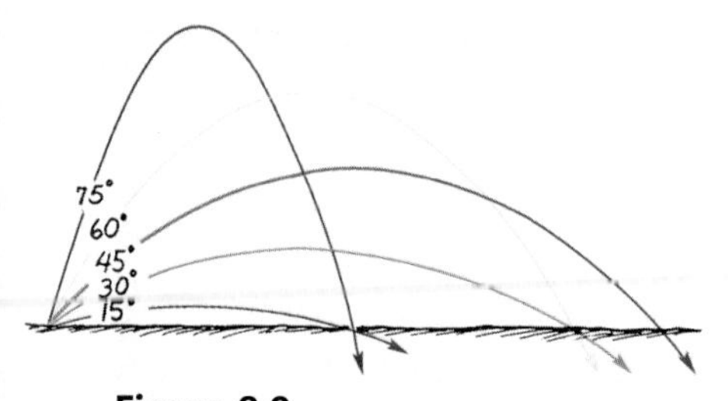

Figure 8.9
The paths of projectiles launched with equal speeds but different projection angles. Note the same range occurs for angles that add to 90 degrees.

8.2 Projectile Altitude and Range

In Figure 8.9, we see the paths of several projectiles in the absence of air drag. All of them have the same initial speed but different projection angles. Notice that these projectiles reach different *altitudes,* or heights above the ground. They also have different *ranges,* or distances traveled horizontally. The remarkable thing to note is that the same range is obtained from two different projection angles—a pair that add up to 90°! An object thrown into the air at an angle of 60°, for example, will have the same range as if it were thrown at the same speed at an angle of 30°. For the smaller angle, of course, the object remains in the air for a shorter time.

When air drag is low enough to be negligible, a projectile will rise to its maximum height in the same time it takes to fall from that height to the ground. This is because the speed it loses while going up is the same as the speed it gains while coming down. So the projectile arrives at the ground with the same speed it had when it was projected from the ground.

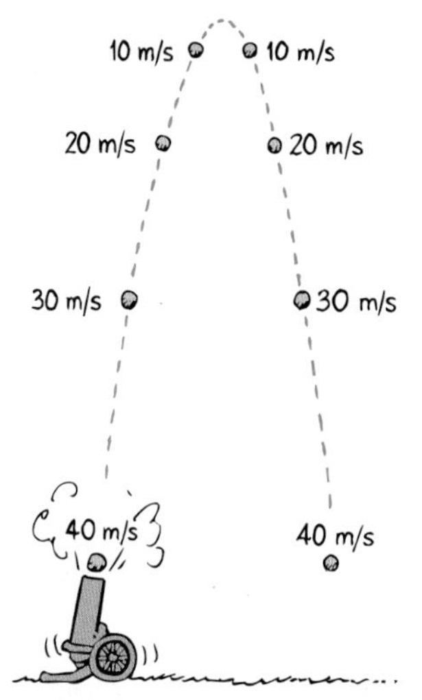

Figure 8.10
Without air drag, speed lost while going up equals speed gained while coming down; time going up equals time coming down.

An athlete or dancer jumping upward is a projectile as soon as the feet leave the ground. The time when feet are off the ground, called *hang time,* depends only on the vertical component of liftoff velocity. Acceleration is that of free fall. How high the jumper goes depends on the vertical component of liftoff velocity. How far the jumper goes horizontally depends only on the horizontal component of velocity and the hang time (the time the jumper is airborne). So when you jump from a skateboard, you reach the same height whether the skateboard is at rest or moving.

In running, however, you *can* jump higher when your foot bounds against the ground during liftoff. When this occurs, you increase hang time. In every case, however, once liftoff is achieved, horizontal and vertical components of velocity are independent of each other. This is an important rule for projectile motion.

The longest hang time for a standing jump is 1 second, for a record 1.25 meters (4 ft) height. Can anyone in your school jump that high? Like with feet 1.25 meters above the ground? Not likely!

8.3 The Effect of Air Drag on Projectiles

We have considered projectile motion without air drag. You can neglect air drag for a ball you toss back and forth with your friends because the speed is small. But higher speed makes a difference. Air drag is a factor for high-speed projectiles. The result of air drag is that both range and altitude are less.

Air drag greatly affects the range of balls batted and thrown in baseball games. Without air drag, a ball normally batted to the middle of center field would be a home run. If baseball were played on the moon (not scheduled in the near future!), the range of balls would be considerably farther—about six times ideal range on Earth. This is because there is no atmosphere on the moon, so air drag on

Calculation Corner

If the boy simply drops a baseball a vertical distance of 5 m it will hit the ground in 1 s. Suppose instead that he throws the ball horizontally as shown. The ball lands 20-m downrange. What is his pitching speed?

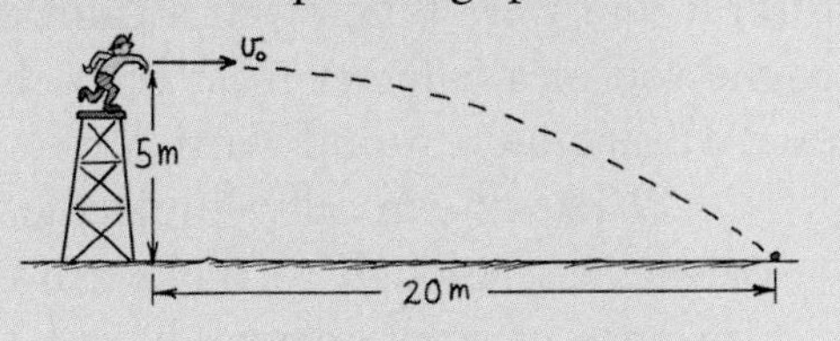

Answer
The ball is thrown horizontally, so the pitching speed is the horizontal distance divided by time. A horizontal distance of 20 m is given. How about the time? Isn't the time along the parabola the same time it takes to vertically fall 5 m? Isn't this time 1 s? So pitching speed $v = d/t = (20\ \text{m})/(1\text{s}) = 20\ \text{m/s}$.

the moon is completely absent. Second, gravity is one-sixth as strong on the moon, which allows higher and longer paths.

Back here on Earth, baseball games normally take place on level ground. Baseballs curve over a flat playing field. The speeds of baseballs are not great enough for the Earth's curvature to affect the ball's path. For very long range projectiles, however, the curvature of the Earth's surface must be taken into account. As we will now see, when an object is projected fast enough, it can fall all the way around the Earth and become a **satellite**.

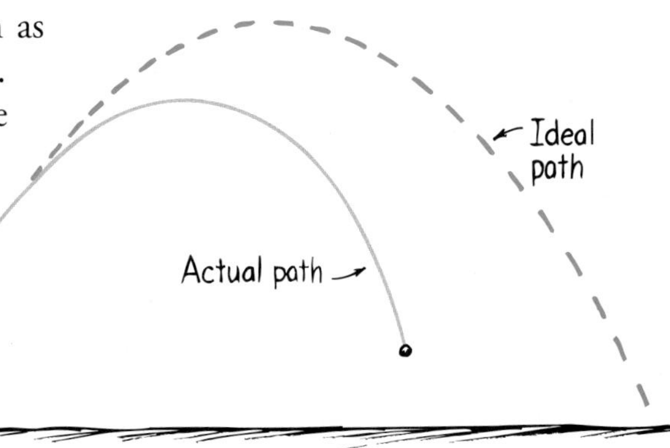

Figure 8.11
In the presence of air resistance, a high-speed projectile falls short of a parabolic path. The dashed line shows an ideal path with no air resistance. The solid line indicates an actual path.

8.4 Fast-Moving Projectiles—Satellites

Suppose a cannon fires a cannonball so fast that its curved path matches the curvature of the Earth. Then without air drag, it would be an Earth satellite! The same would be true if you could throw a stone fast enough. Any satellite is simply a projectile moving fast enough to fall continually around the Earth.

In Figure 8.12, we see the curved paths of a stone thrown horizontally at different speeds. Whatever the pitching speed, in each case the stone drops the same vertical distance in the same time. For a 1-second drop, that distance is 5 meters (perhaps by now you have made use of this fact in lab). So if you simply drop a stone from rest, it will fall 5 meters in 1 second of fall. Toss the stone sideways, and in 1 second it will be 5 meters below where it would have been without gravity. To be an Earth satellite, the stone's horizontal velocity must be great enough for its falling distance to match the Earth's curvature.

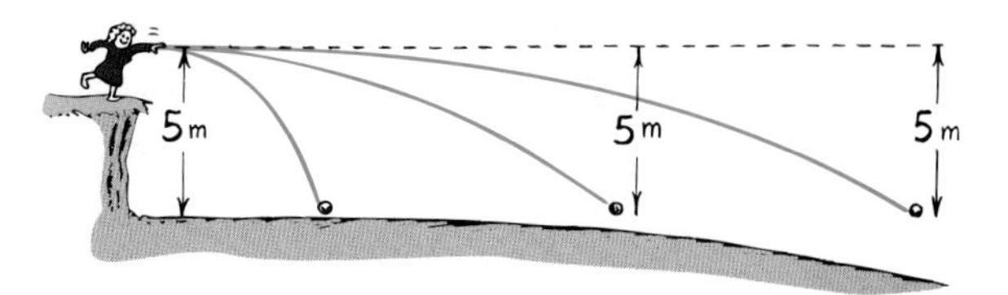

Figure 8.12
Throw a stone at any speed and 1 second later it falls 5 meters below where it would have been if there were no Earth gravity.

8.5 Earth Satellites

It is a geometrical fact that the surface of the Earth drops a vertical distance of 5 meters for every 8000 meters tangent to the surface. A tangent to a circle or to the Earth's surface is a straight line that touches the circle or surface at only one place. (So the tangent is parallel to the circle or sphere at the point of contact). With this amount of Earth's curvature, if you were floating in a calm ocean you would be able to see only the top of a 5-meter mast on a ship 8000 meters (8 kilometers) away. We live on a round Earth.

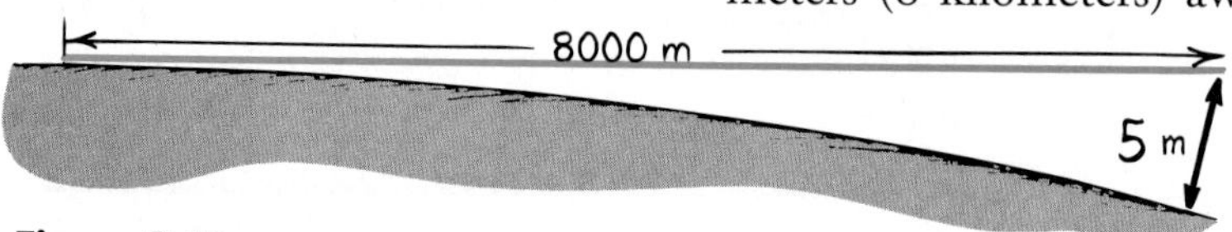

Figure 8.13
The Earth's curvature drops a vertical distance of 5 m for each 8000-m tangent (not to scale).

What do we call a projectile that moves fast enough to travel a horizontal distance of 8 kilometers during 1 second? We call it a satellite. Neglecting air drag, it would follow the curvature of the Earth. A little thought tells you that this speed is 8 kilometers per second. If this doesn't seem fast, convert it to kilometers per hour and you get an impressive 29,000 kilometers per hour (18,000 mi/h). Fast, indeed!

At this speed, atmospheric friction would incinerate the projectile. This happens to grains of sand and other meteorites that graze the Earth's atmosphere, burn up, and appear as "falling stars." That is why satellites like the space shuttles are launched to altitudes higher than 150 kilometers—to be above the atmosphere.

Figure 8.14
If the stone is thrown fast enough so that its curve matches the Earth's curvature, it will be a satellite.

It is a common misconception that satellites orbiting at high altitudes are free from gravity. Nothing could be farther from the truth. The force of gravity on a satellite 150 kilometers above the Earth's surface is nearly as great as at the surface. If there were no gravity, motion would be along a straight-line path instead of curving around the Earth. High altitude puts the satellite beyond the Earth's *atmosphere,* but not beyond Earth's *gravity.* As mentioned in the previous chapter, Earth gravity goes on forever, getting weaker with distance, but never reaching zero.

Satellite motion was understood by Isaac Newton. He reasoned that the moon is simply a projectile circling the Earth under gravitational attraction. This concept is illustrated in Figure 8.15, which is an actual drawing by Newton. He compared the moon's motion to a cannonball fired from the top of a high mountain. He imagined that the mountaintop was above the Earth's atmosphere, so that air drag would not slow the motion of the cannonball. If a cannonball were fired with a low horizontal speed, it would follow a curved path and soon hit the Earth below. If it were fired faster, its path would be wider and it would hit a place on Earth farther away. If the cannonball were fired fast enough, Newton reasoned, the curved path would become a circle and the cannonball would circle the Earth indefinitely. It would be in orbit.

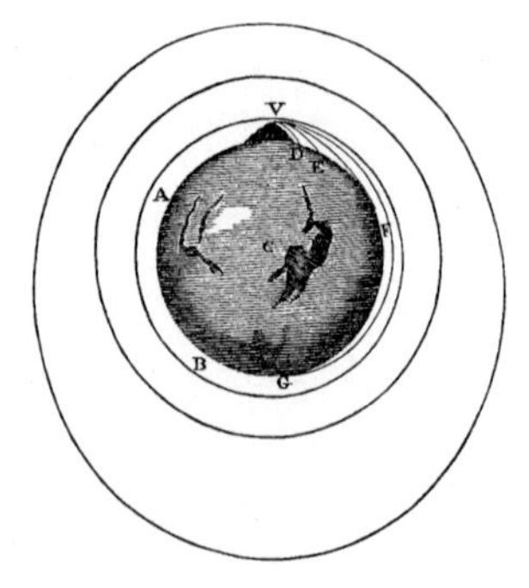

Figure 8.15
A drawing by Newton showing how a faster and faster projectile could circle the Earth and become a satellite.

Newton calculated the speed for circular orbit about the Earth. However, since such a cannon-muzzle velocity was clearly impossible, he did not foresee humans launching satellites. And quite likely he didn't foresee multi-stage rockets.

Both the cannonball and moon have a tangential ("sideways") velocity, parallel to the Earth's surface. This velocity is enough to ensure motion *around* the Earth rather than *into* it. Without air drag to reduce speed, the moon or any Earth satellite "falls" around and around the Earth indefinitely. Similarly with the planets that continually fall around the sun in closed paths.

Why don't the planets crash into the sun? They don't because of their tangential velocities. What would happen if their tangential velocities were reduced to zero? The answer is simple enough: Their motion would be straight toward the sun and they would indeed crash into it. Any objects in the solar system without sufficient tangential velocities have long ago crashed into the sun. What remains is the harmony we observe.

Figure 8.16
The initial thrust of the rocket pushes it above the atmosphere. Another thrust to a tangential speed of at least 8 km/s is needed if it is to fall around rather than into the Earth.

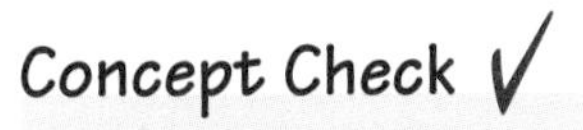

Can we also say a satellite stays in orbit because it is above the Earth's main pull of gravity?

Check Your Answer No, no, no! No satellite is completely "above" the Earth's gravity. If the satellite were not in the grip of Earth's gravity it would not orbit and follow instead a straight-line path.

Satellites are payloads carried above the atmosphere by rockets. Putting a payload into orbit requires control over the speed and direction of the rocket. A rocket initially fired vertically is intentionally tipped from the vertical course as it rises. Then, once above the drag of the atmosphere, it is aimed horizontally, whereupon the payload is given a final thrust to orbital speed (Figure 8.16).

Pretend Corner: A Cannon Shoot

Imagine you could be safely fired from a circus cannon horizontally at 8 kilometers per second. Pretend there is no air drag and the cannon is several meters above the ground. In the first second you'd travel 8 kilometers, fall vertically 5 meters, and still be the same distance above the ground. Your curved path would match the Earth's curvature. After another second you'd be another 8 kilometers down range, but still the same distance above the ground. If you didn't run into any obstacles you'd fall continually while remaining at a constant altitude—you'd be in low-Earth orbit!

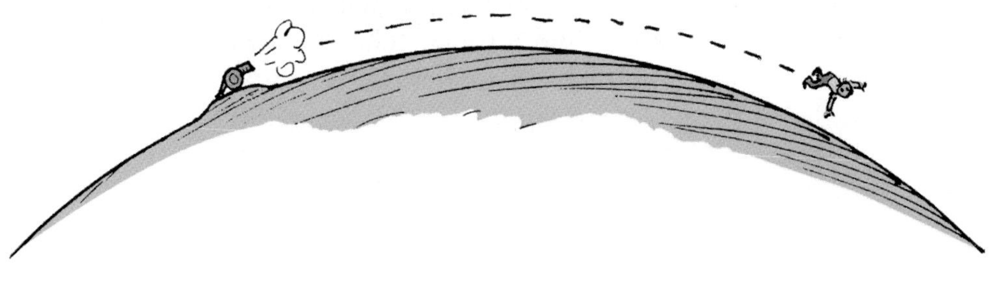

Figure 8.17
If you were shot from a make-believe circus cannon at 8 km/s, with no air drag, you'd be an Earth satellite!

Figure 8.18
Teacher Marshall Ellenstein whirls a bucket of water in a vertical circle and asks his class why water doesn't spill at the top of the swing. How does this relate to satellite motion?

Hands-On Exploration: The Water Bucket Swing

Swing a bucket of water in a vertical circle, as shown in Figure 8.18. If you swing it sufficiently fast, the water won't spill. The explanation is similar to why satellites don't "fall" to Earth. Actually, both water in the bucket and satellites *are* falling. The water doesn't spill at the top of the swing because the bucket swings downward at least as fast as the water falls. Similarly, a satellite doesn't get closer to Earth because it falls a distance that matches the Earth's curvature. Analogies are the way to understand concepts!

Concept Check ✓

What would be the fate of a rocket launched vertically that remains vertical as it rises?

Check Your Answer After the rocket reaches its highest point it would fall back to its launching site—not a good idea!

For a satellite close to the Earth, the period (the time for a complete orbit about the Earth) is about 90 minutes. For higher altitudes, gravitation is less and the orbital speed is less—the period is longer. For example, communication satellites located at an altitude of 5.5 Earth radii have a period of 24 hours. This period matches the period of daily Earth rotation. For an orbit around the equator, these satellites stay above the same point on the ground. The moon is even farther away and has a period of 27.3 days. The higher the orbit of a satellite, the less its speed and the longer its period.

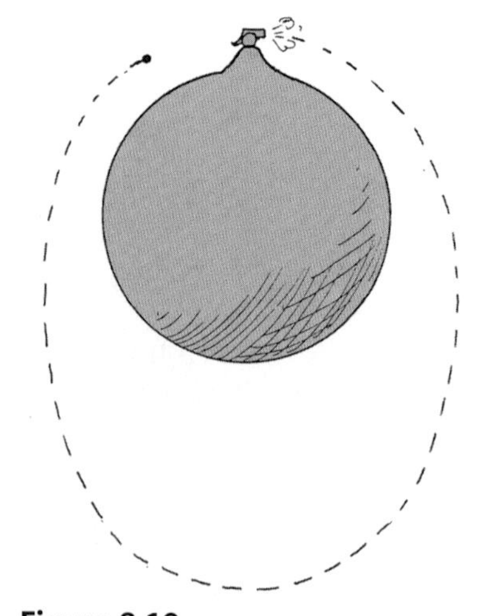

Figure 8.19
A cannonball fired at 9 km/s will overshoot a circular path and follow an ellipse.

8.6 Elliptical Orbits

If a payload above the drag of the atmosphere is given a horizontal speed somewhat greater than 8 kilometers per second, it will overshoot a circular path and trace an elliptical path.

An **ellipse** is a specific curve. It is an oval-like path along which any point has the same sum of distances from two fixed points (called *foci*). For a satellite orbiting Earth, one focus is at the Earth's center; the other focus could be inside or outside the Earth. An ellipse can be easily constructed by using a pair of tacks, one at each focus, a loop of string, and a pencil. The closer the foci are to each other, the closer the ellipse is to a circle. When both foci are together, the ellipse is a circle. So we see that a circle is a special case of an ellipse.

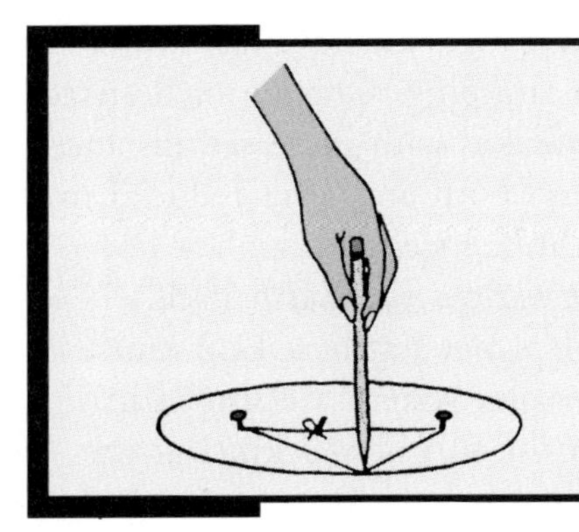

Hands-On Exploration

Construct your own ellipses with string and tacks as shown. Note that the closer your tacks, the more circular your ellipse. When the tacks are farther apart, the ellipse is more *eccentric.*

Unlike the constant speed of a satellite in a circular orbit, speed varies in an elliptical orbit. In circular orbit the satellite path is always parallel to the Earth's surface. Like a bowling ball on a lane, also parallel to the Earth's surface, speed doesn't change. Both the satellite and bowling ball don't go with nor against gravity.

A satellite following an elliptical orbit is different. Half the time the satellite moves away from the Earth, and half the time it moves toward the Earth. When it moves away, against the force of gravity, it loses speed. Like a stone thrown into the air, it slows to a point where it no longer recedes and then begins to fall back toward the Earth. Then it gains speed. The speed lost in receding is regained as it falls back toward the Earth. Then the satellite rejoins its original path with the same speed it had initially. The procedure repeats over and over, and an ellipse is traced each cycle.

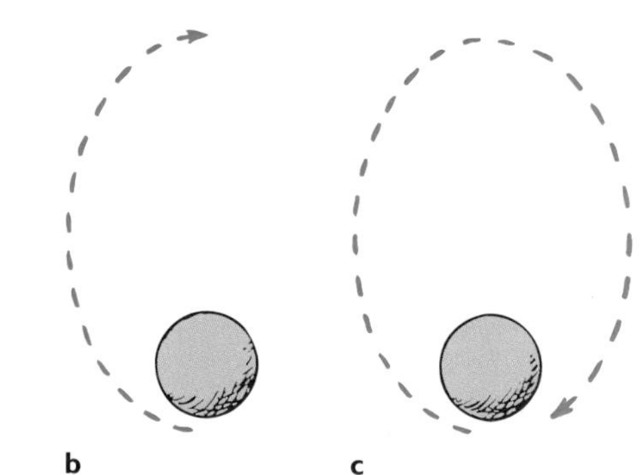

Figure 8.20
The satellite loses speed in receding from the Earth, and regains it when falling back toward the Earth. The cycle is repeated and the satellite remains in an elliptical orbit.

8.7 Escape Speed

We know that a cannonball fired horizontally at 8 kilometers per second from Newton's mountain would be in orbit. But what would happen if the cannonball were instead fired at the same speed *vertically*? It would rise to some maximum height, reverse direction, and then fall back to Earth. Then the old saying "What goes up must come down" would hold true, just as surely as a stone tossed skyward will be returned by gravity (unless, as we shall see, its speed is too great).

Today it is more accurate to say, "What goes up *may* come down," because there is a critical speed at which a projectile can outrun gravity and escape the Earth. This critical speed is called **escape speed** or, if direction is involved, *escape velocity.* From the surface of the Earth, escape speed is 11.2 kilometers per second. A projectile

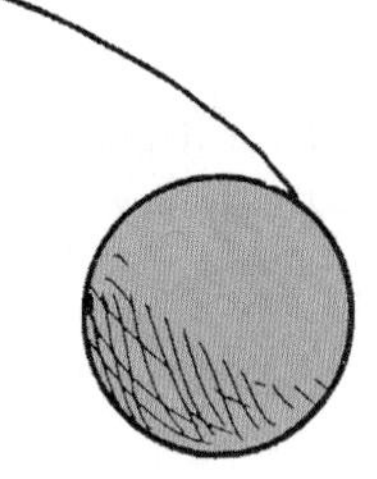

Figure 8.21
Project anything away from the Earth at 11.2 km/s and its trip is a one-way affair. It won't return.

launched at any greater speed will leave the Earth, traveling slower and slower due to Earth's gravity, never stopping. Gravitational attraction to the Earth becomes weaker and weaker with increased distance, and speed becomes less and less—though both are never reduced to zero. The projectile outruns the Earth's influence. Although it never escapes the tug of Earth's gravitation, it escapes the Earth itself.

So escape speed from the surface of planet Earth is 11.2 km/s. Escape speeds from other bodies in the solar system are shown in Table 8.1. Note that escape speed from the sun is 620 km/s at the surface of the sun. Farther from the sun it is less. At a distance equal to the Earth's distance from the sun, escape speed from the sun is 42.5 km/s. This is much greater than escape speed from Earth. An object projected from the Earth at a speed greater than 11.2 km/s but less than 42.5 km/s will escape the Earth, but not the sun. Rather than recede forever, it will take up an orbit around the sun.

Table 8.1

Escape Speeds at the Surface of Bodies in the Solar System

Astronomical Body	Mass (Earth masses)	Radius (Earth radii)	Escape speed (km/s)
Sun	333,000	109	620
Sun (at a distance of the Earth's orbit)		23,500	42.2
Jupiter	318	11	60.2
Saturn	95.2	9.2	36.0
Neptune	17.3	3.47	24.9
Uranus	14.5	3.7	22.3
Earth	1.00	1.00	11.2
Venus	0.82	0.95	10.4
Mars	0.11	0.53	5.0
Mercury	0.055	0.38	4.3
Moon	0.0123	0.27	2.4

Escape speed might well be called the *maximum falling speed.* Here is why. If it takes 11.2 km/s to send a projectile to a place very, very far from Earth, what would be the projectile's speed if it fell back to Earth? As a closer-to-Earth example, suppose you toss a ball at 10 m/s to just barely reach a friend at the top of a building. Whether your friend touches the ball or not, what will be the speed of the ball when it gets back to you? You're correct if you say 10 m/s. The same reasoning applies to a body falling from enormously far away under only the influence of Earth's gravity.

Escape speeds refer to the initial speed given by a *brief* thrust, after which there is no force to assist motion. One could escape the Earth at any *sustained* speed more than zero, given enough time. For example, a slowly moving rocket can reach the moon if its engines continually fire. Interestingly, a rocket doesn't reach its destination by staying on a pre-planned path. And if it strays off course, it makes no attempt to get on any pre-planned path. Instead, the control center in effect asks, "Where is it now relative to where it ought to be? What is the best way to get there from where it now is?" With the aid of high-speed computers, the answers to these questions are used in finding a *new* path. Corrective thrusters put the rocket on this new path. This process is repeated over and over again all the way to the goal.

Perhaps there's a lesson to be learned here. Suppose, for example, in your personal life you find you are "off course." Like the rocket, you may find it better to take a newer course that leads to your goal from where you are now. This may be wiser than trying to get back on the course you plotted from a previous position and time—and perhaps under different circumstances. So many ideas in physical science, it seems, have a moral.

As our space-faring efforts carry us farther into space, we may more and more come to see Earth as our local address—and the entire solar system as our home.

Concept Check ✓

If a flight mechanic drops a wrench from a high-flying airplane, it crashes to Earth. But if an astronaut outside the orbiting space shuttle drops a wrench, it doesn't crash to Earth. Explain.

Check Your Answer If a wrench or anything else is "dropped" from an orbiting space vehicle, it has much the same tangential speed as the vehicle and remains in orbit. If a wrench is dropped from a high-flying airplane, it too has the tangential speed of the airplane. But this speed is too small for the wrench to orbit the Earth. Instead it soon crashes to the Earth's surface.

Mechanics Review

Several positions of a satellite in an elliptical orbit are shown. At which position does the satellite have the greatest

1. speed?
2. velocity?
3. mass?
4. gravitational attraction to Earth?
5. kinetic energy?
6. potential energy?
7. total energy?
8. acceleration? (Let the equation $a = F/m$ guide you.)

Answers

1. *Greatest speed at A.*
2. *Greatest velocity where greatest speed occurs, A.*
3. *Same mass everywhere. Mass does not depend on location.*
4. *Greatest at A, where it is closest to the Earth.*
5. *Greatest at A, where speed is greatest.*
6. *Greatest at C, where distance is greatest and speed least.*
7. *Same everywhere, in accord with energy conservation.*
8. *Greatest at A, where gravitational force is greatest.*

Chapter Review

Key Terms and Matching Definitions

_____ ellipse
_____ escape speed
_____ parabola
_____ projectile
_____ satellite

1. Any object that moves through the air or through space under the influence of gravity.
2. The curved path followed by a projectile near the Earth under the influence of gravity only.
3. A projectile or small body that orbits a larger body.
4. The oval path followed by a satellite. The sum of the distances from any point on the path to two points called foci is a constant. When the foci are together at one point, the ellipse is a circle. As the foci get farther apart, the path gets more "eccentric."
5. The speed that a projectile, space probe, or similar object must reach to escape the gravitational influence of the Earth or celestial body to which it is attracted.

Review Questions

Projectile Motion

1. What exactly is a projectile?
2. How much speed does a freely-falling object gain during each second of fall?
3. With no gravity, a horizontally-moving projectile follows a straight-line path. With gravity, how far below the straight-line path does it fall compared with the distance of free fall?
4. As an object moves horizontally through air (without air drag), how much speed does it gain moving horizontally? How much vertically?
5. A ball is batted upward at an angle. What happens to the vertical component of its velocity as it rises? As it falls?
6. With no air drag, what happens to the horizontal component of velocity for the batted baseball?

Projectile Altitude and Range

7. A projectile is launched upward at an angle of 75 degrees from the horizontal and strikes the ground a certain distance down range. For what other angle of launch at the same speed would this projectile land just as far away?
8. A projectile is launched vertically at 30 m/s. If air drag can be neglected, at what speed will it return to its initial level?
9. What is meant by *hang time?*

The Effect of Air Drag on Projectiles

10. What is the effect of air drag on the height and range of a batted baseball?

Fast-Moving Projectiles—Satellites

11. How can a projectile "fall around the Earth?"

Earth Satellites

12. Why will a projectile that moves horizontally at 8 km/s follow a curve that matches the curvature of the Earth?
13. Why is it important that the projectile in the last question be above the Earth's atmosphere?
14. Are the planets of the solar system simply projectiles falling around and around the sun?
15. Why did Isaac Newton not think humans would one day orbit the Earth?
16. How much time is taken for a complete revolution of a satellite in close orbit about the Earth?
17. For orbits of greater altitude, is the period greater or less?

Elliptical Orbits

18. Why does the speed of a satellite undergo change in an elliptical orbit?

19. At what part of an elliptical orbit does a satellite have the greatest speed? The least speed?

Escape Speed

20. What is the minimum speed for orbiting the Earth in close orbit? The maximum speed? What happens above this speed?

Exercises

1. A heavy crate accidentally falls from a high-flying airplane just as it flies directly above a shiny red sports car parked in a car lot. Relative to the car, where will the crate crash?

2. How does the vertical component of motion for a ball kicked off a high cliff compare with the motion of vertical free fall?

3. In the absence of air drag, why does the horizontal component of the ball's motion not change, while the vertical component does?

4. At what point in its trajectory does a batted baseball have its minimum speed? If air drag can be neglected, how does this compare with the horizontal component of its velocity at other points?

5. Two golfers each hit a ball at the same speed, one at 60° above the horizontal and the other at 30°. Which ball goes farther? Which hits the ground first? (Ignore air resistance.)

6. A park ranger shoots a monkey hanging from a branch of a tree with a tranquilizing dart. The ranger aims directly at the monkey, not realizing that the dart will follow a parabolic path and thus fall below the monkey. The monkey, however, sees the dart leave the gun and lets go of the branch to avoid being hit. Will the monkey be hit anyway? Defend your answer.

7. When you jump upward, your hang time is the time your feet are off the ground. Does hang time depend on your vertical component of velocity when you jump, your horizontal component of velocity, or both? Defend your answer.

8. Since the moon is gravitationally attracted to the Earth, why doesn't it simply crash into the Earth?

9. Which planets have a greater period than 1 Earth year, those closer to the sun than Earth or those farther from the sun than Earth?

10. Does the speed of a falling object depend on its mass? Does the speed of a satellite in orbit depend on its mass? Defend your answers.

11. If you have ever watched the launching of an Earth satellite, you may have noticed that the rocket starts vertically upward, then departs from a vertical course and continues its rise at an angle. Why does it start vertically? Why does it not continue vertically?

12. A satellite can orbit at 5 km above the moon, but not at 5 km above the Earth. Why?

13. If a space shuttle circled the Earth at a distance equal to the Earth-moon distance, how long would it take for it to make a complete orbit? In other words, what would be its period?

14. Consider a high-orbiting spaceship that travels at 7 km/s with respect to the Earth. Suppose it projects a capsule rearward at 7 km/s with respect to the ship. Describe the path of the capsule with respect to the Earth.

15. The orbital velocity of the Earth about the sun is 30 km/s. If the Earth were suddenly stopped in its tracks, it would simply fall directly into the sun. Devise a plan whereby a rocket loaded with radioactive wastes could be fired into the sun for permanent disposal. How fast and in what direction with respect to the Earth's orbit should the rocket be fired?

16. If you stopped an Earth satellite dead in its tracks, it would simply crash into the Earth. Why, then, don't the communication satellites that "hover motionless" above the same spot on Earth crash into the Earth?

17. Escape speed from the surface of the Earth is 11.2 km/s, but a space vehicle could escape from the Earth at half this speed and less. Explain.

18. Suppose a faraway body that is initially at rest falls to the Earth under the influence of Earth's gravity only. What is the maximum possible speed of impact of the object when it hits Earth's surface?

19. If Pluto were somehow stopped short in its orbit, it would fall into rather than around the sun. How fast would it be moving when it hit the sun?

20. Which requires more fuel, a rocket going from the Earth to the moon, or one going from the moon to the Earth? Defend your answer.

Problems

1. Students in a lab roll a steel ball off the edge of a table. They measure the speed of the horizontally-launched ball to be 4.0 m/s. They also know that simply dropping the ball from rest off the edge of the table takes 0.5 seconds to hit the floor. Question: How far from the bottom of the table should they place a small piece of paper so that the ball will hit it when it lands?

2. Calculate the speed in m/s at which the Earth revolves about the sun. You may assume the orbit is nearly circular.

3. The moon is about 3.8×10^5 km from the Earth. Find its average orbital speed about the Earth.

Part 2 Heat

Although the temperature of these sparks exceeds 2000°C, the heat they impart when striking my skin is very small, which tells me that *temperature* and *heat* are very different concepts. Learning to distinguish between closely related concepts is the challenge I face in exploring *Conceptual Physical Science*.

Chapter 9: Thermal Energy

Bits of hot matter from this volcano would burn you if they made contact with you. But when white-hot sparks from a holiday sparkler hit your skin, they don't burn you—why is this so? Can things get colder and colder forever? Hotter and hotter? Or are there limits to hotness and coldness? Why does a bite of the filling of a hot apple pie burn your tongue, while the crust does not? Since both parts of the pie came from the same oven, don't they have the same temperature? Another question . . . why does air in a balloon, the concrete of a sidewalk, and almost everything else expand as it heats up? Why does ice water do the opposite—contract instead of expand as its temperature rises? And why does ice form at the top of a pond, rather than on the bottom? Let's explore!

9.1 Thermal Energy—The Total Energy in a Substance

All matter is made up of constantly jiggling atoms or molecules. When jiggling slowly, the particles form solids. When jiggling faster so they slide over one another, we have liquid. When atoms and molecules move so fast that they disconnect and fly loose, we have a gas. So whether a substance is a solid, liquid, or gas depends on the motion of its particles.

When you strike a penny with a hammer, it becomes warm. Why? Because the hammer's blow causes the coin's atoms to jiggle faster. When you put a flame to a liquid, the liquid becomes warmer. When you rapidly compress air in a tire pump, the air gets warmer. In these cases the molecules are made to move faster. They gain kinetic energy. In general, the warmer an object is, the more kinetic energy its atoms and molecules possess. But that's not all. We can also say that the warmer an object gets, the more thermal energy it contains. The **thermal energy** in a substance is the total energy of all its atoms and molecules. Thermal energy consists both of the potential and kinetic energy of the particles in a substance as they wiggle and jiggle, twist and turn, vibrate, or race back and forth.

9.2 Temperature—Average Kinetic Energy Per Molecule in a Substance

To tell how warm or cold an object is we measure its **temperature.** A common thermometer measures temperature by expansion or contraction of a liquid, usually colored alcohol.

The most common thermometer in the world is the *Celsius thermometer,* named after the Swedish astronomer Anders Celsius (1701–1744). Celsius was the first person to suggest the scale of 100 degrees between the freezing point and boiling point of water. The number 0 represents the temperature at which water freezes and the number 100 represents the temperature at which water boils (at standard atmospheric pressure). In between are 100 equal parts called *degrees.*

In the United States, the number 32 represents the temperature of freezing water and the number 212 for the temperature at which water boils. This temperature scale makes up a Fahrenheit thermometer, named after its originator, the German physicist Gabriel D. Fahrenheit (1686–1736). The Fahrenheit scale is still popular in the United States.

Figure 9.1
Can we trust our sense of hot and cold? Will both fingers feel the same temperature when they are put in the warm water? Try this and see (feel) for yourself.

Arithmetic formulas are used for converting from one temperature scale to the other and are common in classroom exams. Because such arithmetic exercises are not really physical science, we won't be

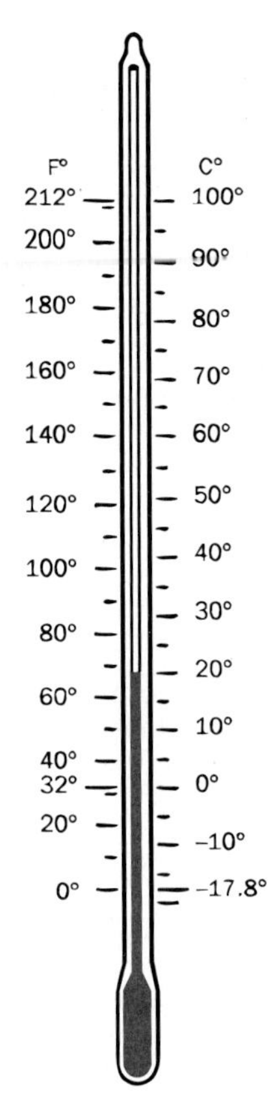

Figure 9.2
Fahrenheit and Celsius scales on a thermometer.

concerned with these conversions. (This may be important in a math class, but not here.) Besides, the conversion between Celsius and Fahrenheit temperatures is closely approximated in the side-by-side scales of Figure 9.2.*

Temperature is related to the random motion of atoms and molecules in a substance. (To be brief, from now on in this chapter, we'll simply say *molecules* to mean *atoms and molecules*.)† Temperature is proportional to the *average* kinetic energy of molecular motion. A substance with a high temperature has molecules with high average kinetic energies.

Interestingly, a thermometer actually registers its own temperature. When a thermometer is in contact with something whose temperature we wish to know, thermal energy flows between the two until their temperatures are equal. At this point, thermal equilibrium is established. So when we look at the temperature of the thermometer, we learn about the temperature of the substance with which it reaches thermal equilibrium.

Concept Check ✓

True or false: Temperature is a measure of the total kinetic energy in a substance.

Check Your Answer False. Temperature is a measure of the *average* (not the *total*!) kinetic energy of the molecules in a substance. For example, there is twice as much total molecular kinetic energy in 2 liters of boiling water as in 1 liter—but the temperatures of the two volumes of water are the same because the *average* kinetic energy per molecule in each is the same.

9.3 Absolute Zero—Nature's Lowest Possible Temperature

In principle, there is no upper limit of temperature. As thermal motion keeps increasing, a solid object melts to a liquid and then evaporates to a gas. Further heating of a gas breaks molecules up into atoms that lose some or all of their electrons. The result of this is a cloud of electrically charged particles—a *plasma*. Plasmas are found in stars, where the temperature is many millions of degrees Celsius. Temperature has no upper limit.

---

* Okay, if you really want to know, the formulas for temperature conversion are: $C = 5/9\ (F - 32)$; $F = 9/5\ C + 32$, where C is the Celsius temperature and F is the corresponding Fahrenheit temperature.

† As we shall see in Chapter 17, a molecule is a particular unit of matter composed of a group of atoms. Atoms make up molecules, and not the other way around.

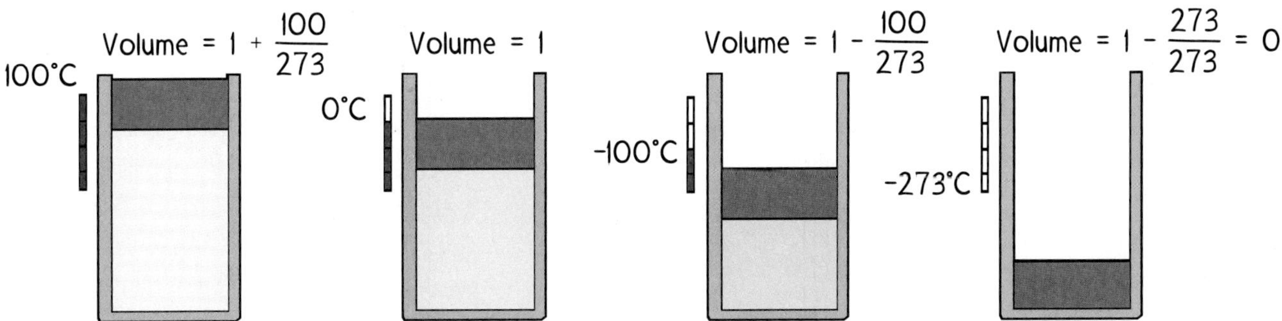

Figure 9.3
When pressure is held constant, the volume of a gas changes by 1/273 of its volume at 0°C with each 1°C change in temperature. At 100°C the volume is 100/273 greater than it is at 0°C. When the temperature is reduced to −100°C, the volume is reduced by 100/273. At −273°C the volume of the gas would be reduced by 273/273 and so would be zero.

In contrast, there is a definite limit at the other end of the temperature scale. Here's how we know that limit. Gases expand when heated and contract when cooled. Experimenters in the 19th century found that all gases shrink by 1/273 of their volume at 0°C for each Celsius degree lowering in temperature. This occurs when the gas pressure is held constant. So if a gas at 0°C were cooled down by 273°C, it would contract 273/273 volumes and be reduced to zero volume. But clearly, we cannot have a substance with zero volume.

The same occurs with pressure. The pressure of a gas of fixed volume goes down by 1/273 for each Celsius degree lowering of temperature. If it is cooled 273°C below zero, it would have no pressure at all. In practice, every gas turns to a liquid before it gets this cold. Nevertheless, these decreases by 1/273 increments suggested the idea of a lowest temperature: −273°C. That's the lower limit of temperature, **absolute zero.** At this temperature molecules have lost all available kinetic energy. No more energy can be taken from a substance at absolute zero. It can't get any colder.

The absolute temperature scale is called the Kelvin scale, named after the famous British physicist, Lord Kelvin. Absolute zero is 0 K (short for "0 kelvin,"; note that the word

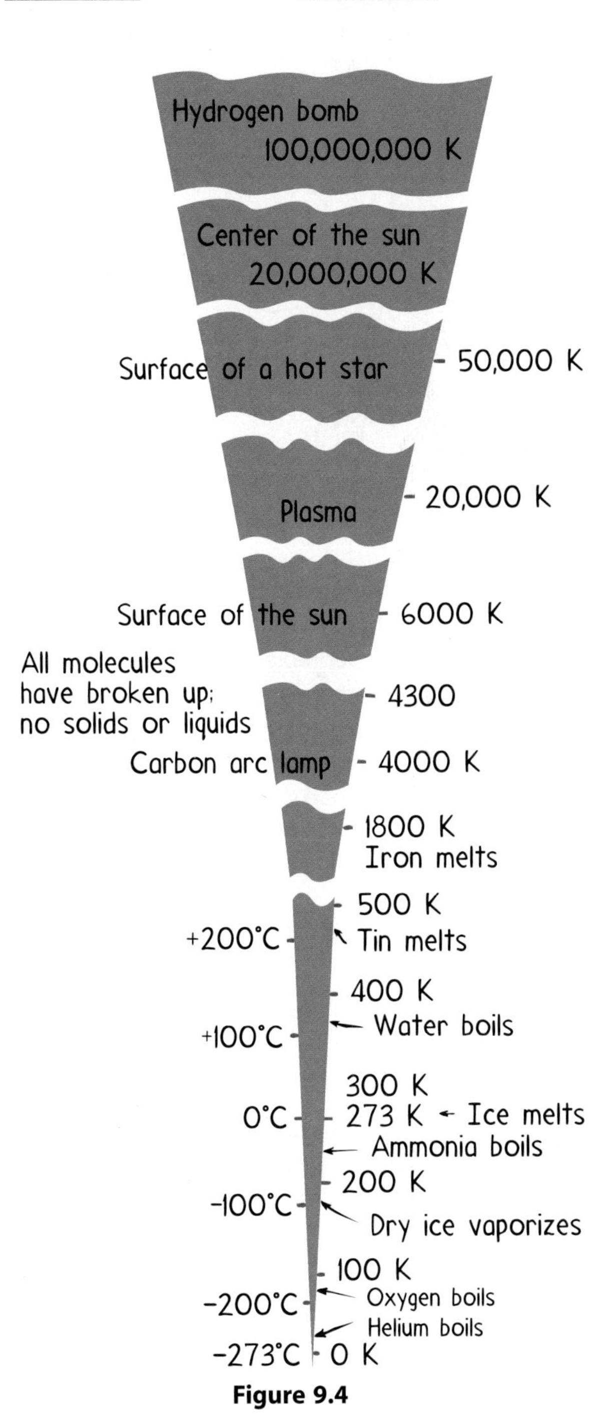

Figure 9.4
Some absolute temperatures.

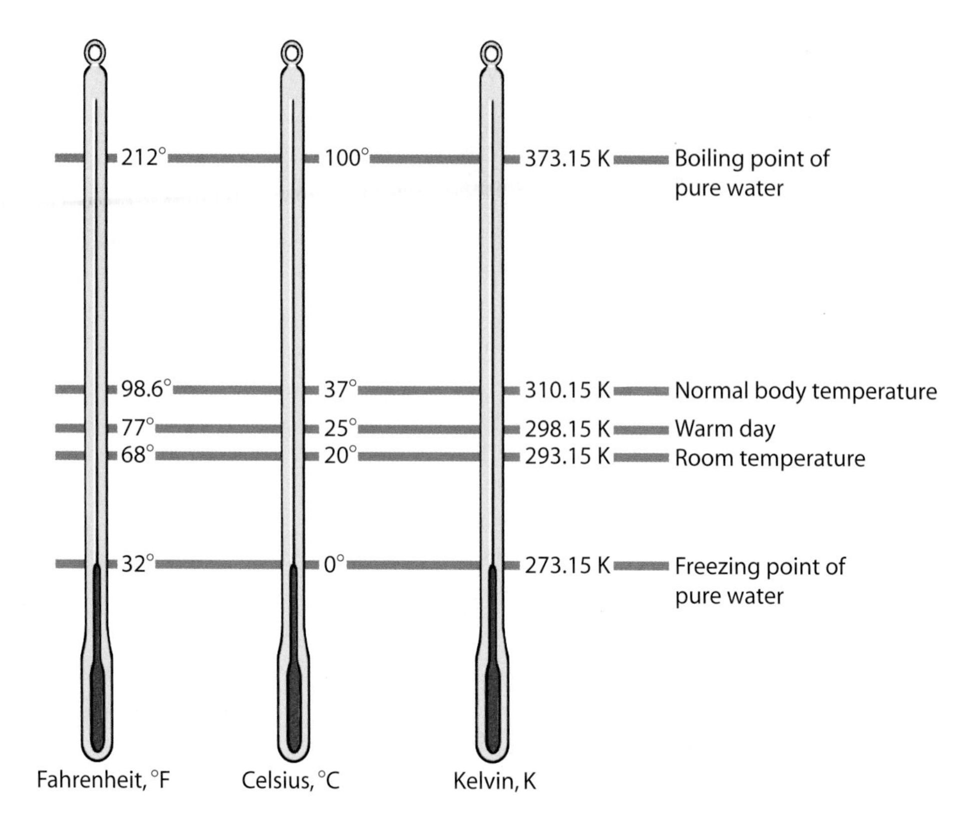

Figure 9.5
Some familiar temperatures measured on the Fahrenheit, Celsius, and Kelvin scales.

Figure 9.6
The temperature of the sparks is very high, about 2000°C. That's a lot of thermal energy per molecule of spark. Because there are only a few molecules per spark, however, the total amount of thermal energy in the sparks is safely small. Temperature is one thing; transfer of thermal energy is another.

"degrees" is not used with Kelvin temperatures). There are no negative numbers on the Kelvin scale. Degrees on it are the same size as divisions on the Celsius scale. Thus the melting point of ice is 273 K, and the boiling point of water is 373 K.

Concept Check ✓

Which is larger, a Celsius degree or a kelvin?

Check Your Answer Neither. They are equal.

9.4 Heat Is the Movement of Thermal Energy

If you touch a hot stove, thermal energy enters your hand because the stove is warmer than your hand. When you touch a piece of ice, however, thermal energy passes out of your hand and into the ice. The direction of thermal energy flow is always from a warmer substance to a cooler one. A scientist defines **heat** as the thermal energy transferred from one substance to another due to a temperature difference between the two substances.

Which has more temperature, a red-hot tack or a cool lake? Which has more thermal energy?

According to this definition, matter does not *contain* heat. Matter contains *thermal energy.* Heat is *thermal energy in transit.* After heat has been transferred to an object or substance, it ceases to be heat. It becomes thermal energy.

For substances in thermal contact, thermal energy flows from the higher-temperature substance into the lower-temperature one until thermal equilibrium is reached. This does not mean it necessarily flows from a substance with more thermal energy into one with less thermal energy. For example, there is more thermal energy in a bowl of warm water than there is in a red-hot thumbtack. If the tack is put into the water, thermal energy doesn't flow from the warm water to the tack. Instead, it flows from the hot tack to the cooler water. Thermal energy never flows by itself from a low-temperature substance into a higher-temperature one.

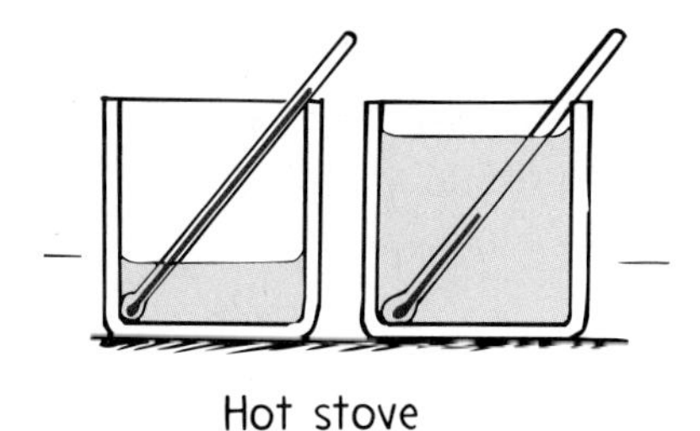

Figure 9.7
The left pot contains 1 liter of water. The right one contains 3 liters. Although both pots absorb the same quantity of heat, the temperature increases three times as much in the pot with the smaller amount of water.

We Know What Heat Is—What Is Cold?

Heat actually exists. Heat is thermal energy that transfers in a direction from hot to cold. But what is cold? Does a cold substance contain something opposite to thermal energy? The answer is no. An object is cold not because it contains something, but because it *lacks* something. It lacks thermal energy. On a near-zero winter day when you're waiting at the bus stop, you feel cold not because something called cold gets to you. You feel cold because you lose heat. Cold is not a thing in itself, but the result of lowered thermal energy.

Concept Check ✓

1. Suppose you apply a flame to 1 L of water and its temperature rises by 3°C. If you apply the same flame for the same length of time to 3 L of water, by how much does its temperature rise?
2. When you touch a cold surface, does cold travel from the surface to your hand or does energy travel from your hand to the cold surface?

Check Your Answers

1. Its temperature rises by only 1°C. This is because there are three times as many molecules in 3 L of water and each molecule receives only one-third as much energy on the average. So the average kinetic energy, and thus the temperature, increases by one-third as much. See Figure 9.7.
2. The direction of energy travel is from hot to cold—from your hand to the cold surface. There is no "cold" that travels in the other direction.

9.5 Heat Units Are Energy Units

Heat is a form of energy and is measured in joules. It takes about 4.2 joules of heat to change 1 gram of water by 1 Celsius degree. A unit of heat still common in the United States is the **calorie.** A calorie is defined as the amount of heat needed to change the temperature of 1 gram of water by 1 Celsius degree. (The relationship between calories and joules is that 1 calorie = 4.18 joules.)

Figure 9.8
To the weight watcher, the peanut contains 10 Calories; to the physicist, it releases 10,000 calories (41,800 joules) of energy when burned or digested.

The energy ratings of foods and fuels are measured by the energy released when they are burned. (Metabolism is really "burning" at a slow rate.) The heat unit for labeling foods is the kilocalorie, which is 1000 calories (the heat needed to change the temperature of 1 kilogram of water by 1°C). To tell the difference between this unit and the smaller calorie, the food unit is usually called a *Calorie* with a capital *C.*

What we've learned thus far about heat and thermal energy is summed up in the laws of thermodynamics. The word thermodynamics stems from Greek words meaning "movement of heat."

Concept Check ✓

Which will raise the temperature of water more, adding 4.18 joules or 1 calorie?

Check Your Answer Both the same. This is like asking which is longer, a 1-mile long track or a 1.6-kilometer long track. They're the same in different units.

9.6 The Laws of Thermodynamics

When thermal energy transfers as heat, it does so without net loss or gain. The energy lost in one place is gained in another. When the conservation of energy, discussed back in Chapter 7, is applied to thermal systems, we have the **first law of thermodynamics.**

> **Whenever heat flows into or out of a system, the gain or loss of thermal energy equals the amount of heat transferred.**

A *system* is any substance or well-defined group of atoms, molecules, or objects. The system may be the steam in a steam engine or it may be the whole Earth's atmosphere. It can even be the body of a living creature. Whether we add heat energy to a steam engine, to the Earth's atmosphere, or to the body of a living creature, we increase the thermal energies of these systems. The added energy enables the system to do work. The first law makes good sense.

The first law is illustrated when you put an airtight can of air on a hot stove and heat it up. The energy put in increases the thermal energy of the enclosed air. So its temperature rises. If the can is fitted with a movable piston, then the heated air can do *mechanical work* as it expands and pushes the piston outward. This ability to do mechanical work is energy that comes from the energy you put in to begin with. The first law says you don't get energy from nothing.

The **second law of thermodynamics** restates what we've learned about the direction of heat flow.

Heat never spontaneously flows from a cold substance to a hot substance.

In winter, heat flows from inside a warm home to the cold air outside. In summer, heat flows from the hot air outside into the cooler interior. The direction of spontaneous heat flow is always from hot to cold. Heat can be made to flow the other way, but only by doing work on the system or by adding energy from another source. This occurs with heat pumps and air conditioners. In these devices, thermal energy is pumped from a cooler to a warmer place. Without external effort, the direction of heat flow is from hot to cold. The second law, like the first, is logical.

The **third law of thermodynamics** restates what we've learned about the lowest limit of temperature.

No system can reach absolute zero.

Figure 9.9
When you push down on the piston you do work on the air inside. What happens to its temperature?

As investigators attempt to reach this lowest temperature, it becomes more difficult to get closer to it. Physicists have been to within less than a millionth of 1 kelvin—but never 0 K.

The laws of thermodynamics were the rage back in the 1800s. At that time, horse and buggies were giving way to steam-driven locomotives. There is the story of the engineer who explains the operation of a steam engine to a peasant. The engineer explains in detail the operation of the steam cycle, how expanding steam drives a piston that in turn rotates the wheels. After some thought, the peasant asked, "Yes, I understand all that, but where's the horse?" This story illustrates how hard it is to abandon our way of thinking about the world when a newer method comes along to replace established ways. Are we any different today?

9.7 Specific Heat Capacity—A Measure of Thermal Inertia

When you're eating, have you noticed that some foods remain hotter much longer than others? For example, just after an apple pie has been taken out of an oven, the filling burns your tongue while the crust doesn't. Or you can take a bite of a piece of hot toast a few seconds after coming from the hot toaster, but you have to wait several minutes before eating soup from a stove as hot as the toaster.

Different substances have different capacities for storing thermal energy. When you heat a pot of water on a stove you find that it takes about 15 minutes to bring it to a boil. If you put an equal mass of iron on the same stove, you'd find it rising through the same temperature range in only about 2 minutes. For silver, the time

Figure 9.10
The filling of hot apple pie may be too hot to eat, even though the crust is not.

would be less than a minute. Different materials require different amounts of thermal energy to raise temperature. This is because different materials absorb energy in different ways. The added energy may increase the jiggling motion of molecules, which raises the temperature. Or added energy may increase the amount of internal vibration or rotation within the molecules and therefore become potential energy. It's mainly the jiggling motion of its atoms and molecules that raises the temperature of a substance.

Each substance has its own characteristic **specific heat capacity***.

The specific heat capacity of any substance is defined as the quantity of heat required to change the temperature of a unit mass of the substance by 1 degree.

Specific heat capacity is a measure of thermal inertia. Recall in our study of Newton's laws that inertia is the property of matter to resist changes in motion. Specific heat capacity is a similar property of matter to resist a change in temperature.

Water has a much higher capacity for storing energy than most all other substances. A lot of heat energy is needed to change the temperature of water. This explains why water is very useful in the cooling system of automobiles and other engines. It absorbs a great quantity of heat for small rises in temperature. Water also takes longer to cool.

Concept Check ✓

Which has a higher specific heat capacity, water or sand? In other words, which takes longer to warm in sunlight (or longer to cool at night)?

Check Your Answer Water has the higher specific heat capacity. In the same sunlight, the temperature of water increases more slowly than the temperature of sand. And water will cool more slowly at night. Sand and soil's low specific heat capacity, as evidenced by how quickly it warms in the morning sun and how quickly it cools at night, affects local climates.

Water's high specific heat capacity changes the world's climate. Look at a world globe and notice the high latitude of Europe. Water's high specific heat keeps climate there milder than regions of the same latitude in northeastern regions of Canada. Both Europe and Canada receive about the same amount of sunlight per square kilometer. What happens is that the Atlantic Ocean current known as the Gulf Stream carries warm water northeast from the Caribbean. It holds much of its thermal energy long enough to reach the North

* If we know the specific heat capacity c of a substance, the formula for the quantity of heat Q involved when a mass m of the substance undergoes a change in temperature ΔT is $Q = cm\Delta T$. In words, heat transferred = specific heat capacity × mass × temperature change.

Atlantic Ocean off the coast of Europe. Then it cools, releasing 2.4 joules of energy for each gram of water that cools 1°C. The released energy is carried by westerly winds over the European continent.

A similar effect occurs in the United States. The winds in North America are mostly westerly. On the West Coast, air moves from the Pacific Ocean to the land. In winter months, the ocean water is warmer than the air. Air blows over the warm water and then moves over the coastal regions. This warms the climate. In summer, the opposite occurs. The water cools the air and the coastal regions are cooled. The East Coast does not benefit from the moderating effects of water because the direction of air is from the land to the Atlantic Ocean. Land, with a lower specific heat capacity, gets hot in the summer but cools rapidly in the winter.

Islands and peninsulas do not have the extremes of temperatures that are common in interior regions of a continent. The high summer and low winter temperatures common in Manitoba and the Dakotas, for example, are largely due to the absence of large bodies of water. Europeans, islanders, and people living near ocean air currents should be glad that water has such a high specific heat capacity. San Franciscans are!

Figure 9.11
Because water has a high specific heat capacity and is transparent, it takes more energy to warm the water than to warm the land. Solar energy striking the land is concentrated at the surface, but solar energy striking the water extends beneath the surface and so is "diluted."

Concept Check ✓

Bermuda is close to North Carolina, but unlike North Carolina, it has a tropical climate year round. Why?

Check Your Answer Bermuda is an island. The surrounding water warms it when it might be too cold, and cools it when it might be too warm.

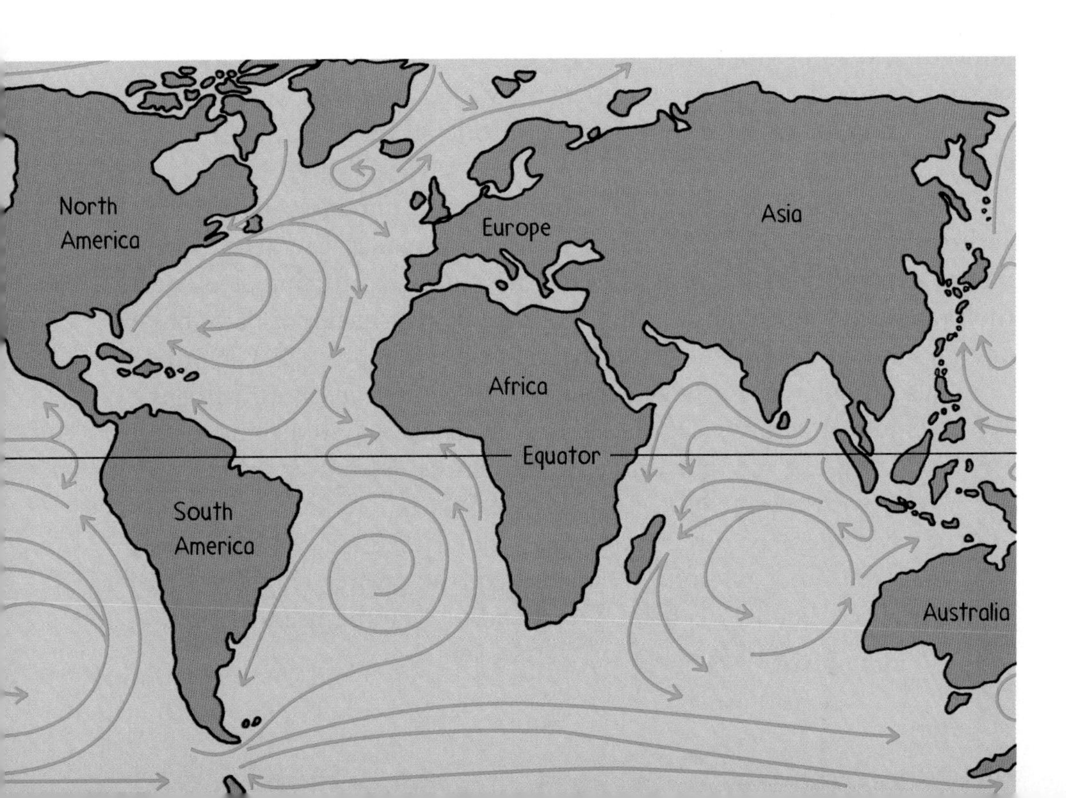

Figure 9.12
Many ocean currents, shown in blue, distribute heat from the warmer equatorial regions to the colder polar regions.

Figure 9.13
Thermal expansion. Extreme heat on a July day caused the buckling of these railroad tracks.

Figure 9.14
One end of the bridge is fixed, but the end shown rides on rockers to allow for thermal expansion.

9.8 Thermal Expansion

Molecules in a hot substance jiggle faster and move farther apart. The result is **thermal expansion.** Most substances expand when heated and contract when cooled. Sometimes the changes aren't noticed, and sometimes they are. Telephone wires are longer and sag more on a hot summer day than in winter. Railroad tracks that are laid on cold winter days expand and buckle in the summer (Figure 9.13).

Thermal expansion must be taken into account in structures and devices of all kinds. A dentist uses filling material with the same rate of expansion as teeth. A civil engineer uses reinforcing steel with the same expansion rate as concrete. A long steel bridge usually has one end fixed while the other rests on rockers (Figure 9.14). Notice also tongue-and-groove gaps called *expansion joints* on bridges (Figure 9.15). We see the effects of expansion all around us.

We can see that different substances expand at different rates with a bimetallic strip (Figure 9.16). This device is made of two strips of different metals welded together, one of brass and the other of iron. When heated, the greater lengthening of the brass bends the strip. This bending may be used to turn a pointer, regulate a valve, or close a switch.

Figure 9.15
This gap in the roadway of a bridge is called an expansion joint; it allows the bridge to expand and contract. (Was this picture taken on a warm or a cold day?)

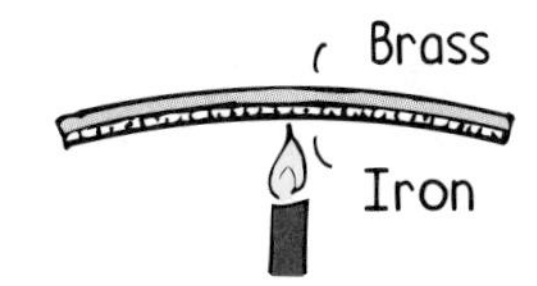

Figure 9.16
A bimetallic strip. Brass expands more when heated than iron does and contracts more when cooled. Because of this behavior, the strip bends as shown.

A practical application of a bimetallic strip wrapped into a coil is the thermostat (Figure 9.17). When a room becomes too cold, the coil bends toward the brass side and activates an electrical switch that turns on the heater. When the room gets too warm, the coil bends toward the iron side, which breaks the electrical circuit and turns off the heater. Bimetallic strips are used in oven thermometers, refrigerators, electric toasters, and various other devices.

Liquids expand more than solids with increases in temperature. We notice this when gasoline overflows from a car's tank on a hot day. If the tank and contents expanded at the same rate, no overflow would occur. This is why a gas tank being filled shouldn't be "topped off" on a hot day.

To furnace

Figure 9.17
A thermostat. When the bimetallic coil expands, the drop of liquid mercury rolls away from the electrical contacts and breaks the electrical circuit. When the coil contracts, the mercury rolls against the contacts and completes the circuit.

Concept Check ✓

1. When you can't loosen a metal lid on a glass jar, how can you use the concept of thermal expansion to rescue the situation?
2. A Concorde supersonic airplane is 20 cm longer when in flight than when parked on the ground. Offer an explanation.

Check Your Answers

1. Hold the lid of the jar under hot water for a few seconds. The metal should expand more than the glass, making it easier to loosen.
2. At cruising speed (faster than the speed of sound), air friction against the Concorde raises its temperature dramatically, resulting in this significant thermal expansion.

Figure 9.18
The six-sided structure of snow crystals is a result of the six-sided ice crystals that make it up.

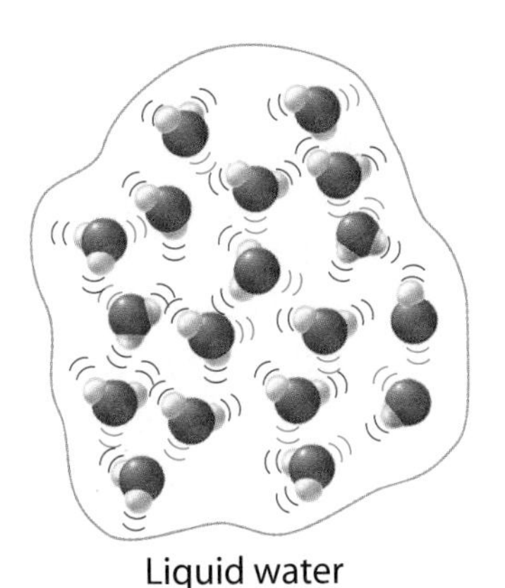

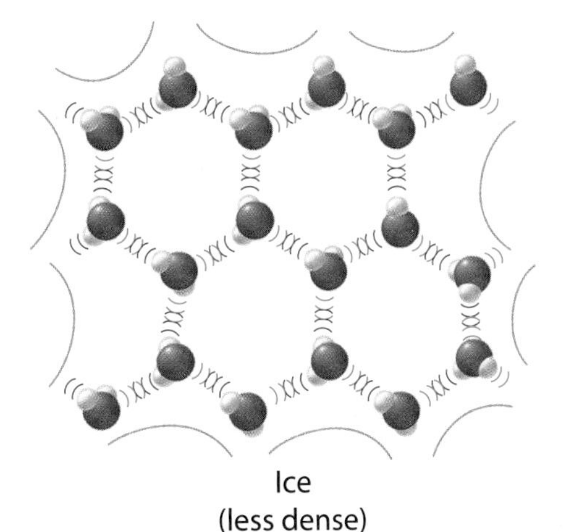

Figure 9.19
Water molecules in a liquid are denser than water molecules frozen in ice, where they have an open crystalline structure.

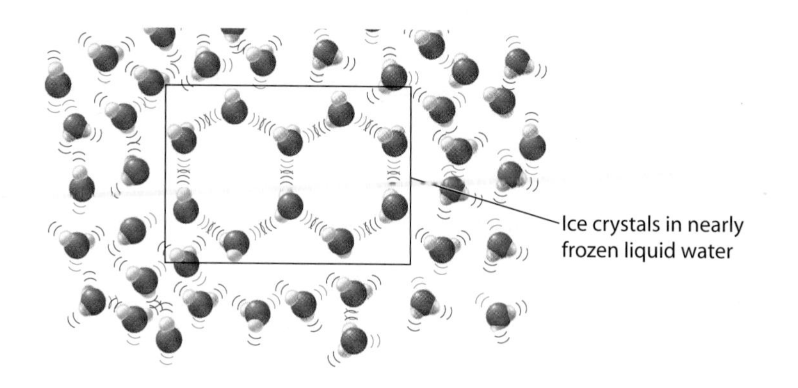

Figure 9.20
Close to 0°C, liquid water contains crystals of ice. The open structure of these crystals increases the volume of water slightly.

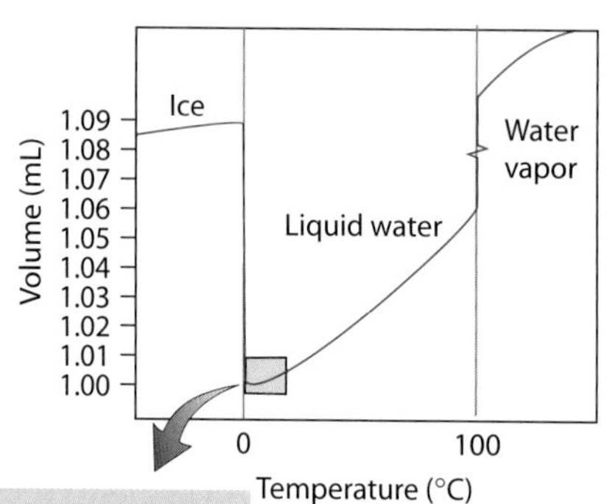

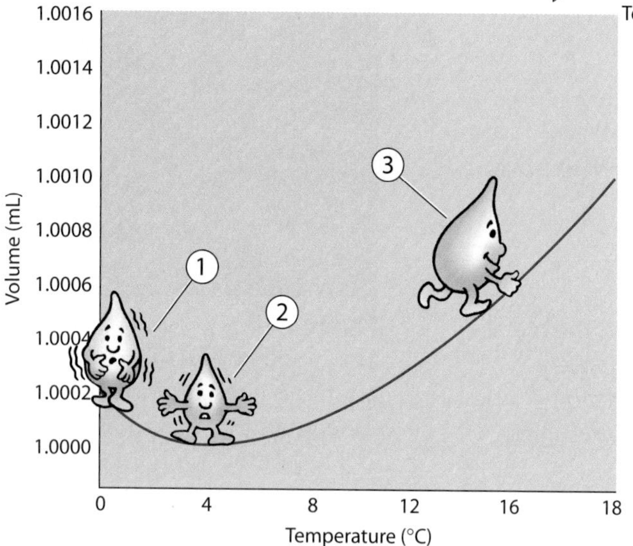

(1) Liquid water below 4°C is bloated with ice crystals.

(2) Upon warming, the crystals collapse, resulting in a smaller volume for the liquid water.

(3) Above 4°C, liquid water expands as it is heated because of greater molecular motion.

Figure 9.21
Between 0°C and 4°C, the volume of liquid water decreases as temperature increases. Above 4°C, water behaves the way other substances do. Its volume increases as its temperature increases. The volumes shown here are for a 1-gram sample.

Expansion of Water

Water, like most other substances, expands when heated. But interestingly, it doesn't expand in the temperature range between 0°C and 4°C. Something quite fascinating happens in this range. Ice has a crystalline structure, with open-structured crystals. Water molecules in this open structure occupy a greater volume than they do in the liquid phase (Figure 9.20). This means that ice is less compact (less dense) than water.

When ice melts, not all the six-sided crystals collapse. Some remain in the ice-water mixture, making up a microscopic slush that slightly "bloats" the water—increases its volume slightly. This results in ice water being less dense than slightly warmer water. As the temperature of water at 0°C is increased, more of the remaining ice crystals collapse. This further decreases the volume of the water. This contraction continues only up to 4°C. That's because two things happen at the same time. Volume becomes less due to ice crystal collapse. Volume becomes more due to greater molecular motion. The collapsing effect dominates until the temperature reaches 4°C. After that, expansion overrides contraction because most of the ice crystals have collapsed. (Figure 9.21).

When ice water freezes to become solid ice, its volume increases tremendously—and its density is much lower. That's why ice floats on water. Like most other substances, solid ice contracts with further cooling.

Concept Check ✓

What's inside the open spaces of the water crystals shown in Figures 9.19 and 9.20 and the cartoon to the right? Is it air, water vapor, or nothing?

Check Your Answer There's nothing at all in the open spaces. It's empty space—a void. If there were air in the spaces, the illustration would have to show air molecules. If there were water vapor in the spaces, H_2O molecules would have to be shown.

This behavior of water is very important in nature. If water were most dense at 0°C it would settle to the bottom of a pond or lake. Water at 0°C is less dense and "floats" at the surface. That's why ice forms at the surface.

So a pond freezes from the surface downward. In a cold winter, the ice will be thicker than in a milder winter. Water at the bottom of an ice-covered pond is 4°C, relatively warm for organisms that live there.

Interestingly, very deep bodies of water are not ice-covered even in the coldest of winters. This is because all the water must be cooled to 4°C before lower temperatures can be reached. For deep water, the winter is not long enough to reduce an entire pond to 4°C. Any 4°C water lies at the bottom. Because of water's high specific heat and poor ability to conduct heat, the bottom of deep bodies of water in cold regions remains at a constant 4°C year round. Fish should be glad that this is so.

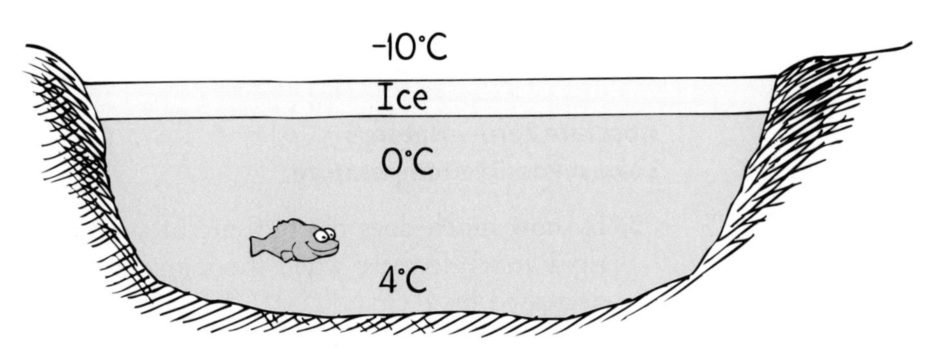

Figure 9.22
As water cools, it sinks until the entire pond is at 4°C. Then, as water at the surface is cooled further, it floats on top and can freeze. Once ice is formed, temperatures lower than 4°C can extend down into the pond.

Concept Check ✓

What was the precise temperature at the bottom of Lake Superior on New Year's Eve in 2000?

Check Your Answer The temperature at the bottom of any body of water with 4°C water in it is 4°C, for the same reason that rocks are at the bottom. Both 4°C water and rocks are more dense than water at any other temperature. Water is a poor heat conductor (next chapter), and so if the body of water is deep and in a region of long winters and short summers, as with Lake Superior, the water at the bottom is 4°C year round.

Chapter Review

Key Terms and Matching Definitions

_____ absolute zero
_____ calorie
_____ first law of thermodynamics
_____ heat
_____ second law of thermodynamics
_____ specific heat capacity
_____ temperature
_____ thermal energy
_____ thermal expansion
_____ third law of thermodynamics

1. The total energy (kinetic plus potential) of atoms and molecules that make up a substance.
2. A measure of the hotness of an object, related to the average kinetic energy per molecule in the object, measured in degrees Celsius, degrees Fahrenheit, or kelvins.
3. The lowest possible temperature; the temperature at which all particles have their minimum kinetic energy.
4. The thermal energy that flows from an object at higher temperature to one at a lower temperature, commonly measured in joules or calories.
5. A unit of thermal energy, or heat. One calorie is the thermal energy required to raise the temperature of one gram of water one Celsius degree (1 cal = 4.18 J). One Calorie (with a capital C) is equal to one thousand calories and is the unit used in describing the energy available from food.
6. A restatement of the law of energy conservation, usually as it applies to thermal systems: Whenever heat flows into or out of a system, the gain or loss of thermal energy equals the amount of heat transferred.
7. Heat never of itself flows from a cold substance to a hot substance.
8. No system can have its absolute temperature reduced to zero.
9. The quantity of heat per unit of mass required to raise the temperature of a substance by 1 degree Celsius.
10. The expansion of a substance due to increased molecular motion in that substance.

Review Questions

Thermal Energy—The Total Energy in a Substance

1. Why does a penny become warmer when struck by a hammer?

Temperature—Average Kinetic Energy Per Molecule in a Substance

2. What are the temperatures for freezing water on the Celsius and Fahrenheit scales? For boiling water at sea level?

3. Is the temperature of an object a measure of the total kinetic energy of molecules in the object, or a measure of the average kinetic energy per molecule in the object?

4. What is meant by the statement, "a thermometer measures its own temperature"?

Absolute Zero—Nature's Lowest Possible Temperature

5. By how much does the pressure of a gas in a rigid vessel decrease when the temperature is decreased by 1°C?

6. What pressure would you expect in a rigid container of 0°C gas if you cooled it by 273°C?

7. What is the temperature of melting ice on the Kelvin scale? Of boiling water at atmospheric pressure?

Heat Is the Movement of Thermal Energy

8. When you touch a cold surface, does "coldness" travel from the surface to your hand or docs thermal energy travel from your hand to the cold surface? Explain.
9. Distinguish between temperature and heat.
10. What determines the direction of heat flow?

Heat Units Are Energy Units

11. How is the energy value of foods determined?
12. Distinguish between a joule and a calorie.

The Laws of Thermodynamics

13. How does the law of the conservation of energy relate to the first law of thermodynamics?
14. How does the second law of thermodynamics relate to the direction of heat flow?

Specific Heat Capacity—A Measure of Thermal Inertia

15. Does a substance that heats up quickly have a high or a low specific heat capacity?
16. Why is the West Coast of the United States warmer in winter than the East Coast?

Thermal Expansion

17. How is a bimetallic strip used to regulate temperature?
18. Which generally expands more for the same increase in temperature—solids or liquids?
19. When the temperature of ice-cold water is increased slightly, does it undergo a net expansion or net contraction?
20. At what temperature do the combined effects of contraction and expansion produce the smallest volume for water?

Exercises

1. In your room there are things such as tables, chairs, other people, and so forth. Which of these things has a temperature (1) lower than, (2) greater than, and (3) equal to the temperature of the air?
2. Which is greater, an increase in temperature of 1°C or one of 1°F?
3. A friend says the temperature inside a certain oven is 500 and the temperature inside a certain star is 50,000. You're unsure about whether your friend means degrees Celsius or kelvins. How much difference does it make in each case?
4. The temperature of the sun's interior is about 10^7 degrees. Does it matter whether this is degrees Celsius or kelvins? Explain.
5. Which has the greater amount of thermal energy, an iceberg or a cup of hot coffee? Explain.
6. On which temperature scale does the average kinetic energy of molecules double when the temperature doubles?
7. Adding the same amount of heat to two different objects does not necessarily produce the same increase in temperature. Why not?
8. Why will a watermelon stay cool for a longer time than sandwiches when both are removed from a cooler on a hot day?
9. Iceland, so named to discourage conquest by expanding empires, is not at all ice-covered like Greenland and parts of Siberia, even though it is nearly on the Arctic Circle. The average winter temperature of Iceland is considerably higher than regions at the same latitude in eastern Greenland and central Siberia. Why is this so?
10. Why does the presence of large bodies of water tend to moderate the climate of nearby land—make it warmer in cold weather, and cooler in hot weather?

11. If the winds at the latitude of San Francisco and Washington, D.C., were from the east rather than from the west, why might San Francisco be able to grow only cherry trees and Washington, D.C., only palm trees?

12. Desert sand is very hot in the day and very cool at night. What does this tell you about its specific heat?

13. Cite an exception to the claim that all substances expand when heated.

14. Creaking noises are often heard in the attic of old houses on cold nights. Give an explanation in terms of thermal expansion.

15. An old remedy for a pair of nested drinking glasses that stick together is to run water at different temperatures into the inner glass and over the surface of the outer glass. Which water should be hot, and which cold?

16. A metal ball is just able to pass through a metal ring. When the ball is heated, however, it will not pass through the ring. What would happen if the ring, rather than the ball, were heated? Does the size of the hole increase, stay the same, or decrease?

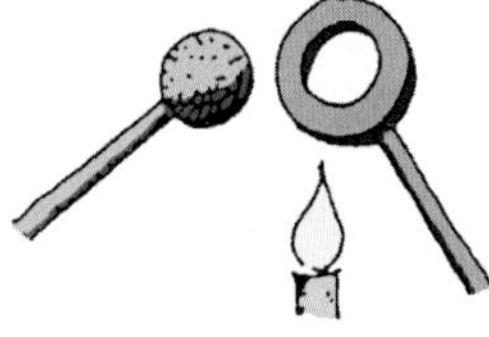

17. Suppose you cut a small gap in a metal ring. If you heat the ring, will the gap become wider or narrower?

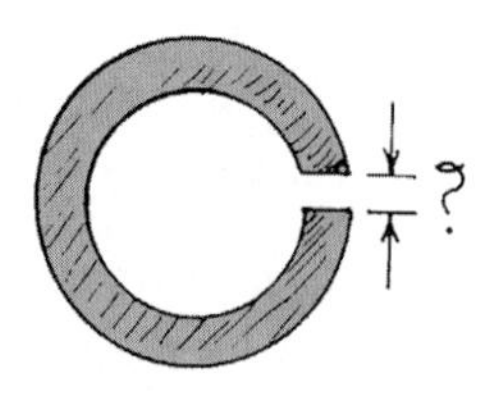

18. State whether water at the following temperatures will expand or contract when warmed a little: 0°C; 4°C; 6°C.

19. Why is it important to protect water pipes so they don't freeze?

20. If a metal object of 0°C is heated until it has twice as much thermal energy, what will its temperature be?

Problems

Quantity of heat, Q, is equal to the specific heat capacity of the substance c multiplied by its mass m and the temperature change ΔT; that is, $Q = cm\Delta T$.

1. The thermal energy of helium gas is directly proportional to its absolute temperature. Consider a flask of helium with a temperature of 10°C. If it is heated until it has twice the thermal energy, what will its temperature be?

2. What would be the final temperature of a mixture of 50 g of 20°C water and 50 g of 40°C water?

3. If you wish to warm 100 kg of water by 20°C for your bath, how much heat is required? (Give your answer in calories and joules.)

4. What would be the final temperature when 100 g of 25°C water is mixed with 75 g of 40°C water? (Hint: Equate the heat gained by the cool water to the heat lost by the warm water.)

5. (Challenge!) Consider a 40,000-km steel pipe that forms a ring to fit snugly all around the circumference of the world. Suppose people along its length breathe on it so as to raise its temperature 1 Celsius degree. The pipe gets longer (because steel expands 11 parts per million for each 1°C increase in temperature). It also is no longer snug. How high does it stand above ground level? (To simplify, consider only the expansion of its radial distance from the center of the Earth, and apply the geometry formula that relates circumference C and radius r, $C = 2\pi r$. The result is surprising!)

Chapter 10: Heat Transfer and Change of Phase

Why can author John Suchocki walk harmlessly with bare feet on red-hot coals? Could he do the same on red-hot pieces of iron? Why does a tile floor feel colder than a carpeted floor when both have the same temperature? Why does air cool when it expands? Is it true that everything continually emits radiation? If so, then why doesn't everything get colder with time? Why are you cooled when water on your skin evaporates? And why are you warmed when it condenses? Why is a steam burn so much more harmful than a burn by boiling water of the same temperature? How can water both boil and freeze at the same time? The answers to these questions are what this chapter is about.

There are three main ways that heat is conducted from one substance to another or from one place to another. These are *conduction, convection,* and *radiation*. We investigate each in turn.

10.1 Conduction—Heat Transfer via Particle Collision

Figure 10.1
The tile floor feels colder than the wooden floor, even though both floors are at the same temperature. This is because tile is a better conductor of heat than wood is, and so heat is more readily conducted out of the foot touching the tile.

When you hold one end of an iron nail in a flame it quickly becomes too hot to hold. Thermal energy at the hot end travels along the nail's entire length. This method of heat transfer is called **conduction.** Heat conduction occurs by means of particles in a material, mainly electrons. Every atom has electrons, and metal atoms have loosely-held electrons that are free to migrate in the metal. We shall see in Chapter 11 that metals are good electrical conductors for the same reason. Heat conduction occurs by electrons colliding and atoms colliding inside the object being heated.

Solids whose atoms or molecules have loosely held electrons are good conductors of heat. Metals have the loosest electrons and are excellent conductors of heat. Silver is the best, copper next, and, among the common metals, aluminum and then iron. Wool, wood, paper, cork, and plastic foam are poor conductors of heat. Molecules in these materials have electrons that are firmly attached to the molecules. Poor conductors are called *insulators.*

Wood is a good insulator and is used for cookware handles. Even when a pot is hot, you can quickly grasp the wooden handle with your bare hand without harm. An iron handle of the same temperature would surely burn your hand. Wood is a good insulator even when it's red hot. This explains how firewalking co-author John Suchocki can walk barefoot on red-hot wooden coals without burning his feet (see the photo at the beginning of this chapter). (CAUTION: Don't try this on your own; even experienced firewalkers sometimes receive bad burns when conditions aren't just right.) The main factor here is the poor conductivity of wood—even red-hot wood. Although its temperature is high, very little heat is conducted to the feet. A firewalker must be careful that no iron nails or other good conductors are among the hot coals. Ouch!

Can you say that a good conductor is a poor insulator? And that a good insulator is a poor conductor?

Air is a very poor conductor. That's why you can briefly put your hand in a hot pizza oven without harm. The hot air doesn't conduct

Figure 10.2
Snow patterns on the roof of a house show areas of conduction and insulation. Bare parts show where heat from inside has conducted through the roof and melted the snow.

heat well. But don't touch the metal in the hot oven. Ouch again! The good insulating properties of such things as wool, fur, and feathers are largely due to the air spaces they contain. Porous substances are also good insulators because of their many small air spaces. Be glad that air is a poor conductor; if it weren't, you'd feel quite chilly on a 20°C (68°F) day!

Snow is a poor conductor of heat. Snowflakes are formed of crystals that trap air and provide insulation. That's why a blanket of snow keeps the ground warm in winter. Animals in the forest find shelter from the cold in snow banks and in holes in the snow. The snow doesn't provide them with thermal energy—it simply slows down the loss of body heat generated by the animals. Then there are the igloos in Arctic dwellings that are shielded from the cold by their snow covering.

Homes are insulated with rock wool or fiberglass. Interestingly, insulation doesn't prevent the flow of heat. Insulation simply slows down the rate at which heat flows. Even a well-insulated warm home gradually cools. Insulation merely delays the rate at which heat conducts from a warmer region to a cooler one. In winter, we wish to slow conduction from inside to outside. But on hot summer days we wish to slow down conduction in the other direction, from outside to inside. Insulation slows conduction in either direction.

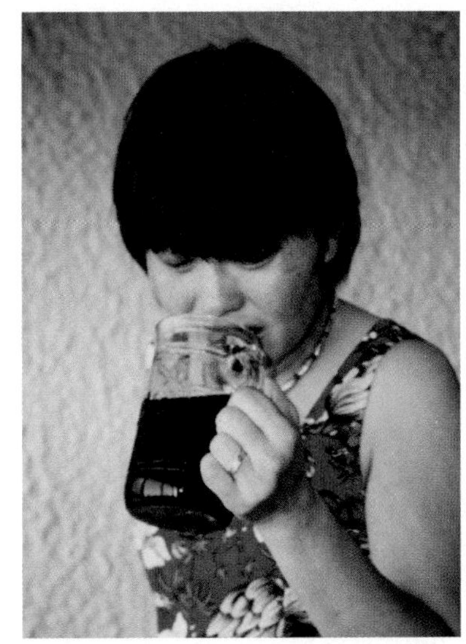

Figure 10.3
Conduction of heat from the hand to the root beer is minimized by the handle of the mug.

Concept Check ✓

In desert regions that are hot in the daytime and cold at night, the walls of houses are often made of mud. Why is it important that the mud walls be thick?

Check Your Answer A wall of correct thickness keeps the house warm at night by slowing heat conduction from warmer inside to cooler outside. In the daytime conduction is slowed from warmer outside to cooler inside. Such a wall has "thermal inertia." Thick walls hold up the roof better, too!

10.2 Convection—Heat Transfer via Movements of Fluid

Liquids and gases transfer heat mainly by **convection,** which is transfer by motion of a fluid—by currents. Convection occurs in all fluids. Whether we heat water in a pot, or warm air in a room, the process is the same (Figure 10.4). As the fluid is heated from below, the molecules at the bottom begin moving faster. They spread apart and become less dense. Then they are buoyed upward. Denser, cooler fluid migrates to the bottom. In this way, convection currents keep the fluid stirred up. Warmer fluid moves away from the heat source and cooler fluid moves toward the heat source and is warmed.

Figure 10.4
Convection currents in a gas (air) and a liquid.

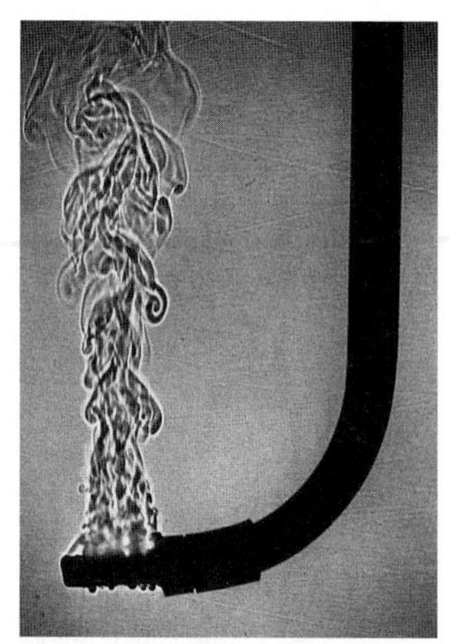

Figure 10.5
A heater at the tip of a heater element submerged in water produces convection currents, which are revealed as shadows (caused by deflections of light in water of different temperatures).

Warm air expands, becomes less dense, and rises in the cooler surrounding air—like a balloon buoyed upward. When the rising air reaches an altitude where air density is the same, it no longer rises. We see this occurring when smoke from a fire rises and then settles off as it cools and its density matches that of the surrounding air. As air expands, it cools.

Cooling by expansion is the opposite of what occurs when air is compressed. If you've ever compressed air with a tire pump, you probably noticed that both air and pump became quite hot.

Convection currents stir the atmosphere and produce winds. Some parts of the Earth's surface absorb heat from the sun more readily than others. This results in uneven heating of the air near the ground. We see this at the seashore, as Figure 10.6 shows. In daytime, the ground warms up more than the water, and air above the ground that is warmed then rises. It is replaced by cooler air that moves in from above the water. The result is a sea breeze. At night, the process reverses because the shore cools off more quickly than the water, and then the warmer air is over the sea. Build a fire on the beach and you'll notice that the smoke sweeps inward during the day and seaward at night.

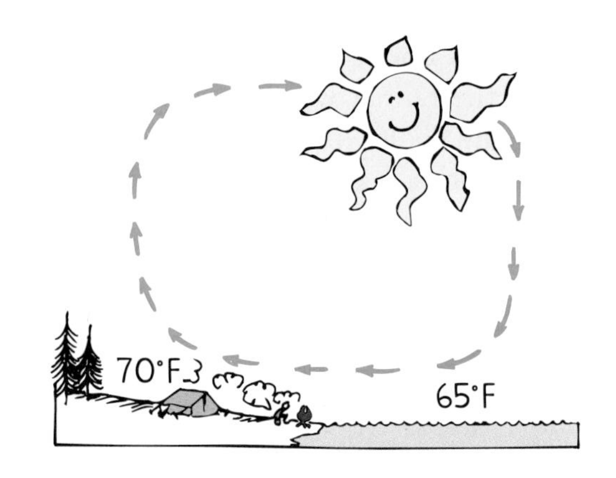

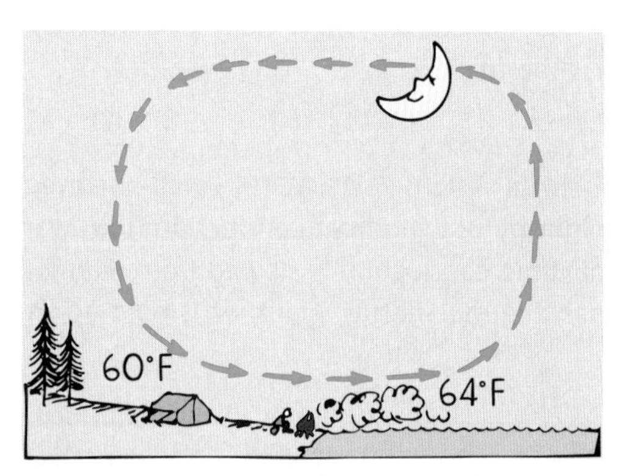

Figure 10.6
Convection currents produced by unequal heating of land and water. During the day, warm air above the land rises, and cooler air over the water moves in to replace it. At night the direction of air flow is reversed because now the water is warmer than the land.

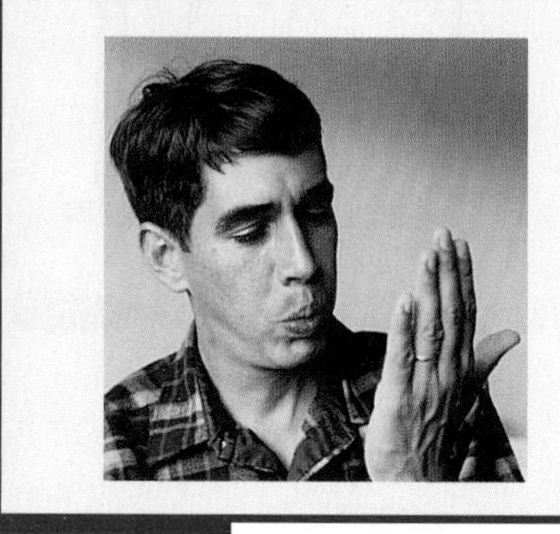

Breath-On Exploration

Do the following experiment right now. With your mouth open, blow on your hand. Your breath is warm. Now repeat, but this time pucker your lips to make a small hole so your breath expands as it leaves your mouth. Note that your breath is appreciably cooler! Expanding air cools.

10.3 Radiation—Heat Transfer via Radiant Energy

Thermal energy from the sun passes through space and then through the atmosphere before it warms the Earth's surface. This heat transfer is not by conduction and convection, for there's no material between the sun and Earth. Heat transfer must be by some other way—by **radiation***. The energy transferred this way is called *radiant energy.*

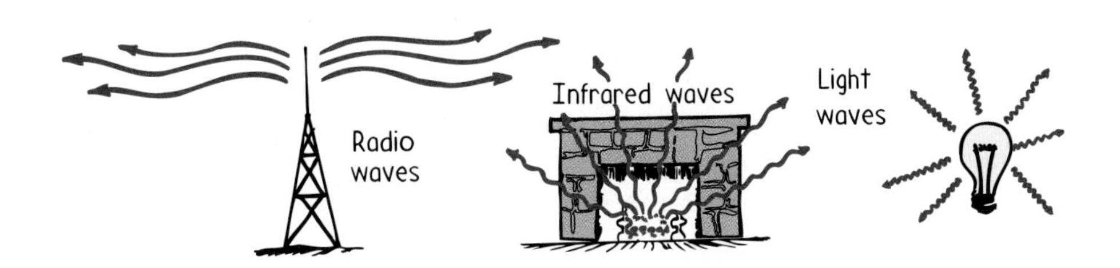

Figure 10.7
Types of radiant energy (electromagnetic waves).

Radiant energy is in the form of *electromagnetic waves.* It includes a wide span of waves that begin with radio waves and infrared waves, continue through to visible-light waves, and end with gamma rays. The lengths of waves differ. The waves of infrared (below-the-red) waves, for instance, are longer than those of visible-light waves. The longest visible wavelengths are for red light, and the shortest are for violet light. Shorter waves can't be seen by the eye. We'll treat waves further in Chapters 13 and 14, and electromagnetic waves in Chapters 12 and 14.

The wavelength of radiation is related to the frequency of radiation. Frequency is the rate of vibration of a wave source. The girl in Figure 10.8 shakes a rope at a low frequency (top), and a higher frequency (bottom). Note that shaking at a low frequency produces a long lazy wave, and the higher-frequency shake produces shorter waves. Likewise with electromagnetic waves. We will see in Chapter 13 that vibrating electrons emit electromagnetic waves. High-frequency vibrations produce short waves and low frequency vibrations produce longer waves.

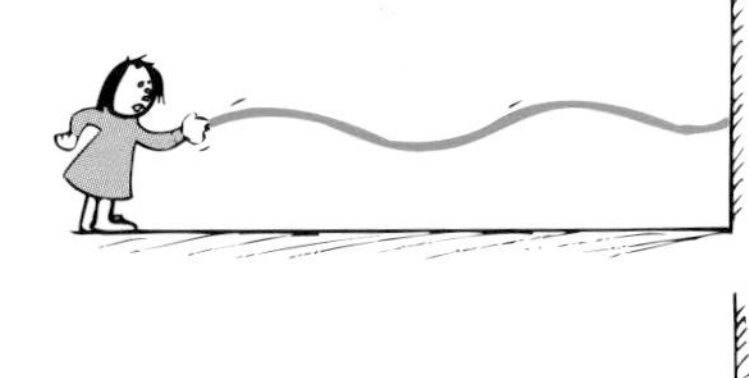

Figure 10.8
A wave of long wavelength is produced when the rope is shaken gently (at a low frequency). When shaken more vigorously (high frequency), a wave of shorter wavelength is produced.

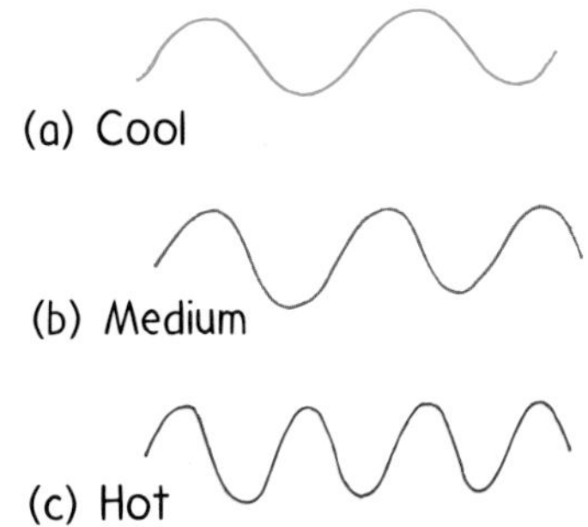

Figure 10.9
(a) A low-temperature (cool) source emits primarily low-frequency, long-wavelength waves. (b) A medium-temperature source emits primarily medium-frequency, long-wavelength waves. (c) A high-temperature source (hot) emits primarily high-frequency, short-wavelength waves.

* Do not confuse radiation with radioactivity, which are reactions that involve the atomic nucleus and are characteristic of nuclear power plants and the like. Radiation here is electromagnetic radiation, "heat" waves of low-frequency light, which we will study in detail in Parts 3 and 4.

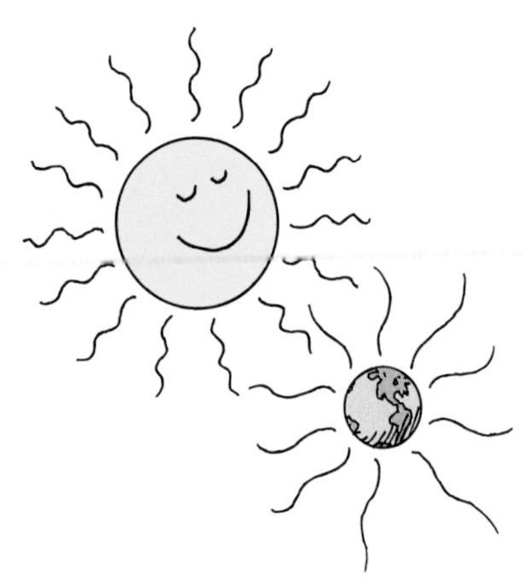

Figure 10.10
Both the sun and the Earth emit the same kind of radiant energy. The sun's glow is visible to the eye; the Earth's glow consists of longer waves and so is not visible to the eye.

Emission of Radiant Energy

All substances at any temperature above absolute zero emit radiant energy. The average frequency f of the radiant energy is directly proportional to the absolute temperature T of the emitter:

$$f \sim T$$

The sun has a very high temperature and therefore emits radiant energy at a high frequency—high enough to stimulate our sense of sight. The Earth, in comparison, is relatively cool. So the radiant energy it emits has a frequency lower than that of visible light. The radiation emitted by the Earth is in the form of infrared waves. Radiant energy emitted by the Earth is called **terrestrial radiation.**

Most people know that the sun glows and emits radiant energy. And many educated people know that the source of the sun's radiant energy involves nuclear reactions in its deep interior. However, relatively few people know that the Earth also "glows" and emits radiant energy of the same nature. If you visit the depths of any mine you'll find it's warm down there—year round. Radioactivity in the Earth's interior warms the Earth. Much of this heat conducts to the surface to become terrestrial radiation. So radiant energy is emitted by both the sun and the Earth, and differs only in the range of frequencies, and the amount. When we study meteorology in Part 7, we'll learn how the atmosphere is transparent to the high-frequency solar radiation, but opaque to much of the lower-frequency terrestrial radiation. This produces a "greenhouse effect" and plays a role in global warming.

All objects—you, your teacher, and everything in your surroundings—continually emit radiant energy in a mixture of frequencies (because temperature corresponds to a mixture of molecular kinetic energies). Objects of everyday temperatures mostly emit low frequency infrared waves. When the higher-frequency infrared waves are absorbed by your skin, you feel the sensation of heat. So it is common to refer to infrared radiation as *heat radiation.*

Common hotter sources that give the sensation of heat are the burning embers in a fireplace, a lamp filament, and the sun. All of these emit both infrared radiation and visible light. When this radiant energy falls on other objects, it is partly reflected and partly absorbed. The part that is absorbed increases the thermal energy of the objects.

Concept Check ✓

Which of the following do *not* give off radiant energy? (a) The sun; (b) Lava from a volcano; (c) Red-hot coals; (d) This book that you're reading.

Check Your Answer Did you answer (d), the book? Sorry, wrong answer. None of the choices is correct. The book, like the other things listed, has temperature—though not as high as the others listed. By the rule $f \sim T$, it therefore emits radiation. Since its temperature is low, the frequency of radiation is also low. Everything with any temperature above absolute zero emits electromagnetic radiation. That's right—*everything*!

Absorption of Radiant Energy

If everything is emitting energy, why doesn't everything finally run out of it? The answer is, all things also *absorb* energy. Good emitters of radiant energy are also good absorbers; poor emitters are poor absorbers. For example, a radio antenna constructed to be a strong emitter of radio waves is also, by its very design, a strong receiver (absorber) of them. A poorly designed transmitting antenna is also a weak receiver. An object that absorbs radiant energy looks dark. If it absorbs all the radiant energy on it, it looks perfectly black.

A dark object that absorbs plenty of radiant energy must emit a lot as well. You can check this out with two metal containers of the same size, one with a light colored shiny surface and the other black (Figure 10.11). Fill them with hot water. A thermometer will show that the black container cools faster. That's because the blackened surface is a better emitter. A hot beverage stays hot longer in a shiny pot than in a dark pot.

Figure 10.11
When the containers are filled with hot (or cold) water, the darker one cools (or warms) faster.

The same experiment can be done in reverse. This time fill each container with ice water and place the containers in front of a fireplace or outside on a sunny day—wherever there is a good source of radiant energy. You'll find that the black container warms up faster. An object that emits well also absorbs well.

Every surface, hot or cold, both absorbs and emits radiant energy. If the surface is hotter than its surroundings the surface will be a net emitter and will cool. If it's colder than its surroundings, it will be a net absorber and will become warmer.

Concept Check ✓

If a good absorber of radiant energy were a poor emitter (instead of a good emitter), how would its temperature compare with the surroundings?

Check Your Answer There would be a net absorption of radiant energy and the temperature would be above the surroundings. Things around us approach a common temperature because good absorbers are, by their very nature, also good emitters.

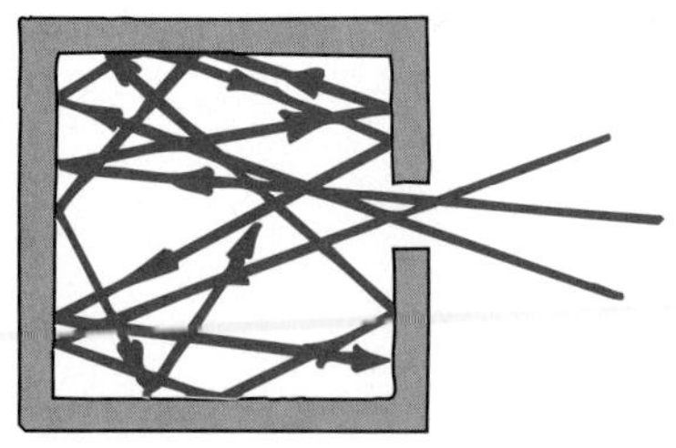

Figure 10.12
Radiation that enters the opening has little chance of leaving because most of it is absorbed. For this reason, the opening to any cavity looks black to us.

Reflection of Radiant Energy

Radiant energy can be reflected. There are no perfect reflectors, so some absorption always occurs. When you look at the open ends of pipes in a stack, the holes appear black. Look at open doorways or windows of distant houses in the daytime, and they, too, look black. Openings appear black because the light that enters them is reflected back and forth on the inside walls many times and is partly absorbed at each reflection. As a result, very little or none of the light returns back out the opening (Figure 10.12).

The pupil of your eye is another example. The pupil is a hole in your iris that allows light to enter with no reflection. That's why it appears black. (An exception occurs in flash photography when pupils appear red or pink, which occurs when very bright light is reflected off the eye's inner pink surface and back through the pupil.)

Light-colored buildings stay cooler in summer because they reflect much of the incoming radiant energy. Light-colored buildings are also poor emitters, and so they retain more of their internal energy than darker buildings and stay warmer in winter. Therefore, paint your house a light color.

Figure 10.13
The hole looks perfectly black and indicates a black interior, when in fact the interior has been painted a bright white.

10.4 Heat Transfer Occurs Whenever Matter Changes Phase

Matter exists in four common phases (states). Ice, for example, is the *solid* phase of water. When thermal energy is added, the increased motion breaks down the rigid molecular structure of water and it becomes the *liquid* phase, water. When more energy is added, the liquid changes to the *gaseous* phase. Add still more energy, and the molecules break into ions and electrons, giving the *plasma* phase. Plasma (not to be confused with blood plasma) is the illuminating gas found in fluorescent and other vapor lamps. The sun, stars, and much of the space in between them is in the plasma phase. When matter changes phase, a transfer of thermal energy is involved.

10.5 Evaporation—A Change of Phase from Liquid to Gas

Water changes to the gaseous phase by the process of **evaporation.** Here's how. Molecules in a liquid move randomly, at a wide variety of speeds. In this random motion they bump into one another. In bumping, some gain kinetic energy while others lose kinetic energy. Molecules at the surface that gain kinetic energy by being bumped from below are the ones to break free of the liquid. They leave the surface and fly into the space above the liquid. In this way they become gas.

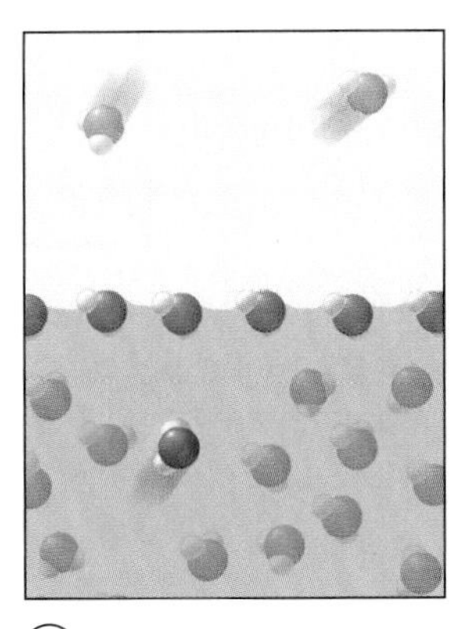

① Liquid water molecule having sufficient kinetic energy to overcome surface hydrogen bonding approaches liquid surface.

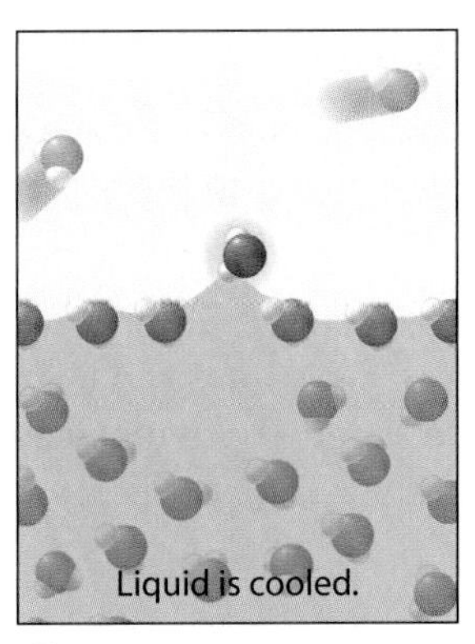

② Liquid water cooled as it loses this high-speed water molecule.

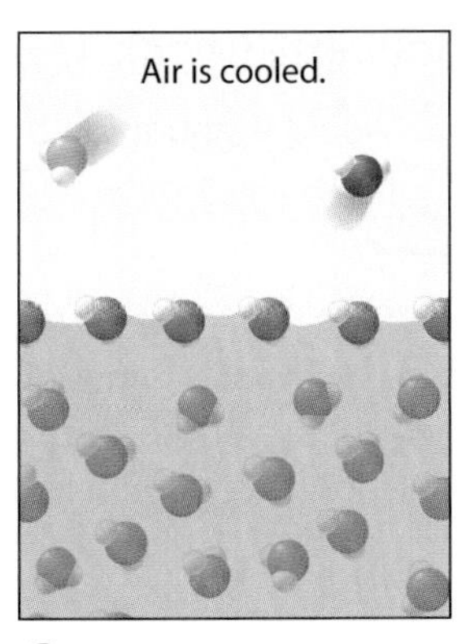

③ Molecule enters gaseous phase, having lost kinetic energy in overcoming hydrogen bonding at the liquid surface. Air is cooled as it collects these slowly moving gaseous particles.

Figure 10.14
Evaporation is a cooling process.

As water is heated, some molecules gain more energy than others. Those with the most energy move fastest and eventually break free of the liquid to become gas. Because the fast-moving molecules leave the liquid, the molecules left behind are the slow-moving ones. As the high-energy molecules leave the liquid, the average kinetic energy of the molecules in the liquid decreases. In other words, the temperature of the liquid goes down.

When our bodies tend to overheat, our sweat glands produce perspiration. This is part of nature's thermostat, for the evaporation of perspiration cools us and helps us maintain a stable body temperature. Many animals do not have sweat glands and must cool themselves by other means (Figures 10.16 and 10.17).

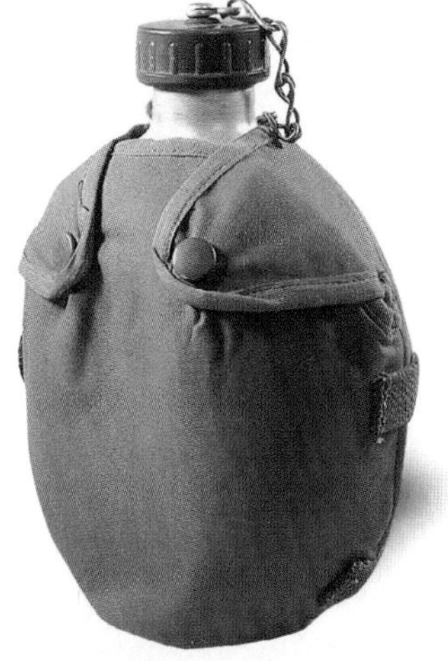

Figure 10.15
When wet, the cloth covering on the canteen promotes cooling. As the faster-moving water molecules evaporate from the wet cloth, its temperature decreases and cools the metal. The metal, in turn, cools the water within. Water in the canteen can become a lot cooler than air temperature.

Figure 10.16
Hanz has no sweat glands (except between the toes). He cools himself by panting. In this way evaporation occurs in the mouth and within the bronchial tract.

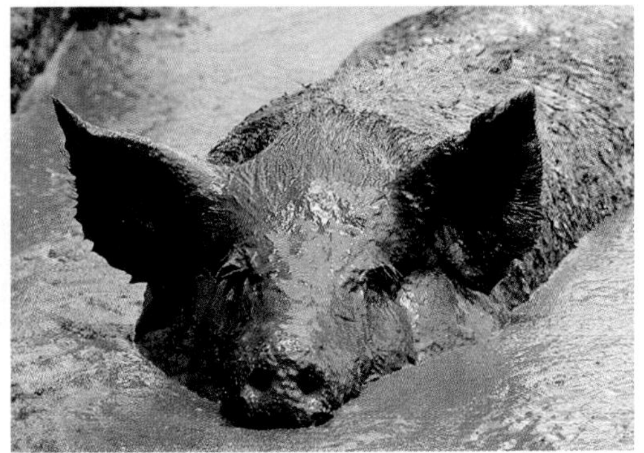

Figure 10.17
Pigs have no sweat glands and therefore cannot cool by the evaporation of perspiration. Instead, they wallow in the mud to cool themselves.

Concept Check ✓

Would evaporation be a cooling process if there were no transfer of molecular kinetic energy from water to the air above?

Check Your Answer No. A liquid cools only when kinetic energy is carried away by evaporating molecules. This is similar to billiard balls that gain speed at the expense of others that lose speed. Those that leave (evaporate) are gainers while losers remain behind and lower the temperature of the water.

Even frozen water "evaporates." In this form of evaporation, called **sublimation,** molecules jump directly from a solid to a gaseous phase. Because water molecules are so tightly held in a solid, frozen water evaporates (sublimes) much slower than liquid water does. Sublimation accounts for the loss of much snow and ice, especially on high sunny mountain tops. Sublimation also explains why ice cubes left in a freezer for a long time get smaller.

10.6 Condensation—A Change of Phase from Gas to Liquid

The opposite of evaporation is **condensation**—the changing of a gas to a liquid. When gas molecules near the surface of a liquid are attracted to the liquid, they strike the surface with increased kinetic energy and become part of the liquid. This kinetic energy is absorbed by the liquid. The result is increased temperature. Condensation is a warming process.

A dramatic example of warming by condensation is the energy given up by steam when it condenses. The steam gives up a lot of energy when it condenses to a liquid and wets the skin. That's why a burn from 100°C steam is much more damaging than a burn from 100°C boiling water. This energy release by condensation is utilized in steam-heating systems.

Figure 10.18
The exchange of molecules at the interface between liquid and gaseous water.

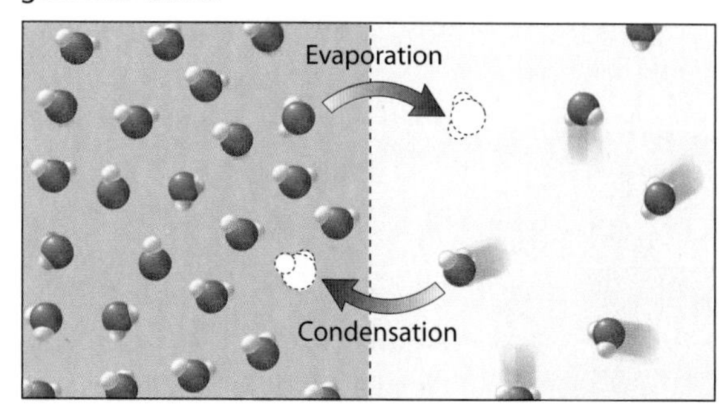

Figure 10.19
Heat is given up by steam when it condenses inside the radiator.

Figure 10.20
If you're chilly outside the shower stall, step back inside and be warmed by the condensation of the excess water vapor there.

Have you noticed when taking a shower that you feel warmer in the moist shower region than outside the shower? You quickly sense the difference when you step outside. Away from the moisture, net evaporation occurs and you feel chilly. But in the shower stall, even with the water off, the warming effect of condensation counteracts the cooling effect of evaporation. When condensation is greater than evaporation, you are warmed. When evaporation is greater than condensation, you are cooled. So now you know why you can dry yourself with a towel much more comfortably if you remain in the shower area. To dry yourself thoroughly, you can finish the job in a less moist area.

You feel cooler on a July afternoon in dry Phoenix or Santa Fe than you do in New York City or New Orleans even though the temperatures are the same. In the dryer cities the rate of evaporation is much greater than the rate of condensation. In humid locations the rate of condensation is greater than the rate of evaporation. You feel the warming effect as vapor in the air condenses on your skin. You are literally being bombarded by the impact of H_2O molecules in the air slamming into you. We will explore condensation in the atmosphere when we study meteorology in Chapter 37.

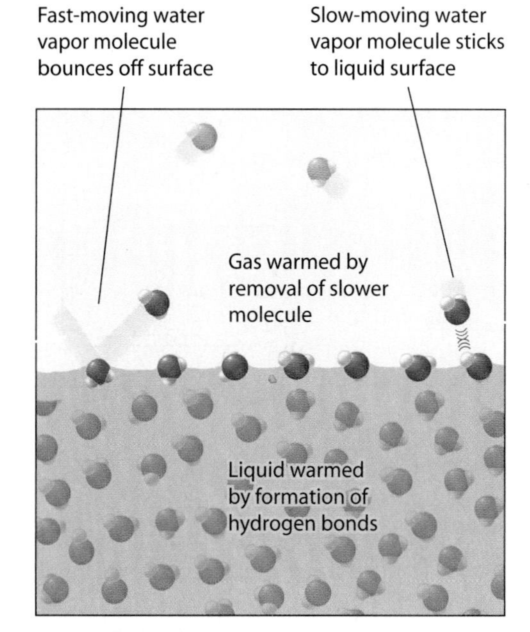

Figure 10.21
Condensation is a warming process.

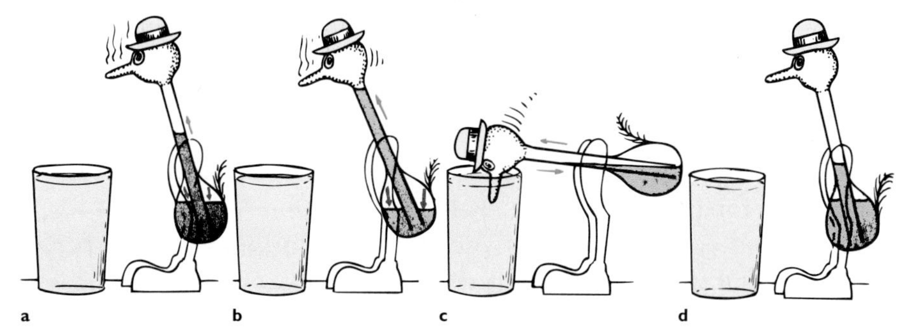

Figure 10.22
The toy drinking bird operates by the evaporation of ether inside its body and by the evaporation of water from the outer surface of its head. The lower body contains liquid ether, which evaporates rapidly at room temperature. As it (a) vaporizes, it (b) creates pressure (inside arrows), which pushes ether up the tube. Ether in the upper part does not vaporize because the head is cooled by the evaporation of water from the outer felt-covered beak and head. When the weight of ether in the head is sufficient, the bird (c) pivots forward, permitting the ether to run back to the body. Each pivot wets the felt surface of the beak and head, and the cycle is repeated.

Concept Check ✓

If the water level in a dish of water remains unchanged from one day to the next, can you conclude that no evaporation or condensation is taking place?

Check Your Answer Not at all, for there is much activity taking place at the molecular level. Both evaporation and condensation occur continuously. The fact that the water level remains constant indicates equal rates of evaporation and condensation. (Unless there is extreme humidity, evaporation is normally greater than condensation.)

10.7 Boiling Is Evaporation within a Liquid

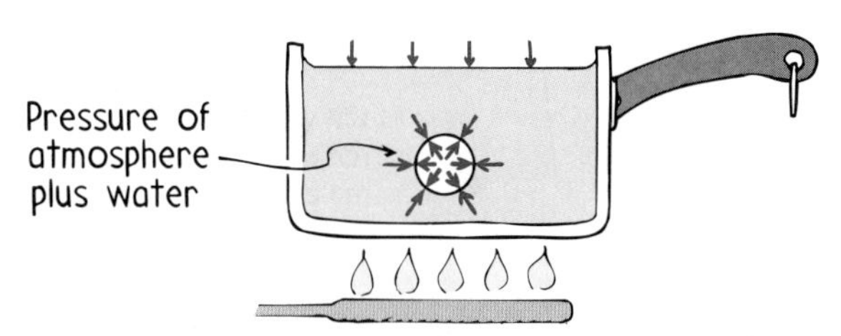

Figure 10.23
The motion of vapor molecules in the bubble of steam (much enlarged) creates a gas pressure (called the *vapor pressure*) that counteracts the atmospheric and water pressure against the bubble.

Evaporation occurs beneath the surface of a liquid in the process called **boiling.** Bubbles of vapor form in the liquid and are buoyed to the surface, where they escape. Bubbles can form only when the pressure of the vapor within them is great enough to resist the pressure exerted by the surrounding water and atmosphere. This occurs at the boiling temperature of the liquid. At lower temperatures the vapor pressure in the bubbles is not enough, and the surrounding pressure collapses any bubbles that might form.

When the pressure on the surface of a liquid increases, boiling is hampered. The temperature needed for boiling rises. The boiling point of a liquid depends on the pressure on the liquid. This is most evident with a pressure cooker (Figure 10.24). Vapor pressure builds up inside and prevents boiling, which results in a higher water temperature. It is important to note that it is the high temperature of the water that cooks the food, not the boiling process itself.

Figure 10.24
The tight lid of a pressure cooker holds pressurized vapor above the water surface, and this inhibits boiling. In this way, the boiling temperature of the water is increased to above 100°C.

Lower atmospheric pressure (as at high altitudes) decreases the boiling temperature. In Denver, Colorado, the "mile-high city," for example, water boils at 95°C, instead of 100°C.* If you try to cook food in boiling water that is cooler than 100°C, you must wait a longer time for proper cooking. A three-minute boiled egg in Denver is yucky. If the temperature of the boiling water were very low, food would not cook at all.

Boiling cools water?

Boiling, like evaporation, is a cooling process. At first thought, this may seem surprising—perhaps because we usually associate boiling with heating. However, heating water is one thing; boiling it is another. When 100°C water

* Mountaineering pioneers in the 19th century, without altimeters, used the boiling point of water to determine their altitudes.

at atmospheric pressure is boiling, it is in thermal equilibrium. It is being cooled by boiling as fast as it is being heated by energy from the heat source (Figure 10.25). If cooling did not take place, continued application of heat to a pot of boiling water would raise its temperature.

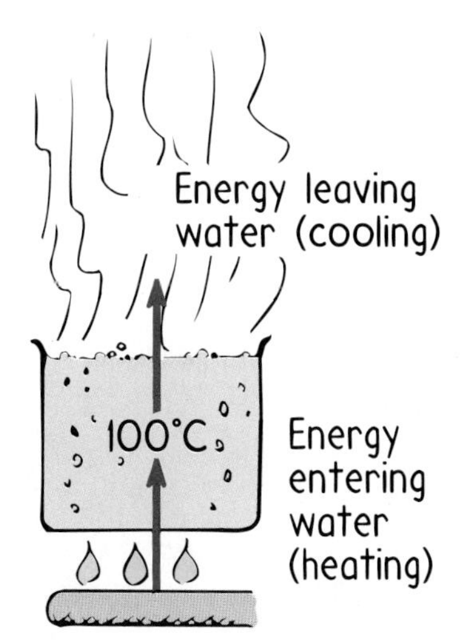

Figure 10.25
Heating warms the water from below, and boiling cools it from above.

Concept Check ✓

Since boiling is a cooling process, would it be a good idea to cool your hot, sticky hands by dipping them into boiling water?

Check Your Answer No, no, no! When we say boiling is a cooling process, we mean that the water (not your hands!) is being cooled relative to the higher temperature it would attain otherwise. Because of the cooling effect of the boiling, the water remains at 100°C instead of getting hotter. A dip in 100°C water would be most uncomfortable for your hands!

A dramatic demonstration of the cooling effect of evaporation and boiling is shown in Figure 10.26. Here we see a shallow dish of room-temperature water in a vacuum jar. When the pressure in the jar is slowly reduced by a vacuum pump, the water starts to boil. The boiling process takes heat away from the water. The water cools. As the pressure is further reduced, more and more of the slower-moving molecules boil away. Continued boiling lowers the temperature until the freezing point of approximately 0°C is reached. Continued cooling by boiling causes ice to form over the surface of the bubbling water. Boiling and freezing take place at the same time! Frozen bubbles of boiling water are a remarkable sight.

Spray some drops of coffee into a vacuum chamber, and they boil until they freeze. Even after they are frozen, the water molecules continue to evaporate into the vacuum until little crystals of coffee solids are left. This is how freeze-dried coffee is made. The low temperature of this process tends to keep the chemical structure of the coffee solids from changing. When hot water is added, more of the original flavor of the coffee is retained.

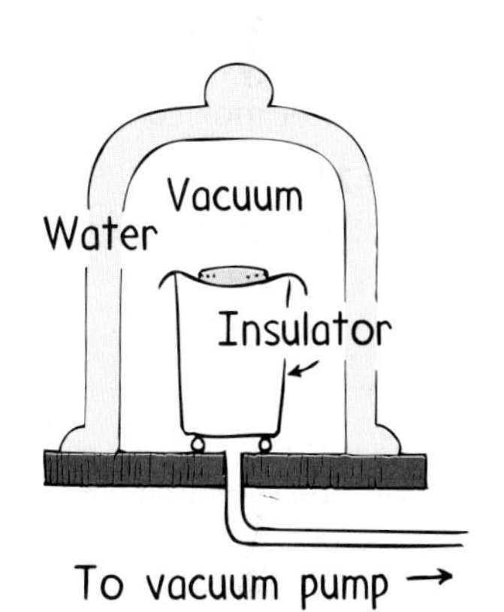

Figure 10.26
Apparatus to demonstrate that in a vacuum water freezes and boils at the same time. A gram or two of water is placed in a dish that is insulated from the base by a polystyrene cup.

10.8 Melting and Freezing—Opposite Directions in Phase Change

Melting occurs when a substance changes from a solid to a liquid. To visualize what happens, pretend you hold hands with someone and then each of you start jumping around. The more violently you jump, the more difficult it is to keep your grasp. If you jump violently enough, keeping your grasp might be impossible. Something like this happens to the molecules of a solid when it is heated. As heat is absorbed, the molecules vibrate more and more violently. If

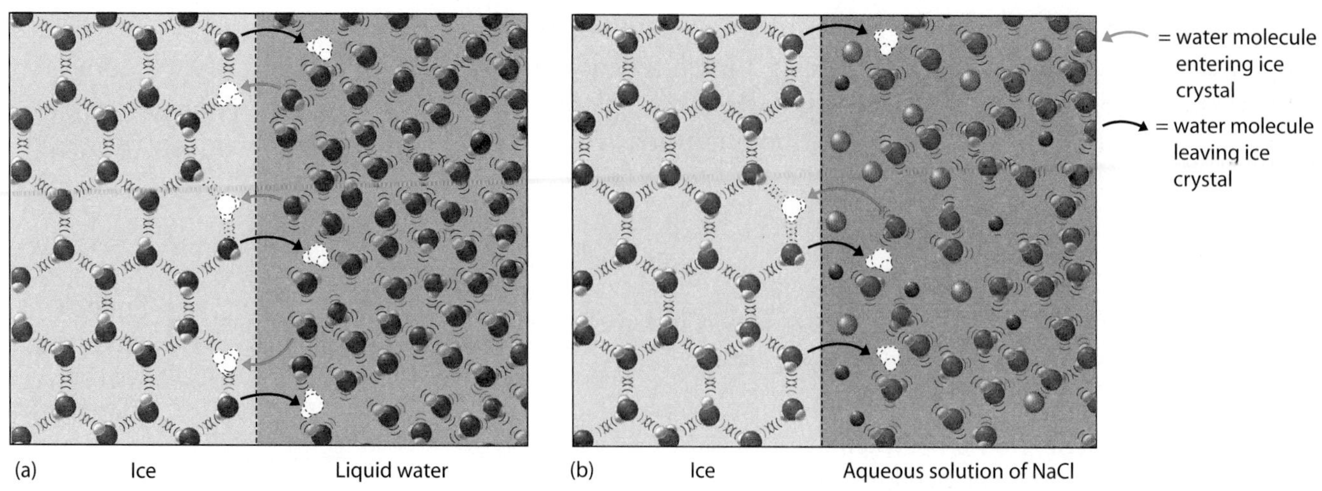

Figure 10.27
(a) In a mixture of ice and water at 0°C, ice crystals gain and lose water molecules at the same time. The ice and water are in thermal equilibrium. (b) When salt is added to the water, there are fewer water molecules entering the ice because there are fewer of them at the interface.

enough heat is absorbed, the attractive forces between the molecules no longer hold them together. The solid melts.

Freezing occurs when a liquid changes to a solid—the opposite of melting. As energy is taken from a liquid, molecular motion slows until molecules move so slowly that attractive forces between them bind them together. The liquid freezes when its molecules vibrate about fixed positions and form a solid.

At atmospheric pressure, ice forms at 0°C. With impurities in the water, the freezing point is lowered. "Foreign" molecules come in between and get in the way of crystal formation. In general, adding anything to water lowers its freezing temperature. Antifreeze is a practical application of this process.

10.9 Energy Is Needed for Changes of Phase

Whenever a substance changes phase, a transfer of energy occurs. Heat must be added to melt ice into water or vaporize water into steam. Heat must be removed to condense steam back into water or freeze water into ice (Figure 10.28).

Figure 10.28
Energy changes with change of phase.

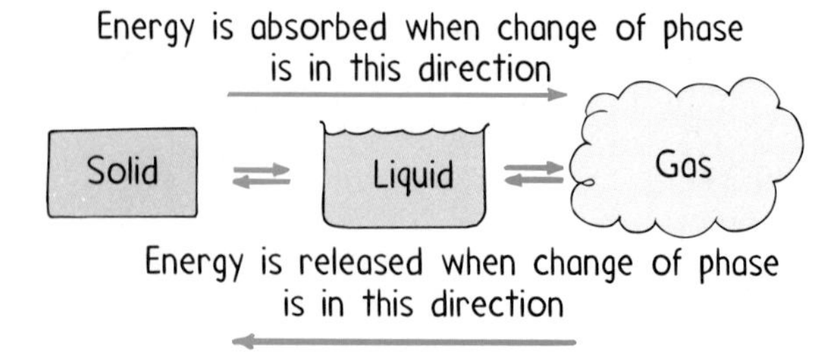

The cooling cycle of a refrigerator nicely illustrates these concepts. A motor pumps a special fluid through the system, where it is made to undergo the cyclic process of vaporization and condensation. In doing so, heat is drawn from things stored inside. The gas with its added energy is directed to outside coils in the back, appropriately called condensation coils. The next time you're near a refrigerator, place your hand near the condensation coils in the back and you'll feel the heat that has been extracted from inside.

An air conditioner uses the same principle and simply pumps heat energy from one part of the unit to another. If the roles of vaporization and condensation were reversed, the air conditioner would become a heater.

Concept Check ✓

In the process of water vapor condensing in the air, the slower-moving molecules are the ones that condense. Does condensation warm or cool the surrounding air?

Check Your Answers As slower-moving molecules are removed from the air, there is an increase in the average kinetic energy of molecules still in the air. Therefore the air is warmed. The same answer is shown in Figure 10.28. The change of phase is from gas to liquid, which releases energy.

The amount of energy needed to change any substance from solid to liquid (and vice versa) is called the **heat of fusion** for the substance. For water, this is 335 joules per gram. The amount of energy required to change any substance from liquid to gas (and vice versa) is called the **heat of vaporization** for the substance. For water this is a whopping 2255 joules per gram. We will see later in Part 6 that these relatively high values are due to the strong forces between water molecules—hydrogen bonds.

Link to Entomology: Life at the Extremes

Some deserts reach surface temperatures of 60°C (140°F). Too hot for life? Not for a species of ant (*Cataglyphis*) that thrives at this searing temperature. The Cataglyphis ant can withstand higher temperatures than any other creatures in the desert. At this extremely high temperature, the desert ants can forage for food without the presence of the lizards who otherwise prey upon them. How they are able to do this is currently being researched. They scavenge the desert surface for corpses of those who did not find cover from the hot sun, touching the hot sand as little as possible while often sprinting on four legs with two high in the air. Although their foraging paths zigzag over the desert floor, their return paths are almost straight lines to their nest holes. They attain speeds of 100 body lengths per second. During an average six-day life, most of these ants retrieve 15 to 20 times their weight in food.

From deserts to glaciers, a variety of creatures have invented ways to survive the harshest corners of the world. A species of worm thrives in the glacial ice in the Arctic. There are insects in the Antarctic ice that pump their bodies full of a substance that acts as a natural antifreeze to prevent being frozen solid. Some fish beneath the ice are able to do the same. Then there are bacteria that thrive in boiling hot springs as a result of heat-resistant proteins in their bodies.

An understanding of how creatures survive at the extremes of temperature can provide clues for practical solutions to physical challenges of humans. Astronauts who venture from our nest, for example, will need all the techniques available for coping with unfamiliar environments.

Chapter Review

Key Terms and Matching Definitions

_____ boiling
_____ condensation
_____ conduction
_____ convection
_____ evaporation
_____ freezing
_____ heat of fusion
_____ heat of vaporization
_____ melting
_____ radiation
_____ sublimation

1. The transfer of heat energy by collisions between the particles in a substance (especially a solid).
2. The transfer of heat energy in a gas or liquid by means of currents in the heated fluid. The fluid moves, carrying energy with it.
3. The transfer of energy by means of electromagnetic waves.
4. The change of phase from liquid to gaseous.
5. The change of phase from solid to gaseous, skipping the liquid phase.
6. The change of phase from gaseous to liquid.
7. Rapid evaporation that takes place within a liquid as well as at its surface.
8. The change of phase from solid to liquid.
9. The change of phase from liquid to solid.
10. The amount of thermal energy needed to change a substance from solid to liquid or from liquid to solid.
11. Amount of thermal energy required to change a substance from liquid to gas or from gas to liquid.

Review Questions

Conduction—Heat Transfer via Particle Collision

1. What is the role of "loose" electrons in heat conductors?

2. Distinguish between a heat conductor and an insulator.

3. In what sense do we say there is no such thing as cold?

Convection—Heat Transfer via Movements of Fluid

4. How is thermal energy transferred from one place to another by convection?

5. What happens to the pressure of air as the air rises? What happens to its volume? Its temperature?

Radiation—Heat Transfer via Radiant Energy

6. How does the frequency of radiant energy vary with the temperature of the radiating source?

7. Which body glows, the sun, the Earth, or both? Explain.

8. Which normally cools faster, a black pot of hot water or a silver pot of hot water? Explain.

9. Is a good absorber of radiation a good emitter or a poor emitter?

Heat Transfer Occurs Whenever Matter Changes Phase

10. What are the four phases of matter?

Evaporation—A Change of Phase from Liquid to Gas

11. Do the molecules in a liquid all have about the same speed or is there a wide variety of speeds?

12. Why is perspiration a cooling process?

Condensation—A Change of Phase from Gas to Liquid

13. What happens to the temperature of a body of water when water molecules condense upon it?

14. Why is a steam burn more damaging than a burn from boiling water at the same temperature?

Boiling Is Evaporation Within a Liquid

15. Distinguish between evaporation and boiling.
16. What condition permits water to boil at a temperature below 100°C?

Melting and Freezing—Opposite Directions in Phase Change

17. Why does increasing the temperature of a solid make it melt?
18. Why does decreasing the temperature of a liquid make it freeze?

Energy Is Needed for Changes of Phase

19. Does a liquid release or absorb energy when it evaporates? When it solidifies?
20. Why does the temperature of boiling water not rise when thermal energy is added?

Exercises

1. If you hold one end of a metal nail against a piece of ice, the end in your hand soon becomes cold. Does cold flow from the ice to your hand? Explain.
2. What is the purpose of a layer or copper or aluminum on the bottom of stainless steel cookware?
3. Many tongues have been injured by licking a piece of metal on a very cold day. Why would no harm result if a piece of wood were licked on the same day even when both have the same temperature?
4. All objects continuously emit radiant energy. Why then doesn't the temperature of all objects continuously decrease?
5. All objects continuously absorb energy from their surroundings. Why then doesn't the temperature of all objects continuously increase?
6. What determines whether an object is a net emitter or a net absorber of radiant energy?
7. Wood is a better insulator than glass. Yet fiberglass is commonly used as an insulator in wooden buildings. Explain.
8. You can comfortably hold your fingers close beside a candle flame, but not very close above the flame. Why?
9. When a hot object is placed in contact with a cooler object, the hot object warms the cooler one. Can you say it loses as much temperature as the cooler one gains? Defend your answer.
10. Consider two equal size rooms connected by an open door. One room is maintained at a higher temperature than the other one. Which room contains more air molecules?
11. In a still room, smoke from a candle will sometimes rise only so far, not reaching the ceiling. Explain why.
12. In a mixture of hydrogen and oxygen gases at the same temperature, which molecules move faster? Why?
13. Turn an incandescent lamp on and off quickly while you are standing near it. You feel its thermal energy but find when you touch the bulb that it is not hot. Explain why you felt thermal energy from it.
14. The thermal energy of volcanoes and natural hot springs comes from trace amounts of radioactive minerals in common rock in the Earth's interior. Why isn't the same kind of rock at the Earth's surface warm to the touch?
15. Why does blowing over the surface of hot soup cool the soup?
16. Why does the temperature of boiling water remain the same as long as the heating and boiling continue?

17. Why does the water in a car radiator sometimes boil explosively when the radiator cap is removed?

18. Why does dew form on a cold soft drink can?

19. Air-conditioning units contain no water whatever, yet it is common to see water dripping from them when they're running on a hot day. Explain.

20. Why does a hot dog pant?

Problems

1. The quantity of heat Q that changes the temperature ΔT of a mass m of a substance is given by $Q = cm\Delta T$, where c is the specific heat capacity of the substance. For example, for H_2O, $c = 1$ cal/g°C. And for a change of phase the quantity of heat Q that changes the phase of a mass m is $Q = mL$, where L is the heat of fusion or heat of vaporization of the substance. For example, for H_2O the heat of fusion is 80 cal/g or 80 kcal/kg, and the heat of vaporization is 540 cal/g or 540 kcal/kg. Use these relationships to determine the number of calories to change
(a) 1 kg of 0°C ice to 0°C ice water;
(b) 1 kg of 0°C ice water to 1 kg of 100°C boiling water;
(c) 1 kg of 100°C boiling water to 1 kg of 100°C steam; and
(d) 1 kg of 0°C ice to 1 kg of 100°C steam.

2. The specific heat capacity of ice is about 0.5 cal/g°C. Supposing that it remains at that value all the way to absolute zero, calculate the number of calories it would take to change a 1-gram ice cube at absolute zero (−273°C) to 1 gram of boiling water. How does this number of calories compare to the number of calories required to change the same gram of 100°C boiling water to 100°C steam?

3. Find the mass of 0°C ice that 10 g of 100°C steam will completely melt.

4. If 50 grams of hot water at 80°C is poured into a cavity in a very large block of ice at 0°C, what will be the final temperature of the water in the cavity? How much ice must melt in order to cool the hot water down to this temperature?

5. A 50-gram chunk of 80°C iron is dropped into a cavity in a very large block of ice at 0°C. How many grams of ice will melt? (The specific heat capacity of iron is 0.11 cal/g°C.)

6. A 0.6 gram peanut is burned beneath 50 grams of water. The water increases in temperature from 22°C to 50°C. (a) Assuming 40% efficiency, what is the food value in calories of the peanut? (b) What is the food value in calories per gram?

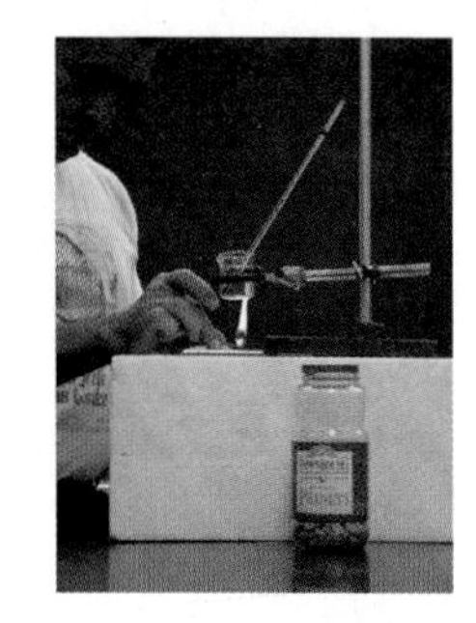

7. Radioactive decay of granite and other rocks in the Earth's interior provides enough energy to keep the interior molten, heat lava, and provide warmth to natural hot springs. This is due to the average release of about 0.03 joules per kilogram each year. How many years are needed for a chunk of thermally insulated granite to increase 500°C in temperature. Assume the specific heat capacity of granite is 800 J/kg·°C.

Part 3 Electricity and Magnetism

How intriguing that this magnet outpulls the whole world when it lifts these nails. The pull between the nails and the Earth I call a **gravitational force**, and the pull between the nails and the magnet I call a **magnetic force**. I can *name* these forces, but I don't yet *understand* them. My learning begins by realizing there's a big difference in *knowing the names* of things and really *understanding* those things.

Chapter 11: Electricity

Why do we sometimes get a shock when we scuff our shoes on the rug? Why does a balloon rubbed on our hair stick to a wall? What causes shock—electric current or electric voltage? And what's the difference between current and voltage? What's the difference between direct current and alternating current? We'll discuss the answers to these questions and more in this chapter.

Electricity is a part of just about everything around us. It's in the lightning from the sky, and in what holds atoms together. Harnessing electricity has enormously changed the world. It's worth serious study. Let's begin with the concept of electric charge.

11.1 Electric Charge Is a Basic Characteristic of Matter

When you rub an inflated rubber balloon on your hair, you charge the balloon. In charging it, you scrape electrons off the atoms of your hair onto the balloon. The rubber apparently has more "grab" on electrons than your hair does. If you do the same with two balloons, you'll find that when you bring them close they repel.

The first rule of electricity is: Like charges repel one another. For historical reasons we say that the charge on an electron is negative. Since there are extra electrons on the balloons rubbed with hair, the balloons are negatively charged. A pair of negatively-charged balloons, or anything else negatively charged, repel. Again, for emphasis:

Rule 1: Like charges repel one another.

If you instead charge one balloon by rubbing it with a piece of kitchen plastic wrap, the plastic grabs electrons from the balloon. The balloon then has an opposite charge to the one rubbed on your hair. When you bring the two oppositely charged balloons together, they attract. Here's the second rule of electricity:

Rule 2: Unlike charges attract one another.

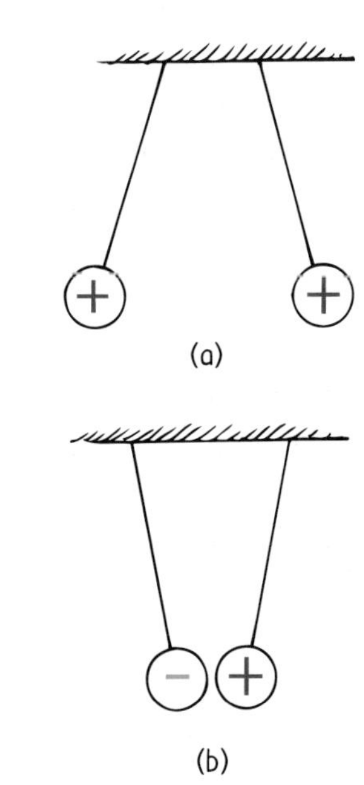

Figure 11.1
(a) Like charges repel.
(b) Unlike charges attract.

In Figure 11.1, we see a pair of like charges attracting, and a pair of unlike charges repelling. In Figure 11.2, we see a simple model of an atom. Particles called *protons* in the nucleus carry the positive charge. The protons attract the whirling electrons and hold them in orbit.

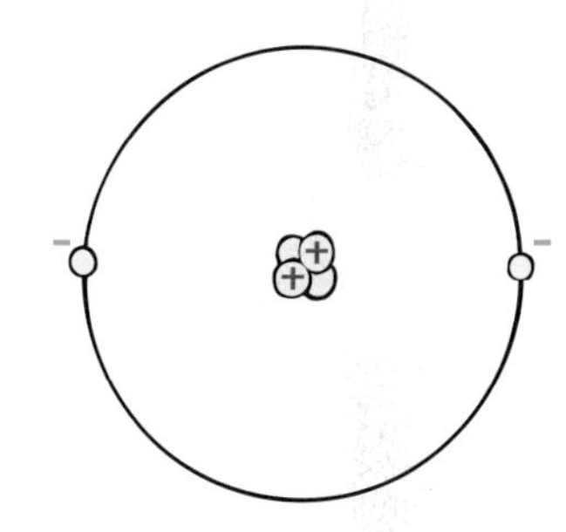

Figure 11.2
Model of a helium atom. The atomic nucleus is made up of two protons and two neutrons. The positively charged protons attract two negative electrons. What is the net charge of this atom?

To understand electric charge, we look at a preview of atomic parts—what we'll learn more about in Part 5. Here are some basic facts about atoms:

1. Every atom has a positively charged *nucleus* surrounded by negatively charged electrons.
2. The electrons of all atoms are identical. Each has the same quantity of negative charge and the same mass.
3. Protons and neutrons make up the nucleus. (The common form of hydrogen has no neutron and is the only exception.) Protons are about 1800 times more massive than electrons, but carry an amount of positive charge equal to the negative charge of electrons. Neutrons have slightly more mass than protons and have no charge.

An atom in its normal state has the same number of electrons as protons. When an atom loses one or more electrons, it has a positive net charge. When it gains one or more electrons, it has a negative net charge. A charged atom is called an *ion.* A *positive ion* has a net positive charge. A *negative ion,* with one or more extra electrons, has a net negative charge.

Negative and *positive* are just the **names** given to opposite charges. The names picked could just as well have been "east and west" or "black and white" or "Fido and Fluffy."

Link to Electronics Technology: Electrostatic Charge

Electric charge can be dangerous. Two hundred years ago, young boys called *powder monkeys* ran below the decks of warships to bring sacks of black gunpowder to the cannons above. It was ship law that this task be done barefoot. Why? Because it was important that no static charge build up on the powder on their bodies as they ran to and fro. Bare feet scuffed the decks much less than shoes and assured no charge buildup that might produce an igniting spark and an explosion.

Static charge is a danger in many industries today. Not because of explosions but because delicate electronic circuits may be destroyed by static charge. Some sensitive circuit components can be "fried" by static electric sparks. Electronics technicians often wear clothing of special fabrics with ground wires between their sleeves and their socks. Some wear special wrist bands that are connected to a grounded surface to prevent static charge build-up—when moving a chair, for example. The smaller the electronic circuit, the more hazardous are sparks that may short-circuit their elements.

Figure 11.3
Electrons are transferred from the fur to the rod. The rod is then negatively charged. Is the fur charged? How much compared to the rod? Positively or negatively?

When you charge the balloon negatively, you also charge your hair positively.

It is important to note that when we charge something, no electrons are created or destroyed. They are simply transferred from one material to another. Charge is *conserved.* In every event, whether on a large-scale or small-scale, the principle of *conservation of charge* has always proved true. No case of the creation or destruction of net electric charge has ever been found. Conservation of charge ranks with conservation of energy and momentum as a significant fundamental principle in physics.

Concept Check ✓

If you walk across a rug and scuff electrons from your feet, are you negatively or positively charged?

Check Your Answer You have fewer electrons after you scuff your feet, and so you are positively charged (and the rug is negatively charged).

11.2 Coulomb's Law—The Force Between Charged Particles

The electrical force has a pattern much like gravitational force. It depends on the quantity of charge and is inversely proportional to the square of the distance between charged particles. This relationship was discovered by Charles Coulomb in the 18th century and so is called **Coulomb's law.** It states that the force between the two charged particles varies directly as the product of their charges and inversely as the square of the separation distance. The force acts along a straight line between the particles. Coulomb's law can be expressed as

$$F = k\frac{q_1 q_2}{d^2}$$

where k is the proportionality constant, q_1 represents the quantity of charge of one particle, q_2 represents the quantity of charge of the other particle, and d is the distance between the charged particles.

The unit of charge is the **coulomb,** abbreviated C. It turns out that a charge of 1 C is the charge on 6.25 billion billion electrons. This might seem like a great number of electrons, but it represents only the amount of charge that passes through a common 100-watt light bulb in little more than a second.

The proportionality constant k in Coulomb's law is similar to G in Newton's law of gravity. Instead of being a very small number like G, k is a very large number, approximately

$$k = 9{,}000{,}000{,}000 \ \mathrm{N{\cdot}m^2/C^2}$$

In scientific notation, $k = 9 \times 10^9 \ \mathrm{N{\cdot}m^2/C^2}$. The unit $\mathrm{N{\cdot}m^2/C^2}$ is not important to learn here. It simply converts the right-hand side of the equation to the unit of force, the newton (N). What is important is the large magnitude of k. If, for example, a pair of like charged particles each carrying a charge of 1 coulomb were 1 meter apart, the force of repulsion between them would be 9 billion newtons. That would be about ten times the weight of a battleship! Obviously, such amounts of net charge do not usually exist in our everyday environment. If such charges were common, we'd see things attracting and repelling quite often. Like charges repel and leave one another before they can build up. The electrical force between even slightly charged objects greatly overwhelms the gravitational force between them.

So Newton's law of gravitation for masses is similar to Coulomb's law for electrically-charged bodies. The most important difference between gravitational and electrical forces is that electrical forces may be either attractive or repulsive, whereas gravitational forces are only attractive. Coulomb's law underlies the bonding forces between molecules that will be covered in chemistry (Part 6).

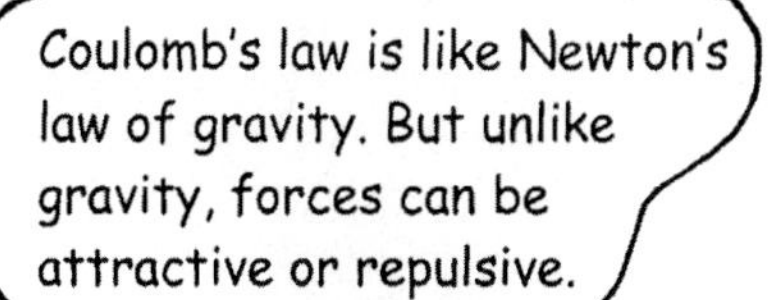

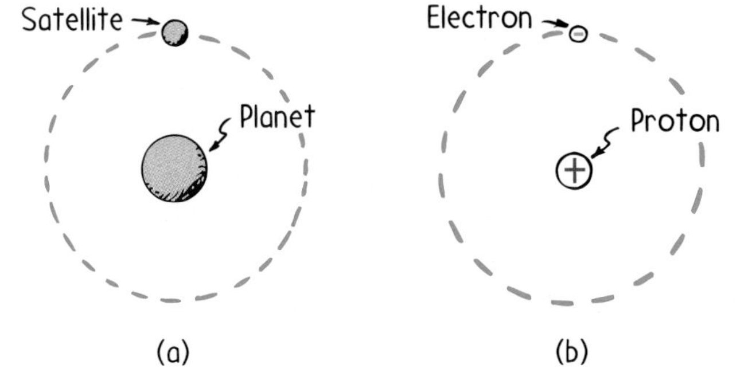

Figure 11.4
(a) A gravitational force holds the satellite in orbit about the planet, and (b) an electrical force holds the electron in orbit about the proton. In both cases the force follows the inverse-square law.

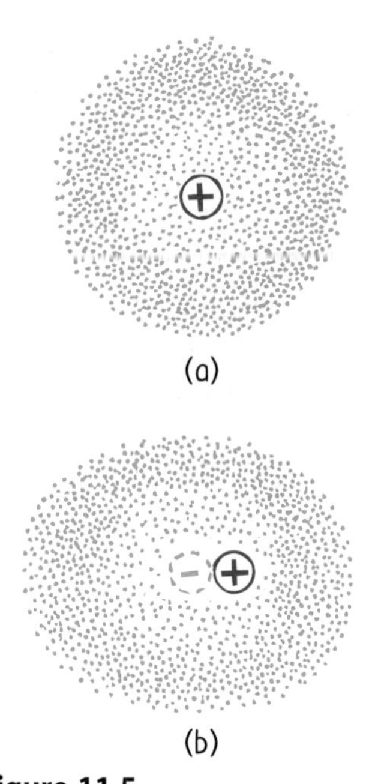

Figure 11.5
(a) The center of the negative "cloud" of electrons coincides with the center of the positive nucleus in an atom.
(b) When an external negative charge is brought nearby to the right, as on a charged balloon, the electron cloud is distorted so the centers of negative and positive charge no longer coincide. The atom is electrically polarized.

Okay, the charged balloon sticks to the wall. If you put your oppositely charged head to the wall, will that stick too?

11.3 Charge Polarization

We began this chapter by discussing an inflated balloon rubbed on your hair. If you place the balloon against a wall you'll see that it sticks. This is because the negative charge on the balloon pulls the positive part of atoms in the wall closer to it. The balloon has the effect of inducing an opposite charge in the wall. Although the atoms in the wall don't move, their "centers of charge" are moved. The positive part of each atom is attracted toward the balloon and the negative part is repelled. We say the distorted atoms (Figure 11.5) are **electrically polarized.**

A polarized object has no net charge. Only the distribution of charge in the material is altered. We will return to electrical polarization in Part 6, and see how it causes the stickiness between many kinds of molecules.

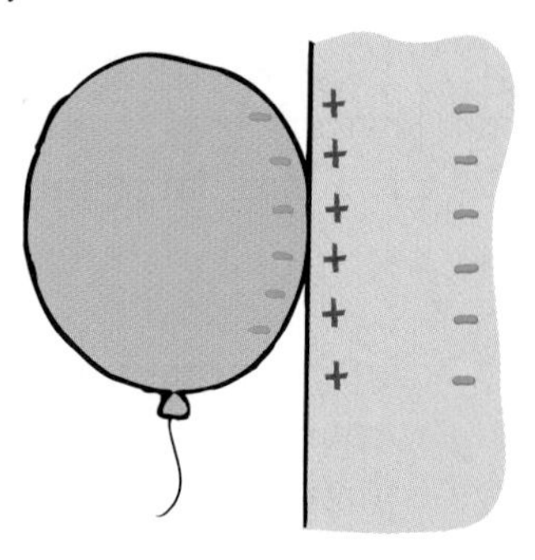

Figure 11.6
Because the negatively charged balloon polarizes atoms in the wall and creates a positively charged surface, the balloon sticks to the wall.

11.4 Electric Current—The Flow of Electric Charge

Recall from Chapter 10 that loose electrons in metals are responsible for the good heat conduction in metals. The same is true for electrical conduction. Loosely-held outer electrons in the atoms of a metal are *conduction electrons.* Protons don't move about in a metal because they are bound inside the atomic nucleus. Conduction electrons, however, can freely migrate through a metal. In Part 6, we'll learn that in addition to electron flow, both positive and negative ions can make up the flow of electric charge in fluids. In this chapter we'll focus on currents made of flowing electrons.

Electrons flow in a way similar to water flow. Just as water current is the flow of H_2O molecules, **electric current** is the flow of electrons. But there are differences between water flow and electron flow. If you buy a water pipe at a hardware store, the clerk doesn't sell you the water to go with it. You provide that yourself. By contrast, when you buy "an electron pipe," an electric wire, you get the electrons too. Every bit of matter, wires included, contains enormous numbers of electrons that swarm about in random directions. When they are set in motion in one direction, a *net* direction, we have an electric current.

The *rate* of electrical flow is measured in *amperes* (abbreviation A). An **ampere** is the rate of flow of 1 coulomb of charge per sec-

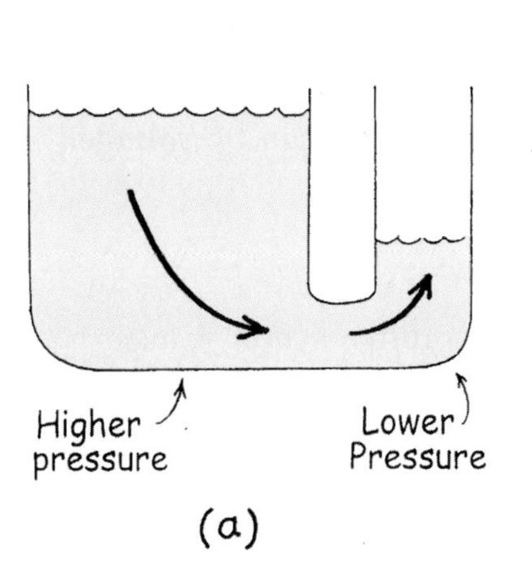

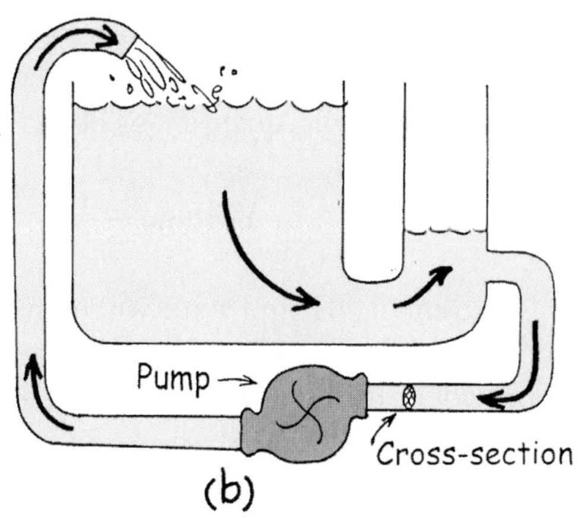

Figure 11.7
(a) Water flows from the reservoir of higher pressure to the reservoir of lower pressure. The flow ceases when the difference in pressure ceases.
(b) Water continues to flow because a difference in pressure is maintained with the pump.

ond. (That's a flow of 6.25 billion billion electrons per second.) In a wire that carries 5 amperes, 5 coulombs of charge pass any cross section in the wire each second. In a wire that carries 10 amperes, twice as many coulombs pass any cross section each second.

11.5 An Electric Current Is Produced by Electrical Pressure—Voltage

When water flows in a pipe, there is more pressure on one end than the other. There must be a pressure difference to keep the water flowing. Also recall from our study of heat in Chapter 10 that heat conduction depends on a temperature difference. Heat flows from zones of high temperature to low temperature. Similarly for electric current. Electrons flow in a wire only when a difference in electrical pressure exists. The name for electrical pressure is *voltage.*

So what, more specifically, is voltage? Voltage is directly proportional to electric potential energy. Recall the concept of potential energy discussed in Chapter 6. Things have energy due to their positions (Figure 11.9). Likewise with electrons and other charged particles. In Figure 11.10, we see that pushing a spring gives it potential energy, for the compressed spring can do work on something when

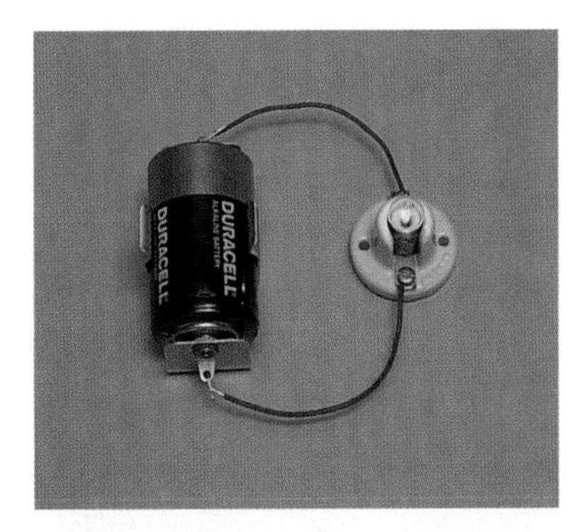

Figure 11.8
Each coulomb of charge that is made to flow in the circuit that connects the ends of this 1.5–V flashlight cell is energized with 1.5 J.

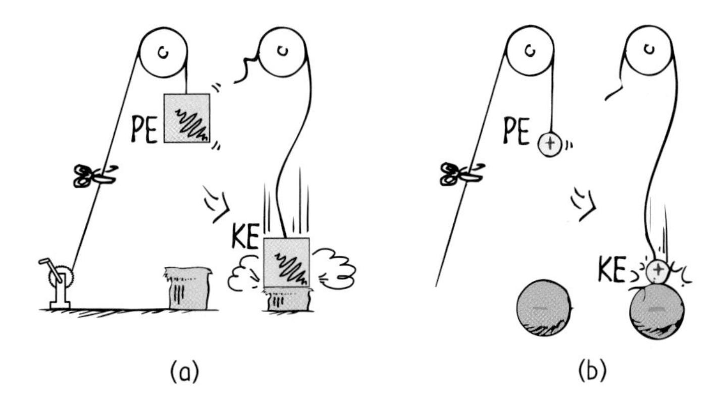

Figure 11.9
(a) Gravitational potential energy converts to kinetic energy.
(b) Similarly, electric potential energy converts to kinetic energy of electrical charge.

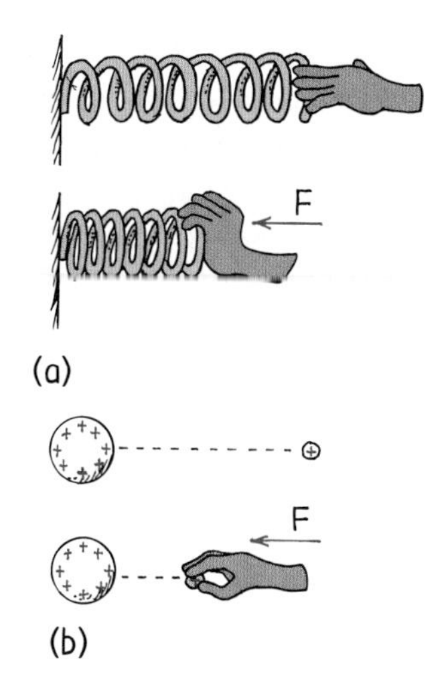

Figure 11.10
(a) The spring has more mechanical PE when compressed. (b) The charged particle similarly has more electrical PE when pushed closer to the charged sphere. In both cases the increased PE is the result of work input.

Figure 11.11
An unusual source of voltage. The electric potential between the head and tail of the electric eel (*Electrophorus electricus*) can be up to 600 V.

released. We also see that pushing a charged particle toward another charge can increase its electrical potential energy. This potential energy compared with the quantity of charge is what we mean by **voltage.**

$$\textbf{Voltage} = \frac{\textbf{potential energy}}{\textbf{charge}}$$

Current flows in a wire when there is a difference in voltage across the ends of the wire. A steady current needs a pumping device to provide a difference in voltage. Chemical batteries or generators are "electrical pumps" that do the job nicely.

A common automobile battery provides a voltage of 12 volts. When each of its terminals are attached to ends of a wire, there is a voltage difference of 12 volts across the wire. That means 12 joules of energy are supplied to each coulomb of charge flowing in the wire. The wire is usually a part of an electric circuit.

There is often some confusion about charge flowing *through* a circuit and voltage *across* a circuit. We can see the difference by thinking of a long pipe filled with water. Water flows *through* the pipe if there is a difference in pressure *across,* or between, its ends; it flows from the high-pressure to the low-pressure end. Only the water flows, not the pressure. Similarly, electrons flow because of a difference in electrical pressure (voltage difference). Electrons flow *through* a circuit because of an applied voltage *across* the circuit. Voltage doesn't flow through a circuit—it doesn't go anywhere, for it is the electrons that flow. Voltage produces current (if there is a complete circuit).

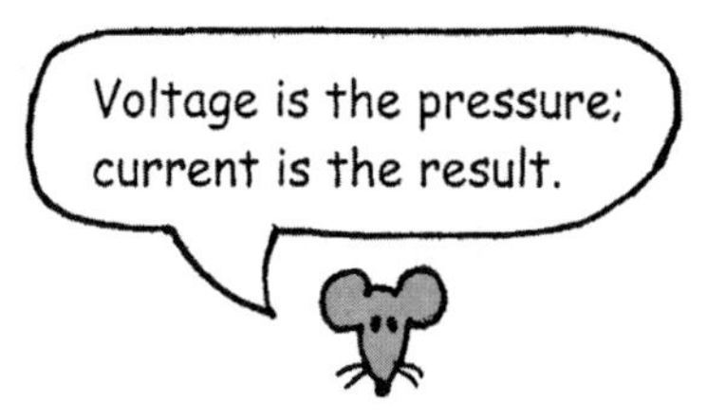

11.6 Electrical Resistance

A battery or generator of some kind moves electrons in a circuit. How much current there is depends on the voltage, and also on the **electrical resistance** of the circuit. Just as narrow pipes resist water flow more than wide pipes, narrow wires resist electrical current more than wider wires. And length contributes to resistance also. Just as long pipes have more resistance than short ones, long wires offer more electrical resistance. And most important is the kind of material. Copper has a low electrical resistance, while a strip of rubber has an enormous resistance. Temperature also affects electrical resistance. The greater the jostling of atoms within a conductor (in other words, the higher the temperature), the greater resistance a conductor has. The resistance of some materials reaches zero at very low temperatures. These are *superconductors.*

Electrical resistance is measured in units called *ohms.* The Greek letter *omega,* Ω, is commonly used as the symbol for the ohm. This

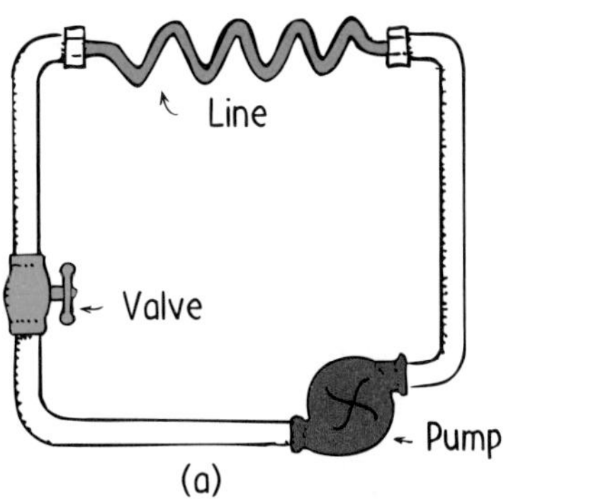

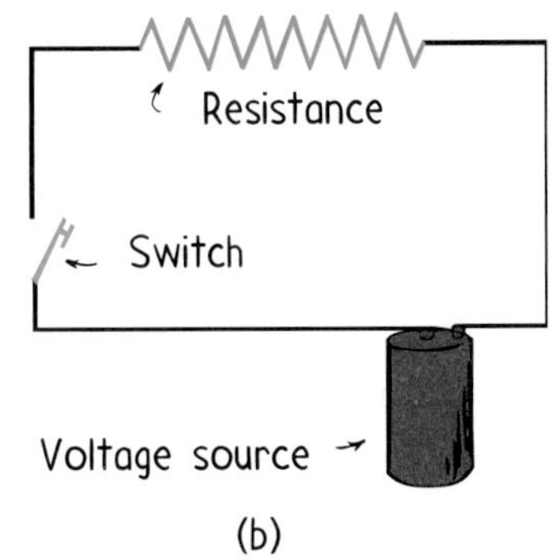

Figure 11.12
(a) In the hydraulic circuit the narrow pipe (green) offers resistance to water flow. (b) In the electric circuit a lamp or other device (shown by the zigzag symbol for resistance) offers resistance to electron flow.

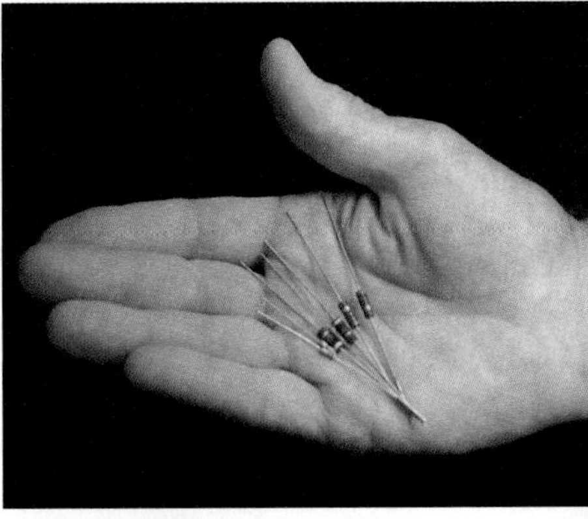

Figure 11.13
Resistors. The symbol of resistance in an electric circuit is -/\/\/-.

unit is named after Georg Simon Ohm, a German physicist who in 1826 discovered a simple and very important relationship among voltage, current, and resistance.

11.7 Ohm's Law—The Relationship Between Current, Voltage, and Resistance

The relationship between voltage, current, and resistance is **Ohm's law.** Ohm discovered that the amount of current in a circuit is directly proportional to the voltage across the circuit and inversely proportional to the resistance of the circuit:

$$\textbf{Current} = \frac{\textbf{voltage}}{\textbf{resistance}}$$

Or, in units form:

$$\textbf{Amperes} = \frac{\textbf{volts}}{\textbf{ohms}}$$

Ohm's law tells us that 1 volt across a circuit with a resistance of 1 ohm produces a current of 1 ampere. If there are 12 volts across the same circuit, the current is 12 amperes. So for a given circuit of constant resistance, current and voltage are proportional to each other.* This means we get twice the current for twice the voltage. The greater the voltage, the greater

Figure 11.14
More water flows through a thick hose than through a thin one connected to a city's water system (same water pressure). Likewise for electric current in thick and thin wires connected across the same voltage.

* Many texts use V for voltage, I for current, and R for resistance, and express Ohm's law as $I = V/R$. It can also be written, $V = IR$, or $R = V/I$, so if any two variables are known, the third can be found. Units are abbreviated V for volts, A for amperes, and Ω for ohms.

the current. But if the resistance is doubled for a circuit, the current is reduced to half. The higher the resistance, the lower the current. Ohm's law makes good sense.

The resistance of a typical lamp cord is much less than 1 ohm, and a typical light bulb has a resistance of more than 100 ohms. An iron or electric toaster has a resistance of 15 to 20 ohms. The current inside these and all other electrical devices is regulated by circuit elements called *resistors*, whose resistance may be a few ohms or millions of ohms.

Concept Check ✓

1. How much current flows through a lamp with a resistance of 60 Ω when the voltage across the lamp is 12 V?
2. What is the resistance of an electric frying pan that draws a current of 12 A when connected to a 120-V circuit?

Check Your Answers

1. This is calculated from Ohm's law: Voltage/resistance = current, so 12 V/60 Ω = 0.2 A.
2. Rearrange Ohm's law to read:

$$\text{Resistance} = \frac{\text{voltage}}{\text{current}} = \frac{120\text{ V}}{12\text{ A}} = 10\ \Omega$$

11.8 Electric Shock

Which causes electric shock in the human body—current or voltage? The damaging effects of shock result from current through the body. From Ohm's law, we see current depends on the voltage applied and also on the body's electrical resistance. A person's resistance ranges

Link to Technology: Superconductors

In ordinary conductors, such as household wiring, the moving electrons that make up current often collide with atomic nuclei in the wire. The colliding electrons transfer their kinetic energy to the wire and the wire heats up. Energy is wasted. However, experiments show that certain metals, when placed in a bath of 4–K liquid helium, lose all their electrical resistance. All of it! The electrons in these extremely cold conductors travel pathways that avoid collisions. Hence the electrons can flow indefinitely. The materials that work this way are called *superconductors.* **Superconductors** have zero electrical resistance to the flow of charge. In superconductivity, no current is lost and no heat is generated.

Various ceramic oxides are superconducting at temperatures above 100 K. Steady currents have been observed to persist for years in some superconductors without any apparent loss in energy. There is presently enormous interest in the technology of superconductors. Imagine the energy-saving potential of superconducting devices! Explanations of superconductivity involve the wave nature of matter (quantum mechanics) and are being vigorously researched.

from about 100 ohms if the body is soaked with salt water to about 500,000 ohms if the skin is very dry. If you touch the two electrodes of a battery with dry fingers, you make up a circuit with a resistance of about 100,000 ohms. You usually cannot feel 12 volts. Even 24 volts just barely tingles. If your skin is moist, 24 volts can be quite uncomfortable. Table 11.1 describes the effects of different amounts of current on the human body.

Table 11.1

Effect of Electric Current on the Body

Current (A)	Effect
0.001	Can be felt
0.005	Is painful
0.010	Causes involuntary muscle contractions (spasms)
0.015	Causes loss of muscle control
0.070	Goes through the heart; serious damage, probably fatal for if current lasts for more than 1 s

Concept Check ✓

1. At 100,000 Ω, how much current will flow through your body if you touch the terminals of a 12-V battery?
2. If your skin is very moist, so that your resistance is only 1000 Ω, and you touch the terminals of a 12-V battery, how much current do you receive?

Check Your Answers:

1. $\frac{12\text{ V}}{100\,000\ \Omega} = 0.00012\text{ A}.$
2. $\frac{12\text{ V}}{1000\ \Omega} = 0.012\text{ A}.$ Ouch!

To receive a shock, there must be a *difference* in voltage between one part of your body and another part. Electron flow will pass along the path of least electrical resistance connecting these two points. Suppose you fell from a bridge and grabbed onto a high-voltage power line, halting your fall. If you touch nothing else of different voltage, you receive no shock. Even if the wire is a few thousand volts and even if you hang by two hands, no significant electron flow will occur between your hands. This is because there is no voltage difference between your hands. If, however, you reach over with one hand and grab onto a wire of different voltage . . . zap! We have all seen birds perched on high-voltage wires. Every part of their bodies is at the same high voltage as the wire, and so they have no problem.

Figure 11.15
The bird can stand harmlessly on one wire of high voltage, but it had better not reach over and grab a neighboring wire! Why not?

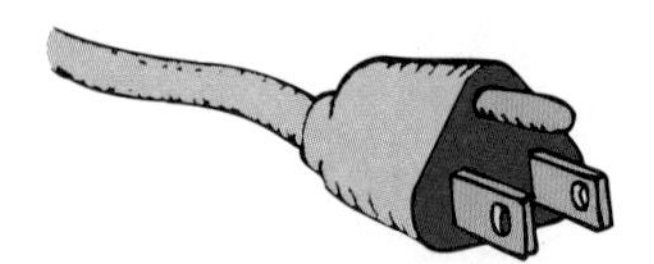

Figure 11.16
The round prong connects the body of the appliance directly to ground (the Earth). Any charge that builds up on an appliance is therefore conducted to the ground—preventing accidental shock.

Most electric plugs and sockets are wired with three connections. The two flat prongs on a plug are for the current-carrying double wire inside the socket. One part is "live" (energized) and the other is neutral. The larger round prong connects to a wire in the electrical system that is grounded—connected directly to the ground (Figure 11.16). If the live wire of the plugged-in appliance accidentally comes in contact with the metal surface of the appliance, and you touch the appliance, you could receive a dangerous shock. This won't occur when the appliance casing is grounded via the ground wire. Then the voltage of the appliance is the same voltage as the ground—relatively speaking, zero. You can't get shocked unless there is a voltage difference.

Concept Check ✓

1. So which causes electric shock—current or voltage?
2. What is the source of electrons that produce a shock in your body?
3. What is the source of a simple battery-powered circuit?

Check Your Answers

1. Electric shock *occurs* when current is produced in the body, which is *caused* by an applied voltage.
2. The source of electrons are those already in your body. Like any conductor, the electrons are already there. A voltage across your body will set them in motion.
3. The circuit elements themselves. Just as a water pump moves water in pipes, a battery supplies energy to move electrons already in a circuit.

Link to Safety: Electric Shock

Many people are killed each year by current from common 120-volt electric circuits. If you touch a faulty 120-volt light fixture with your hand while your feet are on the ground, there may be a 120-volt "electrical pressure" between your hand and the ground. Resistance to current is usually greatest between your feet and the ground, and so the current is usually not enough to do serious harm. But if your feet and the ground are wet, there is a low-resistance electrical path between you and the ground. The 120 volts across this lowered resistance may produce a current greater than your body can stand.

Pure water is not a good conductor. But the ions normally found in water make it a fair conductor. More dissolved materials, especially small amounts of salt, lower the resistance even more. There is usually a layer of salt left from perspiration on your skin, which when wet lowers your skin resistance to a few hundred ohms or less. Handling electrical devices while taking a bath is a definite no-no.

Injury by electric shock comes in three forms: (1) Burning of tissues by heating, (2) muscle contraction, and (3) disruption of cardiac rhythm. These conditions are caused by too much power delivered in critical body regions for too long a time.

Electric shock can upset the nerve center that controls breathing. In rescuing shock victims, the first thing to do is clear them from the electric supply. Use a dry wooden stick or some other nonconductor so that you don't get electrocuted yourself. Then apply artificial respiration. It is important to continue artificial respiration. There have been cases of victims of lightning who did not breathe for several hours, but were eventually revived and completely regained good health.

11.9 Direct Current and Alternating Current

Electric current may be *dc* or *ac*. By *dc*, we mean **direct current.** Direct current is current made up of electrons that flow in *one direction.* A battery produces direct current in a circuit because the terminals of the battery always have the same opposite signs. Electrons move from the repelling negative terminal toward the attracting positive terminal, always moving through the circuit in the same direction.

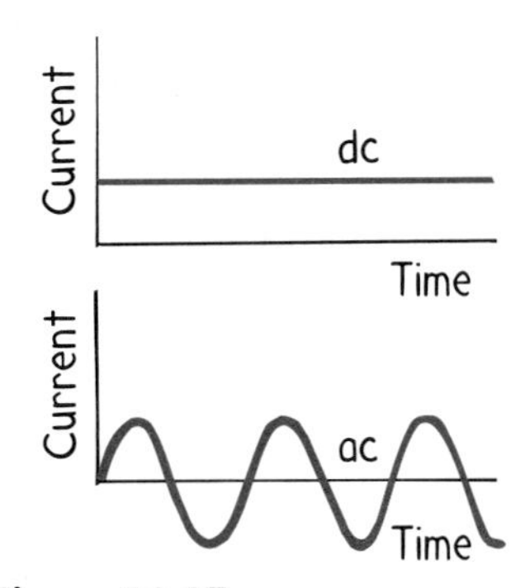

Figure 11.17
Time graphs of dc and ac.

It is interesting to note that the speed of electrons as they drift through a wire is surprisingly slow. This is because electrons continually bump into atoms in the wire. The *drift speed* of electrons in a typical circuit is much less than one centimeter per second. The electric signal, however, travels at nearly the speed of light. That's the speed at which the electric *field* in the wire is established. (An electric field exists around charged particles like a gravitational field exists around massive bodies—both fields travel from one point to another at the speed of light.)

Alternating current (ac) acts as the name implies. Electrons in the circuit flow initially in one direction and then in the opposite direction. This is done by switching the sign at the terminals of the power-station generator (next chapter). The net drift speed of electrons in an ac circuit is zero. They vibrate back and forth about relatively fixed positions and travel nowhere. Nearly all commercial ac circuits involve currents that alternate back and forth at a frequency of 60 cycles per second. This is 60-hertz current (a cycle per second is called a *hertz*).

11.10 Electric Power—The Rate of Doing Work

Moving charges in an electric current can do work. They can heat a circuit or turn a motor. Recall from Chapter 6 how we defined power as the rate of using energy. Power is the energy transformed divided by the elapsed time. Electrical energy may be transformed to mechanical energy (as in a motor), to light (as in a lamp), to

Link to History of Technology: 110 Volts

In the early days of electricity, high voltages burned out electric light filaments, and so low voltages were more practical. The hundreds of power plants built in the U.S. prior to 1900 adopted 110 volts (or 115 or 120 volts) as their standard. Tradition has it that 110 volts was agreed upon because it made bulbs of the day glow as brightly as a gas lamp. By the time electricity became popular in Europe, engineers had figured out how to make light bulbs that would not burn out so fast at higher voltages. Because power transmission is more efficient at higher voltages, Europe adopted 220 volts as their standard. The U.S. remained with 110 volts (today officially 120 volts) because of the installed base of 110-volt equipment.

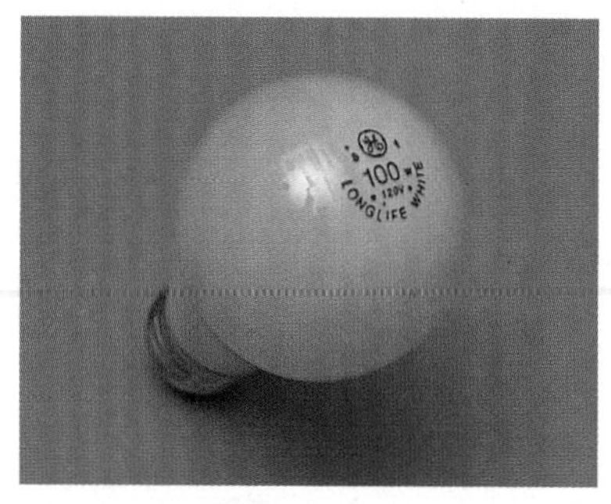

Figure 11.18
The power and voltage on the light bulb read "100 W 120 V." How many amperes will flow through the bulb?

thermal energy (as in a heater), or to other forms. In electrical terms, power is equal to current multiplied by voltage.

Power = current × voltage

When current is in amperes and voltage is in volts, then power is expressed in watts. So in units form,

Watts = amperes × volts.

If a lamp rated at 120 watts operates on a 120-volt line, it draws a current of 1 ampere (120 W = 1 A × 120 V). A 60-watt lamp draws 1/2 ampere on a 120-volt line.

Concept Check ✓

What power is needed to operate a clock radio if it draws a current of 0.05 amperes from your household circuit?

Check Your Answer

Power = current × voltage = 0.05 A × 120 V = 6 W.

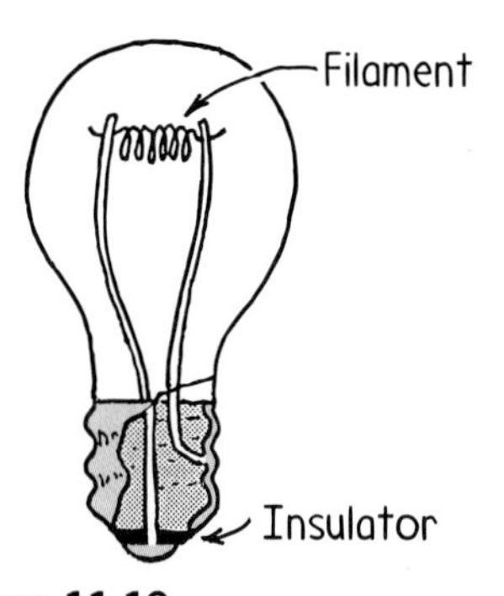

Figure 11.19
The conduction electrons that surge back and forth in the filament of the lamp do not come from the voltage source. They are in the filament to begin with. The voltage source simply provides them with surges of energy.

11.11 Electric Circuits—Series and Parallel

Any path along which electrons can flow is a *circuit.* For a steady current there must be a complete circuit with no gaps. A gap is usually provided by an electric switch that can be opened or closed. Then there is control to either stop or allow energy flow.

Most circuits contain more than one device that receives electric energy in a circuit. It may be several lamps, for example. These devices are commonly connected in one of two ways, *series* or *parallel.* When connected in series, the devices and the wires connecting them form a single pathway for electron flow between the terminals of the battery, generator, or wall socket. When connected in parallel, the devices and wires connecting them form branches, each providing separate paths for electron flow. Series and parallel connections each have their own distinctive characteristics.

Series Circuits

A simple **series circuit** is shown in Figure 11.20, where three lamps are connected in series with a battery. When the switch is closed, the same current exists almost immediately in all three lamps. The current does not "pile up" in any lamp but flows *through* each lamp. Electrons that make up this current leave the negative terminal of the battery, pass through each lamp in turn, and then return to the positive terminal of the battery. Inside the battery the electrons move to the negative terminal again. This means that the amount of

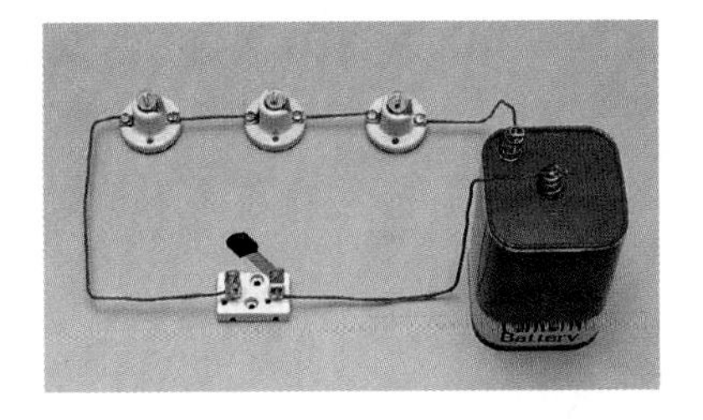

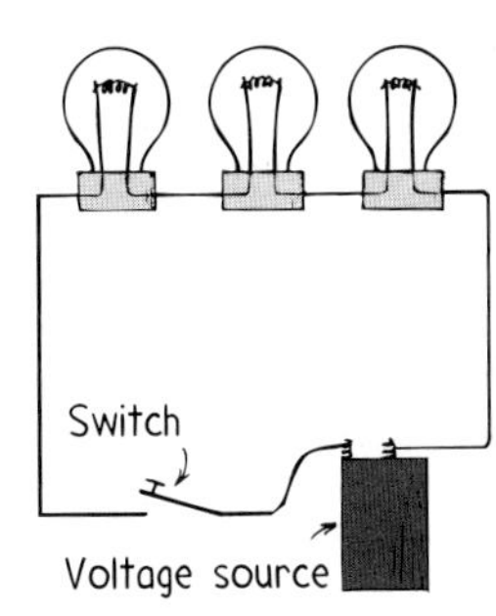

Figure 11.20
A simple series circuit. The 6-V battery provides 2 V across each lamp.

current passing through the battery is the same as the amount passing through the lamps. This is the only path of the electrons through the circuit. If a break occurs anywhere in the path, the flow of electrons stops. Burning out of one of the lamp filaments or simply opening the switch causes such a break. The circuit shown in Figure 11.20 illustrates the following important characteristics of series connections:

1. Electric current has but one pathway. This means that the current is the same in every part of the circuit.
2. This current is resisted by the resistance of the first device, the resistance of the second, and that of the third also. So the total resistance to current in the circuit is the sum of the individual resistances along the circuit path (assuming the resistance of the connecting wires is negligible).
3. The current in the circuit is numerically equal to the voltage supplied by the source divided by the total resistance of the circuit. This is in accord with Ohm's law.
4. The total voltage established across a series circuit divides among the electrical devices in the circuit so that the sum of the "voltage drops" across each device is equal to the total voltage supplied by the source. (This follows from the fact that the amount of energy given to the total current is equal to the sum of energies given to each device.)
5. The voltage drop across each device is proportional to its resistance. (This follows from the fact that more energy is wasted as heat when a current passes through a high-resistance device than through a low-resistance device.)

Concept Check ✓

1. What happens to the current in other light bulbs in a series circuit if one bulb burns out?
2. What happens to the light intensity of each light bulb in a series circuit when more bulbs are added to the circuit?

Check Your Answers

1. If one bulb burns out, the circuit path is broken and current ceases. All lamps go out.
2. The addition of more lamps in a series circuit results in more circuit resistance. This lowers the current in the circuit and therefore in each lamp. That's why the lamps dim. The same amount of energy is divided among more lamps, which means less energy per lamp—less voltage drop across each lamp.

It is easy to see the main disadvantage of a series circuit. If one device fails, current in the entire circuit ceases. Some inexpensive Christmas tree lights are connected in series. When one bulb burns out, it's fun and games (or frustration) trying to locate which bulb to replace.

Most circuits are wired so that it is possible to operate several electrical devices, each independently of the others. In your home, for example, a light switch can be turned on or off without affecting other electrical appliances on the same circuit. This is because these devices are connected not in series, but in parallel.

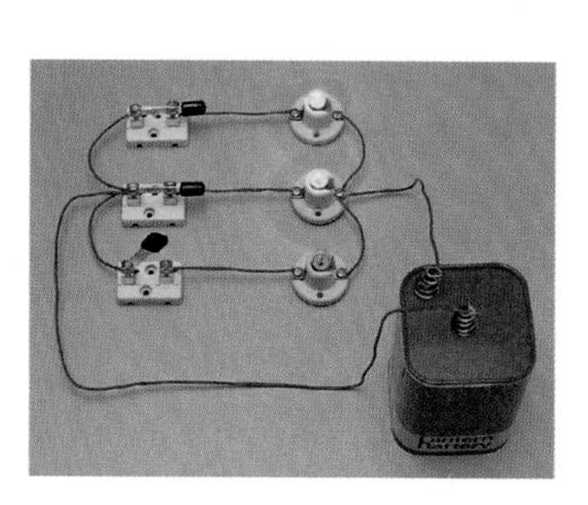

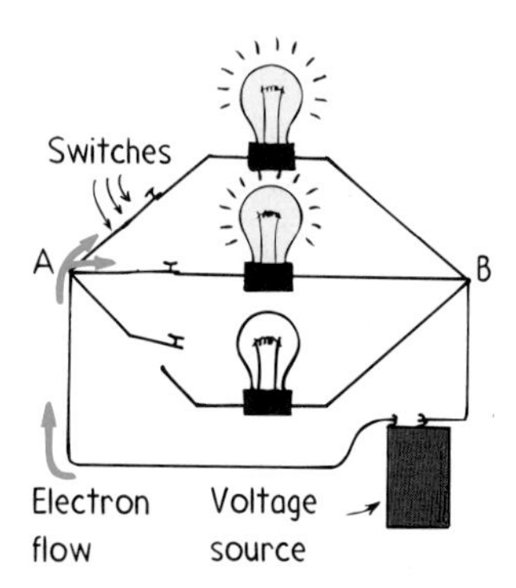

Figure 11.21
A simple parallel circuit. A 6-V battery provides 6 V across each lamp.

Parallel Circuits

A simple **parallel circuit** is shown in Figure 11.21. Three lamps are connected to the same two points A and B. Electrical devices connected to the same two points of an electrical circuit are *connected in parallel.* Electrons leaving the negative battery terminal need to travel through only *one* lamp filament before returning to the positive terminal of the battery. In this case, current branches into three separate pathways from A to B. A break in any one path does not interrupt the flow of charge in the other paths. Each device operates independently of the others.

The circuit shown in Figure 11.21 illustrates the following major characteristics of parallel connections:

1. Each device connects the same two points A and B of the circuit. The voltage is therefore the same across each device.

2. The total current in the circuit divides among the parallel branches. Since the voltage across each branch is the same, the amount of current in each branch is inversely proportional to the resistance of the branch.

More lamps in series increases resistance. More lamps in parallel decreases resistance—like more checkout counters in a store.

3. The total current in the circuit equals the sum of the currents in its parallel branches.

4. As the number of parallel branches is increased, the overall resistance of the circuit is lowered (just as more check-out cashiers at a supermarket lowers people-flow resistance). With each added parallel path, the overall circuit resistance is lowered. This means the overall resistance of the circuit is less than the resistance of any one of the branches.

Concept Check ✓

1. What happens to the current in other light bulbs in a parallel circuit when one bulb burns out?
2. What happens to the light intensity of each light bulb in a parallel circuit when more bulbs are added?

Check Your Answers

1. When one bulb burns out, the others are unaffected. The current in each branch, according to Ohm's law, is equal to voltage/resistance. Since neither voltage nor resistance is affected in the other branches, the current in those branches is unaffected. The total current in the overall circuit (the current through the battery), however, is lowered. It is reduced by an amount equal to the current drawn by the bulb in question before it burned out. But the current in any other single branch is unchanged.
2. The light intensity of each bulb is unchanged as other bulbs are introduced (or removed). Although changes of resistance and current occur for the circuit as a whole, no changes occur in any individual branch in the circuit.

Parallel Circuits and Overloading Electricity is usually fed into a home by way of two wires called *lines.* These lines are very low in resistance and are connected to wall outlets in each room—sometimes through two or more separate circuits. About 110–120 volts are applied across these lines by a transformer in the neighborhood. (A transformer is a device that steps down the higher voltage supplied by the power utility.) As more devices are connected to a circuit, more pathways for current result. This lowers the combined resistance of the circuit. Therefore, more current exists in the circuit, which is sometimes a problem. Circuits that carry more than a safe amount of current are said to be *overloaded.*

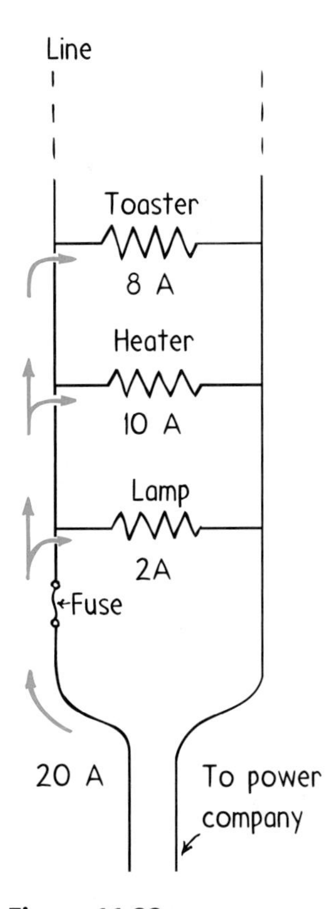

Figure 11.22
Circuit diagram for appliances connected to a household circuit.

We can see how overloading occurs in Figure 11.22. The supply line is connected to an electric toaster that draws 8 amperes, an electric heater that draws 10 amperes, and an electric lamp that draws 2 amperes. When only the toaster is operating and drawing 8 amperes, the total line current is 8 amperes. When the heater is also operating, the total line current increases to 18 amperes (8 amperes

Electrical Energy and Technology

Try to imagine life before electrical energy was something that humans could control. Imagine home life without electric lights, refrigerators, heating and cooling systems, the telephone, and radio and TV. We may romanticize a better life without these, but only if we overlook the hours of daily toil doing laundry, cooking, and heating homes. We'd also have to overlook how difficult it was getting a doctor in times of emergency before the advent of the telephone—when all the doctor had in his bag were laxatives, aspirins, and sugar pills—and when infant death rates were staggering.

We have become so accustomed to the benefits of technology that we are only faintly aware of our dependency on dams, power plants, mass transportation, electrification, modern medicine, and modern agricultural science for our very existence. When we dig into a good meal, we give little thought to the technology that went into growing, harvesting, and delivering the food on our table. When we turn on a light we give little thought to the centrally controlled power grid that links the widely separated power stations by long-distance transmission lines. These lines serve as the productive life force of industry, transportation, and the electrification of civilization. Anyone who thinks of science and technology as "inhuman" fails to grasp the ways in which they make our lives more human.

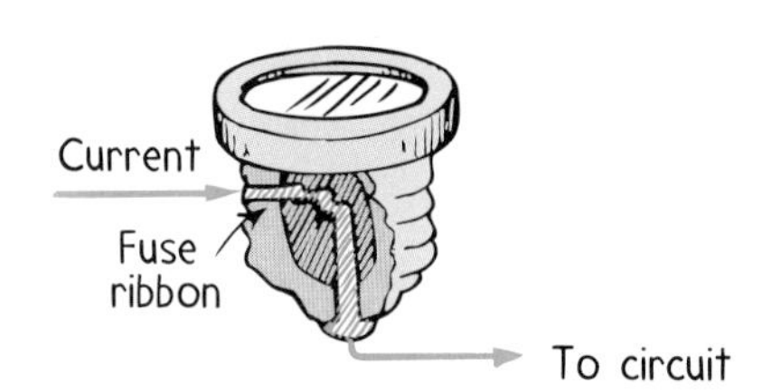

Figure 11.23
A safety fuse.

to the toaster plus 10 amperes to the heater). If you turn on the lamp, the line current increases to 20 amperes. Connecting any more devices increases the current still more. Connecting too many devices into the same circuit results in overheating the wires, which can cause a fire.

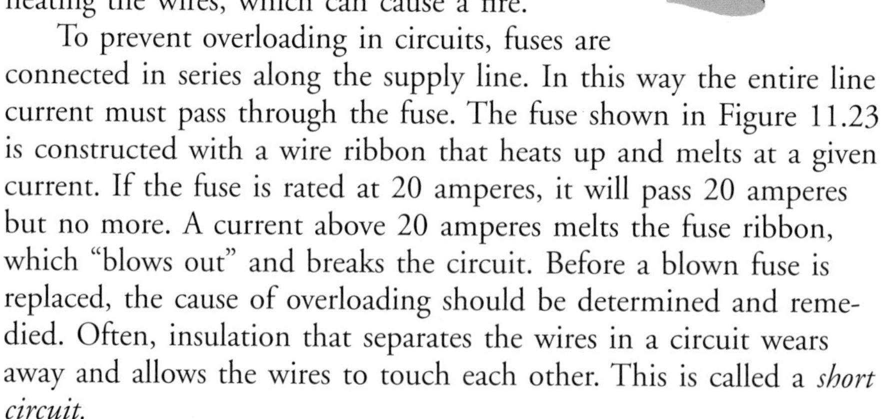

To prevent overloading in circuits, fuses are connected in series along the supply line. In this way the entire line current must pass through the fuse. The fuse shown in Figure 11.23 is constructed with a wire ribbon that heats up and melts at a given current. If the fuse is rated at 20 amperes, it will pass 20 amperes but no more. A current above 20 amperes melts the fuse ribbon, which "blows out" and breaks the circuit. Before a blown fuse is replaced, the cause of overloading should be determined and remedied. Often, insulation that separates the wires in a circuit wears away and allows the wires to touch each other. This is called a *short circuit.*

In modern buildings, fuses have been largely replaced by circuit breakers, which use magnets or bimetallic strips to open a switch when the current is too much. Utility companies use circuit breakers to protect their lines all the way back to the generators.

Chapter Review

Key Terms and Matching Definitions

_____ alternating current (ac)
_____ ampere
_____ coulomb
_____ Coulomb's law
_____ direct current (dc)
_____ electric current
_____ electric power
_____ electrical resistance
_____ electrically polarized
_____ Ohm's law
_____ parallel circuit
_____ series circuit
_____ voltage

1. The electrical force between two charged bodies is directly proportional to the product of the charges and inversely proportional to the square of the distance between them:

   $$F = k\frac{q_1 q_2}{d^2}$$

2. The unit of electrical charge. It is equal in magnitude to the total charge of 6.25×10^{18} electrons.
3. Term applied to an atom or molecule in which the charges are aligned so that one side has a slight excess of positive charge and the other side a slight excess of negative charge.
4. The flow of electric charge that transports energy from one place to another. Measured in amperes.
5. The unit of electric current, equivalent to 1 coulomb per second (the flow of 6.25×10^{18} electrons per second). In symbols, 1 A= 1 C/s.
6. A form of electrical pressure:

   $$\text{Voltage} = \frac{\text{potential energy}}{\text{charge}}.$$

7. The property of a material that resists the flow of charged particles through it. Measured in ohms (Ω).
8. The statement that the current in a circuit varies in direct proportion to the voltage across the circuit and inversely with the circuit's resistance:

   $$\text{Current} = \frac{\text{voltage}}{\text{resistance}}$$

9. Electrically charged particles flowing in one direction only.
10. Electrically charged particles that repeatedly reverse direction, vibrating about relatively fixed positions. In the United States the vibrational rate is 60 Hz.
11. The rate of energy transfer, or rate of doing work. Measured by:

    $$\text{Power} = \text{current} \times \text{voltage}$$

    Measured in watts (or kilowatts), where $1\text{ A} \times 1\text{ V} = 1\text{ W}$.
12. An electric circuit in which electrical devices are connected so that the same electric current exists in all of them.
13. An electric circuit in which electrical devices are connected so that the same voltage acts across each one and any single one completes the circuit independently of all the others.

Review Questions

Electric Charge Is a Basic Characteristic of Matter

1. Which part of an atom is *positively* charged and which part is *negatively* charged?
2. How does the charge of one electron compare with that of another electron?
3. How does the number of protons in the atomic nucleus normally compare with the number of electrons that orbit the nucleus?
4. What is meant by saying charge is *conserved*?

Coulomb's Law—The Force Between Charged Particles

5. How is Coulomb's law similar to Newton's law of gravitation? How is it different?

Charge Polarization

6. How does an electrically *polarized* object differ from an electrically *charged* object?

Electric Current—The Flow of Electric Charge

7. Why do electrons rather than protons make up the flow of charge in a metal wire?

An Electric Current Is Produced by Electrical Pressure—Voltage

8. How much energy is given to each coulomb of charge passing through a 6-V battery?

9. Does electric charge flow *across* a circuit or *through* a circuit? Does voltage *flow* across a circuit or is it *impressed* across a circuit? Explain.

Electrical Resistance

10. Which has the greater resistance, a thick wire or a thin wire of the same length?

Ohm's Law—The Relationship between Current, Voltage, and Resistance

11. When the voltage across the ends of a piece of wire is doubled, what effect does this have on the current in the wire?

12. When the resistance of a circuit is doubled, and no other changes occur, what effect does this have on the current in the circuit?

Electric Shock

13. What is the function of the third prong on the plug of an electric appliance?

Direct Current and Alternating Current

14. Distinguish between *dc* and *ac*.

15. Does a battery produce *dc* or *ac*? Does the generator at a power station produce *dc* or *ac*?

Electric Power—The Rate of Doing Work

16. Which draws more current, a 40-W bulb or a 100-W bulb?

Electric Circuits—Series and Parallel

17. In a circuit consisting of two lamps connected in series, if the current through one lamp is 1 A, what is the current through the other lamp?

18. In a circuit consisting of two lamps connected in parallel, if there is 6 V across one lamp, what is the voltage across the other lamp?

19. How does the total current through the branches of a parallel circuit compare with the current through the voltage source?

20. Are household circuits normally wired in series or in parallel?

Explorations

1. Demonstrate charging by friction and discharging from points with a friend who stands at the far end of a carpeted room. Scuff your shoes across the rug until your noses are close together. This can be a delightfully tingling experience, depending on how dry the air is (and how pointed your noses are).

2. Briskly rub a comb on your hair or a woolen garment and bring it near a small but smooth stream of running water. Is the stream of water charged? (Before you say yes, note the behavior of the stream when an opposite charge is brought nearby.)

3. An electric cell is made by placing two plates made of different materials that have different affinities for electrons in a conducting solution. (A battery is actually a series of cells.) You can make a simple 1.5-V cell by placing a strip of copper and a strip of zinc in a tumbler of salt water. The voltage of a cell depends on the materials used and the solution they are placed in, not on the size of the plates.

An easy cell to construct is the citrus cell. Stick a paper clip and a piece of copper wire into a lemon. Hold the ends of the wire close together, but not touching, and place the ends on your tongue. The slight tingle you feel and the metallic taste you experience result from a slight current of electricity pushed by the citrus cell through the wires when your moist tongue closes the circuit.

Paper clip
Lemon
Copper wire

Exercises

1. When combing your hair, you transfer electrons from your hair onto the comb. Is your hair then positively or negatively charged? How about the comb?
2. If electrons were positive and protons were negative, would Coulomb's law be written the same or differently?
3. The five thousand billion billion freely moving electrons in a penny repel one another. Why don't they fly out of the penny?
4. How does the magnitude of electrical force between a pair of charged objects change when the objects are moved twice as far apart? Three times as far apart?
5. How does the magnitude of electric force compare between a pair of charged particles when they are brought to half their original distance of separation? To one-quarter their original distance? To four times their original distance? (What law guides your answers?)
6. Two equal charges exert equal forces on each other. What if one charge has twice the magnitude of the other. How do the forces they exert on each other compare?
7. Why is a good conductor of electricity also a good conductor of heat?
8. What happens to the brightness of light emitted by a lamp when the current that flows in it increases?
9. Your tutor tells you that an *ampere* and a *volt* really measure the same thing, and the different terms only serve to make a simple concept seem confusing. Why should you consider getting a different tutor?
10. In which of the circuits below does a current exist to light the bulb?

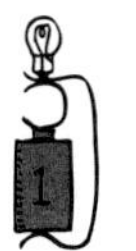

11. Does more current flow out of a battery than into it? Does more current flow into a light bulb than out of it? Explain.
12. Only a small percentage of the electric energy going into a common light bulb is transformed into light. What happens to the rest?
13. Why are thick wires rather than thin wires usually used to carry large currents?
14. What is the effect on current in a wire if both the voltage across it and its resistance are doubled? If both are halved?
15. Will the current in a light bulb connected to a 220-V source be greater or less than when the same bulb is connected to a 110-V source?
16. Which will do less damage—plugging a 110-V appliance into a 220-V circuit or plugging a 220-V appliance into a 110-V circuit? Explain.

17. The damaging effects of electric shock result from the amount of current that flows in the body. Why, then, do we see signs that read "Danger—High Voltage" rather than "Danger—High Current"?

18. If several bulbs are connected in series to a battery, they may feel warm to the touch but not visibly glow. What is your explanation?

19. In the circuit shown, how does the brightness of each identical light bulb compare? Which light bulb draws the most current? What will happen if bulb A is unscrewed? If C is unscrewed?

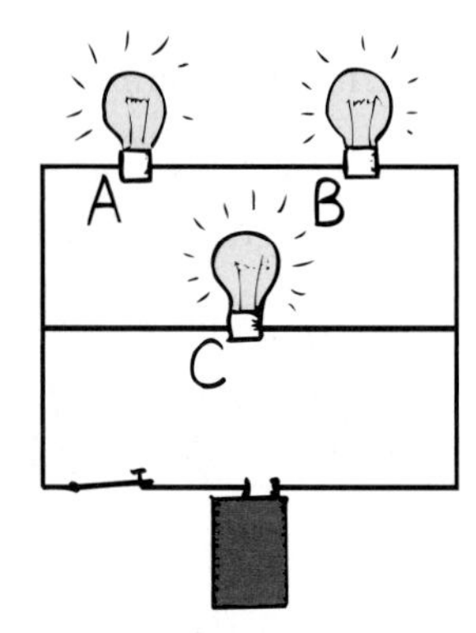

20. As more and more bulbs are connected in series to a flashlight battery, what happens to the brightness of each bulb? Assuming heating inside the battery is negligible, what happens to the brightness of each bulb when more and more bulbs are connected in parallel?

Problems

1. Two point charges are separated by 6 cm. The attractive force between them is 20 N. Find the force between them when they are separated by 12 cm. (Why can you solve this problem without knowing the magnitudes of the charges?)

2. Make use of Coulomb's law here. Suppose you have a pair of electrically charged metal spheres suspended from insulating threads, a certain distance from each other. There is a specific amount of electrical force between them.

a. If the charge on one sphere were doubled, what would happen to the force between them?

b. If the charge on *both* spheres were doubled, what would happen to the force between them?

c. If the distance between the spheres were tripled, what would happen to the force between them?

d. If the distance between them were reduced to one-fourth the original distance, what would happen to the force between them?

e. If the charge on each sphere were doubled and the distance between them were doubled, what would happen to the force between them?

3. The unit of charge is the coulomb (C). The charge on an electron is 1.6×10^{-19} C. How many electrons make a charge of 1 C?

4. A flow of charge of 1.0 C/s is called one *ampere* (1.0 A). How many electrons per second must pass a given point in order to have a current of 2.0 A?

5. How much current flows through a radio speaker that has a resistance of 8 Ω when 12 V is impressed across the speaker?

6. If the circuit shown to the left has a 6-V battery, and the voltage across lamp A is 2 V, what is the voltage across lamp B?

7. Rearrange the equation Current = voltage/resistance to express *resistance* in terms of current and voltage. Then solve the following: A certain device in a 120-V circuit has a current rating of 20 A. What is the resistance of the device (how many ohms)?

8. What is the effect on current through a circuit of steady resistance when the voltage is doubled? What if both voltage and resistance are doubled?

9. Using the definition of power (current $\times$ voltage), how much current flows in a 6-watt clock radio that operates in a common household?

Chapter 12: Magnetism

What is the connection between magnetism and the colorful aurora borealis in the upper atmosphere? Why does a compass needle point north? Is the Earth a giant magnet? What causes magnetism—is it electrical? Will a refrigerator magnet stick to all metal surfaces? A wall? Do electric motors have electromagnets in them? Does the Earth's magnetic field shield us from cosmic rays? These questions and more will be answered in this chapter.

The term *magnetism* comes from the region of Magnesia, a province of Greece. Certain magnetic stones were found by the Greeks who lived there more than 2000 years ago. These stones, called *lodestones,* could attract pieces of iron. When lodestones were rubbed on pieces of iron, magnets were made. Magnets were first made into compasses and used for navigation by the Chinese in the twelfth century. Today much of modern industry—from motors to computers—relies on magnetism.

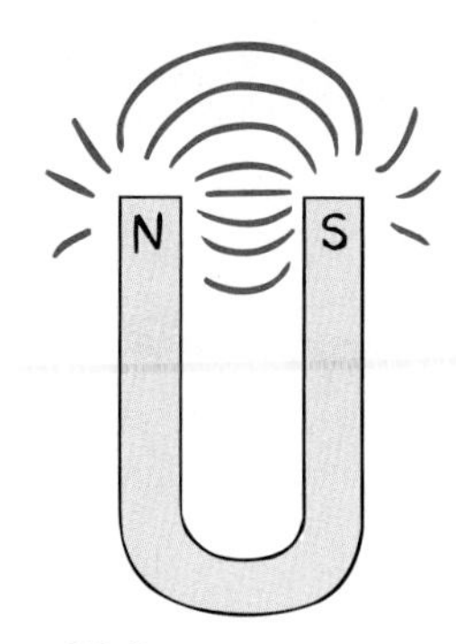

Figure. 12.1
A horseshoe magnet.

12.1 Magnetic Poles—Attraction and Repulsion

If you've played around with magnets, you know they exert forces on one another. **Magnetic forces** are similar to electrical forces, for both kinds of forces cause objects to attract and repel. Also both electrical and magnetic forces act between objects that are not touching. And similar to electrical force, the strength of a magnetic interaction depends on the distance between the two magnets. Whereas electric charges produce electrical forces, regions called *magnetic poles* produce magnetic forces.

Hang a bar magnet at its center by a piece of string and you've got a compass. One end, called the *north-seeking pole,* points northward. The opposite end, called the *south-seeking pole,* point southward. More simply, these are called the *north* and *south poles.* All magnets have both a north and a south pole (some have more than one of each). Refrigerator magnets have narrow strips of alternating north and south poles. These magnets are strong enough to hold sheets of paper against a refrigerator door. But they have a very short range because the north and south poles cancel a short distance from the magnet. In a simple bar magnet the magnetic poles are located at the two ends. A common horseshoe magnet is a bar magnet bent into a U shape. Its poles are also at its two ends.

When the north pole of one magnet is brought near the north pole of another magnet, they repel each other.* The same is true of a south pole near a south pole. If opposite poles are brought together, however, attraction occurs. The rule is:

Like poles repel each other; opposite poles attract.

This rule is similar to the rule for electric charges, where like charges repel one another and unlike charges attract. But there is a very important difference between magnetic poles and electric charges. Electric charges can be isolated, but magnetic poles cannot. Electrons and protons are entities by themselves. A cluster of electrons doesn't need the company of a cluster of protons, and vice versa. But a north magnetic pole never exists without the company of a south pole, and vice versa. The north and south poles of a magnet are like the head and tail of the same coin.

Figure 12.2
Break a magnet in half, and you have two magnets. Break these in half, and you have four magnets, each with a north and south pole. Continue breaking the pieces further and further, and you find the same results. Magnetic poles always exist in pairs.

If you break a bar magnet in half, each half still behaves as a complete magnet. Break the pieces in half again, and you have four complete magnets. You can continue breaking the pieces in half and never isolate a single pole. Even when your piece is one atom thick, there are two poles. This suggests that atoms themselves are magnets.

* The force of interaction between magnetic poles is given by $F \sim p_1 p_2 / d^2$, where p_1 and p_2 represent magnetic pole strengths and d represents the separation distance between the poles. Note the similarity of this relationship to Coulomb's law, and to Newton's law of gravity.

Concept Check ✓

Must every magnet have a north and south pole?

Check Your Answer Yes, just as every coin has two sides, a head and a tail. Some trick magnets may have more than one pair of poles, but nevertheless poles occur in pairs.

Figure 12.3
Top view of iron filings sprinkled around a magnet. The filings trace out a pattern of *magnetic field lines* in the space surrounding the magnet. Interestingly, the magnetic field lines continue inside the magnet (not revealed by the filings) and form closed loops.

12.2 Magnetic Fields—Regions of Magnetic Influence

When you sprinkle some iron filings on a sheet of paper placed on a magnet, you'll see that the filings trace out an orderly pattern of lines that surround the magnet. The space around the magnet contains a **magnetic field.** The shape of the field is shown by the filings. Note that the filings line up with the magnetic field lines that spread out from one pole and return to the other.

The direction of the field outside a magnet is from the north to the south pole. Where the lines are closer together, the field is stronger. The concentration of iron filings at the poles of the magnet in Figure 12.3 shows the magnetic field strength is greater there. If we place another magnet or a small compass anywhere in the field, its poles line up with the magnetic field.

A magnetic field is produced by the motion of electric charge. Where, then, is this motion in a common bar magnet? The answer is, in the electrons of the atoms that make up the magnet. These electrons are in constant motion. Two kinds of electron motion make magnetism: electron spin and electron revolution. Electrons spin about their own axes like tops, and they revolve about the atomic nucleus like planets revolving around the sun. In most common magnets, electron spin is the main contributor to magnetism.

Every spinning electron is a tiny magnet. A pair of electrons spinning in the same direction makes a stronger magnet. A pair of electrons spinning in opposite directions, however, work against each

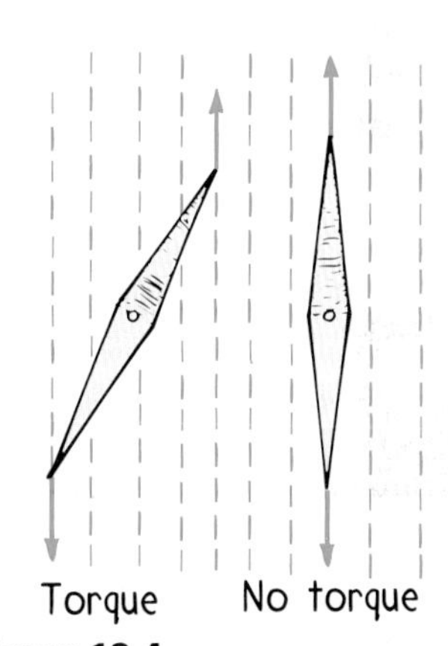

Figure 12.4
(Left) When the compass needle is not aligned with the magnetic field, the oppositely directed forces on the needle produce a pair of torques (twisting forces called a *couple*) that (right) twist the needle into alignment.

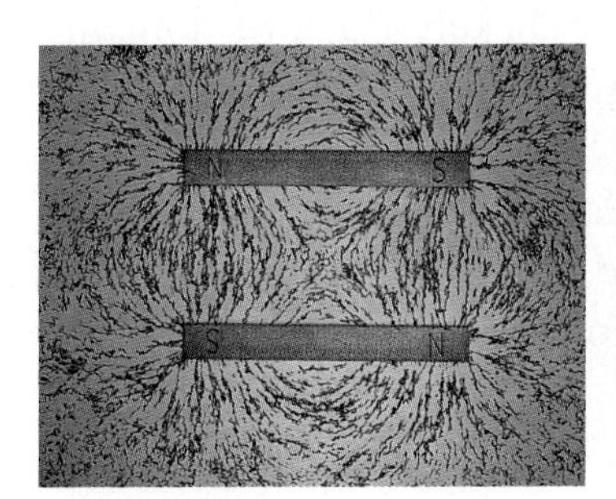

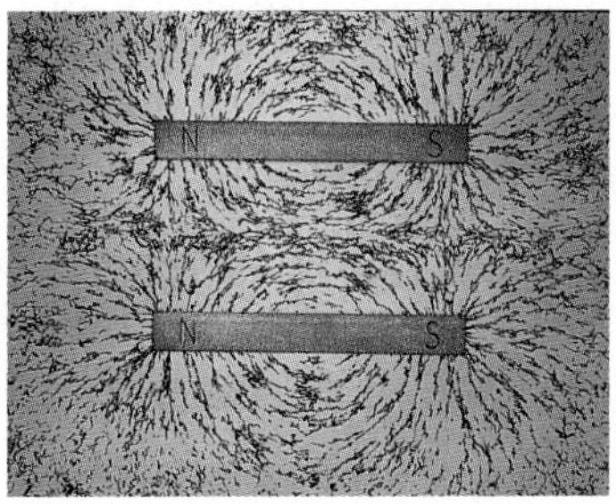

Figure 12.5
The magnetic field patterns for a pair of magnets. (left) Opposite poles are nearest each other, and (right) like poles are nearest each other.

Hands-On Exploration

Most iron objects around you are magnetized to some degree. A filing cabinet, a refrigerator, or even cans of food on your pantry shelf have north and south poles induced by the Earth's magnetic field. Pass a compass from their bottoms to their tops and their poles are easily identified. Turn cans upside down and see how many days it takes for the poles to reverse themselves!

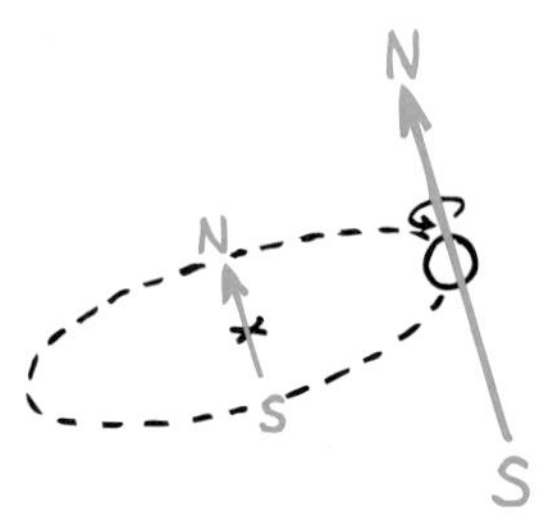

Figure 12.6
Both the spinning motion and the revolving (orbital) motion of every electron in an atom produce magnetic fields. The field due to spin (large vector) combines with the field due to revolution (small vector) to produce the magnetic field of the atom. The resulting field is greater for iron atoms.

other. The magnetic fields cancel. This is why most substances are not magnets. In most atoms, the various fields cancel one another because the electrons spin in opposite directions. In materials such as iron, nickel, and cobalt, however, the fields do not cancel each other entirely. Each iron atom has four electrons whose spin magnetism is uncanceled. Each iron atom, then, is a tiny magnet. The same is true to a smaller extent for the atoms of nickel and cobalt. Most common magnets are therefore made from alloys containing iron, nickel, and cobalt in various proportions.

12.3 Magnetic Domains—Clusters of Aligned Atoms

Large clusters of iron atoms line up with one another. These clusters of aligned atoms are **magnetic domains.** Each domain is made up of billions of aligned atoms. The domains are microscopic (Figure 12.7), and there are many of them in a crystal of iron. Domains themselves tend to align with one another.

Not every piece of iron is a magnet because the domains in ordinary iron are not lined up. In a common iron nail, for example, the domains are randomly oriented. But when you bring a magnet nearby, they can be induced into alignment. (It is interesting to listen with an amplified stethoscope to the clickety-clack of domains aligning in a piece of iron when a strong magnet approaches.) The domains align themselves much as electrical charges in a piece of paper align themselves (become polarized) in the presence of a charged rod. When you remove the nail from the magnet, ordinary thermal motion causes most or all of the domains in the nail to return to a random arrangement.

Figure 12.7
A microscopic view of magnetic domains in a crystal of iron. Each domain consists of billions of aligned iron atoms. In this view the domains are not aligned.

Permanent magnets can be made by placing pieces of iron or other magnetic materials in strong magnetic fields. Iron alloys differ in their ability to become magnetized; soft iron is easier to magnetize than steel. It helps to tap the material to nudge any stubborn domains into alignment. Another way is to stroke the material

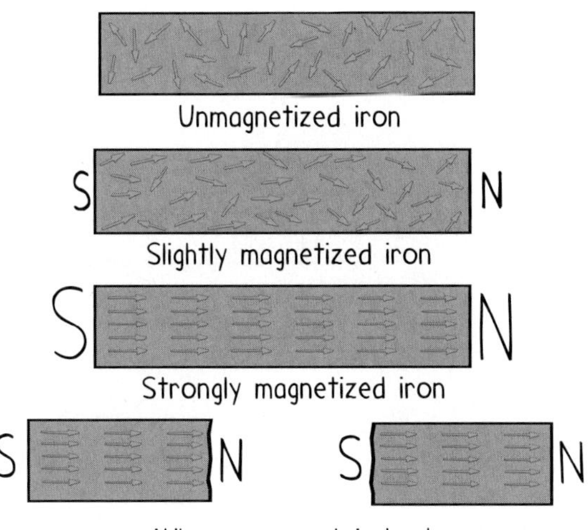

Figure 12.8
Pieces of iron in successive stages of becoming magnetized. The arrows represent domains; each arrowhead is a north pole and each tail a south pole. Poles of neighboring domains neutralize each other's effects, except at the two ends of a piece of iron.

Figure 12.9
The iron nails become induced magnets.

with a magnet. The stroking motion lines up the domains. If a permanent magnet is dropped or heated, some of the domains are jostled out of alignment and the magnet becomes weaker.

Concept Check ✓

1. How can a magnet attract a piece of iron that is not magnetized?
2. Why will a magnet not pick up a penny or a piece of wood?

Check Your Answers

1. Like the compass needle in Figure 12.4, domains in the unmagnetized piece of iron are induced into alignment by the magnetic field of the magnet. One domain pole is attracted to the magnet and the other domain pole is repelled. Does this mean the net force is zero? No, because the force is slightly greater on the domain pole closest to the magnet than on the farther pole. That's why there is a net attraction. In this way a magnet attracts non-magnetized pieces of iron (Figure 12.9).
2. A penny and a piece of wood have no magnetic domains that can be induced into alignment.

12.4 The Interaction Between Electric Currents and Magnetic Fields

A single moving charge produces a magnetic field. A current of charges, then, also produces a magnetic field. The magnetic field that surrounds a current-carrying wire can be demonstrated by arranging an assortment of compasses around the wire (Figure 12.10). The magnetic field about the current-carrying wire makes up a pattern of

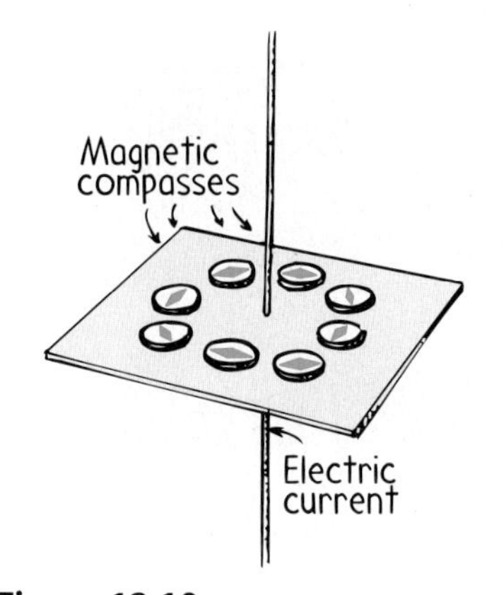

Figure 12.10
The compasses show the circular shape of the magnetic field surrounding the current-carrying wire.

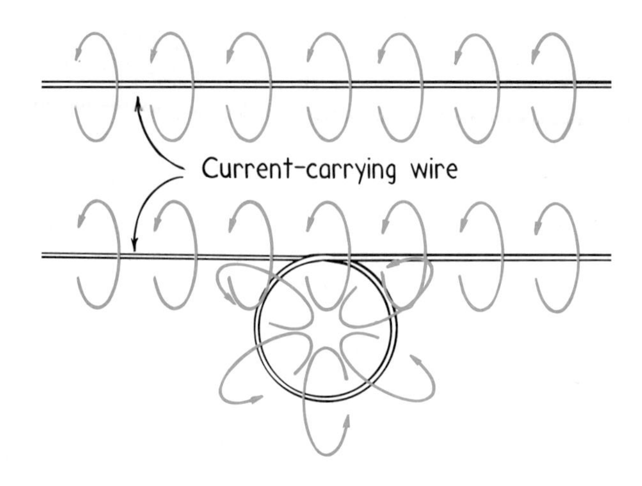

Figure 12.11
Magnetic field lines about a current-carrying wire crowd up when the wire is bent into a loop.

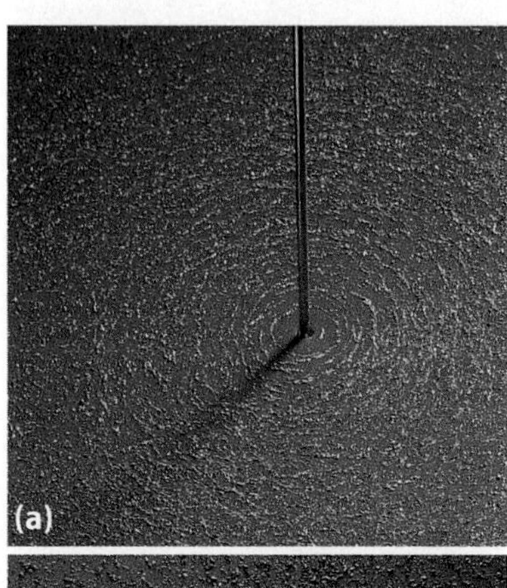

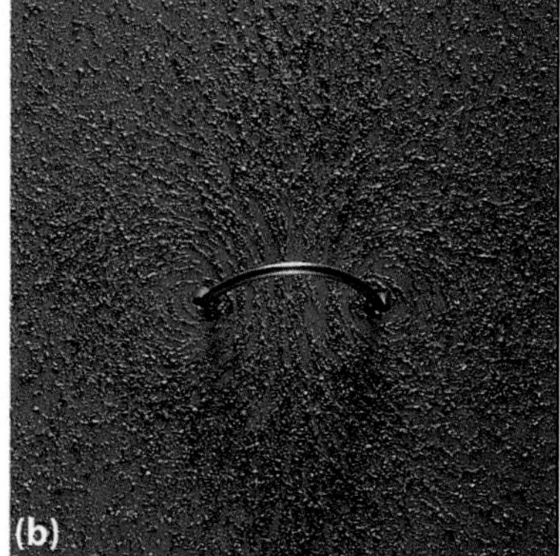

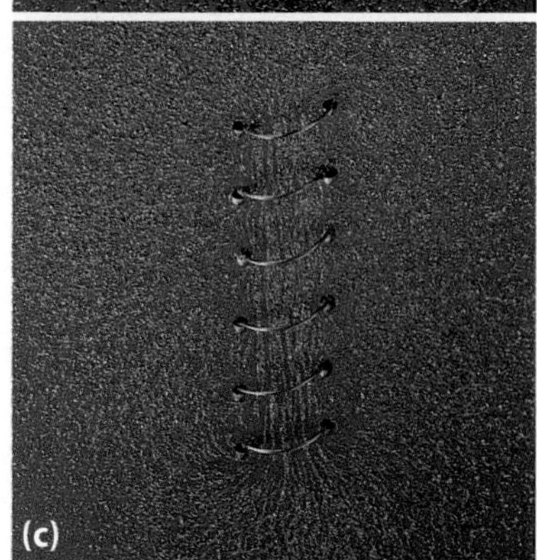

Figure 12.12
Iron filings sprinkled on paper reveal the magnetic field configurations about
(a) a current-carrying wire,
(b) a current-carrying loop, and
(c) a current-carrying coil of loops.

concentric circles. When the current reverses direction, the compass needles turn around, showing that the direction of the magnetic field also changes.*

If the wire is bent into a loop, the magnetic field lines bunch together inside the loop (Figure 12.11). If the wire is bent into another loop, overlapping the first, the bunching of field lines in the loops is doubled. More loops mean more magnetic field intensity. The magnetic field intensity is strong for a current-carrying coil of many loops.

Electromagnets

If a piece of iron is placed in a current-carrying coil of wire, the alignment of magnetic domains in the iron produces a stronger magnet. Then we have an **electromagnet.** Its strength is increased by increasing the current through the coil. Strong electromagnets are used to control charged-particle beams in high-energy accelerators.

Electromagnets powerful enough to lift automobiles are a common sight in junkyards. The strength of these electromagnets is limited mainly by overheating of the current-carrying coils. The most powerful electromagnets omit the iron core and use superconducting coils through which large electrical currents easily flow.

Superconducting Electromagnets

Ceramic superconductors (previous chapter) have the interesting property of expelling magnetic fields. Because magnetic fields cannot penetrate the surface of a superconductor, magnets levitate above them. The reasons for this behavior are beyond the scope of this

---

* Earth scientists think that the Earth's magnetism is the result of electric currents that accompany thermal convection in the molten parts of the Earth's interior. Evidence shows that the Earth's poles periodically reverse places—more than 20 reversals in the past 5 million years. This is perhaps the result of changes in the direction of electric currents within the Earth. More about this in Chapter 32.

Figure 12.13
A permanent magnet levitates above a superconductor because the magnet's magnetic field cannot penetrate the superconducting material.

Figure 12.14
Scale model of a prototype magnetically levitated vehicle—a *magplane.* Whereas conventional trains vibrate as they ride on rails at high speeds, magplanes can travel vibration-free at high speeds because they make no physical contact with the guideway they float above.

book and involve quantum mechanics. One of the hot applications of superconducting electromagnets is the levitation of high-speed trains for transportation. Prototype trains have already been demonstrated in the United States, Japan, and Germany. Watch for the growth of this relatively new technology.

12.5 Magnetic Forces Are Exerted on Moving Charges

A charged particle has to be moving to interact with a magnetic field. Charges at rest don't respond to magnets. But when moving, charged particles experience a deflecting force.* The force is greatest when the particles move at right angles to the magnetic field lines. At other angles, the force is less and becomes zero when the particles move parallel to the field lines. The force is always perpendicular to the magnetic field lines and perpendicular to the velocity of the charged particle (Figure 12.15). So a moving charge is deflected when it crosses through a magnetic field, but when it travels parallel to the field no deflection occurs.

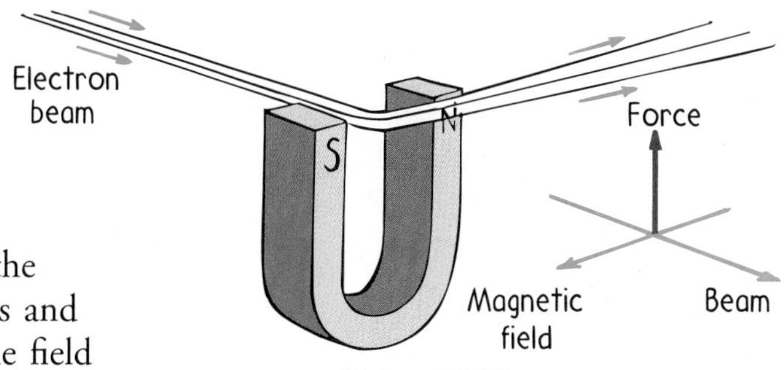

Electron beam
S
N
Force
Magnetic field
Beam

Figure 12.15
A beam of electrons is deflected by a magnetic field.

* When particles of electric charge q and velocity v move perpendicularly into a magnetic field of strength B, the force F on each particle is $F = qvB$. For non perpendicular angles, v in this relationship must be the component of velocity perpendicular to B.

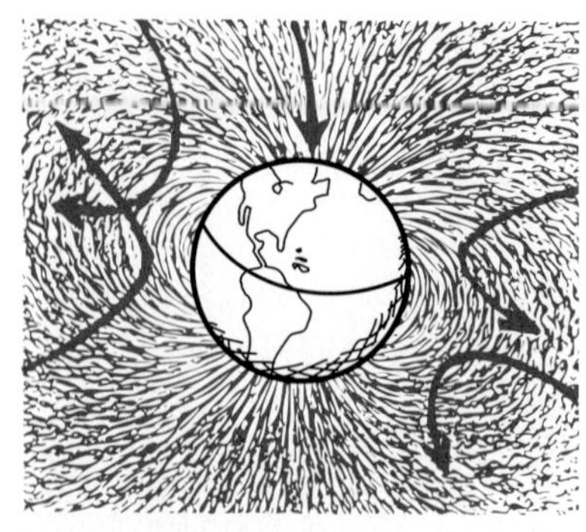

Figure 12.16
The magnetic field of the Earth deflects many charged particles that make up cosmic radiation.

This deflecting force is very different from the forces that occur in other interactions. Gravitation acts in a direction parallel to the line between masses, and electrical force acts in a parallel direction between charges. But magnetic force acts at right angles to the magnetic field and the velocity of the charged particle.

We are fortunate that charged particles are deflected by magnetic fields. This fact is used to guide electrons onto the inner surface of a TV tube and provide a picture. More interesting, charged particles from outer space are deflected by the Earth's magnetic field. The intensity of harmful cosmic rays bombarding the Earth's surface would be more intense otherwise.

Magnetic Force on Current-Carrying Wires

Simple logic tells you that if a charged particle moving through a magnetic field experiences a deflecting force, then a current of charged particles moving through a magnetic field also experiences a deflecting force. If the particles are forced while moving inside a wire, the wire is also forced (Figure 12.17).

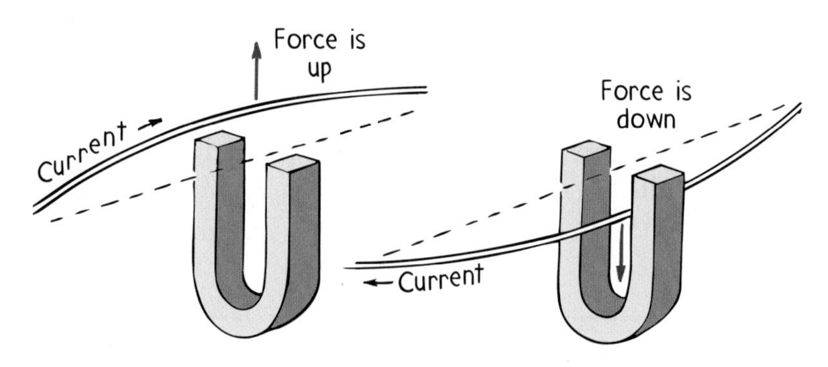

Figure 12.17
A current-carrying wire experiences a force in a magnetic field. (Can you see this is a follow-up of what happens in Figure 12.15?).

If we reverse the direction of current, the deflecting force acts in the opposite direction. This is useful in electric meters and in electric motors.

Electric Meters

The simplest meter that can detect an electric current is a magnetic compass. The next level of complexity is a compass in a coil of wires (Figure 12.18). When an electric current is in the coil, each loop produces its own effect on the needle, and so a very small current can be detected. A current-indicating instrument is called a *galvanometer.*

Figure 12.18
A very simple galvanometer.

A more common galvanometer design is shown in Figure 12.19. It uses more loops of wire and is therefore more sensitive. The coil is free to move, and the magnet is held stationary. The greater the current in its windings, the greater its deflection. The coil turns against a spring, with a needle fixed to show the amount of deflection. A galvanometer may be calibrated to measure current (amperes), in which case it is called an *ammeter.* Or it may be calibrated to measure voltage, in which case it is called a *voltmeter.* *

Figure 12.19
A common galvanometer design.

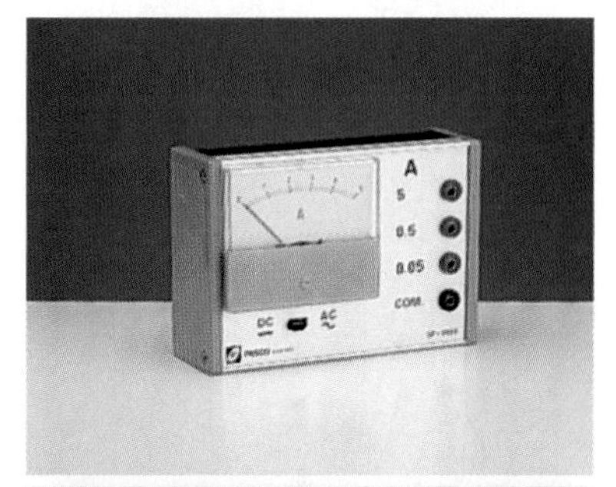

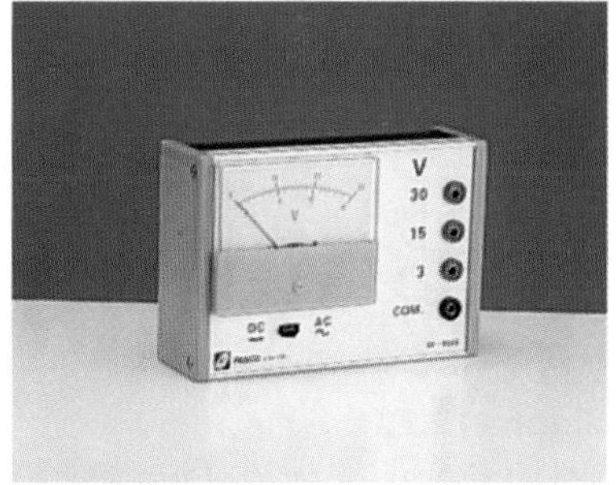

Figure 12.20
Both the ammeter and the voltmeter are basically galvanometers.

Electric Motors

If we change the design of the galvanometer slightly, so that deflection makes a complete turn rather than a partial rotation, we have an **electric motor.** The principal difference is that in a motor the current is made to change direction each time the coil makes a half rotation. This happens in cyclic fashion to produce continuous rotation, which has been used to run clocks, operate gadgets, and lift heavy loads.

In Figure 12.21, we see the principle of the electric motor in bare outline. A permanent magnet produces a magnetic field in a region where a rectangular loop of wire is mounted to turn about the dashed axis shown. Can you see that any current in the loop has one direction in the upper side of the loop and the opposite direction in the lower side? If the upper side of the loop is forced to the left by the magnetic field, the lower side is forced to the right, as if it were a galvanometer. But unlike in a galvanometer, the current in a motor is reversed during each half revolution by stationary contacts on the shaft. The parts of the wire that rotate and brush against these contacts are called *brushes.* In this way, the current in the loop alternates so that the forces on the upper and lower regions do not change directions as the loop rotates. The rotation is continuous as long as current is supplied.

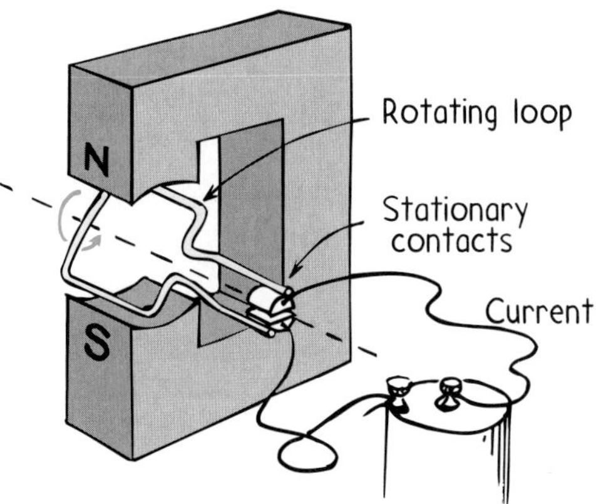

Figure 12.21
A simplified electric motor.

We have described here only a very simple dc motor. Larger motors, dc or ac, are usually made by replacing the permanent magnet by an electromagnet that is energized by some power source. Of course, more than a single loop is used. Many loops of wire are wound about an iron cylinder, called an *armature,* which then rotates when the wire loops carry current.

* To some degree, measuring instruments change what is being measured—ammeters and voltmeters included. Because an ammeter is connected in series with the circuit it measures, its resistance is made very low. That way it doesn't appreciably lower the current it measures. A voltmeter is connected in parallel, so its resistance is made very high so that it draws very little current for its operation.

The advent of electric motors brought to an end much human and animal toil in many parts of the world. Electric motors have greatly changed the way people live.

Concept Check ✓

What is the major similarity between a galvanometer and a simple electric motor? What is the major difference?

Check Your Answers A galvanometer and a motor are similar in that they both use coils positioned in a magnetic field. When a current passes through the coils, forces on the wires rotate the coils. The major difference is that the maximum coil rotation in a galvanometer is one-half turn, whereas in a motor the coil (wrapped on an armature) rotates through many complete turns. This is accomplished by alternating the direction of the current with each half turn of the armature.

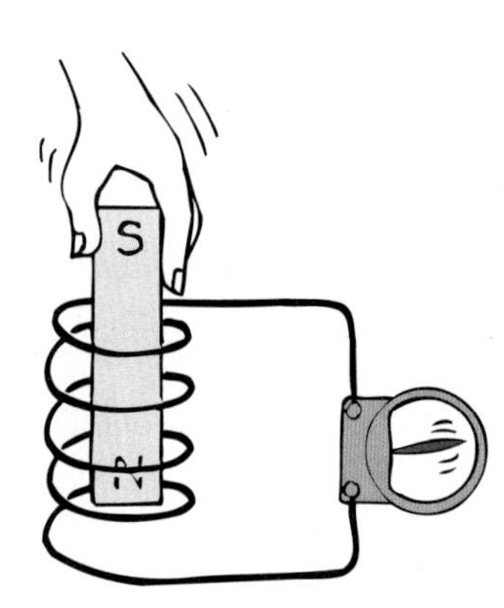

Figure 12.22
When the magnet is plunged into the coil, charges in the coil are set in motion; voltage is induced in the coil.

12.6 Electromagnetic Induction—How Voltage Is Created

In the early 1800s, the only current-producing devices were voltaic cells. These were the first batteries, which produced small currents by dissolving metals in acids. An important question asked was whether electricity could be produced from magnetism. The answer was provided in 1831 by two physicists, Michael Faraday in England and Joseph Henry in the United States—each working without knowledge of the other. Their discovery changed the world. The technology that followed made electricity commonplace, powering industries by day and lighting up cities at night.

Faraday and Henry both discovered that electric current can be produced in a wire simply by moving a magnet in or out of a coiled part of the wire (Figure 12.22). No battery or other voltage source is needed—only the motion of a magnet in a wire coil. They discovered that voltage is caused, or *induced,* by the relative motion between a wire and a magnetic field. Whether the magnetic field moves near a stationary conductor or the conductor moves in a stationary magnetic field—voltage is induced either way (Figure 12.23).

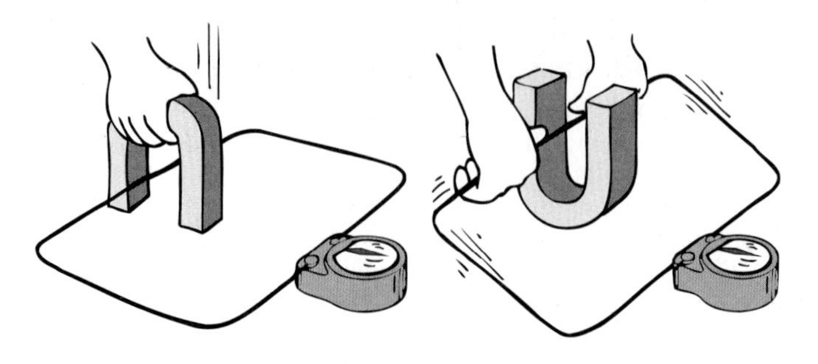

Figure 12.23
Voltage is induced in the wire loop either when the magnetic field moves past the wire or when the wire moves through the magnetic field.

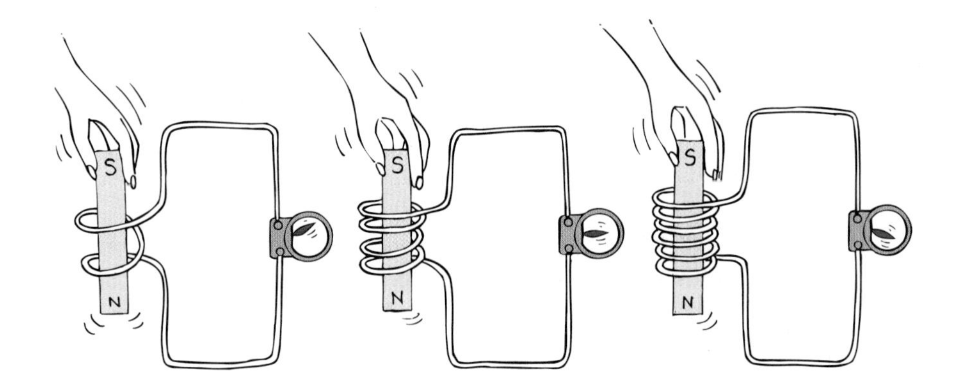

Figure 12.24
When a magnet is plunged into a coil of twice as many loops as another, twice as much voltage is induced. If the magnet is plunged into a coil with three times as many loops, three times as much voltage is induced.

The greater the number of loops of wire moving in a magnetic field, the greater the induced voltage (Figure 12.24). Pushing a magnet into a coil with twice as many loops induces twice as much voltage; pushing into a coil with ten times as many loops induces ten times as much voltage; and so on. It may seem that we get something (energy) for nothing by simply increasing the number of loops in a coil of wire. But we don't: We find it is more difficult to push the magnet into a coil made up of more loops. This is because the induced voltage produces a current, which makes an electromagnet, which repels the magnet in our hand. So we do more work against this "back force" to induce more voltage (Figure 12.25).

The amount of voltage induced depends on how fast the magnetic field lines are entering or leaving the coil. Very slow motion produces hardly any voltage at all. Quick motion induces a greater voltage. This phenomenon of inducing voltage by changing the magnetic field in a coil of wire is **electromagnetic induction.**

Faraday's Law

Electromagnetic induction is summarized by **Faraday's law:**

> **The induced voltage in a coil is proportional to the number of loops multiplied by the rate at which the magnetic field changes within those loops.**

The amount of *current* produced by electromagnetic induction depends on more than the induced voltage. It also depends on the resistance in both the coil and the circuit to which it's connected. For example, we can plunge a magnet in and out of a closed rubber loop and produce no current. Doing the same in a copper loop easily produces current. So the current in each is quite different. The electrons in the rubber sense the same voltage as those in the copper, but their bonding to the fixed atoms prevents the movement of electrons that move so freely in copper.

Figure 12.25
It is more difficult to push the magnet into a coil made up of many loops because the magnetic field of each current loop resists the motion of the magnet.

Concept Check ✓

If you push a magnet into a coil, as shown in Figure 12.25, you'll feel a resistance to your push. Why is this resistance greater in a coil with more loops?

Check Your Answer Simply put, more work is required to provide more energy. You can also look at it this way: When you push a magnet into a coil, you cause the coil to become an electromagnet. The more loops on the coil, the stronger the electromagnet that you produce and the stronger it pushes back against you. (If the coil's electromagnet attracted your magnet instead of repelling it, energy would be created from nothing and the law of energy conservation would be violated. So the coil has to repel your magnet.)

We have mentioned two ways in which voltage can be induced in a loop of wire: by moving the loop near a magnet or by moving a magnet near the loop. There is a third way, by changing a current in a nearby loop. All three cases possess the same essential ingredient—a changing magnetic field in the loop.

We see electromagnetic induction all around us. On the road we see it operate when a car drives over buried coils of wire to activate a nearby traffic light. When iron parts of a car move over the buried coils, the Earth's magnetic field in the coils is changed, inducing a voltage to trigger the changing of the traffic lights. Similarly, when you walk through the upright coils in the security system at an airport, any metal you carry slightly alters the magnetic field in the coils. This change induces voltage, and sounds an alarm. When the magnetic strip on the back of a credit card is scanned, induced voltage pulses identify the card. Similarly with the recording head of a tape recorder. Magnetic domains in the tape are sensed as the tape moves past a current-carrying coil. Electromagnetic induction is everywhere. As we shall see in Chapter 14, it underlies the electromagnetic waves we call light.

12.7 Generators and Alternating Current

A generator is a motor in reverse. The device is much the same, with the roles of input and output reversed. In a motor, electrical energy is the input and mechanical energy the output. In a generator, mechanical energy is the input and electric energy is the output. Both devices simply transform energy from one form to another.

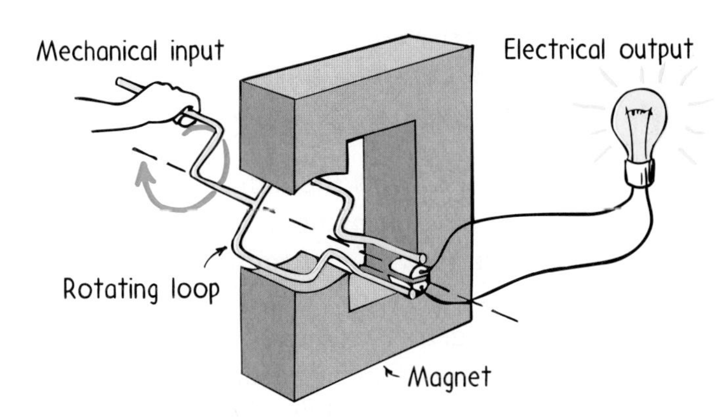

Figure 12.26
A simple generator. Voltage is induced in the loop when it is rotated in the magnetic field.

12.8 Power Production—A Technological Extension of Electromagnetic Induction

Fifty years after Faraday and Henry discovered electromagnetic induction, Nikola Tesla and George Westinghouse put those findings to practical use. They showed the world that electricity could be generated reliably and in sufficient quantities to light entire cities.

Tesla built generators much like those still in use but more complicated than the simple model we have discussed. His generators had armatures made up of bundles of copper wires. The armatures were forced to spin within strong magnetic fields by a turbine, which in turn was spun by the energy of either falling water or steam. The rotating loops of wire in the armature cut through the magnetic field of the surrounding electromagnets. In this way they induced alternating voltage and current.

It's important to know that generators don't produce energy—they simply convert energy from some other form to electric energy. As we discussed in Chapter 6, energy from a source, whether fossil or nuclear fuel or wind or water, is converted to mechanical energy to drive the turbine. The attached generator converts most of this mechanical energy to electrical energy. Some people think that electricity is a primary source of energy. It is not. It is a carrier of energy that must have a source.

Figure 12.27
Steam drives the turbine, which is connected to the armature of the generator.

12.9 The Induction of Fields—Both Electric and Magnetic

Electromagnetic induction explains the induction of voltages and currents. Actually, the more basic *fields* are at the root of both voltages and currents. The modern view of electromagnetic induction states that electric and magnetic fields are induced. These, in turn, produce the voltages we have considered. So induction takes place whether or not a conducting wire or any material medium is present. In this more general sense, Faraday's law states:

An electric field is induced in any region of space in which a magnetic field is changing with time.

There is a second effect, an extension of Faraday's law. It is the same, except that the roles of electric and magnetic fields are interchanged. It is one of nature's many symmetries. This effect was advanced by the British physicist James Clerk Maxwell in about 1860 and is known as **Maxwell's counterpart to Faraday's law:**

A magnetic field is induced in any region of space in which an electric field is changing with time.

In both cases, the strengths of the induced field are proportional to the rates of change of the inducing field. The induced electric and magnetic fields are at right angles to each other.

Maxwell saw the link between electromagnetic waves and light. If electric charges are set into vibration in the range of frequencies that match those of light, waves are produced that *are* light! Maxwell discovered that light is simply electromagnetic waves in the range of frequencies to which the eye is sensitive.

On the eve of his discovery he had a date with a young woman he was later to marry. While walking in a garden, his date remarked about the beauty and wonder of the stars. Maxwell asked how she would feel to know that she was walking with the only person in the world who knew what the starlight really was. For it was true. At that time, James Clerk Maxwell was the only person in the world to know that light of any kind is energy carried in waves of electric and magnetic fields that continually regenerate each other.

Magnetic Therapy

Back in the 18th century a celebrated "magnetizer" from Vienna was Franz Mesmer, who brought his magnets to Paris and established himself as a healer in Parisian society. He healed patients by waving magnetic wands above their heads.

At that time Benjamin Franklin, the world's leading authority on electricity, was visiting Paris as a US representative. He suspected that Mesmer's patients did benefit from his ritual—because it kept them away from the blood-letting practices of other physicians. At the urging of the medical establishment, King Louis XVI appointed a royal commission to investigate Mesmer's claims. The commission included Franklin and Antoine Lavoisier, the founder of modern chemistry. The commissioners designed a series of tests in which some subjects thought they were receiving Mesmer's treatment when they weren't, while others received the treatment but were led to believe they had not. The results of these blind experiments established beyond any doubt that Mesmer's success was due solely to the power of suggestion. To this day the report is a model for clarity and reason. Mesmer's reputation was destroyed and he retired to Austria.

Now two hundred years later, with all that has been learned about magnetism and physiology, hucksters of magnetism are attracting even larger followings. But there is no government commission of Franklins and Lavoisiers to challenge their claims. Instead, magnetic therapy is another of the untested and unregulated "alternative therapies" given official recognition by Congress in 1992.

Although testimonials about the benefits of magnets are many, there is no scientific evidence whatever for magnets boosting body energy or combating aches and pains. None. Yet millions of therapeutic magnets are sold in stores and catalogs. Consumers are buying magnetic bracelets, insoles, wrist and knee bands, back and neck braces, pillows, mattresses, lipstick, and even water. They are told that magnets have powerful effects on the body, mainly increasing blood flow to injured areas. The idea that blood is attracted by a magnet is bunk, for type of iron in blood doesn't respond to a magnet. Furthermore, most therapeutic magnets are the refrigerator type, with a very limited range. To get an idea of how quickly the field of these magnets drops off, see how many sheets of paper one of these magnets will hold on a refrigerator or any iron surface. The magnet will fall off after a few sheets of paper separate it from the iron surface. The field doesn't extend much more than one millimeter and wouldn't penetrate the skin, let alone into muscles. And even if it did, there is no scientific evidence that magnetism has any beneficial effects on the body at all. But again, testimonials are another story.

Sometimes an outrageous claim has some truth to it. For example, the practice of bloodletting in previous centuries was in fact beneficial to a small percentage of men. These men suffered the rare genetic disease *hemochromatosis* (excess iron in the blood)—women were exempt from this benefit due to menstruation. Although the number of men who benefited from bloodletting was small, testimonials of its success prompted the widespread practice that killed many.

No claim is so outrageous that testimonials can't be found to support it. Claims such as those for a flat earth or flying saucers are quite harmless, and may amuse us. Magnetic therapy may likewise be harmless for many ailments, but not when used to treat a serious disorder in place of modern medicine. Pseudoscience may be promoted to intentionally deceive, or it may be the result of flawed and wishful thinking. In either case, pseudoscience is very big business. The market is enormous for therapeutic magnets and other such fruits of unreason.

Scientists must keep open minds, must be prepared to accept new findings, and must be ready to be surprised by new evidence. But scientists also have a responsibility to speak out when the public is being deceived—and in effect robbed—by pseudoscientists whose claims are without substance.

Chapter Review

Key Terms and Matching Definitions

_____ electromagnet
_____ electromagnetic induction
_____ Faraday's law
_____ generator
_____ magnetic domains
_____ magnetic field
_____ magnetic force
_____ Maxwell's counterpart to Faraday's law
_____ motor

1. (1) Between magnets, it is the attraction of unlike magnetic poles for each other and the repulsion between like magnetic poles.
(2) Between a magnetic field and a moving charged particle, it is a deflecting force due to the motion of the particle. It is perpendicular to the velocity of the particle and perpendicular to the magnetic field lines. Also, it is greatest when the particle moves perpendicular to the field lines and zero when the particle moves parallel to the field lines.
2. The region of magnetic influence around either a magnetic pole or a moving charged particle.
3. Clustered regions of aligned magnetic atoms. When these regions are aligned with one another, the substance containing them is a magnet.
4. A magnet whose field is produced by an electric current. Electromagnets are usually in the form of a wire coil with a piece of iron inside the coil.
5. A device employing a current-carrying coil that is forced to rotate in a magnetic field. A motor converts electrical energy to mechanical energy.
6. The induction of voltage when a magnetic field changes with time. If the magnetic field within a closed loop changes in any way, a voltage is induced in the loop:

$$\text{Voltage induced} \sim \text{number of loops} \times \frac{\text{magnetic field change}}{\text{time}}$$

This is a statement of Faraday's law. The induction of voltage is the result of a more fundamental phenomenon: the induction of an electric *field,* as defined for the more general case below.
7. An electric field is induced in any region of space in which a magnetic field is changing with time. The magnitude of the induced electric field is proportional to the rate at which the magnetic field changes.
8. An electromagnetic induction device that produces electric current by rotating a coil within a stationary magnetic field. A generator converts mechanical energy to electrical energy.
9. A magnetic field is induced in any region of space in which an electric field is changing with time. The magnitude of the induced magnetic field is proportional to the rate at which the electric field changes.

Review Questions

Magnetic Poles—Attraction and Repulsion

1. In what way is the rule for the interaction between magnetic poles similar to the rule for the interaction between electric charges?

2. In what way are *magnetic poles very* different from *electric charges*?

Magnetic Fields—Regions of Magnetic Influence

3. An electric field surrounds an electric charge. What additional field surrounds a moving electric charge?

4. What two kinds of motion are exhibited by electrons in an atom?

Magnetic Domains—Clusters of Aligned Atoms

5. Why are some pieces of iron magnets and others not?

6. Why will dropping an iron magnet on a hard floor make it a weaker magnet?

The Interaction Between Electric Currents and Magnetic Fields

7. What is the shape of magnetic field lines about a current-carrying wire?

8. What happens to the direction of the magnetic field about an electric current when the direction of the current is reversed?

Magnetic Forces Are Exerted on Moving Charges

9. In what direction relative to a magnetic field does a charged particle move in order to experience maximum deflecting force? Minimum deflecting force?

10. Both gravitational and electrical forces act along the direction of the force fields. How is the direction of the magnetic force on a moving charge different?

11. Since a magnetic force acts on a moving charged particle, does it make sense that a magnetic force also acts on a current-carrying wire? Defend your answer.

12. What happens to the direction of the force on a wire when the current in it is reversed?

13. What is a galvanometer called when calibrated to read current? Voltage?

Electromagnetic Induction—How Voltage Is Created

14. What must change in order for electromagnetic induction to occur?

15. What are the three ways that voltage can be induced in a wire?

Generators and Alternating Current

16. What is the basic difference between a generator and an electric motor?

17. What is the basic similarity between a generator and an electric motor?

Power Production—A Technological Extension of Electromagnetic Induction

18. What commonly supplies the energy input to a turbine?

The Induction of Fields—Both Electric and Magnetic

19. What is induced by the rapid alternation of a magnetic field?

20. What is induced by the rapid alternation of an electric field?

Explorations

1. Find the direction and dip (slant from the vertical) of the Earth's magnetic field lines in your locality. Magnetize a large steel needle or straight piece of steel wire by stroking it a couple of dozen times with a strong magnet. Run the needle or wire through a cork in such a way that when the cork floats, the magnet remains horizontal (parallel to the water surface). Float the cork in a plastic or wooden container of water. The needle points toward one of the Earth's magnetic poles. Then press unmagnetized common pins into the sides of the cork. Rest the pins on the rims of a pair of drinking glasses so that the needle or wire points toward the magnetic pole. It should dip in line with the Earth's magnetic field.

2. An iron bar can be easily magnetized by aligning it with the magnetic field lines of the Earth and striking it lightly a few times with a hammer. This works best if the bar is tilted down to match the dip of the Earth's field. The hammering jostles the domains so they can better fall into alignment with the Earth's field. The bar can be demagnetized by striking it when it is in an east-west direction.

Exercises

1. Since every iron atom is a tiny magnet, why aren't all iron materials themselves magnets?

2. What is different about the magnetic poles of common refrigerator magnets compared with common bar magnets?

3. What surrounds a stationary electric charge? A moving electric charge?

4. "An electron always experiences a force in an electric field, but not always in a magnetic field." Defend this statement.
5. Why will a magnet attract an ordinary nail or paper clip, but not a wooden pencil?
6. A friend tells you that a refrigerator door, beneath its layer of white painted plastic, is made of aluminum. How could you check to see if this is true (without any scraping)?
7. Why will a magnet placed in front of a television picture tube distort the picture? (*Note:* Do NOT try this with a set that you value. You may magnetize the metal mask in back of the glass screen and have picture distortion even when the magnet is removed!)
8. Magnet A has twice the magnetic field strength of magnet B (at equal distance) and at a certain distance pulls on magnet B with a force of 50 N. With how much force, then, does magnet B pull on magnet A?
9. A strong magnet attracts a paper clip to itself with a certain force. Does the paper clip exert a force on the strong magnet? If not, why not? If so, does it exert as much force on the magnet as the magnet exerts on it? Defend your answers.
10. A common pickup for an electric guitar consists of a coil of wire around a small permanent magnet. The magnetic field of the magnet induces magnetic poles in the nearby guitar string. When the string is plucked, the rhythmic oscillations of the string produce the same rhythmic changes in the magnetic field through the coil, which in turn induce the same rhythmic voltages in the coil, which when amplified and sent to a speaker produce music! Why will this type pickup not work with nylon strings?
11. Why is a generator armature harder to rotate when it is connected to a circuit and supplying electric current?
12. If your metal car moves over a wide, closed loop of wire embedded in a road surface, will the magnetic field of the Earth within the loop be altered? Will this produce a current pulse? Can you think of a practical application for this at a traffic intersection?
13. At the security area of an airport, you walk through a weak ac magnetic field inside a coil of wire. What is the result of a small piece of metal on your person that slightly alters the magnetic field in the coil?
14. A piece of plastic tape coated with iron oxide is magnetized more in some parts than in others. When the tape is moved past a small coil of wire, what happens in the coil? What is a practical application of this?
15. Joseph Henry's wife tearfully sacrificed part of her wedding gown for silk to cover the wires of Joseph's electromagnets. What was the purpose of the silk covering?
16. What is the primary difference between an electric *motor* and an electric *generator*?
17. Your friend says that if you crank the shaft of a dc motor manually, the motor becomes a dc generator. Do you agree or disagree?
18. A length of wire is bent into a closed loop and a magnet is plunged into it, inducing a voltage and, consequently, a current in the wire. A second length of wire, twice as long, is bent into two loops of wire and a magnet is similarly plunged into it. Twice the voltage is induced, but the current is the same as that produced in the single loop. Why?
19. Two separate but similar coils of wire are mounted close to each other, as shown below. The first coil is connected to a battery. The second coil is connected to a galvanometer. How does the galvanometer respond when the switch in the first circuit is closed? After being closed when the current is steady? When the switch is opened?

20. A friend says that changing electric and magnetic fields generate one another, and this gives rise to visible light when the frequency of change matches the frequencies of light. Do you agree? Explain.

Part 4 Waves—Sound and Light

Chapter 13: Waves and Sound

How does the nature of sound differ from the nature of light? Why does sound travel so much slower than light? Is an echo simply sound that's reflected? Does the speed of sound differ in various materials? Can it travel in a vacuum? What's the relationship between vibrations and waves? Can one sound wave cancel another? And how does an airplane create a sonic boom? Let's begin our study of wave motion and find the answers to these questions.

Many things in nature vibrate—the string on a guitar, the reed in a clarinet, and your vocal chords when you speak or sing. When they vibrate in air they make the air vibrate in the same way. When these vibrations reach your ear they are transmitted as impulses to a part of your brain and you hear sound.

13.1 Special Wiggles—Vibrations and Waves

When something moves back and forth, side to side, or up and down, we say it vibrates. A *vibration* is a wiggle. When the wiggle moves through space and time it is a **wave.** A wave extends from one place to another. Light and sound are both vibrations that move through space as waves. But they are two very different kinds of waves. Sound is the movement of vibrations of matter—through solids, liquids, or gases. If there is no matter to vibrate, then no sound is possible. Sound cannot travel in a vacuum. But light can, for light is a vibration of electric and magnetic fields—a vibration of pure energy. Light can pass though many materials, but none is required. It travels quite nicely from the sun and stars to the Earth, for example.

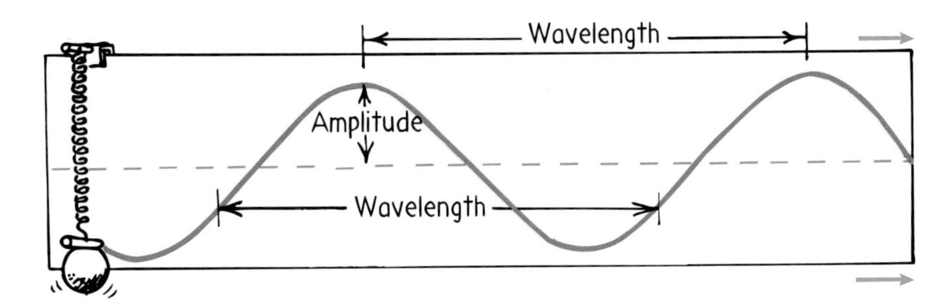

Figure 13.1
When the bob vibrates up and down, a marking pen traces out a sine curve on paper that moves horizontally at constant speed.

You can see the relationship between a vibration and a wave in Figure 13.1. A marking pen on a bob attached to a vertical spring vibrates up and down. Notice that it traces a wave form on a sheet of paper that moves horizontally at constant speed. The wave form is a *sine curve,* with a shape like rolling hills and valleys and a precise mathematical description. As in a water wave, the high points of a sine wave are called *crests* and the low points are *troughs.* The dashed green line shows the "home" position, or midpoint of the vibration. The term **amplitude** refers to the distance from the midpoint to the crest (or trough) of the wave. The amplitude equals the maximum displacement from the home position—from equilibrium.

The **wavelength** of a wave is the distance from one crest to the next one. Equivalently, it is the distance between any two successive identical parts of the wave. The wavelengths of waves at the beach are measured in meters, the wavelengths of ripples in a pond in centimeters, and the wavelengths of light waves in billionths of a meter (nanometers).

How frequently a vibration occurs is described by its **frequency.** The frequency of any vibrating object is the number of to-and-fro vibrations the object makes in a given time (usually one second). If a complete to-and-fro vibration (one cycle) occurs in one second, the frequency is one vibration per second. If two vibrations occur in one second, the frequency is two vibrations per second.

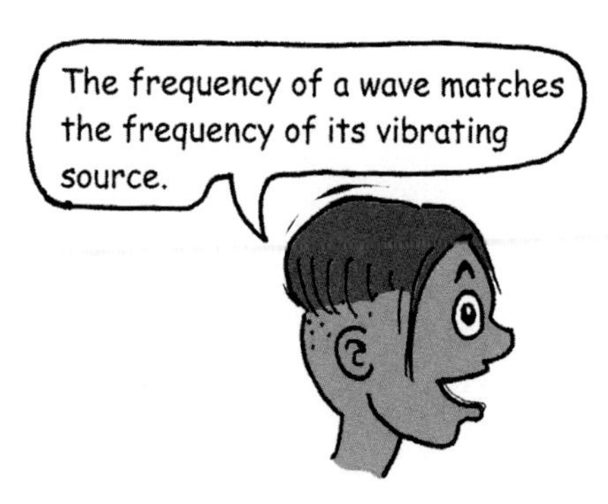

The unit of frequency is the **hertz** (Hz), as was mentioned briefly in the previous chapter. It is named after Heinrich Hertz, who demonstrated the existence of radio waves in 1886. We call one vibration per second 1 hertz; two vibrations per second is 2 hertz, and so on. Higher frequencies are measured in kilohertz (kHz), and still higher frequencies in megahertz (MHz). AM radio waves are usually measured in kilohertz, and FM radio waves are measured in megahertz. A station at 960 kHz on the AM radio dial, for example, broadcasts radio waves that have a frequency of 960,000 vibrations per second. A station at 101.7 MHz on the FM dial broadcasts radio waves that have a frequency of 101,700,000 hertz. These radio-wave frequencies are the frequencies at which electrons in the broadcasting antenna are forced to vibrate.

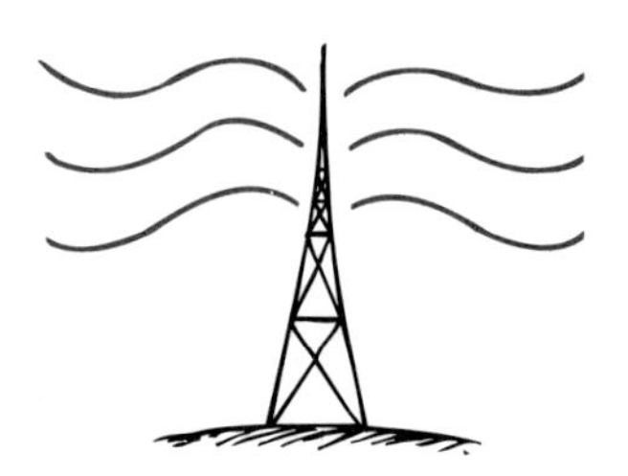

Figure 13.2
Electrons in the transmitting antenna vibrate 940,000 times each second and produce 940-kHz radio waves.

The **period** of a wave or vibration is the time it takes for a complete vibration. Period can be calculated from frequency, and vice versa. Suppose, for example, that a pendulum makes two vibrations in one second. Its frequency is 2 Hz, and the time needed to complete one vibration—that is, the period—is 1/2 second. If the vibration frequency is 3 Hz, then the period is 1/3 second. Frequency and period are the inverse of each other:

$$\textbf{Frequency} = \frac{1}{\textbf{period}} \qquad \textbf{Period} = \frac{1}{\textbf{frequency}}$$

Concept Check ✓

1. An electric toothbrush completes 90 cycles every second. What are (a) its frequency and (b) its period?
2. Gusts of wind cause the Sears Building in Chicago sway back and forth, completing a cycle every ten seconds. What are (a) its frequency and (b) its period?

Check Your Answers

1. (a) 90 cycles per second is 90 vibrations per second or 90 Hz; (b) 1/90 second.
2. (a) 1/10 Hz. (b) 10 s.

13.2 Wave Motion—Transporting Energy

When you drop a stone into a quiet pond, waves travel outward in expanding circles. Energy is carried by the wave, moving from place to place. The water itself goes nowhere. This can be seen by watching a leaf floating in the water. The leaf bobs up and down but doesn't travel with the waves. When you speak, the energy of your voice travels across the room at about 340 meters per second. Wave energy travels through the air. The air itself doesn't travel at this speed. If it did, then speaking would be a windy experience.

Figure 13.3
Water waves.

Wave Speed

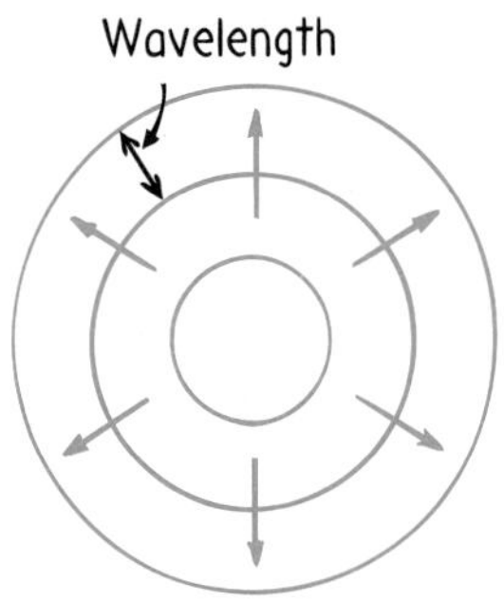

Figure 13.4
A top view of water waves.

The speed of a wave is related to the frequency and wavelength of the waves. You can understand this by considering the simple case of water waves (Figures 13.3 and 13.4). Imagine that you fix your eyes at a stationary point on the surface of water and observe the waves passing by this point. You can measure the amount of time that passes between the arrival of one crest and the arrival of the next one (this time interval is the period). And you can also estimate the distance between crests (the wavelength). You know that speed is defined as distance divided by time. In this case, the distance is one wavelength and the time is one period. That means wave speed = wavelength/period.

For example, if the wavelength is 10 meters and the time between crests at a point on the surface is 0.5 second, the wave moves 10 meters in 0.5 seconds. So its speed is 10 meters divided by 0.5 seconds = 20 meters per second.

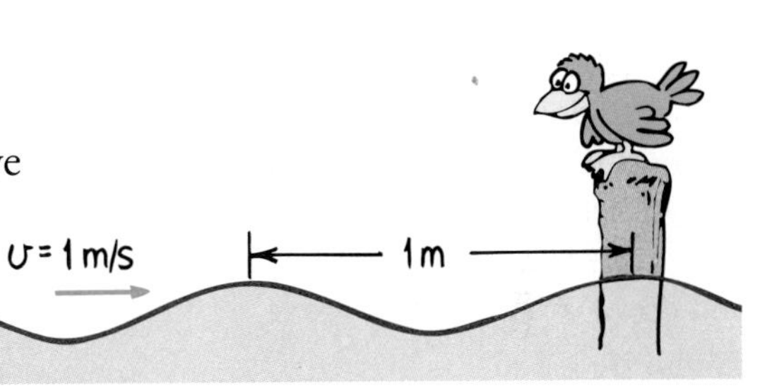

Figure 13.5
If the wavelength is 1 m, and one wavelength per second passes the pole, then the speed of the wave is 1 m/s.

Since period is the inverse of frequency, the formula wave speed = wavelength/period can also be written

Wave speed = frequency × wavelength

This relationship applies to all kinds of waves, whether they are water waves, sound waves, or light waves.

Concept Check ✓

1. If a train of freight cars, each 10 m long, rolls by you at the rate of three cars each second, what is the speed of the train?
2. If a water wave vibrates up and down three times each second and the distance between wave crests is 2 m, what are (a) the wave's frequency? (b) its wavelength? (c) its wave speed?

Check Your Answers

1. 30 m/s. We can see this in two ways. (1) According to the speed definition from Chapter 2, $v = d/t = (3 \times 10\text{ m})/1\text{ s} = 30\text{ m/s}$, because 30 m of train passes you in 1 s. (2) If we compare the train to wave motion, with wavelength corresponding to 10 m and frequency 3 Hz, then Speed = frequency × wavelength = 3 Hz × 10 m = 30 m/s.
2. (a) 3 Hz; (b) 2 m; (c) Wave speed = frequency × wavelength = 3/s × 2 m = 6 m/s. (Note that 3 Hz is 3 vibrations/s, and because "vibrations" has no unit, we write 3 Hz = 3/s.) It is customary to express the equation for wave speed as $v = f\lambda$, where v is wave speed, f is wave frequency, and λ (the Greek letter lambda) is wavelength.

13.3 Two Types of Waves—Transverse and Longitudinal

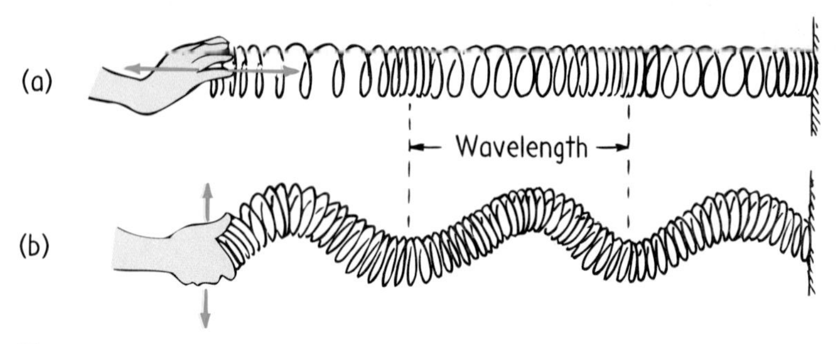

Figure 13.6
Both waves transfer energy from left to right. (a) When the Slinky is pushed and pulled rapidly along its length, a longitudinal wave is produced. (b) When the end of the Slinky is shaken up and down, a transverse wave is produced.

Fasten one end of a Slinky to a wall and hold the free end in your hand. Shake it up and down and you produce vibrations that are at right angles to the direction of wave travel. The right-angled, or sideways, motion is called *transverse motion.* This type of wave—in which the direction of wave travel is perpendicular to the direction of the vibrating source—is called a **transverse wave.** Waves in the stretched strings of musical instruments are transverse. Electromagnetic waves are also transverse.

A **longitudinal wave** is one in which the direction of wave travel is *along* the direction in which the source vibrates. You produce a longitudinal wave with your Slinky when you shake it back and forth along the Slinky's axis (Figure 13.6a). The vibrations are then parallel to the direction of energy transfer. Part of the Slinky is compressed, and a wave of *compression* travels along it. In between successive compressions is a stretched region, called a *rarefaction.* Both compressions and rarefactions travel in the same direction along the Slinky. Together they make up the longitudinal wave.

13.4 Sound Travels in Longitudinal Waves

Figure 13.7
Vibrate a Ping-Pong paddle in the midst of a lot of Ping-Pong balls, and you make the balls vibrate also.

Think of the air molecules in a room as tiny randomly-moving Ping-Pong balls. If you vibrate a Ping-Pong paddle in the midst of the balls, you'll set them vibrating to and fro. The molecules will vibrate in rhythm with your vibrating paddle. In some regions the balls are momentarily bunched up (compressions) and in other regions in between they are momentarily spread out (rarefactions). The vibrating prongs of a tuning fork do the same to air molecules. Vibrations made up of compressions and rarefactions spread from the tuning fork throughout the air. A **sound wave** is produced.

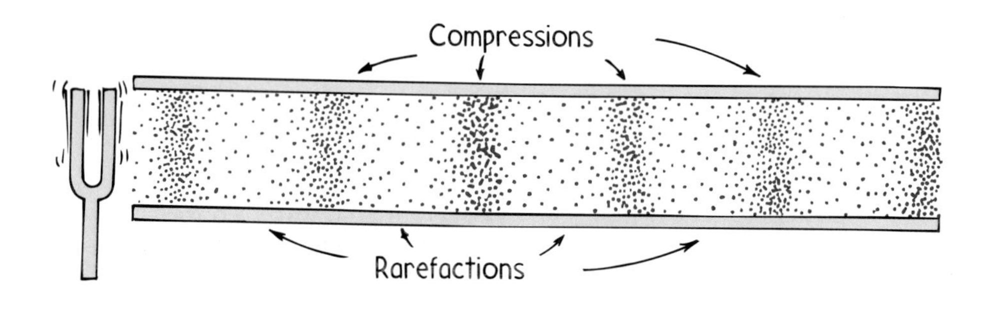

Figure 13.8
Compressions and rarefactions travel (both at the same speed in the same direction) from the tuning fork through the air in the tube.

Sound requires a medium. It can't travel in a vacuum because there's nothing to compress and stretch.

The wavelength of a sound wave is the distance between successive compressions or, equivalently, the distance between successive rarefactions. Each molecule in the air vibrates to-and-fro about some equilibrium position as the waves move by.

Our subjective impression about the frequency of sound is described as **pitch.** A high-pitch sound like that from a tiny bell has a high vibration frequency. Sound from a large bell has a low pitch because its vibrations are of a low frequency.

The human ear can normally hear pitches from sound in a range from about 20 to 20,000 hertz. As we age, this range shrinks. So by the time you can afford to trade in your old sound system for an expensive hi-fi one, you may not be able to tell the difference. Sound waves of frequencies below 20 hertz are called *infrasonic waves,* and those of frequencies above 20,000 hertz are called *ultrasonic waves.* We cannot hear infrasonic or ultrasonic sound waves. But dogs and some other animals can.

Most sound travels through air, but any elastic substance—solid, liquid, or gas—can transmit sound.* Many solids and liquids conduct sound better than air. You can hear the sound of a distant train clearly by placing your ear against the rail. When swimming, have a friend some distance away click two rocks together beneath the water surface while you're submerged. Observe how well water conducts the sound.

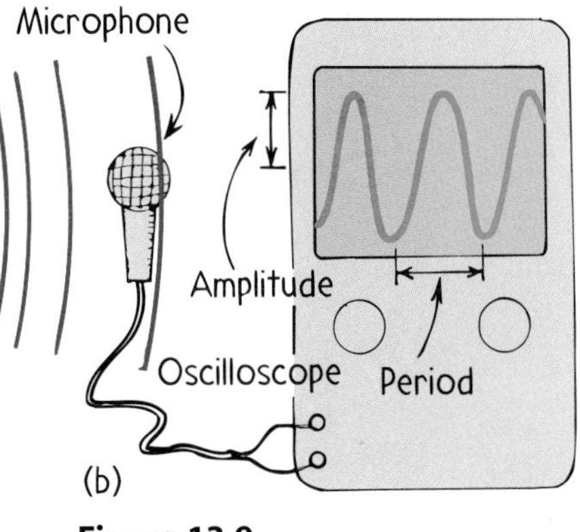

Figure 13.9
The radio loudspeaker is a paper cone that vibrates in rhythm with an electric signal. The sound produced sets up similar vibrations in the microphone, which are displayed on an oscilloscope. The shape of the waveform on the oscilloscope screen reveals information about the sound.

Pause to reflect on the physics of sound while you are quietly listening to your radio sometime. The radio loudspeaker is a paper cone that vibrates in rhythm with an electrical signal. Air molecules next to the vibrating cone are set into vibration. These in turn vibrate against neighboring molecules, which in turn do the same, and so on. As a result, rhythmic patterns of compressed and rarefied air emanate from the loudspeaker and vibrate the air in the whole room. This sets your eardrum into vibration, which in turn sends cascades of rhythmic electrical impulses along the cochlea nerve canal and into your brain. And you listen to the sound of music.

Figure 13.10
Waves of compressed and rarefied air, produced by the vibrating cone of the loudspeaker, make up the pleasing sound of music.

* An elastic substance is "springy," has resilience, and can transmit energy with little loss. Steel, for example, is elastic, whereas lead or putty is not.

Link to Technology: Loudspeakers

The loudspeaker of your radio or other sound-producing systems changes electrical signals into sound waves. The electrical signals pass through a coil wound around the neck of a paper cone. This coil acts as an electromagnet, which is located near a permanent magnet. When the current direction is one way, magnetic force pushes the electromagnet toward the permanent magnet, pulling the cone inward. When the current direction reverses, the cone is pushed outward. Vibrations in the electric signal then cause the cone to vibrate. Vibrations of the cone produce sound waves in the air.

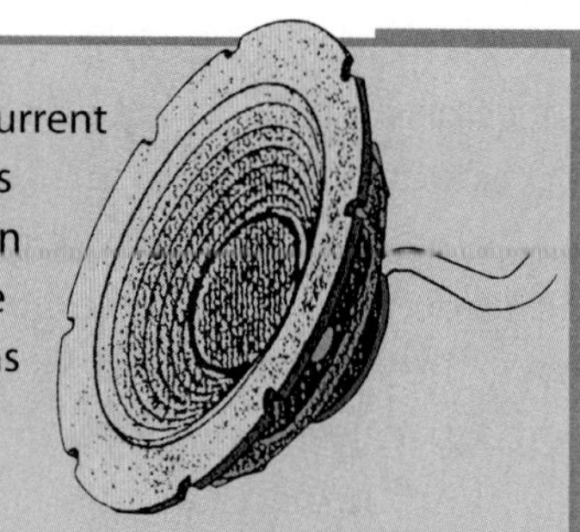

Speed of Sound

If you watch a person at a distance chopping wood or hammering, you can easily see that the blow occurs before you hear it. Likewise, you see a flash of lightning before you hear thunder. Sound takes time to travel from one location to another. The speed of sound depends on wind conditions, temperature, and humidity. But it doesn't depend on the loudness or the frequency of sound. All sounds, loud or soft, high- or low-pitched, travel at the same speed in a given medium. The speed of sound in dry air at 0°C is about 330 meters per second (nearly 1200 km/h). It travels slightly faster if water vapor is in the air. Sound travels faster through warm air than cold air. This makes sense because faster-moving molecules in warm air bump into each other more often and transmit a pulse in less time. For each degree rise in temperature above 0°C, the speed of sound in air increases by 0.6 meter per second. So in air at a normal 20°C room temperature, sound travels at about 340 meters per second. In water, sound speed is about 4 times its speed in air; in steel it's about 15 times.

Concept Check ✓

1. Do compressions and rarefactions in a sound wave travel in the same direction or in opposite directions from one another?
2. What is the approximate distance of a thunderstorm when you note a 3-s delay between the flash of lightning and the sound of thunder? (Use 340 m/s for the speed of sound.)

Check Your Answers

1. They travel in the same direction.
2. In 3 s the sound travels (340 m/s × 3 s) = 1020 m. Since there is no noticeable delay between the time the lightning bolt is created and the time you *see* it, the storm is slightly more than 1 km away.

13.5 Sound Can Be Reflected

We call the reflection of sound an *echo.* A large fraction of sound energy is reflected from a surface that is rigid and smooth. Less sound is reflected if the surface is soft and irregular. Sound energy that is not reflected is either transmitted or absorbed.

Sound reflects from a smooth surface the same way light does—the angle of incidence is equal to the angle of reflection (Figure 13.11). Reflected sound in a room makes it sound lively and full, as you have probably noticed while singing in the shower. Sometimes when the walls, ceiling, and floor of a room are too reflective, the sound becomes garbled. This is due to multiple reflections called *reverberations.* On the other hand, if the reflective surfaces are too absorbent, the sound level is low and the room may sound dull and lifeless. In the design of an auditorium or concert hall, a balance must be achieved between reverberation and absorption. The study of sound properties is called *acoustics.*

In concert halls good acoustics require highly reflective surfaces behind the stage to direct sound out to an audience. Sometimes reflecting surfaces are suspended above the stage. In the San Francisco opera hall the reflective surfaces are large, shiny, plastic plates that also reflect light (Figure 13.12). A listener can look up at these reflectors and see the reflected images of the members of the orchestra (the plastic reflectors are somewhat curved, which increases the field of view). Both sound and light obey the same law of reflection, so if a reflector is oriented so that you can see a particular musical instrument, you'll also hear it. Sound from the instrument follows the line of sight to the reflector and then to you.

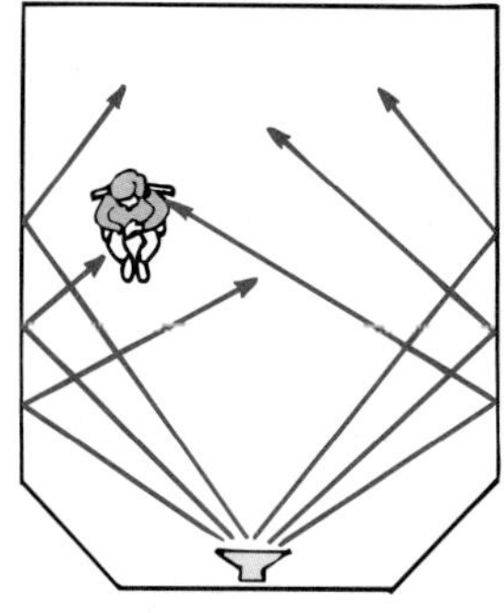

Figure 13.11
The angle of incident sound is equal to the angle of reflected sound.

Figure 13.12
The plastic plates above the orchestra reflect both light and sound. Adjusting them is quite simple: What you see is what you hear.

13.6 Sound Can Be Refracted

Sound waves bend when parts of the waves travel at different speeds. This occurs in uneven winds or when sound is traveling through air of varying temperatures. This bending of sound is called **refraction.** On a warm day, air near the ground may be warmer than air above, and so the speed of sound near the ground increases. Sound waves therefore tend to bend away from the ground, resulting in sound that does not seem to carry well (Figure 13.13).

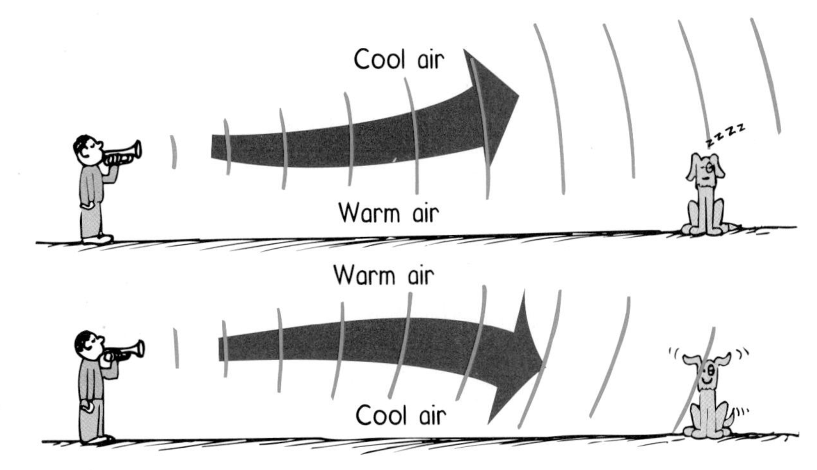

Figure 13.13
Sound waves are bent in air of uneven temperatures.

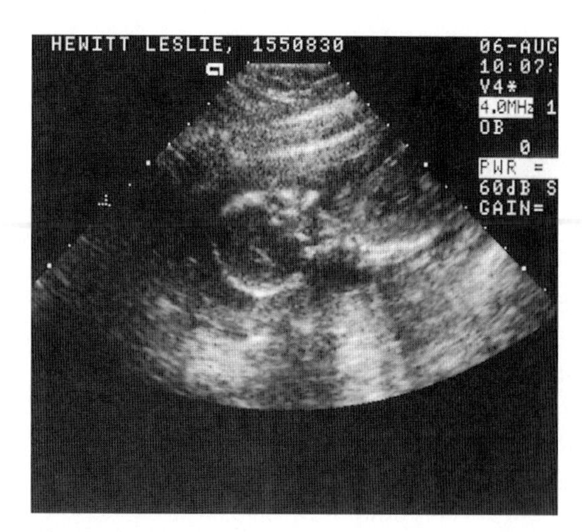

Figure 13.14
The 14-week-old fetus of Megan Hewitt Abrams.

The refraction of sound occurs under water, too, where the speed of sound varies with temperature. This poses a problem for surface vessels that chart the bottom features of an ocean by bouncing ultrasonic waves off the bottom. This is a blessing for submarines that wish to escape detection. Layers of water at different temperatures (thermal gradients) result in refraction of sound that leaves gaps, or "blind spots," in the water. This is where submarines hide. If it weren't for refraction, submarines would be easier to detect.

The multiple reflections and refractions of ultrasonic waves are used by physicians in a technique for harmlessly "seeing" inside the body without the use of X-rays. When high-frequency sound (ultrasound) enters the body, it is reflected more strongly from the outside of organs than from their interior. A picture of the outline of the organs is obtained (Figure 13.14). This ultrasound echo technique has always been used by bats, who emit ultrasonic squeaks and locate objects by their echoes. Dolphins do this and much more.

Figure 13.15
A dolphin emits ultrahigh-frequency sound to locate and identify objects in its environment. Distance is sensed by the time delay between sending sound and receiving its echo, and direction is sensed by differences in time for the echo to reach its two ears. A dolphin's main diet is fish and, since hearing in fish is limited to fairly low frequencies, they are not alerted to the fact they are being hunted.

Concept Check ✓

A depth-sounding vessel surveys the ocean bottom with ultrasonic sound that travels 1530 m/s in seawater. How deep is the water if the time delay of the echo from the ocean floor is 2 s?

Check Your Answer The 2 s delay means it takes 1 s for the sound to reach the bottom (and another 1 s to return). Sound traveling at 1530 m/s for 1 s tells us the bottom is 1530 m deep.

Link to Zoology: Dolphins and Acoustical Imaging

The primary sense of the dolphin is acoustic, for sight is not a very useful sense in the often murky and dark depths of the ocean. Whereas sound is a passive sense for us, it is an active sense for dolphins when they send out sounds and then perceive their surroundings via echoes. The ultrasonic waves emitted by a dolphin enables it to "see" through the bodies of other animals and people. Because skin, muscle, and fat are almost transparent to dolphins, they "see" only a thin outline of the body, but the bones, teeth, and gas-filled cavities are clearly apparent. Physical evidence of cancers, tumors, and heart attacks can all be "seen" by dolphins—as humans have only recently been able to do with ultrasound.

What's more interesting, a dolphin can reproduce the sonic signals that paint the mental image of its surroundings. Thus the dolphin probably communicates its experience to other dolphins by communicating the full acoustic image of what is "seen," placing the image directly in the minds of other dolphins. It needs no word or symbol for "fish," for example, but communicates an image of the real thing. It is quite possible that dolphins highlight portions of the images they send by selective filtering, as we similarly communicate a musical concert to others via various means of sound reproduction. Small wonder that the language of the dolphin is very unlike our own!

13.7 Forced Vibrations and Natural Frequency

If you strike an unmounted tuning fork, its sound is rather faint. Repeat but hold the fork against a table after you strike it, and the sound is louder. This is because the table is forced to vibrate, and with its larger surface it sets more air in motion. The table can be forced into vibration by a fork of any frequency. This is a case of **forced vibration.** The vibration of a factory floor caused by the running of heavy machinery is another example of forced vibration. A more pleasing example is given by the sounding boards of stringed instruments.

Drop a wrench and a baseball bat on a concrete floor, and you can tell the difference in their sounds. This is because each vibrates differently when striking the floor. They are not forced to vibrate at a particular frequency, but instead each vibrates at its own special frequency. When disturbed, any object made of an elastic material vibrates at its own special set of frequencies, which together form its special sound. We speak of an object's **natural frequency,** which depends on factors such as the elasticity and shape of the object. Bells and tuning forks, of course, vibrate at their own special frequencies. And interestingly, most other things, from planets to atoms and almost everything in between, have a springiness to them and vibrate at one or more natural frequencies.

13.8 Resonance and Sympathetic Vibrations

When the frequency of forced vibrations imposed on an object matches the object's natural frequency, a dramatic increase in amplitude occurs. This phenomenon is called **resonance** (which means "resounding" or "sounding again"). Putty doesn't resonate because it

isn't elastic, and a dropped handkerchief is too limp. In order for something to resonate, it needs a force to pull it back to its starting position and enough energy to keep it vibrating.

A common experience illustrating resonance occurs on a swing. When pumping a swing, you pump in rhythm with the natural frequency of the swing. More important than the force with which you pump is the timing. Even small pumps or small pushes from someone else, if delivered in rhythm with the frequency of the swinging motion, produce large amplitudes.

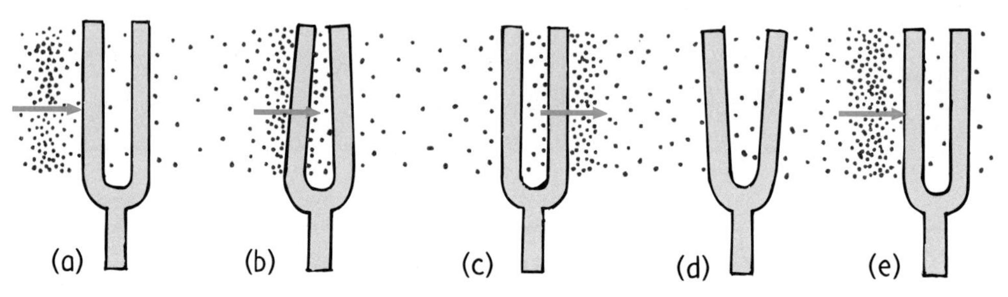

Figure 13.16
Stages of resonance. (a) The first compression meets the prong of the fork and gives it a tiny and momentary push; (b) the prong bends and then (c) returns to its initial position just at the time a rarefaction arrives. Still moving, the prong (d) overshoots in the opposite direction. Just when the prong restores to its initial position (e), the next compression arrives to repeat the cycle. Now it bends farther because it is already moving.

A common classroom demonstration of resonance is illustrated with a pair of tuning forks adjusted to vibrate at the same frequency and spaced a meter or so apart. When one of the forks is struck, it sets the other fork into vibration. This is a small-scale version of pushing a friend on a swing—it's the timing that's important. When a series of sound waves impinge on the fork, each compression gives the prongs a tiny push. Because the frequency of these pushes matches the natural frequency of the fork, the succession of pushes increase the amplitude of vibration. This is because the pushes occur at the right time and in the same direction as the instantaneous motion of the fork. The motion of the second fork is called a *sympathetic vibration.*

If the forks are not adjusted for matched frequencies, the timing of pushes is off and resonance does not occur. When you tune your radio, you adjust the natural frequency of the electronics in the radio to match one of the many surrounding signals in the air. The radio then resonates to one station at a time instead of playing all stations at once.

Resonance occurs whenever successive impulses are applied to a vibrating object in rhythm with its natural frequency. Cavalry troops marching across a footbridge near Manchester, England, in 1831 mistakenly caused the bridge to collapse when they marched in rhythm with the bridge's natural frequency. Since then, it is customary to order troops to "break step" when crossing bridges. A 20th-century bridge disaster was caused by wind-generated resonance (Figure 13.17).

Figure 13.17
In 1940, four months after being completed, the Tacoma Narrows Bridge in the state of Washington was destroyed by wind-generated resonance. A mild gale produced an irregular force in resonance with the natural frequency of the bridge, steadily increasing the amplitude of vibration until the bridge collapsed.

13.9 Interference—The Addition and Subtraction of Waves

One of the most interesting properties of all waves is **interference.** Consider transverse waves. When the crest of one wave overlaps the crest of another, the crests add together. The result is a wave of increased amplitude. This is *constructive interference* (Figure 13.18). When the crest of one wave overlaps the trough of another, the opposite occurs. The high portions of one wave simply fill in the low portions of another. This is *destructive interference.*

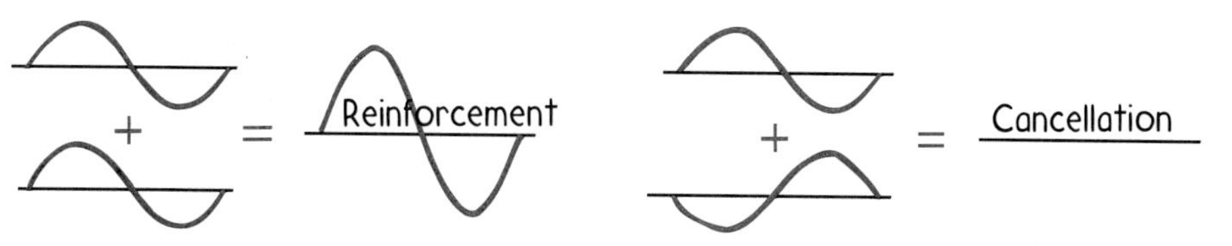

Figure 13.18
Constructive and destructive interference in a transverse wave.

Wave interference is easiest to see in water. Study Figure 13.19 and look at the interference pattern made when two vibrating objects touch the surface of water. Notice the regions where a crest of one

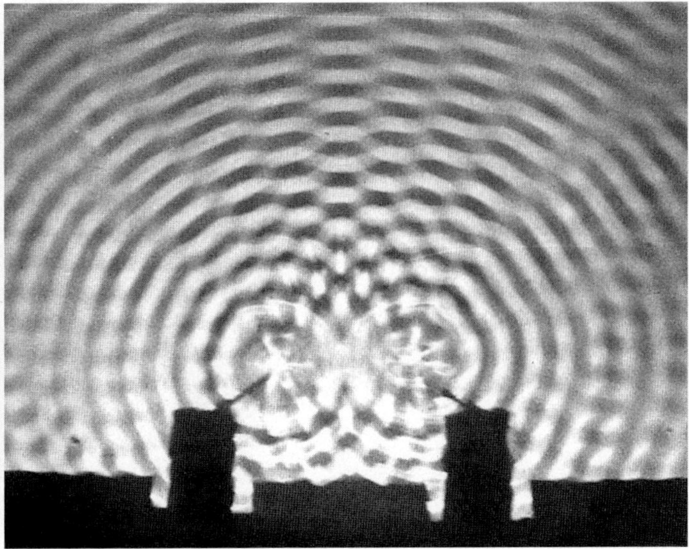

Figure 13.19
Two sets of overlapping water waves produce an interference pattern.

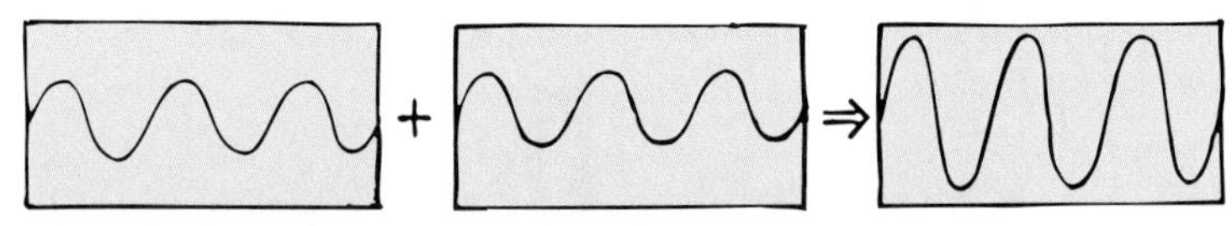

The superposition of two identical transverse waves in phase produces a wave of increased ampitude.

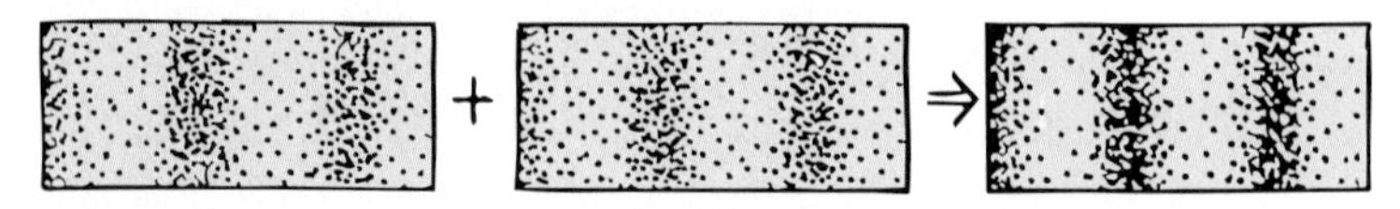

The superposition of two identical longitudinal waves in phase produces a wave of increased intensity.

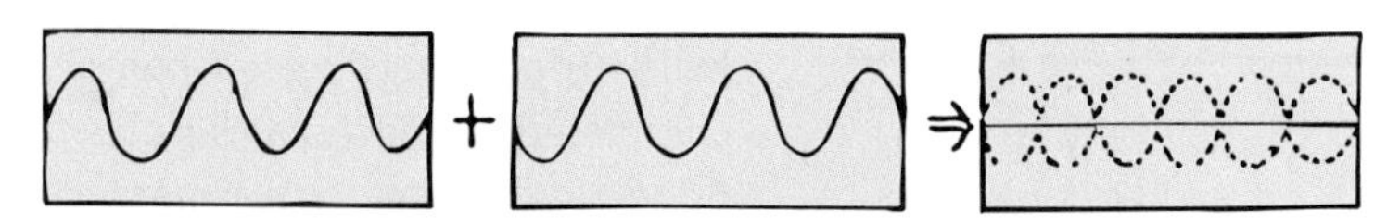

Two identical transverse waves that are out of phase destroy each other when they are superimposed.

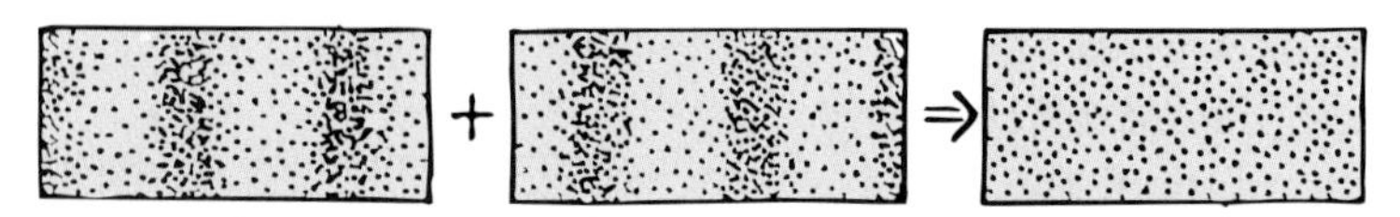

Two identical longitudinal waves that are out of phase destroy each other when they are superimposed.

Figure 13.20
Constructive (top two panels) and destructive (bottom two panels) wave interference in transverse and longitudinal waves.

wave overlaps the trough of another. This results in regions of zero amplitude. At points along these regions, the waves arrive out of step. We say they are *out of phase* with one another.

Interference is a property of all wave motion, whether the waves are water waves, sound waves, or light waves. We see a comparison of interference for both transverse and longitudinal waves in Figure 13.20. In the case of sound, the crest of a wave corresponds to a compression and the trough of a wave corresponds to a rarefaction.

Destructive sound interference is a useful property in *anti-noise technology.* Some noisy devices such as jackhammers are equipped with microphones that send the sound of the device to electronic microchips, which create mirror-image wave patterns of the sound signals. This mirror-image sound signal is fed to earphones worn by the operator. Sound compressions (or rarefactions) from the device are canceled by mirror image rarefactions (or compressions) in the earphones. The combination of signals cancels the noise made by the jackhammer or other device. Watch for this principle applied to electronic mufflers in cars—the anti-noise is blasted through loudspeakers, canceling about 95% of the original noise.

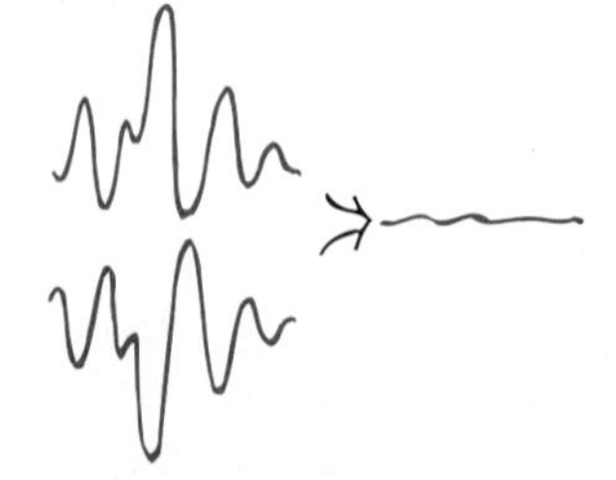

Figure 13.21
When a mirror image of a sound signal combines with original signal, the sound is canceled.

Figure 13.22
Ken Ford tows gliders in quiet comfort when he wears his noise-canceling earphones.

Sound-Off Exploration

Get a sound system with a pair of detachable speakers and set it for monoral sound (not stereo). Reverse the wiring on one of the speakers by switching the positive and negative wire inputs. Your speakers are then out of phase. When both speakers emit the same signal, a compression emitted by one speaker occurs at the same time a rarefaction is being emitted by the other. The resulting sound is not as full and not as loud as from speakers properly connected in phase. That's because the longer waves are being canceled by interference.

Now here's the important step. Unhook the speakers so you can face them toward each other. First hold them apart, with music playing, then bring them closer together. You'll hear weaker sound because shorter waves are being canceled. The sound is tinny. When the pair of speakers are brought face to face and touching each other, you'll hear very little sound! Only the highest frequencies survive cancellation. You must try this to appreciate it.

Beats—An Effect of Sound Interference

A tone is a sound of distinct pitch and duration. When two tones of slightly different frequency are sounded together, a fluctuation in loudness may be heard. The sound is loud, then faint, then loud, then faint, and so on. This periodic variation in loudness is called **beats.** Beats are due to interference.

Beats can occur with any kind of wave and provide a practical way to compare frequencies. To tune a piano, for example, a piano tuner listens for beats produced between a standard tuning fork and those of a particular string on the piano. When the frequencies are identical, the beats disappear. The members of an orchestra tune their instruments by listening for beats between their instruments and a standard tone produced by a piano or some other instrument.

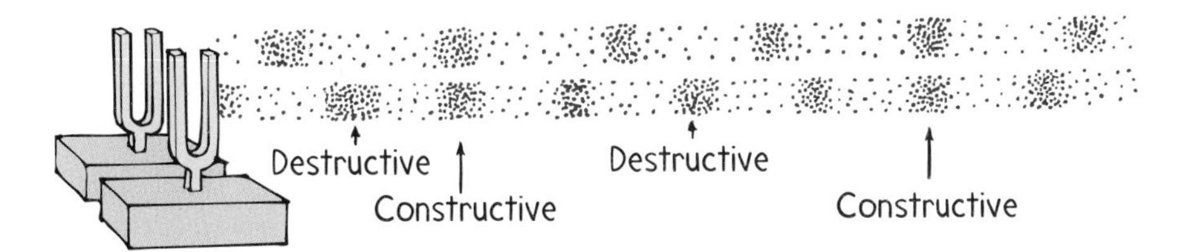

Figure 13.23
The interference of two sound sources of slightly different frequencies produces beats.

Standing Waves—The Effect of Waves Passing Through Each Other

Another interesting effect of interference is *standing waves.* Tie a rope to a wall and shake the free end up and down. (A rubber tube works even better.) Waves that hit the wall are reflected back along the rope. By shaking the rope just right, you can cause the incident and reflected waves to interfere and form a **standing wave,** where parts of

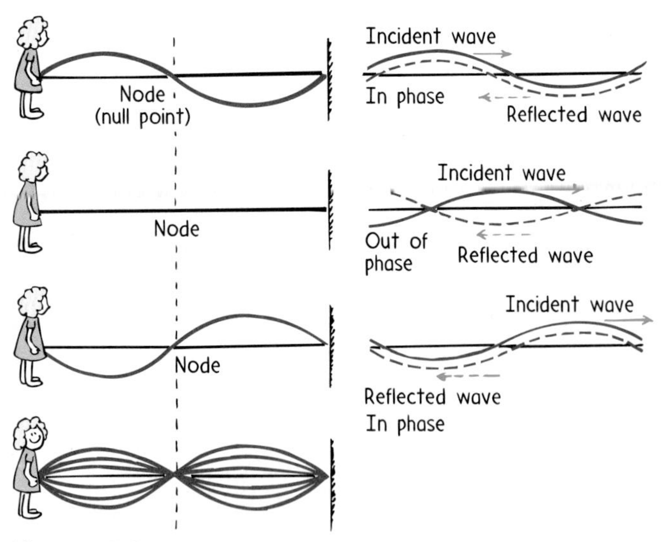

Figure 13.24
The incident and reflected waves interfere to produce a standing wave.

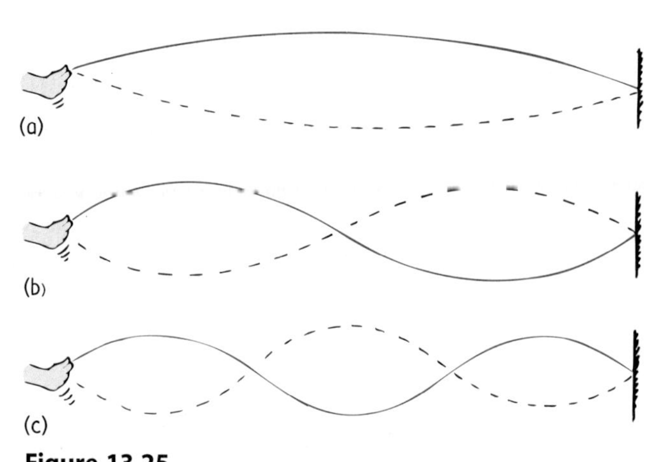

Figure 13.25
(a) Shake the rope until you set up a standing wave of one loop (1/2 wavelength). (b) Shake with twice the frequency and produce a wave having two loops (1 wavelength). (c) Shake with three times the frequency and produce three loops (3/2 wavelengths).

the rope, called the *nodes*, are stationary. You can hold your fingers on either side of the rope at a node, and the rope will not touch them. Thus nodes are positions of zero rope displacement. The distance between nodes is a half wavelength. Two loops make up a complete wave. The positions on a standing wave that have the largest displacements are known as *antinodes* and occur halfway between nodes.

Standing waves are produced when two sets of waves of equal amplitude and wavelength pass through each other in opposite directions. Then the waves are steadily in and out of phase with each other and produce stable regions of constructive and destructive interference (Figure 13.24).

Standing waves are set up in the strings of musical instruments when plucked, bowed, or struck. Standing waves can be set up in a tub of water or a bowl of soup by sloshing it back and forth with the right frequency. They can be produced with either transverse or longitudinal vibrations.

Concept Check ✓

Is it possible for one wave to cancel another so that no amplitude remains?

Check Your Answer

Yes. This is destructive interference. In a standing wave in a rope, for example, parts of the rope—the nodes—have no amplitude because of destructive interference.

13.10 The Doppler Effect—Changes in Frequency Due to Motion

A pattern of water waves produced by a bug jiggling its legs and bobbing up and down in the middle of a quiet puddle is shown in Figure 13.26. The bug is not going anywhere but is merely treading water in a fixed position. The waves it makes are concentric circles because wave speed is the same in all directions. If the bug bobs in the water at a constant frequency, the distance between wave crests (the wavelength) is the same in all directions. Waves encounter point A as frequently as they encounter point B. This means that the frequency of wave motion is the same at points A and B, or anywhere else near the bug. This wave frequency is the same as the bug's bobbing frequency.

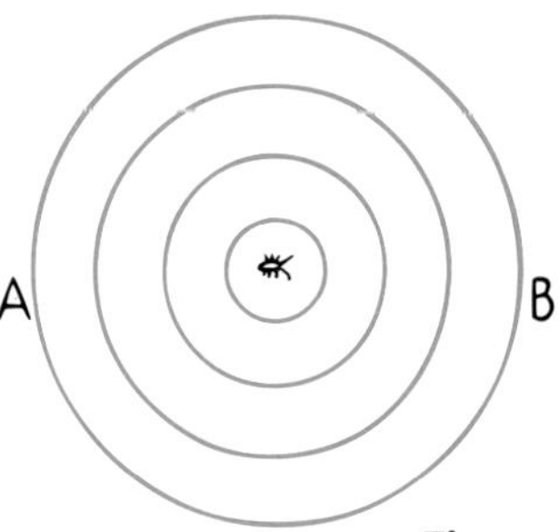

Figure 13.26
Top view of water waves made by a stationary bug jiggling in still water.

Suppose now that the jiggling bug begins moving to the right at a speed less than the wave speed. In effect, the bug chases part of the waves it makes. The circular waves are no longer concentric (Figure 13.27). Each one has its center where the bug was previously. The outermost wave (wave 1 in Figure 13.27) was made when the bug was at the center of that circle. The center of wave 2 was made when the bug was at the center of that circle, and so forth. The centers of the circular waves move in the direction of the swimming bug. Although the bug maintains the same bobbing frequency as before, an observer at B sees the waves coming more frequently. In other words, observer B measures a *higher* frequency. An observer at A, on the other hand, measures a *lower* frequency because of the longer time between wave-crest arrivals. This change in frequency due to the motion of the source (or receiver) is called the **Doppler effect** (after the Austrian scientist Christian Doppler).

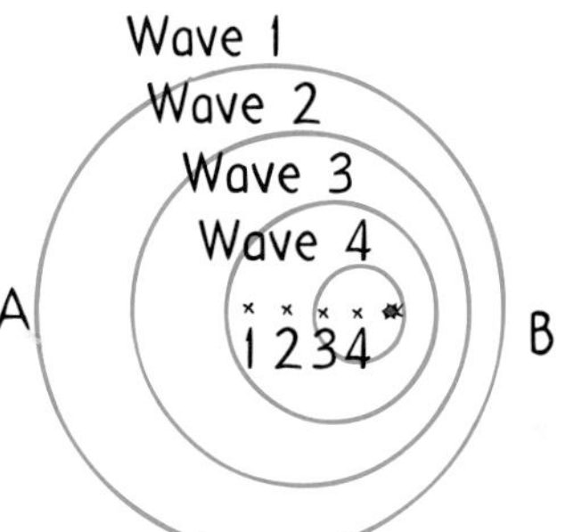

Figure 13.27
Water waves made by a bug swimming in still water toward point B.

Water waves spread over the two-dimensional surface of the water. Sound and light waves, on the other hand, travel in three-dimensional space in all directions like an expanding balloon. Just as circular waves are closer together in front of the swimming bug, spherical sound or light waves ahead of a moving source are closer together and reach a receiver more frequently. The Doppler effect applies to all types of waves.

The Doppler effect is evident when you hear the changing pitch of a siren as the vehicle drives by. When it approaches, the pitch is higher than normal. The wave crests reach your ear more frequently. When the vehicle passes and moves away, you hear a drop in pitch. Then the crests of the waves hit your ear less frequently.

Figure 13.28
The pitch (frequency) of sound increases when the source moves toward you and decreases when the source moves away.

The Doppler effect also occurs when the receiver moves. Move toward a stationary wave source and you encounter its waves more frequently. Move away, and you encounter waves less frequently. The Doppler effect results from relative motion between a wave source and a receiver.

The Doppler effect also occurs with light waves. When a source of light waves approaches, the increase in frequency is called a *blue shift.* That's because the increase is toward the high-frequency, blue end of the color spectrum. A decrease in frequency is called a *red shift,* referring to a shift toward the lower-frequency, red end of the color spectrum. The galaxies, for example, show a red shift in the light they emit. A measurement of this shift permits a calculation of the speeds at which they are receding from the Earth. A rapidly spinning star shows a red shift on the side turning away from us and a relative blue shift on the side turning toward us. This enables astronomers to calculate the star's spin rate.

Concept Check ✓

When a wave source moves toward you, do you measure an increase or decrease in wave speed?

Check Your Answer Neither! It is the *frequency* of the waves that changes for a moving source, not the wave speed. Be clear about the distinction between frequency and speed. How frequently a wave vibrates is altogether different from how fast it moves from one place to another.

13.11 Wave Barriers and Bow Waves

When a moving wave source travels as fast as the waves it produces, a *wave barrier* is produced. Consider the bug in our previous example. When it swims as fast as the wave speed, can you see that it keeps up with the waves it produces? Instead of the waves getting ahead of the bug, they pile up and overlap directly in front of it (Figure 13.29). The bug encounters a wave barrier. Much effort is required of the bug to swim over this barrier—before it can swim faster than wave speed.

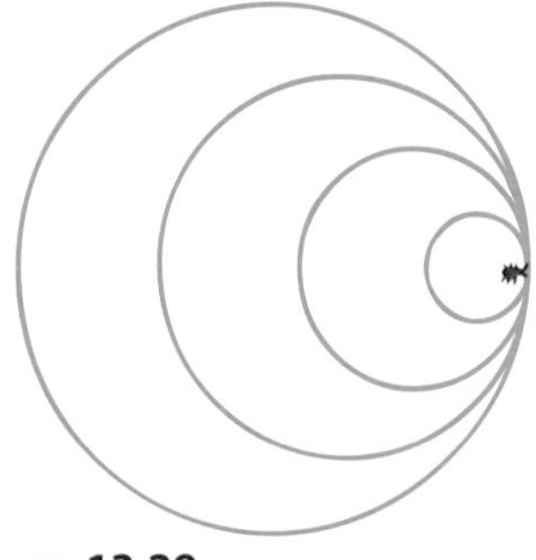

Figure 13.29
Wave pattern made by a bug swimming at wave speed.

The same thing happens when an aircraft travels at the speed of sound. The sound waves produced by the engines overlap to produce a barrier of compressed air on the leading edges of the wings and

Figure 13.30
This aircraft is producing a cloud of water vapor that has just condensed out of the rapidly expanding air in the rarefied region behind the wall of compressed air.

other parts of the craft. Considerable thrust is required for the aircraft to push through this barrier (Figure 13.30). Once through, the craft can fly faster than the speed of sound without similar opposition. It is now *supersonic*. It is like the bug, which once over its wave barrier finds the water ahead relatively smooth and undisturbed.

When the bug swims faster than wave speed, it produces a wave pattern like the one shown in Figure 13.31. It outruns the waves it produces. The overlapping waves form a V shape called a **bow wave,** which appears to be dragging behind the bug. The familiar bow (rhymes with cow) wave generated by a speedboat knifing through the water is produced by the overlapping of many periodic circular waves.

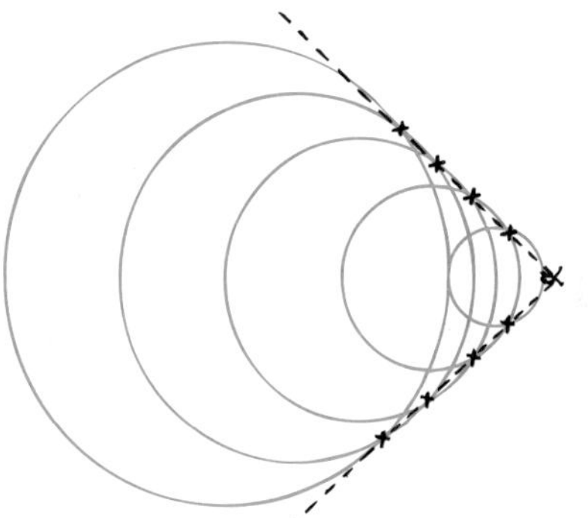

Figure 13.31
A bow wave, the pattern made by a bug swimming faster than wave speed. The points at which adjacent waves overlap (X) produce the V shape.

Some wave patterns made by sources moving at various speeds are shown in Figure 13.32. Note that after the speed of the source exceeds wave speed, increased speeds produce a narrower and narrower V shape.*

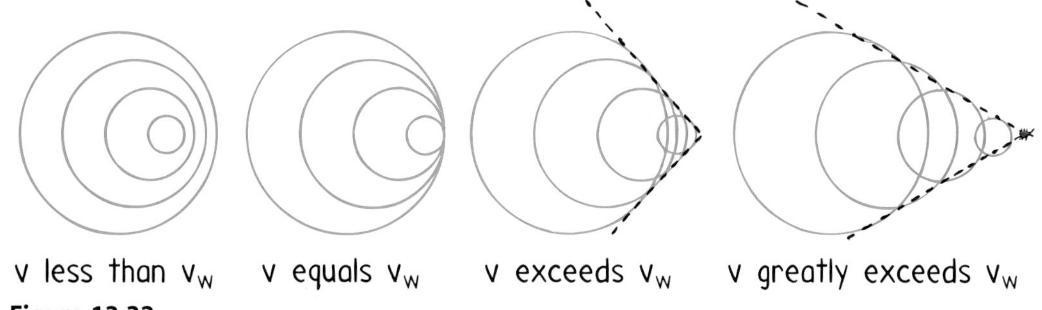

Figure 13.32
Patterns made by a bug swimming at successively greater speeds. Overlapping at the edges occurs only when the bug swims faster than wave speed.

* Bow waves generated by boats in water are more complex than indicated here. Our idealized treatment serves as an analogy for the production of the less complex shock waves in air.

Supersonic has to do with speed—faster than sound. *Ultrasonic* has to do with frequency—higher than we can hear.

13.12 Shock Waves and the Sonic Boom

A speedboat knifing through the water generates a two-dimensional bow wave. A supersonic aircraft similarly generates a three-dimensional **shock wave.** Just as a bow wave is produced by overlapping circles that form a V, a shock wave is produced by overlapping spheres that form a cone. And just as the bow wave of a speedboat spreads until it reaches the shore of a lake, the conical wake generated by a supersonic craft spreads until it reaches the ground.

The bow wave of a speedboat that passes by can splash and douse you if you are at the water's edge. In a sense, you can say that you are hit by a "water boom." In the same way, when the conical shell of compressed air that sweeps behind a supersonic aircraft reaches listeners on the ground, the sharp crack they hear is described as a **sonic boom.**

We don't hear a sonic boom from slower-than-sound, or subsonic, aircraft because the sound waves reach our ears one at a time and make one continuous tone. Only when the craft moves faster than sound do the waves overlap to reach the listener in a single burst. The sudden increase in pressure is much the same in effect as the sudden expansion of air produced by an explosion. Both processes direct a burst of high-pressure air to the listener. The ear normally can't recognize a difference between an explosion and a sonic boom.

A water skier is familiar with the fact that next to the high hump of the bow wave is a V-shaped depression. The same is true of a shock wave. There are two cones: a high-pressure one generated at the bow of the supersonic aircraft and a low-pressure one that follows at the tail. You can see the edges of these cones in the photograph of the supersonic bullet in Figure 13.33. Between these two cones the air pressure rises sharply to above atmospheric pressure, then falls below atmospheric pressure before sharply returning to normal beyond the inner tail cone (Figure 13.34). This overpressure suddenly followed by underpressure intensifies the sonic boom.

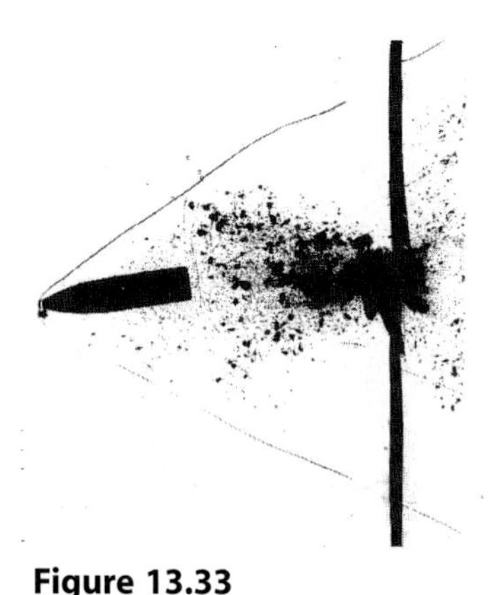

Figure 13.33
Shock wave of a bullet piercing a sheet of Plexiglas. Light is deflected as it passes through the compressed air that makes up the shock wave, making it visible. Look carefully and see the second shock wave originating at the tail of the bullet.

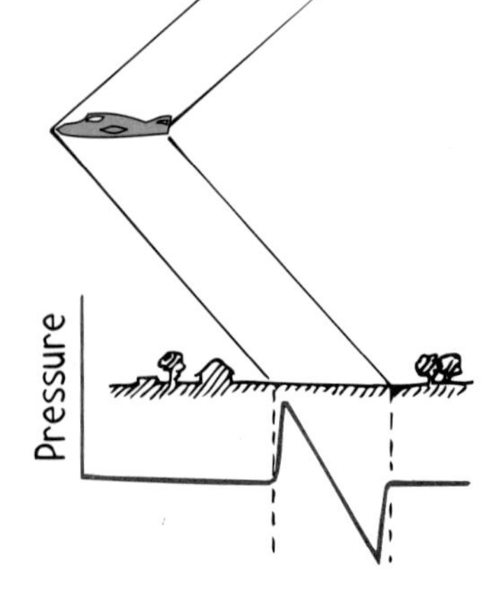

Figure 13.34
A shock wave and the air pressure differences it causes.

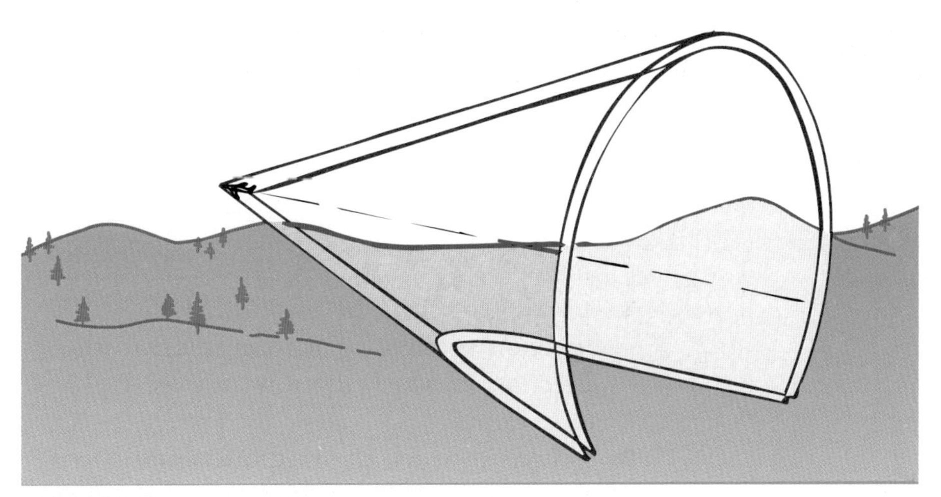

Figure 13.35
A shock wave is made up of two cones—a high-pressure cone with the apex at the bow of the aircraft and a low-pressure cone with the apex at the tail.

Concept Check ✓

Is it correct to say that a sonic boom is produced when an aircraft breaks through the sound barrier?

Check Your Answer This is incorrect, and is similar to saying a boat produces a bow wave when it overtakes its own waves. A shock wave and its resulting sonic boom are swept *continually* behind an aircraft the entire time the craft is traveling faster than sound, just as a bow wave is swept continuously behind a speedboat.

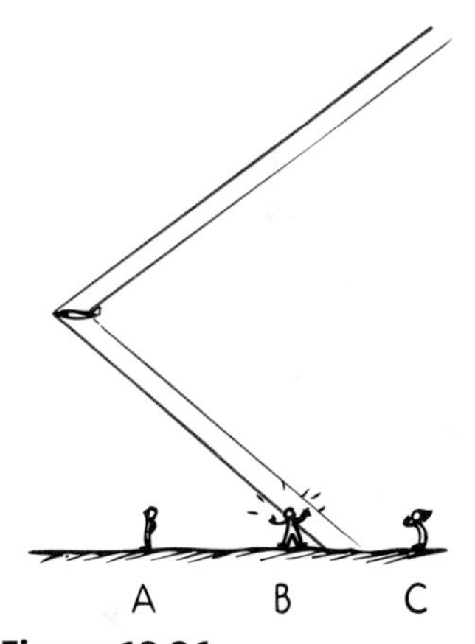

Figure 13.36
The shock wave has not yet reached listener A, but is now reaching listener B and has already reached listener C.

In Figure 13.36, listener B is in the process of hearing a sonic boom. Listener C has already heard it, and listener A will hear it shortly. The aircraft that generated this shock wave may have broken through the sound barrier hours ago!

It is not necessary that the moving source be "noisy" to produce a shock wave. Once an object—even a silent one—is moving faster than the speed of sound, it *makes* sound. A supersonic bullet passing overhead produces a crack, which is a small sonic boom. If it were larger and disturbed more air in its path, the crack would be more boomlike. When a lion tamer cracks a whip, the cracking sound is a sonic boom produced by the tip when it travels faster than the speed of sound. Both the bullet and the whip are not in themselves sound sources. But when traveling at supersonic speeds they produce their own sound as they generate shock waves.

Chapter Review

Key Terms and Matching Definitions

_____ amplitude
_____ beats
_____ bow wave
_____ Doppler effect
_____ forced vibration
_____ frequency
_____ hertz
_____ interference
_____ longitudinal wave
_____ natural frequency
_____ period
_____ refraction
_____ resonance
_____ shock wave
_____ sonic boom
_____ sound wave
_____ standing wave
_____ transverse wave
_____ wave
_____ wave speed
_____ wavelength

1. A disturbance or vibration propagated from point to point in a medium or in space.
2. For a wave or vibration, the maximum displacement on either side of the equilibrium (midpoint) position.
3. The distance between successive crests, troughs, or identical parts of a wave.
4. For a vibrating body or medium, the number of vibrations per unit time. For a wave, the number of crests that pass a particular point per unit time.
5. The SI unit of frequency. It equals one vibration per second.
6. The time required for a vibration or a wave to make a complete cycle; equal to 1/frequency.
7. The speed with which waves pass a particular point:

   $$\text{Wave speed} = \text{frequency} \times \text{wavelength}$$

8. A wave in which the medium vibrates in a direction perpendicular (transverse) to the direction in which the wave travels. Light is an example.
9. A wave in which the medium vibrates in a direction parallel (longitudinal) to the direction in which the wave travels. Sound is an example.
10. A longitudinal vibratory disturbance that travels in a medium, which can be heard in the approximate frequency range 20–20,000 Hertz.
11. The bending of a wave through either a non-uniform medium or from one medium to another, caused by differences in wave speed.
12. The setting up of vibrations in an object by a vibrating force.
13. The pattern formed by superposition of different sets of waves that produces mutual reinforcement in some places and cancellation in others.
14. A stationary wave pattern formed in a medium when two sets of identical waves pass through the medium in opposite directions.
15. The change in frequency of wave motion resulting from motion of the wave source or receiver.
16. The V-shaped wave produced by an object moving across a liquid surface at a speed greater than the wave speed.
17. The cone-shaped wave created by an object moving at supersonic speed through a fluid.
18. The loud sound resulting from the incidence of a shock wave.

19. A frequency at which an elastic object naturally tends to vibrate, so that minimum energy is required to produce a forced vibration or to continue vibrating at that frequency.
20. The response of a body when a forcing frequency matches its natural frequency.
21. A series of alternate reinforcements and cancellations produced by the interference of two waves of slightly different frequency, heard as a throbbing effect in sound waves.

Review Questions

Special Wiggles—Vibrations and Waves

1. What is the source of all waves?

2. How do frequency and period relate to each other?

Wave Motion—Transporting Energy

3. What is it that moves from source to receiver in wave motion?

4. What is the relationship among frequency, wavelength, and wave speed?

Two Types of Waves—Transverse and Longitudinal

5. In a transverse wave, in what direction are the vibrations relative to the direction of wave travel?

6. In a longitudinal wave, in what direction are the vibrations relative to the direction of wave travel?

Sound Travels in Longitudinal Waves

7. Why will sound not travel in a vacuum?

8. How does the speed of sound in water compare with the speed of sound in air? How does the speed in steel compare with the speed in air?

Sound Can Be Reflected

9. How does the angle of incidence compare with the angle of reflection for sound?

10. What is a *reverberation*?

Sound Can Be Refracted

11. What causes refraction?

12. Does sound tend to bend upward or downward when its speed near the ground is greater than its speed at a higher level?

Forced Vibrations and Natural Frequency

13. Why does a struck tuning fork sound louder when it is held against a table?

14. Give three examples of forced vibration.

Resonance and Sympathetic Vibrations

15. Distinguish between *forced vibrations* and *resonance.*

16. When you listen to a radio, why are you able to hear only one station at a time rather than all stations at once?

Interference—The Addition and Subtraction of Waves

17. What kind of waves exhibit interference?

18. Distinguish between *constructive interference* and *destructive interference.*

The Doppler Effect—Changes in Frequency Due to Motion

19. In the Doppler effect, does frequency change? Does wavelength change? Does wave speed change?

20. Can the Doppler effect be observed with longitudinal waves, transverse waves, or both?

Wave Barriers and Bow Waves

21. How does the V shape of a bow wave depend on the speed of the wave source?

Shock Waves and the Sonic Boom

22. How does the V shape of a shock wave depend on the speed of the wave source?

23. True or false: A sonic boom occurs only when an aircraft is breaking through the sound barrier.

24. True or false: In order for an object to produce a sonic boom, it must be a sound source.

Explorations

1. Tie a rubber tube, a spring, or a rope to a fixed support and produce standing waves. See how many nodes you can produce.
2. Test to see which ear has the better hearing by covering one ear and finding the distance away that your open ear can hear the ticking of a clock. Repeat for the other ear. Notice also how the sensitivity of your hearing improves when you cup your ears with your hands.
3. The Doppler shift is nicely heard with a buzzer of any kind that emits a steady tone. Put it in a plastic bag and swing it around your head in a circle. Your friends will hear the frequency shift as the buzzer alternately moves toward and away from them.

Exercises

1. If we double the frequency of a vibrating object, what happens to its period?
2. If the frequency of a sound wave is doubled, what change occurs in its speed? In its wavelength?
3. Red light has a longer wavelength than blue light. Which has the greater frequency?
4. You dip your finger repeatedly into a puddle of water and make waves. What happens to the wavelength if you dip your finger more frequently?
5. How does the frequency of vibration of a small object floating in water compare with the number of waves passing the object each second?
6. Why will marchers at the end of a long parade following a band be out of step with marchers near the front?
7. What two physics mistakes occur in a science fiction movie that shows a distant explosion in outer space, where you see and hear the explosion at the same time?
8. A cat can hear sound frequencies up to 70,000 Hz. Bats send and receive ultrahigh-frequency squeaks up to 120,000 Hz. Which hears shorter wavelengths, cats or bats?
9. At the stands of a racetrack, you notice smoke from the starter's gun before you hear it fire. Explain.
10. Why is it so quiet after a snowfall?
11. Why is the moon described as a "silent planet"?
12. If the speed of sound depended on frequency, how would distant music sound?
13. Why is an echo weaker than the original sound?
14. Would there be a Doppler effect if the source of sound were stationary and the listener in motion? Why or why not? In which direction should the listener move to hear a higher frequency? A lower frequency?
15. When you blow your horn while driving toward a stationary listener, the listener hears an increase in the horn frequency. Would the listener hear an increase in the horn frequency if he were in another car traveling at the same speed in the same direction as you? Explain.

16. Astronomers find that light coming from one edge of the sun has a slightly higher frequency than light from the opposite edge. What do these measurements tell us about the sun's motion?

17. What can you say about the speed of a boat that makes a bow wave? How about the speed of an aircraft that produces a shock wave?

18. What physical principle is used by Manuel when he pumps in rhythm with the natural frequency of the swing?

19. A special device can transmit out-of-phase sound from a noisy jackhammer to earphones worn by its operator. Over the noise of the jackhammer, the operator can easily hear your voice while you are unable to hear his. Explain.

20. If a single disturbance some unknown distance away sends out both transverse and longitudinal waves that travel with distinctly different speeds in the medium, such as in the ground during earthquakes, how could the origin of the disturbance be located?

Problems

1. A weight suspended from a spring bobs up and down over a distance of 20 centimeters twice each second. What is its frequency? Its period? Its amplitude?

2. A rule of thumb for estimating the distance in kilometers between an observer and a lightning stroke is to divide the number of seconds in the interval between the flash and the sound by 3. Is this rule correct?

3. In terms of wavelength, how far does a wave travel during one period?

4. From far away you watch a woman driving nails into her front porch at a regular rate of 1 stroke per second. You hear the sound of the blows exactly synchronized with the blows you see. And then you hear one more blow after you see her cease hammering. How far away is she?

5. A skipper on a boat notices wave crests passing his anchor chain every 5 s. He estimates the distance between wave crests to be 15 m. He also correctly estimates the speed of the waves. What is this speed?

6. An oceanic depth-sounding vessel surveys the ocean bottom with ultrasonic sound that travels 1530 m/s in seawater. How deep is the water if the time delay of the echo from the ocean floor is 6 s?

7. A bat flying in a cave emits a sound and receives its echo 0.1 s later. How far away is the cave wall?

8. What frequency of sound produces a wavelength of 1 meter in room-temperature air?

Chapter 14: Light and Color

Light is the only thing we see, but what exactly *is* light? How does it differ from sound? And why does light travel so fast? How does it get through glass? Does it lose speed when it passes through transparent materials like glass? And why are different things various colors? Why do red, green, and blue light mix to form white; but red, green, and blue paints mix to a muddy brown? Is there a physical science reason for why the sky is blue? Why are sunsets red? Why are clouds white?

Light originates from the accelerated motion of electrons. Light is electromagnetic in nature. It's a tiny portion of a larger whole—the *electromagnetic spectrum*. We begin our study of light by looking at its electromagnetic nature, how it interacts with materials, and how it appears so nicely as color.

14.1 The Electromagnetic Spectrum— A Tiny Bit of Which Is Light

Figure 14.1
Shake an electrically charged object to-and-fro, and you produce an electromagnetic wave.

If you shake the end of a stick back and forth in still water, you'll create waves on the water surface. If you similarly shake an electrically charged rod to-and-fro in empty space, and you create electromagnetic waves in space. This is because the moving charge is an electric current. Recall from Chapter 12 that a magnetic field surrounds an electric current and changes as the current changes. Recall also from Chapter 12 that a changing magnetic field induces an electric field—electromagnetic induction. And what does the changing electric field do? It induces a changing magnetic field. The vibrating electric and magnetic fields regenerate each other to make up an **electromagnetic wave.** In a vacuum, all electromagnetic waves travel at the same speed.

The classification of electromagnetic waves according to frequency is the **electromagnetic spectrum** (Figure 14.3). Electromagnetic waves of frequency as low as 0.01 hertz (Hz) have been detected. Others with frequencies of several thousand hertz (kHz) are classified as low-frequency radio waves. A frequency of one million hertz (1 MHz) lies in the middle of the AM radio band. The very-high-frequency (VHF) television band of waves starts at about 50 million hertz. FM radio ranges from 88 to 108 MHz. Then there are ultra-high frequencies (UHF), followed by microwaves, beyond which are infrared waves, the "heat waves" we studied in Chapter 10. Farther to the right in Figure 14.3 is visible light. Surprisingly, visible light makes up less than a millionth of 1 percent of the electromagnetic spectrum.

The lowest frequency of light that our eyes can see appears red. The highest visible frequencies are nearly twice the frequency of red and appear violet. Still higher frequencies are ultraviolet, which are beyond our range of vision. These higher-frequency waves are more energetic and cause sunburns. Higher frequencies beyond ultraviolet extend into the X-ray and gamma-ray regions. There is no sharp boundary between these regions, which actually overlap each other. The spectrum is broken up into these arbitrary regions merely for classification.

The frequency of an electromagnetic wave as it vibrates through space is identical to the frequency of the vibrating electric charge that

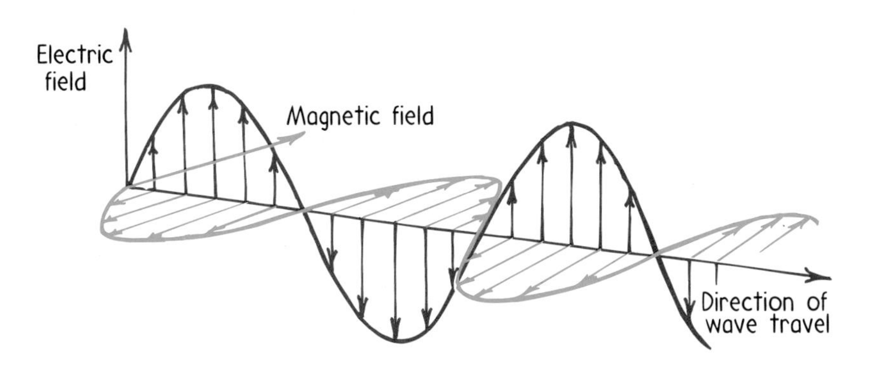

Figure 14.2
The electric and magnetic fields of an electromagnetic wave are perpendicular to each other and to the direction of motion of the wave.

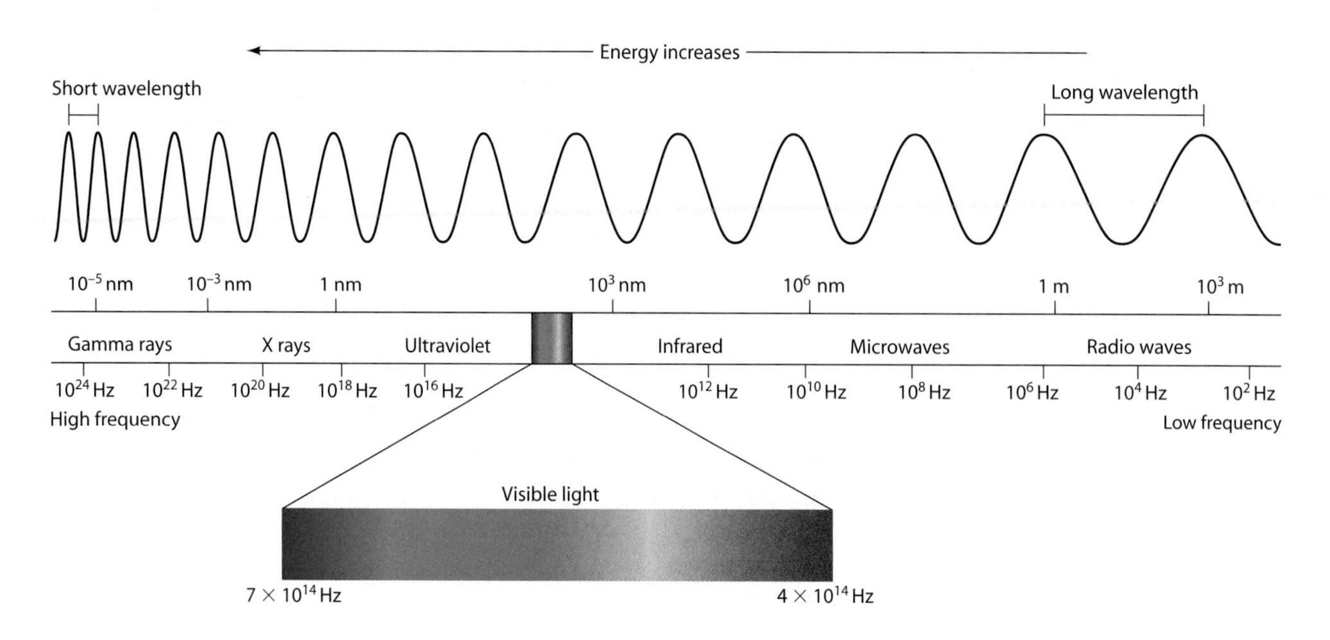

Figure 14.3
The electromagnetic spectrum is a continuous range of waves extending from radio waves to gamma rays. The descriptive names of the sections are merely a historical classification, for all waves are the same in nature. They differ mainly in frequency and wavelength. All travel at the same speed.

generates it. Different frequencies result in different wavelengths—low frequencies produce long wavelengths, and high frequencies produce short wavelengths. The higher the frequency of the vibrating charge, the shorter the wavelength of radiation.*

Concept Check ✓

Is it correct to say that a radio wave is a low-frequency light wave? Is a radio wave also a sound wave?

Check Your Answers Both a radio wave and a light wave are electromagnetic waves, and all electromagnetic waves originate in the vibrations of electrons. Because radio waves have lower frequencies than light waves, a radio wave may be considered a low-frequency light wave (and a light wave a high-frequency radio wave). A sound wave, however, is a *mechanical* vibration of matter and is *not* electromagnetic. A sound wave is fundamentally different from an electromagnetic wave. So a radio wave is definitely not a sound wave. (Don't confuse a radio wave with the sound that a loudspeaker emits.)

Light is energy carried in an electromagnetic wave emitted by vibrating electrons in atoms.

14.2 Why Materials Are Either Transparent or Opaque

When light passes through matter, some of the electrons in the matter are forced into vibration. In this way, vibrations in the emitter are transmitted to vibrations in the receiver. This is similar to the way sound is transmitted (Figure 14.4).

* The relationship is $c = f\lambda$, where c is the waves speed (constant), f is the frequency, and λ is the wavelength. It is common to describe sound waves and radio waves by frequency and light waves by wavelength. In this book, however, we favor the single concept of frequency in describing light.

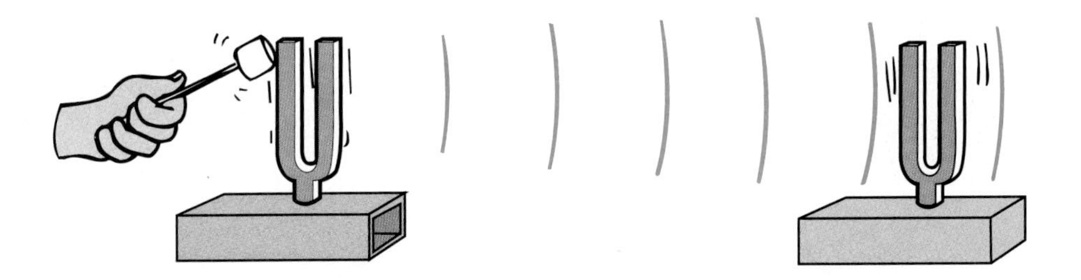

Figure 14.4
Just as a sound wave can force a sound receiver into vibration, a light wave can force electrons in materials into vibration.

Materials such as glass and water allow light to pass through in straight lines. They are **transparent** to light. To understand how light penetrates through glass (or any transparent material), visualize the electrons in the atoms of glass as if they were connected to the nucleus by springs (Figure 14.5). When a light wave meets them, the electrons are set into vibration.

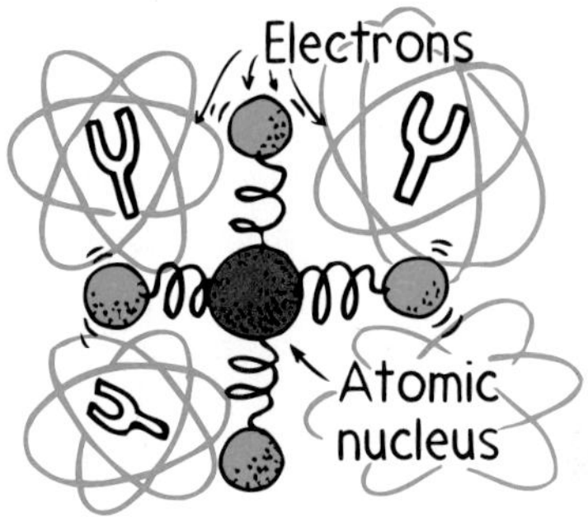

As we learned in the previous chapter, materials that are springy (elastic) respond more to vibrations at particular frequencies than to other frequencies. Bells ring at a particular frequency, tuning forks vibrate at a particular frequency, and so do the electrons of atoms and molecules. The natural vibration frequencies of an electron depend on how strongly it is attached to its atom or molecule. Different atoms and molecules have different "spring strengths." Electrons in atoms of glass vibrate in the ultraviolet range. So when ultraviolet waves shine on glass, resonance occurs. The vibration of electrons builds up to large amplitudes, just as pushing someone at the resonant frequency on a swing builds to a large amplitude. The energy received by a glass atom is either re-emitted or passed on to neighboring atoms by collisions. Resonating atoms in the glass can hold onto the energy of the ultraviolet light for quite a long time (about 100 millionths of a second). During this time the atom makes about 1 million vibrations. This is time enough to collide with neighboring atoms and give up its energy as heat. Glass is therefore not transparent to ultraviolet.

Electrons

Atomic nucleus

Figure 14.5
The electrons of atoms have certain natural frequencies of vibration and can be modeled as particles connected to the atomic nucleus by spring. As a result, atoms and molecules behave somewhat like optical tuning forks.

At lower wave frequencies, such as those of visible light, electrons in the glass atoms are forced into vibration—but at less amplitude. The atoms retain the energy for a briefer time, with less chance of collisions with neighboring atoms. This means that less energy is transformed to heat. Instead, quite nicely, the energy of vibrating electrons is re-emitted as light. Glass is transparent to all the frequencies of visible light. The frequency of the re-emitted light that is passed from atom to atom is identical to the frequency of the light that initially produced the vibration. However, there is a slight time delay between absorption and re-emission.

The "spring model" of the atoms helps us understand how light interacts with matter—even though electrons aren't really connected by springs to the nucleus.

Figure 14.6
A wave of visible light incident upon a pane of glass sets up in the glass atoms vibrations that produce a chain of absorptions and re-emissions, which pass the light energy through the material and out the other side. Because of the time delay between absorptions and re-emissions, the light travels through the glass more slowly than through empty space.

It is this time delay that results in a lower average speed of light through glass (Figure 14.6). Light travels at different average speeds through different materials. We say *average speeds* because the speed of light in a vacuum, whether in interstellar space or in the space between atoms in a piece of glass, is a constant 300,000 kilometers per second.* We call this speed of light c. The speed of light in the atmosphere is slightly less than in a vacuum but is usually rounded off as c. In water, light travels at 75% of its speed in a vacuum, or $0.75c$. In glass, light travels about $0.67c$, depending on the type of glass. In a diamond, light travels at less than half its speed in a vacuum, only $0.41c$. When light emerges from these materials into the air, it again travels at its original speed c.

Infrared waves, which have frequencies lower than those of visible light, vibrate entire molecules in the structure of glass and many other materials. This molecular vibration increases the thermal energy and temperature of the material, which is why infrared waves are often called *heat waves.* Thus glass is not transparent to infrared light.

Concept Check ✓

1. Why is glass transparent to visible light but not to ultraviolet and infrared?
2. Pretend that while you walk across a room you make several short stops along the way to greet people who are "on your wavelength." How is this analogous to visible light traveling through glass?
3. In what way is it not analogous?

Check Your Answers

1. Because the natural vibration frequency for electrons in glass matches the frequency of ultraviolet light, resonance occurs when ultraviolet waves shine on glass. The absorbed energy is passed on to other atoms as heat, instead of being re-emitted as light. So glass is opaque at ultraviolet frequencies. But in the range of visible light, the forced vibrations of electrons in the glass are at smaller amplitudes. Vibrations are smaller

* The presently accepted value is 299,792 km/s, rounded to 300,000 km/s (186,000 mi/s).

and light is re-emitted instead of turning to heat. So the glass is transparent. Lower-frequency infrared light causes entire molecules, rather than electrons, to resonate. So heat is generated and the glass is opaque.

2. Your average speed across the room is less than it would be in an empty room because of the time delays of your stops. In the same way, light takes a longer time to get through glass. The result of interactions of light and atoms it encounters is lower average speed.
3. In walking across the room, it is you who begin and complete the walk. But it is different for light. The light that entered the glass is not the same light that emerges. The frequencies are the same, so like identical twins, they are indistinguishable.

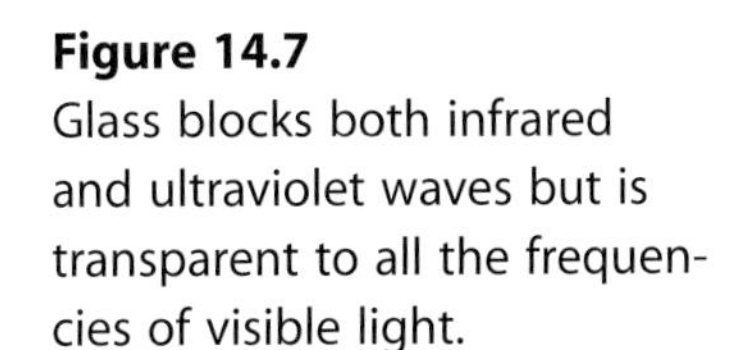

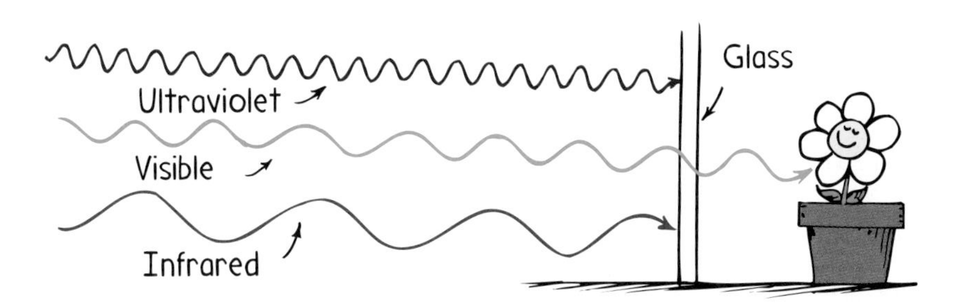

Figure 14.7
Glass blocks both infrared and ultraviolet waves but is transparent to all the frequencies of visible light.

Most things around us are **opaque**—they absorb light without re-emitting it. Books, desks, chairs, and people are opaque. Vibrations given by light to their atoms and molecules are turned into random kinetic energy—into thermal energy. They become slightly warmer.

Metals are opaque. As we learned in Chapter 11, the outer electrons of atoms in metals are not bound to any particular atom. They are loose and free to wander throughout the material (which is why metal conducts electricity and heat so well). When light shines on metal and sets these free electrons into vibration, their energy does not "spring" from atom to atom in the material but is instead reflected. That's why metals are shiny.

The Earth's atmosphere is transparent to some ultraviolet light, all visible light, and some infrared light. But the atmosphere is opaque to high-frequency ultraviolet light. The small amount of ultraviolet that does get through causes sunburns. If all ultraviolet light got through the atmosphere we would be fried to a crisp. Clouds are semitransparent to ultraviolet light, which is why you can get a sunburn on a cloudy day. Ultraviolet light is not only harmful to your skin but is also damaging to tar roofs. Now you know why tarred roofs are often covered with gravel.

Figure 14.8
Metals are shiny because light that shines on them forces free electrons into vibration, and these vibrating electrons then emit their "own" light waves as reflection.

Have you noticed that things look darker when they are wet than when they are dry? Light incident on a dry surface such as sand bounces directly to your eye. But light incident on a wet surface bounces around inside the transparent wet region before it reaches your eye. What happens with each bounce? Absorption! So sand and other things looks darker when wet.

14.3 Color Science

To the scientist, color is not in a material. Color is a physiological experience and is in the eye of the beholder. So when we say that light from a rose is red, in a strict sense we mean that it *appears* red. Many organisms, including people with defective color vision, do not see the rose as red at all.

The colors we see depend on the frequency of the light we see. Different frequencies of light are perceived as different colors; the lowest frequency we detect appears to most people as the color red. The highest frequency appears as violet. In between is the infinite number of hues that make up the color spectrum of the rainbow. By convention these hues are grouped into the seven colors red, orange, yellow, green, blue, indigo, and violet. These colors all blended together appear white. The white light from the sun is a blend of all the visible frequencies.

Figure 14.9
Sunlight passing through a prism separates into a color spectrum. The colors of things depend on the colors of the light that illuminate them.

Selective Reflection

Except for sources of light such as lamps, most things around us reflect rather than emit light. They reflect only part of the light incident upon them. That's the part that gives them their color. A rose, for example, doesn't emit light; it reflects light. If we pass sunlight through a prism and then place a deep-red rose in various parts of the spectrum, the petals appear brown or black in all parts of the spectrum except in the red. In the red part of the spectrum, the petal appears red. But the green stem and leaves look black. This shows that the red petals have the ability to reflect red light but not other colors of light. Likewise, the green leaves have the ability to reflect green light but not other colors. When the rose is held in white light, the petals appear red and the leaves green because the petals reflect the red part of the white light and the leaves reflect the green part. To understand why objects reflect only particular colors of light, we turn again to the atom.

Light reflects from things similar to the way sound "reflects" from a tuning fork when another one sets it into vibration. One tuning fork can cause another to vibrate even when the frequencies are not matched, although at much less amplitude. The same is true of atoms and molecules. Electrons can be forced into vibration by the vibrating electric fields of electromagnetic waves. Once vibrating, these electrons send out their own electromagnetic waves just as vibrating tuning forks send out sound waves.

Usually a material absorbs light of some frequencies and reflects the rest. The color reflected is the color we see. An object that reflects light of all the visible frequencies, for

Figure 14.10
The square on the left *reflects* all the colors illuminating it. In sunlight it is white. When illuminated with blue light, it is blue. The square on the right *absorbs* all the colors illuminating it. In sunlight it is warmer than the white square.

example, the white part of this page, is the same color as the light that shines on it. If a material absorbs all the light that shines on it, it reflects none and is black.

Figure 14.11
The rabbit's dark fur absorbs all the radiant energy in incident sunlight and therefore is black. Light fur on other parts of the body reflects light of all frequencies and therefore is white.

Interestingly, the petals of most yellow flowers, like daffodils, reflect red and green as well as yellow. Yellow daffodils reflect a broad band of frequencies. The reflected colors of most objects are not pure single-frequency colors, but are a mixture of frequencies.

An object can reflect only those frequencies present in the illuminating light. The appearance of a colored object therefore depends on the kind of light that illuminates it. An incandescent lamp, for instance, emits light of lower average frequencies than sunlight, enhancing any reds viewed in this light. In a fabric having only a little bit of red in it, the red is more apparent under an incandescent lamp than under a fluorescent lamp. Fluorescent lamps are richer in the higher frequencies, and so blues are enhanced under them. For this reason, it is difficult to tell the true color of objects viewed in artificial light. What a color looks like depends on the light source (Figure 14.12).

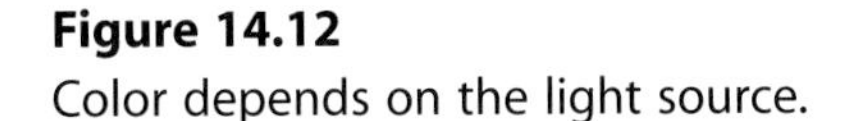

Figure 14.12
Color depends on the light source.

Selective Transmission

The color of a transparent object depends on the color of the light it transmits. A red piece of glass appears red because it absorbs all the colors of white light, except red. So red is transmitted. Similarly, a blue piece of glass appears blue because it transmits primarily blue and absorbs the other colors that illuminate it. These pieces of glass contain dyes or *pigments*—fine particles that selectively absorb light of particular frequencies and selectively transmit others. From an atomic point of view, electrons in the pigment molecules are set into vibration by the illuminating light. Light of some of the frequencies is absorbed by the pigments. The rest is re-emitted from atom to atom in the glass. The energy of the absorbed light increases the kinetic energy of the atoms and the glass is warmed. Ordinary window glass doesn't have a color because it transmits light of all visible frequencies equally well.

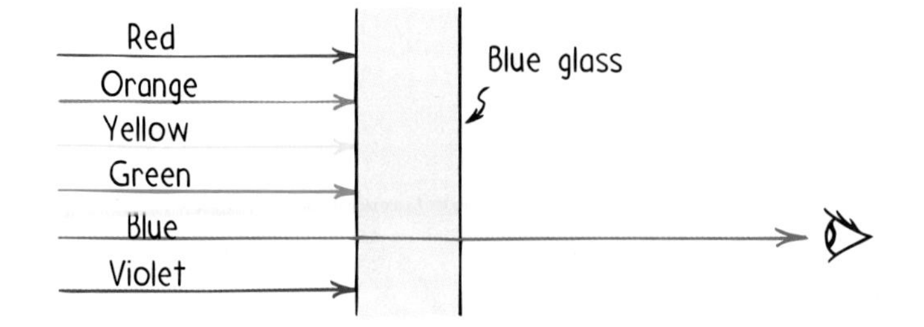

Figure 14.13
Only energy having the frequency of blue light is transmitted; energy of the other frequencies is absorbed and warms the glass.

Concept Check ✓

1. When red light shines on a red rose, why do the leaves become warmer than the petals?
2. When green light shines on a red rose, why do the petals look black?
3. If you hold any small source of white light between you and a piece of red glass, you'll see two reflections from the glass: one from the front surface and one from the back surface. What color is each reflection?

Check Your Answers

1. The leaves don't reflect red light, but absorb it. So the leaves become warmer.
2. The petals don't reflect the green light, but absorb it. Because green is the only color illuminating the rose and because green contains no red to be reflected, the rose reflects no color and appears black.
3. The reflection from the front surface is white because the light doesn't reach far enough into the colored glass to allow absorption of non-red light. Only red light reaches the back surface because the pigments in the glass absorb all the other colors, and so reflection from the back is red.

14.4 Mixing Colored Lights

All the colors added together produce white. All the colors subtracted produce black.

You can see that white light from the sun is composed of all the visible frequencies when you pass sunlight through a prism. The white light is dispersed into a rainbow-colored spectrum. The distribution of solar frequencies (Figure 14.14) is uneven, and is most intense in the yellow-green part of the spectrum. How fascinating that our eyes have evolved to have maximum sensitivity in this range. That's why fire engines are painted yellow-green, particularly at airports, where visibility is vital. Our sensitivity to yellow-green light is also why we see better under the illumination of yellow sodium-vapor lamps at night than under incandescent lamps of the same brightness.

All the colors combined make white. Interestingly, we see white also from the combination of only red, green, and blue light. We can understand this by dividing the solar radiation curve into three

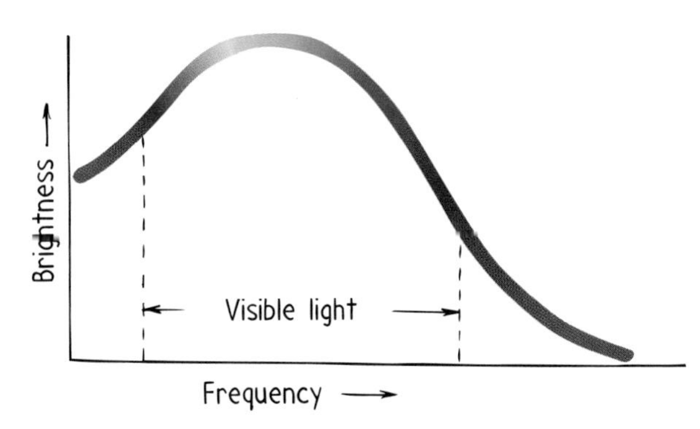

Figure 14.14
The radiation curve of sunlight is a graph of brightness versus frequency. Sunlight is brightest in the yellow-green region, in the middle of the visible range.

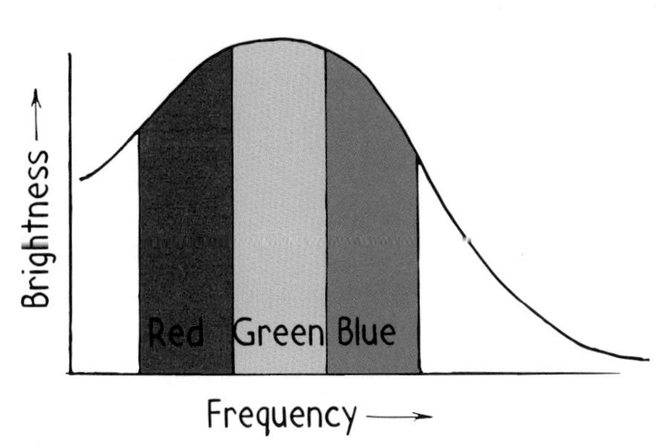

Figure 14.15
Radiation curve of sunlight divided into three regions—red, green, and blue. These are the additive primary colors.

regions, as in Figure 14.15. Three types of cone-shaped receptors in our eyes perceive color. Each is stimulated only by certain frequencies of light. Light of lower visible frequencies stimulates the cones sensitive to low frequencies and appears red. Light of middle frequencies stimulates the mid-frequency-sensitive cones and appears green. Light of higher frequencies stimulates the higher-frequency-sensitive cones and appears blue. When all three types of cones are stimulated equally, we see white.

Additive Primary Colors—Red, Green, and Blue

When red, green, and blue lights are projected on a screen, they overlap to produce white. In the language of scientists, colored lights that overlap *add* to each other. So we say that red, green, and blue light *add to produce white light.* Any two of these colors of light add to produce another color (Figure 14.16). Various amounts of red, green, and blue, produce any color in the spectrum. For this reason, red, green, and blue are called the **additive primary colors.** A close examination of the picture on a TV monitor reveals that the picture is a mixture of tiny spots, each less than a millimeter across. When the screen is lit, some of the spots are red, some green, and some blue. The mixtures of these primary colors at a distance provide a complete range of colors, plus white.*

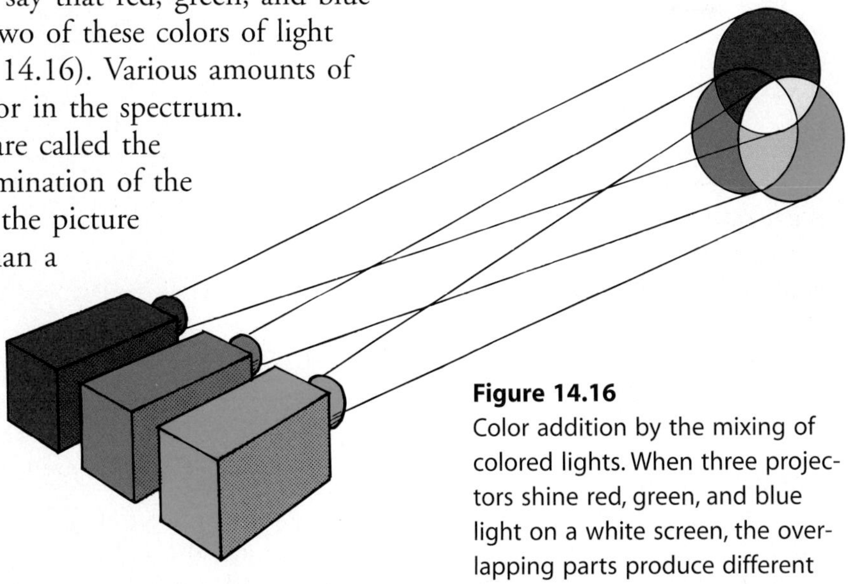

Figure 14.16
Color addition by the mixing of colored lights. When three projectors shine red, green, and blue light on a white screen, the overlapping parts produce different colors. White is produced where all three overlap.

* It's interesting to note that the "black" you see on the darkest scenes on a TV monitor is simply the color of the tube face itself, which is more a light gray than black. Because our eyes are sensitive to the contrast with the illuminated parts of the screen, we see this gray as black.

Complementary Colors

Here's what happens when two of the three additive primary colors are combined:

Red + Blue = Magenta
Red + Green = Yellow
Blue + Green = Cyan

We say that magenta is the opposite of green; cyan is the opposite of red; and yellow is the opposite of blue. When you add any color and its opposite color, you get white.

Magenta + Green = White (= Red + Blue + Green)
Yellow + Blue = White (= Red + Green + Blue)
Cyan + Red = White (= Blue + Green + Red)

When two colors are added together to produce white, they are called **complementary colors.** Every hue has some complementary color that when added to it makes white.

The fact that a color and its complement combine to produce white light is nicely used in lighting stage performances. Blue and yellow lights shining on performers, for example, produce the effect of white light—except where one of the two colors is absent, as in the shadows. The shadow of one lamp, say the blue, is illuminated by the yellow lamp and appear yellow. Similarly, the shadow cast by the yellow lamp appears blue. This is a most interesting effect.

We see this effect in Figure 14.17, where red, green, and blue light shine on the golf ball. Note the shadows cast by the ball. The middle shadow is cast by the green spotlight and is not dark because it is illuminated by the red and blue lights, which make magenta. The shadow cast by the blue light appears yellow because it is illuminated by red and green light. Can you see why the shadow cast by the red light appears cyan?

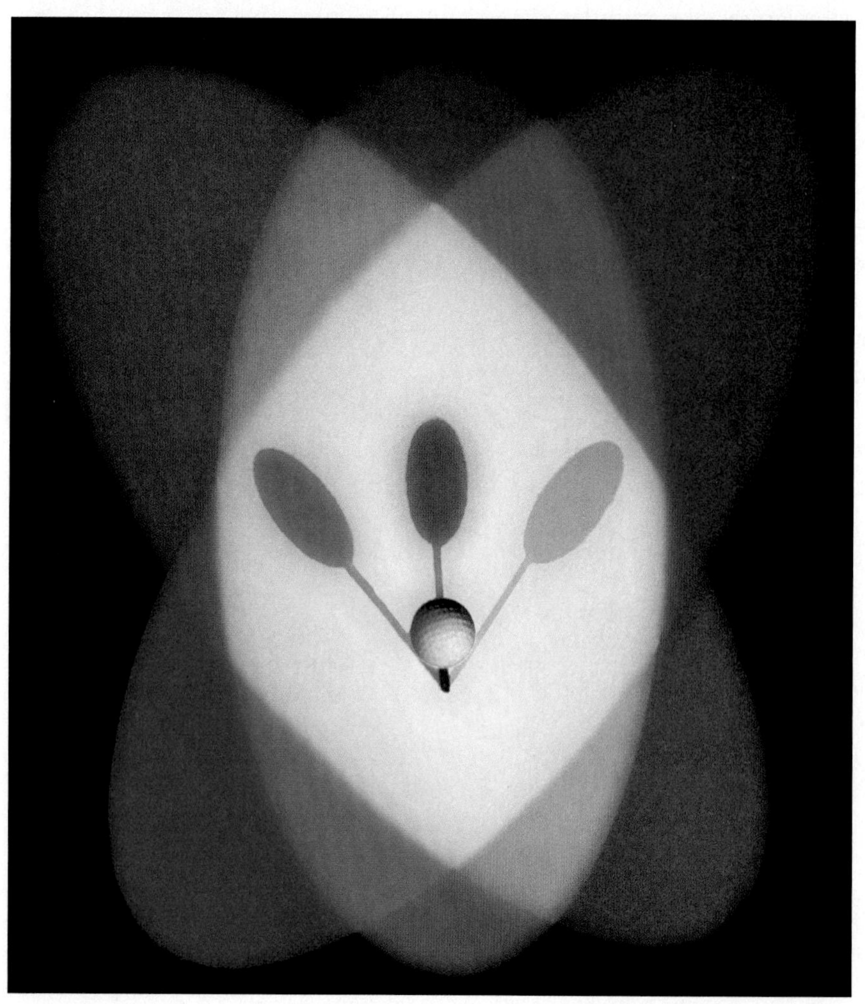

Figure 14.17
The white golf ball appears white when illuminated with red, green, and blue lights of equal intensities. Why are the shadows of the ball cyan, magenta, and yellow?

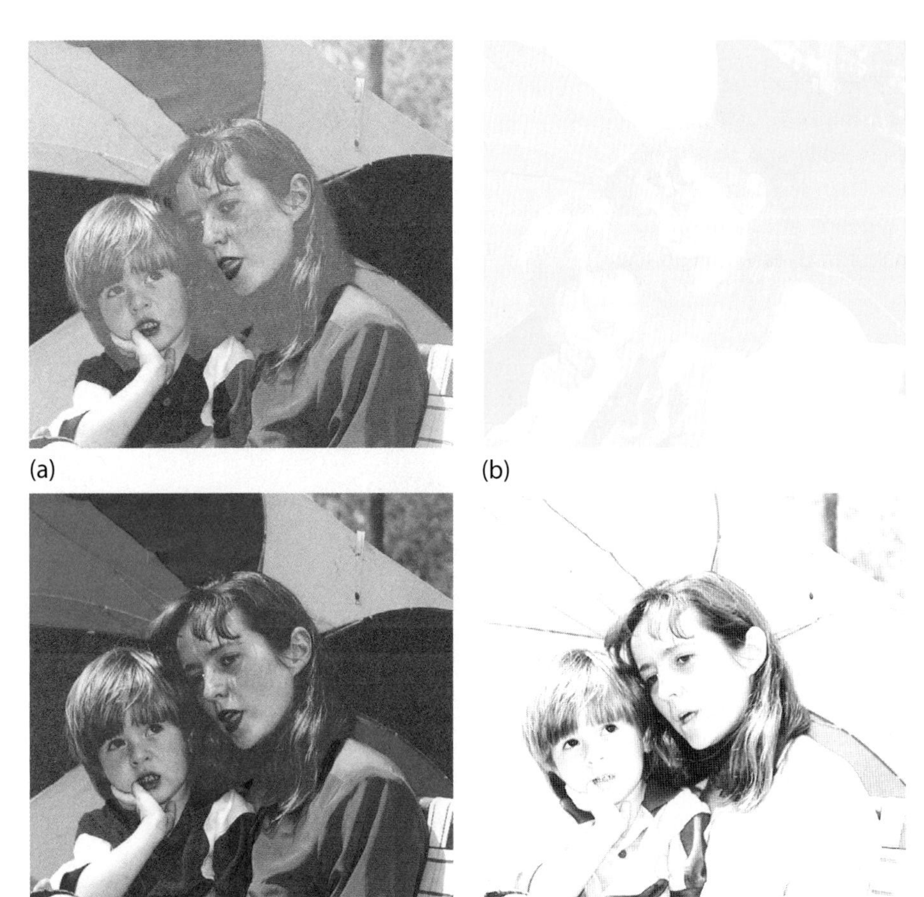

(a) (b) (d) (e)

(c)

(f) **Figure 14.18**
Only four colors of ink are used to print color illustrations and photographs—(a) magenta, (b) yellow, (c) cyan, and black. When magenta, yellow, and cyan are combined, they produce (d). Addition of black (e) produces the finished result, (f).

Concept Check ✓

1. From Figure 14.17, find the complements of cyan, yellow, and red.
2. Red + blue = ______.
3. White − red = ______.
4. White − blue = ______.

Check Your Answers

1. Red, blue, cyan. **2.** Magenta. **3.** Cyan. **4.** Yellow.

14.5 Mixing Colored Pigments

Every artist knows that if you mix red, green, and blue paint, the result is not white but rather a muddy dark brown. Red and green paint certainly do not mix to form yellow, as is the rule for combining colored lights.

Mixing pigments in paints and dyes is entirely different from mixing lights. Pigments are tiny particles that absorb specific colors. For example, pigments that produce the color red absorb the complementary color cyan. So something painted red absorbs cyan, which is

Figure 14.19
Look through a magnifying glass and you will see that the color green on a printed page is made up of blue and yellow dots.

Figure 14.20
Dyes or pigments, as in the three transparencies shown, absorb (subtract) light of some frequencies and transmit only part of the spectrum. When white light passes through overlapping sheets of these colors, all light is blocked (subtracted). Then we have black. Where only yellow and cyan overlap, only green is not subtracted. Various proportions of yellow, cyan, and magenta dyes will produce nearly any color in the spectrum.

Figure 14.21
The rich colors of Sneezlee represent many frequencies of light. The photo, however, is a mixture of only yellow, magenta, cyan, and black.

why it reflects red. In effect, cyan has been *subtracted* from white light. Something painted blue absorbs yellow, and so reflects all the colors except yellow. Take yellow away from white and you've got blue. The colors magenta, cyan, and yellow are the **subtractive primaries.**

The variety of colors you see in the colored photographs in this or any other book are the result of magenta, cyan, and yellow dots. Light illuminates the book, and light of some frequencies is subtracted from the light reflected. The rules of color subtraction differ from the rules of light addition. We leave this topic to the suggested reading.

14.6 Why the Sky Is Blue

Not all colors are the result of the addition or subtraction of light. Some colors, like the blue of the sky, are the result of selective scattering. Consider the analogous case of sound: If a beam of a particular frequency of sound is directed to a tuning fork of similar frequency, the tuning fork is set into vibration and redirects the beam in multiple directions. The tuning fork *scatters* the sound. A similar process occurs with the scattering of light from atoms and particles that are far apart from one another. This is what happens in the atmosphere.

For me, knowing why the sky is blue and why sunsets are red adds to their beauty—knowledge doesn't subtract.

We know that atoms behave like tiny optical tuning forks and re-emit light waves that shine on them. Very tiny particles act the same. The tinier the particle, the higher the frequency of light it will re-emit. This is similar to the way small bells ring with higher notes than larger bells. The nitrogen and oxygen molecules that make up most of the atmosphere are like tiny bells that "ring" with high frequencies when energized by sunlight. Like sound from the bells, the re-emitted light is sent in all directions. When light is re-emitted in all directions, we say the light is *scattered*.

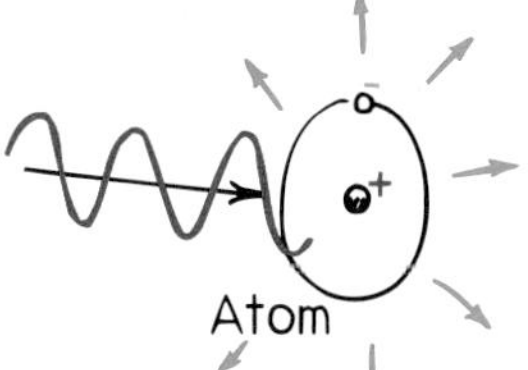

Figure 14.22
A beam of light falls on an atom and increases the vibrational motion of electrons in the atom. The vibrating electrons re-emit the light in various directions. Light is scattered.

Of the visible frequencies of sunlight, violet is scattered the most by nitrogen and oxygen in the atmosphere. Then the other colors are scattered in order; blue, green, yellow, orange, and red. Red is scattered only a tenth as much as violet. Although violet light is scattered more than blue, our eyes are not very sensitive to violet light. Therefore the blue scattered light is what predominates in our vision, so we see a blue sky!

The blue of the sky varies in different places under various conditions. A main factor is the amount of water-vapor content in the atmosphere. On clear dry days the sky is a much deeper blue than on clear humid days. Places where the upper air is exceptionally dry, such as Italy and Greece, have beautiful blue skies that have inspired painters for centuries. Where the atmosphere contains a lot of particles of dust and other particles larger than oxygen and nitrogen molecules, light of the lower frequencies is also scattered strongly. This causes the sky to appear less blue, with a whitish appearance. After a heavy rainstorm when the airborne particles have been washed away, the sky becomes a deeper blue.

The grayish haze in the skies over large cities is the result of particles emitted by car and truck engines and by factories. Even when idling, a typical automobile engine emits more than 100 billion particles per second. Most are invisible but act as tiny centers to which other particles adhere. These are the primary scatterers of lower frequency light. With the largest of these particles, absorption rather than scattering takes place, and a brownish haze is produced. Yuck!

Figure 14.23
In clean air the scattering of high-frequency light gives us a blue sky. When the air is full of particles larger than molecules, lower-frequency light is also scattered, which adds to give a whitish sky.

14.7 Why Sunsets Are Red

Light that isn't scattered is light that is transmitted. Because red, orange, and yellow light are the least scattered by the atmosphere, light of these low frequencies is better transmitted through the air. Red is scattered the least and passes through more atmosphere than any other color. So the thicker the atmosphere through which a beam of sunlight travels, the more time there is to scatter all the higher-frequency parts of the light. This means red light travels through it best. As Figure 14.24 shows, sunlight travels through more atmosphere at sunset, which is why sunsets are red.

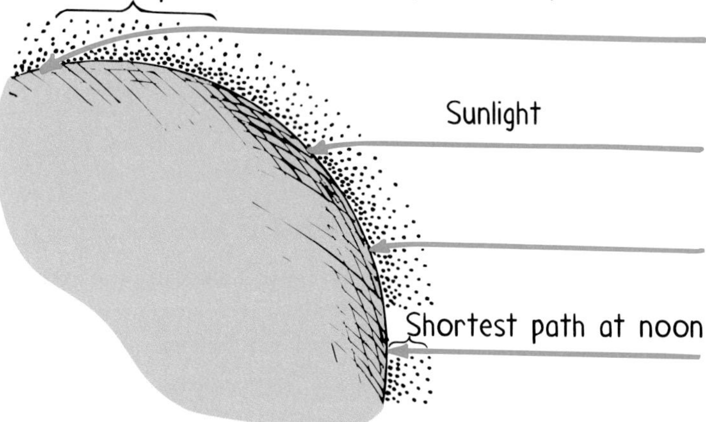

Figure 14.24
A sunbeam must travel through more atmosphere at sunset than at noon. As a result, more blue is scattered from the beam at sunset than at noon. By the time a beam of initially white light gets to the ground, only light of the lower frequencies survives to produce a red sunset.

At noon, sunlight travels through the least amount of atmosphere to reach the Earth's surface. Only a small amount of high-frequency light is scattered from the sunlight, enough to make the sun look yellowish. As the day progresses and the sun descends lower in the sky (Figure 14.24), the path through the atmosphere is longer, and more violet and blue are scattered from the sunlight. The removal of violet and blue leaves the transmitted light redder. The sun becomes progressively redder, going from yellow to orange and finally to a red-orange at sunset. Sunsets and sunrises are unusually colorful following volcanic eruptions, because particles larger than atmospheric molecules are then more abundant in the air.

The colors of the sunset are consistent with our rules for color mixing. When blue is subtracted from white light, the complementary color that remains is yellow. When higher-frequency violet is subtracted, the resulting complementary color is orange. When medium-frequency green is subtracted, magenta is left. The combinations of resulting colors vary with atmospheric conditions, which change daily giving us a variety of sunsets.

Concept Check ✓

If molecules in the sky scattered low-frequency light more than high-frequency light, what color would the sky be? What color would sunsets be?

Check Your Answers If low-frequency light were scattered, the noontime sky would appear reddish-orange. At sunset more reds would be scattered by the longer path of the sunlight, and the sunlight would be predominantly blue and violet. So sunsets would appear blue!

Shine-On Exploration

You can simulate a sunset with a fish tank full of water by adding a few drops of milk. A few drops will do. Then shine a flashlight beam through the water and you'll see that it looks bluish from the side. Milk particles are scattering the higher frequencies of light in the beam. Light emerging from the far end of the tank will have a reddish tinge. That's the light that wasn't scattered.

14.8 Why Clouds Are White

Clouds are made up of clusters of water droplets in a variety of sizes. The different-size clusters result in a variety of scattered colors. The tiniest clusters tend to make blue clouds; slightly larger clusters, green; and still larger clusters, red. The overall result is a white cloud. Electrons close to one another in a cluster vibrate in phase. This results in a greater intensity of scattered light than there would be from the same number of electrons vibrating separately. Hence, clouds are bright!

Larger clusters of droplets absorb much of the light incident upon them, and so the scattered intensity is less. Therefore clouds composed of larger clusters are darker. Further increase in the size of the clusters cause them to fall as raindrops, and we have rain.

The next time you find yourself admiring a crisp blue sky or delighting in the shapes of bright clouds or watching a beautiful sunset, think about all those ultra-tiny optical tuning forks vibrating away. You'll appreciate these everyday wonders of nature even more!

Figure 14.25
A cloud is composed of various sizes of water-droplet clusters. The tiniest clusters scatter blue light, slightly larger ones scatter green light, and still larger ones scatter red light. The result is a white cloud.

Figure 14.26
Water is cyan because it absorbs red light. The froth in the waves is white because, like clouds, it is composed of a variety of tiny clusters of water droplets that scatter all the visible frequencies.

Chapter Review

Key Terms and Matching Definitions

_____ additive primary colors
_____ complementary colors
_____ electromagnetic wave
_____ electromagnetic spectrum
_____ opaque
_____ subtractive primary colors
_____ transparent

1. A wave emitted by vibrating electrical charges (often electrons) and composed of vibrating electric and magnetic fields that regenerate one another.
2. The range of electromagnetic waves extending in frequency from radio waves to gamma rays.
3. The term applied to materials through which light can pass in straight lines.
4. The term applied to materials through which light cannot pass.
5. The three colors—red, blue, and green—that when added in certain proportions produce any other color in the visible-light part of the electromagnetic spectrum.
6. Any two colors that when added produce white light.
7. The three colors of absorbing pigments—magenta, yellow, and cyan—that when mixed in certain proportions reflect any other color in the visible-light part of the electromagnetic spectrum.

Review Questions

The Electromagnetic Spectrum—A Tiny Bit of Which Is Light

1. Does visible light make up a relatively large part or a relatively small part of the electromagnetic spectrum?
2. What is the principal difference between a radio wave and light?
3. How do the speeds of various electromagnetic waves compare?
4. What color do we perceive for the lowest visible frequencies? The highest?
5. How does the frequency of a radio wave compare with the frequency of the vibrating electrons that produces it?

Why Materials Are Either Transparent or Opaque

6. In what region of the electromagnetic spectrum is the resonant frequency of electrons in glass?
7. What is the fate of the energy in ultraviolet light incident on glass?
8. How does the average speed of light in glass compare with its speed in a vacuum?

Color Science

9. Which has the higher frequency, red light or blue light?
10. Distinguish between the white of this page and the black of this ink in terms of what happens to the white light that falls on both.
11. What is the evidence for the statement that white light is a composite of all the colors of the visible part of the electromagnetic spectrum?

Mixing Colored Lights

12. What is the color of the peak frequency of solar radiation?
13. What color light are our eyes most sensitive to?
14. What frequency ranges of the radiation curve do red, green, and blue light occupy?
15. Why are red, green, and blue called the *additive primary colors*?

Mixing Colored Pigments

16. What are the subtractive primary colors? Why are they so called?

17. Why are red and cyan called *complementary colors*?

Why the Sky Is Blue

18. Why does the sky sometimes appear whitish?

Why Sunsets Are Red

19. Why does the sun look reddish at sunrise and sunset but not at noon?

Why Clouds Are White

20. What is the evidence for a variety of particle sizes in a cloud?

Explorations

1. Stare at a piece of colored paper for 45 seconds or so. Then look at a white surface. Because the cones in your retina receptive to the color of the paper have become fatigued, you see an after image of the complementary color when you look at the white area. This is because the fatigued cones send a weaker signal to the brain. All the colors produce white, but all the colors minus one produce the color complementary to the missing one.
2. Stare intently for a minute or so at an American flag. Then turn your view to a white area. What colors do you see in the image of the flag that appears on the wall?

Exercises

1. Which waves have the longest wavelengths: light waves, X-rays, or radio waves? Which have the highest frequencies?
2. What evidence can you cite to support the idea that light can travel in a vacuum?
3. Is glass transparent or opaque to frequencies of light that match its own natural frequencies? Explain.
4. What determines whether a material is transparent or opaque?
5. You can get a sunburn on a cloudy day, but you can't get a sunburn even on a sunny day if you are behind glass. Explain.
6. Suppose that sunlight falls on both a pair of reading glasses and a pair of dark sunglasses. Which pair of glasses would you expect to be warmer in sunlight? Defend your answer.
7. In a clothing shop lit only by fluorescent lighting, a customer during daytime insists on taking garments by the doorway to check their color. Is the customer being reasonable? Explain.
8. What is the common color of tennis balls, and why?
9. What color does red cloth appear when illuminated by sunlight? By red light from a neon sign? By cyan light?
10. Why does a white piece of paper appear white in white light, red in red light, blue in blue light, and so on for every color?
11. A spotlight that has a white-hot filament is coated so that it won't transmit yellow light. What color is the emerging beam of light?
12. Below is a photo of science editor Suzanne with son Tristan wearing red and daughter Simone wearing green. Below that is the negative photo, which shows these colors differently. What is your explanation?

13. Complete the following equations:

Yellow light + blue light = ______ light.

Green light + ______ light = white light.

Magenta light + yellow light + cyan light = ______ light.

14. Check to see if the following three statements are accurate. Then fill in the last statement. (All colors are combined by addition of light.)

Red + green + blue = white.

Red + green = yellow = white − blue.

Red + blue = magenta = white − green.

Green + blue = cyan = white − ______.

15. If the sky were composed of atoms that predominantly scattered orange light rather than blue, what color would sunsets be?

16. Does light travel faster through the lower atmosphere or the upper atmosphere? Defend your answer.

17. Comment on the statement, "Oh, that beautiful red sunset is just the leftover colors that weren't scattered on their way through the atmosphere."

18. Volcanic emissions spew fine ashes in the air that scatter red light. What color does a full moon appear through these ashes?

19. If the atmosphere of the Earth were several times thicker, would ordinary snowfall still seem white or would it be some other color? What color?

20. Tiny particles, like tiny bells, scatter high-frequency waves more than low-frequency waves. Large particles, like large bells, mostly scatter low frequencies. Intermediate-size particles and bells mostly scatter intermediate frequencies. What does this have to do with the whiteness of clouds?

Problems

1. The sun is 1.50×10^{11} meters from the Earth. How long does it take for the sun's light to reach the Earth?

2. In about 1675, the Danish astronomer Olaus Roemer found that light from eclipses of Jupiter's moon took an extra 1000 s to travel 300,000,000 km across the diameter of the Earth's orbit around the sun. Show how this finding provided the first reasonably accurate measurement for the speed of light.

3. In 1969, people had the opportunity to sense the speed of light when on TV they heard a NASA controller in Houston speak to an astronaut on the moon. A longer than normal delay occurred in the back and forth conversation. How long did it take the signals to get to the moon and back?

4. The nearest star beyond the sun is Alpha Centauri, 4.2×10^{16} meters away from the Earth. If we receive a radio message from this star today, how long ago was it sent?

Suggested Reading

Murphy, Pat, and Paul Doherty. *The Color of Nature.* San Francisco: Chronicle Books, 1996.

Chapter 15: Reflection and Refraction

When you view yourself in a mirror, why do your left and right hands appear to reverse places? And why is your image as far behind the mirror as you are in front? Why does light from the cat bend when it goes from air into water? We'll answer these questions and more in this interesting chapter.

When light shines on the surface of a material, the light is either reflected, transmitted, or absorbed. We say light is *reflected* when it is returned into the medium from which it came—the process is **reflection.** When light crosses from one transparent material into another, we say it is *refracted*—the process is **refraction.** Usually some degree of both reflection and refraction occurs when light interacts with matter. Absorption also occurs, but we'll ignore that for now and concentrate on the light that continues to be light after contact with a surface.

15.1 Reflection of Light

When sunlight or a lamp illuminates this page, electrons in the atoms of the paper and ink vibrate more energetically in response to the vibrating electric fields of the light. The energized electrons re-emit the light, which enables you to see the page. When illuminated by white light, the paper appears white, which tells you that the electrons re-emit all the visible frequencies. Very little absorption occurs. The ink is a different story. Except for a bit of reflection, it absorbs all the visible frequencies and therefore appears black.

Law of Reflection

Anyone who has played pool or billiards knows that when a ball bounces from a surface, the angle of rebound is equal to the angle of incidence. Likewise for light. This is the **law of reflection,** and applies to all angles:

The angle of reflection equals the angle of incidence.

The law of reflection is illustrated with arrows that represent light rays in Figure 15.1. Instead of measuring angles from the reflecting surface, it is customary to measure the angles of each ray from a line perpendicular to the reflecting surface. This imaginary line is called the *normal.* The incident ray, the normal, and the reflected ray all lie in the same plane.

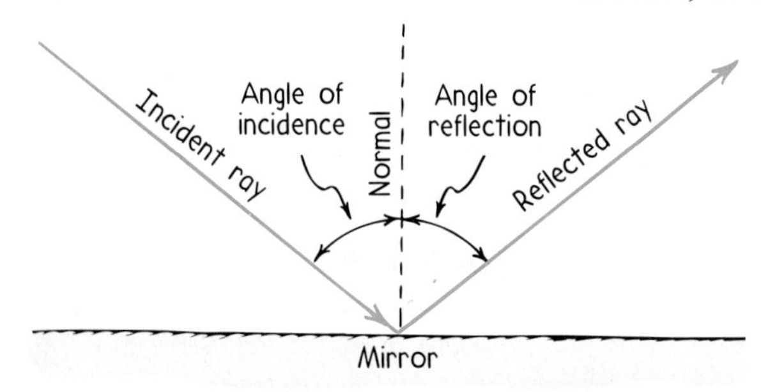

Figure 15.1
The law of reflection.

Look at a candle flame placed in front of a mirror. Rays of light radiate from the flame in all directions. Figure 15.2 shows only four of the infinite number of rays leaving one point on the flame. When these rays meet the mirror, they reflect at angles equal to their angles of incidence. The rays diverge (spread out) from the flame. Notice that they diverge also from the mirror. They appear to emanate from behind the mirror (dashed lines). You would see an image of the candle flame at this point. The image is as far behind the mirror as the flame is in front of the mirror. Both the flame and the image of the flame have the same size.

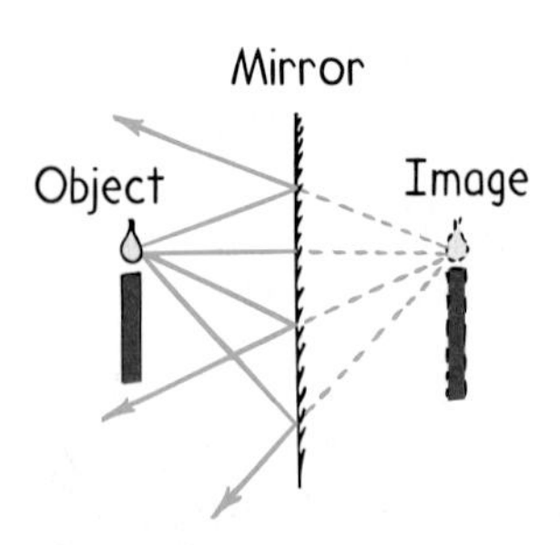

Figure 15.2
A virtual image is formed behind the mirror. It is located at the position where the extended reflected rays (dashed lines) converge.

When you view yourself in a mirror, for example, the size of your image is the same as the size of your twin if located as far behind the mirror as you are in front—as long as the mirror is flat. A flat mirror is called a *plane mirror.*

When the mirror is curved, the sizes and distances of object and image are no longer equal. We shall not study curved mirrors in this book, except to say that the law of reflection still holds. A curved mirror behaves as a succession of flat mirrors, each tilted slightly different from the one next to it. At each point, the angle of incidence is equal to the angle of reflection (Figure 15.4). Note that in a curved mirror, the normals (dashed lines between the solid rays) at different points on the surface are not parallel to one another.

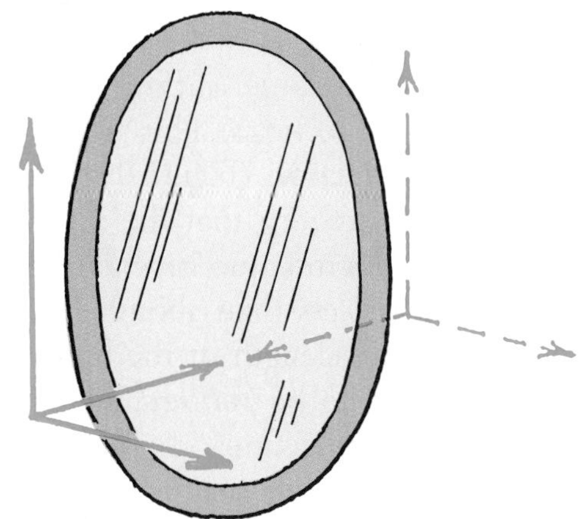

Figure 15.3
Marjorie's image is as far behind the mirror as she is in front. Note that she and her image have the same color of clothing—evidence that light doesn't change frequency upon reflection. Interestingly, her left-right axis is no more reversed than her up-down axis. The axis that *is* reversed, as shown to the right, is front-back. That's why it appears that her left hand faces the right hand of her image.

Whether a mirror is flat or curved, the human eye-brain system cannot ordinarily tell the difference between an object and its reflected image. So we see the illusion that an object exists behind a mirror (or in some cases in front of a concave mirror). Light reaches our eye in the same manner whether it comes from an object or a reflection of the object.

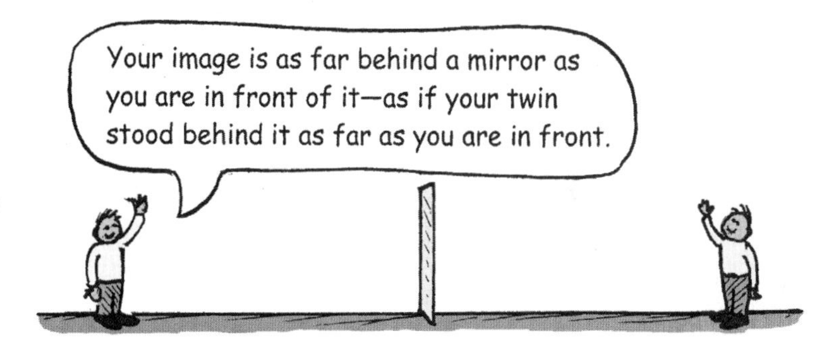

Concept Check ✓

1. What evidence can you cite to support the claim that the frequency of light does not change upon reflection?
2. If you wish to take a picture of your image while standing 2 m in front of a plane mirror, what distance should you set your camera to provide the sharpest focus?

Check Your Answers

1. The color of an image is identical to the color of the object forming the image. Look at yourself in a mirror and the color of your eyes doesn't change. The fact that the color is the same is evidence that the frequency of light doesn't change upon reflection.
2. Set your camera for 4 m; the situation is equivalent to standing 2 m in front of an open window and viewing your twin standing 2 m in back of the window.

Only part of the light that strikes a surface is reflected. On a pane of clear glass, for example, light perpendicular to the surface reflects only about 4 percent from each surface of the pane. On a polished aluminum or silver surface, however, about 90 percent of incident light is reflected.

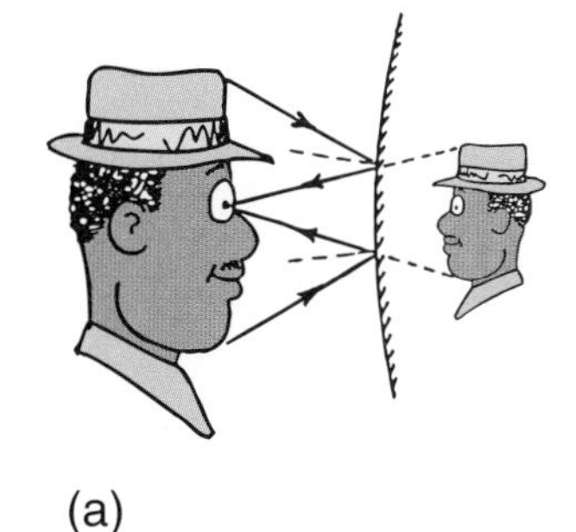

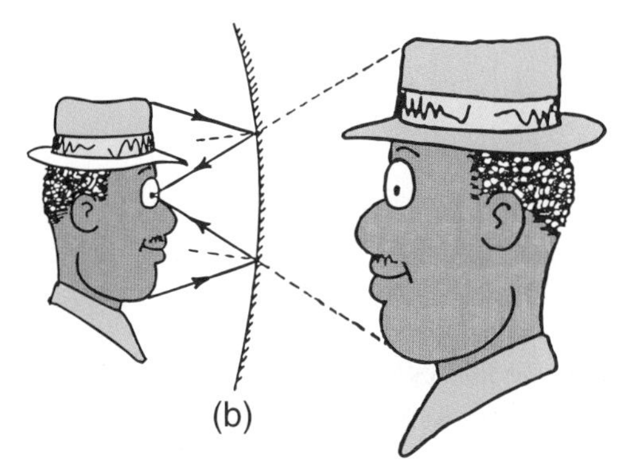

Figure 15.4
(a) The virtual image formed by a *convex* mirror (a mirror that curves outward) is smaller and closer to the mirror than the object. (b) When the object is near a *concave* mirror (a mirror that curves inward like a "cave"), the virtual image is larger and farther away than the object. In either case the law of reflection holds for each ray.

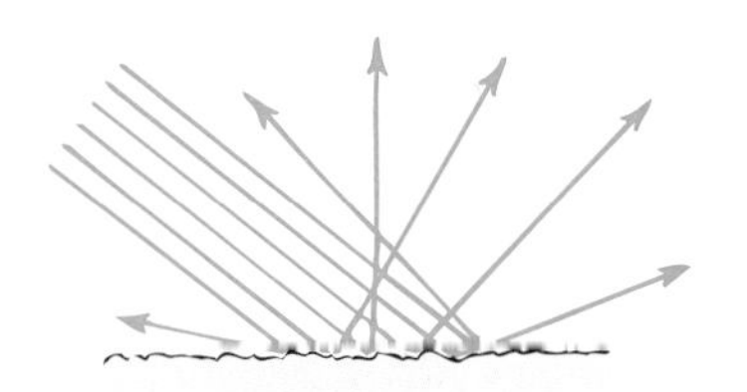

Figure 15.5
Diffuse reflection. Although each ray obeys the law of reflection, the many different surface angles that light rays encounter in striking a rough surface cause reflection in many directions.

Diffuse Reflection

Wanna get rich? Be the first to invent a surface that will reflect 100% of the light incident upon it.

Light incident on a rough surface reflects in many directions. This is called **diffuse reflection** (Figure 15.5). If a surface is so smooth that the distances between microscopic irregularities on the surface are less than about one-eighth the wavelength of the light, the surface is said to be *polished.* There is very little diffuse reflection on a polished surface. Interestingly, an irregular surface for one range of wavelengths may be polished for other ranges. The wire-mesh "dish" shown in Figure 15.6 is very rough for light waves. We certainly don't see it as polished. But for long-wavelength radio waves it is quite polished and is an excellent reflector.

Light reflecting from this page is diffuse. The page may be smooth to a radio wave, but to a light wave it is rough, as Figure 15.7 clearly indicates. Rays of light striking this page encounter millions of tiny flat surfaces facing in all directions. The incident light therefore reflects in all directions. This is good because it lets us see the page without glare. We don't have to hold our head just right to read the page. Similarly for other objects. In an automobile you can see the road ahead at night because of diffuse reflection by the road surface. When the road is wet, however, water provides a more mirrored surface and it is harder to see. Most of our environment is seen by diffuse reflection.

Figure 15.6
The open-mesh parabolic dish is a diffuse reflector for short-wavelength visible light waves but a polished reflector for long-wavelength radio waves.

An undesirable circumstance related to diffuse reflection is the ghost image that occurs on a TV set when the TV signal bounces off buildings and other obstructions. For antenna reception, this difference in path lengths for the direct signal and the reflected signal produces a slight time delay. Multiple reflections may produce multiple ghosts.

Figure 15.7
A magnified view of the surface of ordinary paper.

Concept Check ✓

Why is it more dangerous to drive a car on a rainy night?

Check Your Answer In addition to the less diffuse road surface when wet, as described above, there's another reason. Headlights from oncoming cars reflect from the more mirrored surface full force into your eyes. Glare is more intense.

15.2 Refraction—The Bending of Light Due to Changing Speed

Recall from the previous chapter that light slows down when it enters glass and other transparent materials. Its average speed is lower than in empty space. In other words, light travels at different speeds in various materials. It travels at 300,000 km/s in a vacuum, at a slightly lower speed in air, and at about three-fourths that speed in water. In a diamond, light travels at about 40% of its speed in a vacuum. As mentioned at the beginning of this chapter, when light passes from one medium to another, we call the process *refraction.* When light is not perpendicular to the surface of penetration, bending occurs.

To better understand the bending of light in refraction look at the pair of toy cart wheels in Figure 15.8. The wheels roll from a smooth sidewalk onto a grass lawn. If the wheels meet the grass at an angle, as the figure shows, they are deflected from their straight-line course. Note that the left wheel slows first when it interacts with the grass on the lawn. The right wheel maintains its higher speed while on the sidewalk. It pivots about the slower-moving left wheel because it travels farther in the same time. So the direction of the rolling wheels is bent toward the "normal," the black dashed line perpendicular to the grass-sidewalk border in Figure 15.8.

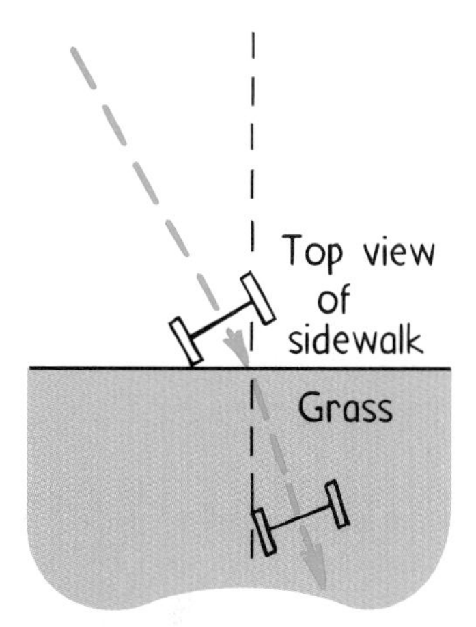

Figure 15.8
The direction of the rolling wheels changes when one slows down before the other does.

Hands-On Explorations: Playing with Mirrors

Stand a pair of pocket-size mirrors on edge with the faces parallel to each other. Place an object such as a coin between the mirrors and look at the reflections in each mirror. Nice?

Now set up the mirrors at right angles and place a coin between them. You'll see four coins. Change the angle of the mirrors and see how many images of the coin you can see.

Look at yourself in the pair of mirrors when they're at right angles to each other. Wink. Notice that you see yourself as others see you. Rotate the mirrors, still at right angles to each other. Does your image rotate also? Now place the mirrors 60° apart so you again see your face. Again rotate the mirrors and see if your image rotates also. Amazing?

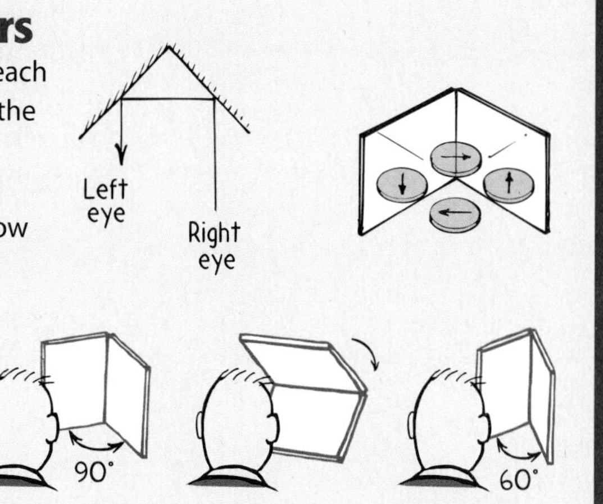

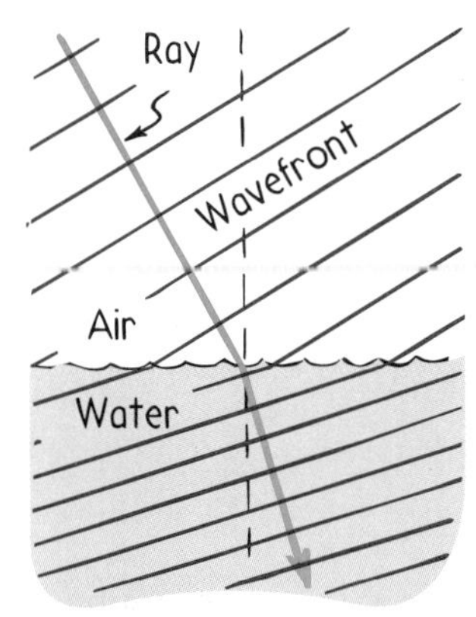

Figure 15.9
When light is refracted the direction of the light waves changes when one part of the wave slows down before the other part.

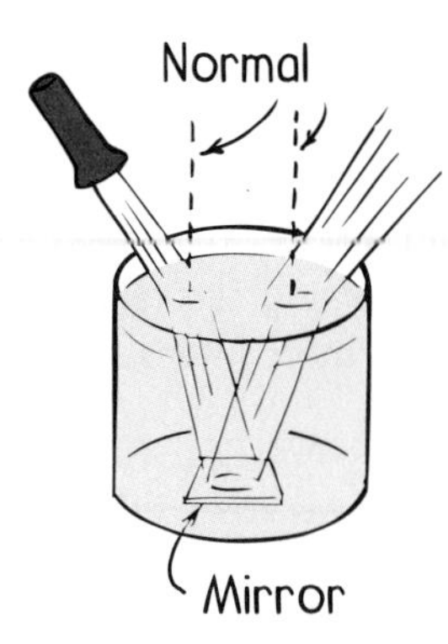

Figure 15.10
Refraction.

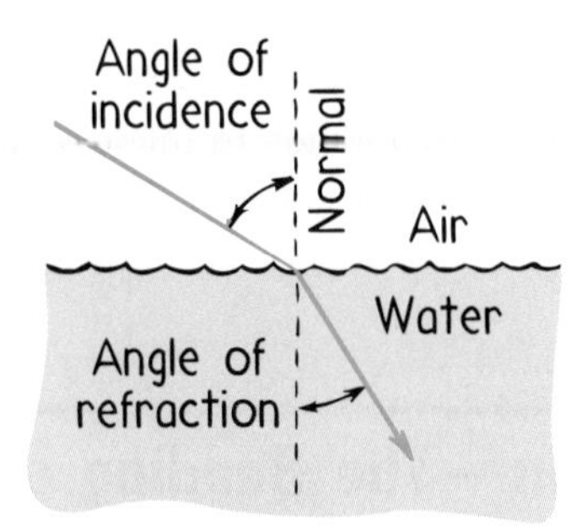

Figure 15.11
When light slows down in going from one medium to another, like going from air to water, it refracts toward the normal. When it speeds up in traveling from one medium to another, like going from water to air, it refracts away from the normal.

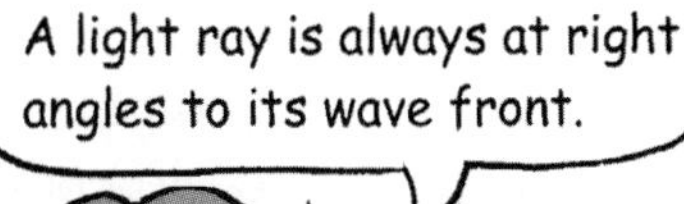

Figure 15.9 shows how a light wave bends in a similar way. Note the direction of light, indicated by the blue arrow (the light ray), and also note the *wave fronts* drawn at right angles to the ray. (If the light source were close, the wave fronts would appear circular; but if the distant sun is the source, the wave fronts are practically straight lines.) The wave fronts are everywhere at right angles to the light rays. In the figure the wave meets the water surface at an angle. This means that the left portion of the wave slows down in the water while the remainder in the air travels at speed *c*. The light ray remains perpendicular to the wave front and therefore bends at the surface. It bends like the wheels bend when they roll from the sidewalk into the grass. In both cases the bending is caused by a change in speed.

Figure 15.11 shows a beam of light entering water at the left and exiting at the right. The path would be the same if the light entered from the right and exited at the left. Light paths are reversible for both reflection and refraction. If you see someone's eyes by means of reflection or refraction, such as with a mirror or a prism, then that person can see you also.

Concept Check ✓

If the speed of light were the same in all media, would refraction still occur when light passes from one medium to another?

Check Your Answer No.

Refraction causes many illusions. One of them is the apparent bending of a stick that is partially in water. The submerged part appears closer to the surface than it actually is. Likewise when you look at a fish in water. The fish appears nearer to the surface and closer than it really is (Figure 15.12). If we look straight down into water, an object submerged 4 meters beneath the surface appears to be only 3 meters deep. Because of refraction, submerged objects appear to be magnified.

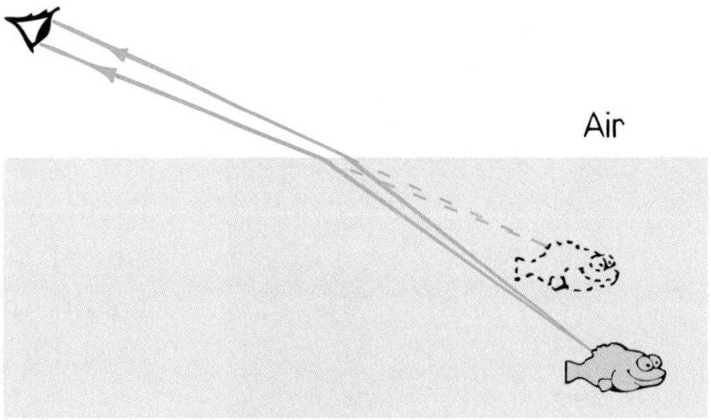

Figure 15.12
Because of refraction, a submerged object appears to be nearer to the surface than it actually is.

15.3 Illusions and Mirages Are Caused by Atmospheric Refraction

Refraction occurs in the Earth's atmosphere. Whenever we watch a sunset, we see the sun for several minutes after it has sunk below the horizon (Figure 15.13). The Earth's atmosphere is thin at the top and dense at the bottom. Because light travels faster in thin air than in dense air, parts of the wavefronts of sunlight higher up travel faster than parts of the wavefronts closer to the ground. Light rays bend. The density of the atmosphere changes gradually, so the light ray bends gradually and follows a curved path. So we gain additional minutes of daylight each day. Furthermore, when the sun (or moon) is near the horizon, the rays from the lower edge are bent more than the rays from the upper edge. This shortens the vertical diameter, causing the sun to appear elliptical (Figure 15.14).

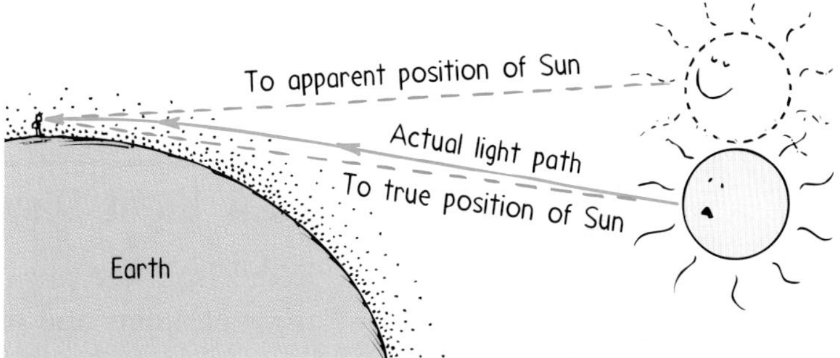

Figure 15.13
The sun's shape is distorted by differential refraction.

Figure 15.14
Because of atmospheric refraction, when the sun is near the horizon it appears higher in the sky.

Mirages are a common sight on a desert. The sky appears to be reflected from water on the distant sand. But when we get there, the sand is dry. Why is this so? The air is very hot just above the surface and cooler above. Light travels faster through the thinner hot air below than through the denser cool air above. So wavefronts near the ground travel faster than above. The result is upward bending (Figure 15.15). So we see an upside-down view as if reflection were occurring from a water surface.

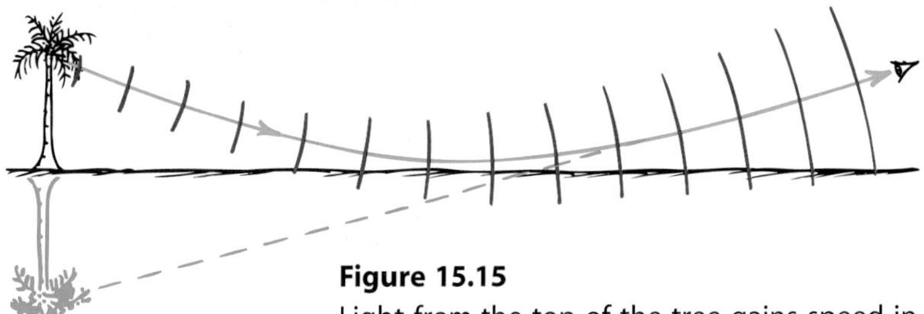

Figure 15.15
Light from the top of the tree gains speed in the warm and less dense air near the ground. When the light grazes the surface and bends upward the observer sees a mirage.

Figure 15.16
A mirage. The apparent wetness of the road is not reflection by water but rather refraction through the warmer and less-dense air near the road surface.

We see a mirage, which is formed by real light and can be photographed (Figure 15.16). A mirage is not, as many people think, a trick of the mind.

When we look at an object over a hot stove or over a hot pavement, we see a wavy, shimmering effect. This is due to varying densities of air caused by changes in temperature. The twinkling of stars results from similar variations in the sky, where light passes through unstable layers in the atmosphere.

Concept Check ✓

If the speed of light were the same in air of various temperatures and densities, would there still be slightly longer daytimes, twinkling stars at night, mirages, and a slightly squashed sun at sunset?

Check Your Answer No.

15.4 Light Dispersion and Rainbows

Recall from the previous chapter that light that resonates with electrons of atoms and molecules in a material are absorbed. Such a material is opaque to light. Also recall that transparency occurs for light of frequencies near the resonant frequencies of the material. The average speed of light is slowed due to the absorption/re-emission sequence. This was shown in Figure 14.6. The grand result is that high-frequency light in a transparent medium travels slower than low-frequency light. Violet light travels about 1 percent slower in ordinary glass than red light. Light of colors between red and violet travel at their own respective speeds in glass.

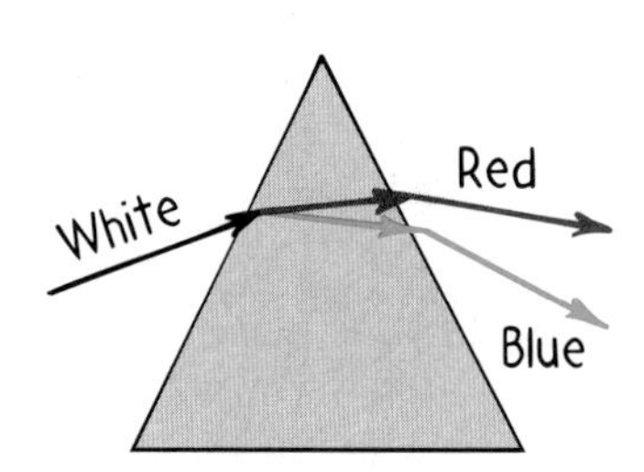

Figure 15.17
Dispersion by a prism makes the components of white light visible.

Because light of various frequencies travel at different speeds in transparent materials, they refract by different amounts. When white light is refracted twice, as in a prism, the separation of colors of light is quite noticeable. This separation of light into colors arranged by frequency is called *dispersion* (Figure 15.17). Because of dispersion, there are rainbows!

To see a rainbow, the sun must shine on water drops in a cloud or in falling rain. The drops act as prisms that disperse light. When you face a rainbow, the sun is behind you, in the opposite part of the sky. Seen from an airplane near midday, the bow forms a complete circle.

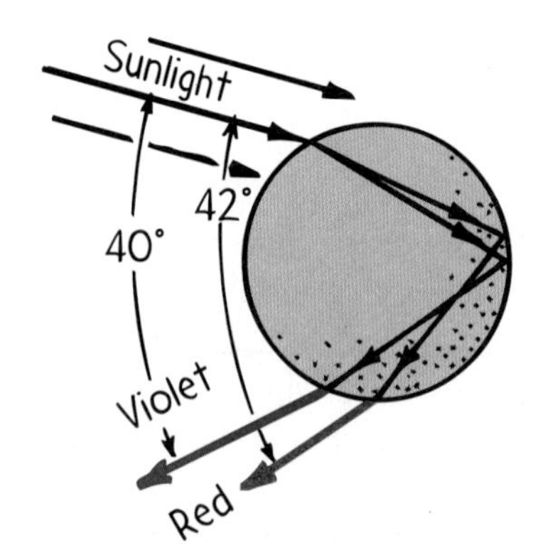

Figure 15.18
Dispersion of sunlight by a single raindrop.

You can see how a raindrop disperses light in Figure 15.18. Follow the ray of sunlight as it enters the drop near its top surface. Some of the light here is reflected (not shown), and the remainder is refracted into the water. At this first refraction, the light is dispersed into its spectrum colors, red being deviated the least and violet the most. When the light reaches the opposite side of the drop, each color is partly refracted out into the air (not shown) and partly reflected back into the water. Arriving at the lower surface of the drop, each color is again partly reflected (not shown) and partly refracted into the air. This refraction at the second surface, like that in a prism, increases the dispersion already produced at the first surface.

All rainbows would be completely round if the ground weren't in the way.

Although each drop disperses a full spectrum of colors, an observer is in a position to see only a single color from any one drop (Figure 15.19). If violet light from a single drop reaches the eye of an observer, red light from the same drop travels a lower path (toward the observer's feet). To see red light, you should look to a higher drop in the sky. The color red is seen when the angle between a beam of sunlight and the dispersed light is 42°. The color violet is seen when the angle between the sunbeams and dispersed light is 40°.

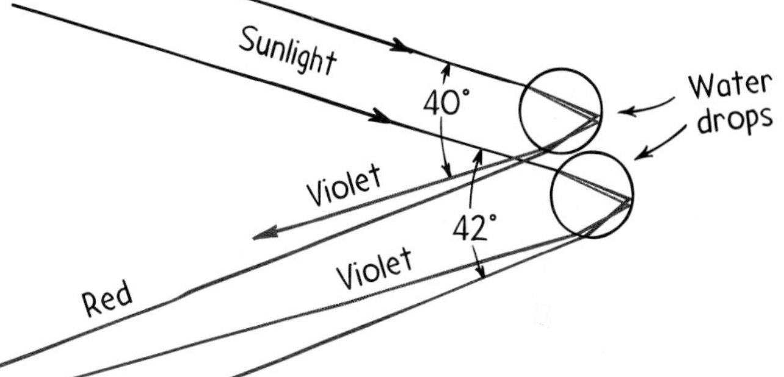

Figure 15.19
Sunlight incident on two raindrops as shown emerges from them as dispersed light. The observer sees the red light from the upper drop and the violet light from the lower drop. Millions of drops produce the whole spectrum of visible light.

A rainbow is not the flat, two-dimensional arc it appears to be. It appears flat for the same reason a spherical burst of fireworks high in the sky look flat—because they're far away. A rainbow is actually part of a three-dimensional cone. The tip of the cone is at your eye. Think of a glass cone. If you held the pointed tip very close to your eye, what would you see? You'd see the glass as a circle. Likewise with a rainbow. All the drops that disperse the rainbow's light toward *you* lie in the shape of a cone—a cone of different layers. Four of the innumerable layers are shown in Figure 15.20. Drops that deflect red to your eye are on the outside, orange is beneath the red, yellow is beneath the orange, and so on all the way to violet on the inner conical surface. The thicker the region containing water drops, the thicker the conical edge you see through, and the more vivid the rainbow.

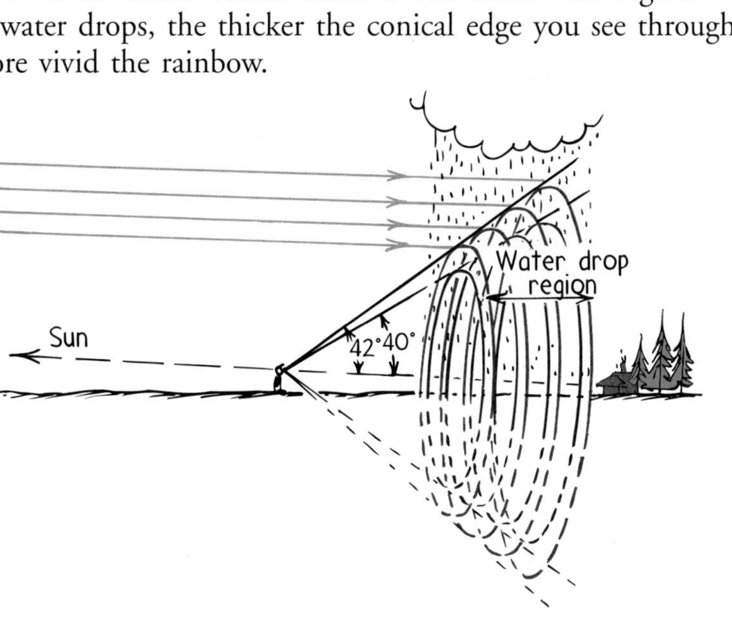

Figure 15.20
When your eye is located between the sun (not shown off to the left) and a water-drop region, the rainbow you see is the edge of a three-dimensional cone that extends through the water-drop region. (Innumerable layers of drops form innumerable two-dimensional arcs like the four suggested here.)

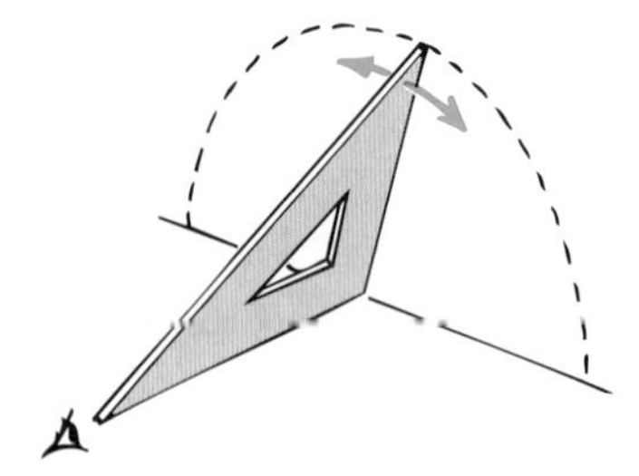

Figure 15.21
Only raindrops along the dashed line disperse red light to the observer at a 42° angle; hence, the light forms a bow.

Your cone of vision intersects the cloud of drops and creates your rainbow. It is ever so slightly different from the rainbow seen by a person nearby. So when a friend says, "Look at the pretty rainbow," you can reply, "Okay, move aside so I can see it, too." Everybody sees his or her own personal rainbow.

Another fact about rainbows: A rainbow always faces you squarely. When you move, your rainbow appears to move with you. So you can never approach the side of a rainbow or see it end-on as in the exaggerated view of Figure 15.20. You *can't* reach its end. Hence the saying "looking for the pot of gold at the end of the rainbow" means pursuing something you can never reach.

Figure 15.22
Two refractions and a reflection in water droplets produce light at all angles up to about 42°, with the intensity concentrated where we see the rainbow at 40° to 42°. No light leaves the water droplet at angles greater than 42° unless it undergoes two or more reflections inside the drop. So the sky is brighter inside the rainbow than outside it. Notice the weak secondary rainbow to the right of the primary.

Snap-Shot Explorations: Making a Pinhole Camera

Make your own historic first camera. Cut out one end of a small cardboard box, and cover the end with semi-transparent tracing or tissue paper. Make a clean-cut pinhole at the other end. (If the cardboard is thick, you can make the pinhole through a piece of tinfoil placed over a larger opening in the cardboard.) Aim the camera at a bright object in a darkened room, and you will see an upside-down image on the tracing paper. The tinier the pinhole, the dimmer and sharper the image. If, in a dark room, you replace the tracing paper with unexposed photographic film, cover the back so that it is light-tight, and cover the pinhole with a removable flap, you have a camera. You're ready to take a picture. Exposure times differ depending mostly on the type of film and amount of light. Try different exposure times, starting with about 3 seconds. Also try boxes of various lengths. The lens on a commercial camera is much bigger than the pinhole and therefore admits more light in less time—hence the name *snapshots.*

Often a larger, secondary rainbow can be seen arching at a greater angle around the primary bow. We won't treat this secondary rainbow except to say that it is formed by similar circumstances and is a result of double reflection within the raindrops (Figure 15.23). Because of this extra reflection (and extra refraction loss), the secondary rainbow is much dimmer and its colors are reversed.

Figure 15.23
Double reflection in a drop produces a secondary bow.

Concept Check ✓

1. Suppose you point to a wall with your arm extended. Then you sweep your arm around, making an angle of about 42° to the wall. If you rotate your arm in a full circle while keeping the same angle, what shape does your arm describe? What shape on the wall does your finger sweep out?

2. If light traveled at the same speed in raindrops as it does in air, would we have rainbows?

Check Your Answers

1. Your arm describes a cone, and your finger sweeps out a circle. Likewise with rainbows.
2. No.

15.5 Lenses Are a Practical Application of Refraction

When you think of a lens, think of a set of glass prisms arranged as shown in Figure 15.24. They refract incoming parallel light rays so that they converge to (or diverge from) a point. The arrangement shown in Figure 15.24a converges the light, and we have a **converging lens.** Notice that it is thicker in the middle. In Figure 15.24b, the middle is thinner than the edges, and this lens diverges the light—and we have a **diverging lens.** Note that the prisms in b diverge the incident rays in a way that makes them appear to originate from a single point in front of the lens.

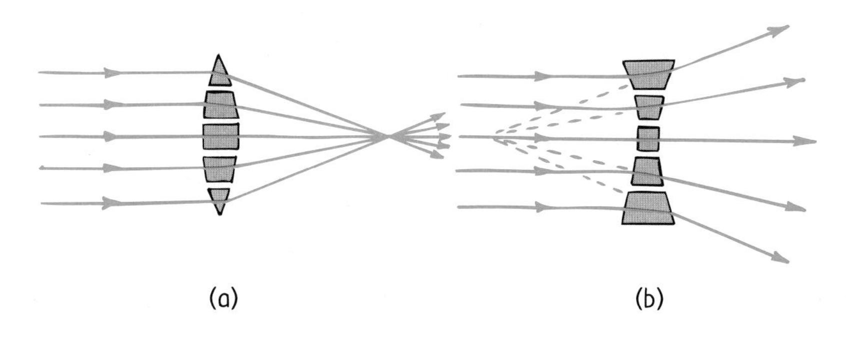

Figure 15.24
Imagine a lens as a set of blocks and prisms.

In both lenses, the greatest deviation of rays occurs at the outermost prisms because they have the greatest angle between the two refracting surfaces. No deviation occurs exactly in the middle, for in that region the two surfaces of the glass are parallel to each other. Real lenses are not made of prisms, of course. They are made of a solid piece of glass with surfaces ground usually to a circular curve. In Figure 15.25 we see how smooth lenses refract waves.

Some key features of lenses are shown for a converging lens in Figure 15.26. The *principal axis* is the line joining the centers of curvatures of the two lens surfaces. The *focal point* is the point of

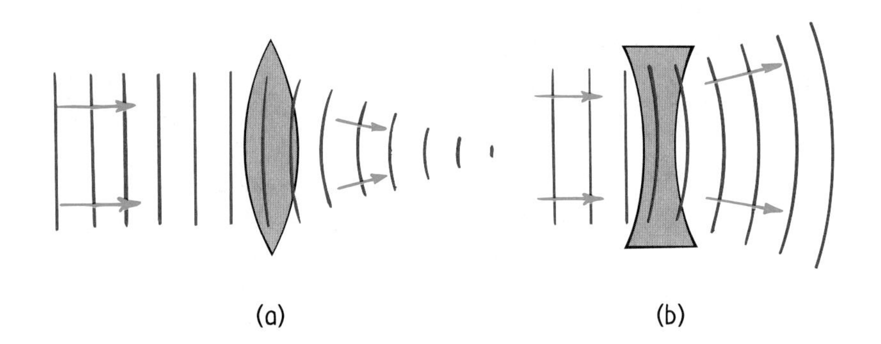

Figure 15.25
Wave fronts travel more slowly in glass than in air. In (a), the waves are retarded more through the center of the lens, and convergence results. In (b), the waves are retarded more at the edges, and divergence results.

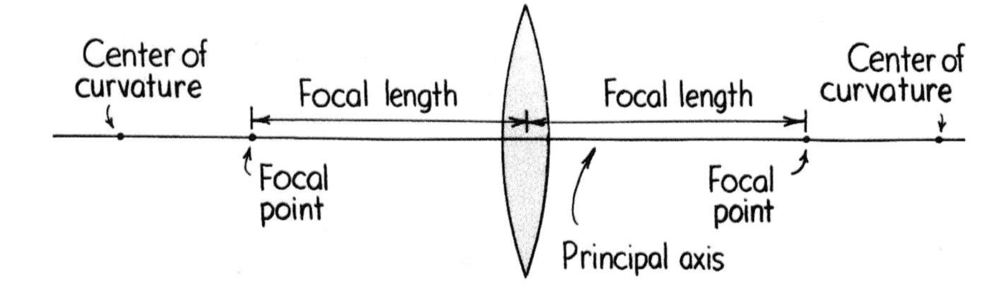

Figure 15.26
Key features of a converging lens.

Hands-On Explorations: Playing Around with Lenses

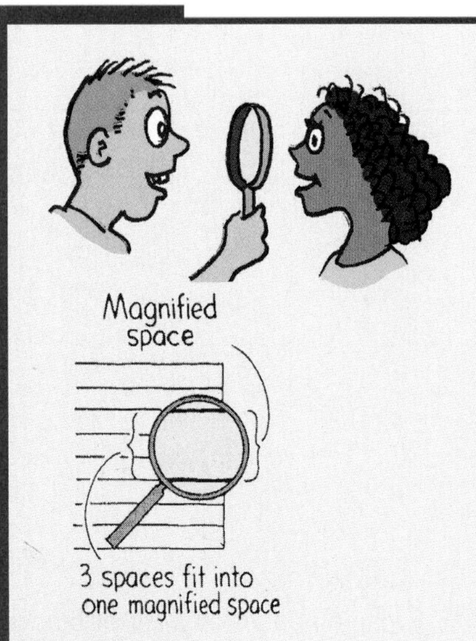

Get your hands on some lenses. Not doing so is like taking swimming lessons away from water. Hold converging lenses in parallel light and note the point where the light converges. Note that different shaped lenses have different focal lengths.

Determine the magnifying power of a lens by focusing on the lines of a ruled piece of paper. Count the spaces between the lines that fit into one magnified space, and you have the magnifying power of the lens. You can do the same with binoculars and a distant brick wall. Hold the binoculars so that only one eye looks at the bricks through the eyepiece while the other eye looks directly at the bricks. The number of bricks seen with the unaided eye that will fit into one magnified brick gives the magnification of the instrument.

convergence for light parallel to the principal axis. Incident beams not parallel to the principal axis focus at points above or below the focal point. All such possible points make up a *focal plane* (not shown). Because a lens has two surfaces, it has two focal points and two focal planes. The *focal length* of the lens is the distance between the center of the lens and either focal point.

In a diverging lens, an incident beam of light parallel to the principal axis is not converged to a point, but is diverged—so the light appears to emerge from a point in front of the lens.

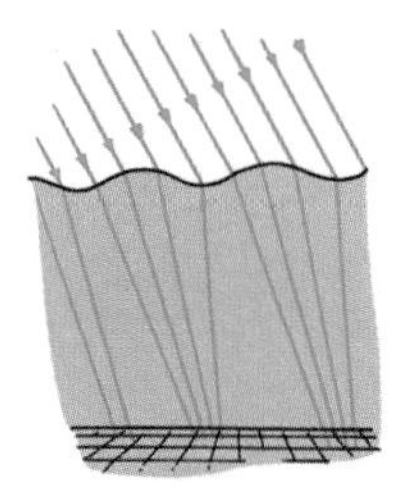

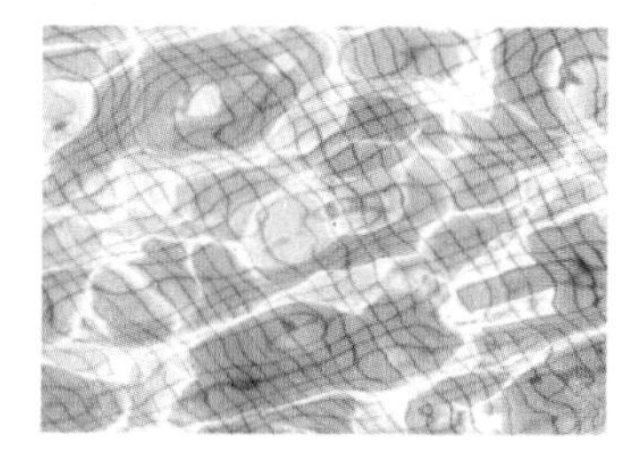

Figure 15.27
The moving patterns of bright and dark areas at the bottom of the pool result from the uneven surface of the water, which behaves like a blanket of undulating lenses. Just as we see the pool bottom shimmering, a fish looking upward at the sun would see it shimmering too. Because of similar irregularities in the atmosphere, we see the stars twinkle.

Link to Physiology: Your Eye

With all of today's technology, the most remarkable optical instrument known is your eye. Light enters through your *cornea,* which does about 70 percent of the necessary bending of the light before it passes through your *pupil* (the aperture—opening—in the iris). Light then passes through your *lens,* which provides the extra bending power needed to focus images of nearby objects on your extremely sensitive *retina.* (Only recently have artificial detectors been made with greater sensitivity to light than the human eye). An image of the visual field outside your eye is spread over the retina. The retina is not uniform. There is a spot in the center of your field of view called the *fovea*—region of most distinct vision. You see greater detail here than at the side parts of your eye. There is also a spot in your retina where the nerves carrying all the information exit; this is your *blind spot.*

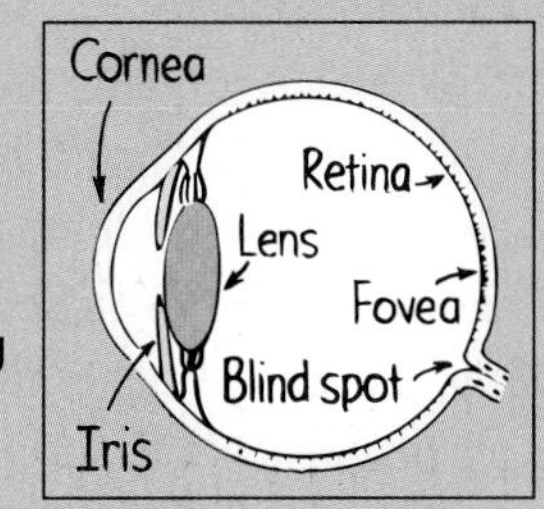

You can demonstrate that you have a blind spot in each eye. Simply hold this book at arm's length, close your left eye, and look at the round dot and X below with your right eye only. You can see both the dot and the X at this distance. Now move the book slowly toward your face, with your right eye fixed upon the dot, and you'll reach a position about 20–25 centimeters from your eye where the X disappears. When both eyes are open, one eye "fills in" the part to which your other eye is blind. Now repeat with only the left eye open, looking this time at the X, and the dot will disappear. But note that your brain fills in the two intersecting lines. Amazingly, your brain fills in the "expected" view even with one eye closed. Instead of seeing nothing, your brain graciously fills in the appropriate background. Repeat this for small objects on various backgrounds. You not only see what's there—you see what's not there!

The light receptors in your retina do not connect directly to your optic nerve but are instead interconnected to many other cells. Through these interconnections a certain amount of information is combined and "digested" in your retina. In this way the light signal is "thought about" before it goes to the optic nerve and then to the main body of your brain. So some brain functioning occurs in your eye. Amazingly, your eye does some of your "thinking."

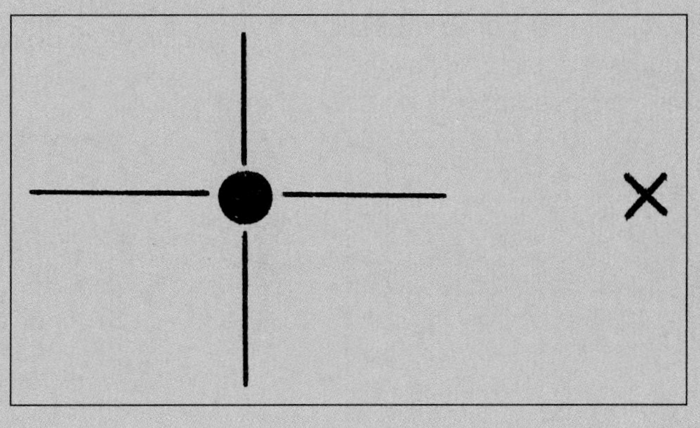

15.6 Image Formation by a Lens

At this moment, light is reflecting from your face onto this page. Light that reflects from your forehead, for example, strikes every part of the page. Likewise for the light that reflects from your chin. Every part of the page is illuminated with reflected light from your forehead, your nose, your chin, and every other part of your face. You don't see an image of your face on the page because there is too much overlapping of light. But place a barrier with a pinhole in it between your face and the page, and the light that reaches the page from your forehead does not overlap the light from your chin. Likewise for the rest of your face. Without this overlapping, an image of your face is formed on the page. It will be very dim, for very little light reflected from your face passes through the pinhole. To see the image, you'd have to shield the page from other light sources. The same is true of the vase and flowers in Figure 15.28b.

The first cameras had no lenses and admitted light through a small pinhole. Long exposure times were required because of the small amount of light admitted by the pinhole. That meant subjects being photographed had to remain still. Motion would produce a blur. If the hole were a bit larger, exposure time would be shorter, but overlapping rays would produce a blurry image. Too large a hole would allow too much overlapping resulting in no image. That's where a converging lens comes in (Figure 15.28c). The lens converges light onto the film without any overlapping of rays. Moving objects can be taken with the lens camera because of the short exposure time. As mentioned earlier, that's why early photographs taken with lens cameras were called snapshots.

The simplest use of a converging lens is a magnifying glass. To understand how it works, think about how you examine objects near and far. With unaided vision, you see a distant object through a relatively narrow angle of view, and a close object through a wider angle of view (Figure 15.29). To see the details of a small object, you want to get as close to it as possible for the widest-angle view. But your

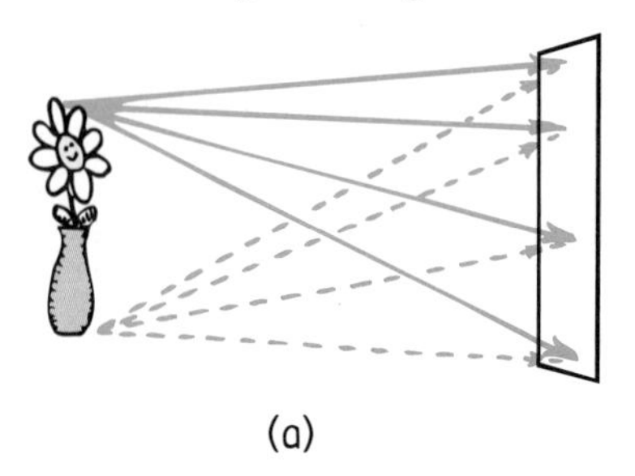

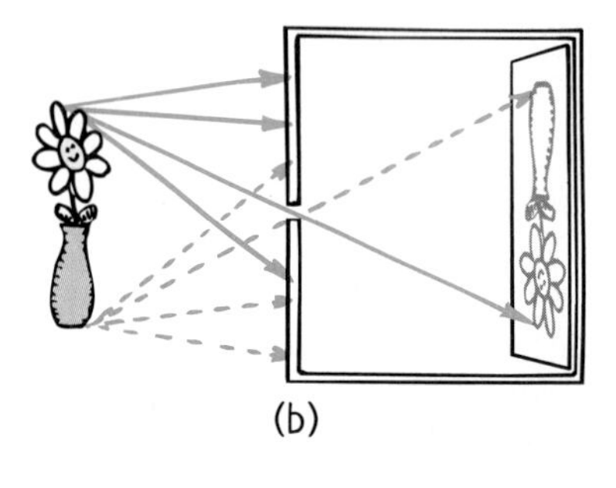

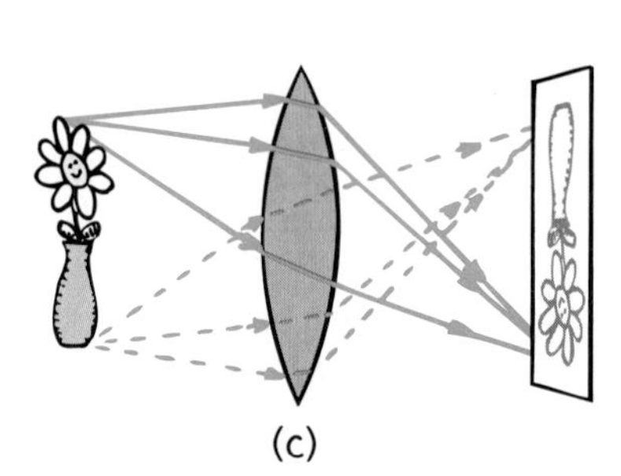

Figure 15.28
Image formation. (a) No image appears on the wall because rays from all parts of the object overlap all portions of the wall. (b) A single small opening in a barrier prevents overlapping rays from reaching the wall; a dim upside-down image is formed. (c) A lens converges the rays upon the wall without overlapping; more light makes a brighter image.

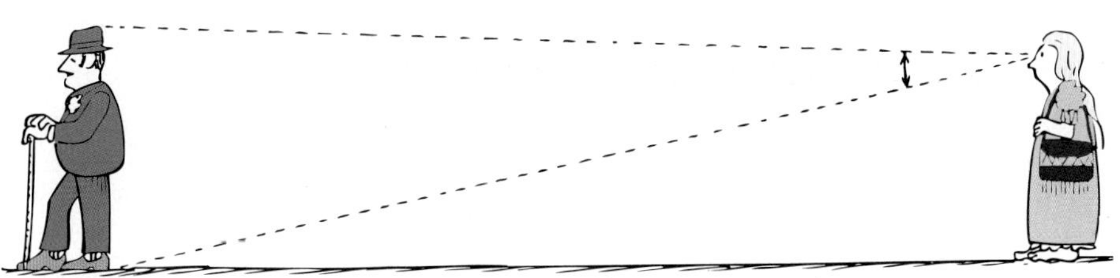

Figure 15.29

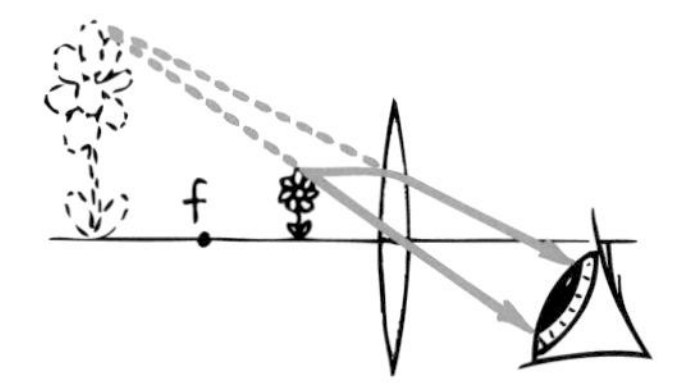

Figure 15.30
When an object is near a converging lens (inside its focal point *f*), the lens acts as a magnifying glass to produce a virtual image. The image appears larger and farther from the lens than the object.

eye can't focus when it's too close to the object. That's where the magnifying glass comes in. When close to the object, the magnifying glass gives you a clear image that would be blurry otherwise.

When you use a magnifying glass, you hold it close to the object you wish to examine. This is because a converging lens provides an enlarged, right-side-up image only when the object is inside the focal point. If a screen is placed at the image distance, no image appears on it because no light is directed to the image position. The rays that reach your eye, however, behave virtually *as if* they originated at the image position. This is called a **virtual image.**

When the object is distant enough to be outside the focal point of a converging lens, a **real image** is formed instead of a virtual image, Figure 15.31 shows this case. A real image is upside-down. Likewise for projecting slides and motion pictures on a screen. To see a right-side image the slide or film is upside-down. The same is true for the image in a camera. Real images formed with a single lens are always upside-down.

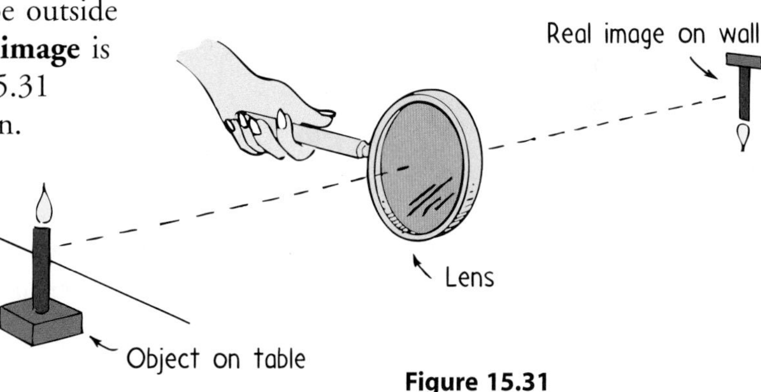

Figure 15.31
When an object is far from a converging lens (beyond its focal point), a real upside-down image is formed.

A diverging lens used alone produces a reduced virtual image. It makes no difference how far or how near the object is. The image is always virtual, right-side up, and smaller than the object. That's why a diverging lens is often used as a "finder" on a camera. When you look at the object to be photographed through such a lens, you see a virtual image that approximates the same proportions as the photograph.

Figure 15.32
A diverging lens forms a virtual, right-side-up image of Jamie and his cat.

Concept Check ✓

Why is the greater part of the photograph in Figure 15.32 out of focus?

Check Your Answer Both Jamie and his cat and the virtual image of Jamie and his cat are "objects" for the lens of the camera that took this photograph. Since the objects are at different distances from the lens, images are at different distances relative to the film in the camera. So only one can be brought into focus. The same is true of your eyes. You cannot focus on near and far objects at the same time.

Lens Defects

No lens forms a perfect image. A distortion in an image is called an *aberration.* Aberrations can be minimized by combining lenses in particular ways. For this reason, most optical instruments use compound lenses instead of using single lenses. Each compound lens consists of several simple lenses.

Spherical aberration results when light passing through the edges of a lens focus at a slightly different place from where light passing through the center of the lens focuses (Figure 15.33). As a result, the image you see is blurred. This can be remedied by covering the edges of the lens. A camera does this with the use of a diaphragm. In good optical instruments, spherical aberration is corrected by a combination of lenses.

Chromatic aberration is the result of different colors having different speeds. This means different refractions in the lens. Different colors of light focus in different places (Figure 15.34). As a result, some colors in the image you see may be in focus, while other colors are out of focus. This is corrected with *achromatic lenses,* which combine simple lenses of assorted glass.

The pupil of your eye changes in size to regulate the amount of light that enters. Vision is sharpest when your pupil is smallest because light passes through only the center of your eye's lens. Through the center, spherical and chromatic aberrations are minimal. Also, the eye then functions like a pinhole camera, needing minimum focusing for a sharp image. Your vision is better in bright light because your pupils are smaller.

Astigmatism of the eye is a defect caused by an irregularity in the curvature of the cornea. It curves more in one direction than another—somewhat like the side of a barrel. Because of this defect, the eye doesn't form sharp images. The remedy is eyeglasses with cylindrical lenses that have more curvature in one direction than in another.

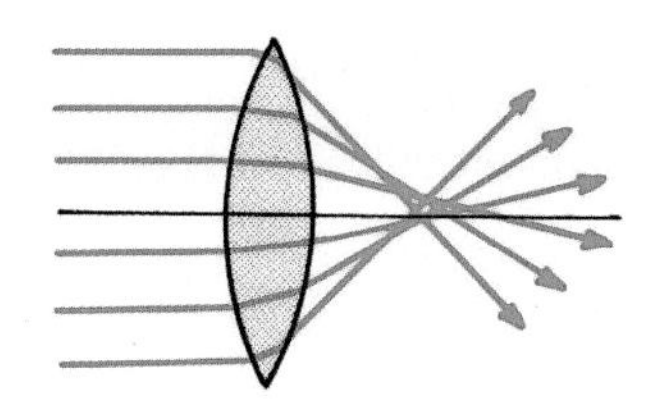

Figure 15.33
Spherical aberration.

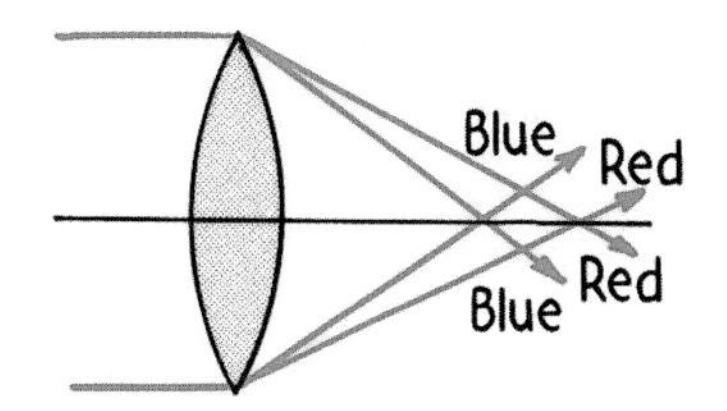

Figure 15.34
Chromatic aberration.

Concept Check ✓

1. If light traveled at the same speed in glass and in air, would glass lenses change the direction of light rays?
2. Why is chromatic aberration associated with a lens but not with a mirror?

Check Your Answers

1. No.
2. Light of different frequencies travel at different speeds in a transparent medium and therefore refract at different angles. This is the cause of chromatic aberration. The angles of reflected light, however, have no relation to frequency. One color reflects the same as any other color. In telescopes, therefore, mirrors are preferred over lenses because of the absence of chromatic aberration for reflection.

To read tiny print clearly, look at it through a pinhole in a piece of paper, close to the print. It works!

Link to Optometry

One option for those with poor sight today is wearing eyeglasses. The advent of eyeglasses probably occurred in Italy in the late 1200s. (Curiously, the telescope wasn't invented until some 300 years later. If, in the meantime, anybody viewed objects through a pair of lenses separated along their axes, such as fixed at the ends of a tube, there is no record of it.) Another present-day option to wearing eyeglasses is contact lenses. And yet another is cornea surgery, where the cornea is shaved to a proper shape for normal vision. Soon the wearing of eyeglasses and contact lenses may be a thing of the past. We really do live in a rapidly changing world.

Chapter Review

Key Terms and Matching Definitions

_____ converging lens
_____ diffuse reflection
_____ diverging lens
_____ law of reflection
_____ reflection
_____ real image
_____ refraction
_____ virtual image

1. The return of light into the medium from which it came.
2. The bending of an oblique ray of light when it passes from one transparent medium to another.
3. The angle of a reflection equals the angle of incidence. The reflected and incident rays lie in a plane that is normal to the reflecting surface.
4. Reflection in irregular directions from an irregular surface.
5. A lens that is thicker in the middle than at the edges and refracts parallel rays passing through it to a focus.
6. A lens that is thinner in the middle than at the edges, causing parallel rays passing through it to diverge as if from a point.
7. An image formed by light rays that do not converge at the location of the image.
8. An image formed by light rays that converge at the location of the image. A real image can be displayed on a screen.

Review Questions

Reflection of Light

1. What does incident light that falls on this page do to the electrons in the atoms of the page?
2. What do the electrons in this illuminated page do when they are energized?
3. What is the law of reflection?
4. Relative to the distance of an object in front of a plane mirror, how far behind the mirror is the image?

Refraction—The Bending of Light Due to Changing Speed

5. How does a change in speed of a ray of light affect its direction when it passes from one medium to another?
6. When a wheel rolls from a smooth sidewalk onto grass, the interaction of the wheel with the blades of grass slows the wheel. What slows light when it passes from air into glass or water?
7. What is the angle between a ray of light and its wave front?

Illusions and Mirages Are Caused by Atmospheric Refraction

8. Why does a setting sun often appear elliptical instead of round?
9. Why do stars twinkle?

Light Dispersion and Rainbows

10. What happens to light of a certain frequency when it resonates with electrons of atoms in a material?
11. Which travels more slowly in glass, red light or violet light?
12. If light of different frequencies has different speeds in a material, does it also refract at different angles in the same material? Explain.
13. What prevents rainbows from being seen as complete circles?
14. Where must the sun be to view a rainbow?
15. Does a viewer see a single color or a spectrum of colors coming from a single faraway raindrop?

16. Why is a secondary rainbow dimmer than a primary bow?

Lenses Are a Practical Application of Refraction

17. Distinguish between a *converging lens* and a *diverging lens.*

18. What is the *focal length* of a lens?

Image Formation by a Lens

19. Distinguish between a *virtual image* and a *real image.*

20. What kind of lens can be used to produce a real image? A virtual image?

Exercises

1. An eye at point P looks into the mirror. Which of the numbered cards is seen reflected in the mirror?

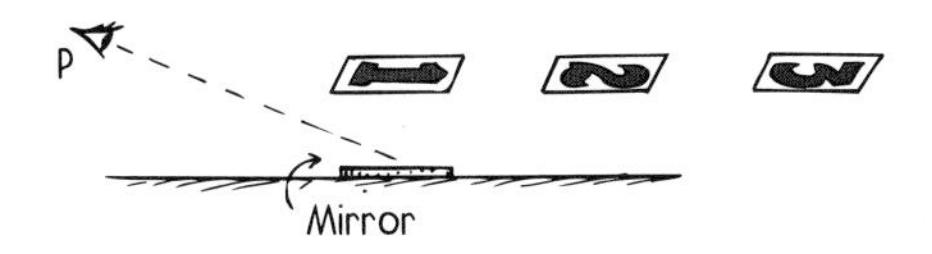

2. What must be the minimum length of a plane mirror in order for you to see a full view of yourself?

3. What effect does your distance from the plane mirror have in the above answer? (Try it and see!)

4. On a steamy mirror wipe away just enough to see your full face. How tall will the wiped area be compared with the vertical dimension of your face?

5. Hold a pocket mirror at almost arm's length from your face and note the amount of your face you can see. To see more of your face, should you hold the mirror closer or farther, or would you have to have a larger mirror? (Try it and see!)

6. The diagram shows a person and her twin at equal distances on opposite sides of a thin wall. Suppose a window is to be cut in the wall so that each twin can see a complete view of the other. Show the size and location of the smallest window that can be cut in the wall to do the job. (*Hint:* Draw rays from the top of each twin's head to the other twin's eyes. Do the same from the feet of each to the eyes of the other.)

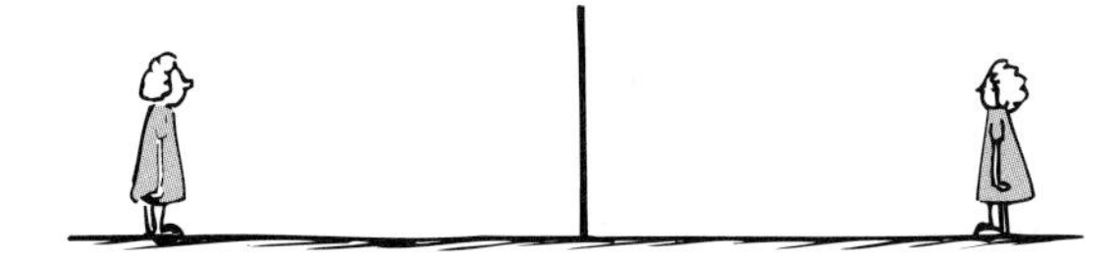

7. Why is the lettering on the front of some vehicles "backwards"?

AMBULANCE (printed mirror-reversed)

8. Which kind of road surface is easier to see when driving at night, a pebbled uneven surface or a mirror-smooth surface? Explain.

9. A person in a dark room looking through a window can clearly see a person outside in the daylight, whereas the person outside cannot see the person inside. Explain.

10. A pair of toy cart wheels are rolled obliquely from a smooth surface onto two plots of grass, a rectangular plot and a triangular plot as shown below. The ground is on a slight incline so that after slowing down in the grass, the wheels speed up again when they emerge on the smooth surface. Finish each sketch by showing some positions of the wheels inside the plots and on the other sides, thereby indicating the direction of travel.

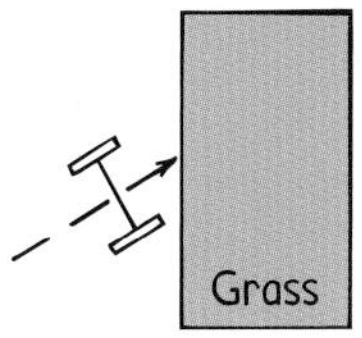

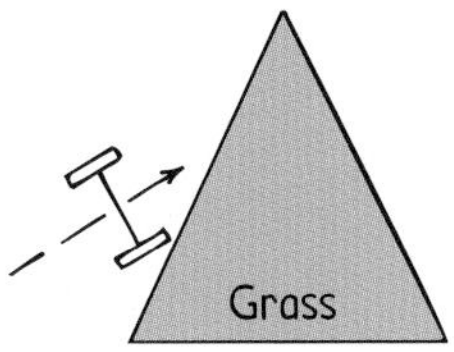

Grass
Grass

11. If light of all frequencies traveled at the same speed in glass, how would white light appear after passing through a prism?

12. If you were in a boat and spearing a fish you see in the water, would you aim above, below, or directly at the fish to make a direct hit? (Assume the fish is stationary in the water.) If you instead used light from a laser as your "spear," would you aim above, below, or directly at the observed fish? Defend your answers.

13. If you were to send a beam of laser light to a space station above the atmosphere and just above the horizon, would you aim the laser above, below, or at the visible space station? Defend your answer.

14. Two observers standing apart from one another do not see the "same" rainbow. Explain.

15. A rainbow viewed from an airplane may form a complete circle. Where will the shadow of the airplane appear? Explain.

16. Transparent plastic swimming-pool covers called *solar heat sheets* have thousands of small lenses made up of air-filled bubbles. The lenses in these sheets are advertised as being able to focus heat from the sun into the water and raise its temperature. Do you think the lenses of such sheets direct a greater amount of solar energy into the water? Defend your answer.

17. Would the average intensity of sunlight measured by a light meter at the bottom of the pool in Figure 15.27 be different if the water were still?

18. In taking a photograph, what would happen to the image if you cover up the bottom half of the lens?

19. Why do you have to put slides into a slide projector upside down?

20. Maps of the moon are upside down. Why?

Problems

1. If you take a photo of your image in a plane mirror, how many meters away should you set your focus for if you are 3 meters in front of the mirror?

2. If you walk toward a mirror at 2 m/s, how fast do you and your image approach each other?

3. No glass is perfectly transparent. Consider a pane of glass that transmits 92% of the light incident upon it. How much light is transmitted through two of these sheets together?

Suggested Reading and Web Sites

Falk, D.S., Brill, D.R., and Stork, D. *Seeing the Light: Optics in Nature.* New York: Harper & Row, 1986.

Murphy, Pat, and Paul Doherty. *The Color of Nature.* San Francisco: Chronicle Books, 1996.

www.exploratorium.edu/exhibits/mix n match/index.html
This has a good adding-light activity from the Exploratorium.

www.exploratorium.edu/snacks/cymk/index.html
This has pigment addition interactions.

Chapter 16: Properties of Light

What causes the colors in soap bubbles—especially big ones, like the one shown here? Does light bend around corners like sound can? Can light waves cancel other light waves? What causes the vivid colors of gasoline spilled on a wet street? Why does the color of some butterfly wings depend on your angle of sight? What's the difference between regular sunglasses and Polaroid sunglasses? Is light a wave or is it a particle? Can it be both at the same time? We'll answer these questions and more in this chapter. Onward!

16.1 Diffraction—The Spreading of Light

When you touch your finger to the surface of still water, circular ripples are produced. When you touch the surface with a straight edge, such as a horizontally held meterstick, you produce a plane wave. You can produce a series of plane waves by successively dipping a meterstick into the surface (Figure 16.1).

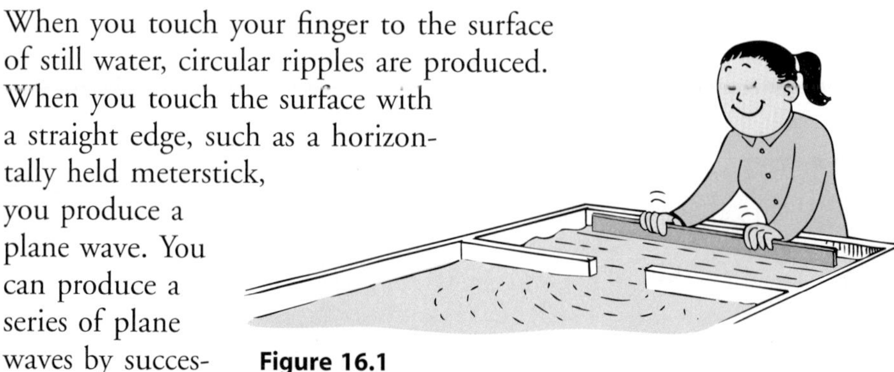

Figure 16.1
The vibrating meterstick makes plane waves in the tank of water. These waves diffract through the opening.

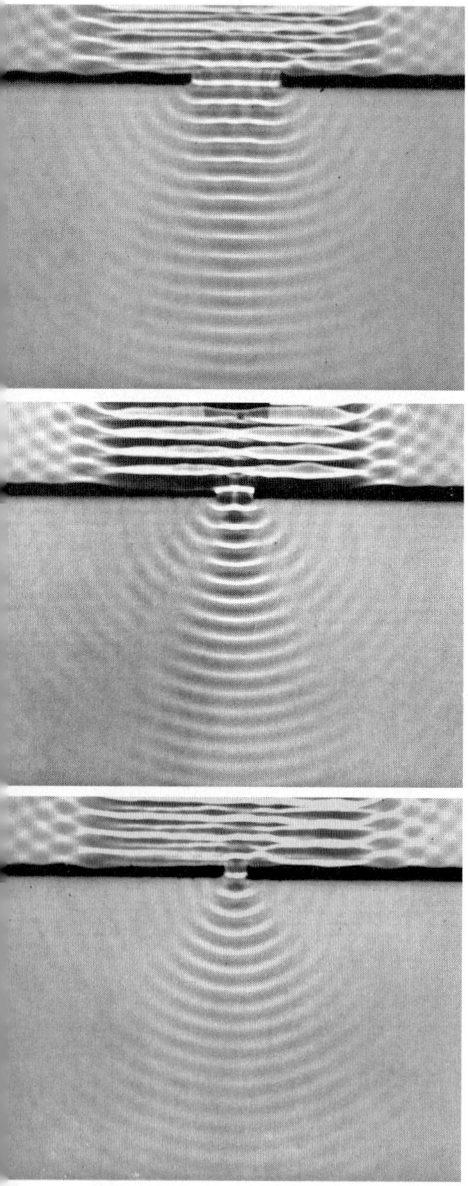

The photographs in Figure 16.2 are top views of water ripples in a shallow glass tank (called a ripple tank). A barrier with an adjustable opening is in the tank. When plane waves meet the barrier, they continue through with some distortion. In the left image, where the opening is wide, the waves continue through the opening almost without change. At the two ends of the opening, however, the waves bend. This bending is called **diffraction.** Any bending of light by means other than reflection and refraction is diffraction. As the width of the opening is narrowed, as in the center image in Figure 16.2, the waves spread more. When the opening is small relative to the wavelength of the incident wave, they spread even more. We see that smaller openings produce more diffraction. Diffraction is a property of all kinds of waves, including sound and light waves.

Diffraction is not confined to narrow slits or to openings. Diffraction occurs around the edges of surfaces and it can be seen with all shadows. On close examination, even the sharpest shadow is blurred slightly at the edge (Figure 16.4).

The amount of diffraction depends on the wavelength of the wave compared with the size of the obstruction that casts the shadow. Long waves are better at filling in shadows, which is why foghorns emit low-frequency sound waves—to fill in any "blind spots." Likewise for radio waves. The wavelength of AM radio waves ranges from 180 to 550 meters. These waves are longer than most objects in their path. They readily bend around buildings and other objects. A long-wavelength radio wave doesn't "see" a relatively small building in its path—but a short-wavelength radio wave does. The radio waves of the FM band range from 2.8 to 3.4 meters and don't bend very well around buildings. This is one of the reasons FM reception is often poor in localities where AM comes in loud and clear. In the case of radio reception, we don't wish to "see" objects in the path of radio waves, and so diffraction is nice.

Figure 16.2
Plane waves passing through openings of various sizes. The smaller the opening, the greater the bending of the waves at the edges—in other words, the greater the diffraction.

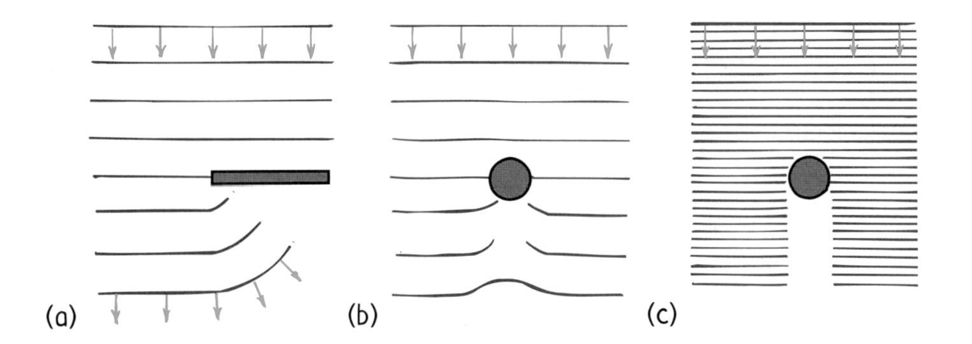

Figure 16.3
(a) Waves tend to spread into the shadow region.
(b) When the wavelength is about the size of the object creating the barrier, the shadow is soon filled in.
(c) When the wavelength is short relative to the object's size, a sharp shadow is cast.

Diffraction is not so nice for viewing very small objects with a microscope. If the size of an object is about the same as the wavelength of light, diffraction blurs the image. If the object is smaller than the wavelength of light, no structure can be seen. The entire image is lost due to diffraction. No amount of magnification or perfection of microscope design can defeat this fundamental diffraction limit.

To minimize this problem, microscopists illuminate tiny objects with electron beams rather than light. Compared with light waves, electron beams have extremely short wavelengths. *Electron microscopes* take advantage of the fact that all matter has wave properties. A beam of electrons has a wavelength smaller than those of visible light. In an electron microscope, electric and magnetic fields, rather than optical lenses, are used to focus and magnify images.

Figure 16.4
Diffraction fringes are evident in the shadows of monochromatic (single-frequency) laser light.

The fact that smaller details can be better seen with smaller wavelengths is useful to dolphins, who scan their environment with ultrasound. The echoes of long-wavelength sound give the dolphin an overall image of objects in its surroundings. To examine more detail, the dolphin emits sound of shorter wavelengths. The dolphin has always done naturally what physicians have only recently been able to do with an ultrasonic imaging devices.

Concept Check ✓

Why does a microscopist use blue light rather than white light to illuminate objects being viewed?

Check Your Answer There is less diffraction with blue light. This allows the microscopist to see more detail (just as a dolphin beautifully investigates fine detail in its environment by the echoes of ultra-short wavelengths of sound).

Figure 16.5
Wave interference.

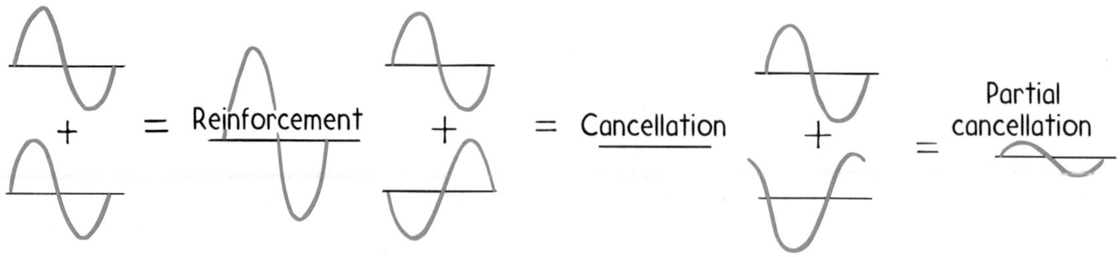

Reinforcement
Cancellation
Partial cancellation

16.2 Interference—Constructive and Destructive

The fringes of brightness and darkness produced when light is diffracted are due to **interference**. Recall our study of interference in Chapter 13, where we learned waves can combine constructively or destructively. Interference is a property of all waves, light included. Constructive and destructive interference is reviewed in Figure 16.5. We see that the adding, or *superposition*, of a pair of identical waves in phase produces a wave of the same frequency but twice the amplitude. If the waves are exactly one-half wavelength out of phase, their superposition results in complete cancellation. If they are out of phase by other amounts, partial cancellation occurs.

Wave interference was convincingly demonstrated by Thomas Young in 1801.* Young shone light through two closely spaced side-by-side pinholes. He found that the light recombines to produce fringes of brightness and darkness on a screen behind. The bright fringes form when a crest from the light wave through one hole

Figure 16.6
Thomas Young's original drawing of a two-source interference pattern.

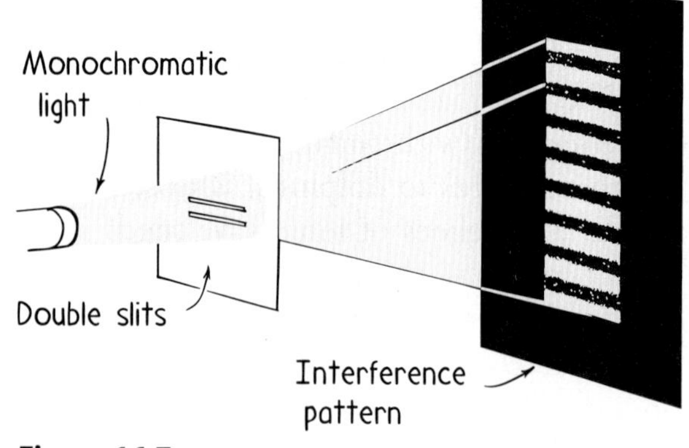

Figure 16.7
When monochromatic light passes through two closely spaced slits, a interference pattern of fringes is produced.

* Thomas Young read fluently at the age of 2. By the age of 4, he had read the Bible twice. When he was 14 he knew eight languages. In his adult life he was a physician and physicist, contributing to an understanding of fluids, work and energy, and the elastic properties of materials. He was the first person to make progress in deciphering Egyptian hieroglyphics. No doubt about it—Thomas Young was a bright guy!

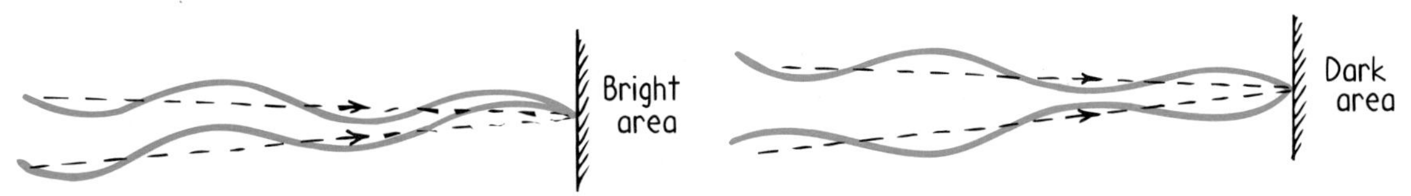

Figure 16.8
A bright area occurs when waves from both slits arrive in phase; a dark area results from the overlapping of waves that are out of phase.

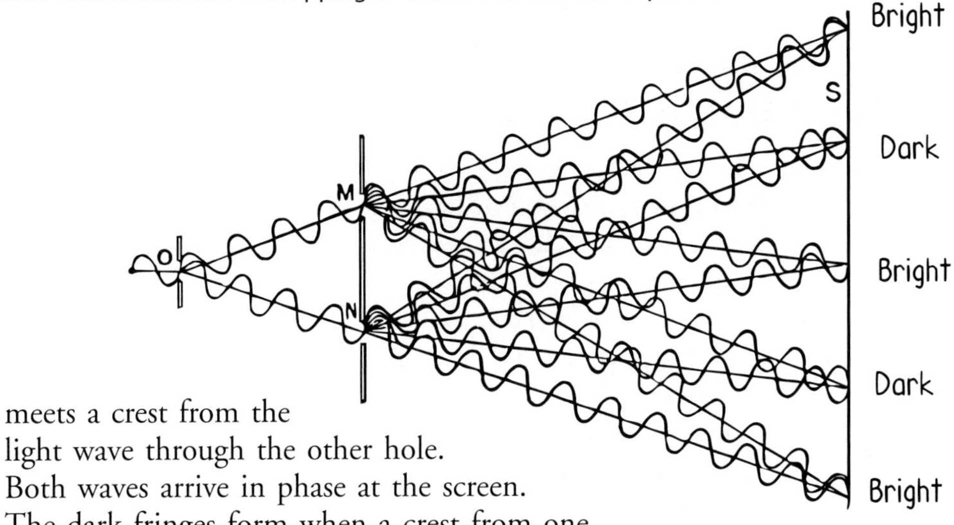

Figure 16.9
Light from O passes through slits M and N and produces an interference pattern on the screen S.

meets a crest from the light wave through the other hole. Both waves arrive in phase at the screen. The dark fringes form when a crest from one wave meets a trough from the other. Figure 16.6 shows Young's drawing of the pattern of superimposed waves from the two sources. When his experiment is done with two closely spaced slits instead of pinholes, the fringe patterns are straight lines (Figure 16.7).

We see in Figures 16.8 and 16.9 how the series of bright and dark fringes result from the different path lengths from slits to screen. For the central bright fringe, the paths from the two slits are the same length and so the waves arrive in phase and reinforce each other. The dark fringes on either side of the central fringe result from one path being longer (or shorter) by one-half wavelength, so waves there arrive half a wavelength out of phase. The other sets of dark fringes occur where the paths differ by odd multiples of one-half wavelength: 3/2, 5/2, and so on.

Concept Check ✓

1. Which fringes would be wider apart, those of red light or blue light, assuming the red and blue are monochromatic (single frequency)?
2. Why is it important that monochromatic light be used?

Check Your Answers

1. Red fringes would be more widely spaced. Can you see that longer waves in Figure 16.9 would produce wider-apart fringes?
2. Light of various wavelengths would result in dark fringes for one wavelength filling in bright fringes for another. The result would be no distinct fringe pattern. If you haven't seen this, perhaps your teacher will demonstrate it.

Link to Optometry: Seeing Star-Shaped Stars

Have you ever wondered why stars are represented with spikes? The stars on the American flag have 5 spikes, and the Jewish Star of David has 6 spikes. All through the ages, stars have been drawn with spikes. Stars do not have spikes, for they are point sources of light in the night sky. The reason for the spikes is poor eyesight.

The surface of your eye, the cornea, becomes scratched by a variety of causes. These scratches make up a sort of diffraction grating. A scratched cornea is not a very good diffraction grating, but its effects are evident. Look at a bright point source against a dark background—like a star in the night sky. Instead of seeing a point of light, you may see a spiky shape. The spikes even shimmer and twinkle if there are some temperature differences in the atmosphere to produce some refraction. And if you live in a windy desert region where sandstorms are frequent, your cornea will be even more scratched and you'll see more vivid star spikes.

Figure 16.10
Because of the interference it causes, a diffraction grating disperses light into colors. It may be used in place of a prism in a spectrometer.

Interference patterns are not limited to one or two slits. A multitude of closely spaced slits makes up a *diffraction grating.* These devices, like prisms, disperse white light into colors. These are used in devices called *spectrometers,* which we shall discuss in Chapter 18. The feathers of some birds act as diffraction gratings and disperse colors. Likewise for the microscopic pits on the reflective surface of compact discs.

16.3 Interference Colors by Reflection from Thin Films

We have all noticed the beautiful spectrum of colors reflected from a soap bubble or from gasoline on a wet street. These colors are produced by the interference of light waves. This phenomenon is often called *iridescence* and is observed in thin transparent films.

A soap bubble appears iridescent in white light when the thickness of the soap film is about the same as the wavelength of light. Light waves reflected from the outer and inner surfaces of the film to your eye travel different distances. When illuminated by white light, the film may be just the right thickness at one place to cause the destructive interference of, say, red light. When red light is subtracted from white light, the mixture left appears as the complementary color—cyan. At another place, where the film is thinner, perhaps blue is cancelled. Then the light seen is the complement of blue—yellow. Whatever color is canceled by interference, the light seen is its complementary color.

This occurs for gasoline on a wet street (Figure 16.11). Light reflects from two surfaces. One is the upper, air-gasoline surface. The other is the lower, gasoline-water surface. If the thickness of the gasoline is such to cancel blue, as the figure suggests, then the gasoline

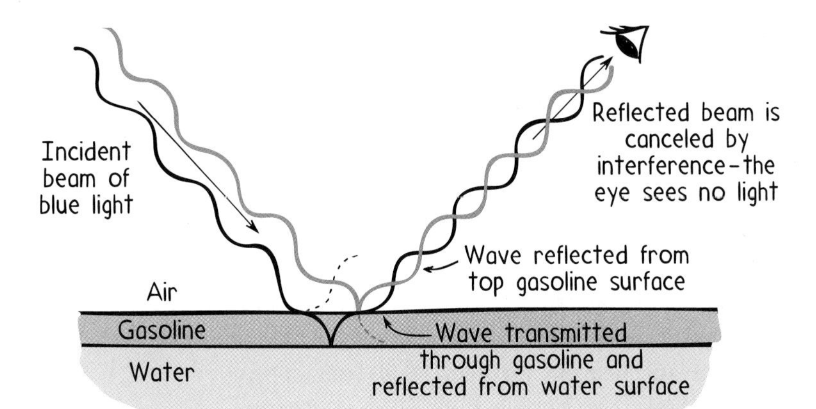

Figure 16.11
The thin film of gasoline is just the right thickness to cancel the reflections of blue light from the top and bottom surfaces. If the film were thinner, perhaps shorter-wavelength violet would be canceled. (One wave is drawn black to show how it is out of phase with the blue wave upon reflection.)

surface appears yellow to the eye. As mentioned earlier, blue subtracted from white leaves yellow. Why are a variety of colors seen in the thin film of gasoline? The answer is that the film thickness is not uniform. Different film thicknesses show a "contour map" of microscopic differences in surface "elevations."

If you view the thin film of gasoline at a lower angle you'll see different colors. That's because light through the film travels a longer path. A longer wave is canceled and a different color is seen. Different wavelengths of light are canceled for different angles.

Dishes washed in soapy water and poorly rinsed have a thin film of soap on them. Hold such a dish up to a light source so that *interference colors* can be seen. Then turn the dish to a new position, keeping your eye on the same part of the dish. Do you notice a change in color? Light reflecting from the bottom surface of the transparent soap film cancels light reflecting from the top surface.

Interference techniques can be used to measure the wavelengths of light and other regions of the electromagnetic spectrum. Interference provides a means of measuring extremely small distances with great accuracy. Instruments called *interferometers* use the principle of interference and are the most accurate instruments known for measuring small distances.

Figure 16.12
Magenta is seen in Lisa's soap bubbles, which is due to the cancellation of green light. What primary color is cancelled to produce cyan?

Hands-On Exploration: Swirling Colors

Dip a dark-colored cup (dark colors make the best background for viewing interference colors) in dishwashing detergent. Then hold the cup sideways and look at the reflected light from the soap film that covers its mouth. Swirling colors appear as the soap runs down to form a wedge that grows thicker at the bottom. The top becomes thinner—so thin that it appears black. This occurs when the film is thinner than 1/4 the wavelength of the shortest waves of visible light. The film soon becomes so thin it pops.

Figure 16.13
A vertically plane-polarized plane wave and a horizontally plane-polarized plane wave.

Concept Check ✓

1. What color appears to be reflected from a soap bubble in sunlight when its thickness is such that green light is canceled?
2. In the left column are the colors of certain objects. In the right column are various ways in which colors are produced. Match the right column to the left.

(a) yellow daffodil	(1) interference
(b) blue sky	(2) diffraction
(c) rainbow	(3) selective reflection
(d) peacock feathers	(4) refraction
(e) soap bubble	(5) scattering

Check Your Answers

1. The composite of all the visible wavelengths except green is the complementary color, magenta. (Go back and see Figures 14.16 and 14.17.)
2. a-3; b-5; c-4; d-2; e-1.

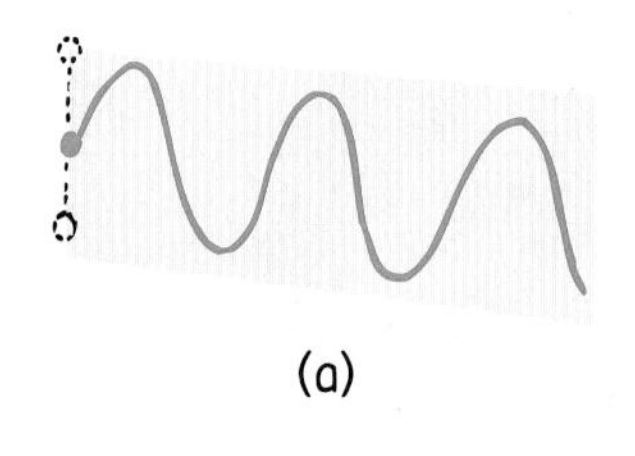

(a)

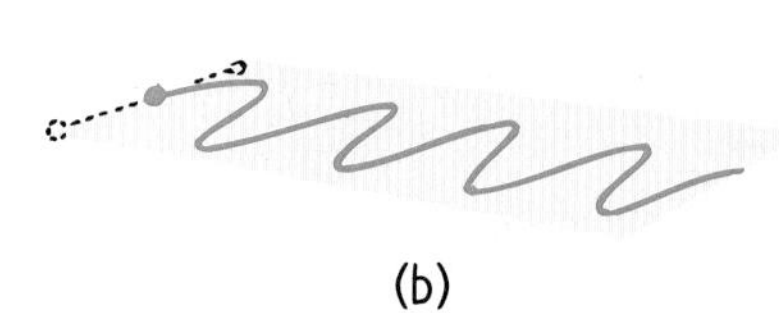

(b)

Figure 16.14
(a) A vertically plane-polarized wave from a charge vibrating vertically.
(b) A horizontally plane-polarized wave from a charge vibrating horizontally.

16.4 Polarization—Evidence for the Transverse Wave Nature of Light

Interference and diffraction provide the best evidence that light is wavelike. As we learned in Chapter 13, waves can be either longitudinal or transverse. Sound waves are longitudinal, which means the vibratory motion of the medium is *along* the direction of wave travel. The fact that light waves exhibit **polarization** demonstrates that they are transverse.

If you shake a rope either up and down or from side to side as shown in Figure 16.13, you'll produce a transverse wave along the rope. The plane of vibration is the same as the plane of the wave. If we shake it up and down, the wave vibrates in a vertical plane. If we shake it back and forth, the wave vibrates in a horizontal plane. We say that such a wave is *plane-polarized*—that the waves traveling along the rope are confined to a single plane. Polarization is a property of transverse waves. (Polarization does not occur among longitudinal waves—there is no such thing as polarized sound.)

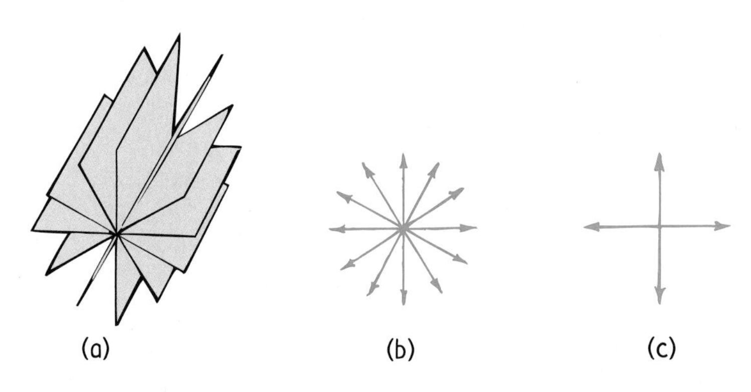

Figure 16.15
Representations of planes of waves. The center and right configurations show electric vectors that make up the electric part of electromagnetic waves.

A single vibrating electron can emit an electromagnetic wave that is plane-polarized. The plane of polarization matches the vibrational direction of the electron. That means a vertically accelerating electron emits light that is vertically polarized. A horizontally accelerating electron emits light that is horizontally polarized (Figure 16.14).

A common light source, such as an incandescent lamp, a fluorescent lamp, or a candle flame, emits light that is unpolarized. This is because electrons that emit the light are vibrating in many random directions. There are as many planes of vibration as the vibrating electrons producing them. A few planes are represented in Figure 16.15a. We can represent all these planes by radial lines, shown in Figure 16.15b. (Or, more simply, the planes can be represented by vectors in two mutually perpendicular directions, Figure 16.15c.) The vertical vector represents all the components of vibration in the vertical direction. The horizontal vector represents all the components of vibration horizontally. The simple model of Figure 16.15c represents unpolarized light. Polarized light would be represented by a single vector.

All transparent crystals having a non-cubic natural shape have the property of polarizing light. These crystals divide unpolarized light into two internal beams polarized at right angles to each other. Some crystals strongly absorb one beam while transmitting the other (Figure 16.16). This makes them excellent polarizers. Herapathite is such a crystal. Microscopic herapathite crystals are aligned and embedded between cellulose sheets. They make up Polaroid filters, popular in sunglasses. Other Polaroid sheets consist of certain aligned molecules rather than tiny crystals.

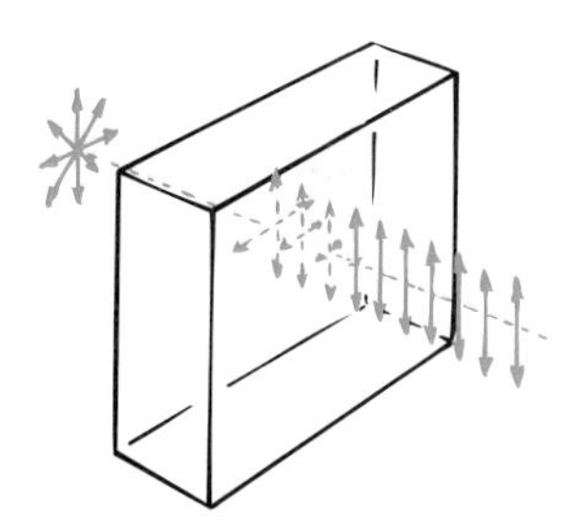

Figure 16.16
One component of the incident unpolarized light is absorbed, resulting in emerging polarized light.

If you look at unpolarized light through a Polaroid filter, you can rotate the filter in any direction and the light appears unchanged. But if the light is polarized, then as you rotate the filter, you progressively cut off more and more of the light until it is blocked out. An ideal Polaroid filter transmits 50 percent of incident unpolarized light. That 50 percent is polarized. When two Polaroid filters are arranged so that their polarization axes are aligned, light can pass through both (Figure 16.17a). If their axes are at right angles to each other,

Hands-On Exploration: Interference Colors with Polaroids

Place a source of white light on a table in front of you. Then place a sheet of Polaroid in front of the source, a bottle of corn syrup in front of the sheet, and a second sheet of Polaroid in front of the bottle. Look through the Polaroid sheets that sandwich the syrup and view spectacular colors as you rotate one of the sheets. (If you can't get sheets, use lenses from Polaroid sunglasses.)

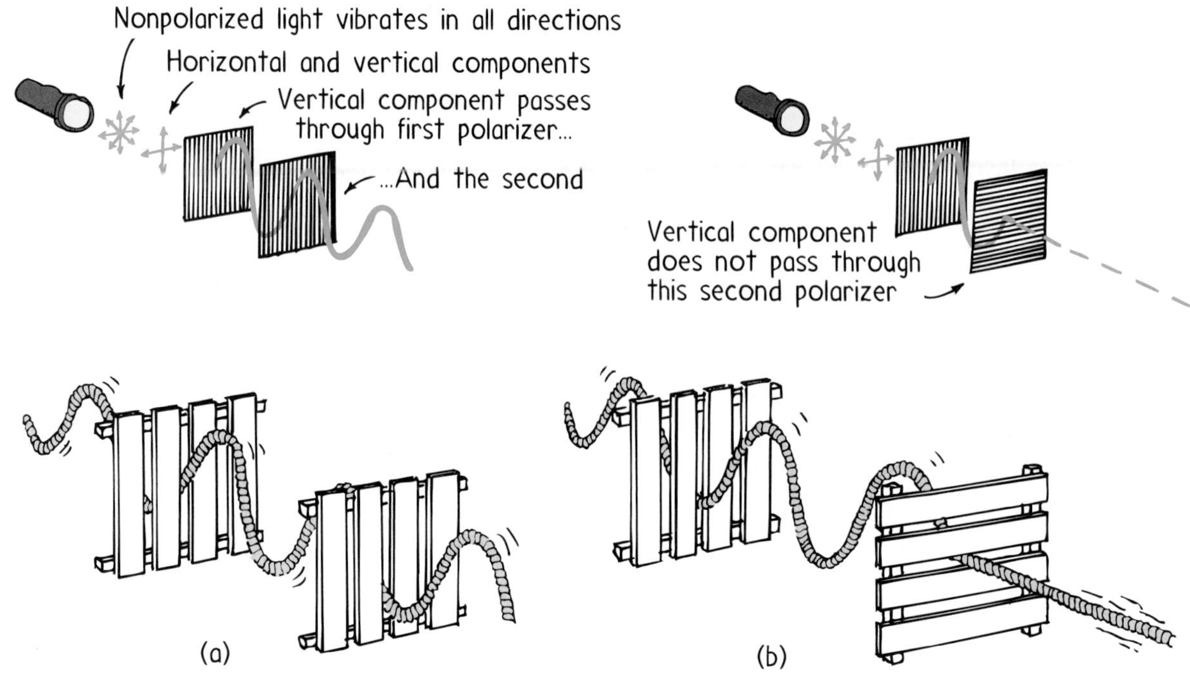

Figure 16.17
(a) Both light and the vibrations of the rope pass through aligned filters.
(b) Neither light nor vibrations of the rope pass through crossed filters. The blue dashed line shows the direction light would travel if the filters weren't crossed.

(in this case we say the filters are *crossed*) almost no light penetrates the pair (Figure 16.17b). (A small amount of shorter wavelengths do get through.) When Polaroid filters are used in pairs like this, the first one is called the *polarizer* and the second one the *analyzer.*

Much of the light reflected from nonmetallic surfaces is polarized. The glare from glass or water is a good example. Except for light that hits vertically, the reflected ray has more vibrations parallel to the reflecting surface. The part of the ray that penetrates the surface has more vibrations at right angles to the surface (Figure 16.19). Skipping flat rocks off the surface of a pond is analogous. When the rocks hit parallel to the surface, they easily reflect. But when they hit with their faces at right angles to the surface, they "refract" into the water. The glare from reflecting surfaces can be dimmed a lot with the use of Polaroid sunglasses. The polarization axes of the lenses are vertical because most glare reflects from horizontal surfaces.

Figure 16.18
Polaroid sunglasses block out horizontally vibrating light. When the lenses overlap at right angles, no light gets through.

Figure 16.19
Most glare from nonmetallic surfaces is polarized. Notice that the components of incident light parallel to the surface are reflected. Also notice that the components perpendicular to the surface pass through the surface into the medium. Because most of the glare we encounter is from horizontal surfaces, the polarization axes of Polaroid sunglasses are vertical.

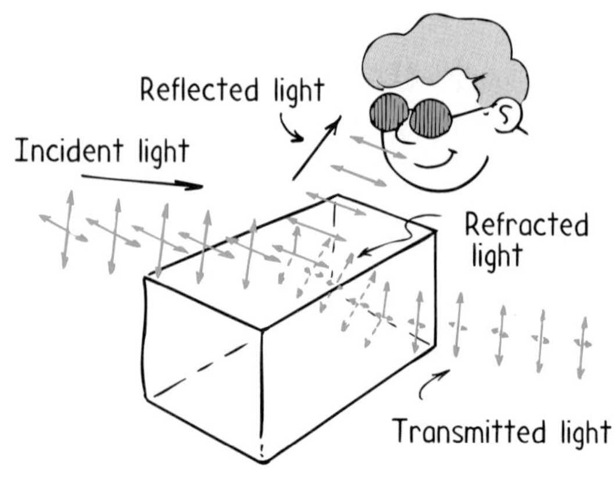

Figure 16.20
Light (a) passes through when the axes of the Polaroid filters are aligned but (b) absorbed when Ludmila rotates one filter so that the axes are at right angles to each other. (c) When she inserts a third Polaroid filter at an angle between the crossed ones, light again passes through. Why? (The answer is in Appendix C.)

Concept Check ✓

Which pair of glasses is best suited for automobile drivers? (The polarization axes are shown by the straight lines.)

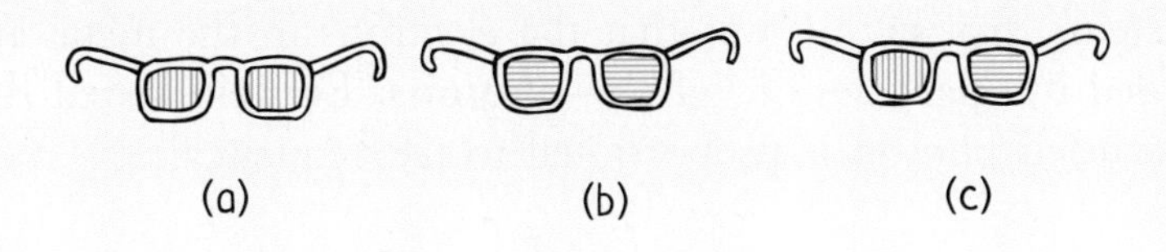

Check Your Answer Pair A is best suited because the vertical axis blocks horizontally polarized light, which makes up much of the glare from horizontal surfaces. Pair C is suited for viewing 3-D movies.

Beautiful colors similar to interference colors can be seen when certain materials are placed between crossed Polaroid filters. Cellophane and transparent tape are wonderful. The explanation for these colors is another story—advanced study.

16.5 Wave-Particle Duality—Light Travels As a Wave and Strikes Like a Particle

In ancient times, Plato and other Greek philosophers hypothesized that light was made up of tiny particles. And in the early 1700s, so did Isaac Newton, who first became famous for his experiments with light. Then a hundred years later, the wave nature of light was demonstrated by Thomas Young in the double-slit experiment. This wave view was validated in 1862 by James C. Maxwell's finding that light is energy carried in the vibrating electric and magnetic fields of electromagnetic waves. The wave view of light was confirmed experimentally by Heinrich Hertz 25 years later. The wave nature of light seemed to be established.

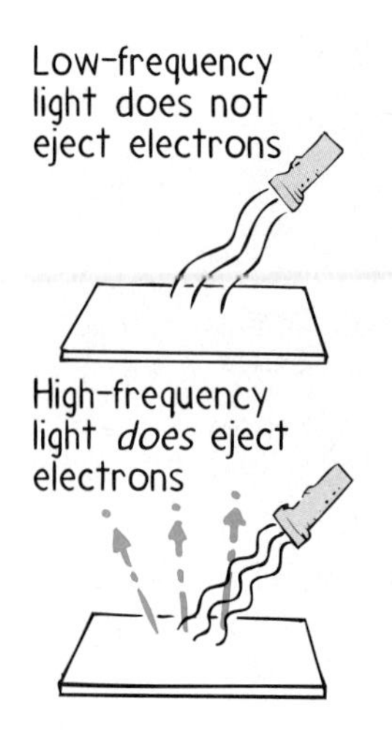

Figure 16.21
The photoelectric effect depends on frequency.

Then in 1905, Albert Einstein published a Nobel Prize–winning paper that challenged the wave theory of light. Einstein stated that, in its interactions with matter, light strikes not as a wave, but as a tiny particle of energy called a *photon.*

So science had come full-circle in its view of light—particle to wave and back to particle. As we shall see, both views are correct! First let's look at Einstein's particle model of light, which explained a mystery to scientists in the early 1900s—the *photoelectric effect.*

When light shines on certain metal surfaces, electrons are ejected from the surfaces. This is the **photoelectric effect,** used in electric eyes, light meters, and some motion-picture sound tracks. What puzzled investigators in the early 1900s was that only ultraviolet and violet light knock electrons from these surfaces. Lower-frequency light, like red, won't do it, even if the red light is very bright with much energy. They discovered that electrons are ejected by high frequency light, not by bright light. Very dim violet light ejects electrons, but a bright red or green light doesn't. If the brightness of violet light is increased, more electrons are ejected—but not at greater energies.

Einstein's explanation was that the electrons in the metal are bombarded by "particles of light"—**photons.** Einstein stated that the energy of each photon is proportional to its frequency:

$$E \sim f$$

So Einstein viewed a beam of light as a hail of photons, each carrying energy proportional to its frequency. One photon is completely absorbed by each electron ejected from the metal.

All attempts to explain the photoelectric effect by waves failed. A light wave is broad and its energy is spread out. For a light wave to eject a single electron from a metal surface, all the light's energy would have to be concentrated on that one electron. This is as unlikely as an ocean wave hitting a beach and knocking only one single seashell far inland with an energy equal to that of the whole wave. The photoelectric effect only make sense by thinking of light as a succession of particle-like photons.

Einstein's idea was experimentally verified 11 years after he announced it. Interestingly, it was verified by an experimenter, the American physicist Robert Millikan, who thought he would discredit it (a not-too-uncommon feature of science). Every aspect of Einstein's concept was confirmed. The photoelectric effect proves convincingly that light has particle properties.

Recall Thomas Young's double-slit interference experiment, which we earlier discussed in terms of waves. When monochromatic light

Figure 16.22
Arrangement for double-slit experiment. The black and white striped tall rectangle is a photograph of the interference pattern. To the far right is a graphic of the pattern.

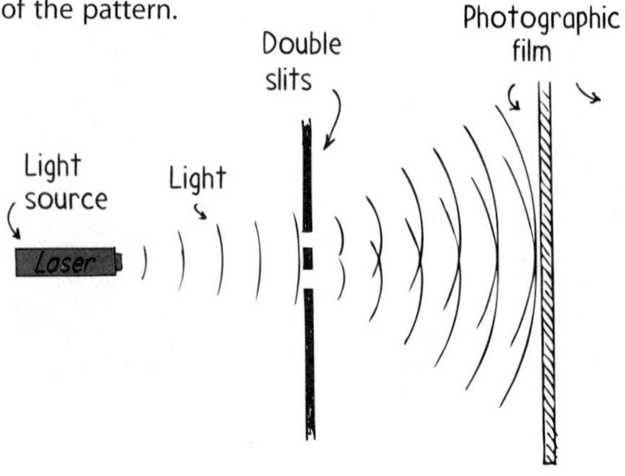

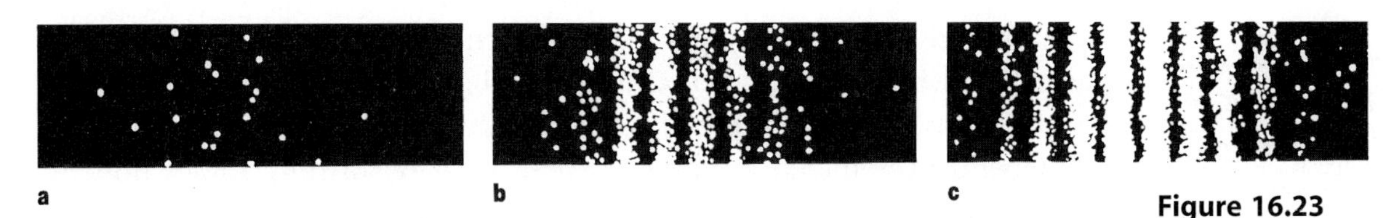

Figure 16.23
Stages of a two-slit interference pattern. The pattern of individual exposed grains progresses from (a) 28 photons to (b) 1000 photons to (c) 10,000 photons. As more photons hit the screen, a pattern of interference fringes appears.

passes through a pair of closely spaced thin slits, an interference pattern is produced on photographic film (Figure 16.22). Now let's consider the experiment in terms of photons. Suppose we dim our light source so that in effect only one photon at a time reaches the thin slits. If the film behind the slits is exposed to the light for a very short time, the film becomes exposed as simulated in Figure 16.23a. Each spot shows where the film has been exposed to a photon. If the light is allowed to expose the film for a longer time, a pattern of fringes begins to emerge as in Figure 16.23b and c. This is quite amazing. We see spots on the film progressing photon by photon to form the same interference pattern characterized by waves!

Evidently, light has both a wave nature and a particle nature—a wave-particle duality. This duality is evident in the formation of optical images. Let's look carefully at the way a photographic image is formed, say on film, photon by photon (Figure 16.24). The photographic film consists of an emulsion that contains grains of silver halide crystal, each grain containing about 10^{10} silver atoms. Each photon that hits the film gives up its energy to a single grain in the emulsion. This energy activates surrounding crystals in the grain and

Figure 16.24
Stages of film exposure reveal the photon-by-photon production of a photograph. The approximate numbers of photons at each stage are (a) 3×10^3; (b) 1.2×10^4; (c) 9.3×10^4; (d) 7.6×10^5; (e) 3.6×10^6; and (f) 2.8×10^7.

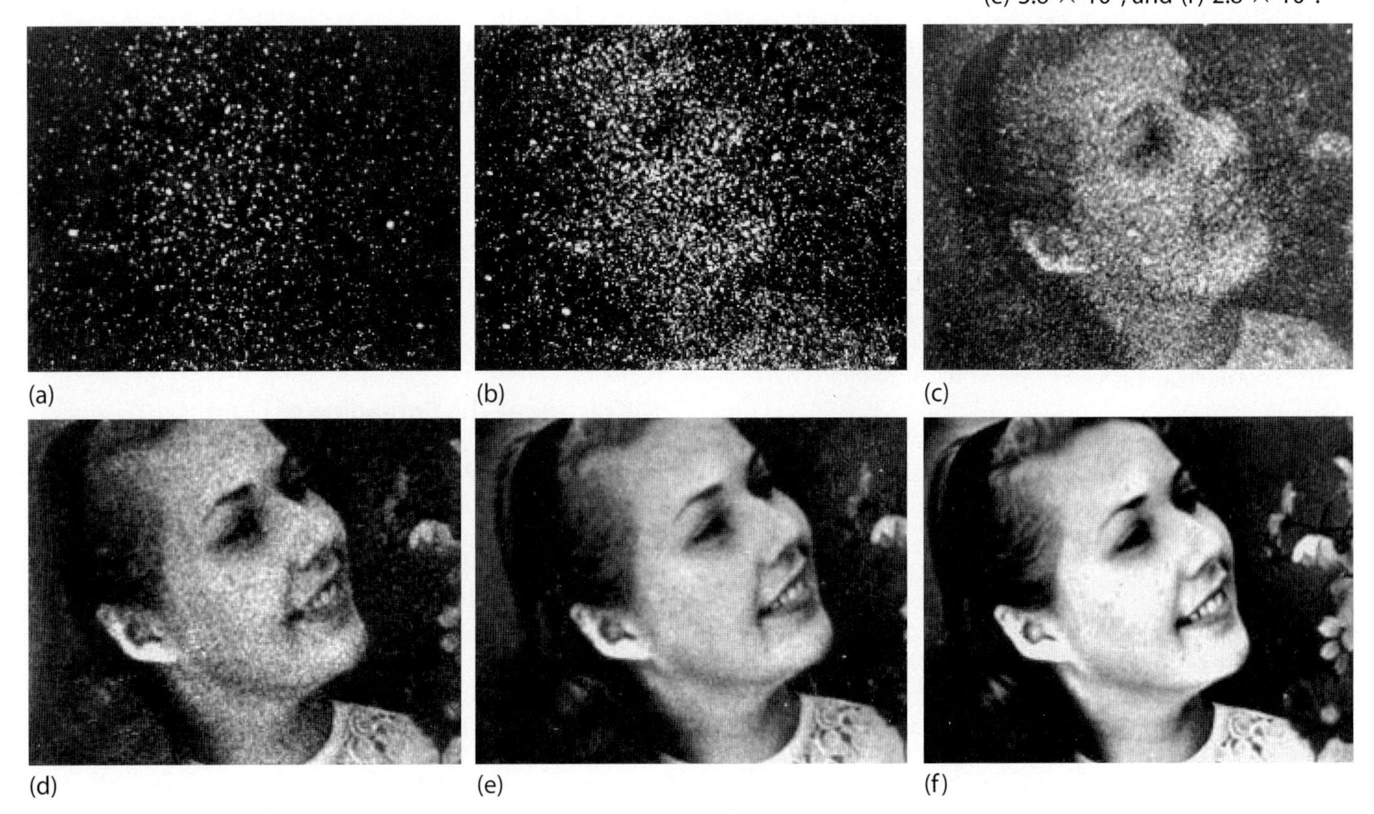

A quantum mechanic.

develops the film. Many photons activating many grains produce the usual photographic exposure. When a photograph is taken with very feeble light, the image is built up by individual photons that arrive independently and at random. Figure 16.24 beautifully shows how an exposure progresses photon by photon.

What all this means is that light has both wave and particle properties. Simply put: *Light behaves as a stream of photons when it interacts with the photographic film or other detectors. But light behaves as a wave in traveling from a source to the place where it is detected.* In interference experiments, photons strike the film at places where we would expect to see constructive interference of waves.

The fact that light shows both wave and particle behavior is one of the most interesting surprises that scientists discovered in the previous century. The finding that light comes in tiny bunches, tiny *quanta* as they are called, led to a whole new way of looking at nature—wave mechanics, or *quantum mechanics.* An outcome of this new mechanics is that just as light has particle properties, particles have wave properties. First electrons were found to have wave properties; a beam of electrons passing through slits exhibits the same type of diffraction pattern as light. Then other particles—even baseballs and orbiting planets—could be described by the new mechanics of waves. Quantum mechanics and Newtonian physics overlap in the macroworld, and both are seen as "correct." But only quantum mechanics, with its emphasis on waves, is wholly accurate in the microworld of the atom. More about this in Part 5. Let's go!

Chapter Review

Key Terms and Matching Definitions

_____ diffraction
_____ interference
_____ photon
_____ photoelectric effect
_____ polarization

1. The bending of light as it passes around an obstacle or through a narrow slit, causing the light to spread and to produce light and dark fringes.
2. The result of superposing two or more waves of the same wavelength.
3. The alignment of the transverse electric vectors that make up electromagnetic radiation.
4. The emission of electrons from a metal surface when light shines on it.
5. A particle of light, or the basic packet of electromagnetic radiation.

Review Questions

Diffraction—The Spreading of Light

1. Is diffraction more pronounced through a small opening or through a large opening?
2. For an opening of a given size, is diffraction more pronounced for a longer wavelength or a shorter wavelength?
3. What are some of the benefits and problems created by diffraction?

Interference—Constructive and Destructive

4. Is interference restricted to only some types of waves or does it occur for all types of waves?
5. What is monochromatic light?

Interference Colors by Reflection from Thin Films

6. What produces iridescence?
7. What causes the variety of colors seen in gasoline splotches on a wet street?
8. What accounts for the variety of colors in a soap bubble?
9. If you look at a soap bubble from different angles so that you're viewing different apparent thicknesses of soap film, do you see different colors? Explain.

Polarization—Evidence for the Transverse Wave Nature of Light

10. What phenomenon distinguishes between longitudinal and transverse waves?
11. How does the direction of polarization of light compare with the direction of vibration of the electrons that produced it?
12. Why does light pass through a pair of Polaroid filters when the axes are aligned but not when the axes are at right angles to each other?
13. How much unpolarized light does an ideal Polaroid filter transmit?
14. When unpolarized light is incident at a grazing angle upon water, what can you say about the reflected light?

Wave-Particle Duality—Light Travels As a Wave and Strikes Like a Particle

15. What evidence can you cite for the wave nature of light? For the particle nature of light?
16. Which are more successful in dislodging electrons from a metal surface, photons of violet light or photons of red light? Why?
17. Why won't a very bright beam of red light impart more energy to an electron than a feeble beam of violet light?
18. Does light behave primarily as a wave or as a particle when it interacts with the crystals of matter in photographic film?
19. Does light travel from one place to another in a wavelike way or a particlelike way?
20. When does light behave as a wave? When does it behave as a particle?

Explorations

1. With a razor blade, cut a slit in a card and look at a light source through it. You can vary the size of the opening by bending the card slightly. See the interference fringes? Try it with two closely spaced slits.
2. Make some slides for a slide projector by sticking some crumpled cellophane onto pieces of slide-sized Polaroid. (Also try strips of cellophane tape overlapped at different angles and experiment with different brands of transparent tape.) Project them onto a large screen or white wall and rotate a second, slightly larger piece of Polaroid in front of the projector lens in rhythm with your favorite music. You'll have your own light show.

Exercises

1. Why do radio waves diffract around buildings but light waves do not?
2. A pattern of fringes is produced when monochromatic light passes through a pair of thin slits. Is such a pattern produced by three parallel thin slits? By thousands of such slits? Give an example to support your answer.

3. The colors of peacocks and hummingbirds are the result not of pigments but of ridges in the surface layers of their feathers. By what physical principle do these ridges produce colors?

4. The colored wings of many butterflies are due to pigmentation, but in some species, such as the Morpho butterfly, the colors do not result from any pigmentation. When the wing is viewed from different angles, the colors change. How are these colors produced?

5. Why do the iridescent colors seen in some seashells, especially abalone shells, change as the shells are viewed from different positions?

6. When dishes are not properly rinsed after washing, different colors are reflected from their surfaces. Explain.

7. If you notice the interference patterns of a thin film of oil or gasoline on water, you'll note that the colors form complete rings. How are these rings similar to the lines of equal elevation on a contour map?

8. Why aren't interference colors seen on films of gasoline on a dry street?

9. Because of wave interference a film of oil on water is seen to be yellow to observers directly above in an airplane. What color does it appear to a scuba diver directly below?

10. Polarized light is a part of nature, but polarized sound is not. Why?

11. Why do Polaroid sunglasses reduce glare, whereas unpolarized sunglasses simply cut down on the total amount of light reaching the eyes?

12. The digital displays of watches and other devices are normally polarized. What problem occurs when you look at them while wearing polarized sunglasses?

13. How can you determine the polarization axis for a single sheet of Polaroid?

14. To reduce the glare of light from a polished floor, should the axis of a Polaroid filter be horizontal or vertical?

15. How can a single sheet of Polaroid film be used to show that the sky is partially polarized? (Interestingly enough, unlike humans, bees and many insects can discern polarized light and use this ability for navigation.)

16. What percentage of light is transmitted by two ideal Polaroid filters atop each other with their polarization axes aligned? With their axes at right angles to each other?

17. A beam of red light and a beam of blue light have exactly the same energy. Which beam contains the greater number of photons?

18. Silver bromide (AgBr) is a light-sensitive substance used in some types of photographic film. In order to be exposed, the film must be illuminated with light having sufficient energy to break apart the AgBr molecules. Why do you suppose this film may be handled without exposure in a darkroom illuminated with red light? How about blue light? How about very bright red light as compared with very dim blue light?

19. Suntanning produces cell damage in the skin. Why is ultraviolet radiation capable of producing this damage, but visible radiation is not?

20. Does the photoelectric effect *prove* that light is made of particles? Do interference experiments *prove* that light is composed of waves? (Is there a distinction between what something *is* and how it *behaves*?)

Suggested Reading

Greenler, Robert. *Chasing the Rainbow—Recurrences in the Life of a Scientist.* Milwaukee: Elton Wolf Publishing, 2000.

Part 5 The Atom

Like everyone, I'm made of atoms — so small and numerous that I inhale billions of trillions with each breath of air. I exhale some of them right away, but other atoms stay for awhile and become part of me, which I may exhale later. Some of my atoms are in each breath you take, and stay to become part of you (and likewise, yours become part of me). There are more atoms in a breath of air than the total number of humans since time zero, so in each breath you inhale, you recycle atoms that once were a part of every person who lived. Hey, in this sense, we're all one!

Chapter 17: Atoms and the Periodic Table

Imagine you are falling off your chair in slow motion. And while you're falling, you're also shrinking in size. By the time you hit the floor, you are the size of an atom, just one billionth of a meter across. What does the world look like from your new perspective? Do the atoms that make up the floor appear solid? Or are they mostly empty space? Do the atoms contain still smaller particles? If so, what do these subparticles look like? Do all the subparticles look the same? In this chapter, we explore the nature of atoms and the amazing chart that tells their story—the periodic table.

17.1 Elements Contain a Single Kind of Atom

You know that atoms make up the matter around you, from stars to steel to chocolate ice cream. You might think that there must be many different kinds of atoms to account for the huge diversity of matter. But the number of different kinds of atoms is surprisingly small. The great variety of substances results from the many ways a few kinds of atoms can be combined. Just as the three colors red, green, and blue can be combined to form any color on a television screen, or the 26 letters of the alphabet make up all the words in a dictionary, only a few kinds of atoms combine in different ways to produce all substances. To date, we know of slightly more than 100 distinct atoms. Of these, about 90 are found in nature. The remaining atoms have been created in the laboratory.

Any material that is made up of only one type of atom is classified as an **element.** A few examples are shown in Figure 17.1. Pure gold, for example, is an element—it contains only gold atoms. Similarly, one of the gases in air is nitrogen, an element. Nitrogen gas is an element because it contains only nitrogen atoms. Likewise, the graphite in your pencil is an element—carbon. Graphite is a made up solely of carbon atoms. All of the elements are listed in a chart called the **periodic table,** shown in Figure 17.2.

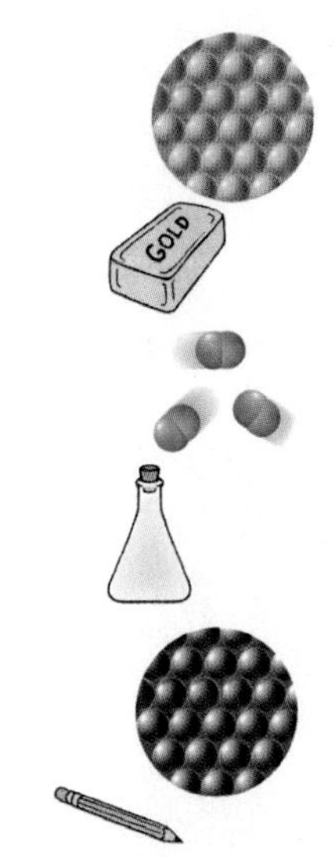

Figure 17.1
Any element consists of only one kind of atom. Gold consists of only gold atoms, a flask of gaseous nitrogen consists of only nitrogen atoms, and the carbon of a graphite pencil consists of only carbon atoms.

1 H																	2 He
3 Li	4 Be											5 B	6 C	7 N	8 O	9 F	10 Ne
11 Na	12 Mg											13 Al	14 Si	15 P	16 S	17 Cl	18 Ar
19 K	20 Ca	21 Sc	22 Ti	23 V	24 Cr	25 Mn	26 Fe	27 Co	28 Ni	29 Cu	30 Zn	31 Ga	32 Ge	33 As	34 Se	35 Br	36 Kr
37 Rb	38 Sr	39 Y	40 Zr	41 Nb	42 Mo	43 Tc	44 Ru	45 Rh	46 Pd	47 Ag	48 Cd	49 In	50 Sn	51 Sb	52 Te	53 I	54 Xe
55 Cs	56 Ba	57 La	72 Hf	73 Ta	74 W	75 Re	76 Os	77 Ir	78 Pt	79 Au	80 Hg	81 Tl	82 Pb	83 Bi	84 Po	85 At	86 Rn
87 Fr	88 Ra	89 Ac	104 Rf	105 Db	106 Sg	107 Bh	108 Hs	109 Mt	110 Uun	111 Uuu	112 Uub		114 Uuq		116 Uuh		118 Uuo

58 Ce	59 Pr	60 Nd	61 Pm	62 Sm	63 Eu	64 Gd	65 Tb	66 Dy	67 Ho	68 Er	69 Tm	70 Yb	71 Lu
90 Th	91 Pa	92 U	93 Np	94 Pu	95 Am	96 Cm	97 Bk	98 Cf	99 Es	100 Fm	101 Md	102 No	103 Lr

Figure 17.2
The periodic table lists all the known elements.

Figure 17.3
A plumb bob is a heavy weight attached to a string and used by carpenters and surveyors to establish a straight vertical line. It gets it name from lead (plumbum, Pb) which is still often used as the weight. Plumbers got their name from working with lead pipes.

As you can see from the periodic table, each element is designated by its **atomic symbol,** which comes from the letters of the element's name. For example, the atomic symbol for carbon is C, and that for chlorine is Cl. In many cases, the atomic symbol is derived from the element's Latin name. Gold has the atomic symbol Au after its Latin name, *aurum.* Lead has the atomic symbol Pb after its Latin name, *plumbum* (Figure 17.3). Elements having symbols derived from Latin names are usually those discovered earliest.

Note that only the first letter of an atomic symbol is capitalized. The symbol for the element cobalt, for instance, is Co. On the other hand, CO is a combination of two elements: carbon, C, and oxygen, O.

Hydrogen was the first element to exist in the universe. And it is still the most common element by far. Today hydrogen makes up more than 90% of all atoms. Hydrogen is very light and has a simple structure. The elements that are heavier than hydrogen have been manufactured in the deep interiors of stars. There, enormous temperatures and pressures fused hydrogen atoms into more complex elements. Most of the elements we find on Earth today were manufactured in stars that exploded long before our solar system came into being.

What else do we know about atoms? One fact that we know is that atoms are very, very old. The birth of most atoms dates back to the origin of the universe. Almost all atoms are older than the sun and the Earth. The atoms in your body have literally existed since the beginning of time. They have been recycling throughout the universe for eons. Atoms combine to form matter of all sorts, living and nonliving. In time, they separate from one another and redistribute.

In this way, you don't "own" the atoms that make up your body—you're simply the present caretaker. The atoms will combine and recombine in countless ways after they leave your body.

Atoms are so small that they can't be seen with visible light. That's because they are even smaller than the wavelengths of visible light. We could stack microscope on top of microscope and never "see" an atom. Photographs of atoms, such as in Figure 17.4, are obtained with a scanning tunneling microscope. This is a non-light imaging device that bypasses light and optics altogether. Even finer detail can be viewed with newer types of imaging devices that are revolutionizing microscopy.

The first direct evidence for atoms was discovered in 1827 by a Scottish botanist, Robert Brown. He was studying grains of pollen in a drop of water under a microscope. He noticed that the grains were continually moving about. At first he thought the grains were moving life forms. But later he found that dust particles and grains of soot moved the same way. This perpetual jiggling of particles—now called *Brownian motion*—results from collisions between visible particles and invisible atoms. Brown's pollen grains were moving because they were constantly being jostled by groups of atoms that make up the water.

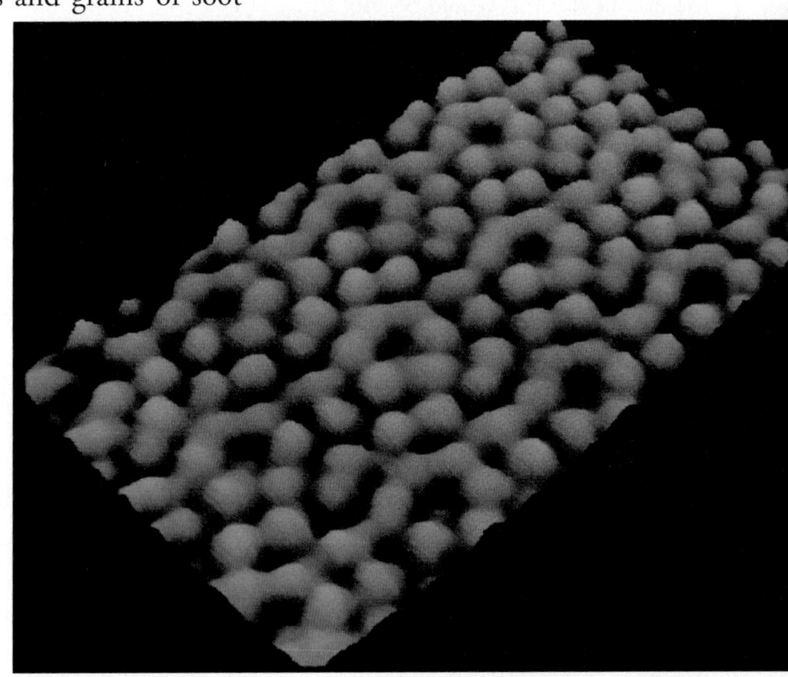

Figure 17.4
An image of carbon atoms obtained with a scanning tunneling microscope.

Since Brown's time, scientific investigation of the atom has been very vigorous and successful. Today we know the atom is made of smaller, subatomic particles—*electrons*, *protons*, and *neutrons*. We also know that atoms differ from one another only in the number of these subatomic particles. Protons and neutrons are bound together at the atom's center to form a larger particle—the **atomic nucleus.** The nucleus is a relatively heavy particle and contains most of an atom's

Link to Physiology: A Breath of Air

Atoms are so small that there are about as many atoms of air in your lungs at any moment as there are breaths of air in the Earth's atmosphere. That's because there are about 10^{22} atoms in a liter of air and about 10^{22} liters of air in the atmosphere. Here's what that means. Exhale a deep breath; the number of atoms exhaled approximately equals the number of breathfuls of air in the Earth's atmosphere. It will take about six years for your breath to become uniformly mixed in the atmosphere. Then anyone, anywhere on Earth, who inhales a breath of air takes in, on average, one of the atoms you exhaled. But you exhale many, many breaths, and so other people breathe in many, many atoms that were once in your lungs—that were once a part of you. And of course, vice versa: With each breath you take in, you recycle atoms that were once a part of everyone who ever lived. Considering that exhaled atoms are part of our bodies, it can be truly said that we are literally breathing one another.

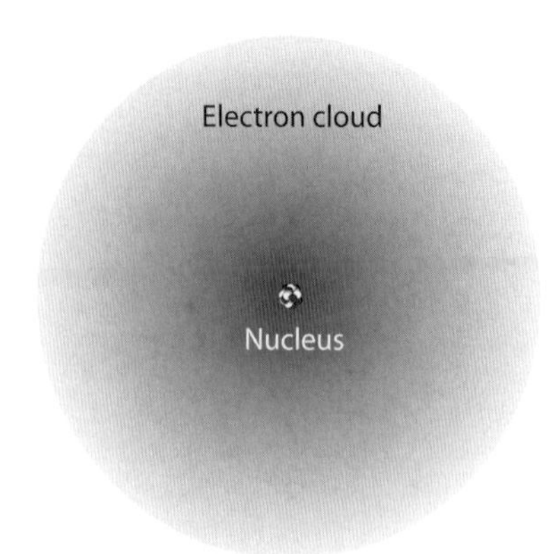

Figure 17.5
Electrons whiz around the atomic nucleus, forming what can be best described as a cloud. If this illustration were drawn to scale, the atomic nucleus would be too small to be seen. An atom is mostly empty space.

mass. Surrounding the nucleus are the tiny **electrons,** as shown in Figure 17.5. When you turn on a light switch you make electrons flow. These make up the current flowing through a light bulb. The electrons that are so important in electricity are the same electrons that dictate chemical reactions. More about this in following chapters.

Concept Check ✓

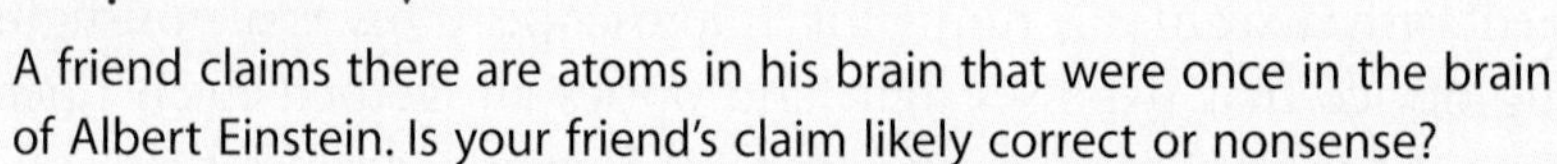

A friend claims there are atoms in his brain that were once in the brain of Albert Einstein. Is your friend's claim likely correct or nonsense?

Check Your Answer Your friend is correct! In addition, there are atoms in your friend's and everyone else's body that were once in Jacqueline Kennedy and everybody else, too! The arrangements of these atoms, however, are now quite different. What's more, the atoms of which you and your friend are composed will live forever in the bodies of all the people on the Earth who are yet to be.

17.2 Atoms Are Mostly Empty Space

You and all materials around you are mostly empty space. How can this be? Electrons move about the nucleus in an atom. But since electrons are very small, and because they are widely separated from the nucleus, atoms are indeed mostly empty space. So why don't atoms simply pass through one another? What keeps us from oozing through the floor we stand on? The answer is: electrical repulsion. Recall from Chapter 11 that like charges repel. In the outer regions of any atom lie the electrons, which repel the electrons of neighboring atoms. Two atoms therefore can get only so close to each other before they start repelling. (As you will see in Chapter 23, however, atoms do make contact when they join in a chemical bond.)

When the atoms of your hand push against the atoms of a wall, electrical repulsions between electrons in your hand and electrons in the wall prevent your hand from passing through the wall. These same electrical repulsions prevent you from falling through the solid floor. They also allow you the sense of touch. Interestingly, when you touch someone, your atoms and those of the other person don't meet. Instead, they get close enough so that you sense the electrical repulsion. There is still a tiny, though imperceptible, gap between the two of you (Figure 17.6).

Figure 17.6
As close as Tracy and Ian are in this photograph, none of their atoms meet. The closeness between us is in our hearts.

17.3 The Atomic Nucleus Is Made of Protons and Neutrons

To get a closer look at the atom, let's investigate the particles found in the atomic nucleus. First, consider protons. A **proton** is a particle found in the atomic nucleus and which carries a positive charge. The proton is heavy—nearly 2000 times as massive as the electron. The proton and electron have the same quantity of charge, but the charges are opposite. The number of protons in the nucleus of any atom is equal to the number of electrons whirling about the nucleus. So the protons' positive charge and electrons' negative charges cancel each other out. Can you see that this guarantees that the atom has an overall electric charge of zero? It is electrically *neutral.* For example, an electrically balanced oxygen atom has eight electrons and eight protons.

How do atoms of one element differ from another element? Each element is identified by its **atomic number.** This is the number of protons contained in each atomic nucleus. Hydrogen, with one proton per atom, has atomic number 1. Helium, with two protons in its nucleus, has atomic number 2; and so on. Look at the periodic table on page 291. Can you see that it lists the elements in order of increasing atomic number?

Concept Check ✓

How many protons are there in an iron atom? (Iron has the chemical symbol Fe and the atomic number 26.)

Check Your Answer The atomic number of an atom and its number of protons are the same. Thus, there are 26 protons in an iron atom. Another way to put this is that all atoms that contain 26 protons are, by definition, iron atoms.

Protons are not alone in the nucleus. The atomic nucleus is made up of other particles as well. Helium, for example, has twice the electric charge of hydrogen but four times the mass. The added mass is due to another subatomic particle in helium's nucleus. What is this other nuclear particle? The neutron. The **neutron** is a nuclear particle with about the same mass as the proton, but with no electric charge. Any object with no net electric charge is said to be electrically neutral (which is where the neutron got its name). We'll have much more to say about the neutron in the following chapter.

Both protons and neutrons are called **nucleons,** a term that states their location in the atomic nucleus. Table 17.1 summarizes the basic facts about our three subatomic particles.

A neutron goes into a restaurant and asks the waiter, "How much for a drink?" The waiter replies, "For you, no charge."

Table 17.1

Subatomic Particles

	Particle	Charge	Mass Compared to Electron	Actual Mass* (kg)
	Electron	−1	1	9.11×10^{-31}†
Nucleons	Proton	+1	1836	1.673×10^{-27}
	Neutron	0	1841	1.675×10^{-27}

* Not measured directly but calculated from experimental data.
† 9.11×10^{-31} kg = 0.000000000000000000000000000000911 kg.

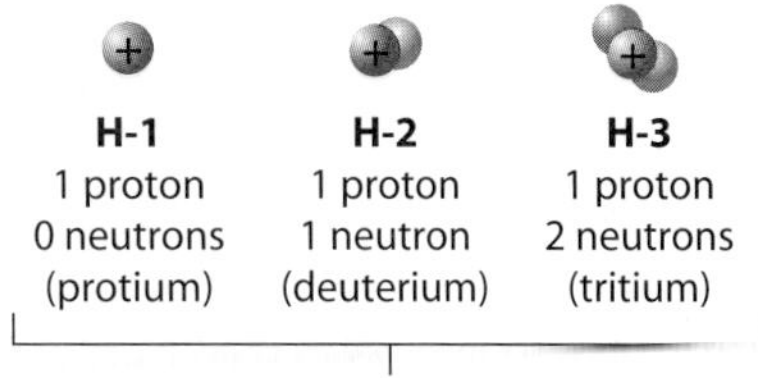

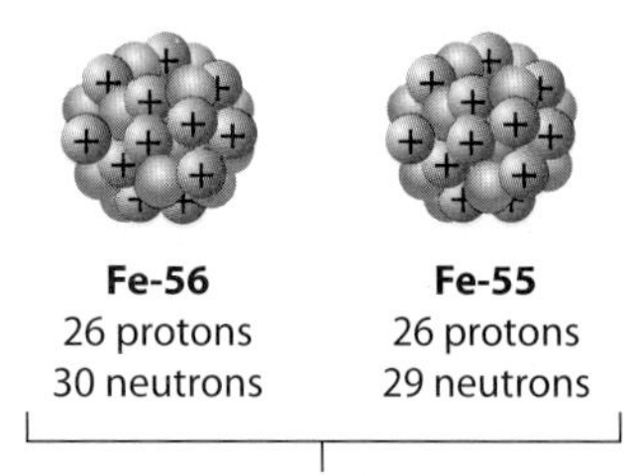

Figure 17.7
Isotopes of an element have the same number of protons but different numbers of neutrons and hence different mass numbers. The three hydrogen isotopes have special names: *protium* for hydrogen-1, *deuterium* for hydrogen-2, and *tritium* for hydrogen-3. Of these three isotopes, hydrogen-1 is most common. For most elements, such as iron, the isotopes have no special names and are indicated merely by mass number.

An element has a definite number of protons, but the number of neutrons it contains may vary. For example, most hydrogen atoms (atomic number 1) have no neutrons. A small percentage, however, have one neutron, and a smaller percentage have two neutrons. Similarly, most iron atoms (atomic number 26) have 30 neutrons, but a small percentage have 29 neutrons. Atoms of the same element that contain different numbers of neutrons are **isotopes.**

17.4 Protons and Neutrons Determine Mass Number and Atomic Mass

How can we tell isotopes apart? We identify isotopes by their **mass number,** which is the total number of protons and neutrons they contain. In other words, mass number is the number of nucleons. As Figure 17.7 shows, a hydrogen isotope with only one proton is called hydrogen-1, where 1 is the mass number. A hydrogen isotope with one proton and one neutron is therefore hydrogen-2, and a hydrogen isotope with one proton and two neutrons is hydrogen-3. Similarly, an iron isotope with 26 protons and 30 neutrons is called iron-56, and one with only 29 neutrons is iron-55.

There is another way to express isotopes. Write the mass number as a superscript and the atomic number as a subscript to the left of the atomic symbol. For example, an iron isotope with a mass number of 56 and atomic number of 26 is written

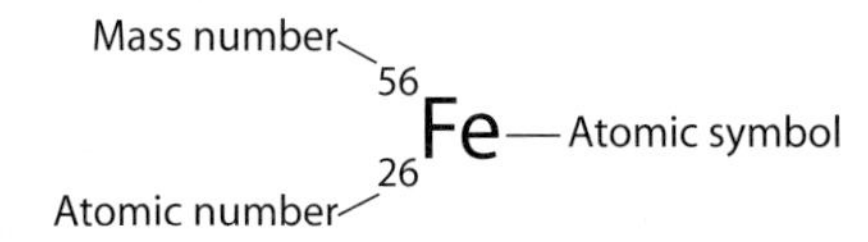

The total number of neutrons in an isotope can be found by subtracting its atomic number from its mass number:

Mass number − atomic number = number of neutrons.

For example, uranium-238 has 238 nucleons. The atomic number of uranium is 92, which tells us that 92 of these 238 nucleons are protons. The remaining 146 nucleons must be neutrons:

238 protons and neutrons − 92 protons = 146 neutrons.

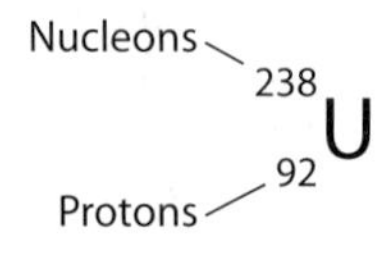

Most of the interactions between atoms are electrical—but not for isotopes. Isotopes can only be distinguished by mass, not by electrical means. Even the mass differences among isotopes are incredibly small. This means that the digestion of sugars containing carbon atoms with seven neutrons is no different than for sugars containing carbon atoms with six neutrons. In fact, about 1% of the carbon we eat is the carbon-13 isotope with seven neutrons per nucleus. The remaining 99% of the carbon in our diet is common carbon-12 isotope, with six neutrons per nucleus. Our bodies don't sense the difference.

The total mass of an atom is called its **atomic mass.** This is the sum of the masses of all the atom's components (electrons, protons, and neutrons). Because electrons are so much less massive than protons and neutrons, their contribution to atomic mass is negligible. A special unit has been developed for atomic masses. This is the *atomic mass unit,* the amu. One atomic mass unit, 1 amu, is slightly less than the mass of a single proton. As shown in Figure 17.8, the atomic masses listed in the periodic table are in atomic mass units.

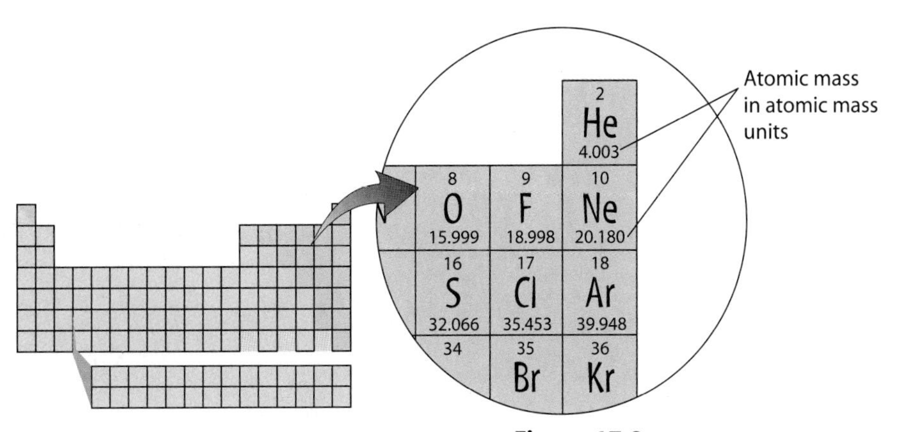

Figure 17.8
Helium, He, has an atomic mass of 4.003 atomic mass units, and neon, Ne, has an atomic mass of 20.18 atomic mass units.

The atomic mass of an element shown in the periodic table is actually the average atomic mass of its various isotopes. This is why atomic masses are not whole numbers. For example, about 99% of all carbon atoms are the isotope carbon-12, and most of the remaining 1% are the heavier isotope carbon-13. This small amount of carbon-13 raises the *average* mass of carbon from 12.0000 amu to the slightly greater value of 12.011 amu.

Concept Check ✓

Distinguish between mass number and atomic mass.

Check Your Answer Both terms include the word mass and so are easily confused. Focus your attention on the second word of each term, however, and you'll get it right every time. Mass number is a count of the number of nucleons in an isotope. An atom's mass number requires no units because it is simply a count. Atomic mass is a measure of the total mass of an atom, which is given in atomic mass units.

17.5 Elements Are Organized in the Periodic Table by Their Properties

So the periodic table is a listing of all the known elements with their atomic masses, atomic numbers, and chemical symbols. But there is much more information in this table. The way the table is organized in groups tells you a lot about the elements' structures and how they behave. Look at how the elements are grouped as metals, nonmetals, and metalloids.

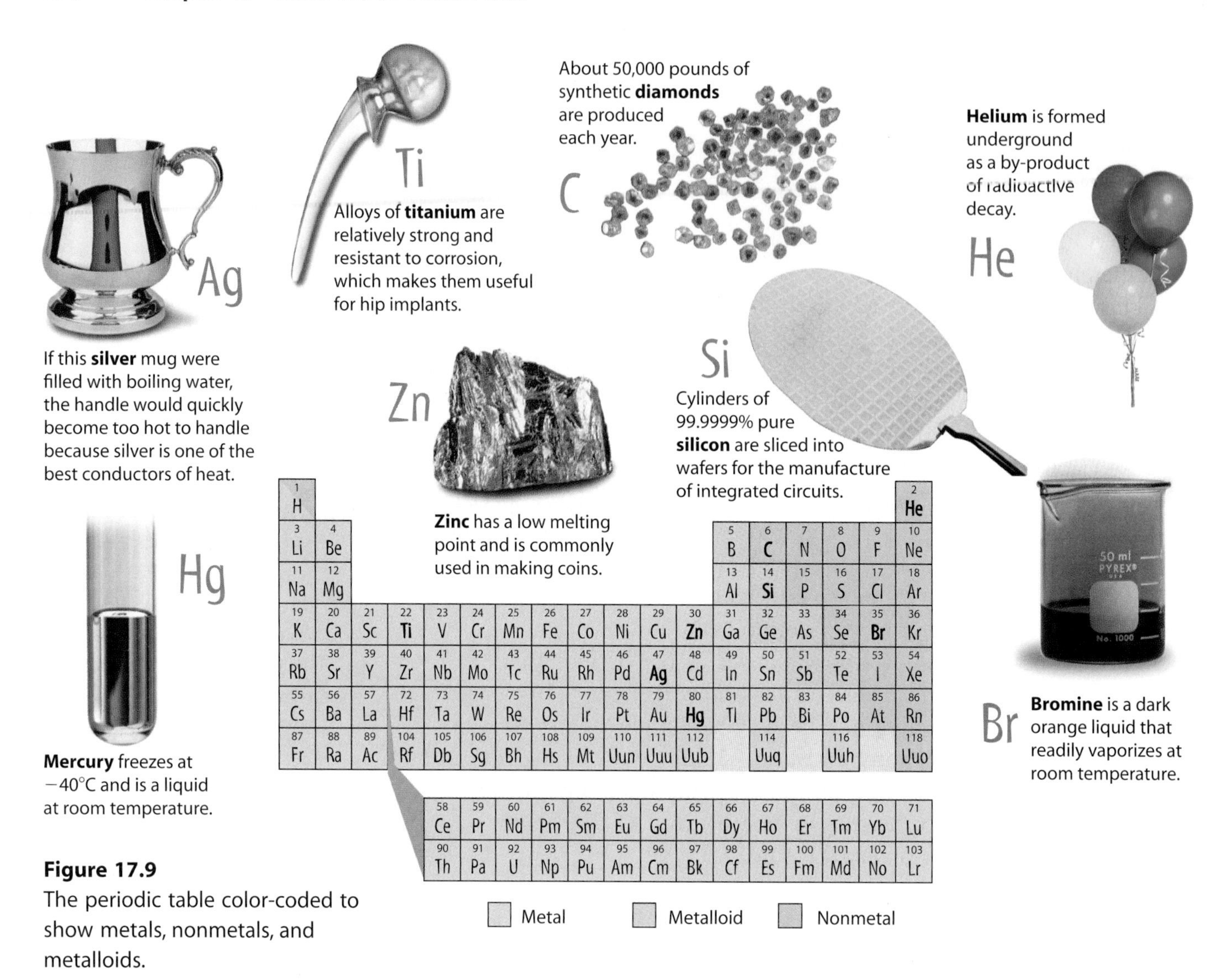

Figure 17.9
The periodic table color-coded to show metals, nonmetals, and metalloids.

Look carefully at Figure 17.9. It shows that metals make up most elements. **Metals** are defined as those elements that are shiny, opaque, and good conductors of electricity and heat. Metals are *malleable*, which means they can be hammered into different shapes or bent without breaking. They are also *ductile*, which means they can be drawn into wires. All but a few metals are solid at room temperature. The exceptions include mercury, Hg; gallium, Ga; cesium, Cs; and francium, Fr. These metals are all liquids at a warm room temperature of 30°C (86°F). Another interesting exception is hydrogen, H. Hydrogen acquires the properties of a liquid metal only at very high pressures (Figure 17.10). Under normal conditions, hydrogen behaves as a nonmetallic gas.

The nonmetallic elements, with the exception of hydrogen, are on the right of the periodic table. **Nonmetals** are very poor conductors of electricity and heat, and they may also be transparent. Solid nonmetals are neither malleable nor ductile. Rather, they are brittle and shatter when hammered. At 30°C (86°F), some nonmetals are solid (such as carbon, C). Other nonmetallic elements are liquid (such as bromine, Br). Still other nonmetals are gaseous (like helium, He).

Six elements are classified as **metalloids**: boron, B; silicon, Si; germanium, Ge; arsenic, As; tin, Sn; and antimony, Sb. You'll see them between the metals and the nonmetals in the periodic table. The metalloids have both metallic and nonmetallic characteristics. For example, they are weak conductors of electricity. This makes them useful as semiconductors in the integrated circuits of computers. Note in the periodic table how germanium, Ge (number 32), is closer to the metals than to the nonmetals. Because of this positioning, we can tell that germanium has more metallic properties than silicon, Si (number 14), and is a slightly better conductor of electricity. So we find that integrated circuits fabricated with germanium operate faster than those fabricated with silicon. Because silicon is much more abundant and less expensive to obtain, however, silicon computer chips remain the industry standard.

Figure 17.10
Geoplanetary models suggest that hydrogen exists as a liquid metal deep beneath the surfaces of Jupiter (shown here) and Saturn. These planets are composed mostly of hydrogen. Inside them, the pressure exceeds 3 million times the Earth's atmospheric pressure. At this tremendously high pressure, hydrogen is pressed to a liquid-metal phase. Back here on the Earth at our relatively low atmospheric pressure, hydrogen exists as a nonmetallic gas of hydrogen molecules, H_2.

17.6 A Period Is a Horizontal Row, a Group Is a Vertical Column

Two other important ways in which the elements are organized in the periodic table are by horizontal rows and vertical columns. Each horizontal row is called a **period,** and each vertical column is called a **group** (or sometimes a *family*). As shown in Figure 17.11, there are 7 periods and 18 groups.

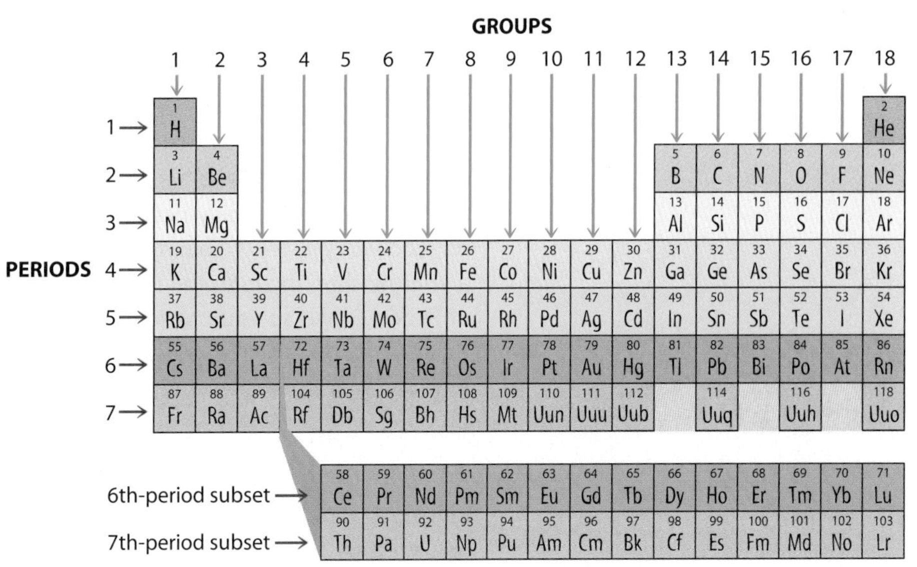

Figure 17.11
The 7 periods (horizontal rows) and 18 groups (vertical columns) of the periodic table. Note that not all periods contain the same number of elements. Also note that, for reasons explained later, the sixth and seventh periods each include a subset of elements, which are listed apart from the main body.

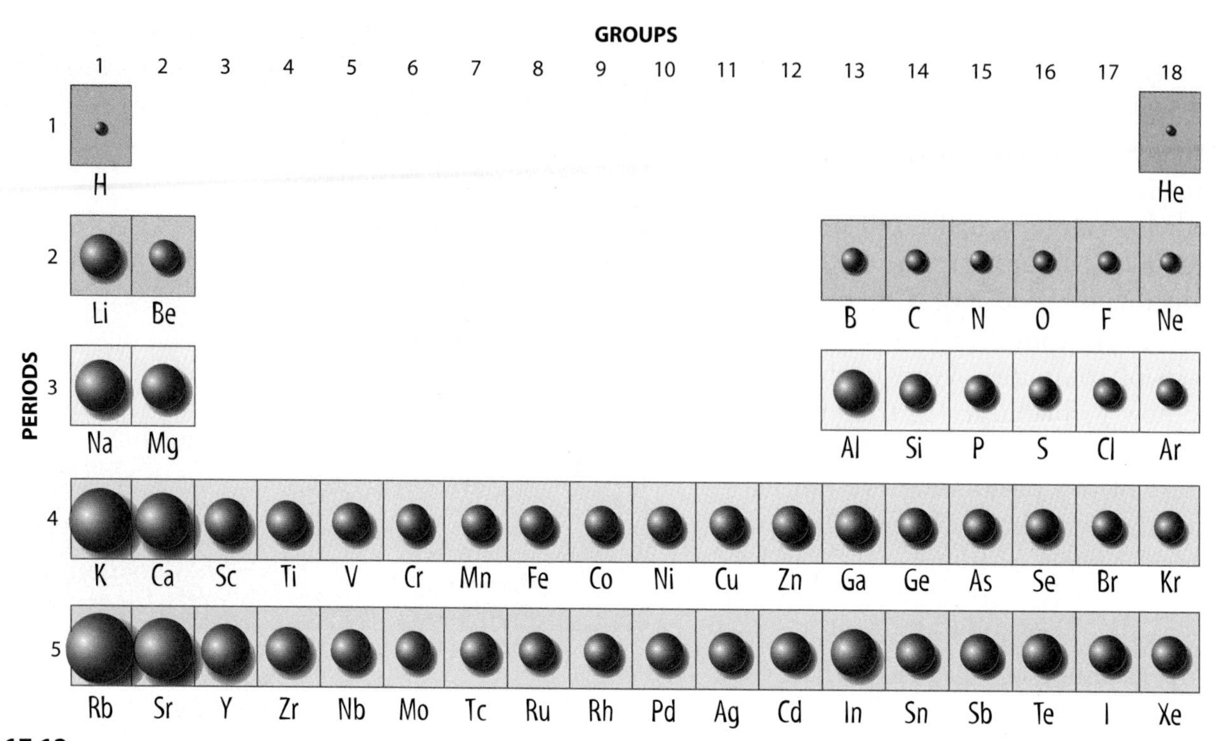

Figure 17.12
The size of atoms gradually decreases in moving from left to right across any period. Atomic size is a periodic (repeating) property.

Across any period, the properties of elements gradually change. This gradual change is called a **periodic trend.** As Figure 17.12 shows, one periodic trend is that atomic size becomes smaller as you move from left to right across any period. Note that the trend repeats from one horizontal row to the next. This repeating of trends is called *periodicity*, a term used to indicate that the trends recur in cycles. Each horizontal row is called a *period* because it corresponds to one full cycle of a trend. There are many other properties of elements that change gradually in moving from left to right across the periodic table.

Concept Check ✓

Which are larger: atoms of cesium, Cs (number 55), or atoms of radon, Rn (number 86)?

Check Your Answer Perhaps you tried looking to Figure 17.12 to answer this question and quickly became frustrated because the 6th period elements are not shown. Well, relax. Look at the trends and you'll see that all atoms to the left are larger than those to the right. Accordingly, cesium is positioned at the far left of Period 6, and so you can reasonably predict that its atoms are larger than those of radon, which is positioned at the far right of Period 6. The periodic table is a road map to understanding the elements.

Moving down any group (vertical column), the properties of elements tend to be remarkably similar. This is why these elements are said to be "grouped" or arranged "in a family." As Figure 17.13

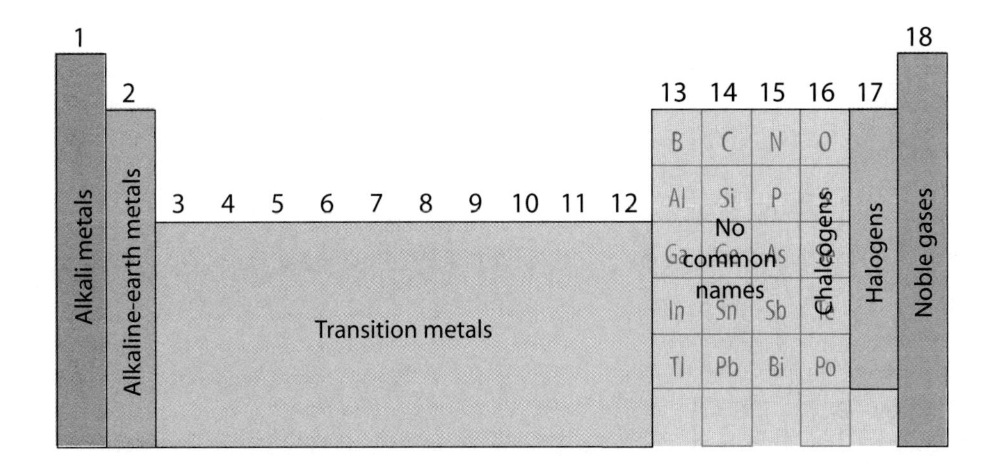

Figure 17.13
The common names for various groups of elements.

shows, several groups have traditional names that describe the properties of their elements. Early in human history, people discovered that ashes mixed with water produce a slippery solution useful for removing grease. By the Middle Ages, such mixtures were described as being *alkaline*, a term derived from the Arabic word for ashes, *al-qali*. Alkaline mixtures found many uses, particularly in the preparation of soaps (Figure 17.14). We now know that alkaline ashes contain compounds of Group 1 elements, most notably potassium carbonate, also known as *potash*. Because of this history, Group 1 elements, which are metals, are called the *alkali metals*.

Figure 17.14
Ashes and water make a slippery alkaline solution once used to clean hands.

Elements of Group 2 also form alkaline solutions when mixed with water. Furthermore, medieval alchemists noted that certain minerals (which we now know are made up of Group 2 elements) do not melt or change when placed in fire. These fire-resistant substances were known to the alchemists as "earth." As a holdover from these ancient times, Group 2 elements are known as the *alkaline-earth metals*.

Toward the right side of the periodic table elements of Group 16 are known as the *chalcogens* ("ore-forming" in Greek) because the top two elements of this group, oxygen and sulfur, are so commonly found in ores. Elements of Group 17 are known as the *halogens* ("salt-forming" in Greek) because of their tendency to form various salts. Interestingly, a small amount of the halogens iodine or bromine inside a lamp allows the tungsten filament of the lamp to glow brighter without burning out so quickly. Such lamps are commonly referred to as halogen lamps. Group 18 elements are all unreactive gases that tend not to combine with other elements. For this reason, they are called the *noble gases* (presumably because people of nobility of earlier times were above interacting with "common folk"!)

The elements of groups 3 through 12 are all metals that do not form alkaline solutions with water. These metals tend to be harder than the alkali metals and less reactive with water. Hence these metals are used for structural purposes. Collectively they are known as the *transition metals*, a name that denotes their central position in the

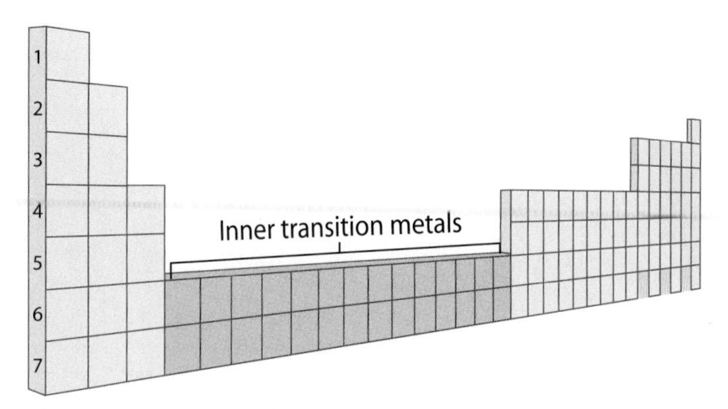

Figure 17.15
Inserting the inner transition metals between atomic groups 3 and 4 results in a periodic table that is not easy to fit on a standard sheet of paper.

periodic table. The transition metals include some of the most familiar and important elements. They are iron, Fe; copper, Cu; nickel, Ni; chromium, Cr; silver, Ag; and gold, Au. They also include many lesser-known elements that are nonetheless important in modern technology. Persons with hip replacements appreciate the transition metals titanium, Ti; molybdenum, Mo; and manganese, Mn, because these noncorrosive metals are used in implant devices.

Concept Check ✓

The elements copper, Cu; silver, Ag; and gold, Au, are three of the few metals that can be found naturally in their elemental state. These three metals have found great use as currency and jewelry for a number of reasons, including their resistance to corrosion and their remarkable colors. How is the fact that these metals have similar properties indicated in the periodic table?

Check Your Answer Copper (number 29), silver (number 47), and gold (number 79) are all in the same group in the periodic table (Group 11). This suggests they should have similar—though not identical—physical and chemical properties.

In the sixth period is a subset of 14 metallic elements (numbers 58 to 71) that are quite unlike any of the other transition metals. A similar subset (numbers 90 to 103) is found in the seventh period. These two subsets are the *inner transition metals.* Inserting the inner transition metals into the main body of the periodic table as in Figure 17.15 results in a long and cumbersome table. So that the table can fit nicely on a standard paper size, these elements are commonly placed below the main body of the table, as shown in Figure 17.16.

The sixth-period inner transition metals are called the *lanthanides* because they follow lanthanum, La. Because of their similar physical and chemical properties, they tend to occur mixed together in the

PERIODS

1	1 H																	2 He
2	3 Li	4 Be											5 B	6 C	7 N	8 O	9 F	10 Ne
3	11 Na	12 Mg											13 Al	14 Si	15 P	16 S	17 Cl	18 Ar
4	19 K	20 Ca	21 Sc	22 Ti	23 V	24 Cr	25 Mn	26 Fe	27 Co	28 Ni	29 Cu	30 Zn	31 Ga	32 Ge	33 As	34 Se	35 Br	36 Kr
5	37 Rb	38 Sr	39 Y	40 Zr	41 Nb	42 Mo	43 Tc	44 Ru	45 Rh	46 Pd	47 Ag	48 Cd	49 In	50 Sn	51 Sb	52 Te	53 I	54 Xe
6	55 Cs	56 Ba	57 La	72 Hf	73 Ta	74 W	75 Re	76 Os	77 Ir	78 Pt	79 Au	80 Hg	81 Tl	82 Pb	83 Bi	84 Po	85 At	86 Rn
7	87 Fr	88 Ra	89 Ac	104 Rf	105 Db	106 Sg	107 Bh	108 Hs	109 Mt	110 Uun	111 Uuu	112 Uub		114 Uuq		116 Uuh		118 Uuo

Inner transition metals

58 Ce	59 Pr	60 Nd	61 Pm	62 Sm	63 Eu	64 Gd	65 Tb	66 Dy	67 Ho	68 Er	69 Tm	70 Yb	71 Lu

← Lanthanides →

90 Th	91 Pa	92 U	93 Np	94 Pu	95 Am	96 Cm	97 Bk	98 Cf	99 Es	100 Fm	101 Md	102 No	103 Lr

← Actinides →

Figure 17.16
The typical display of the inner transition metals. The count of elements in the sixth period goes from lanthanum (La, 57) to cerium (Ce, 58) on through to lutetium (Lu, 71) and then back to hafnium (Hf, 72). A similar jump is made in the seventh period.

same locations in the Earth. Also because of their similarities, lanthanides are unusually difficult to purify. Recently, the commercial use of lanthanides has increased. Several lanthanide elements, for example, are used in the fabrication of the light-emitting diodes (LEDs) of laptop computer monitors.

The seventh-period inner transition metals are called the *actinides* because they follow actinium, Ac. They, too, all have similar properties and hence are not easily purified. The nuclear power industry faces this obstacle because it requires purified samples of two of the most publicized actinides: uranium, U, and plutonium, Pu. Actinides heavier than uranium are not found in nature but are synthesized in the laboratory.

Chapter Review

Key Terms and Matching Definitions

_____ atomic mass
_____ atomic nucleus
_____ atomic number
_____ atomic symbol
_____ electron
_____ element
_____ group
_____ isotope
_____ mass number
_____ metal
_____ metalloid
_____ neutron
_____ nonmetal
_____ nucleon
_____ period
_____ periodic table
_____ periodic trend
_____ proton

1. A fundamental material consisting of only one type of atom.
2. A chart in which all known elements are organized by their properties.
3. An abbreviation for an element or atom.
4. The dense, positively charged center of every atom.
5. An extremely small, negatively charged subatomic particle found outside the atomic nucleus.
6. A positively charged subatomic particle of the atomic nucleus.
7. A count of the number of protons in the atomic nucleus.
8. An electrically neutral subatomic particle of the atomic nucleus.
9. Any subatomic particle found in the atomic nucleus. Another name for either proton or neutron.
10. Atoms of the same element whose nuclei contain the same number of protons but different numbers of neutrons.
11. The number of nucleons (protons and neutrons) in the atomic nucleus. Used primarily to identify isotopes.
12. The mass of an element's atoms listed in the periodic table as an average value based on the relative abundance of the element's isotopes.
13. An element that is shiny, opaque, and able to conduct electricity and heat.
14. An element located toward the upper right of the periodic table and is neither a metal nor a metalloid.
15. An element that exhibits some properties of metals and some properties of nonmetals.
16. A horizontal row in the periodic table.
17. A vertical column in the periodic table, also known as a family of elements.
18. The gradual change of any property in the elements across a period.

Review Questions

Elements Contain a Single Kind of Atom

1. How many types of atoms can you expect to find in a sample of any element?
2. Distinguish between an atom and an element.

Atoms Are Mostly Empty Space

3. If atoms are mostly empty space, why can't we walk through walls?
4. What kind of force prevents atoms from squishing into one another?

The Atomic Nucleus Is Made of Protons and Neutrons

5. How much more massive is a proton compared to an electron?
6. Compare the electric charge on the proton with the electric charge on the electron.
7. What is the definition of atomic number?

8. What role does atomic number play in the periodic table?

Protons and Neutrons Determine Mass Number and Atomic Mass

9. What effect do isotopes of a given element have on the atomic mass calculated for that element?

10. Name two nucleons.

11. Distinguish between atomic number and mass number.

12. Distinguish between mass number and atomic mass.

Elements Are Organized in the Periodic Table by Their Properties

13. How is the periodic table more than just a listing of the known elements?

14. Are most elements metallic or nonmetallic?

15. Why is hydrogen, H, most often considered a nonmetallic element?

16. Where are metalloids located in the periodic table?

A Period Is a Horizontal Row, a Group Is a Vertical Column

17. How many periods are there in the periodic table? How many groups?

18. Why are Group 17 elements called halogens?

19. Which Group of elements are all gases at room temperature?

20. Why are the inner transition metals not listed in the main body of the periodic table?

Exercises

1. A cat strolls across your backyard. An hour later, a dog with its nose to the ground follows the trail of the cat. Explain what is going on in terms of atoms.

2. If all the molecules of a body remained part of that body, would the body have any odor?

3. Which are older, the atoms in the body of an elderly person or those in the body of a baby?

4. In what sense can you truthfully say that you are a part of every person around you?

5. Considering how small atoms are, what are the chances that at least one of the atoms exhaled in your first breath will be in your last breath?

6. Germanium, Ge (number 32), computer chips operate faster than silicon, Si (number 14), computer chips. So how might a gallium, Ga (number 31), chip compare with a germanium chip?

7. Helium, He, is a nonmetallic gas and the second element in the periodic table. Rather than being placed adjacent to hydrogen, H, however, helium is placed on the far right of the table. Why?

8. Name ten elements that you could purchase samples of at a hardware, jewelry, or other store.

9. Strontium, Sr (number 38), is especially dangerous to humans because it tends to accumulate in calcium-dependent bone marrow tissues (calcium, Ca, number 20). How does this fact relate to what you know about the organization of the periodic table?

10. With the periodic table as your guide, describe the element selenium, Se (number 34). Use as many of this chapter's key terms as you can.

11. Which of the following diagrams best represents the size of the atomic nucleus relative to the size of the atom?

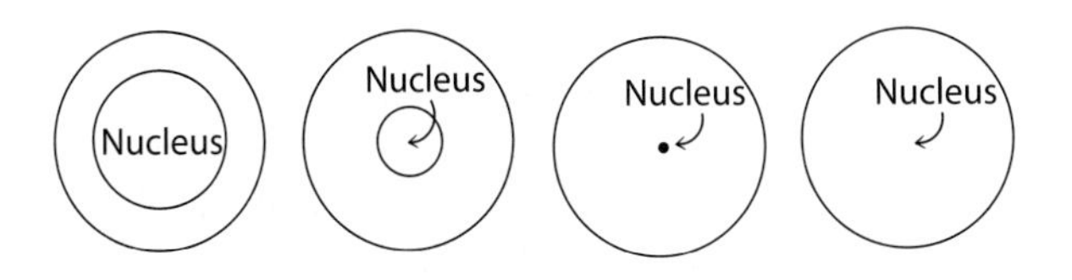

12. If two protons and two neutrons are removed from the nucleus of an oxygen atom, what nucleus remains?

13. You could swallow a capsule of germanium, Ge (atomic number 32), without ill effects. If a proton were added to each germanium nucleus, however, you would not want to swallow the capsule. Why? (Consult a periodic table of the elements.)

14. What happens to the properties of elements across any period of the periodic table?

15. If an atom has 43 electrons, 56 neutrons, and 43 protons, what is its approximate atomic mass? What is the name of this element?

16. The nucleus of an electrically neutral iron atom contains 26 protons. How many electrons does this iron atom have?

17. Evidence for the existence of neutrons did not come until many years after the discoveries of the electron and the proton. Give a possible explanation.

18. Which has more atoms: a 1-gram sample of carbon-12 or a 1-gram sample of carbon-13? Explain.

19. Why are the atomic masses that are listed in the periodic table not whole numbers?

20. Where did the carbon atoms in Leslie's hair originate? (Shown below is a photo of author Leslie at age 16.)

Suggested Readings and Web Sites

The CRC Handbook of Chemistry and Physics. Boca Raton, FL: CRC Press, 1996.
Toward the front of this classic reference book you'll find a section on the history and general properties of each element.

http://clri6c.gsi.de/~demo/wunderland/englisch/Inhalt.html
This is the Web site for the heavy-ion research facility in Darmstadt, Germany, where many of the heaviest but shortest-lived elements are being created.

http://www.shef.ac.uk/~chem/web-elements/
There are a large number of periodic tables posted on the Web, and this is one of the most popular ones.

Chapter 18: Atomic Models

Have you ever wondered how fireworks produce their beautiful displays of color? Was the element helium, common in children's balloons, discovered in the sun before it was found on Earth? If atoms are so small that they're invisible, how can so much be known about them? Do electrons in an atom really orbit the nucleus like planets orbit the sun? And what does whistling down a long tube have in common with the quantum hypothesis? The answers to these questions all involve the structure of the atom and the models we create to understand that structure.

18.1 Models Help Us Visualize the Invisible World of Atoms

Atoms are so small that the number of them in a baseball is roughly equal to the number of Ping-Pong balls that could fit inside a hollow sphere as big as the Earth. This is illustrated in Figure 18.1.

In Chapter 17 you learned that atoms are invisible because they are smaller than the wavelength of visible light. Although we cannot see atoms *directly,* we can generate images of them *indirectly.* In the mid 1980s, researchers developed the *scanning tunneling microscope* (STM). This machine produces images by dragging an ultra-thin needle back and forth over the surface of a sample. Bumps the size of atoms on the surface cause the needle to move up and down. This vertical motion is detected and translated by a computer into a topographical image that corresponds to the positions of atoms on the surface (Figure 18.2). An STM can also be used to push individual atoms into desired positions. This ability opened up the field of nanotechnology. Using nanotechnology, incredibly small electronic circuits and motors can be built atom by atom.

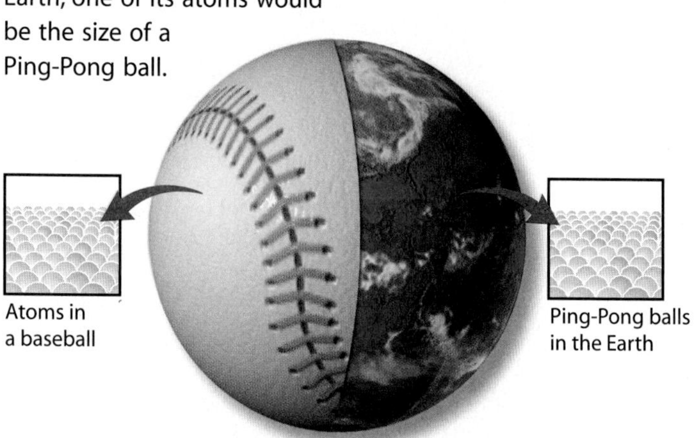

Figure 18.1
If the Earth were filled with nothing but Ping-Pong balls, the number of balls would be roughly equal to the number of atoms in a baseball. Said differently, if a baseball were the size of the Earth, one of its atoms would be the size of a Ping-Pong ball.

Figure 18.2
(a) Scanning tunneling microscopes are relatively simple devices used to create submicroscopic imagery. (b) An image of gallium and arsenic atoms obtained with an STM. (c) Each dot in the world's tiniest map consists of a few thousand gold atoms, each atom moved into its proper place by an STM.

(a)

(b)

(c)

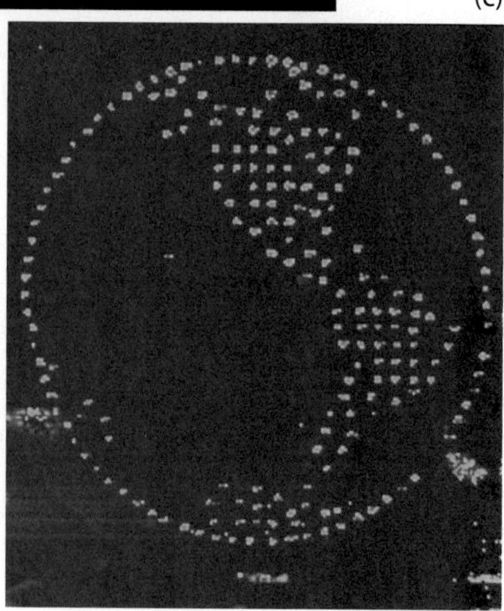

Concept Check ✓

What is the technology that lets us see invisible atoms?

Check Your Answer Scanning tunneling microscopes, for one thing. The atomic images generated by STMs are computer renditions generated from the movements of an ultrathin needle.

The new nanotechnology builds things atom by atom. But trees have been doing that all along!

A very small or very large object that can be seen with microscopes, telescopes or other viewing instruments can be represented with a **physical model.** A physical model is a representation of a very large or small object shown at a convenient scale. Figure 18.3a shows a large-scale physical model of a microorganism typical in a biology class. Making a model of invisible atoms isn't as easy as creating a larger-than-life microorganism however, because there is no way to actually see an atom. The image produced by an STM, for example, merely shows the *positions* of atoms and not actual images of atoms. No detail of the atom is seen. So, instead of describing the atom with a physical model, scientists use a **conceptual model.** A conceptual model is an explanation that treats what is being explained as a *system.* In a system, it is the interactions among component parts that matter most; knowing or seeing the parts themselves does not provide the understanding we are looking for. The accuracy of a conceptual model is determined, not by how closely it matches a physical object, but by how well it predicts the behavior of the system. The weather, for example, is best described using a conceptual model (Figure 18.3b). Such a model shows how the various parts of the system interact with one another. These components of the system include humidity, atmospheric pressure, temperature, electric charge, and the motion of large masses of air.

Other systems described by conceptual models are the economy, population growth, the spread of diseases, and even team sports.

Figure 18.3
(a) This large-scale model of a microorganism is a physical model. (b) Weather forecasters rely on conceptual models such as this one to predict the behavior of weather systems.

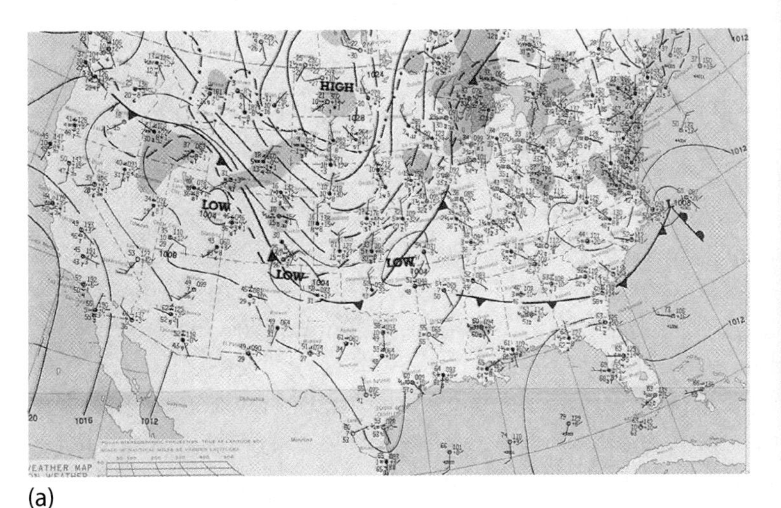

(a)

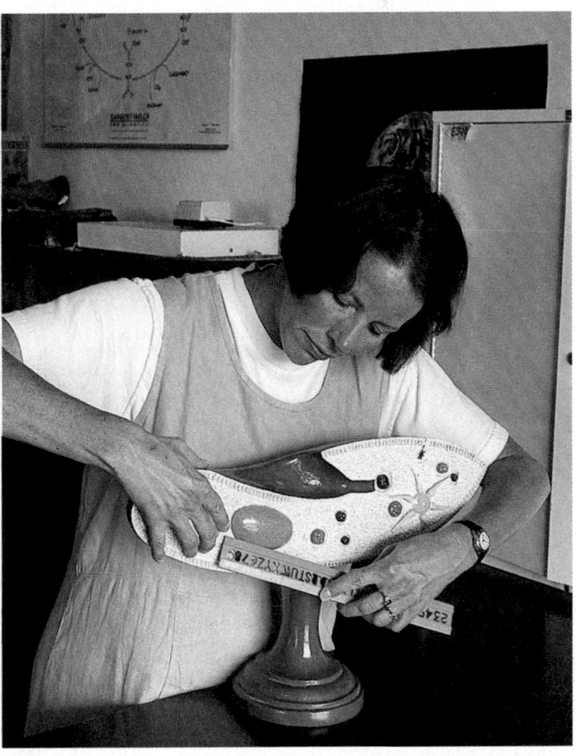

(b)

Concept Check ✓

A basketball coach describes a playing strategy to her team by way of sketches on a game card. Do the illustrations represent a physical model or a conceptual model?

Check Your Answer The sketches are a conceptual model that describes a system (the players on the court) with the hope of predicting an outcome (winning the game).

Atoms are best described with a conceptual model. Don't mistake the model for a visual picture that you'd see with miracle magnification. In Section 18.3, for example, we'll discuss the planetary model of the atom where electrons are shown orbiting the atomic nucleus like planets orbiting the sun. This planetary model is limited. Once it was very popular, but scientists have found that it doesn't explain many properties of atoms. Thus, newer and more accurate (and more complicated) conceptual models of the atom have since been introduced. In these models, electrons appear as a cloud hovering around the atomic nucleus. But even these models have their limitations. Ultimately, the best models of the atom are ones that are purely mathematical.

In this book, our focus is on conceptual atomic models that are easily represented by visual images. These include the planetary atomic model and a model in which electrons are grouped in units called *shells*. Despite their limitations, such images are excellent guides to learning, especially for beginners. These models were developed by scientists to help explain how atoms emit light.

18.2 Atoms Can Be Identified by the Light They Emit

When an atom absorbs energy (from heat or electricity, for example) the atom is temporarily energized. Soon the atom may lose this energy. And as the atom releases its extra energy, something very important happens: The atom emits light. The atoms of a particular element can emit only certain frequencies (or colors) of light, however. As a result, each element produces its own distinctive glow when energized.

Sodium atoms emit bright yellow light. This makes them useful in sodium vapor street lamps, which more efficiently light up areas because our eyes are very sensitive to yellow light. Neon atoms emit a brilliant red-orange light, which is a useful light source in neon signs. Glowing elements are responsible for the colors of fireworks.

As we learned in Chapter 14, we see white light when all frequencies of visible light reach our eyes at the same time. Recall that white light is separated into its color components by a prism

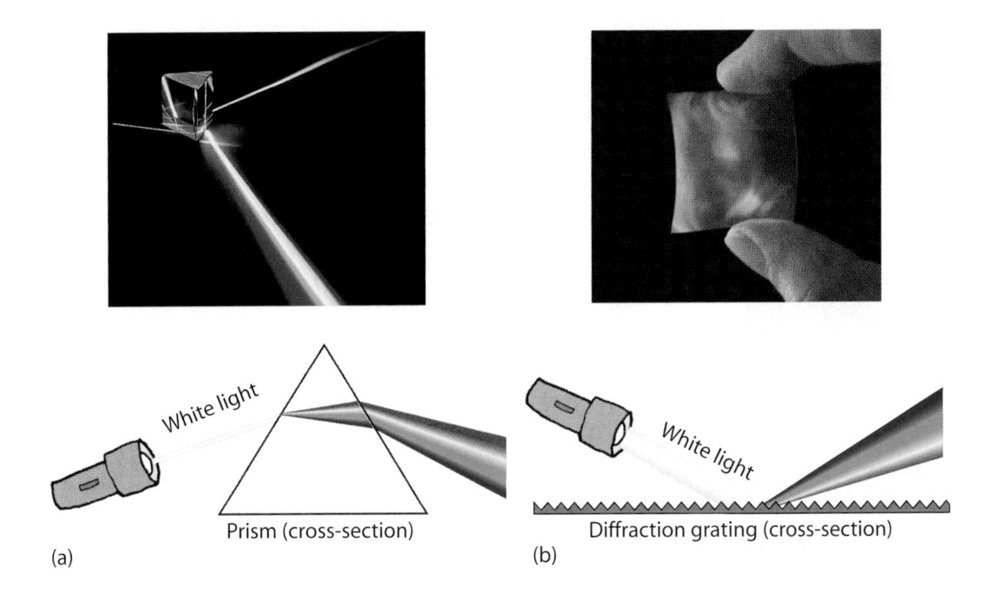

Figure 18.4
White light is separated into its color components by (a) a prism and (b) a diffraction grating.

(or diffraction grating) as it passes through (Figure 18.4). Also recall that each color of visible light corresponds to a different frequency of electromagnetic radiation. A **spectroscope,** shown in Figure 18.5, is an instrument used to observe the color components of any light source. The spectroscope, therefore, allows us to analyze the light emitted by glowing elements.

When we view the light from glowing atoms through a spectroscope, we see that the light consists of a number of discrete (separate from one another) frequencies rather than a continuous spectrum like the one shown in Figure 18.5. The pattern of frequencies formed by a given element is referred to as that element's **atomic spectrum.**

Figure 18.5
(a) In a spectroscope, light emitted by atoms passes through a narrow slit before being separated into particular frequencies by a prism or (as shown here) a diffraction grating.
(b) This is what the eye sees when the slit of a diffraction-grating spectroscope is pointed toward a white-light source. Spectra of colors appear to the left and right of the slit.

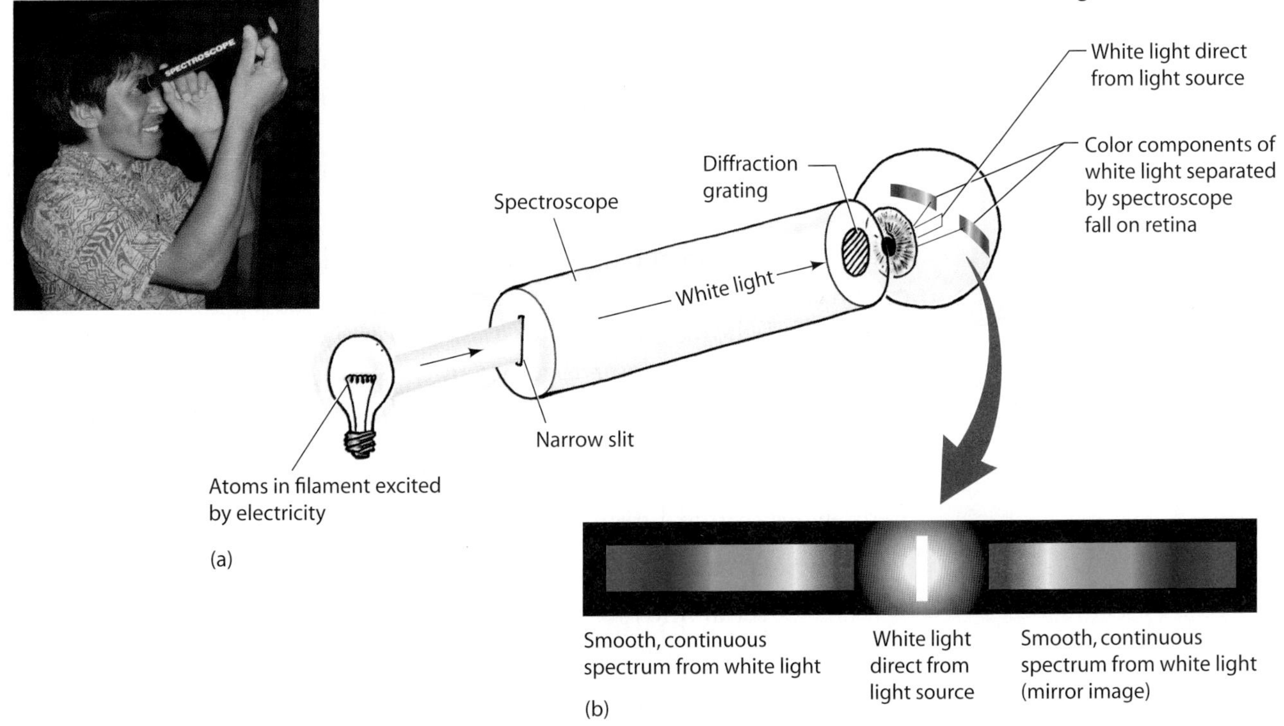

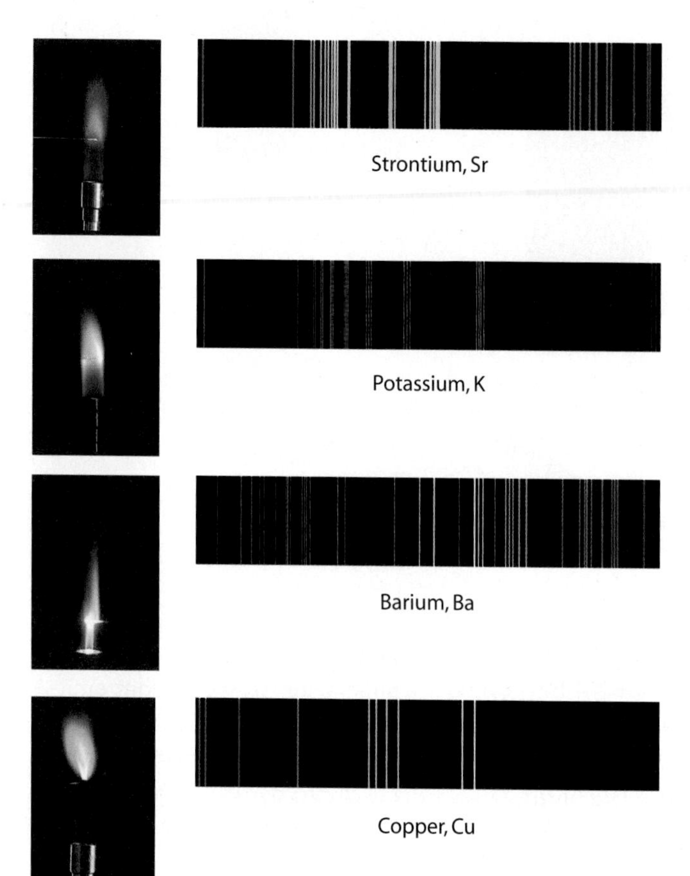

Figure 18.6
Elements heated by a flame glow their characteristic color. This is commonly called a *flame test* and is used to test for the presence of an element in a sample. When viewed through a spectroscope, the color of each element is revealed to consist of a pattern of distinct frequencies known as an *atomic spectrum.*

Look at Figure 18.6 for some examples. The atomic spectrum is an element's "fingerprint." You can identify the elements in a light source by analyzing the light through a spectroscope and looking for characteristic patterns in its atomic spectrum. If you don't have the opportunity to work with a spectroscope in your laboratory, at least focus special attention on the Exploration on page 311.

Concept Check ✓

How might you tell what elements are in a star?

Check Your Answer Aim a well-built spectroscope at the star, and study its spectral patterns. In the late 1800s, this was done with our own star, the sun. Spectral patterns of hydrogen and some other known elements were observed, in addition to one pattern that could not be identified. Scientists concluded that this unidentified pattern belonged to an element not yet discovered on the Earth. They named this element helium after *helios,* the Greek word for "sun."

Figure 18.7
A portion of the atomic spectrum for hydrogen. These frequencies are higher than those of visible light, which is why they are not shown in color.

Researchers in the 1800s noted that the lightest element, hydrogen, has a far more orderly atomic spectrum than the other elements. Figure 18.7 shows a portion of the hydrogen spectrum. Note how successive lines get closer together in a regular way. A Swiss schoolteacher, Johann Balmer, expressed these line positions by a mathematical formula. Another regularity in hydrogen's atomic spectrum was noticed by Johannes Rydberg. Rydberg found that the sum of the frequencies of two lines often equals the frequency of a third line. For example,

First spectral line	1.6×10^{14} Hz
Second spectral line	$+\,4.6 \times 10^{14}$ Hz
Third spectral line	6.2×10^{14} Hz

Specs-On Exploration: Spectral Patterns

Purchase some "rainbow" glasses from a nature, toy, or hobby store. The lenses of these glasses are diffraction gratings. Looking through them, you will see light separated into its color components. Certain light sources, such as the moon or a car's headlights, are separated into a continuous spectrum—in other words, all the colors of the rainbow appear in a continuous sequence from red to violet.

Other light sources, however, emit a distinct number of discontinuous colors. Examples include streetlights, neon signs, sparklers, and fireworks. The spectral patterns you see from these light sources are the atomic spectra of elements heated in the light sources. You'll be able to see the patterns best when you are at least 50 meters from the light source. This distance makes the spectrum appear as a series of dots like those of this chapter's opening photograph and similar to the series of lines shown in Figure 18.8.

The rainbow side of a compact disk can also be used for viewing spectral patterns. Holding the disk at eye level parallel to the ground, look over it at a light source, and observe the rainbow reflection. While focusing on the reflection, bring the disk as close as possible to your eye. Doing so will make the spectral pattern more apparent.

Share your rainbow glasses and disk with a friend on your next "night on the town." You'll find each type of light has its own signature pattern. How many different patterns are you able to observe?

The orderliness of hydrogen's atomic spectrum was most intriguing to Balmer, Rydberg, and other investigators of the time. But they couldn't give a reason why the spectrum was so orderly. These early investigators were unable to formulate any hypothesis that agreed with any accepted atomic model of the day.

18.3 Niels Bohr Used the Quantum Hypothesis to Explain Atomic Spectra

Recall what we learned in Chapter 16: Light travels as a wave, but strikes as a particle. The **quantum hypothesis** states that a beam of light is not a continuous train of waves, but is a stream of zillions of discrete particles (Figure 18.8). We call a particle of light a **photon.** We say the same thing when we say a beam of light is a stream of *quanta.* A **quantum** is the smallest unit of something.

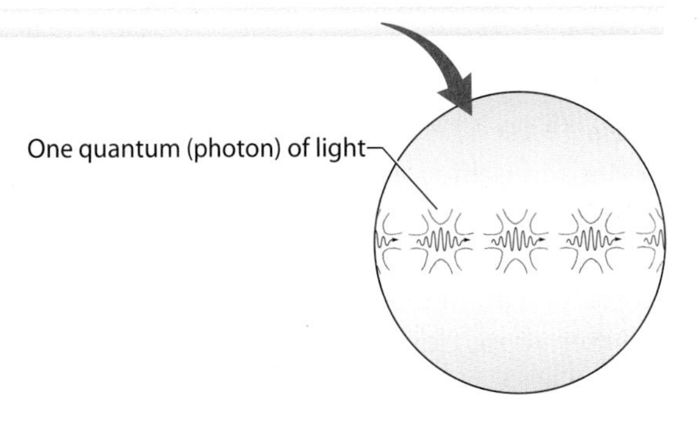

Figure 18.8
Light is quantized, which means it consists of a stream of energy packets. Each packet is called a quantum, also known as a photon.

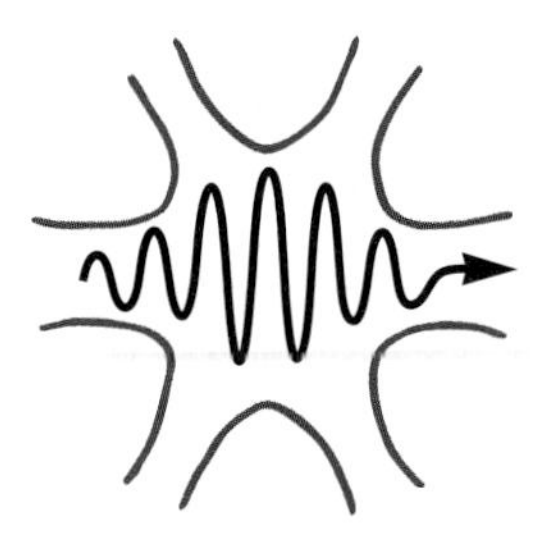

High-frequency,
high-energy
photon

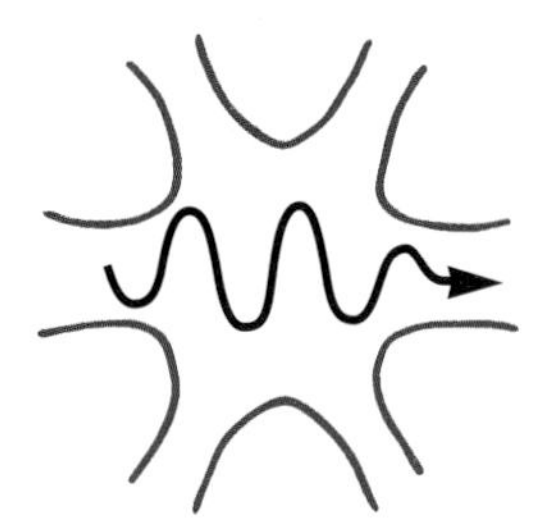

Low-frequency,
low-energy
photon

Figure 18.9
The greater the frequency of a photon of light, the greater the energy packed into that photon.

To better understand the term *quantum,* consider a gold brick. The mass of the brick equals some whole-number multiple of the mass of a single gold atom. Similarly, an electric charge is always some whole-number multiple of the charge on a single electron. Mass and electric charge are therefore said to be *quantized.* They consist of some number of basic units.

As indicated in Figure 18.9, the amount of energy in a photon increases with the frequency of the light. One photon of ultraviolet light, for example, possesses more energy than one photon of infrared light. Why? Because ultraviolet light has higher frequency than infrared. (Recall our discussion of this back in Chapter 16 when we treated the photoelectric effect.)

Using the idea of the quantum, the Danish scientist Niels Bohr (1885–1962) was finally able to explain atomic spectra. Bohr explained that first, the potential energy of an electron in an atom depends on the electron's distance from the nucleus. This is analogous to the potential energy of an object held some distance above the Earth's surface. The object has more potential energy when it is held high above the ground than when it is held close to the ground. Likewise, an electron has more potential energy when it is far from the nucleus than when it is close to the nucleus. Second, when an atom absorbs a photon of light, it absorbs energy. This energy is acquired by one of the electrons surrounding the atom's nucleus. Because of its gained energy, the electron moves away from the nucleus. In other words, absorption of a photon causes a low-potential-energy electron in an atom to become a high-potential-energy electron.

Bohr also realized that the opposite is true: When a high-potential-energy electron in an atom loses some of its energy, the electron moves closer to the nucleus. What happens to the "lost" energy? Energy, of course, is never really lost. The energy lost by the electron as it moves closer to the nucleus is emitted from the atom as a photon of light. Both absorption and emission are illustrated in Figure 18.10.

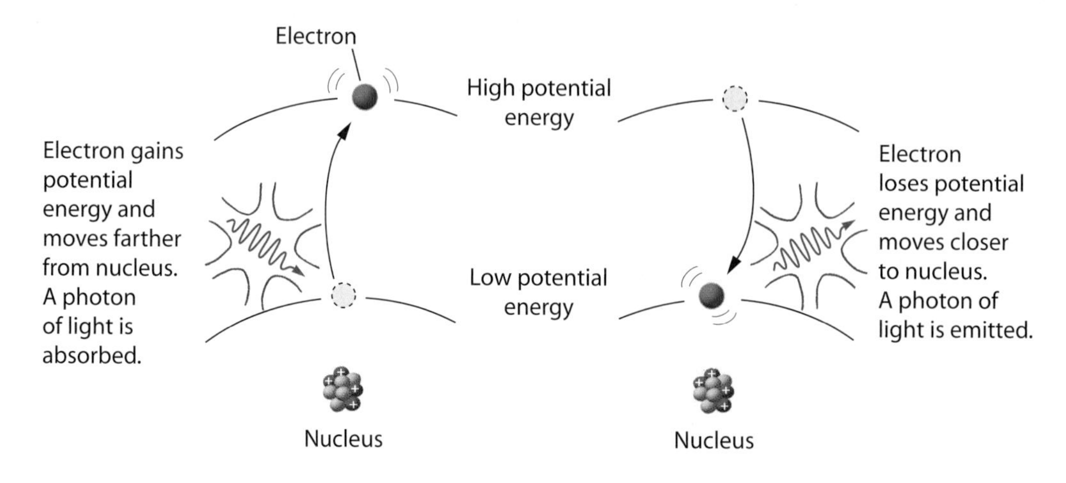

Figure 18.10
An electron is lifted away from the nucleus as the atom it is in absorbs a photon of light. The electron then drops closer to the nucleus as the atom releases a photon of light.

Concept Check ✓

Which has more energy: a photon of red light or a photon of infrared light?

Check Your Answer As was discussed in Chapters 14 and 16, red light has a higher frequency than infrared light, which means a photon of red light has more energy than does a photon of infrared light.

If light energy is quantized, the energy of an electron in an atom must also be quantized. In other words, an electron cannot have just any amount of potential energy. Rather, within the atom there must be a number of distinct energy levels, like steps on a staircase. Your location on a staircase is restricted to where the steps are—you cannot stand at a height that is, say, halfway between two steps. Similarly, there are only a limited number of permitted energy levels in an atom, and an electron can never have an amount of energy between these permitted energy levels. Bohr gave each energy level a **principal quantum number *n*,** where n is always some integer. The lowest energy level has a principal quantum number $n = 1$. An electron for which $n = 1$ is as close to the nucleus as possible, and an electron for which $n = 2$, $n = 3$, and so forth is farther away from the nucleus.

Further, Bohr developed a conceptual model in which an electron moving around the nucleus is restricted to certain distances from the nucleus. These distances determine the amount of energy the electron has. Bohr saw the similarity of this to planets held in orbit around the sun at their respective distances from the sun. The energy levels for any atom, therefore, could be graphically represented as orbits around the nucleus (Figure 18.11). Bohr's quantized model of the atom thus became known as the *planetary model.*

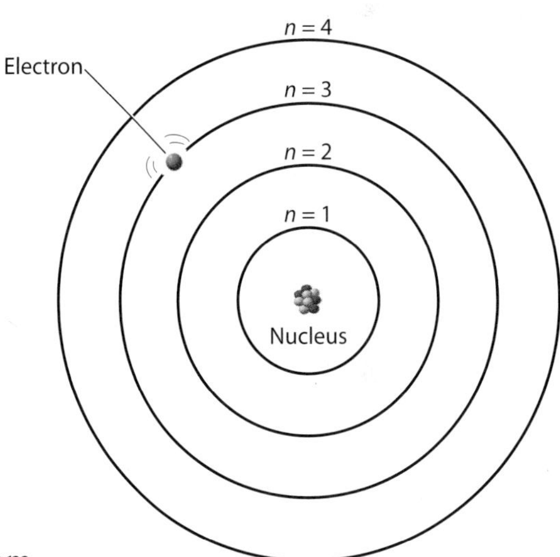

Figure 18.11
Bohr's planetary model of the atom, in which electrons orbit the nucleus much like planets orbit the sun, is a graphical representation. Not to be taken literally, it nevertheless helps us understand how electrons can possess only certain quantities of energy.

Bohr used his planetary model to explain why atomic spectra are made up of bands of different frequency light rather than a continuous spectrum. According to the Bohr's planetary model, photons are emitted by atoms when electrons move from higher-energy, outer orbits to lower-energy, inner orbits. The energy of an emitted photon is equal to the difference in energy between the two orbits. Because an electron can only be found in certain orbits, only certain light frequencies are emitted.

Bohr was also able to explain why the sum of two frequencies of light emitted by an atom often equals a third emitted frequency. If an electron is raised to the third energy level—that is, the third highest orbit, the one for which $n = 3$—it can return to the first

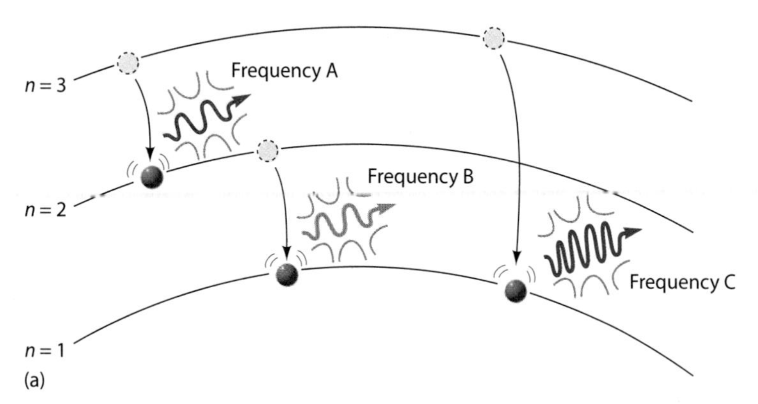

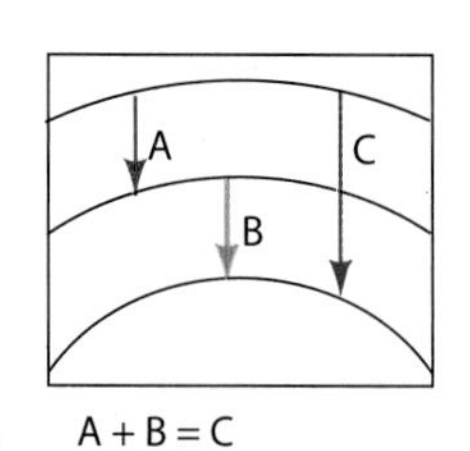

(b) A + B = C

Figure 18.12
(a) The frequency of light emitted (or absorbed) by an atom is proportional to the energy difference between electron orbits. Because the energy differences between orbits are discrete, the frequencies of light emitted (or absorbed) are also discrete. The electron here can emit only three discrete frequencies of light—A, B, and C. The greater the transition, the higher the frequency of the photon emitted. (b) The sum of the energies (and frequencies) for transitions A and B equals the energy (and frequency) of transition C.

orbit by two routes. As shown in Figure 18.12, it can return by a single transition from the third to the first orbit, or it can return by a double transition from the third orbit to the second and then to the first. The single transition emits a photon of frequency C, and the double transition emits two photons, one of frequency A and one of frequency B. These three photons of frequencies A, B, and C are responsible for three spectral lines. Note that the energy transition for A plus B is equal to the energy transition for C. Because frequency is proportional to energy, frequency A plus frequency B equals frequency C.

Concept Check ✓

Suppose the frequency of light emitted in Figure 18.12 is 5 billion hertz along path A and 7 billion hertz along path B. What frequency of light is emitted when an electron makes a transition along path C?

Check Your Answer Add the two known frequencies to get the frequency of path C: 5 billion hertz + 7 billion hertz = 12 billion hertz.

Bohr's planetary atomic model was a tremendous success. By utilizing the quantum hypothesis, Bohr's model solved the mystery of atomic spectra. Despite its successes, though, Bohr's model was limited because it did not explain *why* energy levels in an atom are quantized. Bohr himself was quick to point out that his model was to be interpreted only as a crude beginning, and the picture of electrons whirling about the nucleus like planets about the sun was not to be taken literally (a warning to which popularizers of science paid no heed).

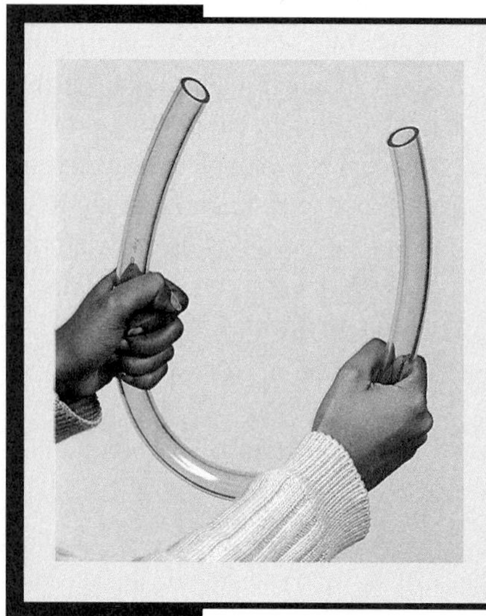

Hands-On Exploration: Quantized Whistle

You can "quantize" your whistle by whistling down a long tube, such as the tube from a roll of wrapping paper. First, without the tube, whistle from a high pitch to a low pitch. Do it in a single breath and as loud as you can. Then repeat while holding the tube to your lips. Ah, ha! Note that some frequencies simply cannot be whistled, no matter how hard you try. These frequencies are forbidden because their wavelengths are not a multiple of the length of the tube.

Try experimenting with tubes of different lengths. To hear yourself more clearly, use a flexible plastic tube and twist the outer end toward your ear.

When your whistle is confined to the tube, the consequence is a quantization of its frequencies. When an electron is confined to an atom, the consequence is a quantization of the electron's energy.

18.4 A Shell Is a Region of Space Where an Electron Can Be Found

The quality of a song depends upon the arrangement of musical notes. In a similar fashion, the properties of an atom depend upon the arrangements of electrons in its atoms. To understand how electrons are arranged in an atom, we turn to a conceptual model known as the *shell model.* This model is similar to Bohr's planetary model in that it is highly simplified.

According to the shell model, electrons behave as though they are arranged in a series of concentric shells. A **shell** is defined as a region of space about the atomic nucleus within which electrons may reside. An important aspect of this model is that each shell can hold only a limited number of electrons. As shown in Figure 18.13, the innermost shell can hold 2, the second and third shells, 8 electrons each; the fourth and fifth shells, 18 each; and the sixth and seventh shells, 32 each.

A series of seven concentric shells accounts for the seven periods of the periodic table. Furthermore, the number of elements in each period is equal to the shell's capacity for electrons. The first shell, for example, has a capacity for only two electrons. That's why we find only two elements, hydrogen and helium, in the first period (Figure 18.14). The second and third shells each have a capacity for eight electrons, and so eight elements are found in both the second and third periods.

The electrons of the outermost occupied shell in any atom are directly exposed to the external environment and are the first to interact with other atoms. Most notably, they are the ones that participate in chemical bonding, as we shall be discussing in Chapter 23. The electrons in the outermost shell, therefore, are quite important. They are called **valence electrons.** The term *valent* refers to the "combining power" of an atom.

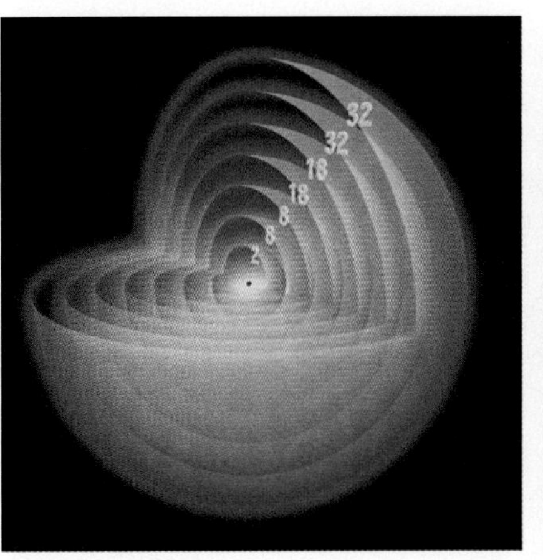

(a)

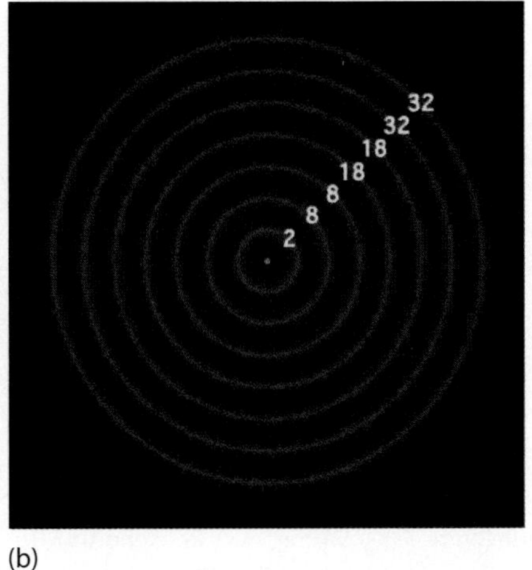

(b)

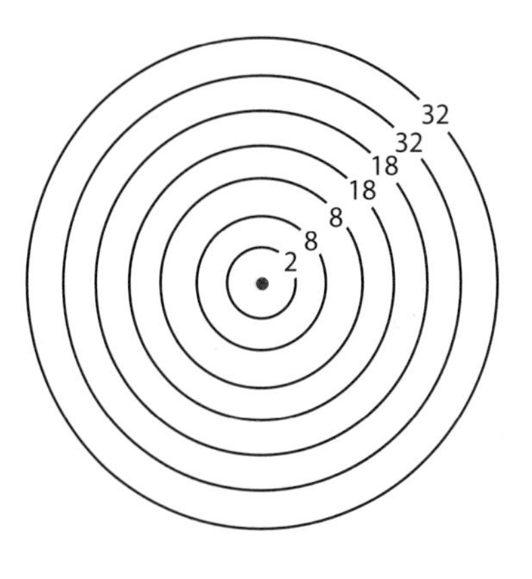

(c)

Figure 18.13
(a) A cutaway view of the seven shells, with the number of electrons each shell can hold indicated.
(b) A two-dimensional, cross-sectional view of the shells.
(c) An easy to draw cross-sectional view that resembles Bohr's planetary model.

Look carefully at Figure 18.14. Can you see that the valence electrons of atoms above and below one another (within the same group) are similarly organized? For example, atoms of the first group, which include hydrogen, lithium, and sodium, each have a single valence electron. The atoms of the second group, including beryllium and magnesium, each have two valence electrons. Similarly, atoms of the last group, including helium, neon, and argon, each have their outermost shells filled to capacity with valence electrons—two for helium, and eight for both neon and argon. In general, the valence electrons of atoms in the same group of the periodic table are similarly organized. This explains why elements of the same group have similar properties—a concept first presented in Section 17.5.

Concept Check ✓

Do atoms really consist of shells that look like those depicted in Figure 18.13?

Check Your Answer The shell model is NOT a depiction of the "appearance of an atom." Rather, it is a conceptual model that allows us to account for observed behavior. An atom, therefore, does not actually contain a series of concentric shells; it merely behaves as though it does.

In this chapter, we have explored a fair amount of detail regarding atomic models, beginning with Bohr's planetary model and ending with the shell model. Remember that these models are not to be interpreted as actual pictures of the atom's physical structure. Rather, they serve as tools to help us understand and predict how atoms behave. In Chapter 23, for example, a simplified shell model, known as *electron-dot structure,* is presented to assist you in understanding chemical bonding. Atomic models are the key to a richer understanding of the world of atoms in which we live.

Figure 18.14
The first three periods of the periodic table according to the shell model. Elements in the same period have electrons in the same shells. Elements in the same period differ from one another by the number of electrons in the outermost shell.

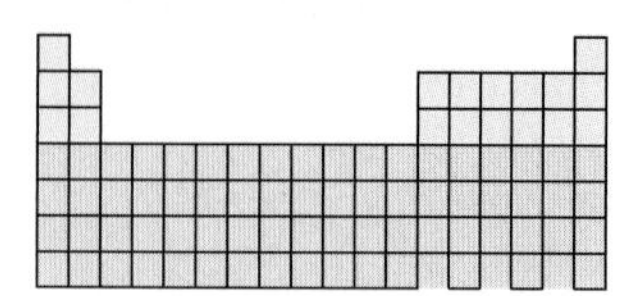

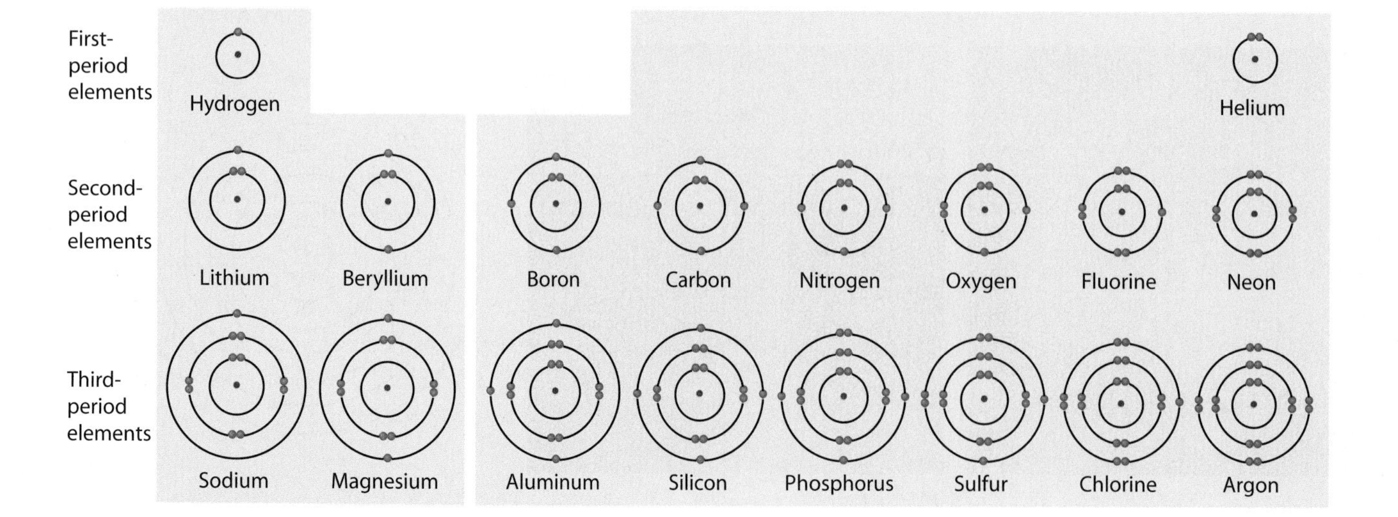

Chapter Review

Key Terms and Matching Definitions

_____ atomic spectrum
_____ conceptual model
_____ physical model
_____ principal quantum number
_____ quantum
_____ quantum hypothesis
_____ shell
_____ spectroscope
_____ valence electrons

1. A representation of an object on a different scale.
2. A representation of a system that helps us predict how the system behaves.
3. A device that uses a prism or diffraction grating to separate light into its color components.
4. The pattern of frequencies of electromagnetic radiation emitted by the atoms of an element, considered to be an element's "fingerprint."
5. The idea that light energy is contained in discrete packets called *quanta.*
6. A small, discrete packet of light energy.
7. An integer that specifies the quantized energy level of an atomic orbital.
8. A region of space in which an electron may be located around the nucleus.
9. The electrons of the outermost shell of an atom.

Review Questions

Models Help Us Visualize the Invisible World of Atoms

1. If a baseball were the size of the Earth, about how large would its atoms be?
2. When we use a scanning tunneling microscope, do we see atoms directly or do we see them only indirectly?
3. Why are atoms invisible to visible light?
4. What is the difference between a physical model and a conceptual model?
5. What is the function of an atomic model?

Atoms Can Be Identified by the Light They Emit

6. What causes an atom to emit light?
7. Why do we say atomic spectra are like fingerprints of the elements?
8. What did Rydberg notice about the atomic spectrum of hydrogen?

Niels Bohr Used the Quantum Hypothesis to Explain Atomic Spectra

9. What was Planck's quantum hypothesis?
10. Which has more potential energy—an electron close to an atomic nucleus or one far from an atomic nucleus?
11. What happens to an electron as it absorbs a photon of light?
12. What is the relationship between the light emitted by an atom and the energies of the electrons in the atom?
13. Did Bohr think of his planetary model as an accurate representation of what an atom looks like?

A Shell Is a Region of Space Where an Electron Can Be Found

14. How is the number of shells an atom of a given element contains related to the row of the periodic table in which that element is found?
15. What is the relationship between the maximum number of electrons each shell can hold and the number of elements in each period of the periodic table?

Explorations

Spectral Patterns

The diffraction gratings used in rainbow glasses have lines etched vertically and horizontally which make the colors appear to the left and right, above and below, and in all corners as well. A compact disk behaves as a diffraction grating because its surface contains many rows of microscopic pits.

To the naked eye, a glowing element appears as only a single color. However, this color is an average of the many different visible frequencies the element is emitting. Only with a device such as a spectroscope are we able to discern the different frequencies. So when you look at an atomic spectrum, don't get confused and think that each frequency of light (color) corresponds to a different element. Instead, remember that what you are looking at is all the frequencies of light emitted by a single element as its electrons make transitions back and forth between energy levels.

Not all elements produce patterns in the visible spectrum. Tungsten, for example, produces the full spectrum of colors (white light), which makes it useful as the glowing component of a car's headlights, as shown in the photograph below. Also, the sunlight reflecting off the moon, also shown below, is so bright and contains the glow of many different elements that it too appears as a broad spectrum.

Quantized Whistle

People watching you perform this activity may not believe that the audible "steps" of your whistling down the tube are not intentional. Explain quantization to them before allowing them to attempt this activity for themselves. Try to count the number of steps in your tubular whistle, understanding that each step is analogous to an energy level in an atom. Does a longer tube create fewer or more steps than a shorter tube? Why is it so difficult to whistle down a garden hose?

If you punch a few holes along the tube, you alter the frequencies of the standing waves that can form in the tube, with the result that different pitches are produced. This is the underlying principle in such musical instruments as flutes and saxophones.

Exercises

1. With scanning tunneling microscopy (STM) technology, we do not see actual atoms. Rather, we see images of them. Explain.

2. Why is it not possible for an STM to make images of the interior of an atom?

3. Would you use a physical model or a conceptual model to describe the following: brain, mind, solar system, birth of universe, stranger, best friend, gold coin, dollar bill, car engine, virus, spread of a cold virus?

4. How might you distinguish a sodium-vapor lamp from a mercury-vapor lamp?

5. How can a hydrogen atom, which has only one electron, create so many spectral lines?

6. Suppose a certain atom has four energy levels. Assuming that all transitions between levels are possible, how many spectral lines will this atom exhibit? Which transition corresponds to the highest-energy light emitted? Which corresponds to the lowest-energy light emitted?

   $n = 4$
   $n = 3$
   $n = 2$
   $n = 1$
   Nucleus

7. An electron drops from the fourth energy level in an atom to the third level and then to the first level. Two frequencies of light are emitted. How does their combined energy compare with the energy of the single frequency that would be emitted if the electron dropped from the fourth level directly to the first level?

8. Figure 18.12 shows three energy-level transitions that produce three spectral lines in a spectroscope. Note that the distance between the $n = 1$ and $n = 2$ levels is greater than the distance between the $n = 2$ and $n = 3$ levels. Would the number of spectral lines produced change if the distance between the $n = 1$ and $n = 2$ levels were exactly the same as the distance between the $n = 2$ and $n = 3$ levels?

9. Which color of light comes from a greater energy transition, red or blue?

10. Place the proper number of electrons in each shell:

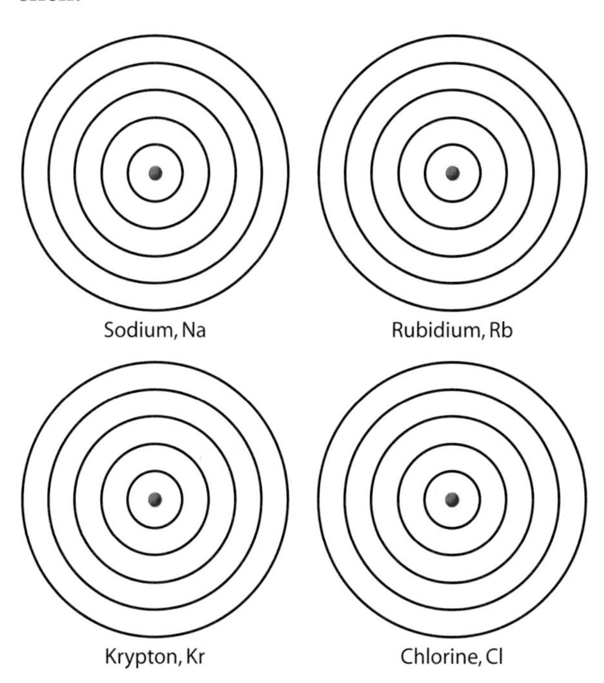

11. Which element is represented in Figure 18.13 if all seven shells are filled to capacity?

12. Does a shell have to contain electrons in order to exist?

Suggested Web Sites

http://www.achilles.net/~jtalbot/
This is a superb site for learning about the spectral patterns of stars and how they are used to study the universe.

http://www.achilles.net/~jtalbot/data/elements/index.html
Here is where you will find high-resolution spectral patterns of a variety of elements.

http://www.physics.purdue.edu/nanophys
Check out the site of the nanoscale physics laboratory of Purdue University, where they look at things that are really, really, really, really, really, really, really small. Lots of pretty pictures.

Chapter 19: Radioactivity

What exactly is radioactivity? How much radioactivity must be present to have an effect on you? How much are we exposed to on a daily basis? And how is radioactivity often used to save lives? Why are some elements radioactive and others not? Is it possible to make gold out of a less expensive element, such as lead? Finally, how do archaeologists use radioactive dating to find the age of ancient artifacts, from prehistoric animal bones to the first human tools?

19.1 Alpha, Beta, and Gamma Radiation Result from Radioactivity

To learn about radioactivity, let's begin with a definition. *Radioactivity* is the process by which certain elements emit particular forms of radiation. We say that any element that emits any of these forms of radiation is **radioactive.** There are three major forms of emitted radiation. These are called: *alpha, beta,* and *gamma* radiation.

Alpha radiation consists of fast-flying positively charged subatomic particles known as *alpha particles.* An **alpha particle** is a combination of two protons and two neutrons. In other words, it is the nucleus of a helium atom, atomic number 2.

Beta radiation consists of fast-flying negatively charged subatomic particles known as *beta particles.* Each **beta particle** is an electron that is ejected by an atomic nucleus.

Gamma radiation is an extremely energetic form of electromagnetic radiation. It lies far to the high end of the electromagnetic spectrum, possessing much more energy than visible light. Unlike alpha and beta radiation, gamma radiation carries no electric charge and has no mass.

Look at Figure 19.1. It shows how the three major types of radiation can be separated from one another. A sample gives off the radiation, which then streams by a magnet. Because alpha particles and beta particles have opposite charge, the magnetic field forces these particles in opposite directions. The gamma rays, which are electrically neutral, are not pushed either way by the field and continue in a straight-line path.

Alpha particles do not easily penetrate solid material. Because of their relatively large size and their double positive charge (+2), solid materials often trap them. However, because alpha particles have a lot of mass and move quickly, they have a great deal of kinetic energy. They can therefore cause significant damage to the surface of a material, especially living tissue.

But the damage that alpha particles can do is limited because of their interactions with electrons. As they travel through air, even through distances as short as a few centimeters, alpha particles pick up electrons, slow down, and become harmless helium. Almost all the Earth's helium atoms, including those in a helium balloon, were once energetic alpha particles ejected from radioactive elements.

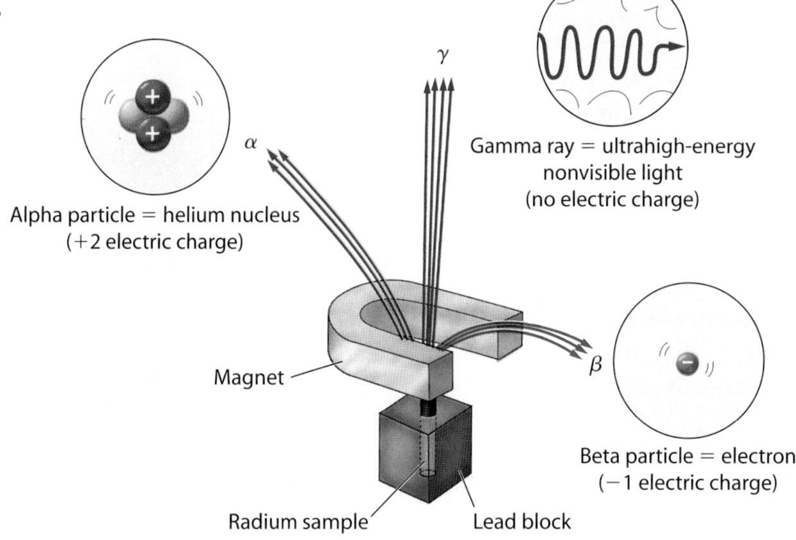

Figure 19.1
The three most common forms of radiation coming from a radioactive substance are called by the first three letters of the Greek alphabet, α, β, γ— *alpha, beta,* and *gamma.* In a magnetic field, alpha rays bend one way, beta rays bend the other way, and gamma rays do not bend at all. Note that the alpha rays bend less than do the beta rays. This happens because the alpha particles have more inertia (because they have more mass) than the beta particles. The source of all three radiations is a radioactive material placed at the bottom of a hole drilled in a lead block.

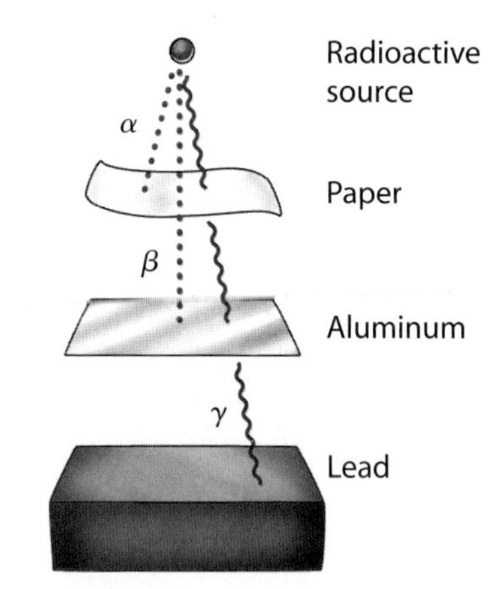

Figure 19.2
Alpha particles are the least penetrating form of radiation and can be stopped by a sheet of paper. Beta particles readily pass through paper but not through a sheet of aluminum. Gamma rays penetrate several centimeters into solid lead.

Beta particles are normally faster than alpha particles and not as easy to stop. For this reason, they are able to penetrate light materials such as paper and clothing. They can penetrate fairly deeply into skin, where they have the potential for harming or killing cells. They are not able to penetrate deeply into denser materials, however, such as aluminum. Beta particles, once stopped, become part of the material they are in, like any other electrons.

Like visible light, a gamma ray is pure energy. The amount of energy in a gamma ray is much greater than the amount of energy in visible light. Because gamma rays have no mass or electric charge and because of their high energies, they can penetrate most materials. However, they cannot penetrate unusually dense materials such as lead. Lead absorbs them. Delicate molecules in cells throughout our bodies that are exposed to gamma rays suffer structural damage. Hence, gamma rays are generally more harmful to us than alpha or beta rays.

Figure 19.2 shows the relative penetrating power of the three types of radiation, and Figure 19.3 shows an interesting practical use for gamma radiation.

Concept Check ✓

Pretend you are given three radioactive rocks—one an alpha emitter, one a beta emitter, and one a gamma emitter. You can throw one away, but you must hold one of the remaining two in your hand and place the other in your pocket. What can you do to minimize your exposure to radiation?

Check Your Answer Ideally, you should get as far from all the rocks as possible. If you must hold one and place one in your pocket, however, hold the alpha emitter because the skin on your hand will shield you. Put the beta emitter in your pocket, because its rays might be stopped by the combined thickness of clothing and skin. Throw away the gamma emitter, because its rays would penetrate deep into your body from either of these places.

Figure 19.3
The shelf life of fresh strawberries and other perishables is increased when the food is subjected to gamma rays from a radioactive source. The strawberries on the right were treated with gamma radiation, which kills the microorganisms that normally lead to spoilage. The food is only a receiver of radiation and is in no way transformed to an emitter of radiation. This can be confirmed with a radiation detector.

19.2 Radioactivity Is a Natural Phenomenon

A common misconception is that radioactivity is new in the environment. Radioactivity, however, has been around far longer than the human race. It is as much a part of our environment as the sun and the rain. It has always occurred in the soil we walk on and in the air we breathe, and it warms the interior of the Earth and makes it molten. The energy released by radioactive substances in the Earth's interior heats the water that spurts from a geyser and the water that wells up from a natural hot spring.

As Figure 19.4 shows, most of the radiation we encounter is natural background radiation that originates in the Earth and in space and was present long before humans arrived. Even the cleanest air we breathe is somewhat radioactive as a result of bombardment by cosmic rays. At sea level, the protective blanket of the atmosphere reduces background radiation, but at higher altitudes radiation is more intense. In Denver, the "Mile-High City," a person receives more than twice as much radiation from cosmic rays as at sea level. A couple of round-trip flights between New York and San Francisco exposes us to as much radiation as we receive in a chest X ray at the physician's office. The air time of airline personnel is limited because of this extra radiation.

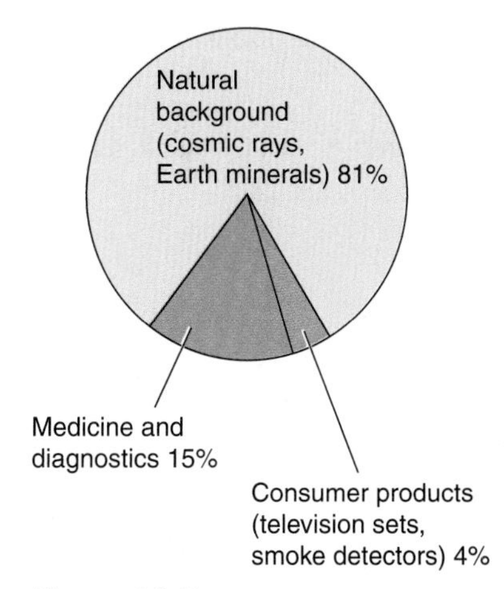

Figure 19.4
Origins of radiation exposure for an average individual in the United States.

Cells are able to repair most kinds of molecular damage caused by radiation if the damage is not too severe. A cell can survive an otherwise lethal dose of radiation if the dose is spread over a long period of time to allow intervals for healing. When a dose of radiation is large enough to kill cells, the dead cells can be replaced by new ones. Sometimes radiation alters the genetic information of a cell by damaging its DNA molecules. New cells arising from the damaged cell retain the altered genetic information, which is called a *mutation.* Usually the effects of a mutation are insignificant, but occasionally the mutation results in cells that do not function as well as unaffected ones, sometimes leading to a cancer. If the damaged DNA is in an individual's reproductive cells, the genetic code of the individual's offspring may retain the mutation.

The capacity for nuclear radiation to cause damage is not just a function of the energy of the radiation. There are different interactions with radiation, some more harmful than others. Suppose you have two arrows of equal

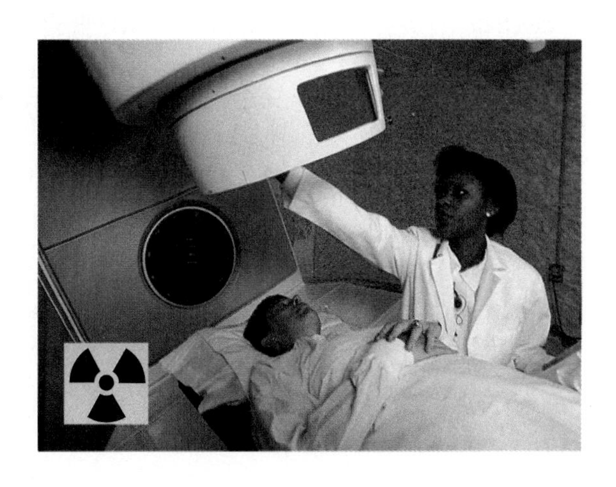

Figure 19.5
Nuclear radiation is focused on harmful tissue, such as a cancerous tumor, to selectively kill or shrink the tissue in a technique known as *radiation therapy.* This application of nuclear radiation has saved millions of lives—a clear-cut example of the benefits of nuclear technology. The inset shows the international symbol indicating an area where radioactive material is being handled or produced.

Figure 19.6
A commercially available radon test kit for the home.

mass, but one with a pointed tip and one with a suction cup at its tip. Aim and release both arrows at an apple. Both have the same amount of kinetic energy. But the one with the pointed tip will do more damage to the apple than the one with the suction cup. Similarly, some forms of radiation cause greater harm than other forms even when we receive the same energy from both forms.

The leading source of naturally occurring radiation is radon-222. Radon-222 is an inert gas arising from uranium deposits. Radon is heavier than air, and therefore tends to accumulate in basements after it seeps up through cracks in the floor. Levels of radon vary from region to region, depending on local geology. You can check the radon level in your home with a radon detection kit like the one shown in Figure 19.6. If levels are abnormally high, corrective measures, such as sealing the basement floor and walls and maintaining adequate ventilation, are recommended. The U.S. Environmental Protection Agency projects that anywhere from 7000 to 30,000 cases of lung cancer each year are attributed to radon exposure. Smokers who inhale the radon that occurs naturally in tobacco smoke are at particularly high risk.

About one-fifth of our annual exposure to radiation comes from non-natural sources, primarily medical procedures. Television sets, fallout from nuclear testing, and the coal and nuclear power industries are minor but significant non-natural sources. Interestingly, the coal industry far outranks the nuclear power industry as a source of radiation. The global combustion of coal annually releases into the

Table 19.1

Annual Radiation Exposure

Source	Typical Relative Amount Received in One Year
Natural Origin	
Cosmic radiation	26
Ground	33
Air (radon-222)	198
Human tissues (potassium-40; radium-226)	35
Human Origin	
Medical procedures	
Diagnostic X-rays	40
Nuclear medicine	15
Television tubes, other consumer products	11
Weapons-test fallout	1

atmosphere about 13,000 tons of radioactive thorium and uranium. Worldwide, the nuclear power industries generate about 10,000 tons of radioactive waste each year. Most of this waste is contained, however, and is *not* released into the environment. As we explore in Chapter 20, where to bury this contained radioactive waste is a heated issue yet to be resolved.

One source of radiation is the human body. This is mainly because of the potassium we ingest. Our bodies contain on average about 200 grams of potassium. Of this quantity, about 20 milligrams is the radioactive isotope potassium-40, a gamma-ray emitter. In a human body, about 5000 potassium-40 atoms emit pulses of radioactivity in the time it takes the heart to beat once. We live with radioactivity—we always have.

19.3 Radioactivity Results from an Imbalance of Forces in the Nucleus

You know that electric charges with the same sign repel one another. Then how is it possible that all the positively charged protons of the nucleus remain clumped together? This question led to the discovery of an attractive force called the **strong nuclear force,** which acts between all nucleons. This force is very strong, but acts only over extremely short distances (about 10^{-15} meter, the diameter of a typical atomic nucleus). Repulsive electrical interactions, on the other hand, are relatively long-ranged. Figure 19.7 compares the strength of these two forces over distance. For protons that are close together, as in a small atomic nucleus, the attractive strong nuclear force is greater than the repulsive electric force. For protons that are far apart, like those on opposite edges of a large nucleus, the attractive strong nuclear force may be weaker than the repulsive electric force.

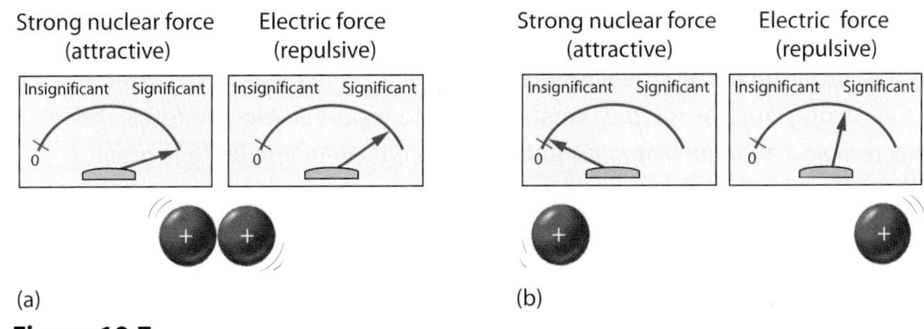

Figure 19.7
(a) Two protons near each other experience both an attractive strong nuclear force and a repulsive electric force. At this tiny separation distance, the strong nuclear force overcomes the electric force, and as a result the protons stay close together. (b) When the two protons are relatively far from each other, the electric force is more significant than the strong nuclear force, and as a result the protons repulse each other. It is this proton-proton repulsion in large atomic nuclei that causes radioactivity.

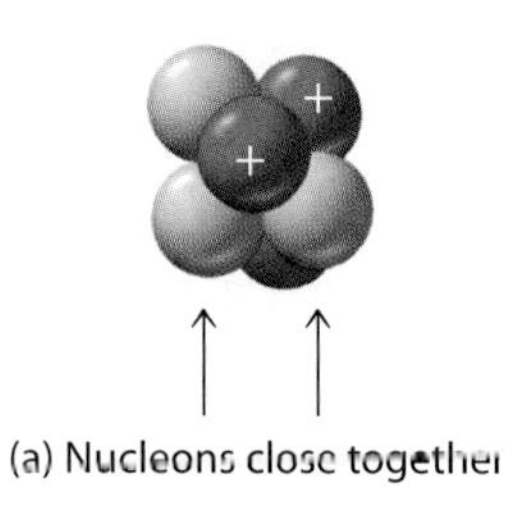

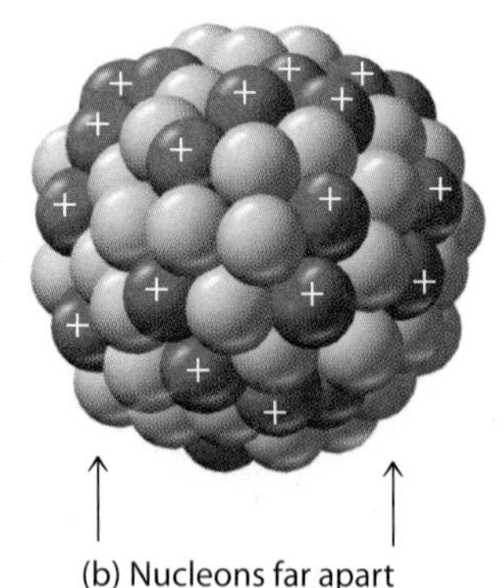

Figure 19.8
(a) All nucleons in a small atomic nucleus are close to one another; hence, they experience an attractive strong nuclear force.
(b) Nucleons on opposite sides of a large nucleus are not as close to one another, and so the attractive strong nuclear forces holding them together are much weaker. The result is that the large nucleus is less stable.

Because the strong nuclear force decreases over distance, a large nucleus is not as stable as a small one (Figure 19.8). In other words, a large atomic nucleus is less stable and decays. Then it emits high-energy particles or gamma radiation. We have *radioactive decay.*

Neutrons act as a "nuclear cement" that holds the atomic nucleus together. We can see why. Protons attract other protons and neutrons alike by the strong nuclear force. The strong nuclear force only attracts. But protons electrically repel other protons. Neutrons, on the other hand, have no electric charge and aren't repelled. By the strong force, neutrons only attract protons and other neutrons. The presence of neutrons therefore adds to the attraction among nucleons and helps hold the nucleus together, as illustrated in Figure 19.9.

The more protons in a nucleus, the more neutrons are needed to balance the repulsive electric forces. For light elements, it is sufficient to have about as many neutrons as protons. The common isotope of carbon-12, for instance, has six protons and six neutrons. For large nuclei, more neutrons than protons are required. Recall that the strong nuclear force diminishes rapidly with increasing distance between nucleons. Nucleons must be practically touching in order for the strong nuclear force to be effective. Nucleons on opposite sides of a large atomic nucleus are not as attracted to one another. The repulsive electric force, however, does not diminish much across the diameter of a large nucleus. So the repulsive electric force can overcome the strong nuclear force. To compensate for the weakening of the strong nuclear force across the diameter of the nucleus, large nuclei have more neutrons than protons. Lead, for example, has about one and a half times as many neutrons as protons.

Concept Check ✓

Two protons in an atomic nucleus repel each other, but they are also attracted to each other. Explain.

Check Your Answer Two protons repel each other by the electric force and at the same time attract each other by the strong nuclear force. When the attractive strong nuclear force is greater than the repulsive electric force, the protons remain together. Under conditions where the electric force is greater than the strong nuclear force, the protons fly apart.

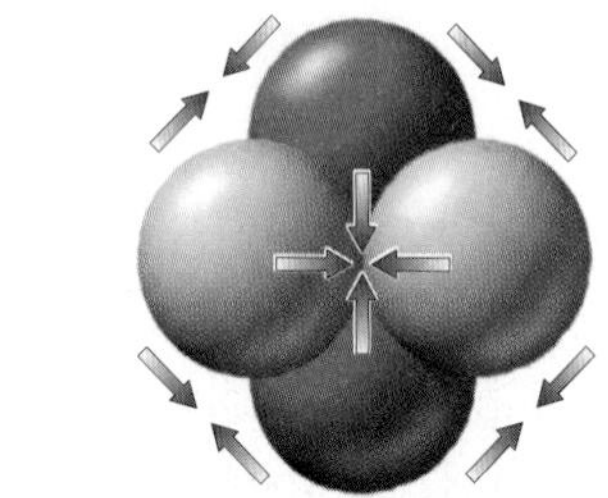
All nucleons, both protons and neutrons, attract one another by the strong nuclear force.

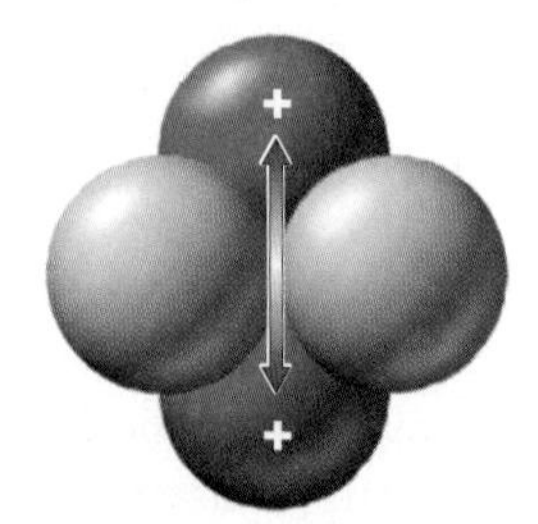
Only protons repel one another by the electric force.

Figure 19.9
The presence of neutrons helps hold the atomic nucleus together by increasing the effect of the attractive strong nuclear force, represented by the single-headed arrows.

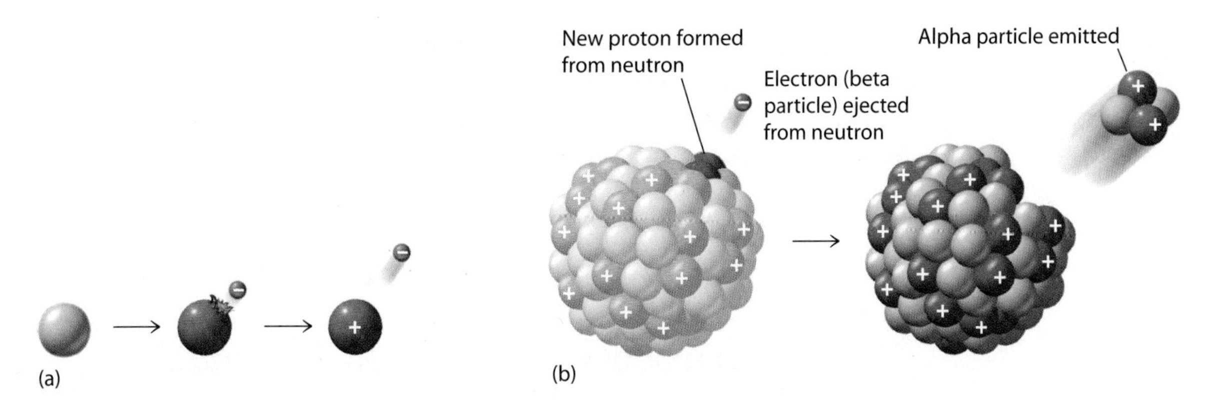

Figure 19.10
(a) A neutron near a proton is stable, but a neutron by itself is unstable and decays to a proton by emitting an electron. (b) Destabilized by an increase in the number of protons, the nucleus begins to shed fragments, such as alpha particles.

An abundance of neutrons helps to stabilize a large nucleus. But a neutron has two drawbacks. First, neutrons are not stable when they are by themselves. A lone neutron will spontaneously transform into a proton and an electron, as shown in Figure 19.10a. A neutron seems to need protons around to prevent this from occurring. After a nucleus reaches a certain size, there are so many more neutrons than protons that there are not enough protons available to prevent the neutrons from changing into protons. As neutrons in a nucleus change to protons, the stability of the nucleus decreases. Why? Because the repulsive electric force becomes more and more significant as we have discussed. The result is that pieces of the nucleus are ejected, as Figure 19.10b shows.

The second reason the stabilizing effect of neutrons is limited is that any proton in the nucleus is attracted by the strong nuclear force only to adjacent protons but is electrically repelled by all other protons in the nucleus. As more and more protons are squeezed into the nucleus, the repulsive electric forces increase substantially. So we find that all nuclei having more than 83 protons are radioactive. Also, the nuclei of the heaviest elements produced in the laboratory are so unstable (radioactive) that they exist for only fractions of a second.

Concept Check ✓

Which is more sensitive to distance: the strong nuclear force or the electric force?

Check Your Answer The strong nuclear force weakens rapidly over relatively short distances, but the electric force remains powerful over such distances.

Small nuclei also have the potential for being radioactive. This generally occurs when a nucleus contains more neutrons than protons. The nucleus of carbon-14, for example, contains eight neutrons but only six protons. With not enough protons to go around, one of the neutrons inevitably transforms to a proton, releasing an electron (beta particle). That's why carbon-14 is a beta emitter.

19.4 A Radioactive Element Can Transmute to a Different Element

When a radioactive nucleus emits an alpha or beta particle, the atomic number of the nucleus is changed. Therefore, the identity of the element also changes. The changing of one element to another is called **transmutation.** Consider a uranium-238 nucleus, which contains 92 protons and 146 neutrons. When an alpha particle is ejected, the nucleus loses two protons and two neutrons. Since an element is defined by the number of protons in its nucleus, the 90 protons and 144 neutrons remaining are no longer uranium. What we have now is the nucleus of a different element—thorium.

This transmutation can be depicted as follows:

Uranium-238 → Thorium-234 + Helium-4

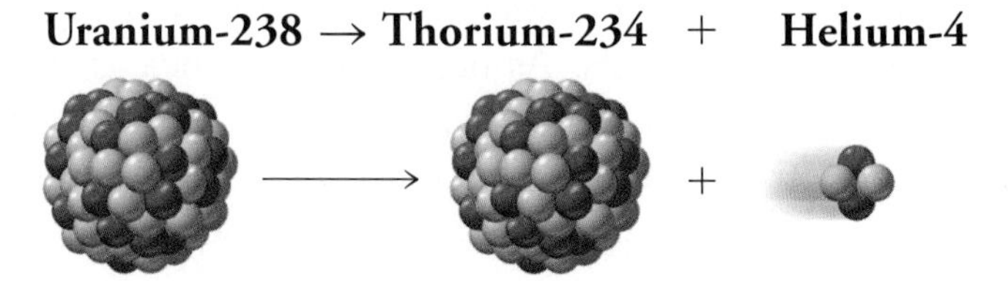

When this transmutation occurs, energy is released. Part of the released energy is in the form of gamma radiation, most is the kinetic energy of the alpha particle, and some is the kinetic energy of the recoiling thorium atom. Notice that in this and all other nuclear equations, the mass numbers balance ($238 = 234 + 4$).

Thorium-234 is radioactive. When it decays, it emits a beta particle. As mentioned previously, a beta particle is an electron emitted by a neutron as the neutron transforms to a proton. Thorium has 90 protons, so beta emission leaves the nucleus with one fewer neutron and one more proton. The new nucleus has 91 protons and is no longer thorium. It becomes the element protactinium. Although the atomic number has increased by 1 in this process, the mass number (protons + neutrons) remains the same. This transmutation can be depicted as follows:

Thorium-234 → Protactinium-234 + Electron

So we see that when an element ejects an alpha particle from its nucleus, the mass number of the remaining atom is decreased by 4 and its atomic number is decreased by 2. The resulting atom is an atom of the element two spaces back in the periodic table because this atom has two fewer protons. When an element ejects a beta particle from its nucleus, the mass of the atom is practically unaffected, meaning there is no change in mass number, but its atomic number increases by 1. The resulting atom is an atom of the element one space forward in the periodic table because it has one more proton.

The decay of uranium-238 to lead-206 is shown in Figure 19.11. Each gray-blue arrow shows an alpha decay, and each red arrow shows a beta decay.

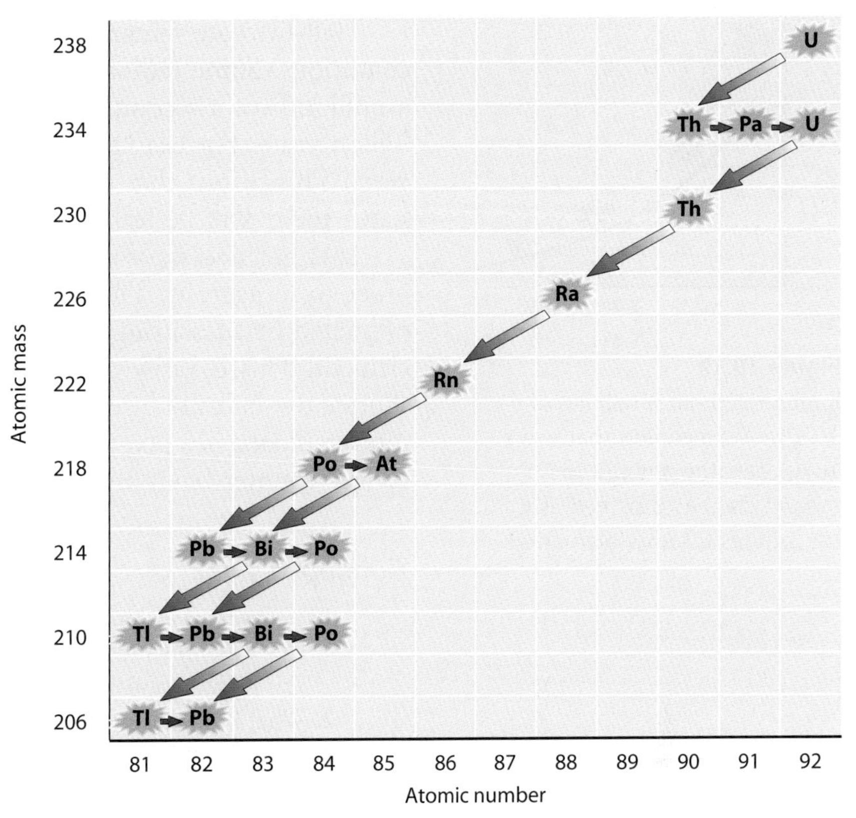

Figure 19.11
Uranium-238 decays to lead-206 through a series of alpha (gray-blue) and beta (red) decays.

Concept Check ✓

What finally becomes of all the uranium that undergoes radioactive decay?

Check Your Answer All uranium ultimately becomes lead. Along the way, it exists as the elements indicated in Figure 19.11.

19.5 The Shorter the Half-Life, the Greater the Radioactivity

Radioactive isotopes decay at different rates. The radioactive decay rate is measured in terms of a characteristic time, the **half-life.** The half-life of a radioactive material is the time needed for half of the radioactive atoms to decay. Radium-226, for example, has a half-life of 1620 years. This means that half of any given specimen of Ra-226 will have decayed by the end of 1620 years. In the next 1620 years, half of the remaining radium decays, leaving only one-fourth the original number of radium atoms. The other three-fourths convert, by a succession of decays, to lead. After 20 half-lives, an initial quantity of radioactive atoms is diminished to about one-millionth of the original quantity.

We can also say that the radioactive half-life of an isotope is the time for its decay rate to reduce to half.

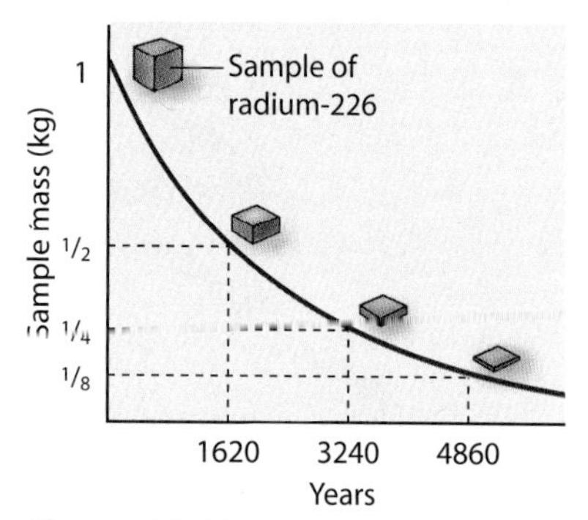

Figure 19.12
Radium-226 has a half-life of 1620 years, meaning that every 1620 years the amount of radium decreases by half as it transmutes to other elements.

Half-lives are remarkably constant and not affected by external conditions. Some radioactive isotopes have half-lives that are less than a millionth of a second, while others have half-lives of more than a billion years. For example, uranium-238 has a half-life of 4.5 billion years. This means that in 4.5 billion years, half the uranium in the Earth today will be lead.

It is not necessary to wait through the duration of a half-life in order to measure it. The half-life of an element can be accurately estimated by measuring the rate of decay of a known quantity of the element. This is easily done using a radiation detector. In general, the shorter the half-life of a substance, the faster it disintegrates and the more radioactivity per minute is detected. Figure 19.13 shows a Geiger counter being used by environmental workers.

Concept Check ✓

1. If you have a sample of a radioactive isotope that has a half-life of one day, how much of the original sample remains at the end of the second day? The third day?
2. What becomes of the decayed atoms of the sample?
3. With equal quantities of material, which results in a higher counting rate on a radiation detector, radioactive material that has a short half-life or radioactive material that has a long half-life?

Check Your Answers

1. At the end of two days, one-fourth of the original sample is left—one-half disappears by the end of the first day, and one-half of that one-half ($1/2 \times 1/2 = 1/4$) disappears by the end of the second day. At the end of three days, one-eighth of the original sample remains.
2. The atoms that decay are now atoms of a different element.
3. The material with the shorter half-life is more active and so produces a higher counting rate.

Figure 19.13
A Geiger counter detects incoming radiation by the way the radiation affects a gas enclosed in the tube that the technician is holding in his right hand.

19.6 Isotopic Dating Measures the Age of a Material

The Earth's atmosphere is continuously bombarded by cosmic rays. This bombardment causes many atoms in the upper atmosphere to transmute. These transmutations result in many protons and neutrons being "sprayed out" into the environment. Most of the protons are stopped as they collide with the atoms of the upper atmosphere. The protons strip electrons from the atoms they collide with and thus become hydrogen atoms. The neutrons, however, continue for longer distances because they have no electric charge and therefore do not interact electrically with matter. Eventually, many of them collide with atomic nuclei in the lower atmosphere. A nitrogen that captures a neutron, for instance, becomes an isotope of carbon by emitting a proton:

Neutron + Nitrogen-14 → Carbon-14 + Proton

This carbon-14 isotope, which makes up less than one-millionth of 1% of the carbon in the atmosphere, is radioactive and has eight neutrons. (The most common isotope, carbon-12, has six neutrons and is not radioactive.) Because both carbon-12 and carbon-14 are forms of carbon, they have the same chemical properties. Both of these isotopes, for example, chemically react with oxygen to form carbon dioxide, which is consumed by plants. This means that all plants contain a tiny quantity of radioactive carbon-14. All animals eat either plants or plant-eating animals, and therefore all animals have a little carbon-14 in them. In short, all living things on the Earth contain some carbon-14.

After carbon-14 emits a beta particle, it later decays back to nitrogen:

Carbon-14 → Nitrogen-14 + Electron

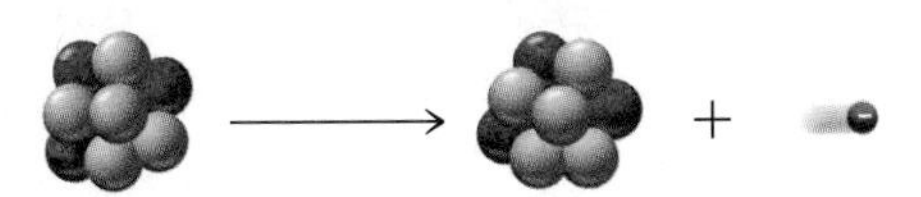

Because plants absorb carbon dioxide as long as they live, any carbon-14 lost to decay is immediately replenished with fresh carbon-14 from the atmosphere. In this way, a radioactive equilibrium is reached where there is a constant ratio of about one carbon-14 atom to every 100 billion carbon-12 atoms. When a plant dies, replenishment of carbon-14 ends. Then the percentage of carbon-14 decreases at a constant rate given by its half-life, but the amount of carbon-12 does not change because this isotope does not undergo

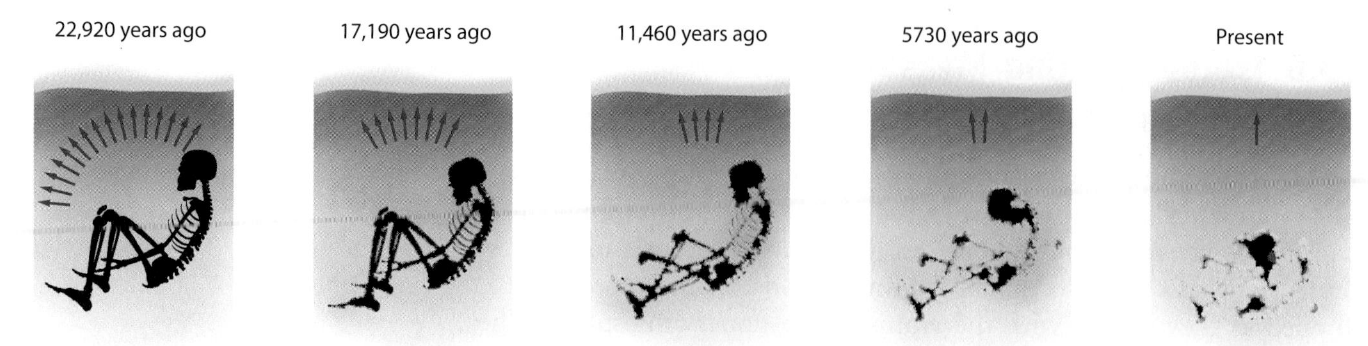

Figure 19.14
The amount of radioactive carbon-14 in the skeleton diminishes by one-half every 5730 years, resulting today with only a fraction of the carbon-14 it originally had. The red arrows symbolize relative amounts of carbon-14.

radioactive decay. The longer a plant or other organism is dead, therefore, the less carbon-14 it contains relative to the constant amount of carbon-12.

The half-life of carbon-14 is about 5730 years. This means that half of the carbon-14 atoms now present in a plant or animal that dies today will decay in the next 5730 years. Half of the remaining carbon-14 atoms will then decay in the following 5730 years, and so on.

With this knowledge, scientists are able to calculate the age of carbon-containing artifacts, such as wooden tools or the skeleton shown in Figure 19.14, by measuring their current level of radioactivity. This process, known as **carbon-14 dating,** enables them to probe as much as 50,000 years into the past. Beyond this time span, there is too little carbon-14 remaining to permit an accurate analysis. (Understanding the local geology is another important tool used by archaeologists in the dating of ancient relics.)

Carbon-14 dating would be an extremely simple and accurate dating method if the amount of radioactive carbon in the atmosphere had been constant over the ages, but it hasn't been. Fluctuations in the magnetic field of the sun and in that of the Earth cause fluctuations in cosmic-ray intensity in the Earth's atmosphere. These ups and downs in cosmic-ray intensity in turn produce fluctuations in the amount of carbon-14 in the atmosphere at any given time. In addition, changes in the Earth's climate affect the amount of carbon dioxide in the atmosphere. The oceans are great reservoirs of carbon dioxide. When the oceans are cold, they release less carbon dioxide into the atmosphere than when they are warm. Because of all these fluctuations in the carbon-14 production rate through the centuries, carbon-14 dating has an uncertainty of about 15%. This means, for example, that the straw of an old adobe brick dated to be 500 years old may really be only 425 years old on the low side or 575 years old on the high side. For many purposes, this is an acceptable level of uncertainty.

Concept Check ✓

Suppose an archaeologist extracts 1.0 g of carbon from an ancient ax handle and finds that it is one-fourth as radioactive as 1.0 g of carbon extracted from a freshly cut tree branch. About how old is the ax handle?

Check Your Answer The ax handle is two half-lives of carbon-14; that's 2 × 5730 years ~11,000 years old.

Scientists use radioactive minerals to date very old nonliving things. The naturally occurring mineral isotopes uranium-238 and uranium-235 decay very slowly and ultimately become lead—but not the common isotope lead-208. Instead, as was shown in Figure 19.11, uranium-238 decays to lead-206. Uranium-235, on the other hand, decays to lead-207. Thus, the lead-206 and lead-207 that now exist in a uranium-bearing rock were at one time uranium. The older the rock, the higher the percentage of these remnant isotopes.

If you know the half-lives of uranium isotopes and the percentage of lead isotopes in some uranium-bearing rock, you can calculate the date the of rock formation. Rocks dated in this manner have been found to be as much as 3.7 *billion* years old. Samples from the moon have been dated at 4.2 billion years, which is close to the estimated age of our solar system: 4.6 billion years.

Chapter Review

Key Terms and Matching Definitions

_____ alpha particle
_____ beta particle
_____ carbon-14 dating
_____ gamma radiation
_____ half-life
_____ radioactive
_____ strong nuclear force
_____ transmutation

1. The tendency of some elements, such as uranium, to emit radiation as a result of changes in the atomic nucleus.
2. A helium atom nucleus, which consists of two neutrons and two protons and is ejected by certain radioactive elements.
3. An electron ejected from an atomic nucleus during the radioactive decay of certain nuclei.
4. High-energy radiation emitted by the nuclei of radioactive atoms.
5. The force of interaction between all nucleons, effective only at very, very, very close distances.
6. The conversion of an atomic nucleus of one element to an atomic nucleus of another element through a loss or gain of protons.
7. The time required for half the atoms in a sample of a radioactive isotope to decay.
8. The process of estimating the age of once-living material by measuring the amount of a radioactive isotope of carbon present in the material.

Review Questions

Alpha, Beta, and Gamma Radiation Result from Radioactivity

1. How do the electric charges of alpha, beta, and gamma rays differ from one another?
2. Which of the three rays has the greatest penetrating power?

Radioactivity Is a Natural Phenomenon

3. What is the origin of most of the radiation you encounter?
4. Is radioactivity on the Earth something relatively new? Defend your answer.

Radioactivity Results From an Imbalance of Forces in the Nucleus

5. How are the strong nuclear force and the electric force different from each other?
6. What role do neutrons play in the atomic nucleus?

A Radioactive Element Can Transmute to a Different Element

7. When thorium, atomic number 90, decays by emitting an alpha particle, what is the atomic number of the resulting nucleus?
8. When thorium decays by emitting a beta particle, what is the atomic number of the resulting nucleus?
9. What change in atomic number occurs when a nucleus emits an alpha particle? A beta particle?
10. What is the long-range fate of all the uranium that exists in the world today?

The Shorter the Half-Life, the Greater the Radioactivity

11. What is meant by *radioactive half-life*?
12. What is the half-life of radium-226?
13. How does the decay rate of an isotope relate to its half-life?

Isotopic Dating Measures the Age of a Material

14. What do cosmic rays have to do with transmutation?
15. How is carbon-14 produced in the atmosphere?
16. Which is radioactive, carbon-12 or carbon-14?
17. Why is there more carbon-14 in living bones than in once-living ancient bones of the same mass?
18. Why is carbon-14 dating useless for dating old coins but not old pieces of cloth?
19. Why is lead found in all deposits of uranium ores?
20. What does the proportion of lead and uranium in rock tell us about the age of the rock?

Exercises

1. Why is a sample of radium always a little warmer than its surroundings?
2. Is it possible for a hydrogen nucleus to emit an alpha particle? Defend your answer.
3. Why are alpha and beta rays bent in opposite directions in a magnetic field? Why is the path of gamma rays unaffected?
4. The alpha particle has twice the electric charge of the beta particle but deflects less in a magnetic field. Why?

5. Which type of radiation—alpha, beta, or gamma—results in the greatest change in mass number? The greatest change in atomic number?

6. Which type of radiation—alpha, beta, or gamma—results in the least change in atomic mass number? The least change in the atomic number?

7. Which type of radiation—alpha, beta, or gamma—predominates in the interior of a high-flying commercial airplane? Why?

8. Why would you expect alpha particles to be less capable of penetrating materials than beta particles?

9. When the isotope bismuth-213 emits an alpha particle, what new element results? What new element results if it instead emits a beta particle?

10. Elements above uranium in the periodic table do not exist in any appreciable amounts in nature, because they have short half-lives. Yet there are several elements below uranium in the table that have equally short half-lives, but do exist in appreciable amounts in nature. How can you account for this?

11. You and a friend journey to the mountain foothills to get closer to nature and escape such things as radioactivity. While bathing in the warmth of a natural hot spring, she wonders aloud how the spring acquires its heat. What do you tell her?

12. A friend checks the local background radiation with a Geiger counter, which ticks audibly. Another friend, who normally fears mostly what isn't understood, makes an effort to keep a distance from the Geiger counter and looks to you for advice. What do you say?

13. Why is carbon-14 dating not accurate for estimating the age of materials more than 50,000 years old?

14. The age of the Dead Sea Scrolls was determined by carbon-14 dating. Could this technique have worked if they had been carved on stone tablets? Explain.

15. A certain radioactive element has a half-life of 1 hour. If you begin with a 1-gram sample of the element at noon, how much remains at 3:00 P.M.? At 6:00 P.M.? At 10:00 P.M.?

Suggested Readings and Web Sites

Waldrop, M. Mitchell. "The Shroud of Turin: An Answer Is at Hand." *Science*, September 30, 1988.

This fascinating article describes the events leading up to the dating of the shroud.

http://www.rw.doe.gov/homejava/homejava.htm

This is the home page for the Office of Civilian Radioactive Waste Management, established in 1982 to develop and manage a federal system for disposing of spent nuclear fuel resulting from atomic energy defense activities. You'll find the official position of the U.S. government regarding Yucca Mountain, Nevada, as a potential nuclear waste repository.

Chapter 20: Nuclear Fission and Fusion

What is nuclear energy? How is it used to produce electricity on one hand and nuclear bombs on the other? How is it that coal-fired power plants emit more radiation into our environment than nuclear power plants? The sun shines with energy, but what is the source of this energy? Are matter and energy really two forms of the same thing? Today, we are making many decisions about the risks and benefits of nuclear technology. In order to make the best possible decisions, we should have a good understanding of nuclear energy and how it arises from nuclear fission and fusion.

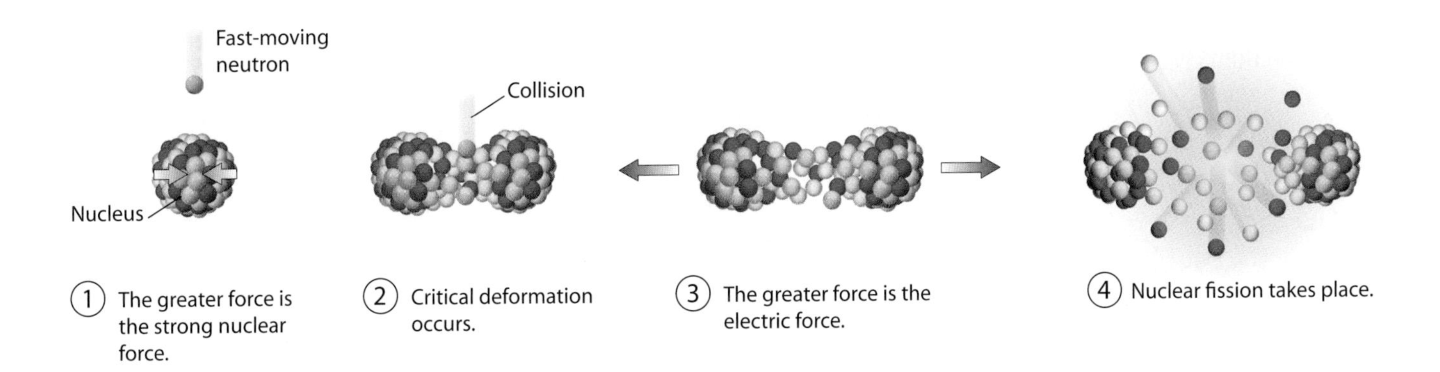

Figure 20.1
Nuclear deformation may result in repulsive electric forces overcoming attractive strong nuclear forces, in which case fission occurs.

20.1 Nuclear Fission Is the Splitting of the Atomic Nucleus

We learned in the previous chapter that there exists a delicate balance of attractive strong nuclear forces and repulsive electric forces in the nucleus of every atom. In all known nuclei, the strong nuclear forces dominate. In many of the heavier nuclei, however, this domination is easily lost and radioactive decay may occur.

For a select number of isotopes, another possibility exists. For example, a uranium-235 nucleus that is hit with a neutron elongates as shown in Figure 20.1. In a nucleus stretched in this manner, the strong nuclear force weakens substantially due to the increased distance between opposite ends. The repulsive electric forces between protons remain strong, however, and these forces may elongate the nucleus even more. If the elongation passes a certain point, the electric forces overwhelm the distance-sensitive strong nuclear forces and the nucleus splits into *fragments.* Typically, there are two large fragments accompanied by several smaller ones. This splitting of a nucleus into fragments is **nuclear fission.**

The energy released by the fission of one uranium-235 nucleus is enormous—about seven million times the energy released by the explosion of one TNT molecule. This energy is mainly in the form of kinetic energy of the fission fragments, which fly apart from one another. A much smaller amount of energy is released as gamma radiation.

Here is the equation for a typical uranium fission reaction:

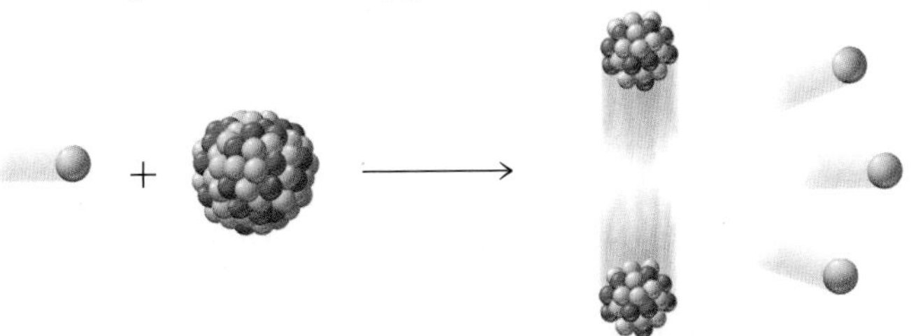

Neutron + Uranium-235 → fission fragments

Note in this reaction that one neutron starts the fission of the uranium nucleus, which in turn produces three neutrons. These new neutrons can cause the fissioning of three other uranium atoms,

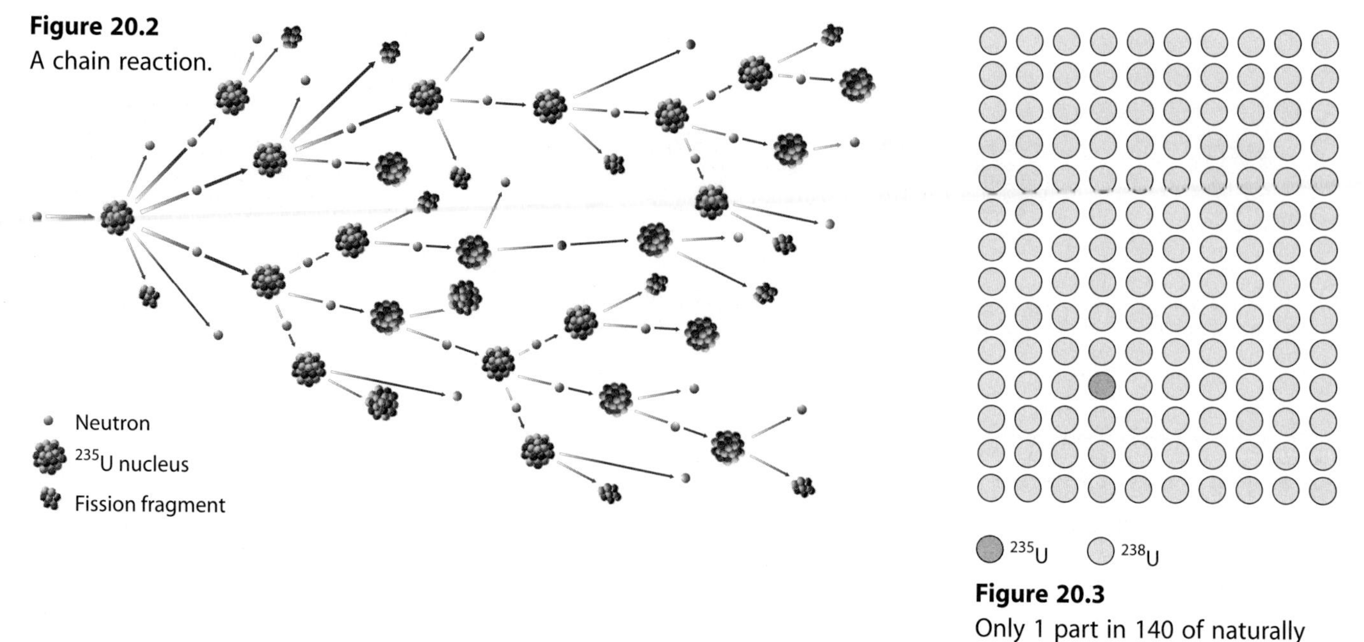

Figure 20.2
A chain reaction.

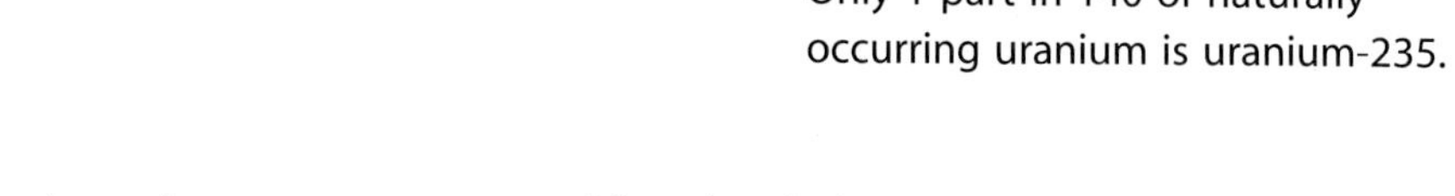

Figure 20.3
Only 1 part in 140 of naturally occurring uranium is uranium-235.

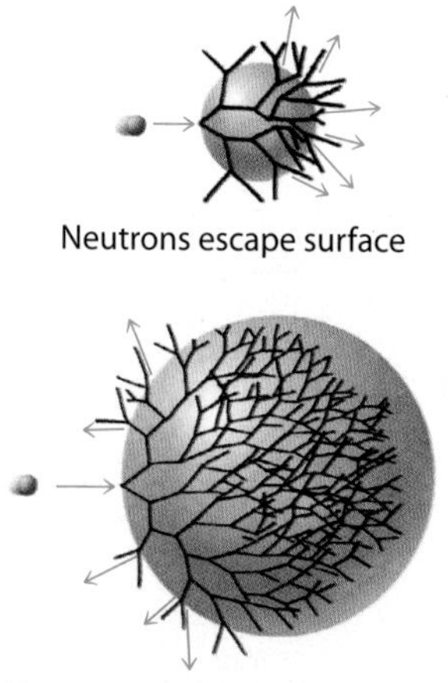

releasing nine more neutrons. If each of these 9 neutrons succeeds in splitting a uranium atom, the next step in the reaction produces 27 neutrons, and so on. Such a sequence, illustrated in Figure 20.2, is called a **chain reaction.** A chain reaction is a self-sustaining reaction in which the products of one reaction event stimulate further reaction events.

Chain reactions do not occur to any great extent in naturally occurring uranium ore because not all uranium atoms fission so easily. Fission occurs mainly in the rare isotope uranium-235. Uranium-235 makes up only 0.7% of the uranium in pure uranium metal (Figure 20.3). When the common isotope uranium-238 absorbs neutrons created by fission of a uranium-235 atom, the uranium-238 typically does not undergo fission. So any chain reaction that might occur in the uranium-235 atoms in an ore sample is snuffed out by the neutron-absorbing uranium-238, as well as by other neutron-absorbing elements in the ore and surrounding rock.

If a chain reaction occurred in a baseball-sized chunk of pure uranium-235, an enormous explosion would result. If the chain reaction were started in a smaller chunk, however, no explosion would occur. This is due to geometry: The ratio of surface area to mass is larger in a small piece than in a large piece. Just as there is more skin on six small potatoes with a combined mass of 1 kilogram than there is on a single 1-kilogram potato, there is more surface area on a bunch of smaller pieces of uranium-235 than on a large piece. In a small piece of uranium-235, therefore, neutrons have a greater chance of reaching the surface and escaping before they cause additional fission events. This is illustrated in Figure 20.4. In a larger piece, the chain reaction builds up to enormous energies before the

Figure 20.4
This exaggerated view shows that a chain reaction in a small piece of pure uranium-235 runs its course before it can cause a large explosion because neutrons leak from the surface too soon. The surface area of the small piece is large relative to the mass. In a larger piece, more uranium and less surface are presented to the neutrons.

neutrons can reach the surface and escape. For masses greater than a certain amount, called the **critical mass,** an explosion of enormous magnitude may occur.

Consider a large quantity of uranium-235 divided into two pieces, each having a mass smaller than critical. The units are *subcritical.* Neutrons in either piece readily reach the surface and escape before a sizable chain reaction builds up. If the pieces are suddenly pushed together, however, the total surface area decreases. If the timing is right and the combined mass is greater than critical, a violent explosion occurs. This is what happens in a nuclear fission bomb (Figure 20.5).

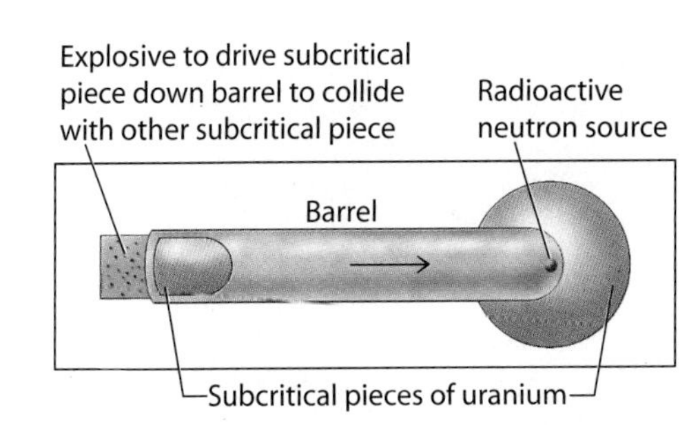

Figure 20.5
Simplified diagram of a uranium fission bomb.

Constructing a fission bomb is a formidable task. The difficulty is in separating enough uranium-235 from the more abundant uranium-238. Scientists took more than two years to extract enough of the 235 isotope from uranium ore to make the bomb that was detonated at Hiroshima in 1945. To this day, uranium isotope separation remains a difficult process.

Concept Check ✓

A 1-kilogram ball of uranium-235 has critical mass, but the same ball broken up into small chunks does not. Explain.

Check Your Answer The small chunks have more combined surface area than the original ball. Neutrons escape via the surface of each small chunk before a sustained chain reaction can develop.

20.2 Nuclear Reactors Convert Nuclear Energy to Electrical Energy

The awesome energy of nuclear fission was introduced to the world in the form of nuclear bombs, and this violent image still colors our thinking about nuclear power. This makes it difficult for many people to recognize the potential usefulness of nuclear fission. Currently, about 20% of electrical energy in the United States is generated by *nuclear fission reactors.* Nuclear fission reactors are simply nuclear boilers, as Figure 20.6 illustrates. Like fossil-fuel furnaces, reactors do

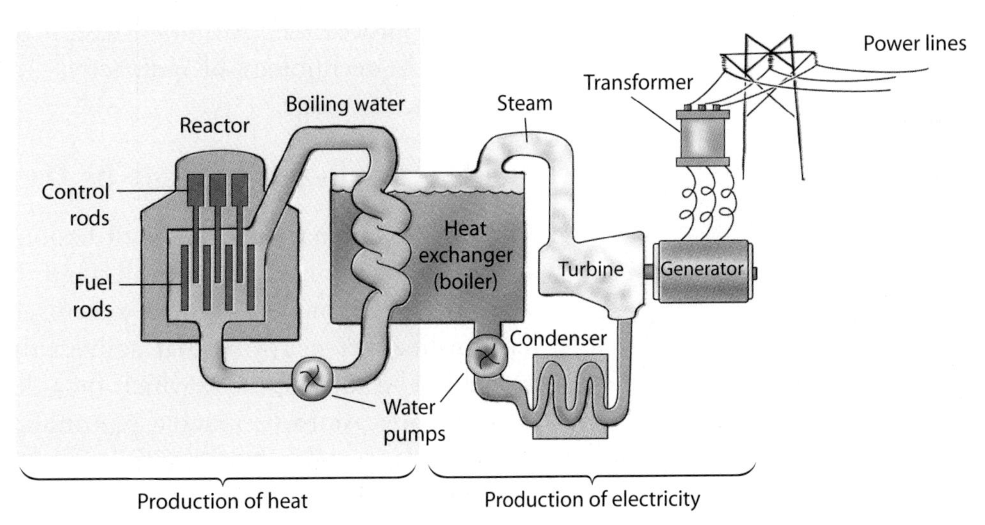

Figure 20.6
Diagram of a nuclear fission power plant. Note that the water in contact with the fuel rods is completely contained, and radioactive materials are not involved directly in the generation of electricity. The details of the production of electricity are covered in Chapter 12.

Figure 20.7
A nuclear reactor is housed within a dome-shaped containment building designed to prevent the release of radioactive isotopes in the event of an accident.

nothing more elegant than boil water to produce steam for a turbine. The greatest practical difference is the amount of fuel involved: A mere 1 kilogram of uranium fuel yields more energy than 30 freight-car loads of coal!

A fission reactor contains three components: nuclear fuel, control rods, and a liquid (usually water) to transfer the heat created by fission from the reactor to the turbine. The nuclear fuel is primarily uranium-238 plus about 3% uranium-235. Because the uranium-235 atoms are so highly diluted with uranium-238 atoms, an explosion like that of a nuclear bomb is not possible. The reaction rate depends on the number of neutrons available to initiate fission of uranium-235 nuclei. The number of neutrons available for fission is controlled by rods inserted into the reactor. The control rods are made of a neutron-absorbing material, usually cadmium or boron.

Water surrounding the nuclear fuel is maintained at high pressure. This keeps the water at a high temperature without boiling. Heated by fission, this water transfers heat to a second, lower-pressure water system, which operates a turbine and an electric generator. Two separate water systems are used so that no radioactivity reaches the turbine, and the entire setup resides inside a building like the one shown in Figure 20.7. The building is designed to prevent the release of radioactive material. Fission power plants don't pollute the environment.

One disadvantage of fission power is that it generates radioactive waste. Smaller atomic nuclei are most stable when they are composed of equal numbers of protons and neutrons. And as we learned earlier, it is mainly heavy nuclei that require more neutrons than protons for stability. For example, there are 143 neutrons but only 92 protons in uranium-235. When this uranium fissions into two medium-sized elements, the extra neutrons in their nuclei make them unstable. These fragments are therefore radioactive. Most of them have very short half-lives, while others have half-lives of thousands of years. Safely disposing of these waste products as well as materials made radioactive in the production of nuclear fuels requires special storage casks and procedures. Although fission power goes back nearly a half-century, the technology of radioactive waste disposal is still in the developmental stage.

The Breeder Reactor Breeds Its Own Fuel

One of the fascinating features of fission power is the breeding of fission fuel from nonfissionable U-238. Breeding occurs when small amounts of fissionable isotopes are mixed with U-238 in a reactor. Fission liberates neutrons that convert the relatively abundant nonfissionable U-238 to U-239, which beta-decays to Np-239, which in turn beta-decays to fissionable plutonium—Pu-239. So in addition to the abundant energy produced, fission fuel is bred from relatively abundant U-238 in the process.

Breeding occurs to some extent in all fission reactors, but a reactor specifically designed to breed more fissionable fuel than is put into it is called a breeder reactor. Using a breeder reactor is like filling your car's gas tank with water, adding some gasoline, then driving the car and having more gasoline after the trip than at the beginning! The basic principle of the breeder reactor is very attractive, for after a few years of operation a breeder-reactor power plant can produce vast amounts of power while at the same time breeding twice as much fuel as its original fuel.

The downside of breeder reactors is the enormous complexity of successful and safe operation. The United States gave up on breeders more than a decade ago, and only France and Germany are still investing in them. Officials in these countries point out that supplies of naturally occurring U-235 are limited. At present rates of consumption, all natural sources of U-235 may be depleted within a century. If countries then decide to turn to breeder reactors, they may well find themselves digging up the radioactive wastes they once buried.

The benefits of fission power are plentiful electricity, conservation of many billions of tons of fossil fuels that every year are literally turned to heat and smoke (which in the long run may be far more precious as sources of organic molecules than sources of heat), and the elimination of the megatons of sulfur oxides and other poisons put into the air each year by the burning of fossil fuels.

20.3 Nuclear Energy Comes from Nuclear Mass and Vice Versa

In the early 1900s, Albert Einstein (1879–1955) discovered that mass is actually "congealed" energy. He realized that mass and energy are two sides of the same coin, as stated in his celebrated equation $E = mc^2$. In this equation, E stands for the energy that any mass has at rest; m stands for mass; and c is the speed of light. This relationship between energy and mass is the key to understanding why and how energy is released in nuclear reactions. Any time a nucleus splits into two smaller nuclei, the combined mass of all nucleons in the smaller nuclei is less than the combined mass of all nucleons in the original nucleus. There is less mass after the splitting than before. The mass "missing" after the fission event is converted to energy and transferred to the surroundings. Let's see how.

From physics we learned that energy is the ability to do work (Section 6.3) and that work is equal to the product of force and distance:

Work = force × distance

How much force is required to pull a nucleon out of a nucleus? A lot—the nucleon has to be pulled hard enough to overcome the

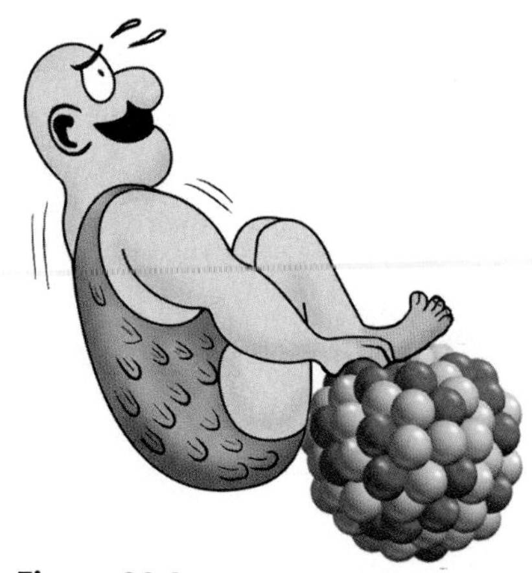

Figure 20.8
Much work is required to pull a nucleon from an atomic nucleus.

attractive strong nuclear force (Figure 20.8). The equation for work reminds us that enormous *force* exerted through a *distance* means that a huge amount of work is required. This work done on the nucleon is energy that is added to the nucleon. It shows itself as increased mass.

Consider a carbon-12 atom. Its nucleus has six protons and six neutrons, with a mass of exactly 12.00000 atomic mass units. Therefore, each proton and each neutron contribute a mass of 1 atomic mass unit. However, outside the nucleus, a proton has a mass of 1.00728 atomic mass units and a neutron has a mass of 1.00867 atomic mass units. Thus we see that the combined mass of six free protons and six free neutrons—$(6 \times 1.00728) + (6 \times 1.00867) = 12.09570$—is greater than the mass of one carbon-12 nucleus. The greater mass of the free nucleons is the form of energy that was required to pull the nucleons apart from one another. Thus, the mass of a nucleon depends on where the nucleon is.

The graph shown in Figure 20.9 results when we plot average mass *per nucleon* for the elements hydrogen through uranium. This graph is the key to understanding the energy released in nuclear processes. To obtain the average mass per nucleon, you divide the total mass of a nucleus by the number of nucleons in the nucleus. (Similarly, if you divide the total mass of a roomful of people by the number of people in the room, you get the average mass per person.)

From Figure 20.9, we can see how energy is released when a uranium nucleus splits into fragment nuclei of lower atomic number. Uranium, represented at the right of the graph, has a relatively large amount of mass per nucleon. When a uranium nucleus splits, note that smaller nuclei of lower atomic numbers are formed. As shown in Figure 20.10, these nuclei are lower on the graph than uranium, which indicates they have less mass per nucleon. Thus, nucleons lose

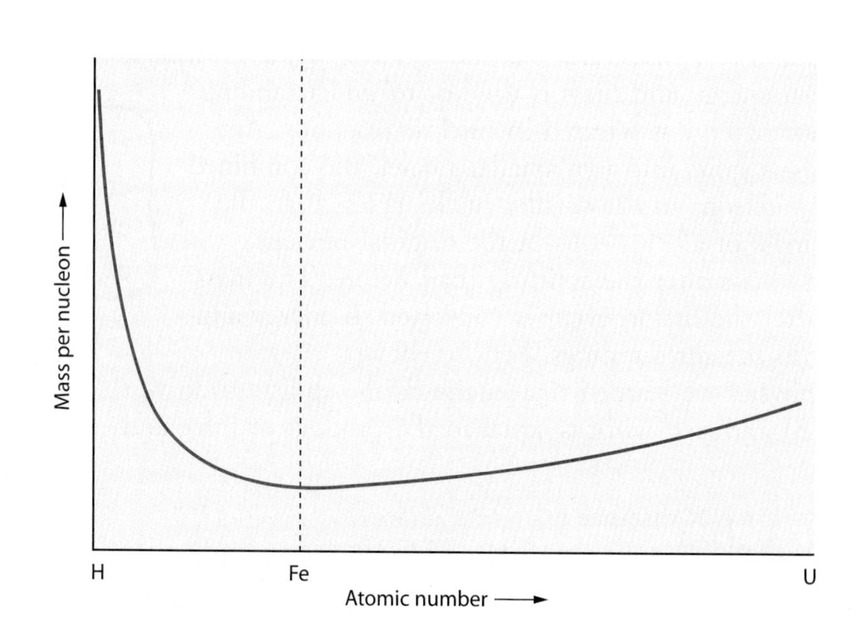

Figure 20.9
This graph shows that the average mass of a nucleon depends on which nucleus it is in. Individual nucleons have the most mass in the lightest nuclei, the least mass in iron, and intermediate mass in the heaviest nuclei.

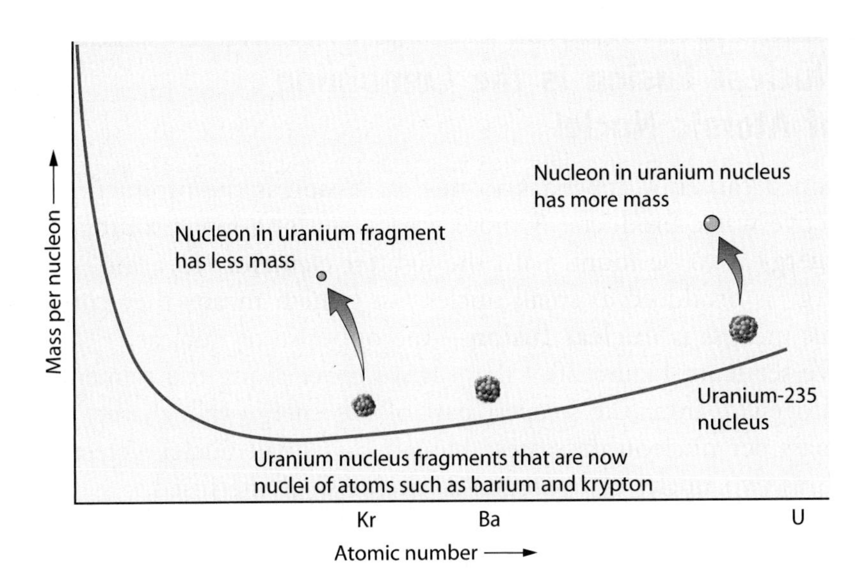

Figure 20.10
The mass of each nucleon in a uranium nucleus is greater than the mass of each nucleon in any one of its fission fragments. This lost mass has been converted to energy, which is the reason nuclear fission is an energy-releasing process.

mass as they convert from a uranium nucleus to one of its fragments. All the mass lost is converted to energy. The harnessing of this energy is referred to as "nuclear power." To calculate exactly how much energy is released in each fission event, use Einstein's equation: Multiply the decrease in mass by the speed of light squared (c^2 in the equation). The product is the amount of energy yielded by each uranium nucleus that undergoes fission.

Interestingly, Einstein's mass/energy relationship applies to chemical reactions as well as to nuclear reactions. For nuclear reactions, the energies involved are so great that the change in mass is measurable, corresponding to about 1 part in a 1000. In chemical reactions, the energy involved is so small that the change in mass, about 1 part in 1,000,000,000, is not detectable.

Concept Check ✓

Correct this statement: When a heavy element undergoes fission, there are fewer nucleons after the reaction than before.

Check Your Answer When a heavy element undergoes fission, there aren't fewer nucleons after the reaction. Instead, there's *less mass* in the same number of nucleons.

We can think of the mass-per-nucleon graph shown in Figures 20.9 and 20.10 as an energy valley that begins at hydrogen (the highest point) and slopes steeply to the lowest point (iron), then slopes gradually up to uranium. Iron is at the bottom of the energy valley and is therefore the most stable nucleus. It is also the most tightly bound nucleus; more energy per nucleon is required to separate nucleons from an iron nucleus than from any other nucleus.

20.4 Nuclear Fusion Is the Combining of Atomic Nuclei

As mentioned earlier, a drawback to nuclear fission is the production of radioactive waste products. A more promising long-range source of nuclear energy is to be found with the lightest elements. In a nutshell, energy is produced as small nuclei *fuse* (which means they combine). This process is **nuclear fusion**—the opposite of nuclear fission. We see from Figure 20.9 that, as we move along the elements from hydrogen to iron (the steepest part of the energy valley), the average mass per nucleon decreases. Thus if two small nuclei were to fuse, such as two nuclei of hydrogen-2, the mass of the fused nucleus, helium-4, would be less than the mass of the two hydrogen-2 nuclei (Figure 20.11). As with fission, the mass lost by the nucleons is converted to useful energy.

If a fusion reaction is to occur, the nuclei must be traveling at extremely high speeds when they collide in order to overcome their mutual electrical repulsion. The required speeds correspond to the extremely high temperatures found in the sun and other stars. Fusion brought about by high temperatures is called **thermonuclear fusion.** In the high temperatures of the sun, approximately 657 million tons of hydrogen is fused to 653 million tons of helium *each second.* That means 4 million tons of nucleon mass each second is converted to radiant energy. How nice—that's our sunshine.

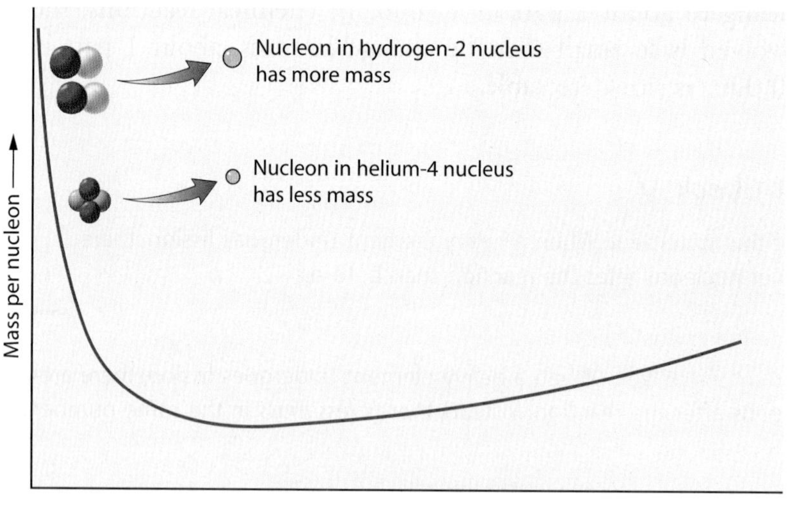

Figure 20.11
The mass of each nucleon in a hydrogen-2 nucleus is greater than the mass of each nucleon in a helium-4, which results from the fusion of two hydrogen-2 nuclei. This lost mass has been converted to energy, which is why nuclear fusion is an energy-releasing process.

Concept Check ✓

To get energy from the element iron, should iron be fissioned or fused?

Check Your Answer Neither, because iron is at the very bottom of the energy-valley curve of Figures 20.9, 20.10, and 20.11. If you fuse two iron nuclei, the product lies somewhere to the right of iron on the curve, which means the product has a higher mass per nucleon. If you split an iron nucleus, the products lie to the left of iron on the curve, which again means a higher mass per nucleon. Because no mass decrease occurs in either reaction, no mass is available to be converted to energy, and as a result no energy is released.

Prior to the development of the atomic bomb, the temperatures required to initiate nuclear fusion on the Earth were unattainable. When researchers found that the temperature inside an exploding atomic bomb is four to five times the temperature at the center of the sun, the thermonuclear bomb was but a step away. This first thermonuclear bomb, a hydrogen bomb, was detonated in 1952. Whereas the critical mass of fissionable material limits the size of a fission bomb (atomic bomb), no such limit is imposed on a fusion bomb (thermonuclear or hydrogen bomb). A typical thermonuclear bomb stockpiled by the United States today, for example, is about 1000 times more destructive than the atomic bomb detonated over Hiroshima, Japan, at the end of World War II.

The hydrogen bomb is another example of a discovery used for destructive rather than constructive purposes. The potential constructive possibility is the controlled release of vast amounts of clean energy, as is discussed in the next section.

20.5 An Important Goal of Nuclear Research Is Controlled Fusion

Carrying out fusion reactions under controlled conditions requires temperatures of millions of degrees. As you can imagine, a big problem is that any reaction vessel being used would melt and vaporize long before these temperatures were reached. The solution to this problem is to confine the reaction in a *nonmaterial container.*

One type of nonmaterial container is a magnetic field. An example is shown in Figure 20.12. Such a magnetic field can exist at any temperature and can exert powerful forces on charged particles in motion. "Magnetic walls" provide a kind of magnetic straitjacket for hot gases called *plasmas.* Plasma is a fourth phase of matter that exists when matter is heated beyond the limits of its gaseous phase. Plasmas

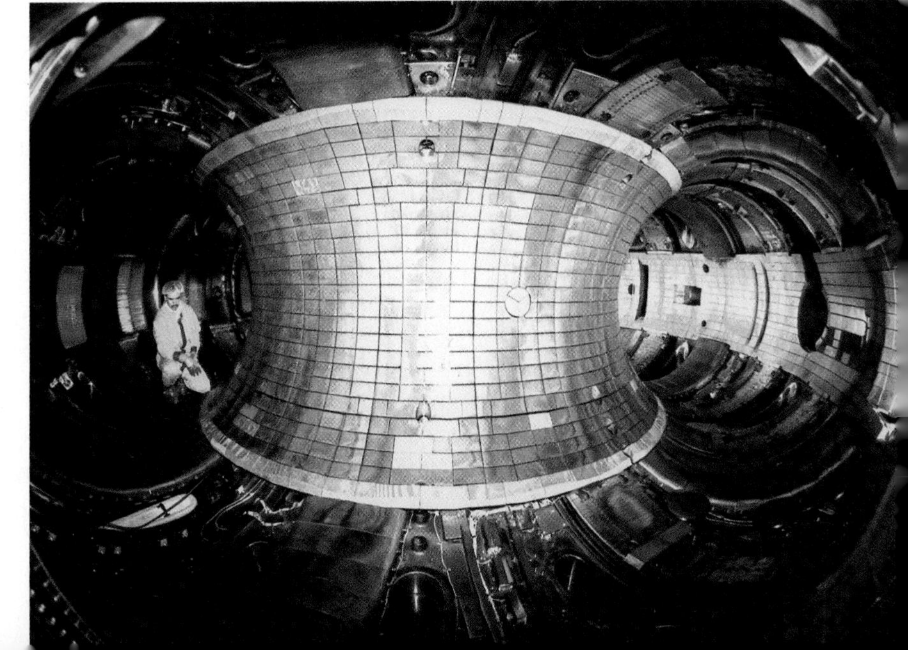

Figure 20.12
An interior view of the former Tokamak Fusion Test Reactor at the Princeton Plasma Physics Laboratory. Magnetic fields confine a fast-moving plasma to a circular path. At a high enough temperature, the atomic nuclei in the confined plasma fuse to produce energy.

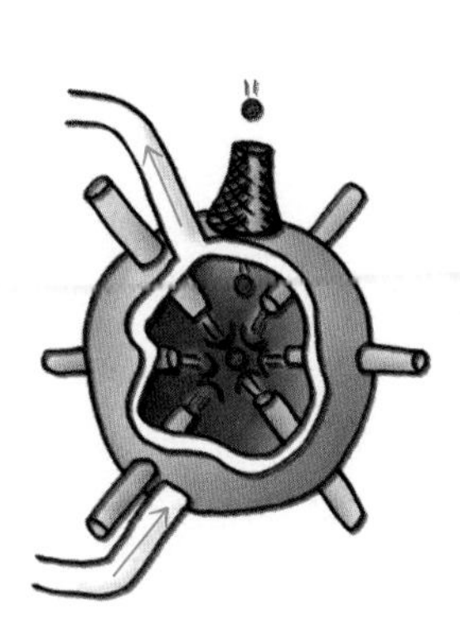

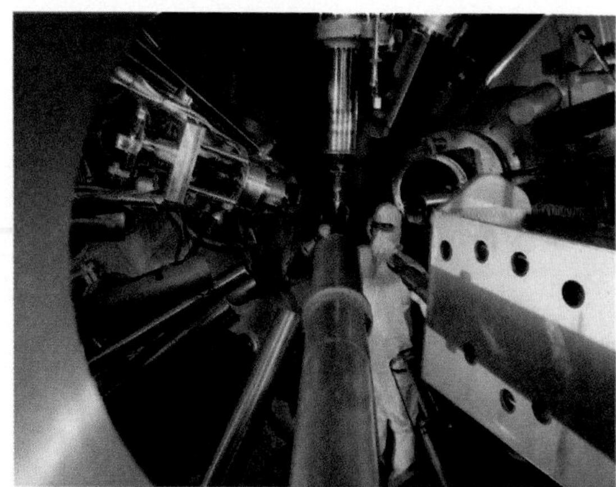

(a) (b)

Figure 20.13
(a) Fusion with multiple laser beams. Pellets of hydrogen isotopes are rhythmically dropped into synchronized laser crossfire in this planned device. The resulting heat is carried off by molten lithium to produce steam. (b) The pellet chamber at Lawrence Livermore Laboratory. The laser source is Nova, which directs ten beams into the target region.

form when matter is heated so much that electrons are stripped away from nuclei and a hot gaslike fluid of charged particles results. The sun is a plasma, as is the interior of a fluorescent lamp when the lamp is operating.

Once a plasma is produced and placed in its magnetic nonmaterial container, magnetic compression further heats the plasmas to fusion temperatures. At about 1 million degrees, some nuclei are moving fast enough to overcome repulsive electric forces, and these nuclei slam together and fuse. The energy released by this fusion, however, is less than the energy used to heat the plasma. Even at 100 million degrees, more energy must be put into the plasma than is given off by fusion. It is not until about 350 million degrees that the fusion reactions produce enough energy to be self-sustaining. At this ignition temperature, all that is needed to produce continuous power is a steady feed of nuclei.

Although several fusion devices are now capable of releasing more energy than they consume, instabilities in the plasma have thus far prevented a sustained reaction. A big problem has been devising a magnetic field system that can hold the plasma steady long enough to allow a sufficient number of nuclei to fuse. Magnetic confinement has been disappointing.

Another approach uses high-energy lasers. One proposed technique is to aim an array of laser beams at a common point and drop solid pellets of hydrogen isotopes through the crossfire, as Figure 20.13 shows. The energy of the multiple beams should crush the pellets to densities 20 times that of lead. Such a fusion could produce several hundred times more energy than the amount delivered by the laser beams. Like the succession of fuel/air explosions in an automobile engine's cylinders that convert to a smooth flow of mechanical power, the successive ignition of pellets in a laser fusion device may similarly produce a steady stream of electric power. A plant equipped with the device could produce 1000 million watts of electric power, enough to supply a city of 600,000 people. High-power lasers that work reliably, however, have yet to be developed.

Concept Check ✓

Fission and fusion are opposite processes, yet each releases energy. Isn't this contradictory?

Check Your Answer No, no, no! As Figure 20.9 shows, only the fusion of light elements and the fission of heavy elements result in a decrease in nucleon mass and therefore a release of energy.

If people are one day to dart about the universe the way we jet about the Earth today, their supply of fuel is assured. The fuel for fusion—hydrogen—is found in every part of the universe, not only in the stars but also in the space between them. About 91% of the atoms in the universe are estimated to be hydrogen. For people of the future, the supply of raw materials is also assured because all the elements known to exist result from the fusing of more and more hydrogen nuclei. Simply stated, if you fuse 8 hydrogen-2 nuclei, you have oxygen; 26, you have iron; and so forth. Future humans might synthesize their own elements and produce energy in the process, just as the stars have always done.

Chapter Review

Key Terms and Matching Definitions

_____ chain reaction
_____ critical mass
_____ nuclear fission
_____ nuclear fusion
_____ thermonuclear fusion

1. The splitting of a heavy nucleus into two lighter nuclei, accompanied by the release of much energy.
2. A self-sustaining reaction in which the products of one fission event stimulate further events.
3. The minimum mass of fissionable material needed in a reactor or nuclear bomb that will sustain a chain reaction.
4. The joining together of light nuclei to form a heavier nucleus, accompanied by the release of much energy.
5. Nuclear fusion produced by high temperature.

Review Questions

Nuclear Fission Is the Splitting of the Atomic Nucleus

1. Why does a chain reaction not occur in uranium mines?
2. Is a chain reaction more likely to occur in two separate pieces of uranium-235 or in the same pieces stuck together?

Nuclear Reactors Convert Nuclear Energy to Electrical Energy

3. How is a nuclear reactor similar to a conventional fossil-fuel power plant? How is it different?
4. What is the function of control rods in a nuclear reactor?

Nuclear Energy Comes from Nuclear Mass and Vice Versa

5. Is work required to pull a nucleon out of an atomic nucleus? Does the nucleon, once outside the nucleus, have more mass than it had inside the nucleus?
6. How does the mass per nucleon in uranium compare with the mass per nucleon in the fission fragments of uranium?
7. If an iron nucleus split in two, would its fission fragments have more mass per nucleon or less mass per nucleon?
8. If a pair of iron nuclei were fused, would the product nucleus have more mass per nucleon or less mass per nucleon?

Nuclear Fusion Is the Combining of Atomic Nuclei

9. When a pair of hydrogen isotopes are fused, is the mass of the product nucleus more or less than the total mass of the hydrogen nuclei?

10. From where does the sun get its energy?

An Important Goal of Nuclear Research Is Controlled Fusion

11. How do the product particles of fusion reactions differ from the product particles of fission reactions?

12. What kind of containers are used to contain multimillion-degree plasmas?

Exercises

1. Why will nuclear fission probably never be used directly for powering automobiles? How could it be used indirectly?

2. Why does a neutron make a better nuclear bullet than a proton or an electron?

3. Does the average distance a neutron travels through fissionable material before escaping increase or decrease when two pieces of fissionable material are assembled into one piece? Does this assembly increase or decrease the probability of an explosion?

4. Why does plutonium not occur in appreciable amounts in natural ore deposits?

5. Uranium-235 releases an average of 2.5 neutrons per fission, while plutonium-239 releases an average of 2.7 neutrons per fission. Which of these elements might you therefore expect to have the smaller critical mass?

6. After a uranium fuel rod reaches the end of its fuel cycle (typically three years), why does most of its energy come from plutonium fission?

7. To predict the approximate energy release of either a fission or a fusion reaction, explain how a physicist uses a table of nuclear masses and the equation $E = mc^2$.

8. Which process would release energy from gold, fission or fusion? From carbon? From iron?

9. Does the mass of a nucleon after it has been pulled from an atomic nucleus depend on which nucleus it was extracted from?

10. If a uranium nucleus were to fission into three segments of approximately equal size instead of two, would more energy or less energy be released? Defend your answer using Figure 20.10.

11. Explain how radioactive decay has always warmed the Earth from the inside and how nuclear fusion has always warmed the Earth from the outside.

12. What percentage of nuclear power plants in operation today are based upon nuclear fusion?

13. Speculate about some worldwide changes likely to follow the advent of successful fusion reactors.

Suggested Web Sites

http://www.iaea.or.at/worldatom

The Web site for the International Atomic Energy Agency, which monitors almost all issues related to nuclear technology. A good starting point for exploring applications of many of the concepts discussed in this chapter.

http://www.iter.org

The Web site for the International Thermonuclear Experimental Reactor project. Explore this site for the latest on the science and politics of this important project.

Part 6 Chemistry

Chapter 21: Elements of Chemistry

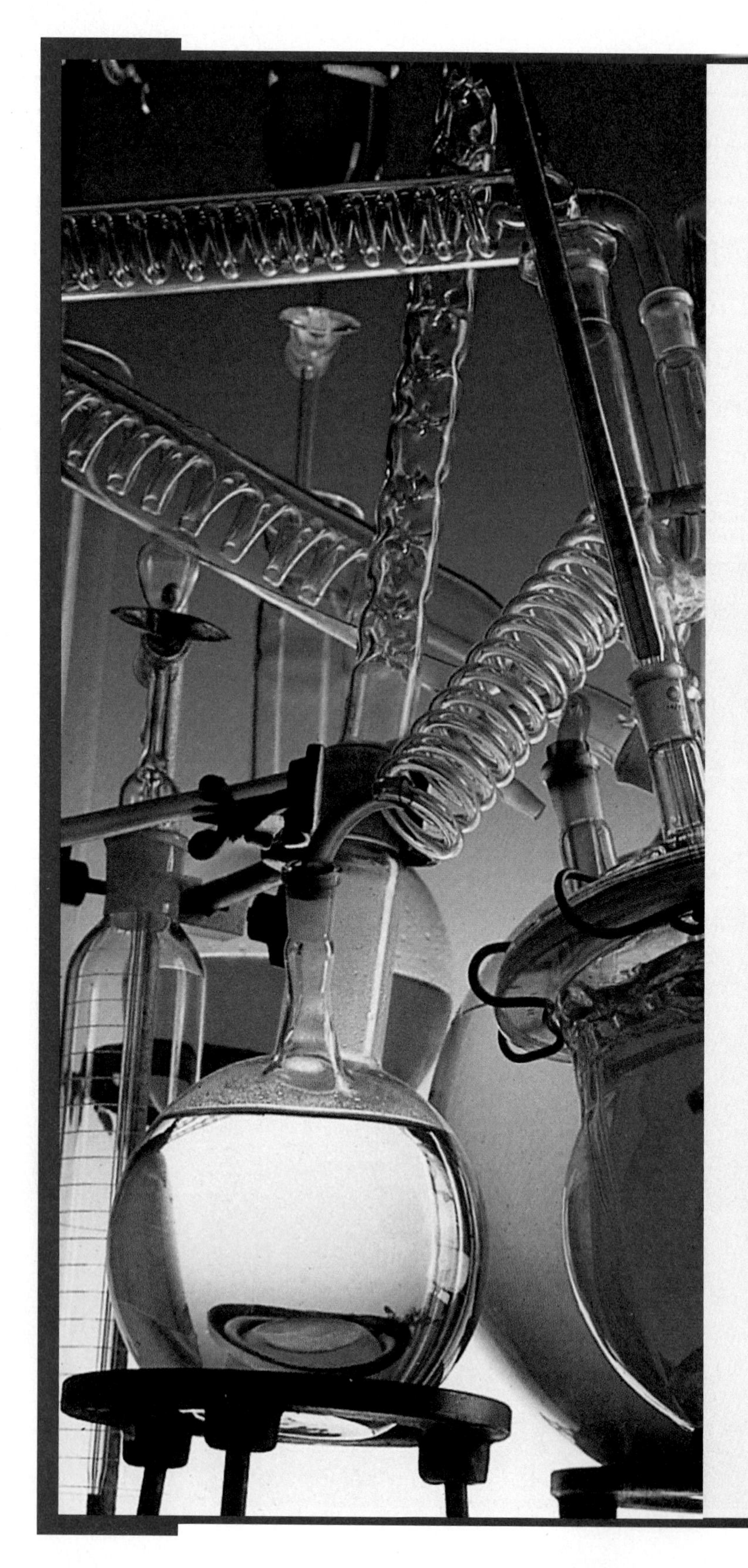

We know that two oxygen atoms joined together make an oxygen molecule—which is good for life. So why does adding another oxygen atom to the pair make a poison? Similarly, we know that common table salt is also good for life. How is it, then, that the two elements composing salt are poisonous by themselves? Why does a strong iron bar eventually become a crumbling pile of rust when left out in the rain? Is rusting the same kind of change that occurs when water freezes or glass breaks? The answers to these questions make up the science of matter: Chemistry!

21.1 Chemistry Is a Central Science Useful to Our Lives

When you wonder what the Earth, sky, or ocean is made of, you are thinking about chemistry. When you wonder how a rain puddle dries up, how a car gets energy from gasoline, or how your body gets energy from the food you eat, you are again thinking about chemistry. By definition, **chemistry** is the study of matter and the transformations it can undergo. **Matter** is anything that occupies space. It is the stuff that makes up all material things—anything you can touch, taste, smell, see, or hear is matter. Chemistry therefore is very broad in scope.

Chemistry is often described as a central science because it touches all the other sciences. It builds up from physics, and serves as the foundation for the most complex science of all—biology. Chemistry is the foundation for the Earth sciences—geology, volcanology, oceanography, meteorology, and archeology. It is also an important component of space science, as described in Figure 21.1. Just as we learned about the origin of the moon from the chemical analysis of moon rocks in the early 1970s, we are now learning about the history of Mars and other planets from the chemical information gathered by space probes.

Figure 21.1
Special materials, such as rocket fuels, metals for the space ships, and fabrics for the space suits, were required to allow the astronauts to reach and explore the surface of the moon. These materials were developed using chemistry, which was also used to study the composition of moon rocks.

Scientific research is activity aimed at discovering and interpreting new knowledge. **Basic research** leads to greater understanding of how the natural world operates. Many scientists focus on basic research. The foundation of knowledge laid down by basic research often leads to useful applications. **Applied research** focuses on developing these applications. While physicists tend to focus on basic research, most chemists focus on applied research. Applied research in chemistry has provided us with medicine, food, water, shelter, and so many of the material goods that characterize modern life. Just a few of a myriad of examples are shown in Figure 21.2.

Figure 21.2
Most of the material items in any modern house are shaped by some human-devised chemical process.

In the past century, we excelled at creating materials to suit our needs. However, mistakes were made in caring for the environment. Waste products were dumped into rivers, buried in the ground, or vented into the air without regard for possible long-term consequences. Many people believed that the Earth was so large that its resources were virtually unlimited and that it could absorb wastes without being significantly harmed. People thought little of the effects of these wastes on human health, as well.

Most nations now recognize this as a dangerous attitude. As a result, government agencies, industries, and concerned citizens are involved in extensive efforts to clean up toxic-waste sites. Such regulations as the international ban on ozone-destroying chlorofluorocarbons (CFCs) have been enacted to protect the environment. Members of the American Chemistry Council, who as a group produce 90% of the chemicals manufactured in the United States, have adopted a program called *Responsible Care,* in which they have pledged to manufacture without causing environmental damage. The Responsible Care program—its emblem shown in Figure 21.3—is based on the understanding that modern technology can be used to both harm and protect the environment. By using chemistry wisely, most waste products can be minimized, recycled, or engineered into sellable ones that are environmentally benign.

Figure 21.3
The Responsible Care symbol of the American Chemistry Council.

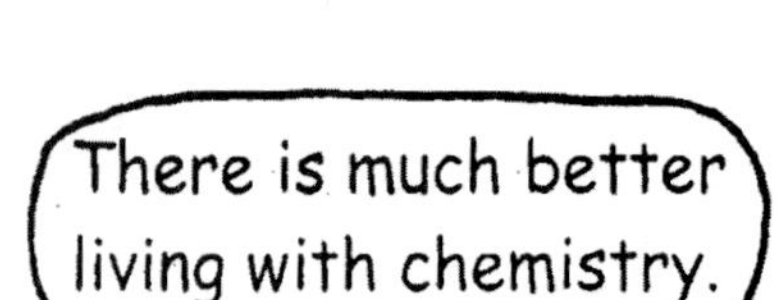

Concept Check ✓

Chemists have learned how to produce aspirin using petroleum as a starting material. Is this an example of basic or applied research?

Check Your Answer This is an example of applied research because the primary goal was to develop a useful commodity. However, the ability to produce aspirin from petroleum depended on an understanding of atoms and molecules, an understanding that came from many years of basic research.

21.2 The Submicroscopic World Is Made of Atoms and Molecules

From afar, a sand dune looks like it's made of a smooth, continuous material. But up close, you can tell the dune is made of tiny particles of sand. Similarly, and as we learned in Chapter 17, everything around us, no matter how smooth it may appear, is made of basic units called *atoms.* Atoms are so small that a single grain of sand contains about 125 million trillion of them. There are roughly a quarter million times more atoms in a single grain of sand than there are grains of sand in the dunes shown in Figure 21.4.

Some atoms link together to form larger but still incredibly small basic units of matter. These are **molecules.** As shown in Figure 21.4,

Figure 21.4
There are far more atoms in a glass of water than there are grains of sand within a mountain-sized sand dune.

two hydrogen atoms and one oxygen atom link to form a single molecule of water, which we know as H_2O. Water molecules are so small that an 8-oz glass of water contains about a trillion trillion of them.

Our world can be studied at different levels of magnification. At the *macroscopic* level, matter is large enough to be seen, measured, and handled. A handful of sand and a glass of water are macroscopic samples of matter. At the *microscopic* level, physical structure is so fine that it can be seen only with a microscope. A biological cell is microscopic, as is the detail on a dragonfly's wing. Beyond the microscopic level is the **submicroscopic**—the realm of atoms and molecules. This realm is an important focus of chemistry.

On the submicroscopic level, solid, liquid, and gaseous phases are distinguished by how the submicroscopic particles hold together. This is illustrated to the right. In solid matter, such as a rock, the attractions between particles are strong enough to hold all the particles together in some fixed three-dimensional arrangement. The particles are able to vibrate about fixed positions, but they cannot move past one another.

Adding heat causes these vibrations to increase until, at a certain temperature, the vibrations are rapid enough to disrupt the fixed arrangements. Rock will melt into lava. Likewise, ice will melt into water. The particles can then slip past one another and tumble around much like a bunch of marbles in a bag. This is the liquid phase of matter, and it is the mobility of the submicroscopic particles that gives rise to the liquid's fluid character—its ability to flow and take on the shape of its container.

Further heating causes the submicroscopic particles in a liquid to move so fast that the attractions they have for one another are unable to hold them together. They then separate from one another forming a gas. For lava, this doesn't easily happen because the particles are so attracted to each other. For water, molecules will separate into a gas at 100°C. For a substance like helium, the submicroscopic particles are already in the gaseous phase at room temperature.

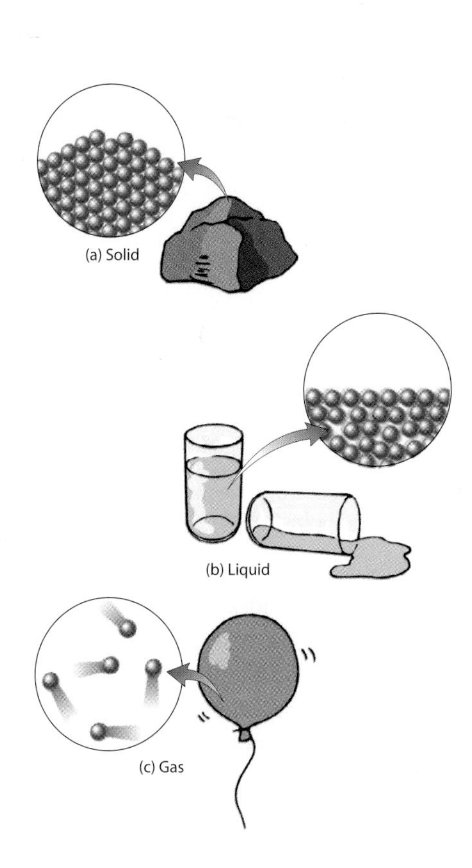

Gold
Opacity: opaque
Color: yellowish
Phase at 25°C: solid
Density: 19.3 g/mL

Diamond
Opacity: transparent
Color: colorless
Phase at 25°C: solid
Density: 3.5 g/mL

Water
Opacity: transparent
Color: colorless
Phase at 25°C: liquid
Density: 1.0 g/mL

Figure 21.5
Gold, diamond, and water can be identified by their physical properties. If a substance has all the physical properties listed under gold, for example, it must be gold.

Moving at an average speed of 500 meters per second (1,100 miles per hour), the particles of a gas are widely separated from one another. Matter in the gaseous phase therefore occupies much more volume than it does in the solid or liquid phase. Applying pressure to a gas squeezes the gas particles closer together, which makes for a smaller volume. Enough air for an underwater diver to breathe for many minutes, for example, can be squeezed (compressed) into a tank small enough to be carried on the diver's back.

21.3 Matter Has Physical and Chemical Properties

Properties that describe the look or feel of a substance, such as color, hardness, density, texture, and phase, are called **physical properties.** Every substance has its own set of characteristic physical properties that we can use to identify that substance (Figure 21.5).

The physical properties of a substance can change when conditions change, but this does not mean a different substance is created. Cooling liquid water to below 0°C causes the water to transform to solid ice, but the substance is still water. H_2O is H_2O no matter which phase it is in. The only difference when water changes phase is how the H_2O molecules are oriented relative to one another. In the liquid, the water molecules tumble around one another, whereas in the ice they vibrate about fixed positions. The freezing of water is an example of what chemists call a **physical change.** During a physical change, a substance changes its phase or some other physical property but *not* its chemical composition, as Figure 21.6 shows.

Concept Check ✓

The melting of gold is a physical change. Why?

Check Your Answer During a physical change, a substance changes only one or more of its physical properties; its identity does not change. Because melted gold is still gold but in a different form, this change is a physical change.

Chemical properties are properties that relate to how one substance reacts with others or how a substance transforms. Figure 21.7 shows three examples. The methane of natural gas has the chemical property of reacting with oxygen to produce carbon dioxide and water, along with lots of heat energy. Similarly, it is a chemical property of baking soda to react with vinegar to produce carbon dioxide and water while absorbing a small amount of heat energy. Copper has the chemical property of reacting with carbon dioxide and water to form a greenish-blue solid known as *patina.* Copper statues exposed to the carbon

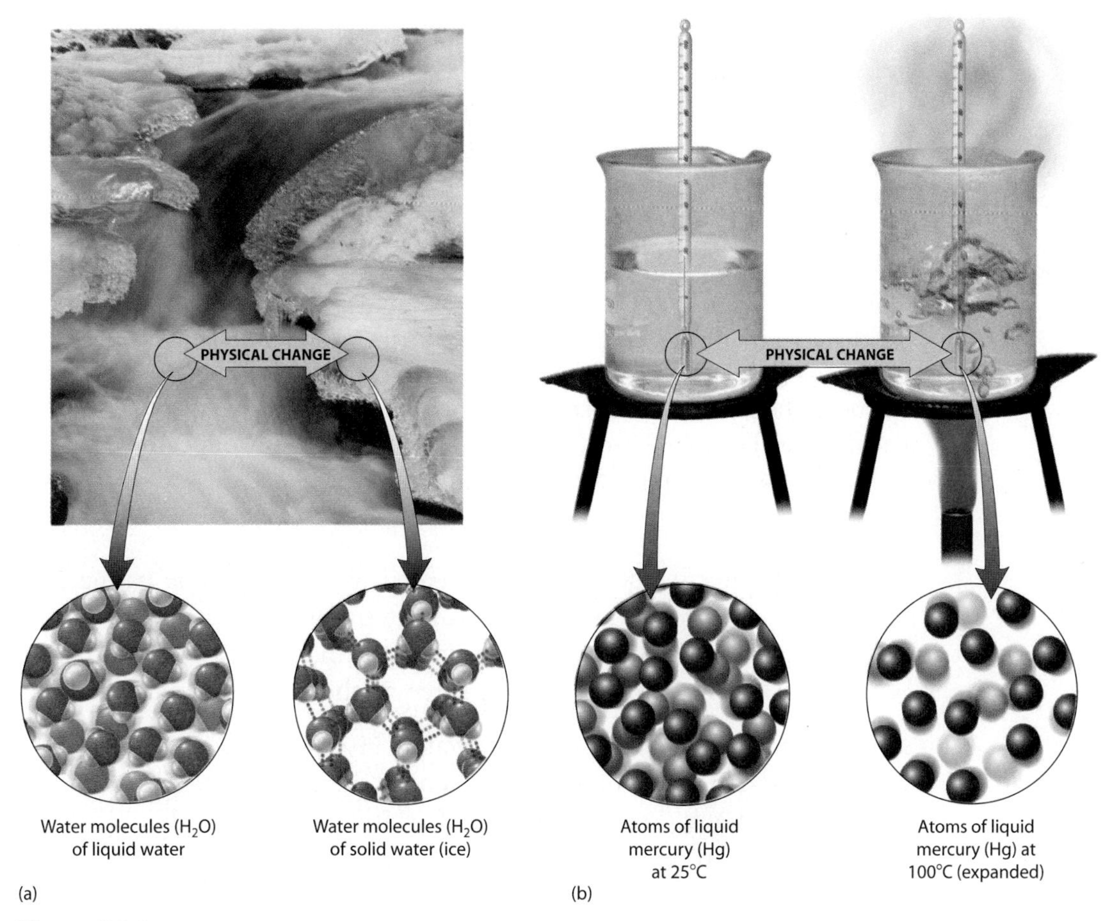

Figure 21.6
Two physical changes. (a) Liquid water and ice might look like different substances, but at the submicroscopic level, it is evident that both consist of water molecules. (b) At 25°C, the atoms in a sample of mercury are a certain distance apart, yielding a density of 13.5 grams per milliliter. At 100°C, the atoms are farther apart. This means that each milliliter now contains fewer atoms than at 25°C, and the density is now 13.3 grams per milliliter. The physical property we call *density* has changed with the temperature, but the identity of the substance remains unchanged: Mercury is mercury.

Methane
Reacts with oxygen to form carbon dioxide and water, giving off lots of heat during the reaction.

Baking soda
Reacts with vinegar to form carbon dioxide and water, absorbing heat during the reaction.

Copper
Reacts with carbon dioxide and water to form the greenish-blue substance called patina.

Figure 21.7
The chemical properties of substances allow them to transform to new substances. Natural gas and baking soda transform to carbon dioxide, water, and heat. Copper transforms to patina.

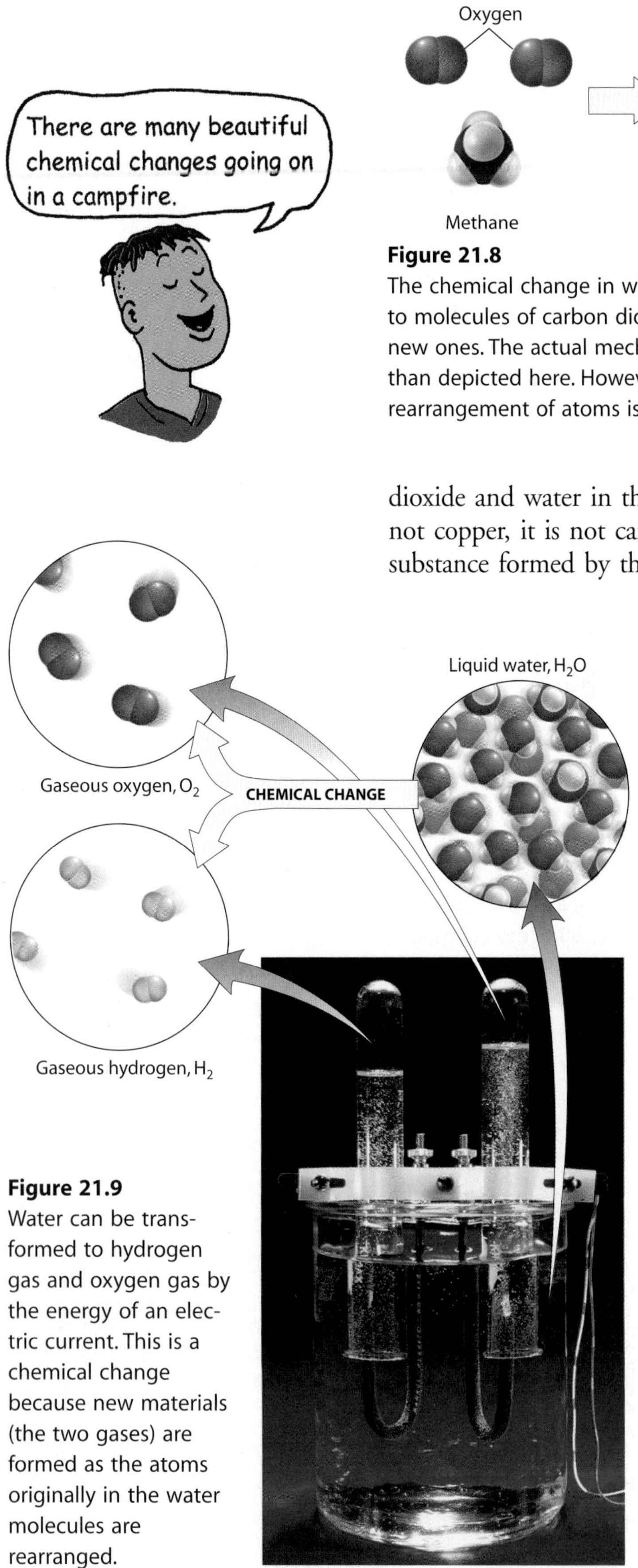

Figure 21.8
The chemical change in which molecules of methane and oxygen transform to molecules of carbon dioxide and water as atoms break old bonds and form new ones. The actual mechanism of this transformation is more complicated than depicted here. However, the idea that new materials are formed by the rearrangement of atoms is accurate.

Figure 21.9
Water can be transformed to hydrogen gas and oxygen gas by the energy of an electric current. This is a chemical change because new materials (the two gases) are formed as the atoms originally in the water molecules are rearranged.

dioxide and water in the air become coated with patina. The patina is not copper, it is not carbon dioxide, and it is not water. It is a new substance formed by the reaction of these chemicals with one another.

All three of these transformations involve a change in the way the atoms in the molecules are *chemically bonded* to one another. A **chemical bond** is the attraction between two atoms that holds them together in a molecule. A methane molecule, for example, is made of a single carbon atom bonded to four hydrogen atoms. An oxygen molecule is made of two oxygen atoms bonded to each other. Figure 21.8 shows the chemical change in which the atoms in a methane molecule and those in two oxygen molecules first pull apart and then form new bonds with different partners, resulting in the formation of molecules of carbon dioxide and water.

Any change in a substance that involves a rearrangement of the way its atoms are bonded is called a **chemical change.** Thus the transformation of methane to carbon dioxide and water is a chemical change, as are the other two transformations shown in Figure 21.7.

The chemical change shown in Figure 21.9 occurs when an electric current is passed through water. The energy of the current causes the water molecules to split into atoms that then form new chemical bonds. Thus, water molecules are changed into molecules of hydrogen and oxygen, two substances that are very different from water. The hydrogen and oxygen are both gases at room temperature, and they can be seen as bubbles rising to the surface.

In the language of chemistry, materials undergoing a chemical change are said to be *reacting*. Methane *reacts* with oxygen to form carbon dioxide and water. Water *reacts* when it's exposed to electricity to form hydrogen gas and oxygen gas. Thus the term *chemical change* means the same thing as *chemical reaction*. During a **chemical reaction,** new materials are formed by a change in the way atoms are bonded together. We shall explore chemical bonds and reactions in which they are formed and broken in later chapters.

Concept Check ✓

Each sphere in the following diagrams represents an atom. Joined spheres represent molecules. One set of diagrams shows a physical change, and the other shows a chemical change. Which is which?

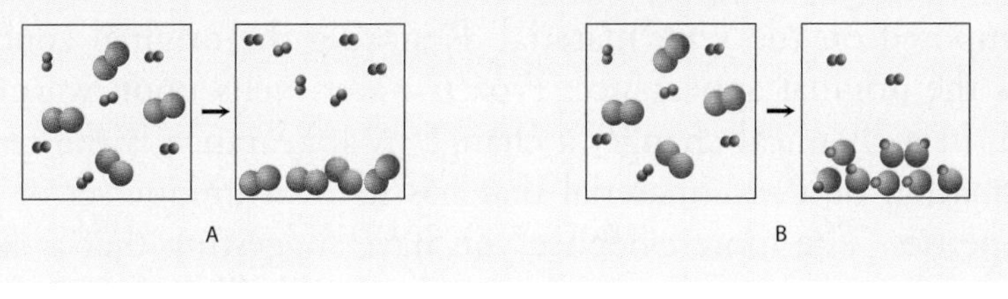

Check Your Answer Remember that a chemical change (also known as a chemical reaction) involves molecules breaking apart so that the atoms are free to form new bonds with new partners. You must be careful to distinguish this breaking apart from a mere change in the relative positions of a group of molecules. In set A, the molecules before and after the change are the same. They differ only in their positions relative to one another. Set A therefore represents a physical change. In set B, new molecules consisting of bonded red and blue spheres appear after the change. These molecules represent a new material, and so B is a chemical change.

How can you tell whether a change you observe is physical or chemical? It can be tricky because in both cases there are changes in physical appearance. Water, for example, looks quite different after it freezes, just as a car looks quite different after it rusts (Figure 21.10). The freezing of water results from a change in how water molecules are oriented relative to one another. This is a physical

Figure 21.10
The transformation of water to ice and the transformation of iron to rust both involve a change in physical appearance. The formation of ice is a physical change, and the formation of rust is a chemical change.

change because liquid water and frozen water are both forms of water. The rusting of a car, by contrast, is the result of the transformation of iron to rust. This is a chemical change because iron and rust are two different materials, each consisting of a different arrangement of atoms. As we shall see in the next two sections, iron is an *element* and rust is a *compound* consisting of iron and oxygen atoms.

By studying this chapter, you can expect to learn the difference between a physical change and a chemical change. However, you cannot expect to have a firm handle on how to categorize an observed change as physical or chemical. Doing so requires a knowledge of the chemical identity of the materials involved as well as an understanding of how their atoms and molecules behave. This sort of insight builds over many years of study and laboratory experience.

There are, however, two powerful guidelines that can assist you in assessing physical and chemical changes. First, in a physical change, a change in appearance is the result of a new set of conditions imposed on the *same* material. Restoring the original conditions restores the original appearance: Frozen water melts upon warming. Second, in a chemical change, a change in appearance is the result of the formation of a *new* material that has its own unique set of physical properties. The more evidence you have suggesting that a different material has been formed, the greater the likelihood that the change is a chemical change. Iron is a material that can be used to build cars. Rust is not. This suggests that the rusting of iron is a chemical change.

Hands-On Exploration: Fire Water

This activity is for those of you with access to a gas stove.

What You Need

gas stove; large pot; water

Safety Note

Tie long hair back and roll up long, loose sleeves. Of course, watch the pot carefully while it is on the burner. Protect your hands with oven mitts if you pick up the pot.

Procedure

Place a large pot of cool water on top of the stove. Set the burner on "High". Observe what happens for several minutes. What product from the combustion of the natural gas do you see condensing on the outside of the pot? Where did it come from? Would more or less of this product form if the pot contained ice water? Where does this product go as the pot gets warmer? What physical and chemical changes can you identify?

Concept Check ✓

Michaela has grown an inch in height over the past year. Is this best described as a physical or a chemical change?

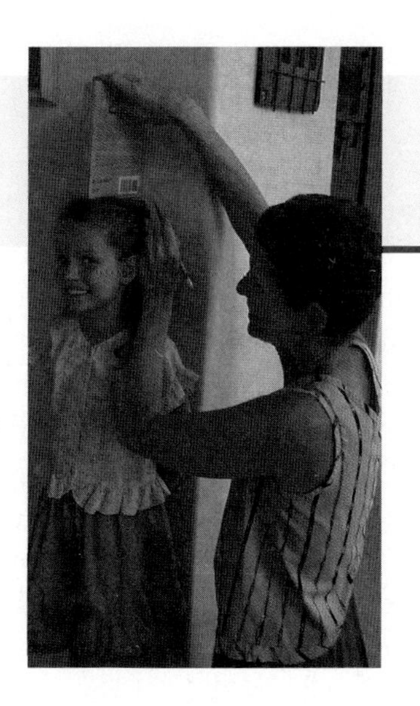

Check Your Answer Are new materials being formed as Michaela grows? Absolutely—created out of the food she eats. Her body is very different from, say, the peanut butter sandwich she ate yesterday. Yet through some very advanced chemistry, her body is able to take the atoms of that peanut butter sandwich and rearrange them into new materials. Biological growth, therefore, is best described as a chemical change.

21.4 An Element Is Made of a Collection of Atoms

The terms *element* and *atom* are often used in a similar context. You might hear, for example, that gold is an element made of gold atoms. Generally, *element* is used in reference to an entire macroscopic or microscopic sample, and *atom* is used when speaking of the submicroscopic particles in the sample. The important distinction is that elements are made of atoms, and not the other way around.

How many atoms are bound together in an element is shown by an **elemental formula.** For elements in which the basic units are individual atoms, the elemental formula is simply the chemical symbol: Au is the elemental formula for gold, and Li is the elemental formula for lithium, to name just two examples. For elements in which the basic units are two or more atoms bonded into molecules, the elemental formula is the chemical symbol followed by a subscript indicating the number of atoms in each molecule. For example, elemental nitrogen, as was shown in Figure 17.1 (page 291), commonly consists of molecules containing two nitrogen atoms per molecule. Thus N_2 is the usual elemental formula given for nitrogen. Similarly, O_2 is the elemental formula for oxygen, and S_8 is the elemental formula for sulfur.

Concept Check ✓

The oxygen we breathe, O_2, is converted to ozone, O_3, in the presence of an electric spark. Is this a physical or chemical change?

Check Your Answer When atoms regroup, the result is an entirely new substance, and that is what happens here. The oxygen we breathe, O_2, is odorless and life-giving. Ozone, O_3, can be toxic and has a pungent smell often associated with electric motors. The conversion of O_2 to O_3 is therefore a chemical change. However, both O_2 and O_3 are elemental forms of oxygen.

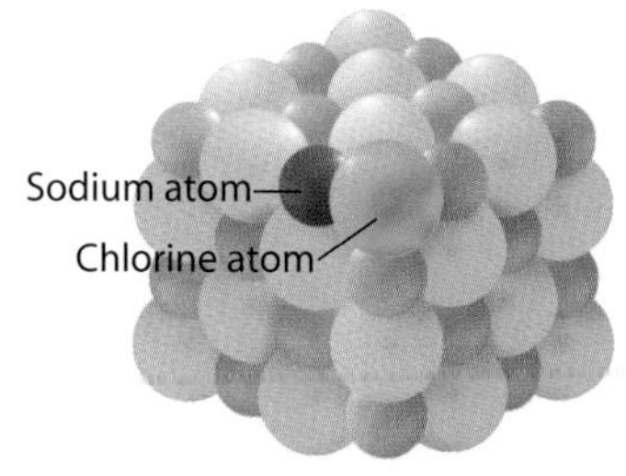

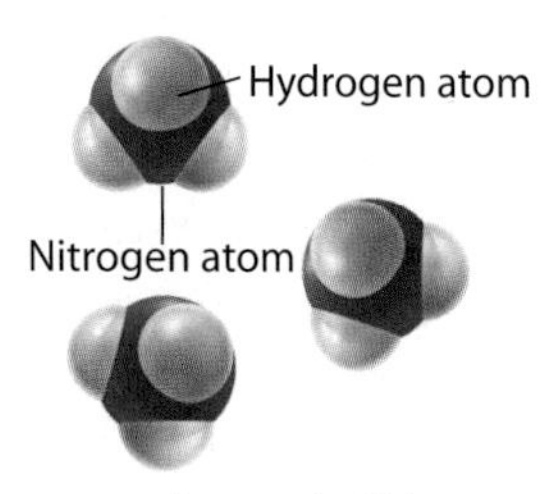

Figure 21.11
The compounds sodium chloride and ammonia are represented by their chemical formulas, NaCl and NH_3. A chemical formula shows the ratio of atoms used to make the compound.

21.5 Elements Can Combine to Form Compounds

When atoms of *different* elements bond to one another, they make a **compound.** Sodium atoms and chlorine atoms, for example, bond to make the compound sodium chloride, commonly known as table salt. Nitrogen atoms and hydrogen atoms join to make the compound ammonia, a common household cleaner.

A compound is represented by its **chemical formula,** in which the symbols for the elements are written together. The chemical formula for sodium chloride is NaCl, and that for ammonia is NH_3. Numerical subscripts indicate the ratio in which the atoms combine. By convention, the subscript 1 is understood and omitted. So the chemical formula NaCl tells us that in the compound sodium chloride there is one sodium for every one chlorine. The chemical formula NH_3 tells us that in the compound ammonia there is one nitrogen atom for every three hydrogen atoms, as Figure 21.11 shows.

Compounds have physical and chemical properties that are different from the properties of their elemental components. Sodium chloride, NaCl, shown in Figure 21.12 is very different from the elemental sodium and elemental chlorine used to form it. Elemental sodium, Na, consists of nothing but sodium atoms, which form a soft, silvery metal that can be cut easily with a knife. Its melting point is 97.5°C, and it reacts violently with water. Elemental chlorine, Cl_2, consists of chlorine molecules. This material, a yellow-green gas at room temperature, is very toxic and was used as a chemical

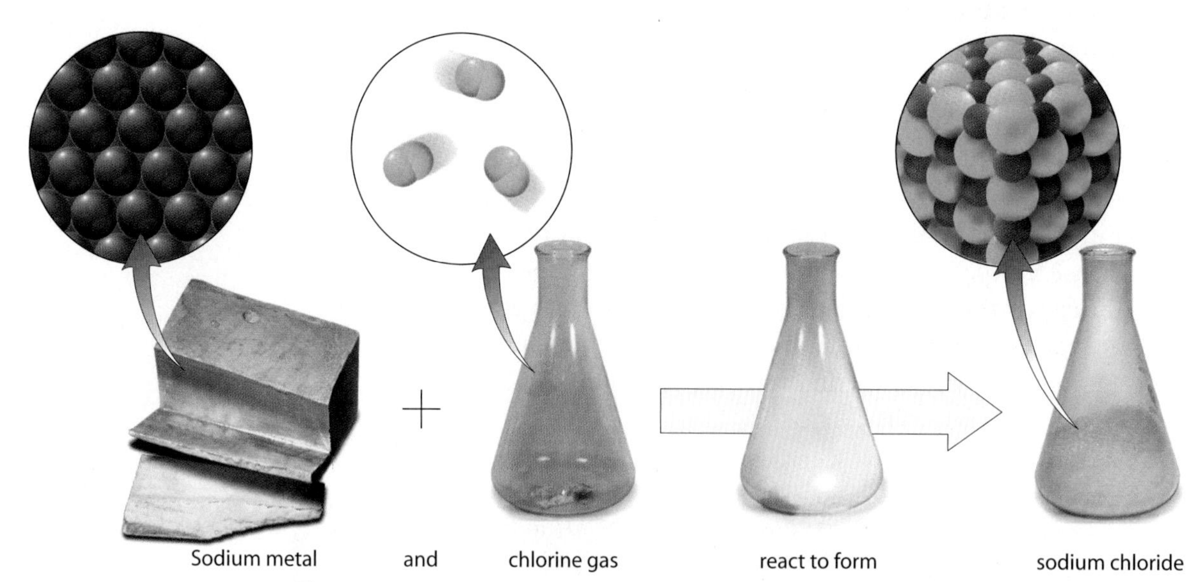

Figure 21.12
A chemical property of sodium metal and chlorine gas is that they react together to form sodium chloride. Although the compound sodium chloride is composed of sodium and chlorine, the physical and chemical properties of sodium chloride are very different from the physical and chemical properties of either sodium metal or chlorine gas.

warfare agent during World War I. Its boiling point is −34°C. The compound sodium chloride, NaCl, is a translucent, brittle, colorless crystal having a melting point of 800°C. Sodium chloride does not chemically react with water the way sodium does, and not only is it not toxic to humans the way chlorine is, but the very opposite is true: It is an essential component of all living organisms. Sodium chloride is not sodium, nor is it chlorine; it is uniquely sodium chloride, a tasty chemical when sprinkled lightly over popcorn.

Concept Check ✓

Hydrogen sulfide, H_2S, is one of the smelliest compounds. Rotten eggs get their characteristic bad smell from the hydrogen sulfide they release. Can you infer from this information that elemental sulfur, S_8, is just as smelly?

Check Your Answer No, you cannot. In fact, the odor of elemental sulfur is negligible compared with that of hydrogen sulfide. Compounds are truly different from the elements from which they are formed. Hydrogen sulfide, H_2S, is as different from elemental sulfur, S_8, as water, H_2O, is from elemental oxygen, O_2.

Hands-On Exploration: Oxygen Bubble Bursts

Compounds can be broken down to their component elements. For example, when you pour a solution of the compound hydrogen peroxide, H_2O_2, over a cut, an enzyme in your blood decomposes it to produce oxygen gas, O_2. You see the oxygen gas as the bubbling that takes place. This oxygen at high concentrations at the site of injury kills off microorganisms and prevents your cut from getting infected. A similar enzyme is found in baker's yeast.

What You Need

Packet of baker's yeast; 3% hydrogen peroxide solution; short, wide drinking glass; tweezers; matches

Safety Note

Wear safety glasses, and remove all materials that could burn (such as paper towels, etc.) from the area where you are working. Keep your fingers well away from the flame because the flame will glow brighter as it is exposed to the oxygen.

Procedure

1. Pour the yeast into the glass. Add a couple of capfuls of the hydrogen peroxide and watch the oxygen bubbles form.
2. Test for the presence of oxygen. Hold a lighted match with the tweezers and put the flame near the bubbles. Look for the flame to glow brighter as the escaping oxygen passes over it.

Describe oxygen's physical and chemical properties.

21.6 Chemical Reactions Are Represented by Chemical Equations

During a chemical reaction, atoms rearrange to create one or more new compounds. This activity is neatly summed up in written form as a **chemical equation.** A chemical equation shows the reacting substances, called **reactants,** to the left of an arrow. On the right side of the arrow lie the products, the newly formed substances that result from the chemical reaction. The arrow always points away from the reactants and toward the products.

reactants → products

Typically, reactants and products are represented by their elemental or chemical formulas. Sometimes molecular models or simply names may be used instead. Phases are also often shown: *(s)* for solid, *(ℓ)* for liquid, and *(g)* for gas. Compounds dissolved in water are designated *(aq)* for aqueous. Lastly, numbers are placed in front of the reactants or products to show the ratio in which they either combine or form. These numbers are called *coefficients,* and they represent numbers of individual atoms and molecules. For instance, to represent the chemical reaction in which coal (solid carbon) burns in the presence of oxygen to form gaseous carbon dioxide, we write the chemical equation using coefficients of 1.

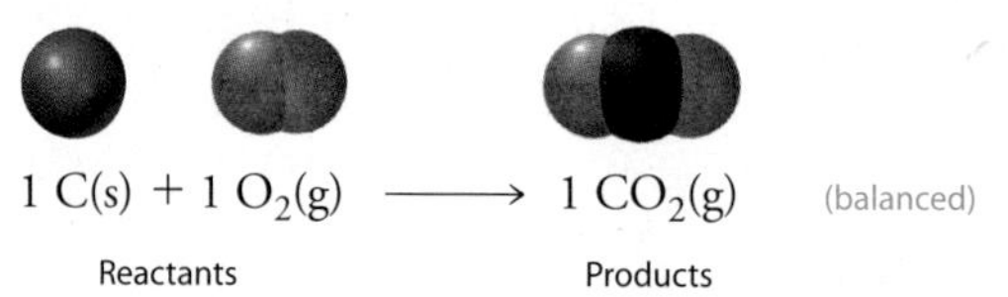

One of the most important principles of chemistry is the **law of mass conservation.** The law of mass conservation states that matter is neither created nor destroyed during a chemical reaction. The atoms present at the beginning of a reaction merely rearrange to form new molecules. This means that no atoms are lost or gained during any reaction. The chemical equation must therefore be *balanced.* In a balanced equation, each atom must appear on both sides of the arrow the same number of times. The equation for the formation of carbon dioxide is balanced because each side shows one carbon atom and two oxygen atoms. You can count the number of atoms in the models to see this for yourself.

In another chemical reaction, two hydrogen gas molecules, H_2, react with one oxygen gas molecule, O_2, to produce two molecules of water, H_2O, in the gaseous phase:

$$2\ H_2(g) + 1\ O_2(g) \longrightarrow 2\ H_2O(g) \quad \text{(balanced)}$$

This equation for the formation of water is also balanced—there are four hydrogen and two oxygen atoms before and after the arrow.

A coefficient in front of a chemical formula tells us the number of times that element or compound must be counted. For example, 2 H_2O indicates two water molecules, which contain a total of four hydrogen atoms and two oxygen atoms.

By convention, the coefficient 1 is omitted so that the above chemical equations are typically written

$$C(s) + O_2(g) \rightarrow CO_2(g) \quad \text{(balanced)}$$

$$2H_2(g) + O_2(g) \rightarrow 2H_2O(g) \quad \text{(balanced)}$$

Concept Check ✓

How many oxygen atoms are indicated by the following balanced equation?

$$3O_2(g) \rightarrow 2O_3(g)$$

Check Your Answer Before the reaction, these six oxygen atoms are found in three O_2 molecules. After the reaction, these same six atoms are found in two O_3 molecules.

You Can Balance Unbalanced Equations

An unbalanced chemical equation shows the reactants and products without the correct coefficients. For example, the equation

$$NO(g) \rightarrow N_2O(g) + NO_2(g) \quad \text{(not balanced)}$$

is not balanced because there is one nitrogen atom and one oxygen atom before the arrow, but three nitrogen atoms and three oxygen atoms after the arrow.

You can balance unbalanced equations by adding or changing coefficients to produce correct ratios. (It's important ***not to change subscripts,*** however, because to do so changes the compound's identity—H_2O is water, but H_2O_2 is hydrogen peroxide!) For example, to balance the above equation, add a 3 before the NO:

$$3NO(g) \rightarrow N_2O(g) + NO_2(g) \quad \text{(balanced)}$$

Now there are three nitrogen atoms and three oxygen atoms on each side of the arrow, and the law of mass conservation is not violated.

There are many methods of balancing equations. For example, consider the following equation in which aluminum oxide, Al_2O_3, and carbon, C, react to form elemental aluminum, Al, and carbon dioxide, CO_2. Here is an unbalanced equation for this reaction:

$$\_\_\_\ Al_2O_3(s) + \_\_\_\ C(s) \rightarrow \_\_\_\ Al(s) + \_\_\_\ CO_2(g) \quad \text{(not balanced)}$$

Balancing an equation usually proceeds most efficiently when you balance one element at a time, starting with elements in the most complex reactant. For this example, therefore, we can start by balancing the aluminum. This element can be balanced by placing a 2 in front of the product Al, so that there are two aluminum atoms before the arrow and two after.

(aluminum balanced)

$$\_\_\_\ Al_2O_3(s) + \_\_\_\ C(s) \rightarrow 2Al(s) + \_\_\_\ CO_2(g)$$

(oxygen not balanced)
(carbon balanced)

The oxygen can then be balanced by placing a 2 in front of the Al_2O_3 and a 3 in front of the CO_2:

(aluminum not balanced)

$$2Al_2O_3(s) + \_\_\_\ C(s) \rightarrow 2Al(s) + 3CO_2(g)$$

(oxygen balanced)
(carbon not balanced)

Doing this gives six oxygen atoms before and after the arrow. Ignore the fact that adding these coefficients upsets the balance of aluminum and carbon atoms. It is best to focus on one element at a time, and for the above our focus was on oxygen.

We've now worked with all the elements of the most complex reactant, Al_2O_3, and so it's time to balance the carbon. This can be done by placing a 3 in front of its symbol:

(aluminum not balanced)

$$2Al_2O_3(s) + 3C(s) \rightarrow 2Al(s) + 3CO_2(g)$$

(oxygen balanced)
(carbon balanced)

Go through the equation again, focusing on one element at a time to make sure each is balanced. The number of aluminum atoms, for example, can be balanced by changing the coefficient on Al to 4:

(aluminum balanced)

$$2Al_2O_3(s) + 3C(s) \rightarrow 4Al(s) + 3CO_2(g)$$

(oxygen balanced)
(carbon balanced)

Knowing *why* chemical equations need to be balanced is more important than being able to balance them!

As you will discover, you can follow many paths to balance a chemical equation. For the above example, you could have started by balancing the carbon first, though it's usually wisest to start with elements in the most complex reactant. If the coefficients start getting very large (beyond 12), you've likely chosen a path that is looping you around to no end. In such an event, start over.

Here is a summary of the steps used in our example:

1. Balance one element at a time. Modify the coefficients to make this element appear the same number of times on both sides of the arrow. Start with the reactant having the most complex formula and finish with the reactant having the simplest formula.

2. If you incidentally unbalance an element that you worked with previously, leave it alone and come back to it only after you have worked with all other elements.

3. After you have worked with each element, make another pass through the equation, changing coefficients as needed.
4. Repeat Step 3 until all elements are balanced.
5. If necessary, minimize the coefficients by dividing by the lowest common denominator. The coefficients 2:4:2, for example, should be reduced to 1:2:1 by dividing by the lowest common denominator, which is 2.

Helpful Hints: Never, ever, EVER alter a subscript. Remember that a coefficient must appear before *a chemical compound, not within it. Use a pencil so that you can erase coefficients as needed.*

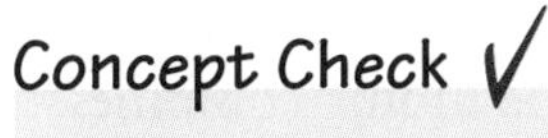

Write a balanced equation for the reaction showing hydrogen gas and nitrogen gas forming ammonia gas below:

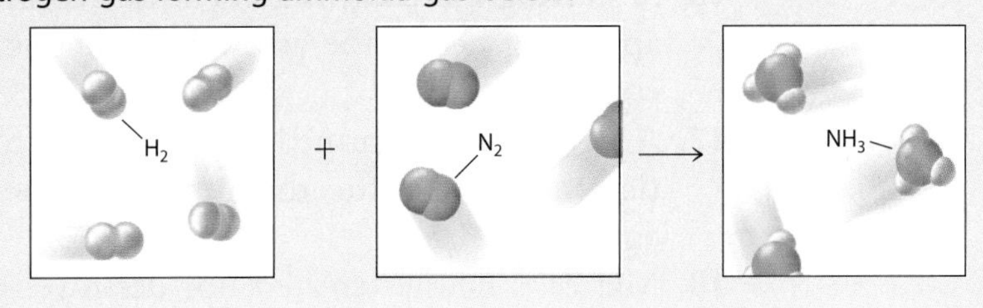

Check Your Answer Start by writing the equation without the coefficients:

$$\_\_\_ H_2(g) + \_\_\_ N_2(g) \rightarrow \_\_\_ NH_3(g)$$

Then go through the steps outlined in the text. The balanced equation is

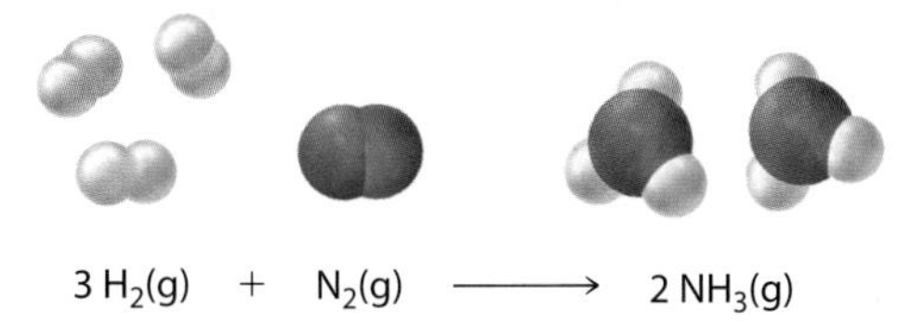

$$3\,H_2(g) + N_2(g) \longrightarrow 2\,NH_3(g)$$

You can see that there are equal numbers of each kind of atom before and after the arrow. For more practice balancing equations, see the Exercises at the end of this chapter.

Practicing chemists develop a skill for balancing equations. This skill involves creative energy and, like other skills, improves with experience. There are some useful tricks of the trade for balancing equations, and maybe your teacher will share some with you. For brevity, however, this text introduces only the basics. More important than being an expert at balancing equations is knowing why they need to be balanced. And the reason is the law of mass conservation, which tells us that atoms are neither created nor destroyed in a chemical reaction—they are simply rearranged. So every atom present before the reaction must be present after the reaction, even though the groupings of atoms are different.

Chapter Review

Key Terms and Matching Definitions

_____ applied research
_____ basic research
_____ chemical change
_____ chemical equation
_____ chemical formula
_____ chemical properties
_____ chemical reaction
_____ chemistry
_____ compound
_____ elemental formula
_____ law of mass conservation
_____ matter
_____ molecules
_____ physical change
_____ physical properties
_____ products
_____ reactants
_____ submicroscopic

1. The study of matter and the transformations it can undergo.
2. Anything that occupies space.
3. Research dedicated to the discovery of the fundamental workings of nature.
4. Research dedicated to the development of useful products and processes.
5. A group of atoms that collectively make the fundamental unit of a material, such as water.
6. The realm of atoms and molecules, where objects are smaller than can be detected by optical microscopes.
7. Any physical attribute of a substance, such as color, density, or hardness.
8. A change in which a substance changes its physical properties without changing its chemical identity.
9. A property that relates to how a substance changes its chemical identity.
10. During this kind of change, atoms in a substance are rearranged to give a new substance having a new chemical identity.
11. Synonymous with chemical change.
12. A notation that uses the atomic symbol and (sometimes) a numerical subscript to denote how atoms are bonded in an element.
13. A material in which atoms of different elements are bonded to one another.
14. A notation used to indicate the composition of a compound, consisting of the atomic symbols for the different elements of the compound and numerical subscripts indicating the ratio in which the atoms combine.
15. A representation of a chemical reaction.
16. A starting material in a chemical reaction, appearing before the arrow in a chemical equation.
17. A new material formed in a chemical reaction, appearing after the arrow in a chemical equation.
18. Matter is neither created nor destroyed during a chemical reaction.

Review Questions

Chemistry Is a Central Science Useful to Our Lives

1. What is the difference between basic research and applied research?
2. Why is chemistry often called the *central science*?
3. What do members of the Chemical Manufacturers Association pledge in the Responsible Care program?

The Submicroscopic World Is Made of Atoms and Molecules

4. Are atoms made of molecules, or are molecules made of atoms?
5. Which is smaller: the microscopic or the submicroscopic?

Matter Has Physical and Chemical Properties

6. What is a physical property?
7. What is a chemical property?
8. What doesn't change during a physical change?
9. Why is it sometimes difficult to decide whether an observed change is physical or chemical?
10. What are some of the clues that help us determine whether an observed change is physical or chemical?

An Element Is Made of a Collection of Atoms

11. Is it possible for an element to have more than one atomic formula?
12. How many atoms are in a sulfur molecule that has the elemental formula S_8?

Elements Can Combine to Form Compounds

13. What is the difference between an element and a compound?
14. How many atoms are there in one molecule of H_3PO_4? How many atoms of each element are there in one molecule of H_3PO_4?
15. Are the physical and chemical properties of a compound necessarily similar to those of the elements from which it is composed?

Chemical Reactions Are Represented by Chemical Equations

16. What is the purpose of coefficients in a chemical equation?
17. How many chromium atoms and how many oxygen atoms are indicated on the right side of this balanced chemical equation:

$$4\ Cr(s) + 3\ O_2(g) \rightarrow 2\ Cr_2O_3(g)$$

18. What do the letters (s), (l), (g), and (aq) stand for in a chemical equation?
19. Why is it important that a chemical equation be balanced?
20. Why is it important never to change a subscript in a chemical formula when balancing a chemical equation?
21. Which equations are balanced?
    a. $Mg(s) + 2HCl(aq) \rightarrow MgCl_2(aq) + H_2(g)$
    b. $3Al(s) + 3Br_2(l) \rightarrow Al_2Br_3(s)$
    c. $2HgO(s) \rightarrow 2Hg(l) + O_2(g)$

Explorations

Fire Water

As you can see in Figure 21.8, the two primary products when natural gas burns are carbon dioxide and water. Because of the heat generated by the burning, the water is released as water vapor. When it comes into contact with the relatively cool sides of the pot, this water vapor condenses to the liquid phase and is seen as "sweat." If the pot contained ice water, more vapor would condense, enough to form drops that roll off the bottom edge. As the pot gets warmer, this liquid water is heated and returns to the gaseous phase.

The only chemical change is the conversion of natural gas to carbon dioxide and water vapor. There are two physical changes—condensation of the water vapor created in the methane combustion and evaporation of this water once the pot gets sufficiently hot. (Of course, the evaporation of the water in the pot is another physical change.)

Oxygen Bubble Bursts

Hydrogen peroxide, H_2O_2, is a relatively unstable compound. In water, it slowly decomposes, producing oxygen gas. In describing oxygen's physical properties, you should have noted that it is an invisible gas having no odor detectable over that of the yeast. Oxygen is light enough to rise out of the glass once it is released from the bubbles. What is your evidence of this? A chemical property of oxygen is that it intensifies burning.

Exercises

1. Why is it important to work through the Review Questions before attempting the Exercises?

2. In what sense is a color computer monitor or television screen similar to our view of matter? Place a drop (and only a drop) of water on your computer monitor or television screen for a closer look.

3. Of the three sciences physics, chemistry, and biology, which is the most complex? Explain.

4. Is chemistry the study of the submicroscopic, the microscopic, the macroscopic, or all three? Defend your answer.

5. Each night you measure your height just before going to bed. When you arise each morning, you measure your height again and consistently find that you are 1 inch taller than you were the night before but only as tall as you were 24 hours ago! Is what happens to your body in this instance best described as a physical change or a chemical change? Be sure to try this activity if you haven't already.

6. Classify the following changes as physical or chemical. Even if you are incorrect in your assessment, you should be able to defend why you chose as you did.
a. grape juice turns to wine
b. wood burns to ashes
c. water begins to boil
d. a broken leg mends itself
e. grass grows
f. an infant gains 10 pounds
g. a rock is crushed to powder

7. Is the following transformation representative of a physical change or a chemical change?

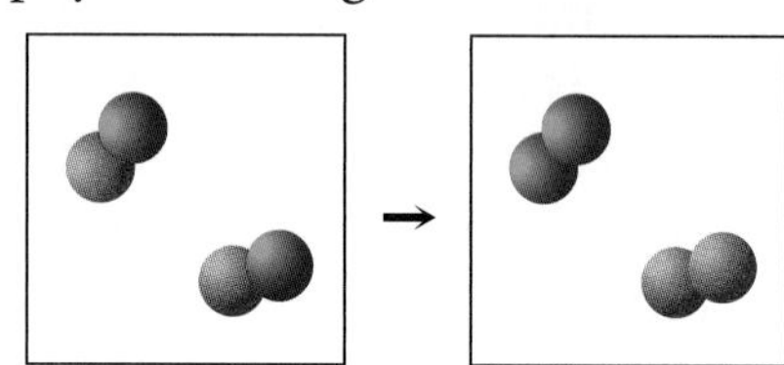

8. Each sphere in the diagrams below represents an atom. Joined spheres represent molecules. Which box contains a liquid phase? Why can you not assume that box B represents a lower temperature?

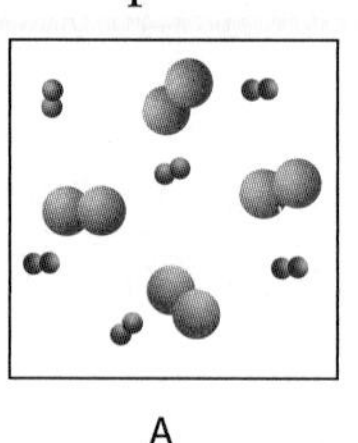
A

B

9. Based on the information given in the following diagrams, which substance has the higher boiling point?

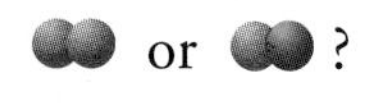

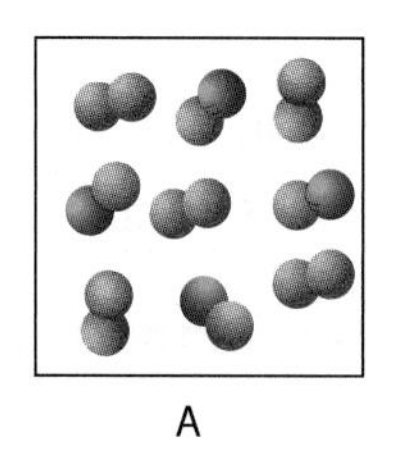
A

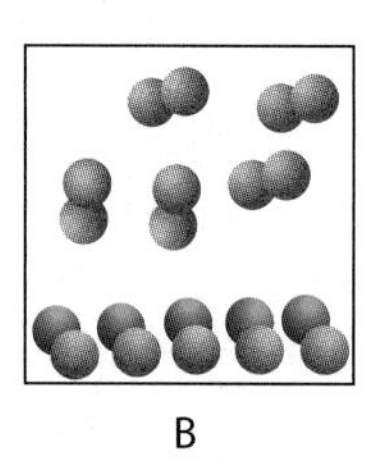
B

10. What physical and chemical changes occur when a wax candle burns?

11. Which elements are some of the oldest known? What is your evidence?

12. Oxygen atoms are used to make water molecules. Does this mean that oxygen, O_2, and water, H_2O, have similar properties? Why do we drown when we breathe in water despite all the oxygen atoms present in this material?

13. Is this chemical equation balanced?
$2C_4H_{10}(g) + 13O_2(g)$
$\rightarrow 8CO_2(g) + 10H_2O(l)$

14. Balance these equations:
a. ___ $Fe(s)$ + ___ $O_2(g) \rightarrow$ ___ $Fe_2O_3(s)$
b. ___ $H_2(g)$ + ___ $N_2(g) \rightarrow$ ___ $NH_3(g)$
c. ___ $Cl_2(g)$ + ___ $KBr(aq) \rightarrow$
___ $Br_2(l)$ + ___ $KCl(aq)$
d. ___ $CH_4(g)$ + ___ $O_2(g) \rightarrow$
___ $CO_2(g)$ + ___ $H_2O(l)$

15. Balance these equations:

a. ___ $Fe(s)$ + ___ $S(s) \rightarrow$ ___ $Fe_2S_3(s)$

b. ___ $P_4(s)$ + ___ $H_2(g) \rightarrow$ ___ $PH_3(g)$

c. ___ $NO(g)$ + ___ $Cl_2(g) \rightarrow$ ___ $NOCl(g)$

d. ___ $SiCl_4(l)$ + ___ $Mg(s) \rightarrow$ ___ $Si(s)$ + ___ $MgCl_2(s)$

Suggested Readings and Web Sites

Breslow, Ronald. *Chemistry Today and Tomorrow: The Central, Useful, and Creative Science.* Sudbury, MA: Jones and Bartlett, 1997.

Written by a past president of the American Chemical Society, this paperback analyzes the role chemistry has played in the development of our modern society and the role it is sure to play in our future.

Hoffmann, Roald. *The Same and Not the Same.* New York: Columbia University Press, 1995.

Written by a noted Nobel Prize winner, researcher, and teacher, this book seeks to explain the workings of chemistry to the general public, with an emphasis on the beauty of molecular design.

http://www.chemcenter.org

Maintained by the American Chemical Society, this site is an excellent starting point for searching out such chemistry-related information as current events or the status of a particular avenue of research.

http://www.csicop.org/

Home page for the Committee for the Scientific Investigation of Claims of the Paranormal. This organization of Nobel laureates and other respected scientists takes on the claims of pseudoscience with all the rigor required of any scientific claim.

http://www.rsc.org/lap/rsccom/wcc/wccindex.htm

Home page for the Women Chemists Committee of The Royal Society of Chemistry (UK). This organization promotes the entry and re-entry of women to the profession of chemistry and collects and disseminates information about women in chemistry.

http://www.sciencenews.org/sn_arch

Archives of ScienceNews, a widely read weekly magazine covering current developments in science.

Chapter 22: Mixtures

Why is it impossible for anything, from drinking water to the air to copper that occurs naturally, to be truly 100% pure? When tap water is left boiling on the stove too long, it evaporates completely but leaves a chalky, hard-to-clean-up residue in the pot. What is this residue and where did it come from? When you stir sugar into water, the sugar crystals disappear. Where do they go? What are clouds made of and what do they have in common with the blood that runs through our veins? Is it true that a fish can drown in water? The answers to these sorts of questions can be found with an understanding of mixtures.

22.1 Most Materials Are Mixtures

A **mixture** is a combination of two or more substances in which each substance retains its properties. Most materials around us are mixtures—mixtures of elements, mixtures of compounds, or mixtures of elements and compounds. Stainless steel, for example, is a mixture of the elements iron, chromium, nickel, and carbon. Seltzer water is a mixture of the liquid compound water and the gaseous compound carbon dioxide. Our atmosphere, as Figure 22.1 illustrates, is a mixture of the elements nitrogen, oxygen, and argon plus small percentages of such compounds as carbon dioxide and water vapor.

Tap water is a mixture containing mostly water but also many other compounds. Depending on your location, your water may include compounds of calcium, magnesium, fluorine, iron, and potassium; chlorine disinfectants; trace amounts of compounds of lead, mercury, and cadmium; organic compounds; and dissolved oxygen, nitrogen, and carbon dioxide. While it is surely important to minimize any toxic components in your drinking water, it is unnecessary, undesirable, and impossible to remove all other substances from it. Some of the dissolved solids and gases give water its characteristic taste, and many of them promote human health. Fluoride compounds protect teeth, chlorine destroys harmful bacteria, and as much as 10% of our daily requirement for iron, potassium, calcium, and magnesium is obtained from drinking water (Figures 22.2 and 22.3).

Figure 22.1
The Earth's atmosphere is a mixture of gaseous elements and compounds. Some of them are shown here.

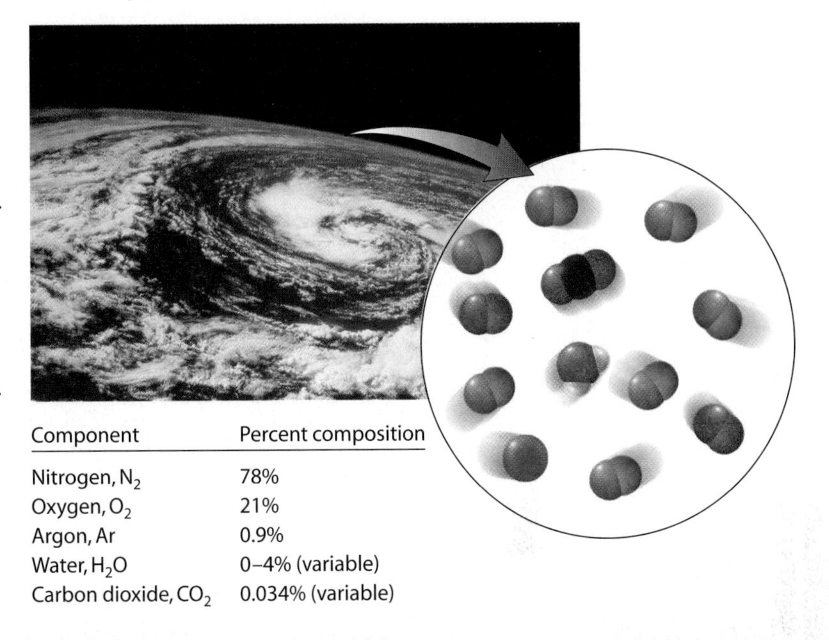

Component	Percent composition
Nitrogen, N_2	78%
Oxygen, O_2	21%
Argon, Ar	0.9%
Water, H_2O	0–4% (variable)
Carbon dioxide, CO_2	0.034% (variable)

Figure 22.2
Tap water provides us with water as well as a large number of other compounds, many of which are flavorful and healthful. Bottoms up!

Figure 22.3
Most of the oxygen in the air bubbles produced by an aquarium aerator escapes into the atmosphere. Some of the oxygen, however, mixes with the water. It is this oxygen the fish depend on to survive. Without this dissolved oxygen, which they extract with their gills, the fish would promptly drown. So fish don't "breathe" water. They breathe the oxygen, O_2, dissolved in the water.

Concept Check ✓

So far, you have learned about three kinds of matter: elements, compounds, and mixtures. Which box below contains only an element? Which contains only a compound? Which contains a mixture?

A B C

Check Your Answer The molecules in box A each contain two different types of atoms and so are representative of a compound. The molecules in box B each consist of the same atoms and so are representative of an element. Box C is a mixture of the compound and the element.

Note how the molecules of the compound and those of the element remain intact in the mixture. That is, upon the formation of the mixture, there is no exchange of atoms between the components.

There is a difference between the way substances—either elements or compounds—combine to form mixtures and the way elements combine to form compounds. Each substance in a mixture retains its chemical identity. The sugar molecules in the teaspoon of sugar in Figure 22.4, for example, are identical to the sugar molecules already in the tea. The only difference is that the sugar molecules in the tea are mixed with other substances, mostly water. The formation of a mixture, therefore, is a physical change. As was discussed in Section 21.5, when elements join to form compounds, there is a change in

Figure 22.4
Table sugar is a compound consisting of only sugar molecules. Once these molecules are mixed into hot tea, they become interspersed among the water and tea molecules and form a sugar-tea-water mixture. No new compounds are formed, and so this is an example of a physical change.

chemical identity. Sodium chloride is not a mixture of sodium and chlorine atoms. Instead, sodium chloride is a compound. It is entirely different from the elements used to make it. The formation of a compound is therefore a chemical change.

22.2 Mixtures Can Be Separated by Physical Means

The components of mixtures can be separated from one another by taking advantage of differences in the components' physical properties. A mixture of solids and liquids, for example, can be separated using filter paper through which the liquids pass but the solids do not. This is how coffee is often made. The caffeine and flavor molecules in the hot water pass through the filter and into the coffee pot while the solid coffee grinds remain behind. This method of separating a solid-liquid mixture is called *filtration*. Filtration is a common technique used by chemists.

Mixtures can also be separated by using differences in boiling or melting points. Seawater is a mixture of water and a variety of compounds, mostly sodium chloride. Whereas pure water boils at 100°C, sodium chloride doesn't even *melt* until 800°C. One way to separate pure water from the mixture seawater, therefore, is to heat the seawater to about 100°C. At this temperature, the liquid water readily transforms to water vapor, but the sodium chloride stays behind, dissolved in the remaining water. As the water vapor rises, it can be channeled into a cooler container, where it condenses to a liquid without the dissolved solids. This process of collecting a vaporized substance, called *distillation*, is illustrated in Figure 22.5. After all the water has been distilled from seawater, what remains are dry solids.

Figure 22.5
(a) A simple distillation setup used to separate one component—water—from the mixture we call *seawater*. The seawater is boiled in the flask on the left. The rising water vapor is channeled into a downward-slanting tube kept cool by cold water flowing across its outer surface. The water vapor inside the cool tube condenses and collects in the flask on the right. (b) A whiskey still works on the same principle. A mixture containing alcohol is heated to the point where the alcohol, some flavoring molecules, and some water are vaporized. These vapors travel through the copper coils, where they condense to a liquid ready for collection.

These solids, also a mixture of compounds, contain a variety of valuable materials, including sodium chloride, potassium bromide, and a small amount of gold! Further separation of the components of this mixture is of significant commercial interest (Figure 22.6).

Figure 22.6
At the southern end of San Francisco Bay are areas where the seawater has been partitioned off. These are evaporation ponds where the water is allowed to evaporate, leaving behind the solids that were dissolved in the seawater. These solids are further refined for commercial sale. The remarkable color of the ponds results from suspended particles of iron oxide and other minerals, which are easily removed during refining.

Hands-On Exploration: Bottoms Up and Bubbles Out

What's in a glass of water? Separate the components of your tap water to find out.

What You Need

Tap water, sparkling clean cooking pot, stove, metal spoon

Safety Note

Wear safety glasses for step 1 because some splattering may occur.

Procedure

1. Put on your safety glasses and add the tap water to the cooking pot. Boil the water to dryness. (Turn off the burner before the water is all gone. The heat from the pot will finish the evaporation.)
2. Examine the resulting residue by scraping it with the side of the spoon. These are the solids you ingest with every glass of water you drink.
3. To see the gases dissolved in your water, fill a clean cooking pot with water and let it stand at room temperature for several hours. Note the bubbles that stick to the inner sides of the pot.

Where did the bubbles of Step 3 come from? What do you suppose they contain?

22.3 Chemists Classify Matter as Pure or Impure

If a material is **pure,** it consists of only a single element or a single compound. In pure gold, for example, there is nothing but the element gold. In pure table salt, there is nothing but the compound sodium chloride. If a material is **impure,** it is a mixture and contains two or more elements or compounds. This classification scheme is shown in Figure 22.7.

Since atoms and molecules are so small, it is impractical to prepare a sample that is truly pure—that is, truly 100% of a single material. For example, if just one atom or molecule out of a trillion trillion were different than the others, then the 100% pure status would be lost. Samples can be "purified" by various methods, however, such as distillation. When we say *pure,* it is understood to be a relative term. Comparing the purity of two samples, the purer one contains fewer impurities. A sample of water that is 99.9% pure has a greater proportion of impurities than does a sample of water that is 99.9999% pure.

Orange juice may be 100% natural, but never 100% pure.

Sometimes naturally occurring mixtures are labeled as being pure, as in "pure orange juice." Such a statement merely means that nothing artificial has been added. According to a chemist's definition, however, orange juice is anything but pure. It contains a wide variety of materials, including water, pulp, flavorings, vitamins, and sugars.

Mixtures may be heterogeneous or homogeneous. In a **heterogeneous mixture,** the different components can be seen as individual substances. For example, orange juice, sand in water, and salad dressing are heterogeneous mixtures. You can see the pulp in orange juice, sand particles in water, or oil globules dispersed in vinegar. The different components are visible. On the other hand, **homogeneous mixtures** have the same composition throughout. Any one region of the mixture has the same ratio of substances as does any other region. The components of a homogeneous mixture do not appear separated. The distinction is shown in Figure 22.8.

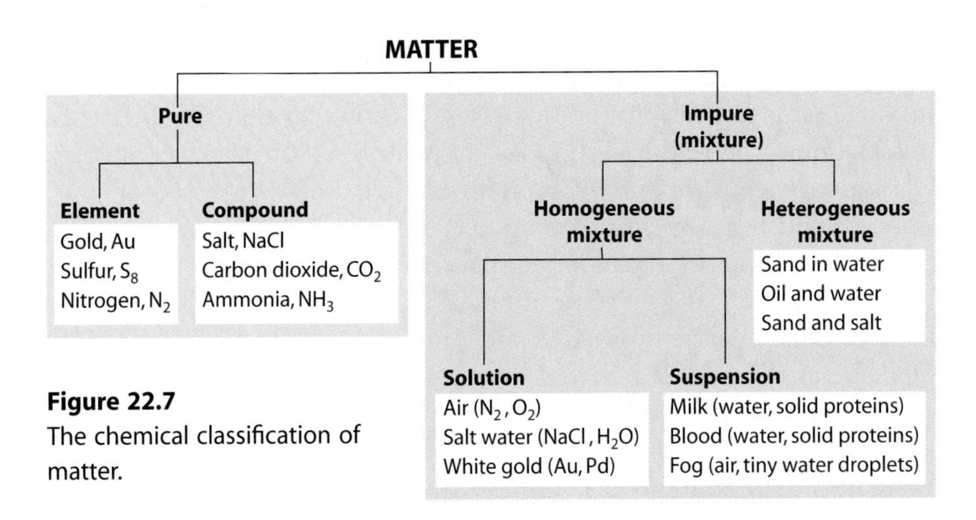

Figure 22.7
The chemical classification of matter.

Figure 22.8
(a) In heterogeneous mixtures, the different components can be seen with the naked eye. (b) In homogeneous mixtures, the different components are mixed at a much finer level and so are not readily distinguished.

A homogeneous mixture may be either a solution or a suspension. In a **solution,** all components are in the same phase. The atmosphere we breathe is a gaseous solution consisting of the gaseous elements nitrogen and oxygen as well as minor amounts of other gaseous materials. Salt water is a liquid solution because both the water and the dissolved sodium chloride are found in a single liquid phase. An example of a solid solution is white gold, which is a homogeneous mixture of the elements gold and palladium. We shall be discussing solutions in more detail in the next section.

Figure 22.9
The path of light becomes visible when the light passes through a suspension.

A **suspension** is a homogeneous mixture in which the different components are in different phases, such as solids in liquids or liquids in gases. In a suspension, the mixing is so thorough that the different phases cannot be readily distinguished. Milk is a suspension because it is a homogeneous mixture of proteins and fats finely dispersed in water. Blood is a suspension composed of finely dispersed blood cells in water. Another example of a suspension is clouds, which are homogeneous mixtures of tiny water droplets suspended in air. Shining a light through a suspension, as is done in Figure 22.9, results in a visible cone as the light is reflected by the suspended components.

The easiest way to distinguish a suspension from a solution in the laboratory is to spin a sample in a centrifuge. This device, spinning at thousands of revolutions per minute, separates the components of suspensions but not those of solutions, as Figure 22.10 shows.

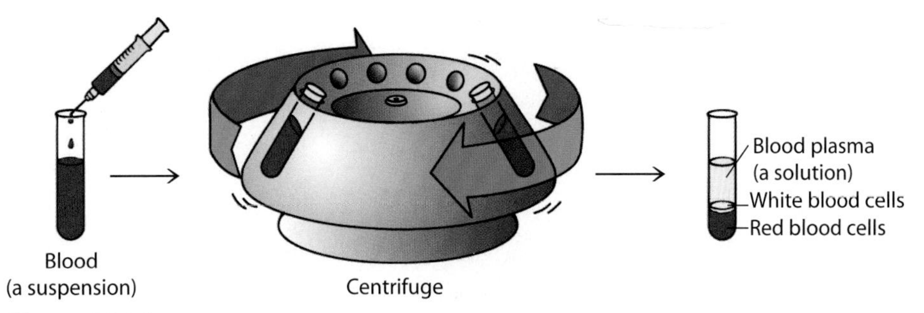

Figure 22.10
Blood, because it is a suspension, can be centrifuged into its components, which include the blood plasma (a yellowish solution) and white and red blood cells. The components of the plasma cannot be separated from one another here because a centrifuge has no effect on solutions.

Concept Check ✓

Impure water can be purified by

a. removing the impure water molecules.
b. removing everything that is not water.
c. breaking down the water to its simplest components.
d. adding some disinfectant such as chlorine.

Check Your Answer Water, H_2O, is a compound made of the elements hydrogen and oxygen in a 2-to-1 ratio. Every H_2O molecule is exactly the same as every other, and there's no such thing as an impure H_2O molecule. Just about anything, including you, beach balls, rubber ducks, dust particles, and bacteria, can be found in water. When something other than water is found in water, we say that the water is impure. It is important to see that the impurities are *in* the water and not part *of* the water, which means that it is possible to remove them by a variety of physical means, such as filtration or distillation. The answer to this Concept Check is b.

22.4 A Solution Is a Single-Phase Homogeneous Mixture

What happens when table sugar, known chemically as *sucrose,* is stirred into water? Is the sucrose destroyed? We know it isn't because it sweetens the water. Does the sucrose disappear because it somehow ceases to occupy space or because it fits within the nooks and crannies of the water? Not so, for the addition of sucrose changes the volume. This may not be noticeable at first, but continue to add sucrose to a glass of water and you'll see that the water level rises just as it would if you were adding sand.

Sucrose stirred into water loses its crystalline form. Each sucrose crystal consists of billions upon billions of sucrose molecules packed neatly together. When the crystal is exposed to water, as was first shown in Figure 22.4 and again here in Figure 22.11, an even greater number of water molecules pull on the sucrose molecules. With a little stirring, the sucrose molecules soon mix throughout the water.

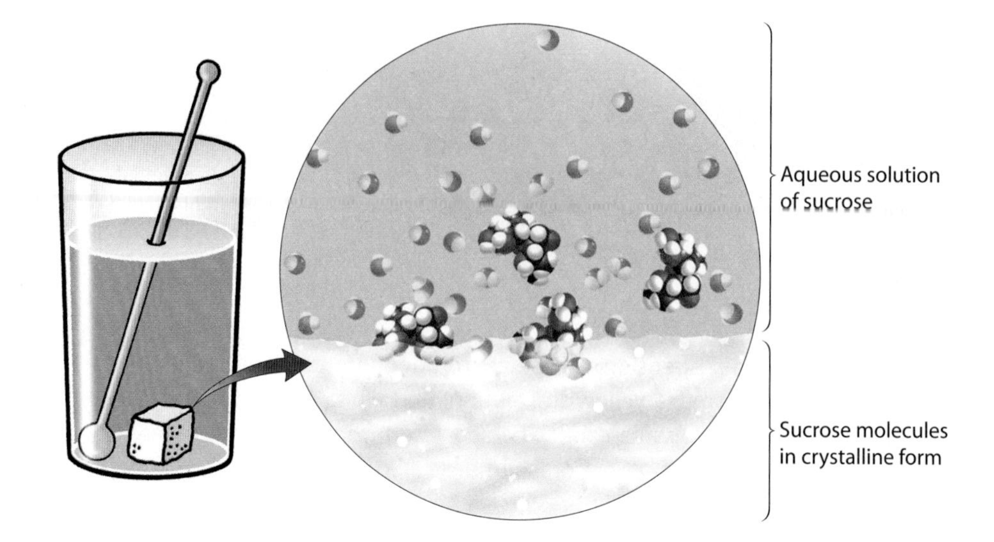

Figure 22.11
Water molecules pull the sucrose molecules in a sucrose crystal away from one another. They do not, however, affect the covalent bonds within each sucrose molecule. This is why each dissolved sucrose molecule remains intact as a single molecule.

In place of sucrose crystals and water, we have a homogeneous mixture of sucrose molecules in water. Recall, homogeneous means that a sample taken from one part of a mixture is the same as a sample taken from any other part of the mixture. In our sucrose example, this means that the sweetness of the first sip of the solution is the same as the sweetness of the last sip.

As discussed in the previous section, a homogeneous mixture consisting of a single phase is called a solution. Sugar in water is a solution in the liquid phase. Solutions aren't always liquids, however. They can also be solid or gaseous, as Figure 22.12 shows. Gem stones are solid solutions. A ruby, for example, is a solid solution of trace quantities of red chromium compounds in transparent

(a)

(b)

(c)

Figure 22.12
Solutions may occur in (a) the solid phase, (b) the liquid phase, or (c) the gaseous phase.

aluminum oxide. A blue sapphire is a solid solution of trace quantities of light green iron compounds and blue titanium compounds in aluminum oxide. Another important example of solid solutions is metal alloys, which are mixtures of different metallic elements. The alloy known as brass is a solid solution of copper and zinc, for instance, and the alloy stainless steel is a solid solution of iron, chromium, nickel, and carbon.

An example of a gaseous solution is the air we breathe. By volume, this solution is 78% nitrogen gas, 21% oxygen gas, and 1% other gaseous materials, including water vapor and carbon dioxide. The air we exhale is a gaseous solution of 75% nitrogen, 14% oxygen, 5% carbon dioxide, and around 6% water vapor.

In describing solutions, it is usual to call the component that is present in the largest amount the **solvent.** The other components is the **solute.** For example, when a teaspoon of table sugar is mixed with 1 liter of water, we identify the sugar as the solute and the water as the solvent. Often, a solution has more than one solute.

The process of a solute mixing in a solvent is called **dissolving.** To make a solution, a solute must dissolve in a solvent. In other words, when solute and solvent form a homogeneous mixture, the solute has dissolved.

To most people solutions mean finding the answers. But to chemists, solutions are things that are still all mixed up.

Concept Check ✓

What is the solvent in the gaseous solution we call air?

Check Your Answer Nitrogen is the solvent because it is the component present in the greatest quantity.

There can be a limit to how much of a given solute can dissolve in a given solvent, as Figure 22.13 illustrates. When you add table sugar to a glass of water, for example, the sugar rapidly dissolves. As you continue to add sugar, however, there comes a point when it no longer dissolves. Instead, it collects at the bottom of the glass, even after stirring. At this point, the water is *saturated* with sugar. The water cannot accept any more sugar. When this happens, we have what is called a **saturated solution,** defined as a solution in which no more solute can dissolve. A solution that has not reached the limit of solute that will dissolve is called an **unsaturated solution.**

The quantity of solute dissolved in a solution is described in mathematical terms by the solution's **concentration,** which is the amount of solute dissolved per amount of solution:

$$\text{concentration of solution} = \frac{\text{amount of solute}}{\text{amount of solution}}$$

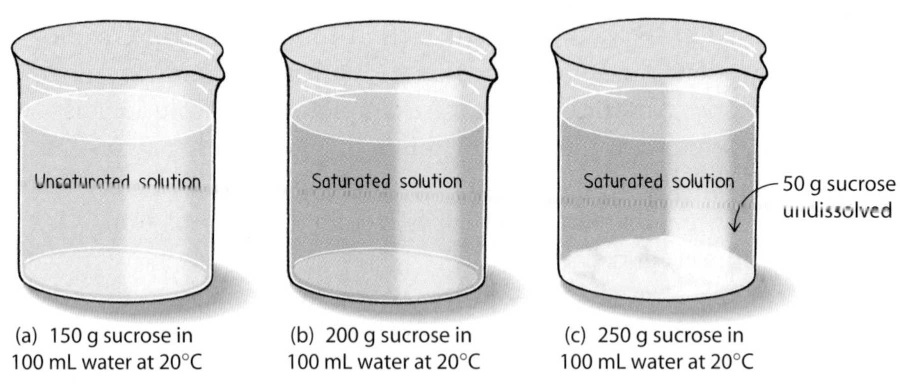

Figure 22.13
A maximum of 200 grams of sucrose dissolves in 100 milliliters of water at 20°C. (a) Mixing 150 grams of sucrose in 100 milliliters of water at 20°C produces an unsaturated solution. (b) Mixing 200 grams of sucrose in 100 milliliters of water at 20°C produces a saturated solution. (c) If 250 grams of sucrose is mixed with 100 milliliters of water at 20°C, 50 grams of sucrose remains undissolved. (As we discuss later, the concentration of a saturated solution varies with temperature.)

For example, a sucrose-water solution may have a concentration of 1 gram of sucrose for every liter of solution. This can be compared with concentrations of other solutions. A sucrose-water solution containing 2 grams of sucrose per liter of solution, for example, is more *concentrated,* and one containing only 0.5 gram of sucrose per liter of solution is less concentrated, or more *dilute.*

Chemists are often more interested in the number of solute particles in a solution rather than the number of grams of solute. Submicroscopic particles, however, are so very small that the number of them in any observable sample is incredibly large. To get around having to use cumbersome numbers, scientists use a unit called the mole. One **mole** of any type of particle is, by definition, 6.02×10^{23} particles (this superlarge number is about 602 billion trillion):

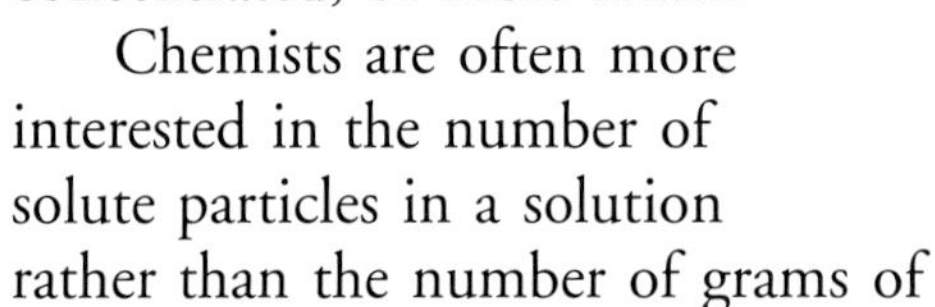

$$
\begin{aligned}
1 \text{ mole} &= 6.02 \times 10^{23} \text{ particles} \\
&= 602{,}000{,}000{,}000{,}000{,}000{,}000{,}000 \text{ particles}
\end{aligned}
$$

One mole of pennies, for example, is 6.02×10^{23} pennies, 1 mole of marbles is 6.02×10^{23} marbles, and 1 mole of sucrose molecules is 6.02×10^{23} sucrose molecules.

Even though you've probably never heard the term *mole* in your life before now, you are already familiar with the basic idea. "One mole" is just a shorthand way of saying "six point oh two times ten to the twenty-third." Just as "a couple of" means 2 of something and "a dozen of" means 12 of something, "a mole of" means 6.02×10^{23} of something. It's as simple as that:

- a couple of coconuts = 2 coconuts
- a dozen of donuts = 12 donuts
- a mole of mints = 6.02×10^{23} mints
- a mole of molecules = 6.02×10^{23} molecules

A stack containing 1 mole of pennies would reach a height of 903 quadrillion kilometers, which is roughly equal to the diameter of our Milky Way galaxy. A mole of marbles would be enough to cover the entire land area of the 50 United States to a depth greater than 4 meters. Sucrose molecules are so small, however, that there are 6.02×10^{23} of them in only 342 grams of sucrose, which is about a cupful. Thus because 342 grams of sucrose contains 6.02×10^{23} molecules of sucrose, we can use our shorthand wording and say that 342 grams of sucrose contains 1 mole of sucrose. As Figure 22.14 shows, therefore, an aqueous solution that has a concentration of 342 grams of sucrose per liter of solution also has a concentration of 6.02×10^{23} sucrose molecules per liter of solution or, by definition, a concentration of 1 mole of sucrose per liter of solution.

Figure 22.14
An aqueous solution of sucrose that has a concentration of 1 mole of sucrose per liter of solution contains 6.02×10^{23} sucrose molecules (342 grams) in every liter of solution.

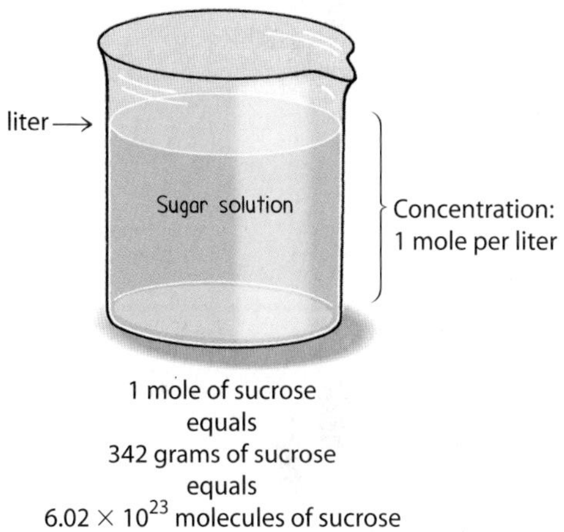

The number of grams tells you the mass of solute in a given solution, and the number of moles tells you the actual number of molecules. Interestingly, the term *mole* is derived from the Latin word for "pile." A mole of marbles would be one amazingly large pile, now wouldn't it!

A common unit of concentration used by chemists is **molarity.** This concentration unit expresses the concentration of a solution moles of solute per liter of solution:

molarity = number of moles of solute/liters of solution

A solution that contains 1 mole of solute per liter of solution has a concentration of 1 molar, which is often abbreviated 1 *M*. A more concentrated, 2-molar (2 *M*) solution contains 2 moles of solute per liter of solution.

The difference between referring to the number of molecules of solute and referring to the number of grams of solute can be illustrated by the following question. A saturated aqueous solution of sucrose contains 200 grams of sucrose and 100 grams of water. Which is the solvent: sucrose or water?

As shown in Figure 22.15, there are 3.5×10^{23} molecules of sucrose in 200 grams of sucrose but almost 10 times as many molecules of water in 100 grams of water—3.3×10^{24} molecules. As defined earlier, the solvent is the component present in the largest amount, but what do we mean by amount? If amount means number of molecules, then water is the solvent. If amount means mass,

Component	Mass	Number of molecules
Sucrose	200 g	3.5×10^{23}
Water	100 g	3.3×10^{24}

Figure 22.15
Although 200 grams of sucrose is twice as massive as 100 grams of water, there are about 10 times as many water molecules in 100 grams of water as there are sucrose molecules in 200 grams of sucrose. How can this be? Each water molecule is about 20 times less massive (and smaller) than each sucrose molecule, which means that about 10 times as many water molecules can fit within half the mass.

then sucrose is the solvent. So, the answer depends on how you look at it. From a chemist's point of view, "amount" typically means the number of molecules. So water is the solvent in this case.

Concept Check ✓

1. How much sucrose, in moles, is there in 0.5 liter of a 2-molar solution? How many molecules of sucrose is this?
2. Does 1 liter of a 1-molar solution of sucrose in water contain 1 liter of water, less than 1 liter of water, or more than 1 liter of water?

Hands-On Exploration: Overflowing Sweetness

Just because a solid dissolves in a liquid doesn't mean the solid no longer occupies space.

What You Need

Tall glass, warm water, container larger than the tall glass,
4 tablespoons of table sugar

Procedure

1. Fill the glass to its brim with the warm water, and then carefully pour all the water into the larger container.
2. Add the sugar to the empty glass.
3. Return half of the warm water to the glass and stir to dissolve all the solid.
4. Return the remaining water, and as you get close to the top, ask a friend to predict whether the water level will be less than before, about the same as before, or more than before so that the water spills over the edge of the glass. What do you observe?

If your friend doesn't understand the result, ask him or her what would happen if you had added the sugar to the glass when the glass was full of water.

Check Your Answers

1. First you need to understand that "2-molar" means *2 moles of sucrose per liter of solution*. Then you should multiply concentration by volume to obtain amount of solute:

   (2 moles/L)(0.5 L) = 1 mole

   which is the same as 6.02×10^{23} molecules.
2. The definition of *molarity* refers to the number of liters of solution, not liters of solvent. When sucrose is added to a given volume of water, the volume of the solution increases. So, if 1 mole of sucrose is added to 1 liter of water, the result is more than 1 liter of solution. Therefore, 1 liter of a 1-molar solution requires less than 1 liter of water.

Chapter Review

Key Terms and Matching Definitions

_____ concentration
_____ dissolving
_____ heterogeneous mixture
_____ homogeneous mixture
_____ impure
_____ mixture
_____ molarity
_____ mole
_____ pure
_____ saturated solution
_____ solute
_____ solution
_____ solvent
_____ suspension
_____ unsaturated solution

1. A combination of two or more substances in which each substance retains its properties.
2. The state of a material that consists of a single element or compound.
3. The state of a material that is a mixture of more than one element or compound.
4. A mixture in which the various components can be seen as individual substances.
5. A mixture in which the components are so finely mixed that the composition is the same throughout.
6. A homogeneous mixture in which all components are in the same phase.
7. A homogeneous mixture in which the various components are in different phases.
8. The component in a solution present in the largest amount.
9. Any component in a solution that is not the solvent.
10. The process of mixing a solute in a solvent.
11. A solution containing the maximum amount of solute that will dissolve.
12. A solution that will dissolve additional solute if it is added.
13. A quantitative measure of the amount of solute in a solution.
14. 6.02×10^{23} of anything.
15. A unit of concentration equal to the number of moles of a solute per liter of solution.

Review Questions

Most Materials Are Mixtures

1. What defines a material as being a mixture?
2. Why is drinking water considered a mixture?
3. Why is sodium chloride not considered a mixture?

Mixtures Can Be Separated by Physical Means

4. How can the components of a mixture be separated from one another?
5. How does distillation separate the components of a mixture? What is the phase of nitrogen, N_2, at this temperature?

6. The boiling point of oxygen, O_2, is 90 K (−183°C) and that of nitrogen, N_2, is 77 K (−196°C). What is the phase of oxygen, O_2, at 80 K (−193°C)? What is the phase of nitrogen, N_2, at this temperature?

Chemists Classify Matter as Pure or Impure

7. Why is it not practical to have a macroscopic sample that is 100% pure?

8. Classify the following as (a) homogeneous mixture, (b) heterogeneous mixture, (c) element, or (d) compound:

milk	steel
ocean water	blood
sodium	Planet Earth

9. How is a solution different from a suspension?

10. How can a solution be distinguished from a suspension?

A Solution Is a Single-Phase Homogeneous Mixture

11. What happens to the volume of a sugar solution as more sugar is dissolved in it?

12. Why is a ruby gemstone considered to be a solution?

13. Distinguish between a solute and a solvent.

14. What does it mean to say a solution is concentrated?

15. Distinguish between a saturated solution and an unsaturated solution.

16. How is the amount of solute in a solution calculated?

17. Is 1 mole of particles a very large number or a very small number of particles?

Exercises

1. A sample of water that is 99.9999% pure contains 0.0001% impurities. Consider from Chapter 19 that a glass of water contains on the order of a trillion trillion (1.3×10^{24}) molecules. If 0.0001% of these molecules were the molecules of some impurity, about how many impurity molecules would this be?
   a. 1,000 (one thousand: 1.3×10^{3})
   b. 1,000,000 (one million: 1.3×10^{6})
   c. 1,000,000,000 (one billion: 1.3×10^{9})
   d. 1,000,000,000,000,000,000 (one million trillion: 1.3×10^{18})

   How does your answer make you feel about drinking water that is 99.9999% free of some poison, such as a pesticide? (See Appendix C for a discussion of scientific notation.)

2. Read carefully: Twice as much as one million trillion is two million trillion. One thousand times as much is 1,000 million trillion. One million times as much is 1,000,000 million trillion, which is the same as one trillion trillion. Thus, one trillion trillion is one million times greater than a million trillion. Got that? So how many more water molecules than impurity molecules are there in a glass of water that is 99.9999% pure?

3. Someone argues that he or she doesn't drink tap water because it contains thousands of molecules of some impurity in each glass. How would you respond in defense of the water's purity, if it indeed does contain thousands of molecules of some impurity per glass?

4. Explain what chicken noodle soup and garden soil have in common without using the phrase *heterogeneous mixture.*

5. Classify the following as element, compound, or mixture, and justify your classifications: salt, stainless steel, tap water, sugar, vanilla extract, butter, maple syrup, aluminum, ice, milk, cherry-flavored cough drops.

6. If you eat metallic sodium or inhale chlorine gas, you stand a strong chance of dying. Let these two elements react with each other, however, and you can safely sprinkle the compound on your popcorn for better taste. What is going on?

7. Which of the following boxes contains an elemental material? A compound? A mixture? How many different types of molecules are shown altogether in all three boxes?

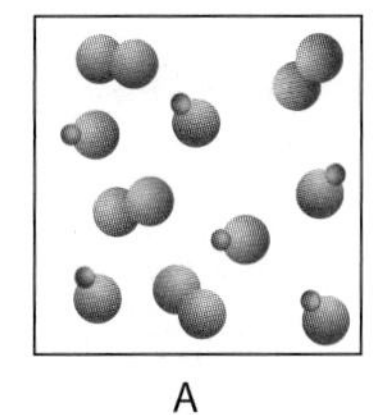
A

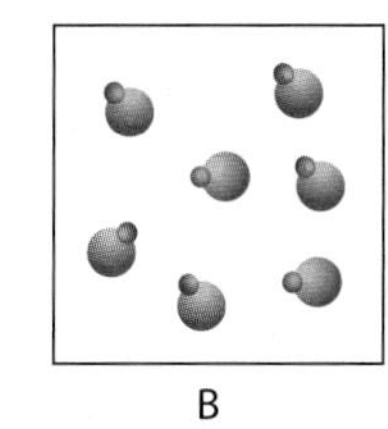
B

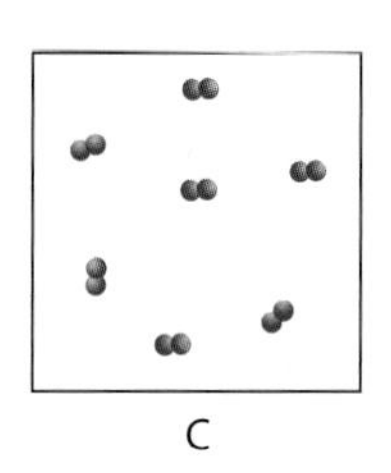
C

8. Common names of chemical compounds are generally much shorter than the corresponding systematic names. The systematic names for water, ammonia, and methane, for example, are dihydrogen monoxide, H_2O; trihydrogen nitride, NH_3; and tetrahydrogen carbide, CH_4. For these compounds, which would you rather use: common names or systematic names? Which do you find more descriptive?

9. What is the difference between a compound and a mixture?

10. How might you separate a mixture of sand and salt? How about a mixture of iron and sand?

11. Mixtures can be separated into their components by taking advantage of differences in the chemical properties of the components. Why might this separation method be less convenient than taking advantage of differences in the physical properties of the components?

12. Why can't the elements of a compound be separated from one another by physical means?

13. Is the air in your house a homogeneous or heterogeneous mixture? What evidence have you seen?

14. Many dry cereals are fortified with iron, which is added to the cereal in the form of small iron particles. How might these particles be separated from the cereal?

15. Why is half-frozen fruit punch always sweeter than the same fruit punch completely melted?

16. Describe two ways to tell whether a sugar solution is saturated or not.

Suggested Web Site

http://www.sugar.org/scoop/refine.html

This page, sponsored by the Sugar Association, takes you on a tour that follows sugar from cane field to the table.

Chapter 23: Chemical Bonding

Have you ever looked closely at table salt crystals and noticed that they are all cubic in shape (except for some rounded edges here and there)? Why does water remain a liquid when other molecules of the same molecular mass are gases? Why is water so different? And why don't oil and water mix? What does the way that molecules bond have to do with the answers to these questions? Let's explore the connections atoms make with one another—chemical bonding.

23.1 An Atomic Model Is Needed to Understand How Atoms Bond

In Chapter 18, we discussed how electrons are arranged around an atomic nucleus. As discussed in Section 18.4, electrons can be described as occupying a series of shells. Recall from Figure 18.14 that there are seven shells available to the electrons in any atom, and that electrons occupy these shells in order, from the innermost shell to the outermost shell. Furthermore, the maximum number of electrons allowed in the first shell is 2, and for the second and third shells it is 8. The fourth and fifth shells can each hold 18 electrons, and the sixth and seventh shells can each hold 32 electrons. These numbers match the number of elements in each period (horizontal row) of the periodic table. Figure 23.1 shows how this model applies to the first four elements of Group 18.

Electrons in the outermost shell of any atom give that atom its chemical properties, including its ability to form chemical bonds. These important outermost electrons are called *valence electrons.* The shell they occupy is called the atom's **valence shell.** We represent the valence electrons of an atom with a special notation called an *electron-dot diagram* or *electron-dot structure.* An electron-dot structure shows the valence electrons of an atom distributed around its atomic symbol. Significantly, some of these electrons are in pairs while others are unpaired. Figure 23.2 shows the electron-dot structures for the atoms important to our discussion of chemical bonds.

When you look at the electron-dot structure of an atom, you immediately know two important things about that element. You know how many valence electrons it has and how many of these are

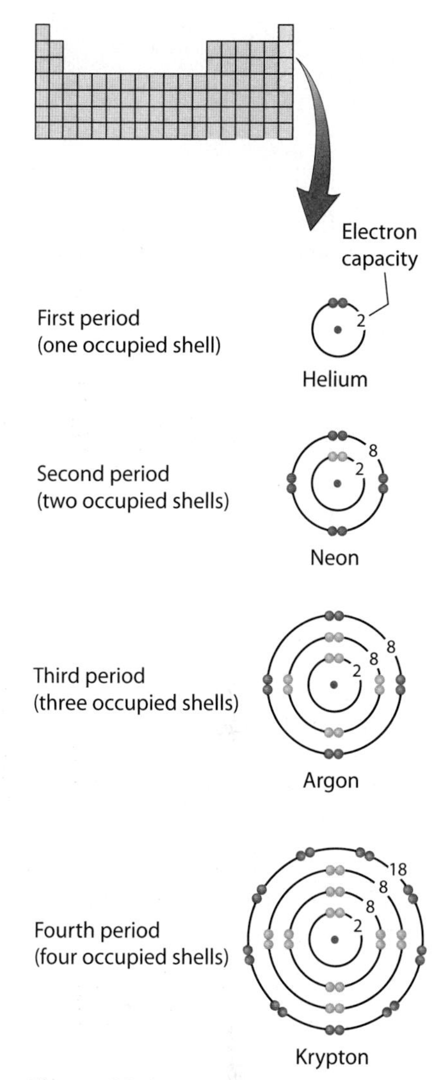

Figure 23.1
Occupied shells in the Group 18 elements helium through krypton. Each of these elements has a filled outermost occupied shell. Also, the number of electrons in each outermost occupied shell corresponds to the number of elements in the period to which a particular Group 18 element belongs.

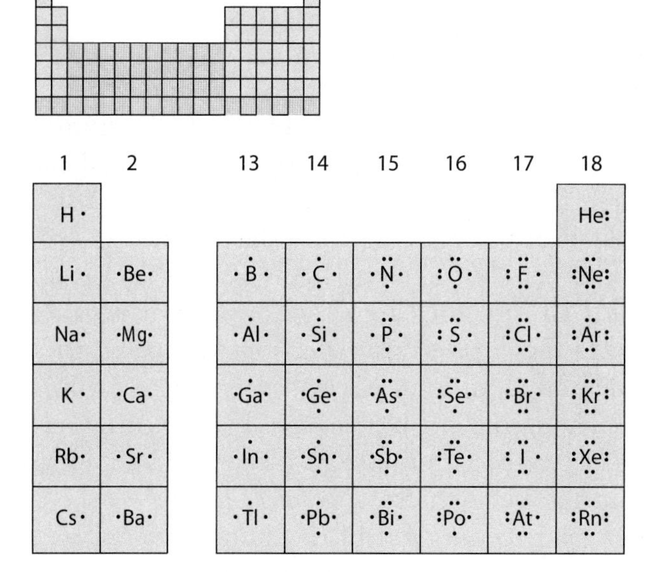

Figure 23.2
The valence electrons of an atom are shown in its electron-dot structure. Note that the first three periods here parallel Figure 18.15.

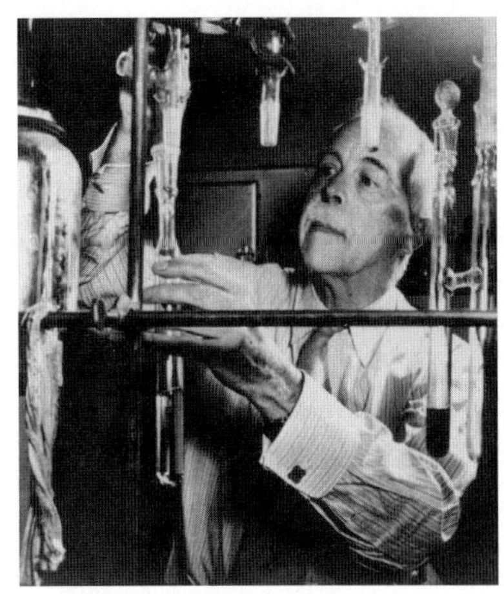

Figure 23.3
Gilbert Newton Lewis (1875–1946) revolutionized chemistry with his theory of chemical bonding, which he published in 1916. He worked most of his life in the chemistry department of the University of California at Berkeley, where he was not only a productive researcher but also an exceptional teacher. Among his teaching innovations was the idea of providing students with problem sets as a follow-up to lectures and readings!

paired. Chlorine, for example, has three sets of paired electrons and one unpaired electron. Carbon has four unpaired electrons:

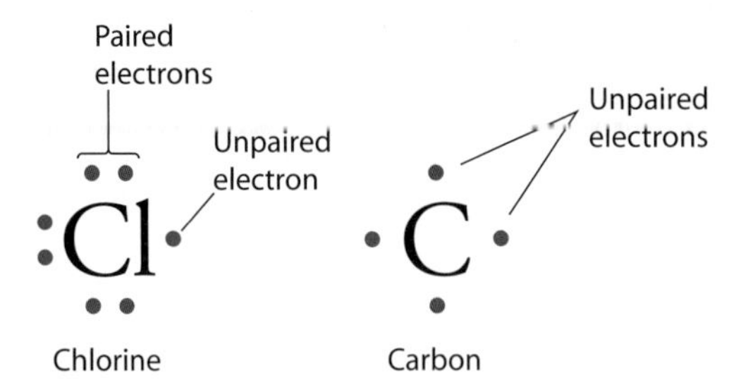

Paired valence electrons are relatively stable. In other words, they usually do not form chemical bonds with other atoms. For this reason, electron pairs in an electron-dot structure are called **nonbonding pairs.**

By contrast, valence electrons that are unpaired strongly tend to bond. When an unpaired electron does form a bond, it does so by pairing up with an electron from another atom. The chemical bonds discussed in this chapter form in either one of two ways: 1) A bond forms when one atom gives an unpaired electron to another atom which also has an unpaired valence electron, or 2) two atoms, both of which have an unpaired electron in their valence shell share their valence electrons to form a single shared valence pair.

Concept Check ✓

What is a valence electron?

Check Your Answer A valence electron is any electron in the outermost occupied shell of an atom that is available to form a chemical bond with another atom.

23.2 Atoms Can Lose or Gain Valence Electrons to Become Ions

When the number of protons in an atomic nucleus equals the number of surrounding electrons in the atom, the charges balance and the atom is electrically neutral. If one or more valence electrons are lost or gained, as illustrated in Figures 23.4 and 23.5, the balance is upset. The atom then has a net electric charge. Any atom that has a net electric charge is called an **ion.** Ions are either positive or negative, depending on whether they lose or gain electrons compared to their neutral state. If a neutral atom loses electrons, its protons will outnumber its electrons. The result is a *positive ion.* If the neutral atom gains electrons, its electrons outnumber its protons. The ion's net charge is negative and the ion is called a *negative ion.*

Chemists use a superscript to the right of the atomic symbol to show the magnitude and sign of an ion's charge. Examples are shown in Figures 23.4 and 23.5. The positive ion formed from the sodium atom is written Na^{1+} while the negative ion formed from the fluorine atom is written F^{1-}. Usually the numeral 1 is omitted when indicating either a 1+ or 1− charge. Hence, these two ions are most frequently written Na^{+} and F^{-}.

To give two more examples, a calcium atom that loses two electrons is written Ca^{2+}, and an oxygen atom that gains two electrons is written O^{2-}. (Note that the convention is to write the numeral before the sign, not after it: 2+, not +2.)

The key to the shell game is ending up as close as you can to a filled shell.

We can use the shell model to tell the type of ion an atom tends to form. This model gives us an important rule:

Atoms tend to lose or gain electrons so that they end up with an outermost occupied shell that is filled to capacity.

Let's take a moment to consider this rule, looking to Figures 23.4 and 23.5 as visual guides.

If an atom has only one or a few electrons in its valence shell, it tends to lose these electrons so that the next shell inward, which is already filled, becomes the outermost occupied shell. The sodium atom of Figure 23.4, for example, has one electron in its valence

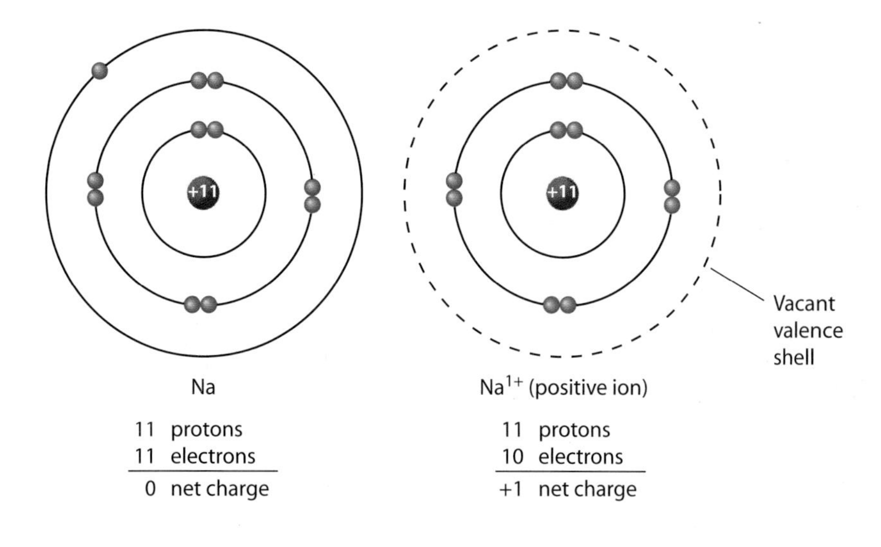

Figure 23.4
An electrically neutral sodium atom contains 11 negatively charged electrons surrounding the 11 positively charged protons of the nucleus. When this atom loses an electron, the result is a positive ion.

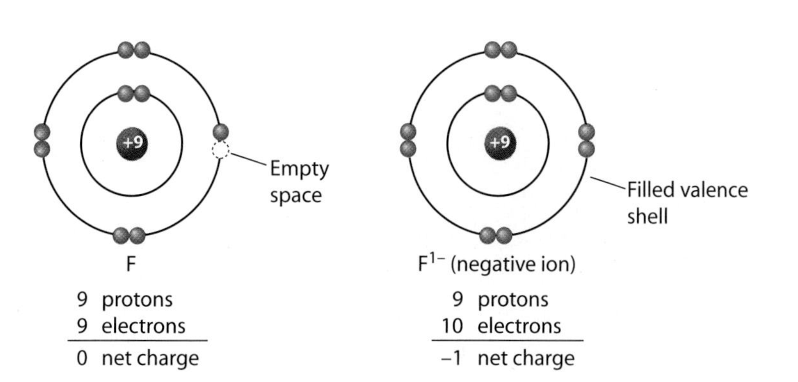

Figure 23.5
An electrically neutral fluorine atom contains nine protons and nine electrons. When this atom gains an electron, the result is a negative ion.

shell, which is the third shell. In forming an ion, the sodium atom loses this electron, thereby making the second shell, which is already filled to capacity, the outermost occupied shell. Because the sodium atom has only one valence electron to lose, it tends to form the 1+ ion.

If the valence shell of an atom is almost filled, that atom attracts electrons from another atom and so forms a negative ion. The fluorine atom of Figure 23.5, for example, has one space available in its valence shell for an additional electron. After this additional electron is gained, the fluorine achieves a filled valence shell. Fluorine therefore tends to form the 1− ion.

You can use the periodic table as a quick reference when determining the type of ion that an atom tends to form. As Figure 23.6 shows, each atom of any Group 1 element, for example, has only one valence electron. So each of the elements in this group tend to form the 1+ ion. Each atom of any Group 17 element has room for one additional electron in its valence shell. Group 17 elements thus tend to form the 1− ion. Atoms of the noble-gas elements tend not to form any type of ion because their valence shells are already filled to capacity.

Concept Check ✓

What type of ion does the magnesium atom, Mg, tend to form?

Check Your Answer The magnesium atom (atomic number 12) is found in Group 2 and has two valence electrons to lose (see Figure 23.2). It therefore tends to form the 2+ ion.

As is indicated in Figure 23.6, the attraction that an atom's nucleus has for its valence electrons is weakest for elements on the left in the periodic table and strongest for elements on the right.

Figure 23.6
The periodic table is your guide to the types of ions atoms tend to form.

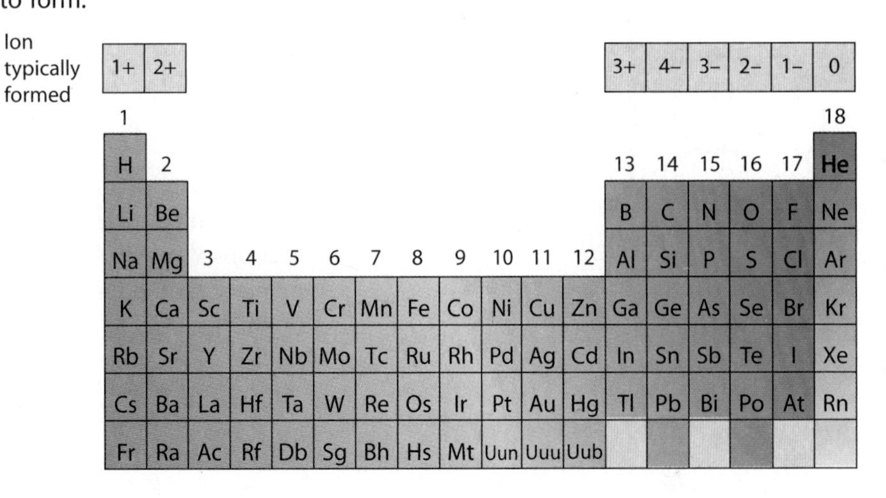

From sodium's position in the table, we see that a sodium atom's single valence electron is not held very strongly. This is why this electron is so easily lost.

At the other side of the periodic table, the nucleus of a fluorine atom holds on strongly to its valence electrons. This explains why the fluorine atom tends *not* to lose any electrons to form a positive ion. Instead, fluorine's nuclear pull on the valence electrons is strong enough to accommodate even an additional electron "imported" from some other atom.

The nucleus of a noble-gas atom pulls so strongly on its valence electrons that they are very difficult to lose. Because there is no room left in the valence shell of a noble-gas atom, no additional electrons are gained. Thus, a noble-gas atom tends not to form an ion of any sort.

Concept Check ✓

Why does the magnesium atom tend to form the 2+ ion?

Check Your Answer Magnesium is on the left in the periodic table, and so atoms of this element do not hold on to the two valence electrons very strongly.

We can use the shell model to explain ion formation for Groups 1 and 2 and 13 through 18. This model is too simplified to work well for the transition metals of Groups 3 through 12, however. It is also too simple to account for how the inner transition metals form ions. In general, these metal atoms tend to form positive ions, but the number of electrons lost varies. Depending on conditions, for example, an iron atom may lose two electrons to form the Fe^{2+} ion, or it may lose three electrons to form the Fe^{3+} ion. Other complexities are beyond the scope of this book. All you need to know is that the positive charge on any ion formed from an atom of a transition metal or inner transition metal is equal to the number of electrons lost.

23.3 Ionic Bonds Result from a Transfer of Valence Electrons

When an atom that tends to lose electrons comes in contact with an atom that tends to gain them, an electron transfer occurs. The result of the transfer is that two oppositely charged ions are formed. The combination of sodium and chlorine is a good example. As shown in

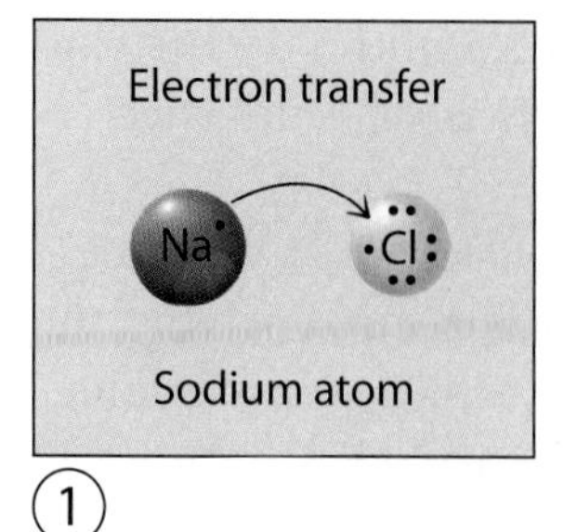

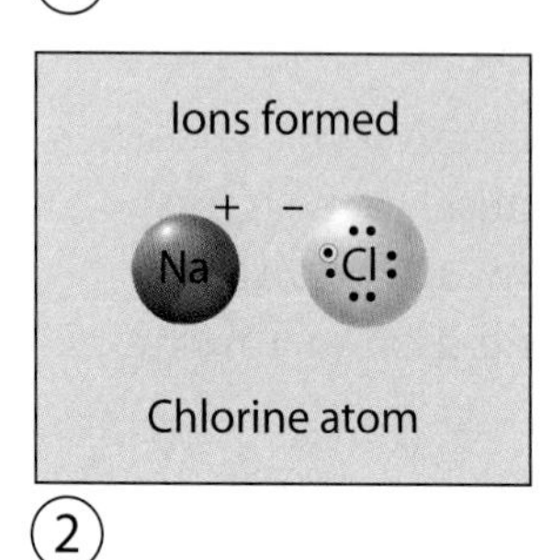

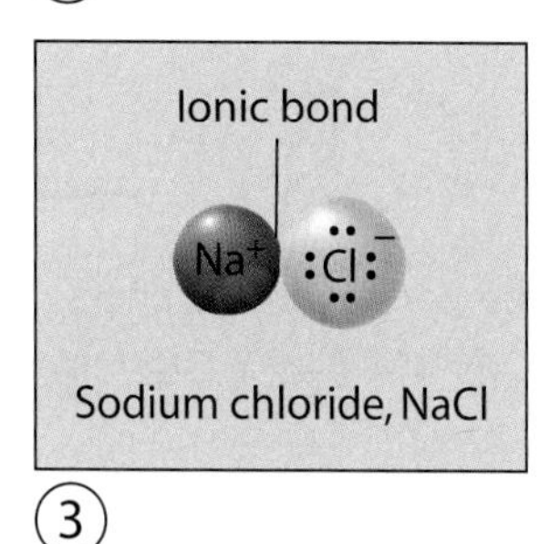

Figure 23.7
1. An electrically neutral sodium atom loses its valence electron to an electrically neutral chlorine atom. 2. This electron transfer results in two oppositely charged ions. 3. The ions are then held together by an ionic bond. The spheres drawn around the electron-dot structures here and in subsequent illustrations indicate the relative sizes of the atoms and ions. Note that the sodium ion is smaller than the sodium atom because the lone electron in the third shell is gone once the ion forms, leaving the ion with only two occupied shells. The chloride ion is larger than the chlorine atom because adding that one electron to the third shell makes the shell expand as a result of the repulsions among the electrons.

Figure 23.7, the sodium atom loses one electron to the chlorine atom, forming a positive sodium ion and a negative chloride ion. The two oppositely charged ions are then attracted to each other by electric force, which holds them close together. This electric force of attraction between two oppositely charged ions is called an **ionic bond.**

A sodium ion and a chloride ion together make the chemical compound sodium chloride, commonly known as table salt. This and all other chemical compounds containing ions are referred to as **ionic compounds.** All ionic compounds are completely different from the elements from which they are made. As you know, sodium chloride is not sodium, nor is it chlorine. Rather, it is a collection of sodium and chloride ions that form a unique material having its own physical and chemical properties.

Concept Check ✓

Is the transfer of an electron from a sodium atom to a chlorine atom a physical change or a chemical change?

Check Your Answer Recall from Chapter 21 that only a chemical change involves the formation of new material. Thus this or any other electron transfer, because it results in the formation of a new substance, is a chemical change.

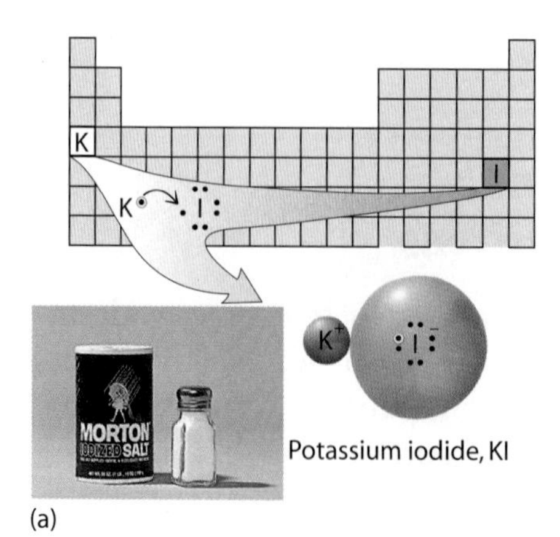

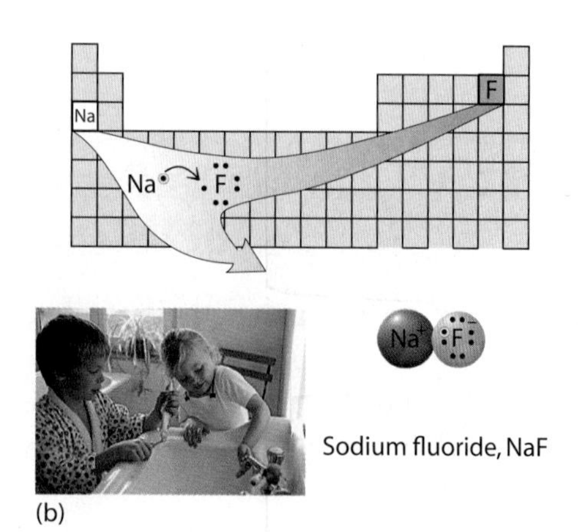

Figure 23.8
(a) The ionic compound potassium iodide, KI, is added in minute quantities to commercial salt because the iodide ion, I^-, it contains is an essential dietary mineral. (b) The ionic compound sodium fluoride, NaF, is often added to municipal water supplies and toothpastes because it is a good source of the tooth-strengthening fluoride ion, F^-.

As Figure 23.8 on page 394 shows, ionic compounds typically consist of elements found on opposite sides of the periodic table. Also, because of how the metals and nonmetals are organized in the periodic table, positive ions are generally made from metallic elements and negative ions are generally made from nonmetallic elements.

For all ionic compounds, positive and negative charges must balance. In sodium chloride, for example, there is one sodium 1+ ion for every chloride 1− ion. Charges must also balance in compounds containing ions that carry multiple charges. The calcium ion, for example, carries a charge of 2+, but the fluoride ion carries a charge of only 1−. Because two fluoride ions are needed to balance each calcium ion, the formula for calcium fluoride is CaF_2, as Figure 23.9 illustrates. Calcium fluoride occurs naturally in the drinking water of some communities. It is a good source of the tooth-strengthening fluoride ion, F−.

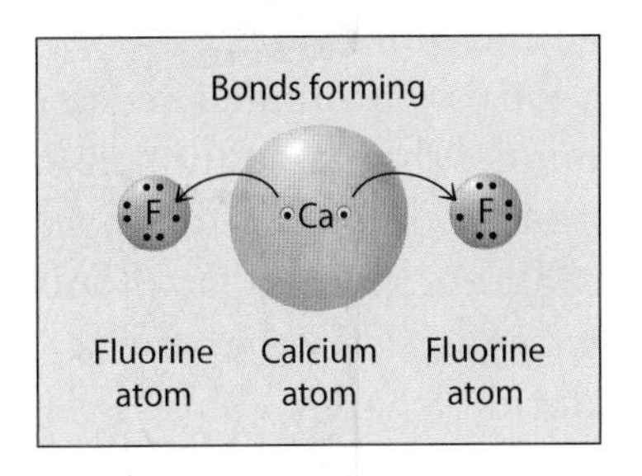

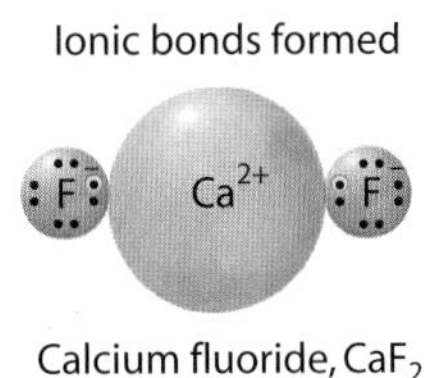

Fluorite

Figure 23.9
A calcium atom loses two electrons to form a calcium ion, Ca^{2+}. These two electrons may be picked up by two fluorine atoms, transforming the atoms to two fluoride ions. Calcium and fluoride ions then join to form the ionic compound calcium fluoride, CaF_2, which occurs naturally as the mineral fluorite.

An aluminum ion carries a 3+ charge, and an oxide ion carries a 2+ charge. Together, these ions make the ionic compound aluminum oxide, Al_2O_3. Aluminum oxide is the main component of such gemstones as rubies and sapphires. Figure 23.10 illustrates the formation of aluminum oxide. The three oxide ions in Al_2O_3 carry a total charge of 6−, which balances the total 6+ charge of the two aluminum ions. Interestingly, rubies and sapphires differ in color

Figure 23.10
Two aluminum atoms lose a total of six electrons to form two aluminum ions, Al^{3+}. These six electrons may be picked up by three oxygen atoms, transforming the atoms to three oxide ions.

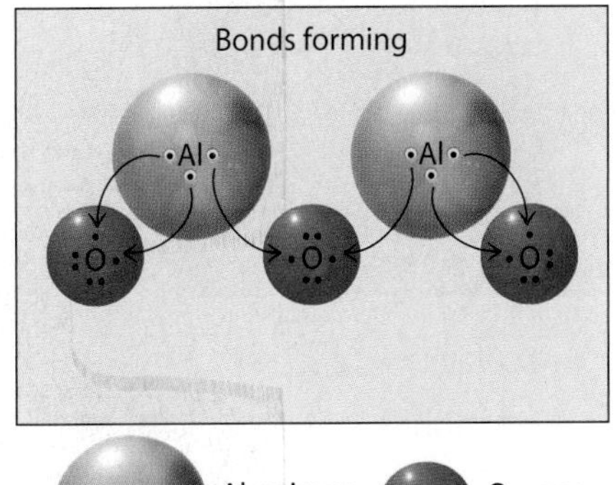

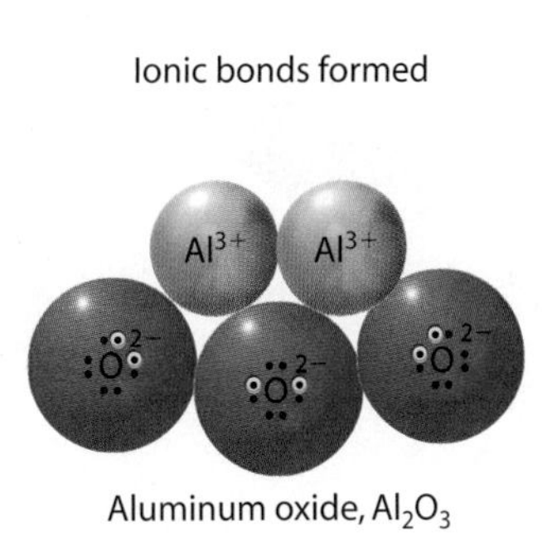

Sapphire

because of the impurities they contain. Rubies are red because of minor amounts of chromium ions, and sapphires are blue because of minor amounts of iron and titanium ions.

Concept Check ✓

What is the chemical formula for the ionic compound magnesium oxide?

Check Your Answer Because magnesium is a Group 2 element, you know a magnesium atom must lose two electrons to form an Mg^{2+} ion. Because oxygen is a Group 16 element, an oxygen atom gains two electrons to form an O^{2-} ion. These charges balance in a one-to-one ratio, and so the formula for magnesium oxide is MgO.

An ionic compound typically contains a multitude of ions grouped together in a highly ordered three-dimensional array. In sodium chloride, for example, each sodium ion is surrounded by six chloride ions and each chloride ion is surrounded by six sodium ions, as shown in Figure 23.11. Overall, there is one sodium ion for each chloride ion, but there are no identifiable sodium-chloride pairs. Such an orderly array of ions is known as an *ionic crystal.* On the atomic level, the crystalline structure of sodium chloride is cubic. This is why macroscopic crystals of table salt are also cubic. Smash a large cubic sodium chloride crystal with a hammer, and what do you get? Smaller cubic sodium chloride crystals!

Similarly, the crystalline structures of other ionic compounds, such as calcium fluoride and aluminum oxide, are a consequence of how the ions pack together.

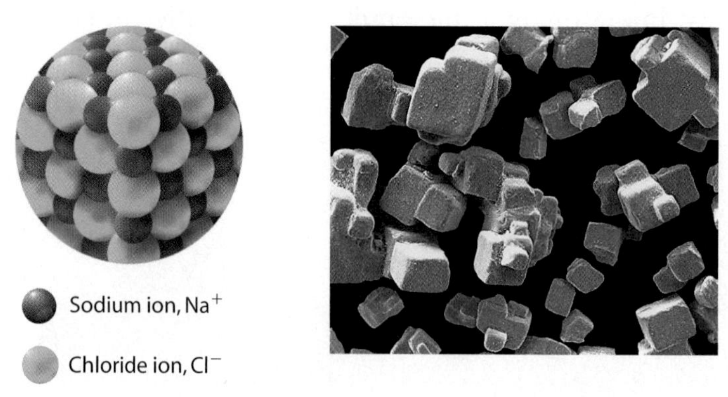

Figure 23.11
Sodium chloride, as well as other ionic compounds, forms ionic crystals in which every internal ion is surrounded by ions of the opposite charge. (For simplicity, only a three-by-three ion array is shown here. A typical NaCl crystal involves millions and millions of ions.) A view of crystals of table salt through a microscope shows their cubic structure. The cubic shape is a consequence of the cubic arrangement of sodium and chloride ions.

Hands-On Exploration: Up Close with Crystals

What You Need

table salt (NaCl); sodium-free salt (KCl); microscope or magnifying glass; metal spoon

Procedure

View crystals of table salt with a magnifying glass or, better yet, with a microscope if you have one. If you do have a microscope, crush the crystals with a spoon and examine the resulting powder. Next, examine some sodium-free salt, which is potassium chloride, KCl. Examine these ionic crystals, both intact and crushed.

Sodium chloride and potassium chloride both form cubic crystals, but there are significant differences. What are they?

23.4 Covalent Bonds Result from a Sharing of Valence Electrons

Imagine two children playing together and sharing their toys. A force that keeps the children together is their mutual attraction to the toys they share. In a similar fashion, two atoms can be held together by their mutual attraction for electrons they share. A fluorine atom, for example, has a strong attraction for one additional electron to fill its outermost occupied shell. As shown in Figure 23.12, a fluorine atom can obtain an additional electron by holding on to the unpaired valence electron of another fluorine atom. This results in a situation in which the two fluorine atoms are mutually attracted to the same two electrons. This type of electrical attraction in which atoms are

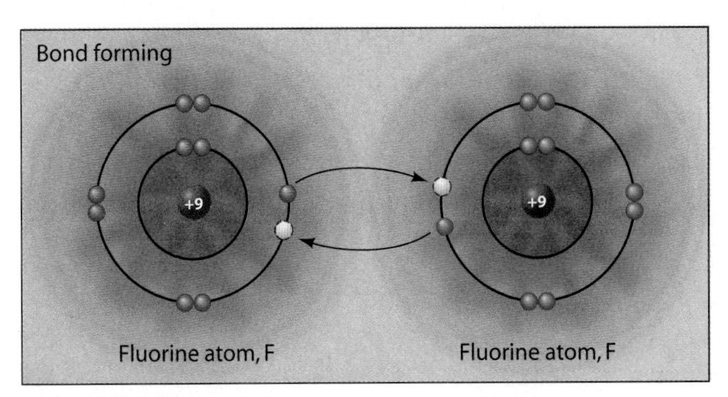

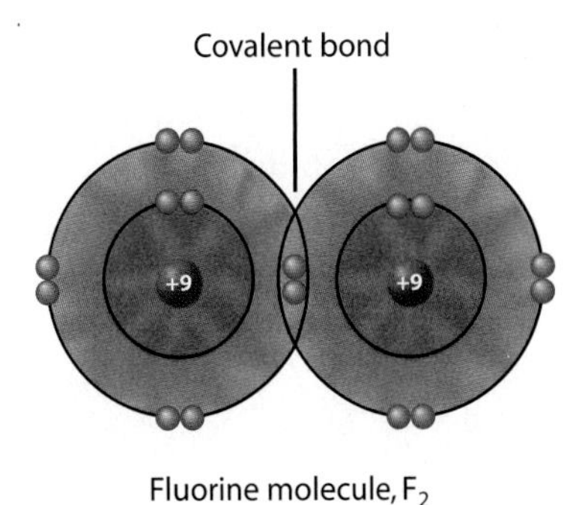

Figure 23.12
The effect of the positive nuclear charge (represented by red shading) of a fluorine atom extends beyond the atom's outermost occupied shell. This positive charge can cause the fluorine atom to become attracted to the unpaired valence electron of a neighboring fluorine atom. Then the two atoms are held together in a fluorine molecule by the attraction they both have for the two shared electrons. Each fluorine atom achieves a filled valence shell.

held together by their mutual attraction for shared electrons is called a **covalent bond.** In the term *covalent, co-* signifies sharing and *-valent* refers to the valence electrons being shared.

A substance that is made up of atoms which are held together by covalent bonds is a **covalent compound.** Molecules are held together by covalent bonds, so a substance made of molecules (as opposed to ions) is a covalent compound.

Now that you know the definition of a covalent bond, you are ready for the true scientific definition of a molecule. A **molecule** is any group of atoms held together by covalent bonds. Figure 23.13 uses the element fluorine to illustrate this principle.

When writing electron-dot structures for covalent compounds, chemists often use a straight line to represent the two electrons involved in a covalent bond. In some representations, the nonbonding electron pairs are left out. This is done in instances where these electrons play no significant role in the process being illustrated. Here are two frequently used ways of showing the electron-dot structure for a fluorine molecule without using spheres to represent the atoms:

$$:\ddot{\underset{..}{F}}-\ddot{\underset{..}{F}}: \qquad F-F$$

Remember: The straight line in both versions represents two electrons, *one from each atom.* Thus we now have two types of electron pairs to keep track of. The term *nonbonding pair* refers to any pair that exists in the electron-dot structure of an individual atom, and the term *bonding pair* refers to any pair that results from formation of a covalent bond. In a nonbonding pair, both electrons come from the same atom. In a bonding pair, one electron comes from one of the atoms taking part in the covalent bond and the other electron comes from the other atom taking part in the bond.

Recall from Section 23.3 that an ionic bond is formed when an atom that tends to lose electrons finds itself in contact with an atom that tends to gain them. A covalent bond, by contrast, is formed when two atoms that tend to gain electrons come into contact with each other. Atoms that tend to form covalent bonds are therefore primarily atoms of the nonmetallic elements in the upper right corner of the periodic table. (The noble-gas elements, which are very stable and tend not to form bonds, are the exception to this rule.)

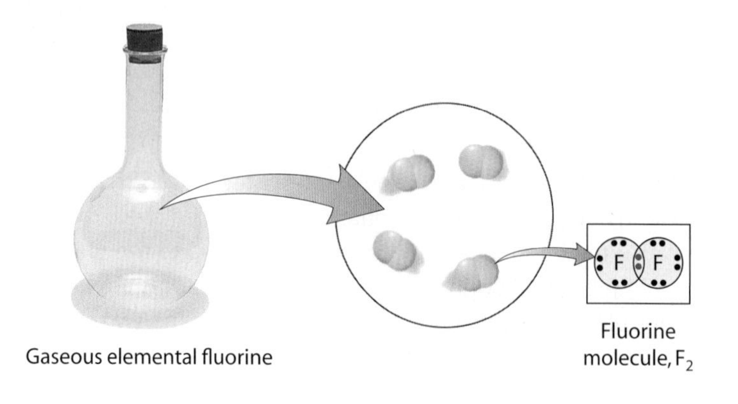

Figure 23.13
Molecules are the fundamental units of the gaseous covalent compound fluorine, F_2. Notice that in this model of a fluorine molecule, the spheres overlap, whereas the spheres shown earlier for ionic compounds do not. Now you know that this difference in representation is because of the difference in bond types.

Hydrogen tends to form covalent bonds because, unlike the other Group 1 elements, it has a fairly strong attraction for an additional electron. Two hydrogen atoms, for example, covalently bond to form a hydrogen molecule, H_2, as shown in Figure 23.14.

The number of covalent bonds an atom can form is equal to the number of additional electrons it can attract. This is the number it needs to fill its valence shell. Hydrogen attracts only one additional electron, and so it forms only one covalent bond. Oxygen, which attracts two additional electrons, finds them when it encounters two hydrogen atoms and reacts with them to form water, H_2O, as Figure 23.15 shows. In water, not only does the oxygen atom have access to two additional electrons by covalently bonding to two hydrogen atoms, but each hydrogen atom has access to an additional electron by bonding to the oxygen atom. Each atom thus achieves a filled valence shell.

Before bonding

H• •H

Hydrogen atom Hydrogen atom

Covalent bond formed

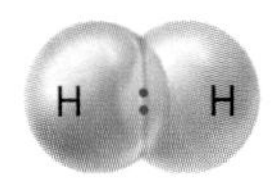

Hydrogen molecule, H_2

Figure 23.14
Two hydrogen atoms form a covalent bond as they share their unpaired electrons.

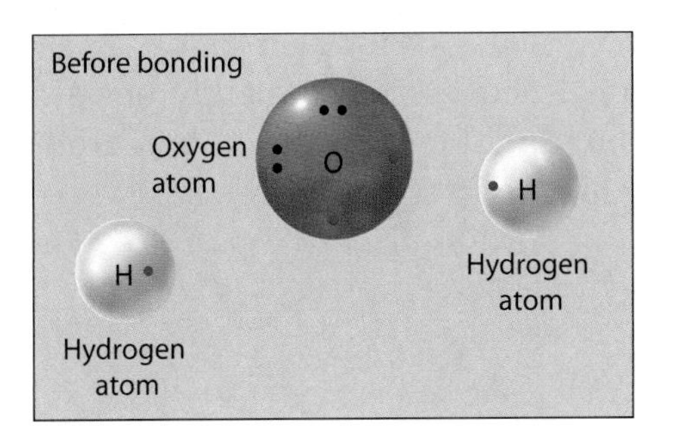

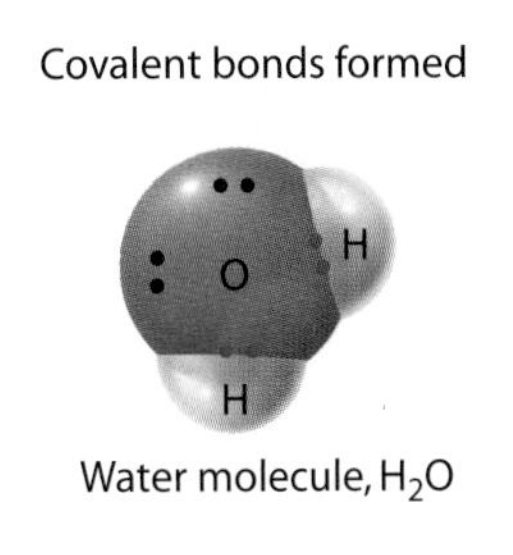

Figure 23.15
The two unpaired valence electrons of oxygen pair with the unpaired valence electrons of two hydrogen atoms to form the covalent compound water.

Nitrogen attracts three additional electrons and can form three covalent bonds, as occurs in ammonia, NH_3, shown in Figure 23.16. Likewise, a carbon atom can attract four additional electrons and is able to form four covalent bonds, as occurs in methane, CH_4. Note that the number of covalent bonds formed by these and other nonmetal elements matches the type of negative ions they tend to

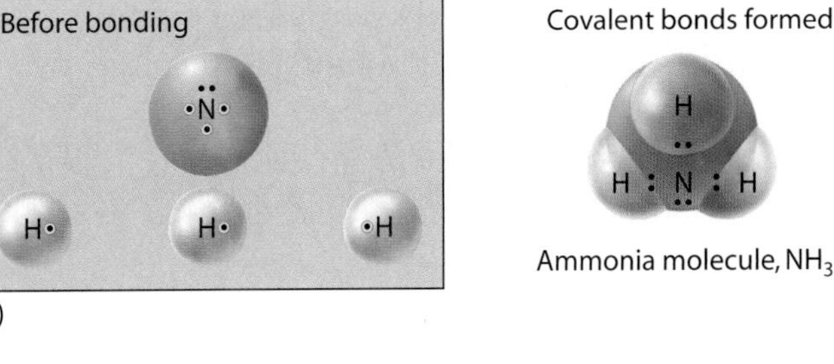
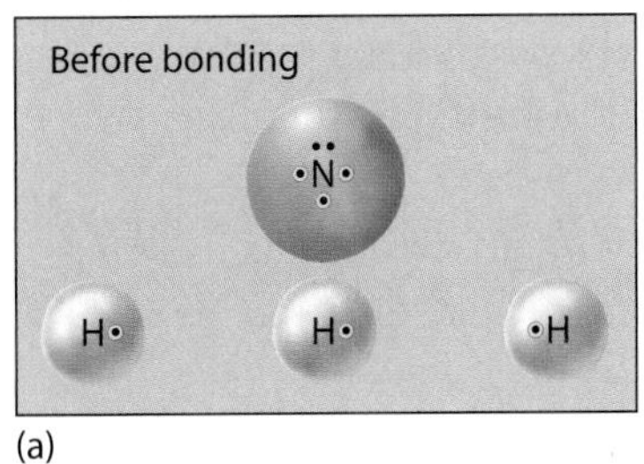

(a)

Covalent bonds formed

H

H : N : H

Ammonia molecule, NH_3

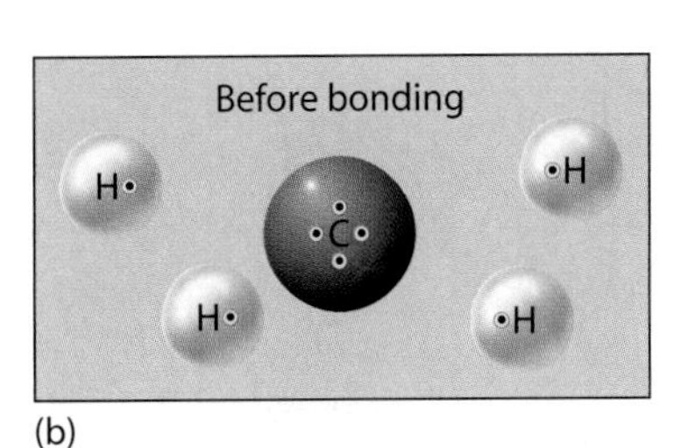

(b)

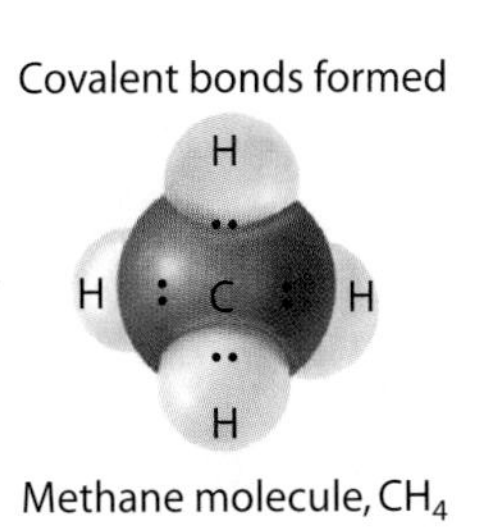

Figure 23.16
(a) A nitrogen atom attracts the three electrons in three hydrogen atoms to form ammonia, NH_3, a gas that can dissolve in water to make an effective cleanser. (b) A carbon atom attracts the four electrons in four hydrogen atoms to form methane, CH_4, the primary component of natural gas. In these and most other cases of covalent bond formation, the result is a filled valence shell for all the atoms involved.

Figure 23.17
The crystalline structure of diamond is best illustrated by using sticks to represent the covalent bonds. It is the molecular nature of diamond that is responsible for this material's unusual properties, such as its extreme hardness.

form (see Figure 23.16). This makes sense because covalent bond formation and negative ion formation are both applications of the same concept: Nonmetallic atoms tend to gain electrons until their valence shells are filled.

Diamond is a most unusual covalent compound made up of carbon atoms covalently bonded to one another in four directions. The result is a *covalent crystal,* which, as shown in Figure 23.17, is a highly ordered three-dimensional network of covalently bonded atoms. The geometry of the bonding between the carbon atoms in diamond forms a very strong and rigid structure. This is why diamonds are so hard.

Concept Check ✓

How many electrons make up a covalent bond?

Check Your Answer Two—one from each participating atom.

It is possible to have more than two electrons shared between two atoms, and Figure 23.18 shows a few examples. Molecular oxygen, O_2, consists of two oxygen atoms connected by four shared electrons. This arrangement is called a *double covalent bond* or, for short, a *double bond.* As another example, the covalent compound carbon dioxide, CO_2, consists of two double bonds connecting two oxygen atoms to a central carbon atom.

Some atoms can form *triple covalent bonds,* in which six electrons—three from each atom—are shared. One example is molecular nitrogen, N_2.

Any double or triple bond is often referred to as a *multiple covalent bond.* Multiple bonds higher than these, such as the quadruple covalent bond, are not commonly observed.

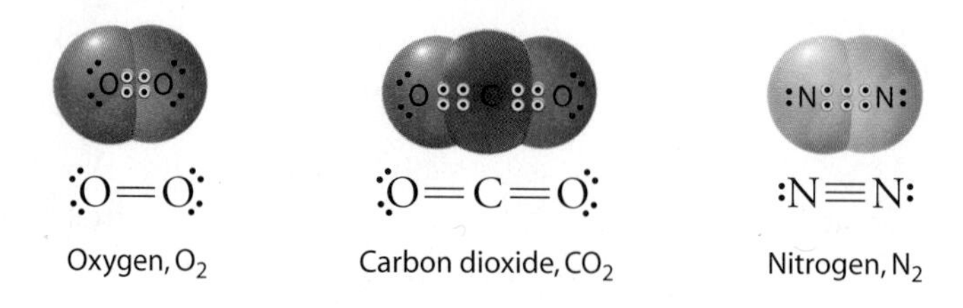

Figure 23.18
Two representations of multiple covalent bonds in molecules of oxygen, O_2, and carbon dioxide, CO_2, both double covalent bonds, and nitrogen, N_2, a triple covalent bond.

23.5 Polar Covalent Bonds Result from an Uneven Sharing of Electrons

If the two atoms in a covalent bond are identical, their nuclei have the same positive charge, and therefore the electrons are shared *evenly.* We can represent these electrons as being centrally located between the two hydrogens as shown below to the left. A different way to show how the electrons are shared is to draw the various positions of the electrons over time as a series of tiny dots. According to this drawing, the electrons are most likely to be found where the dots are grouped closest together, as shown below to the right.

H : H   H H

In a covalent bond between nonidentical atoms, the nuclear charges are different. Consequently the bonding electrons may be shared *unevenly.* This occurs in a hydrogen–fluorine bond, where electrons are more attracted to fluorine's greater nuclear charge:

H : F   H F

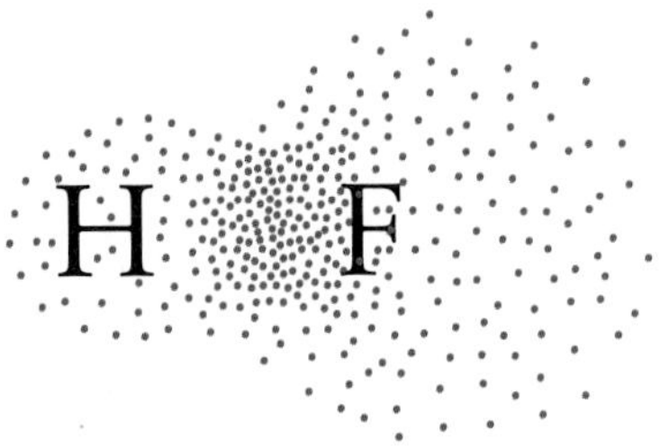

The bonding electrons spend more time around the fluorine atom. For this reason, the fluorine side of the bond is slightly negative and, because the bonding electrons have been drawn away from the hydrogen atom, the hydrogen side of the bond is slightly positive. The centers of charge in the atom are separated (recall electrical polarization in Chapter 11, particularly Figure 11.5b). This separation of charge is called a **dipole** (pronounced *die*-pole) and is represented either by the characters $\delta-$ and $\delta+$, read "slightly negative" and "slightly positive," respectively. Charge separation can also be represented by a crossed arrow pointing to the negative side of the bond:

$$\overset{\delta+}{H}-\overset{\delta-}{F} \qquad \overset{\longmapsto}{H-F}$$

So, a chemical bond is a tug-of-war between atoms for electrons. How strongly an atom is able to tug on bonding electrons has been measured experimentally and quantified as the atom's

H 2.2																	He —
Li 0.98	Be 1.57											B 2.04	C 2.55	N 3.04	O 3.44	F 3.98	Ne —
Na 0.93	Mg 1.31											Al 1.61	Si 1.9	P 2.19	S 2.58	Cl 3.16	Ar —
K 0.82	Ca 1.0	Sc 1.36	Ti 1.54	V 1.63	Cr 1.66	Mn 1.55	Fe 1.83	Co 1.88	Ni 1.91	Cu 1.90	Zn 1.65	Ga 1.81	Ge 2.01	As 2.18	Se 2.55	Br 2.96	Kr —
Rb 0.82	Sr 0.95	Y 1.22	Zr 1.33	Nb 1.6	Mo 2.16	Tc 1.9	Ru 2.2	Rh 2.28	Pd 2.20	Ag 1.93	Cd 1.69	In 1.78	Sn 1.96	Sb 2.05	Te 2.1	I 2.66	Xe —
Cs 0.79	Ba 0.89	La 1.10	Hf 1.3	Ta 1.5	W 2.36	Re 1.9	Os 2.2	Ir 2.20	Pt 2.8	Au 2.54	Hg 2.00	Tl 2.04	Pb 2.33	Bi 2.02	Po 2.0	At 2.2	Rn —
Fr 0.7	Ra 0.9	Ac 1.1	Rf —	Db —	Sg —	Bh —	Hs —	Mt —	Uun —	Uuu —	Uub —						

Figure 23.19
The experimentally measured electronegativities of elements. The greater the electronegativity of an element the greater its ability to pull on bonding electrons.

electronegativity. The range of electronegativities runs from 0.7 to 3.98, as Figure 23.19 shows. The greater an atom's electronegativity, the greater its ability to pull electrons toward itself when bonded. Thus, in hydrogen fluoride, fluorine has a greater electronegativity, or *pulling power,* than does hydrogen.

Electronegativity is greatest for elements at the upper right of the periodic table and lowest for elements at the lower left. Noble gases are not considered in electronegativity discussions because, with only a few exceptions, they do not participate in chemical bonding.

When the two atoms have the same electronegativity, no dipole is formed (as is the case with H_2) and the bond is classified as a **nonpolar bond.** When the electronegativity of the atoms differs, a dipole may form (as with HF) and the bond is classified as a **polar bond.** Just how polar a bond is depends on the difference between the electronegativity values of the two atoms. The greater the difference, the more polar the bond, which means that there is a greater dipole (separation of charge).

As can be seen in Figure 23.19, the farther apart two atoms are in the periodic table, the greater the difference in their electronegativities, and hence the greater the polarity of the bond between them. So a chemist need not even read the electronegativities to predict which bonds are more polar than others. To find out, he or she need only look at the relative positions of the atoms in the periodic table—the farther apart they are, especially when one is at the lower left and one is at the upper right, the greater the polarity of the bond between the atoms.

Concept Check ✓

List these bonds in order of increasing polarity: P—F, S—F, Ga—F, Ge—F (F, fluorine, atomic number 9; P, phosphorus, atomic number 15; S, sulfur, atomic number 16; Ga, gallium, atomic number 31; Ge, germanium, atomic number 32):

(least polar) ________ < ________ < ________ < ________ (most polar)

Check Your Answer *If you answered the question, or attempted to, before reading this answer, hooray for you! You're doing more than reading the text—you're learning physical science.* The greater the *difference* in electronegativities between two bonded atoms, the greater the polarity of the bond, and so the order of increasing polarity is S—F, P—F, Ge—F, Ga—F.

Note that this answer can be obtained by looking only at the relative positions of these elements in the periodic table, rather than by calculating their differences in electronegativities.

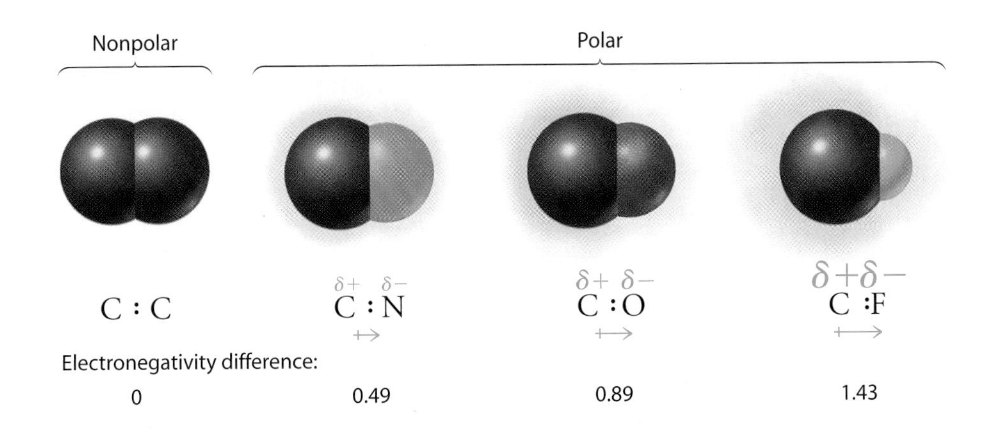

Figure 23.20
These bonds are in order of increasing polarity from left to right, a trend indicated by the larger and larger crossed arrows and δ+/δ− symbols. Which of these pairs of elements are farthest apart in the periodic table?

The magnitude of bond polarity is sometimes indicated by the size of the crossed arrow or δ+/δ− symbol used to depict a dipole, as shown in Figure 23.20.

What is important to understand here is that there is no black-and-white distinction between ionic and covalent bonds. Rather, there is a gradual change from one to the other as the atoms that bond are located farther and farther apart in the periodic table. This continuum is illustrated in Figure 23.21. Atoms on opposite sides of the table have great differences in electronegativity, and hence the bonds between them are highly polar—in other words, ionic. Nonmetallic atoms of the same type have the same electronegativities, and so their bonds are nonpolar covalent. The polar covalent bond with its uneven *sharing* of electrons and slightly *charged* atoms is between these two extremes.

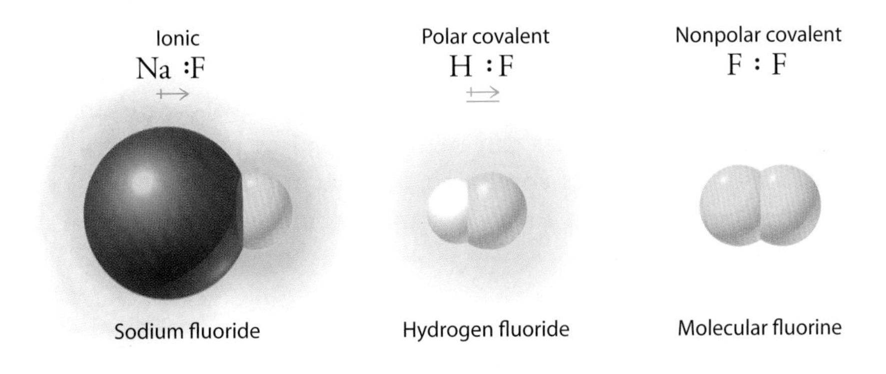

Figure 23.21
The ionic bond and the nonpolar covalent bond represent the two extremes of chemical bonding. The ionic bond involves a *transfer* of one or more electrons, and the nonpolar covalent bond involves the equitable *sharing* of electrons. The character of a polar covalent bond falls between these two extremes.

23.6 Molecular Polarity Results from an Uneven Distribution of Electrons

If all the bonds in a molecule are nonpolar, the molecule as a whole is also nonpolar. This is the case with H_2, O_2, and N_2. If a molecule consists of only two atoms and the bond between them is polar, the polarity of the molecule is the same as the polarity of the bond. This happens in HF, HCl, and ClF.

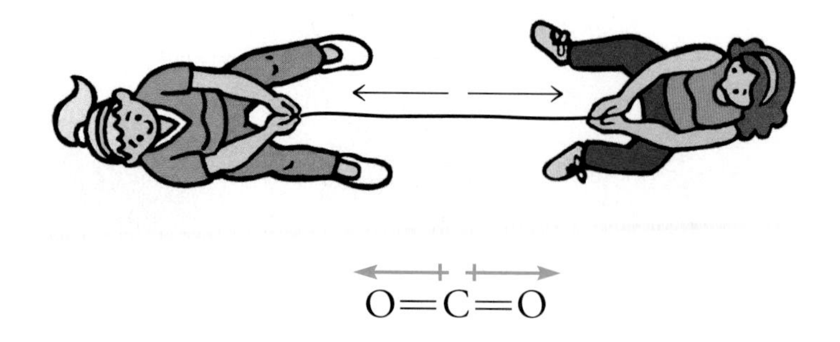

Figure 23.22
There is no net dipole in a carbon dioxide molecule, and so the molecule is nonpolar. This is analogous to two people in a tug-of-war. As long as they pull with equal forces but in opposite directions, the rope remains stationary.

Complexities arise when trying to determine the polarity of a molecule made up of more than two atoms. Look at carbon dioxide, CO_2, in Figure 23.22. The cause of the dipole in either one of the carbon—oxygen bonds is oxygen's greater pull (because oxygen is more electronegative than carbon) on the bonding electrons. The second oxygen atom found on the opposite side of the carbon pulls those electrons back to the carbon, however. The net result is an even distribution of bonding electrons around the whole molecule. So, dipoles that are of equal strength but pull in opposite directions in a molecule effectively cancel each other, with the result that the molecule as a whole is nonpolar.

There are many instances in which the dipoles of different bonds in a molecule combine to make a stronger dipole. Perhaps the most relevant example is water, H_2O. Each hydrogen–oxygen covalent bond is a relatively large dipole. Because of the bent shape of the molecule, however, the two dipoles, shown in blue in Figure 23.23, do not cancel each other the way the C=O dipoles in Figure 23.22 do. Instead, the dipoles in the water molecule work together to give an overall dipole, shown in purple, for the molecule.

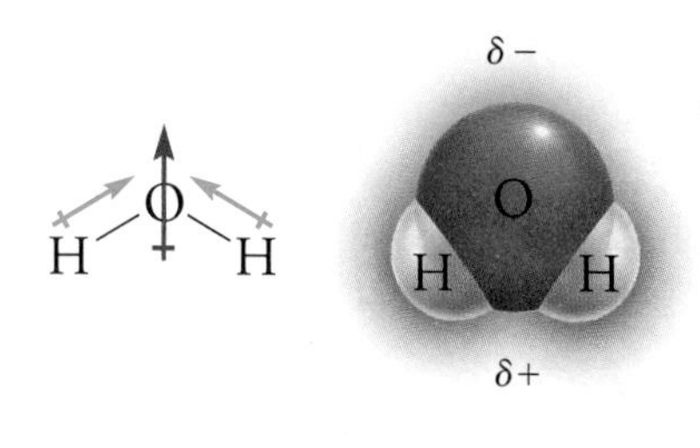

Figure 23.23
(a) The individual dipoles in a water molecule add together to give a large overall dipole for the whole molecule, shown in purple. (b) The region around the oxygen atom is therefore slightly negative, and the region around the two hydrogens is slightly positive.

Figure 23.24 illustrates how polar molecules electrically attract one another and as a result are relatively difficult to separate. In other words, polar molecules can be thought of as being "sticky." For this reason, substances composed of polar molecules typically have higher boiling points than substances composed of nonpolar

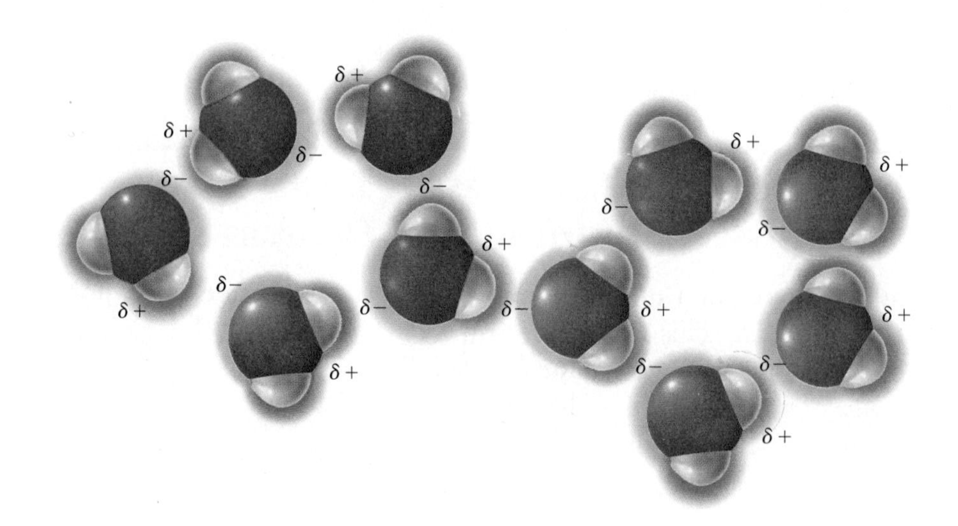

Figure 23.24
Water molecules attract one another because each contains a slightly positive side and a slightly negative side. The molecules position themselves such that the positive side of one faces the negative side of a neighbor.

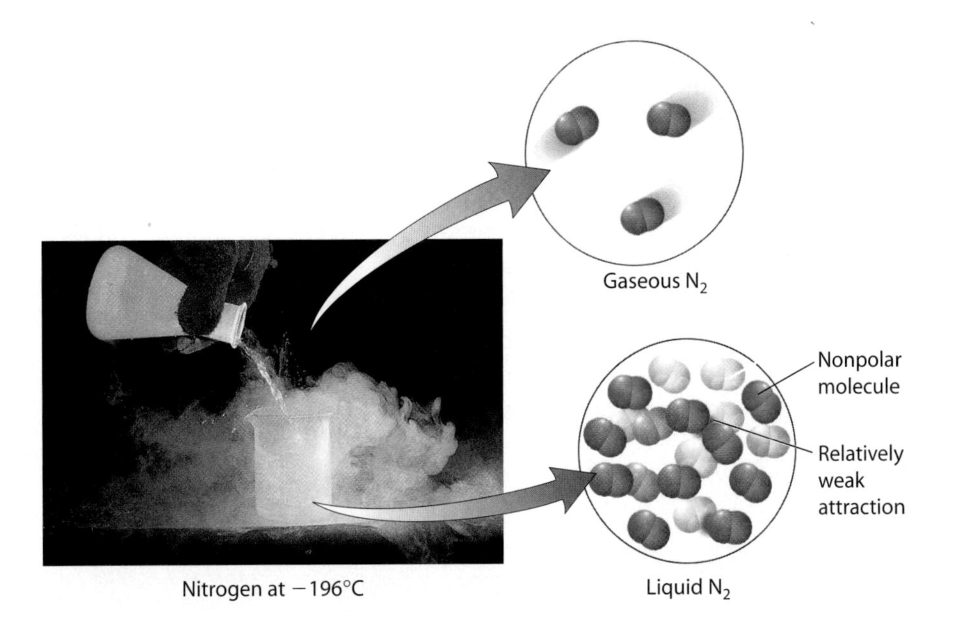

Figure 23.25
Nitrogen is a liquid at temperatures below its chilly boiling point of 2196°C. Nitrogen molecules are not very attracted to one another because they are nonpolar. As a result, the small amount of heat energy available at 2196°C is enough to separate them and allow them to enter the gaseous phase.

molecules. That's because it takes more heat energy to separate the polar molecules into the gaseous phase. Water, for example, boils at 100°C, whereas carbon dioxide boils at −79°C. This 179°C difference is quite dramatic when you consider that a carbon dioxide molecule is more than twice as massive as a water molecule.

Because molecular "stickiness" can play a lead role in determining a substance's macroscopic properties, molecular polarity is a central concept of chemistry. Figure 23.25 shows how nitrogen, N_2, has such a low boiling point because nitrogen molecules do not stick very well to one another. Figure 23.26 describes how molecular stickiness is responsible for the fact that oil and water don't mix. In the next chapter, we explore many other examples of the consequence of the stickiness among molecules.

Figure 23.26
Oil and water are difficult to mix, as is evident from this 1989 oil spill of the Exxon Valdez oil tanker in Alaska's Prince William Sound. It's not, however, that oil and water repel each other. Rather, water molecules are so attracted to themselves because of their polarity that they pull themselves together. The nonpolar oil molecules are thus excluded and left to themselves. Being less dense than water, oil floats on the surface, where it poses great danger to wildlife.

Concept Check ✓

A substance made of polar molecules tends to have a higher boiling point than one made of nonpolar molecules. Why?

Check Your Answer Polar molecules are attracted to one another, and so a lot of heat energy must be added to separate them into the gaseous phase. Therefore a substance made of polar molecules has a relatively high boiling point. There is less electrical attraction between nonpolar molecules, and so less heat energy is needed to separate them into the gaseous phase. Therefore a substance made of nonpolar molecules has a relatively low boiling point.

Chapter Review

Key Terms and Matching Definitions

_____ covalent bond
_____ covalent compound
_____ dipole
_____ electron-dot structure
_____ electronegativity
_____ ion
_____ ionic bond
_____ ionic compound
_____ molecule
_____ nonbonding pairs
_____ nonpolar bond
_____ polar bond
_____ valence shell

1. The outermost occupied shell of an atom that contains electrons that can participate in chemical bonding by either being lost, gained, or shared.
2. A shorthand notation of the shell model of the atom in which valence electrons are shown around an atomic symbol.
3. Two paired valence electrons that don't participate in a chemical bond.
4. An electrically charged particle created when an atom either loses or gains one or more electrons.
5. A chemical bond in which an attractive electric force holds ions of opposite charge together.
6. Any chemical compound containing ions.
7. A chemical bond in which atoms are held together by their mutual attraction for two or more electrons they share.
8. An element or chemical compound in which atoms are held together by covalent bonds.
9. A group of atoms held tightly together by covalent bonds.
10. A separation of electric charge that occurs in a chemical bond because of differences in the electronegativities of the bonded atoms.
11. The ability of an atom to attract a bonding pair of electrons to itself when bonded to another atom.
12. A chemical bond in which bonding electrons are distributed evenly between bonding atoms. Such a bond has no dipole.
13. A chemical bond in which bonding electrons are distributed unevenly between bonding atoms. Such a bond has a dipole. The more uneven the sharing, the greater the dipole.

Review Questions

An Atomic Model Is Needed to Understand How Atoms Bond

1. How many shells are needed to account for the seven periods of the periodic table?
2. How many electrons can fit in the first shell? How many in the second shell?
3. How many shells are completely filled in an argon atom, Ar (atomic number 18)?
4. Which electrons are represented by an electron-dot structure?
5. How do the electron-dot structures of elements in the same group in the periodic table compare with one another?
6. How many nonbonding pairs are there in an oxygen atom? How many unpaired valence electrons?

Atoms Can Lose or Gain Valence Electrons to Become Ions

7. How does an ion differ from an atom?
8. To become a negative ion, does an atom lose or gain electrons?
9. Do metals more readily gain or lose electrons?
10. How many electrons does the calcium atom tend to lose?
11. Why does the fluorine atom tend to gain only one additional electron?

Ionic Bonds Result from a Transfer of Valence Electrons

12. Which elements tend to form ionic bonds?
13. What is the electric charge on the calcium ion in the compound calcium chloride, $CaCl_2$?
14. What is the electric charge on the calcium ion in the compound calcium oxide, CaO?
15. Suppose an oxygen atom gains two electrons to become an oxygen ion. What is its electric charge?
16. What is an ionic crystal?

Covalent Bonds Result from a Sharing of Valence Electrons

17. Which elements tend to form covalent bonds?
18. What force holds two atoms together in a covalent bond?
19. How many electrons are shared in a double covalent bond?
20. How many electrons are shared in a triple covalent bond?
21. How many additional electrons is an oxygen atom able to attract?
22. How many covalent bonds is an oxygen atom able to form?

Polar Covalent Bonds Result from an Uneven Sharing of Electrons

23. What is a dipole?
24. Which element of the periodic table has the greatest electronegativity? Which has the smallest?
25. Which is more polar: a carbon–oxygen bond or a carbon–nitrogen bond?
26. How is a polar covalent bond similar to an ionic bond?

Molecular Polarity Results from an Uneven Distribution of Electrons

27. How can a molecule be nonpolar when it consists of atoms that have different electronegativities?
28. Why do nonpolar substances tend to boil at relatively low temperatures?

29. Why don't oil and water mix?

30. Which would you describe as "stickier": a polar molecule or a nonpolar one?

Exploration

Up Close with Crystals

One thing you probably noticed under the magnifying glass when you compared uncrushed crystals was sharp, angular edges in NaCl and rounded edges in KCl. Then you probably found it easier to grind the KCl crystals to powder. These differences have the same origin: A potassium ion, K^+, is larger than a sodium ion, Na^+.This means the positive and negative charges of the ionic bond in potassium chloride are farther apart than in sodium chloride:

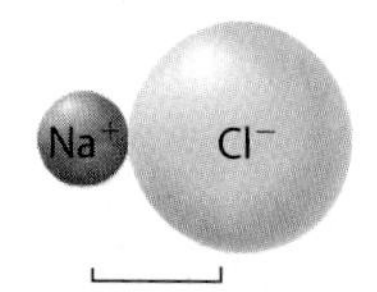

Shorter distance between positive and negative charges

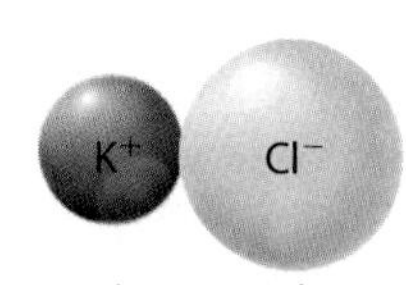

Longer distance between positive and negative charges

Because the electric force weakens with increasing distance between the opposite charges, the KCl ionic bond is weaker than the NaCl ionic bond. Weaker ionic bonds mean that KCl crystals are less resilient to stress and impact than are NaCl crystals. This accounts for the rounder edges you observed in the KCl crystals and for the fact that it was easier to grind the KCl to a powder.

Exercises

1. An atom loses an electron to another atom. Is this an example of a physical change or a chemical change?
2. Why is it so easy for a magnesium atom to lose two electrons?
3. Why doesn't the sodium atom gain seven electrons so that its third shell becomes the filled outermost shell?
4. Magnesium ions carry a 2+ charge, and chloride ions carry a 1− charge. What is the chemical formula for the ionic compound magnesium chloride?
5. Barium ions carry a 2+ charge, and nitrogen ions carry a 3− charge. What is the chemical formula for the ionic compound barium nitride?
6. Does an ionic bond have a dipole?
7. Why doesn't a neon atom tend to gain electrons?
8. Why doesn't a neon atom tend to lose electrons?
9. Why doesn't a hydrogen atom form more than one covalent bond?
10. What drives an atom to form a covalent bond: its nuclear charge or the need to have a filled outer shell? Explain.
11. Is there an abrupt change or a gradual change between ionic and covalent bonds? Explain.
12. Classify the following bonds as ionic, polar covalent, or nonpolar covalent (O, atomic number 8; F, atomic number 9; Na, atomic number 11; Cl, atomic number 17; Ca, atomic number 20; U, atomic number 92):
    O with F
    Ca with Cl
    Na with Na
    U with Cl
13. Nonmetal atoms form covalent bonds, but they can also form ionic bonds. How is this possible?
14. Metal atoms can form ionic bonds, but they are not very good at forming covalent bonds. Why?

15. Phosphine is a covalent compound of phosphorus, P, and hydrogen, H. What is its chemical formula?

16. Which bond is most polar: H—N, N—C, C—O, C—C, O—H, C—H?

17. Which molecule is most polar: S=C=S, O=C=O, O=C=S?

18. In each molecule, which atom carries the greater positive charge: H—Cl, Br—F, C—O, Br—Br?

19. List these bonds in order of increasing polarity: N—N, N—F, N—O, H—F

________, ________, ________, ________

(least polar) (most polar)

20. Which is more polar: a sulfur–bromine bond, S—Br, or a selenium–chlorine bond, Se—Cl?

21. An individual carbon–oxygen bond is polar. Yet carbon dioxide, CO_2, which has two carbon–oxygen bonds, is nonpolar. Explain.

22. Water, H_2O, has less than half as heavy as carbon dioxide, CO_2. Why then is the boiling point of water so much higher than that of carbon dioxide?

Suggested Web Sites

http://www.ada.org/topics/fluoride.html

Fluoride page of the American Dental Association, with many links to information regarding fluorides and fluoridation of drinking water and toothpastes.

http://www.saltinstitute.org/idd.htm

Numerous reports in the literature demonstrate the effectiveness of iodized salt in controlling the medical condition called *goiter.* Check out this site for historical case studies that first pointed to this conclusion.

http://www.soils.wisc.edu/virtual_museum/

Home page of the Virtual Museum of Minerals and Molecules, curated by Phillip Barak of the University of Minnesota and Ed Nater of the University of Wisconsin. Through this site, you will find molecular models that you can manipulate in three dimensions. To do so, your browser will need to be equipped with the Chime plug-in, which you may download by following the hyperlinks to http://www.mdli.com/download/chimedown.html.

Chapter 24: Molecular Mixing

How does oxygen stay in water? Where does sugar disappear to when it is dissolved in water? Does the water level in a cup of tea rise when sugar is dissolved into the tea? Why does a soda drink taste more bubbly when it is warm than when it is cold? What is soap and how does it work? Why does plastic wrap stick to glass? Let's find out!

24.1 Submicroscopic Particles Electrically Attract One Another

What is a *polar* molecule? You probably remember from Chapter 23 that a *polar* molecule is one in which the bonding electrons are unevenly distributed. One side of the molecule carries a slight negative charge, and the opposite side carries a slight positive charge. This separation of charge is a *dipole.*

So what happens to polar molecules, such as water molecules, when they are near an ionic compound, such as sodium chloride? The opposite charges electrically attract one another. The positive sodium ions attract the negative side of the water molecules. And the negative chloride ions attract the positive side of the water molecules. This is illustrated in Figure 24.1. Such an attraction between an ion and the dipole of a polar molecule is called an *ion–dipole attraction.*

Ion–dipole attractions are much weaker than ionic bonds. However, a large number of ion-dipole attractions can act collectively to disrupt an ionic bond. This is what happens to sodium chloride in water. Attractions exerted by the water molecules break the ionic bonds and pull the ions apart. The result, represented in Figure 24.2, is a solution of sodium chloride in water. (A solution in water is called an *aqueous solution.*)

A water molecule is a natural dipole—a bit positive on one end and negative on the other. What's the net charge of a dipole?

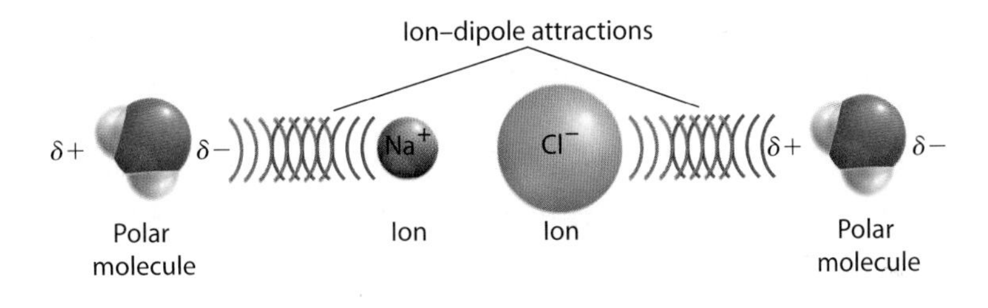

Figure 24.1
An ion–dipole attraction occurs between the slightly negative side (oxygen side) of a water molecule and the positively charged sodium ion of sodium chloride. Another ion–dipole attraction occurs between the slightly positive side (hydrogen side) of a water molecule and the negatively charged chloride ion. Electrical attractions are shown as a series of overlapping arcs. The blue arcs indicate negative charge, and the red arcs indicate positive charge.

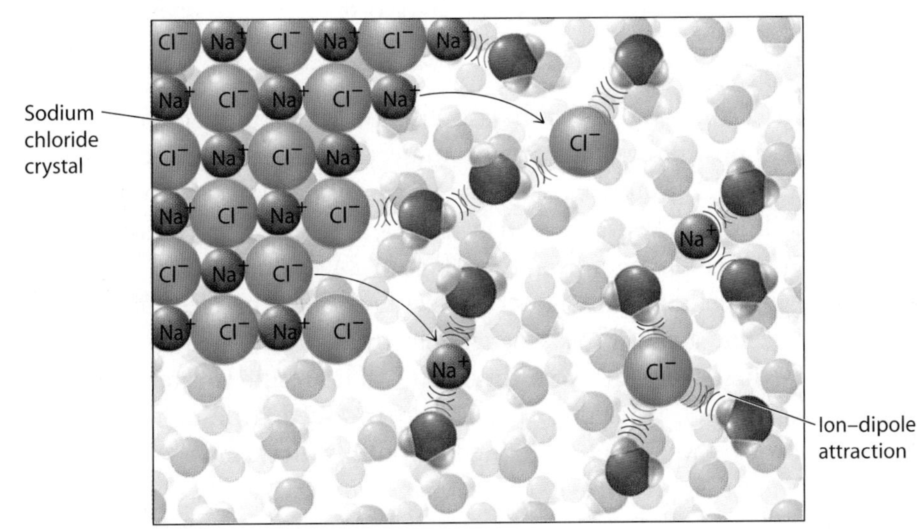

Figure 24.2
Sodium and chloride ions tightly bound in a crystal lattice are separated from one another by the collective attraction exerted by many water molecules to form an aqueous solution of sodium chloride.

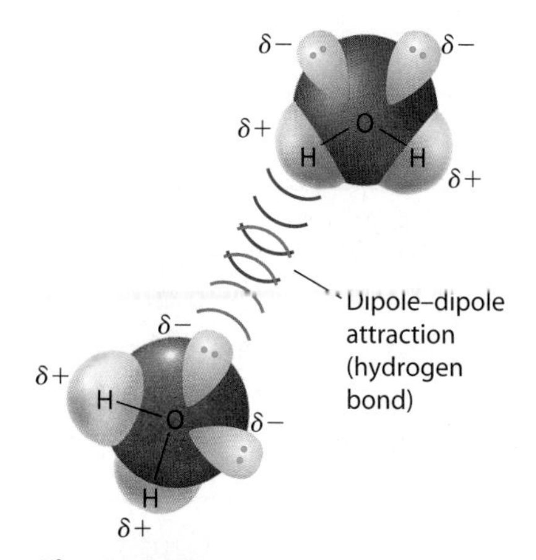

Figure 24.3
The dipole–dipole attraction between two water molecules is a hydrogen bond because it involves hydrogen atoms bonded to highly electronegative oxygen atoms.

Polar Molecules Attract Other Polar Molecules

An attraction between two polar molecules is called a *dipole–dipole attraction.* An unusually strong dipole–dipole attraction is the **hydrogen bond.** This attraction occurs between molecules that have a hydrogen atom covalently bonded to a highly electronegative atom, usually nitrogen, oxygen, or fluorine. Recall from Chapter 23 that the electronegativity of an atom describes how well that atom is able to pull bonding electrons towards itself. The greater the atom's electronegativity, the better it is able to gain electrons and thus the more negative is its charge.

Look at Figure 24.3 to see how hydrogen bonding works. The hydrogen side of a polar molecule (water in this example) has a positive charge, as the more electronegative atom hydrogen is bonded to tugs on hydrogen's electron. This hydrogen is therefore electrically attracted to a pair of nonbonding electrons on the negatively charged atom of another molecule. This mutual attraction between hydrogen and the negatively charged atom of another molecule is a hydrogen bond.

The strength of a hydrogen bond depends on two things: (1) the strength of the dipoles involved (this in turn depends on the differences in electronegativity of the atoms within the polar molecules) and (2) how strongly nonbonding electrons on one molecule can attract a hydrogen atom on a nearby molecule.

Even though the hydrogen bond is much weaker than true chemical bonds between atoms (covalent or ionic bonds), the effects of hydrogen bonding can be very pronounced. For example, water owes many of its properties to hydrogen bonds. The hydrogen bond is also of great importance in the chemistry of large molecules, such as DNA and proteins, found in living organisms.

In the previous chapter we talked about how molecules form. In this chapter we're looking at how molecules mix together.

24.2 Polar Molecules Can Induce Dipoles in Nonpolar Molecules

In many molecules, the electrons are distributed evenly. So in these cases there is no dipole. The oxygen molecule, O_2, is one such nonpolar molecule. A nonpolar molecule can be induced to become a temporary dipole, however, when it is brought close to a water molecule or any other polar molecule, as Figure 24.4 illustrates. The slightly negative side of the water molecule pushes the electrons in the oxygen molecule away. Thus oxygen's electrons are pushed to the side which is farthest from the water molecule. The result is a temporary uneven distribution of electrons called an **induced dipole.** The resulting attraction between the permanent dipole (water) and the induced dipole (oxygen) is a *dipole–induced dipole attraction.*

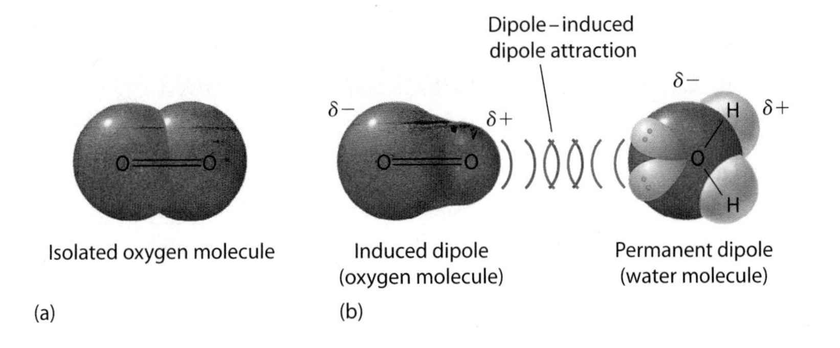

Figure 24.4
(a) An isolated oxygen molecule has no dipole; its electrons are distributed evenly. (b) An adjacent water molecule induces a redistribution of electrons in the oxygen molecule. (The slightly negative side of the oxygen molecule is shown larger than the slightly positive side because the slightly negative side contains more electrons.)

Concept Check ✓

How does the electron distribution in an oxygen molecule change when the hydrogen side of a water molecule is nearby?

Check Your Answer Because the hydrogen side of the water molecule is slightly positive, the electrons in the oxygen molecule are pulled *toward* the water molecule.

Remember: Induced dipoles are only temporary. If the water molecule in Figure 24.4b were removed, the oxygen molecule would return to its normal, nonpolar state. As a consequence, dipole–induced dipole attractions are weaker than dipole–dipole attractions. They are strong enough to hold relatively small quantities of oxygen dissolved in water, however. This attraction between water and molecular oxygen is vital for fish and other forms of aquatic life that rely on molecular oxygen mixed in water.

Dipole–induced dipole attractions also occur between molecules of carbon dioxide, which are nonpolar, and water. It is these attractions that help to keep carbonated beverages (which are mixtures of carbon dioxide in water) from losing their fizz too quickly after they've been opened. Dipole–induced dipole attractions are also responsible for holding plastic wrap to glass, as shown in Figure 24.5. These wraps are made of very long nonpolar molecules that are induced to have dipoles when placed in contact with glass, which is highly polar.

Figure 24.5
Temporary dipoles induced in normally nonpolar molecules in plastic wrap makes it stick to glass.

Concept Check ✓

Distinguish between a dipole–dipole attraction and a dipole–induced dipole attraction.

Check Your Answer The dipole–dipole attraction is stronger and involves two permanent dipoles. The dipole–induced dipole attraction is weaker and involves a permanent dipole and a temporary one.

Hands-On Exploration: Circular Rainbows

Black ink contains pigments of many different colors. Acting together, these pigments absorb all the frequencies of visible light. Because no light is reflected, the ink appears black. We can use electrical attractions to separate the color components of black ink with a special technique called *paper chromatography.*

What You Need

Black felt-tip pen or black water-soluble marker; piece of porous paper, such as paper towel, table napkin, or coffee filter; solvent, such as water, acetone (fingernail polish remover), rubbing alcohol, or white vinegar.

Procedure

1. Place a concentrated dot of ink at the center of the piece of porous paper.
2. Carefully place one drop of solvent on top of the dot, and watch the ink spread radially with the solvent. Because the different components of the ink have different affinities for the solvent (based on the electrical attractions between component molecules and solvent molecules), they travel with the solvent at different rates.
3. Just after the drop of solvent is completely absorbed, add a second drop at the same place you put the first one, then a third, and so on until the ink components have separated to your satisfaction.

How the components separate depends on several factors, including your choice of solvent and your technique. It's also interesting to watch the leading edge of the moving ink under a strong magnifying glass or microscope.

24.3 Solubility Is a Measure of How Well a Solute Dissolves

Recall from Section 22.4 that a *solvent* is the major component of a mixture, and the *solute* is the minor component. For example, in a glass of sugar water, the water is the solvent while the sugar is the solute.

The **solubility** of a solute is its ability to dissolve in a solvent. This ability depends in great part on the submicroscopic attractions between solute particles and solvent particles. Sugar, for example, is said to have good solubility in water. If a solute has any solubility in a solvent, then we say that solute is **soluble** in that solvent.

Solubility also depends on attractions between solute particles and attractions between solvent particles. As shown in Figure 24.6, for example, there are many polar hydrogen–oxygen bonds in a sugar molecule. Sugar molecules, therefore, can form multiple hydrogen bonds with each other. These hydrogen bonds are strong enough to make sucrose a solid at room temperature and give it a relatively high melting point of 185°C. In order for sucrose to dissolve in water, the water molecules must first pull sucrose mol cules away from one another. This puts a limit on the amount of sucrose that can dissolve in water—eventually a point is reached where there are not enough water molecules to separate the sucrose molecules from one another. As was discussed in Section 22.4, this is the point of saturation. Any additional sucrose added to the solution does not dissolve.

Figure 24.6
A sucrose molecule contains many hydrogen–oxygen covalent bonds in which the hydrogen atoms are slightly positive and the oxygen atoms are slightly negative. These dipoles in any given sucrose molecule result in the formation of hydrogen bonds with neighboring sucrose molecules.

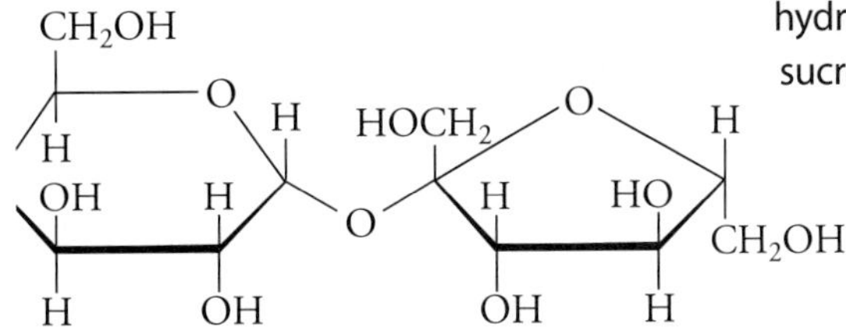

Sucrose

When the molecule-to-molecule attractions among solute molecules are comparable to the molecule-to-molecule attractions among solvent molecules, the result is that there is no saturation point. As shown in Figure 24.7, for example, the hydrogen-bonds among water molecules are about as strong as those among ethanol molecules. These two liquids therefore mix together quite well and in just about any proportion. We can even add ethanol to water until the ethanol rather than the water may be considered the solvent. In fact, ethanol and water stick so well to each other that, even after distillation (Section 22.2) of a water-ethanol solution, the purest ethanol we can get is 95%. To get 100% ethanol requires special procedures.

A solute that has no practical point of saturation in a given solvent is said to be *infinitely soluble* in that solvent. Ethanol, for example, is infinitely soluble in water. Also, gases are generally infinitely soluble in each other because they can be mixed together in just about any proportion.

Let's now look at the other extreme of solubility, where a solute has very little solubility in a given solvent. An example is oxygen, O_2 in water. In contrast to sucrose, which has a solubility of 200 grams per 100 milliliters of water, only 0.004 gram of oxygen can dissolve in 100 milliliters of water. We can account for oxygen's low solubility in

Grease is soluble in paint thinner, which is why paint thinner can be used to clean one's hands of grease. But body oils are also soluble in paint thinner, which is why hands cleaned with paint thinner feel dry and chapped.

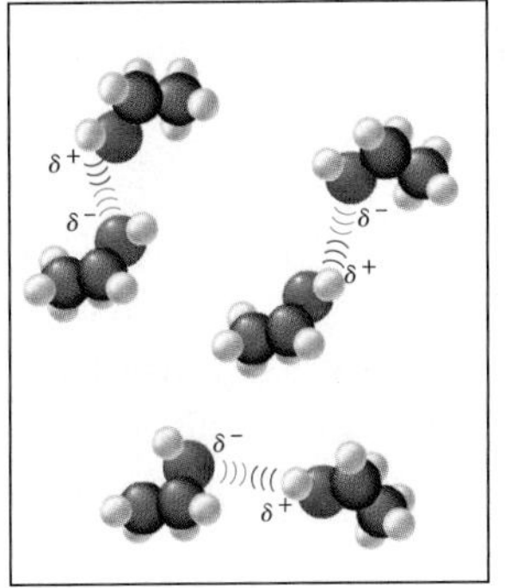

Ethanol

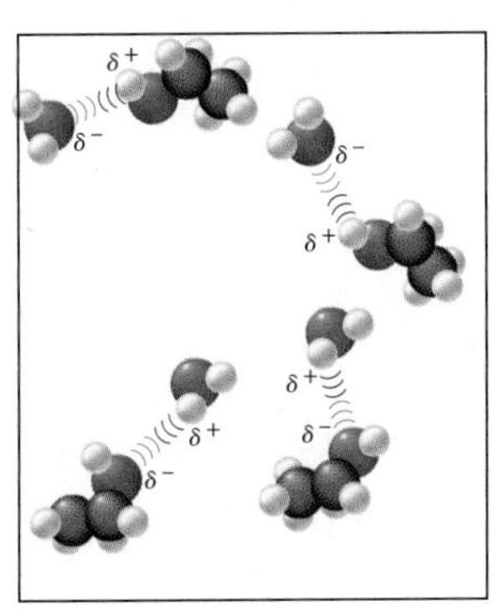

Ethanol and water

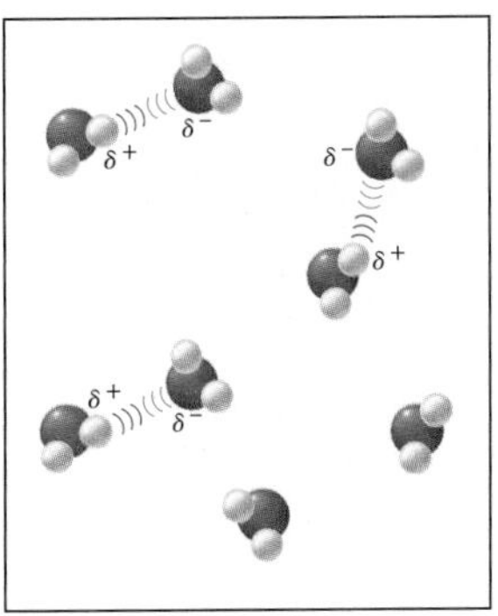

Water

Figure 24.7
Ethanol and water molecules are about the same size, and they both form hydrogen bonds. As a result, ethanol and water readily mix with each other.

Figure 24.8
Glass is frosted by dissolving its outer surface in hydrofluoric acid.

water by noting that the only electrical attractions that occur between oxygen molecules and water molecules are relatively weak dipole–induced dipole attractions. More important, however, is the fact that the stronger attractions of water molecules for one another—through the hydrogen bonds the water molecules form with one another—effectively exclude oxygen molecules from intermingling.

A material that does not dissolve in a solvent to any appreciable extent is said to be **insoluble** in that solvent. There are many substances we consider to be insoluble in water, including sand and glass. Just because a material is not soluble in one solvent, however, does not mean it won't dissolve in another. Sand and glass, for example, are soluble in hydrofluoric acid, HF, which is used to give glass the decorative frosted look shown in Figure 24.8. Also, although Styrofoam® is insoluble in water, it is soluble in acetone, a solvent used in fingernail polish remover. Pour a little acetone into a Styrofoam cup, and the acetone soon dissolves the Styrofoam, as you can see in Figure 24.9.

Concept Check ✓

Why isn't sucrose infinitely soluble in water?

Check Your Answer The attraction between two sucrose molecules is much stronger than the attraction between a sucrose molecule and a water molecule. Because of this, sucrose dissolves in water only so long as the number of water molecules far exceeds the number of sucrose molecules. When there are too few water molecules to dissolve any additional sucrose, the solution is saturated.

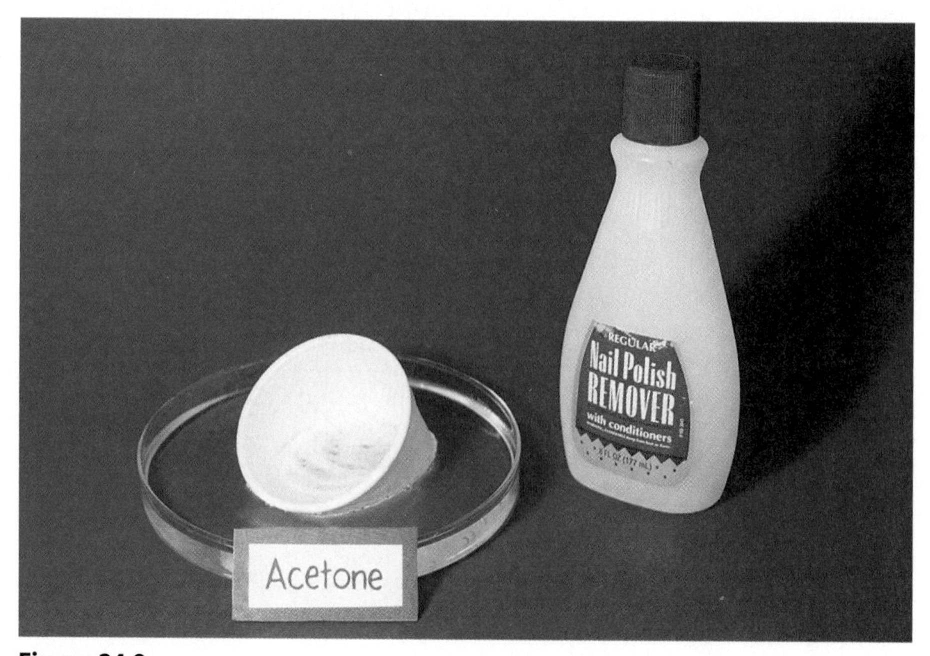

Figure 24.9
Is this cup melting or dissolving?

24.4 Solubility Changes with Temperature and Pressure

Maybe you know from experience that water-soluble solids usually dissolve better in hot water than in cold water. A highly concentrated solution of sucrose in water, for example, can be made by heating the solution almost to the boiling point. This is how syrups and hard candy are made.

Solubility increases with increasing temperature because hot water molecules have greater kinetic energy and therefore are able to collide with the solid solute more vigorously. The vigorous collisions disrupt the electrical particle-to-particle attractions in the solid.

The solubilities of some solid solutes—sucrose, to name just one example—are greatly affected by temperature changes. However, the solubilities of other solid solutes, such as sodium chloride, are only mildly affected. This is shown in Figure 24.10. This difference has to do with a number of factors, including the strength of the chemical bonds in the solute molecules and the way those molecules are packed together.

When a solution saturated at a high temperature cools, some of the solute usually comes out of solution and forms what is called a **precipitate.** When this happens, we say the solute has *precipitated* from the solution. For example, at 100°C the solubility of sodium nitrate, $NaNO_3$, in water is 180 grams per 100 milliliters of water. As we cool this solution, the solubility of $NaNO_3$ decreases as shown in Figure 24.10. This change in solubility causes some of the dissolved $NaNO_3$ to precipitate (come out of solution). At 20°C, the solubility of $NaNO_3$ is only 87 grams per 100 milliliters of water. So if we cool the 100°C solution to 20°C, 93 grams (180 grams − 87 grams) precipitates, as shown in Figure 24.11.

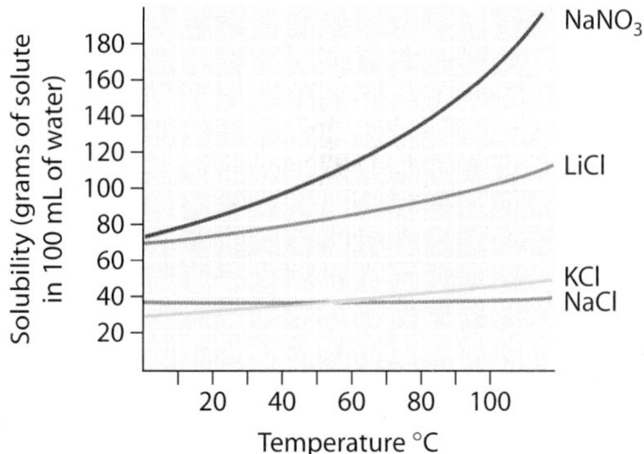

Figure 24.10
The solubility of many water-soluble solids increases with temperature, while the solubility of others is only very slightly affected by temperature.

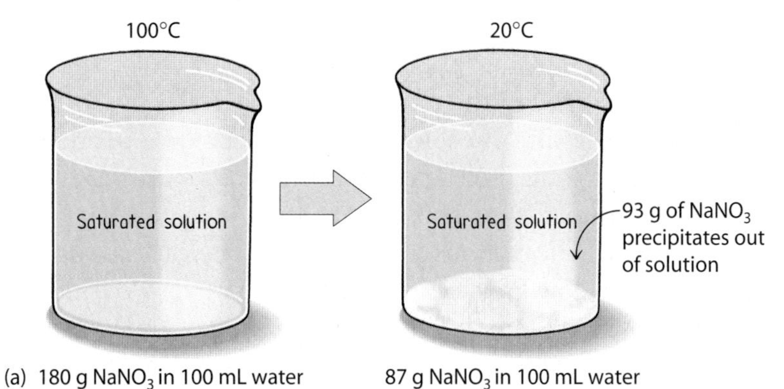

Figure 24.11
The solubility of sodium nitrate is 180 grams per 100 milliliters of water at 100°C but only 87 grams per 100 milliliters at 20°C. Cooling a 100°C saturated solution of $NaNO_3$ to 20°C causes 93 grams of the solute to precipitate.

Table 24.1

Temperature-Dependent Solubility of Oxygen Gas in Water at a Pressure of 1 Atmosphere

Temperature (°C)	O_2 Solubility (g O_2/L H_2O)
0	0.0141
10	0.0109
20	0.0092
25	0.0083
30	0.0077
35	0.0070
40	0.0065

In contrast to the solubilities of most solids, the solubilities of gases in liquids *decrease* with increasing temperature, as Table 24.1 shows. This is because an increase in temperature means the solvent molecules have more kinetic energy. This makes it more difficult for a gaseous solute to stay in solution because the solute molecules are kicked out by the high-energy solvent molecules.

Have you noticed that warm carbonated beverages go flat faster than cold ones? The higher temperature causes the molecules of carbon dioxide gas to leave the liquid solvent at a higher rate.

The solubility of a gas in a liquid also depends on the pressure of the gas immediately above the liquid. In general, a higher gas pressure above the liquid means more of the gas dissolves. A gas at a high pressure has many, many gas particles crammed into a given volume. The "empty" space in an unopened soft drink bottle, for example, is crammed with carbon dioxide molecules in the gaseous phase. With nowhere else to go, many of these molecules dissolve in the liquid, as shown in Figure 24.12. Alternatively, we might say that the great pressure forces the carbon dioxide molecules into solution. When the bottle is opened, the "head" of highly pressurized carbon dioxide gas escapes. Now the gas pressure above the liquid is lower than it was. As a result, the solubility of the carbon dioxide drops and the carbon dioxide molecules once squeezed into the solution begin to escape into the air above the liquid.

The rate at which carbon dioxide molecules leave an opened soft drink is relatively slow. You can increase the rate by pouring in granulated sugar, salt, or sand. The microscopic nooks and crannies on the surface of the grains serve as *nucleation sites* where carbon dioxide bubbles can form rapidly and then escape by buoyant forces. Shaking the beverage also increases the surface area of the liquid-to-gas interface, making it easier for the carbon dioxide to escape from the solution. Once the solution is shaken, the rate at which carbon dioxide

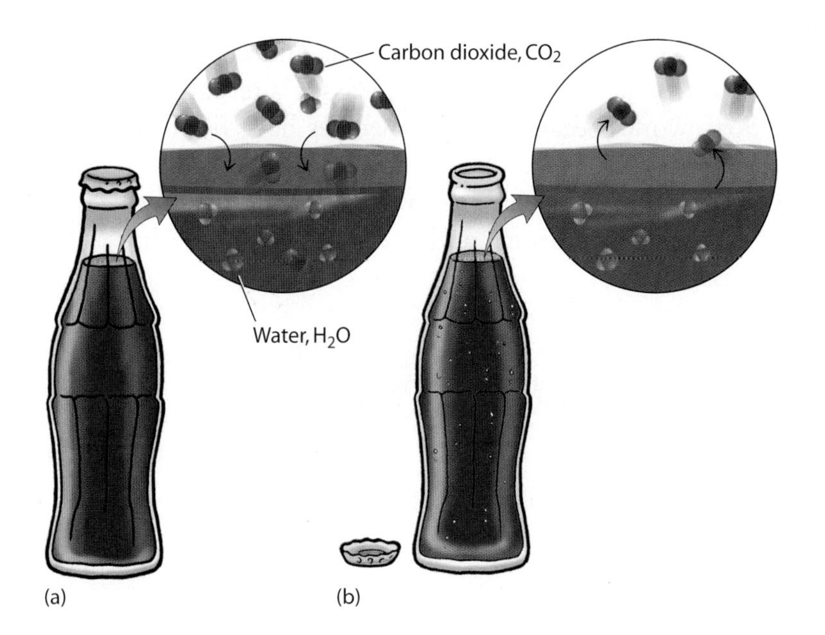

Figure 24.12
(a) The carbon dioxide gas above the liquid in an unopened soft drink bottle consists of many tightly packed carbon dioxide molecules that are forced by pressure into solution. (b) When the bottle is opened, the pressure is released and carbon dioxide molecules originally dissolved in the liquid can escape into the air.

escapes becomes so great that the beverage froths over. You also increase the rate at which carbon dioxide escapes when you pour the beverage into your mouth, which abounds in nucleation sites. You can feel the resulting tingly sensation.

Concept Check ✓

You open two cans of soft drinks: one from a warm kitchen shelf, the other from the coldest depths of your refrigerator. Which provides more bubbles in the first gulp you take and why?

Check Your Answer The solubility of carbon dioxide in water decreases with increasing temperature. The warm drink will therefore fizz in your mouth more than the cold one.

24.5 Soap Works by Being Both Polar and Nonpolar

Dirt and grease together make grime. Because grime contains many nonpolar components, grime is difficult to remove from hands or clothing using just water. To remove most grime, we can use a nonpolar solvent such as turpentine or trichloroethane. Such solvents dissolve grime because of strong induced dipole attractions. Turpentine, also known as a paint thinner, is good for removing the grime left on hands after such activities as changing a car's motor oil. Trichloroethane is the solvent used to "dry clean" clothes. In dry cleaning, dirty clothes are churned in a container full of this nonpolar solvent, which removes the toughest nonpolar stains without the use of water.

Rather than washing our dirty hands and clothes with nonpolar solvents, however, we have a more pleasant alternative—soap and water. Soap works because soap molecules have both nonpolar and polar properties.

A typical soap molecule has two parts: a long *nonpolar tail* of carbon and hydrogen atoms and a *polar head* containing at least one ionic bond:

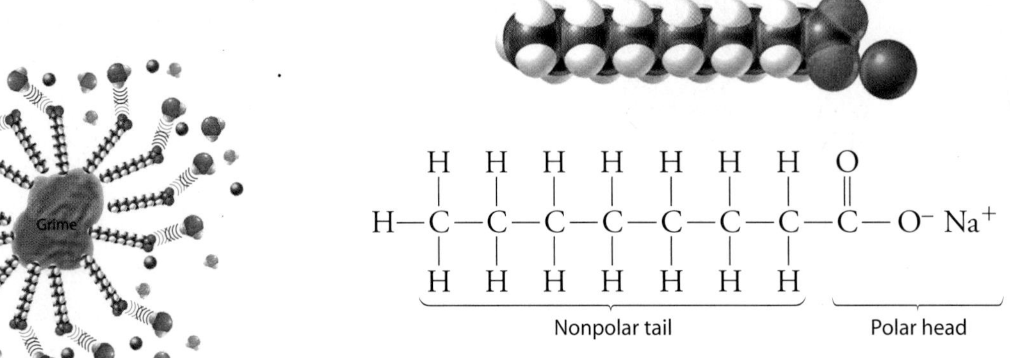

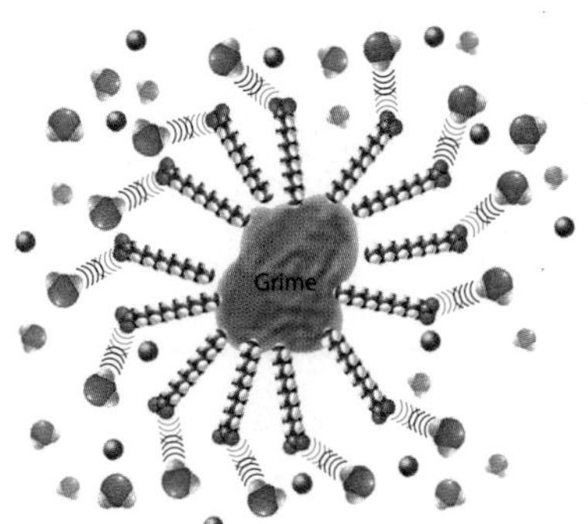

Figure 24.13
Nonpolar grime attracts and is surrounded by the nonpolar tails of soap molecules. The polar heads of the soap molecules are attracted by ion–dipole attractions to water molecules, which carry the soap–grime combination away.

Because most of a soap molecule is nonpolar, it attracts nonpolar grime molecules, as Figure 24.13 illustrates. In fact, grime quickly finds itself surrounded in three dimensions by the nonpolar tails of soap molecules. This attraction is usually enough to lift the grime away from the surface being cleaned. With the nonpolar tails facing inward toward the grime, the polar heads are all directed outward, where they are attracted to water molecules by relatively strong ion-dipole attractions. If the water is flowing, the whole conglomeration of grime and soap molecules flows with it, away from your hands or clothes and down the drain.

For the past several centuries, soaps have been prepared by treating animal fats with sodium hydroxide, NaOH, also known as caustic lye. In this reaction, which is still used today, each fat molecule is broken down into three *fatty acid* soap molecules and one glycerol molecule:

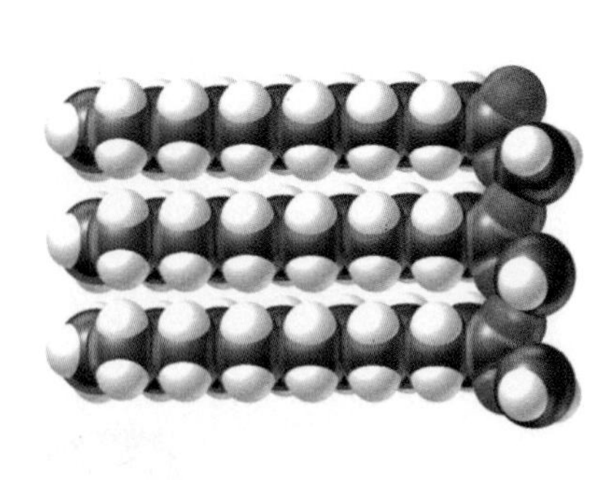

Treat with NaOH →

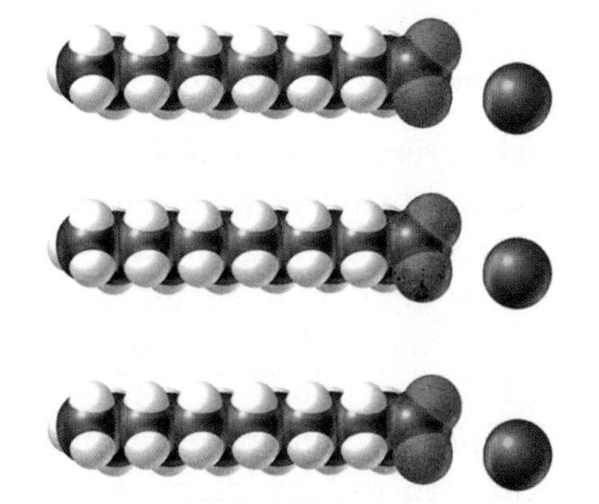

Fat molecule

Three fatty acid soap molecules

Glycerol molecule

Detergents Are Synthetic Soaps

In the 1940s, chemists began developing a class of synthetic soaps, known as *detergents.* Detergents offer several advantages over soaps, such as stronger grease penetration and lower price.

The chemical structure of detergent molecules is similar to that of soap molecules in that both possess a polar head attached to a nonpolar tail. The polar head in a detergent molecule, however, typically consists of either a sulfate, $—OSO_3^-$, or a sulfonate, $—SO_3^-$, group. The nonpolar tail can have an assortment of structures.

One of the most common sulfate detergents is sodium lauryl sulfate, a main ingredient of many toothpastes. A common sulfonate detergent is sodium dodecyl benzenesulfonate, also known as a linear alkylsulfonate, or LAS. You'll often find this compound in dishwashing liquids. Both these detergents are biodegradable, which means microorganisms can break down the molecules once they are released into the environment.

$$CH_3CH_2CH_2CH_2CH_2CH_2CH_2CH_2CH_2CH_2CH_2CH_2—O—SO_2—O^-\ Na^+$$

Sodium lauryl sulfate

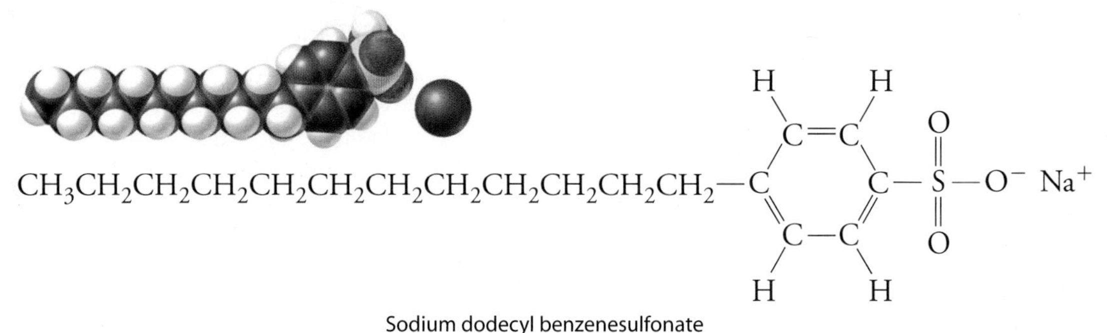

$$CH_3CH_2CH_2CH_2CH_2CH_2CH_2CH_2CH_2CH_2CH_2CH_2—C_6H_4—SO_2—O^-\ Na^+$$

Sodium dodecyl benzenesulfonate

Concept Check ✓

What type of molecular attraction makes it most possible for water to rinse soap off of your hands?

Check Your Answer Three possible molecular attractions discussed in this chapter include: ion–dipole, dipole–dipole, and dipole–induced dipole. If the answer isn't immediately apparent to you, why not back up and re-read Section 24.1. It's there that you find the discussion of the correct answer, which is ion-dipole attractions.

Hands-On Exploration: Crystal Crazy

If a hot saturated solution is allowed to cool slowly and without disturbance, the solute may stay in solution. The result is a *supersaturated* solution. Supersaturated solutions of sucrose (table sugar) are fairly easy to make. From this solution it is easy to grow large crystals of sugar.

What You Need

Cooking pot, stove, water, table sugar, butter knife or chopstick, string, weight (a nut or bolt works well).

Safety

Wear safety glasses to protect eyes from any hot liquid that may splatter. Never perform experiments on your own. Some times the best safety measure is the assistance you receive from others. Involve classmates, siblings, and especially parents so that they too can enjoy the experiments.

Procedure

1. Fill the pot no more than 1 inch deep with water and heat the water to boiling.
2. Lower the heat to medium—low. Slowly pour in sugar while carefully stirring to avoid splattering. Because sugar is very soluble in hot water, be prepared to add a volume of sugar equal to or greater than the volume of water you began with. Continue to add sugar until no more will dissolve even with persistent stirring.
3. Allow the solution to come back to a boil while stirring carefully. This should help dissolve any excess sugar added in Step 2. Do not set the burner on *High* because doing so may make the sugar solution froth up and spill out of the pot. If sugar still doesn't fully dissolve after the solution is brought to a slow boil, add more water 1 teaspoon at a time. If the sugar dissolves after being brought to a slow boil, add more sugar 1 tablespoon at a time. Ideally, you want a boiling-hot sugar solution that is just below saturation. This may be difficult for you to assess if you do not have prior experience.

4. Remove the clear (no undissolved sugar) boiling sugar solution from the heat. Tie some string to the weight and lower the weight into the hot solution. Support the string with a butter knife or chopstick set across the rim of the pot so that the weight does not touch the bottom.
5. Leave the mixture undisturbed for about a week, but check it periodically. You will see large sugar crystals, also known as rock candy, form on the string and also along the sides of the pot. The longer you wait, the larger the crystals.

Interesting crystals can also be made from supersaturated solutions of Epsom salts ($MgSO_4 \cdot 7\ H_2O$) and alum ($KAl(SO_4)_2 \cdot 12\ H_2O$), which is used for pickling and is available in the spice sections of some grocery stores. Crystal shape directly relates to how the ions or molecules of a substance pack together. In fact, substances are often characterized by the shape of the crystals they form. *Crystallography* is the study of mineral crystals and their shapes and structure.

Chapter Review

Key Terms and Matching Definitions

_____ hydrogen bond
_____ induced dipole
_____ insoluble
_____ precipitate
_____ solubility
_____ soluble

1. A strong dipole–dipole attraction between a slightly positive hydrogen atom on one molecule and a pair of nonbonding electrons on another molecule.
2. A dipole temporarily created in an otherwise nonpolar molecule, induced by a neighboring charge.
3. The ability of a solute to dissolve in a given solvent.
4. Capable of dissolving to an appreciable extent in a given solvent.
5. Not capable of dissolving to any appreciable extent in a given solvent.
6. A solute that has come out of solution.

Review Questions

Submicroscopic Particles Electrically Attract One Another

1. What is the primary difference between a chemical bond and an attraction between two molecules?
2. Which is stronger, the ion–dipole attraction or the induced dipole–induced dipole attraction?
3. Why are water molecules attracted to sodium chloride?
4. How are ion–dipole attractions able to break apart ionic bonds, which are relatively strong?
5. Are electrons distributed evenly or unevenly in a polar molecule?
6. What is a hydrogen bond?
7. How are oxygen molecules attracted to water molecules?
8. Are induced dipoles permanent?

Solubility Is a Measure of How Well a Solute Dissolves

9. Why does oxygen have such a low solubility in water?
10. By what means are ethanol and water molecules attracted to each other?
11. What does it mean to say that two materials are infinitely soluble in each other?
12. What kind of electrical attraction is responsible for oxygen dissolving in water?

Solubility Changes with Temperature and Pressure

13. What effect does temperature have on the solubility of a solid solute in a liquid solvent?
14. What effect does temperature have on the solubility of a gas solute in a liquid solvent?
15. What is the relationship between a precipitate and a solute?
16. Why does the solubility of a gas solute in a liquid solvent decrease with increasing temperature?

Soap Works by Being Both Polar and Nonpolar

17. Which portion of a soap molecule is nonpolar?
18. Water and soap are attracted to each other by what type of electrical attraction?
19. How many soap molecules can be made from a single fat molecule?
20. What is the difference between a soap and a detergent?

Exercises

1. Why are ion–dipole attractions stronger than dipole–dipole attractions?

2. Explain why, for these three substances, the solubility in 20°C water goes down as the molecules get larger but the boiling point goes up:

Substance	**Boiling point/ Solubility**
$CH_3{-}O{-}H$	65°C infinite
$CH_3CH_2CH_2CH_2{-}O{-}H$	117°C 8 g/100 mL
$CH_3CH_2CH_2CH_2CH_2{-}O{-}H$	138°C 2.3 g/100 mL

3. The boiling point of 1,4-butanediol is 230°C. Would you expect this compound to be soluble or insoluble in room-temperature water? Explain.

4. Why are noble gases infinitely soluble in other noble gases?

5. Which solute in Figure 24.10 has a solubility in water that changes the least with increasing temperature?

6. At 10°C, which is more concentrated: a saturated solution of sodium nitrate, $NaNO_3$, or a saturated solution of sodium chloride, NaCl? (See Figure 24.10.)

7. Suggest why salt is insoluble in gasoline. Consider the electrical attractions.

8. Which would you expect to have a higher melting point: sodium chloride, NaCl, or aluminum oxide, Al_2O_3? Why?

9. Hydrogen chloride, HCl, is a gas at room temperature. Would you expect this material to be very soluble or not very soluble in water?

10. Would you expect to find more dissolved oxygen in ocean water around the North Pole or tropical ocean water close to the equator? Why?

11. What is the boiling point of a single water molecule? Why does this question not make sense?

12. Account for the observation that ethanol, C_2H_5OH, dissolves readily in water but dimethyl ether, CH_3OCH_3, which has the same number and kinds of atoms, does not.

```
  H H H H H H H H H H H H H H H H
  | | | | | | | | | | | | | | | |
H-C-C-C-C-C-C-C-C-C-C-C-C-C-C-C-C-H
  | | | | | | | | | | | | | | | |
  H H H H H H H H H H H H H H H H
```

Structure A

```
  H H H H H H H H
  | | | | | | | |
H-C-C-C-C-C-C-C-C-H
  | | | | | | | |
  H H H H H H H H
```

Structure B

13. Why are the melting points of most ionic compounds far higher than the melting points of most covalent compounds?

14. How necessary is soap for removing salt from your hands? Why?

15. When you set a pot of tap water on the stove to boil, you'll often see bubbles start to form well before boiling temperature is reached. Explain this observation.

16. Fish don't live very long in water that has been boiled and brought back to room temperature. Why?

Suggested Web Sites

http://www.med.umich.edu/liquid/Research.html

Use *perfluorocarbon* as a Web-search keyword. A perfluorocarbon is an unusual solvent in that it readily dissolves gases such as oxygen and carbon dioxide. The site listed here is that of the Liquid Ventilation Program at the University of Michigan. Scroll to the bottom of the home page for a list of useful links.

http://www.sugar.org/scoop/refine.html

This page, sponsored by the Sugar Association, takes you on a tour that follows sugar from cane field to table.

Chapter 25: Acids and Bases

What force shaped the huge caverns of Carlsbad Caverns National Park? Was it earthquakes? Volcanic eruptions? No to both. Would you believe it was acid rain? Where does acid rain come from and why is it a perfectly natural phenomenon? What happens to the oceans when rained upon by acid rain? Is the pH changed? But wait . . . What is pH anyway? How does pH measure the level of acidity of a solution? Why are all acidic foods sour? How are ashes used to make soap? If an otherwise inviting swimming pool has a pH of 10, would this be good or bad news? Let's learn about the acids and bases that are found throughout our environment.

25.1 Acids Donate Protons, Bases Accept Them

The term *acid* comes from the Latin *acidus,* which means "sour." The sour taste of vinegar and citrus fruits is due to the presence of acids. Food is digested in the stomach with the help of acids, and acids are also essential in the chemical industry. Today, for instance, more than 85 billion pounds of sulfuric acid is produced annually in the United States, making this the number-one manufactured chemical. Sulfuric acid is used in fertilizers, detergents, paint dyes, plastics, pharmaceuticals, storage batteries, iron, and steel. It is so important in the manufacturing of goods that its production is considered a standard measure of a nation's industrial strength. Figure 25.1 shows only a very few of the acids we commonly encounter.

Bases are on the opposite end of the pH scale from acids. Bases are characterized by their bitter taste and slippery feel. Interestingly, bases themselves are not slippery. Rather, they cause skin oils to transform into slippery solutions of soap. Most commercial preparations for unclogging drains are composed of sodium hydroxide, NaOH (also known as *lye*). Sodium hydroxide is extremely basic and hazardous when concentrated. Bases are also heavily used in industry. Each year in the United States about 25 billion pounds of sodium hydroxide are manufactured for industrial use. This vast amount of sodium hydroxide is mainly used in the production of various chemicals and in the pulp

(a) (b) (c) (d)

Figure 25.1
Examples of acids. (a) Vinegar contains acetic acid, $C_2H_4O_2$, and can be used to preserve foods. (b) All carbonated beverages contain carbonic acid, H_2CO_3, while many also contain phosphoric acid, H_3PO_4. (c) Many toilet bowl cleaners are formulated with hydrochloric acid, HCl. (d) Citrus fruits contain many types of acids, including ascorbic acid, $C_6H_8O_8$, which is vitamin C.

and paper industry. As you learned in Section 17.5, solutions containing bases are often called *alkaline,* a term derived from the Arabic word for ashes (*al-qali*). Ashes are slippery when wet because of the presence of the base potassium carbonate, K_2CO_3. Figure 25.2 shows some common bases with which you are probably familiar.

Acids and bases may be defined in several ways. Simply said, an **acid** is any chemical that donates a hydrogen ion, H+, to another substance. A **base** is any chemical that accepts a hydrogen ion into its own structure. A hydrogen ion is simply a bare proton, or equivalently, a hydrogen atom with its electron removed. Therefore we can also say that an acid is a chemical that donates a proton and a base is a chemical that accepts a proton.

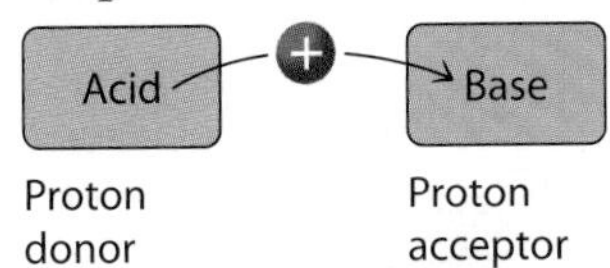

Consider what happens when hydrogen chloride is mixed into water:

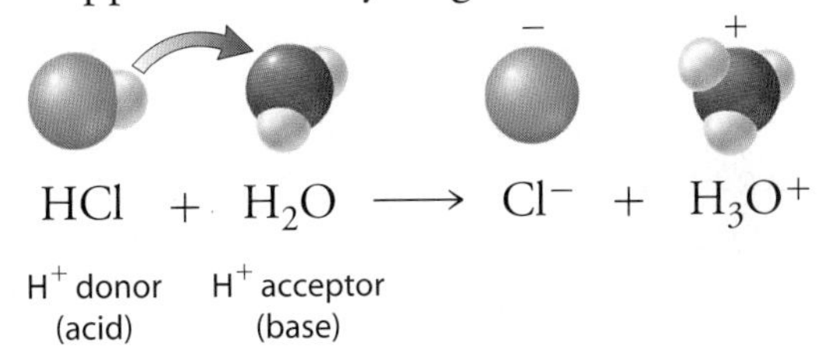

Figure 25.2
Examples of bases. (a) Reactions involving sodium bicarbonate, $NaHCO_3$, make baked goods rise. (b) Ashes contain potassium carbonate, K_2CO_3. (c) Soap is made by reacting bases with animal or vegetable oils. The soap itself, then, is slightly alkaline. (d) Powerful bases, such as sodium hydroxide, NaOH, are used in drain cleaners.

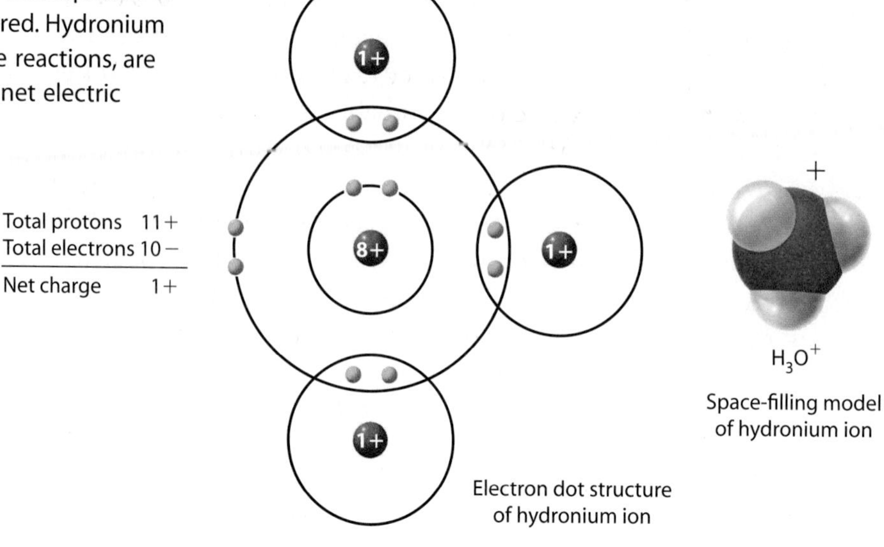

Figure 25.3
The hydronium ion's positive charge is a consequence of the extra proton this molecule has acquired. Hydronium ions, which play a part in many acid-base reactions, are *polyatomic ions*—molecules that carry a net electric charge.

Hydrogen chloride donates a hydrogen ion to one of the nonbonding electron pairs of a water molecule. The result is a third hydrogen bonded to the oxygen. In this case, hydrogen chloride behaves as an acid (proton donor) and water behaves as a base (proton acceptor). The products of this reaction are one chloride ion and one **hydronium ion,** H_3O+. As Figure 25.3 shows, a hydronium ion is a water molecule with an extra proton.

When added to water, ammonia behaves as a base by accepting a hydrogen ion from water, which, in this case, behaves as an acid:

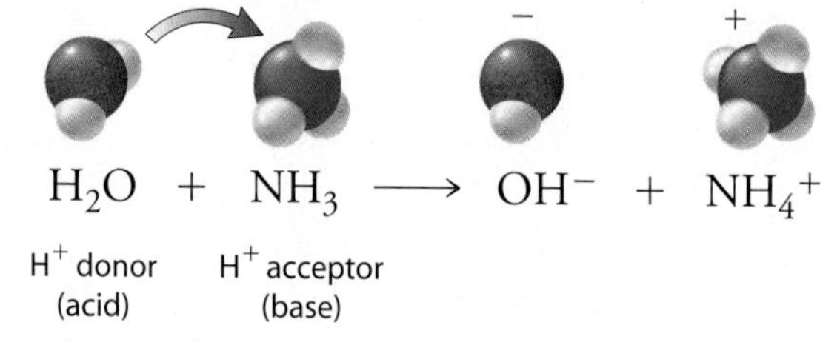

This reaction results in the formation of an ammonium ion and a **hydroxide ion.** As shown in Figure 25.4, a hydroxide ion is a water molecule without the nucleus of one of the hydrogen atoms.

Total protons 9+
Total electrons 10−
Net charge 1−

8+ 1+

−

OH^-

Electron dot structure of hydroxide ion

Space-filling model of hydroxide ion

Figure 25.4
Hydroxide ions have a net negative charge, which is a consequence of having lost a proton. Like hydronium ions, they play a part in many acid-base reactions.

Our definition of acids and bases recognizes the *behavior* of acids and bases. We say, for example, that hydrogen chloride *behaves* as an acid when mixed with water, which *behaves* as a base. Similarly, ammonia *behaves* as a base when mixed with water, which under this circumstance *behaves* as an acid. Because acid-base is seen as a behavior, there is really no contradiction when a chemical like water behaves as a base in one instance but as an acid in another instance. By analogy, consider yourself. You are who you are, but your behavior changes depending on whom you are with. Likewise, it is a chemical property of water to behave as a base (accept H+) when mixed with hydrogen chloride and as an acid (donate H+) when mixed with ammonia.

The products of an acid-base reaction can also behave as acids or bases. An ammonium ion, for example, may donate a hydrogen ion back to a hydroxide ion to re-form ammonia and water:

$$H_2O + NH_3 \longleftarrow OH^- + NH_4^+$$

OH^-: H^+ acceptor (base); NH_4^+: H^+ donor (acid)

Forward and reverse acid-base reactions proceed simultaneously. We represent them as occurring at the same time by using two oppositely facing arrows:

$$H_2O + NH_3 \rightleftarrows OH^- + NH_4^+$$

H_2O: H^+ donor (acid); NH_3: H^+ acceptor (base); OH^-: H^+ acceptor (base); NH_4^+: H^+ donor (acid)

How we behave depends on who we're with. Likewise for chemicals.

When the equation is viewed from left to right, the ammonia behaves as a base because it accepts a hydrogen ion from the water. The water therefore acts as an acid. Viewed in the reverse direction, the equation shows that the ammonium ion behaves as an acid because it donates a hydrogen ion to the hydroxide ion. The hydroxide ion, by accepting the hydrogen ion, behaves as a base.

Concept Check ✓

Identify the acid or base behavior of each participant in the reaction.

$$H_2PO_4^- + H_3O^+ \rightleftarrows H_3PO_4 + H_2O$$

Check Your Answer Moving left to right, $H_2PO_4^-$ gains a hydrogen ion to become H_3PO_4. In accepting the hydrogen ion, $H_2PO_4^-$ is behaving as a base. It gets the hydrogen ion from the H_3O^+, which is behaving as an acid. From right to left, H_3PO_4 becomes $H_2PO_4^-$ and is thus behaving as an acid. The hydrogen ion recipient is the H_2O, which is behaving as a base as it transforms to H_3O^+.

Figure 25.5
"Salt-free" table-salt substitutes contain potassium chloride in place of sodium chloride. Caution is advised in using these products, however, because excessive quantities of potassium salts can lead to serious illness. Furthermore, sodium ions are a vital component of our diet and should never be totally excluded. For a good balance of these two important ions, you might inquire about commercially available half-and-half mixtures of sodium chloride and potassium chloride such as the one shown here.

A Salt Is the Ionic Product of an Acid–Base Reaction

In everyday language, when people say "salt" they are referring to sodium chloride, NaCl, table salt. In the language of chemistry, however, a **salt** is any ionic compound formed from the reaction between an acid and a base. Hydrogen chloride and sodium hydroxide, for example, react to produce the salt sodium chloride plus water:

HCl	+	NaOH →	NaCl	+	H_2O
Hydrogen chloride (acid)		Sodium hydroxide (base)	Sodium choride (salt)		Water

Similarly, the reaction between hydrogen chloride and potassium hydroxide yields the salt potassium chloride and water:

HCl	+	KOH →	KCl	+	H_2O
Hydrogen chloride (acid)		Potassium hydroxide (base)	Potassium choride (salt)		Water

Potassium chloride is the main ingredient in "salt-free" table salt, as noted in Figure 25.5.

Salts are generally far less corrosive than the acids and bases from which they are formed. A *corrosive* chemical has the power to disintegrate a material or wear away its surface. Hydrogen chloride is a remarkably corrosive acid. This makes it useful for cleaning toilet bowls and etching metal surfaces. Sodium hydroxide is a very corrosive base used for unclogging drains. Mixing hydrogen chloride and sodium hydroxide together in equal portions, however, produces an aqueous solution of sodium chloride—salt water, which is not nearly as destructive as either starting material.

There are as many salts as there are acids and bases. Sodium cyanide, NaCN, is a deadly poison. "Salt peter," which is potassium nitrate, KNO_3, is useful as a fertilizer and in the formulation of gun powder. Calcium chloride, $CaCl_2$, is commonly used to de-ice roads, and sodium fluoride, NaF, prevents tooth decay. The acid-base reactions forming these salts are shown in Table 25.1.

Table 25.1

Acid-base Reactions and The Salts Formed

Acid		Base		Salt		Water
HCN Hyrogen cyanide	+	NaOH Sodium hydroxide	→	NaCN Sodium cyanide	+	H_20
HNO_3 Nitric acid	+	KOH Potassium hydroxide	→	KNO_3 Potassium nitrate	+	H_20
2 HCL Hyrogen chloride	+	$Ca(OH)_2$ Calcium hydroxide	→	$CaCl_2$ Calcium cholide	+	2 H_20
HF Hyrogen flouride	+	NaOH Sodium hydroxide	→	NaF Sodium flouride	+	H_20

The reaction between an acid and a base is called a **neutralization** reaction. As can be seen in the color-coding of the neutralization reactions in Table 25.1, the positive ion of a salt comes from the base and the negative ion comes from the acid. The remaining hydrogen and hydroxide ions join to form water.

Not all neutralization reactions result in the formation of water. In the presence of hydrogen chloride, for example, the drug cocaine behaves as

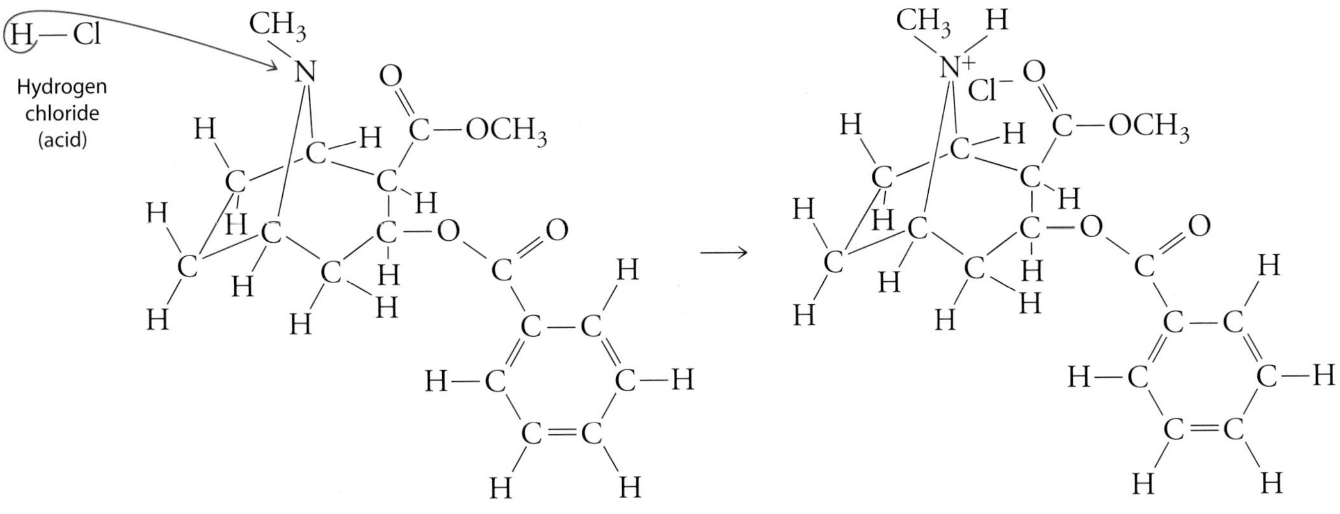

H—Cl

Hydrogen chloride (acid)

Cocaine (base)

→

Cocaine hydrochloride (salt)

Figure 25.6
Hydrogen chloride and cocaine react to form the salt cocaine hydrochloride, which, because of its solubility in water, is readily absorbed into the body through moist membranes.

a base by accepting H^+ from a hydrogen chloride. The negative Cl^- then attaches to form the salt cocaine hydrochloride, shown in Figure 25.6. This salt of cocaine is soluble in water and can be absorbed through the moist membranes of the nasal passages or mouth. The nonsalt form of cocaine, also known as "free-base cocaine" or "crack cocaine," is a nonpolar material that vaporizes easily when heated. Its vapors are inhaled directly into the lungs, resulting in dangerously high concentrations of cocaine in the bloodstream. We shall return to the actions of various drugs in Chapter 29.

Concept Check ✓

Is a neutralization reaction best described as a physical change or a chemical change?

Check Your Answer New chemicals are formed during a neutralization reaction, meaning the reaction is a chemical change.

25.2 Some Acids and Bases Are Stronger Than Others

In general, the stronger an acid, the more readily it donates hydrogen ions. Likewise, the stronger a base, the more readily it accepts hydrogen ions. An example of a strong acid is hydrogen chloride, HCl, and an example of a strong base is sodium hydroxide, NaOH. The corrosiveness of these materials is a result of their strength.

One way to assess the strength of an acid or base is to measure how much of it remains after it has been added to water. If little remains, the acid or base is strong. If a lot remains, the acid or base is weak. To illustrate this concept, consider what happens when the strong acid

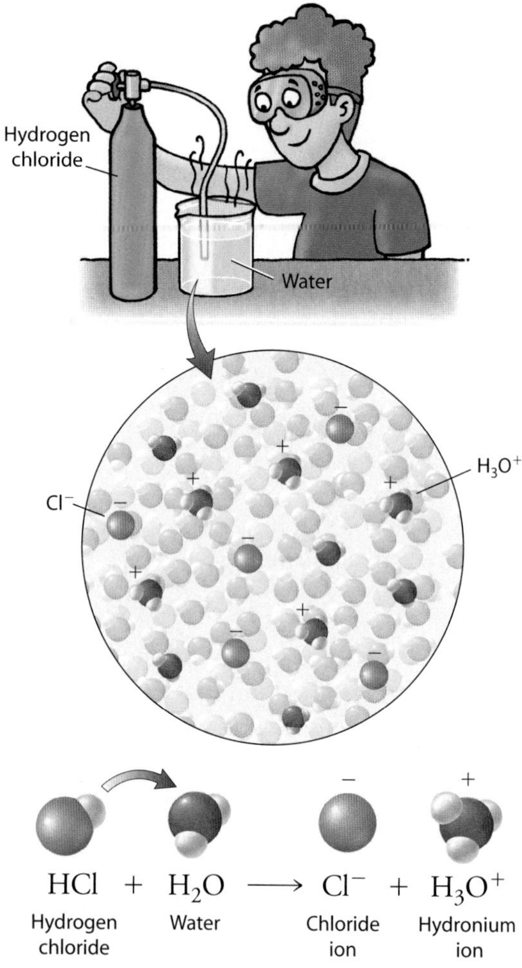

Figure 25.7
Immediately after hydrogen chloride, which is a gaseous substance, is added to water, it reacts with the water to form hydronium ions and chloride ions. That very little HCl remains (none shown here) tells us that HCl is a strong acid.

hydrogen chloride is added to water and what happens when the weak acid acetic acid, $C_2H_4O_2$ (the active ingredient of vinegar), is added to water.

Being an acid, hydrogen chloride donates hydrogen ions to water, forming chloride ions and hydronium ions. Because HCl is such a strong acid, nearly all of it is converted to these ions, as is shown in Figure 25.7.

Because acetic acid is a weak acid, it has much less tendency to donate hydrogen ions to water. When this acid is dissolved in water, only a small portion of the acetic acid molecules are converted to ions. This occurs as the polar O—H bonds are broken (the C—H bonds of acetic acid are unaffected by the water because of their nonpolarity). The majority of acetic acid molecules remain intact in their original nonionized form, as shown in Figure 25.8.

Figures 25.7 and 25.8 show the submicroscopic behavior of strong and weak acids in water. However, molecules and ions are too small to see. How then does a chemist measure the strength of an acid? One way is by measuring a solution's ability to conduct an electric current, as Figure 25.9 illustrates. In pure water there are practically no ions to conduct electricity. When a strong acid is dissolved in water many ions are generated, as indicated in Figure 25.7. The presence of these ions allows for the flow of a large electric current. A weak acid dissolved in water generates only a few ions, as indicated in Figure 25.8. The presence of fewer ions means there can be only a small electric current.

This same trend is seen with strong and weak bases. Strong bases, for example, tend to accept hydrogen ions more readily than weak bases. In solution, a strong base allows the flow of a large electric current and a weak base allows the flow of a small electric current.

Concept Check ✓

According to the aqueous solutions illustrated here, which is the stronger base, NH_3 or NaOH?

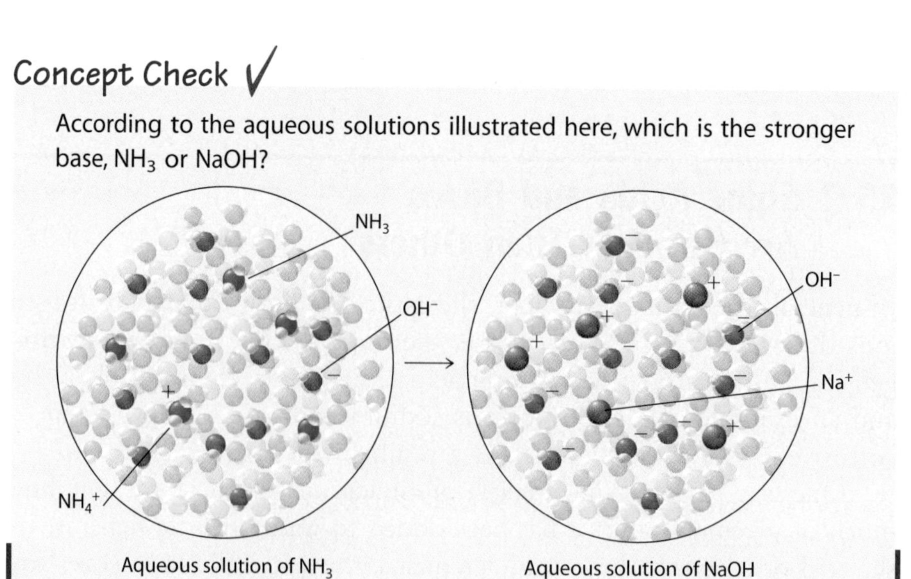

Check Your Answer The solution on the right contains the greater number of ions, meaning sodium hydroxide, NaOH, is the stronger base. Ammonia, NH_3, is the weaker base, indicated by the relatively few ions in the solution on the left.

Just because an acid or base is strong doesn't mean a solution of that acid or base is corrosive. The corrosive action of an acidic solution is caused by the hydronium ions rather than by the acid that generated those hydronium ions. Similarly, the corrosive action of a basic solution results from the hydroxide ions it contains, regardless of the base that generated those hydroxide ions. A *very* dilute solution of a strong acid or a strong base may have little corrosive action because in such solutions there are only a few hydronium or hydroxide ions. (Almost all the molecules of the strong acid or base break up into ions, but, because the solution is dilute, there are only a few acid or base molecules to begin with. As a result, there are only a few hydronium or hydroxide ions.) You shouldn't be too alarmed, therefore, when you discover that some toothpastes are formulated with small amounts of sodium hydroxide, one of the strongest bases known.

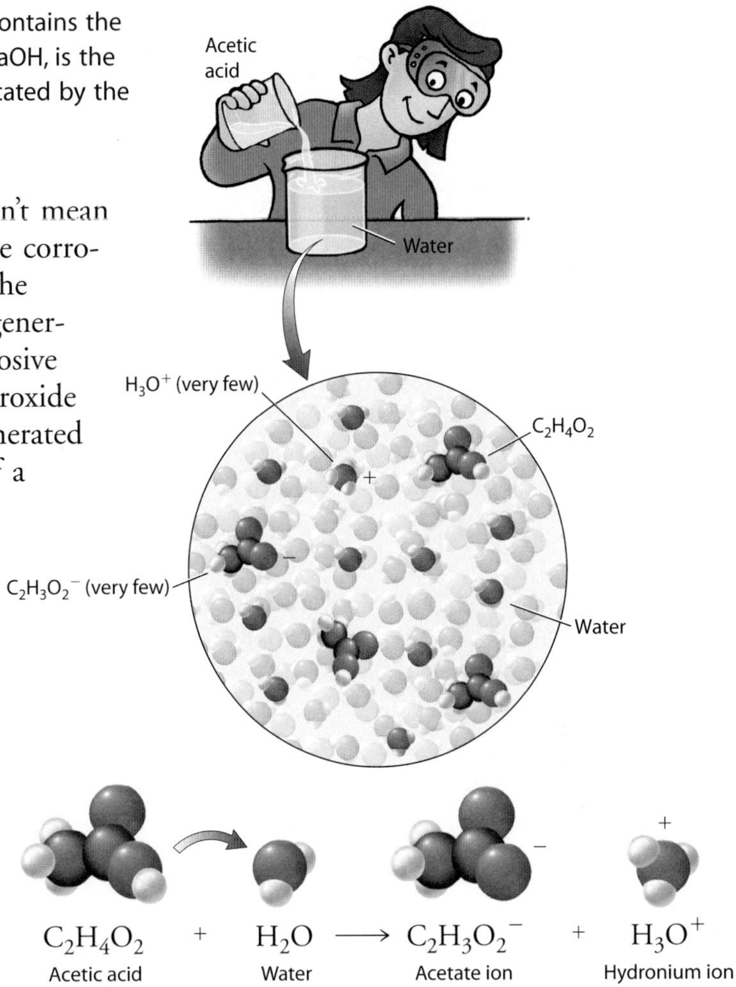

$$C_2H_4O_2 + H_2O \longrightarrow C_2H_3O_2^- + H_3O^+$$

Acetic acid, Water, Acetate ion, Hydronium ion

Figure 25.8
When liquid acetic acid is added to water, only a few acetic acid molecules react with water to form ions. The majority of the acetic acid molecules remain in their nonionized form. This tells us that acetic acid is a weak acid.

(a)

(b)

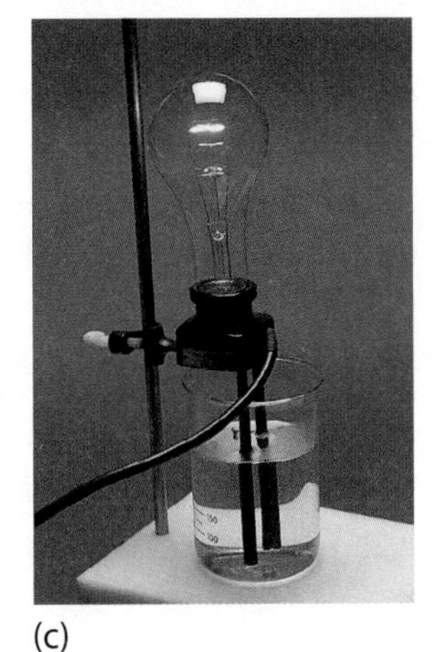
(c)

Figure 25.9
(a) The pure water in this circuit is unable to conduct electricity because it contains practically no ions. The light bulb in the circuit therefore remains unlit. (b) Because HCl is a strong acid, nearly all of its molecules break apart in water, giving a high concentration of ions. The ions are able to conduct an electric current that lights the bulb. (c) Acetic acid, $C_2H_4O_2$, is a weak acid, and in water only a small portion of its molecules break up into ions. Because fewer ions are generated, only a weak current exists and the bulb is dimmer.

On the other hand, a concentrated solution of a weak acid, such as acetic acid, may be just as corrosive as or even more corrosive than a dilute solution of a strong acid, such as hydrogen chloride. Concentrated solutions of acetic acid, for example, are corrosive enough that they are sometimes used to remove warts! The relative strengths of two acids in solution or two bases in solution, therefore, can be compared only when the two solutions have the same concentration.

25.3 Solutions Can Be Acidic, Basic, or Neutral

A substance whose ability to behave as an acid is about the same as its ability to behave as a base is said to be **amphoteric.** Water is a good example. Because it is amphoteric, water has the ability to react with itself. In behaving as an acid, a water molecule donates a hydrogen ion to a neighboring water molecule, which in accepting the hydrogen ion is behaving as a base. This reaction produces a hydroxide ion and a hydronium ion, which react together to re-form the water:

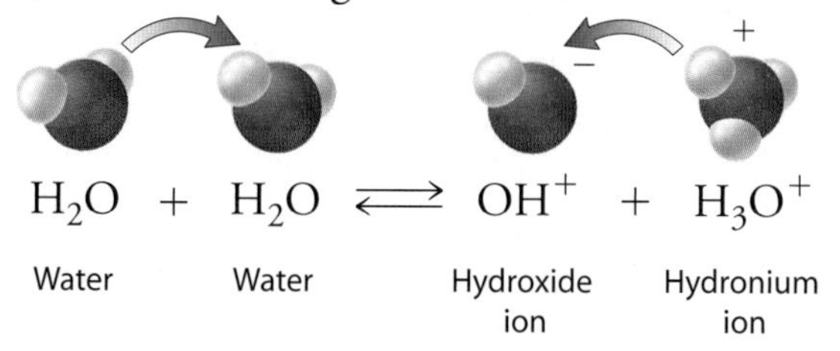

From this reaction we can see that, in order for a water molecule to gain a hydrogen ion, a second water molecule must lose a hydrogen ion. This means that for every one hydronium ion formed, there is also one hydroxide ion formed. In pure water, therefore, the total number of hydronium ions must be the same as the total number of hydroxide ions. Experiments reveal that the concentration of hydronium and hydroxide ions in pure water is extremely low—about 0.0000001 *M* for each, where *M* stands for molarity or moles per liter (Section 22.4). Water by itself, therefore, is a very weak acid as well as a very weak base, as evidenced by the unlit light bulb in Figure 25.9a.

In an **acidic** solution, $[H_3O^+] > [OH^-]$.

In a **basic** solution, $[H_3O^+] < [OH^-]$.

In a **neutral** solution, $[H_3O^+] = [OH^-]$.

Figure 25.10
The relative concentrations of hydronium and hydroxide ions determine whether a solution is acidic, basic, or neutral.

Concept Check ✓

Do water molecules react with one another?

Check Your Answer Yes, but not to any large extent. When they do react, they form hydronium and hydroxide ions. (Note: Make sure you understand this point because it serves as a basis for most of the rest of the chapter.)

An aqueous solution can be described as acidic, basic, or neutral, as Figure 25.10 summarizes. An **acidic solution** is one in which the hydronium ion concentration is higher than the hydroxide ion concentration. An acidic solution is made by adding an acid to water. The effect of this addition is to increase the concentration of hydronium ions. A **basic solution** is one in which the hydroxide ion concentration is higher than the hydronium ion concentration. A basic solution is made by adding a base to water. This addition increases the concentration of hydroxide ions. A **neutral solution** is one in which the hydronium ion concentration equals the hydroxide ion concentration. Pure water is an example of a neutral solution—not because it contains so few hydronium and hydroxide ions but because it contains equal numbers of them. A neutral solution is also obtained when equal quantities of acid and base are combined, which is why acids and bases are said to *neutralize* each other.

Concept Check ✓

How does adding ammonia, NH_3, to water make a basic solution when there are no hydroxide ions in the formula for ammonia?

Check Your Answer Ammonia indirectly increases the hydroxide ion concentration by reacting with water:

$$NH_3 + H_2O \rightarrow NH_4^+ + OH^-$$

This reaction raises the hydroxide ion concentration, which has the effect of lowering the hydronium ion concentration. With the hydroxide ion concentration now higher than the hydronium ion concentration, the solution is basic.

The pH Scale Is Used to Describe Acidity

The *pH scale* is a numeric scale used to express the acidity of a solution. Mathematically, **pH** is equal to the negative logarithm of the hydronium ion concentration:

$$pH = -\log[H_3O^+]$$

A few words about this definition. First, note that the brackets are used to represent concentration. So the H_3O^+ set in brackets as $[H_3O^+]$ can be read "concentration of hydronium ions." Are you also asking, "What does 'log' mean?" The term "log" is short for *logarithm.* A logarithm is a fancy way of finding the power to which 10 is raised to make a certain number. The logarithm of 10^2, for example, is 2 because that is the power to which 10 is raised. If you know that 10^2 is equal to 100, then you'll understand that the logarithm of 100 also is 2. The logarithm of 10^3, which is the same as 1000, is 3, and so forth.

The logarithm of a number can be found on any scientific calculator by typing in the number and pressing the [log] button. What the calculator does is find the power to which 10 is raised to give this number.

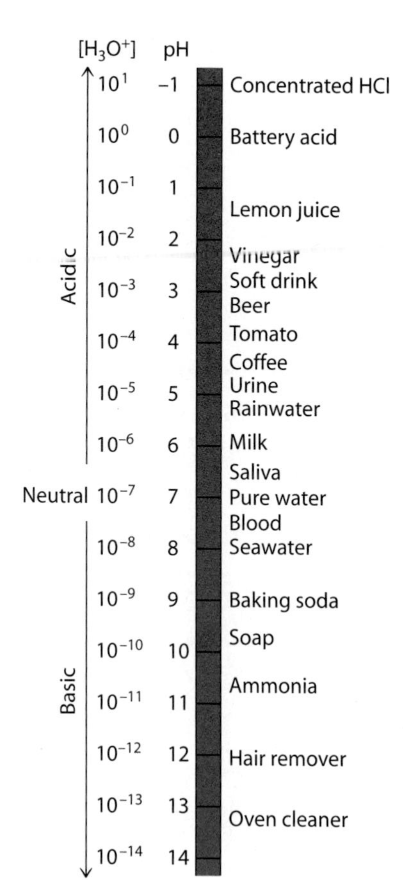

Figure 25.11
The pH values of some common solutions.

(a)

(b)

Figure 25.12
(a) The pH of a solution can be measured electronically using a pH meter. (b) A rough estimate of the pH of a solution can be obtained with litmus paper. The litmus paper is coated with a dye that changes color with pH.

Consider a neutral solution that has a hydronium ion concentration of 10^{-7} *M*. To find the pH of this solution, we first take the logarithm of this value, which is −7. The pH is by definition the *negative* of this value, which means $-(-7) = +7$. Hence, in a neutral solution, where the hydronium ion concentration equals 10^{-7} *M*, the pH is 7.

Acidic solutions have pH values less than 7. For an acidic solution in which the hydronium ion concentration is 10^{-4} *M*, for example, $\text{pH} = -\log(10^{-4}) = 4$. The more acidic a solution is, the greater its hydronium ion concentration and the lower its pH.

Basic solutions have pH values greater than 7. For a basic solution in which the hydronium ion concentration is 10^{-9} *M*, for example, $\text{pH} = -\log(10^{-9}) = 9$. The more basic a solution is, the smaller its hydronium ion concentration and the higher its pH.

Figure 25.11 shows typical pH values of some familiar solutions, and Figure 25.12 shows two common ways of determining pH values.

25.4 Rainwater Is Acidic and Ocean Water Is Basic

Rainwater is naturally acidic. One source of this acidity is carbon dioxide, the same gas that gives fizz to soda drinks. There are 670 billion tons of CO_2 in the atmosphere, most of it from such natural sources as volcanoes and decaying organic matter but a growing amount from human activities.

Hands-On Exploration: Rainbow Cabbage

The pH of a solution can be approximated with a *pH indicator.* A pH indicator is any chemical whose color changes with pH. Many pH indicators are found in plants; the pigment of red cabbage is a good example. This pigment is red at low pH values (1 to 5), light purple around neutral pH values (6 to 7), light green at moderately alkaline pH values (8 to 11), and dark green at very alkaline pH values (12 to 14).

Safety Note

Wear safety glasses. Test only solutions provided for you by your teacher or parent. Do not use any bleach products as they can be dangerous.

What You Need

Head of red cabbage; small pot; water; four colorless plastic cups or drinking glasses; toilet-bowl cleaner; vinegar; baking soda; ammonia cleanser

Procedure

1. Shred about a quarter of the head of red cabbage and boil the shredded cabbage in 2 cups of water for about 5 minutes. Strain and collect the broth, which contains the pH-indicating pigment.
2. Pour one-fourth of the broth into each cup. (If the cups are plastic, either allow the broth to cool before pouring or dilute with cold water.)
3. Add less than half a teaspoon of toilet-bowl cleaner to the first cup, about a teaspoon of vinegar to the second cup, less than half a teaspoon of baking soda to the third, and a teaspoon of ammonia solution to the fourth.
4. Use the different colors to estimate the pH of each solution.
5. Mix some of the acidic and basic solutions together and note the rapid change in pH (indicated by the change in color).

Water in the atmosphere reacts with carbon dioxide to form *carbonic acid:*

$$\underset{\text{Carbon dioxide}}{CO_2\ (g)} + \underset{\text{Water}}{H_2O\ (l)} \rightarrow \underset{\text{Carbonic acid}}{H_2CO_3\ (aq)}$$

Carbonic acid, as its name implies, behaves as an acid and lowers the pH of water. The CO_2 in the atmosphere brings the pH of rainwater to about 5.6—noticeably below the neutral pH value of 7. Because of local fluctuations, the normal pH of rainwater varies between 5 and 7. This natural acidity of rainwater may accelerate the erosion of land and, under the right circumstances, can lead to the formation of underground caves, as was discussed in this chapter's introduction.

By convention, *acid rain* is a term used for rain having a pH lower than 5. Acid rain is created when airborne pollutants such as

(a)

(b)

Figure 25.13
The two photographs in (a) show the same obelisk before and after the effects of acid rain. (b) Many forests downwind from heavily industrialized areas, such as in the northeastern United States and in Europe, have been noticeably hard hit by acid rain.

sulfur dioxide are absorbed by atmospheric moisture. Sulfur dioxide is readily converted to sulfur trioxide, which reacts with water to form *sulfuric acid:*

$$2\ SO_2\ (g) + O_2\ (g) \rightarrow 2\ SO_3\ (g)$$

Sulfur Dioxide — Oxygen — Sulfur trioxide

$$SO_3\ (g) + H_2O\ (l) \rightarrow H_2SO_4\ (aq)$$

Sulfur Trioxide — Water — Sulfuric acid

Interestingly, it was sulfuric acid-laced groundwater that helped create the great chambers of Carlsbad Caverns, as shown in the opening photograph of this chapter. The sulfuric acid was generated from sulfur dioxide (and hydrogen sulfide) from subterranean fossil-fuel deposits. When we burn these fossil fuels, the reactants that produce sulfuric acid are emitted into the atmosphere. Each year, for example, about 20 million tons of SO_2 is released into the atmosphere by the combustion of sulfur-containing coal and oil. Sulfuric acid is much stronger than carbonic acid, and as a result rain laced with sulfuric acid eventually corrodes metal, paint, and other exposed substances. Each year the damage costs billions of dollars. The cost to the environment is also high. Many rivers and lakes receiving acid rain become less capable of sustaining life. Much vegetation that receives acid rain doesn't survive. This is particularly evident in heavily industrialized regions.

Concept Check ✓

When sulfuric acid, H_2SO_4, is added to water, what makes the resulting aqueous solution corrosive?

Check Your Answer Because H_2SO_4 is a strong acid, it readily forms hydronium ions when dissolved in water. Hydronium ions are responsible for the corrosive action.

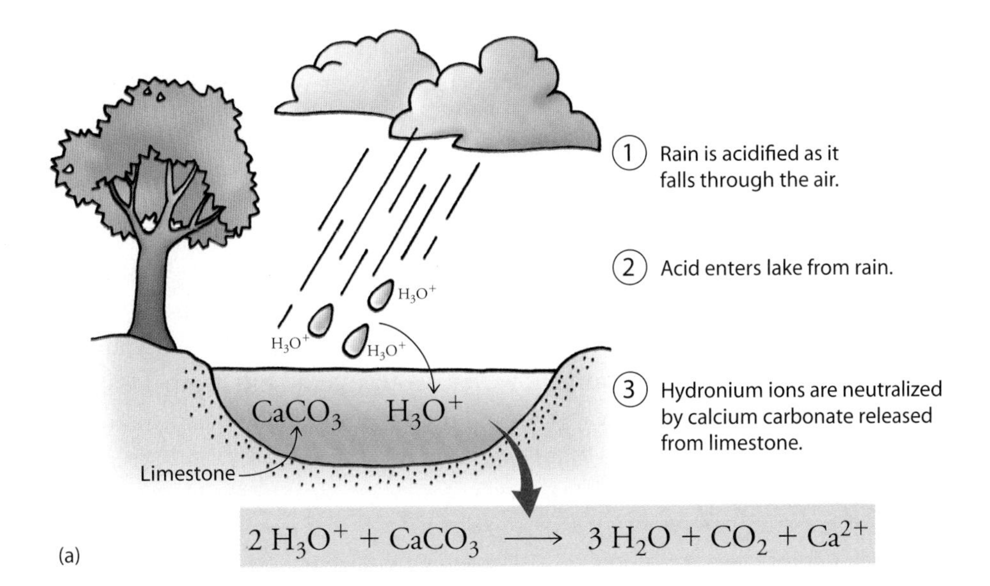

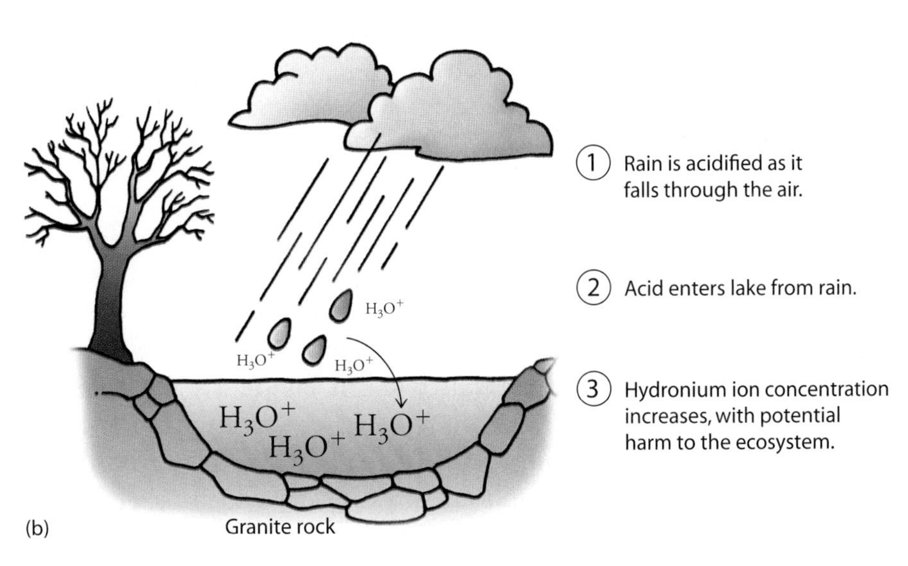

Figure 25.14
(a) The damaging effects of acid rain do not appear in bodies of fresh water lined with calcium carbonate, which neutralizes any acidity. (b) Lakes and rivers lined with inert materials are not protected.

Figure 25.15
Most chalks are made from calcium carbonate, which is the same chemical found in limestone. The addition of even a weak acid, such as the acetic acid of vinegar, produces hydronium ions that react with the calcium carbonate to form several products. The most notable is carbon dioxide, which rapidly bubbles out of solution. Try this for yourself! If the bubbling is not as vigorous as shown here, then the chalk is made of other mineral components.

The environmental impact of acid rain depends on local geology, as Figure 25.14 illustrates. In certain regions, such as the midwestern United States, the ground contains significant quantities of the alkaline compound calcium carbonate (limestone). The calcium carbonate was deposited when these lands were submerged under oceans 200 million years ago. Acid rain pouring into these regions is often neutralized by the calcium carbonate before any damage is done. (Figure 25.15 shows calcium carbonate neutralizing an acid.) In the northeastern United States and many other regions, however, the ground contains very little calcium carbonate and is composed primarily of less chemically reactive materials, such as granite. In these regions, the effect of acid rain on lakes and rivers accumulates.

One demonstrated solution to this problem is to raise the pH of acidified lakes and rivers by adding calcium carbonate—a process known as *liming*. The cost of transporting the calcium carbonate coupled with the need to monitor treated water systems closely limits liming to only a small fraction of the vast number of water systems already affected. Furthermore, as acid rain continues to pour into these regions, the need to lime also continues.

A longer-term solution to acid rain is to prevent most of the generated sulfur dioxide and other pollutants from entering the atmosphere in the first place. Toward this end, smokestacks have been designed or retrofitted to minimize the quantities of pollutants released. Though costly, the positive effects of these adjustments have been demonstrated. An ultimate long-term solution, however, would be a shift from fossil fuels to cleaner energy sources, such as nuclear and solar energy.

Concept Check ✓

What kind of lakes are protected against the negative effects of acid rain?

Check Your Answer Lakes that have a floor consisting of basic minerals, such as limestone, are more resistant to acid rain because the chemicals of the limestone (mostly calcium carbonate, $CaCO_3$) neutralize any incoming acid.

It should come as no surprise that the amount of carbon dioxide put into the atmosphere by human activities is growing. What is surprising, however, is that studies indicate that the atmospheric concentration of CO_2 is not increasing proportionately. A likely explanation has to do with the oceans and is illustrated in Figure 25.16. When atmospheric CO_2 dissolves in any body of water—a raindrop, a lake, or the ocean—it forms carbonic acid. In fresh water, this carbonic acid transforms back to water and carbon dioxide, which is released back into the atmosphere. Carbonic acid in the ocean, however, is quickly neutralized by dissolved alkaline substances such as calcium carbonate (the ocean is alkaline, pH $>$ 8.2). The products of this

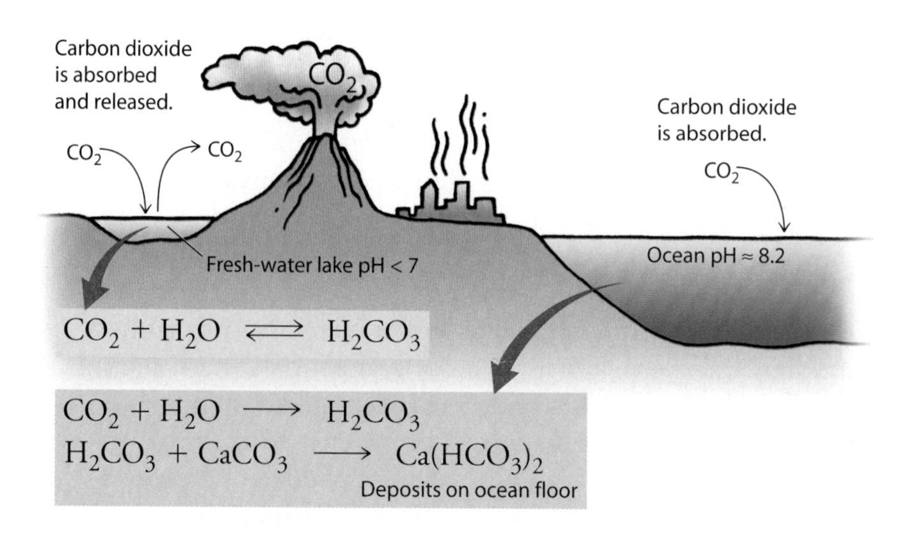

Figure 25.16
Carbon dioxide forms carbonic acid upon entering any body of water. In fresh water, this reaction is reversible, and the carbon dioxide is released back into the atmosphere. In the alkaline ocean, the carbonic acid is neutralized to compounds such as calcium bicarbonate, $Ca(HCO_3)_2$, which precipitate to the ocean floor. As a result, most of the atmospheric carbon dioxide that enters our oceans stays there.

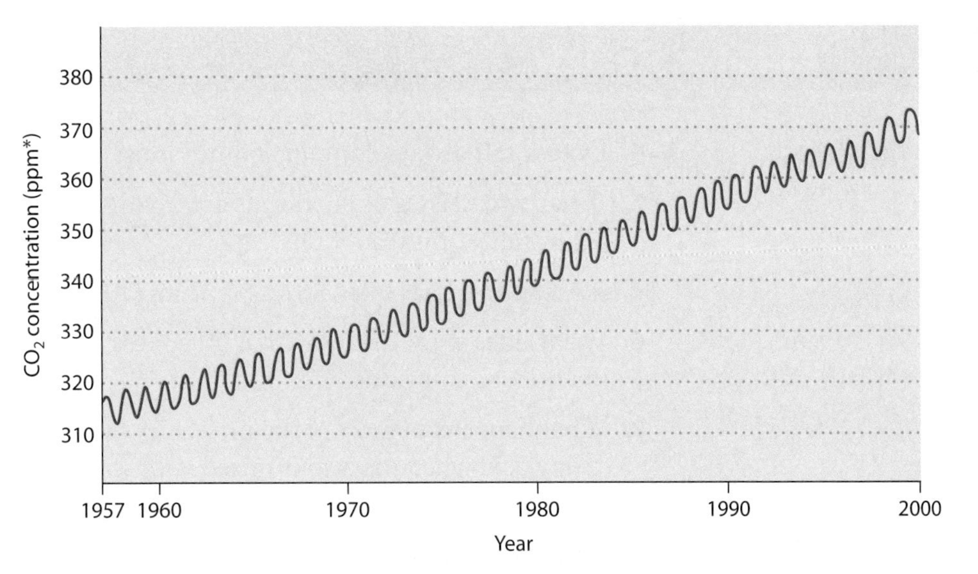

Figure 25.17
Researchers at the Mauna Loa Weather Observatory in Hawaii have recorded increasing concentrations of atmospheric carbon dioxide since they began collecting data in the 1950s. The oscillations of this graph reflect seasonal changes in CO_2 levels.

neutralization eventually end up on the ocean floor as insoluble solids. Thus carbonic acid neutralization in the ocean prevents CO_2 from being released back into the atmosphere. The ocean therefore is a carbon dioxide *sink*—most of the CO_2 that goes in doesn't come out. So, pushing more CO_2 into our atmosphere means pushing more of it into our vast oceans. This is another of the many ways in which the oceans regulate our global environment.

Nevertheless, as Figure 25.17 shows, the concentration of atmospheric CO_2 *is* increasing. Carbon dioxide is being produced faster than the ocean can absorb it, and this may alter the Earth's environment. Carbon dioxide is a *greenhouse gas.* It helps keep the surface of the Earth warm by preventing infrared radiation from escaping into outer space. Without greenhouse gases in the atmosphere, the Earth's surface would average a frigid −18°C. However, with increasing concentration of CO_2 in the atmosphere, we might experience higher average temperatures. Higher temperatures may significantly alter global weather patterns. Also, higher global temperatures could raise the average sea level as the polar ice caps melt and the volume of seawater increases (because of thermal expansion). Global warming is explored in more detail in Chapter 37.

So we find that the pH of rain depends, in great part, on the concentration of atmospheric CO_2, which depends on the pH of the oceans. These systems are interconnected with global temperatures, which naturally connect to the countless living systems on the Earth. How true it is—all the parts of our environment are intricately connected, down to the level of atoms and molecules!

Chapter Review

Key Terms and Matching Definitions

_____ acid
_____ acidic solution
_____ amphoteric
_____ base
_____ basic solution
_____ hydronium ion
_____ hydroxide ion
_____ neutral solution
_____ neutralization
_____ pH
_____ salt

1. A substance that donates hydrogen ions.
2. A substance that accepts hydrogen ions.
3. A water molecule after it accepts a hydrogen ion.
4. A water molecule after it loses a hydrogen ion.
5. An ionic compound formed from the reaction between an acid and a base.
6. A reaction in which an acid and base combine to form a salt.
7. A description of a substance that can behave as either an acid or a base.
8. A solution in which the hydronium ion concentration is higher than the hydroxide ion concentration.
9. A solution in which the hydroxide ion concentration is higher than the hydronium ion concentration.
10. A solution in which the hydronium and hydroxide ion concentrations are equal.
11. A measure of the acidity of a solution, equal to the negative of the base-10 logarithm of the hydronium ion concentration.

Review Questions

Acids Accept Protons, Bases Donate Them

1. What is a hydrogen ion?
2. When an acid is dissolved in water, what ion does the water form?
3. When a chemical loses a hydrogen ion, is it behaving as an acid or a base?
4. Does a salt always contain sodium ions?
5. What two classes of chemicals are involved in a neutralization reaction?

Some Acids and Bases Are Stronger Than Others

6. What does it mean to say that an acid is strong in aqueous solution?
7. What happens to most of the molecules of a strong acid when the acid is mixed with water?
8. Why does a solution of a strong acid conduct electricity better than a solution of a weak acid having the same concentration?
9. Which has a greater ability to accept hydrogen ions: a strong base or a weak base?
10. When can a solution of a weak base be more corrosive than a solution of a strong base?

Solutions Can Be Acidic, Basic, or Neutral

11. Is it possible for a chemical to behave as an acid in one instance and as a base in another instance?
12. Is water a strong acid or a weak acid?
13. What does the pH of a solution indicate?
14. As the hydronium ion concentration of a solution increases, does the pH of the solution increase or decrease?

Rainwater Is Acidic and Ocean Water Is Basic

15. What is the product of the reaction between carbon dioxide and water?
16. How can rain be acidic and yet not qualify as acid rain?
17. What does sulfur dioxide have to do with acid rain?
18. How do humans generate the air pollutant sulfur dioxide?
19. How does one lime a lake?
20. Why aren't atmospheric levels of carbon dioxide rising as rapidly as might be expected based on the increased output of carbon dioxide resulting from human activities?

Exercises

1. Suggest an explanation for why people once washed their hands with ashes.
2. What is the relationship between a hydroxide ion and a water molecule?
3. An acid and a base react to form a salt, which consists of positive and negative ions. Which forms the positive ions: the acid or the base? Which forms the negative ions?
4. Water is formed from the reaction between an acid and a base. Why is water not classified as a salt?
5. Identify the acid or base behavior of each substance in these reactions:
   a. $H_3O^+ + Cl^- \rightleftarrows H_2O + HCl$
   b. $H_2PO_4 + H_2O \rightleftarrows H_3O^+ + HPO_4^-$
   c. $HSO_4^- + H_2O \rightleftarrows H_3O^+ + SO_4^{2-}$
6. Identify the acid or base behavior of each substance in these reactions:
   a. $HSO_4^- + H_2O \rightleftarrows OH^- + H_2SO_4$
   b. $O^{2-} + H_2O \rightleftarrows OH^- + OH^-$
7. Sodium hydroxide, NaOH, is a strong base, which means it readily accepts hydrogen ions. What products are formed when sodium hydroxide accepts a hydrogen ion from a water molecule?
8. What happens to the corrosive properties of an acid and a base after they neutralize each other? Why?
9. What is true about the relative concentrations of hydronium and hydroxide ions in an acidic solution? How about a neutral solution? A basic solution?
10. Why do we use the pH scale to indicate the acidity of a solution rather than simply stating the concentration of hydronium ions?
11. When the hydronium ion concentration of a solution equals 1 mole per liter, what is the pH of the solution? Is the solution acidic or basic?
12. When the hydronium ion concentration of a solution equals 10 moles per liter, what is the pH of the solution? Is the solution acidic or basic?
13. What is the concentration of hydronium ions in a solution that has a pH of −3? Why is such a solution impossible to prepare?
14. When the pH of a solution decreases by 1, say from pH = 4 to pH = 3, by what factor does the hydronium ion concentration increase?
15. What happens to the pH of an acidic solution as pure water is added?
16. Why might a small piece of chalk reduce acid indigestion?
17. How might you tell whether or not your toothpaste contained calcium carbonate, $CaCO_3$, or perhaps baking soda, $NaHCO_3$, without looking at the ingredients label?
18. Why do lakes lying in granite basins tend to become acidified by acid rain more readily than lakes lying in limestone basins?
19. Cutting back on the pollutants that cause acid rain is one solution to the problem of acidified lakes. Suggest another.
20. How might warmer oceans accelerate global warming?

Suggested Web Sites

http://www.nps.gov/cave/
http://www.carlsbad.caverns.national-park.com/info.htm
http://www.nps.gov/maca/
http://www.mammoth.cave.national-park.com/info.htm

Check these official and unofficial sites for Carlsbad Caverns National Park and Mammoth Cave National Park for details on how these underground landmarks formed. Ample travel information is included.

http://www.epa.gov

Go to this home page for the Environmental Protection Agency and use *acid rain* as a keyword in their search engine to find numerous articles on this subject.

http://mloserv.mlo.hawaii.gov/mloinfo/program.htm

This address itemizes the atmospheric projects of the Climate Monitoring and Diagnostic Laboratory of the Mauna Loa Weather Observatory. Links to the Network for the Detection of Stratospheric Changes are included.

Chapter 26: Oxidation and Reduction

What do our bodies have in common with the burning of a campfire or the rusting of old farm equipment? Why does silver tarnish? How can aluminum restore tarnished silver? Why is it unwise for people with fillings in their teeth to bite down on aluminum foil? How do batteries work and what is the source of their energy? Why is hydrogen the ultimate fuel of the future? The answers to all these questions involve the transfer of electrons from one substance to another. These kinds of chemical reactions are the main focus of this chapter.

26.1 Oxidation Is the Loss of Electrons and Reduction Is the Gain of Electrons

As we learned in Chapter 21, two chemicals that react with one another are called *reactants.* After the two reactants react, they form new chemicals known as *products.* In an acid base reaction, a proton is transferred between the two reactants. In this chapter we look at a class of reactions in which an electron is transferred between the two reactants. These types of reactions are called *oxidation-reduction reactions.*

Oxidation is the process whereby a reactant loses one or more electrons. **Reduction** is the opposite process whereby a reactant gains one or more electrons. Oxidation and reduction are complementary processes that occur at the same time. They always occur together; you cannot have one without the other. The electrons lost by one chemical in an oxidation reaction don't simply disappear; they are gained by another chemical in a reduction reaction.

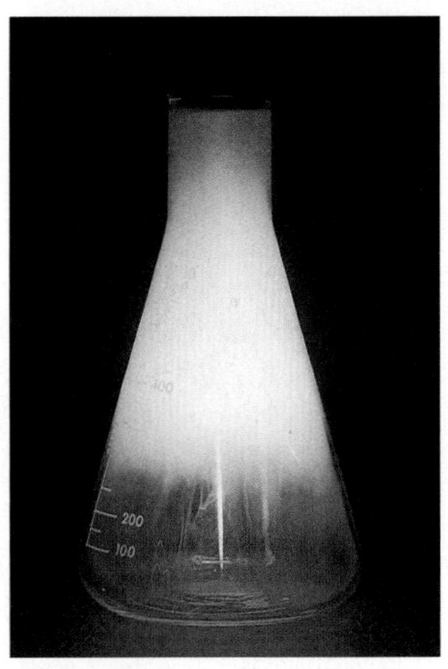

Figure 26.1
In the exothermic formation of sodium chloride, sodium metal is oxidized by chlorine gas, and chlorine gas is reduced by sodium metal.

An oxidation–reduction reaction occurs when sodium and chlorine react to form sodium chloride, as shown in Figure 26.1. The equation for this reaction is:

$$2\ Na + Cl_2 \rightarrow 2\ NaCl$$

To see how electrons transfer in this reaction, look at each reactant individually. Each electrically neutral sodium atom changes to a positively charged ion. We can also say each atom loses an electron and is therefore oxidized:

$$2\ Na \rightarrow 2Na^+ + 2e^- \quad \text{Oxidation}$$

Each electrically neutral chlorine molecule changes to two negatively charged ions. Each of these atoms gains an electron and is therefore reduced:

$$Cl_2 + 2\ e^- \rightarrow 2\ Cl^- \quad \text{Reduction}$$

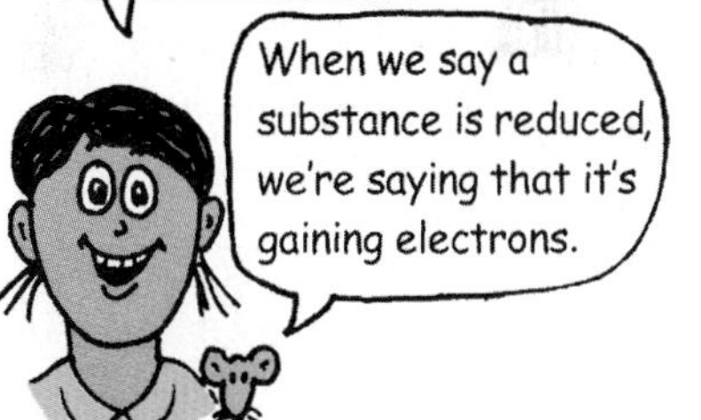

The net result is that the two electrons lost by the sodium atoms are transferred to the chlorine atoms. Therefore, the two equations shown above actually represent one process, called a **half reaction.** In other words, an electron won't be lost from a sodium atom without there being a chlorine atom available to pick up that electron. Both half reactions are required to represent the *whole* oxidation–reduction process. Half reactions are useful for showing which reactant loses electrons and which reactant gains them, which is why half reactions are used throughout this chapter.

Because the sodium causes reduction of the chlorine, the sodium acts as a *reducing agent.* A reducing agent is any reactant that causes another reactant to be reduced. Note that sodium is oxidized when it behaves as a reducing agent—it loses electrons. Conversely, the chlorine causes oxidation of the sodium and also acts as an *oxidizing agent.* Since it gains electrons in the process, an oxidizing agent is

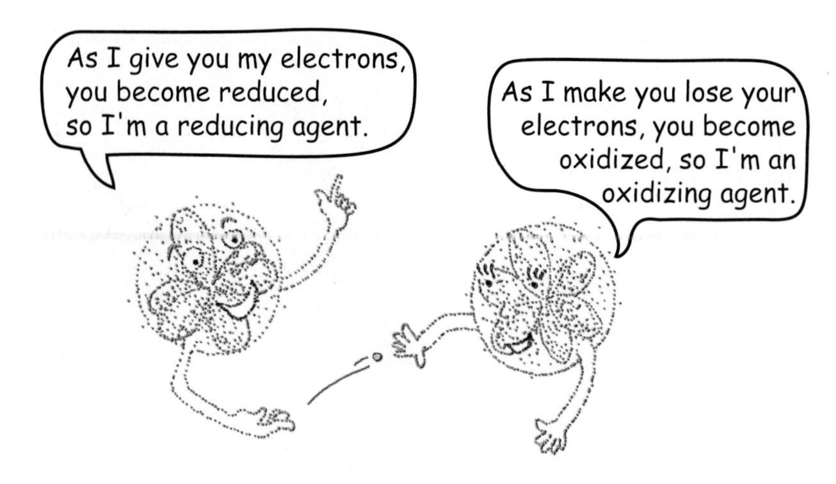

reduced. Just remember that **l**oss of **e**lectrons is **o**xidation, and **g**ain of **e**lectrons is **r**eduction. Here is a helpful mnemonic adapted from a once-popular children's story: **Leo** the lion went "**ger.**"

Different elements have different oxidation and reduction tendencies. Some lose electrons more readily, while others gain electrons more readily (Figure 26.2). In general, elements toward the upper right of the periodic table tend to gain electrons. They thus behave as oxidizing agents. Elements toward the lower left tend to lose electrons and thus behave as reducing agents. The exception to this periodic trend are the nobel gas elements of Group 18, which tend to behave as neither.

Concept Check ✓

True or false:

1. Reducing agents are oxidized in oxidation–reduction reactions.
2. Oxidizing agents are reduced in oxidation–reduction reactions.

Check Your Answer Both statements are true.

Have you ever seen antique (or just plain old) silverware that hasn't been polished? Over time, silverware and other silver objects get tarnished. Tarnish is a blackish coating of silver sulfide, Ag_2S, an ionic compound that consists of two silver ions, Ag^+, and one sulfide ion, S^{2-}. Tarnishing is a common oxidation-reduction reaction.

Tarnishing begins when silver atoms in the silverware come into contact with airborne hydrogen sulfide, H_2S, a smelly gas produced by the digestion of food in mammals and other organisms. The half reaction for the silver and hydrogen sulfide is

$$4\,Ag + 2\,H_2S \rightarrow 4\,Ag^+ + 4\,H^+ + 2\,S^{2-} + 4\,e^- \quad \text{Oxidation}$$

The silver ions and sulfide ions combine to form blackish silver sulfide, while at the same time the hydrogen ions and electrons combine with atmospheric oxygen to form water:

$$4\,H^+ + 4\,e^- + O_2 \rightarrow 2\,H_2O \quad \text{Reduction}$$

The balanced chemical equation for the tarnishing of silver is the combination of these two half reactions:

$$4\,Ag + 2\,H_2S + O_2 \rightarrow 2\,Ag_2S + 2\,H_2O$$

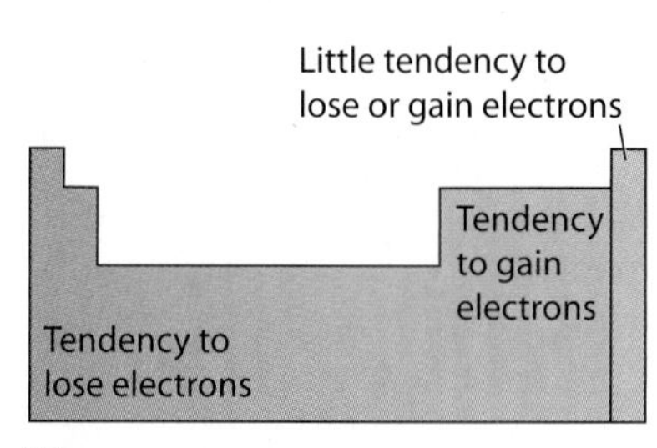

More likely to behave as oxidizing agent (be reduced)

More likely to behave as reducing agent (be oxidized)

Figure 26.2
The ability of an atom to gain or lose electrons is a function of its position in the periodic table. Those at the upper right tend to gain electrons, and those at the lower left tend to lose them.

Hand-On Exploration: The Silver Lining

What You Need

Very clean aluminum pot (or non-aluminum pot and aluminum foil), water, baking soda, piece of tarnished silver. Hot plate provided for by your teacher, or use your kitchen stove if you are performing this activity at home with your parents.

Procedure

1. Put about a liter of water and several heaping tablespoons of baking soda in the aluminum pot or the non-aluminum pot containing aluminum foil.
2. Bring the water to boiling and then remove the pot from the heat source.
3. Slowly immerse the tarnished silver; you'll see an immediate effect as the silver and aluminum make contact. (Add more baking soda if you don't.) Also, as the silver ions accept electrons from the aluminum and are thereby reduced to shiny silver atoms, the sulfide ions are free to re-form hydrogen sulfide gas, which is released back into the air. You may smell it!

The baking soda serves as a conductive ionic solution that permits electrons to move from the aluminum atoms to the silver ions. What is the advantage of this approach over polishing the silver with an abrasive paste?

From these equations we see that the hydrogen sulfide causes the silver to lose electrons to oxygen. To restore the silver to its shiny elemental state, we need to return the electrons it lost. The oxygen won't relinquish electrons back to silver. But there is a way to do it, as you can discover in the activity shown above.

26.2 The Energy of Flowing Electrons Can Be Harnessed

Electrochemistry is the study of the relationship between electrical energy and chemical change. It makes use of an oxidation–reduction reaction to produce an electric current or an electric current to produce an oxidation–reduction reaction.

To understand how an oxidation–reduction reaction generates an electric current, consider what happens when a reducing agent is placed in direct contact with an oxidizing agent: Electrons flow from the reducing agent to the oxidizing agent. This flow of electrons is an electric current—a flow of kinetic energy that can be harnessed for useful purposes.

Iron, Fe, for example, is a better reducing agent than the copper ion Cu^{2+}. So when a piece of iron metal and a solution containing copper ions are placed in contact with each other, electrons flow

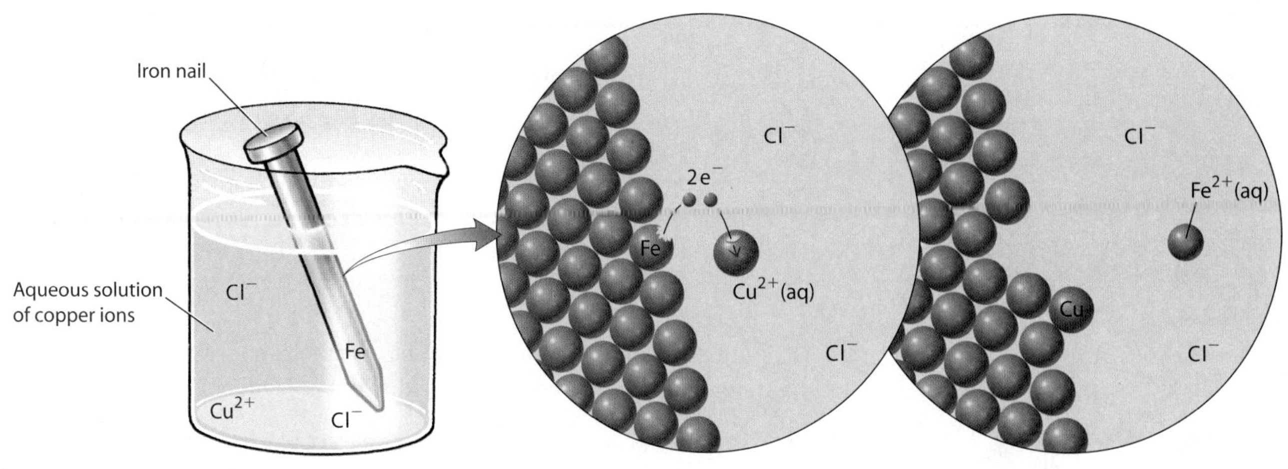

Figure 26.3
A nail made of iron placed in a solution of Cu^{2+} ions oxidizes to Fe^{2+} ions, which dissolve in the water. At the same time, copper ions are reduced to metallic copper, which coats the nail. (Negatively charged ions, such as chloride ions, Cl^-, must also be present to balance these positively charged ions in solution.)

Oxidation $Fe \longrightarrow Fe^{2+} + 2\,e^-$

Reduction $Cu^{2+} + 2\,e^- \longrightarrow Cu$

from the iron to the copper ions, as Figure 26.3 illustrates. The result is the oxidation of iron and the reduction of copper ions.

Iron and copper ions don't have to be in physical contact for electrons to flow between them. When in separate containers but bridged by a conducting wire, the electrons flow from the iron to the copper ions through the wire. The resulting electric current in the wire can be attached to some useful device, such as a light bulb. But alas, an electric current won't keep flowing by this arrangement.

The reason the electric current is not sustained is shown in Figure 26.4. An initial flow of electrons through the wire immediately results

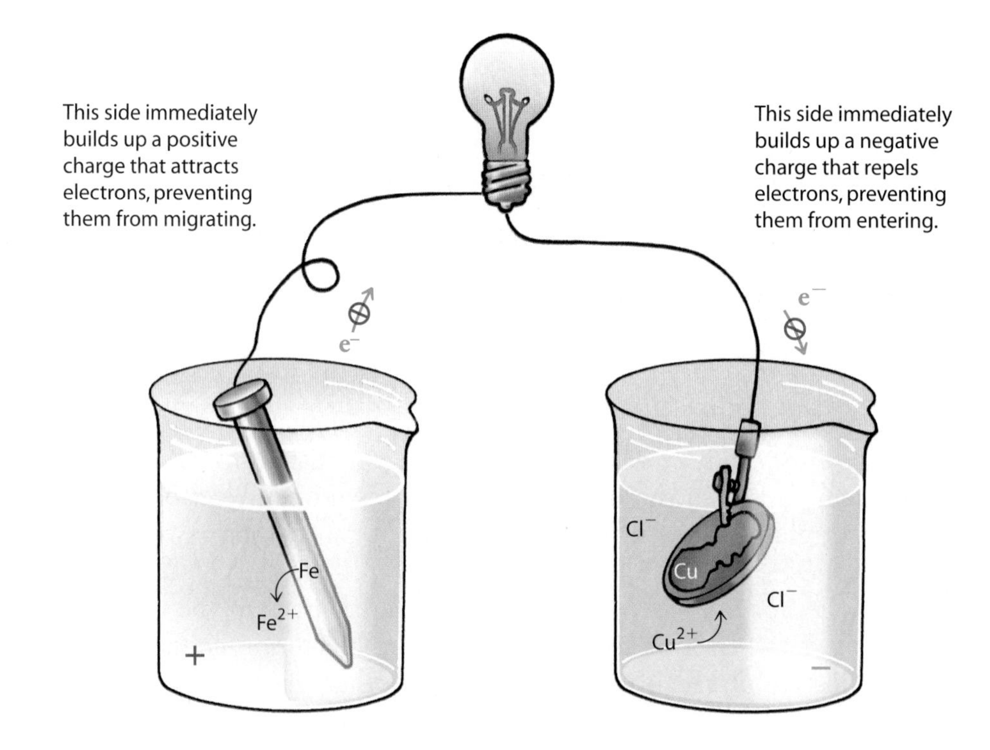

Figure 26.4
An iron nail is placed in water and connected by a conducting wire to a solution of copper ions. Nothing happens because this arrangement results in a buildup of charge that prevents the further flow of electrons.

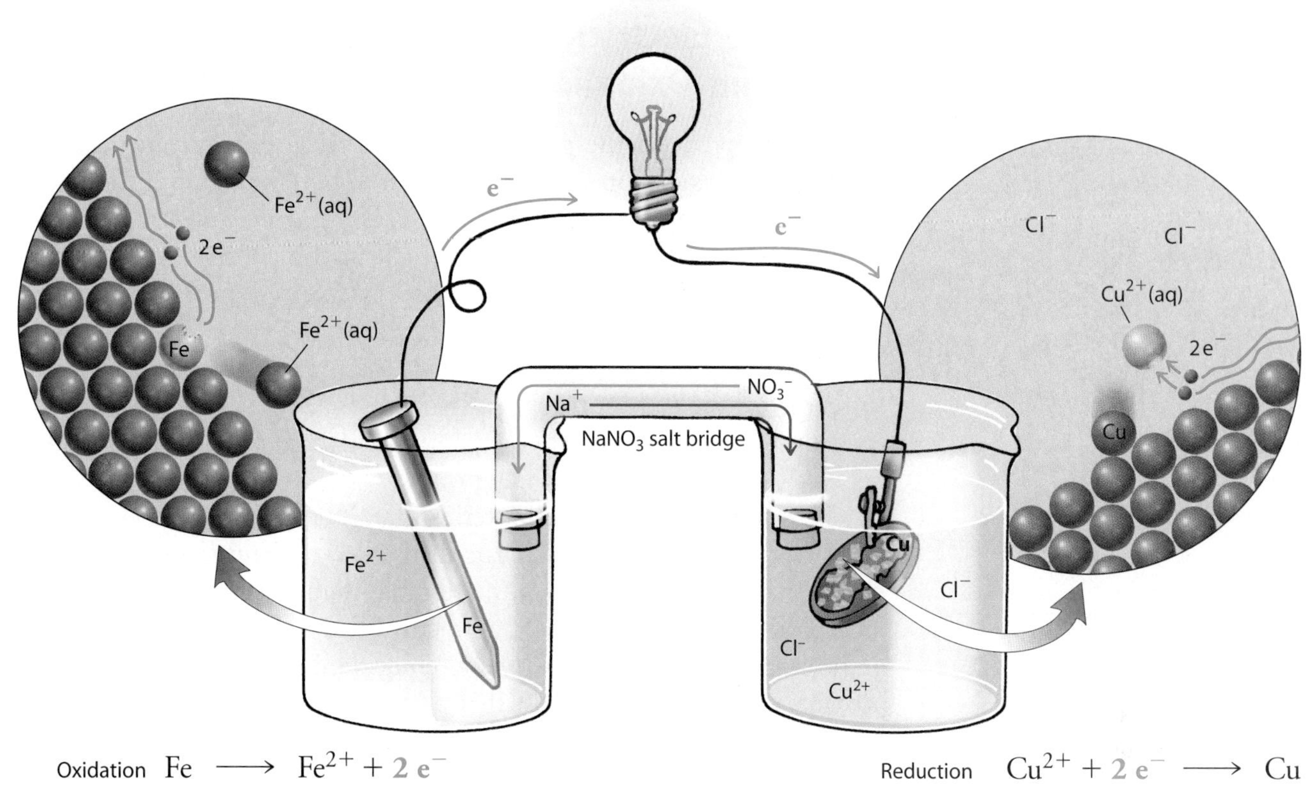

Oxidation $Fe \longrightarrow Fe^{2+} + 2\ e^-$ Reduction $Cu^{2+} + 2\ e^- \longrightarrow Cu$

Figure 26.5
The salt bridge completes the electric circuit. Electrons freed as the iron is oxidized pass through the wire to the container on the right. Nitrate ions, NO_3^-, from the salt bridge flow into the left container to balance the positive charges of the Fe^{2+} ions that form and prevent any buildup of positive charge. Meanwhile, Na^+ ions from the salt bridge enter the right container to balance the Cl^- ions "abandoned" by the Cu^{2+} ions as the Cu^{2+} ions pick up electrons to become metallic copper.

in a buildup of electric charge in both containers. The container on the left builds up positive charge as it accumulates Fe^{2+} ions from the nail. The container on the right builds up negative charge as electrons accumulate there. This situation prevents any further migration of electrons through the wire. Since electrons are negative, they are repelled by the negative charge in the right container and attracted to the positive charge in the left container. The net result is no electron flow through the wire, and the bulb remains unlit.

The solution to this problem is to allow ions to migrate into either container so that neither builds up any positive or negative charge. This is accomplished with a *salt bridge.* A salt bridge may be a U-shaped tube filled with a salt, such as sodium nitrate, $NaNO_3$, and closed with semiporous plugs. Figure 26.5 shows how a salt bridge allows the ions it holds to enter either container. This permits the flow of electrons through the conducting wire and creating a complete electric circuit.

26.3 The Electricity of a Battery Comes from Oxidation–Reduction Reactions

So we see that with the proper setup, electrical energy can be harnessed in an oxidation–reduction reaction. The apparatus shown in Figure 26.5 is one example. Such devices are called *voltaic cells.* Instead of two containers, a voltaic cell can be an all-in-one, self-contained unit. In this case it is called a *battery.* Batteries are either disposable or rechargeable, and here we explore some examples of each. Although the two types differ in design and composition, they function by the same principle: Two materials that oxidize and reduce each other are connected by a medium through which ions travel to balance an external flow of electrons.

Let's look at disposable batteries first. The common *dry-cell battery* was invented in the 1860s. It is still used today and is probably the cheapest disposable energy source for flashlights, toys, and the like. The basic design consists of a zinc cup filled with a thick paste of ammonium chloride, NH_4Cl, zinc chloride, $ZnCl_2$, and manganese dioxide, MnO_2. Immersed in this paste is a porous stick of graphite that projects to the top of the battery, as shown in Figure 26.6.

Graphite is a good conductor of electricity. Chemicals in the paste receive electrons at the graphite stick and so are reduced. The reaction for the ammonium ions is:

$$2\,NH_4^+\,(aq) + 2\,e^- \rightarrow 2\,NH_3\,(g) + H_2\,(g) \quad \text{Reduction}$$

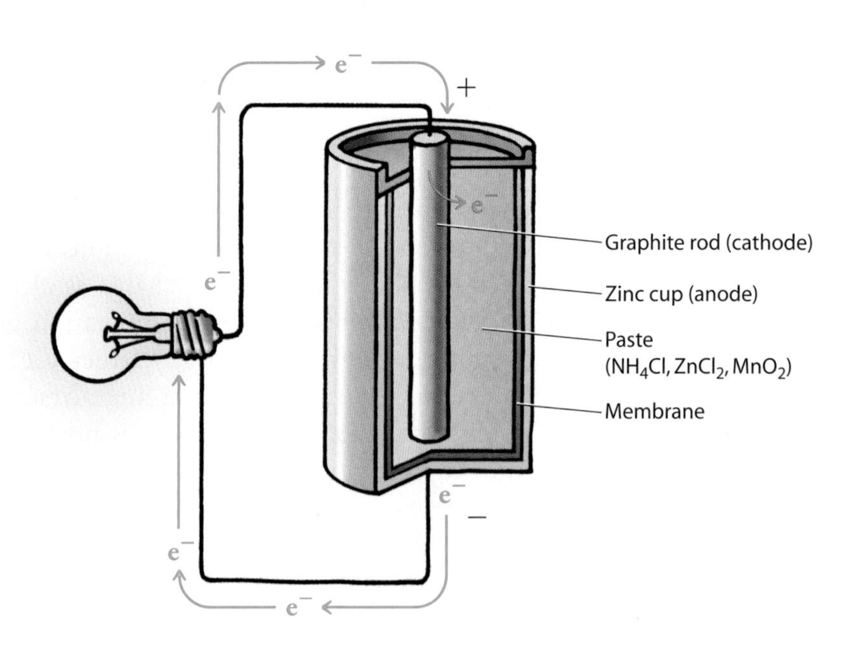

Figure 26.6
A common dry-cell battery with a graphite rod immersed in a paste of ammonium chloride, manganese dioxide, and zinc chloride.

An **electrode** is any material that conducts electrons into or out of a medium in which electrochemical reactions are occurring. The electrode where chemicals are reduced is called a **cathode.** For any battery, such as the one shown in Figure 26.6, the cathode is always positive (+), which indicates that electrons are naturally attracted to this location. The electrons gained by chemicals at the cathode originate at the **anode,** which is the electrode where chemicals are oxidized. For any battery, the anode is always negative (−), which indicates that electrons stream away from this location. The anode in Figure 26.6 is the zinc cup, where zinc atoms lose electrons to form zinc ions:

$$Zn\ (s) \rightarrow Zn^{2+}\ (aq) + 2\ e^- \quad \text{Oxidation}$$

The reduction of ammonium ions in a dry-cell battery produces two gases—ammonia, NH_3, and hydrogen, H_2—that need to be removed to avoid a pressure buildup and a potential explosion. Removal is accomplished by having the ammonia and hydrogen react with the zinc chloride and manganese dioxide:

$$ZnCl_2\ (aq) + 2\ NH_3\ (g) \rightarrow Zn(NH_3)_2Cl_2\ (s) \quad \text{Oxidation}$$

$$2\ MnO_2\ (s) + H_2\ (g) \rightarrow Mn_2O_3(s) + H_2O\ (l) \quad \text{Reduction}$$

The life of a dry-cell battery is relatively short. Oxidation causes the zinc cup to deteriorate, and eventually the contents leak out. Even while the battery is not operating, the zinc corrodes as it reacts with ammonium ions. This zinc corrosion can be inhibited by storing the battery in a refrigerator, which increases the life of the battery.

Another type of disposable battery is the more expensive *alkaline battery,* shown in Figure 26.7. This battery avoids many of the problems of dry-cell batteries by operating in a strongly alkaline paste. In the presence of hydroxide ions, the zinc oxidizes to insoluble zinc oxide:

$$Zn\ (s) + 2\ OH^-\ (aq) \rightarrow ZnO(s) + H_2O(l) + 2\ e^- \quad \text{Oxidation}$$

while at the same time manganese dioxide is reduced:

$$2\ MnO_2\ (s) + H_2O\ (l) + 2\ e^- \rightarrow Mn_2O_3(s) + 2\ OH^-\ (aq) \quad \text{Reduction}$$

Note how these two reactions avoid the use of the zinc-corroding ammonium ion (which means alkaline batteries last a lot longer than dry-cell batteries) and the formation of any gaseous products. Furthermore, these reactions maintain a given voltage for longer periods of operation.

Figure 26.7
Alkaline batteries last a lot longer than dry-cell batteries and give a steadier voltage, but they are expensive.

The small mercury and lithium disposable batteries used for calculators and cameras are variations of the alkaline battery. In the mercury battery, mercuric oxide, HgO, is reduced rather than manganese dioxide. Manufacturers are phasing out these batteries because of the environmental hazard posed by mercury, which is poisonous. In the lithium battery, lithium metal is used as the source of electrons rather than zinc. Not only is lithium able to maintain a higher voltage than zinc, it is about 13 times less dense, which allows for a lighter battery.

Disposable batteries have relatively short lives because electron-producing chemicals are consumed. The main feature of rechargeable batteries is the reversibility of the oxidation and reduction reactions. In your car's rechargeable lead storage battery, for example, electrical energy is produced as lead dioxide, lead, and sulfuric acid are consumed to form lead sulfate and water. The lead is oxidized to Pb^{2+}, and the oxygen in the lead dioxide is reduced from the O^{-} state to the O^{2-} state. Combining the two half reactions gives the complete oxidation–reduction reaction:

$$PbO_2 + Pb + 2\,H_2SO_4 \rightarrow PbSO_4 + 2\,H_2O + \text{electrical energy}$$

This reaction can be reversed by supplying electrical energy, as Figure 26.8 shows. This is the task of the car's alternator, which is powered by the engine:

$$\text{electrical energy} + 2\,PbSO_4 + 2\,H_2O \rightarrow PbO_2 + Pb + 2\,H_2SO_4$$

So running the engine maintains concentrations of lead dioxide, lead, and sulfuric acid in the battery. With the engine turned off, these reactants stand ready to supply electric power as needed to start the engine, operate the emergency blinkers, or play the radio.

Concept Check ✓

What is recharged in a car battery?

Check Your Answer When the battery is being recharged, electrical energy from a source (the alternator) outside the battery is used to regenerate reactants. These are the reactants that were earlier transformed to products during the oxidation–reduction reaction that produced the electrical energy needed to start the engine. The reactants being regenerated are lead dioxide, elemental lead, and sulfuric acid.

Many rechargeable batteries smaller than car batteries are made of compounds of nickel and cadmium (ni–cad batteries). As with the lead storage battery, ni–cad reactants are replenished by supplying electrical energy from some external source, such as an electrical wall outlet. Like mercury batteries, ni–cad batteries pose an environmental hazard because cadmium is toxic to humans and other organisms. For this reason, alkaline batteries designed to be rechargeable are rapidly gaining a place in the market.

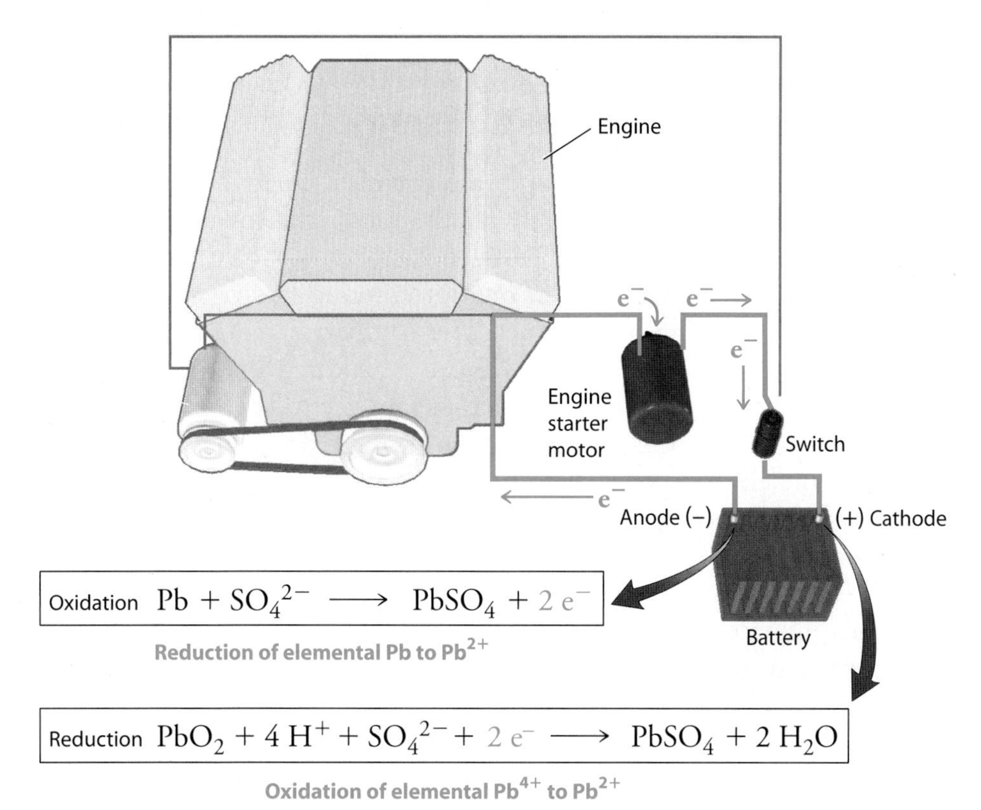

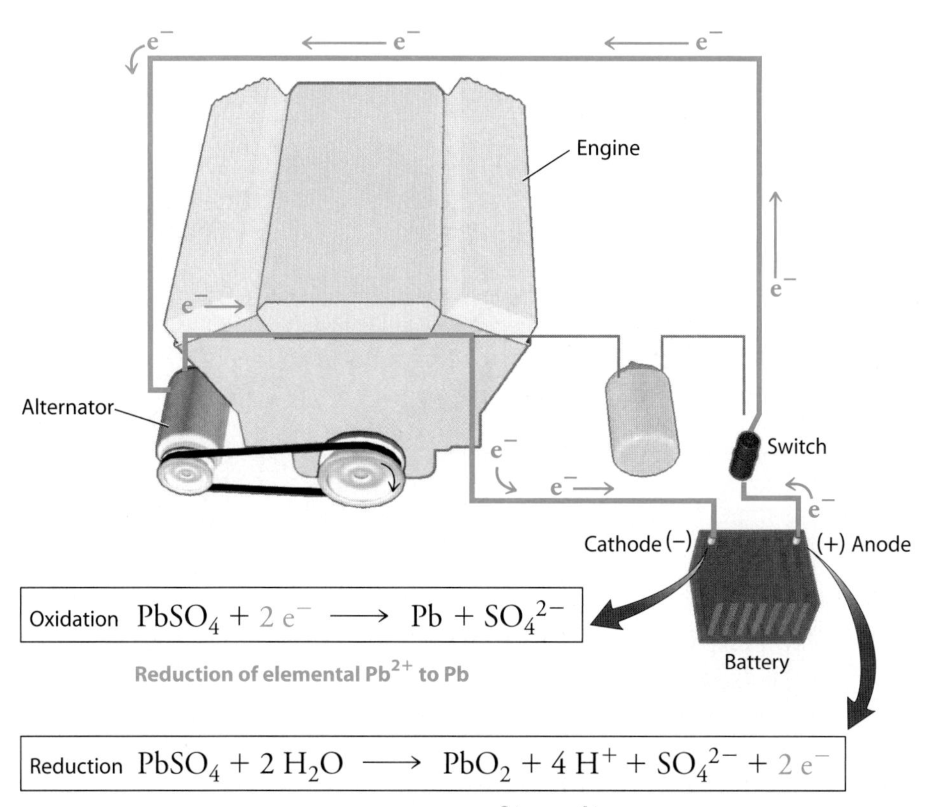

Figure 26.8
(a) Electrical energy from the battery forces the starter motor to start the engine. (b) The combustion of fuel keeps the engine running and provides energy to spin the alternator, which recharges the battery. Note that the battery has a reversed cathode–anode orientation during recharging.

26.4 Fuel Cells Are Highly Efficient Sources of Electrical Energy

A *fuel cell* is a device that changes the chemical energy of a fuel to electrical energy. Fuel cells are by far the most efficient means of generating electricity. A hydrogen–oxygen fuel cell is shown in Figure 26.9. It has two compartments, one for entering hydrogen fuel and the other for entering oxygen fuel. The compartments are separated by a set of porous electrodes. Hydrogen is oxidized upon contact with hydroxide ions at the hydrogen-facing electrode (the anode). The electrons from this oxidation flow through an external circuit and provide electric power before meeting up with oxygen at the oxygen-facing electrode (the cathode). The oxygen readily picks up the electrons (in other words, the oxygen is reduced) and reacts with water to form hydroxide ions. To complete the circuit, these hydroxide ions migrate across the porous electrodes. They flow through an ionic paste of potassium hydroxide, KOH, to meet up with hydrogen at the hydrogen-facing electrode.

As the oxidation equation shown at the top of Figure 26.9 demonstrates, the hydrogen and hydroxide ions react to produce steam as well as electrons. This steam may be used for heating or to generate electricity in a steam turbine. Furthermore, the water that condenses from the steam is pure water, suitable for drinking!

Figure 26.9
The hydrogen–oxygen fuel cell.

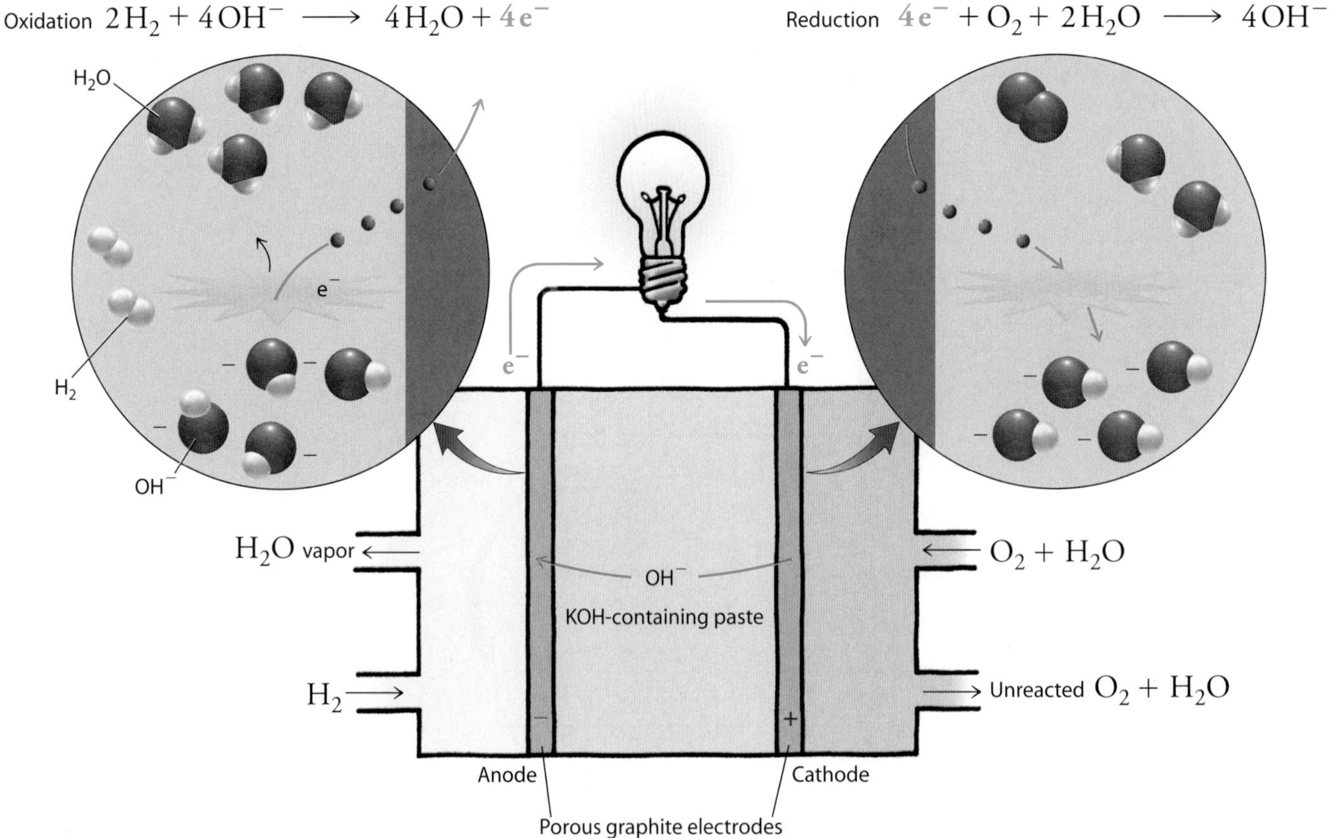

Figure 26.10
Because this bus is powered by a fuel cell, its tail pipe emits mostly water vapor.

Fuel cells are similar to dry-cell batteries, but as long as fuel is supplied, fuel cells don't run down. The space shuttle uses hydrogen–oxygen fuel cells to meet its electrical needs. The cells also produce more than 100 gallons of drinking water for the astronauts during a typical week-long mission. Back on the Earth, researchers are developing fuel cells for buses and automobiles. As shown in Figure 26.10, experimental fuel-cell buses are already operating in several cities, such as Vancouver, British Columbia, and Chicago, Illinois. These vehicles produce very few pollutants and can run much more efficiently than those that run on fossil fuels.

In the future, commercial buildings as well as individual homes may be outfitted with fuel cells as an alternative to receiving electricity (and heat) from regional power stations. Researchers are also working on miniature fuel cells that could replace the batteries used for portable electronic devices, such as cellular phones and laptop computers. Such devices could operate for extended periods of time on a single "ampule" of fuel available at your local supermarket.

Amazingly, a car powered by a hydrogen–oxygen fuel cell requires only about 3 kilograms of hydrogen to travel 500 kilometers. However, this much hydrogen gas at room temperature and atmospheric pressure would occupy a volume of about 36,000 liters, the volume of about four midsize cars! Thus the major hurdle to the development of fuel-cell technology lies not with the cell but with the fuel. This volume of gas could be compressed to a much smaller volume, as it is on the experimental buses in Vancouver.

Compressing a gas takes energy, however, and as a consequence the efficiency of the fuel cell is lowered. Chilling hydrogen to its liquid phase, which occupies much less volume, poses similar problems. Instead, researchers are looking for novel ways of providing fuel cells with hydrogen. In one design, hydrogen is generated within the fuel cell from chemical reactions involving liquid hydrocarbons, such as methanol, CH_3OH. Alternatively, certain porous materials, including the recently developed carbon nanofibers shown in Figure 26.11, can hold large volumes of hydrogen on their surfaces. They behave like hydrogen "sponges." The hydrogen is "squeezed" out of these materials on demand by controlling the temperature. The warmer the material, the more hydrogen released.

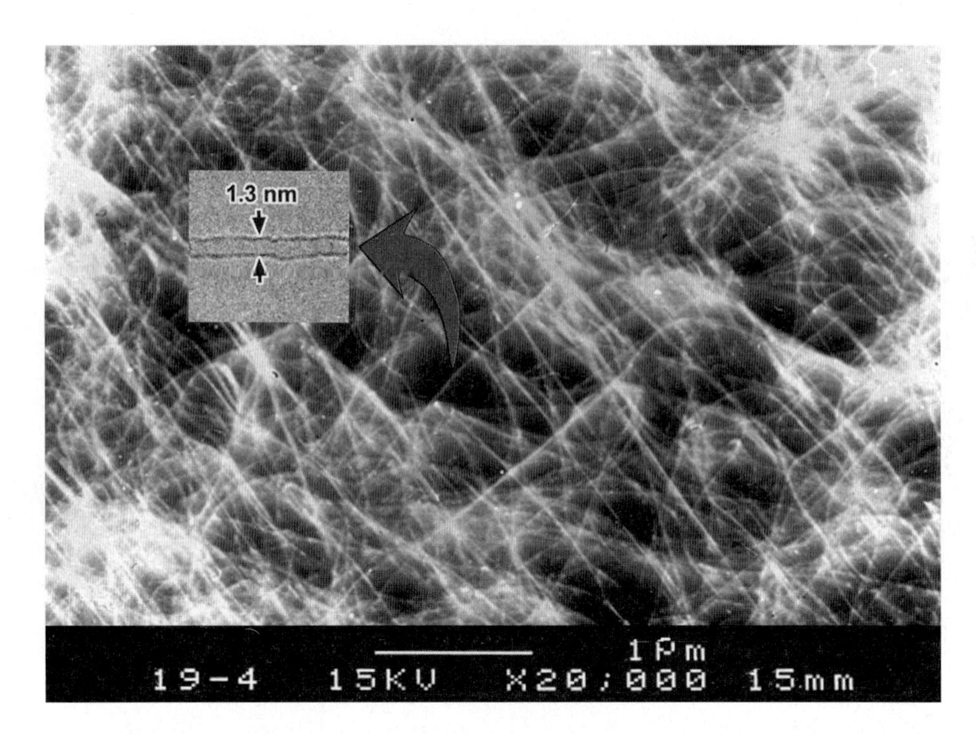

Figure 26.11
Carbon nanofibers consist of near-submicroscopic tubes of carbon atoms. They outclass most all other known materials in their ability to absorb hydrogen molecules. With carbon nanofibers, for example, a volume of 36,000 liters of hydrogen can be reduced to a mere 35 liters. Carbon nanofibers are a recent discovery, however, and much research is still required to confirm their applicability to hydrogen storage and to develop the technology.

Concept Check ✓

As long as fuel is available to it, a given fuel cell can supply electrical energy indefinitely. Why can't batteries do the same?

Check Your Answer Batteries generate electricity by reducing and oxidizing chemical reactants. Once these reactants are consumed, the battery can no longer generate electricity. A rechargeable battery can be made to operate again, but only after the energy flow is interrupted so that the reactants can be replenished.

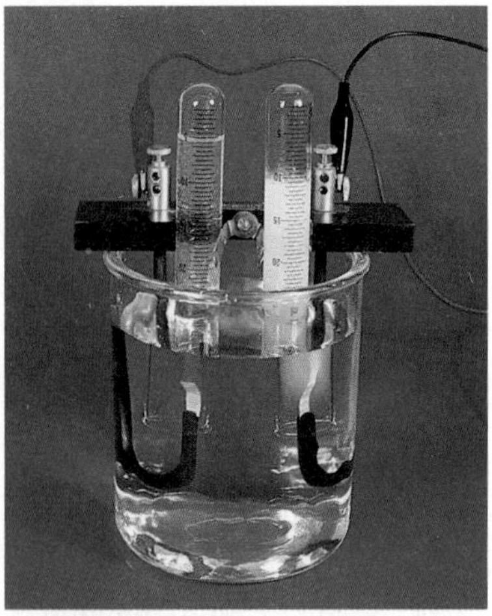

26.5 Electrical Energy Can Produce Chemical Change

Electrolysis is the use of electrical energy to produce chemical change. The recharging of a car battery is an example of electrolysis. Another, shown in Figure 26.12, is passing an electric current through water. This process breaks the water down into its elemental components:

$$\text{electrical energy} + 2\ H_2O \rightarrow 2\ H_2\ (g) + O_2\ (g)$$

Electrolysis is used to purify metals from metal ores. An example is aluminum, the third most abundant element in the Earth's crust. Aluminum occurs naturally bonded to oxygen in an ore called *bauxite*. Aluminum metal wasn't known until about 1827, when it was prepared by reacting bauxite with hydrochloric acid. This gave the aluminum ion, Al^{3+}, which was reduced to aluminum metal with sodium metal acting as the reducing agent:

$$Al^{3+} + 3\ Na^{+} \rightarrow Al + 3\ Na^{+}$$

Figure 26.12
The electrolysis of water produces hydrogen gas and oxygen gas in a 2:1 ratio by volume, which is in accordance with the chemical formula for water: H_2O. For this process to work, ions must be dissolved in the water so that the electricity can be conducted between the electrodes.

This chemical process was expensive. The price of aluminum at that time was about $100,000 per pound, and it was considered a rare and precious metal. In 1855, aluminum dinnerware and other items were exhibited in Paris with the crown jewels of France. Then, in 1886, two men working independently, Charles Hall (1863–1914) in the United States and Paul Heroult (1863–1914) in France, almost simultaneously discovered a process whereby aluminum could be produced from aluminum oxide, Al_2O_3, a main component of bauxite. In what is now known as the Hall–Heroult process, shown in Figure 26.13, a strong electric current is passed through a molten mixture of aluminum oxide and cryolite, Na_3AlF_6, a naturally occurring mineral. The fluoride ions of the cryolite react with the aluminum oxide to form various aluminum fluoride ions, such as $AlOF_3^{2-}$. These are then oxidized to the aluminum hexafluoride ion, AlF_6^{3-}. The Al^{3+} in this ion is then reduced to elemental aluminum, which collects at the bottom of the reaction chamber. This process, which is still in use today, greatly facilitated mass production of aluminum metal, and by 1890 the price of aluminum had dropped to about $2 per pound.

Today, worldwide production of aluminum is about 16 million tons annually. For each ton produced from ore, about 16,000 kilowatt-hours of electrical energy is required, as much as a typical American household consumes in 18 months. Processing recycled aluminum, on the other hand, consumes only about 700 kilowatt-hours for every ton. Thus recycling aluminum not only reduces litter but also helps reduce the load on power companies, which in turn reduces air pollution.

For a nerve-wracking experience involving the oxidation of elemental aluminum, bite a piece of aluminum foil with a tooth filled with dental amalgam. (If you don't have any dental fillings, hooray for you! You'll need to ask a less fortunate friend what this activity is

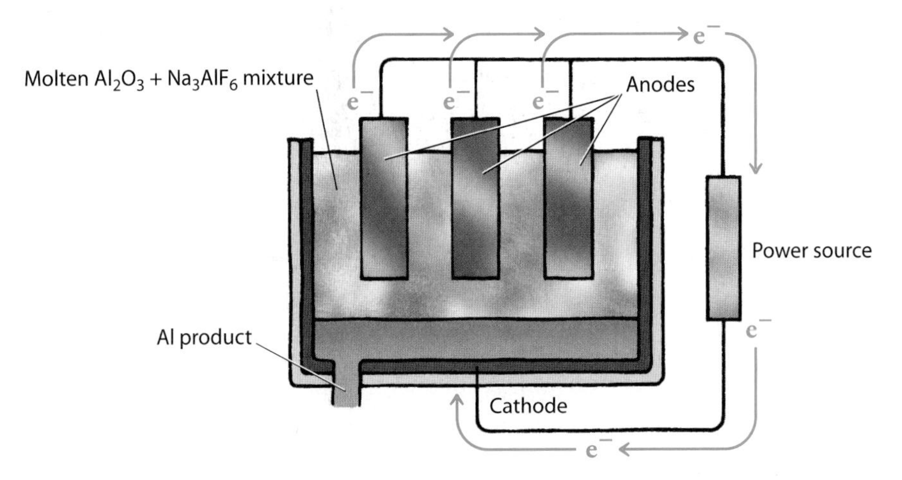

Figure 26.13
The melting point of aluminum oxide (2030°C) is too high for it to be efficiently electrolyzed to aluminum metal. When the oxide is mixed with the mineral cryolite, the melting point of the oxide drops to a more reasonable 980°C. A strong electric current passed through the molten aluminum oxide–cryolite mixture generates aluminum metal at the cathode, where aluminum ions pick up electrons and so are reduced to elemental aluminum.

like.) The aluminum behaves as an anode and releases electrons to the amalgam (a mix of silver, tin, and mercury). The amalgam behaves as a cathode by transferring these electrons to oxygen, which then combines with hydrogen ions to form water. The slight current that results produces a jolt of . . . ouch . . . pain.

Concept Check ✓

Is the reaction that goes on in a hydrogen–oxygen fuel cell an example of electrolysis?

Check Your Answer During electrolysis, electrical energy is used to produce chemical change. In the hydrogen–oxygen fuel cell, chemical change is used to produce electrical energy. Therefore, the answer to the question is no.

26.6 Oxygen Is Responsible for Corrosion and Combustion

Look to the upper right of the periodic table, and you will find one of the most common oxidizing agents—oxygen. In fact, if you haven't guessed already, the term *oxidation* is derived from this element. Oxygen is able to pluck electrons from many other elements, typically those that lie at the lower left of the periodic table. Two common oxidation–reduction reactions involving oxygen as the oxidizing agent are *corrosion* and *combustion.*

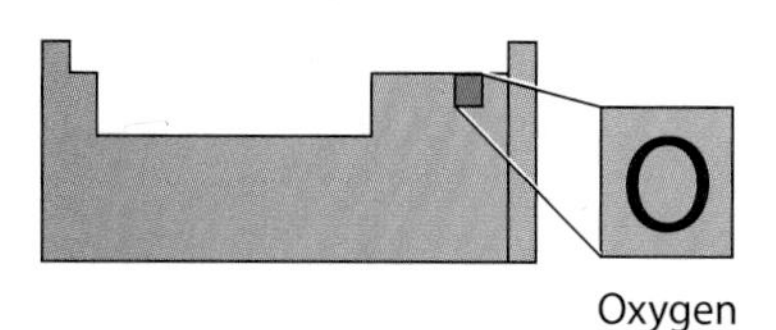

Oxygen

Concept Check ✓

Oxygen is a good oxidizing agent, but so is chlorine. What does this tell you about their relative positions in the periodic table?

Hands-On Exploration: Splitting Water

You Will Need

9-volt battery; glass of fresh water; glass of water with a tablespoon of salt added

Procedure

Immerse the top of a disposable 9-volt battery in the fresh water. Note that not much occurs. Next, immerse the top of the battery into the salt water. The bubbles that form contain hydrogen gas produced as the water decomposes.

Why does this activity work better with salt water than with tap water? Why does this activity quickly ruin the battery (which should therefore not be used again)?

Check Your Answer Chlorine and oxygen must lie in the same area of the periodic table. Both have strong tendencies to attract additional elements, so both are strong oxidizing agents.

Figure 26.14
Rust itself is not harmful to the iron structures on which it forms. Rather it is the loss of metallic iron that ruins the structural integrity.

Corrosion is the process whereby a metal deteriorates. Corrosion caused by atmospheric oxygen is a widespread and costly problem. About one-quarter of the steel produced in the United States, for example, goes into replacing corroded iron at a cost of billions of dollars annually. Iron corrodes when it reacts with atmospheric oxygen and water to form iron oxide trihydrate—the naturally occurring reddish-brown substance you know as *rust*, shown in Figure 26.14.

We can better understand rusting by considering the half equations shown in Figure 26.15. (1) Iron loses electrons to form the Fe^{2+} ion. (2) Oxygen accepts these electrons and then reacts with water to form hydroxide ions, OH^-. (3) Iron ions and hydroxide ions combine to form iron hydroxide, $Fe(OH)_2$, which is further oxidized by oxygen to form rust, $Fe_2O_3 \cdot 3\,H_2O$.

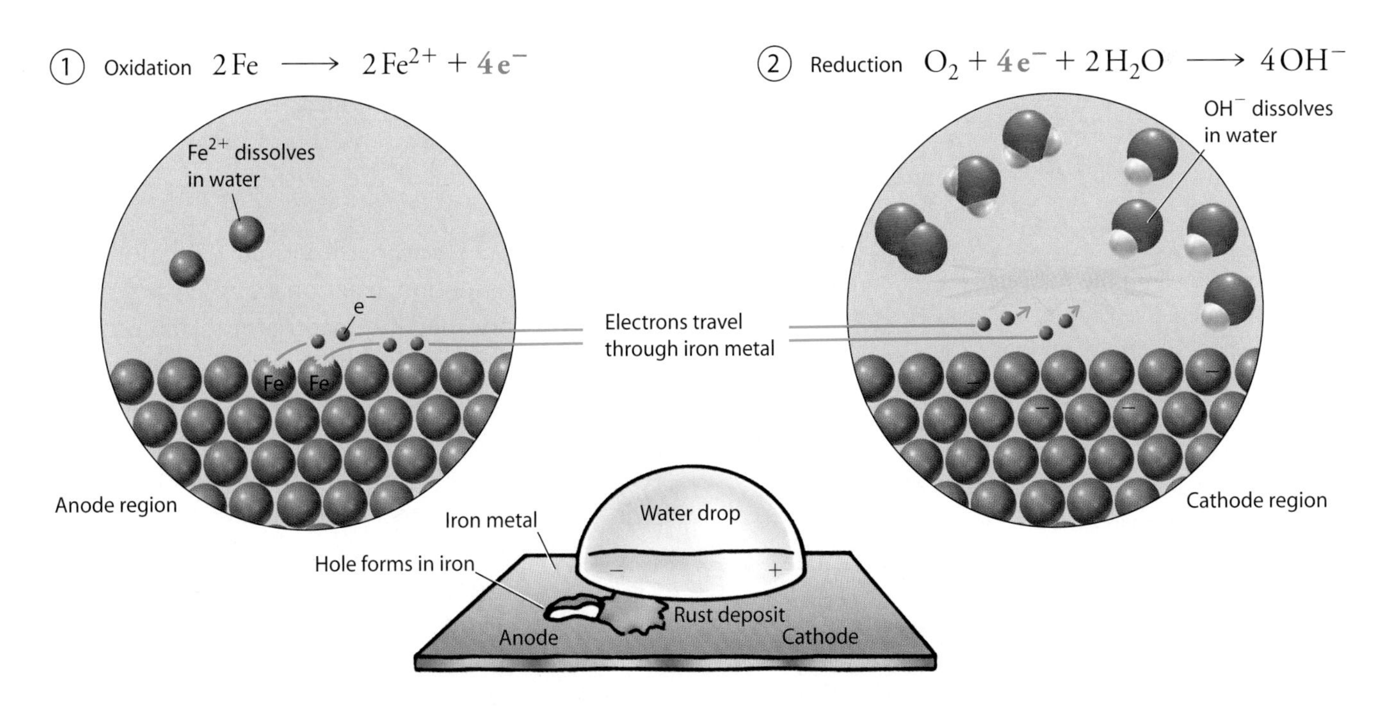

Figure 26.15
Rusting begins when iron atoms lose electrons to form Fe^{2+} ions. These electrons are lost to oxygen atoms, which are thereby reduced to hydroxide ions, OH^-. One region of the iron behaves as the anode while another region behaves as the cathode. Rust forms only in the region of the anode, where iron atoms lose electrons. The loss of elemental iron in this region causes a hole to form in the metal. The formation of rust, however, is not as much of a problem as is the loss of metallic iron, which results in a loss of structural integrity.

Another common metal oxidized by oxygen is aluminum. The product of aluminum oxidation is aluminum oxide, Al_2O_3, which is not soluble in water. Because of its water insolubility, aluminum oxide forms a protective coat that shields aluminum from further oxidation. This coat is so thin that it's transparent, which is why aluminum maintains its metallic shine.

A protective water-insoluble oxidized coat is the principle underlying a process called *galvanization*. Zinc has a slightly greater tendency to oxidize than does iron. For this reason, many iron articles, such as the nails pictured in Figure 26.16, are *galvanized* by coating them with a thin layer of zinc. The zinc oxidizes to zinc oxide, an inert, insoluble substance that protects the inner iron from rusting.

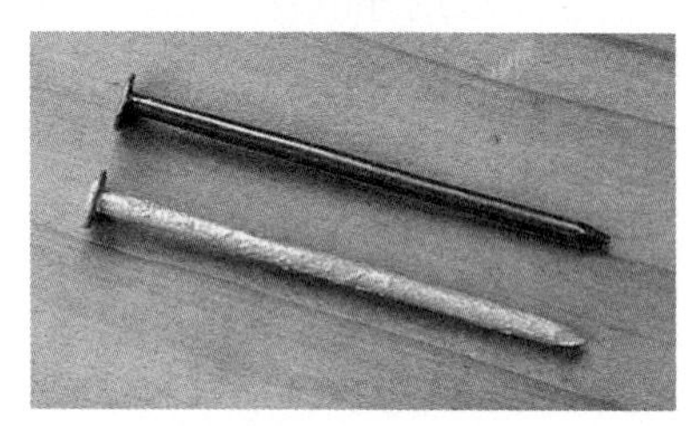

Figure 26.16
The galvanized nail (bottom) is protected from rusting by the sacrificial oxidation of zinc.

Yet another way to protect iron and other metals from oxidation is to coat them with a corrosion-resistant metal, such as chromium, platinum, or gold. *Electroplating* is the operation of coating one metal with another by electrolysis, and it is illustrated in Figure 26.17. The object to be electroplated is connected to a negative battery terminal and then submerged in a solution containing ions of the metal to be used as the coating. The positive terminal of the battery is connected to an electrode made of the coating metal. The circuit is completed when this electrode is submerged in the solution. Dissolved metal ions are attracted to the negatively charged object, where they pick up electrons and are deposited as metal atoms. The ions in solution are replenished by the forced oxidation of the coating metal at the positive electrode.

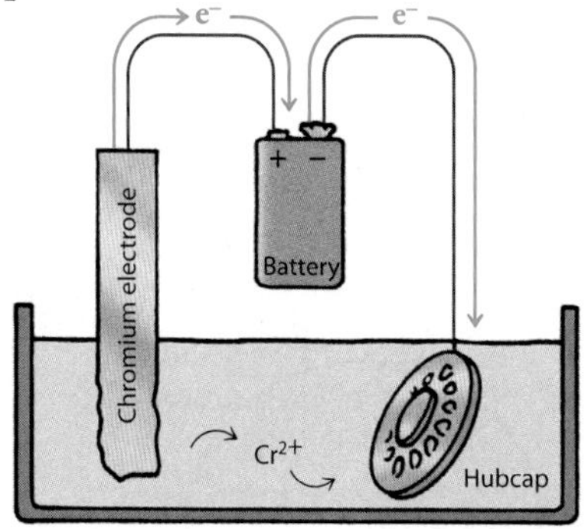

Figure 26.17
As electrons flow into the hubcap and give it a negative charge, positively charged chromium ions move from the solution to the hubcap and are reduced to chromium metal, which deposits as a coating on the hubcap. The solution is kept supplied with ions as chromium atoms in the cathode are oxidized to Cr^{2+} ions.

Combustion is an oxidation–reduction reaction between a nonmetallic material and molecular oxygen. Combustion reactions characteristically release energy. A violent combustion reaction is the formation of water from hydrogen and oxygen. The energy from this reaction is used to power rockets into space. More common examples of combustion include the burning of wood and fossil fuels. The combustion of these and other carbon-based chemicals forms carbon dioxide and water. Consider, for example, the combustion of methane, the major component of natural gas:

$$\underset{\text{Methane}}{CH_4} + \underset{\text{Oxygen}}{2\,O_2} \rightarrow \underset{\text{Carbon dioxide}}{CO_2} + \underset{\text{water}}{2\,H_2O} + \text{energy}$$

In combustion, electrons are transferred when polar covalent bonds are formed from nonpolar covalent bonds, or vice versa. This is in contrast to the other examples of oxidation–reduction reactions

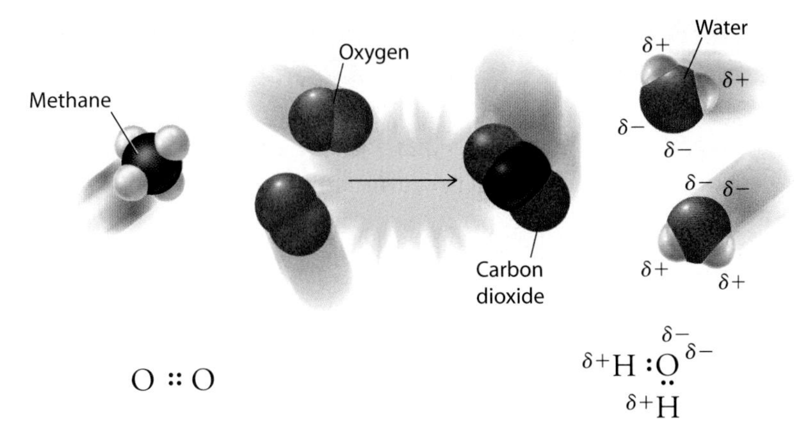

(a) Reactant oxygen atoms share electrons equally in O_2 molecules.

(b) Product oxygen atoms pull electrons away from H atoms in H_2O molecules and are reduced.

Figure 26.18
(a) Neither atom in an oxygen molecule is able to preferentially attract the bonding electrons.
(b) The oxygen atom of a water molecule pulls the bonding electrons away from the hydrogen atoms on the water molecule, making the oxygen slightly negative and the two hydrogens slightly positive.

presented in this chapter, which involve the formation of ions from atoms or, conversely, atoms from ions. This concept is illustrated in Figure 26.19, which compares the electronic structures of the combustion starting material molecular oxygen and the combustion product water. Molecular oxygen is a nonpolar covalent compound. Although each oxygen atom in the molecule has a fairly strong electronegativity, the four bonding electrons are pulled equally by both atoms and thus are unable to gather on one side or the other. After combustion, however, the electrons are shared between the oxygen and hydrogen atoms in a water molecule and are pulled to the oxygen. This gives the oxygen a slight negative charge, which is another way of saying it has gained electrons and has thus been reduced. At the same time, the hydrogen atoms in the water molecule develop a slight positive charge, which is another way of saying they have lost electrons and have thus been oxidized. This gain of electrons by oxygen and loss of electrons by hydrogen is an energy-releasing process. Typically, the energy is released either as molecular kinetic energy (heat) or as light (the flame).

Interestingly, the same sort of oxidation–reduction reaction occurs in your body. You can visualize a simplified model of your metabolism by reviewing Figure 26.18 and substituting a food molecule for the methane. Food molecules relinquish their electrons to the oxygen molecules you inhale. The products are carbon dioxide, water vapor, and energy. You exhale both the carbon dioxide and water vapor. Much of the energy from the reaction is used to keep your body warm and to drive the many other biochemical reactions necessary for living.

Chapter Review

Key Terms and Matching Definitions

_____ anode
_____ cathode
_____ combustion
_____ corrosion
_____ electrochemistry
_____ electrode
_____ electrolysis
_____ half reaction
_____ oxidation
_____ reduction

1. The process whereby a reactant loses one or more electrons.
2. The process whereby a reactant gains one or more electrons.
3. One portion of an oxidation–reduction reaction, represented by an equation showing electrons as either reactants or products.
4. The branch of chemistry concerned with the relationship between electrical energy and chemical change.
5. Any material that conducts electrons into or out of a medium in which electrochemical reactions are occurring.
6. The electrode where reduction occurs.
7. The electrode where oxidation occurs.
8. The use of electrical energy to produce chemical change.
9. The deterioration of a metal, typically caused by atmospheric oxygen.
10. An exothermic oxidation–reduction reaction between a nonmetallic material and molecular oxygen.

Review Questions

Oxidation Is the Loss of Electrons and Reduction Is the Gain of Electrons

1. Which elements have the greatest tendency to behave as oxidizing agents?
2. Write an equation for the half reaction in which a potassium atom, K, is oxidized.
3. Write an equation for the half reaction in which a bromine atom, Br, is reduced.
4. What happens to a reducing agent as it reduces?

The Energy of Flowing Electrons Can Be Harnessed

5. What is electrochemistry?
6. What is the purpose of the salt bridge shown in Figure 26.5?

The Electricity of a Battery Comes from Oxidation–Reduction Reactions

7. What is the purpose of the manganese dioxide in a dry-cell battery?
8. What chemical reaction is forced to occur while a car battery is being recharged?
9. Why don't the electrodes of a fuel cell deteriorate the way the electrodes of a battery do?

Fuel Cells Are Highly Efficient Sources of Electrical Energy

10. What are the chemical emissions of a hydrogen/oxygen fuel cell?
11. What is a major technological hurdle to the development of the fuel cell?

Electrical Energy Can Produce Chemical Change

12. What is electrolysis, and how does it differ from what goes on inside a battery?
13. What ore is used in the manufacture of aluminum?

Oxygen Is Responsible for Corrosion and Combustion

14. Why is oxygen such a good oxidizing agent?
15. What do the oxidation of zinc and the oxidation of aluminum have in common?
16. What is electroplating, and how is it accomplished?

17. What are some differences between corrosion and combustion?

18. What are some similarities between corrosion and combustion?

Explorations

Silver Lining

This is one of the better party tricks you can perform for any willing dinner host who has a cabinet full of tarnished silver pieces. Be forewarned, however, that many pieces coming out of the treatment will still be in need of some buffing with silver polish. Lively conversation is guaranteed.

Polishing with an abrasive paste removes both the thin layer of tarnish and some silver atoms. Silver-plated pieces are therefore susceptible to losing their thin coating of silver. The aluminum method, by contrast, restores the silver lost to the tarnishing.

For pieces too large to fit in the pot, try rubbing lightly with a paste of baking soda and water, using aluminum foil as your rubbing cloth.

Splitting Water

Try this activity with tap water instead of salt water to see the difference dissolved ions can make—the ions are needed to conduct electricity between the two electrodes.

The primary reaction occurs at the negative electrode, where water molecules accept electrons to form hydrogen gas and hydroxide ions. Recall from Chapter 25 that an increase in hydroxide ion concentration causes the pH of the solution to rise. You can track the production of hydroxide ions by adding a pH indicator to the solution. The indicator of choice is phenolphthalein, which you might obtain from your teacher. Alternatively, you might use the red cabbage extract discussed in Chapter 25. Whichever indicator you use, note the swirls of color forming at the negative electrode as hydroxide ions are generated.

The battery is quickly ruined because placing it in the conducting liquid short-circuits the terminals, which results in a large drain on the battery.

You may be wondering why oxygen gas is not generated along with the hydrogen gas. For reasons beyond the scope of this text, oxygen gas is generated only when the positive electrode is made of certain metals, such as gold or platinum. The steel electrode of the 9-volt battery does not suffice.

Exercises

1. Which atom is oxidized, ● or ●?

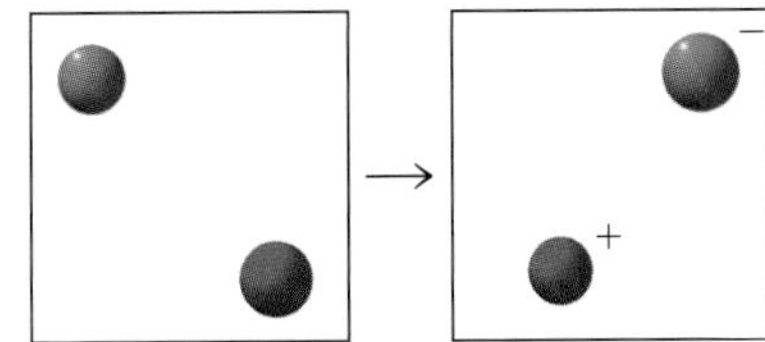

2. In the previous exercise, which atom behaves as the oxidizing agent, ● or ●?

3. What correlation might you expect between an element's electronegativity (Section 23.5) and its ability to behave as an oxidizing agent? How about its ability to behave as a reducing agent?

4. Based on their relative positions in the periodic table, which might you expect to be a stronger oxidizing agent, chlorine or fluorine? Why?

5. Iron is a better reducing agent than the copper ion Cu^{2+}. In which direction do electrons flow when an iron nail is submerged in a solution of Cu^{2+} ions?

6. Why is the anode of a battery indicated with a minus sign?

7. Is sodium metal oxidized or reduced in the production of aluminum?

8. Why is the formation of iron hydroxide, $Fe(OH)_2$, from Fe^{2+} and OH^- not considered an oxidation–reduction reaction?

9. Your car lights were left on while you were shopping, and now your car battery is dead. Has the pH of the battery fluid increased or decreased?

10. Jewelry is often manufactured by electroplating an expensive metal such as gold over a cheaper metal. Sketch a setup for this process.

11. Some car batteries require the periodic addition of water. Does adding the water increase or decrease the battery's ability to provide electric power to start the car? Explain.

12. Why does a battery that has thick zinc walls last longer than one that has thin zinc walls?

13. The oxidation of iron to rust is a problem structural engineers need to be concerned about, but the oxidation of aluminum to aluminum oxide is not. Why?

14. Why do combustion reactions generally release energy?

15. Which element is closer to the upper right corner of the periodic table, ● or ●?

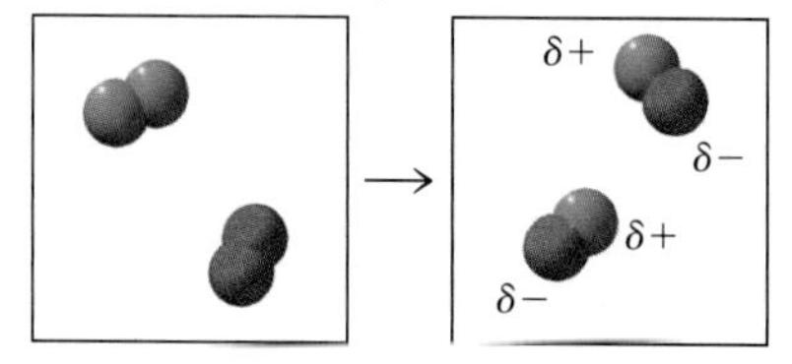

16. Water is 88.88% oxygen by mass. Oxygen is exactly what a fire needs to grow brighter and stronger. So why doesn't a fire grow brighter and stronger when water is added to it?

17. Clorox® is a laundry bleaching agent used to remove stains from white clothes. Suggest why the name begins with *Clor-* and ends with *-ox*.

18. Iron has a greater tendency to oxidize than does copper. Is this good news or bad news for a home in which much of the plumbing consists of iron and copper pipes connected together? Explain.

19. Copper has a greater tendency to be reduced than iron does. Was this good news or bad news for the Statue of Liberty, whose copper exterior was originally held together by steel rivets?

20. One of the products of combustion is water. Why doesn't this water extinguish the combustion?

Suggested Web Sites

http://www.aluminum.org/

The Web site of the Aluminum Association, Inc., where you will find basic facts about the aluminum industry, recycling efforts, and the impact of our aluminum use on the environment.

http://www.duracell.com/Fun_Learning/index.html

An excellent site to explore the chemistry and history of the battery. Also included is a Battery IQ Test you should take. Any questions you miss are prompted by a link to help you find the correct answers. Remember: **Leo** the lion went "**ger.**"

http://www.internationalfuelcells.com/
http://www.fuelcellworld.org/

Use *fuel cells* as a search keyword and you will find a number of private companies and organizations, such as the two named here, that are dedicated to improving the efficiency of fuel cells and publicizing their use. Fuel cells are certainly a wave of the future.

Chapter 27: Organic Compounds

Tetramethylpyrazine

Vanillin

What makes chocolate taste like chocolate and vanilla like vanilla? How is gasoline made from the thick black oil that comes from the ground? What is the difference between the alcohol of alcoholic beverages and the rubbing alcohol used to disinfect skin? Why does rotten fish smell so bad? And why does a scented flower smell so good? In this chapter, we explore how carbon atoms are unique in their ability to form a limitless number of chemical compounds, including the ones of which we ourselves are made.

27.1 Organic Chemistry Is the Study of Carbon Compounds

Carbon atoms can join to form large molecules made up of many carbon atoms. Add to this the fact that any of the carbon atoms so linked can also bond with atoms of other elements, and we have an endless number of different carbon-based molecules.

Each molecule has its own unique set of physical, chemical, and biological properties. The flavor of vanilla, for example, is perceived when the compound *vanillin* is absorbed by the sensory organs in the mouth and nose. A model of vanillin is shown in the chapter opening photograph. Note how it consists of a ring of carbon atoms (in black) with oxygen atoms (in red) attached in a particular fashion. Vanillin is the essential ingredient in anything that has the flavor of vanilla. Without vanillin, there is no vanilla flavor.

The flavor of chocolate, on the other hand, is not generated by a single molecule. You taste chocolate when, not just one, but a wide assortment of carbon-based molecules are absorbed in the mouth and nose. One of the more significant of these molecules is *tetramethylpyrazine,* which is also shown along with the chapter opening photograph. Note how tetramethylpyrazine has a ring of nitrogen (in blue) and carbon atoms (in black) attached in a particular fashion.

Life is based on carbon's ability to form diverse structures. Reflecting this fact, the branch of chemistry that is the study of carbon-containing compounds has come to be known as **organic chemistry.** The term *organic* is derived from *organism.* It is not necessarily related to the environment-friendly form of farming known as *organic farming.* Today, more than 13 million organic compounds are known, and about 100,000 new ones are added to the list each year. This includes those discovered in nature and those synthesized in the laboratory. (By contrast, there are only 200,000 to 300,000 known *inorganic* compounds. Inorganic compounds are those based on elements other than carbon.)

Because organic compounds are so closely tied to living organisms and because they have many applications—from flavorings to fuels, polymers, medicines, agriculture, and more—it is important to have a basic understanding of them. We begin with the simplest organic compounds. The simplest organic compounds are those consisting of only carbon and hydrogen.

27.2 Hydrocarbons Contain Only Carbon and Hydrogen

Methane, CH_4

Organic compounds that contain only carbon and hydrogen are **hydrocarbons.** Hydrocarbons differ from one another by the number of carbon and hydrogen atoms they contain. The simplest hydrocarbon is methane, CH_4. As the formula shows, it has only one carbon atom per molecule. Methane is the main component of natural gas. The hydrocarbon octane, C_8H_{18}, has eight carbons per molecule and is a component of gasoline. The hydrocarbon polyethylene contains hundreds of carbon and hydrogen atoms per molecule. Polyethylene is a plastic used to make many items, including milk containers and plastic bags.

Octane, C_8H_{18}

Hydrocarbons also differ from one another in the way the carbon atoms connect to each other. Figure 27.1 shows the three hydrocarbons *n*-pentane, *iso*-pentane, and *neo*-pentane. These hydrocarbons all have the same molecular formula, C_5H_{12}, but they are structurally different from one another. The carbon framework of *n*-pentane is a chain of five carbon atoms. In *iso*-pentane, the carbon chain branches, so that the framework is a *four*-carbon chain branched at the second carbon. In *neo*-pentane, a central carbon atom is bonded to four surrounding carbon atoms.

Polyethylene

We can see the different structural features of *n*-pentane, *iso*-pentane, and *neo*-pentane more clearly by drawing the molecules in two dimensions, as shown in the middle row of Figure 27.1. Another way to represent them is by the *stick structures* shown in the bottom row. A stick structure is a commonly used, shorthand notation for representing an organic molecule. Each line (stick) represents a covalent bond, and carbon atoms are understood to be

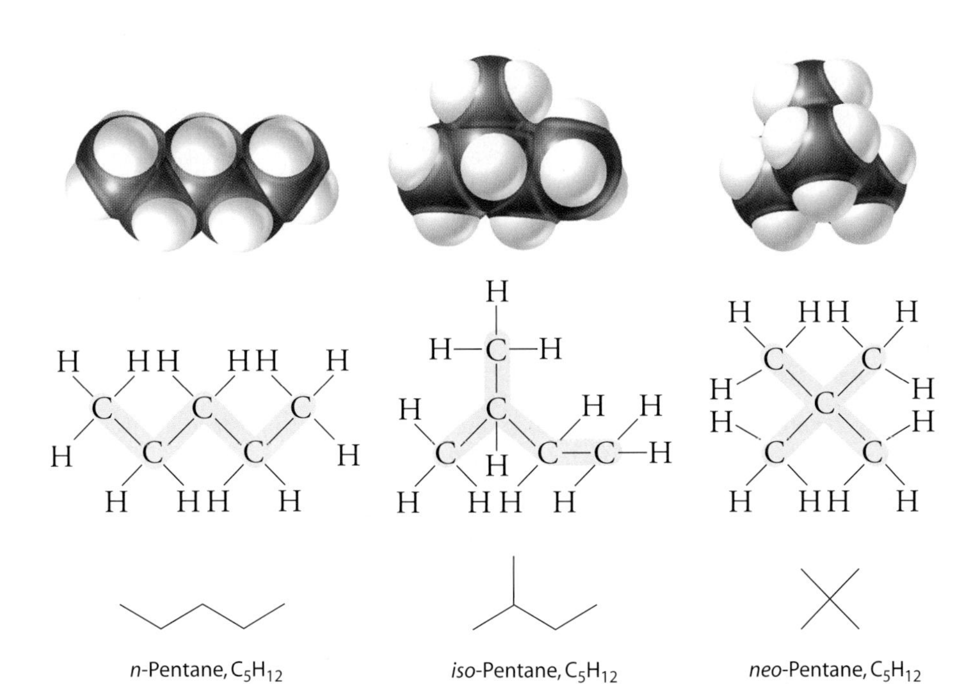

Figure 27.1
These three hydrocarbons all have the same molecular formula. We can see their different structural features by highlighting the carbon framework in two dimensions. Easy-to-draw stick structures that use lines for all carbon-carbon covalent bonds can also be used.

wherever two or more straight lines meet and at the end of any line (unless another type of atom is drawn at the end of the line). Any hydrogen atoms bonded to the carbons are also typically not shown. Instead, their presence is implied so that the focus can remain on the skeletal structure formed by the carbon atoms.

When every carbon atom in a hydrocarbon except the two end ones is bonded to only two other carbon atoms, the molecule is called a *straight-chain hydrocarbon.* (Do not take this name literally, for, as the *n*-pentane structures in Figure 27.1 show, this is a straight-chain hydrocarbon despite the zigzag nature of the drawings representing it.) When one or more carbon atoms in a hydrocarbon is bonded to either three or four carbon atoms, the molecule is a *branched hydrocarbon.* Both *iso*-pentane and *neo*-pentane are branched hydrocarbons.

Molecules such as *n*-pentane, *iso*-pentane, and *neo*-pentane, which have the same molecular formula but different structures, are known as **structural isomers.** Structural isomers have different physical and chemical properties. For example, *n*-pentane has a boiling point of 36°C, *iso*-pentane's boiling point is 30°C, and *neo*-pentane's is 10°C.

The number of possible structural isomers for a chemical formula increases rapidly as the number of carbon atoms increases. There are 3 structural isomers for compounds having the formula C_5H_{12}; 18 for C_8H_{18}; 75 for $C_{10}H_{22}$; and a whopping 366,319 for $C_{20}H_{42}$!

Carbon-based molecules can have different spatial orientations called **conformations.** Flex your wrist, elbow, and shoulder joints, and you'll find your arm passing through a range of conformations. Likewise, organic molecules can twist and turn about their carbon-carbon single bonds and thus have a range of conformations. The structures in Figure 27.2, for example, are different conformations of *n*-pentane.

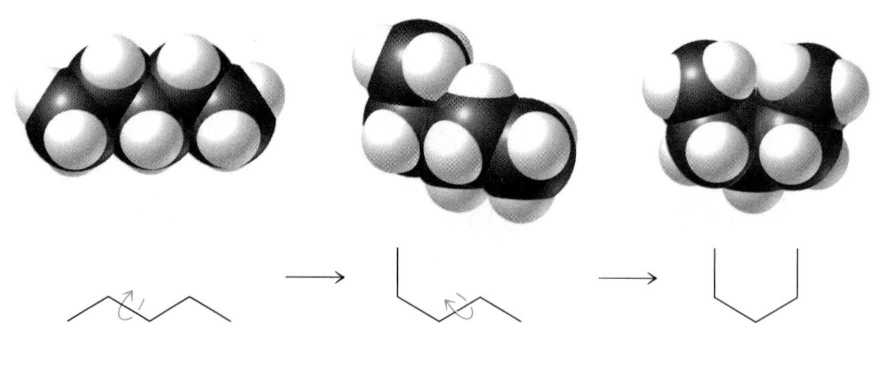

Figure 27.2
Three conformations for a molecule of *n*-pentane. The molecule looks different in each conformation, but the five-carbon framework is the same in all three conformations. In a sample of liquid *n*-pentane, the molecules are found in all conformations—not unlike a bucket of worms.

Concept Check ✓

Which carbon—carbon bond was rotated to go from the "before" conformation of *iso*-pentane to the "after" conformation:

b a c d — Before

b a c d — After

Check Your Answer The best way to answer any question about the conformation of a molecule is to play around with molecular models that you can hold in your hand. In this case, bond c rotates in such a way that the carbon at the right end of bond d comes up out of the plane of the page, momentarily points straight at you, and then plops back into the plane of the page below bond c. This rotation is similar to that of the arm of an arm wrestler who, her arm just above the table as she is on the brink of losing, suddenly gets a surge of strength. She swings her opponent's arm (and her own) through a half-circle arc and wins.

Before After

Hydrocarbons are obtained primarily from coal and petroleum. Coal and petroleum are both formed when plant and animal matter decays in the absence of oxygen. Most of the coal and petroleum that exist today were formed between 280 and 395 million years ago. At that time, there were extensive swamps that, because they were close to sea level, periodically became submerged. The organic matter of the swamps was buried beneath layers of marine sediments and was eventually transformed to either coal or petroleum.

Coal is a solid material containing many large, complex hydrocarbon molecules. Most of the coal mined today is used for the production of steel and for generating electricity at coal-burning power plants.

Petroleum, also called *crude oil,* is a liquid that can easily be separated into its hydrocarbon components through the process of *fractional distillation,* shown in Figure 27.3. The crude oil is heated in a pipe still to a temperature high enough to vaporize most of the components. The hot vapor flows into the bottom of a fractionating tower, which is warmer at the bottom than at the top. As the vapor rises in the tower and cools, the various components begin to condense. Hydrocarbons that have high boiling points, such as tar and lubricating stocks, condense first at warmer temperatures. Hydrocarbons that have

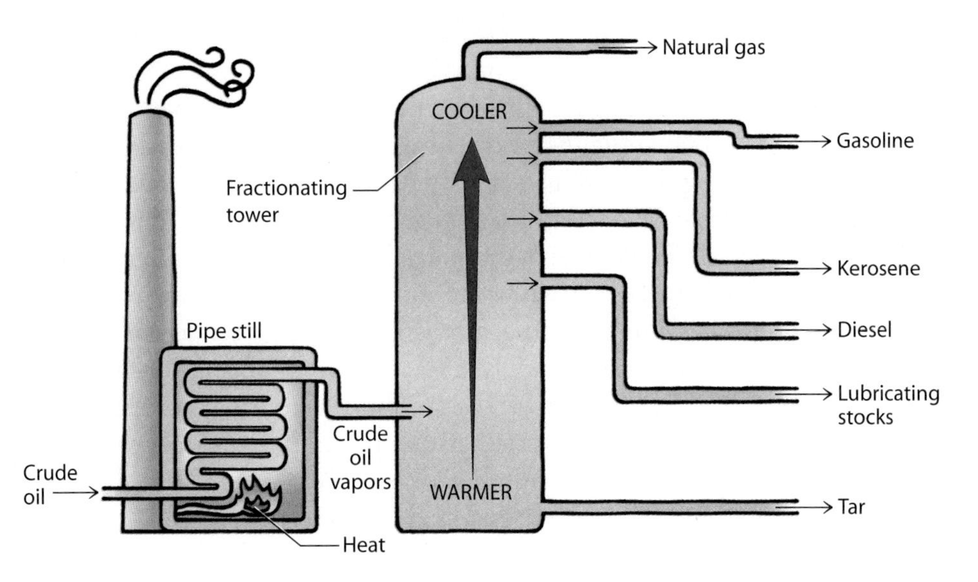

Figure 27.3
A schematic for the fractional distillation of petroleum into its useful hydrocarbon components.

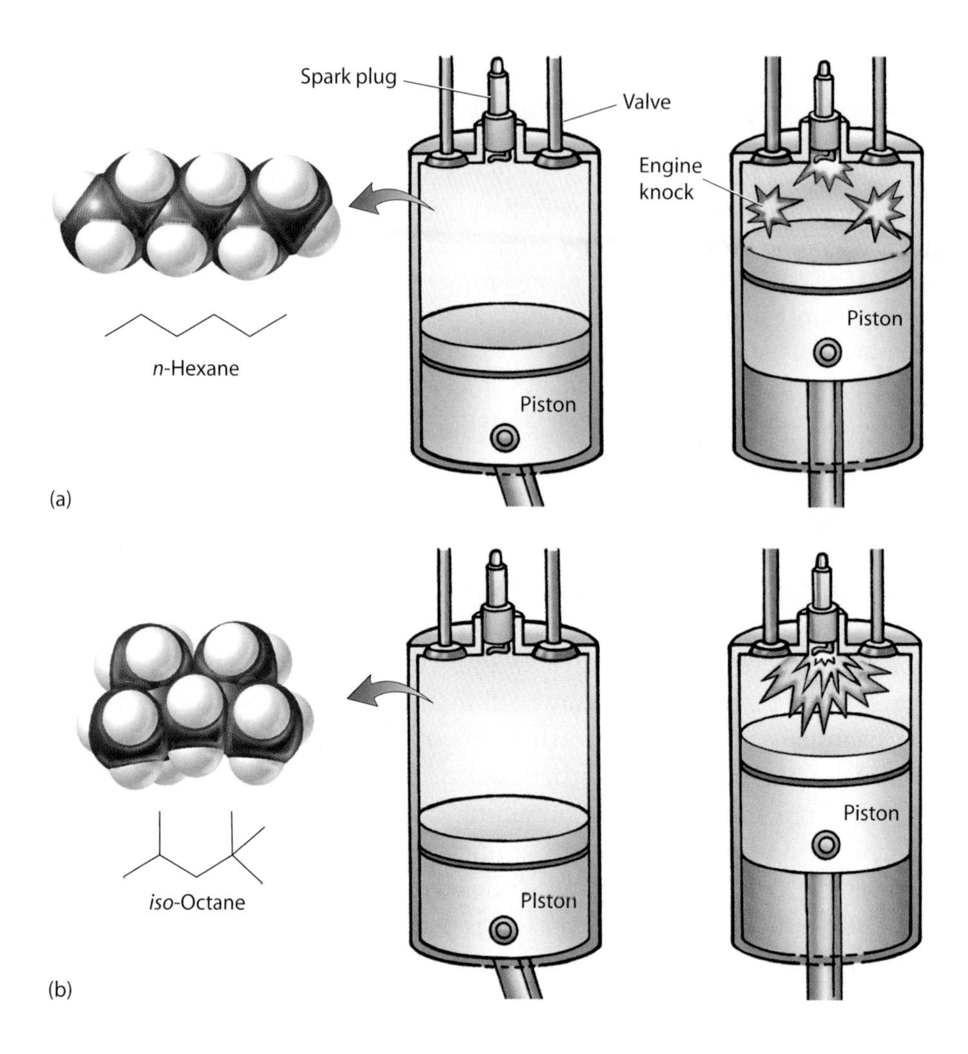

Figure 27.4
(a) A straight-chain hydrocarbon, such as *n*-hexane, can be ignited from the heat generated as gasoline is compressed by the piston—before the spark plug fires. This upsets the timing of the engine cycle, giving rise to a knocking sound. (b) Branched hydrocarbons, such as *iso*-octane, burn less readily and are ignited not by compression alone but only when the spark plug fires.

low boiling points, such as gasoline, travel to the cooler regions at the top of the tower before condensing. Pipes drain the various liquid hydrocarbon fractions from the tower. Natural gas, which is primarily methane, does not condense. It remains a gas and is collected at the top of the tower.

The gasoline obtained from the fractional distillation of petroleum consists of a wide variety of hydrocarbons that have similar boiling points. Some of these components burn more efficiently than others in a car engine. The straight-chain hydrocarbons, such as *n*-hexane, tend to burn too quickly. This causes what is called *engine knock,* as illustrated in Figure 27.4. Gasoline hydrocarbons that have more branching, such as *iso*-octane, burn slowly, and as a result the engine runs more smoothly. These two compounds, *n*-hexane and *iso*-octane, are used as standards in assigning *octane ratings* to gasoline. An octane number of 100 is arbitrarily assigned to *iso*-octane, and *n*-hexane is assigned an octane number of 0. The antiknock performance of a particular gasoline is compared with that of various mixtures of *iso*-octane and *n*-hexane, and an octane number is assigned. Figure 27.5 shows octane information on a typical gasoline pump.

Figure 27.5
Octane ratings are posted on gasoline pumps.

Concept Check ✓

Which structural isomer in Figure 27.1 should have the highest octane rating?

Check Your Answer The structural isomer with the greatest amount of branching in the carbon framework will likely have the highest octane rating, making *neo*-pentane the clear winner. Just for the record, the ratings are

Compound	Octane rating
n-Pentane	61.7
iso-Pentane	92.3
neo-Pentane	116

27.3 Unsaturated Hydrocarbons Contain Multiple Bonds

Recall from Section 23.1 that carbon has four unpaired valence electrons. As shown in Figure 27.6, each of these electrons is available for pairing with an electron from another atom, such as hydrogen, to form a covalent bond.

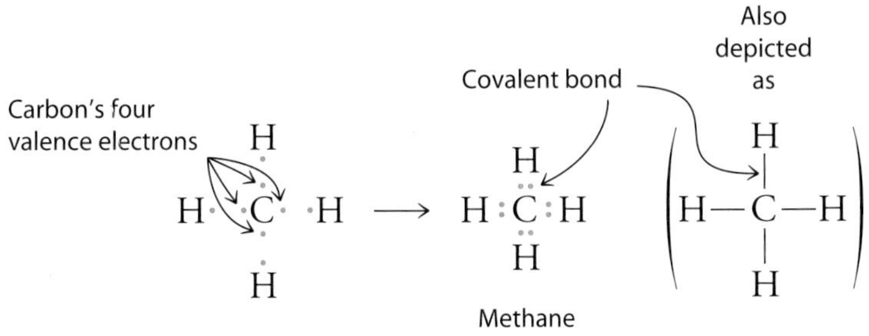

Carbon's four valence electrons

Covalent bond

Also depicted as

Methane

Figure 27.6
Carbon has four valence electrons. Each electron pairs with an electron from a hydrogen atom in the four covalent bonds of methane.

In all the hydrocarbons discussed so far, including the methane shown in Figure 27.6, each carbon atom is bonded to four neighboring atoms by four single covalent bonds. Such hydrocarbons are known as **saturated hydrocarbons.** The term *saturated* means that each carbon has as many atoms bonded to it as possible. We now explore cases where one or more carbon atoms in a hydrocarbon are bonded to fewer than four neighboring atoms. This occurs when at least one of the bonds between a carbon and a neighboring atom is a multiple bond.

A hydrocarbon with a multiple bond—either double or triple—is known as an **unsaturated hydrocarbon.** Two of the carbons are bonded to fewer than four other atoms because of the multiple bond. These carbons are thus said to be *unsaturated.*

Figure 27.7 compares the saturated hydrocarbon *n*-butane with the unsaturated hydrocarbon 2-butene. The number of atoms bonded to each of the two middle carbons of *n*-butane is four, whereas each of the two middle carbons of 2-butene is bonded to only three other atoms—a hydrogen and two carbons.

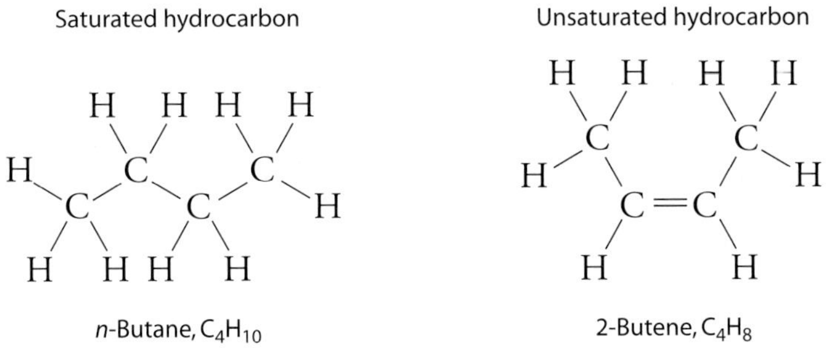

Saturated hydrocarbon

n-Butane, C_4H_{10}

Unsaturated hydrocarbon

2-Butene, C_4H_8

Figure 27.7
The carbons of the hydrocarbon *n*-butane are *saturated,* each being bonded to four other atoms. Because of the double bond, two of the carbons of the unsaturated hydrocarbon 2-butene are bonded to only three other atoms, which makes the molecule an unsaturated hydrocarbon.

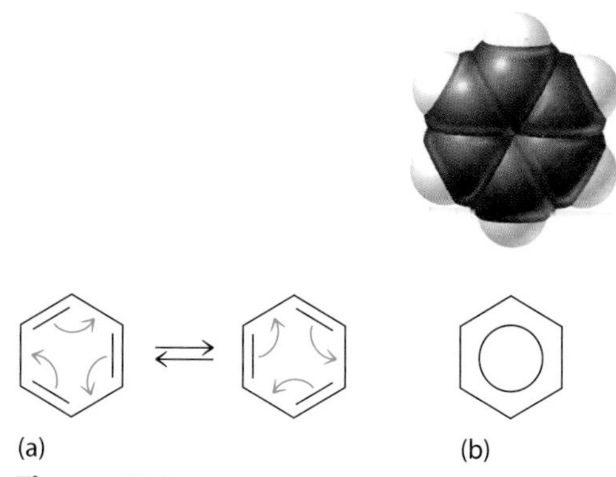

Figure 27.8
(a) The double bonds of benzene, C_6H_6, are able to migrate around the ring. (b) For this reason, they are often represented by a circle within the ring.

An important unsaturated hydrocarbon is benzene, C_6H_6, which may be drawn as three double bonds contained within a flat hexagonal ring, as is shown in Figure 27.8a. Unlike the double-bond electrons in most other unsaturated hydrocarbons, however, the electrons of the double bonds in benzene are not fixed between any two carbon atoms. Instead, these electrons are able to move freely around the ring. This is commonly represented by drawing a circle within the ring, as shown in Figure 27.8b, rather than the individual double bonds.

Many organic compounds contain one or more benzene rings in their structure. Because many of these compounds are fragrant, any organic molecule containing a benzene ring is classified as an **aromatic compound** (even if it is not particularly fragrant). Figure 27.9 shows a few examples. Toluene, a common solvent used as paint thinner, is toxic and gives airplane glue its distinctive odor. Some aromatic compounds, such as naphthalene, contain two or more benzene rings fused together. At one time, mothballs were made of naphthalene. Most mothballs sold today, however, are made of the less toxic 1,4-dichlorobenzene.

An example of an unsaturated hydrocarbon containing a triple bond is acetylene, C_2H_2. A confined flame of acetylene burning in oxygen is hot enough to melt iron, which makes acetylene a choice fuel for the welding shown in Figure 27.10.

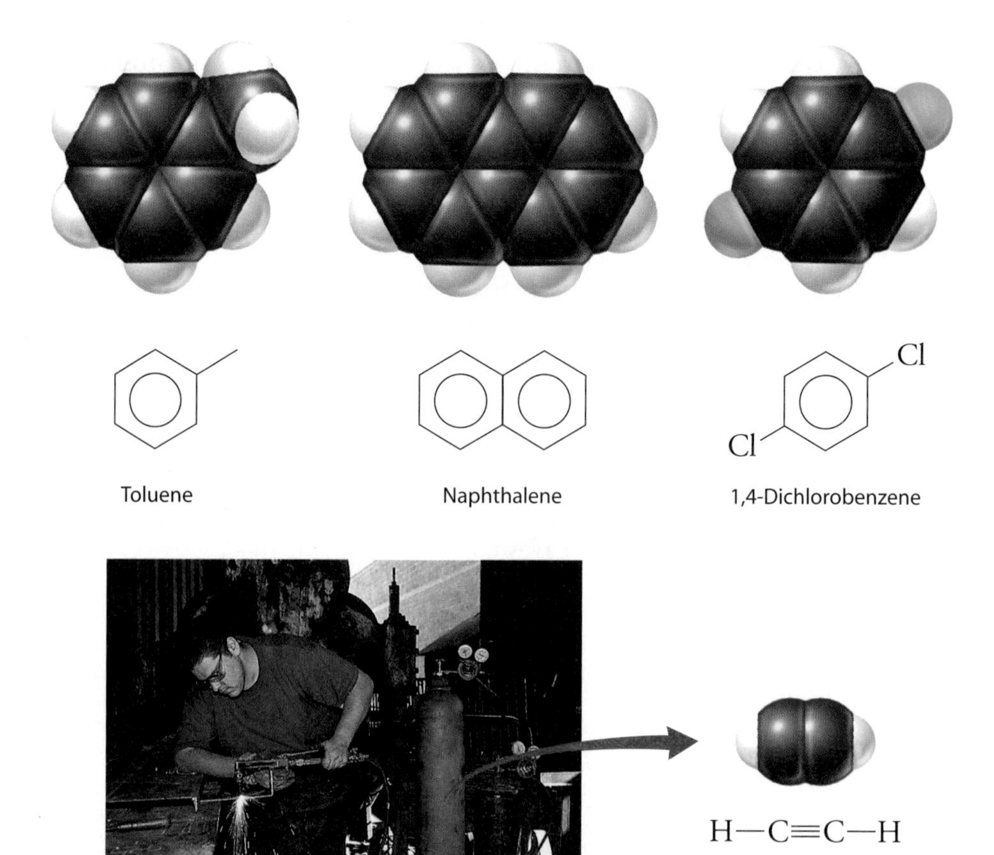

Figure 27.9
The structures for three odoriferous organic compounds containing one or more benzene rings: toluene, naphthalene, and 1,4-dichlorobenzene.

Figure 27.10
The unsaturated hydrocarbon acetylene, C_2H_2, burned in this torch produces a flame hot enough to melt iron.

Concept Check ✓

Prolonged exposure to benzene increases the risk of certain cancers. The structure of aspirin contains a benzene ring. Does this indicate that prolonged exposure to aspirin increases a person's risk of developing cancer?

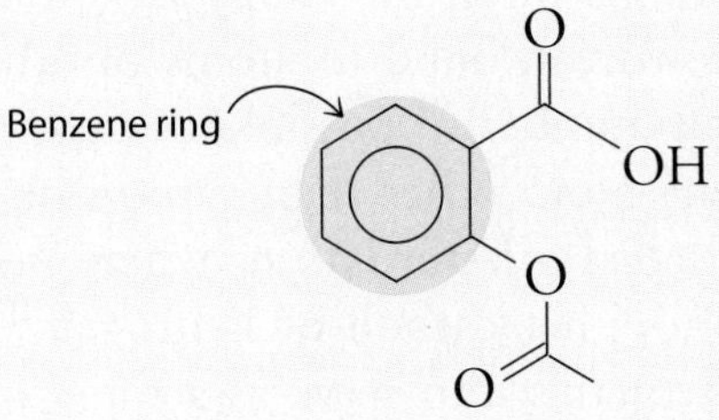

Benzene ring
O
OH
O
O

Check Your Answer No. Although benzene and aspirin both contain a benzene ring, these two molecules have different overall structures. Therefore the properties of one are quite different from the properties of the other. Each carbon-containing organic compound has its own set of unique physical, chemical, and biological properties. While benzene may cause cancer, aspirin is a safe remedy for headaches.

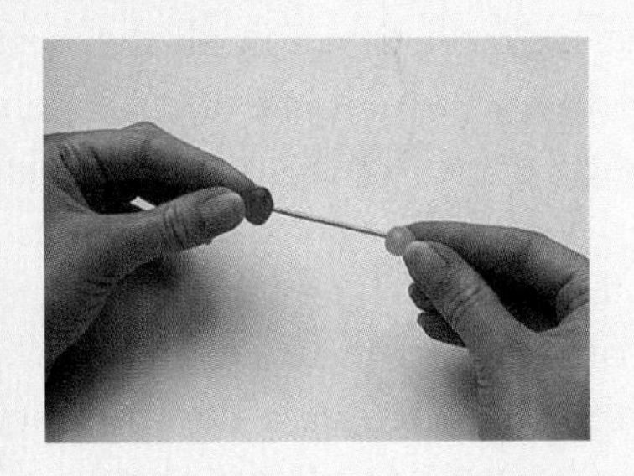

Hands-On Exploration: Twisting Jellybeans

Two carbon atoms connected by a single bond can rotate relative to each other. As we discussed in Section 27.2, this ability to rotate can give rise to numerous conformations (spatial orientations) of an organic molecule. Is it also possible for carbon atoms connected by a double bond to rotate relative to each other? Perform this quick activity to see for yourself.

What You Need

Jellybeans (or gumdrops), round toothpicks.

Procedure

1. Attach one jellybean to each end of a single toothpick. Hold one of the jellybeans firmly with one hand while rotating the second jellybean with your other hand. Observe how there is no restriction on the different orientations of the two jellybeans relative to each other.

2. Hold two toothpicks side by side and attach one jellybean to each end such that each jellybean has both toothpicks poked into it. As before, hold one jellybean while rotating the other. What kind of rotations are possible now?

Relate what you observe to the carbon–carbon double bond. Which structure of Figure 27.7 do you suppose has more possible conformations: *n*-butane or 2-butene? What do you suppose is true about the ability of atoms connected by a carbon–carbon triple bond to twist relative to each other?

27.4 Organic Molecules Are Classified by Functional Group

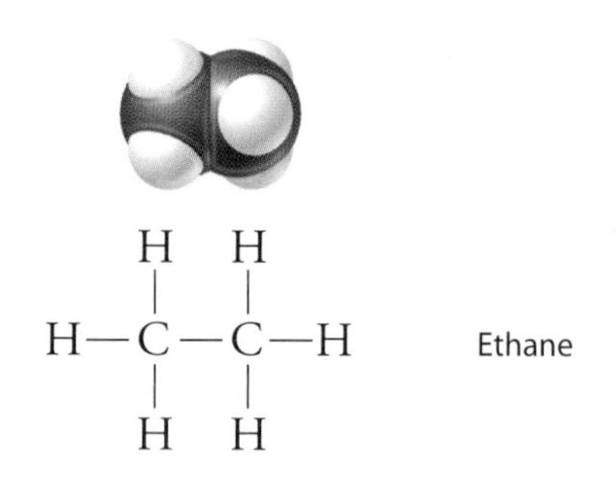

Ethane

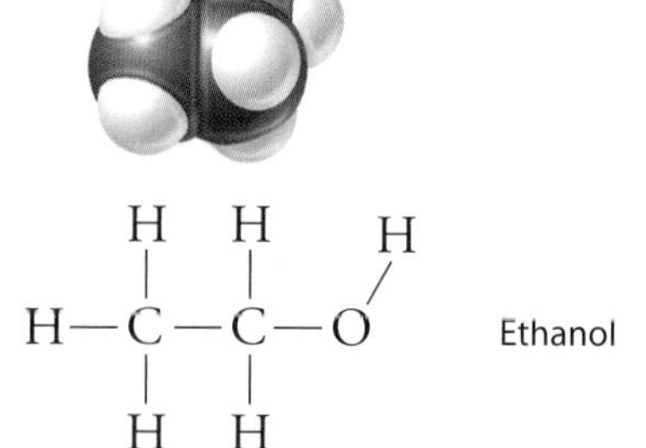

Ethanol

Ethylamine

Carbon atoms can bond to one another and to hydrogen atoms in many ways, which results in an incredibly large number of hydrocarbons. But carbon atoms can bond to atoms of other elements as well, further increasing the number of possible organic molecules. In organic chemistry, any atom other than carbon or hydrogen in an organic molecule is called a **heteroatom,** where *hetero-* in this case means "different from either carbon or hydrogen."

A hydrocarbon structure can serve as a framework to which various heteroatoms can be attached. This is analogous to a Christmas tree serving as the scaffolding on which ornaments are hung. Just as the ornaments give character to the tree, so do heteroatoms give character to an organic molecule. In other words, heteroatoms can have profound effects on the properties of an organic molecule.

Consider ethane, C_2H_6, and ethanol, C_2H_6O, which differ from each other by only a single oxygen atom. Ethane has a boiling point of 288°C, making it a gas at room temperature, and it does not dissolve in water very well. Ethanol, by contrast, has a boiling point of 178°C, making it a liquid at room temperature. It is infinitely soluble in water and is the active ingredient of alcoholic beverages. Consider further ethylamine, C_2H_7N, which has a nitrogen atom on the same basic two-carbon framework. This compound is a corrosive, pungent, highly toxic gas—most unlike either ethane or ethanol.

Organic molecules are classified according to the functional groups they contain. A **functional group** is defined as a combination of atoms that behave as a unit. Most functional groups are distinguished by the heteroatoms they contain, and some common groups are listed in Table 27.1.

The remainder of this chapter introduces the classes of organic molecules shown in Table 27.1. The role heteroatoms play in determining the properties of each class is the underlying theme. As you study this material, focus on understanding the chemical and physical properties of the various classes of compounds. Doing so will give you a greater appreciation of the remarkable diversity of organic molecules and their many applications.

Concept Check ✓

What is the significance of heteroatoms in an organic molecule?

Check Your Answer Heteroatoms largely determine an organic molecule's "personality."

Table 27.1

Functional Groups in Organic Molecules

General Structure	Name	Class
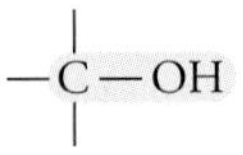 —C—OH	Hydroxyl group	Alcohols
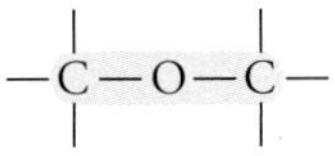 —C—O—C—	Ether group	Ethers
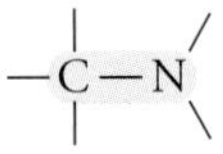 —C—N	Amine group	Amines
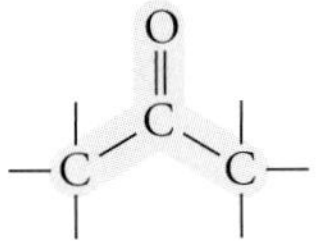 —C—C(=O)—C—	Ketone group	Ketones
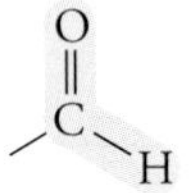 —C(=O)—H	Aldehyde group	Aldehydes
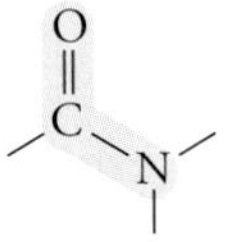 —C(=O)—N—	Amide group	Amides
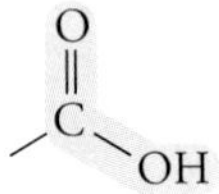 —C(=O)—OH	Carboxyl group	Carboxylic acids
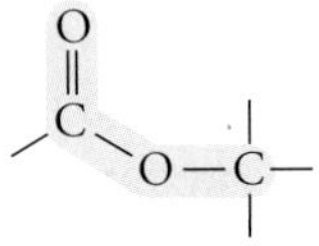 —C(=O)—O—C—	Ester group	Esters

Alcohols Contain the Hydroxyl Group

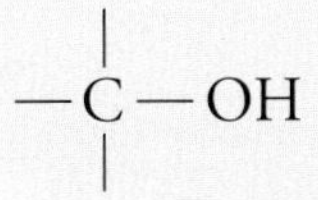

Alcohols are organic molecules in which a *hydroxyl group* is bonded to a saturated carbon. The hydroxyl group consists of an oxygen bonded to a hydrogen. Because of the polarity of the oxygen-hydrogen bond, low-formula-mass alcohols are often soluble in water, which is itself very polar. Some common alcohols and their melting and boiling points are listed in Table 27.2.

—C—OH

Hydroxyl group

More than 11 billion pounds of methanol, CH_3OH, are produced annually in the United States. Most of it is used for making formaldehyde and acetic acid, which are important starting materials in the production of plastics. In addition, methanol is used as a solvent, an octane booster, and an anti-icing agent in gasoline. Sometimes called *wood alcohol* because it can be obtained from wood,

Table 27.2

Some Simple Alcohols

Structure	Scientific Name	Common Name	Melting Point (°C)	Boiling Point (°C)
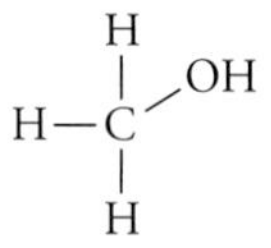 H—C(H)(H)—OH	Methanol	Methyl alcohol	−97	65
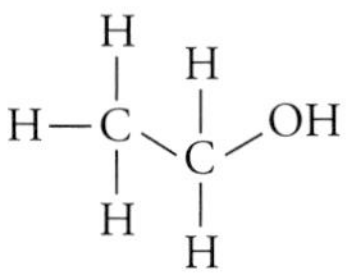 H—C(H)(H)—C(H)(H)—OH	Ethanol	Ethyl alcohol	−115	78
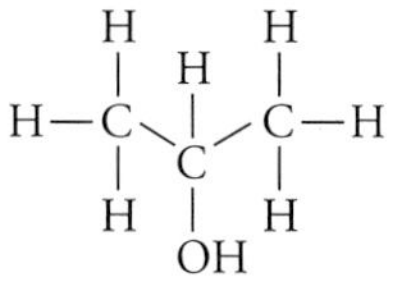 H—C(H)(H)—C(H)(OH)—C(H)(H)—H	2-Propanol	Isopropyl alcohol	−126	97

methanol should never be ingested. In the body it is metabolized to formaldehyde and formic acid. Formaldehyde is harmful to the eyes, can lead to blindness, and was once used to preserve dead biological specimens. Formic acid, the active ingredient in an ant bite, can lower the pH of the blood to dangerous levels. Ingesting only about 15 milliliters of methanol may lead to blindness, and about 30 milliliters can cause death.

Ethanol, C_2H_5OH, is one of the oldest chemicals manufactured by humans. The "alcohol" of alcoholic beverages, ethanol is prepared by feeding the sugars of various plants to certain yeasts, which produce ethanol through a biological process known as *fermentation.* Ethanol is widely used as an industrial solvent. For many years, ethanol intended for this purpose was made by fermentation. But today industrial-grade ethanol is more cheaply manufactured from petroleum byproducts, such as ethene. Figure 27.11 illustrates this.

The liquid produced by fermentation has an ethanol concentration no greater than about 12% because at this concentration the yeast begin to die. This is why most wines have an alcohol content of 11% or 12%—they are produced solely by fermentation. To attain the higher ethanol concentrations found in such "hard"

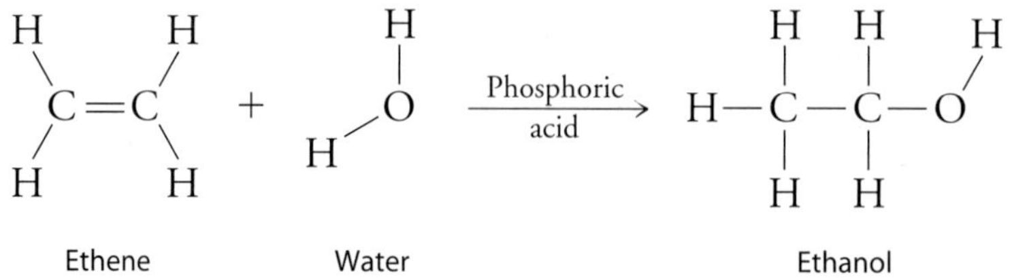

$$\underset{\text{Ethene}}{H_2C{=}CH_2} + \underset{\text{Water}}{H{-}O{-}H} \xrightarrow[\text{acid}]{\text{Phosphoric}} \underset{\text{Ethanol}}{H{-}CH_2{-}CH_2{-}O{-}H}$$

Figure 27.11
Ethanol can be synthesized from the unsaturated hydrocarbon ethene, with phosphoric acid as a catalyst.

alcoholic beverages as gin and vodka, the fermented liquid must be distilled. In the United States, the ethanol content of alcoholic beverages is measured as *proof,* which is twice the percent of ethanol. An 86-proof whiskey, for example, is 43% ethanol by volume. The term *proof* evolved from a crude method once employed to test alcohol content. Gunpowder was wetted with a beverage of suspect alcohol content. If the beverage was primarily water, the powder would not ignite. If the beverage contained a significant amount of ethanol, the powder would burn, thus providing "proof" of the beverage's worth.

A third well-known alcohol is isopropyl alcohol, also called 2-propanol. This is the rubbing alcohol you buy at the drugstore. Although 2-propanol has a relatively high boiling point, it readily evaporates, leading to a pronounced cooling effect when applied to skin—an effect once used to reduce fevers. (Isopropyl alcohol is very toxic if ingested. Washcloths wetted with cold water are nearly as effective in reducing fever and far safer.) You are probably most familiar with the use of isopropyl alcohol as a topical disinfectant.

The Oxygen of an Ether Group Is Bonded to Two Carbon Atoms

Ethers are organic compounds structurally related to alcohols. The oxygen atom in an ether group, however, is bonded not to a carbon and a hydrogen but rather to two carbons. As we see in Figure 27.12, ethanol and dimethyl ether have the same chemical formula, C_2H_6O, but their physical properties are vastly different. Whereas ethanol is a liquid at room temperature (boiling point 78°C) and mixes quite well with water, dimethyl ether is a gas at room temperature (boiling point 225°C) and is much less soluble in water.

—C—O—C—

Ether group

Ethers are not very soluble in water because, without the hydroxyl group, they are unable to form strong hydrogen bonds with water (Section 24.1). Furthermore, without the polar hydroxyl group, the molecular attractions among ether molecules are relatively weak. As a result, it does not take much energy to separate ether molecules

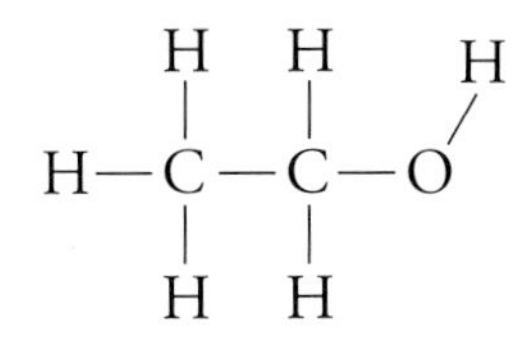

Ethanol: Soluble in water, boiling point 78°C

Dimethyl ether: Insoluble in water, boiling point −25°C

Figure 27.12
The oxygen in an alcohol such as ethanol is bonded to one carbon atom and one hydrogen atom. The oxygen in an ether such as dimethyl ether is bonded to two carbon atoms. Because of this difference, alcohols and ethers of similar molecular mass have vastly different physical properties.

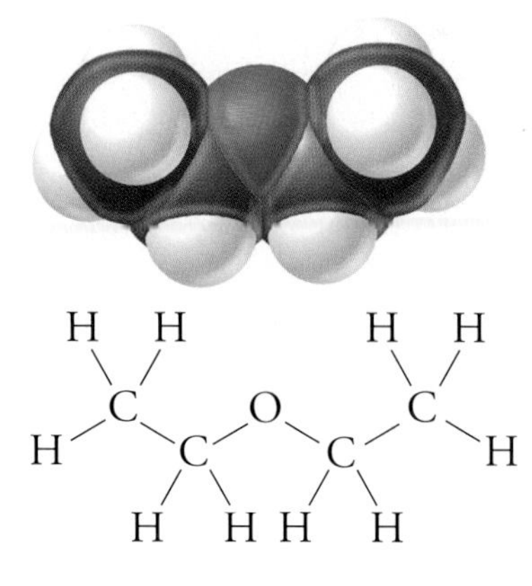

Diethyl ether,
boiling point 35°C

Figure 27.13
Diethyl ether is the systematic name for the "ether" historically used as an anesthetic.

from one another. This is why ethers have relatively low boiling points and evaporate so readily.

Diethyl ether, shown in Figure 27.13, was one of the first anesthetics. When the anesthetic properties of this compound were discovered in the early 1800s, the practice of surgery was revolutionized. Because of its high volatility at room temperature, inhaled diethyl ether rapidly enters the bloodstream. But since this ether has low solubility in water and high volatility, it quickly leaves the bloodstream once introduced. These physical properties allow a doctor to bring a surgical patient in and out of anesthesia (a state of unconsciousness) simply by regulating the gases breathed. Modern-day gaseous anesthetics have fewer side effects than diethyl ether but work on the same principle.

Amines Form Alkaline Solutions

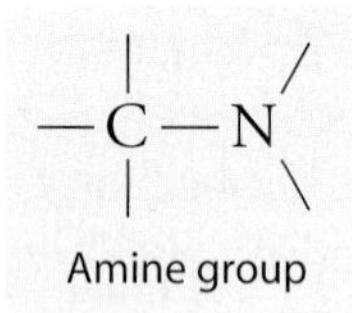

Amines are organic compounds that contain the amine group—a nitrogen atom bonded to one, two, or three saturated carbons. Amines are typically less soluble in water than are alcohols because the nitrogen-hydrogen bond is not quite as polar as the oxygen-hydrogen bond. The lower polarity of amines also means their boiling points are typically somewhat lower than those of alcohols of similar formula mass. Table 27.3 lists three simple amines.

Table 27.3

Three Simple Amines

Structure	Name	Melting Point (°C)	Boiling Point (°C)
$CH_3CH_2NH_2$	Ethylamine	281	17
$CH_3CH_2NHCH_2CH_3$	Diethylamine	250	55
$(CH_3CH_2)_3N$	Triethylamine	27	89

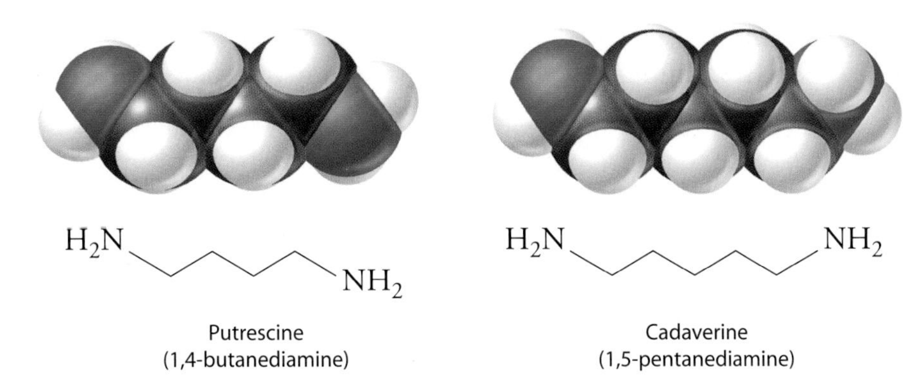

Putrescine (1,4-butanediamine)

Cadaverine (1,5-pentanediamine)

Figure 27.14
Low-formula-mass amines like these tend to have an offensive odor.

One of the most notable physical properties of many low-formula-mass amines is their offensive odor. Figure 27.14 shows two appropriately named amines, putrescine and cadaverine, responsible for the odor of decaying flesh.

Amines are typically alkaline because the nitrogen atom readily accepts a hydrogen ion from water, as Figure 27.15 illustrates.

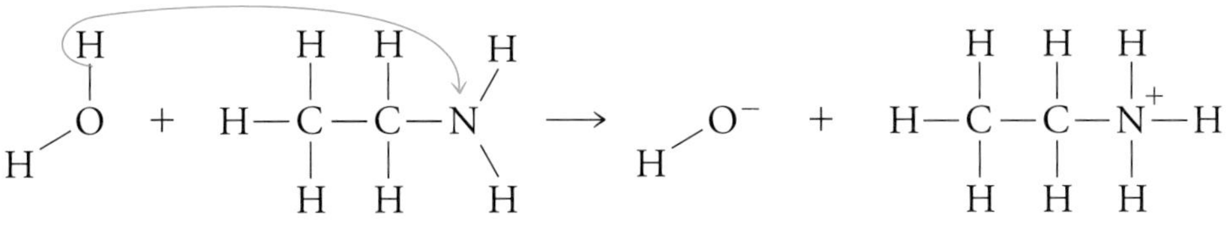

$$H_2O + CH_3CH_2NH_2 \longrightarrow OH^- + CH_3CH_2NH_3^+$$

Water (acid) | Ethylamine (base) | Hydroxide ion | Ethylammonium ion

Figure 27.15
Ethylamine acts as a base and accepts a hydrogen ion from water to become the ethylammonium ion. This reaction generates a hydroxide ion, which increases the pH of the solution.

A group of naturally occurring complex molecules that are alkaline because they contain nitrogen atoms are often called *alkaloids.* Because many alkaloids have medicinal value, there is great interest in isolating these compounds from plants or marine organisms containing them. As shown in Figure 27.16, an alkaloid reacts with an acid to form a salt that is usually quite soluble in water. This is in contrast to the nonionized form of the alkaloid, known as a *free base* and typically insoluble in water.

Most alkaloids exist in nature not in their free-base form but rather as the salt of naturally occurring acids known as *tannins,* a group of organic compounds that have complex structures and the ability to donate hydrogen ions. The alkaloid salts of these acids are usually much more soluble in hot water than in cold water. The caffeine in coffee and tea exists in the form of the tannin salt, which is

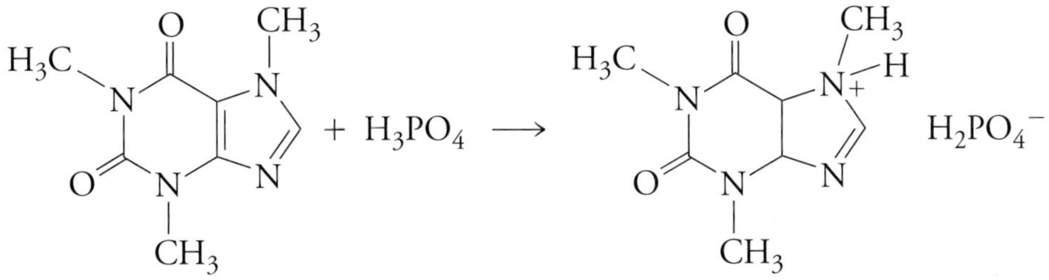

$$\text{Caffeine} + H_3PO_4 \longrightarrow \text{Caffeine-}H^+ \; H_2PO_4^-$$

Caffeine, free-base form (water-insoluble) | Phosphoric acid | Caffeine–phosphoric acid salt (water-soluble)

Figure 27.16
All alkaloids are bases that react with acids to form salts. An example is the alkaloid caffeine, shown here reacting with phosphoric acid.

Figure 27.17
Tannins are responsible for the brown stains in coffee mugs or on a coffee drinker's teeth. Because tannins are acidic, they can be readily removed with an alkaline cleanser. Use a little laundry bleach on the mug, and brush your teeth with baking soda.

why coffee and tea are more effectively brewed in hot water. As Figure 27.17 relates, tannins are also responsible for the stains caused by these beverages.

Concept Check ✓

Why do most caffeinated soft drinks also contain phosphoric acid?

Check Your Answer The phosphoric acid, as shown in Figure 27.16, reacts with the caffeine to form the caffeine phosphoric acid salt, which is much more soluble in cold water than the naturally occurring tannin salt.

Ketones, Aldehydes, Amides, Carboxylic Acids, and Esters All Contain a Carbonyl Group

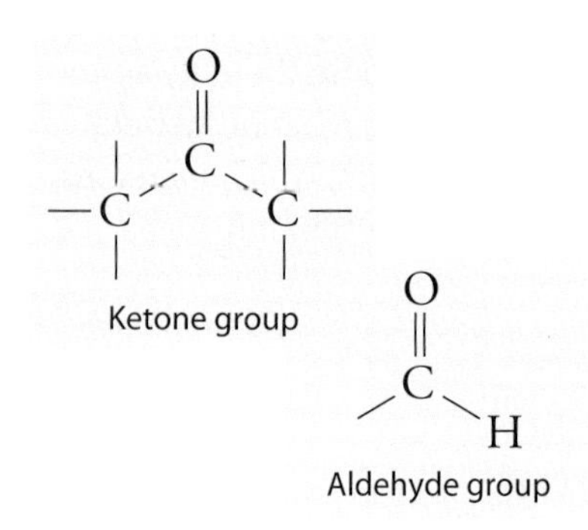

Ketones, aldehydes, amides, carboxylic acids, and esters all contain a **carbonyl group.** The carbonyl group consists of a carbon atom double-bonded to an oxygen atom.

A **ketone** is a carbonyl-containing organic molecule in which the carbonyl carbon is bonded to two carbon atoms. A familiar example of a ketone is *acetone,* which is often used in fingernail polish remover and is shown in Figure 27.18a. In an **aldehyde,** the carbonyl carbon is bonded either to one carbon atom and one hydrogen atom, as in Figure 27.18b, or, in the special case of formaldehyde, to two hydrogen atoms.

Many aldehydes are particularly fragrant. A number of flowers, for example, owe their pleasant odor to the presence of simple aldehydes. The smells of lemons, cinnamon, and almonds are due to the aldehydes citral, cinnamaldehyde, and benzaldehyde, respectively. The structures of these three aldehydes are shown in Figure 27.19. The aldehyde vanillin, introduced at the beginning of this chapter, is the key flavoring molecule derived from the vanilla orchid. You may have noticed that vanilla beans and vanilla extract are fairly expensive.

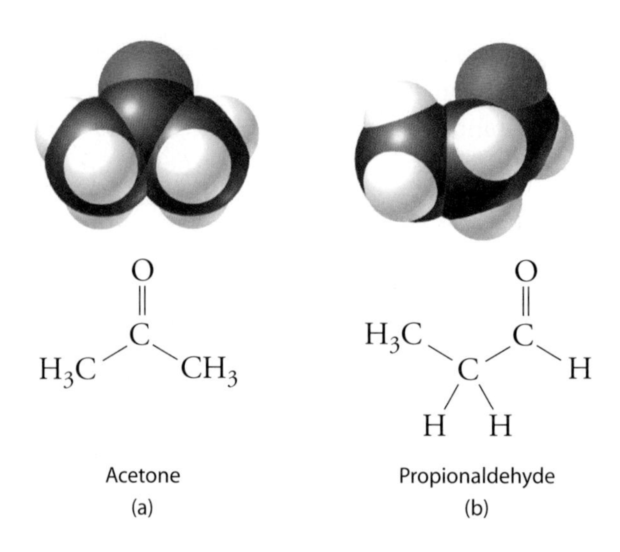

Figure 27.18
(a) When the carbon of a carbonyl group is bonded to two carbon atoms, the result is a ketone. An example is acetone.
(b) When the carbon of a carbonyl group is bonded to at least one hydrogen atom, the result is an aldehyde. An example is propionaldehyde.

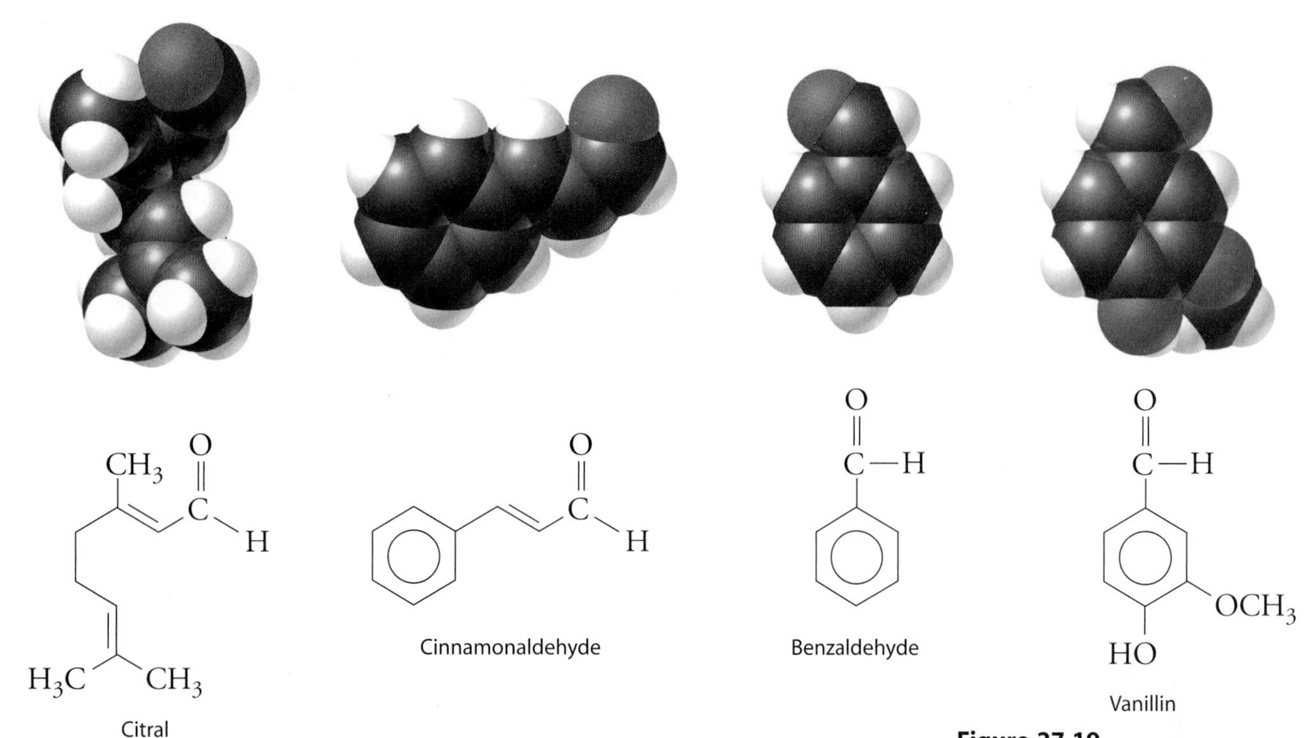

Figure 27.19
Aldehydes are responsible for many familiar fragrances.

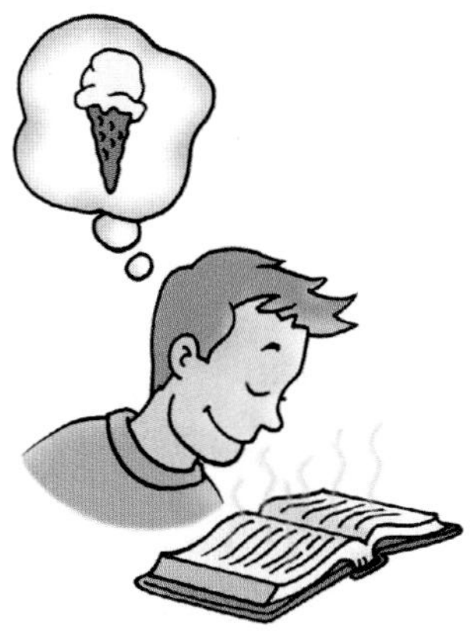

Imitation vanilla flavoring is less expensive because it is merely a solution of the compound vanillin, which is economically synthesized from the waste chemicals of the wood pulp industry. Imitation vanilla does not taste the same as natural vanilla extract, however, because in addition to vanillin many other flavorful molecules contribute to the complex taste of natural vanilla. Many books made in the days before "acid-free" paper smell of vanilla because of the vanillin formed and released as the paper ages, a process that is accelerated by the acids the paper contains.

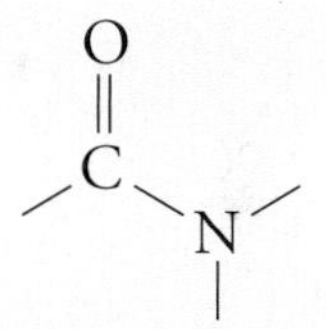

Amide group

An **amide** is a carbonyl-containing organic molecule in which the carbonyl carbon is bonded to a nitrogen atom. The active ingredient of most mosquito repellents is an amide whose chemical name is *N,N*-diethyl-*m*-toluamide but is commercially known as DEET, shown in Figure 27.20. This compound is actually not an insecticide. Rather, it causes certain insects, especially mosquitoes, to lose their sense of direction, which effectively protects DEET wearers from being bitten.

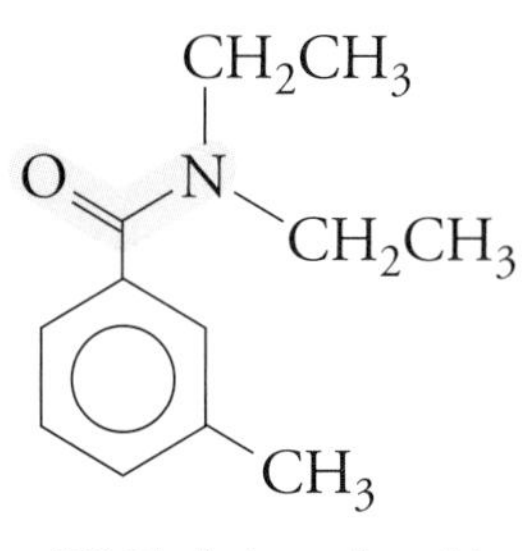

N,N-Diethyl-*m*-toluamide (DEET)

Figure 27.20
N,N-diethyl-*m*-toluamide is an example of an amide. Amides contain the amide group, shown highlighted in blue.

A **carboxylic acid** is a carbonyl-containing organic molecule in which the carbonyl carbon is bonded to a hydroxyl group. As its name implies, this functional group is able to donate hydrogen ions, and as a result organic molecules containing it are acidic. An example is acetic acid, $C_2H_4O_2$, the main ingredient of vinegar. You may recall that this organic compound was used as an example of a weak acid back in Chapter 25. An interesting example of an organic

Figure 27.21
Aspirin, acetylsalicylic acid, has both the carboxylic acid and ester functional groups.

Carboxyl group

Ester

Aspirin
(acetylsalicylic acid)

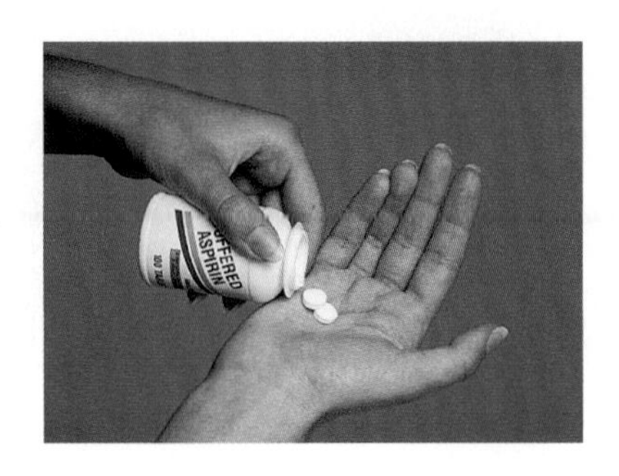

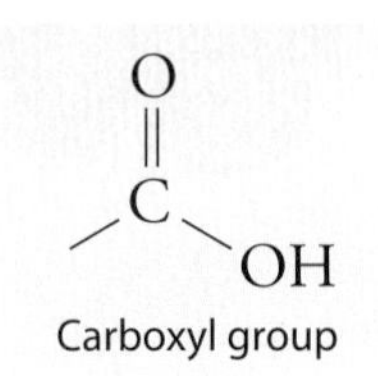

Carboxyl group

Ester group

compound that contains a carboxylic acid is acetylsalicylic acid, also known as aspirin (Figure 27.21).

An **ester** is an organic molecule similar to a carboxylic acid except that in the ester the hydroxyl hydrogen is replaced by a carbon. Unlike carboxylic acids, esters are not acidic because they lack the hydrogen of the hydroxyl group. Like aldehydes, many simple esters have notable fragrances and are used as flavorings. Some familiar ones are listed in Table 27.4.

Table 27.4

Some Esters and Their Flavors

Structure	Name	Flavor
$H_3C-C(=O)-O-CH_2CH_2-CH(CH_3)_2$	Isopentyl acetate	Banana
$CH_3CH_2CH_2-C(=O)-O-CH_2CH_3$	Ethyl butyrate	Pineapple
$H-C(=O)-O-CH_2-CH(CH_3)_2$	Isobutyl formate	Raspberry
$C_6H_4(OH)-C(=O)-O-CH_3$	Methyl salicylate	Wintergreen

Concept Check ✓

Identify all the functional groups in these four molecules (ignore the sulfur group in penicillin G):

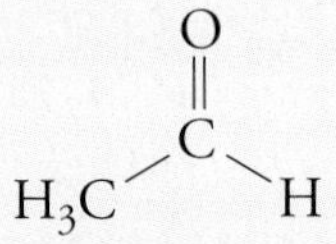

O
C
H_3C H
Acetaldehyde

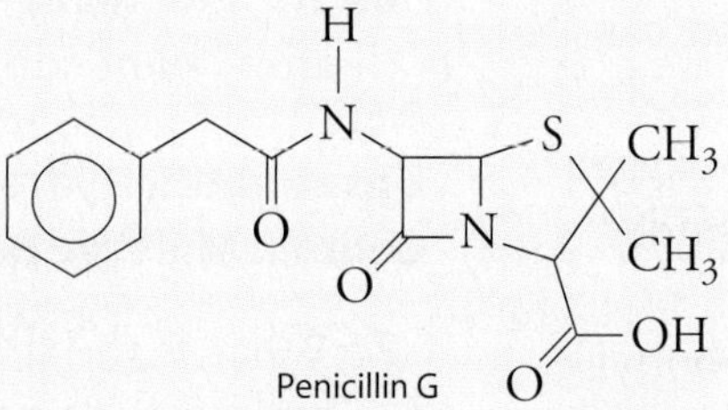

H N S CH_3 O N CH_3 O OH O
Penicillin G

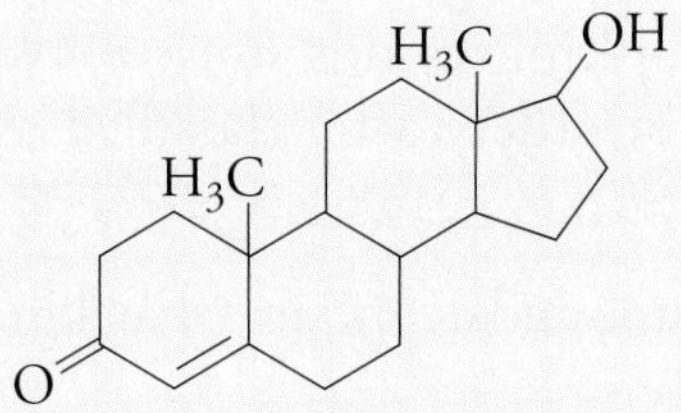

H_3C OH H_3C O
Testosterone

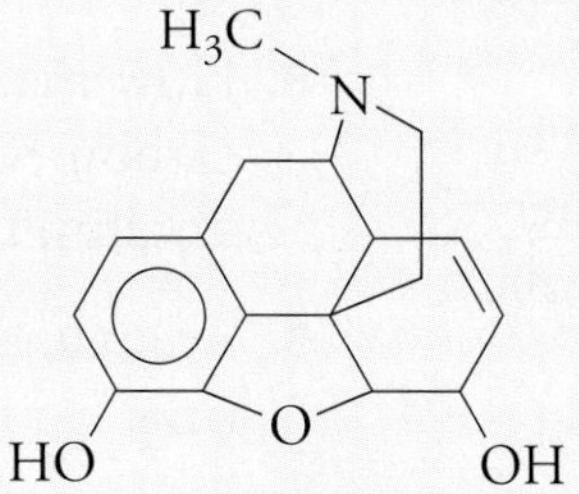

H_3C N HO O OH
Morphine

Check Your Answer Acetaldehyde: aldehyde; penicillin G: amide (two amide groups), carboxylic acid; testosterone: alcohol and ketone; morphine: alcohol, phenol, ether, and amine.

Chapter Review

Key Terms and Matching Definitions

_____ organic chemistry
_____ hydrocarbons
_____ structural isomers
_____ conformations
_____ saturated hydrocarbons
_____ unsaturated hydrocarbons
_____ aromatic compound
_____ heteroatom
_____ functional group
_____ alcohol
_____ ethers
_____ amines
_____ carbonyl group
_____ ketone
_____ aldehyde
_____ amide
_____ carboxylic acid
_____ ester

1. The study of carbon-containing compounds.
2. A chemical compound containing only carbon and hydrogen atoms.
3. Molecules that have the same molecular formula but different chemical structures.
4. One of the possible spatial orientations of a molecule.
5. A hydrocarbon containing no multiple covalent bonds, with each carbon atom bonded to four other atoms.
6. A hydrocarbon containing at least one multiple covalent bond.
7. Any organic molecule containing a benzene ring.
8. Any atom other than carbon or hydrogen in an organic molecule.
9. A specific combination of atoms that behave as a unit in an organic molecule.
10. An organic molecule that contains a hydroxyl group bonded to a saturated carbon.

11. An organic molecule containing an oxygen atom bonded to two carbon atoms.
12. An organic molecule containing a nitrogen atom bonded to one or more saturated carbon atoms.
13. A carbon atom double-bonded to an oxygen atom, found in ketones, aldehydes, amides, carboxylic acids, and esters.
14. An organic molecule containing a carbonyl group the carbon of which is bonded to two carbon atoms.
15. An organic molecule containing a carbonyl group the carbon of which is bonded either to one carbon atom and one hydrogen atom or to two hydrogen atoms.
16. An organic molecule containing a carbonyl group the carbon of which is bonded to a nitrogen atom.
17. An organic molecule containing a carbonyl group the carbon of which is bonded to a hydroxyl group.
18. An organic molecule containing a carbonyl group the carbon of which is bonded to one carbon atom and one oxygen atom bonded to another carbon atom.

Review Questions

Organic Chemistry Is the Study of Carbon Compounds

1. Why is carbon able to form a limitless number of compounds?
2. What organic molecule gives rise to the flavor of vanilla?

Hydrocarbons Contain Only Carbon and Hydrogen

3. How do two structural isomers differ from each other?
4. How are two structural isomers similar to each other?
5. What physical property of hydrocarbons is used in fractional distillation?
6. To how many atoms is a saturated carbon atom bonded?

Unsaturated Hydrocarbons Contain Multiple Bonds

7. What is the difference between a saturated hydrocarbon and an unsaturated hydrocarbon?
8. How many multiple bonds must a hydrocarbon have in order to be classified as unsaturated?
9. Aromatic compounds contain what kind of ring?

Organic Molecules Are Classified by Functional Group

10. What is a heteroatom?
11. Why are low-formula-mass alcohols soluble in water?
12. What distinguishes an alcohol from an ether?
13. Why do ethers typically have lower boiling points than alcohols?
14. Which heteroatom is characteristic of an amine?
15. Do amines tend to be acidic, neutral, or basic?
16. Are alkaloids found in nature?
17. What are some examples of alkaloids?
18. Which elements make up the carbonyl group?
19. How are ketones and aldehydes related to each other? How are they different from each other?
20. How are amides and carboxylic acids related to each other? How are they different from each other?

Exploration

Twisting Jellybeans

What you should discover in this activity is that the carbon-carbon bond greatly restricts the number of possible conformations for an organic molecule. While *n*-butane, for instance, can twist like a snake, 2-butene is restricted to one of two possible conformations. (Check back to Figure 27.7 for the structures of these two molecules.) In one conformation, the two end carbons are on the same side of the double bond—this is called the *cis* conformation. In the second conformation, the two end carbons are on opposite sides of the double bond—the *trans* conformation:

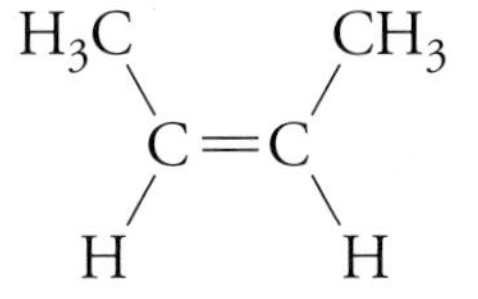

```
H3C       CH3        H3C        H
   \     /              \      /
    C=C                  C=C
   /     \              /     \
  H       H            H       CH3
```

cis-2-Butene 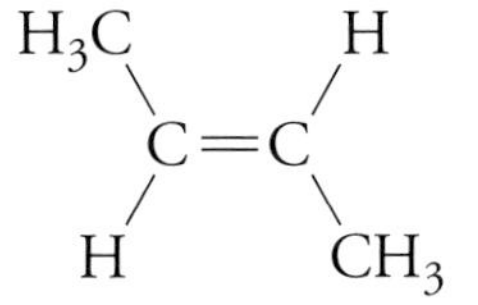   *trans*-2-Butene

Because the double bond cannot rotate, the *cis* and *trans* conformations cannot be made from each other. They therefore represent two different molecules (structural isomers), each having its own unique set of properties. The melting point of *cis*-2-butene, for example, is −139°C, while that of *trans*-2-butene is a warmer −106°C.

Exercises

1. Which contains more hydrogen atoms: a five-carbon saturated hydrocarbon molecule or a five-carbon unsaturated hydrocarbon molecule?

2. Why does the melting point of hydrocarbons increase as the number of carbon atoms per molecule increases?

3. Draw all the structural isomers for hydrocarbons having the molecular formula C_4H_{10}.

4. Draw all the structural isomers for hydrocarbons having the molecular formula C_6H_{14}.

5. How many structural isomers are shown here?

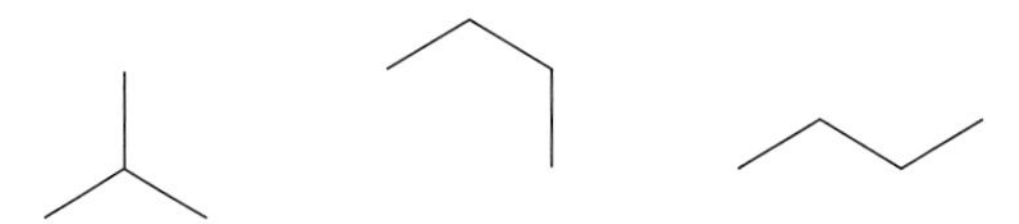

6. Which two of these four structures are of the same structural isomer?

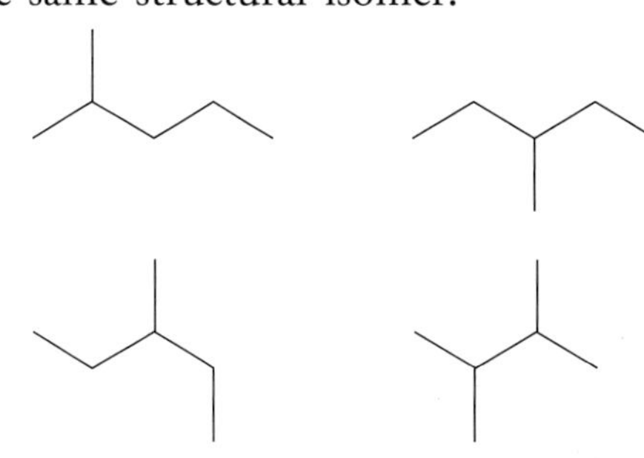

7. The temperatures in a fractionating tower at an oil refinery are important, but so are the pressures. Where might the pressure in a fractionating tower be greatest, at the bottom or at the top? Defend your answer.

8. Heteroatoms make a difference in the physical and chemical properties of an organic molecule because;
   a. they add extra mass to the hydrocarbon structure.
   b. each heteroatom has its own characteristic chemistry.
   c. they can enhance the polarity of the organic molecule.
   d. all of the above.

9. What is the percent volume of water in 80-proof vodka?

10. How does ingested methanol lead to the damaging of a person's eyes?

11. The phosphoric acid salt of caffeine has the structure

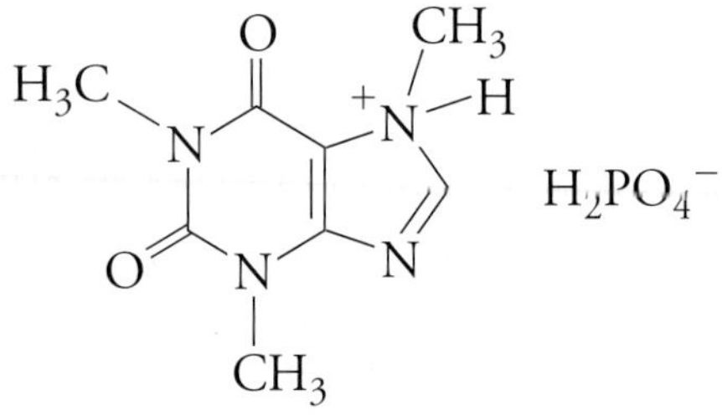

$H_2PO_4^-$

Caffeine–phosphoric acid salt

This molecule behaves as an acid in that it can donate a hydrogen ion, created from the hydrogen atom bonded to the positively charged nitrogen atom. What are all the products formed when 1 mole of this salt reacts with 1 mole of sodium hydroxide, NaOH, a strong base?

12. The solvent diethyl ether can be mixed with water but only by shaking the two liquids together. After the shaking is stopped, the liquids separate into two layers, much like oil and vinegar. The free-base form of the alkaloid caffeine is readily soluble in diethyl ether but not in water. Suggest what might happen to the caffeine of a caffeinated beverage if the beverage was first made alkaline with sodium hydroxide and then shaken with some diethyl ether.

13. Alkaloid salts are not very soluble in the organic solvent diethyl ether. What might happen to the free-base form of caffeine dissolved in diethyl ether if gaseous hydrogen chloride, HCl, were bubbled into the solution?

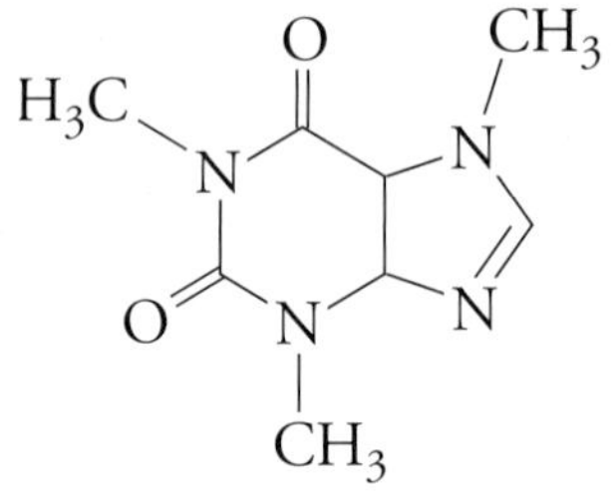

Caffeine
(free base)

14. Draw all the structural isomers for amines having the molecular formula C_3H_9N.

15. Explain why caprylic acid, $CH_3(CH_2)_6COOH$, dissolves in a 5% aqueous solution of sodium hydroxide but caprylaldehyde, $CH_3(CH_2)_6CHO$, does not.

16. In water, does the molecule

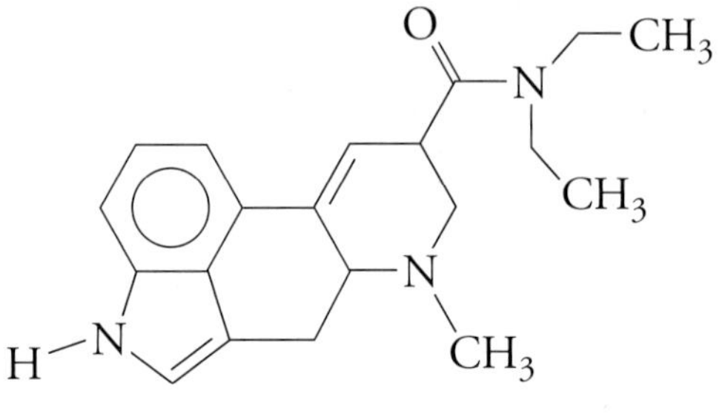

Lysergic acid diethylamide

act as an acid, a base, neither, or both?

17. If you saw the label phenylephrine · HCl on a decongestant, would you worry that consuming it would expose you to the strong acid hydrochloric acid? Explain.

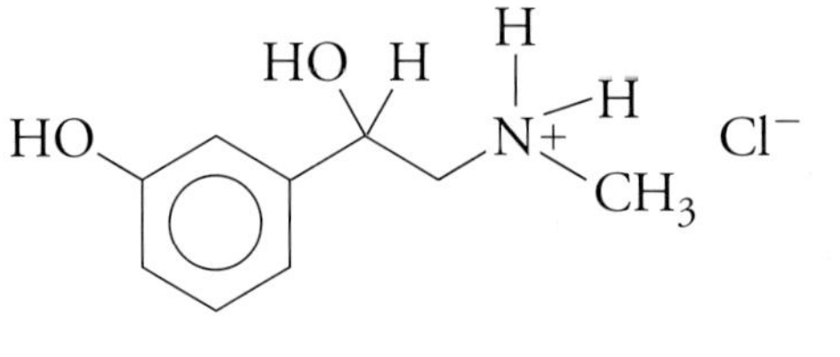

Cl^-

Phenylephrine

18. Suggest an explanation for why aspirin has a sour taste.

19. An amino acid is an organic molecule that contains both an amine group and a carboxyl group. At an acidic pH, which structure is most likely?

(a) 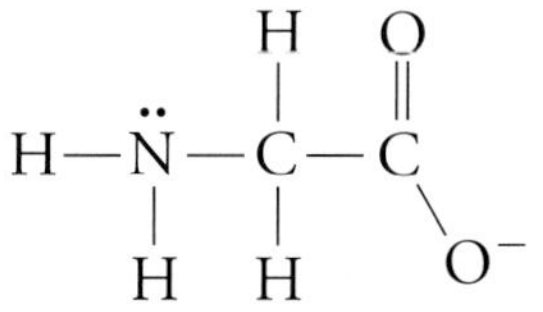 $H-\ddot{N}(H)-C(H)(H)-C(=O)-O^-$

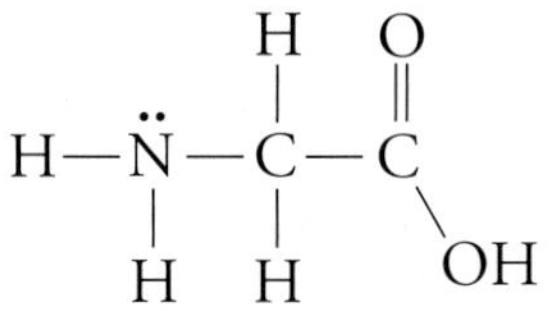

(b) $H-\ddot{N}(H)-C(H)(H)-C(=O)-OH$

(c) $H-N^+(H)(H)-C(H)(H)-C(=O)-OH$

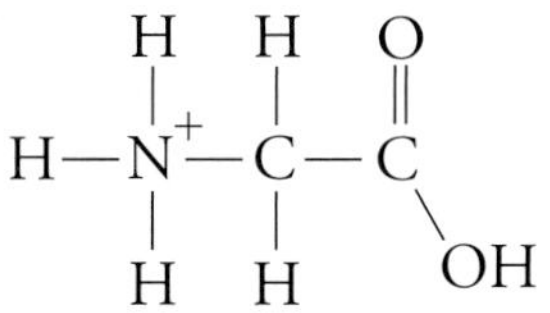

Explain your answer.

20. Identify the following functional groups in this organic molecule—amide, ester, ketone, ether, alcohol, aldehyde, amine:

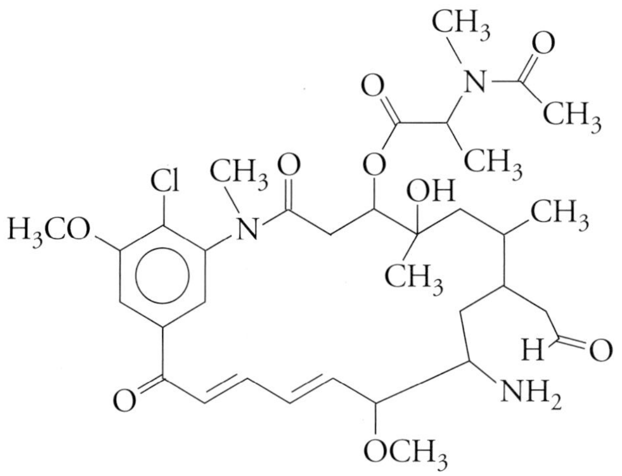

Suggested Readings and Web Sites

P. W. Atkins. *Molecules.* New York: W. H. Freeman, 1987.
An enchanting account of some of the more important organic molecules of nature as well as those produced by chemists. Written for the general public, the dialogue is warm, intriguing, and accompanied by spectacular photographs.

http://www.icco.org/
The home page of the International Cocoa Organization. Through this site, you can find answers to many of the questions you may have regarding the chemistry of chocolate and its path from the cocoa tree to your mouth.

http://www.chevron.com/explore/index.html
Web address for the Learning Zone of the Chevron Corporation, where you can find information about crude oil and the refining process.

Chapter 28: The Chemistry of Drugs

Where do most drugs come from? How do chemists design drugs in the laboratory? How do drugs cure us of infections? Why do some drugs keep us awake, while others make us drowsy? How is caffeine removed from coffee and tea? Why do cigarette smokers usually smoke after eating a meal? Why is nicotine many times more addicting than heroin? And how do drugs stop us from experiencing pain? For the answers, we turn to the chemistry of drugs.

28.1 There Are Several Ways to Classify Drugs

What is a drug? Loosely defined, a *drug* is any substance other than food or water that affects how the body functions. A drug that can improve a person's health is also called a *medicine.* Drugs, some legal and others illegal, are also used for nonmedical purposes. Legal non-medical drugs include alcohol, caffeine, and nicotine. Illegal nonmedical drugs include LSD and cocaine.

There are a variety of ways to classify drugs. As Table 28.1 shows, the U.S. Drug Enforcement Agency (DEA) classifies them according to safety and social acceptability.

Drugs can also be classified according to how they are derived, as is done in Table 28.2. Drugs that are natural products come directly

Table 28.1

U.S. Drug Enforcement Agency Classification of Drugs

Classification	Description	Examples
Over-the-counter (OTC) drugs	Available to anyone	Aspirin, cough medicines
Permitted nonmedical drugs	Available in food, beverages, and tobacco products	Alcohol, caffeine, nicotine
Prescription drugs	Requires physician authorization	Antibiotics, birth control pills
Controlled substances		
Schedule 1	No medical use, high abuse potential	Heroin, LSD, mescaline, marijuana
Schedule 2	Some medical use, high abuse potential	Amphetamines, cocaine, morphine
Schedule 3	Prescription drugs, abuse potential	Barbiturates, Valium

Table 28.2

The Origin of Some Common Drugs

Origin	Drug	Biological Effect
Natural product	Caffeine	Nerve stimulant
	Reserpine	Hypertension reducer
	Vincristine	Anticancer agent
	Penicillin	Antibiotic
	Morphine	Analgesic
Chemical derivative of natural product	Prednisone	Antirheumatic
	Ampicillin	Antibiotic
	LSD	Hallucinogenic
	Chloroquinine	Antimalarial
	Ethynodiol diacetate	Contraceptive
Synthetic	Valium	Antidepressant
	Benadryl	Antihistamine
	Allobarbital	Sedative-hypnotic
	Phencyclidine	Veterinary anesthetic
	Methadone	Analgesic

Most all drugs have a multitude of effects on the body.

from terrestrial or marine plants or animals. Drugs that are chemical derivatives are natural products that have been chemically modified to increase potency or decrease side effects. Synthetic drugs are those that originate in the laboratory.

Perhaps the most common way to classify drugs is according to their primary biological effect. It must be noted, however, that most drugs exhibit a broad spectrum of activity. Therefore they may fall under several classifications. Aspirin, for example, relieves pain, but it also reduces fever and inflammation, thins the blood, and causes ringing in the ears. Morphine relieves pain, but it also constipates and suppresses the urge to cough.

At times, the multiple effects of a drug are desirable. Aspirin's pain-reducing and fever-reducing properties work well together in treating flu symptoms in adults, for instance. Additionally, aspirin's blood-thinning ability helps prevent heart disease. Morphine was widely used during the American Civil War both for relieving the pain of battle wounds and for controlling diarrhea. Often, however, the side effects of a drug are less desirable. Ringing in the ears and upset stomach are a few of the negative side effects of aspirin, and a major side effect of morphine is its addictiveness. A main goal of drug research, therefore, is to find drugs that are specific in their action and have minimal side effects.

Although two drugs that are taken together may have different primary activities, they may share a common secondary activity. The effect that both drugs share can be amplified when the two drugs are taken together. One drug enhancing the action of another is called the **synergistic effect.** A synergistic effect is often more powerful than the sum of the activities of the two drugs taken separately. One of the great challenges of physicians and pharmacists is keeping track of all the possible combinations of drugs and potential synergistic effects.

The synergism that results from mixing drugs that have the same primary effect is particularly hazardous. For example, a moderate dose of a sedative that makes you sleepy combined with a moderate amount of alcohol may be lethal. In fact, most drug overdoses are the result of a combination of drugs rather than the abuse of a single drug.

Concept Check ✓

Distinguish between a drug and a medicine.

Check Your Answer A drug is any substance administered to affect body function. A medicine is any drug administered for its therapeutic effect. All medicines are drugs, but not all drugs are medicines.

28.2 The Lock-and-Key Model Guides Chemists in Creating New Drugs

To find new and more effective medicines, chemists use various models that describe how drugs work. By far, the most useful model of drug action is the **lock-and-key model.** The basis of this model is that there is a connection between a drug's chemical structure and its biological effect. For example, morphine and all related pain-relieving opioids, such as codeine and heroin, have the T-shaped structure shown in Figure 28.1.

Many drugs, though not all, are thought to work by the lock-and-key mechanism.

According to the lock-and-key model, illustrated in Figure 28.2, biologically active molecules function by fitting into *receptor sites* in the body, where they are held by molecular attractions, such as hydrogen bonding. When a drug molecule fits into a receptor site the way a key fits into a lock, a particular biological event is triggered, such as a nerve impulse or even a chemical reaction. In order for a

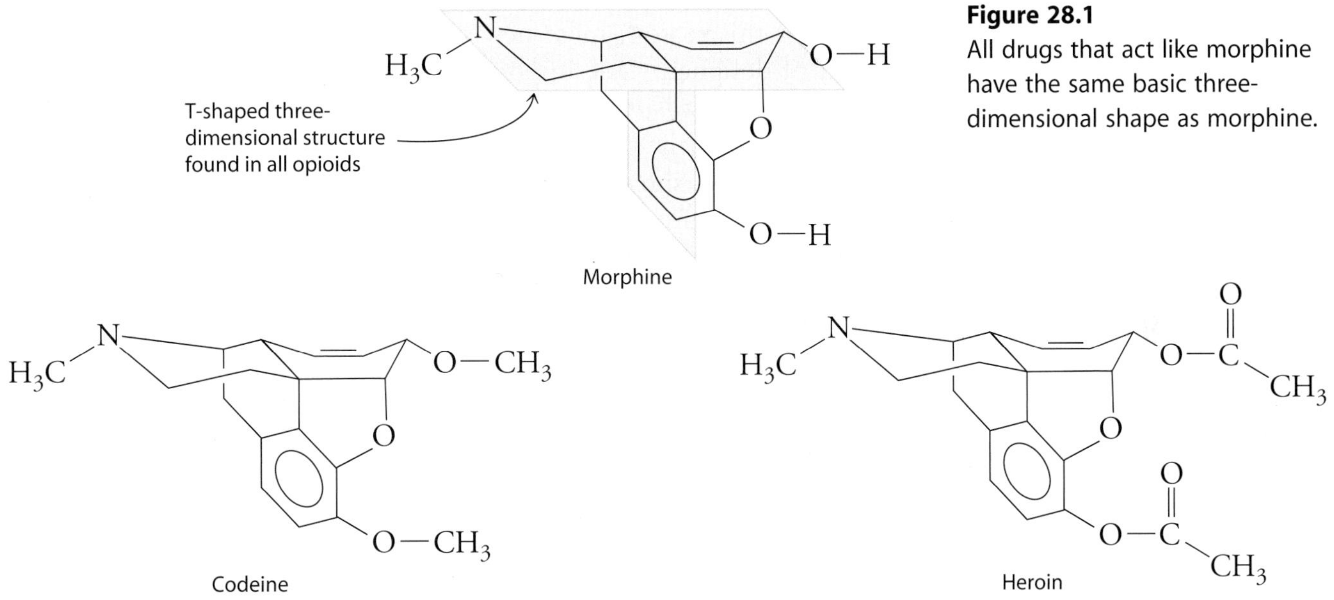

T-shaped three-dimensional structure found in all opioids

Morphine

Codeine

Heroin

Figure 28.1
All drugs that act like morphine have the same basic three-dimensional shape as morphine.

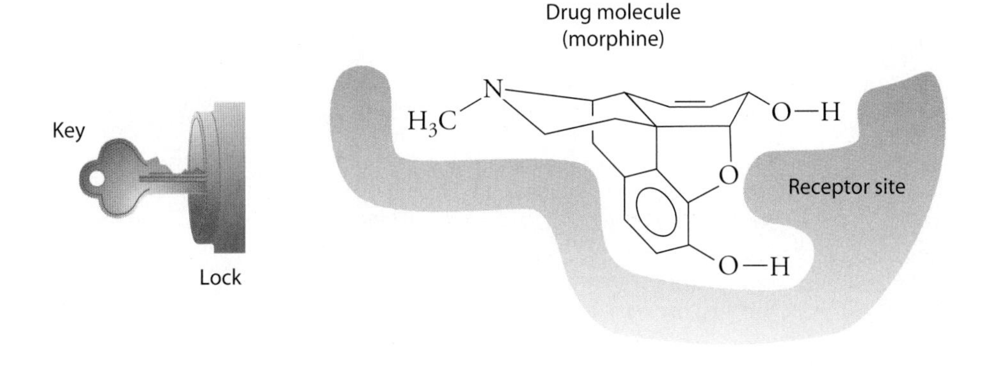

Figure 28.2
Many drugs act by fitting into receptor sites on molecules found in the body, much as a key fits in a lock.

molecule to fit into a particular receptor site, however, it must have the proper shape, just as a key must have properly shaped notches in order to fit a lock.

Another facet of this model is that the molecular attractions holding a drug to a receptor site are easily broken. (Recall from Chapter 24 that most molecular attractions are many times weaker than chemical bonds.) A drug is therefore held to a receptor site only temporarily. Once the drug is removed from the receptor site, body chemistry destroys the drug's chemical structure and the effects of the drug are said to have worn off.

Using this model, we can understand why some drugs are more potent than others. Heroin, for example, is a more potent pain-killer than is morphine because the chemical structure of heroin allows for tighter binding to its receptor sites.

The lock-and-key model has developed into one of the most important tools of pharmaceutical study. Knowing the precise shape of a target receptor site allows chemists to design molecules that have an optimum fit and a specific biological effect.

Biochemical systems are so complex, however, that our knowledge is still limited, as is our capacity to design effective medicinal drugs. For this reason, most new medicinal drugs are still discovered instead of designed. One important avenue for drug discovery is *ethnobotany*. *An ethnobotanist is* a researcher who learns about the medicinal plants used in indigenous cultures, such as the root of the Bobgunnua tree, shown in Figure 28.3. Today there are hundreds of clinically useful prescription drugs derived from plants. About three quarters of these came to the attention of the pharmaceutical industry as a result of their use in folk medicine.

Another important method of drug discovery is the random screening of vast numbers of compounds. Each year, for example, the National Cancer Institute screens some 20,000 compounds for anticancer activity. One successful hit was the compound Taxol, shown in Figure 28.4. This compound has significant activity against several forms of cancer, especially ovarian cancer.

Figure 28.3
Ethnobotanists directed natural-products chemists to the yellow coating on the root of the African Bobgunnua tree. Indigenous people have known for many generations that this coating has medicinal properties. From extracts of the coating, the chemists isolated a compound that is highly effective in treating fungal infections. This compound, produced by the tree to protect itself from root rot, shows much promise in the treatment of the opportunistic fungal infections that plague those suffering from AIDS.

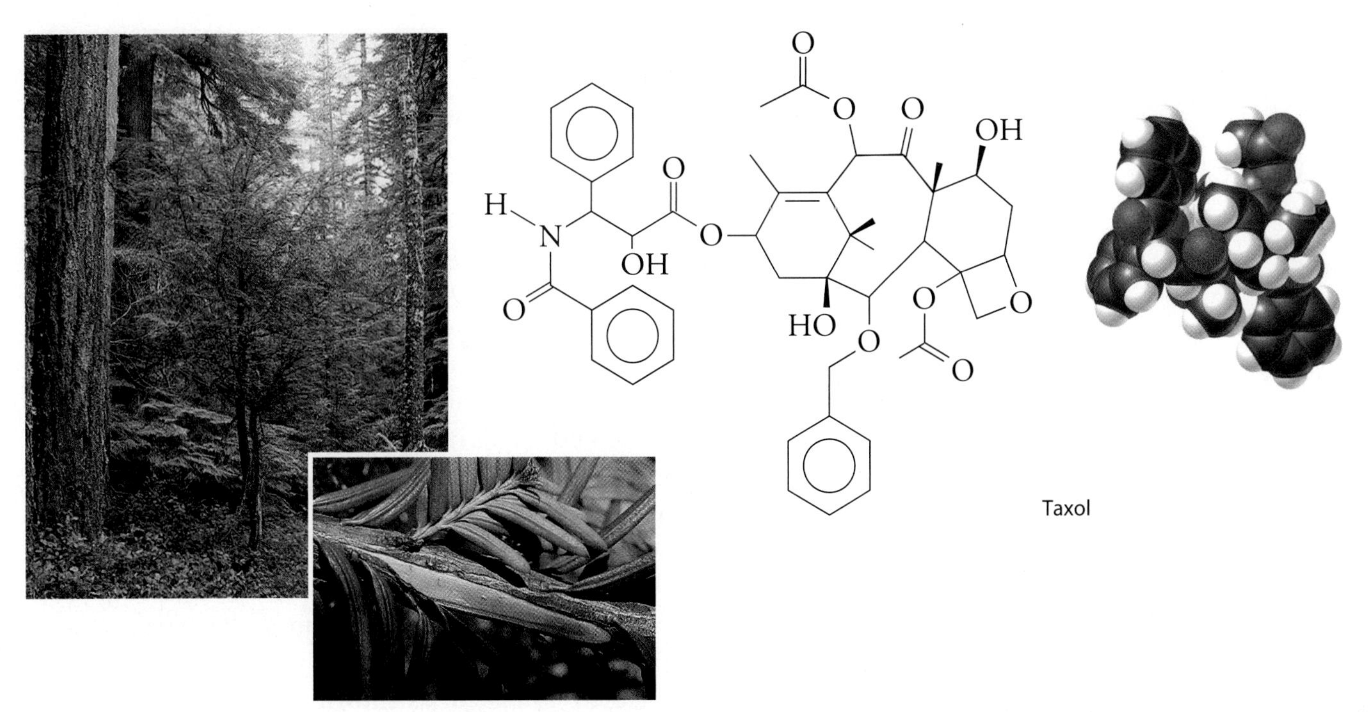

Figure 28.4
Originally isolated from the bark of the Pacific yew tree, Taxol is a complex natural product useful in the treatment of various forms of cancer.

A drug isolated from a natural source is not necessarily better or more gentle than one produced in the laboratory. Aspirin, for example, is a human-made chemical derivative, and it is certainly more gentle than cocaine, which is 100% natural. The main advantage of natural products is their great *diversity.* Each year, more than 3000 new chemical compounds are discovered from plants. Many of these compounds are biologically active, serving the plant as a chemical defense against disease or predators. Nicotine, for example, is a naturally occurring insecticide produced by the tobacco plant to protect itself from insects.

It has been estimated that only 5000 plant species have been studied exhaustively for possible medical applications. This is a minor fraction of the estimated 250,000 to 300,000 plant species on our planet, most of which are located in tropical rainforests. That we know little or nothing about much of the plant kingdom has raised justified and well-publicized concern. For as rain forests are being destroyed, also being destroyed are plant species that might yield useful medicines.

Medicines are just one of many reasons that our planet's rainforests need to be preserved.

Concept Check ✓

Why are chemicals from natural sources so suitable for making drugs?

Check Your Answer Because there are so many naturally derived chemicals with biological effects. Their vast diversity permits the manufacture of the many different types of medicines needed to combat the many different types of illnesses humans are subject to.

28.3 Chemotherapy Cures the Host By Killing the Disease

The use of drugs that destroy disease-causing agents without destroying the animal host is known as **chemotherapy.** This approach is effective in the treatment of many diseases, including bacterial infections. It works by taking advantage of the ways a disease-causing agent, also known as a *pathogen,* is different from a host.

Sulfa drugs are synthetic drugs first used to treat bacterial infections in the 1930s. They work by taking advantage of a striking difference between humans and bacteria. Both humans and bacteria must have the nutrient *folic acid* in order to remain healthy. We humans can obtain folic acid from what we eat. But bacteria cannot absorb folic acid from outside sources. Instead, bacteria must make their own supply of folic acid. For this, they possess receptor sites that help make folic acid from a simpler molecule found in all bacteria, para-aminobenzoic acid (PABA). The PABA attaches to the specific receptor site and is converted to folic acid, as shown in Figure 28.5.

Figure 28.5
Bacterial enzymes use para-aminobenzoic acid (PABA) to synthesize folic acid.

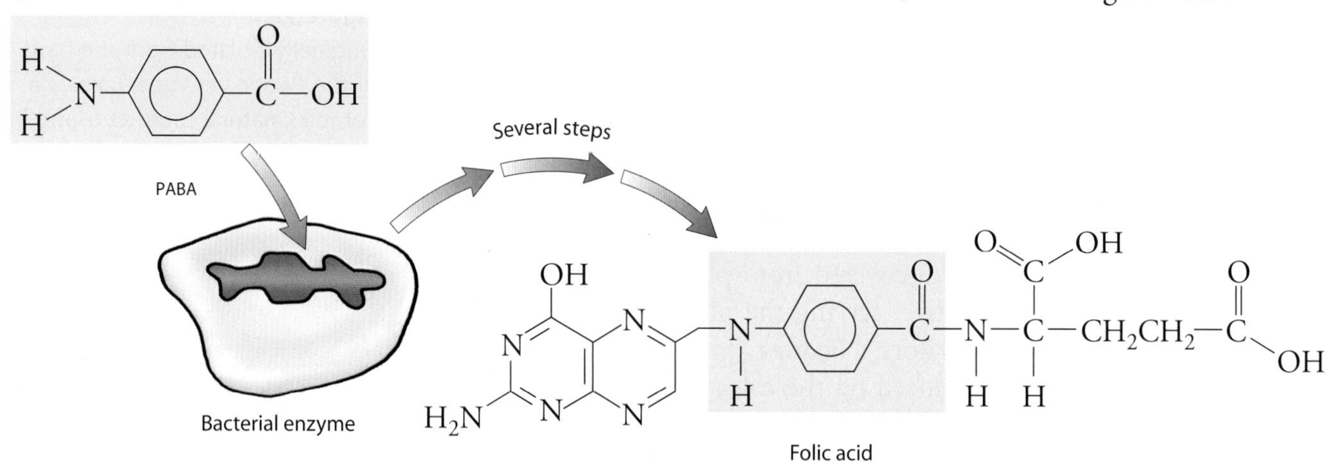

Sulfa drugs have a close structural resemblance to PABA. When taken by a person suffering from a bacterial infection, a sulfa drug is transformed by the body into the compound *sulfanilamide,* which attaches to the bacterial receptor sites designed for PABA, as shown in Figure 28.6. This prevents the bacteria from synthesizing folic acid. Without folic acid, the bacteria soon die. The patient, however, lives on because he or she receives folic acid from the diet.

Concept Check ✓

How is sulfanilamide poisonous to bacteria but not to humans?

Check Your Answer Sulfanilamide is poisonous to bacteria because it prevents them from synthesizing the folic acid they need to survive. Humans utilize folic acid from their diet and so they are not bothered by sulfanilamide's ability to disrupt the synthesis of folic acid.

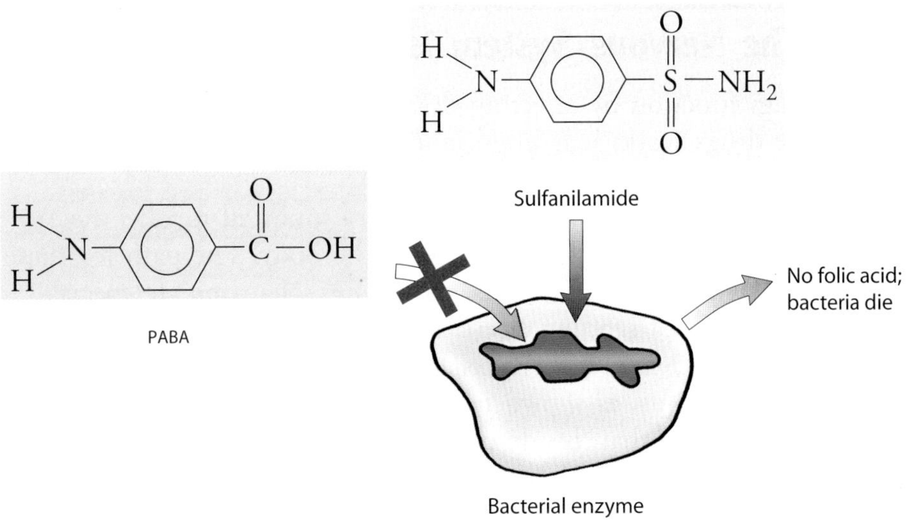

Sulfanilamide

PABA

No folic acid; bacteria die

Bacterial enzyme

Figure 28.6
In the body, sulfa drugs are transformed into sulfanilamide, which binds to the bacterial receptor sites and keeps them from doing their job.

Antibiotics are chemicals that prevent the growth of bacteria. They are produced by such microorganisms as molds, other fungi, and even bacteria. The first antibiotic discovered was penicillin. Many derivatives of penicillin, such as the penicillin G shown in Figure 28.7, have since been isolated from microorganisms as well as prepared in the laboratory. Penicillins and the closely related compounds known as cephalosporins, also shown in Figure 28.7, kill bacteria by inactivating receptor sites responsible for strengthening the bacterial cell wall. With this receptor site inactivated, bacterial cell walls grow weak and eventually burst.

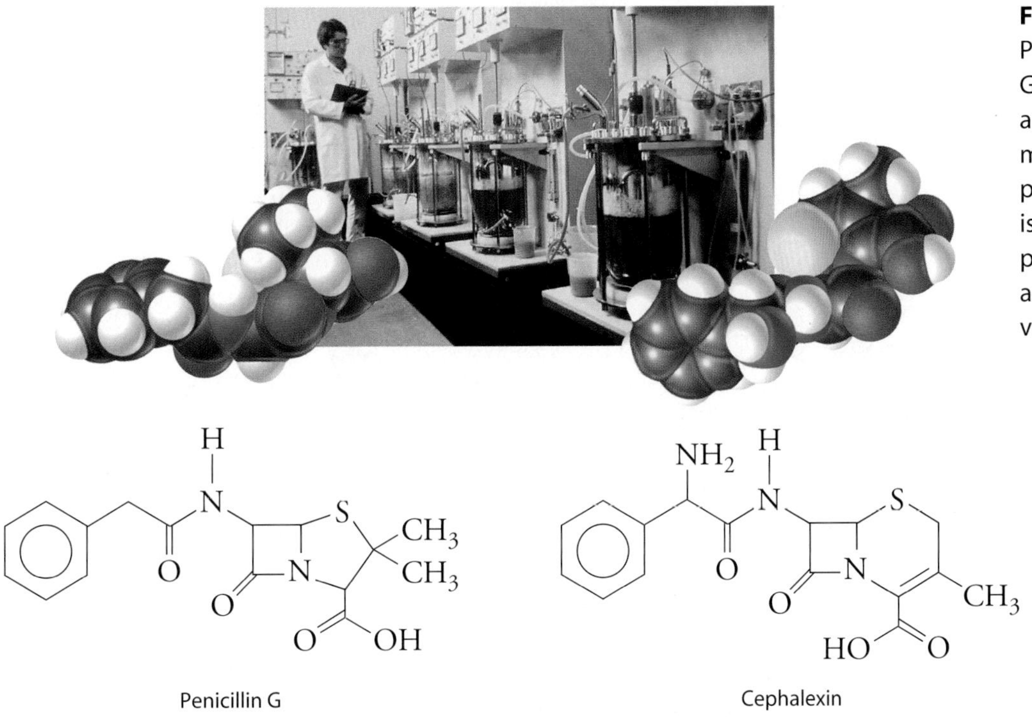

Penicillin G

Cephalexin

Figure 28.7
Penicillins, such as penicillin G, and cephalosporins, such as cephalexin, as well as most other antibiotics, are produced by microorganisms that can be mass-produced in large vats. The antibiotics are then harvested and purified.

28.4 The Nervous System Is a Network of Neurons

Many drugs function by affecting the nervous system. To understand how these drugs work, it is important to know the basic structure and functions of the nervous system.

Thoughts, physical actions, and sensory input all involve the transmission of electrical signals through the body. The path for these signals is a network of nerve cells, or *neurons.* **Neurons** are specialized cells capable of sending electrical impulses. First, in what is called the *resting phase,* a nerve cell primes itself for an impulse by ejecting sodium ions, as shown in Figure 28.8a. More sodium ions outside the neuron than inside creates a separation of charge. And the separation of charge gives rise to an electric potential of around −70 millivolts across the cell membrane. As shown in Figure 28.8b, a nerve impulse is a reversal in this electric potential that travels down the length of the neuron to the *synaptic terminals.* The reversal of the electric potential within an impulse occurs as sodium ions flush back into the neuron.

After the impulse passes a given point along the neuron, the cell again ejects sodium ions at that point to re-establish the original distribution of ions and the −70 millivolt potential.

Unlike the wires in an electric circuit, most neurons are not physically connected to one another. Nor are they connected to the

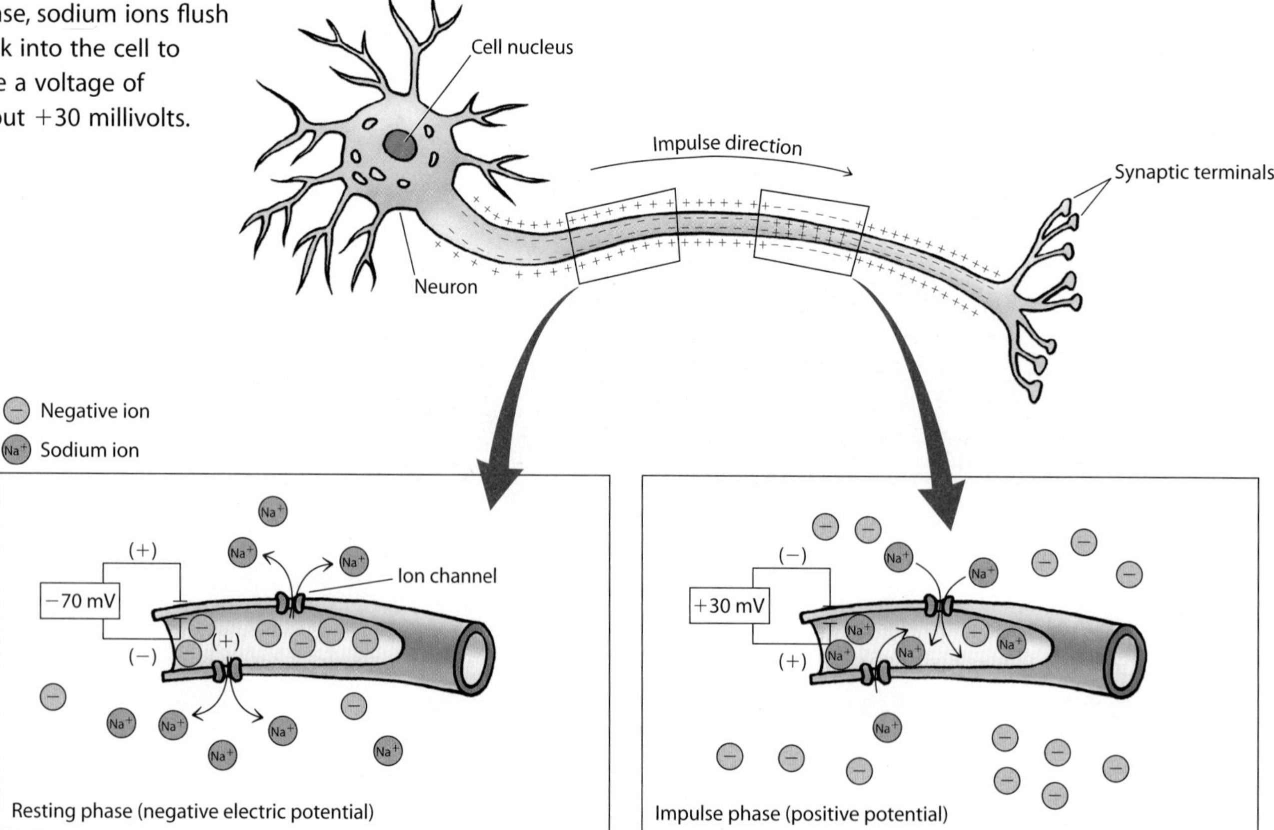

Figure 28.8
(a) The resting phase of a neuron maintains a greater concentration of sodium ions outside the cell. This results in a voltage of about −70 millivolts. (b) In the impulse phase, sodium ions flush back into the cell to give a voltage of about +30 millivolts.

muscles or glands on which they act. Rather, as Figure 28.9 shows, they are separated from one another or from a muscle or gland by a narrow gap known as the **synaptic cleft.**

A nerve impulse reaching a synaptic cleft causes bubble-like compartments in the terminal, called *vessels,* to release neurotransmitters into the cleft. **Neurotransmitters** are organic compounds released by a neuron capable of activating receptor sites.

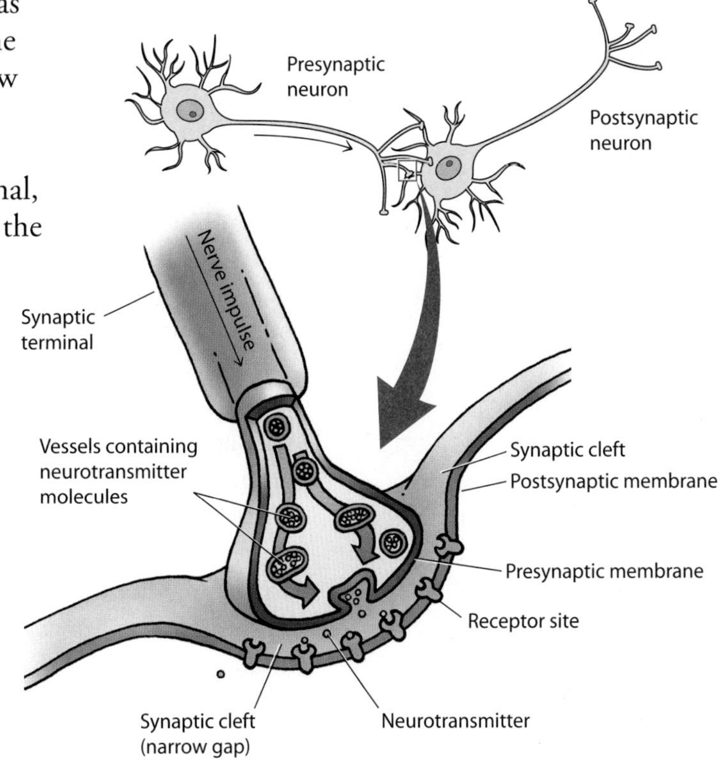

Figure 28.9
The passage of neurotransmitters across a synaptic cleft.

A neurotransmitter, once released into the synaptic cleft, migrates across the cleft to receptor sites on the opposite side. If the receptor sites are located on a *postsynaptic neuron,* as shown in Figure 28.9, the binding of the neurotransmitter may start a nerve impulse in that neuron. If the receptor sites are located on a muscle or organ, then binding of the neurotransmitter may start a bodily response, such as muscle contraction or the release of hormones.

Two important classes of neurons include the *stress* and *maintenance* neurons. Both types are always firing. But in times of stress, as when facing an angry bear or giving a speech, the stress neurons are more active than the maintenance neurons. This condition is the *fight-or-flight* response, during which fear causes stress neurons to trigger rapid bodily changes to help defend against impending danger: The mind becomes alert, air passages in the nose and lungs open to bring in more oxygen, the heart beats faster to spread the oxygenated blood throughout the body, and nonessential activities such as digestion are temporarily stopped. In times of relaxation, such as sitting down in front of the television with a bowl of potato chips, the maintenance neurons are more active than the stress neurons. Under these conditions, digestive juices are secreted, intestinal muscles push food through the gut, the pupils constrict to sharpen vision, and the heart pulses at a minimal rate.

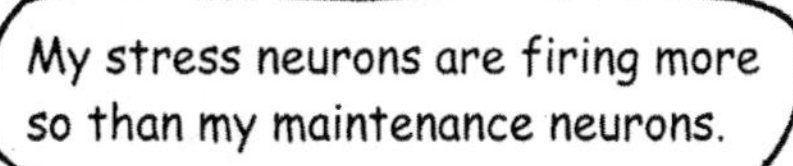

Concept Check ✓

What is a neurotransmitter?

Check Your Answer A neurotransmitter is a small organic molecule released by a neuron. It influences neighboring tissues, such as nerve membranes, by binding to receptor sites.

Norepinephrine

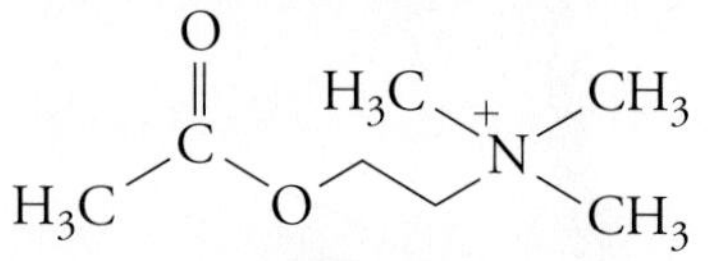

Acetylcholine

Figure 28.10
The chemical structures of the stress neurotransmitter norepinephrine and the maintenance neurotransmitter acetylcholine.

Neurotransmitters Include Norepinephrine, Acetylcholine, Dopamine, Serotonin, and GABA

On the chemical level, stress and maintenance neurons can be distinguished by the types of neurotransmitters they use. The primary neurotransmitter for stress neurons is *norepinephrine.* The primary neurotransmitter for maintenance neurons is *acetylcholine,* both shown in Figure 28.10. As we shall see in the following sections, many drugs function by altering the balance of stress and maintenance neuron activity.

In addition to norepinephrine and acetylcholine, a host of other neurotransmitters contribute to a broad range of effects. Three examples are the neurotransmitters dopamine, serotonin, and gamma aminobutyric acid, shown in Figure 28.11.

Dopamine plays a significant role in activating the brain's reward center. The brain's reward center is located in the hypothalamus. This is an area at the lower middle of the brain, as illustrated in Figure 28.12. The hypothalamus is the main control center for the involuntary part of the peripheral nervous system and for emotional response and behavior. Stimulation of the reward center by dopamine results in a pleasurable sense of *euphoria,* which is an exaggerated sense of well-being.

The control of physical responses ultimately allows us to perform such complex tasks as driving a car or playing the piano. The control of emotional responses allows us to refine our behavior, such as overcoming anxiety in tense social interactions or remaining calm in an emergency. The brain controls both physical and emotional responses by inhibiting the transmission of nerve impulses. The neurotransmitter responsible for this inhibition—*gamma aminobutyric acid* (GABA)—is *the* major inhibitory neurotransmitter of the brain. Without it, coordinated movements and emotional skills would not be possible.

Serotonin is the neurotransmitter used by the brain to block unneeded nerve impulses. To make sense of the world, the frontal lobes of the brain selectively block out a multitude of signals coming from the lower brain and from the peripheral nervous system. We are not born with this ability to selectively block out information. In order to have an appropriate focus on the world, newborns must learn from experience which lights, sounds, smells, and feelings

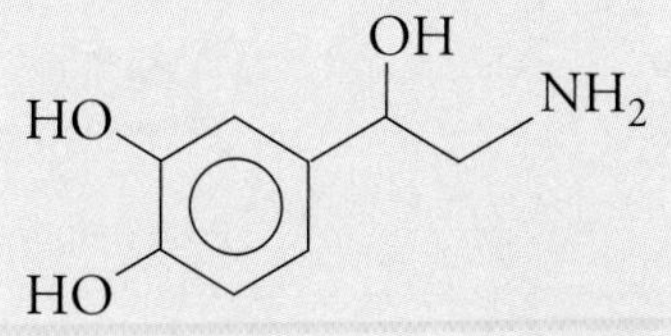

Figure 28.11
The chemical structures of three neurotransmitters important to the central nervous system.

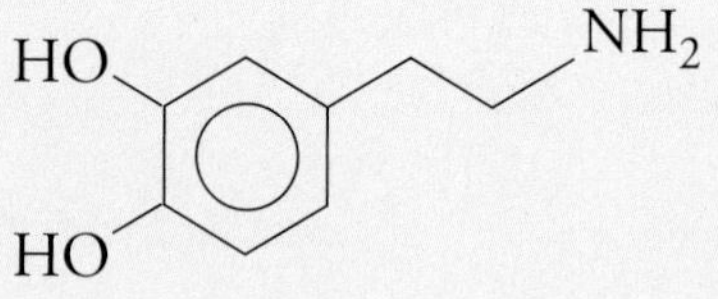

Dopamine

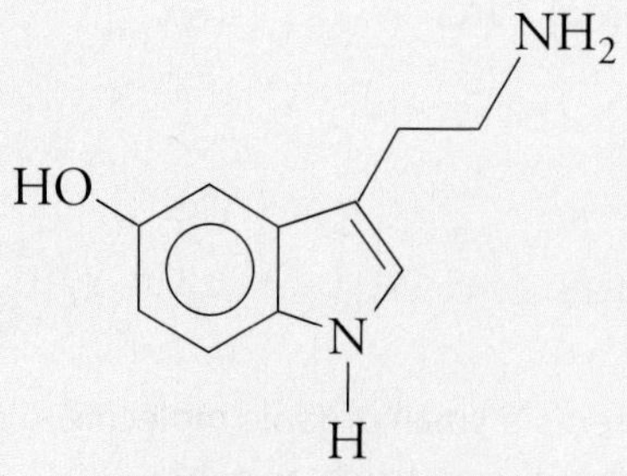

Serotonin

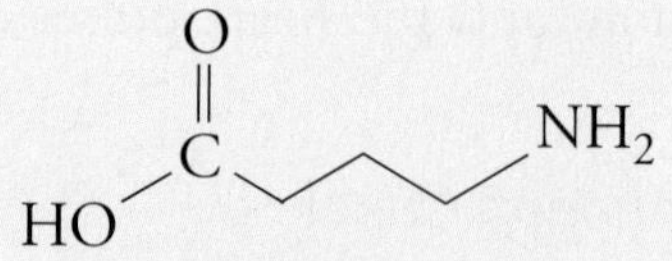

Gamma aminobutyric acid (GABA)

outside and inside their bodies must be dampened. A healthy, mature brain is one in which serotonin successfully suppresses lower-brain nerve signals. Information that does make it to the higher brain can then be sorted efficiently.

Drugs such as LSD, that modify the action of serotonin, alter the brain's ability to sort information, and this alters perception. While hallucinating, for example, an LSD user rarely sees something that isn't there. Rather, the user has an altered perception of something that does exist.

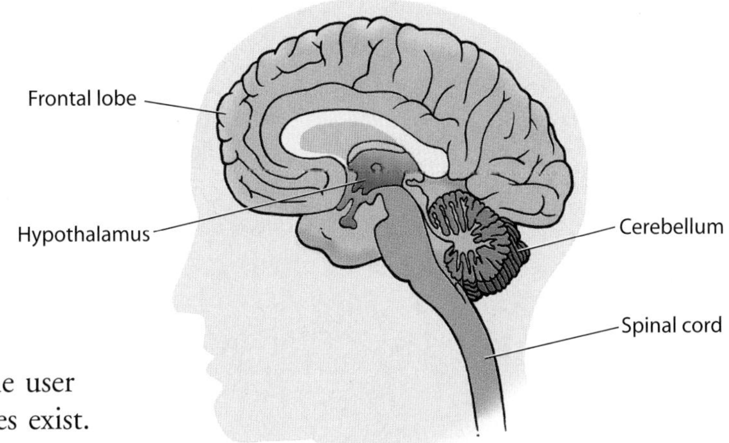

Figure 28.12
The human brain.

Concept Check ✓

Match the neurotransmitter to its primary function:

____ norepinephrine	**a.** inhibits nerve transmission
____ acetylcholine	**b.** stimulates reward center
____ dopamine	**c.** selectively blocks nerve impulses
____ serotonin	**d.** maintains stressed state
____ GABA	**e.** maintains relaxed state

Check Your Answers d, e, b, c, a.

28.5 Psychoactive Drugs Alter the Mind or Behavior

Any drug that affects the mind or behavior is classified as **psychoactive.** In this chapter we focus on two classes of psychoactive drugs: stimulants and depressants.

Stimulants Activate the Stress Neurons

By enhancing the intensity of our reactions to stimuli, *stimulants* cause brief periods of heightened awareness, quick thinking, and elevated mood. They exert this effect by activating the stress neurons. Four widely recognized stimulants are amphetamines, cocaine, caffeine, and nicotine.

Amphetamines are a family of stimulants that include the parent compound *amphetamine* (also known as "speed") and such derivatives as methamphetamine and pseudoephedrine. As you can see by comparing Figure 28.13 on page 500 with Figures 28.10 and 28.11, these drugs are structurally similar to the neurotransmitters norepinephrine and dopamine. Amphetamines bind to receptor sites for these neurotransmitters. So amphetamines mimic many of the effects of norepinephrine and dopamine on the stress neurons, including the fight-or-flight response and the ability to give a person a sense of euphoria.

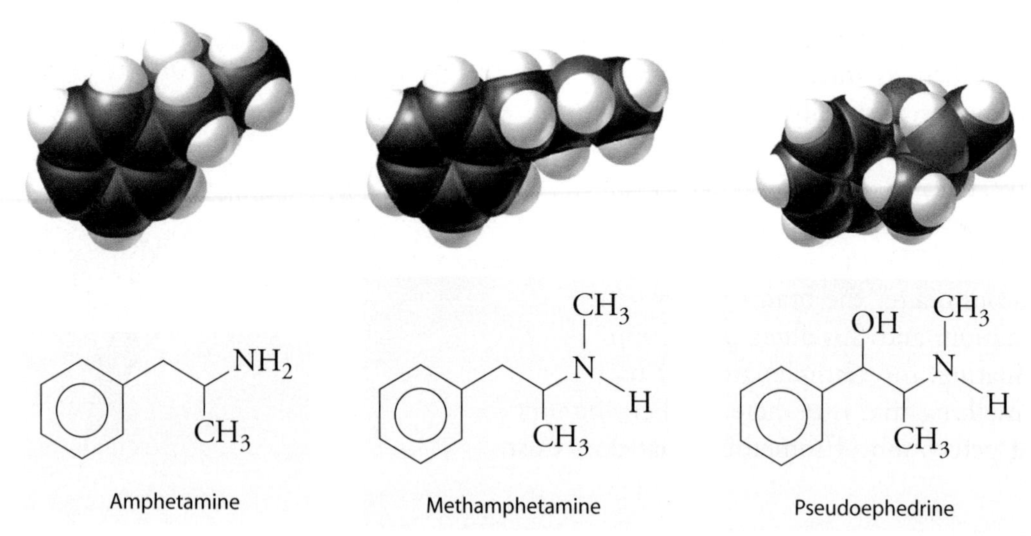

Figure 28.13
Amphetamines are a family of compounds structurally related to the neurotransmitters norepinephrine and dopamine.

The stimulating and mood-altering effects of amphetamines give them a high abuse potential. Side effects include insomnia, irritability, loss of appetite, and paranoia. Amphetamines take a particularly hard toll on the heart. Hyperactive heart muscles are prone to tearing. Subsequent scarring of tissue ultimately leads to a weaker heart. Furthermore, amphetamines cause blood vessels to constrict and blood pressure to rise. And these conditions increase the likelihood of heart attack or stroke, especially for the one person out of four whose blood pressure is already high.

Drug addiction is not completely understood, but scientists do know that it involves both physical and psychological dependence. **Physical dependence** is the need to continue taking the drug to avoid withdrawal symptoms. For amphetamines, drug withdrawal symptoms include depression, fatigue, and a strong desire to eat. **Psychological dependence** is the *craving* to continue drug use. This craving may be the most serious and deep-rooted aspect of addiction. It can persist even after withdrawal from physical dependence, frequently leading to renewed drug-seeking behavior.

One of the more notorious and abused stimulants is *cocaine,* a natural product isolated from the South American coca plant, shown in Figure 28.14. Once in the bloodstream, cocaine produces a sense of euphoria and increased stamina. It is also a powerful local anesthetic when applied topically. Within a few decades of its first isolation from plant material in 1860, cocaine was used as a local anesthetic for eye surgery and dentistry. This practice was stopped once safer local anesthetics were discovered in the early 1900s.

Cocaine and amphetamines share a similar profile of addictiveness, though cocaine's addictive properties are more intense. The cocaine that is inhaled nasally is the hydrochloride salt. The free-base form of cocaine, called *crack cocaine,* is also abused. As with the street drug "ice," which is the free-base form of methamphetamine, crack cocaine is volatile. It may be smoked for what is an intense but profoundly dangerous and addictive high.

Figure 28.14
The South American coca plant has been used by indigenous cultures for many years in religious ceremonies and as an aid to staying awake on long hunting trips. Leaves are either chewed or ground to a powder that is inhaled nasally.

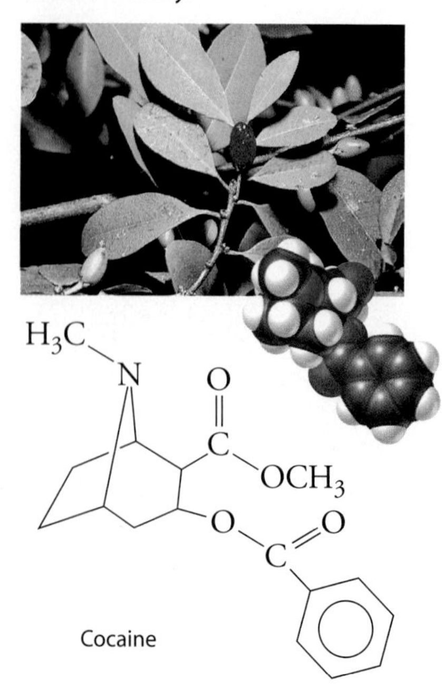

Long-term cocaine or amphetamine abuse leads to a deterioration of the nervous system. The body recognizes the excessive stimulatory actions produced by these drugs. To deal with the overstimulation, the body creates more depressant receptor sites for neurotransmitters that inhibit nerve transmission. A tolerance for the drugs therefore develops. Then, to receive the same stimulatory effect, the abuser is forced to increase the dose. And this induces the body to create even more depressant receptor sites. The end result over the long term is that the abuser's natural levels of dopamine and norepinephrine are insufficient to compensate for the excessive number of depressant sites. Lasting personality changes are thus often observed. Addicts, even when recovered, often report feelings of psychological depression.

Concept Check ✓

How do amphetamines and cocaine exert their effects?

Check Your Answer Amphetamines and cocaine in the synaptic cleft both cause an overstimulation of receptor sites for norepinephrine and dopamine.

A much milder and legal stimulant is *caffeine,* depicted in Figure 28.15. A number of mechanisms have been proposed for caffeine's stimulatory effects. Perhaps the most straightforward mechanism is that caffeine facilitates the release of norepinephrine into synaptic clefts. Caffeine also exerts many other effects on the body, such as dilation of arteries, relaxation of bronchial and gastrointestinal muscles, and stimulation of stomach-acid secretion.

The caffeine people ingest comes from various natural sources, including coffee beans, teas, kolanuts, and cocoa beans. Kolanut extracts are used for making cola drinks, and cocoa beans (not to be confused with the cocaine-producing coca plant) are roasted and then ground to a paste used for making chocolate. Caffeine is relatively easy to remove from these natural products using high-pressure carbon dioxide, which selectively dissolves the caffeine. This allows for the economical production of "decaffeinated" beverages, many of which, however, still contain small amounts of caffeine. Interestingly, cola drink manufacturers use decaffeinated kolanut extract in their

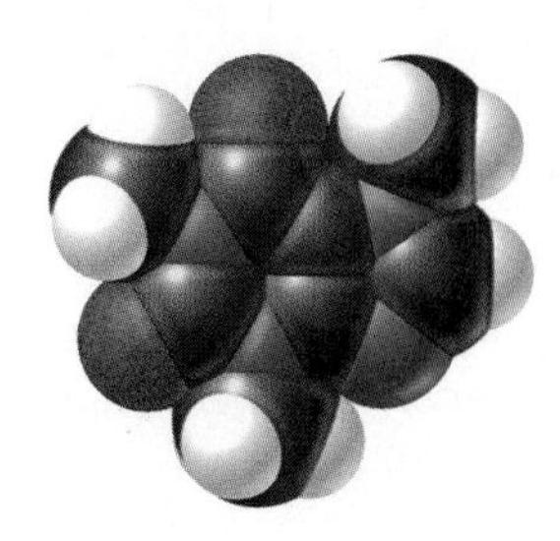

Figure 28.15
A coffee plant with its ripening caffeine-containing beans.

Table 28.3

Approximate Caffeine Content of Various Products

Product	Caffeine Content
Brewed coffee	100–150 mg/cup
Instant coffee	50–100 mg/cup
Decaffeinated coffee	2–10 mg/cup
Black tea	50–150 mg/cup
Cola drink	35–55 mg/12 oz
Chocolate bar	1–2 mg/oz
Over-the-counter stimulant	100 mg/dose
Over-the-counter analgesic	30–60 mg/dose

beverages. The caffeine is added in a separate step to guarantee a particular caffeine concentration. In the United States, about two million pounds of caffeine is added to soft drinks each year. Table 28.3 shows the caffeine content of various commercial products. For comparison, the maximum daily dose of caffeine tolerable by most adults is about 1500 milligrams.

Another legal, but far more toxic, stimulant is *nicotine.* As noted earlier, tobacco plants produce nicotine as a chemical defense against insects. This compound is so potent that a lethal dose in humans is only about 60 milligrams. A single cigarette may contain up to 5 milligrams of nicotine. Most of it is destroyed by the heat of the burning embers, however, so that less than 1 milligram is typically inhaled by the smoker.

Nicotine and the neurotransmitter acetylcholine, which acts on maintenance neurons, have similar structures, as Figure 28.16 illustrates. Nicotine molecules are therefore able to bind to acetylcholine receptor sites and trigger many of acetylcholine's effects, including relaxation and increased digestion. This explains the tendency of smokers to smoke after eating meals. In addition, acetylcholine is used for muscle contraction, and so the smoker may also experience some muscle stimulation immediately after smoking. After these initial responses, however, nicotine molecules remain bound to the acetylcholine receptor sites. This blocks acetylcholine molecules from binding. The result is that the activity of these neurons is depressed.

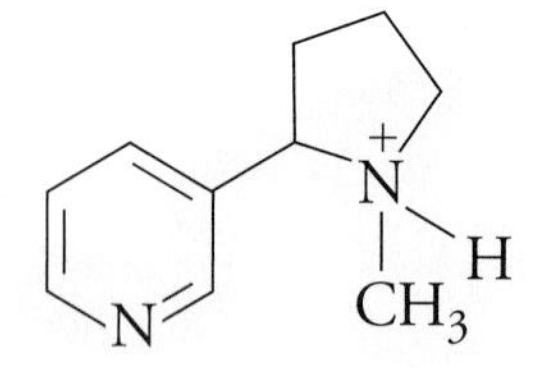

Nicotine

Acetylcholine

Figure 28.16
Nicotine is able to bind to receptor sites for acetylcholine because of structural similarities.

Recall that maintenance neurons and stress neurons are both always working. Thus inhibiting the activity of one type increases the activity of the other type. So, as nicotine depresses the maintenance neurons, it favors the stress neurons. This raises the smoker's blood pressure and stresses the heart.

Animal studies show inhaled nicotine to be about six times more addictive than injected heroin. Because nicotine leaves the body quickly, withdrawal symptoms begin about 1 hour after a cigarette is smoked, which means the smoker is inclined to light up frequently.

Tobacco field

Tobacco curing on racks

Cigarette manufacture

User

Blackened lungs

Figure 28.17
The path of tobacco from the field to a smoker's lungs. About 46 million Americans smoke despite an awareness of the dangers of this habit.

Figure 28.17 shows what a smoker's lungs look like. In the United States, about 450,000 individuals die each year from such tobacco-related health problems as emphysema, heart failure, and various forms of cancer, especially lung cancer, which is brought on primarily by tobacco's tar component. Some relief from the addiction can be obtained with nicotine chewing gum and nicotine skin patches. In order for any method to be effective, however, the smoker must first genuinely desire to quit smoking.

Concept Check ✓

Caffeine and nicotine both add stress to the nervous system, but they do so by different means. Briefly describe the difference.

Check Your Answer Caffeine stimulates the release of the stress neurotransmitter norepinephrine, and nicotine both depresses the action of the maintenance neurotransmitter acetylcholine and enhances the release of norepinephrine.

Figure 28.18
One of the initial effects of alcohol is a depression of social inhibitions, which can serve to bolster mood. Alcohol is not a stimulant, however. From the first sip to the last, body systems are being depressed.

$CH_3CH_2—OH$

Ethanol

Depressants Inhibit the Ability of Neurons to Conduct Impulses

Depressants are a class of drugs that inhibit the ability of neurons to conduct impulses. Two commonly used depressants are ethanol and benzodiazepines.

Ethanol, also known simply as *alcohol,* is by far the most widely used depressant. Its structure is shown in Figure 28.18. In the United States, about a third of the population, or about 100 million people, drink alcohol. It is well established that alcohol consumption leads to about 150,000 deaths each year in the United States. The causes of these deaths are overdoses of alcohol alone, overdoses of alcohol combined with other depressants, alcohol-induced violent crime, cirrhosis of the liver, and alcohol-related traffic accidents.

Benzodiazepines are a potent class of antianxiety agents. Compared to many other types of depressants, benzodiazepines are relatively safe and rarely produce cardiovascular and respiratory depression. Their antianxiety effects were identified in 1957 by chance. During a routine laboratory clean-up, a synthesized compound that had been sitting on the shelf for two years was submitted for routine testing despite the fact that compounds thought to have similar structures had shown no promising pharmacologic activity. This particular compound, however, shown in Figure 28.19 and now known as chlordiazepoxide, contained an unexpected seven-membered ring. Chlordiazepoxide showed a significant calming effect in humans, and by 1960 was marketed under the trade name Librium® as an antianxiety agent. Shortly thereafter, a derivative, Diazepam®, was found to be five to ten times more potent than Librium. In 1963 Diazepam hit the market under the trade name Valium.

A primary way in which alcohol and benzodiazepines exert their depressant effect is by enhancing the action of GABA. As shown in Figure 28.20, GABA keeps electrical impulses from passing

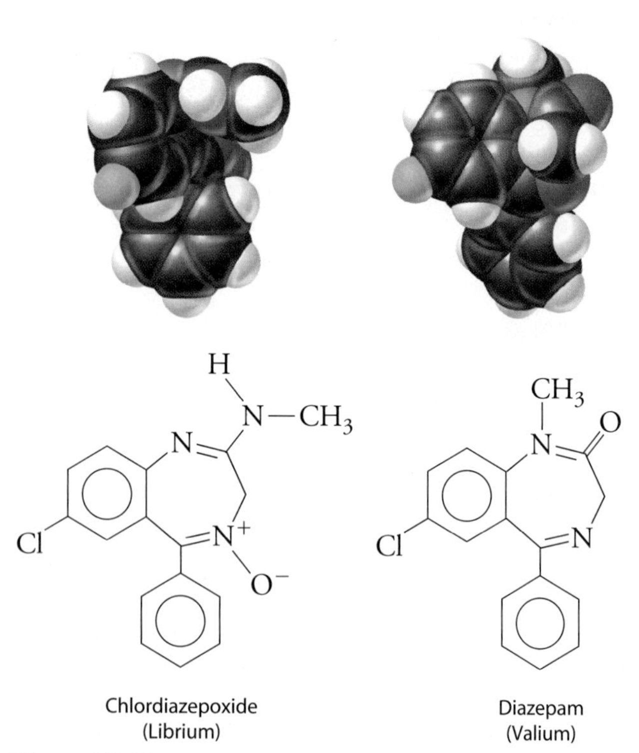

Figure 28.19
The benzodiazepines Librium and Valium.

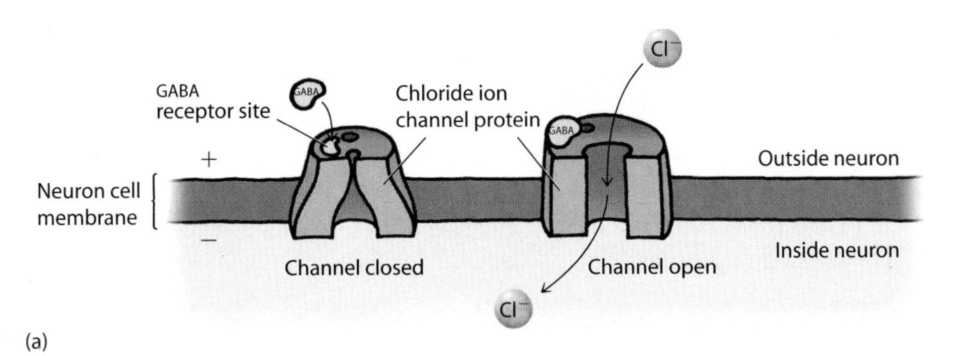

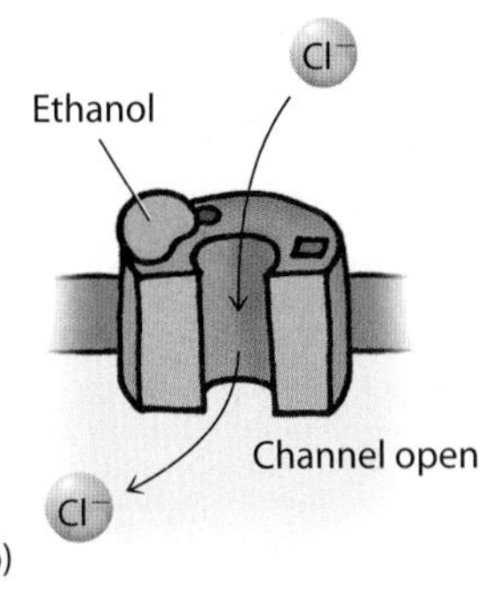

Figure 28.20
(a) When GABA binds to its receptor site, a channel opens to allow negatively charged chloride ions into the neuron. The high concentration of negative ions inside the neuron prevents the electric potential from reversing from negative to positive. Because that reversal is necessary if an impulse is to travel through a neuron, no impulse can move through the neuron. (b) Ethanol mimics GABA by binding to GABA receptor sites.

through a neuron by binding to a receptor site on a channel that penetrates the cell membrane of the neuron. Figure 28.20a shows that when GABA binds to the receptor site, the channel opens, allowing chloride ions to migrate into the neuron. The resulting negative charge buildup in the neuron maintains the negative electric potential across the cell membrane. This inhibits a reversal to a positive potential and prevents an impulse from traveling along the neuron. (If you are confused, go back and review Figure 28.8 and the text describing it. Perhaps you now know why doctors spend many years hitting the books in their medical training.)

Ethanol mimics the effect of GABA by binding to GABA receptor sites. This allows chloride ions to enter the neuron, as shown in Figure 28.20b. The effect of alcohol is dose-dependent, which means that the greater the amount drunk, the greater the effect. At small concentrations, few chloride ions are permitted into the neuron; these low concentrations of ions decrease inhibitions, alter judgment, and impair muscle control. As the person continues to drink and the chloride ion concentration inside the neuron rises, both reflexes and consciousness diminish, eventually to the point of coma and then death.

Figure 28.21 illustrates how benzodiazepines exert their depressant effects by binding to receptor sites located adjacent to GABA receptor sites. Benzodiazepine binding merely helps GABA bind. Because benzodiazepine doesn't directly open chloride-ion channels, overdoses of this

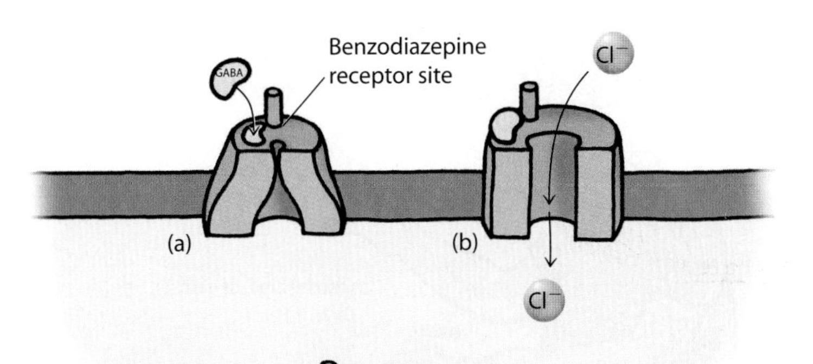

Figure 28.21
The receptor sites for benzodiazepines are adjacent to GABA receptor sites. (a) Benzodiazepines cannot open up the chloride channel on their own. (b) Rather, benzodiazepines help GABA in its channel-opening task.

compound are less hazardous than alternatives. This makes the benzodiazepines the drugs of choice for treating symptoms of anxiety.

Concept Check ✓

Does the activity of a neuron increase or decrease as chloride ions are allowed to pass into it?

Check Your Answer Chloride ions inside the neuron help to maintain the negative electric potential. This inhibits the neuron from being able to conduct an impulse (See Figure 28.8) Chloride ions, therefore, decrease the activity of a neuron.

28.6 Pain Relievers Inhibit the Transmission or Perception of Pain

Physical pain is a complex body response to injury. On the cellular level, pain-inducing biochemicals are rapidly synthesized at the site of injury, where they initiate swelling, inflammation, and other responses that get your body's attention. These pain signals are sent through the nervous system to the brain, where the pain is perceived. To alleviate pain, drugs act at various stages of this process, as shown in Figure 28.22.

Anesthetics prevent neurons from transmitting sensations to the brain. *Local anesthetics* are applied either topically to numb the skin or by injection to numb deeper tissues. These mild anesthetics are useful for minor surgical or dental procedures. As described earlier, cocaine was the first medically used local anesthetic. Others having fewer side effects soon followed, such as the ones shown in Figure 28.23.

A *general anesthetic* blocks out pain by rendering the patient unconscious. As discussed in Section 27.4, diethyl ether was one of the first general anesthetics. Two of the more popular gaseous general anesthetics used by anesthesiologists today are those shown in Figure 28.24, sevoflurane and nitrous oxide. When inhaled, these

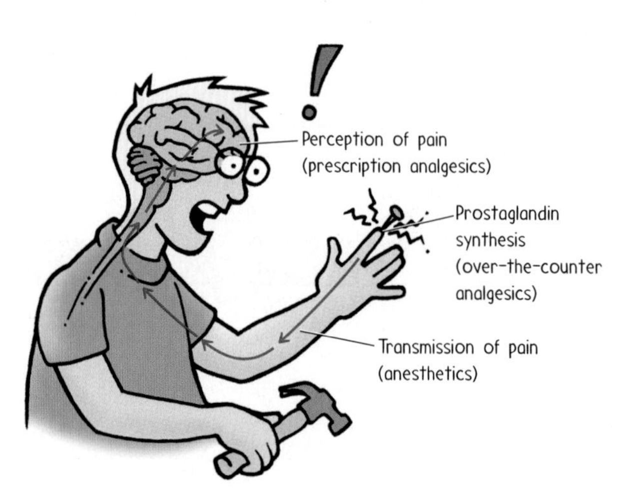

Figure 28.22
Injury to tissue causes the transmission of pain signals to the brain. Pain relievers prevent this transmission, inhibit the inflammation response, or dampen the brain's ability to perceive the pain.

Benzocaine

Procaine (Novocaine)

Tetracaine

Lidocaine (Xylocaine)

Cocaine

Aromatic ring | Intermediate chain | Amine group

Figure 28.23
Local anesthetics have similar structural features, including an aromatic ring, an intermediate chain, and an amine group. Ask your dentist which ones he or she uses for your treatment.

compounds enter the bloodstream and are distributed throughout the body. At certain blood concentrations, general anesthetics render the individual unconscious, which is useful for invasive surgery. General anesthesia must be monitored very carefully, however, so as to avoid a major shutdown of the nervous system and subsequent death.

Analgesics are a class of drugs that enhance our ability to tolerate pain without abolishing nerve sensations. Over-the-counter analgesics, such as aspirin, ibuprofen, naproxen, and acetaminophen, inhibit the formation of *prostaglandins.* As Figure 28.25 on page 508 illustrates, prostaglandins are biochemicals the body quickly synthesizes to generate pain signals. These analgesics also reduce fever because of the role prostaglandins play in raising body temperature.

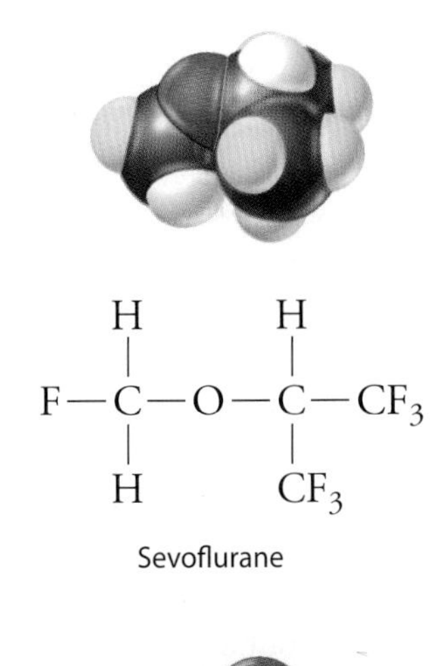

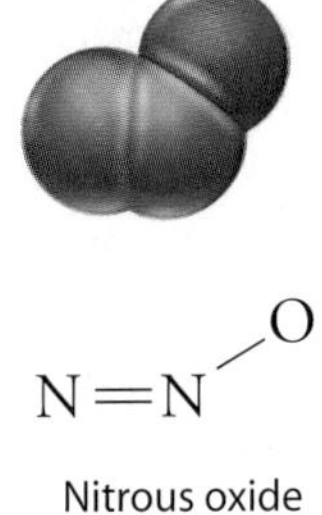

Figure 28.24
The chemical structures of sevoflurane and nitrous oxide.

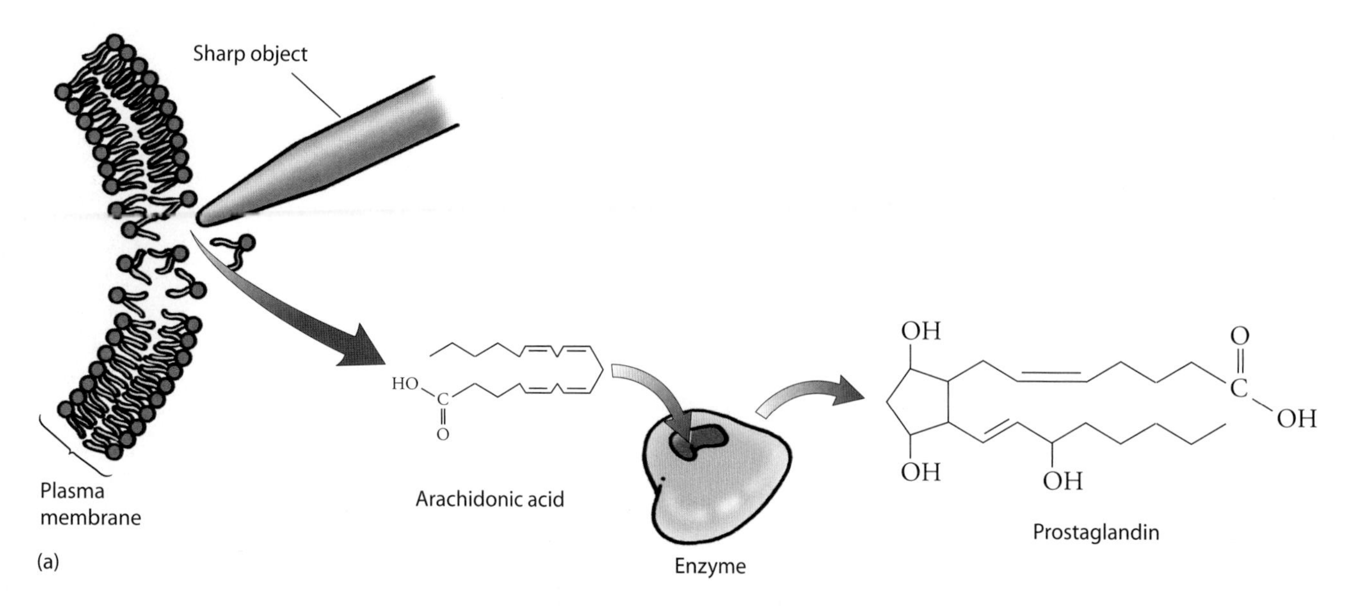

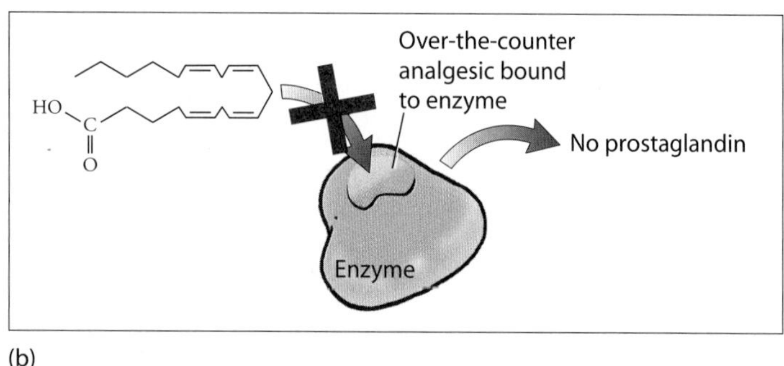

Figure 28.25
(a) Prostaglandins, which cause pain signals to be sent to the brain, are synthesized by the body in response to injury. The starting material for all prostaglandins is arachidonic acid, which is found in the membrane of all cells. Arachidonic acid is transformed to prostaglandins with the help of a receptor site. There are a variety of prostaglandins, each having its own effect, but all have a chemical structure resembling the one shown here. (b) Analgesics inhibit the synthesis of prostaglandins by binding to the arachidonic acid receptor site. With no prostaglandins, no pain signals are generated.

In addition to reducing pain and fever, aspirin, ibuprofen, and naproxen act as anti-inflammatory agents because they block the formation of a certain type of prostaglandin responsible for inflammation. Acetaminophen does not act on inflammation. These four analgesics are shown in Figure 28.26.

The more potent opioid analgesics—morphine, codeine, and heroin (Figure 28.1)—moderate the brain's perception of pain by binding to receptor sites on neurons in the central nervous system, which includes the brain and spinal column. Initial discovery of these receptor sites raised the question of why they exist. Perhaps, it was hypothesized, opioids mimic the action of a naturally occurring brain chemical. *Endorphins,* a group of large biomolecules that have strong opioid activity, were subsequently isolated from brain tissue. It has been suggested that endorphins evolved as a means of suppressing awareness of pain that would otherwise be incapacitating in life-threatening situations. The "runner's high" experienced by many athletes after a vigorous workout is caused by endorphins.

Endorphins are also implicated in the *placebo effect,* in which patients experience a reduction in pain after taking what they believe is a drug but is actually a sugar pill. (A *placebo* is any inactive substance used as a control in a scientific experiment.) Through the placebo effect, it is the patients' belief in the effectiveness of a medicine rather than the medicine itself that leads to pain relief. The

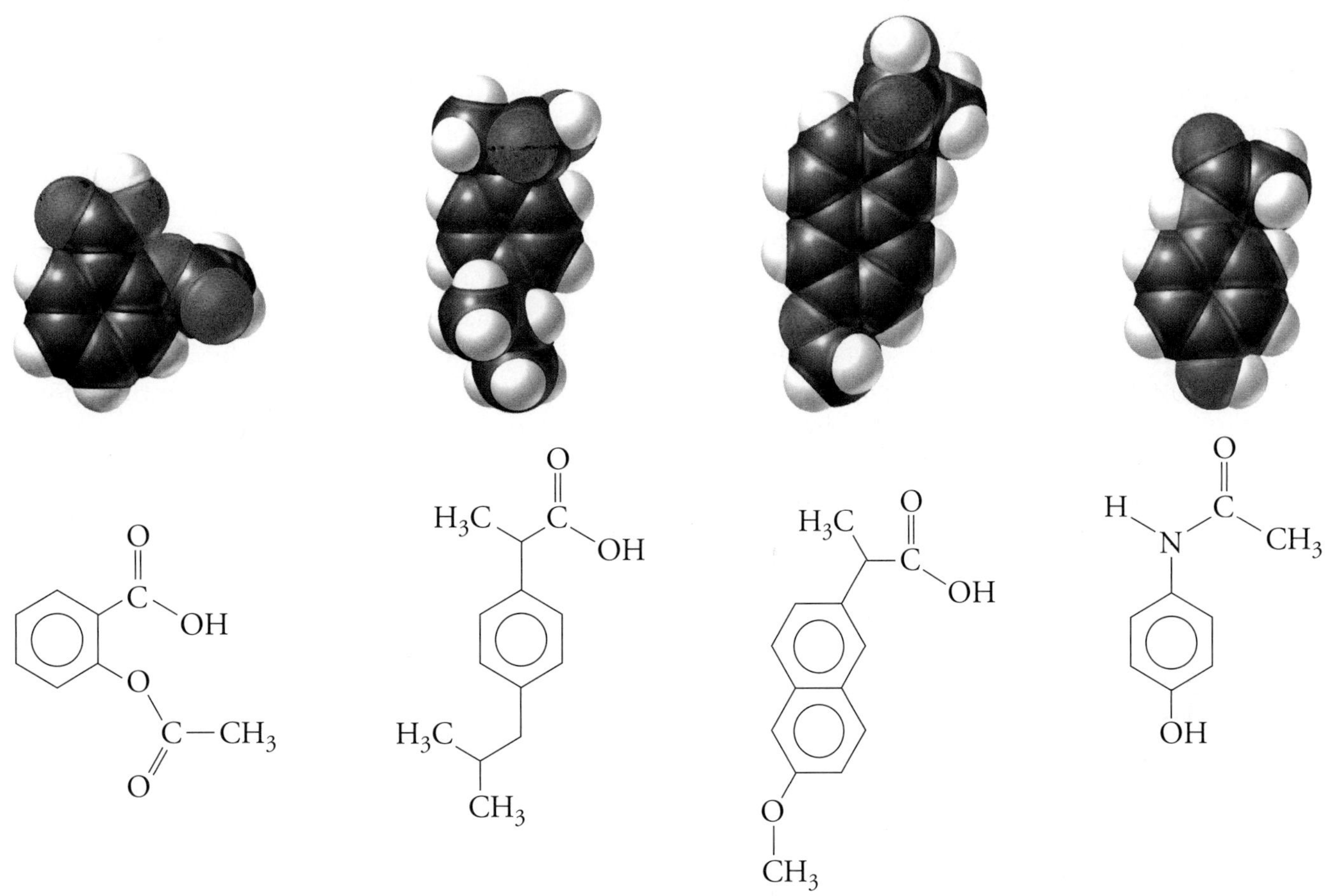

Figure 28.26
Aspirin, ibuprofen, and naproxen block the formation of prostaglandins responsible for pain, fever, and inflammation. Acetaminophen blocks the formation only of prostaglandins responsible for pain and fever.

involvement of endorphins in the placebo effect has been demonstrated by replacing the sugar pills with drugs that block opioids or endorphins from binding to their receptor sites. Under these circumstances, the placebo effect vanishes.

In addition to acting as analgesics, opioids can induce euphoria, which is why they are so frequently abused. With repeated use, the body develops a tolerance to these drugs: More and more must be administered to achieve the same effect. Abusers also become physically dependent on opioids, which means they must continue to take the opioids to avoid severe withdrawal symptoms, such as chills, sweating, stiffness, abdominal cramps, vomiting, weight loss, and anxiety. Interestingly, when opioids are used primarily for pain relief rather than for pleasure, the withdrawal symptoms are much less dramatic—especially when the patient does not know he or she has been on these drugs.

Concept Check ✓

Distinguish between an anesthetic and an analgesic.

Check Your Answer An anesthetic blocks pain signals from reaching the brain. An analgesic facilitates the ability to manage pain signals once they are received by the brain.

Perhaps nowhere is the impact of chemistry on society more evident than in the development of drugs. On the whole, drugs have increased our life-span and improved our quality of living. They have also presented us with a number of ethical and social questions. How do we care for an expanding elderly population? What drugs, if any, should be permissible for recreational use? How do we deal with drug addiction—as a crime, as a disease, or both? As we continue to learn more about ourselves and our ills, we can be sure that more powerful drugs will become available. All drugs, however, carry certain risks that we should be aware of. As most physicians would point out, drugs offer many benefits, but they are no substitute for a healthy lifestyle and preventative approaches to medicine.

Chapter Review

Key Terms and Matching Definitions

_____ analgesic
_____ anesthetic
_____ chemotherapy
_____ lock-and-key model
_____ neuron
_____ neurotransmitters
_____ physical dependence
_____ psychological dependence
_____ synaptic cleft
_____ synergistic effect

1. One drug enhancing the effect of another.
2. A model that explains how drugs interact with receptor sites.
3. The use of drugs to destroy pathogens without destroying the animal host.
4. A specialized cell capable of receiving and sending electrical impulses.
5. A narrow gap across which neurotransmitters pass either from one neuron to the next or from a neuron to a muscle or gland.
6. An organic compound capable of activating receptor sites on proteins embedded in the membrane of a neuron.
7. A dependence characterized by the need to continue taking a drug to avoid withdrawal symptoms.
8. A deep-rooted craving for a drug.
9. A drug that prevents neurons from transmitting sensations to the central nervous system.
10. A drug that enhances the ability to tolerate pain without abolishing nerve sensations.

Review Questions

There Are Several Ways to Classify Drugs

1. What are the three origins of drugs?
2. Are a drug's side effects necessarily bad?
3. What is the synergistic effect?

The Lock-and-Key Model Guides Chemists in Creating New Drugs

4. In the lock-and-key model, is a drug viewed as the lock or the key?
5. What holds a drug to its receptor site?

Chemotherapy Cures the Host by Killing the Disease

6. Why do bacteria need PABA but humans can do without it?
7. How does penicillin G cure bacterial infections?

The Nervous System Is a Network of Neurons

8. How does a neuron maintain an electric potential difference across its membrane?
9. What are the symptoms that a person's stress neurons have been activated?

10. What are some of the things going on in the body when maintenance neurons are more active than stress neurons?

11. What neurotransmitter functions most in the brain's reward center?

12. What is the role of GABA in the nervous system?

Psychoactive Drugs Alter the Mind or Behavior

13. How is psychological dependence distinguished from physical dependence?

14. What is one mechanism for how caffeine stimulates the nervous system?

15. What neurotransmitter does nicotine mimic?

16. What drugs enhance the action of GABA?

Pain Relievers Inhibit the Transmission or Perception of Pain

17. What is an anesthetic?

18. What is an analgesic?

19. Where are the major opioid receptor sites located?

20. What biochemical is thought to be responsible for the placebo effect?

Exercises

1. Aspirin can cure a headache, but when you pop an aspirin tablet, how does the aspirin know to go to your head rather than your big toe?

2. Which is better for you: a drug that is a natural product or one that is synthetic?

3. Would formulating a sulfa drug with PABA be likely to increase or decrease its antibacterial properties?

4. What is an advantage of synaptic clefts between neurons rather than direct connections?

5. How is a drug addict's addiction similar to our need for food? How is it different?

6. Nicotine solutions are available from lawn and garden stores as an insecticide. Why must gardeners handle this product with extreme care?

7. A variety of gaseous compounds behave as general anesthetics even though their structures have very little in common. Does this support the role of a receptor site for their mode of action?

8. How might the structure of benzocaine be modified to create a compound having greater anesthetic properties?

9. Which is the more appropriate statement: Opioids have endorphin activity, or endorphins have opioid activity? Explain your answer.

10. A person may feel more relaxed after smoking a cigarette, but his or her heart is actually stressed. Why?

Discussion Topics

1. Alcohol-free and caffeine-free beverages have been quite successful in the marketplace, while nicotine-free tobacco products have yet to be introduced. Speculate about possible reasons.

2. What would be the advantages and disadvantages of making the production, sale, and consumption of tobacco products illegal?

3. Historically, drug abuse has been viewed as criminal behavior. Advances in medicine, however, show the complexity of drug abuse, leading some people to believe it should be viewed and treated as a disease. This newer view suggests that education and medical treatment, not fines and jail sentences, should be the major weapons combating drug abuse. Assume this newer view is the predominant legal one, and prioritize where you think resources should be spent to help alleviate drug abuse. Under what circumstances do you think a drug user should be incarcerated?

4. Research suggests that skin loses its flexibility as we get older because of sugar molecules that bind to the skin protein called collagen. New classes of drugs that prevent this from happening and are even able to reverse the process may hit the market within a decade, promising a long-lasting youthful look for everyone. Should such drugs be sold as prescription or over-the-counter medicine? What if there is a minor side effect, such as increased risk of infections, or a major side effect, such as a weakening of heart tissue? What if there are no significant side effects except for a doubling of the average human-life span?

Suggested Web Sites

http://www.usdoj.gov/dea/

Home page for the Drug Enforcement Administration, the lead federal agency for the enforcement of narcotics and controlled-substance laws and regulations. The agency's priority is the long-term immobilization of major drug trafficking organizations through the jailing of their leaders, termination of their trafficking networks, and seizure of their assets.

http://www.nida.nih.gov/

A primary mission of The National Institute on Drug Abuse is to ensure that science, not ideology or anecdote, forms the foundation for our nation's drug abuse policies. This Web site is designed to ensure the rapid and effective transfer of scientific data to policy makers, health care practitioners, and the general public.

http://www.lungusa.org/

This site for the American Lung Association can lead you to information about all kinds of lung disease, the air you breathe, and events and services in your area. Since 1904, the Association has worked to fight lung disease by helping people quit smoking, funding research, improving indoor and outdoor air quality, and educating millions about asthma.

Chapter 29: Plastics

Where do plastics come from? How are they made? How long ago did the United States start making more plastic than steel? How many pounds of polyethylene do you use each year? Who first invented a plastic, and what were some of the first applications? Why does the inside of a ping pong ball smell like athletic muscle rub? Would the Allied forces have won World War II if it were not for the invention of plastics? How was Teflon used in the development of the atomic bomb? In this chapter, we explore some of the profound impacts that plastics have had on modern civilization.

29.1 Organic Molecules Can Link To Form Polymers

Many of the molecules that make up living organisms are **polymers.** These include DNA, proteins, the cellulose of plants, and the complex carbohydrates of starchy foods. Polymers are exceedingly long molecules that consist of repeating molecular units called **monomers,** as Figure 29.1 illustrates. Monomers have relatively simple structures consisting of anywhere from 4 to 100 atoms per molecule. When chained together, they can form polymers consisting of hundreds of thousands of atoms per molecule. These large molecules are still too small to be seen with the unaided eye. They are, however, giants in the world of the submicroscopic—if a typical polymer molecule were as thick as a kite string, it would be 1 kilometer long.

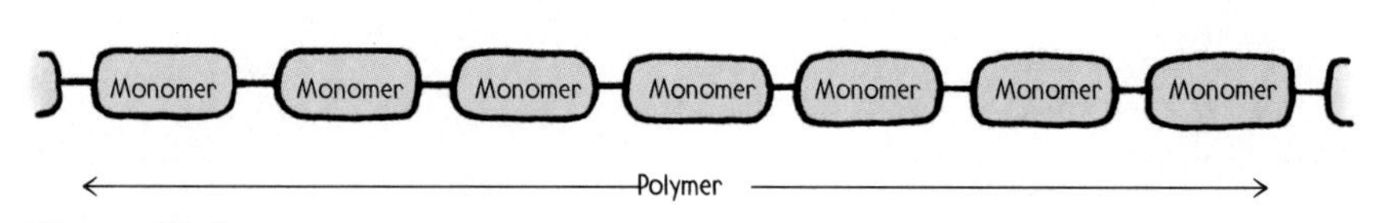

Figure 29.1
A polymer is a long molecule consisting of many smaller monomer molecules linked together.

We leave a discussion of these important biological molecules to a course in the life sciences. For now, we focus on the human-made polymers, also known as synthetic polymers. These make up the class of materials commonly known as **plastics.**

We begin by exploring the two major types of synthetic polymers in use today—*addition polymers and condensation polymers.* As shown in Table 29.1 (on page 514 and 515), addition and condensation polymers have a wide variety of uses. Solely the product of human design, these polymers are important to modern living. In the United States, for example, synthetic polymers have surpassed steel as the most widely used material.

29.2 Addition Polymers Result from the Joining Together of Monomers

When monomer units join one another, the result is what is called **addition polymers.** For this to occur, each monomer must contain at least one double bond. As shown in Figure 29.2, polymerization occurs when electrons from each monomer jump towards neighboring monomer molecules. The result are covalent bonds that hold the monomer units together. During this process, no atoms are lost. This means that the total mass of the polymer is equal to the sum of the masses of all the monomers.

Nearly 12 million tons of polyethylene is produced annually in the United States; that's about 90 pounds per U.S. citizen. The monomer from which it is synthesized, ethylene, is an unsaturated hydrocarbon produced in large quantities from petroleum.

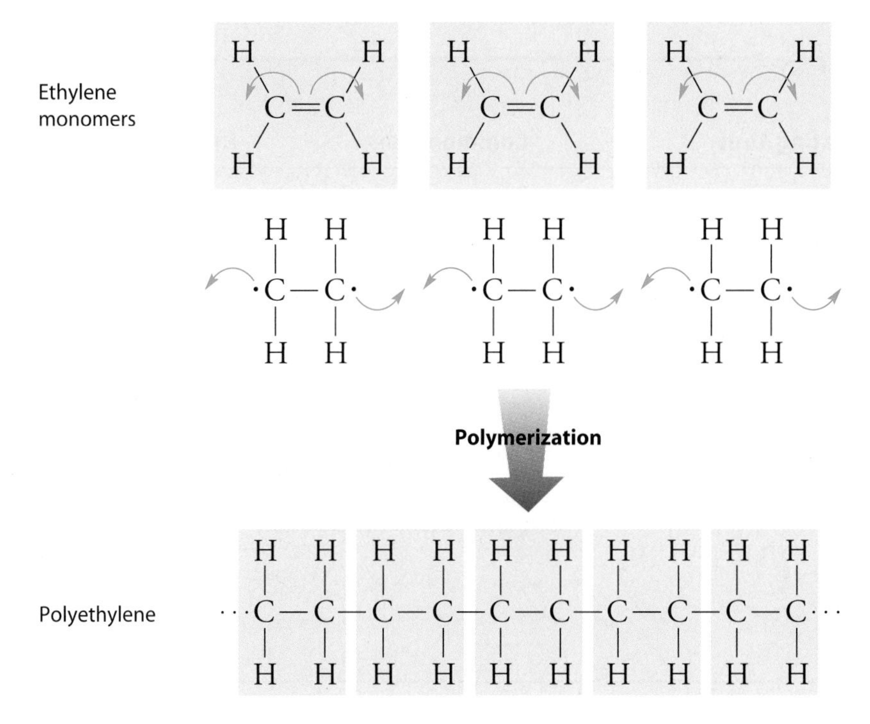

Figure 29.2
The addition polymer polyethylene is formed as electrons from the double bonds of ethylene monomer molecules split away and become unpaired valence electrons. Each unpaired electron then joins with an unpaired electron of a neighboring carbon atom to form a new covalent bond that links two monomer units together.

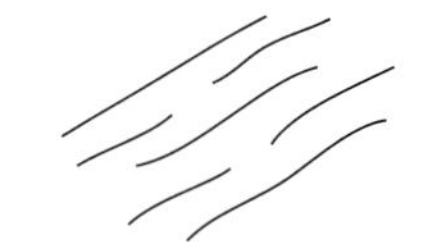

(a) Molecular strands of HDPE

(b) Molecular strands of LDPE

Figure 29.3
(a) The polyethylene strands of HDPE are able to pack closely together, much like strands of uncooked spaghetti. (b) The polyethylene strands of LDPE are branched, which prevents the strands from packing well.

Plastic bottles and milk jugs are made from a high-density polyethylene (HDPE), shown schematically in Figure 29.3a. Note how it consists of long strands of straight-chain molecules packed closely together. The tight alignment of neighboring strands makes HDPE a relatively rigid, tough plastic. Low-density polyethylene (LDPE), shown in Figure 29.3b, is used for plastic bags, photographic film, and electrical wire insulation. Note the strands of highly branched chains. This is an architecture that prevents the strands from packing closely together. This makes LDPE more bendable than HDPE and gives it a lower melting point. LDPE will hold its shape in boiling water, while HDPE deforms.

For pipes, hard-shelled suitcases, and appliance parts, other addition polymers are created with different monomers—ones that have a double bond. The monomer propylene, for example, yields polypropylene, as shown in Figure 29.4. Fibers of polypropylene are used for upholstery, indoor–outdoor carpets, and even thermal underwear.

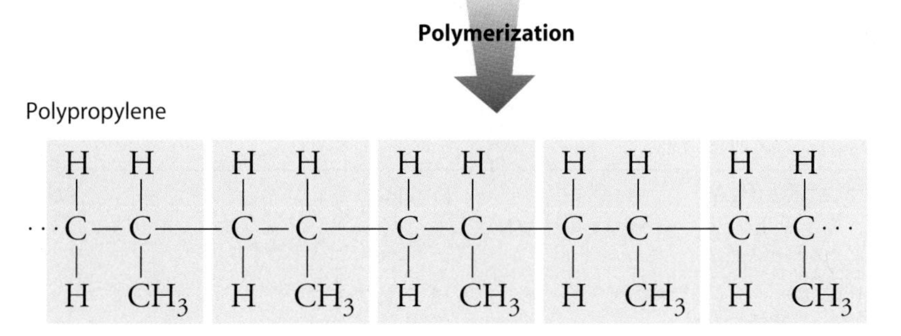

Figure 29.4
Propylene monomers polymerize to form polypropylene.

Table 29.1

Addition and Condensation Polymers

Addition Polymers	Repeating Unit	Common Uses	Recycling Code
Polyethylene (PE)	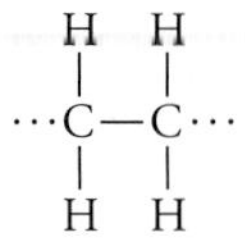 $\cdots\mathrm{CH_2{-}CH_2}\cdots$	Plastic bags, bottles	2 HDPE, 4 LDPE
Polypropylene (PP)	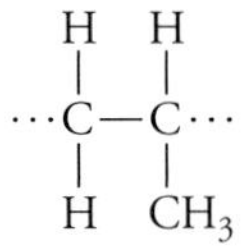 $\cdots\mathrm{CH_2{-}CH(CH_3)}\cdots$	Indoor–outdoor carpets	5 PP
Polystyrene (PS)	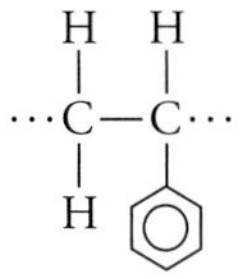 $\cdots\mathrm{CH_2{-}CH(C_6H_5)}\cdots$	Plastic utensils, insulation	6 PS
Polyvinyl chloride (PVC)	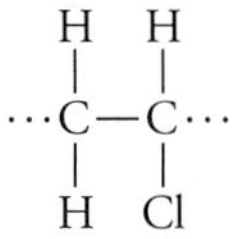 $\cdots\mathrm{CH_2{-}CHCl}\cdots$	Shower curtains, tubing	3 V
Polyvinylidene chloride (Saran)	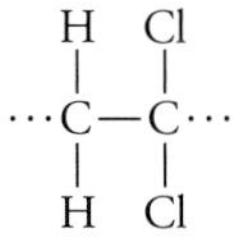 $\cdots\mathrm{CH_2{-}CCl_2}\cdots$	Plastic wrap	—
Polytetrafluoroethylene (Teflon)	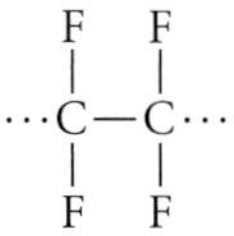 $\cdots\mathrm{CF_2{-}CF_2}\cdots$	Nonstick coating	—
Polyacrylonitrile (Orlon)	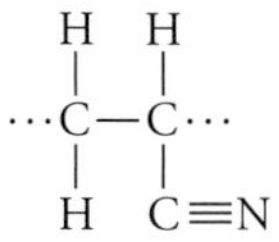 $\cdots\mathrm{CH_2{-}CH(C{\equiv}N)}\cdots$	Yarn, paints	—
Polymethyl methacrylate (Lucite, Plexiglas)	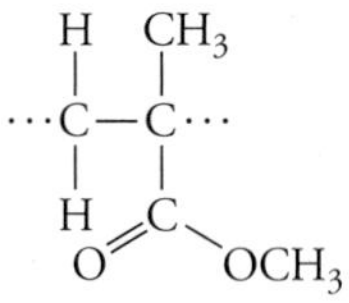 $\cdots\mathrm{CH_2{-}C(CH_3)(C({=}O)OCH_3)}\cdots$	Windows, bowling balls	—
Polyvinyl acetate (PVA)	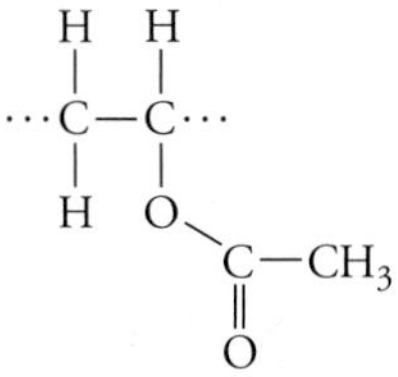 $\cdots\mathrm{CH_2{-}CH(O{-}C({=}O){-}CH_3)}\cdots$	Adhesives, chewing gum	—

(continued)

Table 29.1

Addition and Condensation Polymers

Condensation Polymers	Repeating Unit	Common Uses	Recycling Code
Nylon	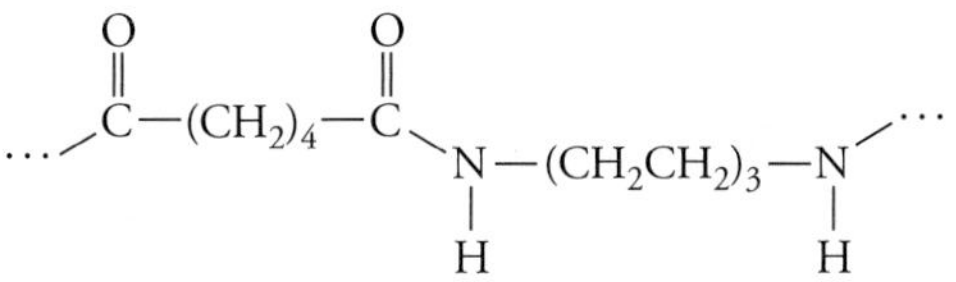 $\cdots C(=O)-(CH_2)_4-C(=O)-N(H)-(CH_2CH_2)_3-N(H)\cdots$	Carpeting, clothing	—
Polyethylene terephthalate (Dacron, Mylar)	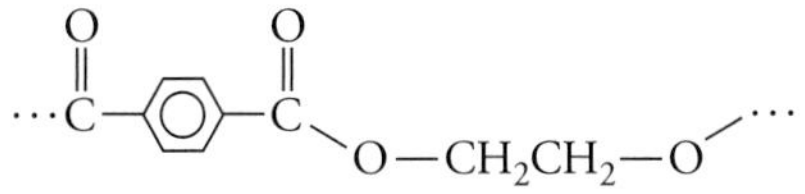 $\cdots C(=O)-C_6H_4-C(=O)-O-CH_2CH_2-O\cdots$	Clothing, plastic bottles	1 PET
Melamine–formaldehyde resin (Melmac, Formica)	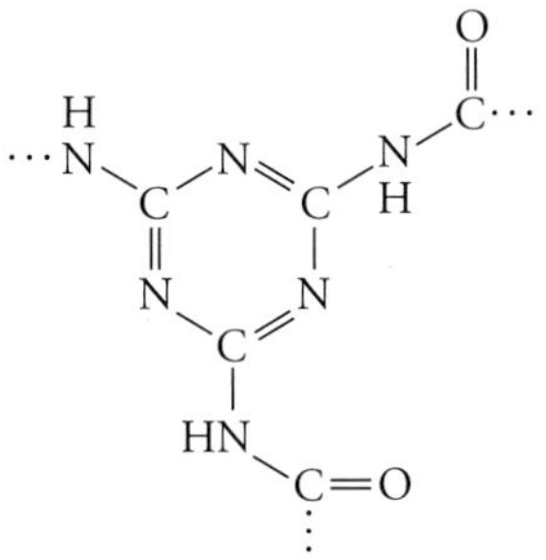 $\cdots N(H)-C_3N_3(-N(H)-C(=O)\cdots)(-HN-C(=O)\cdots)$	Dishes, countertops	—

Blowing gas into liquid polystyrene generates Styrofoam, widely used for coffee cups, packing material, and insulation. Figure 29.5 shows that the use of styrene as the monomer yields polystyrene. Transparent plastic cups are made of polystyrene, as are thousands of other household items.

Another important addition polymer is polyvinylchloride (PVC), which is tough and easily molded. Floor tiles, shower curtains, and pipes are most often made of PVC, shown in Figure 29.6.

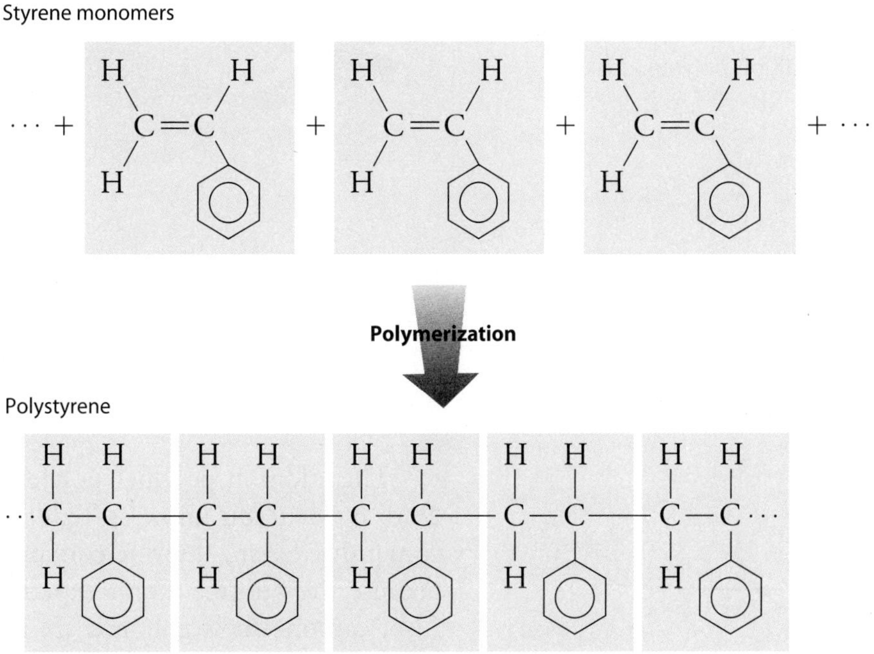

Styrene monomers

Polymerization

Polystyrene

Figure 29.5
Styrene monomers polymerize to form polystyrene.

Figure 29.6
PVC is tough and easily molded, which is why it is used to fabricate many household items.

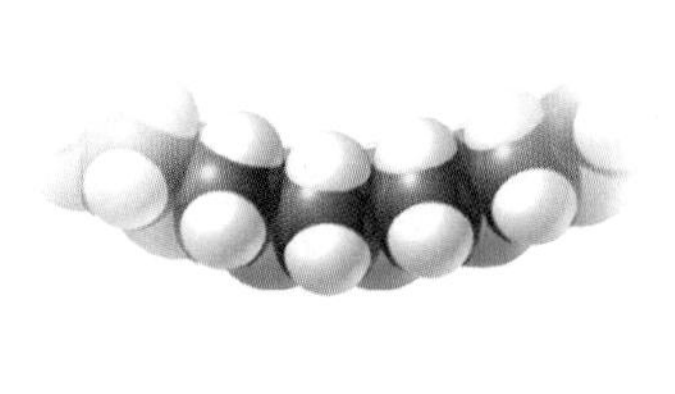

Polyvinyl chloride (PVC)

```
      H  H   H  H   H  H   H  H   H  H
      |  |   |  |   |  |   |  |   |  |
   ···C—C—  C—C—C—C—C—C—C—C···
      |  |   |  |   |  |   |  |   |  |
      H  Cl  H  Cl  H  Cl  H  Cl  H  Cl
```

The addition polymer polyvinylidene chloride (trade name Saran®), shown in Figure 29.7, is used as plastic wrap for food. The large chlorine atoms in this polymer help it stick to surfaces such as glass by dipole type attractions, as we saw in Section 24.1.

Figure 29.7
The large chlorine atoms in polyvinylidene chloride make this addition polymer sticky.

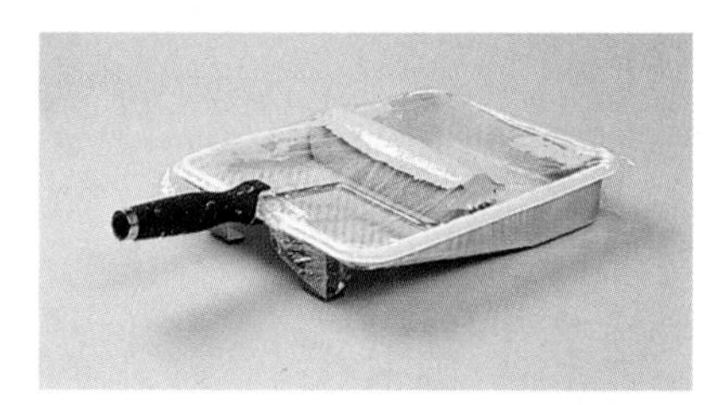

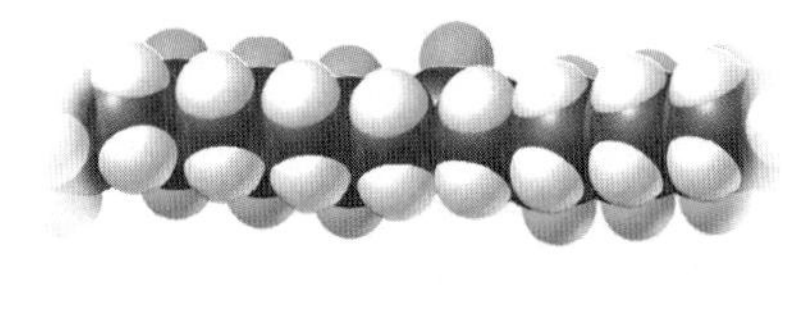

Polyvinylidene chloride (Saran)

```
      H  Cl  H  Cl  H  Cl  H  Cl  H  Cl
      |  |   |  |   |  |   |  |   |  |
   ···C—C—C—C—C—C—C—C—C—C···
      |  |   |  |   |  |   |  |   |  |
      H  Cl  H  Cl  H  Cl  H  Cl  H  Cl
```

The addition polymer polytetrafluoroethylene, shown in Figure 29.8, is what you know as Teflon®. In contrast to the chlorine-containing Saran, fluorine-containing Teflon has a nonstick surface because the fluorine atoms experience very few molecular attractions. Also, carbon–fluorine bonds are unusually strong, which means Teflon can be heated to high temperatures before decomposing. These properties make Teflon an ideal coating for cooking surfaces.

Figure 29.8
The fluorine atoms in polytetrafluoroethylene tend not to experience molecular attractions, which is why this addition polymer is used as a nonstick coating and lubricant.

Polytetrafluoroethylene (Teflon)

```
       F   F   F   F   F   F   F   F   F   F
       |   |   |   |   |   |   |   |   |   |
    ...C — C — C — C — C — C — C — C — C — C...
       |   |   |   |   |   |   |   |   |   |
       F   F   F   F   F   F   F   F   F   F
```

Concept Check ✓

What do all monomers of addition polymers have in common?

Check Your Answer A double covalent bond between two carbon atoms.

29.3 Condensation Polymers Form with the Loss of Small Molecules

A **condensation polymer** is one formed when the joining of monomer units is accompanied by the loss of a small molecule like water, H_2O, or hydrogen chloride, HCl. Any monomer capable of becoming part of a condensation polymer must have a functional group on each end. When two such monomers come together to form a condensation polymer, one functional group of the first monomer links up with one functional group of the other monomer. The result is a two-monomer unit that has two terminal functional groups, one from each of the two original monomers. Each of these terminal functional groups in the two-monomer unit is now free to link up with one of the functional groups of a third monomer, and then a fourth, and so on. In this way a polymer chain is built.

Figure 29.9 shows this process for the condensation polymer called *nylon*, created in 1937. This polymer is composed of two different monomers, as shown in Figure 29.9. One monomer is adipic acid, which contains two reactive end groups, both carboxyl groups. The second monomer is hexamethylenediamine, in which two amine groups are the reactive end groups. One end of an adipic acid molecule and one end of a hexamethylamine molecule can be made to react

with each other, splitting off a water molecule in the process. After two monomers have joined, reactive ends still remain for further reactions, which leads to a growing polymer chain.

Concept Check ✓

The structure of 6-aminohexanoic acid is

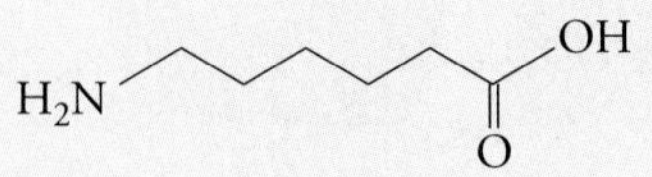

H_2N OH O

Is this compound a suitable monomer for forming a condensation polymer? If so, what is the structure of the polymer formed, and what small molecule is split off during the condensation?

Check Your Answer Yes, because the molecule has two reactive ends. You know both ends are reactive because they are the ends shown in Figure 29.9. The only difference here is that both types of reactive ends are on the same molecule. Monomers of 6-aminohexanoic acid combine by splitting off water molecules to form the polymer known as nylon-6:

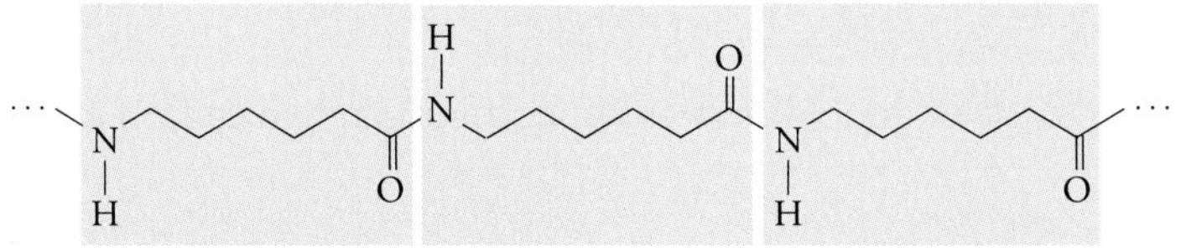

... N H O H N O N H O ...

Figure 29.9
Adipic acid and hexamethylenediamine polymerize to form the condensation copolymer nylon.

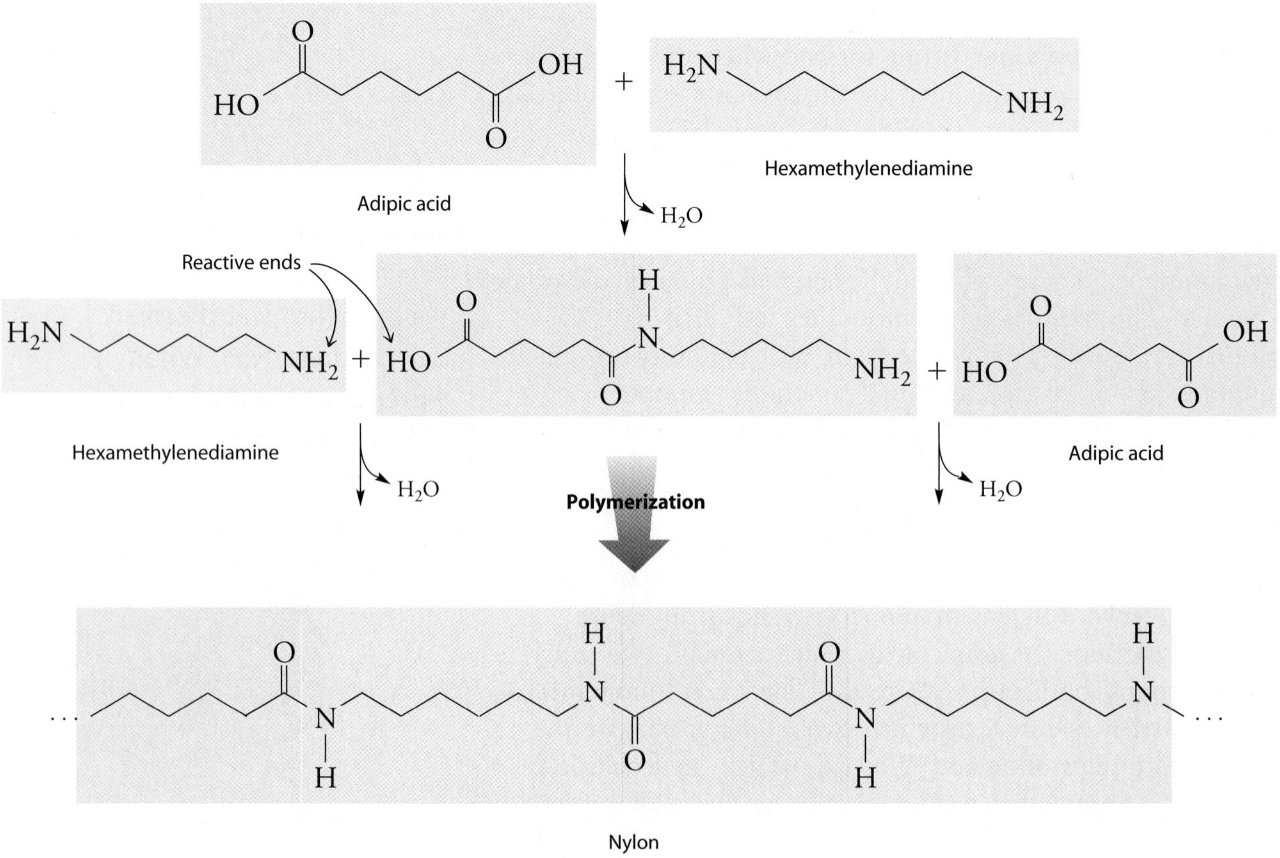

O HO OH O + H_2N NH_2

Adipic acid

Hexamethylenediamine

H_2O

Reactive ends

H_2N NH_2 + HO O H N O NH_2 + HO O OH O

Hexamethylenediamine

H_2O

Polymerization

H_2O

Adipic acid

... O N H H N O O N H H N ...

Nylon

Another widely used condensation polymer is polyethylene terephthalate (PET), formed from the copolymerization of ethylene glycol and terephthalic acid, as shown in Figure 29.10. Plastic soda bottles are made from this polymer. Also, PET fibers are sold as Dacron® polyester, used in clothing and stuffing for pillows and sleeping bags. Thin films of PET are called Mylar® and can be coated with metal particles to make magnetic recording tape or those metallic-looking balloons you see for sale at most grocery store check-out counters.

Figure 29.10
Terephthalic acid and ethylene glycol polymerize to form the condensation copolymer polyethylene terephthalate.

Terephthalic acid

Ethylene glycol

H_2O H_2O H_2O

Polymerization

Polyethylene terephthalate (PET)

Hands-On Exploration: Racing Water Drops

The chemical composition of a polymer has a significant effect on its macroscopic properties. See this for yourself.

What You Need

plastic sandwich bag; plastic food wrap; drop of water

Procedure

Place a drop of water on a new plastic sandwich bag, and then tilt the bag vertically so that the drop races off. Observe the behavior of the water carefully. Now race a drop of water off a freshly pulled strop of plastic food wrap. How does the behavior of the drop on the wrap compare with the behavior of the drop on the sandwich bag?

Most brands of sandwich bags are made of polyethylene chloride. Look carefully at the chemical composition of these polymers, shown in Table 29.1. Which contains larger atoms? Which might be involved in stronger dipole-induced dipole interactions with water? Need help with these questions? Refer back to Section 24.1.

Monomers that contain three reactive functional groups can also form polymer chains. These chains become interlocked in a rigid three-dimensional network that lends considerable strength and durability to the polymer. Once formed, these condensation polymers cannot be remelted or reshaped, which makes them hard-set, or *thermoset,* polymers. A good example is the thermoset polymer shown in Figure 29.11, formed from the reaction of formaldehyde with melamine. Hard plastic dishes (Melmac®) and countertops (Formica®) are made of this material. A similar polymer, Bakelite®, made from formaldehyde and phenol, is used to bind plywood and particle board. Bakelite was synthesized in the early 1900s, and it was the first widely used polymer.

Figure 29.11
The three reactive groups of melamine allow it to polymerize with formaldehyde to form a three-dimensional network.

Formaldehyde + Melamine

Three reactive ends

Polymerization

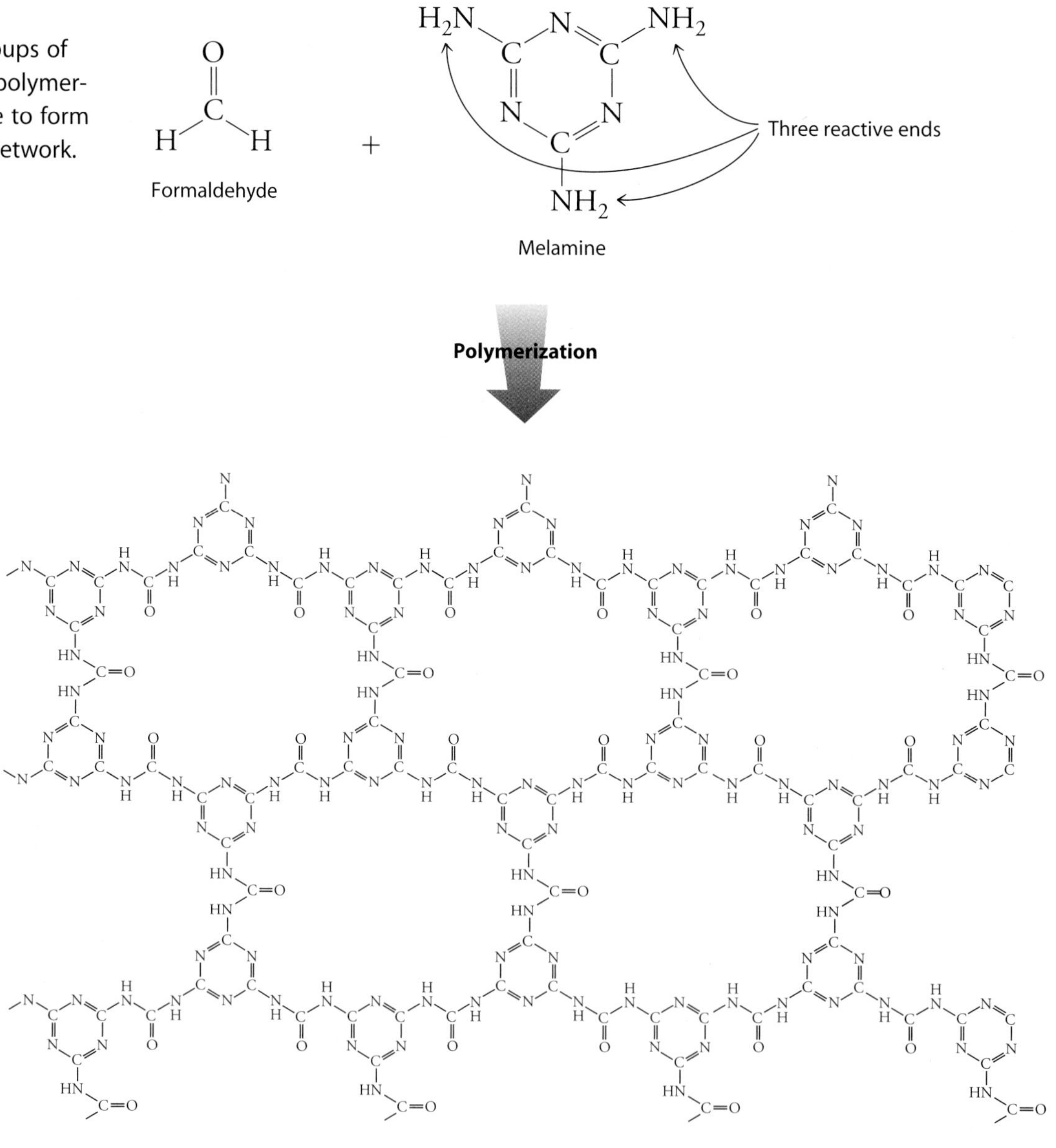

Melmac

The synthetic-polymers industry has grown remarkably over the past 50 years. Annual production of polymers in the United States alone has grown from 3 billion pounds in 1950 to 100 billion pounds in 2000. Today, it is a challenge to find any consumer item that does *not* contain a plastic of one sort or another. Try this yourself.

29.4 The Development of Plastics Involved Experiments and Discovery

This section covers the inventors who developed synthetic polymers and the effect these new materials have had on society. The successes and failures in the story of plastics show that chemistry is a process of experimentation and discovery.

The search for a lightweight, nonbreakable, moldable material began with the invention of vulcanized rubber. This material is derived from natural rubber, which is a semisolid, elastic, natural polymer. The fundamental chemical unit of natural rubber is polyisoprene, which plants produce from isoprene molecules, as shown in Figure 29.12. In the 1700s, natural rubber was noted for its ability to rub off pencil marks. (This is the origin of the term *rubber*.) Natural rubber has few other uses, however, because it turns gooey at warm temperatures and brittle at cold temperatures.

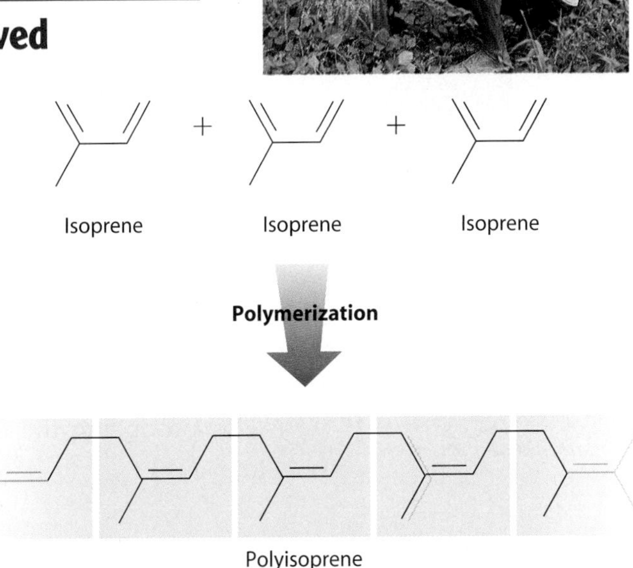

Figure 29.12
Isoprene molecules react with one another to form polyisoprene, the fundamental chemical unit of natural rubber, which comes from rubber trees.

In 1839, an American inventor, Charles Goodyear, discovered *rubber vulcanization*, a process in which natural rubber and sulfur are heated together. (His discovery occurred after he accidentally tipped an open jar of sulfur into a pot of heated natural rubber.) The product, vulcanized rubber, is harder than natural rubber and retains its elastic properties over a wide range of temperatures. This is the result of *disulfide cross-linking* between polymer chains, as illustrated in Figure 29.13.

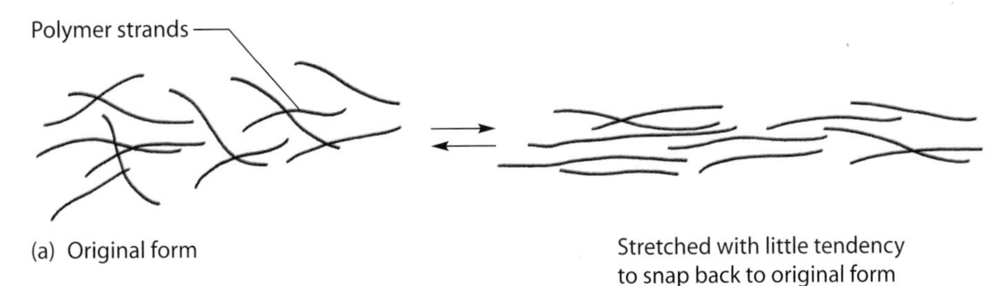

Polymer strands

(b) Original form with disulfide cross-links

Stretched with great tendency to snap back because of cross-links

Figure 29.13
(a) When stretched, the individual polyisoprene strands in natural rubber slip past one another and the rubber stays stretched. (b) When vulcanized rubber is stretched, the sulfur cross-links hold the strands together, allowing the rubber to return to its original shape.

Being at the right place at the right time is not enough for making a great scientific discovery. Curiosity and hard work are also important.

Vulcanized rubber has found innumerable applications, from tires to raingear, and has grown into a multibillion dollar industry. To help supply the need for vulcanized rubber, natural rubber (polyisoprene) is now a byproduct of petroleum. Goodyear unfortunately reaped very few rewards from his discovery. He was a man of ill health who died in jail serving time for debts he was unable to pay. The present-day Goodyear Corporation was founded not by Goodyear but by others who sought to pay tribute to his name 15 years after he died.

In 1845, as vulcanized rubber was becoming popular, the Swiss chemistry professor Christian Schobein wiped up a spilled mixture of nitric and sulfuric acids with a cotton rag that he then hung up to dry. Within a few minutes, the rag burst into flames and then vanished, leaving only a tiny bit of ash. Schobein had discovered nitrocellulose, in which most of the hydroxyl groups in cellulose are bonded to nitrate groups, as Figure 29.14 illustrates. Schobein's attempts to market nitrocellulose as a smokeless gunpowder (*guncotton*) were unsuccessful, mainly because of a number of lethal explosions at plants producing the material.

Nitrate group

Nitrocellulose (cellulose nitrate)

Figure 29.14
Nitrocellulose, also known as cellulose nitrate, is highly combustible because of its many nitrate groups, which facilitate oxidation.

Collodion and Celluloid Begin with Nitrocellulose

While Schobein failed at marketing guncotton, researchers in France discovered that solvents such as diethyl ether and alcohol transformed nitrocellulose to a gel that could be molded into various shapes. Furthermore, spread thin on a flat surface, the gel dried to a tough, clear, transparent film. This workable nitrocellulose material was dubbed *collodion*, and its first application was as a medical dressing for cuts.

In 1855, the moldable features of collodion were exploited by the British inventor and chemist Alexander Parkes, who marketed the material as Parkesine. Combs, earrings, buttons, bracelets, billiard balls, and even false teeth were manufactured in his factories. Parkes chose to focus more on quantity than on quality, however. Because he used low-grade cotton and cheap but unsuitable solvents, many of his products lacked durability, which led to commercial failure. In 1870, John Hyatt, a young inventor from Albany, New York, discovered that collodion's moldable properties were vastly improved by using camphor as a solvent. Hyatt's brother Isaiah named this camphor-based nitrocellulose material *celluloid.* Because of its greater workability, celluloid became the plastic of choice for the manufacture of many household items. In addition, thin transparent films of celluloid made excellent supports for photosensitive emulsions, a boon to the photography industry and a first step in the development of motion pictures.

As wonderful as celluloid was, it still had the major drawback of being highly flammable. Today, one of the few commercially available products made of celluloid is Ping-Pong balls, shown in Figure 29.15.

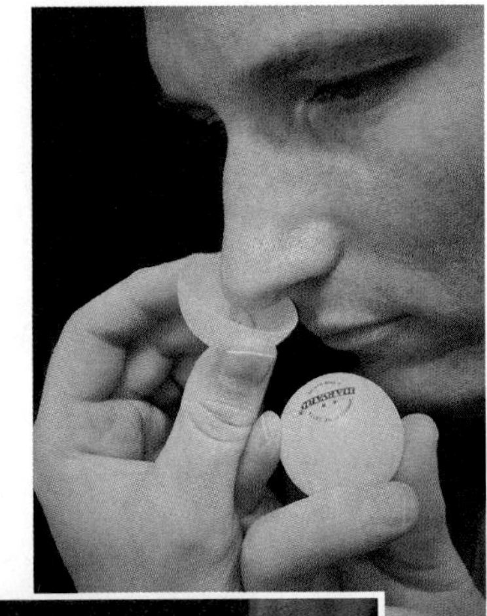

Figure 29.15
(a) Smell a freshly cut Ping-Pong ball, and you will note the distinct odor of camphor, which is the same smell that arises from heat cream for sore muscles. This camphor comes from the celluloid from which the ball is made. (b) Ping-Pong balls burn rapidly because they are made of nitrocellulose.

Bakelite Was the First Widely Used Plastic

About 1899, Leo Baekeland, a chemist who had immigrated to the United States from Belgium, developed an emulsion for photographic paper that was exceptional in its sensitivity to light. He sold his invention to George Eastman, who had made a fortune selling celluloid-based photographic film along with his portable Kodak® camera. Expecting no more than $50,000 for his invention, Baekeland was shocked at Eastman's initial offer of $750,000 (in today's dollars, that would be about $25 million). Suddenly a very wealthy man, Baekeland was free to pursue his chemical interests.

He decided that the material the world needed most was a synthetic shellac to replace the natural shellac produced from the resinous secretions of the lac beetle native to southeastern Asia. At the time, shellac was the optimal insulator for electrical wires. Ever since Edison's 1879 invention of the incandescent light bulb, miles of shellac-coated metal wire were being stretched across the land. The supply of shellac, however, was unable to keep up with demand.

Baekeland explored a tarlike solid once produced in the laboratories of Alfred von Bayer, the German chemist who played a role in the development of aspirin. Whereas Bayer had dismissed the solid as worthless, Baekeland saw it as a virtual gold mine. After several years, he produced a resin that, when poured into a mold and then heated under pressure, solidified into a transparent positive of the mold. Baekeland's resin was a mixture of formaldehyde and a phenol that polymerized into the complex network shown in Figure 29.16.

The solidified material, which he called *Bakelite,* was impervious to harsh acids or bases, wide temperature extremes, and just about any solvent. Bakelite quickly replaced celluloid as a molding medium, finding a wide variety of uses for several decades. It wasn't until the 1930s that alternative thermoset polymers, such as Melmac and Formica began to challenge Bakelite's dominance in the evolving plastics industry.

Figure 29.16
The molecular network of Bakelite shown in two dimensions. The actual structure projects in all three dimensions. The first handset telephones were made of Bakelite.

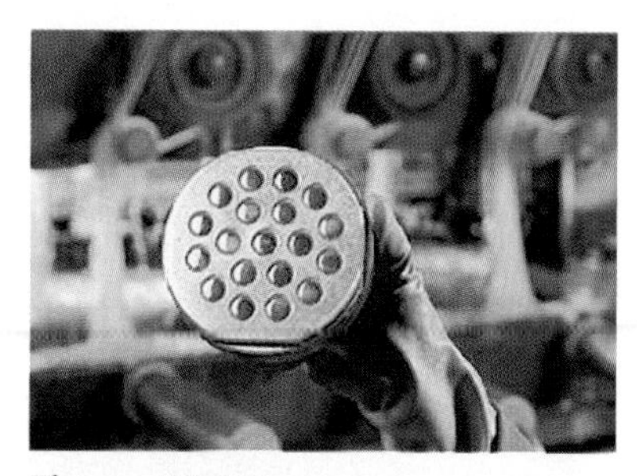

Figure 29.17
Viscose is still used today in the manufacture of fibers used to make the synthetic fabric called *rayon*. Shown here is the die through which the viscose is passed in order to transform it into threads.

The First Plastic Wrap Was Cellophane

Cellophane had its beginnings in 1892, when Charles Cross and Edward Beven of England found that treating cellulose with concentrated sodium hydroxide followed by carbon disulfide created a thick, syrupy, yellow liquid they called *viscose*. Extruding the viscose into an acidic solution, shown in Figure 29.17, generated a tough cellulose filament that could be used to make a synthetic silky cloth today called *rayon*.

In 1904, Jacques Brandenberger, a Swiss textile chemist, observed restaurant workers discarding fine tablecloths that had only slight stains on them. Working with viscose at the time, he had the idea of extruding it not as a fiber but as a thin, transparent sheet that might be adhered to tablecloths and provide an easy-to-clean surface. By 1913, Brandenberger had perfected the manufacture of a viscose-derived, thin, transparent sheet of cellulose, which he named *cellophane*. After failing to form an adequate adhesion between cellophane and cloth, Brandenberger investigated cellophane's possible use as a film support for photography and motion pictures. This idea didn't work because of the cellophane's tendency to warp when heated. From these failures, Brandenberger began to realize that the most likely utility of his newly created cellophane was as a wrapping material.

Within several years, the DuPont Corporation bought the rights to cellophane. After producing several batches, investigators discovered that cellophane did not provide an effective barrier to water vapor and hence did little to keep foods from drying out. By 1926, DuPont chemists had solved this problem by incorporating small amounts of nitrocellulose and wax. Vaporproof cellophane gained wide popularity as a wrapping for such products as chocolates, cigarettes, cigars, and bakery goods. Hermetically sealed by cellophane, a product could be kept free of dust and germs. And unlike paper or tin foil—the alternatives of the day—cellophane was transparent and thus allowed the consumer to view the packaged contents, as seen in Figure 29.18. With properties like these, cellophane played a great role in the success of supermarkets, which first appeared in the 1930s. Perhaps, cellophane's greatest appeal to the consumer, however, was its shine. As marketing people soon discovered, nearly any product—soaps, canned goods, or golf balls—would sell faster when wrapped in cellophane.

Figure 29.18
Cellophane transformed the way foods and other items were marketed.

29.5 Polymers Win in World War II

In the 1930s, more than 90% of the natural rubber used in the United States came from Malaysia. Then came Pearl Harbor and World War II. Within days Japan captured Malaysia. That was when the United States faced its first natural resource crisis. The military implications were devastating because without rubber for tires, military airplanes and jeeps were useless. Petroleum-based synthetic rubber had been developed in 1930 by the chemist Wallace Carothers but was not widely used because it was much more expensive than natural rubber. With Malaysian rubber impossible to get and a war on, cost was no longer an issue. Synthetic rubber factories were constructed across the nation, and within a few years, the annual production of synthetic rubber rose from 2000 tons to about 800,000 tons.

Also in the 1930s, British scientists developed radar as a way to track thunderstorms. With war approaching, these scientists turned their attention to the idea that radar could be used to detect enemy aircraft. Their equipment was massive, however. A series of ground-based radar stations could be built, but placing massive radar equipment on aircraft was not feasible. The great mass of the equipment was due to the large coils of wire needed to generate the intense radio waves. The scientists knew that if they could coat the wires with a thin, flexible electrical insulator, they would be able to design a radar device that was much less massive. Fortunately, the recently developed polymer polyethylene turned out to be an ideal electrical insulator. This permitted British radar scientists to construct equipment light enough to be carried by airplanes. These planes were slow, but flying at night or in poor weather, they could detect, intercept, and destroy enemy aircraft. Midway through the war, the Germans developed radar themselves, but without polyethylene, their radar equipment was inferior, and the tactical advantage stayed with the Allied forces.

As mentioned earlier, the polymer nylon was invented in 1937, which was just prior to World War II. Soon after its invention, and before the start of the war, executives at DuPont, where it had been invented, sought a market for this polymer. They knew this polymer was remarkable for its ability to form strong, silk-like fibers. Stockings made of the polymer were comfortable and resistant to the tears and runs common to silk stockings. The stockings were also easy to manufacture and could be sold at prices much cheaper than silk stockings. Tremendous demand ensued, and commercial success was immediate. To reflect the benefits of this new polymer, the folks at DuPont considered marketing it under the trade name "norun." That didn't succeed, and so they spelled it backwards to come up with "nuron." When that didn't succeed either, they changed the name to "nylon."

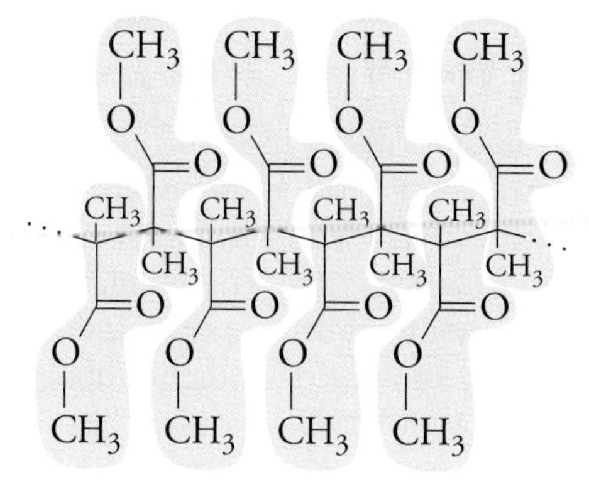

Figure 29.19
The bulky side groups in poly(methyl methacrylate) prevent the polymer chains from aligning with one another. This makes it easy for light to pass through the material which is tough, transparent, lightweight, and moldable. (Plexiglas® is a registered trademark belonging to ATOFINA.)

During World War II, nylon stockings were a scarce but hot item. It's said that spies even used nylon stockings to bribe enemy agents for information.

Aside from its use in hosiery, nylon also found great use in the manufacture of parachutes, important to the U.S. military. Up to that time, parachutes were made mainly of silk. The world's foremost supplier of silk, however, was Japan. By the time World War II began, Japan had stopped exporting silk to the United States. The U.S., however, now had nylon, which in many regards was better than silk. Over the course of the war practically all the nylon that DuPont could produce went to the military for the manufacture of a wide variety of nylon-based commodities suited for military purposes, such as parachutes, ropes, and clothing.

Four other polymers that had a significant impact on the outcome of World War II were Plexiglas®, polyvinyl chloride, Saran, and Teflon. Plexiglas, shown in Figure 29.19, is a polymer known to chemists as poly(methyl methacrylate). This glasslike but moldable and lightweight material made excellent domes for the gunner's nests on fighter planes and bombers. Although both Allied and German chemists had developed poly(methyl methacrylate), only the Allied chemists learned how small amounts of this polymer in solution could prevent oil or hydraulic fluid from becoming too thick at low temperatures. Equipped with only a few gallons of a poly(methyl methacrylate) solution, Soviet forces were able to keep their tanks operational in the Battle of Stalingrad during the winter of 1943. While Nazi equipment halted in the bitter cold, Soviet tanks and artillery functioned perfectly, resulting in victory and an important turning point in the war.

Polyvinyl chloride (PVC) had been developed by a number of chemical companies in the 1920s. The problem with this material, however, was that it lost resiliency when heated. In 1929, Waldo Semon, a chemist at BFGoodrich, found that PVC could be made into a workable material by the addition of a plasticizer. Semon got the idea of using plasticized PVC as a shower curtain when he

Figure 29.20
The now-familiar plastic food wrap carton with a cutting edge was first introduced in 1954 by Dow Chemical for its brand of Saran Wrap.

observed his wife sewing together a shower curtain made of rubberized cotton. Other uses for PVC were slow to appear, however, and it wasn't until World War II that this material became recognized as an ideal waterproof material for tents and rain gear. After the war, PVC replaced Bakelite as the medium for making phonograph records.

Originally designed as a covering to protect theater seats from chewing gum, Saran found great use in World War II as a protective wrapping for artillery equipment during sea voyages. (Before Saran, the standard operating procedure had been to disassemble and grease the artillery to avoid corrosion.) After the war, the polymer was reformulated to eliminate the original formula's unpleasant odor. This improvement over cellophane resulted in Saran Wrap becoming the most popular food wrap of all time (Figure 29.20).

In the late 1930s, the DuPont researcher Roy Plunkett and his colleagues, shown in Figure 29.21, filled a pressurized cylinder with a gas called tetrafluoroethylene. The next morning they were surprised to discover that the cylinder appeared empty. Not believing that the contents had simply vanished, they hacksawed the cylinder apart and discovered a white solid coating the inner surface. Driven by curiosity, they continued to investigate the material, which eventually came to be known and marketed as the polymer Teflon.

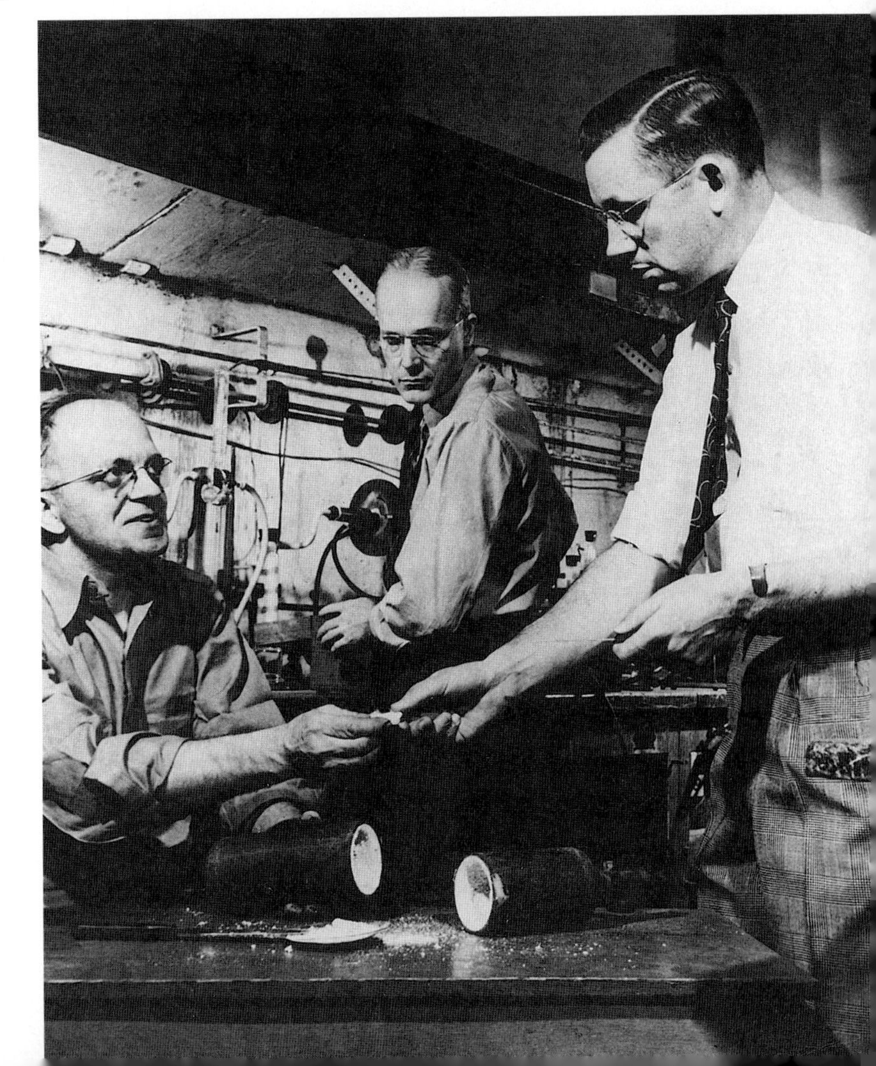

Figure 29.21
Roy Plunkett (right) and his colleagues pose for this reenactment photograph of their discovery of Teflon. Their success was due in great part to their curiosity.

Initially, the discoverers of Teflon were impressed by the long list of things this new material would *not* do. It would not burn and it would not completely melt. Instead, at 620°F, it congealed into a gel that could be conveniently molded. It would not conduct electricity and it was impervious to attack by mold or fungus. No solvent, acid, or base could dissolve or corrode it. And most remarkably, nothing would stick to it, not even chewing gum.

Because of all the things Teflon would not do, DuPont was not quite sure what to do with it. Then, in 1944 they were approached by governmental researchers who were in desperate need of a highly inert material to line the valves and ducts of an apparatus being built to isolate uranium-235 in the manufacture of the first nuclear bomb. Thus Teflon found its first application, and one year later, World War II came to a close with the nuclear bombing of Japan.

Concept Check ✓

Name seven polymers that had a significant impact on the effectiveness of the Allied forces in World War II.

Check Your Answer Synthetic rubber, polyethylene, nylon, Plexiglas, polyvinyl chloride, Saran, and Teflon.

29.6 Attitudes About Plastics Have Changed

With a record of wartime successes, plastics were readily embraced in the postwar years. In the 1950s, polyester was introduced as a substitute for wool. Also, the 1950s were the decade during which the entrepreneur Earl Tupper created a line of polyethylene food containers known as Tupperware.

By the 1960s, a decade of environmental awakening, many people began to recognize the negative attributes of plastics. Being cheap, disposable, and non-biodegradable, plastic readily accumulated as litter and as landfill. With petroleum so readily available and inexpensive, however, and with a growing population of plastic-dependent baby boomers, little stood in the way of an ever-expanding array of plastic consumer products. By 1977, plastics surpassed steel as the number one material produced in the United States. Environmental

Figure 29.22
Flexible and flat video displays can now be fabricated from polymers.

concerns also continued to grow, and in the 1980s, plastics-recycling programs began to appear. Although the efficiency of plastics recycling still holds room for improvement, we now live in a time when sports jackets made of recycled plastic bottles are a valued commodity.

In the past 50 years, there have been a number of significant technological advances in plastics. Polymers that emit light, for example, can be used to build display monitors that roll up like a newspaper or can be rolled onto walls like wallpaper. One interesting application is shown in Figure 29.22. We have polymers that conduct electricity, replace body parts, and are stronger but much lighter than steel. Imagine synthetic polymers that mimic photosynthesis by transforming solar energy to chemical energy or synthetic polymers that efficiently separate fresh water from the oceans. These are not dreams. They are realities that chemists have already demonstrated in the laboratory. Polymers have played a significant role in our past, and they hold a clear promise for our future. Let's work to ensure that the petroleum starting materials from which we fabricate most polymers are not exhausted before this promise is realized.

Chapter Review

Key Terms and Matching Definitions

_____ addition polymer
_____ condensation polymer
_____ monomer
_____ plastic
_____ polymer

1. A long organic molecule made of many repeating units.
2. The small molecular unit from which a polymer is formed.
3. Another name for a moldable material made out of organic polymers.
4. A polymer formed by the joining together of monomer units with no atoms lost as the polymer forms.
5. A polymer formed by the joining together of monomer units accompanied by the loss of a small molecule, such as water.

Review Questions

Organic Molecules Can Link to Form Polymers

1. Are monomers made of polymers or are polymers made of monomers?
2. Are all polymers made by humans?

Addition Polymers Result from the Joining Together of Monomers

3. What happens to the double bond of a monomer participating in the formation of an addition polymer?
4. Which has more branching in its polymer chains: HDPE or LDPE?
5. Which has more chlorine atoms per polymer stand: polyvinyl chloride, or polyvinylidene chloride?
6. Why is polyvinylidene chloride a stickier plastic than polyethylene?

Condensation Polymers Form with the Loss of Small Molecules

7. What is released in the formation of a condensation polymer?
8. What is a copolymer?
9. What is a thermoset polymer?

The Development of Plastics Involved Experimentation and Discovery

10. How did Schobein discover nitrocellulose?
11. What chemical is used to make celluloid a workable material?
12. What is one of the major drawbacks of celluloid?
13. Who provided Baekeland with the financial resources to develop Bakelite?
14. What prompted Brandenberger to seek a way to make thin sheets of viscose?
15. How did chemists transform cellophane into a vaporproof wrap?

Polymers Win in World War II

16. What was one of the prime motivations for the Japanese to invade Malaysia at the beginning of World War II?
17. What polymer proved useful in the development of radar equipment lightweight enough to be carried on airplanes?
18. What role did Teflon play in World War II?

Attitudes About Plastics Have Changed

19. What is an environmental drawback of plastics?
20. What raw material are plastic made from?

Exploration

Racing Drops

You may need to play around with the drops for a while in order to see the differing affinities that the bag and wrap have for water. One way to do this is to tape the polymers side by side stretched out on a sturdy piece of cardboard. Tilt the cardboard to various angles, testing for the speed with which water drops roll down the incline on the two surfaces. Ultimately, you should find that the drops roll more slowly on the wrap (polyvinylidene chloride) than on the bag (polyethylene terephthalate). The source of this greater "stickiness" in the wrap is the fairly large chlorine atoms of the polyvinylidene chloride. The larger the atom, the greater its potential for forming induced dipole molecular attractions.

The greater stickiness of the wrap is also apparent when you try to glide one sheet of wrap over another.

Exercises

1. Would you expect polypropylene to be denser or less dense than low-density polyethylene? Why?
2. Many polymers emit toxic fumes when burning. Which polymer in Table 29.1 produces hydrogen cyanide, HCN? Which two produce toxic hydrogen chloride, HCl, gas?
3. One solution to the problem of our overflowing landfills is to burn plastic objects instead of burying them. What would be some of the advantages and disadvantages of this practice?
4. Which would you expect to be more viscous, a polymer made of long molecular strands or one made of short molecular stands? Why?
5. Hydrocarbons release a lot of energy when ignited. Where does this energy come from?
6. What type of polymer would be best to use in the manufacture of stain-resistant carpets?
7. As noted in the Concept Check of Section 29.3, the compound 6-aminohexanoic acid is used to make the condensation polymer nylon-6. Polymerization is not always successful, however, because of a competing reaction. What is this reaction? Would polymerization be more likely in a dilute solution of this monomer or in a concentrated solution? Why?
8. Why does a Styrofoam cup insulate better than a transparent plastic cup, even though they are both made of the same material, which is polystyrene?
9. What role did chance discovery play in the history of polymers? Cite some examples.
10. What is the difference between collodion and celluloid?
11. What is the chemical difference between celluloid and cellophane?
12. Why does a freshly cut Ping-Pong ball smell of camphor?
13. Why are Ping-Pong balls so highly flammable?
14. How are the chemical structures of Bakelite and Melmac similar to each other? How are they different?
15. List these plastics in order of the year in which they were developed: cellophane, celluloid, collodion, nylon, parkesine, PVC, teflon, viscose yarn.

Discussion Topics

1. What value is there in learning about the history of something as "boring" as plastic? Are plastics really "boring"?
2. Should the government require that plastics be recycled? If so, how should this requirement be enforced?
3. What are some of the obstacles people face when trying to recycle materials? How might these obstacles be overcome in your community?
4. You are given the choice of shopping either from a modern catalog or from one from the 1930s. Which catalog offers cheaper prices? Which catalog offers more goods? Which catalog would you choose and why?

Suggested Reading and Web Sites

Fenichell, Stephen. *Plastic: The Making of a Synthetic Century.* New York: Harper Collins, 1997.
This engaging history tells how the development of plastics has had a profound effect on our society.

http://www.invent.org/
Home page for the National Inventors Hall of Fame. Use this site to find additional information about many of the people referred to in this chapter. To search by name, follow the links to www.invent.org/book/index.html.

http://www.socplas.org/
Home page for the Society of the Plastics Industry. Follow the link "about the industry" to learn more about the history of plastics and their social and economic impact on society.

http://www.npcm.plastics.com
Home page for the National Plastics Center and Museum. Be sure to explore their "experiment of the month" feature.

Part 7 Earth Science

This hunk of igneous rock fascinates me. Before being erupted by a volcano it was part of the Earth's interior, kept hot by radioactive decay. Over the vast span of geologic time volcanoes have spewed out not only molten rock, but gases and water vapor to form much of our atmosphere. When condensed, water vapor, together with cometary debris from outer space, formed our oceans. Ah, the wonders of *Conceptual Physical Science!*

Chapter 30: Minerals and Their Formation

Most people know that rocks are made of minerals. But what are minerals? In what ways are minerals different from one another? Why are some minerals found in beautiful geometric forms, while others have no distinct shape? Why are some minerals hard while others are soft? Why do minerals occur in so many different colors? Which minerals are common in everyday life and which are rare? And how are minerals used in the world today? To answer these questions, we enter the realm of Earth science—a rocky place.

30.1 Minerals Can Be Identified by Their Properties

A **mineral** is a naturally occurring, solid substance made up of a single element or compound. Minerals include familiar substances such as halite (table salt), copper, gold, and silver. There are many strange and rare minerals as well. All minerals are inorganic—they are never formed by living or once-living things. They have specific chemical compositions. Finally, minerals have crystalline structures that reflect the internal, orderly arrangement of their atoms.

Geologists classify minerals into different groups. The classification is based on their chemical composition (which elements are present) and their crystal structure (how the elements are arranged). It's often easy to identify and classify minerals by observing their physical properties. The physical properties of minerals include crystal form, hardness, cleavage, luster, color, streak, and specific gravity. Let's discuss these in order.

Crystal Form Expresses the Arrangement of Atoms in a Mineral

The orderly arrangement of atoms in a crystal is expressed in its shape, or **crystal form.** When you look at a crystal you are seeing the actual arrangement of atoms! Every mineral has its own crystal form. In the mineral pyrite (fool's gold), for example, you can see its intergrown cubes. When you look at quartz, you can see six-sided prisms that end in a point (Figure 30.1). Asbestos minerals often look like narrow, threadlike fibers. The mineral hematite often has a globular form that resembles a bunch of grapes (Figure 30.2). Unfortunately, well-shaped crystals are rare in nature. That's because crystals usually grow in cramped spaces.

(a)

(b)

Figure 30.1
The unique crystal form of each mineral is the external expression of the mineral's internal arrangement of atoms. (a) The intergrown cubes of pyrite. (b) The six-sided prisms of quartz.

(a)

(b)

Figure 30.2
Well-shaped crystal forms do not develop when growing occurs in a confined space. Nevertheless, the distinctive growth patterns of many minerals are apparent. (a) The narrow, threadlike fibers of the asbestos group minerals. (b) The grape-cluster shape of hematite.

Sometimes, two or more minerals contain the same elements in the same proportions but their elements are arranged differently. As a result, their crystalline structure and properties are different. Such minerals are **polymorphs** of each other. Graphite and diamond are polymorphs because they both are made of the same element, carbon.

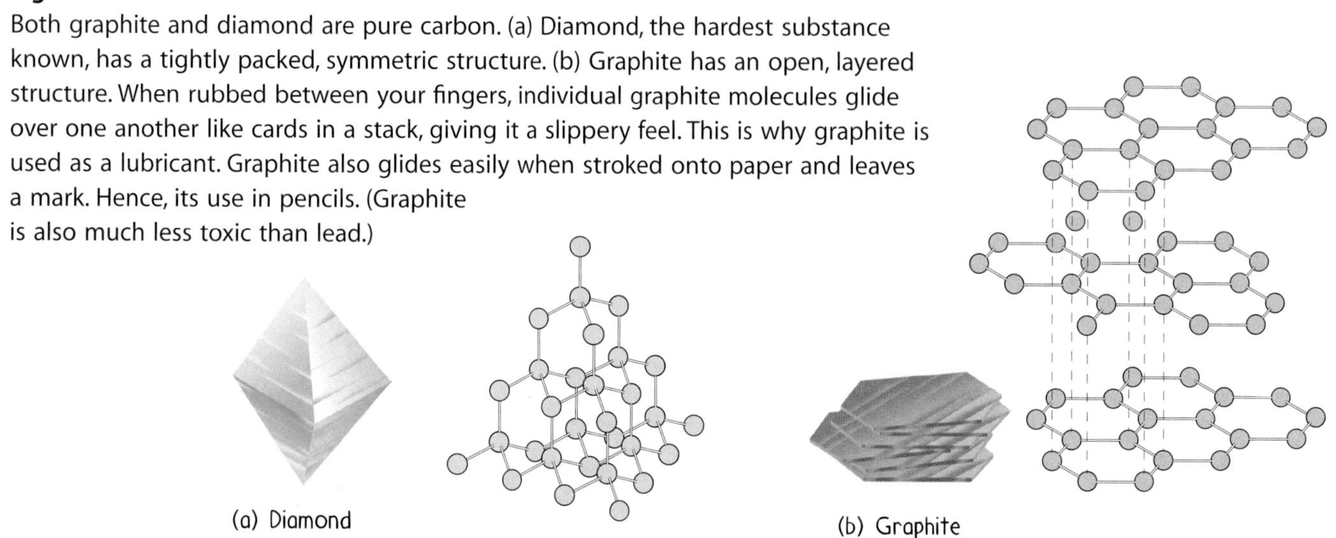

Figure 30.3
Both graphite and diamond are pure carbon. (a) Diamond, the hardest substance known, has a tightly packed, symmetric structure. (b) Graphite has an open, layered structure. When rubbed between your fingers, individual graphite molecules glide over one another like cards in a stack, giving it a slippery feel. This is why graphite is used as a lubricant. Graphite also glides easily when stroked onto paper and leaves a mark. Hence, its use in pencils. (Graphite is also much less toxic than lead.)

Crystal Power

We're most familiar with crystals in jewelry. In times past when superstition was in its heyday, crystals were valued for their alleged healing benefits and powers. Crystals were thought to channel good "energy" and ward off bad "energy." It was thought that crystals carry "vibrations" that resonate with healing "frequencies" and produce a beneficial body balance. Amazingly, these beliefs are still around. Today people are told that certain crystals provide protection against electromagnetic forces emitted by power lines, cellular phones, computer monitors, microwave ovens, and other people. Some of those who promote crystals say they are "medically proven" to heal and to protect, and the power to do so is "based on Nobel Prize–winning physics."

Crystals *do* give off energy—as every other object does. As we've learned in Chapter 10, *everything* radiates energy, and everything also absorbs energy. If a crystal or any material radiates more energy than it absorbs, its temperature drops. Atoms in crystals *do* vibrate, and *do* resonate with matching frequencies of external vibrations—just as molecules in gases and liquids do. But if promoters of crystal power are talking about some kind of energy special to crystals, or to life, no scientific evidence supports this (the discoverer of such would become world famous in a short time). Of course evidence for a new kind of energy could one day be found. But this is not what crystal power advocates say when citing scientific evidence to support their wares.

Their evidence for crystal power is not experimental. They persuade instead with *testimonials.* As advertising illustrates, more people are persuaded by testimonials than by experimental evidence. Testimonials of people convinced of personal benefits are numerous. Being convinced by scientific evidence is one thing. But another is wishful thinking. No claims for the special powers of crystals are backed up by scientific evidence.

Wearing crystal pendants seems to give some people a good *feeling,* and even a feeling of protection. This, and the physical beauty of crystals, are their virtues. Some people feel they bring good luck, just as carrying a rabbit's foot in your pocket does. The difference between crystal power and rabbit's feet, however, is that the benefits of crystals are stated in scientific terms, whereas claims for carrying a rabbit's foot are not. Hence the advocates of crystal power are into full-fledged pseudoscience.

But the carbon atoms are arranged differently. As a result, graphite and diamond show vastly different properties (Figure 30.3). Because the formation of these similar-yet-different minerals depends on particular temperatures and pressures, polymorphs are good indicators of the geological conditions at their sites of formation.

Concept Check ✓

Many minerals can be identified by their physical properties—crystal form, hardness, cleavage, luster, color, streak, and specific gravity. Why is identifying a mineral by its crystal form usually difficult?

Check Your Answer Well-shaped crystals are rare in nature because minerals typically grow in cramped spaces.

Hardness Is the Resistance of a Mineral to Scratching

Diamond is one of the minerals that can scratch glass. Diamond can do this because it is harder than glass. Similarly, a quartz crystal can scratch a feldspar crystal because quartz is harder than feldspar. The ability of one mineral to scratch another and the resistance of a mineral to being scratched are measures of hardness. We use the **Mohs scale of hardness** (Table 30.1) to compare the hardness of different minerals. You'll see that diamond is at the top of the scale with a rating of 10. Look at gypsum in the table. Would you consider it to be a hard or soft mineral?

Table 30.1

Mohs Scale of Hardness

Mineral	Scale Number	Common objects with similar hardness
Diamond	10	
Corundum	9	
Topaz	8	
Quartz	7	Steel file
Feldspar	6	Window glass
Apatite	5	Pocket knife
Fluorite	4	
Calcite	3	Copper wire or coin
Gypsum	2	Fingernail
Talc	1	

(a)

(b)

Figure 30.4
A mineral's cleavage is very useful in its identification.
(a) Muscovite (mica) has perfect cleavage in one direction.
(b) Calcite has perfect cleavage in three directions.

Why are some minerals harder than others? Hardness depends on the strength of a mineral's chemical bonds—the stronger its bonds, the harder the mineral. The things that influence bond strength are ionic charge, atom or ion size, and packing. So these things influence hardness. Strong bonds are generally found between highly charged ions—the greater the attraction, the stronger the bond. Size affects bond strength because small atoms and ions can generally pack closer together than large atoms and ions. Closely packed atoms and ions have a smaller distance between one another, and thus form stronger bonds because they attract one another with more force. Gold, with its large atoms, is soft. Its atoms are rather loosely packed and loosely bonded. Diamond, with its small carbon atoms and tightly packed structure, is hard.

Cleavage and Fracture Are Ways in Which Minerals Break

If you shatter a sample of the mineral calcite with a hammer, it will tend to break along its *planes of weakness*—planes along which chemical bonds are weak or little in number. We call this property of a mineral to break along planes of weakness **cleavage.** Planes of weakness are determined by crystal structure and chemical bond strength.

Some minerals show more tendencies toward cleavage than others do. In general, minerals that have strong bonds between flat (planar) crystal surfaces show poor cleavage while those with weak bonds along planar surfaces show more developed cleavage. Muscovite and calcite both have very distinct cleavage. Muscovite has perfect cleavage in one direction. It breaks apart to form thin, flat sheets (Figure 30.4a). Calcite has perfect cleavage in three directions. It breaks to produce rhombohedral faces (Figure 30.4b). Garnet, whose crystal structure has strong bonds in all directions, shows no cleavage.

Sometimes a mineral breaks, but the break is not along a cleavage plane. This is called a **fracture.** A fracture that is smooth and curved so that it resembles broken glass is called *conchoidal.* Quartz and olivine display smooth conchoidal fractures. Some minerals, such

Hands-On Exploration: Salt Crystals

Look at some crystals of table salt under a microscope or magnifying glass and observe their generally cubic shapes. There's no machine at the salt factory specifically designed to give salt crystals these cubic shapes, as opposed to round or triangular ones. The cubic shape occurs naturally and is a reflection of how the atoms of salt are organized—cubically. Smash a few of these salt cubes and then look at them again carefully. What you'll see are smaller salt cubes! Use the cleavage properties of crystals to explain these results.

as hematite and serpentine, break into splinters or fibers. But most minerals fracture irregularly. The degree and type of cleavage or fracture are useful guides for identifying minerals.

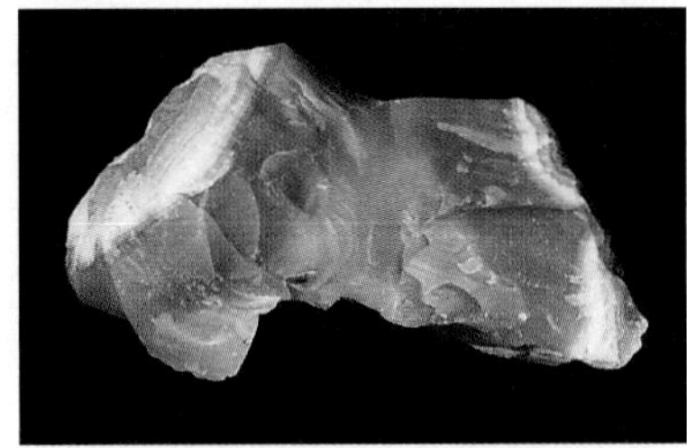

Figure 30.5
This rock, which is composed of quartz, does not exhibit cleavage. When it breaks it instead develops a conchoidal fracture—a curved, smooth surface that resembles broken glass.

Concept Check ✓

When pieces of calcite and fluorite are scraped together, which scratches which?

Check Your Answer Looking at Table 30.1, we see that fluorite is harder than calcite. So fluorite scratches calcite.

Luster Is the Appearance of a Mineral's Surface in Reflected Light

The **luster** of a mineral is the way its surface looks when it reflects light. Luster does not depend on color. Minerals of the same color may have different lusters, and minerals of the same luster may have different colors. Mineral lusters are listed in Table 30.2.

A Mineral's Color May Vary, but Its Streak Is Always the Same

Although color is an easily observable feature of a mineral, it is not a very reliable means of identification. Some minerals—copper and turquoise are two examples—have a distinctive color. But the majority of minerals occur in a variety of colors or can be colorless.

Chemical impurities in a mineral affect color. For example, the common mineral quartz, SiO_2, can be found in many colors, depending on slight impurities. It can be clear and colorless if it has no impurities. It can be milky white from tiny fluid inclusions. Rose-colored quartz results from small amounts of titanium, violet quartz

Figure 30.6
The mineral corundum (Al_2O_3) comes in a variety of colors as a result of chemical impurities. The addition of small amounts of chromium in place of aluminum produces the red gemstone *ruby*. With the addition of small amounts of iron and titanium, the result is the blue gemstone *sapphire*.

Table 30.2

Mineral Lusters

Mineral Luster	Appearance
Metallic	Strong reflection; polished or dull
Vitreous	Bright, glassy
Resinous	Waxy
Greasy	Like oily glass, also may feel greasy
Pearly	Pearly iridescence
Silky	Sheen of silk
Adamantine	Diamond, brilliant

Figure 30.7
The streak test can be used to identify minerals that have a metallic or semimetallic luster.

(amethyst) results from small amounts of iron, and smoky-gray to black quartz results from radiation damage. The color of the mineral corundum, Al_2O_3, is commonly white or grayish. But impurities in corundum give us rubies and sapphires.

Streak, the name given to the color of a mineral in its powdered form, is an important characteristic for identifying minerals that have a metallic or semimetallic luster. When rubbed across an unglazed porcelain plate, all minerals leave behind a thin layer of powder—a streak. Although different samples of the same mineral often have different colors, the color of the streak is always the same from sample to sample (Figure 30.7). For example, the mineral hematite is normally reddish-brown to black but always makes a streak that is reddish-brown. Magnetite can be gray or brown to black, but always makes a black streak. Limonite is normally yellowish-brown to dark brown and always streaks yellowish-brown. Minerals that do not have a metallic luster generally leave behind a white streak. A white streak cannot be used to identify minerals.

Specific Gravity Is a Ratio of Densities

Density is a property of all forms of matter, minerals included. In practical terms, the density of a mineral tells us how heavy a mineral feels for its size. One standard measure of density is **specific gravity**—the ratio between the weight of a substance and the weight of an equal volume of water. For example, if 1 cubic centimeter of a mineral weighs three times as much as 1 cubic centimeter of water, its specific gravity is 3. So, we see that specific gravity is simply a ratio of densities. The specific gravities of some minerals are shown in Table 30.3.

Gold's particularly high specific gravity of 19.3 is nicely taken advantage of by miners panning for gold. Fine gold pieces hidden in a mixture of mud and sand settle to the bottom of the pan when the

Table 30.3

Specific Gravity of Various Minerals

Borax	1.7	Pyrite	5.0
Quartz	2.65	Hematite	5.26
Talc	2.8	Copper	8.9
Mica	3.0	Silver	10.5
Chromite	4.6	Gold	19.3

mixture is swirled in water. Water and less dense materials spill out. After a succession of dunks and swirls, only the substance with the highest specific gravity remains—gold!

Concept Check ✓

Why are there no units for specific gravity?

Check Your Answer Remember, specific gravity is a ratio of densities. Density units divided by density units cancel out. For example, the density of the mineral hematite, Fe_2O_3, is 5.26 g/cm^3, and that of water is 1.0 g/cm^3. Therefore the specific gravity of hematite is (5.26 g/cm^3) ÷ (1.0 g/cm^3) = 5.26.

Asbestos: Friend and Foe

What do you think when you hear the word *asbestos*? Perhaps you have heard of the lung disease it causes, *asbestosis.* Or maybe images of workers in "space suits" removing asbestos from buildings come to your mind. Today, when we think of asbestos we first think of its hazards. But it hasn't always been this way.

The first known reference to asbestos goes back to the time of Aristotle. That was when asbestos was discovered to have fireproof qualities. Since then, the incombustibility and low heat conductivity of asbestos, plus its fibrous, flexible nature, have proven useful in many ways. For example, asbestos has been woven into fabrics and used in theater curtains and fireproof suits. It has been utilized in building materials as fireproof insulation and as a flame retardant in plaster, ceiling, and floor tile. It has also been used in automobile brake shoes and clutch facings, air and water filters, military gas masks, and toothpaste! In the 1970s, the commercial use of asbestos reached an all-time high. But then asbestos was linked to lung disease. The fibrous nature that makes asbestos so flexible also allows it to easily penetrate body tissues, particularly the lungs. The history of asbestos is one of bitter paradox because the unique qualities that allowed it to save lives have also been found to endanger lives.

Asbestos is not really a single mineral. Actually, asbestos is a family of silicate minerals known for their fibrous structures. There are six types of asbestos minerals. Two of these are of commercial importance—chrysotile and crocidolite. The asbestos mineral chrysotile accounts for 95% of asbestos production worldwide, and crocidolite accounts for the remaining 5%. Chrysotile has a sheet-like silicate structure that makes it soft and flexible (Figure 30.9). Because of its softness, chrysotile is easily broken down in the body, and produces no apparent damage. This form of asbestos, leached from the ground, is present in many reservoirs of quite-safe drinking water. Recent medical evidence indicates that people exposed for long periods to moderate amounts of chrysotile show no lung ailments.

Crocidolite is a different story, however. This type of asbestos has a double-chain silicate structure (Figure 30.9). Its structure makes it strong and stiff, and thus more dangerous in the body. People exposed either to high levels of crocidolite for a short time or to moderate levels over a prolonged period of time have been found to develop lung disease. Thus it is crocidolite that is the principal culprit in asbestos-related lung diseases. Despite this knowledge, many reports on asbestos health hazards fail to make a distinction between the various types.

Asbestos is deeply feared by people all around the world. Yet, much of this fear is based on a failure to distinguish between harmful and harmless forms of asbestos. It is by far the most expensive pollutant to regulate and remove. The removal of asbestos-containing materials from schools, hospitals, and other public buildings has cost billions of dollars over the past 20 years. But since only a fraction of asbestos in use today poses a health problem, many scientists question the practice of eliminating all forms. Just like electricity in the 1800s, gasoline-powered vehicles in the early 1900s, and radioactivity in the late 1900s, public fears about asbestos will likely persist longer than the actual threat exists. Ultimately, however, we may view asbestos as both friend (chrysotile) and foe (crocidolite).

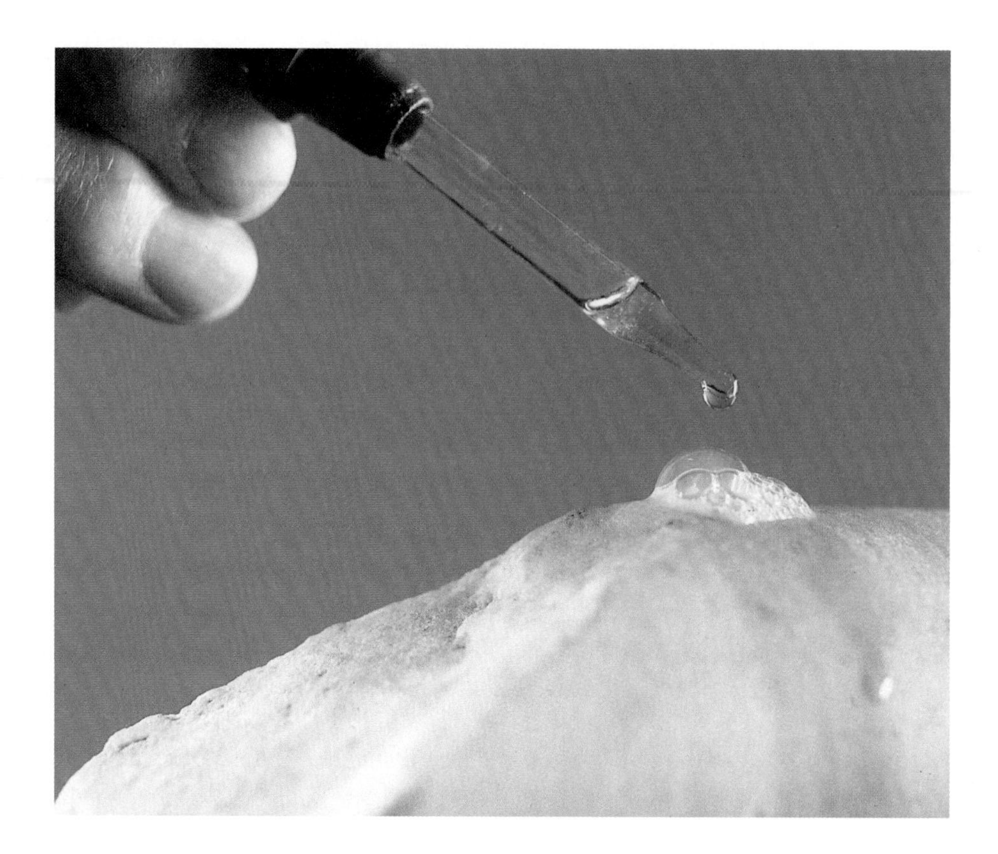

Figure 30.8
Bubbles of carbon dioxide effervesce when carbonate minerals make contact with HCl.

Chemical Properties—The Taste Test and the Acid Test

Sometimes chemical properties are used to identify a mineral. Two simple chemical tests for identifying minerals are the taste test and the fizz test. The taste test is commonly used to identify the mineral halite, NaCl (common table salt), which has a distinctive taste—salty. (But you should never do the taste test on a mineral without getting the okay from your teacher first. Some minerals are poisonous!) Another common test is the fizz test or acid test, which is used on carbonate minerals. Carbonate minerals fizz (effervesce) in dilute hydrochloric acid (HCl). That is, carbonate minerals give off bubbles of carbon dioxide gas when they chemically react with HCl (Figure 30.8). Some carbonate minerals react more easily with HCl than others, however.

30.2 Minerals That Form Rock Fall into Five Main Groups

Minerals are the building blocks from which rocks are made. Like letters of the alphabet, minerals combine in various ways to create the huge diversity of rocks found on Earth. But minerals themselves are of course made of smaller building blocks, the chemical elements. Minerals are sometimes made of single elements, but more often,

Table 30.4

Most Common Chemical Elements in the Earth's Crust

Element	Symbol	Percent by Mass	Percent by Volume
Oxygen	O	46.60	93.8
Silicon	Si	27.72	0.9
Aluminum	Al	8.13	0.5
Iron	Fe	5.00	0.4
Calcium	Ca	3.63	1.0
Sodium	Na	2.83	1.3
Potassium	K	2.59	1.8
Magnesium	Mg	2.09	0.3
	Total	98.59	100.0

Source: *Principles of Geochemistry* by Brian Mason and Carleton B. Moore.

are composed of elements joined as compounds. Of the 115 known elements, 88 occur naturally in the Earth's crust. These 88 elements combine to make up the more than 3400 types of minerals.

Although there are thousands of types of minerals on the Earth, only about two dozen are common. And the two dozen common minerals are chiefly composed of just eight elements. Table 30.4 lists them. These eight elements represent about 98% of the mass of the Earth's crust. And almost half of this mass is the element oxygen. What does this tell you about the amount of oxygen found in the Earth's crust?

The oxygen in the Earth's crust is found in such common groups of minerals as the *silicates, oxides,* and *carbonates.* Other groups of rock-forming minerals are *sulfides* and *sulfates.* With few exceptions, all rock-forming minerals are members of these five groups.

Silicates Make Up Nearly 90% of the Earth's Crust

After oxygen, the second most abundant element in the Earth's crust is silicon. The tendency of silicon to bond with oxygen is so strong that silicon is never found in nature as a pure element. Silicon is always combined with oxygen. Oxygen and silicon combine to form the largest mineral group, the *silicates.* The silicate known as quartz is the second most common mineral in the Earth's crust. Quartz is composed only of oxygen and silicon (SiO_2). Most other silicates contain elements in addition to oxygen and silicon. Feldspar, the most common and abundant mineral, is a silicate that also contains aluminum, sodium, potassium, and calcium.

Silcate Mineral		Typical Formula	Cleavage	Silicate Structure
Olivine		$(Mg, Fe)_2SiO_4$	None	Single tetrahedron
Pyroxene		$(Mg, Fe)SiO_3$	Two planes at right angles	Chains
Amphibole		$(Ca_2Mg_5)Si_8O_{22}(OH)_2$	Two planes at 60° and 120°	Double chains
Micas	Muscovite	$KAl_3Si_3O_{10}(OH)_2$	One plane	Sheets
	Biotite	$K(Mg, Fe)_3Si_3O_{10}(OH)_2$		
Feldspars	Orthoclase	$KAlSi_3O_8$	Two planes at 90°	Three-dimensional networks
	Plagioclase	$(Ca, Na) AlSi_3O_8$		
Quartz		SiO_2	None	

Figure 30.9
As silicate tetrahedra link to one another, they polymerize to form chains, sheets, and various network patterns. The complexity of the silicate structure increases down the chart.

All silicates have the same basic structure, which is the silicon-oxygen tetrahedron shown in Figure 30.10. These tetrahedra are not chemically stable until they form polymers by chemically bonding to one another (Figure 30.9).

Oxides Are Important Ore Minerals

The oxides are minerals that contain oxygen combined with one or more metals. These metals include iron (which forms hematite and magnetite), chromium (which forms chromite), manganese (which forms pyrolusite), tin (which forms cassiterite), and uranium (which forms uraninite). Oxides are of great economic importance. Many of the ores necessary for industrial and technological manufacturing are oxides.

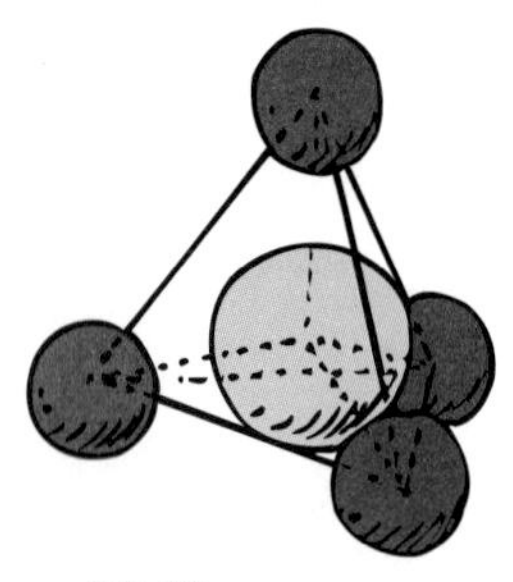

Figure 30.10
The silicon-oxygen tetrahedron is four oxygen atoms surrounding a central silicon atom.

Carbonate Minerals Make Limestone

The carbonate minerals are much simpler in structure than the silicates. Carbonate structure is triangular. It has a central carbon atom bonded to three oxygen atoms. The structure of the carbonate ion, CO_3^{2-}, is shown in Figure 30.11. Two common carbonate minerals are calcite and dolomite. Calcite is the chemical compound calcium carbonate, $CaCO_3$. Dolomite is a mixture of calcium carbonate and magnesium carbonate, $CaMg(CO_3)_2$. Calcite and dolomite are the main minerals found in the group of rocks called *limestone.*

Figure 30.11
The carbonate ion CO_3^{2-} has a structure that features a central carbon atom bonded to three oxygen atoms.

Sulfides and Sulfates Are Also Important Ore Minerals

As the names imply, the sulfide and sulfate minerals have sulfur as a main element. In the sulfide group, sulfide ions combine with metallic elements. This is the reason most sulfide minerals look like metals. In fact, the sulfides form an important class of minerals because they, in addition to oxides, are very important ores for modern society. The most common sulfide mineral is pyrite (fool's gold), FeS_4.

The sulfates contain sulfur in the form of a sulfate ion, SO_4^{2-}. The sulfate ion is shaped like a tetrahedron with one sulfur atom and four oxygen atoms. Gypsum, a calcium sulfate, is one of the most abundant minerals of the sulfate group. It is used to make plaster of Paris.

30.3 On the Way to Rocks

Rocks are made of many different minerals. So how do all these different minerals form? Minerals are formed by the process of **crystallization.** Crystallization is simply the growth of a solid from a material whose atoms come together in specific chemical compositions and crystalline arrangements. Crystallization of minerals commonly comes from two different sources—from **magma** (molten rock from the Earth's interior) and from water solutions. As we will see in the following chapter, *igneous* rocks are formed from magma and certain *sedimentary* rocks are formed from water solutions.

Minerals and Rock Formed from Magma

Just as there is water and ice, there is magma and rock. Magma cools and solidifies to become minerals and rock. Minerals and rock that are heated melt to become magma. Just as ice melts at the same temperature that water freezes, the temperature at which a solid mineral melts is the same temperature at which the same mineral in molten form solidifies. So when we discuss the melting temperature of a mineral, we imply that the magma form of that mineral solidifies at the same temperature.

For minerals and rock to melt, the temperature needs to be very high, in a range from 750°C to 1000°C. The Earth's temperature increases with depth. It increases about 30°C for each kilometer (Figure 30.12). Thus all minerals and rock melt if deep enough inside the Earth. There are two basic keys to understanding the different types of minerals and rock that are formed from magma. When minerals and rock melt, new magma is produced. As magma cools, mineral crystals begin to form and new rock is produced.

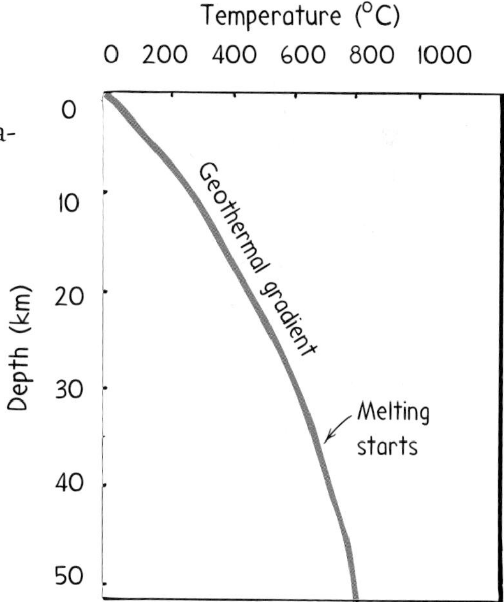

Figure 30.12
Temperature inside the Earth increases about 30°C for each kilometer of depth from the surface. This increase of temperature with depth is known as the *geothermal gradient.* Very deep in the Earth, temperatures become hot enough to melt rock.

Melting Keep in mind that rock is composed of many minerals. As rock is heated, the first minerals to melt are those with the lowest melting points. Therefore the melting of rock into magma occurs over a broad temperature range. When conditions are such that all minerals within a rock can melt, the composition of the resulting magma is the same as the composition of the original rock. Most often, however, melting is not complete; we have **partial melting.** Magma resulting from the partial melting of rock is made from only those minerals that have melted, the ones with the lowest melting points. This results in

Figure 30.13
Chemical sediments:
(a) calcite,
(b) dolomite,
(c) gypsum, and
(d) halite.

magmas of many different compositions. What happens when magmas of different compositions cool and crystallize? The answer is that many different kinds of igneous minerals and rocks are formed.

Magma is classified by the amount of silicon it contains. Minerals with a high silicon content tend to have lower melting temperatures (they melt more easily). Partial melting thus produces magma with a higher silicon content than the parent rock because the high silicon content minerals are the first to melt. And minerals in the rock with a lower silicon content are left unmelted.

There are three major types of magma—*basaltic, andesitic,* and *granitic. Basaltic magma* is low in silicon content. Basaltic magma that has solidified is the dark igneous rock—basalt—that makes up the Hawaiian Islands. *Andesitic magma,* which is medium in silicon content, is produced from partial melting of basaltic oceanic crust. The rock andesite, produced from andesitic magma, gets its name from the Andes Mountains in South America, where it is very common. When andesitic rocks undergo partial melting, *granitic magma,* which is high in silicon content, is produced. This magma, when solidified, forms granite and other granitic rocks.

Crystallization The process of crystallization brings us minerals and rock from magma. It is like the process of partial melting, which brings us magma, but the changes of state occur in reverse order. Crystallization begins when magma starts to cool. Solid crystals form out of the liquid magma. Just as rock that completely melts produces a magma of identical composition, a magma that cools "completely" produces rock of identical composition. As with melting, this seldom happens. Minerals that have the highest melting points—those that have the lowest silicon content—crystallize first, followed by minerals with lower melting points (those with larger amounts of silicon).

In general, as minerals crystallize in a cooling magma, they settle out of the magma. Additionally, portions of the remaining magma may escape from the region where crystallization is occurring. The result is that newly formed crystals become separated from the liquid that is still molten. So as crystallization proceeds, the composition of the liquid that the crystals formed in changes continuously. It becomes *depleted* in the constituents of minerals that have already crystallized and *enriched* in the constituents of minerals yet to crystallize. Can you see that this process enriches the remaining liquid in silicon? The crystallization process allows a single magma to generate several different types of igneous minerals and rock.

Concept Check ✓

How is the partial melting of rock similar to the crystallization of magma that produces igneous rock?

Check Your Answer Both produce temperature-dependent materials but by opposite processes. Partial melting (solid to liquid) produces *magmas* of various compositions that depend on the melting temperatures of the minerals making up the rock that is melting. Crystallization (liquid to solid) produces *crystals* of various compositions that depend on the solidification temperatures of the minerals that form from the magma that is solidifying. In both processes, materials separate depending on the temperature.

Minerals and Rock Formed from a Water Solution

The process of crystallization is not unique to magma. Crystallization also occurs in water solutions. **Chemical sediments** are formed when minerals precipitate from water in which they are dissolved. Chemical sediments fall into two categories: *carbonates* and *evaporites.*

Carbonates are minerals and rocks composed mostly of calcite, $CaCO_3$, or dolomite, $CaMg(CO_3)_2$. Calcium ions and carbonate ions in seawater are used by living organisms to grow hard, protective shells. The shells eventually crystallize to become calcite. Thus, *limestone,* the most abundant carbonate rock, is generally formed as a result of biologic activity. Here's how it works: When organisms with calcium carbonate shells die, their shells accumulate on the sea floor. The shells begin to dissolve, forming a non-crystalline *ooze* of calcium carbonate. This ooze eventually crystallizes into calcite, which forms limestone. Because of compaction and how easily calcium carbonate dissolves, the original textures and structures of the seashells are sometimes obliterated. Dolomite forms when magnesium chemically replaces some of the calcium. Some carbonates, however, form without the aid of organisms. Cave dripstones, such as stalactites and stalagmites, provide an interesting example of calcium carbonate precipitating inorganically from dripping water (Figure 30.14).

Evaporites are minerals and rocks that are precipitated when a restricted body of seawater or the water of a salty lake evaporates. Examples are gypsum, anhydrite, and halite. These names apply both to individual minerals and to rocks made of a single type of evaporite mineral. As evaporation proceeds, the minerals that are the most difficult to dissolve, such as gypsum, precipitate first, followed by the minerals that dissolve more easily, such as anhydrite and then halite. Although carbonates make up the bulk of chemical sediments, evaporites are a small but important group.

Figure 30.14
Calcium carbonate precipitating from dripping water in a cave forms icicle-shaped stalactites hanging down from the ceiling and cone-shaped stalagmites protruding upward from the ground.

Minerals—The Link to Rocks

Now that we have seen what minerals are and the different ways that minerals form, we can begin to learn about the combination of minerals called *rock.* As we will see, minerals formed from the crystallization of magma make up igneous rock; and minerals formed from the precipitation or evaporation of water make up some sedimentary rock. We will also see that igneous rock breaks down to form sedimentary rock, and that a third rock type—metamorphic rock—is also formed from rock that already existed. So, minerals, in their many forms, are the building blocks of the many different rocks on Earth.

Chapter Review

Key Terms and Matching Definitions

_____ chemical sediments
_____ cleavage
_____ crystal form
_____ crystallization
_____ fracture
_____ luster
_____ magma
_____ mineral
_____ Mohs scale of hardness
_____ partial melting
_____ polymorphs
_____ specific gravity
_____ streak

1. A naturally formed, inorganic solid composed of an ordered array of atoms chemically bonded to form a particular crystalline structure.
2. The orderly internal arrangement of atoms in a crystal.
3. Two or more minerals that contain the same elements in the same proportions but have different crystal structures.
4. A ranking of the relative hardness of minerals.
5. The tendency of a mineral to break along planes of weakness.
6. A break that is not along a plane of weakness.
7. The appearance of a mineral's surface when it reflects light.
8. The name given to the color of a mineral in its powdered form.
9. The ratio between the weight of a substance and the weight of an equal volume of water.
10. The growth of a solid from a material whose constituent atoms can come together in the proper chemical proportions and crystalline arrangement.
11. The incomplete melting of rocks, resulting in magmas of different compositions.
12. Formed by the precipitation of minerals from water.
13. Molten rock from the Earth's interior.

Review Questions

Minerals Can Be Identified by Their Properties

1. What is a mineral?
2. What physical properties are used in the identification of minerals?
3. All minerals are defined by an orderly internal arrangement of their atoms—the crystal form. Yet, most mineral samples do not display their crystal form. Why not?
4. What are two of the classifications for mineral luster?
5. Although color is an obvious feature of a mineral, it is not a very reliable means of identification. Why not?
6. Will the mineral topaz scratch quartz, or will quartz scratch topaz? Why?
7. The minerals calcite, halite, and gypsum are all nonmetallic, light and softer than glass, and all have three directions of cleavage. In what ways can they be distinguished from one another?
8. Silver has a density of 10.5 g/cm^3. What is its specific gravity?
9. What is the relationship of density to specific gravity?

Minerals That Form Rock Fall into Five Groups

10. What is the most abundant element in the Earth's crust? What is the second most abundant element?
11. What is the most abundant mineral in the Earth's crust? What is the second most abundant mineral?
12. Name the five most common mineral groups found in most rocks.

On the Way to Rocks

13. What is meant by *partial melting*?
14. How are partial melting and crystallization similar?
15. If a rock contains mineral A and mineral B, which would melt first, mineral A with 30% silicon, or mineral B with 25% silicon?
16. What are three common chemical sediments?

17. When water evaporates from a body of water, what type of sediment is left behind?
18. What is the most common chemical sediment?
19. What is the most abundant carbonate rock?
20. How are most of the carbonate minerals and rocks formed?

Exploration

A physical property of any material is its density—its mass per volume. Pennies made after 1982 contain both copper and zinc. Pennies made before this date are pure copper. Zinc is less dense than copper, so post-1982 pennies are less dense and have less mass than pre-1982 pennies. Dig into your penny collection and find 20 pre-1982 and 20 post-1982 pennies. Measure their masses on a sensitive scale, such as a home postage scale. Alternatively, hold the pennies in opposite hands to see if you can feel the difference in their masses. How few pennies can you hold and still feel the difference? Try holding single pennies on your left and right index fingers. Can you tell the difference with your eyes closed? Try this with a friend.

Exercises

1. Clearly distinguish between physical and chemical properties of minerals, and give examples of each.
2. Describe the difference between a mineral and an element.
3. Why is color not always the best way to identify a mineral?
4. What does a colorless streak tell you about a mineral?
5. What makes gold so soft (easily scratched) while quartz and diamond are so much harder?
6. Imagine that we have a liquid with a specific gravity of 3.5. Knowing that objects of higher specific gravity will sink in the liquid, will a piece of quartz sink or float in the liquid? How about a piece of chromite?
7. Does the fact that carbon dioxide is given off in the fizz when HCl touches a mineral mean that the mineral contains carbon?
8. Is cleavage the same as crystal form? Why or why not?
9. Silicon is essential for the computer industry in making microchips. Can silicon be mined directly from the Earth?
10. What mineral groups provide most of the ore that we, as a society, need?
11. What two minerals make up most of the sand in the world?
12. What class of minerals does Galena (PbS) belong to?
13. What class of minerals does Anglesite ($PbSO_4$) belong to?
14. Is it possible for crystallization to enrich a magma in more than just silicon?
15. If high silicon-content magma is the last to cool, why aren't high silicon rocks the last to melt?
16. Why is halite the last mineral to precipitate from evaporating seawater?
17. Would you expect to find any fossils in limestone? Why or why not?
18. How do chemical sediments produce rock? Name two rock types that form by chemical sedimentation.
19. Why is asbestos in drinking water not particularly harmful to humans, whereas asbestos particles in air are very harmful?
20. Make an argument that the removal of asbestos products such as ceiling tile and pipe coverings may be more hazardous to humans than simply covering them up.

Suggested Web Sites

http://www.usgs.gov/education

http://minerals.usgs.gov/

http://webmineral.com/

http://edtech.kennesaw.edu/web/rocks.html

Chapter 31: Rocks

Geology is the study of rocks. "Geo" means "Earth."

What kinds of rock exist on our planet? Where and how do rocks form? Are rocks at the Earth's surface different from rocks deep inside the Earth? Are rocks at the bottom of the ocean different from rocks on land? Do volcanoes produce rock? Are there different kinds of volcanoes? And different kinds of lava? Do most rocks form quickly, or slowly over long periods of time? The study of rocks—*geology*—helps to answer these questions.

In Chapter 30, we learned that the rocks making up the Earth's crust are formed from crystals called minerals. Each mineral has its own chemical formula and its own atomic structure. Rocks differ from one another because they are mixtures of different kinds or proportions of minerals. So we see that there are many types of minerals and many types of rocks.

31.1 Rocks Are Divided into Three Main Groups

The Earth's surface is not static and unchanging. Instead, there is constant destruction, rebuilding, and rearrangement. As land and rock are formed in one area, they are destroyed in another. These processes of change are summed up in the theory of *plate tectonics,* which we will discuss in Chapter 33. Plate tectonics tells us that the Earth's surface is broken into several large, rigid plates. These plates shift because of very slow movements in the Earth's interior. The boundaries of these rigid plates are places of intense geological activity. Earthquakes, volcanoes, and young mountain ranges all tend to cluster there. Many rocks are either created, changed, or destroyed at plate boundaries.

The rocks of the Earth are classified into three types according to their origin: *igneous, sedimentary,* and *metamorphic.*

Igneous rocks are formed by the cooling and crystallization of hot, molten rock called *magma.* The word *igneous* means "formed by fire." Igneous rocks make up about 95% of the entire Earth's crust. Basalt and granite are common igneous rocks.

Sedimentary rocks are formed from pieces of other rocks (sediments) carried by water, wind, or ice. Sedimentary rocks are the most common rocks in the uppermost part of the Earth's crust. In fact, sedimentary rocks cover more than two thirds of the Earth's surface. Sandstone, shale, and limestone are common sedimentary rocks.

Metamorphic rocks are formed from older, preexisting rocks (igneous, sedimentary, or metamorphic) that are transformed by high temperature, high pressure, or both—without melting. The word *metamorphic* means "changed in form." Marble and slate are common metamorphic rocks.

Figure 31.1
The three main types of rock. (a) Basalt and granite are igneous rocks. (b) Sandstone and limestone are sedimentary rocks. (c) Marble and slate are metamorphic rocks.

31.2 Igneous Rocks Form When Magma Cools

Most of the Earth's crust is made up of different kinds of igneous rock. On the continents, the most common igneous rocks are granite and andesite. On the ocean floor, basalt is the most common kind of rock. All igneous rock originated as magma, which is molten rock from the Earth's interior.

Some Igneous Rocks Form at the Earth's Surface

Igneous rocks may form either at or below the Earth's surface. Igneous rocks that form at the Earth's surface are called **extrusive** rocks. The word "extrusive" means "pushed out of."

As discussed in Chapter 30, partial melting and crystallization produce a variety of magmas. These magmas differ in the amount of silicon they contain. Magma with higher silicon content flows more slowly because it is thicker and more gooey—*viscous*—than magma with lower silicon content. (This is like a spilled milkshake that flows slower than spilled milk.) Basaltic magma is one important example of a low-silicon, fast-flowing, magma. The temperature of magma also affects its ability to flow. Hotter magma flows more easily than cooler magma.

Magma that moves upward from inside the Earth and flows onto the surface is called **lava.** The term *lava* refers to both the molten rock itself and to solid rocks that form from it. Lava may be extruded through cracks and fractures in the Earth's surface, or through a central vent—a **volcano.** Eruptions from a volcano are familiar to us because they are very exciting to see. But the outpourings of magma from fissures are much more common than volcanic eruptions.

Figure 31.2
The flood basalts that produced the Columbia Plateau covered more than 200,000 km^2 of the preexisting land surface.

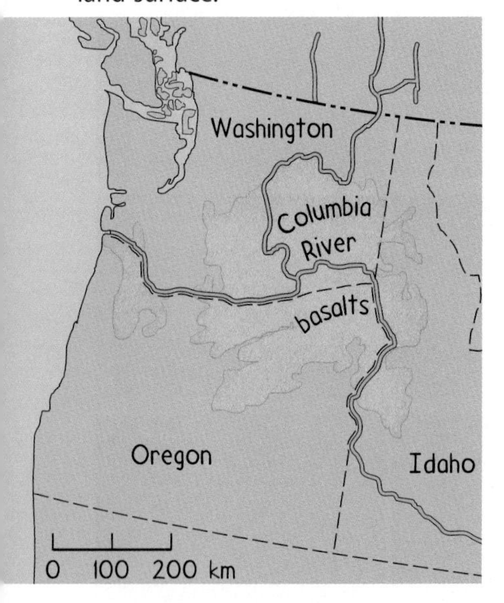

Fissure Eruptions Occur Under Water and on Land

Most fissure eruptions occur when fast-flowing basaltic lava erupts at the bottom of the ocean. These underwater fissure eruptions form the ocean floors. Fissure eruptions also occur on land. Lava outpourings known as *flood basalts* have flooded large areas and created extensive lava plains. The Columbia Plateau in the Pacific Northwest is the result of extensive flood basalts (Figure 31.2), as is the Deccan Plateau in India.

Concept Check ✓

Is it correct to say that lava and magma with a high silicon content are more viscous than those with lower silicon content?

Check Your Answer Yes. Higher silicon content in a lava or magma causes it to be gooier and it flows slower. Viscosity is the property of a liquid that describes how well it flows. Molasses flows very slowly (is more viscous) than water (which is less viscous). Therefore it is correct to say that magmas and lavas with a high silicon content are more viscous than those with lower silicon content.

Link To Mythology

According to Greco-Roman mythology, volcanic activity can be traced to Vulcan, the Roman god of volcanic fire and metalworking. The word *volcano* comes from the island of Vulcano off the coast of southern Italy. In ancient times, Vulcano was believed to be the metal workshop of Vulcan. The people surrounding Vulcano believed that the lava fragments and glowing ash that erupted from the island were a result of Vulcan's work as he forged thunderbolts for Jupiter, king of the gods, and weapons for Mars, god of war. The people of Polynesia, who attribute volcanic activity to Pele, goddess of volcanoes, tell a similar story.

Volcanoes Come in a Variety of Shapes and Sizes

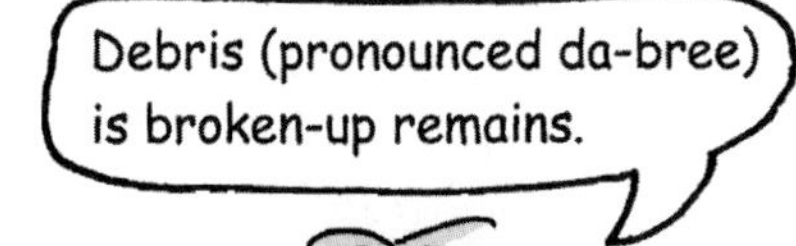

All volcanoes are basically vents—holes—where magma can rise to the Earth's surface. Volcanoes come in many shapes and sizes. Volcanoes that are built by a steady supply of easily flowing basaltic lava have a broad, gently sloping cone that resembles a shield. These are *shield volcanoes.* They are built from many lava flows that pour out in all directions to cool as thin, gently sloping sheets. Some of the largest volcanoes in the world are shield volcanoes. The enormous size of Mauna Loa in Hawaii is the result of the building up of many individual lava flows, each one only a few meters thick. Mauna Loa is the largest volcano on Earth, standing 4145 meters above sea level and more than 9750 meters above the deep ocean floor.

Cinder cones are common in many volcanically active areas. They are very steep but rarely rise more than 300 meters or so above ground level. They are formed from the piling up of ash, cinders, and rocks that have been explosively erupted from a single vent. As debris showers down, the larger fragments pile up near the top of the cone to form a symmetrical, steep-sided cone around the vent. The smaller particles fall farther from the vent to form gentle slopes at the base. Two well-known examples of cinder cones are Sunset Crater in Arizona (Figure 31.3b) and Parícutin in Mexico.

When a volcano erupts both lava and ash, a *composite cone* of alternating layers of lava, ash, and mud is produced (the word "composite" means mixture). The layers build up to form a volcano with a steep-sided summit and gently sloping lower flanks. Composite cones are usually bigger than cinder cones because the mixture of lava and ash helps to hold the cone together. Mount Fujiyama (Figure 31.3c) is a classic example of a majestic composite cone.

(a)

(b)

(c)

Figure 31.3
The three types of volcanoes. (a) Shield volcanoes, such as Mauna Loa, have broad, gentle slopes that average between 1° and 10° (from the horizontal). (b) Cinder cones, such as Sunset Crater, generally have smooth steep slopes of 25°–40° and bowl-shaped summit craters. (c) Composite cones, such as picturesque Mount Fujiyama, are also very steep. On average, the slope of a composite cone starts out at 30° at the summit and gradually flattens to 10° at the base.

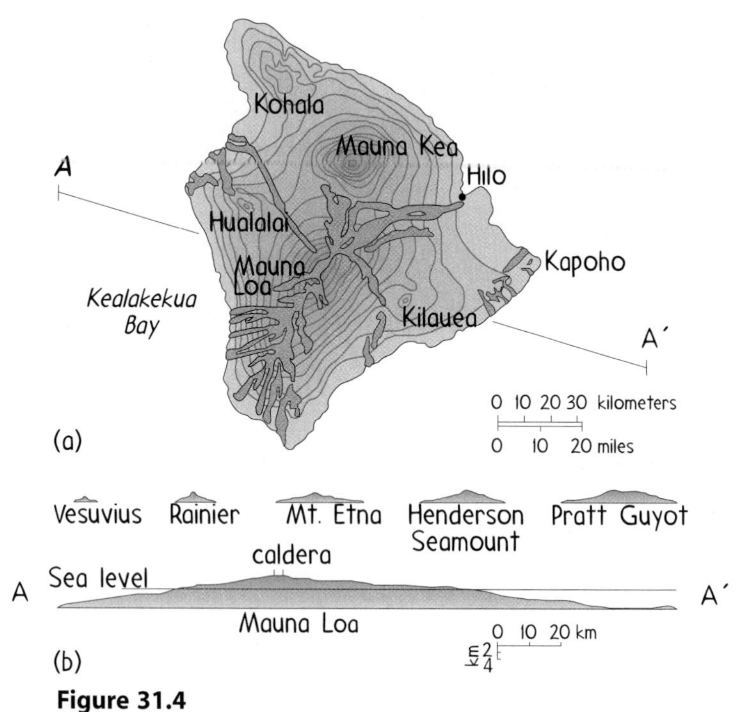

Figure 31.4
(a) Mauna Loa, a shield volcano on the Island of Hawaii, is the largest volcano on Earth.
(b) When compared to other large volcanoes, its immense size and volume is dramatic.

Composite cones tend to erupt explosively because their magmas and lavas usually do not flow easily. This thicker magma traps volcanic gases, which increases the pressure inside the volcano. We can compare the gases in magma to gases in a bottle of carbonated soda. If we cover the top of the bottle and shake vigorously, the gases separate from the liquid and form bubbles. When we remove the cover, pressure is released and gases and liquid explode from the bottle. The gases in magma behave in much the same way. In a volcanic blast, the pressure and temperature increases, and the whole mass of gooey magma and overlying rock explodes into dust and rubble. When mixed with volcanic ash, this mixture can expand and destroy everything in its path. Examples of this kind of volcanic activity are Mount Vesuvius in 79 A.D., Mt. Pelee in 1902, Mount St. Helens in 1980, and Mount Pinatubo in 1991.

Concept Check ✓

How are the lavas that form shield volcanoes different from those that form composite cones?

Check Your Answer Lavas that make up shield volcanoes generally have a low silicon content, so they tend to easily flow outwards from the vent. This creates a volcano that is wide and gently sloping. Lavas that make up composite cones generally have a higher silicon content, so they do not flow easily out from the vent. This causes explosive eruptions of thick lava and ash that do not flow away, but rather build a steep composite cone.

Figure 31.5
Craters form in the central vent of volcanoes. This crater, in pre-1980 Mount St. Helens, shows a rising steam plume and a lava dome.

A *crater* is commonly formed above the central vent of an erupting volcano (Figure 31.5). During an eruption, the upward moving lava overflows the crater walls and then sinks back into the vent as the eruption ends. The walls of a crater often collapse after an eruption, making the area of the central vent larger.

A crater can grow to more than a kilometer in diameter. A very large crater is referred to as a *caldera.* Calderas range from 5 to 30 kilometers in diameter. Most calderas are formed when the central part of a composite cone simply collapses into the empty space inside the volcano where magma used to be (the magma chamber). However, a few calderas have been formed by explosive eruptions in which the top of a volcano was blown out. The volcanic eruption 7000 years ago of Mount Mazama in Oregon was one of these rare events. This eruption blasted ash throughout the northwestern United States. After the eruption, most of the cone collapsed into the emptied magma chamber. The caldera, filled with rainwater, resulted in Oregon's famous Crater Lake, which is 9 kilometers wide and 590 meters deep.

Figure 31.6
Crater Lake in Oregon is a remnant of the eruption of Mount Mazama 7000 years ago.

Yellowstone National Park is located in one of the most geologically active regions of the Rocky Mountains. It is considered a "hotspot" in the Earth's crust. Situated in a caldera 70 kilometers long and 45 kilometers wide, Yellowstone Park is the remnant of an ancient volcano that violently erupted about 600,000 years ago. Most of the hot springs, bubbling muds, steaming pools, and spouting geysers that the park is famous for lie within the caldera. Heat from the enormous amount of magma that produced the massive eruption still remains close to the Earth's surface. This heat keeps the present volcanic activity going. Although no eruption has occurred in recorded human time, some magma remains so close to the surface that the possibility of an eruption in the future cannot be ignored. Yellowstone is an example of geology in action, where the eruption of scalding water may be but a "prequel" to more violent activity in the near future.

Some Igneous Rocks Form Beneath the Earth's Surface

Igneous rocks that form beneath the Earth's surface are called **intrusive** rocks (the word "intrusive" means *pushed into*). Large intrusive igneous rock bodies are called **plutons.** They occur in a great variety of shapes and sizes, ranging from thin slabs to wide, shapeless blobs (Figure 31.7). As you might have guessed, intrusive rocks can only be studied after they are exposed at the Earth's surface.

A common type of pluton is a *dike*—formed by the intrusion of magma into fractures that cut across the layers of existing rock. Dikes are old channel ways for rising magma and often occur near volcanic vents.

Figure 31.7
Intrusive igneous features in cross-sectional view.

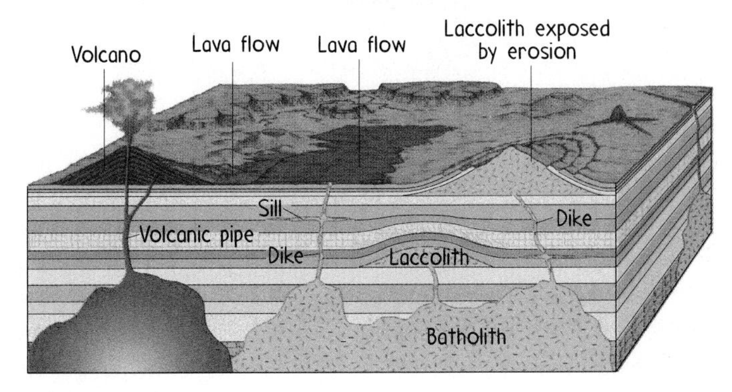

Figure 31.8
Shiprock, New Mexico. Radiating dikes surround the eroded remains of a volcanic vent.

A spectacular example of this is the radiating dikes around the exposed volcanic neck at Shiprock, New Mexico (Figure 31.8). At Shiprock, the volcanic neck and the dikes are more difficult to erode than the softer rock they intruded. Erosion has acted to remove the surrounding softer rock, leaving the harder, volcanic rocks behind as a peak (the volcanic neck) and wall-like ridges (the dikes).

Another type of pluton is a *sill.* A sill is formed by the intrusion of magma into fractures that are parallel to the layering of existing rock. Most sills are formed by the intrusion of more-fluid basaltic magmas at shallow depth. A variation of a sill is a *laccolith,* a body of rock that is created when slow-moving, less-fluid (more gooey) magma rising upward in the Earth's crust spreads to form a mushroom shape. Unlike sills, laccoliths push the overlying layers upward in domelike fashion.

Batholiths are the largest of the plutons and are defined as having more than 100 square kilometers of surface exposure. A batholith is usually not generated by a single intrusion. Instead, numerous intrusive events over millions of years create a massive batholith. Batholiths form the cores of many major mountain systems of the world. Additionally, many modern mountains are actually the exposed cores of the batholiths of larger mountains that have long since eroded away. Some of the largest batholiths in North America include the Coast Range batholith and the Sierra Nevada batholith. These batholiths continue to push upward, increasing the height of the mountain range (Figure 31.9). We all know that mountains wear down due to erosion. Quite interestingly, the Sierra Nevada mountains are presently gaining height with time because the upward push of the batholith is faster than erosion can wear it down.

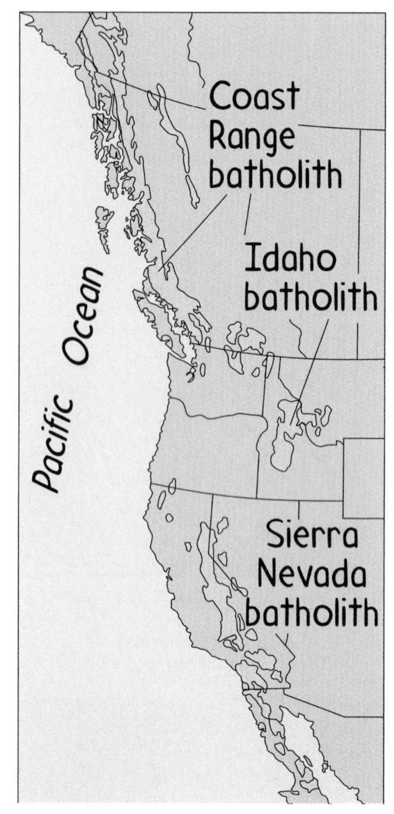

Figure 31.9
Some of the largest batholiths in North America include the Coast Range batholith and the Sierra Nevada batholith.

Concept Check ✓

1. Why is it incorrect to say that igneous rocks may form from the intrusion of lava?
2. Is it correct to say that igneous rocks may form from the extrusion of lava?

Check Your Answers

1. The terminology in the statement is used incorrectly. The term *intrusion* refers to solidification that occurs in the Earth's interior and therefore has nothing to do with lava. Lava is not a synonym for magma, but is the term for magma that has been extruded at the Earth's surface in molten form. By definition, there is no lava beneath the Earth's surface. Magma is intruded and lava is extruded to form igneous rocks.
2. Yes, once magma is extruded from the Earth it is called *lava,* which when solidified becomes igneous rock.

31.3 Sedimentary Rocks Blanket Most of the Earth's Surface

Sedimentary rocks are the most common rocks in the uppermost part of the crust. They form a thin blanket over igneous and metamorphic rocks and cover two thirds of the Earth's surface. As you will see, sedimentary rocks are the remains of older rocks. For this reason, they provide information about geological events that have occurred over time at the Earth's surface.

From Weathering to Sedimentation—Making Sedimentary Rock

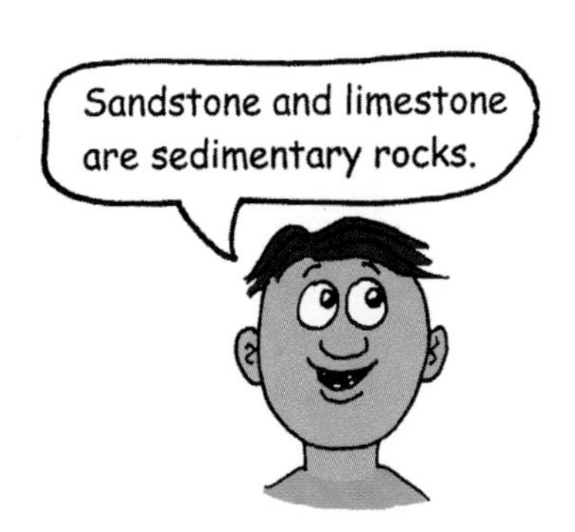

While the process of volcanism is always generating new rock material at the Earth's surface, the opposing process of *weathering* breaks rock down. There are two kinds of weathering, *mechanical* and *chemical.* Both kinds of weathering produce *sediment,* small pieces of rock. **Mechanical weathering** physically breaks rocks into smaller and smaller pieces. **Chemical weathering** consists of chemical reactions that involve water and decompose rock into smaller pieces. As rock is weathered, it erodes. **Erosion** is the process by which weathered rock particles are removed and transported away by water, wind, or ice.

Sediments composed of small fragments of other rocks are called *clastic* sediments. On the other hand, sediments produced by chemical means are called *chemical* sediments. When clastic particles are first produced, they are normally quite angular and jagged. But as the small pieces are transported (usually by water), they collide and break. This decreases their size and rounds off their sharp edges. When transportation stops, *deposition* and *sedimentation* begin.

Deposition occurs when a particle that is being transported stops and is deposited. The larger a sediment grain is, the stronger a water current must be to carry it. Usually, water currents get weaker as they move away from their source area. As the water slows down, the larger grains are the first to be deposited, while smaller grains are able to stay with the flow. In this way, grains tend to be sorted according to size as they are deposited. A deposit that contains grains of very similar sizes is called a *well-sorted* deposit. A deposit that contains grains of various sizes is called a *poorly sorted* deposit.

Figure 31.10
(a) Well-sorted sediment grains.
(b) Poorly-sorted sediment grains.

During transportation, sediment is continuously sorted and worn down. So the size, shape, and sorting of grains in a sedimentary deposit provide clues to the distance of transport. In general, poorly sorted, angular grains of various shapes traveled only a short distance, whereas well-sorted, well-rounded grains imply a greater transportation distance. We can also get clues to the method of transportation. Glacial deposits, for example, tend to be very poorly sorted and angular because they are moved by ice. On the other hand, wind-blown deposits tend to be very well sorted and fine-grained.

Figure 31.11
The red and orange in the sedimentary rocks at Bryce Canyon are caused by the presence of iron oxide.

In the process of *sedimentation,* sediment particles are deposited one layer at a time in a horizontal manner. As the deposited sediment accumulates, it begins to change into *sedimentary rock.* The transformation of sediments to sedimentary rock occurs in two ways: *compaction* and *cementation.* As the weight of overlying sediments presses down upon deeper layers, sediment grains are squeezed and compacted together. This compaction squeezes water out of the spaces (called *pores*) between sediment grains. This water often contains dissolved compounds such as silicon dioxide, calcium carbonate, and iron oxide. These compounds, when chemically precipitated from solution, partially fill the pore spaces with mineral matter. This glues the grains together and acts as a cementing agent. Silicon dioxide cement, the most durable, produces some of the hardest and most resistant sedimentary rocks. When iron oxide acts as a cementing agent, it produces the red or orange stain of many sedimentary rocks. The colors of Bryce Canyon provide a beautiful example of iron oxide stain.

Clastic Sediments Are Classified by Particle Size

Clastic sedimentary rocks are classified by particle size (Table 31.1). The three most abundant clastic sedimentary rocks are *shale, sandstone,* and *conglomerate.* Shale is composed of very fine grains that are too small to be visible with a magnifying hand lens. Sandstone is composed of medium sized grains such as those found in typical beach sand. *Conglomerate* is composed of a variety of larger particles, ranging from pebbles to boulders.

Shale is formed by the compaction of silt and clay-sized particles. It is finely layered. The extremely fine grain size of shale suggests that the particles it contains were deposited in quiet waters, such as deep ocean basins, flood plains, deltas, lakes, or lagoons. The color of shale

Table 31.1

Particle Size Classification of Clastic Sediments and Sedimentary Rocks

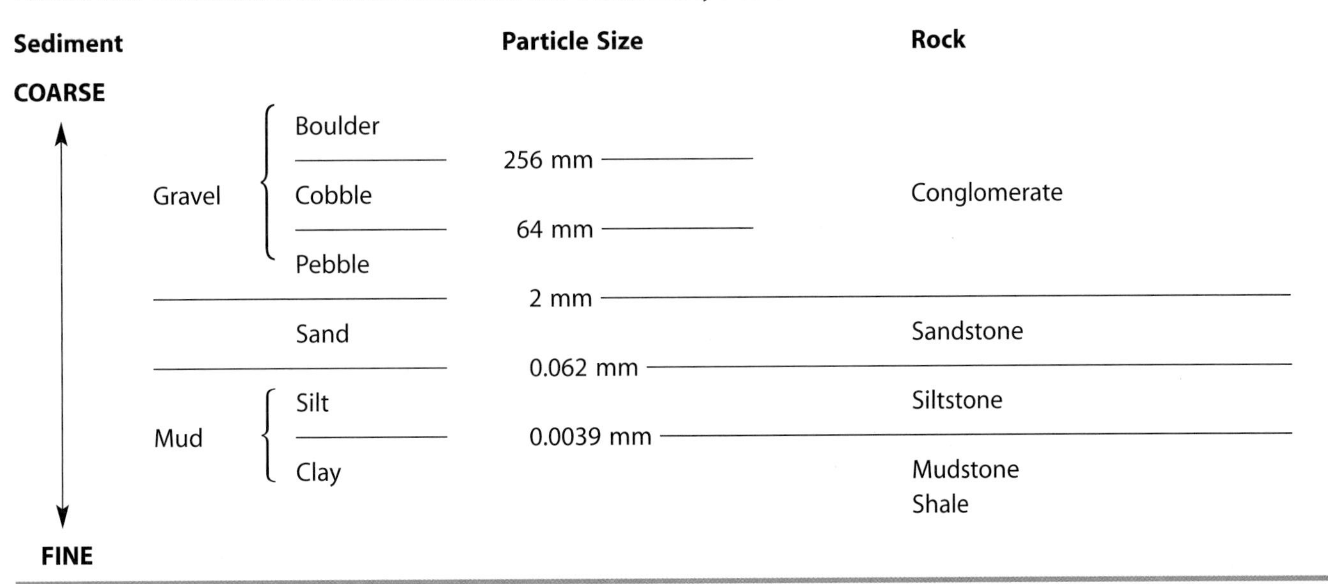

Sediment		Particle Size	Rock
COARSE			
Gravel	Boulder		Conglomerate
		256 mm	
	Cobble		
		64 mm	
	Pebble		
		2 mm	
	Sand		Sandstone
		0.062 mm	
Mud	Silt		Siltstone
		0.0039 mm	
	Clay		Mudstone Shale
FINE			

ranges from gray to black and from red to brown to green. The colors of shale tell us about the environment in which it formed. Gray to black shale indicates organic matter in the shale. Organic matter can be preserved only in a swampy environment with little or no oxygen. Black shale is the main source rock for crude oil.

There are three kinds of sandstone, each with a different mineral make-up. When quartz is the primary mineral, the rock is simply called *quartz sandstone,* which is composed of well-sorted, well-rounded quartz grains. Sandstone that contains considerable amounts of the mineral feldspar is called *arkose.* The grains in arkose tend to be poorly rounded and not as well sorted as those in quartz sandstone. Sandstone made of a mixture of quartz, feldspar, and rock fragments is called *graywacke* (pronounced gray-wack-ee). Sandstones form in a variety of environments, including dunes, beaches, marine sandbars, river channels, and land and underwater canyons.

Conglomerates are composed of gravels and rounded rock fragments. The rock fragments are usually large enough for easy identification, which gives us useful information about the areas from which the sediments were eroded. Water currents strong enough to carry them must have transported the larger rock fragments. Because these strong currents round out the rock fragments, the roundness of their edges and corners are good guides to the distance they traveled. Conglomerates are often found in river channels and in rapidly eroding coastlines.

(a)

(b)

(c)

Figure 31.12
Sedimentary rocks. (a) Shale, composed of fine sized particles; (b) sandstone, composed of medium-sized particles; and (c) conglomerate, made up of a wide variety of large, rounded particles.

Limestones and Evaporites Are Sedimentary Rocks

Recall from Chapter 30 that chemical sediments are formed by the precipitation of minerals from water. Carbonate rocks, like limestone for example, are formed by the precipitation of calcium carbonate. Evaporites, like halite for example, are formed by the evaporation of salty waters. In general, chemical sedimentary rocks form where there are no clastic sediments.

Warm climates favor carbonate deposition because carbonates dissolve more easily in cold water than in warm water. Evaporite deposits require a dry climate that promotes evaporation of lake or sea water. As the water dries out, evaporite minerals precipitate and are left behind. Modern-day as well as ancient evaporites are created in desert basins, tidal flats, and restricted sea basins. Vast carbonate and evaporite deposits on the continents are evidence that expansive, shallow seas have periodically covered the land surfaces in the past.

Fossils, Clues to Life in the Past

Because sedimentary rocks (clastic and chemical) are formed at the Earth's surface, they often contain the remains of preexisting life forms—fossils. Fossils give us important information for interpreting the Earth's geologic past. As we shall see in Chapter 36, fossils play an important role as indicators of where and when sediments were deposited. They are also useful in the matching up of rocks from different places that are of similar geologic age. Some fossils are made of whole organisms, but most fossils are just parts of organisms. Other fossils are simply an impression, or print, made in the rock before it hardened. Plants commonly leave their impression as a thin film of carbon. There are many methods of fossilization (Figure 31.13).

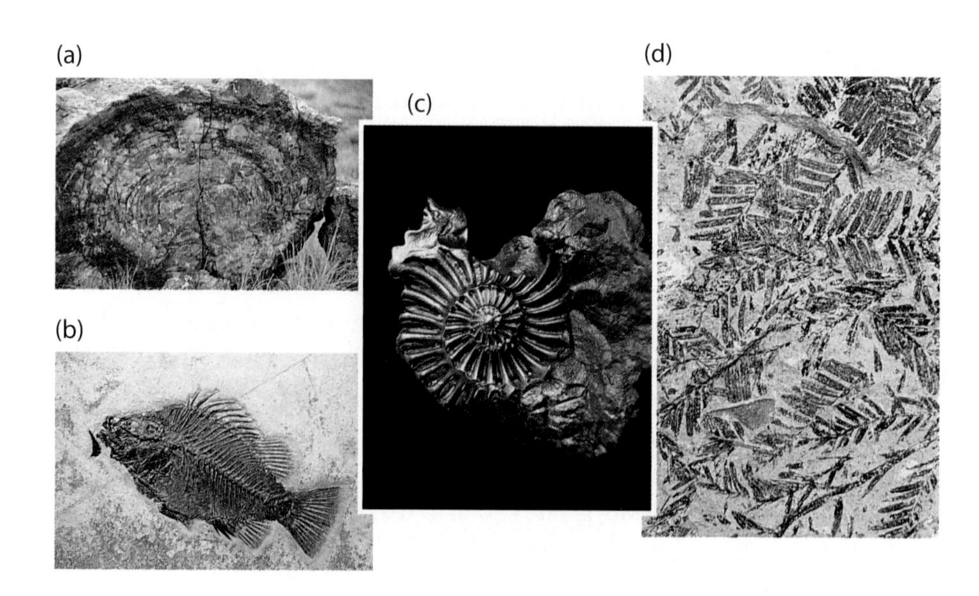

Figure 31.13
Some of the many methods of fossilization.
(a) Permineralization occurs when the porous remains of an organism become filled with water that is rich in dissolved minerals—like petrified wood.
(b) Impression is made by an organism buried quickly, before it can decompose, thereby preserving its impression.
(c) Replacement occurs when the remains of organisms are replaced by a mineral. Pyrite has replaced the original shell in this specimen.
(d) Carbonization occurs when a plant or an organism is preserved as a thin film of carbon.

Fossil Fuels

When ancient plants and animals died, most of the organic matter they were made of was quickly decomposed by bacteria and converted to nutrients consumed by other organisms. Material that escaped bacterial decay was either stored as sparsely distributed organic matter or converted to coal, oil, or gas.

Coal, oil, and gas are all fossils in the sense that they are the remains of past organisms. However, these remains have been so changed over time that the forms and even the composition of the accumulated organisms are beyond recognition.

The source of oil and gas is fossilized organic matter found in buried sediments. When buried sediment with a lot of plant or animal material is heated over a long enough period of time, chemical changes take place that create oil. Under pressure of the overlying sediments, tiny oil droplets are squeezed out of the source rocks and into overlying porous rocks. The porous rocks, commonly sandstone, become oil reservoirs. Just as in the metamorphism of rocks, deeper burial results in higher temperatures. If the temperature gets high enough, natural gas is generated rather than oil.

Coal is formed from plants that do not completely decay but are so altered that the original structure is destroyed. Coal, oil, and natural gas are the primary fuels of our modern economy.

Concept Check ✓

1. Why is it incorrect to say that clastic sediments are precipitated out of water?
2. Is it correct to say that evaporites are deposited by water?

Check Your Answers

1. Clastic sediments are made of small fragments of other rocks. Precipitation is a specific chemical reaction in which a solid forms out of a solution. The solid left behind is chemical sediment, not clastic sediment. It is correct to say that clastic sediments are deposited by water, air or ice.
2. Yes, because the word *deposit* is a general term that is applied to all types of sediments.

31.4 Metamorphic Rocks Are Changed Rocks

Existing rocks, whether igneous, sedimentary, or metamorphic, can undergo change called **metamorphism.** Metamorphism takes place if rocks are heated or compressed for a very long time. An everyday example of metamorphism is potter's clay. Potter's clay is soft at room temperature. But when heated, it turns to a hard ceramic. Similarly, limestone subjected to enough heat and pressure becomes marble. Shale is metamorphosed to slate. Rocks may also be drastically stretched or compressed. It is important to note that during metamorphism, no minerals are melted. Change instead occurs by means of *recrystallization* of preexisting minerals and by *mechanical deformation* of rock.

Recrystallization occurs when the minerals in a rock change because the rock has been exposed to high temperatures and pressures.

Marble and slate are metamorphic rocks.

Mechanical deformation occurs when a rock is subjected to physical stress, and may or may not involve high temperatures. Such physical stress occurs deep in the Earth's crust.

Concept Check ✓

Is recrystallization the opposite of partial melting?

Check Your Answer No. The process of igneous crystallization is the opposite of partial melting. Recrystallization is a chemical change that occurs within a rock because of exposure to high temperatures or pressures, without partial melting.

Two Kinds of Metamorphism: Contact and Regional

The most common types of metamorphism are *contact* and *regional.* Each type of metamorphism is characterized by differences in mechanical deformation and recrystallization.

Contact metamorphism occurs when a body of rock is intruded by magma (Figure 31.14). The high temperature of the magma produces a zone of alteration that surrounds the intrusion. The alteration is greatest at the *contact,* which is the surface that separates the intrusive rock from the surrounding rock. The width of the altered zone may range from a few centimeters to several hundred meters. One of the most common changes found with contact metamorphism is an increase in the grain size of the rock due to recrystallization. The grain size is greatest at the contact, and decreases with distance from that point. The water content of the rock also changes with distance from the contact. At the contact, where temperature is high, water content is low. So we find dry, high-temperature minerals such as garnet and pyroxene at the contact. Farther away, we find water-rich, low-temperature minerals such as muscovite and chlorite.

Regional metamorphism is the other common type of metamorphism. This is the kind of change that rock undergoes when it is subjected to physical stress and heat over a large region. It is more than just recrystallization combined with mechanical deformation.

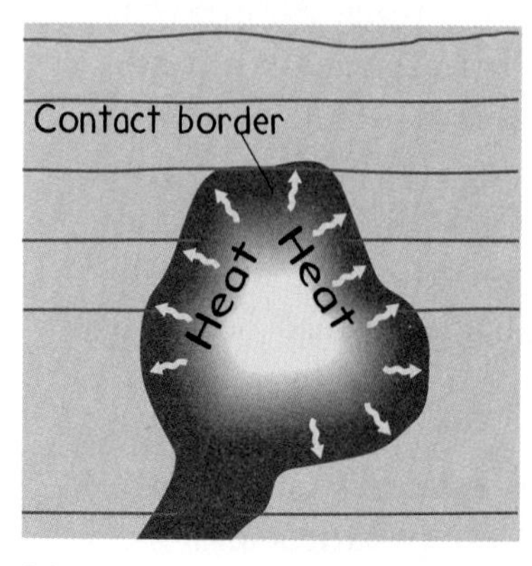

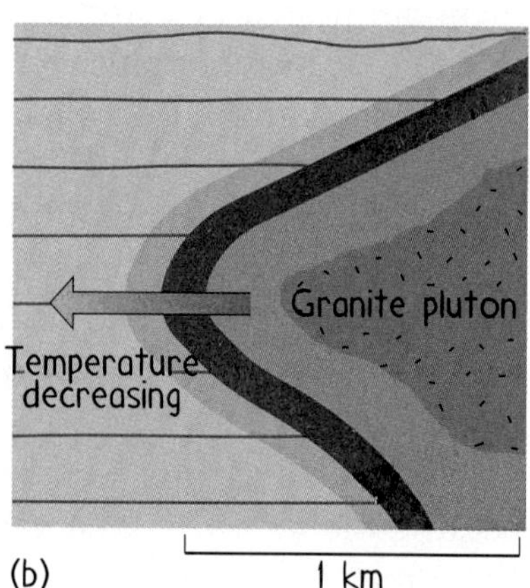

Figure 31.14
(a) Contact metamorphism is the result of rising molten magma that intrudes a rock body.
(b) Surrounding the solidified intrusive rock is a zone of alteration. Alteration is greatest at the contact and decreases away from the contact area.

Figure 31.15
Satellite photo of regional-scale folding in the Appalachian Mountains of Central Pennsylvania.

Regionally metamorphosed rocks are found in all the major mountain belts of the world. During the process of mountain building, the Earth's crust is severely compressed into a mass of highly deformed rock. You can see this deformation in the folded and faulted rock layers in many mountain ranges. The effects of regional metamorphism are most pronounced in the cores of deformed mountains. Rocks develop a distinctly "layered" appearance and zoned sequences of minerals. Because of the large-scale nature of regional metamorphism, these zones tend to be broad and extensive. Such areas are the hunting grounds of gem prospectors, because the heat and pressure that accompany regional metamorphism can produce beautiful minerals.

Metamorphic rocks are defined by their appearance and the minerals they contain. For classification and identification, metamorphic rocks can be divided into two groups: *foliated* and *nonfoliated.*

Foliated Metamorphic Rocks Have a Layered Appearance

As rock is subjected to an increase in temperature and pressure, some of its minerals realign into parallel planes as they recrystallize. The face of each of these parallel planes is perpendicular to the main direction of the compressive force. This leads to a layered appearance called *foliation.* Foliation is a prominent visual feature of regionally metamorphosed rocks and it is very different than the layering seen in sedimentary rock. For example, sheet-structured minerals such as the micas start to grow and orient themselves so that their sheets are perpendicular to the direction of maximum pressure (Figure 31.16). The new rock, which now has parallel flakes, or plates, of mica, is said to be foliated. The most common foliated metamorphic rocks—slate, schist, and gneiss—are derived from fine-grained sedimentary rocks that have the appropriate chemical composition for micas to form.

Slate is the "lowest-grade" foliated metamorphic rock. This means that it was formed under relatively low temperature and pressure. Slate, which is metamorphosed shale, is a very fine-grained foliated rock composed of tiny mica flakes. The most obvious characteristic of slate is its excellent rock cleavage, which allows it to be split into thin slabs. The best pool tables and chalkboards are made from slate. Slate is also used as roof and floor tile.

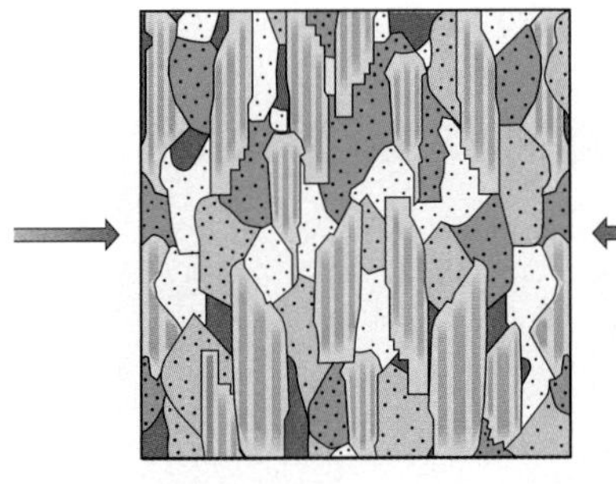

Figure 31.16
As compressive forces squeeze platy and sheet structured minerals, the grains align themselves perpendicular to the main direction of force. Arrows indicate the direction of compressive force.

Schist is one of the most easily recognizable kinds of metamorphic rock because it is shiny. Schist forms under higher temperature and pressure conditions than slate. This causes the minerals in schist to grow large enough to be identified with the naked eye. Schists usually contain 50% platy minerals, most commonly muscovite and biotite. The larger mica flakes give the rock a shiny surface that is quite striking. Schists are named according to the major minerals in the rock (biotite schist, staurolite-garnet schist, and so on).

(a)

(b)

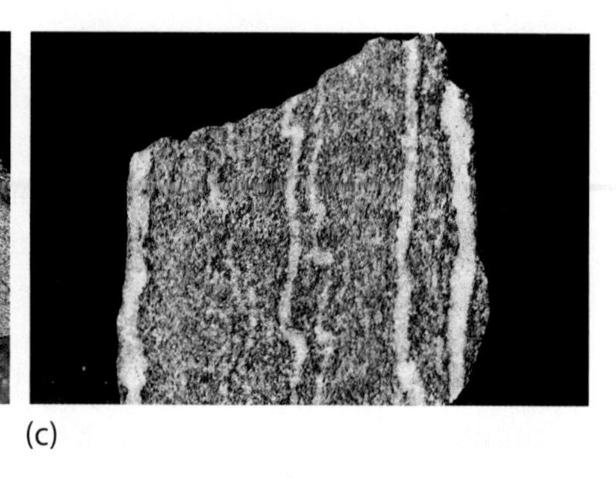
(c)

Figure 31.17
Common foliated metamorphic rocks: (a) slate, (b) schist, (c) gneiss.

Gneiss (pronounced "nice") is a foliated metamorphic rock that contains alternating layers of dark platy minerals and lighter granular minerals, which gives it a characteristic striped appearance. This change in appearance is caused by even greater temperature and pressure conditions than those for schists. The most common granular minerals found in gneisses are quartz and feldspar, which are also the most common granular minerals in granite. In fact, some gneisses are actually metamorphosed granites.

Nonfoliated Metamorphic Rocks Have a Smoother Appearance

Nonfoliated metamorphic rocks can form either in areas of increased temperature and pressure, or in areas where only the temperature increased. Even under high pressure, foliation cannot develop if the rock doesn't have the right chemical composition for platy minerals, such as mica, to form. If the chemical composition is correct, but the pressure is not high enough, such as in contact metamorphism, foliation cannot develop. Two common nonfoliated rocks that generally do not have the potential for micas to grow are marble and quartzite.

Marble (Figure 31.18a) is a coarse-grained, crystalline, metamorphosed limestone. Pure marble is white and virtually 100% calcite. Because of its color and relative softness (hardness 3) marble is a popular building stone. Often the limestone from which marble forms contains impurities that produce various colors. Thus marble can range from pink to gray, green, or even black.

Quartzite (Figure 31.18b) is metamorphosed quartz sandstone and is therefore very hard (hardness 7). The recrystallization of quartzite is so complete that the rock splits across the original quartz grains when broken, rather than between them. Although pure quartzite is white, it commonly contains impurities and thus can be a variety of colors, such as pink, green, and light gray.

Figure 31.18
Nonfoliated metamorphic rocks: (a) marble, (b) quartzite.

(a)

(b)

31.5 The Rock Cycle, a Descriptive Key

We have seen that the igneous, sedimentary, and metamorphic rocks of the Earth's crust have different origins. Although formed by different processes, the three rock types are related. This relationship is graphically shown in the model of the rock cycle (Figure 31.19). By following the different pathways in the model, we can illustrate the origin of the three basic rock types and the different geologic processes that change one rock type into another. The figure helps to summarize this chapter.

We have learned that igneous rock is formed when the molten magma beneath the Earth's crust cools and crystallizes. Magma can crystallize into many different kinds of igneous rock. Although most of the Earth's crust is either igneous or derived from rock that was initially igneous, the rock we see at the surface is mainly sedimentary.

We have learned that sedimentary rock is the result of the decomposition and disintegration of other rocks by weathering and erosion. When sedimentary rock is buried deep within the Earth or is caught up in mountain building, we have seen that great pressures and heat can transform it into metamorphic rock. When subjected to still greater heat and pressure, metamorphic rock melts and turns to magma, which eventually solidifies as igneous rock to complete the rock cycle.

The rock cycle varies in its paths. Igneous rock, for example, may be subjected to heat and pressure far below the Earth's surface to become metamorphic rock. Or metamorphic or sedimentary rocks at the Earth's surface may decompose to become sediment that becomes new sedimentary rock. Cycles within cycles occur in the recycling of the Earth's crust. Whatever the path, the Earth's crust is formed when molten rock rises from the depths of the Earth, cools, and solidifies to form a crust. Over eons, rocks of the Earth's crust are reworked by shifting and erosion, and can eventually be returned to the interior. There, they may be melted and once again become magma.

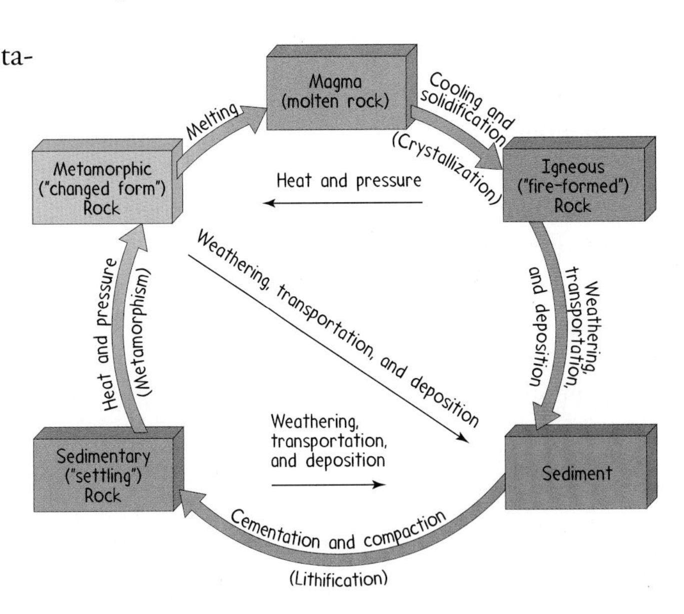

Figure 31.19
The rock cycle: Igneous rock subjected to heat and pressure far below the Earth's surface may become metamorphic rock. Or metamorphic or sedimentary rocks at the Earth's surface may decompose to become sediment that becomes new sedimentary rock. Whatever the variety of routes, molten rock rises from the depths of the Earth, cools, and solidifies to form a crust that over eons is reworked by shifting and erosion, only to eventually return to become magma in the Earth's interior.

Chapter Review

Key Terms and Matching Definitions

_____ chemical weathering
_____ erosion
_____ extrusive rocks
_____ igneous rocks
_____ intrusive rocks
_____ lava
_____ mechanical deformation
_____ mechanical weathering
_____ metamorphic rocks
_____ metamorphism
_____ pluton
_____ recrystallization
_____ rock cycle
_____ sedimentary rocks
_____ volcano

1. Rocks formed by the cooling and crystallization of hot, molten rock material called magma.
2. Rocks formed from the accumulation of weathered material (sediments) carried by water, wind, or ice.
3. Rocks formed from preexisting rocks that have been changed or transformed by high temperature, high pressure, or both.
4. Igneous rocks that form at the Earth's surface.
5. Igneous rocks that crystallize below the Earth's surface.
6. Magma once it reaches the Earth's surface.
7. A central vent through which lava, gases, and ash erupt and flow.
8. A very large intrusive body formed below the Earth's surface.
9. The breakdown of rocks on the Earth's surface by physical means.
10. The breakdown of rocks on the Earth's surface by chemical means.
11. The process by which rock particles are transported away by water, wind, or ice.
12. The process of changing one kind of rock into another kind as a result of high temperature, high pressure, or both.
13. Metamorphism caused by high temperatures.
14. Metamorphism caused by stress, such as increased pressure.
15. A sequence of events involving the formation, destruction, alteration, and reformation of rocks as a result of the generation and movement of magma, the weathering, erosion, transportation, and deposition of sediment, and the metamorphism of preexisting rocks.

Review Questions

Rocks Are Divided into Three Main Groups

1. Name the three major types of rocks, and describe the conditions of their origin.

Igneous Rocks Form When Magma Cools

2. What are the most common igneous rocks and where do they generally occur?
3. What percentage of the Earth's crust is composed of igneous rocks?
4. Where on the Earth's surface are lava flows most common?
5. What are the three major types of volcanoes?
6. Which type of volcano produces the most violent eruptions? Which type produces the quietest eruptions?
7. What are three common types of plutons?
8. Are the Sierra Nevada mountains presently losing or gaining height?

Sedimentary Rocks Blanket Most of the Earth's Surface

9. How does weathering produce sediment? Distinguish between weathering and erosion.

10. What does roundness tell us about sediment grains?

11. Relate the shape and sorting of sand grains to the way they were probably transported.

12. What is a clastic sedimentary rock?

13. In what two ways does sediment turn into sedimentary rock?

14. What are the three most common clastic sedimentary rocks?

15. What is a fossil? How are they used in the study of geology?

Metamorphic Rocks Are Changed Rocks

16. What is metamorphism? What causes metamorphism?

17. What are the two processes by which rock is changed?

18. What patterns of alteration are characteristic of contact metamorphism?

19. What changes are characteristic of regional metamorphism?

The Rock Cycle, a Descriptive Key

20. List the different cycles of rock formation.

Exercises

1. Name at least one other national park, besides Yellowstone National Park, that was formed by volcanic or plutonic activity.

2. What type of rock is formed when magma rises slowly and solidifies before reaching the Earth's surface? Give an example.

3. Where on the Earth's crust do we find the two most common igneous rocks, basalt and granite?

4. Are the Hawaiian Islands primarily made up of igneous, sedimentary, or metamorphic rock?

5. Can metamorphic rocks exist on an island of purely volcanic origin? Explain.

6. What accounts for the differences in lava composition of the two volcanoes that have erupted in recent times, Mauna Loa in Hawaii, and Mount St. Helens in Washington?

7. Can a sequence of plutonic rocks be both intrusive and extrusive?

8. What types of minerals are clastic sedimentary rocks commonly made of?

9. What general rock feature does a geologist look for in a sedimentary rock to determine the distance the rock has traveled from its place of origin?

10. What feature of clastic sedimentary rock enables the flow of oil after it has been formed?

11. Of the rocks (a) granite (b) sandstone (c) limestone (d) halite, which is the first to weather in a humid (wet) climate? Which is the last to weather? Why?

12. Which type of rock is most sought by petroleum prospectors—igneous, sedimentary, or metamorphic? Why?

13. What kind of weathering is imposed on a rock when it is smashed into small pieces? When dissolved in acid?

14. What type of rock is made of previously existing rock whose formation does not involve high temperatures and pressure?

15. What properties of slate make it good roofing material?

16. Name two mica minerals that can give a metamorphic rock its foliation.

17. How is foliation different than sedimentary layering?

18. Can metamorphism, caused solely by elevated temperature occur without the presence of magma? Why or why not?

19. Each of the following statements describes one or more characteristics of a particular metamorphic rock. For each statement, name the metamorphic rock being described.

(a) Foliated rock derived from granite.

(b) Hard, nonfoliated, single mineral rock, formed under high to moderate metamorphism.

(c) Foliated rock possessing excellent rock cleavage. Generally used in making blackboards.

(d) Nonfoliated rock composed of carbonate minerals.

(e) Foliated rock containing about 50% platy minerals; named according to the major minerals in the rock.

20. Explain the different cycles of rock formation.

Suggested Readings and Web Sites

Press, Frank and Siever, Raymond. *Understanding Earth.* New York: W.H. Freeman, 1994.

Wenkam, Robert. *The Edge of Fire.* San Francisco: Sierra Club Books, 1987.

http://sln.fi.edu/fellows/fellow1/oct98/create/index.html

http://wrgis.wr.usgs.gov/docs/parks/rxmin/content.html

http://hvo.wr.usgs.gov/

Chapter 32: The Architecture of the Earth

Did you ever think about digging a hole to China? The idea is intriguing, for why go all the way around the Earth if you can take a shortcut straight through? Digging such a hole, unfortunately, is not a very realistic possibility. If such a hole were possible, what would we find in the Earth's interior? Is the Earth's interior different than its surface? If so, how so? And how do we know that it is different?

Because scientists can't dig straight through the Earth, they look for clues about the Earth's interior in things they can see and measure. For example, surface rocks are all around us, and they can tell us a great deal about the inside of our Earth. Earthquakes and volcanic eruptions are a direct link to the inner workings of the planet. Rocks, earthquakes, and volcanic eruptions are external expressions of the Earth's internal processes.

32.1 Earthquakes Make Seismic Waves

All earthquakes create waves that travel through the Earth's interior. Such earthquake-generated waves are called *seismic waves.* The way these waves travel has provided Earth scientists with a view into the Earth's interior and revealed a planet that is layered. The major layers of the Earth are the *crust, mantle, outer core,* and *inner core* (Figure 32.1). Before examining the composition of the Earth, however, we will first learn how seismic waves reveal the Earth's hidden mysteries.

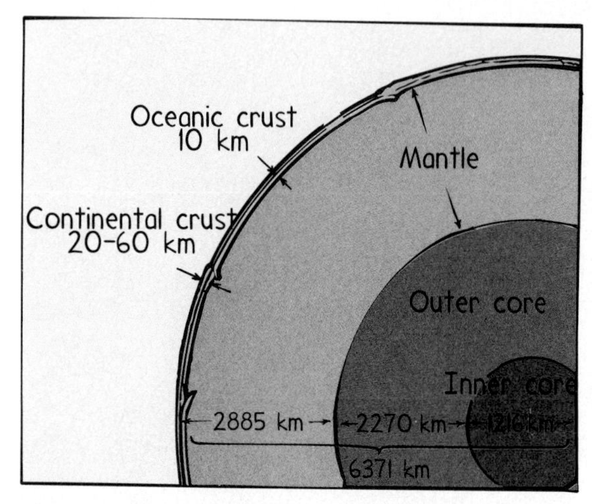

Figure 32.1
Cross section of the Earth's interior showing the four major layers and their approximate thicknesses.

Recall from Chapter 13 that a wave's speed depends on the medium through which it travels. We learned that the sound waves made by two rocks clicking together travel faster through water than through air. And sound waves travel even faster through a solid than a liquid. Just like sound waves, the speed of seismic waves depends on the elasticity of the material through which they are traveling. The elasticity of a material is controlled by its density and stiffness. So when we measure the speeds of seismic waves, we can learn a lot about the composition of the Earth.

During an earthquake, energy is released within the Earth's interior and radiates in all directions. This energy travels to the Earth's surface in the form of seismic waves. Seismic waves from an earthquake cause the ground to shake and move. This ground movement is recorded on a machine called a *seismograph.* Seismographic records provide a map of the Earth's interior.

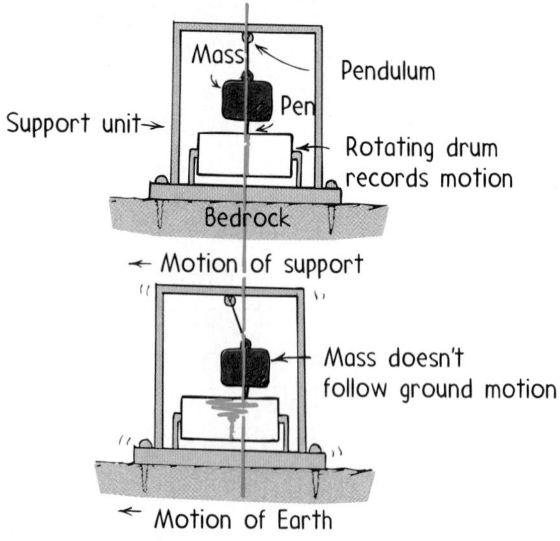

Figure 32.2
Diagram of a seismograph. When the Earth moves, the support unit attached to the ground also moves. However, because of inertia, the mass at the end of the pendulum tends to stay in place. A pen attached to the mass marks the relative displacement on the slowly rotating drum beneath. In this way, ground movement is recorded.

There are two types of seismic waves: **body waves,** which travel through the Earth's interior, and **surface waves,** which travel on the Earth's surface.

Body waves are further classified as either **primary waves** (P-waves) or **secondary waves** (S-waves). Primary waves, like sound waves, are longitudinal—they compress and expand the rock as they move through it. Like vibrations in a bell, primary waves move out in all directions from their source. They are the fastest of all seismic waves and so they are the first to reach a seismograph. P-waves travel through any type of material—solid granite or magma or water or air. But secondary waves are transverse, like the waves produced by a vibrating violin string. They vibrate the particles of their medium up and down and side to side. S-waves travel more slowly than P-waves and are the second to register on a seismograph. S-waves cannot move through fluids. S-waves can travel only through solids.

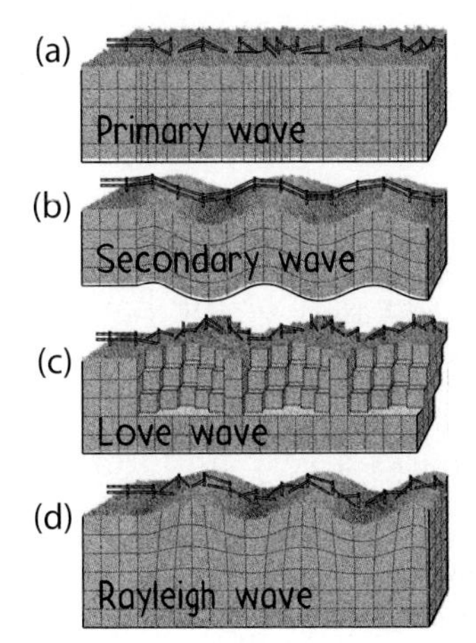

Figure 32.3
Block diagrams show the effects of seismic waves. The yellow portion on the left side of each diagram represents the undisturbed area. (a) Primary body waves alternately compress and expand the Earth's crust, as shown by the different spacing between the vertical lines, similar to the action of a spring. (b) Secondary body waves cause the crust to oscillate up and down and side to side. (c) Love surface waves whip back and forth in horizontal motion. (d) Rayleigh surface waves operate much like secondary body waves but affect only the surface of the Earth.

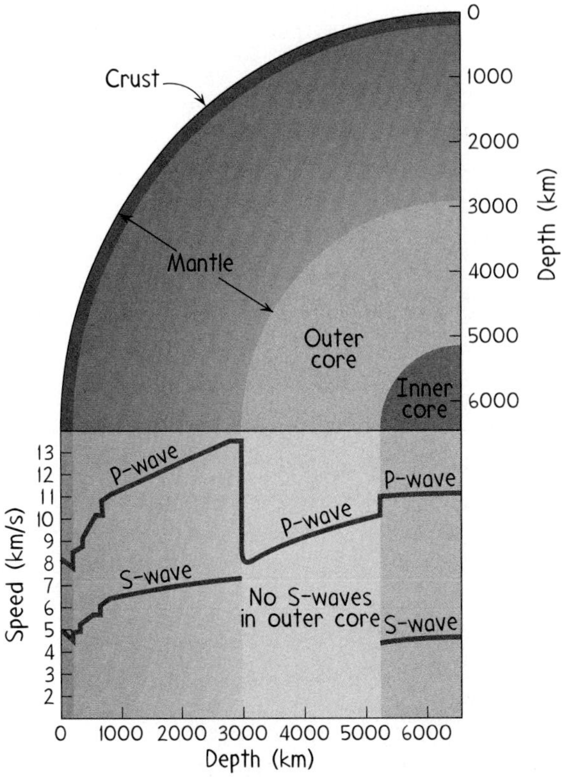

Figure 32.4
Cross section of the Earth's internal layers, showing the increases and decreases of P- and S-wave velocity in the different layers.

There are also two types of surface waves: *Rayleigh waves* and *Love waves.* Rayleigh waves move in an up-and-down motion, and Love waves move in a side-to-side, whiplike motion. Both of these surface-wave types move slower than P- and S-waves. So they are the last to register on a seismograph.

Like any kind of wave, seismic waves reflect and refract from surfaces. Just as medical people use ultrasound to scan the inside of a body, geoscientists do much the same with seismic waves. With this technique they scan the insides of the Earth. Seismic wave research tells us that the Earth's interior is layered (Figure 32.4).

32.2 Seismic Waves Show the Earth Has Layers

In 1909, the Croatian seismologist Andrija Mohorovičić presented the first convincing evidence that the Earth's interior is layered. While studying seismographic data from a recent earthquake, he discovered that the seismic waves generated by the quake suddenly picked up speed at a certain depth below the surface. Knowing that the speeds of these waves depend on the elasticity of the material they pass through, Mohorovičić concluded that the speed increase he observed was due to variations in the density of the Earth. The seismographic data had literally drawn a map of the upper boundary of the Earth's mantle, a layer of dense rock underlying the less dense crust. This boundary, known as the **Mohorovičić discontinuity** (called the "Moho" for short), separates the Earth's crust from the rocks of different composition in the mantle below.

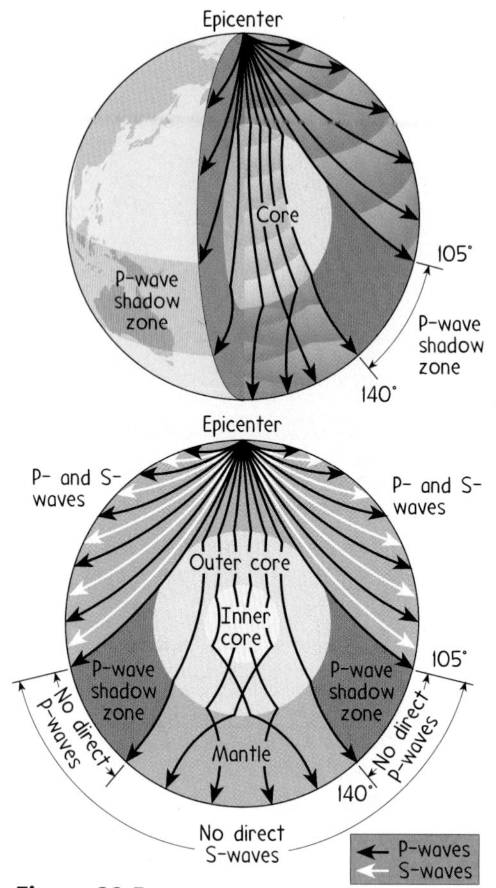

Figure 32.5
Cut-away and cross-sectional diagrams that show the change in wave paths at the major internal boundaries and the P-wave shadow. The P-wave shadow lies in an arc between 105° and 140° from an earthquake's epicenter. The P-wave shadow is caused by the refraction of the P-waves at the core-mantle boundary. Note that any location that is more than 105° from an earthquake's epicenter does not receive S-waves because the liquid outer core does not transmit S-waves.

Two years after the discovery of the Moho, the mantle-core boundary was detected. Both P- and S-waves are strongly influenced by a pronounced boundary 2900 kilometers deep. When P-waves reach that boundary, they are reflected and refracted so strongly that the boundary actually casts a P-wave shadow over part of the Earth (Figure 32.5). In the area of the P-wave shadow, there are no seismic P-waves. Because the boundary is so distinct in the seismographic record, we infer that it marks a very important change in the density of the materials present. Both the overall density* of the Earth and the speed that seismic waves travel through the core suggest that the core is composed of metallic iron, a material that is much denser than the silicate rocks that make up the mantle.

The sharp boundary between the mantle and core casts an S-wave shadow that is even more extensive than the P-wave shadow. As a result we find S-waves are unable to pass through the core. Because S-waves, being transverse, cannot travel through liquids, we infer that the outer portion of the core is liquid.

In 1936, the discovery that P-waves are reflected from a boundary within the core showed there was yet another layer in the Earth. The P-waves passing through the inner portion of the core traveled faster than P-waves passing through the outer core. This change in wave speed tells us that the inner core must be solid. Do you suppose these layers in the Earth's interior influence the geologic changes our planet experiences? The answer is yes, as you will now see.

Concept Check ✓

What evidence supports the theory that the Earth's inner core is solid and the outer core is liquid?

Check Your Answer The evidence is the differences between how P- and S-waves move through the Earth's interior. As the waves encounter the boundary at 2900 km, a very pronounced wave shadow develops. P-waves are both reflected and refracted at the boundary, but S-waves are only reflected. S-waves cannot travel through liquids, implying a liquid outer core. As P-waves move through the outer core, there is a depth at which there is a sudden increase in speed. Knowing that waves travel faster in solids, we infer a solid inner core.

* The density of rocks at the Earth's surface is 2.7–3.0 g/cm^3, whereas the average density of the Earth as a whole is 5.5 g/cm^3. Thus surface rocks are not representative of the planet's interior. To account for the Earth's high average density, the density of the core must be at least 10 g/cm^3. This and other reasons suggest that the core is composed of iron and nickel, the most abundant of the heavier elements.

The Core Has Two Parts: A Solid Inner Core, and a Liquid Outer Core

The **core** is composed mostly of iron and nickel. In the inner core, the iron and nickel are solid. Although the inner core is very hot, intense pressure from the weight of the rest of the Earth prevents the material of the inner core from becoming a liquid (like a pressure cooker prevents high-temperature water from boiling to a gas). Because less weight is exerted on the outer core, it is made of liquefied iron and nickel. The molten outer core flows at the very slow rate of several kilometers per year. This motion sets up the electric current that generates the Earth's magnetic field. This magnetic field is not stable but has changed throughout geologic time. Recall from Chapter 12 that there have been times when the Earth's magnetic field has diminished to zero, only to build up again with the poles reversed. These magnetic pole reversals probably result from changes in the direction of flow in the molten outer core of the Earth.

Concept Check ✓

Iron's normal melting point is 1535°C, yet the Earth's inner core temperature is about 4300°C. Why doesn't the solid inner core melt?

Check Your Answer The intense pressure from the weight of the Earth above crushes atoms so tightly that even the high temperature cannot budge them. In this way melting is prevented.

The Mantle Is Dynamic

Surrounding the core of the planet is the **mantle,** a rocky layer that is about 3000 kilometers thick. Composed of hot, iron-rich silicate rocks, the mantle behaves like a *plastic.* In other words, the mantle is not as rigid as a solid but not able to flow as freely as a true fluid. The mantle itself is further subdivided. The lower mantle extends from Earth's outer core most of the way to Earth's surface. The other part of the mantle is the upper mantle, which extends from the crust-mantle boundary down to a depth of about 350 kilometers. The upper mantle is itself divided. Scientists classify the upper mantle into two zones, as Figure 32.6 shows. The lower part of the upper mantle, called the **asthenosphere,** behaves like a plastic but flows much more easily than the rest of the mantle.

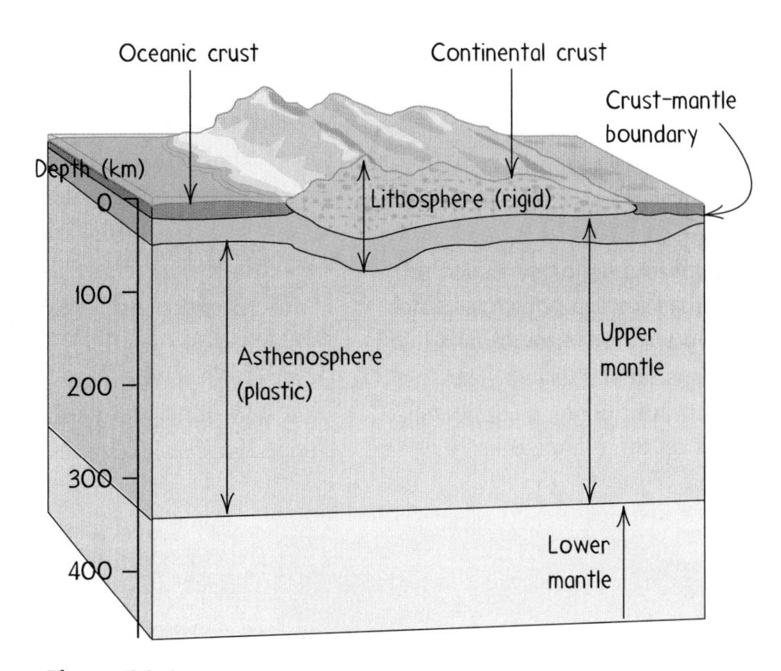

Figure 32.6
The bottom portion of the Earth's upper mantle is the plastic asthenosphere. The top portion of the upper mantle plus the crust form the rigid layer called the lithosphere.

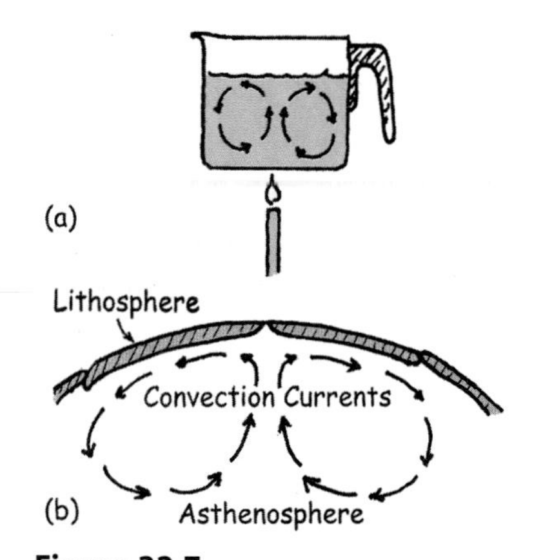

Figure 32.7
(a) A familiar example of convection is seen when water is heated in a pan. (b) A simple model showing convection currents in the asthenosphere.

This asthenosphere flows gradually because convection currents carry warmer material upward as colder, denser material sinks. (Figure 32.7). The constant flowing movements in the asthenosphere greatly affect the surface features of our planet.

Above the asthenosphere is the **lithosphere.** The lithosphere is about 100 km thick. The lithosphere includes the entire crust and the uppermost part of the mantle (Figure 32.6). Unlike the asthenosphere, the lithosphere is more rigid and brittle and does not flow. The lithosphere is, in a sense, floating on top of the asthenosphere like a raft on a pond. The lithosphere is carried along by the motions of the material beneath it in the asthenosphere. The motions in the asthenosphere are not uniform. Because of this, as we shall see in the next chapter, the brittle lithosphere is broken into many individual pieces called *plates.*

The lithospheric plates are always in motion. They float on the circulating asthenosphere. Although asthenosphere currents move at a leisurely pace—taking hundreds of millions of years to complete one loop—they are powerful enough to move continents and reshape many of our surface features. The movement of the lithospheric plates causes earthquakes, volcanic activity, and the crunching of large masses of rock to create mountains.

The Crustal Surface Makes Up the Ocean Floor and Continental Land

The top part of the lithosphere is the **crust.** We live on the continental crust while the oceanic crust makes up the ocean floor. The crust that makes up the deep ocean floor is very different than the crust that makes up the continents. First of all, the crust of the ocean floor is thin. It's only about 10 kilometers thick. Also, it's made of dense, basaltic rocks. The part of the crust we know as continents is between 20 and 60 kilometers thick. Continental masses are mostly composed of granitic rocks. These are less dense than the basaltic rock of the ocean floor.

If continental crust is so much thicker than oceanic crust, then why are the ocean floors underwater and the continents high and dry? The answer is found in their density differences and buoyancy. The less-dense continental crust always floats higher than the more-dense oceanic crust, even if the continental crust has more mass. It may be hard to visualize this idea. It might help to know that the bottom of the less-dense continental crust extends deeper into the mantle than the more-dense oceanic crust, as Figure 32.8 shows, forming a deep "root." The thicker the crust, the higher it floats on the mantle, but the further it extends into the mantle.

Figure 32.8
Continental crust is thicker and less dense than oceanic crust. As such, continental crust floats higher on the mantle than oceanic crust.

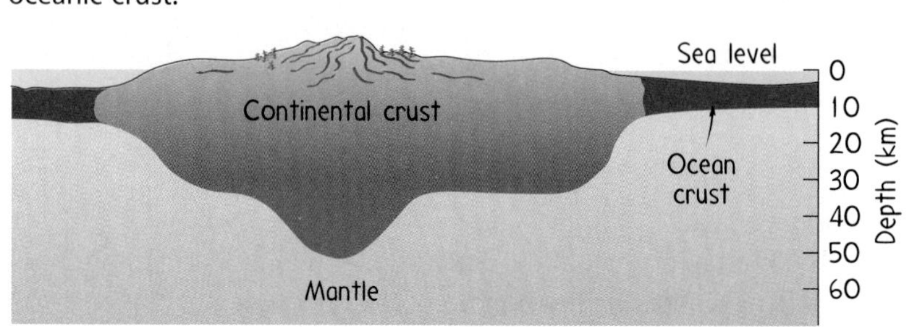

Concept Check ✓

If you wished to drill the shortest hole to the mantle, would you drill in western Colorado or in Florida?

Check Your Answer Put the question another way: If you wanted to drill the shortest hole through ice to the water below, would you drill atop an iceberg or through a slab of ice that hardly extends out of the water? You would drill your hole in the slab, of course, and likewise you should drill through the thinner crust of mountain-free Florida. In mountainous western Colorado, the crust is much thicker. (If you really want the shortest hole, you should drill through the ocean floor—exactly what scientists have done in Project MoHole, in the East Pacific Ocean.)

If Earth were the size of a cue ball, it would be just as smooth! Its highest mountains are insignificant compared with the Earth's radius.

32.3 Folds, Faults, and Earthquakes

The slow-moving mantle causes the overlying lithospheric plates to move slowly but constantly. As the mantle's motion pushes and pulls on the plates in several directions, the plates undergo stress. Rocks subjected to stress begin to deform. They form intricate and broad *folds.* If enough stress is applied, rocks break. The rocks then can move along *faults.* Faults come in sizes ranging from small and almost impossible to see to large and loaded with the potential to devastate. Are there any large folds or faults where you live?

Rock Folds, an Expression of Compressive Force

Compressive stresses push rocks together. The rocks then begin to buckle and fold. The term *fold* has a precise geological meaning. To see what a fold is, suppose you had a throw rug on your floor, with a friend standing on one end. If you push the rug toward your friend while keeping it on the floor, the rug begins to tilt away from your hand, and a series of ripples, or **folds,** develop. This is what happens to the Earth's crust when it is subjected to compressive stress.

We know from Chapter 31 that sediments settling from water in an ocean or bay are deposited in horizontal layers. The layer at the bottom, naturally, is deposited first. This bottom layer is therefore the oldest in the sequence of deposited layers. Each new layer is deposited on top of the previous layer, with the youngest at the top.

As initially flat sedimentary rock layers are subjected to compressive stress, they tilt and become folded, just as the throw rug became tilted and folded. Each high point and low point in a series of folds creates an axis. Imagine such an axis as a plane extending downward into the Earth, as Figure 32.9 shows. When the layers tilt

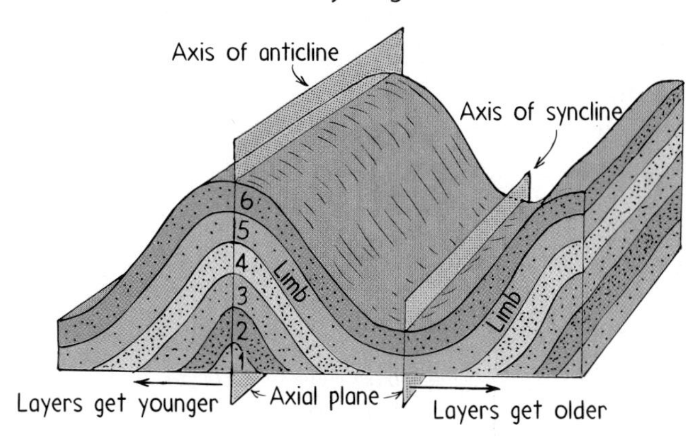

Figure 32.9
Anticline and syncline folds. Layer 1 is the oldest rock, and layer 6 is the youngest. The limbs of an anticline tilt away from the axis of the fold (a marble would roll away from the axis). Rock layers are oldest at the core of the fold. The limbs of a syncline tilt toward the axis of the fold (a marble would roll toward the axis). Rocks are youngest at the core.

in toward the fold axis, so that if you put a marble on the rock it would roll toward the axis, the fold is called a **syncline.** The rocks at the center, or *core,* of a syncline are the youngest. As you move horizontally away from the axis, the rocks get older and older. If the fold layers tilt away from the axis, so that if you put a marble on the rock it would roll away from the axis, the fold is called an **anticline.** The rocks in an anticline are oldest at the core, and as you move horizontally away from the axis, the rocks get younger. Another way to think about it is that anticlines are pushed upward and synclines are pushed downward.

Figure 32.10
The terms *footwall* and *hanging wall* were used by miners because one could hang a lamp on a hanging wall and stand on a footwall.

Concept Check ✓

Why are rocks at the core of a syncline younger than those farther out from the core while the opposite is true for an anticline?

Check Your Answer Think of the rug example. Assume the top surface of the rug is younger than the lower surface. When you push the rug it can (1) fold upward (like the letter "A"), or (2) fold downward (like a sag). In the first case, the bottom surface makes up the core—an anticline. In the second case, the top surface makes up the core—a syncline. Makes sense!

Faults Are Made by the Forces of Compression or Tension

When compressional stress is stronger than rock, the rock fractures into two parts. If one part then moves relative to the other part, the fracture is called a **fault.**

Note the angle of the fault in Figure 32.10 relative to the horizontal ground surface. Imagine you could pull the block diagram apart at the fault, as is done in the lower drawing. The half containing the fault surface where someone could stand is the *footwall* block. The fault surface of the other half is inclined to make standing impossible; this is the *hanging wall* block. These terms were coined by miners because one could hang a lamp on a hanging wall and could stand on a footwall.

Once compressional forces have created a fault, these forces cause rocks in the hanging wall to be pushed upward along the fault plane relative to rocks in the footwall, as Figure 32.11 shows. This type of fault is called a *reverse fault.* The Rocky Mountain foreland, the Canadian Rockies, and the Appalachian Mountains, to name a few, were formed in part by reverse faulting.

In addition to compression, stress in rocks can also occur by tension. Opposite of compression forces that push, tension forces pull at the rocks. Tension causes rocks in a hanging wall to drop downward along the fault plane relative to those in the adjacent footwall, producing a *normal fault* (Figure 32.12). Virtually the entire state of

Figure 32.11
A reverse fault. In a zone of compressional faulting, rocks in the hanging wall are pushed up relative to rocks in the footwall. (a) A reverse fault before erosion (b) the same reverse fault after erosion.

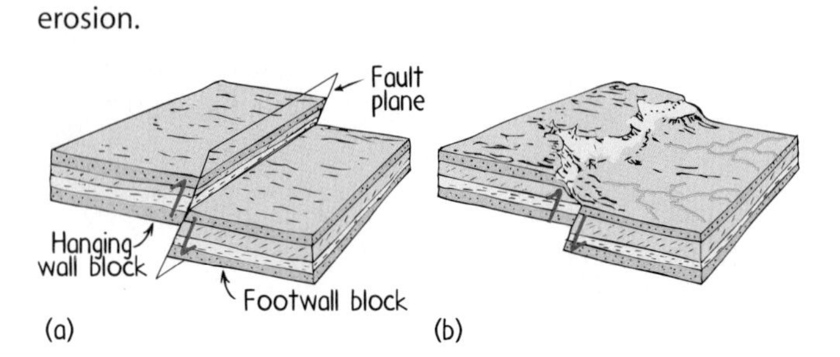

Nevada, eastern California, southern Oregon, southern Idaho, and western Utah are greatly affected by normal faulting.

The faults described so far have mostly vertical (up and down) motion. But some faults have almost no vertical motion; their motion is horizontal. These are called *strike-slip* faults. Some of the world's most famous faults, such as the San Andreas Fault in California, are strike-slip faults.

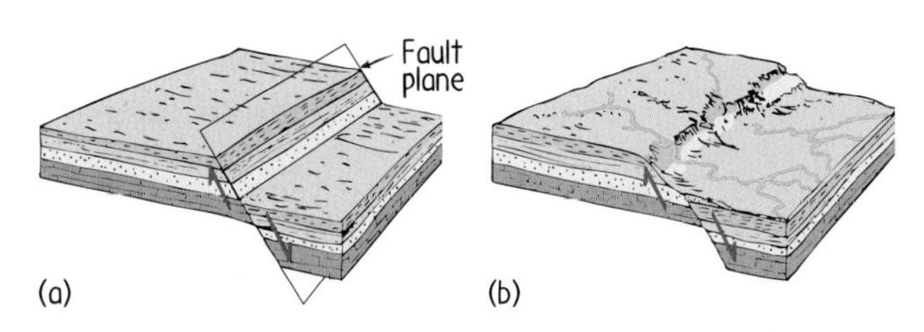

Figure 32.12
A normal fault. In a zone of tensional faulting, rocks in the hanging wall drop down relative to those in the footwall, forming a normal fault. (a) A normal fault before erosion (b) the same normal fault after erosion.

Devastating earthquakes can occur with all three types of faults: strike-slip, reverse, or normal. The Great San Francisco Earthquake and Fire of 1906 registered near 8.3 on the Richter scale (a measure of ground motion caused by earthquakes). It caused 700 deaths and extensive fire damage. The 1989 Loma Prieta earthquake near Santa Cruz, California, registered 7.1 on the Richter scale. It sadly caused 62 deaths and more than $6 billion in damage. The 1999 earthquake in Turkey registered 7.6 on the Richter scale and caused 17,000 deaths. Strike-slip faulting caused all three of these earthquakes. Still larger and more catastrophic earthquakes have occurred along reverse faults. The 1964 earthquake in Anchorage, Alaska registered 8.5 on the Richter scale, and caused 131 deaths and $300 million in damage.* The 2001 earthquake in India registered 7.7 on the Richter scale and caused an unimaginable 20,000 deaths! Normal faults have also had their share of tragic earthquakes. The 2001 earthquake in El Salvador registered 7.7 on the Richter scale and led to 844 deaths.

Understanding earthquakes is obviously of major importance to society. Unfortunately, earthquakes and fault movement occur with little or no warning and thus are very difficult to predict.

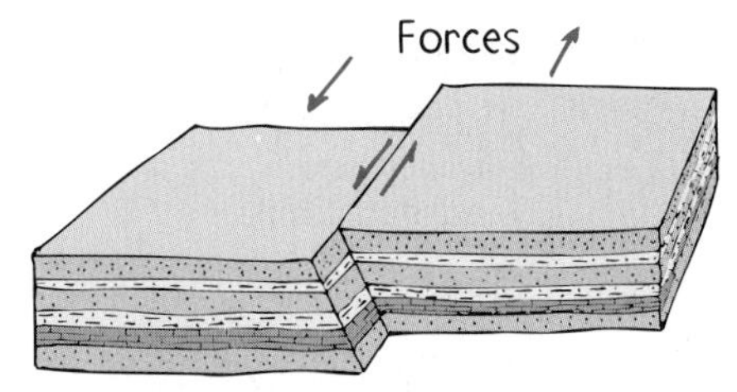

Figure 32.13
The relative movement of a strike-slip fault is horizontal.

Figure 32.14
Offset orchard rows in an orange grove that straddles the San Andreas Fault. The rows in the background have moved to the right relative to the rows in the foreground.

* The death toll was largely due to great seismic sea waves, or *tsunamis*. A tsunami is generated from the displacement of water as a result of an earthquake, submarine landslide, or an underwater volcanic eruption.

Table 32.1 lists some of the world's most notable earthquakes according to their impact on society. The 1906 San Francisco earthquake is notable because of its damage to the city and its inhabitants. On the other hand, in the winter of 1811–1812, a much greater earthquake

Table 32.1

Some Notable Earthquakes

Year	Location	Magnitude	Estimated Deaths	Comments
1556	Shensei, China		830,000	Possibly the greatest natural disaster
1811	New Madrid, Missouri		few	
1906	San Francisco, California	8.25	700	Fires caused extensive damage
1908	Messina, Italy	7.5	120,000	
1920	Kansu, China	8.5	180,000	
1923	Tokyo, Japan	8.2	150,000	Fire caused extensive destruction
1960	Southern Chile	8.7	5700	The largest earthquake ever recorded
1964	Anchorage, Alaska	8.5	131	More than $300 million in damage
1970	Peru	7.8	66,000	Great rockslide
1971	San Fernando, California	6.5	65	More than $5 billion in damage
1975	Liaoning, China	7.5	few	First major earthquake to be predicted
1976	Tangshan, China	7.6	500,000	
1985	Mexico City, Mexico	8.5	7,000	
1989	San Francisco, California	7.1	62	More than $6 billion in damage
1994	Northridge, California	6.7	57	More than $25 billion in damage
1995	Kobe, Japan	7.2	5,500	Between $95 to $147 billion in damage

Table 32.2

Richter Magnitude

Magnitude	Maximum Number per Year	Mercalli Intensity	Characteristic Effects
<3.4	800,000	I	Recorded only by seismographs.
3.4–4.4	30,000	II and III	Felt by some people in the area.
4.4–4.8	4,800	IV	Felt by many people in the area.
4.8–5.4	1,400	V	Felt by everyone in the area.
5.4–6.0	500	VI and VII	Slight building damage.
6.0–7.0	100	VIII and IX	Much building damage.
7.0–7.4	15	X	Serious damage, bridges twisted, walls fractured.
7.4–8.0	4	XI	Great damage, buildings collapse.
>8.0	one every 5–10 years	XII	Total damage, waves seen on ground, objects thrown in air.

Source: From B. Gutenberg, 1950

on the New Madrid fault in Missouri changed the landscape beyond recognition as it shifted the direction and course of the Mississippi River. Fortunately, because the quake occurred in a remote region where there were few settlers, human injuries and deaths were few. Going much, much, further back in time, the presently inactive Appalachian Mountains and parts of the Rocky Mountains were once zones of intense earthquake activity. You can imagine that they were once much like the places listed in Table 32.1 that have recently endured the awesome power of the quaking Earth.

Earthquake Measurements—Mercalli and Richter Scales

Every year hundreds of thousands of earthquakes occur. Although most are small and go undetected, the danger of large earthquakes certainly exists. Earthquake-prone regions experience large earthquakes about every 50–100 years.

The Mercalli scale measures quake intensity in terms of the effects the quake produces on the local environment. The scale ranges from an intensity of I, barely detectable, to an intensity of XII, total destruction. The Mercalli intensity at any location depends (1) on how far that location is from where the earthquake occurred, and (2) on the nature of the subsurface materials at the location. (For example: Is the subsurface material solid rock or loose sediment?)

Based purely on observation, the Mercalli scale, though a valuable yardstick, does not provide a precise measurement of quake size. For this reason seismologists developed a more precise way to estimate the energy released in an earthquake. The Richter scale, which is a magnitude scale, measures quake severity in terms of the amount of energy released and the amount of ground-shaking at a standard distance from the location of the quake. The Richter magnitude scale, which measures the amplitude of seismic waves recorded by a seismograph, is logarithmic. This means that each increase of 1 unit on the scale is equivalent to an increase of 10 in the amplitude of ground-shaking. Because seismic waves occur over a range of frequencies, this tenfold increase in amplitude translates into 30 times more energy released! For example, the 1985 Mexico City earthquake, which had a magnitude of 8.5, released about 30 times as much energy, and had 10 times more ground-shaking than the 1908 Messina earthquake, which had a magnitude of 7.5. The magnitude 8.5 Mexico City quake released 900 times as much energy as and produced 100 times more ground shaking than the 1971 magnitude 6.5 quake in San Fernando, California.

The Mercalli Scale of Intensity

I Not felt except by a very few under especially favorable circumstances.

II Felt only by a few persons at rest, especially on upper floors of buildings.

III Felt quite noticeably indoors, especially on upper floors of buildings, but many people do not recognize this as an earthquake.

IV Most people feel it indoors, a few outdoors. Dishes, windows, doors rattle.

V Felt by nearly everyone. Disturbances of trees, poles, and other tall objects.

VI Felt by all; many frightened and run outdoors. Some heavy furniture moved; a few instances of fallen plaster or damaged chimneys. Damage slight.

VII Everybody runs outdoors. Damage negligible in buildings of good design and construction; slight to moderate in well-built structures; considerable in poorly built structures.

VIII Damage slight in specially designed structures; considerable in ordinary substantial buildings with partial collapse; great in poorly built structures (fall of chimneys, factory stacks, columns, monuments, walls).

IX Damage considerable in specially designed structures. Buildings shifted off foundations. Ground conspicuously cracked.

X Some structures destroyed. Most masonry and frame structures destroyed with foundations. Ground badly cracked.

XI Few, if any, structures (masonry) remain standing. Bridges destroyed. Broad fissures in ground.

XII Damage total. Waves seen on ground surfaces. Objects thrown up in air.

Source: U.S. Coast Guard and Geodetic Survey

Chapter Review

Key Terms and Matching Definitions

_____ anticline
_____ asthenosphere
_____ core
_____ crust
_____ fault
_____ lithosphere
_____ mantle
_____ Mohorovičić discontinuity (Moho)
_____ primary waves
_____ secondary waves
_____ syncline

1. A transverse body wave; cannot travel through liquids and so does not travel through the Earth's outer core.
2. The central layer in the Earth's interior, divided into an outer liquid core and an inner solid core.
3. A subdivision of the upper mantle situated below the lithosphere, a zone of plastic, easily deformed rock.
4. The Earth's outermost layer.
5. A fold in strata that has relatively young rocks at its core, with rock age increasing as you move horizontally away from the fold core.
6. A fracture along which visible movement can be detected on one side relative to the other.
7. A longitudinal body wave; travels through solids, liquids, and gases and is the fastest seismic wave.
8. The crust-mantle boundary, marking the depth at which the speed of P-waves traveling toward the Earth's center increases.
9. The middle layer in the Earth's interior, between crust and core.
10. The entire crust plus the portion of the mantle above the asthenosphere.
11. A fold in strata that has relatively old rocks at its core, with rock age decreasing as you move horizontally away from the fold core.

Review Questions

Earthquakes Make Seismic Waves

1. P-waves and S-waves move through the Earth's interior in two ways. What is the difference in their mode of propagation?
2. Can S-waves travel through liquids? Explain.
3. Name the two types of surface waves and describe the motion of each.

Seismic Waves Show the Earth Has Layers

4. What was Andrija Mohorovičić's major contribution to geology?
5. List the different properties of the 4 types of seismic waves. How did seismic waves contribute to the discovery of the Earth's internal boundaries?
6. What does the wave shadow that develops 105–140° from the origin of an earthquake tell us about the Earth's composition?
7. What is the evidence for the solidity of the Earth's inner core?
8. Even though the inner and outer cores are both composed predominantly of iron and nickel, the inner core is solid and the outer core is liquid. Why?
9. What is the evidence that the Earth's outer core is liquid?
10. Describe the asthenosphere and the lithosphere. In what way are they different from each other?
11. What convectional movement is responsible for the motion of lithospheric plates?
12. How does continental crust differ from oceanic crust?
13. Why does continental crust float higher on the Earth's surface than oceanic crust?

Folds, Faults, and Earthquakes

14. What are folds?

15. Are folded rocks the result of compressional or tensional forces?
16. Distinguish between anticlines and synclines.
17. Two major types of faults are reverse faults and normal faults. What is the difference between them?
18. How are strike-slip faults different from normal or reverse faults?
19. What type of fault is associated with the 1964 earthquake in Alaska?
20. The Mercalli Scale measures earthquake intensity. The Richter Scale measures earthquake magnitude. Which scale is the more precise measurement? Why?

Exercises

1. Compare the relative speeds of primary and secondary seismic waves, and relate speeds of travel to the medium in which they travel.
2. Explain how seismic waves indicate whether regions within the Earth are solid or liquid.
3. How do seismic waves indicate layering of materials in the Earth's interior?
4. What is the evidence for the Earth's central core being solid?
5. What is the evidence for distinguishing a boundary at the base of the crust?
6. Speculate on why the lithosphere is rigid and the asthenosphere plastic, even though they are both part of the mantle.
7. If the Earth's mantle is composed of rock, how can we say that the crust floats on the mantle?
8. Is the lithosphere part of the mantle? Explain the relationships of the crust to the lithosphere and the lithosphere to the mantle.
9. Which extends farther into the mantle, the continental or the oceanic crust? Why?
10. How does erosion and wearing away of a mountain affect the depth to which the crust right below a mountain extends into the lithosphere?
11. Describe how the presence of faults and folds supports the idea that lithospheric plates are in motion.
12. What kind of force creates reverse faults? Where in the United States do we find evidence of reverse faults?
13. What kind of force creates normal faults? Where in the United States do we find evidence of normal faults?
14. Strike-slip faults show horizontal motion. Where in the United States do we find strike-slip faulting?
15. If you found folded beds of sedimentary rock in the field, what detail would you need to know in order to tell if the fold was an anticline or a syncline?
16. Does the fact that the mantle is beneath the crust necessarily mean that the mantle is denser than the crust? Explain.
17. The weight of ocean floor bearing down upon the lithosphere is increased by the weight of ocean water. Relative to the weight of the 10-km-thick basaltic ocean crust (specific gravity 3), how much weight does the 3-km-deep ocean (specific gravity 1) contribute? Express your answer as a percent of the crust's weight.
18. The Richter scale is logarithmic, meaning that each increase of 1 on the Richter scale corresponds to an increase of 10 in the amplitude of the seismic waves created by an earthquake. An earthquake that measures 8 on the Richter scale has how many times more ground-shaking ability than a quake that measures 6 on the Richter scale?
19. What is a very likely cause for the existence of the Earth's magnetic field?

Suggested Web Sites

http://pubs.usgs.gov/publications/text/inside.html

http://scign.jpl.nasa.gov/learn/plate1.htm

http://www.tinynet.com/faults.html

http://earth.leeds.ac.uk/faults/

Chapter 33: Our Restless Planet

Why do the great majority of earthquakes and volcanic eruptions occur in narrow zones, with very few occurring outside these zones? Does the answer have anything to do with the formation of mountains? Where does granite occur, and why? Where do we find basalt? How did the Atlantic Ocean form? In this chapter we apply the scientific method to investigate why rocks fold or break, and why earthquakes and volcanoes are found in certain locations.

33.1 The Theory of Continental Drift

Scientists of the early 20th century believed that oceans and continents were geographically fixed. They thought the surface of the planet is an unchanging, static skin spread over a molten, but gradually cooling interior. They believed that the cooling of the planet resulted in its contraction (the Shrinking Earth Theory). These early scientists assumed that shrinking caused the outer skin to contort and wrinkle into mountains and valleys.

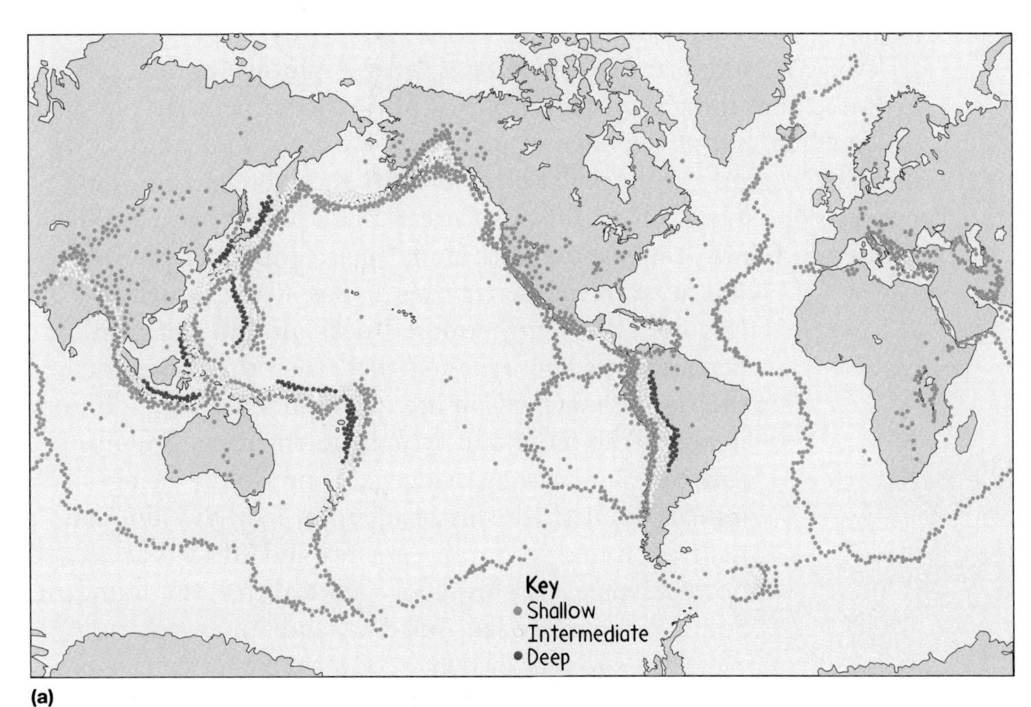

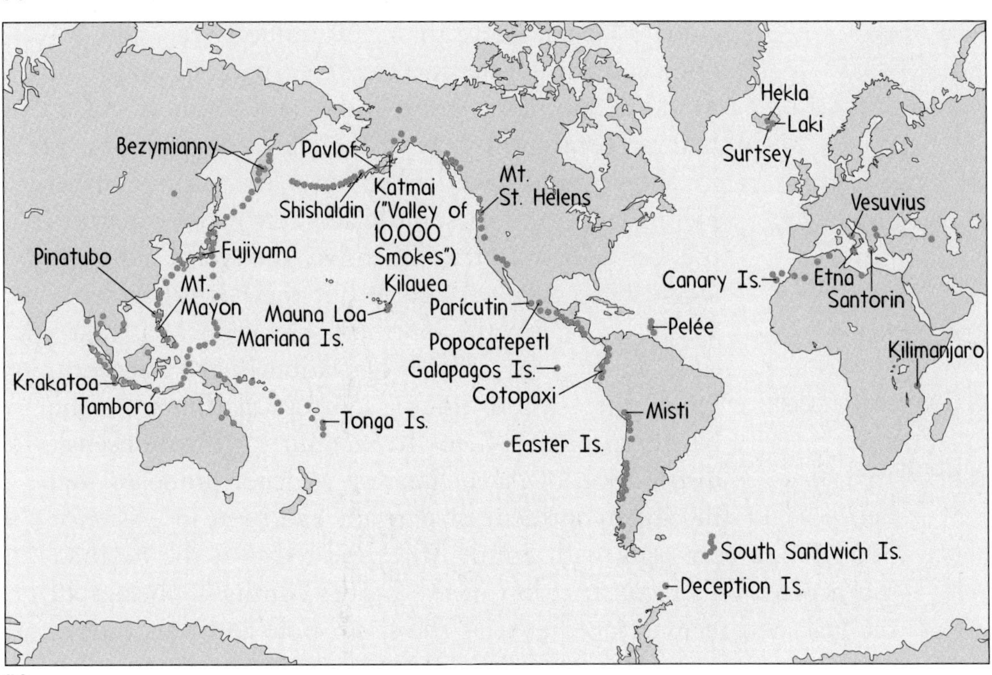

Figure 33.1
(a) The global distribution of earthquakes. (b) The global distribution of recent volcanic centers. (Source: NOAA)

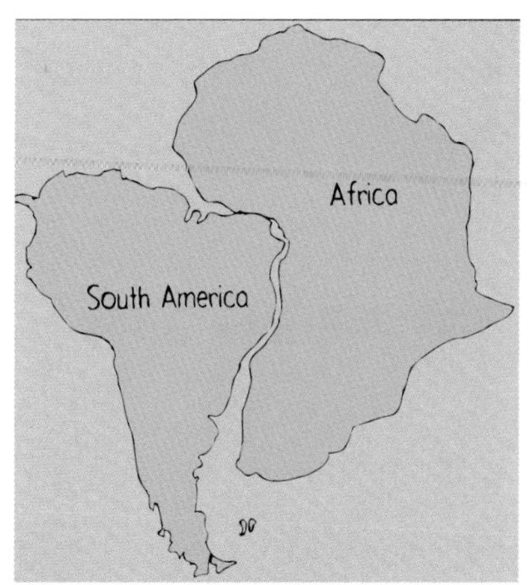

Figure 33.2
When you align the shorelines of South America and Africa, the continents fit together like pieces in a jigsaw puzzle.

Many people had noticed, however, that the eastern shorelines of South America and the western shoreline of Africa seemed to fit together like a jigsaw puzzle (Figure 33.2). An Earth scientist named Alfred Wegener took this observation seriously. Wegener saw the Earth as a dynamic planet with the continents in slow, but constant motion. He believed that all the continents had once been joined together into one great supercontinent he called **Pangaea** (pronounced Pan-gee-ah), meaning "all land". His hypothesis was that Pangaea had fractured into a number of large pieces, and that South America and Africa had indeed once been joined together as part of a larger land mass.

Wegener supported his hypothesis with impressive evidence. He proposed that the geological boundary of each continent lay not at its shoreline, but at the edge of its *continental shelf.* The continental shelf is the gently sloping platform between the shoreline and the steeper slope that leads to the deep ocean floor. When Wegener fit Africa and South America together along their continental shelves, the fit was even better than it was at the shorelines (Figure 33.3). Furthermore, rocks on different continents that are brought together when the continental shelves are matched up are nearly identical. In addition, many of the mountain systems in Africa and South America show strong evidence of previously being joined. Similarly, fossils of identical land-dwelling animals are found in South America and Africa but nowhere else. And fossils of identical trees are found in South America, India, Australia, and Antarctica.

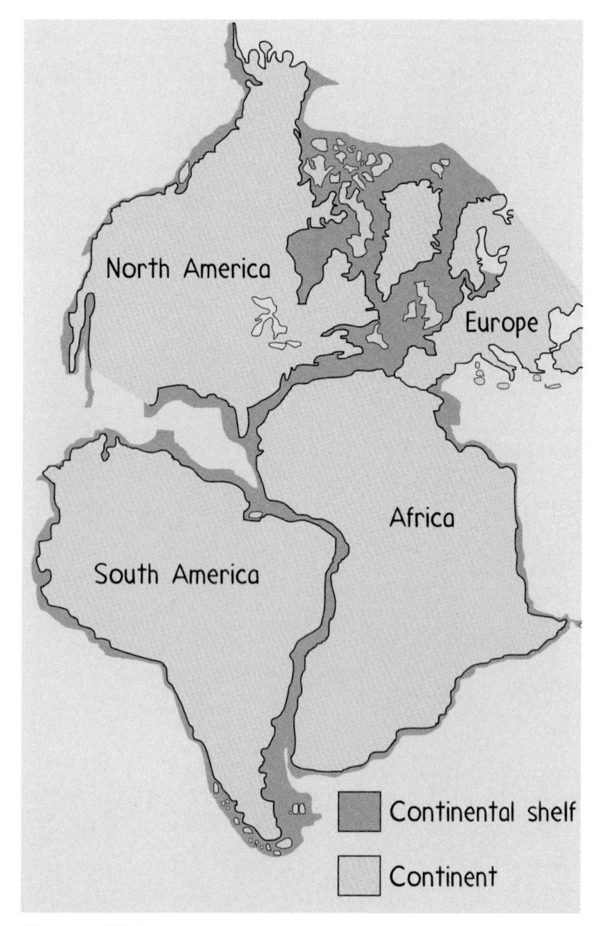

Figure 33.3
The jigsaw-puzzle fit between continents is even better at the continental shelves than at the shorelines.

Even stronger evidence for a supercontinent comes from paleoclimatic (*paleoclimatic* means "ancient climate") data. More than 300 million years ago, a huge continental ice sheet covered parts of South America, southern Africa, India, and southern Australia. If these continents were in their present positions, the ice sheet would have had to cover the entire Southern Hemisphere, and in some places, cross the equator! If the ice sheet were that extensive, the world climate would have been very cold. But there is no evidence of a cold climate in the Northern Hemisphere at that time. In fact, the time of glaciation in the Southern Hemisphere was a time of subtropical climate in the Northern Hemisphere. To account for this nonsensical distribution of paleoclimates, Wegener proposed that the supercontinent of Pangaea had been in existence 300 million years ago, with South Africa located over the Earth's south pole. This reconstruction of the Earth's continents brings all the glaciated regions together near the south pole and puts today's northern continents closer to the tropics.

Wegner was neither the first nor the last worthy investigator to be discredited by his fellow scientists. His idea was simply too far from the thinking of the day. Without a convincing explanation for *why* the plates move, the early twentieth-century scientific community was not ready to believe the moving-plate hypothesis. Recent discoveries now confirm Wegener's ideas. The concept of continental drift is very much accepted today.

A Scientific Revolution

One of the first key discoveries that support continental drift came from studies of the Earth's magnetic field. We know from Chapter 12 that the Earth is a huge magnet. Its magnetic north and south pole are near the geographic poles. Because certain minerals align themselves with the magnetic field when a rock is formed, many rocks have a preserved record of the changes in the Earth's magnetism over the expanse of geologic time. These changes include times when the magnetic north and south poles were reversed. This magnetism from the geologic past is known as **paleomagnetism.**

Three important bits of information are contained in the preserved magnetic record: (1) the polarity of the Earth's magnetic field at the time the rock was formed (whether the magnetic north pole corresponds to the geographic north pole), (2) the direction to the magnetic pole from the rock's location at the time the rock was formed, and (3) the magnetic latitude of the rock's location at the time the rock was formed. Once the magnetic latitude of a rock and the direction of the magnetic poles are known, the position of the magnetic pole at the time of formation can be determined. During the 1950s, a plot of the position of the magnetic north pole through time revealed that, over the past 500 million years, the position of the pole had wandered all over the world (Figure 33.4)! It seemed that either the magnetic poles migrated through time or the continents had drifted. Because the apparent path of polar movement was different from continent to continent, it was more reasonable that the continents had moved. Thus the hypothesis of continental drift was revived, but a mechanism to explain how the movement occurred was still missing.

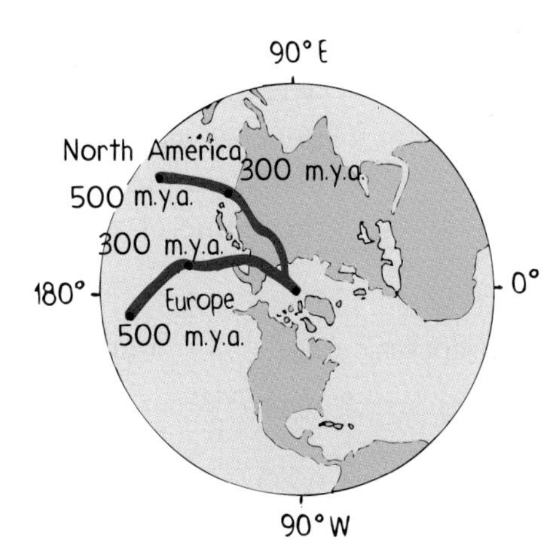

Figure 33.4
The path of the magnetic north pole during the last 500 million years. (The unit m.y.a. stands for "millions of years ago.") The lower red line is derived from evidence collected in Europe, and the upper red line is derived from evidence collected in North America. One would expect that these two lines would overlie each other. Thus either the magnetic pole wanders erratically, or the continents have moved. But how could the pole be in more than one place at the same time? This question strongly suggests that the continents have indeed moved relative to each other.

Figure 33.5
A detailed map of the ocean floor reveals enormous mountain ranges (brown regions) in the middle of the oceans and deep ocean trenches near some continental landmasses.

Like tape from a tape recorder, history of the ocean bottom is preserved in a magnetic record.

The 1950s were a time of extensive and detailed mapping of the ocean floors. Huge mountain ranges running down the middle of the Atlantic, Pacific, and Indian Oceans were discovered. Deep ocean trenches were found near some of the continental landmasses, particularly around the edges of the Pacific (Figure 33.5). It turned out that some of the deepest parts of the ocean are actually near some of the continents, and out in the middle of the oceans the water is relatively shallow because of the underwater mountains. Volcanic activity and lots of heat were discovered near the underwater mountains.

With this new information, H. H. Hess, an American geologist, presented the hypothesis of **seafloor spreading.** Hess proposed that the seafloor is not permanent. Instead, it is constantly being renewed. He theorized that the underwater mountains (mid-ocean ridges) are located above upwelling *convection cells* in the mantle. As rising material from the mantle oozes upward, new lithosphere is formed at the midoceanic ridge. The old lithosphere is simultaneously destroyed in the deep ocean trenches near the edges of continents. In a conveyor belt fashion, new lithosphere forms at a mid-ocean ridge (also known as a *spreading center*). Older lithosphere is pushed from the ridge crest to

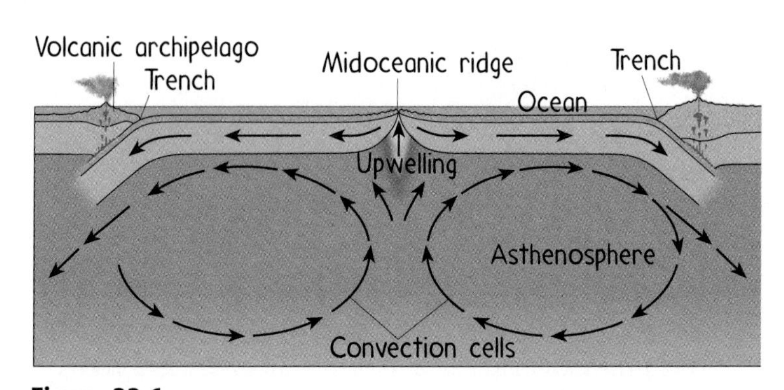

Figure 33.6
In conveyor-belt fashion, new lithosphere is formed at the mid-ocean ridges (also called "spreading centers") as old lithosphere is recycled back into the asthenosphere at a deep ocean trench.

be eventually recycled back into the mantle at a deep ocean trench (Figure 33.6).

Support for this theory came from paleomagnetic studies of the ocean floor. As new basalt is extruded at a mid-ocean ridge, it is magnetized according to the existing magnetic field. The magnetic surveys obtained from mapping the ocean's floor showed alternating strips of normal and reversed polarity, paralleling either side of the ridge areas (Figure 33.7). As in a very slow magnetic tape recording, the magnetic history of the Earth is recorded in the spreading ocean floors. In this way, we have a continuous record of the movement of the seafloors. Since the dates of pole reversals can be independently determined, the magnetic pattern of the spreading seafloor tells us both the age of the seafloor and the rate at which it spreads. The oceanic crust was found to be thin and young near the central ridge region and progressively thicker and older away from the ridge.

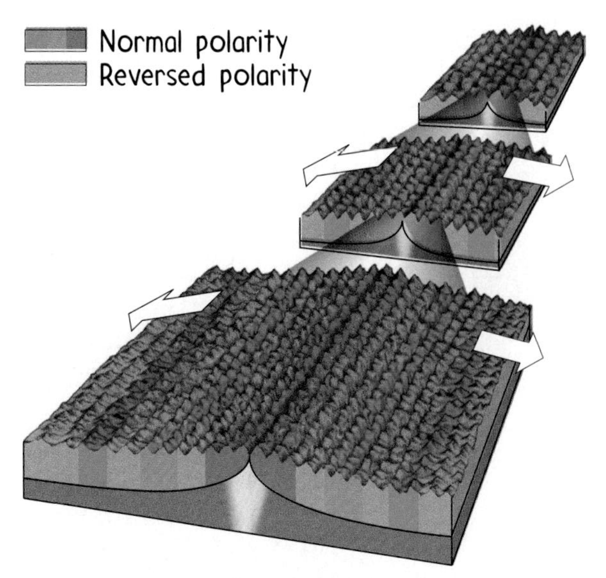

Figure 33.7
As new material is extruded at an oceanic ridge (spreading center), it is magnetized according to the existing magnetic field. Magnetic surveys show alternating strips of normal and reversed polarity paralleling both sides of the rift area. Like a very slow magnetic tape recording, the magnetic history of the Earth is thus recorded in the spreading ocean floors.

The theory of seafloor spreading provided a mechanism for continental drift. The time was right for the revolutionary concepts to follow. The tide of scientific opinion had indeed switched in favor of an ever-changing Earth.

Concept Check ✓

Why was Wegener's theory of continental drift not taken more seriously in the early part of this century?

Check Your Answer Wegener failed to produce a suitable mechanism that could cause the continents to move. Even if he had proposed the role of the convective interior, though, we can only guess how quickly the scientific community would have accepted his hypothesis. Scientists, like all other human beings, tend to identify with the ideas of their time. Do advances in knowledge, scientific or otherwise, occur because they are accepted by the status quo or because holders of the status quo eventually die off? Knowledge that is radical and unacceptable to the old guard is often easily accepted by newcomers who use it to push the knowledge frontier further. Hooray for the young (and the young-at-heart)!

33.2 The Theory of Plate Tectonics

Wegner's ideas have evolved to the present *theory of plate tectonics.* This theory describes the forces within the Earth that create the continents, ocean basins, mountain ranges, earthquake belts, and other large-scale features of the Earth's surface. The **theory of plate tectonics** states that the Earth's outer shell, the lithosphere, is divided into eight relatively large plates and a number of smaller ones

(Figure 33.8). These lithospheric plates ride atop the plastic asthenosphere below. Because each plate moves as a single unit in relation to the other plates, the interiors of the plates are generally stable geologically. All major interactions between plates occur along the plate boundaries. Thus most of the Earth's earthquakes, volcanoes, and mountains occur along these active margins. In fact, the creation and destruction of lithosphere described in Figure 33.6 takes place at such margins.

33.3 There Are Three Types of Plate Boundaries

Moving plates interact at their boundaries. Two types of plate boundaries mark the edges of lithospheric formation and destruction. The third type of plate boundary simply accommodates plate movement (with no formation or destruction of lithosphere).

Hot Spots and Lasers—A Measurement of Tectonic Plate Motion

Motion is relative. Whenever we discuss the motion of something, all we describe is its motion relative to something else. We call this place from which motion is observed and measured a *reference frame.* Living in a world where everything is in motion, how can we measure rates of plate motion? What do we choose for our reference frame?

The Canadian geophysicist J. Tuzo Wilson suggested that one measure of Pacific seafloor movement could be found in the ages of the volcanic Hawaiian Islands. Wilson postulated that the islands are the tips of huge volcanoes that formed as the Pacific seafloor moved over a fixed *hot spot*—a magma source rising from the Earth's interior. As we learned in Chapters 10 and 19, radioactive decay in the Earth's interior keeps the Earth hot. The concept of a concentrated or fixed hot spot provides a stationary reference point on the surface of the Earth against which plate motion can be determined. There are about 100 such hot spots around the world, and they are used to determine rates of plate movements.

There is, however, some debate as to whether the hot spots are truly stationary. A fascinating and more precise way to measure Earth movement is by a laser beam reflecting off "mirrors" in outer space. Lasers are used to detect the broad movements of tectonic plates. Laser pulses beamed to the reference point in space from a pair of ground stations located on opposite sides of a plate boundary or a fault start timers that run until the reflected pulses are received back at the stations. A computer combines this elapsed time with the known position of the reference point to determine the exact position of the ground station. Any movement of the ground station is registered as a change in the elapsed time.

Measurements can also be done with radio signals. Radio telescopes are keyed to a reference point in outer space (a quasar or a satellite) from which relative positions of points on Earth can be plotted. Local fault movement can be measured by placing the mirror on the opposite side of the fault from the laser, instead of in outer space. Neat-O!

Quasar

Radio telescopes

Longer path than before as measured by different arrival times of quasar radio signals

(a)

Mirror

Laser

Longer path as measured by difference in arrival time for reflected light laser

(b)

Earth movements are measured by radio telescopes or lasers keyed to a stationary reference point. (a) Broad movements are detected using satellites and quasars as reference points. (b) Fault movements are measured by laser beams shot from opposite sides of a fault. The laser flashes a beam off a reflector; the bounce-back time is recorded. Because of light's constant speed, the bounce-back time changes with Earth movement.

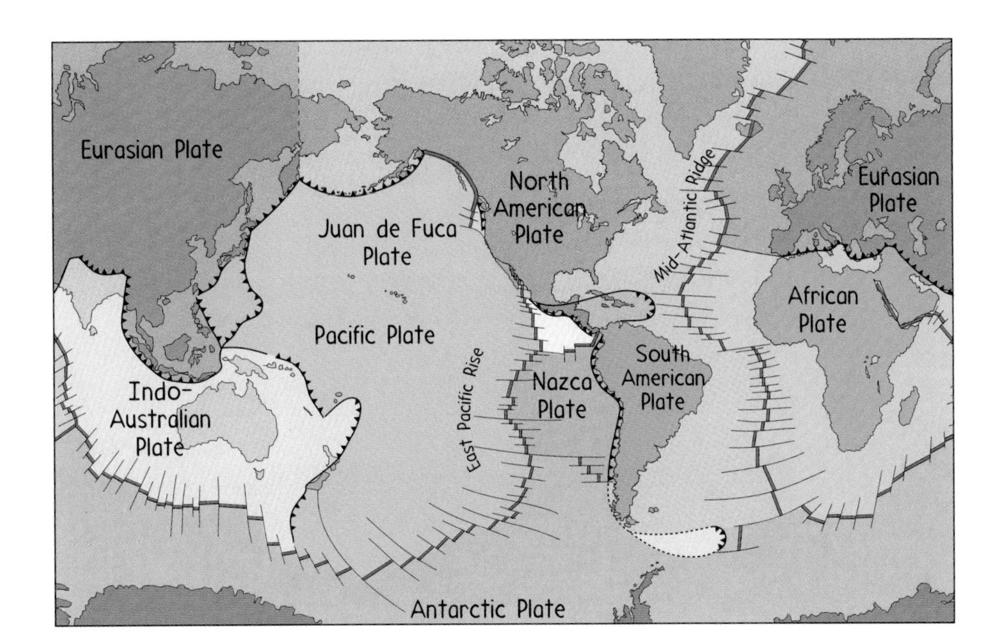

Figure 33.8
The lithosphere is divided into eight large plates and a number of smaller ones.

Divergent Plate Boundaries

Where two plates are moving apart, as in Figure 33.9a, tension stretches the lithosphere and creates a spreading center. Hot, molten rock from the Earth's asthenosphere buoyantly rises up to form new lithosphere. The lithosphere near the spreading edge is thin and has a relatively low density due to the heat and expansion of the rising magma. As it moves away from the spreading center, the new lithosphere cools, contracts, and becomes denser.

The Mid-Atlantic Ridge is a spreading center that has been producing the Atlantic Ocean floor for 160 million years. The spreading is accompanied by almost continuous earthquake activity. The earthquakes are not noticed because they are mild and harmless to humans. Continents on opposite sides of the ridge drift apart as the ocean floor grows. The lithosphere is younger near the central ridge region and progressively older away from the ridge. The spreading rate has been slow, about 2 centimeters each year (about as fast as fingernails grow). This rate over 160 million years adds up to 3200 kilometers—the width of the Atlantic Ocean. So, it has taken only 160 million years for a fracture in an ancient continent to turn into the Atlantic Ocean!

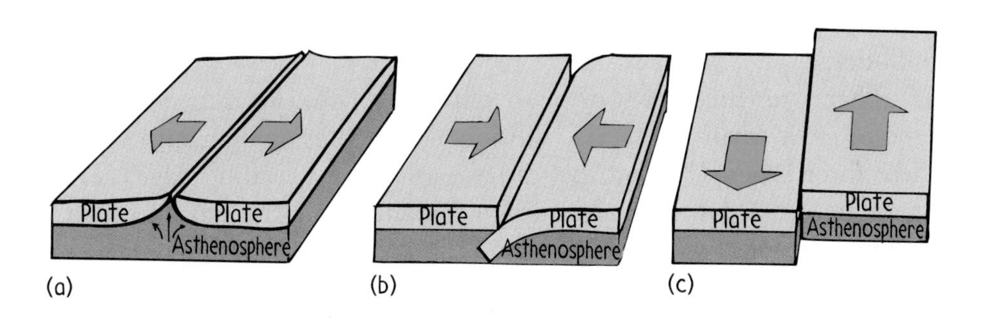

Figure 33.9
Plate boundaries are often sites of lithospheric formation and destruction. Named for the movement they accommodate, three types of plate boundaries are (a) divergent, (b) convergent, and (c) transform-fault boundaries.

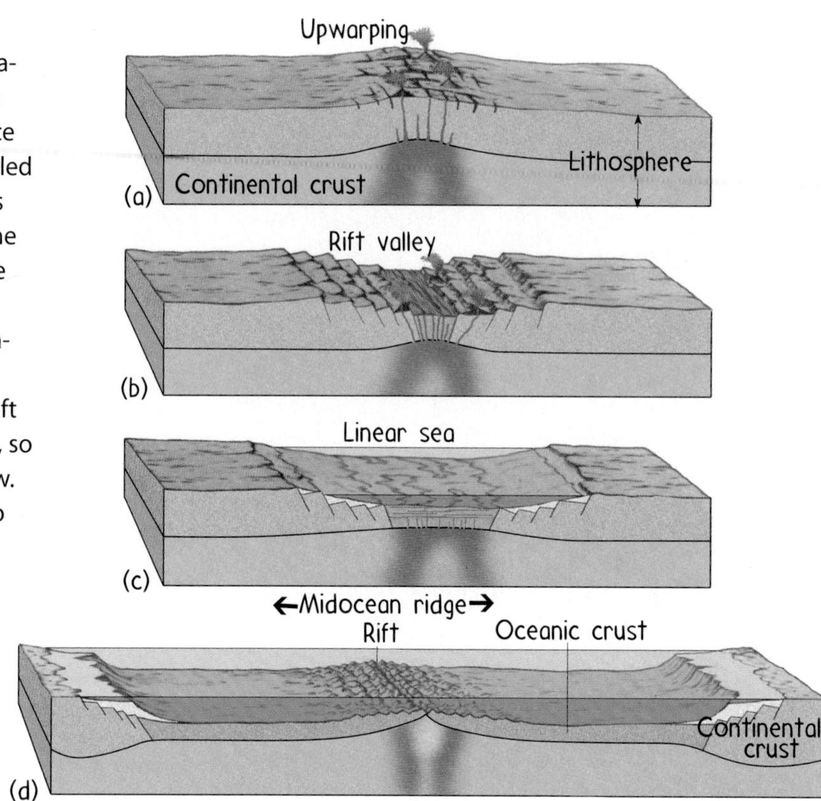

Figure 33.10
Formation of a rift valley and its transformation into an ocean basin. (a) Rising magma uplifts continental crust, causing the surface to crack. (b) Rift valley forms as crust is pulled apart. It is in this stage that we find Africa's Great Rift Valley today. (The two sides of the valley move away from each other because they happen to be located above mantle convection cells that have the same circulation pattern as the cells in Figure 33.6.) (c) Water from the ocean drains in as the rift drops below sea level, forming a linear sea, so called because it is usually long and narrow. (d) Over millions of years, the rift widens to become an ocean basin.

Spreading centers are not restricted to the ocean floors. They also develop on land. Hot, molten material in the Earth's interior rising beneath continental landmasses causes the Earth's crust to bend upward (this is called *upwarping*). Gaps in the crust are produced, and large slabs of rock slide and sink down into these gaps. The large down-faulted valleys generated by this process are called either **rifts** or **rift valleys** (Figure 33.10). The Great Rift Valley of East Africa is an excellent example of such a feature. If spreading continues there, it may be the beginning of a new ocean basin.

Convergent Plate Boundaries

Convergent boundaries occur where plates come together. When motion due to convection cells pushes two plates toward each other, one of two things may happen. Compression either pushes the lithosphere of one plate downward under the other plate. Or else the compression of the crust shortens the lithosphere by folding and faulting. The regions of plate collisions are regions of great mountain building.

How fascinating to see a time-lapse view of mountain building!

There are three types of plate collisions. They are classified by the type of crust involved in the collision. Plate collisions fall into one of these three types: (1) Both plates have an oceanic leading edge, (2) one plate has a continental leading edge and the other has an oceanic leading edge, and (3) both plates have a continental leading edge (Figure 33.11).

Oceanic-Oceanic Convergence Collision, or *convergence,* between two oceanic plates results in one plate bending and descending beneath the other. This is *subduction,* which forms a deep ocean trench. The deepest trench known is the Marianas Trench in the western Pacific Ocean. Here, the seafloor is as much as 11 kilometers below sea level. Partial melting of the subducted oceanic crust produces andesitic magma* that buoyantly migrates upward, eventually erupting onto the ocean floor to produce volcanoes that create a series of islands called a volcanic *island arc.* The size and elevation of the island arc increase over time because of continued volcanic activity. Such island arcs have formed the Aleutian, Marianas, and Tonga Islands and the island arc systems of the Alaskan Peninsula, the Philippines, and Japan.

Deep ocean trenches mark the active subduction zones that border island arcs. Earthquakes occur along the subduction zones as the downgoing plate grinds against the overriding plate. The earthquakes get deeper and deeper in the direction of subduction (Figure 33.12).

Oceanic-Continental Convergence When an oceanic plate and a continental plate come together (converge), the denser oceanic plate is subducted beneath the less dense continental plate, and a deep ocean trench is formed. The downward moving oceanic crust carries a mixture of oceanic basaltic

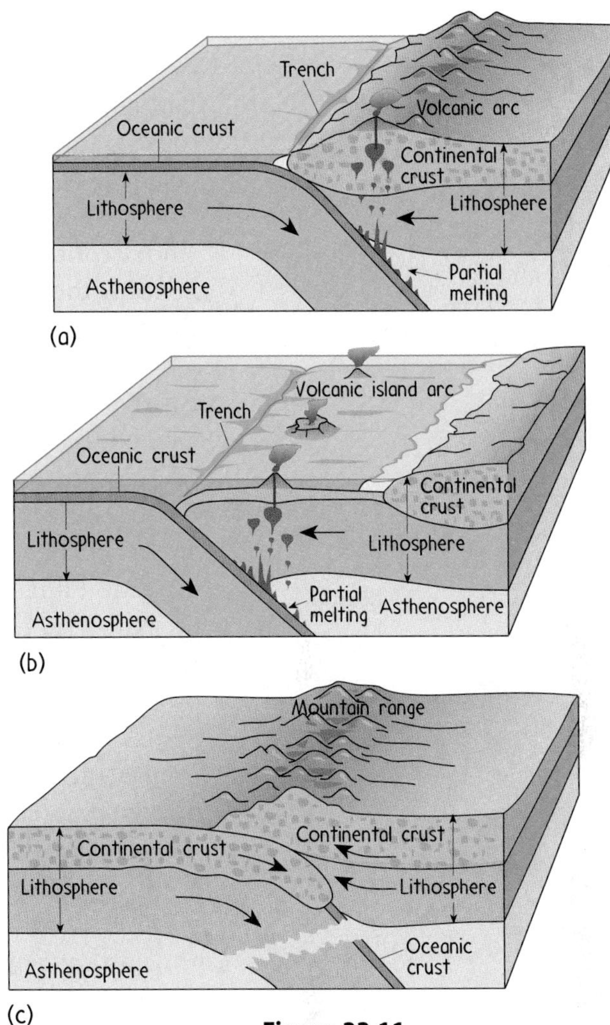

Figure 33.11
The three types of convergent margins: (a) oceanic-oceanic, b) oceanic-continental, and (c) continental-continental.

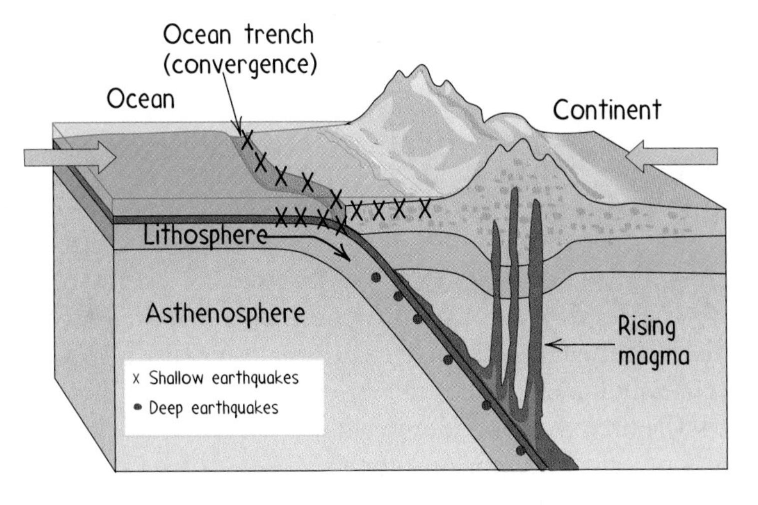

Figure 33.12
Earthquakes at a subduction zone get deeper and deeper in the direction of subduction.

* Recall from Chapter 30 that andesitic rock, or magma, has a higher silicon content than basaltic rock (or magma). Silicon is relatively light, so, the higher the silicon content the lower the density of the rock, or magma.

rock and sediment eroded from the overriding continental plate. Andesitic magma is produced and rises upward as the oceanic crust partially melts. Some of the magma crystallizes below the surface to form granite, and some reaches the surface where it erupts to form a chain of volcanic mountains on the overriding continental plate. The Andes mountains of western South America formed in this way. The Andes continue to grow higher as subduction of the Nazca Plate beneath the South American Plate causes sediments accumulated on the Nazca Plate to be scraped off onto the granitic roots of the Andes. This scraped material becomes permanently attached to the South American Plate and adds thickness and buoyancy to the mountains. The additional thickness and buoyancy causes the Andes mountains to rise upward more rapidly than they are eroded by wind and rain.

In the western United States, examples of such volcanic activity are found in the Sierra Nevada, an ancient volcanic range, and the Cascade Range, which is currently active. The Sierra Nevada was produced by subduction of the ancient Farallon Plate beneath the North American Plate. The Sierra Nevada batholith is a remnant of the original volcanic range, while the California Coast Range has remnants of the sediments that accumulated in the trench. The Cascade Range, produced from the subduction of the Juan de Fuca Plate (a piece of the Farallon Plate) beneath the North American Plate, includes the volcanoes Mounts Rainier, Shasta, and St. Helens. The eruption of Mount St. Helens gives testimony that the Cascade Range is still quite active.

Earthquakes, similar to those in oceanic-oceanic convergence, are also characteristic of oceanic-continental convergence.

Concept Check ✓

Erosion wears mountains down, and yet the Andes mountains grow taller each year. Why?

Check Your Answer Subduction is still occurring which causes the uplift of the Andes mountains. Because the rate of uplift is greater than the erosion rate, the mountains continue to grow.

Continental-Continental Convergence The collision between two continental landmasses is always preceded by oceanic-continental convergence. Since continental crust is light and buoyant, it doesn't undergo significant subduction. The result is a head-on collision of continental plates (Figure 33.11c). Compression causes the plates to break and fold upon each other, making the crust very thick. The region where the plates meet is defined by intensely compressed and metamorphosed rock. In contrast with convergence of oceanic plates or one continental and one oceanic plate, volcanic activity is not characteristic of continental-continental collisions. But earthquakes are!

The collision between continental plates has produced some of the most famous mountain ranges. One majestic example is the snow-capped Himalayas, the highest mountain range in the world. This chain of towering peaks is still being thrust upward as India continues crunching against Asia (Figure 33.13). The European Alps were formed in a similar fashion when part of the African Plate collided with the Eurasian Plate 80 million years ago. Relentless pressure between the two plates continues and is slowly closing up the Mediterranean Sea. In North America, the Appalachian Mountains were produced from a continental-continental collision that ultimately resulted in the formation of the supercontinent Pangaea.

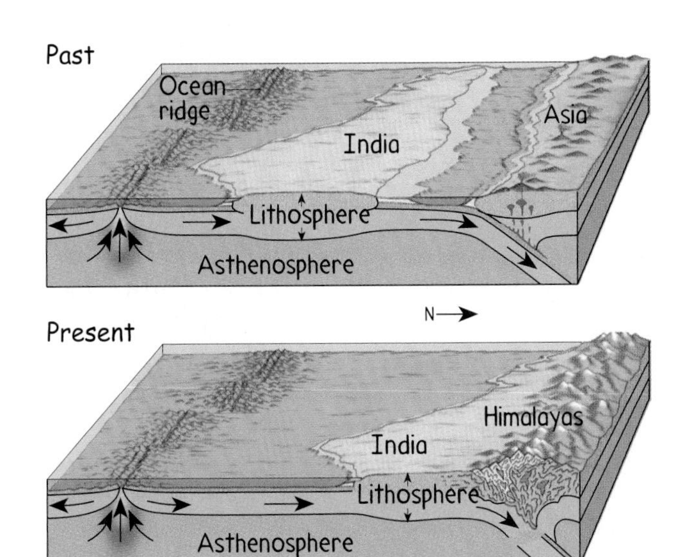

Figure 33.13
The continent-to-continent collision of India with Asia produced—and is still producing—the Himalayas.

Transform-Fault Plate Boundaries

A **transform fault** is a plate boundary that occurs where two plates are neither colliding nor pulling apart. Rather, they are sliding horizontally past each other (Figure 33.14). The fault "transforms" the motion from one ridge segment to the other. Because there is no tension or compression between the plates, there is no creation or destruction of the lithosphere. A transform fault is a zone of horizontal accommodation of plate movement, with neither side of the fault moving up or down relative to the other. Transform faults have the same type of motion as *strike-slip faults*.

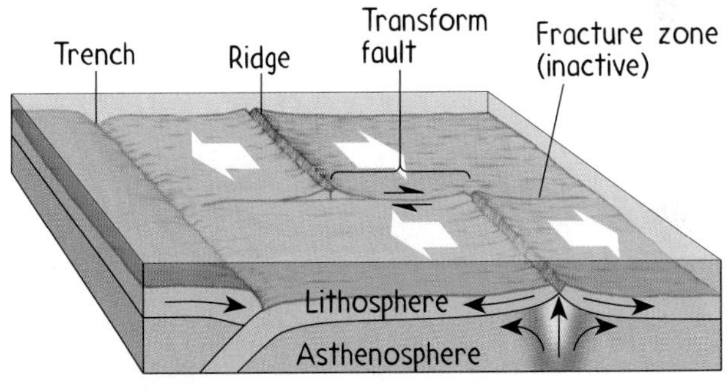

Figure 33.14
Transform faults allow two plates to slide past one another where ridge segments are offset.

The San Andreas Fault is one of the most famous transform faults. It stretches for 1500 kilometers from Cape Mendocino in northern California to the East Pacific Rise in the Gulf of California. The Pacific Plate is moving northwest at a rate of about 5.0 centimeters per year relative to the North American Plate. The San Andreas Fault accommodates about 70% of this motion, or about 3.5 centimeters per year. The rest of the motion occurs along other faults (such as the Hayward Fault). Grinding and crushing take place as the two plates move past each other. When sections of the plates become locked together, stress builds up until it is relieved in the form of an earthquake. On April 18, 1906, the Pacific Plate lurched about 6 meters northward over a 434-kilometer stretch of the fault, releasing the built-up stress, and causing the catastrophic San Francisco earthquake.

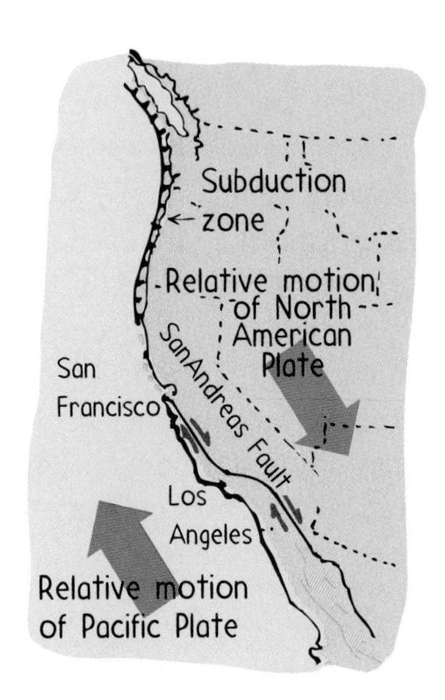

Figure 33.15
The San Andreas Fault shows horizontal offset, where the Pacific Plate slides northward past the adjacent North American Plate.

33.4 The Theory That Explains Much

Before the theory of plate tectonics was proposed, processes such as mountain building, folding, and faulting were poorly understood. Plate tectonics offers explanations as to the where and why of many geologic processes.

Why are the Appalachian Mountains located where they are? What about the Sierra Nevada? The Rocky Mountains? The Alps? The plate tectonic model gives an answer: All the mountain-building events take place near convergent plate boundaries.

Calculation Corner

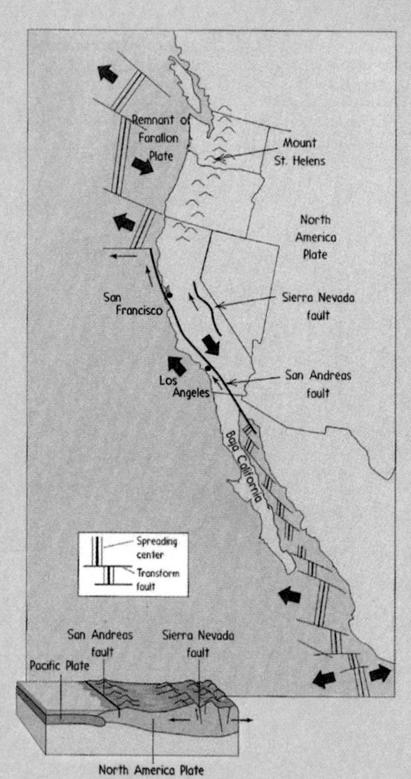

Moving Faults

Knowing the rate of movement along a fault, it's easy to calculate the amount of *offset* over a period of time. Movement along the San Andreas Fault is about 3.5 centimeters per year. If a fence was built across the fault in 1990, we can figure out how far apart the two sections of the now-broken fence will be in 2005. The period of time is

$$2005 - 1990 = 15 \text{ years.}$$

Thus the two parts of the fence will be separated by

$$\frac{3.5 \text{ cm}}{\cancel{\text{yr}}} \times 15 \cancel{\text{yr}} = 52.5 \text{ cm}$$

If you want to know how many feet that is, we can convert by knowing that there are 2.54 centimeters in an inch and 12 inches in a foot.

$$52.5 \cancel{\text{cm}} \times \frac{1 \cancel{\text{in}}}{2.54 \cancel{\text{cm}}} \times \frac{1 \text{ ft}}{12 \cancel{\text{in}}} = 1.7224 \text{ ft}$$

which rounds off to 1.7 feet.

Example

Two trees were planted on opposite sides of the San Andreas Fault in 1907. How far apart were they in 2001?

Answer

The period of time is 2001 − 1907 = 94 years.

$$94 \cancel{\text{yr}} \times \frac{3.5 \text{ cm}}{\cancel{\text{yr}}} = 329 \text{ cm}$$

Your Turn

1. The distance between Los Angeles and San Francisco is 600 kilometers. What is the distance between these two cities in centimeters?
2. The San Andreas Fault separates the northwest moving Pacific Plate, on which Los Angeles sits, from the North American Plate, on which San Francisco sits. How long will it take the two cities to form one large city?

We can also relate the formation of the three different rock types to the plate tectonic theory. All rocks are tied to plate interaction in some manner or another.

The intense heat and pressure caused by subduction and continental collisions result in the metamorphism of preexisting rocks. Here is where regional metamorphism occurs. As the subducting slab is heated, it eventually begins to melt. Recall from Chapter 30 that when rock melts, the liquid contains more silicon than the parent rock (partial melting). Thus andesite forms from basaltic material (subducted ocean floor), and we find belts of composite cone volcanoes, like those in the Andes Mountains and the Cascade Range.

What about granite? Where does most of that come from? As the ocean floor melts, it can produce large volumes of andesitic magma. This magma doesn't all erupt at once, but accumulates in the Earth's crust. As the magma bodies cool, they undergo crystallization, resulting in a liquid containing more silicon than the original. When this silicon-enriched magma cools, it forms granite. Where in nature has this occurred? The Sierra Nevada is a largely granitic mountain range in California that formed in such a manner. The large batholiths of granitic rocks are the "roots," or solidified magma bodies, from a once-extensive volcanic belt formed as a result of subduction.

We can also explain the creation of basaltic rocks using the plate tectonic model. Where plates are diverging, such as at a mid-ocean ridge, mantle rocks partially melt to form new lithosphere that is capped by oceanic crust—basalt.

What about sedimentary rocks? As mountains grow as a result of plate collisions, they also begin to weather and erode. The clastic sediments produced are transported downslope. There, they accumulate, layer upon layer, and eventually become sedimentary rock.

Virtually all earthquake and volcanic activity can be tied directly to plate tectonics. These energetic responses to plate interactions are almost always found where plates interact. Earthquakes are found at all types of plate boundaries, and volcanoes are concentrated where plates either collide or pull apart.

So we see that the tectonic interaction between lithospheric plates, which occurs mostly at their boundaries, explains the origin of mountain chains, the development and destruction of the ocean floors, the three types of rocks found on Earth, and the global distribution of earthquakes and volcanoes. The internal motions that change the Earth's surface do so in a cycle. In Chapter 36, we shall see the effects of plate tectonic interaction through time. The study of geology uses processes that occur today to understand what may have occurred in the past. This concept is commonly stated as "the present is the key to the past." But what has happened in the past provides clues as to what may happen in the future. The Earth is indeed a dynamic planet.

Chapter Review

Key Terms and Matching Definitions

_____ Alfred Wegener
_____ convergent boundary
_____ divergent boundary
_____ H.H. Hess
_____ Pangaea
_____ paleomagnetism
_____ seafloor spreading
_____ transform fault boundary

1. The study of natural magnetization in a rock that is used to determine the intensity and direction of the Earth's magnetic field at the time of the rock's formation.
2. Earth scientist who hypothesized that the Earth's continents moved in slow, but constant motion. He believed that all the continents had, at one time, been joined together into one supercontinent.
3. A single, large landmass that existed in the geologic past and was composed of all the present-day continents.
4. Geologist who proposed that the seafloor is constantly being renewed; new lithosphere is created at spreading centers and old lithosphere is destroyed at subduction zones.
5. Plate boundary where two plates are moving apart from one another.
6. The moving apart of two oceanic plates to form a rift in the seafloor.
7. Plate boundary where two plates are being pushed toward one other.
8. A plate boundary formed by two plates that are sliding horizontally past each other.

Review Questions

The Theory of Continental Drift

1. What key evidence did Alfred Wegener use to support his idea of continental drift?
2. How does evidence of prehistoric glaciation found in parts of South America, southern Africa, India, and southern Australia support the concept of a supercontinent?
3. What was the stated reason for the scientific community rejecting Wegener's idea of Continental Drift?
4. What information can be learned from a rock's magnetic record?
5. What role did paleomagnetism play in supporting continental drift?
6. Where are the deepest parts of the ocean?
7. What major discovery at the bottom of the ocean did H. H. Hess make?
8. How is the ocean floor similar to a gigantic slow-moving tape recorder?
9. What does the Earth's crust have in common with a conveyor belt?
10. In what way does seafloor spreading support continental drift?

The Theory of Plate Tectonics

11. Why does most geologic activity occur near plate boundaries?
12. What are hot spots? How are they useful for understanding plate movements?

There Are Three Types of Plate Boundaries

13. Describe the three types of plate boundaries.
14. What is a rift?
15. What kind of boundary separates the South American Plate from the African Plate?
16. Describe the three types of plate collisions that occur at convergent boundaries.
17. The Appalachian Mountains were produced at what type of plate boundary?
18. What is a transform fault?
19. What kind of plate boundary separates the North American Plate from the Pacific Plate?

The Theory That Explains Much

20. At what type of plate boundary is regional metamorphism found?

21. Do you think it is likely to find andesitic rocks close to regionally metamorphosed rocks, or should these two rock types be widely separated? Why?

Exercises

1. Describe how the different paths of polar wandering helped establish that continents move over geologic time.
2. Why are most earthquakes generated near plate boundaries?
3. Why do mountains tend to form in long narrow ranges?
4. Relate the formation of metamorphic rocks to plate tectonics. Would you expect to find metamorphic rocks at all three types of plate boundaries? Why or why not?
5. Cite one line of evidence that suggests subduction once occurred off the coast of California.
6. Distinguish between continental drift and plate tectonics.
7. Are the present ocean basins a permanent feature on our planet? Why or why not?
8. Are the present continents a permanent feature on our planet? Discuss why or why not.
9. Why is it that the most ancient rocks are found on the continents, and not on the ocean floor?
10. Upon crystallization, certain minerals (the most important being magnetite) align themselves in the direction of the surrounding magnetic field, providing a magnetic fossil imprint. How does the Earth's magnetic record support the theory of continental drift?
11. How is the theory of seafloor spreading supported by paleomagnetic data?
12. What kind of boundaries are associated with seafloor spreading centers?
13. What is meant by magnetic pole reversals? What useful information do they tell us about the Earth's history?
14. Using a photocopy of Figure 33.8, mark the different boundaries of plate interaction. Draw arrows showing direction of plate movement for convergent, divergent and transform fault boundaries.
15. Earthquakes, the result of sudden motion in the Earth caused by abrupt release of slowly accumulated stress, cause rocks to fracture or fault. Relate such faulting to horizontal movement of plates. Where does this type of movement occur?
16. Lithospheric material is continuously created and destroyed. Where does this creation and destruction take place? Do the rates of the two processes balance each other?
17. Subduction is the process of one lithospheric plate descending beneath another. Why does the oceanic portion of the lithosphere undergo subduction while the continental portion does not?
18. What geologic features are explained by plate tectonics?
19. In 1964, a large tsunami struck the Hawaiian Islands without warning, devastating the coastal town of Hilo, Hawaii. Since that time a tsunami warning station has been established for the coastal areas of the Pacific. Why do you think these stations are located around the Pacific rim?
20. Where is the world's longest mountain range located?
21. If the mid-Atlantic ocean is spreading at 2 cm per year, how many years has it taken for it to reach its present width of 5000 km?

Suggested Readings and Web Sites

McPhee, John. *Assembling California.* New York: Farrar, Strauss & Giroux, 1993.

http://wrgis.wr.usgs.gov/docs/usgsnps/pltec/pltec1.html

http://www.ucmp.berkeley.edu/geology/tectonics.html

http://greenwood.cr.usgs.gov/pub/open-file-reports/ofr-99-0132/

http://kids.earth.nasa.gov/archive/pangaea/

Chapter 34: Water on Our World

Where is most of the Earth's water? Are there such things as underground rivers? How does water move underground? Why do some rivers move slowly, while others have lots of rapids? What kinds of water pollution threaten our water resources? And what can be done to save our planet's most precious resource?

This chapter begins with a look at the Earth's fresh water supply—underground water, river and lake water, and water frozen in glaciers and ice caps. We end the chapter with an investigation of the Earth's oceans. If you "thirst" for knowledge about water, read on!

34.1 The Hydrologic Cycle

Have you ever witnessed the power of a waterfall? Or been caught in a sudden rainstorm? Have you ever been sailing on a lake? Or gone fishing in a rushing stream? If so, it may seem to you that the supply of fresh water on the Earth is unlimited. The Earth's population, however, has grown to more than 6 billion people. For such a large population, fresh water is in fact a limited resource.

A view of the Earth from space shows our planet to be a vast expanse of water interrupted here and there by island-like continents. Indeed, about 70% of the Earth's surface is covered with water. Slightly more than 97% of all the Earth's water is in the oceans. Just over 2% is frozen in the polar icecaps and glaciers. The remainder of the Earth's water, less than 1%, consists of water vapor in the atmosphere, water in the ground, and water in rivers and lakes.

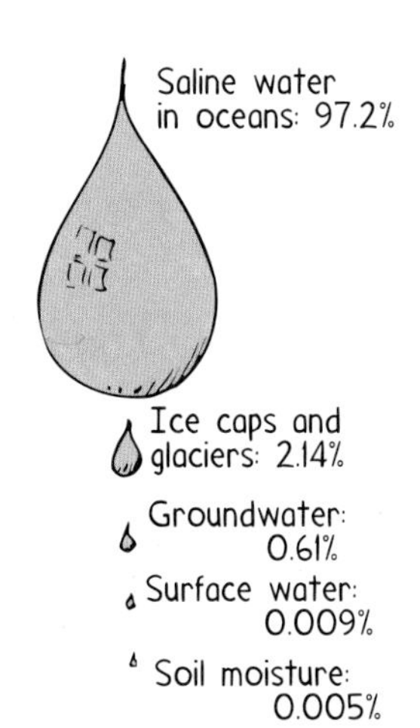

Figure 34.1
Distribution of the Earth's water supply.

Water on the Earth is constantly circulating, driven by the heat of the sun and the force of gravity. As the sun's energy evaporates ocean water, a cycle begins (Figure 34.2). Evaporation moves water molecules from the Earth's surface to the atmosphere. The resulting moist air may be transported great distances by wind. Some of the water molecules condense to form clouds and then precipitate as rain or snow. The total amount of water vapor in the atmosphere remains relatively constant. Therefore, evaporation and precipitation balance each other.

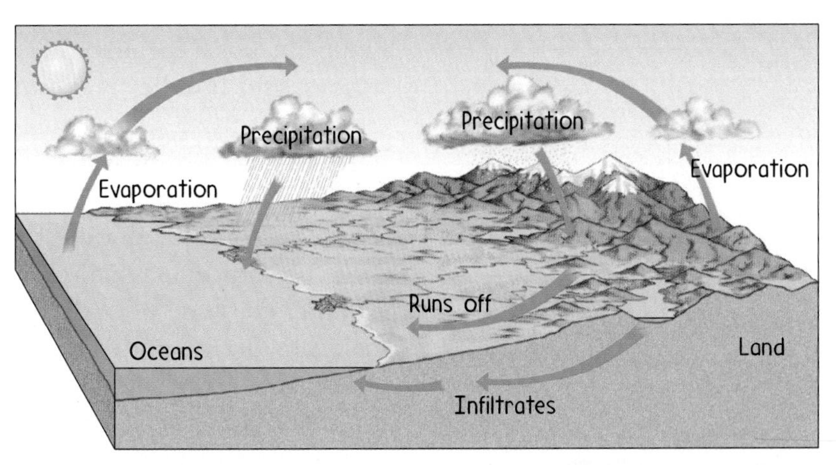

Figure 34.2
The hydrologic cycle. Water evaporated at the Earth's surface enters the atmosphere as water vapor, condenses into clouds, precipitates as rain or snow, and falls back to the surface, only to evaporate again and go through the cycle yet another time.

If precipitation falls on the ocean, the cycle is complete—from ocean back to ocean. A longer cycle occurs when precipitation falls on land, for water may drain to streams, then to rivers, and then journey back into the ocean. Or it may soak deep into the ground, or evaporate back into the atmosphere before reaching the ocean. Also, water falling on land may become part of a snow pack or glacier. Although snow or ice may lock water up for many years, it eventually melts or evaporates, returning the water to the cycle. This natural circulation of water from the oceans to the air, to the ground, then to the oceans and then back to the atmosphere is called the **hydrologic cycle.***

* This key concept is another conservation principle, of the type we first saw in Chapters 5 and 6 when we studied conservation of momentum and energy. In Chapter 11, we saw it as conservation of electric charge; and in Chapter 21, as conservation of nucleons. So now we learn that the amount of water on the Earth is conserved. A lack of it in one place means an abundance someplace else, which in most instances is the ocean.

The rain or snow that falls on the continents is the Earth's only natural supply of fresh water. More than three-quarters of the Earth's fresh water is locked up in the polar icecaps and glaciers. It is surprising to note that the rest of the fresh water is not only in lakes and rivers but is mostly beneath the ground. As rain falls and sinks into the ground, it flows downward to fill the open pore spaces between sediment grains. This water is **groundwater.**

Concept Check ✓

What percent of the Earth's water supply is fresh water?

Check Your Answer Less than 3%, as you can see by adding the freshwater values in Figure 34.1: 2.14% (glaciers) + 0.61% (groundwater) + 0.009% (surface water) + 0.005% (soil moisture) = 2.764%.

34.2 Water Below the Surface

A pond or lake is simply a place where the land surface dips below the water table.

Except in polar regions, the water in lakes, ponds, rivers, streams, springs, and puddles is the only fresh water that meets our eye. Interestingly, all these water sources together hold only about 1.5% of the Earth's non-ice fresh water. The other 98.5% resides in porous regions beneath the Earth's surface.

Have you ever noticed how, during a rainstorm, sandy ground soaks up rain like a sponge? The water quickly disappears into the ground. The type of surface material influences how much water will soak into the ground. Some soils, like sand, easily soak up water. Other soils, like clay, do not easily soak up water and cause runoff. Rocky surfaces with little or no soil are the worst water absorbers, with water penetration occurring only through cracks in the rock.

Water that does soak into the Earth to fill all the open pore spaces becomes groundwater. If the open pore spaces are not filled with water, but still have air in them, then the water is called *soil moisture.* The amount of groundwater at any location depends on the *porosity* of the soil or rock. **Porosity** is a measure of the volume of open space compared with the total volume of the soil or rock. Porosity depends on the size and shape of the soil or rock particles and on how closely these particles are packed together. For example, a soil composed of rounded particles that are all about the same size has a higher porosity than a soil composed of rounded particles that are many sizes. This is because the smaller particles fill up the pores formed by the larger particles, thereby reducing the overall porosity. In addition, angular particles can fill in pores because of their irregular shape.

The ability of a material to transmit fluid is its **permeability.** If the spaces between particles are extremely small and poorly connected (as with flattened clay particles), water may barely move at all. So although clay may have a large porosity, it is difficult for water to move through the pores. In other words, clay is practically impermeable. In contrast, sand and gravel have well-connected, large, open pore spaces, and water moves freely from one pore space to the next. Thus sand and gravels are highly porous and highly permeable (Figure 34.3).

Porosity and permeability of surface and subsurface material are very important to the storage and movement of groundwater.

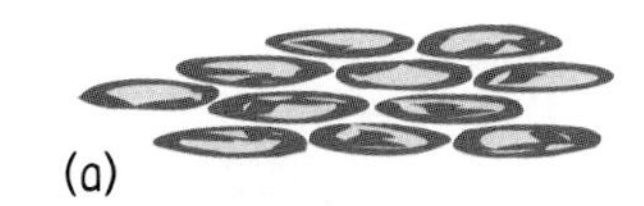
(a)

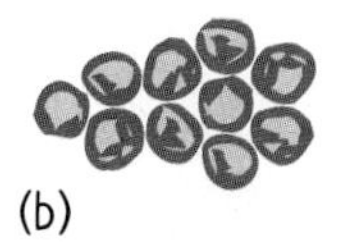
(b)

Figure 34.3
Porosity and Permeability
(a) The sediment particles in clay are small, flat, and tightly packed. Because the sediment particles are flat, the many small pore spaces are poorly connected. Thus clays have high porosity but low permeability.
(b) Sediment particles in sand or gravel are relatively uniform in size and shape, with large and well-connected pore spaces. This allows water to flow freely. So sands and gravels can have both high porosity and high permeability.

Concept Check ✓

Why is a sandy soil better for water flow than a soil composed of clay?

Check Your Answer Water flows easily through sandy soil because sandy soil is usually made up of rounded particles with large and well-connected pore spaces. Thus sandy soil has a high permeability. Water doesn't flow very easily through clay soils because clay is made up of flattened particles with small, poorly connected pore spaces between them. Thus clay soils have a low permeability. Water flow is easier in more permeable soils.

The Water Table

If you were to dig a hole into the ground, you'd find that the wetness of the soil varies with depth. Just below the surface is an *unsaturated zone*, or *zone of aeration*. Here, pore spaces are filled with air and water and we have soil moisture. As you descend farther, you enter the *zone of saturation*. In this zone, water flowing down from the surface has filled all pore spaces and we have groundwater. A hole would fill with water if dug to this zone. The upper boundary of the saturated zone is called the **water table.**

The depth of the water table beneath the Earth's surface varies with precipitation and climate. It ranges from zero depth in marshes and swamps to hundreds of meters below the ground in some parts of the deserts. The water table also tends to rise and fall with the surface topography (Figure 34.4). At lakes and streams that flow year round, the water table is above the land surface.

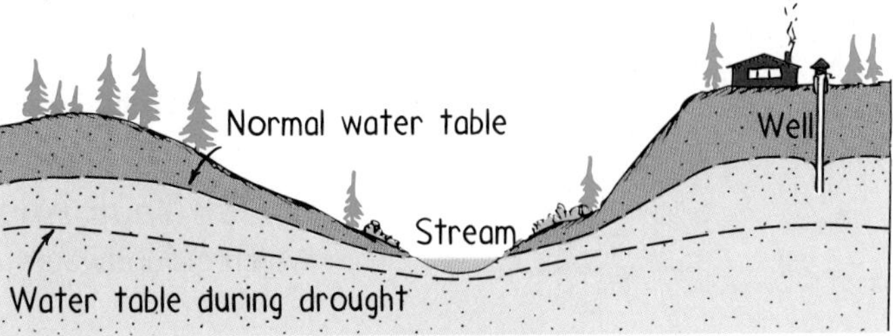

Figure 34.4
The water table roughly parallels the ground surface. In times of drought, the water table falls, reducing stream flow and drying up wells. The water table also falls if the rate at which water is pumped out of a well exceeds the rate at which the groundwater is replaced.

Concept Check ✓

If you dig a hole in the ground and pour water into it, the water seeps out. Why doesn't water seep out of ponds and lakes?

Check Your Answer It does! Water does seep in and out of ponds, but because the surrounding ground (to the sides and below the pond) is already soaking wet, seepage in equals seepage out. There is no *net* seepage.

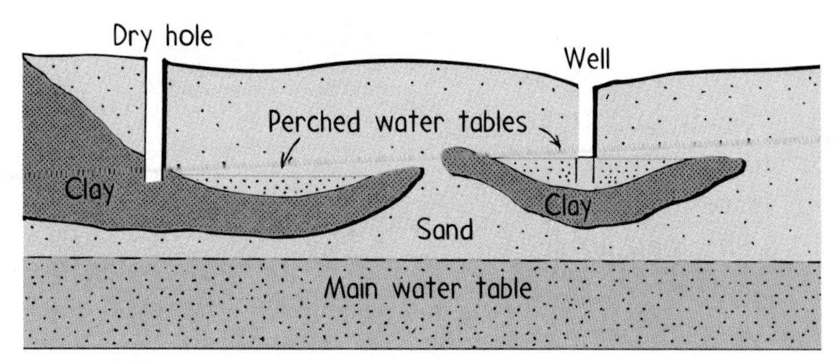

Figure 34.5
A perched water table is separated from the main water table by a low-permeability layer—clay, in this case.

Aquifers and Springs

Any water-bearing underground layer that groundwater can flow through is called an *aquifer*. These reservoirs of groundwater underlie the land in many places. Aquifers contain an enormous amount of water—much more than is found in freshwater lakes, rivers, and streams. More than half the land area in the United States is underlain by aquifers. One such aquifer is the Ogallala aquifer, which stretches from South Dakota to Texas and from Colorado to Arkansas!

The flow of groundwater in an aquifer can be affected by low-permeability beds of rock or soil that slow or prevent groundwater movement.* Low-permeability layers can intercept water above the main water table; when this happens, a *perched* water table is created.

Where the water table meets the land surface, groundwater emerges from an aquifer as either a spring, stream, or lake. Springs can generally be found where the water table meets the surface along a slope, such as on a hillside or coastal cliff. Because water tends to leak out of the ground through breaks and cracks in a rock, springs are often associated with faults. In fact, field geologists can often locate faults by looking for springs.

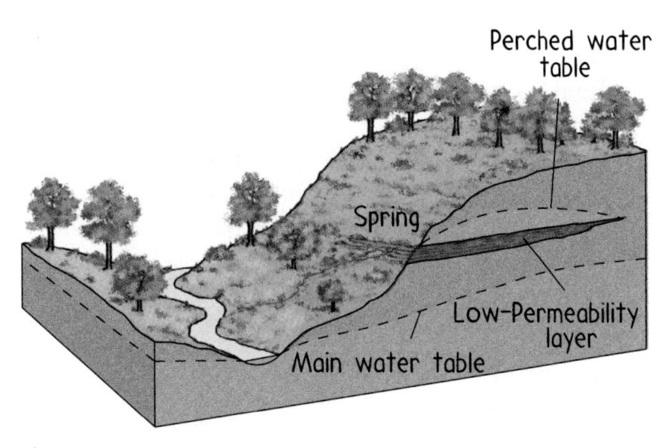

Figure 34.6
When the water table intersects the land surface, groundwater is released. From the perched water table, water is released via a spring; from the main water table, water is released by or into a stream.

Concept Check ✓

What is an aquifer?

Check Your Answer An aquifer is a body of rock or sediment through which groundwater moves easily.

Groundwater Movement

The rate of groundwater flow through an aquifer is directly proportional to the aquifer's permeability. But there's another factor that affects flow rate, too: the slope of the water table. (Geologists call the slope a *hydraulic gradient*).

Due to gravity, groundwater flows "downhill" underground, but the path it takes depends on differences in slope of the water table rather than differences in the slope of the ground. In shallow

* Geologists call these beds *aquicludes* or *aquitards*.

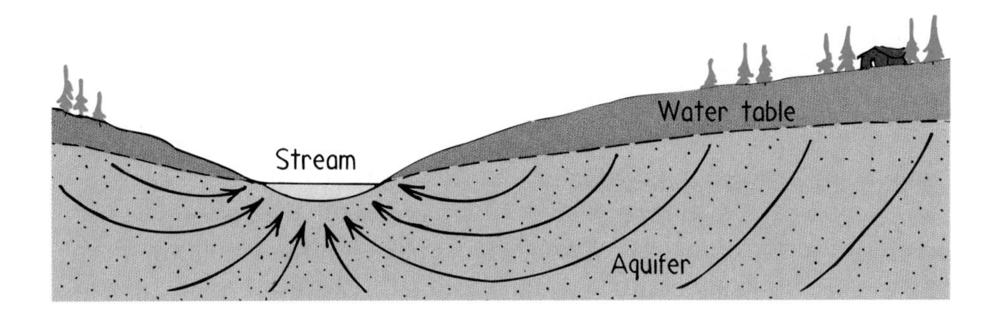

Figure 34.7
Groundwater flows from an area where the water table is high, such as beneath a hill, to an area where the water table is low, such as beneath a stream valley. The curved arrows indicate flow, which show the stream is fed from below.

aquifers, the water table tends to rise and fall with topography, as mentioned earlier. So, responding to the force of gravity, water moves from regions where the water table is high to regions where the water table is low (Figure 34.7).

The speed of groundwater movement is generally very slow compared with the speed of water in rivers and streams. The more permeable the aquifer, the faster the flow. Also, the greater the hydraulic gradient, the faster the flow. How do we know? The speed and route of groundwater flow can be measured by introducing dye into a well and noting the time it takes to travel to the next well. In most aquifers, groundwater speed is only a few centimeters per day, enough to keep underground reservoirs full.

34.3 Streams Come in Different Shapes and Sizes

Streams are all made up of flowing surface water. A stream may be as small as a brook near your school or as big as the mighty Mississippi River. Large or small, streams carve out the landscape. They affect both the surface of the land and the people who live on that land. They also provide energy, irrigation, and a means of transportation for people.

Stream Speed Changes as a Stream Moves

As rain falls on land, it begins a complex journey back to the ocean. Some of the rain soaks into the ground, some evaporates back into the atmosphere, and some runs off into streams.

Streams come in a variety of forms—straight or curved, fast or slow. At their headwaters (the place where a stream originates), stream channels are narrow and water flows quickly through deep, V-shaped mountain valleys. Farther downstream, channels widen so that water flows into and along broad, low valleys.

There are three variables that influence how fast the water in a stream flows. These three factors are: stream gradient; stream discharge; and channel geometry. The *gradient* of a stream is a measure of its slope (the ratio of the vertical drop to the horizontal distance

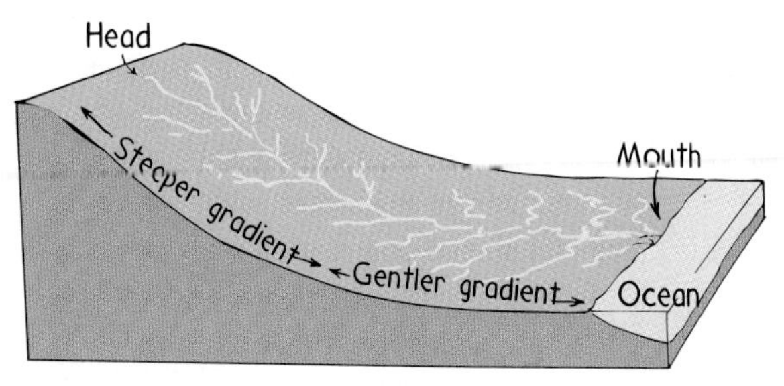

Figure 34.8
The long profile of a stream. At a stream's headwaters, the gradient is high, the channels narrow and shallow, and the stream flow is rapid. As the stream progresses downslope, the gradient decreases, the channel widens, and discharge increases.

for that drop). If we look at a long profile of a stream (Figure 34.8), we see that the gradient is steep near the stream's head and gentler, almost horizontal, near its mouth. Because of gravity, stream speed tends to be greater where the stream gradient is steep.

Discharge is the volume of water that passes a given location in a channel in a certain amount of time. It is directly related to the width and depth (cross-sectional area) of the channel and to the *average* stream speed.

Channel geometry, the shape and dimensions of a stream channel, greatly influences stream speed. Imagine two streams that have the same discharge. If one of the stream channels has a larger cross-sectional area, it will have a slower stream speed than a smaller stream.

The shape of the channel also affects stream speed. Imagine two streams that have the same cross-sectional area. Water flowing in the channel touches the channel bottom and sides. Friction between the water and the channel slows water speed. The more water that touches the bottom and sides, the more friction there is (Figure 34.9). If the stream channel is rounded and deep, as opposed to flat-bottomed and relatively shallow, the stream speed will be faster because there is less water touching the bottom and sides.

When either channel geometry or discharge changes, so does stream speed. In the headwaters, the gradient is high, and channels are narrow, so stream speed is high. In fact, these upland sections of a stream are often called "rapids." As the stream moves downslope, the gradient gradually decreases, the channel typically widens, and as other streams feed into the main branch, discharge increases. Even though discharge increases, water speed can either increase or decrease because stream channels usually become larger as the stream moves downslope. A large channel can carry the extra water without causing the water to move faster. For example, imagine a stream whose discharge increases a little bit downslope but the channel stays the same size. For this stream, the speed increases as we move downslope. Now think about a different stream whose discharge also increases a little bit downslope but the channel gets very wide and deep. The speed of this stream will get slower as we move downslope.

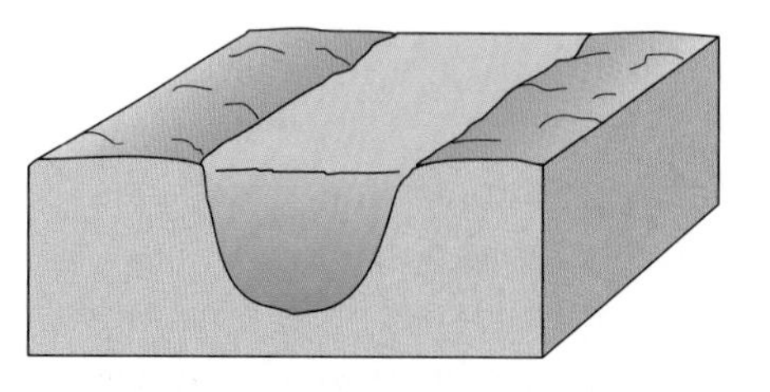

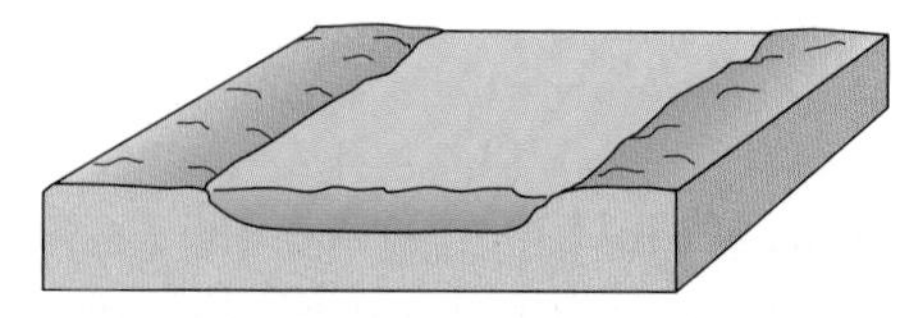

Figure 34.9
(a) In a rounded, deep channel, the water flow speed is relatively high because there is relatively less water in contact with the channel (there is less friction). (b) Wide, shallow channels tend to have slower flows because more water is in contact with the channel (there is more friction).

Concept Check ✓

Stream discharge usually increases downstream as more and more tributary streams feed into the main channel. Does stream speed always increase downstream?

Check Your Answer Not always! If the stream channel widens enough to compensate for the additional water, the speed may not change or may even decrease.

Drainage Networks Are Made Up of Many Streams

A stream is one small piece of a much larger system called a *drainage basin.* A drainage basin is defined as the total area that feeds water to a given stream. A drainage basin can cover a vast area or be as small as 1 square kilometer. Drainage basins are separated from one another by *divides,* lines tracing out the highest ground between streams. A divide can be either very long, if it separates two enormous drainage basins, or a divide can be a mere ridge separating two small gullies. The *Continental Divide,* a continuous line running north to south down the length of North America, separates the Pacific basin on the west from the Atlantic basin on the east. Water west of the Divide eventually flows to the Pacific Ocean, and water east of it flows to the Atlantic Ocean (Figure 34.10).

Streams merge with other streams as they flow downhill, becoming larger and larger. The entire assembly of streams draining a region is called a *drainage network* and can be characterized by the

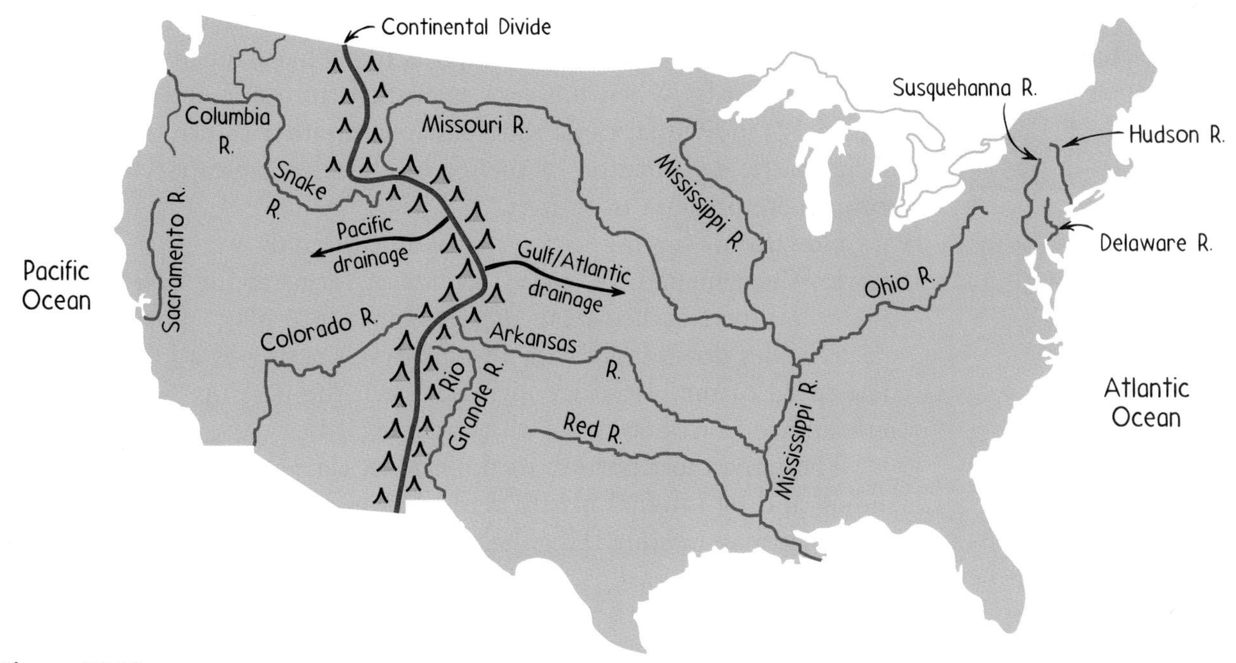

Figure 34.10
The Continental Divide in North America separates the Pacific basin on the west from the Atlantic basin on the east.

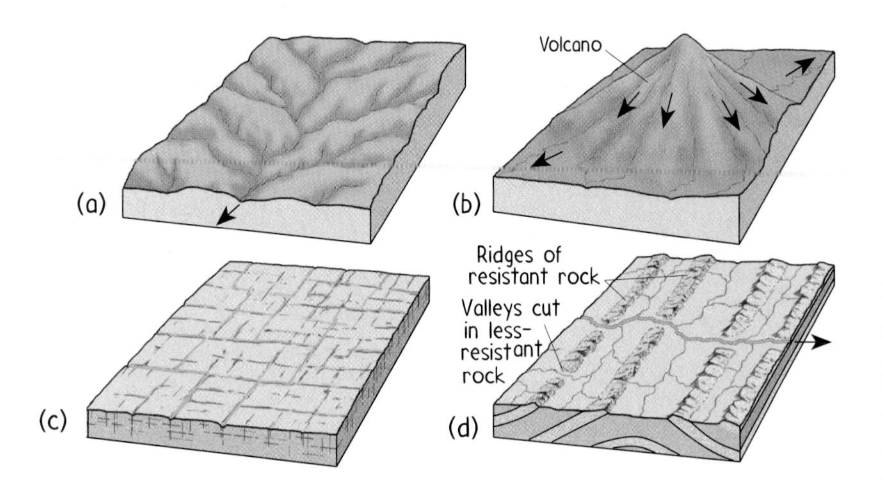

Figure 34.11
Different drainage patterns develop according to surface material and surface structure: (a) dentritic, (b) radial, (c) rectangular, (d) trellis.

branching pattern formed by the streams (Figure 34.11). Because streams erode the land surface and hence erode the rocks and rock material on the land, a drainage pattern is greatly influenced by the rock type and rock material eroded.

34.4 Glaciers Are Flowing Ice

The mightiest rivers on the Earth are frozen solid. As such they normally flow a sluggish few centimeters per day. These great icy rivers are called **glaciers.**

Glacier Formation and Movement

The ice of a glacier is formed from recrystallized snow. This ice does not become a true glacier, however, until it moves under its own weight. This occurs when the ice mass reaches a critical thickness of about 50 meters. Then the weight of the overlying ice causes the ice crystals at the base of the glacier to deform *plastically* and flow downslope. This plastic deformation is like what happens to a deck of playing cards. When the deck is pushed from one end, as in Figure 34.12, individual cards slide past one another shifting the entire deck. Plastic deformation of the ice in a glacier is greatest at the bottom, where pressure is greatest.

Figure 34.12
When a deck of cards is pushed from one side, the individual playing cards slide past one another, thus shifting the whole deck.

Flow from slipping ice crystals is not the only way that glaciers move. When melted ice, called *meltwater*, forms at the base of the glacier, a process called *basal sliding*, comes into play.* The entire glacier slides downslope, with the meltwater acting as a lubricant. The speed of the glacial ice increases from the base up. So the glacier's surface moves fastest of all (Figure 34.13).

The uppermost portion of the glacier, carried along both by basal sliding and by internal plastic deformation, behaves like a stiff, brittle mass that may fracture. Huge, gaping cracks called *crevasses* sometimes

* Meltwater may result from pressure of the overlying ice (the melting point of ice decreases as pressure increases), from the internal heat of the Earth, or from the generation of heat from frictional drag as the glacier moves. No matter what caused it to form, however, meltwater contributes to movement of the glacier.

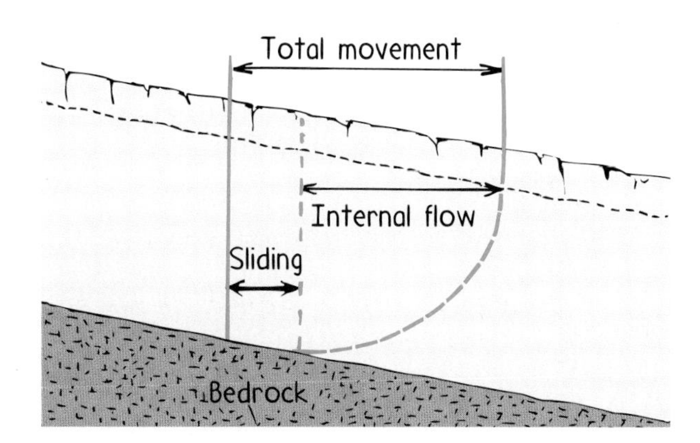

Figure 34.13
Cross section of a glacier. Glacial movement has two components: Internal flow and sliding resulting from lubrication by meltwater. Movement is slowest at the base because of frictional drag and fastest at the surface. The upper parts of the glacier are carried along in piggyback fashion by plastic flow within the ice.

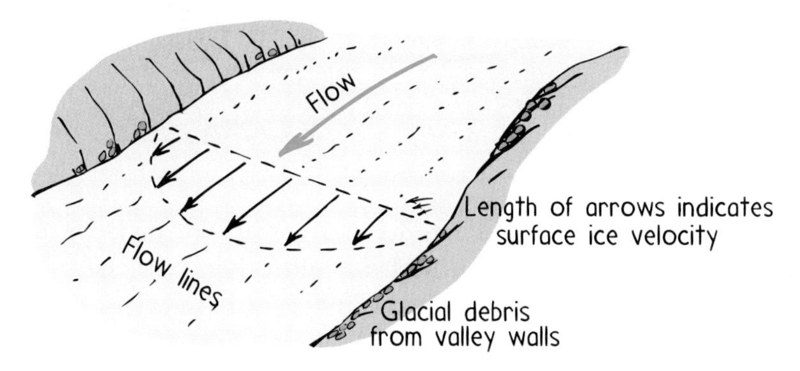

Figure 34.14
Top view of a glacier. Movement is fastest at the center and gradually decreases along the edges because of friction.

develop in surface ice. Crevasses can extend to great depths and can therefore be quite dangerous for people attempting to cross a glacier.

Average glacier speed varies from glacier to glacier and can range from only a few centimeters to a few hundred centimeters per day. Ice moves fastest in the center and slower toward the edges because of frictional drag (Figure 34.14). Some glaciers experience surges, periods of much more rapid movement. These surges are probably caused by periodic melting of the base and sudden redistribution of mass. The flow rate in these relatively brief surges can be 100 times faster than the normal rate. Viewed by air, flow bands of rock debris and ice normally have a parallel pattern, but during a surge the flow bands become intricately folded (Figure 34.15).

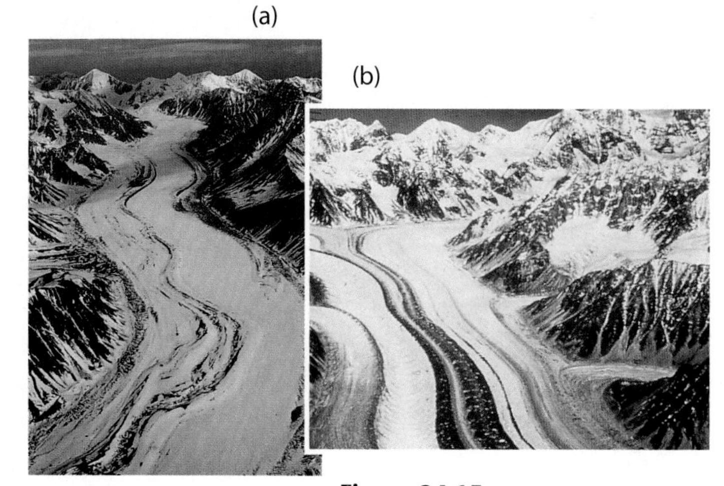

Figure 34.15
Glacial flows:
(a) surge flow
(b) normal flow.

Glacial Mass Balance

From season to season, and over longer periods of time, the mass of a glacier changes. Typically, a glacier grows in the winter as snow accumulates on its surface. The amount of snow added to a glacier annually is aptly termed **accumulation.**

As ice accumulates and begins to flow downhill, it may move to an altitude where temperatures are warmer. Then some ice melts and the glacier loses some of its mass. A glacier may also lose mass as it

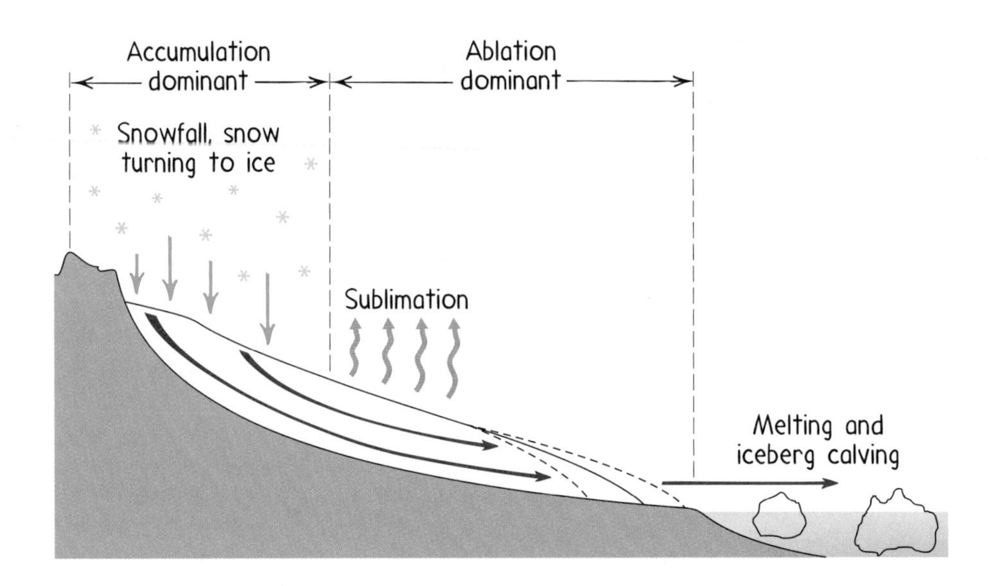

Figure 34.16
Accumulation on a glacier takes place as snow falls on the glacier and turns to ice at high elevations. Ablation takes place as ice melts and/or calves into icebergs at lower, warmer elevations, or as ice is lost through sublimation.

moves downslope to a shoreline. There ice may break off, or *calve*, to form icebergs that float away to sea. Melting and calving are the two primary mechanisms by which glaciers lose mass. Although less noticeable, glaciers may also lose mass as the ice *sublimates* to water vapor. By whatever means, the total amount of ice lost annually is called **ablation** (Figure 34.16).

When accumulation equals ablation, the size of the glacier remains constant. For example, in a mountain glacier, accumulation occurs with winter snowfall in the farther-back, higher-elevation parts of the glacier. Ablation occurs in the lower portions, where spring and summer meltings are greatest. When the accumulation rate equals the ablation rate, the melting of the lower portions is offset by the downslope flow of ice from higher portions. As a result, the location of the front edge of the glacier does not change. When accumulation exceeds ablation, the glacier advances—it grows. When ablation exceeds accumulation, the glacier retreats—it shrinks. Naturally, in all these cases, the ice of the glacier is always flowing downslope.

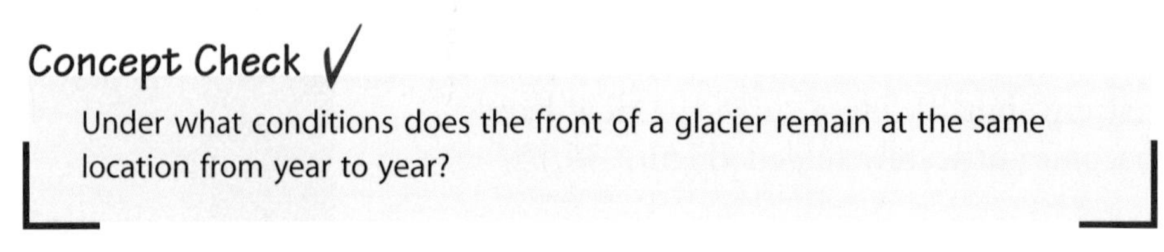

Concept Check ✓

Under what conditions does the front of a glacier remain at the same location from year to year?

Figure 34.17
Map of the ocean floor showing variation in topography.
(a) Atlantic profile
(b) Pacific profile.

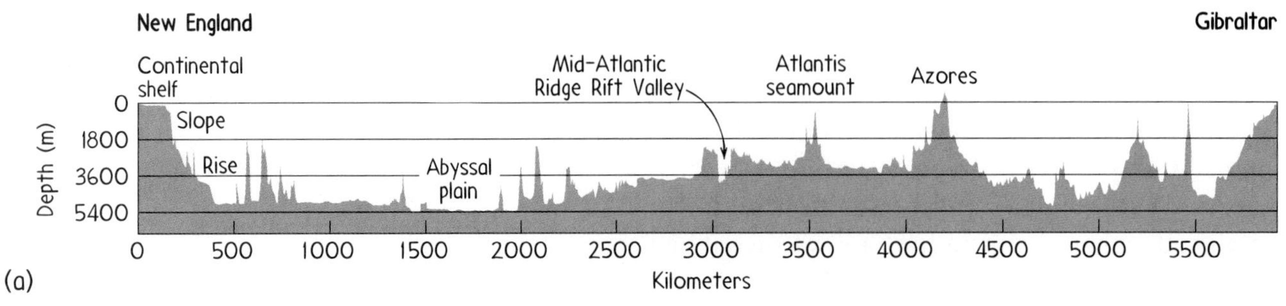

Check Your Answer The front of a glacier remains at the same location when the rate of growth (accumulation) for the year equals the rate of shrinking (ablation). In the spring, as ice at the glaciers front melts away, the glacier retreats upslope. At the same time, the increased mass from the past winter's accumulation causes the glacier to move forward. When the rate at which this forward movement matches the rate of melting, the location of the front edge remains constant.

34.5 Most of the Earth's Water Is in the Oceans

If we could drain the water from the Earth's oceans, we'd see enormous mountain ranges in the middle of the ocean basins and deep trenches bordering many continents. These features of the ocean bottom are very pronounced. In fact, land rises an average of about 840 meters above sea level while the ocean bottom averages about 3800 meters below sea level. If we were to compare the height of Mount Everest in the Himalayas, a majestic 8848 meters above sea level, to the Marianas Trench in the Pacific Ocean, an unfathomable 11,035 meters below sea level, we would see that the oceans are much deeper than land mountains are high!

The **continental margin** is the boundary between the continents and the ocean. As Figure 34.18 shows, the continental margin consists of a *continental shelf* (the submerged upper portion of the margin), a *continental slope* (the break point where the shelf steepens as it descends to depths of 2–3 kilometers), and a *continental rise* (the area from the base of the slope seaward to the deep ocean floor). Continental margins are formed by the deposition of continental sediments.

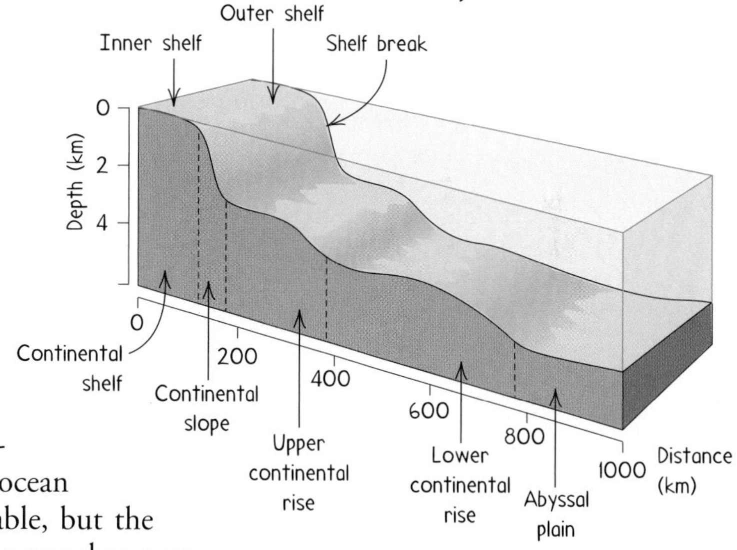

Figure 34.18
Profile of the continental margin going from land to the deep ocean bottom. The vertical dimension is exaggerated for clarity.

Between continental margins, the topography of the ocean floor varies greatly. The midocean ridges that encircle the globe are tall and variable, but the sediment-covered ocean bottom is flat. Seafloor trenches near continental margins are very deep (Figure 34.17).

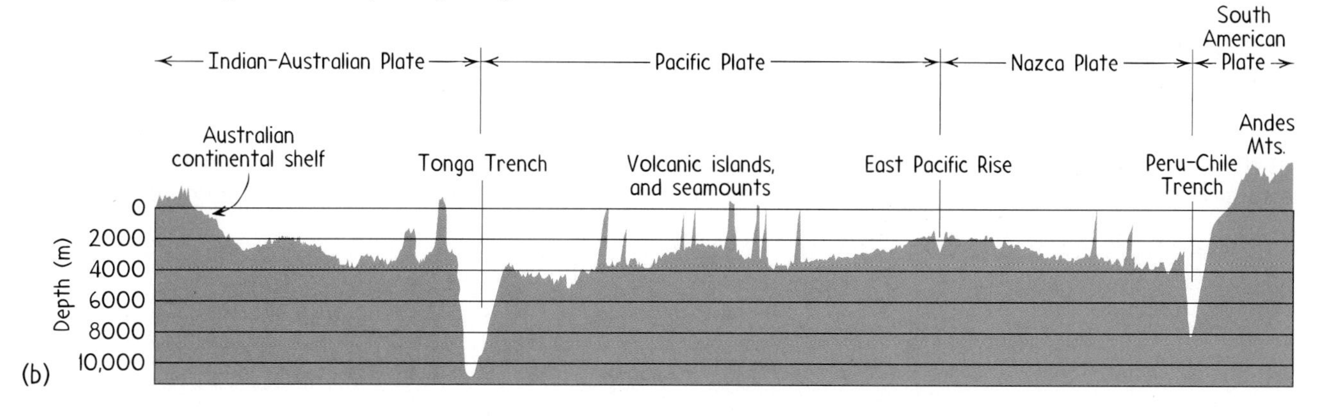

Ocean Waves

Ocean waves come in a variety of sizes and shapes, from tiny ripples to the gigantic waves in hurricanes, to the huge tides as we saw in Chapter 7. Water waves, like all other waves, begin by some disturbance. The most common disturbance that causes ocean waves is the wind. Blow on a bowl filled with water and you'll see a succession of small ripples moving across the water surface.

The generation of ripples in the ocean is similar. As wind speed increases, the ripples grow to full-sized waves; as stronger winds blow, larger waves are created. As waves travel away from their origin, they develop into regular patterns of smooth, rounded waves called *swells*—the mature undulations of the open ocean.

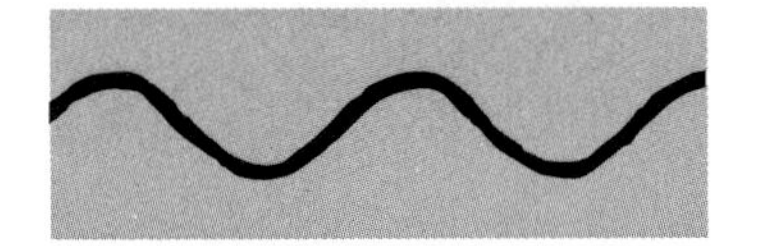

Figure 34.19
Ocean waves have characteristics of simple sine waves.

Recall from our study of waves in Chapter 13 that wave motion can be described in terms of a sine curve (Figure 34.19). It is the *disturbance* that is carried by a wave, not the material through which the wave is moving. An ocean wave travels across the ocean, while the water making up the wave remains for the most part in one place.

However, ocean waves have both transverse and circular motion, so ocean waves are more complicated than the simple transverse waves discussed in Chapter 13. As a water wave passes a given point, the water particles at that point move in circular paths. This circular motion can be seen by observing the behavior of a floating piece of wood on the ocean surface. The wood sways to and fro while bobbing up and down, actually tracing a circle during each wave cycle. This circular motion occurs near the water surface and decreases gradually with depth (Figure 34.20). At a depth of about one-half wavelength, the circular part of the wave is negligible. For this reason, we can say with reasonable accuracy that water waves occur mainly at the surface.

When a wave approaches the shore, where the water gets shallow, the water's circular motion is interrupted by the ocean bottom. When water is only as deep as half of a wave's wavelength, the bottom of the circular path flattens. This slows the wave. The wave period remains

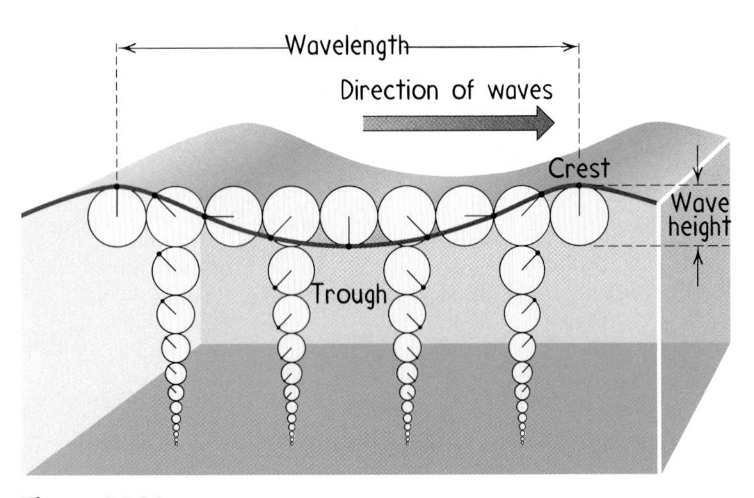

Figure 34.20
Movement of water particles with the passage of a wave. The particles move in a circular orbit. Orbital motion is greatest at the surface and gradually decreases with depth. At depths greater than half a wavelength, orbital motion is negligible.

unchanged because the swells from deeper waters continue to advance. As a result, incoming waves gain on leading slower waves and the distance between the waves decreases. This bunching up of waves in a narrower zone produces higher, steeper waves. When wave height gets so steep that water can no longer support itself, the wave overturns, breaking with a crash. This breakwater is called *surf*—the area of wave activity between the line of breakers and the shore (Figure 34.21).

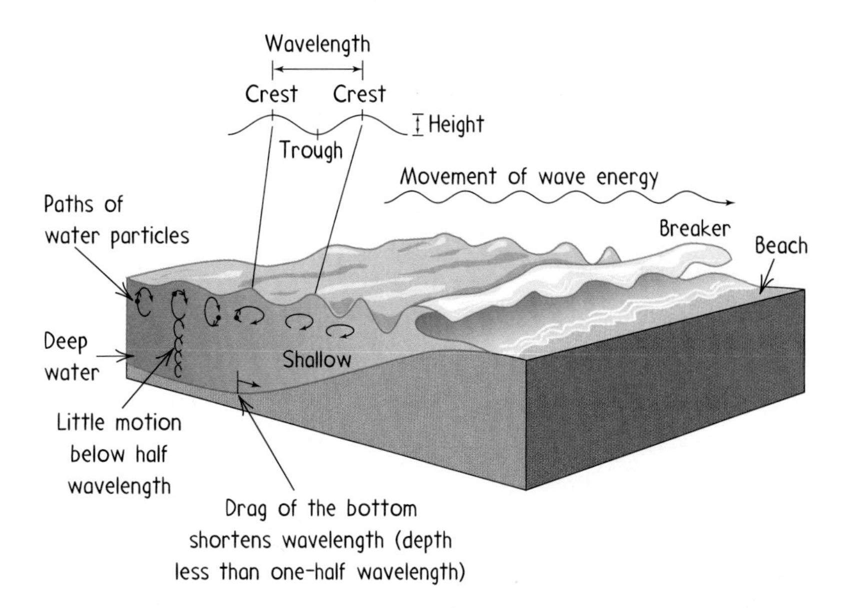

Figure 34.21
Waves change form as they travel from deep water through shallow water to shore. In deep water, orbital motion is circular. In shallow water orbital motion becomes elliptical as a result of contact with the bottom. This change decreases the wave speed. As incoming waves continue to advance, the distance between waves decreases, causing wave height to increase. When waves reach a critical height, they break and crash into the surf zone.

Wave Refraction Occurs When Waves Encounter Obstacles

Wave refraction (bending) occurs in two different settings. As waves enter shallow water, their direction of approach changes as they refract to a direction more parallel to the shore. This refraction occurs whenever waves approach a shoreline at an angle. As the portion of the wave closest to the shore begins to touch the ocean bottom, it slows and begins to lag behind the portions of the wave still in deeper water. As the next portion of the wave touches the bottom, it too slows. Thus in a continuous fashion, the line of the wave crest bends as it moves into shallower water, becoming more nearly parallel to the coastline (Figure 34.22). This bending is not always complete and some waves approach the shoreline at a small angle. The oblique approach of waves causes a *longshore current* that flows parallel to shore.

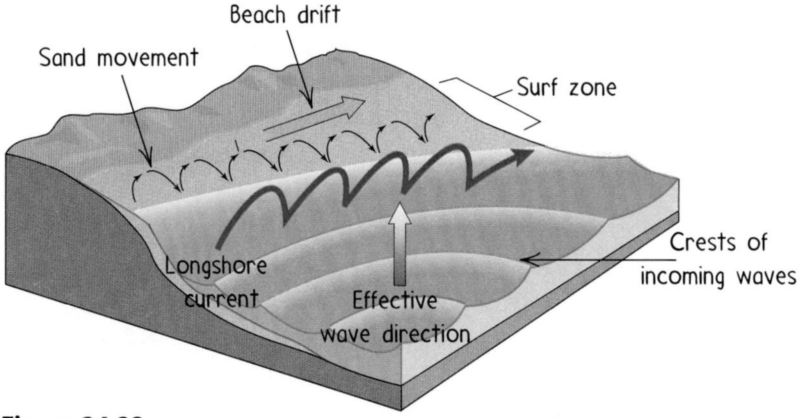

Figure 34.22
When waves approach a shoreline, they refract (bend) so that the crests of the approaching waves become more parallel to the shore as they move into shallower water. Because the overall direction of wave movement is oblique to the shore, a longshore current forms, which causes water and sand to move parallel to the shoreline.

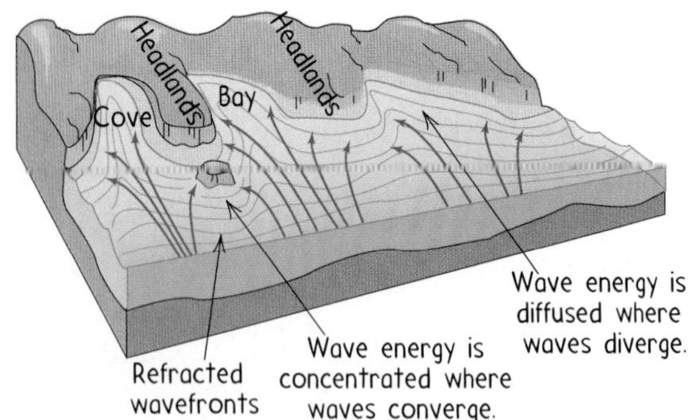

Figure 34.23
On irregular coastlines, wave energy is concentrated as it converges on headlands and diffused as it diverges in coves and bays.

Wave refraction also greatly impacts irregular shorelines, mainly those where there are protruding headlands and small bays. Refraction causes wave energy to be unevenly distributed. Wave energy is concentrated at headland areas, where shorelines project into the water, and diluted in adjacent bays (Figure 34.23).

34.6 Can We Drink the Water?

Water occurs in many environments. The quality of the water we drink and use for industry, recreation, and agriculture is an important factor in the quality of our lives. Rainwater is used as the standard of water purity. Most of our water supply is of good quality—as good as rainwater. There are important variations though.

If rain falls through clean air and soaks into a bed of quartz sand, the water quality after filtering through the sand will be about the same as before filtration. On the other hand, rain soaking into a bed of carbonate rocks (like limestone) can become "hard" from dissolved calcium carbonate. Dissolved substances can affect the taste of water as well as its usefulness. For example, hard water does not rinse away detergents well and therefore is not very useful for washing. So we see that the quality of groundwater depends very much on the type of soil and rock through which it flows.

As you know, natural drinking water is never pure. It is a solution, containing dissolved substances from rock and soil. The quantity of dissolved substances in drinking water is generally very small. How can we measure the purity of drinking water? Often we measure the concentration of dissolved substances in units of parts-per-million (ppm). If there are 100 grams of a certain substance in a total of one million grams of solution, the concentration of the solution is 100 parts-per-million, or 100 ppm.

Good-quality water averages 150 parts-per-million for total dissolved substances, with an upper limit of 1000 parts-per-million. The taste of water depends on the type of dissolved substances. Water containing 1000 parts-per-million of dissolved calcium tastes fine, but water containing 200 parts-per-million of sodium chloride tastes salty.

Several other dissolved substances, many introduced by human activities, have a strong effect on the quality of water. Some of these ingredients are beneficial to health, while others can be quite dangerous. Added fluoride, for example, helps reduce tooth decay. Zeolite minerals in water filters soften hard water by chemically exchanging dissolved calcium for sodium. In contrast, lead and arsenic, two naturally occurring minerals, can make water unsafe to drink even if present in small amounts. Bacteria from sewage also contaminate water.

Figure 34.24
In hard water, dissolved calcium prevents soap from developing a sudsy lather and also from rinsing clean. Bathtub rings are more prevalent in areas that have hard water.

Concept Check ✓

How is hard water softened?

Answer Hard water is softened by removal of dissolved calcium. This is accomplished by passing it through a zeolite filter. The filter absorbs calcium ions while releasing an equivalent number of sodium ions. Unlike hard water, soft water allows the formation of soap suds that can readily be rinsed off.

Water-Supply Contamination

The primary source of water-supply contamination is human activity. As rivers and streams receive discharge from factories, sewage, and chemical spills, surface water is polluted. Surface waters are linked to groundwaters, so what affects one can easily affect the other.

The most common source of groundwater contamination is sewage. Sewage includes drainage from septic tanks, inadequate or broken sewer lines, and barnyard wastes. Sewage water contains bacteria that, if untreated, can cause waterborne diseases such as typhoid, cholera, and infectious hepatitis.

Agricultural areas where a lot of nitrate fertilizers are used also contribute to groundwater contamination. Nitrate fertilizer spread over the land is used by plants, but the unused nitrate flows down into the groundwater as a contaminant. Nitrate levels in groundwater have to be closely monitored because these compounds are toxic to humans in amounts as small as 15 parts-per-million. Fertilizers in surface waters can cause damage to the environment as well.

As a population grows, so does its garbage. The most common means of waste disposal is burial in a landfill. Even radioactive, toxic, and hazardous wastes are buried. The location of underground storage sites is tricky to decide. For a site to be considered safe, it must be located where waste products and their containers cannot be affected chemically by water, physically by Earth movements, or accidentally by people. Precipitation infiltrating the site may dissolve a variety of compounds from the solid waste. The resulting liquid, known as **leachate,** can move vertically downward from the landfill to the water table and contaminate the groundwater. When leachate

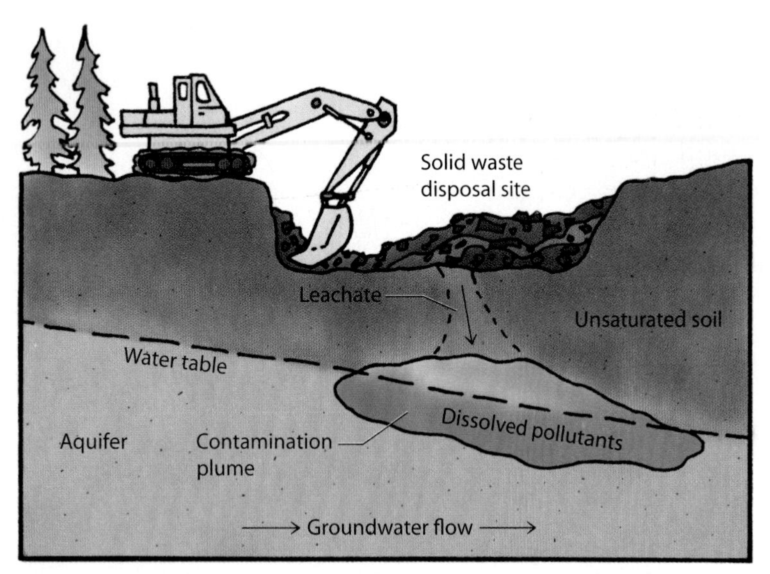

Figure 34.25
A contaminant plume of leachate spreads in the direction of groundwater flow.

mixes with groundwater, it forms a plume (an area of contamination) that spreads in the direction of the flowing groundwater (Figure 34.25).

Groundwater contamination also occurs as a result of spills and leaks of toxic and hazardous chemicals. These discharges can be sudden, as in a train or tanker truck accident, or as a result of slow leakage from a holding container. As we can see in Figure 34.26, groundwater is susceptible to a wide variety of contamination sources.

Luckily, the human-caused contamination described above is not without hope. Ways to prevent such contamination are being used more and more. New technologies are constantly being developed to prevent and clean up contaminated water. Our water supply is precious. We, as a society, must do everything we can to protect and conserve it.

How Can We Conserve Water?

Clean, fresh water is the most valuable resource our planet has to offer. Clearly, each of us must be mindful of fresh water's true value. We must practice conservation measures whenever possible.

Toward this goal, please consider the following guidelines adapted from the American Water and Energy Savers Association.

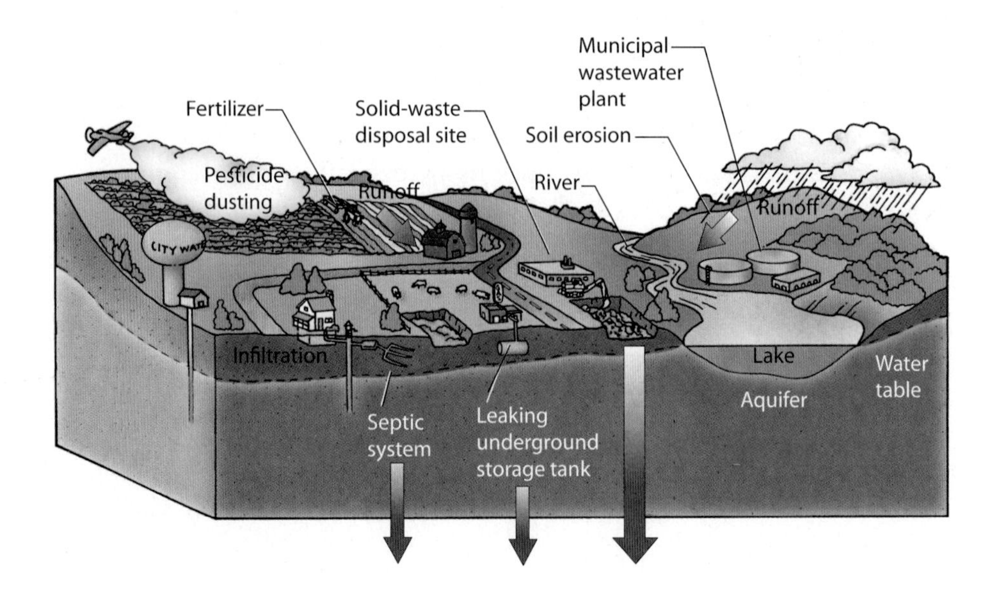

Figure 34.26
Groundwater can be contaminated by a wide variety of sources. The arrows indicate some major sources of contamination.

10 Ways to Save Water

1. Read your water meter before and after a 2-hour period when no water is being used. If the reading has changed, there is a leak somewhere in your water system. Get the leak fixed.
2. Fix dripping faucets by replacing the washers in them. If a faucet is dripping at the rate of one drop per second, you can expect to waste 11,000 liters per year!
3. Check for toilet tank leaks by adding food coloring to the tank. If the toilet is leaking, color will appear in the bowl within 30 minutes.
4. Avoid flushing the toilet unnecessarily.
5. Install a composting toilet, which uses no water. Rather, it allows human wastes to decompose *aerobically* as air is vented over the waste, which is buried in peat moss. Dried odor-free compost is removed every few months and is useful as a garden fertilizer.
6. Take shorter showers and don't let the water run unnecessarily while you brush your teeth.
7. When washing dishes by hand, do so in a sink or large pot filled with soapy water. Quickly rinse under a small stream from the faucet.
8. Store drinking water in the refrigerator rather than letting the tap run every time you want a cool glass of water.
9. Always follow all water conservation and water shortage rules and restrictions in effect in your area.
10. Encourage family, friends, and neighbors to be part of a water-conscious community.

Chapter Review

Key Terms and Matching Definitions

_____ ablation
_____ accumulation
_____ continental margin
_____ glacier
_____ groundwater
_____ hydrologic cycle
_____ leachate
_____ permeability
_____ porosity
_____ water table

1. The natural circulation of water from ocean to atmosphere to ground, back to ocean.
2. Subsurface water in the zone of saturation.
3. The volume of open space in a rock or sediment compared to the total volume of solids plus open space.
4. The ability of a material to transmit fluid.
5. The upper boundary of the zone of saturation, below which every pore space is filled with water.

6. A solution formed by water that has percolated through soil containing water-soluble substances.
7. A large mass of ice formed by the compaction and recrystallization of snow, moving downslope under its own weight.
8. The amount of snow added to a glacier in a year.
9. The amount of ice a glacier loses in a year.
10. The boundary between continental land and deep ocean basins, consisting of continental shelf, continental slope, and continental rise.

Review Questions

The Hydrologic Cycle

1. Describe the hydrologic cycle.

Water Below the Surface

2. Distinguish between *porosity* and *permeability.*
3. What soil, or soil type, can have high porosity but low permeability?
4. A kitchen table is usually flat, but the water table is generally not flat. Why?
5. Compare and contrast the zone of aeration with the zone of saturation.
6. What type of soils allow for the greatest infiltration of rainfall?
7. How does an aquitard differ from an aquifer?
8. What factors affect the rate of groundwater movement?

Streams Come in Different Shapes and Sizes

9. What is meant by stream gradient, and how does it affect stream velocity?
10. What happens when (a) the discharge of a stream increases, and (b) the speed of a stream increases?
11. How does the shape of a stream channel affect flow?
12. What is the significance of the Continental Divide in North America with respect to water flow to the Atlantic and Pacific Oceans?

Glaciers Are Flowing Ice

13. What conditions are necessary for a glacier to form?
14. What distinguishes a huge block of ice from a glacier?
15. What two main ways do glaciers flow?
16. Why do crevasses form on the surface of glaciers?
17. Under what conditions does a glacier front advance?
18. Under what conditions does a glacier front retreat?

Most of the Earth's Water Is in the Oceans

19. Why do waves become taller as they approach shore?
20. What is wave refraction? Why does it happen in ocean waves?

Can We Drink the Water?

21. List three ways our water supply is being contaminated.

Exercises

1. Look at a map of any part of the world and you'll see the locations of older cities where rivers occur, or have occurred. What is your explanation?
2. How much of the Earth's supply of water is fresh water, and where is most of it located?

3. Where does most rainfall on Earth finally end up before becoming rain again?

4. If the water table at location X is lower than the water table at location Y, does groundwater flow from X to Y or from Y to X?

5. Is the infiltration of water into the ground greatest on steep, rocky slopes or on gentle, sandy slopes? Defend your answer.

6. How is the local hydrologic cycle affected by the practice of drawing drinking water from a river, then returning sewage to the same river?

7. Why is pollution of groundwater a greater environmental hazard than pollution of surface waters?

8. Some metals can be extremely dangerous to water supplies. Aluminum has been linked to Alzheimer's and Parkinson's disease; cadmium is known to cause liver damage; and lead affects the circulatory, reproductive, nervous, and kidney systems. What are the likely ways these metals can get into our water supply?

9. Is water in the unsaturated zone called groundwater? Why or why not?

10. By what means can Earth scientists predict the discharge of a stream after a rainstorm?

11. As a population increases, so does the amount of garbage produced by that population. In many areas the way to deal with increasing wastes is by burial in a landfill or underground storage facilities. What principal factors must be considered in the planning and building of such sites?

12. What effect does a dam have on the water table in the vicinity of the dam?

13. What effect does the accumulation of sediments behind a dam have on its capacity for storing water?

14. In an aquifer, if the water table next to a stream is lower than the water level in the stream, does groundwater flow into the stream or does stream water flow into the ground? Explain.

15. As runoff into streams increases, what variables of stream flow (as discussed in the text) increase?

16. How is a glacier formed?

17. How does "frictional drag" play a role in the external movement of a glacier? How about the internal movement?

18. As waves approach shallow water, those with longer wavelengths slow down before those with shorter wavelengths. Why?

19. How does leachate form?

20. What is a continental divide?

Suggested Readings and Web Sites

Reisner, Marc. *Cadillac Desert.* New York: Penguin Books, 1993.

http://www.groundwater.org/GWBasics/hydro.htm

http://www.epa.gov/kids/water.htm

http://water.usgs.gov/education.html

http://www.clemson.edu/waterquality/educate.htm

Chapter 35: Our Natural Landscape

We have learned that the Earth is a planet that is ever-changing. For example, internal forces cause the tectonic plates to move. We have also learned that water is constantly circulating on and in the Earth's crust. So, what work does water do as it circulates? How does it change the appearance of the landscape around us? Is liquid water the only agent of change? How are sediments moved from one place to another? Let's explore the forces of change that alter our natural landscape!

35.1 The Work of Air

Water is the most important force that changes our natural landscape. But air plays a role too. If you've ever been in a wind storm or at the beach on a windy day, you may have felt the "sandblasting" effect of the sand carried by wind. Once in the air, particles of sediment can be carried great distances by the wind. Red dust from the Sahara is found on glaciers in the Swiss Alps, and on islands in the Caribbean Ocean. Fine grains of quartz from central Asia blow onto the Hawaiian Islands!

Figure 35.1
Generated by blowing winds, ripple marks are narrow ridges of sand separated by wider troughs. They are small, elongated sand dunes.

In the desert, winds move over dry sand surfaces. The wind picks up the small, easily transported particles but leaves the large, harder-to-move particles behind. The small particles bounce across the desert floor. In doing this, they knock more particles into the air. Tiny sand dunes, called *ripple marks*, form (Figure 35.1). Ripple marks also form as water currents move sand grains. You can see these ripple marks in shallow streams or under the waves at beaches.

Sand dunes begin to form when the flow of air is blocked by an obstacle, such as a rock or clump of vegetation. As the wind sweeps over and around the obstacle, wind speed slows. Sand grains fall out of the air into the wind shadow (the place where wind is blocked). As more sand falls, a mound forms which blocks the flow of air even more. With more sand and more wind, the mound becomes a dune. As a dune grows, the whole mound starts moving. Where? It moves downwind as sand grains on the windward slope move up and over the crest of the dune to fall on the leeward slope (Figure 35.2). Over time, this continuous process moves the entire dune.

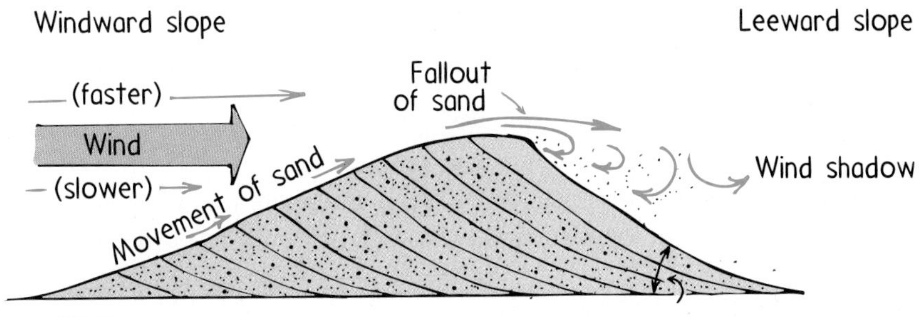

Figure 35.2
Formation of a sand dune. When air flow is obstructed, air speed drops and as a result sand grains settle in the wind shadow. With more wind, more sand settles, and a dune is formed. As a dune grows, sand grains on the windward slope move up and over the crest to fall on the leeward slope. This results in motion of the whole mound downwind.

35.2 The Work of Groundwater

Flowing groundwater—no matter how slow—can cause large changes in landscapes. These impacts usually occur without any human interference. Sometimes, however, our actions can drastically alter our landscape.

Pumping Can Cause Land Subsidence

Most wells are drilled to pump groundwater out of the ground. In areas where groundwater withdrawal has been extreme, the land surface is lowered—it *subsides.* The problem of land subsidence is most noticeable where the ground and underlying subsurface is made of thick layers of loosely packed sediments. These thick layers of sediments usually have many layers of easily compressed, water-bearing clays sandwiched between a series of sandy aquifers. Recall from Chapter 34 that clay has a very low permeability. Nevertheless, as water is pumped from the aquifers, water slowly leaks out of the clay layers to replenish the aquifers, which usually continue to be pumped. As the clays lose water, they shrink, and the land surface subsides.

The best known example of land subsidence is the Leaning Tower of Pisa in Italy. The tilting tower was built on loose sediments from the Arno River. Over the years, as groundwater has been withdrawn to supply the growing city, the tilt of the tower has increased (Figure 35.3). Another region where land subsidence can be seen is Mexico City, which is built in the middle of an ancient shallow lake. The withdrawal of groundwater beneath this city now finds many "street-level" buildings at basement level. Some areas in Mexico City have subsided by as much as 6–7 meters. In the United States, large amounts of groundwater pumping for irrigation in the San Joaquin Valley of California has caused the water table to drop 75 meters in 20 years. And this has lowered the land surface by as much as 9 meters. Because water for irrigation is now provided by canals, the sandy aquifers are slowly recharging (refilling with water). Unfortunately, though, most of the land subsidence from the shrunken clay layers cannot be reversed.

Figure 35.3
The Leaning Tower of Pisa. Construction began in about 1173 and was suspended when builders realized the foundation was inadequate. Work was later resumed, however, and the 60-m tower was completed 200 years later. Deviation from the vertical is about 4.6 m. The tower's foundation has been recently stabilized by groundwater withdrawal management, and so the tower should remain stable for years to come.

Concept Check ✓

Why is land subsidence most evident in regions where the underlying geology is a series of low-permeability clay layers sandwiched between sandy aquifers?

Check Your Answer Clay layers lose water and shrink as water is pumped from the adjacent aquifers. Shrinkage causes the land to subside. When the underlying ground surface is made up of mostly sandy layers, however, the ground compresses only a little when pumped. That's why subsidence is more noticeable in regions underlain by alternating layers of loosely packed clays and sands.

Some Rocks Are Dissolved by Groundwater

The vast carbonate deposits (like limestone) that underlie millions of square kilometers of the Earth's surface provide storage areas for groundwater. The effect of groundwater on limestone is unique. It creates some very interesting landforms. Recall from Chapters 30 and 31 that limestone is made of the mineral calcite (calcium carbonate or $CaCO_3$). On its way to becoming groundwater, rainwater naturally reacts with carbon dioxide in the air and soil to produce carbonic acid. When this slightly acidic water comes in contact with limestone, the carbonic acid partially dissolves the rocks. As groundwater steadily dissolves the limestone, it creates unusual erosional features, like sinkholes and caverns.

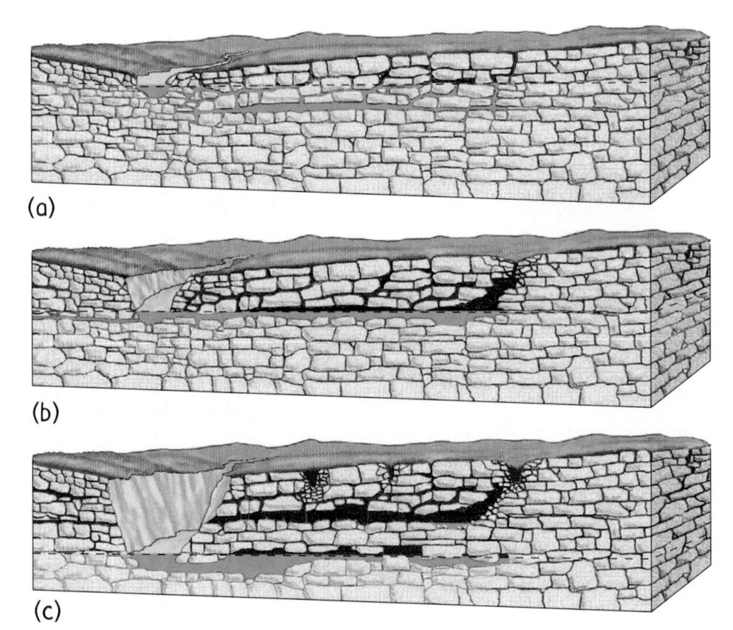

Figure 35.4
The formation of a cave begins with a layer of carbonate rock, mildly acidic groundwater, and an enormous span of time. (a) Groundwater makes its way toward a stream. (b) As the stream valley deepens because of erosion, the water table is lowered. The carbonate rock is eaten away as acidified water erodes and enlarges the existing fractures into small caves. (c) Further deepening of the stream valley causes the water table to drop even lower. Water in the cave seeps downward, leaving an empty cave above a lowered groundwater level.

Caverns and Caves The dissolving action of underground water has carved out magnificent caverns and caves. Groundwater flow in limestone aquifers occurs mostly through fractures in the rock, rather than through pores. It is in limestone that we find the only true underground rivers. (In other rocks and soils, underground water is found only in pore spaces, not in large, open channels.)

As rainwater (enriched in carbonic acid) soaks into limestone, it moves through rock fractures. As the groundwater flows toward a stream, as shown in Figure 35.4a, the slightly acidic water dissolves the surrounding limestone. Fractures become larger. This eventually creates an underground channel. The stream that the groundwater is flowing toward also dissolves and erodes its stream channel. Then the water level in the stream and the level of the water table drop. Lowering the water table drains the water from the underground channel, forming caves. The dropping water table also causes water in the main cave channel to seep downward to begin a new level of fractures and caves (Figure 35.4c).

Water dripping down from the cave ceiling, now rich in dissolved calcium carbonate, creates icicle-shaped stalactites as water evaporates and the calcium carbonate precipitates. Some of the water solution drips off the end of the stalactites to build corresponding cone-shaped stalagmites on the floor. As caves grow and develop into interconnecting chambers, they are called *caverns.* One of the most impressive caverns in the United States is Carlsbad Caverns in southeastern New Mexico (Figure 35.5). The cavern descends to a

Figure 35.5
Cave dripstone formations at Carlsbad Cavern near Hobbs, New Mexico.

maximum depth of 486 meters. Other famous caves and caverns include Mammoth Cave in Kentucky, Adelsberg Cave in Austria, and Good Luck Cave in Borneo.

Sinkholes Sinkholes are funnel-shaped cavities in the ground that are open to the sky. They are formed in much the same way as caves. Groundwater dissolves limestone and eventually the surface collapses in on itself. Some sinkholes are caves whose roofs have collapsed. Some sinkholes are formed by drought conditions or excessive groundwater pumping.

Karst Regions When sinkholes, caves, and caverns define the land surface, the terrain is called *karst topography*, named after the Karst region of Yugoslavia. There, weathering and erosion of highly soluble limestone characterizes the landscape. The pattern of streams in this type of landscape is very irregular; streams and rivers disappear into the ground and reappear as springs. Some karst areas appear as soft, rolling hills with large depressions that dot the landscape. The depressions are old sinkholes now covered with vegetation (Figure 35.6). In general, karst areas have sharp, rugged surfaces and very thin soils as a result of high runoff and dissolution of surface material.

Karst regions can be found throughout the world: in the Mediterranean basin; in sections of the Alps and the Pyrenees; in southern China; and in Kentucky, Missouri, Florida, and Tennessee. The beauty of southern China's karst landscape is depicted in Figure 35.7.

35.3 The Work of Surface Water

Water on the surface of the Earth plays two very important yet contrasting roles in shaping the landscape. One is erosion, and the other is deposition of sediments. Surface water is both a destroyer and creator of sediments and sedimentary rocks. We learned in Chapter 31

Figure 35.6
Karst topography covered by vegetation makes up the rolling hills in south central Kentucky.

Figure 35.7
The karst landscape of China has been an inspiration to classical Chinese brush artists for centuries.

that weathering and erosion create and move sediment. Erosion by water is the most common way clastic sediments are transported away from the site of their creation. After the sediments are deposited, they eventually become sedimentary rock. The most dominant feature of sedimentary rocks is the way the particles of sediment are laid down, layer upon layer. These layers are referred to as *beds.* Varying in both thickness and area, each bed represents one episode of deposition. For example, flooding in a particular year might produce a layer of sediment next to a river. A flood any time after that produces an overlying layer. The deposition and erosion of clastic sediments occurs in many different environments, including rivers and streams, deserts, deltas, and shorelines.

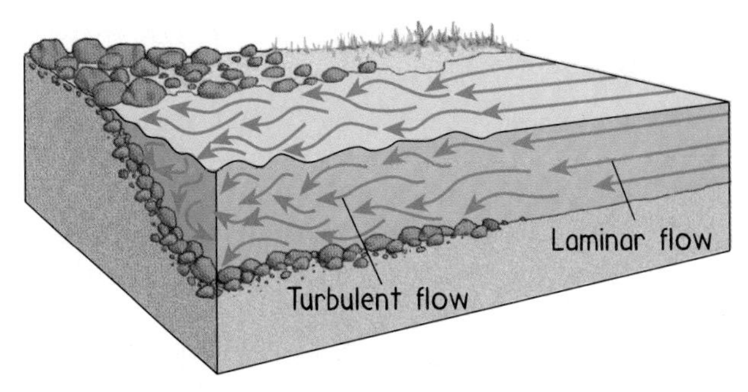

Figure 35.8
Laminar flow is slow and steady, with no mixing of sediment in the channel. Turbulent flow is fast and jumbled, stirring up everything in the flow.

Before we learn about the different erosional and depositional environments, we need to understand some characteristics of surface water flow. The pattern of flow can have a large effect on how the water changes the landscape. The flow pattern of moving water is of two types—turbulent and laminar (Figure 35.8). When water moves erratically downstream, stirring everything it comes in contact with, the flow is **turbulent.** When water flows steadily downstream with no mixing of sediment, the flow is **laminar.** In general, slow, shallow flows tend to be laminar and faster moving flows tend to be turbulent. Whether a flow is laminar or turbulent depends on the nature and geometry of the stream channel and the speed of the flow.

Erosion, Transport, and Deposition of Sediment

Surface water erodes sediment and rocks, transports them downstream and eventually deposits them in another place. In such a manner, surface water reshapes our landscape. Moving water erodes stream channels in several different ways. First of all, stream water contains many dissolved substances that chemically weather the rocks they encounter. Another powerful mechanism for erosion is hydraulic action—the sheer force of running water. Swiftly flowing streams and streams that are flooding have great erosive power as they break up and loosen large quantities of sediment and rock. The most powerful type of erosion, however, is abrasion, where sediments and particles actually scour a channel, much like sandpaper on wood. When powered by turbulent spiraling water, rock particles can rotate like drill bits to carve out deep potholes (Figure 35.9). The faster the current, the greater the turbulence, and the greater the erosion.

Figure 35.9
When powered by turbulent circular currents of water, rock particles rotate like drill bits and carve out deep potholes.

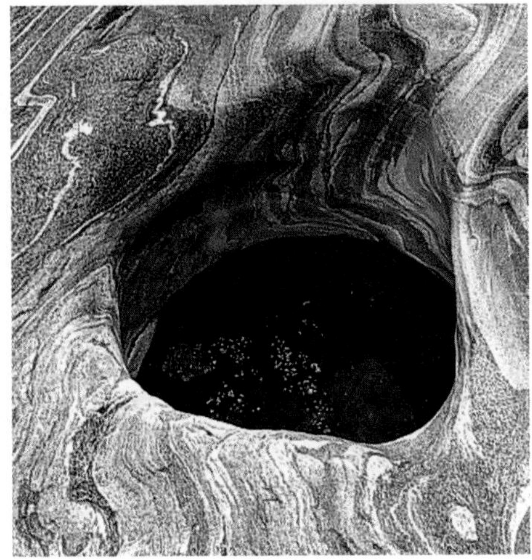

Erosion is only the beginning of the story of how surface water changes the Earth's surface. Streams carry much more than just water—they transport large amounts of sediment from one place to another. In general, laminar flows can lift and carry only the very smallest and lightest particles.

A turbulent flow, however, depending on its speed, can move and carry a range of particle sizes—from the smallest particles of clay to large pebbles and cobbles. A turbulent current gathers and moves particles downstream mainly by lifting them into the flow or by rolling and sliding them along the channel bottom. The smaller, finer particles are easily lifted into the flow and remain suspended to make the water murky. As we would expect, the faster the current, the larger the particles that can be carried. Also, the larger the volume of water, the larger the amount of sediment that can be carried.

So, streams that have a higher discharge can carry larger amounts of sediment. Streams in which the water is moving fast can carry larger sizes of sediment. The continuous abrasion of sediment in the stream channel breaks up the sediments and thus contributes to the overall decrease in particle size as we move downstream. At the river's mouth, only finer particles of sand, silt, and clay remain. As we shall soon see, these tiny particles are deposited to form deltas when a stream loses speed as it enters the sea.

Eventually, particles that are being transported by surface water drop out of suspension—they are deposited. This happens when the water loses energy and slows. When the water has less energy than the force of gravity, the particles are deposited. As a river gradually loses energy, larger particles are deposited first and then smaller ones. In this manner, surface water deposits tend to be well sorted. Deposition occurs in all surface water bodies; rivers, lakes, puddles, and the ocean.

Now that we have learned about the role of surface water in erosion, transport and deposition, let's explore some specific environments in which these processes occur. But first. . .

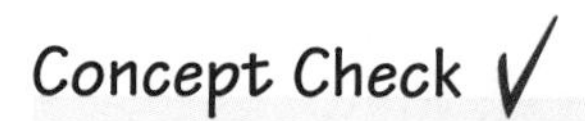

Concept Check ✓

Which is more effective in transporting sediment, laminar or turbulent flow? Why?

Check Your Answer Turbulent flow, because the water motion is irregular and sediments have a greater tendency to remain in suspension. Turbulent flow carries more energy in its churning water, and therefore carries more sediments. In laminar flow, water moves steadily in a straight-line path with no significant mixing of sediment in the channel.

Stream Valleys and Floodplains

As rainfall hits the ground, it loosens soil and washes it away. As more and more rain falls and the ground continues to lose soil, a gully forms. Once water and soil particles funnel into such a gully, a stream channel is created. This erosive action may be extremely rapid, as in the erosion of loose sediments, or very slow, as in the erosion of solid rock. Water's erosive power enables a stream to widen and deepen its channel, transport sediment away, and, in time, create a valley. In high mountain areas, the erosive action of a stream cuts down into the underlying rock to form a narrow V-shaped valley. Because the valley is narrow, the stream channel takes up the whole valley bottom. Fast-moving rapids and beautiful waterfalls are characteristic of V-shaped mountain stream valleys (Figure 35.10).

Figure 35.10
At a stream's headwaters, high gradients contribute to fast-moving rapids. When there is an abrupt change in gradient, we see a beautiful cascading waterfall.

Stream speed plays an important role in both erosion and deposition. In Chapter 34 we learned about the variations in stream speed that occur as we follow it downstream. But water speed also varies at a given location within a channel. Flow speed is slower along the stream bed, where water is in contact with the channel creating friction. Flow speed is faster near the water surface. In a large stream flowing in a straight channel, the maximum flow speed is found in the middle of the stream (Figure 35.11b). In a stream running through a bending, looping channel, the maximum flow speed is found toward the outside of each bend (Figure 35.11a and c).

As a stream flows downhill and its gradient becomes gentler and its speed slows, the focus of its energy changes from eroding downward (deepening the channel) to eroding laterally in a side-to-side motion. As a result of this lateral action, the stream develops a more curvy form (Figure 35.12). As the stream bends and curves, the flow speed in the channel shifts so that the maximum speed is always toward the outside of each bend (Figure 35.11). Rapidly moving water effectively erodes material from the outside of the bend. The result is a steep bank called a *cut bank.* Material eroded from the cut bank is transported downstream, where it may eventually be deposited in areas where stream speed decreases. Sandy *point bars* form on the insides of bends by this process.

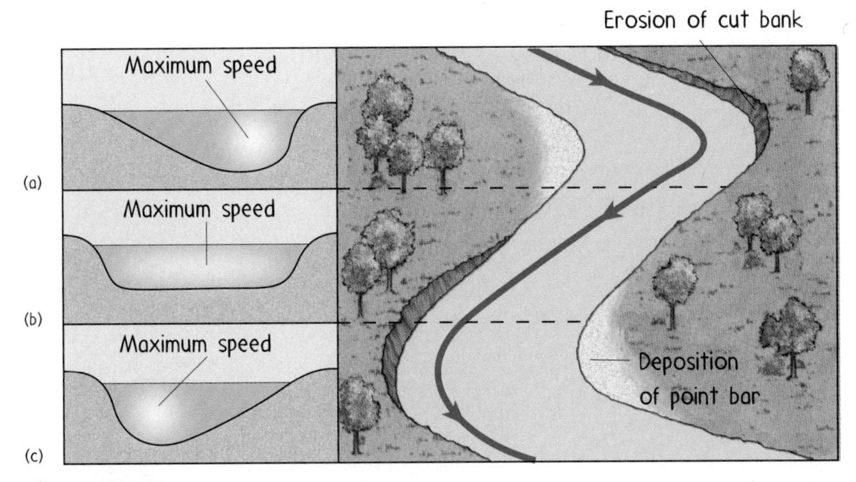

Figure 35.11
In a stream that bends (a and c), maximum flow speed is toward the outside of each bend and slightly below the surface. In a straight-channel stream (b), maximum speed is midchannel and near the water's surface. Erosion of the stream channel occurs where stream speed is greatest (cut bank); deposition occurs where stream flow slows (point bar).

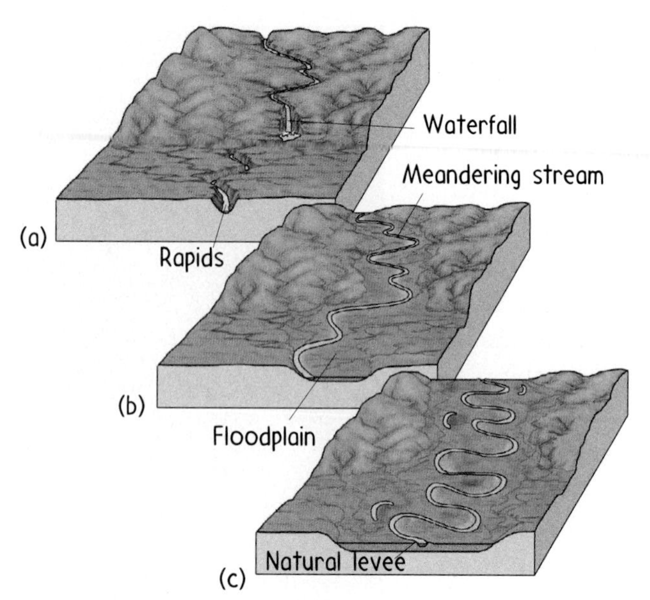

Figure 35.12
The evolution of a stream valley and development of a floodplain. (a) At the headwaters, the V-shaped stream valley is characterized by steep gradients and fast-moving water that cuts down into the stream channel. Features in this area include cascading rapids and waterfalls. (b) Downstream, with reduced gradient, the stream focuses its erosive action in a side-to-side sinuous manner, thereby widening the stream valley. (c) Farther downstream, meandering increases and further widens the stream valley to form a large floodplain.

As the stream continues to change its profile (shape) and channel, it also widens the stream's valley. Farther downstream, the curviness increases and the stream develops a wandering (meandering) pattern that winds back and forth further widening the valley into a broad, low-lying area called a *floodplain* (Figure 35.12b and c).

Floodplains are so named because they are the sites of periodic flooding. In a flood, as discharge and flow speed increase, so does the stream's ability to carry sediment. Thus when a stream overflows its banks, sediment-rich water spills out onto the floodplain. Because the speed of the water quickly decreases as it spreads out over the large, flat, floodplain, a sequence of coarse to fine particles is deposited. As expected, larger, coarse-grained sediments are deposited along the edges of the channel and smaller, fine-grained sediments are deposited farther away from the stream channel on the floodplain. The coarse materials deposited close to the stream channel act as *natural levees* that help to confine future floodwaters (Figure 35.13).

Concept Check ✓

Floodplains are often prime agricultural areas. Why would people want to work and live in areas so prone to flooding?

Check Your Answer People live and work in floodplain areas because such plains are next to rivers that provide the residents easy access to water, food, and a means of transportation. Also, because of periodic flooding, floodplain soils are often extremely fertile and thus serve as prime farmland.

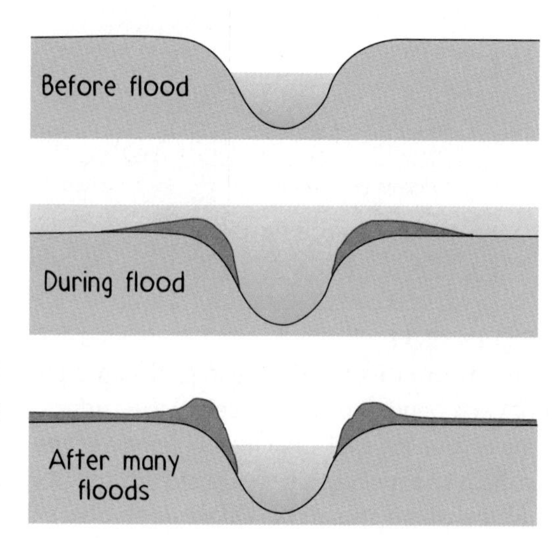

Figure 35.13
In a flood, increased discharge and flow speed help a stream to carry not only a large sediment load but also larger particles. Larger, coarse-grained sediment deposited close to the stream channel forms natural levees that confine the stream between flood stages. Successive floods increase the height of the levees and may even raise the overall elevation of the channel bed so that it is higher than the surrounding floodplain.

Deltas Are the End of the Line for a River

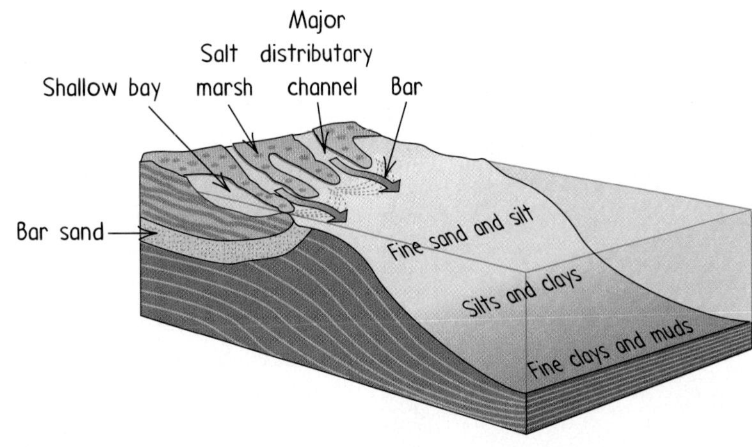

Figure 35.14
Deltas are areas of land generation. As streams flow to the sea, they carry sediment. These sediments are deposited in order of decreasing weight, with heavy, coarse particles settling at or near the shoreline and light, fine particles settling farther offshore. Layer upon layer, the depositional platform called a delta takes form.

As a stream flows into a standing body of water, such as a sea, bay, or lake, the moving water gradually loses its forward momentum. With reduced energy, the moving water slows down and the stream loses its ability to carry sediment. In short, the stream dumps the sediment load it has been transporting. In this way, the mouth of the stream and the area immediately offshore become filled with sediment—the dumped sediment is referred to as a **delta.** A delta is a fan-shaped deposit formed when a stream loses its ability to transport sediment. Sediment is deposited in order of decreasing weight, with heavy, coarse particles deposited first, at and near the shoreline. Light, fine particles are deposited farther offshore. With the continual addition of incoming sediment, the delta builds itself outward as an extension of land into the body of water.

As the main stream channel becomes choked with sediment, smaller stream channels (called *distributaries*) form to carry the moving water off to the standing body of water. As the delta continues to grow outward, the distance the water in the distributaries has to travel to reach the body of standing water becomes so great that the stream shifts its course and begins to cut new, shorter pathways to the standing water. In this way the distributaries take on the appearance of branching fingers. When the fingers get too long, the branching process begins again. As streams continue to flow to the standing water and as sediment layers are deposited one on top of the other, the delta continues to build itself outward (Figure 35.14). Thus delta environments are areas where new land is continuously created.

Some of the world's greatest rivers have huge deltas at their mouth. Millions of years ago, the mouth of the Mississippi River was where Cairo, Illinois, is today (Figure 35.15). Since that time, the delta has extended 1600 kilometers south to the city of New Orleans. Less than 5000 years ago, the site of New Orleans was underwater in the Gulf of Mexico!

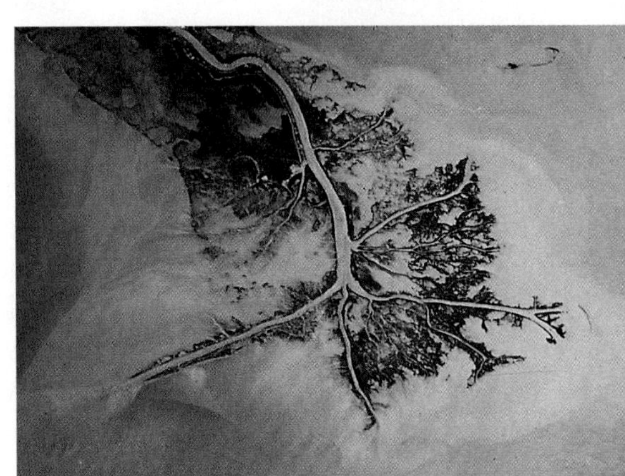

Figure 35.15
Satellite image of the Mississippi Delta. Note how smaller streams are formed branching off from the main river.

Even Dry Places Are Affected by Surface Water

We now shift our discussion to the arid desert environment, with its angular hills, shear canyon walls, and sand dunes. Because such an environment lacks moisture, mechanical

Figure 35.16
Desert erosion on a cliff face. A desert has many extremes—scorching daytime heat, chilling night air, and strong winds. Mechanical weathering physically breaks down the rocks to smaller and smaller pieces.

weathering predominates (Figure 35.16). Interestingly, although the desert lacks moisture, water is the main cause of erosion and transportation of sediments.

Rare as water is in the desert, when a heavy rain falls, the water does not have time to soak into the ground and so it causes flash floods. These flash floods transport and then deposit great quantities of debris and sediment at the bases of mountain slopes and on the floors of wide valleys and basins. Evidence of alternating wet and dry conditions can be seen in the unique sedimentary feature called *mudcracks,* found in desert basins (Figure 35.17). Flash floods can create temporary lakes that deposit mud as they disappear. When exposed to air, mud dries out and shrinks, producing cracks. Mudcracks are also associated with shallow lakes and tidal flats.

The desert environment is also ideal for the formation of evaporites (like halite). Evaporite deposits require a dry climate that is favorable to the evaporation of lake water or seawater. As the water dries out, evaporite minerals precipitate. Modern-day and ancient evaporites are found in desert basins, tidal flats, and restricted sea basins.

Concept Check ✓

What is the ultimate destination of all water flow and hence the eventual site of deposition of most sediments?

Check Your Answer Water flows eventually to the ocean, and sediments to the ocean floor.

Figure 35.17
Mudcracks, a feature unique to sedimentary rocks, are evidence of alternating wet and dry conditions in the desert.

35.4 The Work of Glaciers

Glacial Erosion and Erosional Landforms

The icy currents called glaciers are powerful agents of erosion. In many ways, a glacier is like a plow as it scrapes and plucks up rock and sediment. It is also like a sled as it carries its heavy load to distant places. As it moves across the Earth's surface, a glacier loosens and lifts up blocks of rock, incorporating them into the ice. The large rock fragments carried at the bottom of a glacier scrape the underlying bedrock and leave long, parallel scratches (like sled tracks) aligned in the direction of ice flow (Figure 35.18). These are called *striations.*

Figure 35.18
Striations mark the presence of a former glacier.

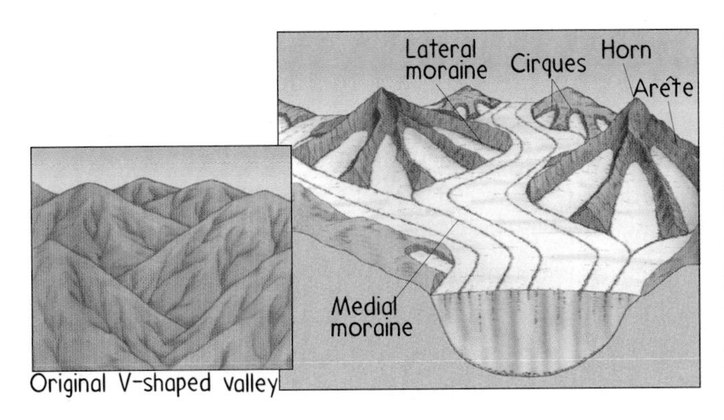

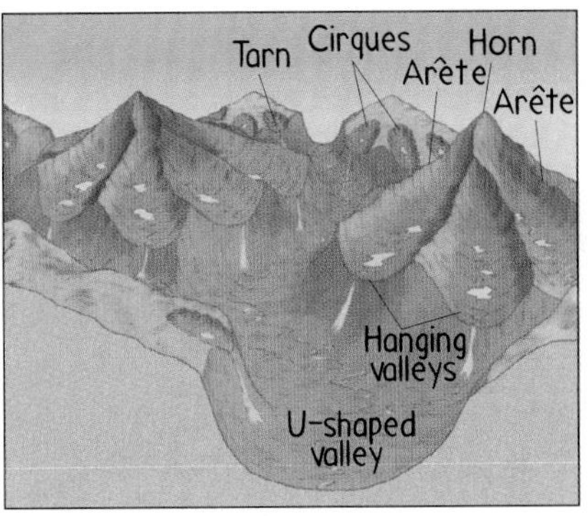

Figure 35.19 The many erosional features of alpine glaciation.

The two main types of glaciers, *alpine* and *continental*, have different erosional effects and produce different landforms. Alpine glaciers develop in mountainous areas and are often confined to individual valleys, while continental glaciers cover much larger areas. Alpine glaciers occur in most tall mountain chains in the world, like the Cascades, Rockies, Andes, and the Himalayas. The erosional features of alpine glaciation are depicted in Figure 35.19.

Continental glaciers spread over the land surface, smoothing and rounding the underlying land. Although striations are produced by both alpine and continental glaciers, they have played a larger role in the study of continental glaciers. Since continental glaciers scour very large tracts of land, they tend to lack obvious valleys (making it difficult to understand its direction of flow). By mapping striations

Geologist Bob Abrams observes the grandeur of the Juneau Ice Field, Alaska.

The Matterhorn—named for its characteristic "horn" feature.

Hanging valleys are a spectacular feature found in areas that have been shaped by alpine glacial erosion. Bridalveil Falls in Yosemite National Park spills out of a hanging valley into the larger valley that was once occupied by the main glacier.

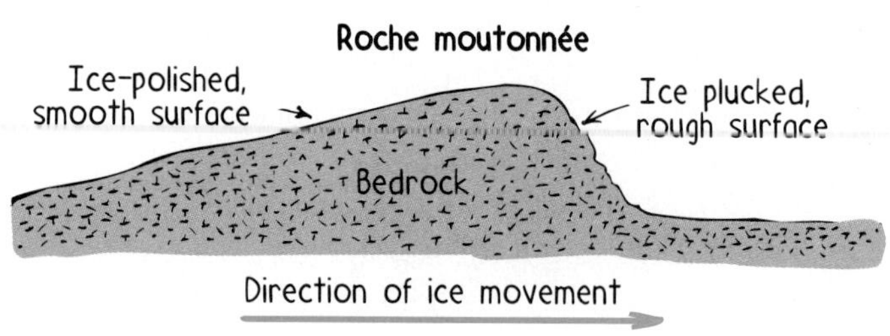

Figure 35.20
Small asymmetrical hills called roches moutonnées show the direction of continental glacial movement. On the side of the hill facing the approaching glacier, the slope is smooth and gentle. On the side facing away from the approaching glacier, the slope is rough and steep with a plucked appearance.

on land once covered by continental glaciers, geologists can decipher the flow direction of the ice. The direction of ice flow is also indicated by small, asymmetrical hills called *roches moutonnées* (Figure 35.20). In the direction of ice flow, the hill's slope is smooth and striated from the abrasion of ice on bedrock. On the down-flow side, the slope is rough and steep because the ice plucked away rock fragments from cracks in the bedrock.

Glacial Sedimentation and Depositional Landforms

As a glacier advances across the land, it picks up and transports great quantities of debris. When the glacier retreats, this debris is left behind because it is melted out of the ice. Glacial deposits are collectively called **drift.** This term dates back to the nineteenth century, when it was thought that all such debris had been "drifted in" by the great Biblical Flood.

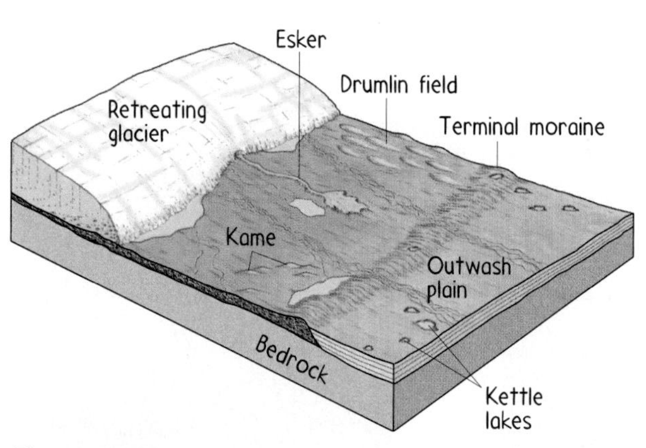

Figure 35.21
Glacial depositional landforms. Of special importance is the terminal moraine which marks the farthest point of a glaciers advance.

Drift is deposited in two main ways. When glacial sediment is released into meltwater, it is carried and deposited like any other waterborne sediment. Thus it is well sorted. This type of drift is called *outwash.* Material deposited directly by melting ice—an unsorted mixture of rock debris containing boulders and clay—is called *till.* Many of the old stone walls and fences of New England are found in areas where the surface material is glacial till. Settlers who tried to farm this land had to remove all the larger boulders before they could plow, and piled them along the edges of their fields.

The most common landform created out of till is the *moraine,* a ridge-shaped landform that marks the boundaries of ice flow. Of all the different types of moraines probably the most important is the *terminal moraine,* as it marks the farthest point of a glacier's advance (Figure 35.21). Another distinctive landform consisting of till is the *drumlin,* an elongated hill shaped like the back of a whale. Formed by continental glaciation and lined up in the direction of ice flow, drumlins have a steep, blunt end in the direction from which the ice came and a tapered gentle slope on the down flow side (Figure 35.22). The most famous drumlin is Bunker Hill in Massachusetts.

As continental glaciers of the past retreated northward, large blocks of ice were sometimes left behind. They often became incorporated into moraine deposits or were partially buried by outwash. As the chunks of ice melted, drift material caved in around them. Large dish-shaped hollows called *kettles* were created. Such kettles dot

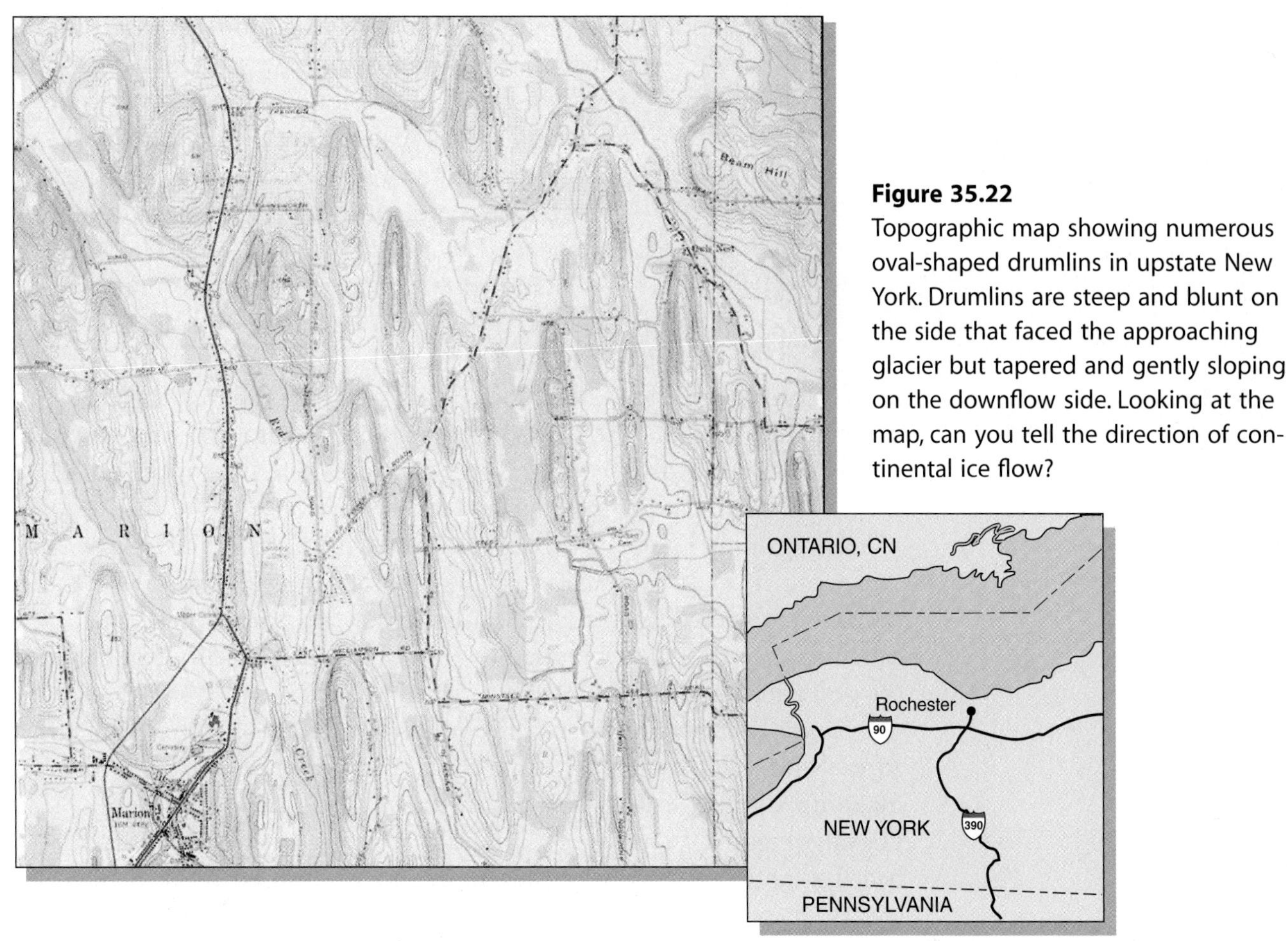

Figure 35.22
Topographic map showing numerous oval-shaped drumlins in upstate New York. Drumlins are steep and blunt on the side that faced the approaching glacier but tapered and gently sloping on the downflow side. Looking at the map, can you tell the direction of continental ice flow?

the surface of the northern United States by the tens of thousands. Today, filled with water, these kettle lakes make up the "10,000 Lakes" of Minnesota.

Concept Check ✓

What land surface forms can be used to tell the direction of glacial flow?

Check Your Answer The direction of glacial flow can be known by striations (long, parallel scratches aligned in the direction of ice flow), *roches moutonnées* (small, asymmetrical hills), and *drumlins* (elongated hills shaped like the back of a whale).

35.5 The Work of Oceans

Shoreline environments are dominated by beaches and barrier islands. In Chapter 34 we learned that winds blowing across the ocean surface generate waves. As the waves approach shallow water near land, they become higher and steeper until they finally collapse, or break. This is the *surf zone,* where wave activity moves sediment back and forth, shoreward and seaward. Because the amount of surf at a

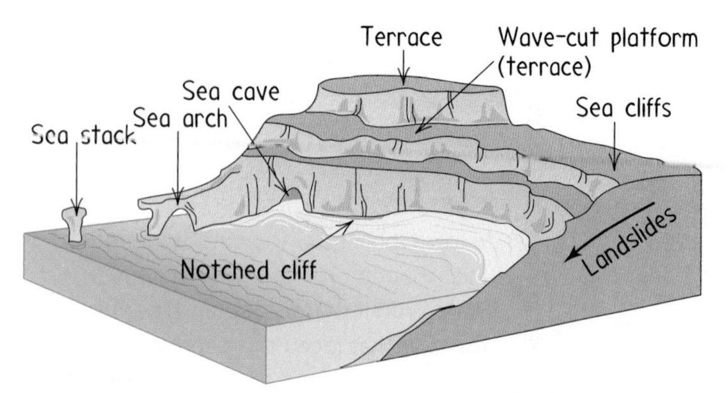

Figure 35.23
Characteristic coastal erosional landforms.

shoreline varies with time and because the rocks at any shoreline have different degrees of resistance to erosion, surf can form many different erosional features. Soft rocks and highly fractured rocks erode fastest, whereas hard rocks and unfractured ones erode more slowly.

Along shorelines consisting of hard rock, the pounding surf cuts into and notches the base of the land. As erosion proceeds, the notch deepens and the rocks above begin to jut out over the empty space at the base. As the overhanging rocks fall into the surf, the cliff progressively retreats. In time, waves cut into the cliff to form a relatively flat surface known as a *wave cut platform.*

Along some rocky coastlines, *sea caves* form along with cliffs. *Sea arches* can form if two caves, usually on opposite sides of a headland, become connected. When an arch collapses, an isolated remnant called a *sea stack* is left behind. In time, wave action also erodes away the sea stack (Figure 35.23).

Sandy beaches are the result of the turbulent motion of the surf zone. Sand-sized fragments from coral reefs and carbonate platforms make up the white-sand beaches in many island areas, such as Hawaii (Figure 35.24). Look carefully at the sand in such tropical beaches, and you'll see it is predominantly composed of shell fragments. In contrast, the sand on the beaches of the continents is mostly composed of silicate minerals. So, whereas sand in Hawaii is organic in origin, beach sand along the American western coast is largely inorganic.

Beaches are the most common shoreline depositional feature. Beaches tend to be elongated by longshore currents that form when waves approach the shoreline at an oblique angle. For example, waves approaching a north-south coastline from the northwest would cause a longshore current southward down the beach. These currents move sand down the length of the coast. Where the currents deposit sand, we have the formation of *spits.* Spits begin as submerged ridges of

(a)

(b)

Figure 35.24
(a) The white-sand beaches of California are composed of silicate minerals and are classified as inorganic. (b) The white-sand beaches of Hawaii are composed of carbonate minerals—the sediment remains of tiny shells—and are classified as organic.

Figure 35.25
Spit forming on northern tip of Cape Cod, Massachusetts.

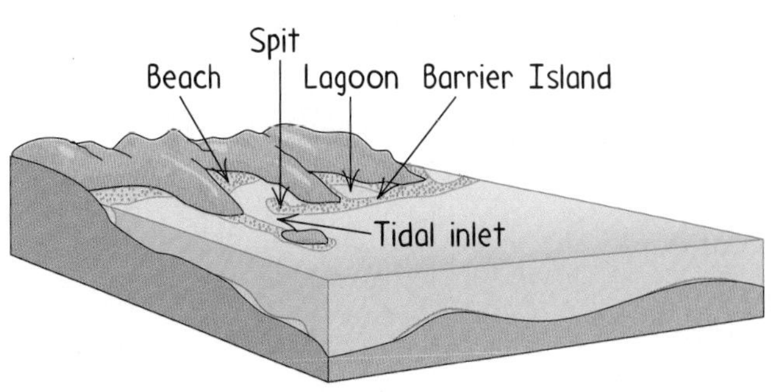

Figure 35.26
Characteristic coastal depositional landforms.

sand. As sand accumulates, the spit rises above the surface and projects from the coast into open water as a continuation of the beach, frequently as a fingerlike piece of land (Figure 35.25).

When sand ridges form parallel to the coast, they eventually grow into *barrier islands.* Barrier islands form where ridges of sand break the surface of the water for a long enough period of time that vegetation begins to take hold. Vegetation allows the new barrier island to become resistant to surf and storm erosion. With continued safeguards, the barrier island grows even more. Separated from the coast by tidal flats or shallow lagoons, barrier islands form a barricade between the coast and the open ocean (Figure 35.26). The Gulf Coast of the United States and much of the eastern shore south of New York City have abundant barrier islands. Since the lagoons separating these narrow islands from the shore are zones of relatively quiet water, small boats often use the lagoons as a "freeway" between Florida and New York, thus avoiding the potentially rough waters of the open Atlantic.

Geologic features are best viewed from an airplane. Next time you are flying in an airplane, request a window seat and enjoy the geology below.

Chapter Review

Key Terms and Matching Definitions

_____ cavern
_____ cut-bank
_____ delta
_____ drift
_____ laminar flow
_____ point bar
_____ sand dune
_____ sinkhole
_____ subsidence
_____ turbulent flow

1. Landform created when air flow is blocked by an obstacle, slowing air speed and therefore depositing sand.
2. When groundwater is severely removed from an area, the land surface is lowered.
3. A large cave.
4. A funnel-shaped hole, caused by the dissolution of groundwater into the subsurface, that is open to the sky.
5. Water flowing erratically in a jumbled manner, stirring up everything it touches.
6. Water flowing smoothly in straight lines with no mixing of sediment.
7. An accumulation of sediments, commonly forming a triangular or fan-shaped plain, deposited where a stream flows into a standing body of water.
8. Steep bank on the outside bend of a river channel. An area of erosion.
9. Sandy, gentle bank on the inside bend of a river channel. An area of deposition.
10. General term for all glacial deposits. Waterborne glacial deposits are known as *outwash.* Material deposited directly by melting ice is *till.*

Review Questions

The Work of Air

1. How are sand dunes formed?
2. How are ripple marks formed?

The Work of Groundwater

3. What are the consequences of overpumping groundwater?
4. How does rainwater become acidic? How does this affect groundwater?
5. What is karst topography? Where on Earth is it found?
6. How does a stalactite form? How does a stalagmite form?

The Work of Surface Water

7. What is the greater transporter of sediment, a laminar flow or a turbulent flow? Why?
8. Name three ways the movement of water erodes the stream channel. Which one creates potholes?
9. What factors are responsible for the formation of a stream valley?
10. Under what conditions do curvy, meandering rivers form along a floodplain?
11. What type of streams and stream valleys do we generally find in high mountainous regions?
12. What is a delta?
13. Deserts are generally dry areas. Why is water still a major factor of erosion in the desert environment?

The Work of Glaciers

14. What are striations? What is their significance?

15. What erosional features might you find in an area of alpine glaciation? See Figure 35.19.

16. What are the two types of glacial deposits? Discuss their depositional features.

The Work of Oceans

17. Describe how a sea stack forms.

18. What types of land features are associated with transport of sand from a longshore current?

Exercises

1. In the formation of a river delta, why is it that coarser material is deposited first, followed by medium and finer material farther out? Defend your answer.

2. What causes the formation of branches off the main channel of a river delta?

3. What is a sinkhole? What factors contribute to its formation?

4. The Mississippi Delta has moved south from near Cairo, Illinois, to its present location in Louisiana. Other than the length of time, why has the delta moved so far?

5. Which of the three agents of transportation (wind, water, or ice) transports the largest boulders? Why?

6. Which of the three agents of transportation is limited to transporting small size rocks? Why?

7. In what way does a glaciated mountain valley differ from a nonglaciated mountain valley? See Figure 35.19.

8. Are underground rivers ever found in nature? Explain.

9. Describe the formation of caves and caverns in limestone.

10. Do you think a stream in which the flow is laminar can become turbulent without increasing the volume of water in the stream?

11. Why is it that water is the main cause of erosion in the desert, even though there is very little water?

12. Why is it that the sand of some beaches is composed of small pieces of sea shells?

13. Describe the formation of stalactites.

14. Why do we say that surface water is both a creator and destroyer of sediments and sedimentary rocks?

15. Why do point bars form on the inside bends of meandering streams?

16. Name three environments that favor evaporite deposits. What, if anything, do they have in common?

17. What can we learn from glacial striations? In the context of modern continental glaciation, is there any other way to get the same information about the direction of glacial movement? Would that method work to learn about ancient continental glaciation?

18. How is a roche moutonnée different from a drumlin?

Suggested Readings

Abbey, Edward. *Desert Solitaire.* New York: Ballantine Books, 1968.

Roadside Geology Series. Missoula, Montana: Mountain Press.

Chapter 36: A Brief History of the Earth

How long has Planet Earth been around? Has the Earth always been as we see it today? Was the climate on Earth ever much warmer than it is today? When did sharks appear in the ocean? Do you know why the dinosaurs became extinct? When were the American and African continents connected? Are they still moving apart? In this chapter we'll learn about the many ways in which the Earth is continually changing.

36.1 The Geologic Clock

Geologic time is difficult to comprehend. We can begin to see how long geologic time is in comparison to the time spans we can imagine with the following thought exercise. Let's imagine that we can compress 4.5 billion years, the age of the Earth, into a single year. Then, Planet Earth would have begun forming from matter surrounding the sun on January 1st. The oldest known Earth rocks would appear at the end of February. Simple bacterial life would appear in the sea at the end of March, and more complex plants and animals would not emerge until late October or early November. Dinosaurs would rule the Earth in mid-December and disappear by December 26. *Homo sapiens* (humans) would appear at 11:50 P.M. on the evening of December 31. All of recorded human history would occur in the last minute of New Year's Eve!

The Earth's history is recorded in the rocks of its crust. Scientists use an assumption called *uniformitarianism* to relate what we know about present-day processes to past events. The present is the key to the past. Simply put, uniformitarianism states that the natural laws (like the laws of physics) we know about today have been constant over the geologic past. The rock record is like a long and detailed diary, containing the history of Earth-shaping events. The book is incomplete, however. Many pages, especially in the early part, are missing, and many others are tattered, torn, and difficult to read. But enough pages are preserved to give an account of the remarkable events of the Earth's four and a half billion years of history.

36.2 Relative Dating—The Placement of Rocks in Order

Sedimentary rock layers and lava flows provide good evidence of relative rock ages. This is because the rock layers were deposited one atop the other. Lower layers were formed before top layers and so are older than the top layers. Perhaps the world's most spectacular display of the rock record is the Grand Canyon of the Colorado River in Arizona. The many layers of rock exposed in the canyon walls and the thickness of these layers are testimony to great geologic activity over millions of years. The conditions under which the sedimentary layers were deposited varied widely, changing from season to season and year to year. Some layers reveal climatic cycles that span centuries, other layers indicate times when the land surface became submerged beneath a shallow sea, while still other layers show periods of increased rainfall accompanied by gradual uplift of the entire area. Millions of years after the top layer was deposited, erosion from the Colorado River cut through all these accumulated layers of sedimentary rock like a knife into a layer cake, forming the canyon!

Figure 36.1
The lowermost layers of the Grand Canyon are older than the uppermost layers—the principle of superposition.

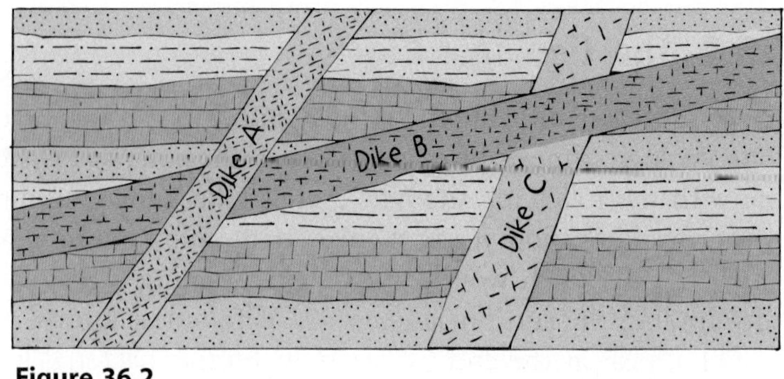

Figure 36.2
Dikes cutting into a rock body are younger than the rock they cut into. In the diagram, dike A cuts into dike B, and dike B cuts into dike C. From the principle of cross-cutting relationships, A is the youngest dike, B the next youngest, and C the oldest of the three. The horizontal layers, which are cut by all three dikes, are all older than C.

In the Grand Canyon and elsewhere, Earth scientists use five common-sense principles to determine the relative ages of rocks. The principles are:

1. **Original horizontality** Layers of sediment are deposited evenly, with each new layer laid down nearly horizontally over older sediment. Layers that are inclined at any angle—from very slight to very steep—indicate they were moved into that position by crustal disturbances after deposition.
2. **Superposition** In an undeformed (flat) sequence of sedimentary rocks, each layer is older than the one above and younger than the one below. Like the layers of a huge wedding cake, the rock record was formed from the bottom layer to the top. Upper layers are younger than lower layers.
3. **Cross-cutting** An igneous intrusion or fault that cuts through preexisting rock is younger than the rock through which it cuts (Figure 36.2).
4. **Inclusion** Inclusions are pieces of one rock type contained within another. Any inclusion is older than the rock containing it; just as pieces of rock that make up a slab of concrete were formed before the concrete was formed (Figure 36.3).
5. **Faunal succession** The evolution of life is recorded in the rock record in the form of fossils. Fossil organisms follow one another in a definite, irreversible time sequence. Fossils are a great tool for matching up rocks of similar age in different regions because any time period can be recognized by the fossils it contains.

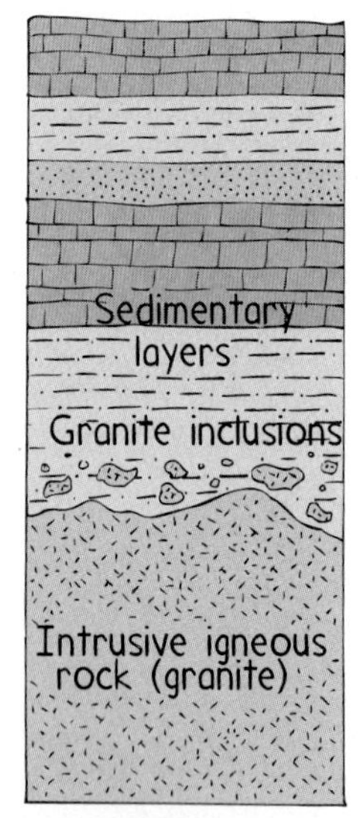

Figure 36.3
The rocks locked in the sedimentary layer existed before the sedimentary layer formed—the principle of inclusion.

Figure 36.4
Hunting for fossils can be a lot of fun. Finding one is a delightful experience, as the author indicates.

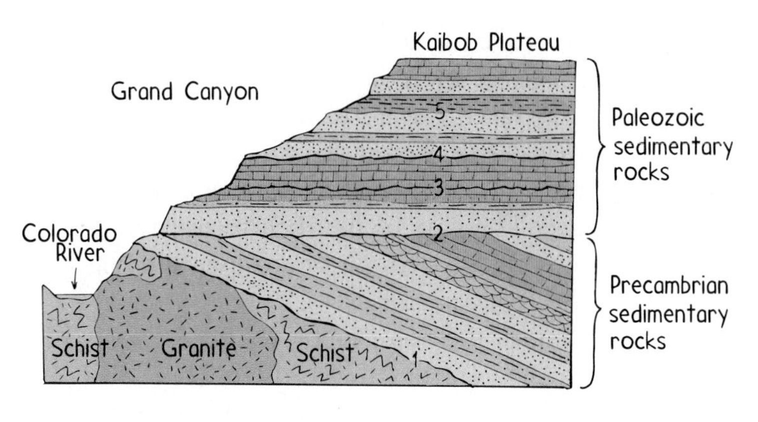

Figure 36.5
The age of the Grand Canyon can be deciphered by its sequence of rock layers. As in other places, the sequence is not continuous and there are time gaps. (1) A nonconformity separating older metamorphic rocks from sedimentary layers. (2) An angular unconformity separating older tilted layers from horizontal layers. Time gaps are also represented between horizontal sedimentary layers. The unconformities (3)–(5) are difficult to identify and often require a good eye and a knowledge of fossils.

It always comes as a surprise to find the fossil of an extinct sea animal encased in rock high above sea level. Such fossils are evidence that many of today's land surfaces were yesterday's sea bottoms. Finding fossils is a delight to first-time and experienced fossil hunters alike.

We know that most rock layers were deposited without interruption. However, a continuous sequence of rock layers that began at the Earth's formation and leads up to the present time has not survived anywhere on Earth. Weathering and erosion, crustal uplifts, and other geologic processes interrupt the normal sequence of deposition. So there are breaks or gaps in the rock record as Figure 36.5 shows. We can find these gaps, called **unconformities,** by observing the relationships of layers and fossils.

The most easily recognized of all unconformities is an **angular unconformity.** In an angular unconformity, tilted or folded sedimentary rocks are overlain by younger, relatively horizontal rock layers. They are easy to recognize because rock layers on one side of the unconformity are at an angle (not parallel) to the rock layers on the other side of the unconformity. An angular unconformity forms when older, previously horizontal rock layers are uplifted and tilted by the Earth's movements (Figure 36.6). During and after the uplift, erosion wears down the tilted layers so

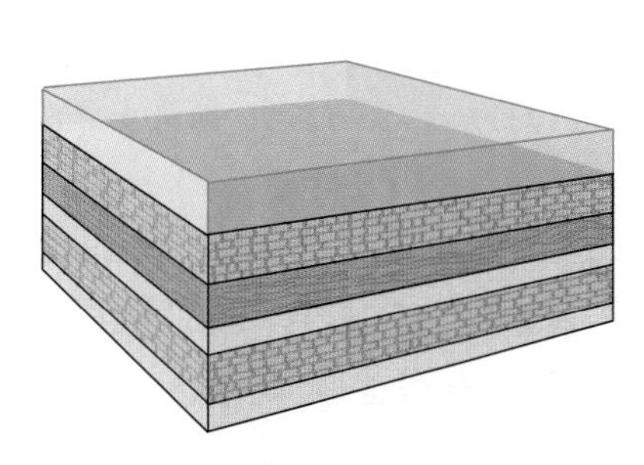

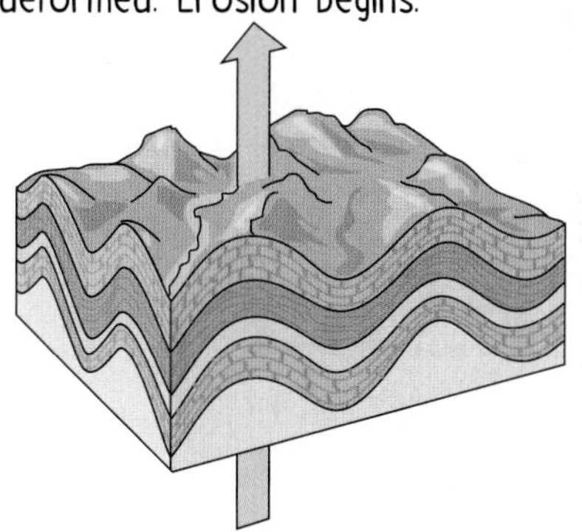

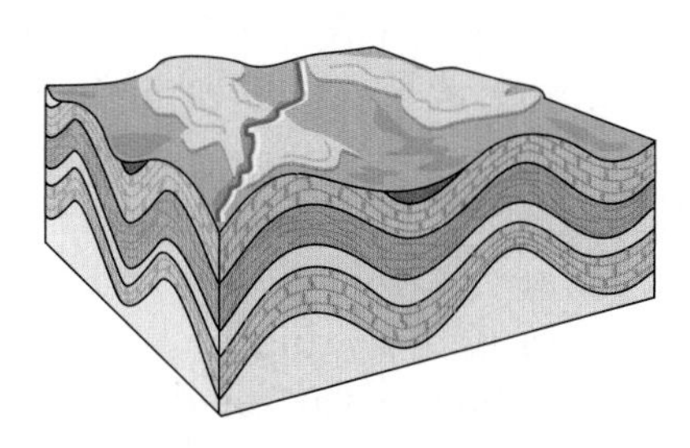

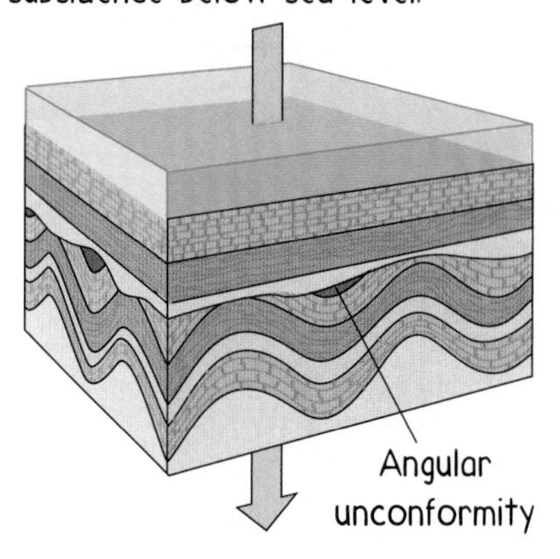

Figure 36.6
Sequence of events in the formation of an angular unconformity.

that rocks at the surface are eroded to a more or less even plane. After the period of erosion is over, more layers are deposited over the tilted ones, and these younger layers are horizontal. The angular unconformity is the "surface" that separates the tilted layers from the horizontal layers. It represents the long interval of time during which uplift and erosion took place. The part of the rock record representing this long interval is now missing because of erosion, and the unconformity is the evidence that remains.

When overlying sedimentary rocks are found on an eroded surface of metamorphic or intrusive igneous rocks, the unconformity is called a **nonconformity.** The older intrusive igneous or metamorphic rocks formed deep beneath the Earth's surface but were present at the Earth's surface when the overlying sedimentary rocks were deposited on top of them. Therefore, a nonconformity shows that a great deal of uplift and erosion occurred before the sedimentary layers were deposited, with a large stretch of time "missing" from the rock record.

Concept Check ✓

If a granitic intrusion, a dike for example, cuts into or across sedimentary layers, which is older: the granite or the sedimentary layers?

Check Your Answer The intrusion is new rock in the making. Therefore the sedimentary layers are older than the intrusions that cut into them.

36.3 Radiometric Dating Reveals the Actual Time of Rock Formation

Relative dating tells us which parts of the Earth's crust are older or younger. However, it doesn't tell us the actual age of rock—the amount of time that has passed since the rock was formed. The actual age of a rock can be determined by **radiometric dating.** This process measures the ratio of radioactive isotopes to their decay products.

Recall from Chapter 17 that atoms of the same element that contain different numbers of neutrons are isotopes. And recall our discussion of isotopic dating in Chapter 19. Some of the common radioactive isotopes frequently used for dating and estimates of geologic time are shown in Table 36.1.

In a uranium-bearing rock, there are two naturally occurring radioactive *isotopes* used for dating. Uranium-238 decays to its stable daughter isotope, lead-206. Uranium-235 decays to the stable isotope lead-207. Neither uranium isotope decays to the common isotope of lead (lead-208). Therefore, lead-206 and lead-207 found in a rock today were at one time uranium. If, for example, a sample contains equal numbers of uranium-235 and lead-207 atoms, the age of the sample is one uranium-235 half-life—704 million years. (We assume

Table 36.1

Isotopes Most Commonly Used for Radiometric Dating

Radioactive Parent	Stable Daughter Product	Half-life Value
uranium-238	lead-206	4.5 billion years
uranium-235	lead-207	704 million years
potassium-40	argon-40	1.3 billion years
carbon-14	nitrogen-14	5730 years

all the lead-207 at one time was uranium-235.) If uranium ore contains only a relatively small amount of lead-207, it is relatively young.

Radiometric dating has shown that the oldest known mineral ever found on Earth is an astounding 4.4 billion years old! The oldest whole rock found so far on the Earth is 3.8 billion years old. Radiometric dating involves some uncertainty due to the procedures used and the random nature of radioactive decay.

Organic matter uses carbon-14 for dating. Because of its short half-life (5730 years), this isotope is useful only for geologically recent events, within the last 50,000 years or so. Less than one-millionth of 1% of the carbon in the atmosphere is carbon-14. Some of this tiny amount enters plants via photosynthesis. Because all animals either eat plants or plant-eating animals, all living things have a little carbon-14 in them. Carbon-14 decays to nitrogen-14. But because living organisms continuously take in carbon, this decay is accompanied by a replenishment of carbon-14. For this reason, the amount of carbon-14 in a living organism remains constant. When the organism dies, however, the replenishment stops. So the amount of carbon-14 remaining in a fossil tells us the amount of time that has elapsed since the time of the organism's death.

The *Geologic Time Scale* was developed through the use of relative dating, and specific dates have been applied to it with radiometric dating. Now, through these techniques, we have a fascinating history

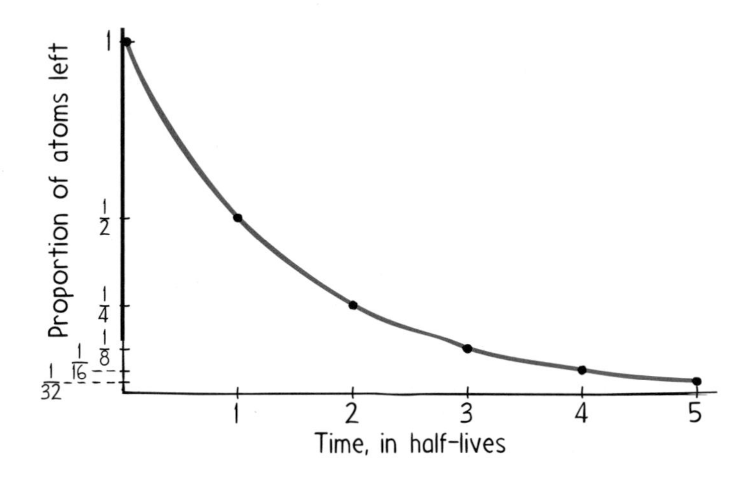

Figure 36.7
Amount of parent material versus number of half-lives as the radioactive parent decays.

THE GEOLOGIC TIME SCALE

ERA	PERIOD	EPOCH	
Cenozoic	Quaternary	Holocene	0.01 Years
		Pleistocene	1.8
	Tertiary	Pliocene	5
		Miocene	24
		Oligocene	37
		Eocene	58
		Paleocene	65
Mesozoic	Cretaceous		144
	Jurassic (first bird)		208
	Triassic		245
Paleozoic	Permian (first reptiles)		286
	Carboniferous: Pennsylvanian		325
	Carboniferous: Mississippian		360
	Devonian (first amphibian)		410
	Silurian (first insect fossils)		440
	Ordovician (first vertebrate fossils)		505
	Cambrian (first plant fossils)		544
Precambrian	3956 million years		
Precambrian			4500

of our planet. Earth scientists divide the geologic time scale into three eras, the **Paleozoic** era, the **Mesozoic** era and the **Cenozoic** era. These eras are listed in chronological order: *Paleozoic* means "time of ancient life" *Mesozoic* means "time of middle life," and *Cenozoic* means "time of recent life." Each of the three eras is further divided into periods, which are still further divided into epochs. The largest span of time, the time period preceding the Paleozoic, is known as the **Precambrian** ("the time of hidden life").

Concept Check ✓

Could carbon-14 be used for dating rocks from Precambrian time?

Check Your Answer No. Carbon-14 has a half-life of 5730 years and can be used to date only relatively young rocks. Any carbon-14 in Precambrian carbonaceous material would have long since been reduced to insignificant amounts.

36.4 The Precambrian Era, the Time of Hidden Life

The Precambrian era ranges from about 4.5 billion years ago, when the Earth formed, to about 544 million years ago, when abundant macroscopic life appeared. The Precambrian—the time of which we know the least—makes up 85% of Earth's history! Most of the rocks that formed in this early part of the Earth's history have been eroded away, metamorphosed, or recycled into the Earth's interior. Relatively few fossils are preserved in Precambrian rocks because organisms of that period didn't have the easily fossilized hard body parts that later organisms developed.

The beginning of the Precambrian era was likely a time of considerable volcanic activity and frequent meteorite* impact. Let's imagine the Earth as it was at that time: an oceanless planet covered with countless volcanoes belching forth gases and steam from its scorching interior. Huge holes and gashes left by falling meteorites scarred its surface. Intense convection in the mantle, and severe heat escaping from the interior left the surface of the Earth's early crust in turmoil. The earliest crustal formations were short-lived, ever-changing small lithospheric plates. After about 4 billion years, the Earth's heat slowly dissipated, large meteorite impacts decreased, and crustal blocks began to survive. All continents were completely devoid of life during this violent time.

* A meteorite is any solid rock object from interplanetary space that has fallen to the Earth's surface without being vaporized during its passage through the atmosphere. We shall learn more about these objects in Chapter 39 when we study the formation of the solar system.

Gases brought to the surface of the Earth by volcanic processes eventually created both a primitive atmosphere and an ocean. The early atmosphere was rich in water vapor but very poor in free oxygen. The first simple organisms for which fossils have been found are dated at 3.5 billion years old. These fossils, known as *stromatolites*, are the remains of wavy layers of algae that lived in shallow seas.

During the middle of the Precambrian era, organisms such as stromatolites and blue-green algae developed a simple version of photosynthesis. Photosynthetic organisms require CO_2 to use the sun's energy. They keep the carbon and expel oxygen. With the release of free oxygen, a primitive ozone layer began to develop above the Earth's surface. The ozone layer reduced the amount of harmful ultraviolet radiation reaching the Earth. This protection and the accumulation of free oxygen in the Earth's atmosphere permitted the emergence of new life.

Figure 36.8
Primitive stromatolites found in western Australia are dated as old as 3.5 billion years. They are very similar in structure to the present-day stromatolites pictured here. Although the first stromatolites evolved in an anaerobic (oxygen-poor) environment, in time they developed the ability to use sunlight to convert carbon dioxide to food, generating oxygen as a waste product. With the production of oxygen, many anaerobic life forms became poisoned by oxygen while adaptive stromatolites continued to flourish. Stromatolites thus changed the Earth's history as its atmosphere became oxygen-rich.

The primitive blue-green algae and bacteria that lived during this time were composed of simple cells without a nucleus. Reproduction was by simple cell division. The first evidence of single-celled organisms with a nucleus (green algae) occurs in rocks dated at approximately 1.5 billion years ago. The discovery of multicellular plants and animals, dated at approximately 700 million years ago, shows evidence of major evolutionary changes that began during the later half of the Precambrian. Some rocks in southern Australia contain diverse fossils of soft-bodied animals, ranging from jellyfish to wormlike forms. This area provides us with the first evidence of an animal community that lived in shallow marine waters.

Precambrian Tectonics

Crustal plates began to form during the Precambrian era. Evidence from folded and faulted rocks and radiometric dating indicates that the first significant continental crust movements took place 2.5 billion years ago. Continents then began to form as small landmasses came together. Scientists speculate that about a billion and a half years ago Siberia merged into the western edge of North America while Europe was converging with the eastern region of North America. Other continents were converging from the south to form the first documented supercontinent (long before Pangaea).

36.5 The Paleozoic Era, a Time of Life Diversification

The Paleozoic era is better known than the Precambrian era. However, the Paleozoic era was comparatively short. The Paleozoic era began about 544 million years ago and lasted about 300 million years. During this time, sea levels rose and fell several times worldwide. This allowed shallow seas to cover the continents and marine life to flourish. Changing sea levels greatly influenced the progression and diversification of life forms—from marine invertebrates to fishes, amphibians, and reptiles. An important event in the Paleozoic era was the development of shelled organisms. In fact, it is because of shelled organisms that we know so much more about the Paleozoic than the Precambrian era. Shelled organisms have "hard parts" and can be preserved as fossils. The Paleozoic era is divided into six periods, each characterized by changes in life forms and changes in tectonics.

The Cambrian Period, an Explosion of Life Forms

Figure 36.9
The trilobite was the dominant fossil of the Cambrian period.

The Cambrian period marks the beginning of the Paleozoic era. Almost all major groups of marine organisms came into existence during this time, as shown by abundant fossil evidence. A most important event in the Cambrian was the development of organisms having the ability to secrete calcium carbonate and calcium phosphate for the formation of outer skeletons, or shells. This ability helped organisms become less vulnerable to predators and provided protection against ultraviolet rays. Organisms could then move into shallower habitats. In addition, the support provided by a skeleton allowed organisms to grow larger.

The fossil record of the Cambrian is dominated by the skeletons of shallow marine organisms. A variety of these organisms flourished, including the *trilobite*, the armored "cockroaches" of the Cambrian sea.

The Ordovician Period, the Explosion of Life Continues

Figure 36.10
The hagfish is a descendant of the *agnatha*, a primitive jawless fish that first appeared in the Cambrian and flourished in the Ordovician.

Fossil records show that the Ordovician period was a time of great diversity and abundant marine life. The Ordovician period marks the earliest unquestionable appearance of vertebrates with the jawless fish known as the *agnatha*. The end of the Ordovician brought many extinctions. The extinctions were likely a result of widespread cooling and glaciation. Tropical shallow-water marine groups were the most affected, while high-latitude and deep-water organisms were relatively unaffected.

The Silurian Period, Life Begins to Emerge on Land

During the Silurian period much of what is now the North American continent was at or above sea level. Thick gypsum and other evaporite minerals accumulated in the vanishing shallow seas. The Silurian period brought the emergence of terrestrial (land-based) life—plants. The earliest known land plants with a well-developed circulatory system (vascular plants) appeared during the Silurian. These plants were closely tied to their water origins and inhabited only low wetlands. As plants moved ashore, so did other terrestrial organisms. Air-breathing scorpions and millipedes were common land animals during this period.

Figure 36.11
Life in the Devonian sea. In the front center, a nautiloid, which is related to the modern squid, is attacking a trilobite. The colorful organisms on the left are corals.

The Devonian Period, the Age of the Fishes

By the Devonian period, known as the "age of fishes," many dramatic changes had taken place. Plants had spread over the land surfaces. Lowland forests of seed ferns, scale trees, and true ferns flourished. In the seas, fishes diversified into many new groups. Some well adapted groups, such as the shark and bony fishes, are still present today. In the bony fish group, the lobe-finned fishes are of particular importance because they led the way for the development of land animals. Some lobe-finned fishes developed internal nostrils, which enabled them to breathe air. Today, the lungfishes and the *coelacanth* (pronounced see-la-kanth), a "living fossil," have such internal nostrils and breathe in a similar way.* Another important characteristic of the lobe-finned fishes is that their fins were lobed and muscular with jointed appendages that enabled the animals to walk. Eventually, animal life moved to land. Descended from the lobe-finned fishes, the first amphibians made their appearance during the late Devonian. The arrival of amphibians was of enormous importance to the evolutionary chain of air-breathing vertebrate land animals. Amphibians, although able to live on land, need to return to water to lay their eggs.

Figure 36.12
Warm, moist climatic conditions contributed to the lush vegetation and swampy coal forests of the Carboniferous period. These forests produced most of the world's coal deposits.

The Carboniferous Period, a Time of Great Swampy Forests

The Carboniferous period includes both the Mississippian and the Pennsylvanian periods.† Warm, moist climatic conditions contributed to lush vegetation and dense swampy forests. These swamps were the source of the extensive coal beds that now lie under North America, Europe, and northern China. In the Carboniferous period, insects underwent rapid changes that led to such diverse forms as giant

---

* The coelacanth was thought to have become extinct after the Mesozoic era. However, in 1938, the first living specimen was caught off the coast of East Africa. Since then other specimens have been discovered in the Madagascar area. The coelacanth is now considered a "living fossil."

† The term *Carboniferous period* originated in England but is now used around the world. In North America, the terms *Mississippian period,* named for the Mississippi River valley, and *Pennsylvanian period,* named for the state of Pennsylvania, refer to localities where rocks of these periods are well exposed. Whatever the name, rocks from this time period are known for their coal beds.

cockroaches and dragonflies with wingspans of 80 centimeters. The evolution of the first reptiles took place with the development of the amniote egg. The amniote egg features a porous shell that contains a membrane that provided a completely self-contained environment for an embryo. The shell protected the embryo from drying out. And this allowed animals to complete the transition, begun by amphibians in the Devonian period, from aquatic environments to land. Thanks to the amniote egg, reptiles do not need to lay their eggs in water the way amphibians do.

The Permian Period, the Beginning of the Age of Reptiles

The evolution of reptiles continued in the Permian period. The reptiles must have been well suited to their environment, for they ruled the Earth for 200 million years. (By comparison, modern humans have inhabited the Earth for less than 100,000 years.) Two major groups of reptiles appeared during the Permian time: the *diapsids* and the *synapsids*. The synapsids, which include ancestors of the earliest mammals, dominated the Permian. The most famous of the synapsids are the fin-backed *pelycosaurs* (Figure 36.13), whose fins may have

Figure 36.13
The fin-backed pelycosaurs, a famous member of the reptile group *synapsid*.

served to regulate body temperature. The diapsids were less dominant than the synapsids in the Permian period. But it was the diapsids that eventually gave rise to the dinosaurs early in the Mesozoic era.

At the end of the Permian period one of the greatest extinctions of animals in the Earth's history occurred. Marine invertebrates were affected more than terrestrial life forms were. Half of all animal families, up to 95% of all marine species, and 70% of all land species became extinct. The cause of this mass extinction is not well understood. One hypothesis is that worldwide global cooling resulted in glaciation with a lowering of sea level. Climatic extremes ranging from glaciers to deserts are clearly recorded in the rocks of this time. The long duration of low sea level, about 20–25 million years, undoubtedly put much stress on the environments of marine organisms. Yet this alone cannot account for the large marine extinction. Whatever happened took a less drastic toll on terrestrial life. Terrestrial life, although affected, continued to evolve and expanded rapidly as new land habitats appeared, likely due to the lowered sea level. As we shall see in the next section, one likely explanation for the Permian extinction is the tectonic activity that accompanied the formation of Pangaea.

Paleozoic Tectonics

The breakup of the Precambrian supercontinent began about 600 million years ago (latest Precambrian) and continued into the Cambrian period (earliest Paleozoic era). This was a time of active seafloor spreading, with the North American and European Plates diverging from each other. This active seafloor spreading opened up new ocean basins, and resulted in the first of several major worldwide rises in Paleozoic sea level.

Later in the Paleozoic (Devonian to Permian periods), the collision of all major land masses resulted in the supercontinent of *Pangaea* (Figure 36.14). Mountain-building activity continued and was widespread throughout the Appalachian Mountains in North America, the Hercynian and Caledonian Mountains in Europe, and the Ural Mountains in Russia. Crustal disturbances were so great that they affected not only the continental margins but also the inner regions of continents. The ancestral Rocky Mountains, for example, owe their formation to the dramatic collision.

The southern climate of Pangaea was dominated by widespread glaciation due to the close proximity to the south pole. Paleomagnetic evidence suggests that Pangaea was drifting as a unit across the south pole, accounting for the shift in centers of glaciation. Being very large, Pangaea greatly influenced the climate belts and the evolution of land life.

Figure 36.14
With the dramatic collision of continental landmasses, the supercontinent Pangaea was formed.

36.6 The Mesozoic Era, When Dinosaurs Ruled the Earth

The Mesozoic era, known as "the age of reptiles," is made up of three periods: *Triassic, Jurassic,* and *Cretaceous.* Reptiles that survived the Permian extinction at the end of the Paleozoic era evolved to become the rulers of the world. The most significant event of the Mesozoic era was the rise of the dinosaurs. Mammals evolved from reptiles early in the Mesozoic, but were relatively small and insignificant compared to the dinosaurs.

Land plants greatly diversified during the Mesozoic era. True pines and redwoods appeared and rapidly spread throughout the land. Flowering plants arose in the Cretaceous period and diversified so quickly that by the end of the period they were the dominant types of plants. The emergence of the flowering plants also accelerated the evolution and specialization of insects.

The end of the Cretaceous, 65 million years ago, was another time of great extinction. The dinosaurs, flying reptiles, and marine reptiles were completely wiped out, as were many nonreptiles, both on land and in the seas.

The cause of the extinction is still a source of some debate among scientists. Perhaps the best-documented hypothesis comes from Luis and Walter Alvarez. They hypothesize that the extinction was caused by the impact of a very large meteorite. Support for their hypothesis comes from an abundance of the element iridium found in sediments that mark the boundary between the Cretaceous and Tertiary periods. The concentration of iridium in a meteorite is higher than the iridium concentration in the Earth's crust. Yet all over the world the concentration of iridium at the Cretaceous-Tertiary boundary is much greater than in sediments above or below the boundary. This indicates that iridium was probably spread worldwide by the impact event. The Cretaceous-Tertiary boundary layer was deposited about 65 million years ago—the time of the great dinosaur extinction.

The Alvarez team have suggested that a large meteorite hit the Earth with such force that a gigantic light-blocking cloud of dust developed—a "nuclear winter." The dust cloud lasted for months, perhaps even longer. The huge cloud stopped photosynthesis, terminated the food supply, and chilled the Earth. Nuclear winter ended when the dust settled, and deposited the layer of iridium-enriched sediment. Other killing mechanisms associated with a meteorite impact of this size would include acid rain, tsunamis, wildfires, and a delayed greenhouse effect. According to recent research, the site of the impact crater is located in the Mexican Yucatan peninsula.

An alternative to the Alvarez hypothesis suggests that the iridium layer may have been generated from massive volcanic eruptions.

The ash and debris from these eruptions also could have blocked out the sun. A third possibility is that large-scale volcanic eruptions could have been caused by the impact of an extraterrestrial object.

The Cretaceous extinction marked the close of the Mesozoic era.

Mesozoic Tectonics

The Mesozoic era witnessed the initial breakup of Pangaea (Figure 36.15). The breakup began at the end of the Triassic period with the eruption of extensive basalt flows associated with two major rift zones. One of these rift zones initiated the separation of North America from Pangaea, thus forming the central Atlantic Ocean basin. During the Jurassic period, India started on a northward journey, while South America/Africa separated from Australia/Antarctica. The south Atlantic Ocean formed after the split of South America and Africa during the Cretaceous period. The breakup of Pangaea occurred during the entire Mesozoic era, which probably caused a worldwide rise in sea level in the Cretaceous. The plate movements that initiated the breakup continue today. Of all the former continental unions that existed in Paleozoic time, only that of Europe and Asia has survived to the present time.

Subduction of the Farallon Plate (see Section 33.3) and related tectonic *accretions* to the North American continent began no later than the Triassic period. Accretions are pieces of one plate that eventually become part of another. They often occur as one plate is subducted beneath another. This activity produced deformation and widespread volcanism in both the North American and the Andean mountain belts. Granitic batholiths of the Andes and the Sierra Nevada are the remnants left behind from the numerous volcanic arcs that rimmed the eastern Pacific basin.

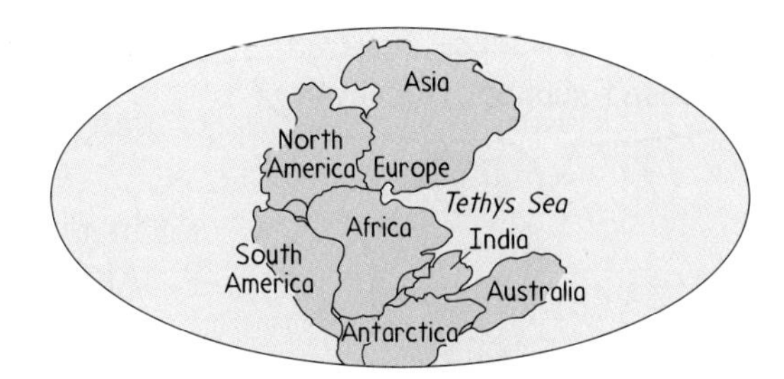

200 million years ago
Mesozoic Era

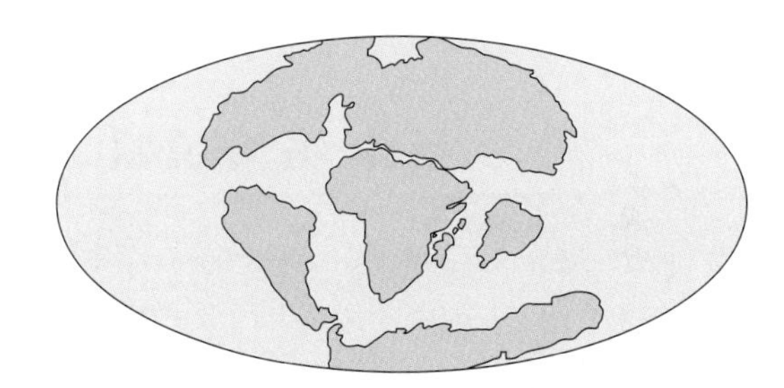

65 million years ago
Cenozoic Era

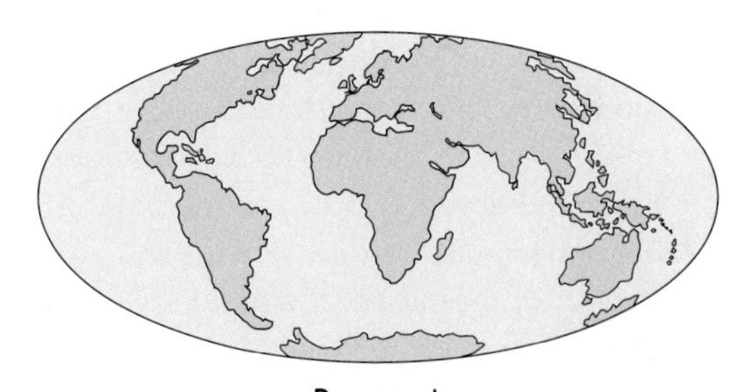

Present

Figure 36.15
Stages in the breakup of Pangaea.

36.7 The Cenozoic Era, the Time of the Mammal

The Cenozoic era, known as the "age of the mammals," is made up of two periods—*Tertiary* and *Quaternary*. From oldest to youngest, these two periods are broken up into the *Paleocene*, *Eocene*, *Oligocene*, *Miocene*, and *Pliocene* epochs for the Tertiary period; and the *Pleistocene* and *Holocene* epochs for the Quaternary period. We are currently in the Holocene epoch.

After the mass extinctions at the end of the Mesozoic era, many environmental niches were left vacant. This allowed the relatively rapid evolution of mammals in habitats formerly occupied by their extinct predecessors.* Flying bats, some large land mammals, and marine animals such as whales and dolphins began to occupy niches left vacant by the extinct Mesozoic reptiles.

Climates cooled during much of the Cenozoic era, culminating in the widespread glaciation that characterized the Pleistocene. Although this *ice age* continues today, there have been many alternations between glacial and interglacial conditions (as the following box shows). During the glacial episodes, as much as one-third of the present land area was covered by great thicknesses of ice. The huge continental glaciers were very heavy, depressing the land by their weight and altering the courses of many streams and rivers. The

Link to Global Thermodynamics: Is It Cold Outside?

Yes it is, relatively speaking. For 90% of the Earth's history there were no glaciers of continental magnitude anywhere. In fact, because such glaciers exist today, mainly in the polar ice caps and Greenland, we are technically now in an *ice age*. Since continental-scale glaciers are now restricted to the polar regions, we are in what is known as an *interglacial* period of an ice age.

Ice ages have occurred five times over the course of Earth's history. Evidence shows the first one occurred more than 2 billion years ago. Another began about 840 million years ago and lasted an incredible 240 million years! There were two ice ages during the Paleozoic era, but none in the Mesozoic. For the first 50 million years or so of the Cenozoic era, there were also no ice ages. The present ice age actually began 8–10 million years ago, but the extensive glaciation that characterized the Pleistocene epoch began about 1 million years ago.

So what causes ice ages? There most likely is no single explanation. But most scientists agree that the global-scale cooling that leads to ice ages is caused by the right combination of three things: (1) the arrangement of continents around the globe, (2) the amount of sunlight reflected back into space, and (3) the geometry of the Earth's rotation on its axis and revolution around the sun.

The arrangement of continents greatly influences ocean and atmospheric currents. These currents in turn distribute ocean and atmospheric heat around the globe. Continents grouped together in one location are easier to warm, because warm waters from the equator flow with less obstruction toward the cooler poles. When continents are spread out around the globe, as they are today, heat distribution is less efficient.

When sea level is lower, for whatever reason, more land area is exposed. This increased amount of land area increases the amount of sunlight reflected back into space. This reflection results in cooler temperatures globally. Cloud cover and/or dust in the atmosphere also causes sunlight to be reflected back into space, reducing the absorption of solar radiation.

Other factors include variations in the angle of the Earth's rotation axis (currently tilted about 23.5°), wobbling of the Earth's rotation axis, and variations in the eccentricity ("ovalness") of the Earth's orbit around the sun. Combinations of these factors can reduce solar radiation at high northern latitudes during the summer. If the reduction is great enough, much of the snow from the preceding winter doesn't melt and, over many years, can form continental-scale glaciers.

So what's next? Large-scale glaciation or global warming? Only further research can provide hope of finding an answer. Anything less is pure speculation, at best.

* Biologists refer to this phenomenon as *adaptive radiation* because organisms begin to adapt to new environments and radiate, or diversify, away from a smaller set of ancestors.

glaciers eroded and scratched the land in some places and deposited huge moraines in others, leaving behind abundant evidence of the extent of their former existence.

The Cenozoic era also brought about the evolution of humans. The extensive glaciation of the Pleistocene caused sea level to drop because a great deal of water was tied up in glaciers. Even though the distribution of landmasses was essentially the same as it is today, the lowered sea level resulted in "land bridge" connections between landmasses that are now separated by water. One of these land bridges existed across the present-day Bering Strait. It provided the route for the human migration from Asia to North America. The expansion of humans, not only into North America but also throughout the world, coincided with a period of extinction that occurred during the Pleistocene. The Pleistocene extinctions primarily involved large terrestrial mammals, while marine animals were for the most part unaffected. In North America, many large mammals became extinct after humans arrived. And in Africa, mammalian extinctions can be related to the appearance of the Stone Age hunters.

The cause of the Pleistocene extinction is a much-debated issue. The extreme climatic variation that existed at the time could have been partly responsible. Even though large-scale glaciation was occurring in some regions, the climate in many areas was relatively mild. This leads some scientists to believe that harsh climate was just a small factor in the Pleistocene extinctions.

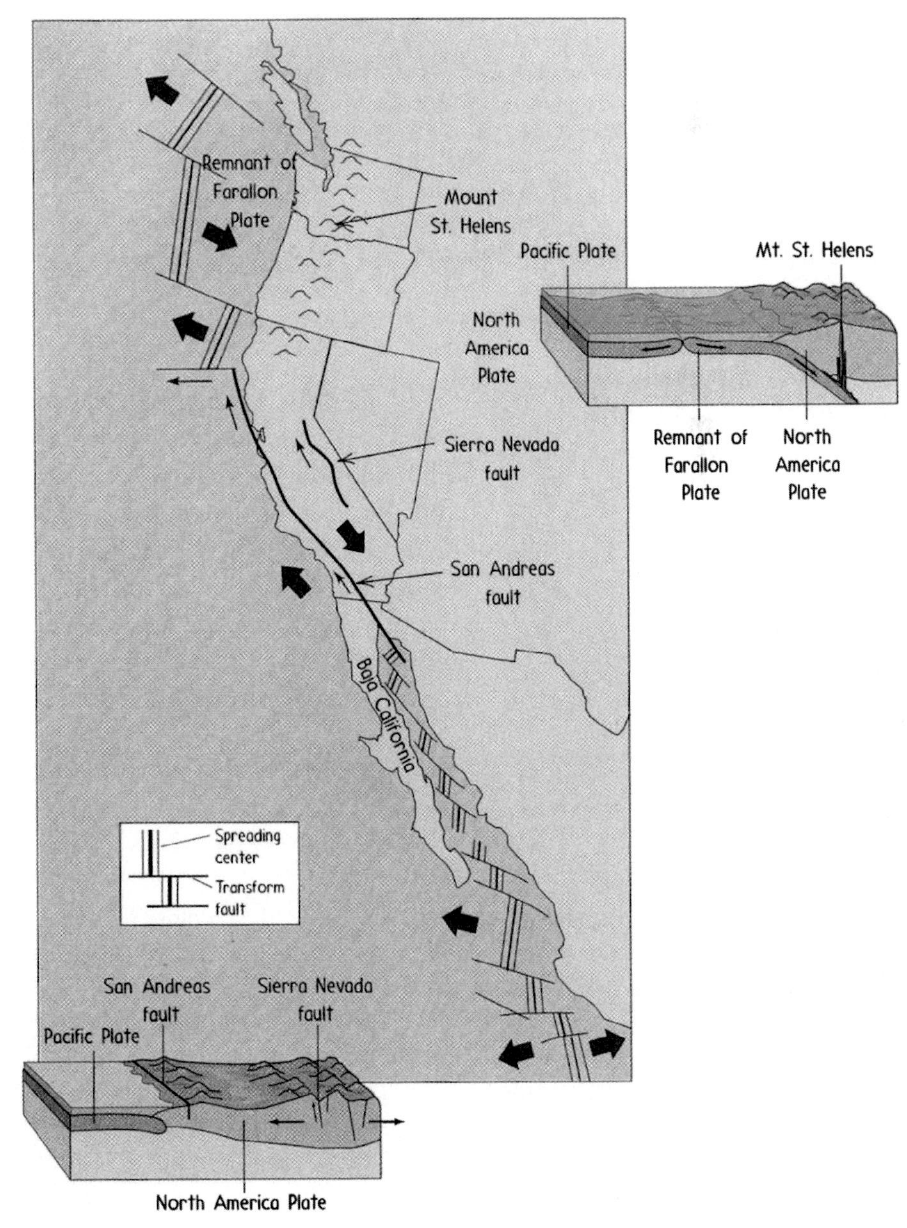

Figure 36.16
The San Andreas Fault is the result of an encounter between the North American Plate and the Pacific Plate. As the fault grew longer, Baja California was torn from the continental margin.

Cenozoic Tectonics

Enormous tectonic disturbances occurred rapidly throughout the world during the Tertiary period. In the early Tertiary period, there was a spreading center off the western margin of North America, with the Pacific Plate on the west and the Farallon Plate on the east (Figure 36.16 and

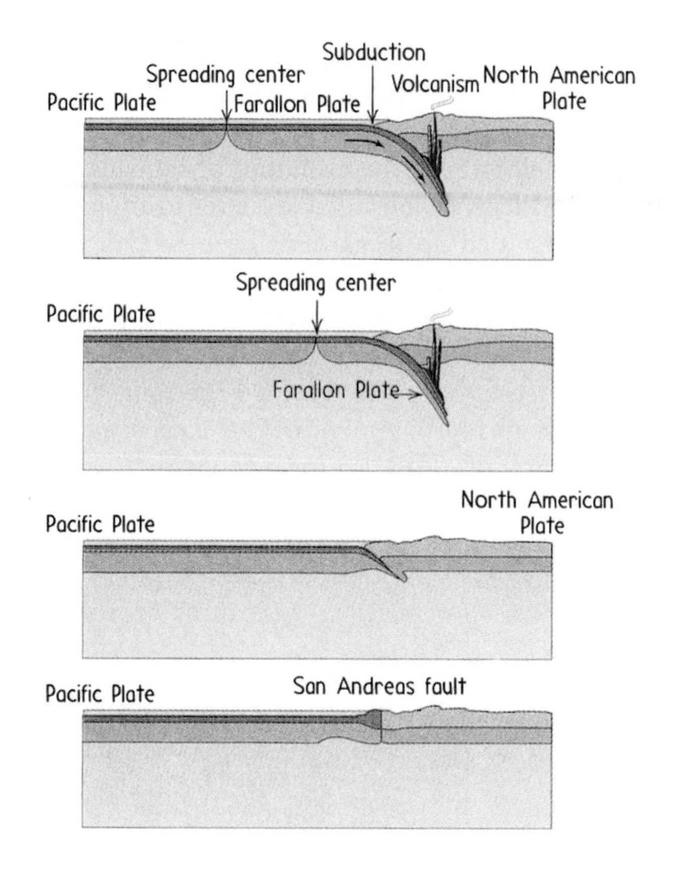

Figure 36.17
Subduction sequence of the Farallon Plate beneath the North American Plate. As the spreading center and the continental margin approached each other, the San Andreas Fault formed as a transform fault between the Pacific and North American Plates.

36.17). As the Farallon Plate subducted beneath North America, the spreading center approached the North American continental margin. The collision between the westward moving North American Plate and the Pacific ridge system occurred about 30 million years ago, giving birth to the San Andreas Fault. Baja California was torn away from the Mexican mainland, and the Gulf of California was created. Because the plates are still moving, western California and Baja California will eventually either become completely detached from the mainland or will find themselves joined to western Canada.

Human Geologic Force

Although the "human age" amounts to only a brief 0.002% of geologic time, we are almost certainly the most clever and adaptable organism to have evolved on the planet. All life forms alter their environment. Humans do it more, as we manipulate it to meet our needs. We have but to look at the irrigation systems of Mesopotamia, the cultivation of the Nile, the plowing of the prairies in the Great Plains, the invention of machines to further utilize the land, and the dams and locks on the Mississippi, Missouri, and Colorado rivers to illustrate the human role in geologic changes. These geologic changes also include such problems as deterioration of the ozone layer, hydrocarbon pollution, and global warming. Because we have the capacity to affect geologic change, it is imperative that we take care of our terrestrial home. It's the only one we've got!

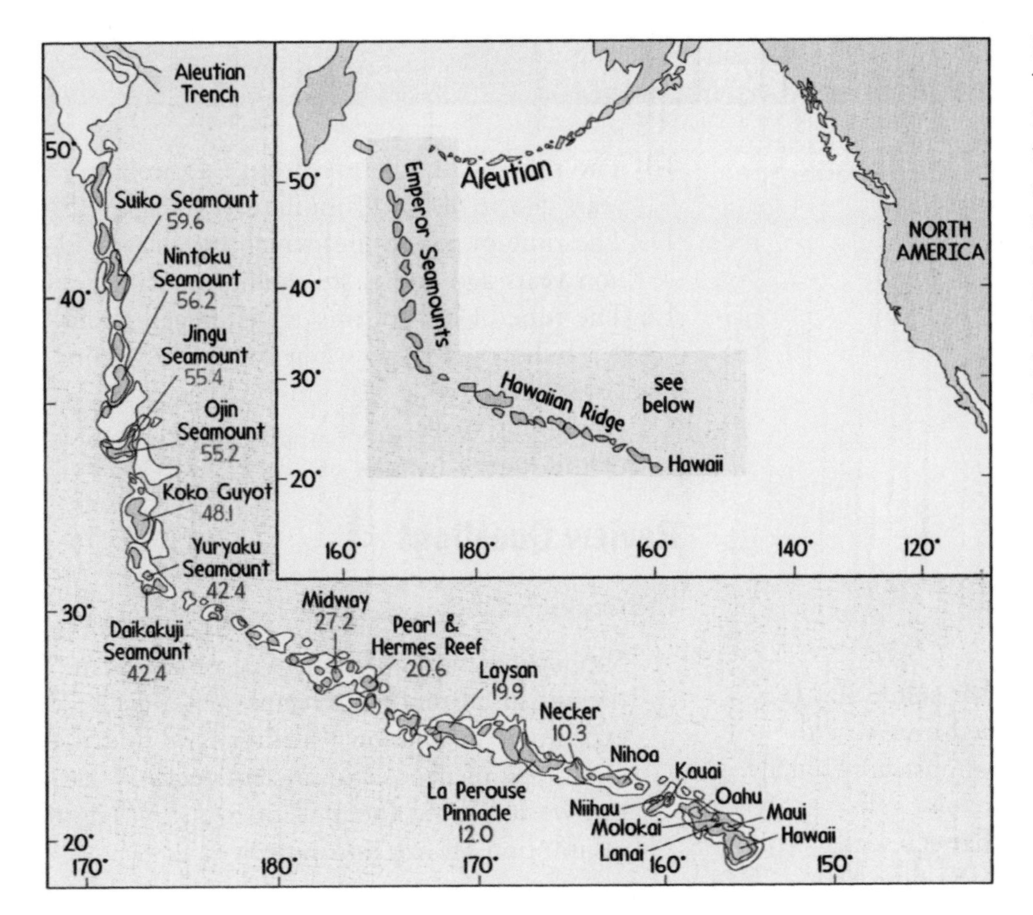

Figure 36.18
The Hawaiian Island/Emperor Seamount Chain. The bend in the chain shows the change in direction of the Pacific Plate as a result of the collision of the North American Plate with the Pacific ridge. The red numbers indicate the age (in millions of years) of the individual islands and seamounts.

The Hawaiian Island/Emperor Seamount chain (Figure 36.18) gives evidence of another significant Tertiary tectonic disturbance: the change in direction of the Pacific Plate. The bend in the chain of islands occurred between 30 and 40 million years ago (mid-Tertiary) when plate motion changed from nearly due north to northwesterly. The change in direction occurred at about the same time in Earth history as the collision of the North American Plate with the Pacific ridge system.

Chapter Review

Key Terms and Matching Definitions

_____ angular unconformity
_____ Cenozoic
_____ cross-cutting
_____ faunal succession
_____ inclusions
_____ Mesozoic
_____ original horizontality
_____ Paleozoic
_____ Precambrian
_____ radiometric dating
_____ superposition
_____ unconformity

1. Relative dating principle that states that layers of sediment are deposited evenly, with each new layer laid down almost horizontally over the older sediment.
2. Relative dating principle that states that in an undeformed sequence of sedimentary rocks, each bed or layer is older than the one above and younger than the one below.
3. Relative dating principle that states that when an igneous intrusion or fault cuts through other rocks, the intrusion or fault is younger than the rock it cuts.
4. Relative dating principle that states that any inclusion (pieces of one rock type contained within another) is older than the rock containing it.
5. Relative dating principle that states that fossil organisms succeed one another in a definite, irreversible order.
6. A break or gap in the geologic record, caused by an interruption in the sequence of deposition or by erosion of preexisting rock.
7. An unconformity in which older tilted strata are overlain by horizontal younger beds.
8. A method of calculating the age of geologic materials based on the nuclear decay of naturally occurring radioactive isotopes.
9. The time of ancient life, from 544 million years ago to 245 million years ago.
10. The time of middle life, from 245 million years ago to about 65 million years ago.
11. The time of recent life, which began 65 million years ago and is still going on.
12. The time of hidden life, which began about 4.5 billion years ago when the Earth formed, lasted until about 544 million years ago (beginning of Paleozoic), and makes up 85% of the Earth's history.

Review Questions

The Geologic Clock

1. Suppose we find a certain type of sediment deposit in all modern streams. On a geologic expedition in unknown territory, we find the same type of deposit in ancient rocks. What can we say about the ancient rocks? What assumption are we making?

Relative Dating—The Placement of Rocks in Order

2. What five principles are used in relative dating? Describe each one.
3. When a granitic dike is found in a bed of sandstone, what can be said about the relative ages of the dike and the age of the sandstone? What is this principle called?
4. Why aren't all rock formations found with a continuous sequence from the beginning of time to the present?
5. Explain how fossils of fishes and other marine animals occur at high elevations such as the Himalayas.
6. In an undeformed sequence of rocks, fossil X is found in a limestone layer at the bottom of the formation, and fossil Y is found in a shale layer at the top of the formation. What can we say about the ages of fossils X and Y?

Radiometric Dating Reveals the Actual Time of Rock Formation

7. What is the definition of half-life?
8. What isotope is best for dating very old rocks?
9. What isotope is commonly used for dating sediments or organic material from the Pleistocene?

The Precambrian Era, the Time of Hidden Life

10. Which of the geologic time units spans the greatest length of time?
11. How old is the Earth?

The Paleozoic Era, a Time of Life Diversification

12. The Paleozoic era experienced several fluctuations in sea level. What effect did this have on life forms?
13. What is the Silurian period best known for?
14. The Devonian is known as "the age of fishes." What were some of the Devonian life forms?
15. Why do many geologists consider the lobe-finned fishes to be especially significant?
16. During what time period were most coal deposits laid down? Why was this period unique?
17. What group evolved from the amphibians with the development of the amniote egg?

The Mesozoic Era, When Dinosaurs Ruled the Earth

18. What is the Mesozoic era known as?
19. What is the most likely cause of the Cretaceous extinction that wiped out the dinosaurs?
20. What Pangaean landmass survives to this day?

The Cenozoic Era, the Time of the Mammal

21. What geologic event allowed the development of many mammals in the early Cenozoic?
22. What role did tectonic activity play in the formation of the San Andreas Fault?

Exercises

1. Suppose you see a group of sedimentary rock layers overlain by a basalt flow. A fault displaces the bedding of the sedimentary rock but does not intersect the basalt flow. Relate the fault to the ages of the two rock types in the formation.
2. If a sedimentary rock contains inclusions of metamorphic rock, which rock is older?
3. Refer to the following figure. Using the principles of relative dating, determine the relative ages of the rock bodies and other lettered features. Start with the question: What was there first?

Sequence of events

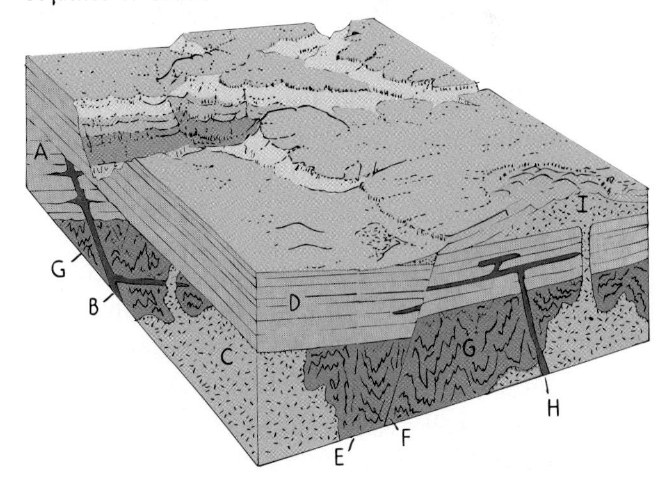

4. Which isotopes are most appropriate for dating rocks from the following ages: (a) the early Precambrian; (b) the Mesozoic; (c) the late Pleistocene?

5. Has the amount of uranium in the Earth increased over geologic time? Has the amount of lead increased? Explain.
6. Granitic pebbles within a sedimentary rock have a radiometric age of 300 million years. What can you say about the age of the sedimentary rock? Nearby, a dike having a radiometric age of 200 million years intrudes an outcrop of the same sedimentary rock. What can we say about the age of the sedimentary rock?
7. Geologists often refer to the early Paleozoic as the "Cambrian Explosion". What do you think is meant by this phrase?
8. What is the difference between a nonconformity and an angular unconformity?
9. What key developments in life occurred during the Precambrian era?
10. What factors are believed to have contributed to the generation of free oxygen during the early Precambrian? In what way did the increase in oxygen affect our planet?
11. What evidence do we have of Precambrian life?
12. Why can we find Paleozoic marine sedimentary rocks, such as limestone and dolomite, widely distributed in the continental interiors?
13. Coal beds are formed from the accumulation of plant material that becomes trapped in swamp floors. Yet coal deposits are found on the continent of Antarctica, where no swamps or vegetation exists. How can this be?
14. A radiometric date is determined from mica that has been removed from a rock. What does the date signify if the mica is found in granite? What does the date signify if the mica is found in a sedimentary rock?
15. How does iridium relate to the time of the extinction of the dinosaurs?
16. During the Earth's long history, life has emerged and life has perished. Briefly discuss the emergence of life and the extinction of life for each era.
17. In what ways could sea level be lowered? What effect might this have on existing life forms?
18. What could cause a rise in sea level? Is this likely to happen in the future? Why or why not?
19. What general assumption must be made to understand the processes that occurred throughout the Earth's history?
20. If fine muds were laid down at a rate of 1 cm per 1000 years, how long would it take to accumulate a sequence 1 km thick?

Suggested Readings and Web Sites

Gould, Stephen Jay. *Wonderful Life—The Burgess Shale and the Nature of History.* New York: Norton, 1989.

http://www.scotese.com/earth.htm

http://seaborg.nmu.edu/earth/

http://www.ucmp.berkeley.edu/help/timeform.html

http://www.enchantedlearning.com/subjects/Geologictime.html

Chapter 37: The Atmosphere, the Oceans, and Their Interactions

How did Earth get its atmosphere? Where did the oceans come from? How do the atmosphere and ocean interact? Is it true that without an atmosphere, the ocean would boil away? What causes seasons? Is it true that North America is warmer when it is farthest from the sun? What causes wind? Can ocean water freeze over and become ice? How salty are the oceans? How the atmosphere interacts with the oceans affects all of us. Let's explore and learn why!

37.1 Earth's Atmosphere and Oceans

As seen from space, Planet Earth is blue with wisps of silver. It is blue because of the oceans. The wispy silver is clouds in the atmosphere. Seventy percent of the Earth's surface is covered by water (Figure 37.1). The remaining 30% is land, most of which is located in the Northern Hemisphere (Figure 37.2). Although the many oceans are named for their various locations, they are connected. All of Earth's oceans are actually a single continuous ocean.

Earth's Oceans Moderate Land Temperatures

As we learned in Section 34.1, the oceans are the reservoir from which water evaporates into the atmosphere to later precipitate as rain and snow. Oceans play a major role in moderating the Earth's temperature and climate. Recall from Chapter 6 that water has a large heat capacity. Because of this, Earth is slow to heat up or cool down. For water to cool down, it must transfer large amounts of heat to its surroundings. On the other hand, water absorbs a great deal of heat before its temperature increases. The large heat capacity is the reason that land bordering the oceans experience moderate temperatures. The moderating influence of the oceans can be seen when we look at seasonal

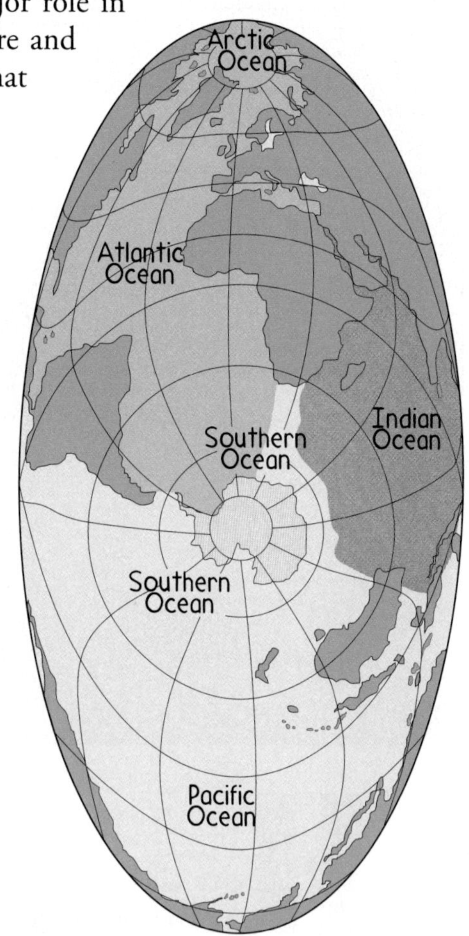

Figure 37.2
When a map is centered over Antarctica, the expanse of the world ocean can be seen. In terms of size and volume, the Pacific Ocean accounts for more than half of the world ocean and is thus the largest ocean. In fact, the Atlantic and Indian Oceans combined would easily fit into the space occupied by the Pacific.

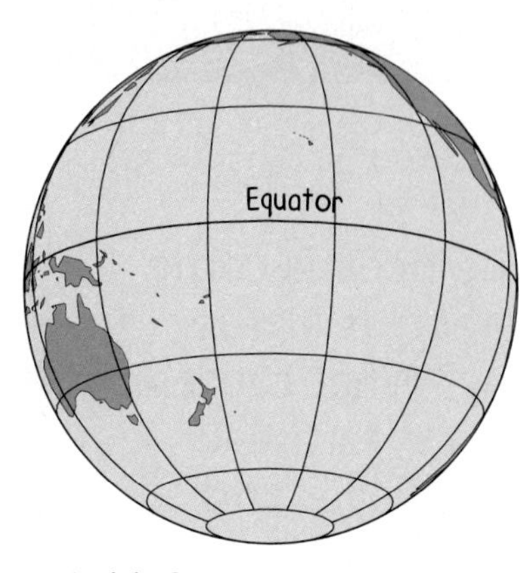

Figure 37.1
Most of the Earth's surface is covered by water. We can divide the Earth into (a) an ocean-dominated hemisphere, and (b) a land-dominated hemisphere.

temperature variations for two cities at the same latitude: coastal San Francisco, California and continental Wichita, Kansas (Figure 37.3). Whereas temperatures in San Francisco tend to have small seasonal variations, temperatures in Wichita show strong seasonal fluctuations—cold winters and hot summers. The oceans do a great job of both making summers cooler, and winters warmer.

Figure 37.3
Comparison of seasonal temperature ranges for coastal San Francisco, California, and continental Wichita, Kansas.

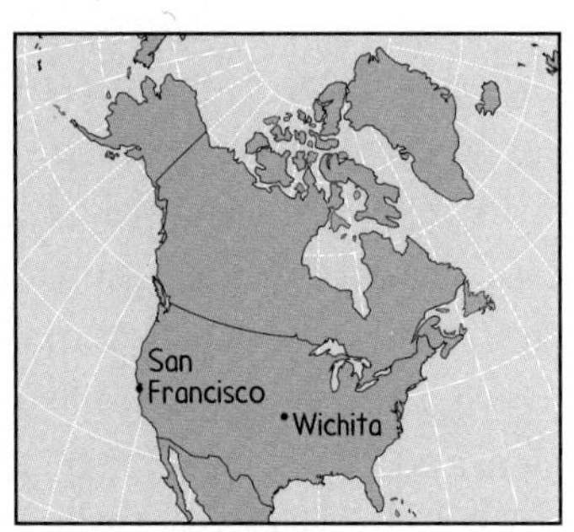

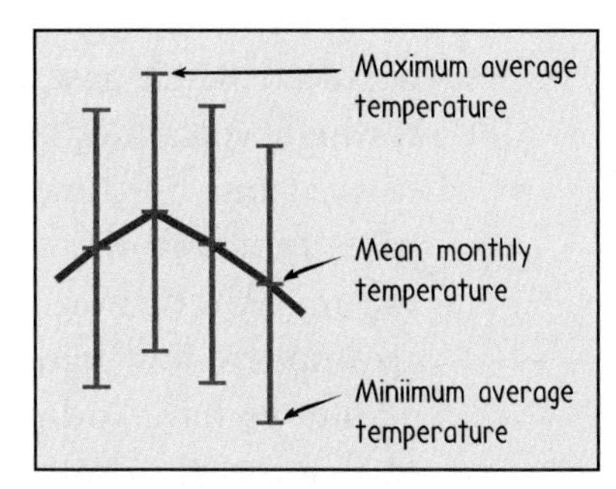

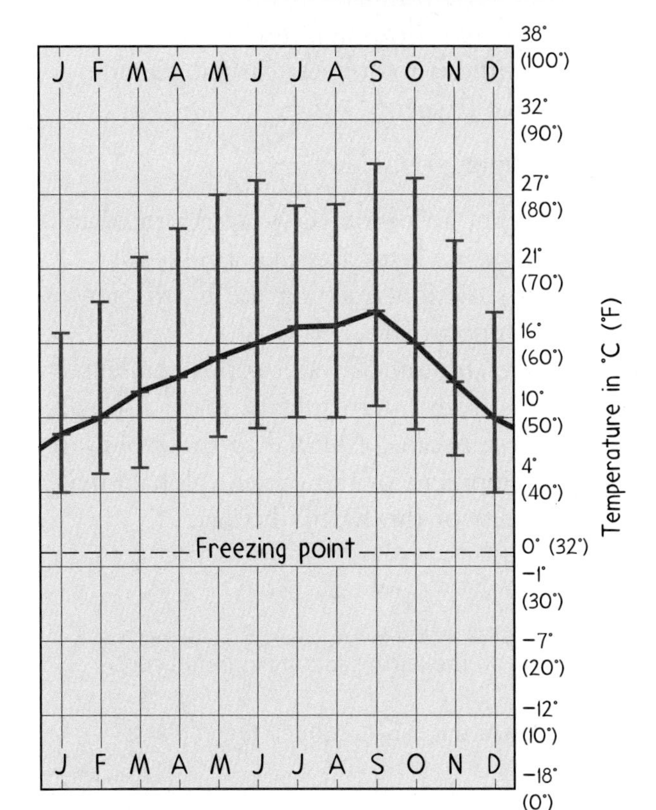

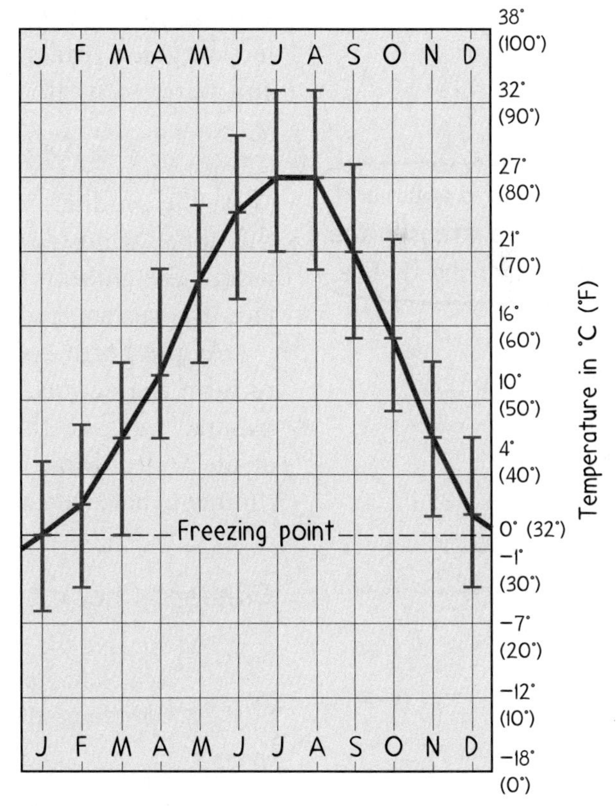

Station: San Francisco, California
Latitude/longtitude: 37°37′ N, 122°23′ W
Average annual temperature: 14°C (57.2°F)
Total annual precipitation: 47.5 cm (18.7 in.)
Elevation: 5 m (16.4 ft)
Population: 750,000
Annual temperature range: 9°C (16.2°F)

Station: Wichita, Kansas
Latitude/longtitude: 37°39′ N, 97°25′ W
Average annual temperature: 13.7°C (56.6°F)
Total annual precipitation: 72.2 cm (28.4 in.)
Elevation: 402.6 m (1321 ft)
Population: 280,000
Annual temperature range: 27°C (48.6°F)

Evolution of the Earth's Atmosphere and Oceans

The Earth probably had an atmosphere before the sun was fully formed. This first atmosphere was possibly composed of only hydrogen and helium, the two most abundant gases in the universe, along with a trace amount of ammonia and methane. But no oxygen could be found in the early atmosphere. When the sun was born, the blast from the sun's formation must have produced a strong outflow of charged particles—an outflow strong enough to sweep the Earth of its earliest atmosphere.

The next stage in the formation of the atmosphere probably occurred when gases trapped in the Earth's hot interior escaped through volcanoes and fissures at the Earth's surface. The gases spewed out by these early eruptions were probably much like the gases found in the volcanic eruptions of today—about 85% water vapor, 10% carbon dioxide, and 5% nitrogen, by mass. The early atmosphere still had no free oxygen and could not support the type of life we have today. As we learned in Chapter 36, the production of free oxygen did not occur until the primitive plants known as stromatolites and green algae appeared. Stromatolites and green algae, like all green plants, use photosynthesis to convert carbon dioxide and water to hydrocarbon and free oxygen.

$$CO_2 + H_2O + \text{light} \rightarrow CH_2O + O_2$$

With the production of free oxygen, an ozone (O_3) layer formed in the upper atmosphere. Since the ozone layer acts like a filter to reduce the amount of ultraviolet radiation reaching the Earth's surface, the surface became able to support life.

As the Earth cooled, the huge amount of water vapor condensed to form the oceans. Comet debris from interplanetary space also contributed water to the oceans. These oceans, essential to the evolution of life and ultimately to the development of the present global environment, have remained for the rest of the Earth's history.

Concept Check ✓

1. Why are the hottest climates on the Earth typically found in continental interiors?
2. Did the ozone layer exist before the Earth acquired green plants?

Check Your Answers

1. The large heat capacity of water tends to keep coastal areas from experiencing extreme temperatures. Therefore very hot climates are usually some distance away from the ocean.
2. No. The formation of ozone, O_3, was preceded by the introduction of free oxygen, which came from photosynthesizing plants.

37.2 Components of the Earth's Atmosphere

If gas molecules in the atmosphere were not always moving, gravity would force them to lie on the ground like popcorn at the bottom of a popcorn machine. But add heat to the popcorn, or to atmospheric gas, and both will bumble their way up to higher altitudes. Popcorn attains speeds of maybe 1 meter per second and can rise a meter or two. Air molecules move at speeds of about 450 meters per second and some rise to an altitude of more than 50 kilometers. If there were no gravity, gas molecules in the atmosphere would fly off into outer space.

If you have ever gone mountain climbing, you probably noticed that the air grows cooler and thinner with increasing elevation. At sea level, the air is generally warmer and denser. The greater density near the Earth's surface is due to gravity. Like a deep pile of feathers, the density is greatest at the bottom and least at the top. More than half the atmosphere's mass lies below an altitude of 5.6 kilometers, and about 99% lies below an altitude of 30 kilometers. Unlike a pile of feathers, the atmosphere doesn't have a distinct top. It gradually thins to the near vacuum of outer space.

Since air has weight, it exerts pressure on the Earth's surface. This pressure is known as *atmospheric pressure* or simply *air pressure.* Like density, air pressure also decreases with increasing height above the Earth's surface. Interestingly, the weight of the air on the ocean surface keeps the ocean from boiling away. Recall in Chapter 10 that water will boil at 0°C when no air pressure acts on it. So not only birds, but also fish, should appreciate the atmosphere.

Table 37.1 shows that the Earth's present-day atmosphere is a mixture of various gases—primarily nitrogen and oxygen with small percentages of water vapor, argon, and carbon dioxide*, and trace amounts of other elements and compounds.

Table 37.1

Composition of the Atmosphere

Gas	Symbol	Percent by Volume
Permanent Gases		
Nitrogen	N_2	78%
Oxygen	O_2	21%
Argon	Ar	0.9%
Neon	Ne	0.0018%
Helium	He	0.0005%
Methane	CH_4	0.0001%
Hydrogen	H_2	0.00005%
Variable Gases		
Water Vapor	H_2O	0 to 4
Carbon dioxide	CO_2	0.035
Ozone	O_3	0.000004*
Carbon monoxide	CO	0.00002*
Sulfur dioxide	SO_2	0.000001*
Nitrogen dioxide	NO_2	0.000001*
Particles (dust, pollen)		0.00001*

* Average value in polluted air.

* Carbon dioxide is a minor constituent in the Earth's atmosphere because much of the carbon dioxide gases spewed into the air from the Earth's interior easily dissolve in ocean water. After carbon dioxide is dissolved in our oceans, it undergoes various chemical reactions, most of which lead to the formation of calcium carbonate precipitates as we saw in Chapters 30 and 31.

Vertical Structure of the Atmosphere

The atmosphere is classified in layers, each distinct in its own characteristics (Figure 37.4). The lowest layer, the **troposphere,** is where weather occurs. The troposphere extends to a height of 16 kilometers over the equatorial region and 8 kilometers over the polar regions. Commercial jets generally fly at the top of the troposphere to minimize the bumpiness caused by weather disturbances. Even though the troposphere is the thinnest atmosphere layer, it contains 90% of the atmosphere's mass and almost all of its water vapor and clouds. Temperature in the troposphere decreases steadily (6°C per kilometer) with increasing altitude. At the top of the troposphere, temperature averages a freezing −50°C.

Above the troposphere is the **stratosphere,** which reaches a height of 50 kilometers. Ultraviolet radiation from the sun is absorbed by

Calculation Corner

Dense as Air

Knowing the density of air (1.25 kilograms/cubic meter), it's a straight-forward calculation to find the mass of air for any given volume—simply multiply air's density by the volume. The volume of an average-sized room is assumed to be 4.00 meters × 4.00 meters × 3.00 meters = 48.0 cubic meters. Thus the mass of the air in the room is

$$\frac{1.25 \text{ kg}}{\text{m}^3} \times 48.0 \text{m}^3 = 60.0 \text{ kg}$$

If you're curious to know how many pounds this is, multiply by the conversion factor 2.20 pounds/1 kilogram.

$$60.0 \text{ kg} \times \frac{2.20 \text{ lb}}{\text{kg}} = 132 \text{ lb}$$

Example

What is the mass in kilograms of the air in a classroom that has a volume of 796 cubic meters?

Answer

Each cubic meter of air has a mass of 1.25 kilograms, and so

$$796 \text{ m}^3 \times \frac{1.25 \text{ kg}}{\text{m}^3} = 995 \text{ kg}$$

which is as much as the combined mass of 17 students having a mass of about 60 kilograms each.

Your Turn

1. What is the mass in kilograms of the air in an "empty" nonpressurized scuba tank that has an internal volume of 0.0100 cubic meter?
2. What is the mass in kilograms of the air in a scuba tank that has an internal volume of 0.0100 cubic meter and is pressurized so that the density of the air in the tank is 240 kilograms/cubic meter?

the ozone layer in the stratosphere. This causes the temperature of the stratosphere to rise from about −50°C at the bottom to about 0°C at the top. Above the stratosphere, the **mesosphere** extends upward to about 80 kilometers. The gases that make up the mesosphere absorb very little of the sun's radiation. As a result, temperature decreases again from about 0°C at the bottom of the layer to about −90°C at the top.

The **thermosphere** is a layer above the mesosphere that extends upward to 500 kilometers. It has so little air that temperature variations in this layer have very little significance.

The **ionosphere** is an ion-rich region within the thermosphere and uppermost mesosphere. The ions in it are produced from the interaction between high-frequency solar radiation and atoms in the atmosphere. The incoming solar rays strip electrons from nitrogen and oxygen atoms, producing a large concentration of free electrons and positively charged ions in this layer. The degree of ionization in the ionosphere depends on air density and on the amount of solar radiation. Ionization is greatest in the upper part of the ionosphere where air density is low and solar radiation is high.

Ions in the ionosphere cast a faint glow that prevents moonless nights from becoming stark black. Near the Earth's magnetic poles, fiery light displays called *auroras* occur as the solar wind (high-speed charged particles ejected by the sun) stirs up the ionosphere (Figure 37.5). These auroral displays are particularly spectacular during times of solar flares.

Finally, above 500 kilometers, in the **exosphere,** the thinning atmosphere gradually yields to the radiation belts and magnetic fields of interplanetary space.

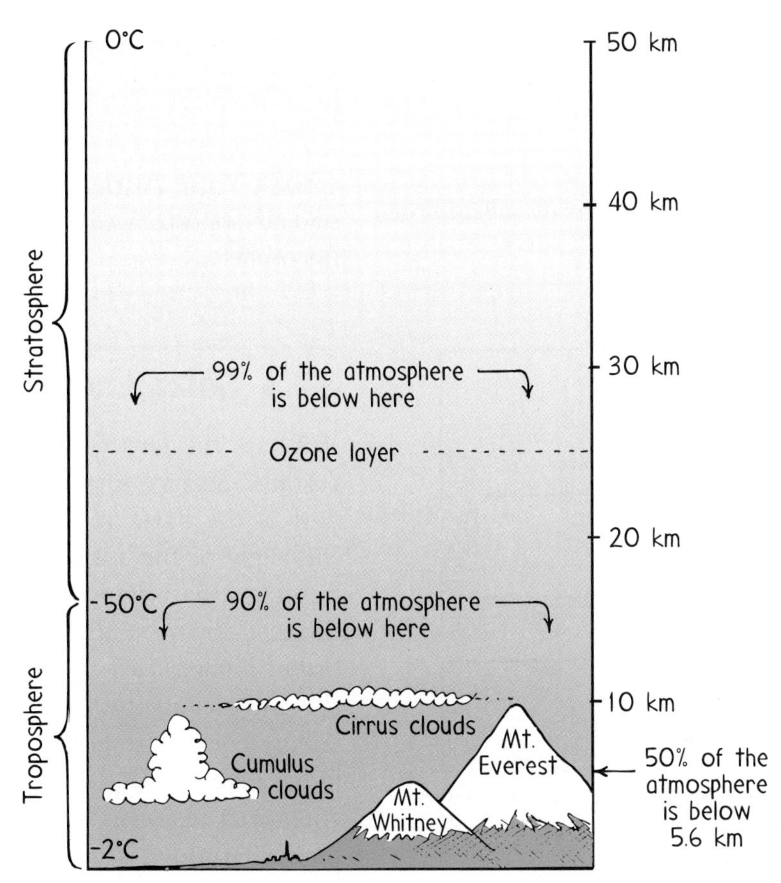

Figure 37.4
The two lowest atmospheric layers, the troposphere and stratosphere.

Figure 37.5
The aurora borealis over Alaska is created by solar-charged particles that strike the upper atmosphere and light up the sky (just as similar particles on a smaller scale light up a fluorescent lamp).

Concept Check ✓

Why do commercial airliners tend to fly at the top of the troposphere?

Check Your Answer They fly high because the captain wants to have a smooth ride! Most weather disturbances don't extend past the top of the troposphere.

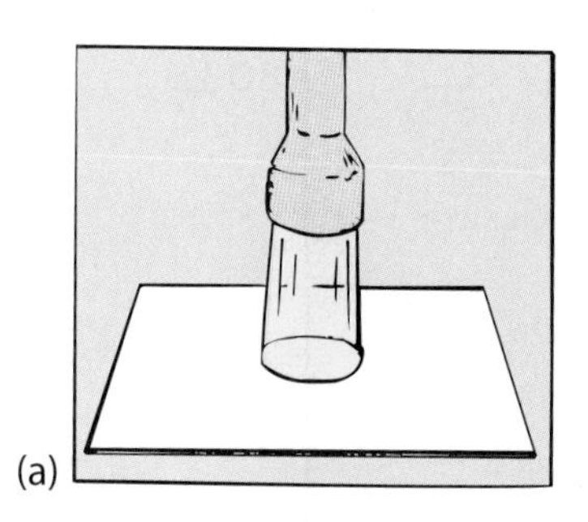

(a)

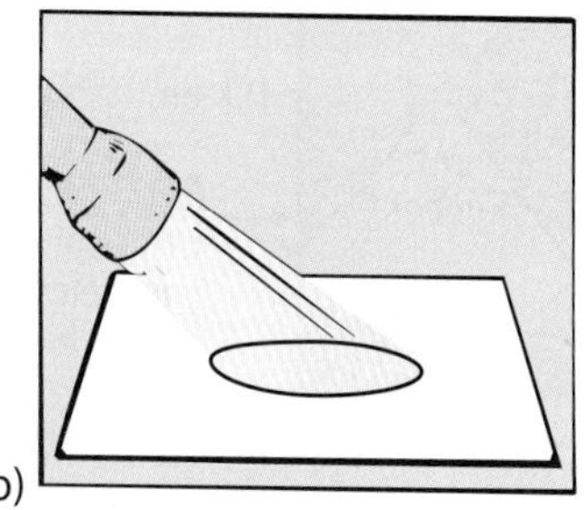

(b)

Figure 37.6
(a) When the flashlight is held directly above at a right angle to the surface, the beam of light produces a bright circle. (b) When the light is shone at an angle, the light beam is dispersed over a larger area and is therefore less intense.

37.3 Solar Energy

Why are the Earth's equatorial regions always warmer than the polar regions? Surface temperatures on Earth depend on the energy each part of the Earth receives from the sun each day. This depends on the angle of the sun's rays to the Earth's surface. You can see this by holding a flashlight vertically over a table, and shining the light directly down on the flat surface (Figure 37.6a). The light produces a bright circle. Now tip the flashlight at various angles and notice that the circle elongates into ellipses, spreading the same amount of energy over more area and therefore decreasing the intensity of the light. Likewise for sunlight on the Earth's surface. High noon in equatorial regions is like the vertically held flashlight; high noon at higher latitudes is like the flashlight held at an angle.

The Seasons

The northern United States and Canada, both temperate regions, have distinct summer and winter seasons because of variations in the angle of the sun's rays striking the Earth's surface. Figure 37.7 shows the tilt of the Earth and how the corresponding different arrays of solar radiation produce the yearly cycle of seasons. When the sun's rays are closest to perpendicular at any spot on the Earth, that region experiences summer. Six months later the rays fall upon the same region more obliquely, and we have winter. In between are fall and spring.

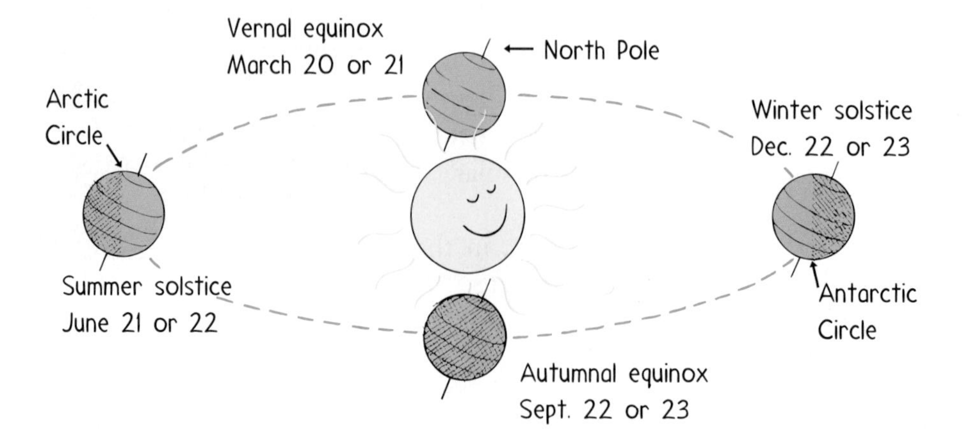

Figure 37.7
The tilt of the Earth and the corresponding different spreading of solar radiation produce the yearly cycle of seasons.

It is interesting to note that because the Earth follows an elliptical path around the sun, the Earth is farthest from the sun when the Northern Hemisphere experiences summer. So the angle of the sun's rays, not the distance from the sun, is most responsible for Earth's surface temperatures.

Another effect of the tilting rays is the length of daylight each day. Can you see in Figure 37.7 that a location in summer has more daylight per daily rotation of the Earth than the same location when the Earth is on the opposite side of the sun in winter? If you have trouble visualizing this, take a look at the high latitudes near the poles. Consider the special latitude where daylight lasts nearly 24 hours during the summer solstice (around June 21), and night lasts about 24 hours at the winter solstice (around December 21). This latitude is called the Arctic Circle in the Northern Hemisphere and the Antarctic Circle in the Southern Hemisphere. During the summer solstice, the north pole leans towards the sun and the south pole leans away from the sun. (Summer and winter are reversed, of course, in the two hemispheres.)

Halfway between the peaks of the winter and summer solstice, around mid-September and mid-March, the hours of daylight and night are of equal length. These are called the equinoxes (Latin for "equal nights"). The equal hours of day and night during the equinoxes are not restricted to high latitudes, they occur all over the world.

Another interesting thing happens as you travel north of the Arctic Circle (or south of the Antarctic Circle). There are more summer days with the sun always above the horizon and more winter days with the sun always below the horizon! At the poles there is a full six months of continuous sunlight followed by a full six months of continuous night! These 24-hour-long "days" in the polar regions are never very bright because the sun is never very far above the horizon. Likewise, the 24-hour-long "nights" aren't all that dark because the sun never sinks very far below the horizon. The polar regions can be eerie places.

Terrestrial Radiation

Solar radiation covers a wide spectrum of wavelengths, mostly in the visible short-wavelength part of the spectrum. The Earth absorbs this energy, and in turn, reradiates part of it back to space. As we learned in Chapter 10, this is *terrestrial radiation,* emitted from the Earth's surface (Figure 37.8). Terrestrial radiation is emitted in the infrared long-wavelength part of the spectrum.

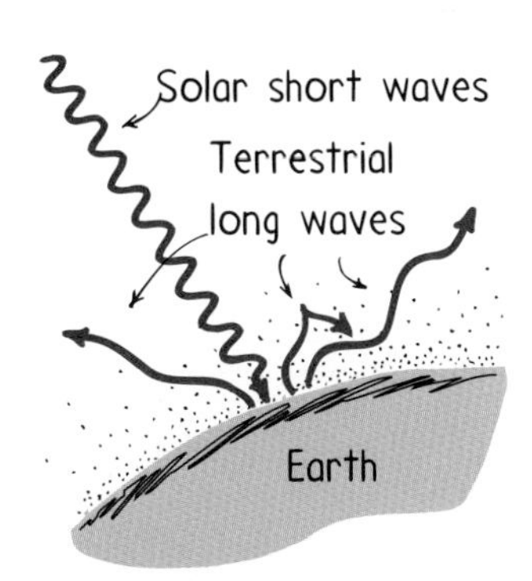

Figure 37.8
The hot sun emits short waves, and the cool Earth re-emits long waves. Radiation emitted from the Earth is called terrestrial radiation.

Interestingly, it is terrestrial radiation rather than solar radiation that directly warms the lower atmosphere. This explains why air close to the ground is so much warmer than air at higher elevations. The temperature of the Earth's surface depends on the amount of solar radiation coming in compared with the amount of terrestrial

radiation going out. In direct sunlight, the net effect is warming because the Earth's surface absorbs more energy from the sun than it emits. At night, the net effect is cooling because the Earth's surface emits more energy than it absorbs. Cloud cover blocks either incoming solar radiation or outgoing terrestrial radiation. Hence cooler cloudy days and warmer cloudy nights.

The Greenhouse Effect and Global Warming

The Earth absorbs short-wavelength radiation from the sun and reradiates it as long-wavelength terrestrial radiation. Incoming short-wavelength solar radiation easily penetrates the atmosphere to reach and warm the Earth's surface, but outgoing long-wavelength terrestrial radiation cannot penetrate the atmosphere to escape into space. Instead, atmospheric gases (mainly water vapor and carbon dioxide) absorb the long-wave terrestrial radiation. As a result, this long-wave radiation ends up keeping the Earth's surface warmer than it would be if there were no atmosphere. This process is very nice, for the Earth would be a frigid −18°C otherwise! Our present environmental concern, however, is that increased levels of carbon dioxide and other gases in the atmosphere may make the Earth *too* warm.

Similar to the panes of glass in a greenhouse, atmospheric gases trap long-wave terrestrial radiation, thereby warming the lower atmosphere. This warming of the lower atmosphere is called the **greenhouse effect** and plays a significant role in global warming. Gases released primarily by volcanic eruptions, but also by the burning of fossil fuels (coal, oil, and gas) and from agricultural and manufacturing industries, add carbon dioxide and other greenhouse gases to the atmosphere, changing its composition. This compositional change, in turn, affects atmospheric absorption of both solar and terrestrial energy.

Of all the greenhouse gases, water vapor plays the largest role in confining the Earth's heat. As part of the Earth's natural hydrologic cycle, water vapor levels have remained relatively constant throughout time. Like water vapor, carbon dioxide occurs naturally in the Earth's atmosphere. Unlike the case with water vapor, however, carbon dioxide levels are on the rise. Since the Industrial Revolution of the 1800s, atmospheric carbon dioxide levels have been steadily increasing and before the end of this century will probably double pre-industrial levels. This increase may account for the warming of the Earth's surface by about 0.6°C since 1850. Some scientists and policy makers believe that further warming will likely occur if carbon dioxide emissions are not held in check. Other gases, such as methane, nitrous oxides, and CFCs, are also on the increase and as such, they too may play a role in changing the Earth's atmosphere.

The effects of warming the Earth's surface are not known. One concern is that warming will cause polar ice caps to melt, raising sea level and flooding low-lying coastal lands. Warming would also likely change rainfall patterns, seriously affecting agricultural industries. Grain-growing regions of North America and Asia might shift northward as local climates warmed and growing seasons lengthened. On the other hand, deserts in the interior of continents might spread to cover much larger areas. We don't know. What we do know is that the Earth has experienced warmer and colder periods in times past and that global scale climatic changes may have contributed to many of the extinctions discussed in Chapter 36. More research is needed to fully understand the impacts of global warming.

Concept Check ✓

1. What does it mean to say the greenhouse effect is like a one-way valve?
2. Which gas in the atmosphere is the greatest contributor to the greenhouse effect?
3. What is the primary contributor to greenhouse gases in the Earth's atmosphere?

Check Your Answers

1. The transparent material—atmosphere for the Earth and glass for the florist's greenhouse—passes only incoming short waves and blocks outgoing long waves. In other words, radiation travels only one way.
2. Water vapor.
3. Volcanic eruptions. Interestingly, the volcanic eruption of Mount Pinatubo in 1991 spewed more chlorine into the atmosphere than the combined leakage of CFCs over a century.

37.4 Driving Forces of Air Motion

As warm air rises, it expands and cools. As the air rises, cooler air sinks to occupy the region left vacant by the rising warm air. Such motion constitutes a convection cycle and thermal circulation of the air—in other words, a *convection current.* As convection currents stir the atmosphere, the result is *wind*—defined as air in nearly horizontal motion. Wind is generated in response to pressure differences in the atmosphere, which are largely the result of temperature differences.

A difference in pressure between two different locations is called a *pressure gradient,* and forces caused by changes in pressure are called *pressure-gradient forces.*

The underlying cause of general air circulation is the unequal heating of the Earth's surface. On a global level, equatorial regions receive optimum radiant energy from the sun and as a result have higher average temperatures than other regions. As air heated by the hot ground or ocean at the equator rises, it moves toward the polar regions, cooling gradually in the upper atmosphere. This cooled air

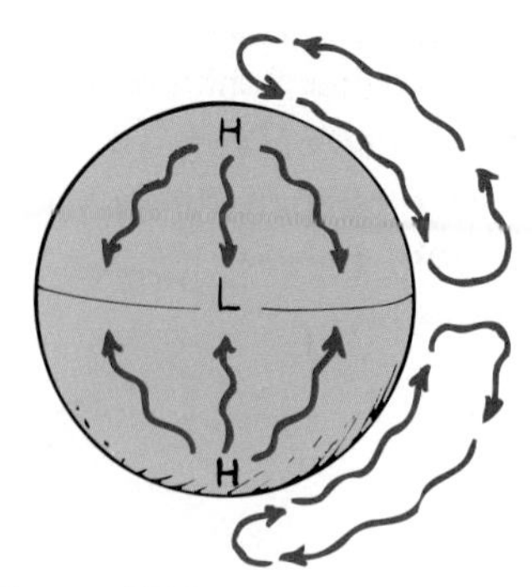

Figure 37.9
If the Earth were simply a nonrotating sphere, air circulation would be in a single Northern Hemisphere cell and a single Southern Hemisphere cell. In each cell, heated air would rise at the equator and move toward the polar regions, where it would cool, sink, and be drawn back to the warmer regions of the equator.

then sinks at the poles and is drawn back to the warmer regions of the equator. If we assume the Earth to be a nonrotating sphere, the effect is one simple single-cell circulation pattern in the Northern Hemisphere and another in the Southern Hemisphere, as shown in Figure 37.9.

But the Earth rotates, which greatly affects the path of moving air.* Think of the Earth as a large merry-go-round rotating in a counterclockwise direction (the same direction the Earth spins as viewed from above the north pole). Pretend you and a friend are playing catch on this merry-go-round. When you throw the ball to your friend, the circular movement of the merry-go-round affects the direction the ball appears to travel. Although the ball travels in a straight-line path, it appears to curve to the right, as shown in Figure 37.10. (The ball travels straight, but your friend never catches it because the movement of the merry-go-round causes his position to change.) This apparent curving is similar to what happens on the Earth. As the Earth spins, all free-moving objects—air and water, aircraft and ballistic missiles, and even snowballs to a small extent—appear to deviate from their straight-line paths as the Earth rotates under them. This apparent deflection due to the rotation of the Earth is called the **Coriolis effect.**

A significant result of the Coriolis effect is the apparent deflection of winds toward the right in the Northern Hemisphere and toward the left in the Southern Hemisphere (Figure 37.11). The impact of the Coriolis effect varies according to the speed of the

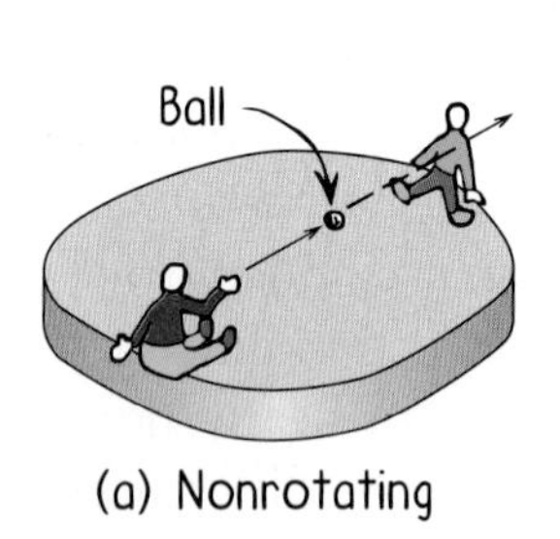

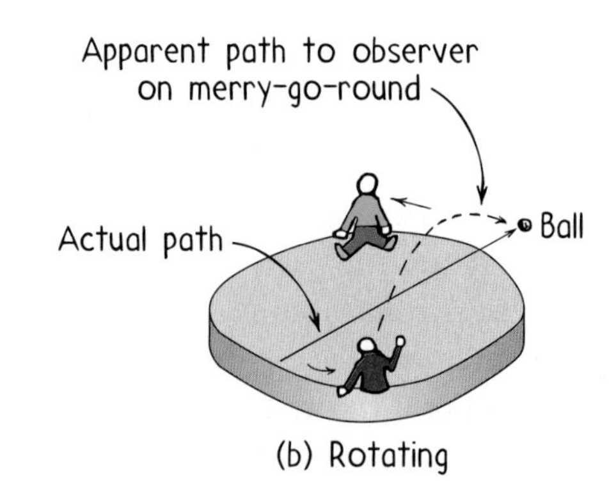

Figure 37.10
(a) On the nonrotating merry-go-round, a thrown ball travels in a straight line. (b) On the counterclockwise rotating merry-go-round, the ball moves in a straight line. However, because the merry-go-round is rotating, the ball appears to deflect to the right of its intended path.

* As air circulates over the oceans, it causes surface water to drift along with it. A severe storm in 1990 gave scientists an unusual tool for studying the currents of the Pacific Ocean. Five cargo containers of athletic shoes were washed overboard from freighters that ran into stormy seas en route from South Korea to the Pacific Northwest. Since then, in a confirmation of theories about currents in the Northwest Pacific, thousands of sneakers, hiking boots, children's sandals and other shoes were picked up along beaches from British Columbia to Oregon and as far into the Mid-Pacific as Hawaii. Despite months at sea, most shoes were wearable after washing. The problem, though, is that the shoes were not tied together. Beachcombers formed "swap meets" to search for mates of found shoes!

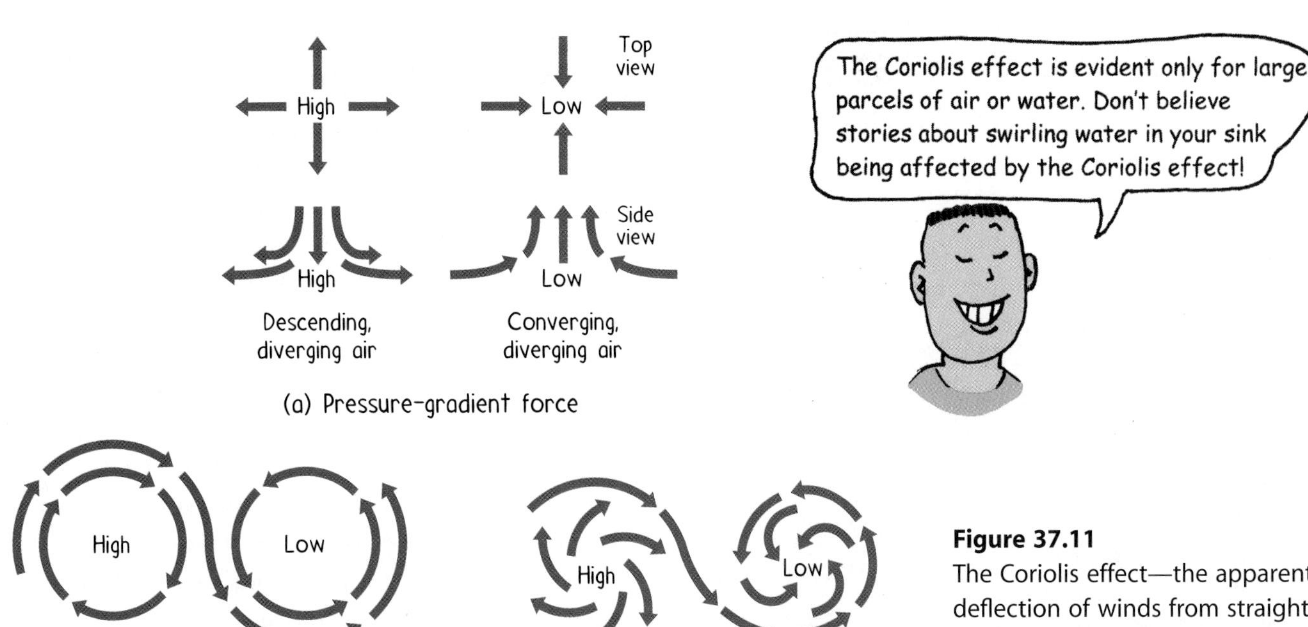

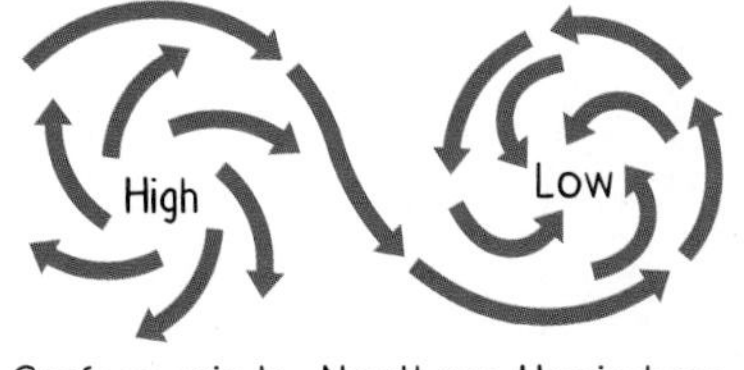

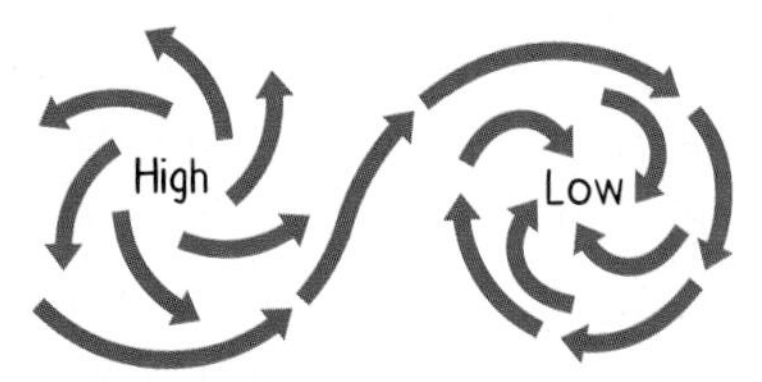

High

Low

Upper-level winds—Southern Hemisphere

(b) Coriolis effect

(c) Frictional force at the ground

Figure 37.11
The Coriolis effect—the apparent deflection of winds from straight-line paths by the Earth's rotation, is a principal force in the production of wind. It is, however, not the only force. (a) First of all, air moves due to pressure differences—the pressure gradient force. (b) Once the air is moving, it is affected by the Earth's rotation—the Coriolis effect. (c) As air moves close to the ground it slows due to frictional force.

wind. The faster the wind, the greater the deflection. Latitude also influences the degree of deflection. Deflection is greatest at the poles and decreases to zero at the equator (Figure 37.12).

Air moving close to the Earth's surface encounters a *frictional force.* The rougher the surface, the greater the friction and so the greater the drag. Because surface friction reduces wind speed, it reduces the effect of the Coriolis force. This causes winds in the Northern Hemisphere to spiral out clockwise from a high-pressure region and spiral counterclockwise into a low-pressure region (top part of Figure 37.11c). In the Southern Hemisphere, these circulation patterns are reversed (bottom part of Figure 37.11c).

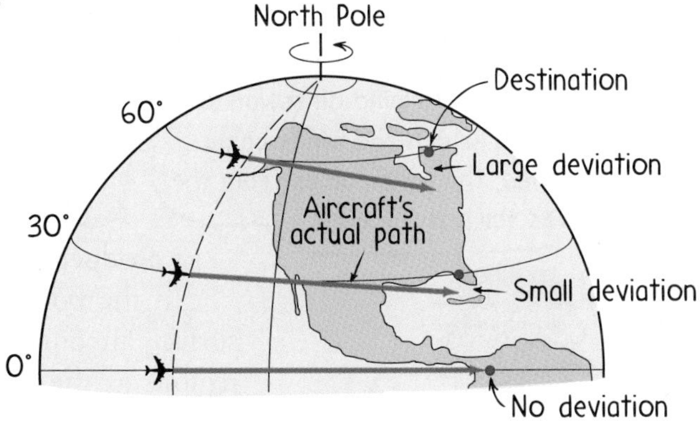

Figure 37.12
Latitude influences the apparent deflection resulting from the Coriolis effect. A free-moving object heading east (or west) appears to deviate from its straight-line path as the Earth rotates beneath it. Deflection is greatest at the poles and decreases to zero at the equator.

37.5 Global Circulation Patterns

Cell-like circulation patterns are responsible for the redistribution of heat across the Earth's surface and the global winds (Figure 37.13). At the equator, warmed air flows straight up with very little horizontal movement, resulting in a vast low-pressure zone. This rising motion creates a narrow, windless realm of air that is still, hot, and stagnant. Seamen of long ago cursed the equatorial seas as their ships floated listlessly for lack of wind, and referred to the area as the *doldrums.* When the moist air from the doldrums rises, it cools and releases torrents of rain. When over land areas, these frequent rains give rise to the tropical rain forests that characterize the equatorial region.

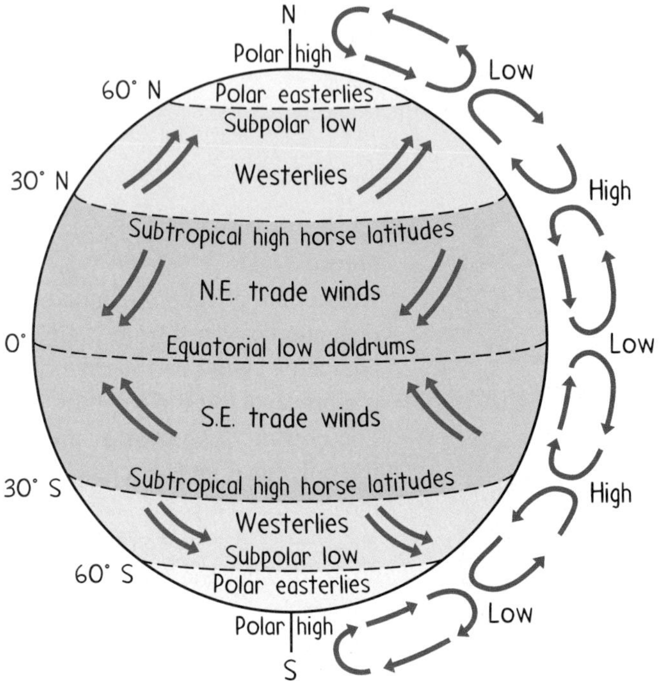

Figure 37.13
Global winds are the result of several cell-like circulation patterns, brought about by unequal heating of the Earth's surface and compounded by effects of the Earth's rotation

The air of the sweltering doldrums rises to the boundary between the troposphere and stratosphere, where it divides and spreads out either to the north or south. (Very little wind crosses the equator into the neighboring hemisphere.) By the time it has reached about 30° N and 30° S latitudes, this air has cooled enough to descend toward the surface. The descending air warms as it is compressed. A resulting high-pressure zone girdles the Earth, creating a belt of hot, dry surface air. On land, these high-pressure zones account for the world's great deserts—the Sahara in Africa, the Arabian Desert in the Middle East, the Mojave Desert in the United States, and the Great Victoria Desert in Australia. At sea, the hot, descending air produces very weak winds. According to legend, early sailing ships were frequently stalled at these latitudes, both north and south. As food and water supplies dwindled, horses on board were either eaten or cast overboard to conserve fresh water. As a result, this region is now known as the *horse latitudes.* The thermal convection cycle that starts at the equator is completed when air flowing southward from the horse latitudes in the Northern Hemisphere and northward in the Southern Hemisphere is deflected westward to produce the *trade winds.* Air that flows northward from the horse latitudes in the Northern Hemisphere and southward in the Southern Hemisphere is deflected eastward to produce the prevailing *westerlies.*

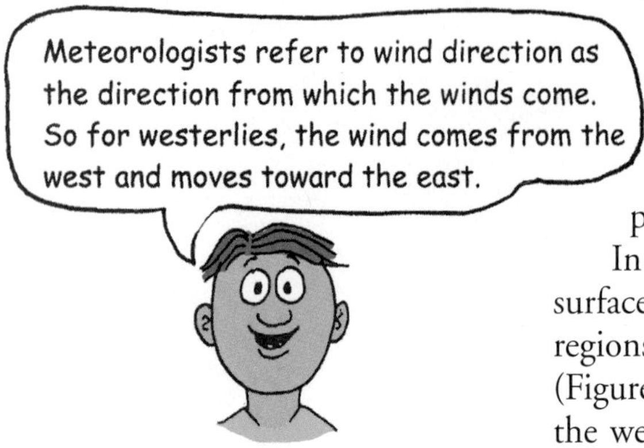

Meteorologists refer to wind direction as the direction from which the winds come. So for westerlies, the wind comes from the west and moves toward the east.

In the polar regions, frigid air continuously sinks, pushing the surface air outward. The Coriolis effect is quite evident in the polar regions as the wind deflects to the west to create the *polar easterlies* (Figure 37.13). The cool dry polar air meets the warm moist air of the westerlies at latitudes 60° N and 60° S. This boundary, called the *polar front,* is a zone of low pressure where contrasting air masses converge, often resulting in storms.

The midlatitudes are noted for their unpredictable weather. Although the winds tend to be westerlies, they are often quite changeable as the temperature and pressure differences between the subtropical and polar air masses at the polar front produce powerful winds. As air moves from regions of high pressure, where air is denser, toward regions of low pressure, the result is a *cyclone* effect.

Irregularities in the Earth's surface also influence wind behavior. Mountains, valleys, deserts, forests, and great bodies of water all play a part in determining which way the wind blows.

Upper Atmospheric Circulation

In the upper troposphere, "rivers" of rapidly moving air meander around the Earth at altitudes of 9–14 kilometers. These high-speed winds are the *jet streams.* With wind speeds averaging between 95 and 190 kilometers per hour, the jet streams play an essential role in the global transfer of heat from the equator to the poles.

The two most important jet streams, the *polar jet stream* and the *subtropical jet stream* form in both the Northern and Southern Hemispheres. The formation of polar jet streams is a result of a temperature gradient at the polar front—at about 60° N and 60° S latitude—where cool polar air meets warm tropical air. This temperature gradient causes a steep pressure gradient that increases the wind speed. During the winter, the polar jet is strong and extensive as it migrates to lower latitudes, bringing strong winter storms and blizzards to the United States. In summer, the jet stream is weaker and migrates to higher latitudes.

The subtropical jet stream is generated as warm air is carried from the equator to the poles, producing a sharp temperature gradient along the subtropical front—about 30° N and 30° S latitude. Once again a pressure gradient caused by the temperature gradient generates strong winds.

The subtropical jet stream above Southeast Asia, India, and Africa merits special mention (Figure 37.14). The formation of this jet stream is related to the warming of the air above the Tibetan highlands. During the summer, the air above the continental highlands is warmer than the air above the ocean to the south. Thus temperature and pressure gradients generate strong on-shore winds that contribute to the region's *monsoon* (rainy) climate. During winter, the winds change direction to produce a dry season.

This cycle of winds characterizes the climates of much of Southeast Asia. The predictable rain-bearing summer wind from the sea that moves over the heated land is called the *summer monsoon*; the prevailing wind from land to sea in winter is called the *winter monsoon.*

Figure 37.14
Winds over Southeast Asia. (a) During the summer months, air over the oceans is cooler than the air over land. The summer monsoon brings heavy rains as the winds blow from sea to land. (b) During the winter months, air over continents is cooler than air over oceans. The winter monsoon generally has clear skies and winds that blow from land to sea.

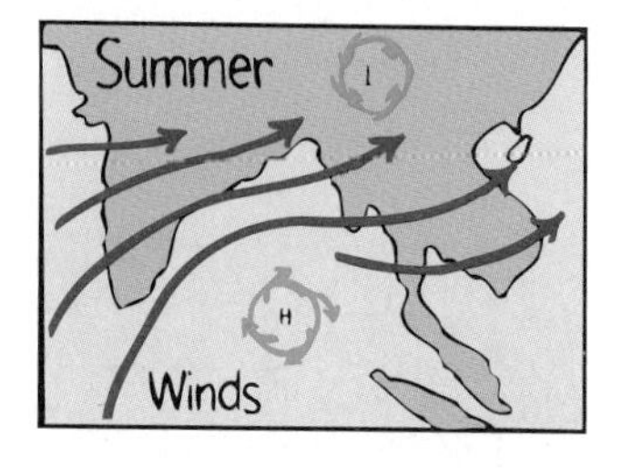

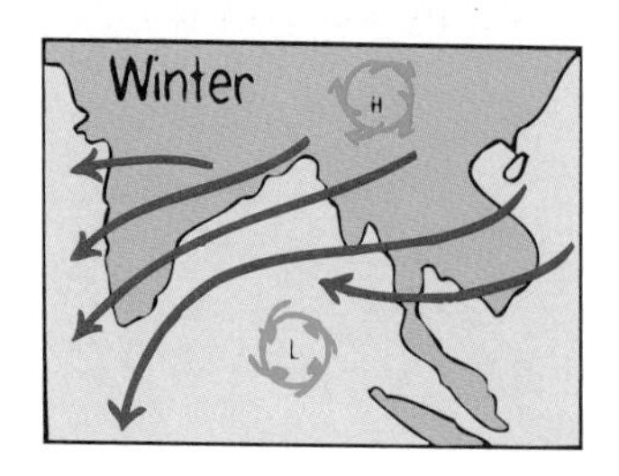

Concept Check ✓

1. What are the main causes of the trade winds, jet streams, monsoons, and ultimately their bearing on the world's climates?
2. In the midlatitudes, airlines schedule shorter flight times for planes traveling west to east and longer flight times for planes traveling east to west. Why are eastbound planes faster?

Check Your Answers

1. Simply enough, the unequal heating of the Earth's surface coupled with the Earth's rotation.
2. The upper-level westerly moving winds of the jet stream account for faster-moving eastbound aircraft. As the jet stream moves from west to east it carries along everything in its path. To save time and fuel, air pilots seek the jet stream when traveling west to east and avoid it when traveling east to west.

Oceanic Circulation

The forces that drive the winds also affect the movement of seawater. In the open ocean, the major movement of seawater results from two types of currents: wind-driven surface currents and density-driven deep-water currents. Near coastlines, water movement is affected not only by surface and deep-water currents but also by coastal boundaries. Density is controlled by two things—temperature and salinity.

Salts make up 99.7% of the ocean's dissolved materials. The amount of dissolved salts in seawater is measured as **salinity**—the mass of salts dissolved in 1000 grams of seawater. Although salinity varies from one part of the ocean to another, the overall composition of seawater is about the same from place to place—a mixture of about 96.5% water and 3.5% salt. Variation is, of course, influenced by factors that increase or decrease supplies of fresh water. Fresh water enters the ocean in three ways: runoff from streams and rivers, precipitation, and melting of glacial ice. Fresh water leaves the ocean in two ways: evaporation and formation of ice. Overall balance is maintained when evaporation is offset by precipitation and runoff and ice formation is offset by ice melting.

Like the atmosphere, the ocean can be divided into several vertical layers—the surface zone, a transition zone, and the deep zone. Scuba divers notice an increase in water pressure when swimming to lower depths. The deeper you descend, the greater the water pressure. The pressure is simply the weight of the water above you pushing against you. Another factor that generally changes as you descend is temperature. Deeper waters are cooler. So, in addition to variations in salinity, seawater also varies in temperature and pressure. Because cold water is denser than warm water, cold seawater sinks below warmer

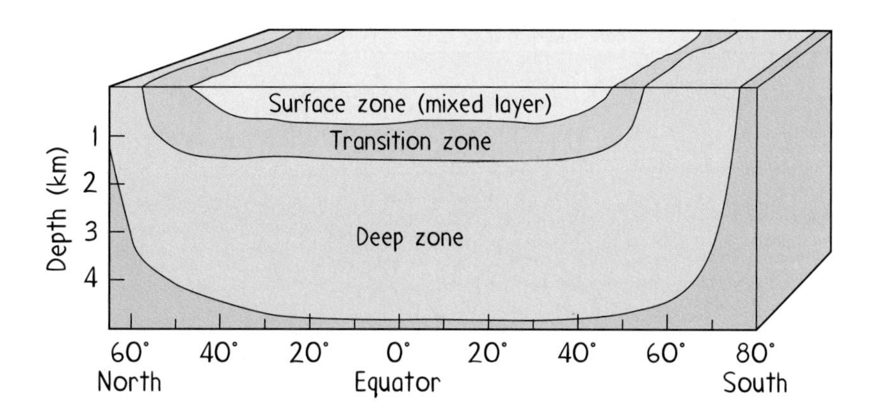

Figure 37.15
The ocean's vertical structure. In the surface zone, water is well mixed as it moves vertically in response to temperature and density changes, and horizontally in response to wind. Water in the transition zone moves along density surfaces. Water in the deep zone is density-driven as it circulates from cold polar regions to warmer equatorial regions.

seawater. Salinity also affects density: The greater the salinity, the greater the density. These variations are best illustrated when we look at the ocean's vertical structure (Figure 37.15).

Concept Check ✓

Would you expect the pressure 100 m deep in the equatorial Pacific to be the same as it is 100 m deep in the northern Pacific?

Check Your Answer Probably not. If salinity is the same at both locations (a good assumption if we are away from coastal areas), the pressure will be slightly higher in the cold, northern Pacific. Remember, cold water is denser than warm water, and so a volume of cold water weighs more than an equal volume of warm water. Pressure is the weight per unit of area.

Surface Currents As winds blow across the ocean, frictional forces set surface waters into motion. If distances are short, the surface waters move in the same direction as the wind. For longer distances, however, other factors come into play. One such factor is the deflective Coriolis effect, which causes surface waters to spiral in a circular whirl pattern called a **gyre.** The circular motion is clockwise in the Northern Hemisphere and counterclockwise in the Southern Hemisphere.

In the tropics, the trade winds drive equatorial ocean currents westward. When the westward flow is blocked by a continental shoreline, the current splits, with some flow going north and some going south. At temperate latitudes, the prevailing westerlies take over to drive the surface currents eastward. In the Northern Hemisphere, huge gyres are created as eastward moving water encounters land boundaries. In the Southern Hemisphere, with fewer land obstructions, eastward flow is able to encircle the globe (Figure 37.16).

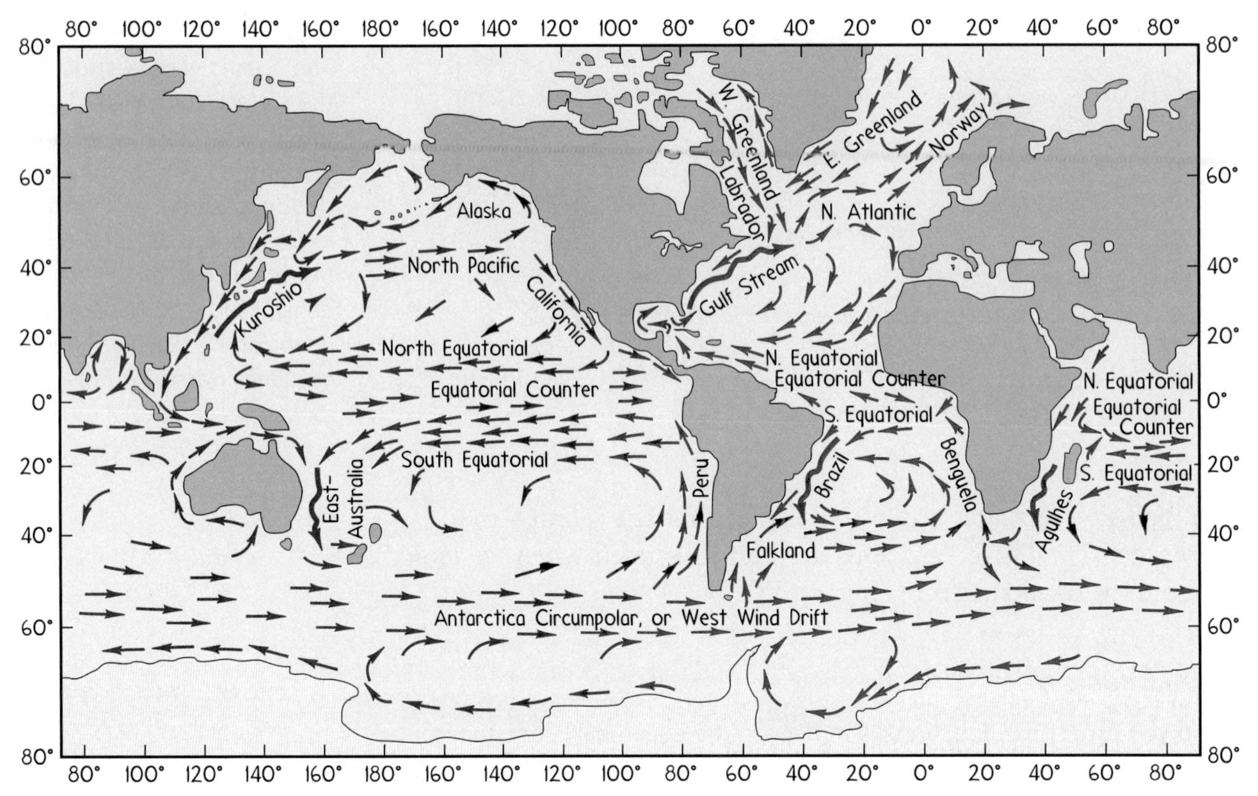

Figure 37.16
Circulation of the ocean's surface waters. The names of the major currents are indicated.

An important consequence of these large gyres is the transport of heat from equatorial regions to higher latitudes. In the North Atlantic Ocean, for example, warm equatorial water flows westward into and around the Gulf of Mexico then northward along the eastern coast of the United States. This warm-water current is called the *Gulf Stream.* As the Gulf Stream flows northward along the North American coast, the prevailing westerlies steer the warm current eastward toward Europe (Figure 37.16). Great Britain and Norway benefit from the warm waters in the Gulf Stream, for lands at this northern latitude would be much colder without being bathed by the heat of warm water from the Gulf of Mexico. As the warm current encounters Europe, it is turned southward toward the equator where it is once again picked up by the trade winds to move westward into the Gulf of Mexico and once again become part of the Gulf Stream.

Oceanic circulation in the North Pacific is similar to that in the North Atlantic. The Pacific counterpart of the Gulf Stream is the warm, northward-flowing current known as the *Kuroshio.* In the Southern Hemisphere, surface oceanic circulation (with the exception of the Antarctica Circumpolar Current) is similar except that the gyres move counterclockwise.

Deep-Water Currents Surface waters are driven by winds, but deeper waters are driven by gravity. In essence, deep water flows because dense water sinks. Although deep water flows more slowly than surface water, the volume of deep-water flow is like a large global conveyor belt (Figure 37.17).

In the high latitudes, where seawater in the deep zone interacts with seawater in the surface zone, a very slow worldwide, north-south circulation pattern develops. To understand how this pattern develops, we need to look at what happens when seawater begins to freeze.

First of all, seawater does not freeze easily. When it does, however, only the water freezes, and the salt is left behind. Thus the seawater that does not freeze experiences an increase in salinity, which in turn causes an increase in density. The cold, denser, saltier seawater sinks, which sets up a pattern of vertical movement. There is also horizontal movement as the dense water that sinks in the polar regions flows along the bottom to the deeper parts of the ocean floor.

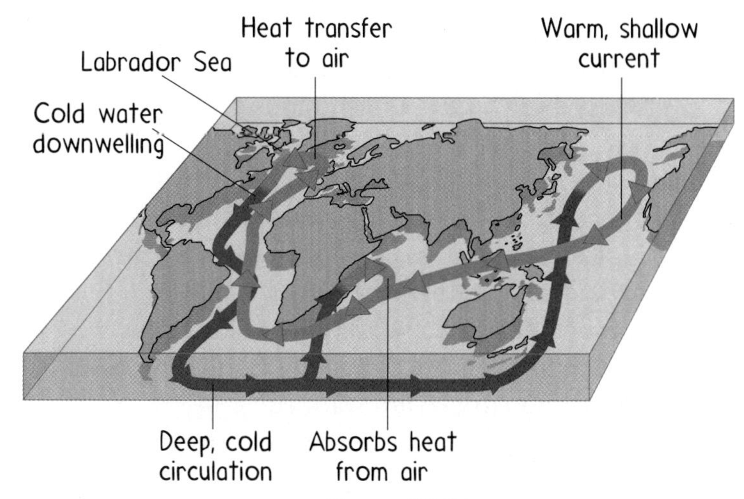

Figure 37.17
Deep-ocean currents act like a conveyor belt, transporting cold water from the North Atlantic to the equator and onto the Antarctic. From the Antarctic, water flows eastward then northward into the Pacific and Indian Oceans.

Thus conveyor-belt circulation begins in the North Atlantic as dense, cold, salty seawater around Greenland and Iceland sinks and flows along the ocean bottom toward the equator then onto the Antarctic Ocean. Once near Antarctica, the water flows eastward around the continent, then northward into the Pacific and Indian Oceans. Thus deep-water currents flow in a north-south circulation pattern.

Chapter Review

Key Terms and Matching Definitions

_____ Coriolis effect
_____ greenhouse effect
_____ gyre
_____ ionosphere
_____ mesosphere
_____ salinity
_____ stratosphere
_____ thermosphere
_____ troposphere

1. The atmospheric layer closest to the Earth's surface, 16 km high over the equator and 8 km high over the poles. It contains 90% of the atmosphere's mass and essentially all its water vapor and clouds.
2. The second atmospheric layer above the Earth's surface, extending from the top of the troposphere up to 50 km.
3. The third atmospheric layer above the Earth's surface, extending from the top of the stratosphere to 80 km.
4. The fourth atmospheric layer above the Earth's surface, extending from the top of the mesosphere to 500 km.

5. An electrified region within the thermosphere and uppermost mesosphere where fairly large concentrations of ions and free electrons exist.
6. Warming caused by short-wavelength radiant energy from the sun that easily enters the atmosphere and is absorbed by the Earth. This energy is then reradiated at longer wavelengths that cannot easily escape the Earth's atmosphere. Thus the Earth's atmosphere and surface are warmed.
7. The mass of salts dissolved in 1000 g of seawater.
8. The apparent deflection from a straight-line path observed in any body moving near the Earth's surface, caused by the Earth's rotation.
9. Circular or spiral whirl pattern, usually applied to very large current systems in the open ocean.

Review Questions

Earth's Atmosphere and Oceans

1. What were the main components of the Earth's first atmosphere? What happened to this atmosphere?
2. The Earth's present atmosphere likely developed from gases that escaped from the interior of the Earth during volcanic eruptions. What were the three principal atmospheric gases produced by these eruptions?
3. Explain the importance of photosynthesis in the evolution of the atmosphere.

Components of the Earth's Atmosphere

4. What elements make up today's atmosphere?
5. Being that our atmosphere developed as a result of volcanic eruptions, why aren't there higher traces of atmospheric carbon dioxide, one of the principal volcanic gases?
6. Does temperature increase or decrease as one moves upward in the troposphere? As one moves upward in the stratosphere?
7. What causes the fiery displays of light called the *auroras*?

Solar Energy

8. What does the angle at which the sun strikes the Earth have to do with the temperate and polar regions?
9. What does the tilt of the Earth have to do with the change of seasons?
10. Why are the hours of daylight equal all around the world on the two equinoxes?
11. How does radiation emitted from the Earth differ from that emitted by the sun?
12. How is the atmosphere near the Earth's surface heated from below?

Driving Forces of Air Motion

13. What are the three main driving forces of air motion?
14. What is the underlying cause of air motion?
15. How does the Coriolis effect determine the general path of air circulation?

Global Circulation Patterns

16. What is the characteristic climate of the doldrums and why does it occur?
17. In summer, Southeast Asia, India, and Africa experience heavy flooding. Why?
18. What factors set surface ocean currents into motion?
19. How does the Coriolis effect influence the movement of surface waters?
20. Explain the circulation pattern of the Gulf Stream.
21. How does the density of seawater vary with changes in temperature? How does density change with salinity?

Exercises

1. If it is true that a gas fills all the space available to it, why doesn't the atmosphere go off into space?
2. Why do your ears pop when you ascend to higher altitudes? Explain.
3. The Earth is closest to the sun in January, but January is cold in the Northern Hemisphere. Why?
4. How do the total number of hours of sunlight in a year compare for tropical regions and polar regions of the Earth? Why are polar regions so much colder?
5. How do the wavelengths of radiant energy vary with the temperature of the radiating source? How does this affect solar and terrestrial radiation?
6. How is global warming affected by the relative transparencies of the atmosphere to long- and short-wavelength electromagnetic radiation?
7. Why is it important that mountain climbers wear sunglasses and use sunblock even when the temperature is below freezing?
8. If there were no water on the Earth's surface, would weather occur? Defend your answer.
9. If the Earth were not spinning, what direction would the surface winds blow where you live? What direction does it blow on the real Earth at 15° S latitude and why?
10. What is the relationship between global atmospheric circulation and ocean currents? Relate oceanic gyres to patterns of subtropical high pressure.
11. Relate the jet stream to upper-air circulation. How does this circulation pattern relate to airline schedules from New York to San Francisco and the return trip to New York?
12. What are the jet streams and how do they form?
13. Why are temperature fluctuations greater over land than water? Explain.
14. How does the ocean influence weather on land?
15. What happens to the salinity of seawater when evaporation at the ocean surface exceeds precipitation? When precipitation exceeds evaporation? Explain.
16. Water denser than surrounding water sinks. With respect to the densities of deeper water, how far does it sink?
17. What effect does the formation of sea ice in polar regions have on the density of seawater? Explain.
18. Why are most of the world's deserts found in the area known as the horse latitudes?
19. As a volume of seawater freezes, the salinity of the surrounding water increases. Explain.
20. Why do the temperate zones have unpredictable weather?

Suggested Web Sites

http://www.ncar.ucar.edu/ncar/

http://www.noaa.gov/

http://response.restoration.noaa.gov/kids/kids.html

http://www.geo.nsf.gov/atm/atmkids.htm

Chapter 38: Weather

We've all seen weather reports on the news. But how do meteorologists figure out whether it will rain or not? What causes air to warm up? What makes air cool down? Why do deserts occur on one side of a tall mountain range? Which side is the desert on? And what do most people talk about in idle conversation? That's right, what this chapter is about—the weather!

38.1 Water in the Atmosphere

Water. . . it is vital to life on Earth as we know. But it also has a huge importance in shaping the Earth's surface and determining its weather. There is always some water vapor in the air. *Humidity* is a measure of the water vapor in the air. Specifically, humidity is the mass of water per volume of air.

When you hear a TV newsperson describing humidity, however, they are probably speaking of **relative humidity.** Relative humidity is the ratio of the amount of water vapor currently in the air compared to the largest amount of water vapor the air can hold at that temperature. A relative humidity of 50%, for example, means that the water content in the air is half the amount that air can hold at that temperature.*

When air contains as much water vapor as it can possibly hold, the air is *saturated.* Saturation occurs when the air temperature drops and water vapor molecules in the air begin to condense to form droplets. Because slower moving molecules characterize lower air temperatures, saturation and condensation are more likely to occur in cool air than in warm air (Figure 38.1). Warm air can contain more water vapor than cold air.

As air rises, it expands. The expansion occurs because air moves to a region of lower air pressure. As we learned in Chapter 10, air cools when it expands. As the air cools, water molecules move slower and condensation occurs. If there are larger and slower-moving

Fast-moving H_2O molecules rebound upon collision

Slow-moving H_2O molecules condense upon collision

Figure 38.1
Condensation of water molecules.

* Relative humidity is a good indicator of comfort. For most people, conditions are ideal when the temperature is about 20°C and the relative humidity 50–60%. When the relative humidity is too high, moist air feels "muggy" as condensation counteracts the evaporation of perspiration. Cold air that has a high relative humidity feels colder than dry air of the same temperature because of increased conduction of heat from the body. When the relative humidity is high, hot weather feels hotter, and cold weather feels colder.

particles or ions present in the air, water vapor condenses on these particles. And this creates a cloud. As the size of the cloud droplets grow, they fall to the Earth and we have rain. Rain is one form of precipitation. Other familiar forms of precipitation are mist, hail, snow, and sleet. Precipitation comes from water vapor in the air that condenses to make clouds then falls as liquid water or ice.

Water vapor in the air can condense close to the ground as well. When condensation in the air occurs near the Earth's surface, we call it *dew*, *frost*, or *fog*. On cool, clear nights, objects near the ground cool down more rapidly than the surrounding air. As the air cools below a certain temperature, called the *dew point*, the air cannot hold as much water vapor as when the air was warmer. Water from the now-saturated air condenses on any available surface. This may be a twig, a blade of grass, or the windshield of a car. We often call this type of condensation *early-morning dew*. When the dew point is at or below freezing, we have frost. When a large mass of air cools and reaches its dew point, the relative humidity approaches 100%. And this produces a cloud near the ground—fog.

Figure 38.2
San Francisco is well known for its summer fog.

Concept Check ✓

What is the major difference between fog and a cloud?

Check Your Answer Altitude.

38.2 Air Masses—Movement and Temperature Changes

Air pressure, temperature, and density are three key variables that control weather. To understand and predict the weather, we must understand all three. First, consider air pressure. Air is a mixture of molecules that move randomly and collide with one another like billiard balls. When a molecule bumps into something, it exerts a small push on whatever it hits. This push (force) spreads over the area that is hit as pressure. The name for any force compared to the area over which that force is exerted is *pressure*, as you recall from earlier chapters. In the case of the force exerted by air molecules, we use the term *air pressure.*

The faster that air molecules are moving, the greater their kinetic energy. The greater the kinetic energy, the harder the molecular collisions and the higher the air pressure. Fast-moving air molecules—in other words, warm air—has higher air pressure than cooler air.

Another factor, besides temperature, that affects air pressure is density. The denser the air, the more molecular collisions take place, and the greater the air pressure. How air behaves depends on its pressure, temperature, and density.

Hands-On Exploration: Atmospheric Can-Crusher

When water vapor condenses in a closed container, very low pressure is created within that container. The atmospheric pressure on the outside of the container is then able to crush the container. In this activity you will see how this works when water vapor condenses on the inside of an aluminum soda can.

What You Need

Water, aluminum soda can, saucepan, tongs, stove or other heat source.

Safety Note

Wear safety goggles, and avoid touching the steam produced in this experiment—steam burns can be severe.

Procedure

1. Fill the saucepan with water and set it aside.
2. Put about a tablespoon of water in the can and heat it on the stove until steam comes out. The steam shows that the air has been driven out of the can and has been replaced by water vapor, which is also leaving the can.
3. Quickly grasp the can with the tongs, invert it, and dip it into the water in the saucepan just enough to place the can's opening under water. Crunch! The can is crushed by atmospheric pressure! Why?

Adiabatic Processes in Air

The concept of *heat exchange* shows us that air pressure, temperature, and density are interrelated. When heat is added to an air mass, its temperature, its pressure, or both increase. Heat can be added to air by solar radiation, by moisture condensation, or by contact with warm ground. When heat is subtracted from an air mass, the temperature or the pressure of the air falls. Heat can be subtracted from air by radiation to space, by evaporation of rain falling through dry air, or by contact with cold surfaces.

However, atmospheric processes—even those that involve temperature change—do not always involve significant heat exchange. When heat transfer is zero, or nearly zero, we call the atmospheric process *adiabatic*. For example, when air is suddenly compressed or allowed to expand, its temperature changes as its pressure changes. The air was warmed or cooled even though no heat exchange took place.

Large bodies of air in the atmosphere often follow adiabatic processes. To illustrate how bodies of air behave, let us imagine a body of air enclosed in a very thin plastic garment bag—an air *parcel*. Like a free-floating balloon, the parcel can expand and contract freely without heat exchange with air outside the parcel. Air pressure always decreases with increasing height. So, as an air parcel flows up the side of a mountain, air pressure within the parcel decreases. The air parcel can expand and cool without any heat exchange.

Wind is like the air, only pushier.

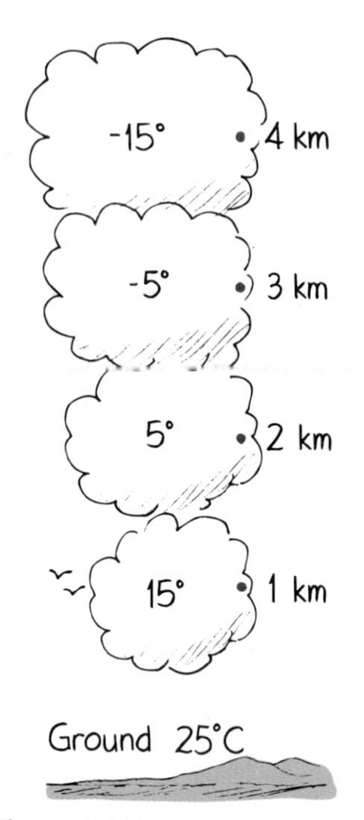

Figure 38.3
The temperature of a parcel of dry air that expands adiabatically changes by about 10°C for each kilometer of elevation.

With adiabatic expansion, the temperature of a dry air parcel decreases about 10°C for each kilometer it rises (Figure 38.3). This rate of cooling for dry air is called the *dry adiabatic rate.* Adiabatic processes are not restricted to dry air. As rising air cools, its ability to hold water vapor decreases so the relative humidity of the rising air increases. If the air cools to its dew point, the water vapor condenses and a cloud forms. Because condensation releases heat, the surrounding air is warmed. This added heat offsets the cooling due to expansion, making the air cool at a lesser rate—a *moist adiabatic rate.* Although the moist adiabatic rate varies according to temperature and moisture content, on average a moist air parcel cools 6°C for every kilometer it rises.

Atmospheric Stability

Stable air is air that resists upward movement. When a parcel of rising air is cooler than its surroundings, it is also denser than its surroundings. The denser air tends to sink. The two effects—rising and sinking—balance each other and the air is stable. Stable air that is forced to rise spreads out horizontally. When clouds develop in stable air, they too spread out into thin horizontal layers having flat tops and bottoms.

On the other hand, when a rising air parcel is warmer than its surroundings, it is less dense and continues to move upward until its temperature equals the temperature of its surroundings. In this case the air is unstable and favors upward movement. Rising dry air cools at the dry adiabatic rate while air at the surface warms up. When the rising air is moist, billowy, towering clouds develop.

A dramatic example of adiabatic warming is the *Chinook*—a dry wind that blows down from the Rocky Mountains across the Great Plains (Figure 38.4). Cold air moving down a mountain slope is compressed as it moves to lower elevations (where air pressure is greater than at higher elevations) and becomes much warmer. The effect of expansion or compression of gases is quite impressive.*

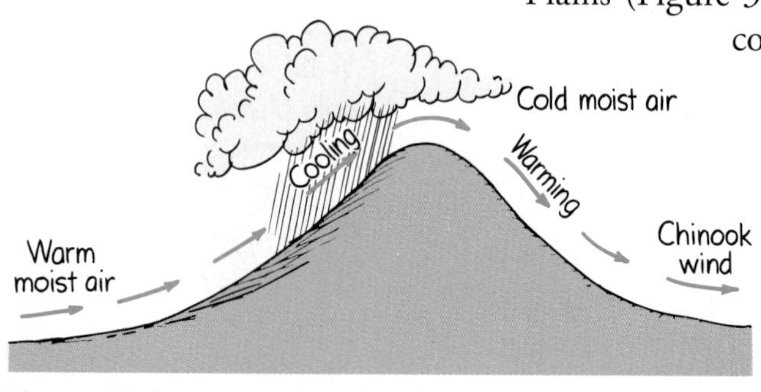

Figure 38.4
Chinooks—which are warm, dry winds—occur when high-altitude air descends and is adiabatically warmed.

A rising parcel of air continues to rise as long as it is warmer and less dense than the surrounding air. If it gets cooler and denser than its surroundings, it sinks. Under some conditions, large parcels of cold air sink and remain at a low elevation. This results in air above that is warmer. When the upper regions of the atmosphere are warmer than the lower regions, which is the opposite of what normally occurs, we have a **temperature inversion.** Then most rising air can't pass through the upper layer of warmer air.

* When you're flying at high altitudes where outside air temperature is typically −35°C, you're quite comfortable in your warm cabin—but not because of heaters. The process of compressing outside air to a cabin pressure nearly that of sea level normally heats the air to a roasting 55°C (131°F). So air conditioners must be used to extract heat from the pressurized air.

Evidence of this is commonly seen over a cold lake when visible gas and other small particles such as smoke spread out in a flat layer above the lake rather than rising and dissipating higher in the atmosphere (Figure 38.5).

The smog of Los Angeles is trapped by such an inversion, caused by cold air from the ocean capped by a layer of hot air moving westward over the mountains from the Mojave Desert. The west-facing side of the mountains helps hold the trapped air (Figure 38.6). The mountains on the edge of Denver play a similar role in trapping smog beneath a temperature inversion.

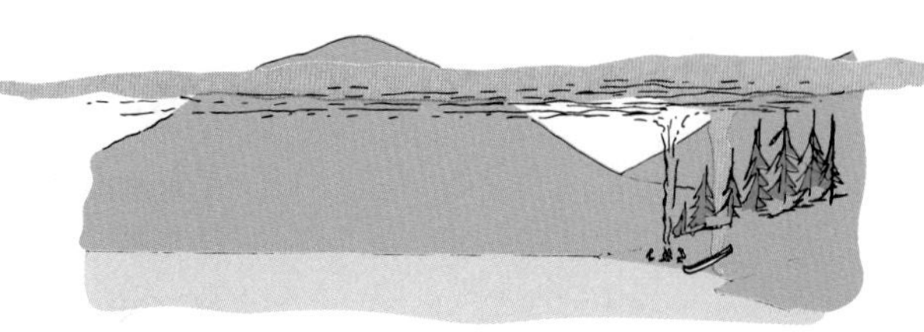

Figure 38.5
The layer of campfire smoke over the lake indicates a temperature inversion. The air above the smoke is warmer than the smoke, and the air below is cooler.

Concept Check ✓

1. If a parcel of dry air initially at 0°C expands adiabatically while flowing upward alongside a mountain, what is its temperature when it has risen 2 km? 5 km?
2. What happens to the air temperature in a valley when dry, cold air blowing across the mountains descends into the valley?
3. Imagine a gigantic dry-cleaner's garment bag full of dry, −10°C air floating 6 km above the ground like a balloon with a string hanging from it. If you yank it suddenly to the ground, what will its temperature be?

Check Your Answers

1. The air cools at the dry adiabatic rate of 10°C for each kilometer it rises. When the parcel rises to an elevation of 2 km, its temperature is −20°C. At an elevation of 5 km, its temperature is −50°C.
2. The air is adiabatically compressed, and so its temperature increases. Residents of some valley towns in the Rocky Mountains, such as Salida, Colorado, benefit from this adiabatic compression and enjoy "banana belt" weather in midwinter.
3. If the bag of air is pulled down quickly and heat conduction is negligible, the atmosphere adiabatically compresses the air and its temperature rises to a piping hot 50°C.

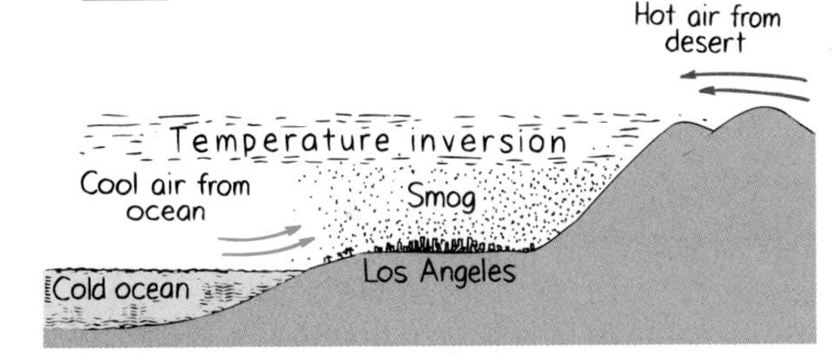

Figure 38.6
Smog in Los Angeles is trapped by the mountains and a temperature inversion caused by warm air from the Mohave Desert overlying cool air from the Pacific Ocean.

Hands-On Exploration: Adiabatic Expansion

Repeat the experiment that you might have done back in Chapter 10. Open your mouth and blow on your hand. You can feel that your breath is warm. Do the same again, only this time pucker your lips to make a small hole so your breath expands as it leaves your mouth. Note that your breath is appreciably cooler! Adiabatic expansion, hooray!

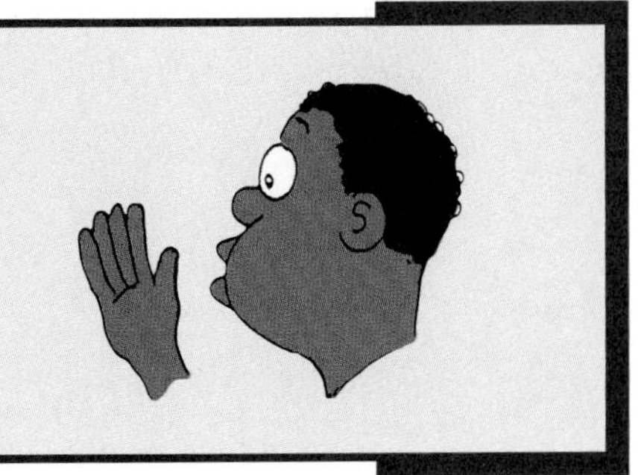

38.3 There Are Many Different Clouds

Clouds are a mixture of suspended water droplets and ice crystals. They are formed from rising, warm, moist air. As the warm air rises, it cools and therefore becomes less able to hold water vapor. The water vapor condenses into tiny droplets, and clouds are formed.

Clouds are generally classified according to their altitude and shape. There are ten principal cloud forms, each of which belongs to one of four major groups (Table 38.1).

High Clouds

High clouds are clouds that form at altitudes above 6000 meters. High clouds are denoted by the prefix *cirro-*. The air at this elevation is quite cold and dry, and so clouds this high are made up almost entirely of ice crystals.

The most common high clouds are thin, wispy *cirrus* clouds. Cirrus clouds are blown by high winds into the well-known wispy shapes, such as the classic "mare's tail." They usually indicate fair weather, but may also indicate approaching rain.

Cirrocumulus clouds are the familiar rounded white puffs. They are found in patches, and seldom cover more than a small portion of the sky. Small ripples and a wavy appearance make the cirrocumulus clouds look like fish scales. Hence cirrocumulus clouds are often said to make up a *mackerel sky.*

Cirrostratus clouds are thin and sheetlike, and often cover the whole sky. The ice crystals in these clouds refract light and produce a halo around the sun or the moon. When cirrostratus clouds thicken, they give the sky a white, glary appearance—an indication of coming rain or snow.

Middle Clouds

Middle clouds form between 2000 and 6000 meters high in the atmosphere. Middle clouds are denoted by the prefix *alto-*. These clouds are made up of water droplets and, when temperature allows, ice crystals.

Altostratus clouds are gray to blue-gray, and often cover the sky for hundreds of square kilometers. Altostratus clouds are often so thick that they diffuse incoming sunlight to the extent that objects on the ground don't produce shadows. Altostratus clouds often form before a storm. Look on the ground next time you're planning a picnic. If you don't see your shadow, cancel!

Altocumulus clouds appear as gray, puffy masses in parallel waves or bands. The individual puffs are much larger than those found in cirrocumulus clouds, and the color is also much darker. The appearance of altocumulus clouds on a warm, humid summer morning often indicates thunderstorms by late afternoon.

Figure 38.7
The four cloud groups. (a) High clouds: cirrus, cirrostratus, cirrocumulus. (b) Middle clouds: altostratus, altocumulus. (c) Low clouds: stratus, stratocumulus, nimbostratus. (d) Clouds with vertical development: cumulus, cumulonimbus.

Table 38.1

The Four Major Cloud Groups

1. **High clouds (above 6000 m)**	2. **Middle clouds (2000–6000 m)**	3. **Low clouds (below 2000 m)**	4. **Clouds having vertical development**
Cirrus	Altostratus	Stratus	Cumulus
Cirrostratus	Altocumulus	Stratocumulus	Cumulonimbus
Cirrocumulus	Nimbostratus		

Low Clouds

Low clouds ranging from the surface up to 2000 meters are called *stratus* clouds. They are almost always made up of water droplets. But in cold weather low clouds may also contain ice crystals and snow. *Stratus* clouds are uniformly gray and often cover the whole sky. They are very common in winter and give the sky a hazy, gray look. They resemble a high fog that doesn't touch the ground. Although stratus clouds are not directly associated with falling precipitation, they sometimes generate a light drizzle or mist.

Stratocumulus clouds either form a low, lumpy layer that grows in horizontal rows or patches or, with weak rising motion, appear as rounded masses. The color is generally light to dark gray. To tell the difference between altocumulus clouds and stratocumulus clouds, hold your hand at arm's length and point toward the cloud in question. An altocumulus cloud commonly appears to be the size of your thumbnail; a stratocumulus cloud appears to be about the size of your fist. Rain and snow do not usually fall from stratocumulus clouds.

Nimbostratus clouds are dark and foreboding. They are a wet-looking cloud layer associated with light to moderate rain or snow.

Clouds That Have Vertical Development

Rising air currents produce vertical clouds. These are *cumulus* clouds, which are the most familiar of the many cloud types. Cumulus clouds resemble pieces of floating cotton. They have sharp outlines and a flat base. They are white to light gray and generally occur about 1000 meters above the ground. The tops of cumulus clouds are often in the form of rising towers, showing the upward limit of the rising air. These are the clouds childhood daydreams are made of. Remember the horses, dragons, and magic palaces you saw in them?

Since clouds are denser than the surrounding air, why don't they fall from the sky? Ah, they do! They fall as fast as the air below rises. So without updrafts, we'd have no clouds.

When cumulus clouds turn dark and are accompanied by precipitation, they are referred to as *cumulonimbus* clouds. In this case, they indicate a coming storm. As we shall see, cumulonimbus clouds often become *thunderheads.*

Water vapor is less dense than air, though cloud droplets are considerably denser than air. The gravitational force pulling it down is enough to make the droplet fall. So why don't all the water droplets in clouds fall to the ground? The answer has to do with *updrafts.* An updraft is an upward movement of air. A typical cumulus cloud has an updraft speed of at least 1 meter per second, which is faster than the droplets can fall. So the droplets are supported by the upward-rising air. Without updrafts, the droplets drift so slowly out of the bottom of the cloud and evaporate so quickly that they have no chance to reach the ground. They are replaced by new droplets forming above.

Raindrops, on the other hand, are huge compared with typical cloud droplets. They fall faster than most updrafts can push upwards. And raindrops evaporate slowly enough that they can easily reach the ground.

38.4 Air Masses, Fronts, and Storms

An *air mass* is a volume of air much larger than the parcels of air we have discussed. Various distinct air masses cover large portions of the Earth's surface. Each has its own characteristics. An air mass formed over water in the tropics is different from one formed over land in the polar regions. Air masses are divided into six general categories according to what type of land or water they form over and the latitude where formation occurs (Table 38.2 and Figure 38.8). The type of surface an air mass forms over is designated by a lowercase letter (m for maritime, c for continental). The source region where an air mass forms is designated by an uppercase letter (A for arctic, P for polar, T for tropical.)

Each air mass has its own characteristics. Continental polar (cP) and continental arctic (cA) air masses generally produce very cold, dry weather in winter and cool, pleasant weather in summer. Maritime polar (mP) and maritime arctic (mA) air masses, picking up moisture as they travel across the oceans, generally bring cool, moist weather to a region. Continental tropical (cT) air masses are generally responsible for the hot, dry weather of summer, and warm, humid conditions are due to maritime tropical (mT) air masses.

When two different air masses meet, a variety of different weather conditions can develop.

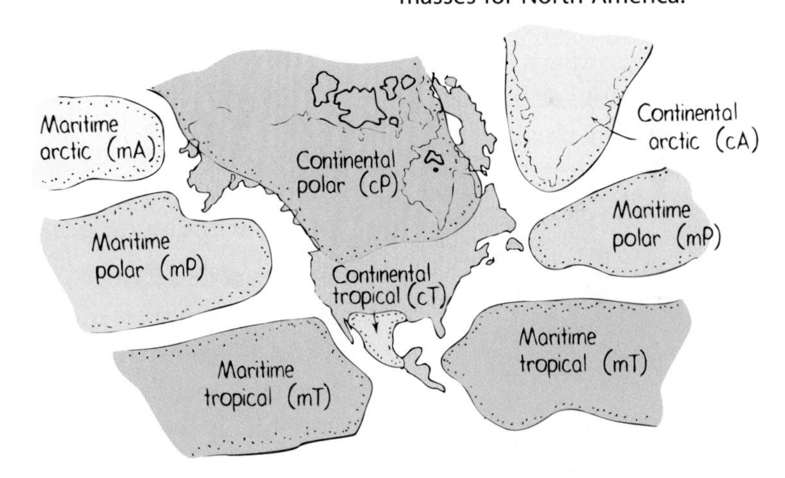

Figure 38.8
Typical source regions of air masses for North America.

Table 38.2

Classification of Air Masses and Their Characteristics

Typical Source Region	Classification	Symbol	Characteristics
Arctic	maritime arctic	mA	cool, moist, unstable
Greenland	continental arctic	cA	cold, dry, stable
North Atlantic, Pacific Oceans	maritime polar	mP	cool, moist, unstable
Alaska, Canada	continental polar	cP	cold, dry stable
Caribbean Sea, Gulf of Mexico	maritime tropical	mT	warm, moist; usually unstable
Mexico, Southwestern United States	continental tropical	cT	hot, dry, stable—aloft; unstable—surface

Figure 38.9
Cumulus clouds are often found as individual towering white clouds separated from each other by expanses of blue sky.

Atmospheric Lifting Creates Clouds

Clouds are great indicators of weather. For clouds to form, air must be lifted. The three principal lifting mechanisms in the atmosphere are convectional lifting, orographic lifting, and frontal lifting.

Convectional Lifting The Earth's surface is heated unequally. Some areas are better absorbers of solar radiation than others and so they heat up more quickly. The air that touches these surface "hot spots" becomes warmer than the surrounding air. This warmed air rises, expands, and cools. The rising air is accompanied by the sinking of cooler air from above. This circulatory motion produces **convectional lifting.**

If cooling occurs close to the air's saturation temperature, the condensing moisture forms a cumulus cloud. Air movement within the cumulus cloud moves in a cycle: Warm air rises, cool air descends. Because descending cool air inhibits the expansion of warm air beneath it, small cumulus clouds usually have a great deal of blue sky between them (Figure 38.9).

Cumulus clouds often remain in the same place that they formed, dissipating and reforming many times. As they grow, they shade the ground from the sun. This slows surface heating and inhibits the upward convection of warm air. Without a continuous supply of rising air, the cloud begins to dissipate. Once the cloud is gone, the ground reheats, allowing the air above it to warm and rise. Thus convectional lifting begins again, and another cumulus cloud begins at the same place.

No way can *all* air rise. Some has to come back down. Where air rises we see clouds; where it descends we see blue sky between the clouds.

Orographic Lifting An air mass that is pushed upward over an obstacle such as a mountain range undergoes **orographic lifting.** The rising air cools. If the air is humid, clouds form. The type of cloud formed depends on the air's stability and moisture content. If the air is stable, a layer of stratus clouds may form. If the air is unstable, cumulus clouds may form. As the air mass moves down the other side of the mountain (the leeward slope), it warms adiabatically. This descending air is dry because most of its moisture was removed in the form of clouds and precipitation on the windward (upslope) side of the mountain. Because the dry leeward (downslope) sides of mountain ranges are sheltered from rain and moisture, they are often referred to as regions of *rain shadow* (Figure 38.10).*

* Just as a "regular" shadow is a place where no light falls because some obstacle blocks the light, a rain shadow is a place where no rain falls because some obstacle (such as a mountain) blocks precipitation.

Frontal Lifting In weather reports we often hear about *fronts.* A front is the contact zone between two different air masses. When two air masses make contact, differences in temperature, moisture, and pressure can cause one air mass to ride over the other. When this occurs, we have **frontal lifting.** If a cold air mass moves into an area occupied by a non-moving warm air mass, the contact zone between them is called a *cold front,* and if warm air moves into an area occupied by a non-moving mass of cold air, the zone of contact is called a *warm front.* If neither of the air masses is moving, the contact zone is called a *stationary front.* Fronts are usually accompanied by wind, clouds, rain, and storms.

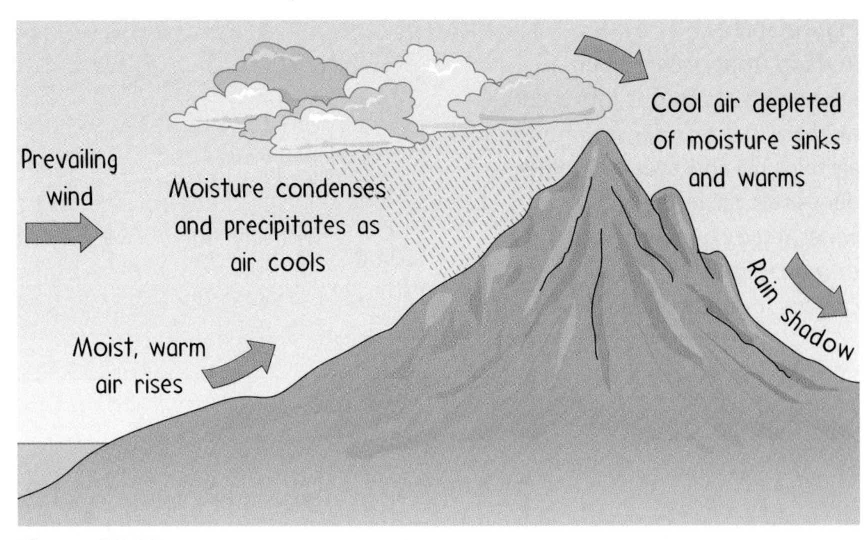

Figure 38.10
A mountain range may produce a rain shadow on its leeward slope. As warm, moist air rises on the windward slope, the air cools and precipitation develops. By the time it reaches the leeward slope, the air is depleted of moisture, so that the leeward side is dry. It lies in a rain shadow.

Meteorologists and other observers of the sky can often tell when a cold front is approaching by observing high cirrus clouds, a shift in wind direction, a drop in temperature, and a drop in air pressure. As cold air moves into a warm air mass, forming a cold front, the warm air is forced upward (Figure 38.11). As it rises, it cools, and water vapor condenses into a series of cumulonimbus clouds. The advancing wall of clouds at the front develops into thunderstorms with heavy showers and gusty winds. After the front passes, the air cools and sinks, pressure rises, and rain ceases. Except for a few fair-weather cumulus clouds, the skies clear and we have the calm after the storm.

When warm air moves into a cold air mass, forming a warm front, the less-dense warmer air gradually rides up and over the

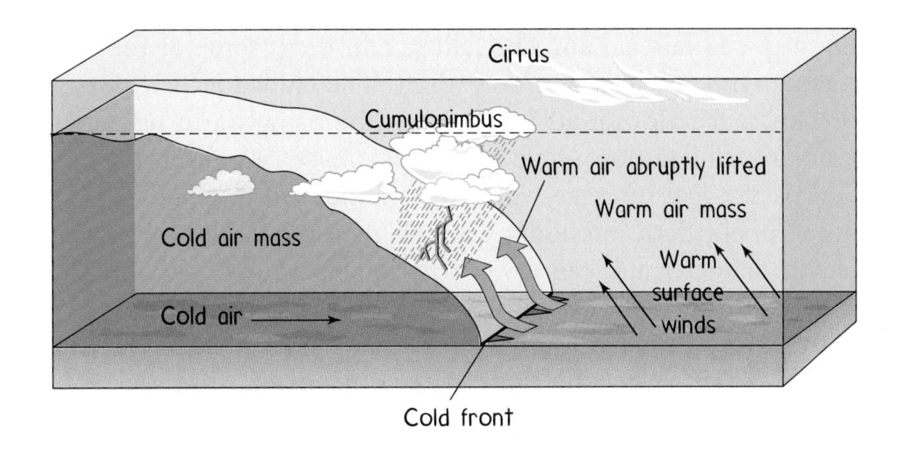

Figure 38.11
A cold front occurs when a cold air mass moves into a warm air mass. The cold air forces the warm air upward, where it condenses to form clouds. If the warmer air is moist and unstable, heavy rainfall and gusty winds develop.

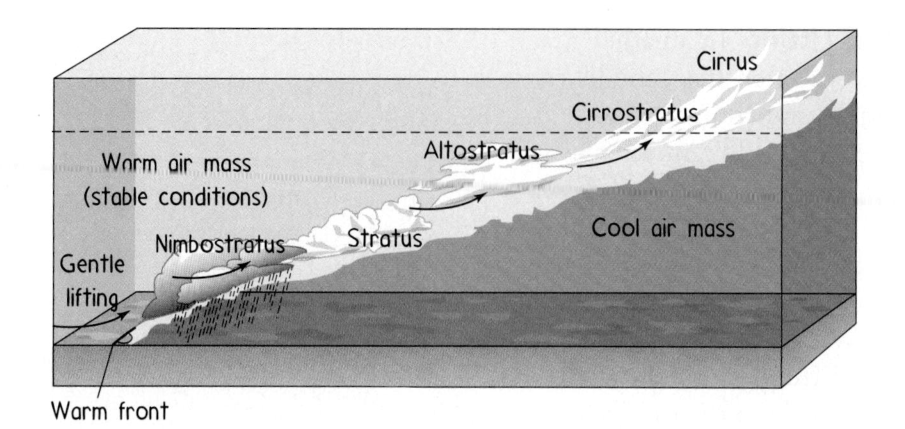

Figure 38.12
A warm front occurs when a warm air mass moves into a cold air mass. The less-dense warmer air rides up and over the colder, denser air, resulting in widespread cloudiness and light to moderate precipitation that can cover great distances.

colder, denser air (Figure 38.12). The approach of a warm front, although less obvious and more gradual than the approach of a cold front, is also indicated by cirrus clouds. Ahead of the front, the cirrus clouds descend and thicken into altocumulus and altostratus clouds that turn the sky an overcast gray. Moving still closer to the front, light to moderate rain or snow develops, and winds become brisk. At the front, air gradually warms, and the rain or snow turns to drizzle. Behind the front, the air is warm and the clouds scatter.

38.5 Weather Can Be Violent

The three types of lifting just discussed bring about many different weather conditions. Weather resulting from air masses in contact depends on the conditions in their source regions. Weather changes can occur slowly or very quickly. The most rapid changes, and the most violent ones occur with three major types of storms: thunderstorms, tornadoes, and hurricanes.

Thunderstorms

A thunderstorm begins with humid air rising, cooling, and condensing into a single cumulus cloud. This cloud builds and grows upward as long as it is fed by an updraft of rising warm air from below. Cloud droplets grow larger and heavier within the cloud until they eventually begin to fall as rain. The falling rain drags some of the cool air along with it, creating a downdraft. The chilled air is colder and denser than the air around it. Together, the rising warm updraft and the sinking chilled downdraft make up a *storm cell* within the cloud. This is the mature stage, where the thunderstorm cloud appears as a lonely, dark, brooding giant. It typically has a base several kilometers in diameter and can tower to altitudes up to 12 kilometers. At these high altitudes, horizontal winds and lower temperatures flatten and stretch the thunderhead crown into a characteristic anvil shape (Figure 38.13). After the thunderstorm dissipates, it leaves behind the cirrus anvil as a reminder of its once mighty presence.

Figure 38.13
The mature stage of a thunderstorm cloud appears as a towering cumulonimbus cloud that reaches up to about 12 km. Strong horizontal winds and icy temperatures flatten and distend the cloud's crown into a characteristic anvil shape.

Figure 38.14
Time exposure of cloud-to-ground lightning during an intense thunderstorm. (Can you see that a diffraction grating was in front of the camera lens?)

At any given time, there are about 1800 thunderstorms in progress in the Earth's atmosphere. Wherever thunderstorms occur, there is lightning and thunder. As water droplets in the cloud bump into and rub against one another, the cloud becomes electrically charged—usually positively charged at the top and negatively charged at the base. As electrical stress between the oppositely charged regions builds up, the charge becomes great enough that electrical energy is released and passed to other points of opposite charge, which quite often means the ground. The electrical energy flow from cloud to ground is lightning (Figure 38.14). As lightning heats up the air, the air expands and we hear lightning's noisy companion, thunder. Lightning strikes the Earth some 100 times every second. Some bolts have an electric potential of as much as 100 million volts. Lightning claims more than 200 victims per year in the United States alone.

Tornadoes

A revolving object, such as a whirling ball on a string, speeds up when pulled toward its axis of revolution, thus conserving its angular momentum. Similarly, winds slowly rotating over a large area speed up when the radius of rotation decreases. This increase in speed produces a *tornado.* A tornado is a funnel-shaped cloud that extends downward from a large cumulonimbus cloud. The funnel cloud is

Figure 38.15
Like a gigantic vacuum cleaner, the strong wind of a tornado can pick up and obliterate everything in its path.

called a tornado only after it touches the ground. The winds of a tornado travel at speeds of up to 800 kilometers per hour in a counterclockwise direction (clockwise in the Southern Hemisphere). As a tornado moves across the land, moving at speeds from 45 to 95 kilometers per hour, it may bounce and skip, rising briefly from the ground and then touching down again. A tornado acts like a gigantic vacuum cleaner, picking up everything in its path. It wreaks havoc not only by suction but also by the battering power of its whirling winds. In its wake, an explosive trail of flying dirt and debris is left behind (Figure 38.15).

Tornadoes occur in many parts of the world. In the flat central plains of the United States, a tornado zone extends from northern Texas through Oklahoma, Kansas, and Missouri. In this area, more than 300 tornadoes touch down each year. Hence the name for this area: Tornado Alley. Tornadoes are so frequent in this part of the country that some homes are built with underground shelters. The power of a tornado is terrifying and devastating.

Hurricanes

In the steamy tropics where the sun warms the oceans, the transfer of heat to the atmosphere by evaporation and conduction is so thorough that air and water temperatures are about equal. The high humidity in this part of the world favors the development of cumulus clouds and afternoon thunderstorms. Most of the individual storms are not severe. However, as the moisture content and temperature of the air increase and surface winds collide, a strong vertical wind shear can cause the rising warm, moist air to tilt inward and spiral. This condition produces a more violent storm—a *hurricane*—with wind speeds up to nearly 300 kilometers per hour. Gaining energy from its source area, a hurricane grows as more air rises. Increasing winds rotate around a central relatively calm, low-pressure zone—the eye of the storm.

Concept Check ✓

Would storms occur if all parts of the Earth's surface were heated evenly?

Check Your Answer No. The principal factor in the formation of storms is contact between warm air and cool air.

38.6 The Weather—Number One Topic of Conversation

Meteorologists have the important job of forecasting hurricanes and other storms. Weather forecasting is in part a matter of determining air-mass characteristics, predicting how and why the characteristics might change, and in what direction an air mass might move. In the case of hurricanes and tornadoes, such predictions are life saving. Meteorologists have a long and remarkable record of reducing property loss and saving many lives.

Weather forecasting involves great quantities of data from all over the world. Before the 1960s, most of these data were assembled, analyzed, and plotted on weather maps and charts by hand. This took thousands of calculations, a large work force, and long hours. Now, with computers, great quantities of data from around the world can be processed in a matter of minutes. Computers not only plot and analyze data, they also predict the weather. The computer draws a map of the projected weather conditions. The weather forecaster then uses these projections as a guide to predicting the weather. Even so, the many variables involved are not exactly predictable, and so it may unexpectedly rain on your parade!

Figure 38.16
On August 24, 1992, Hurricane Andrew struck the southern peninsula of Florida with gusts measured at 200 km per hour. The storm was catastrophic, with 45 deaths and more than 180,000 people left homeless. Without the timely hurricane watch issued by the National Weather Service, casualty losses could have been much more severe.

Weather Maps

The weather forecaster's primary tool is the surface weather map or chart. A weather map is essentially a representation of the frontal systems and the high-pressure and low-pressure systems that overlie the areas outlined in the map. Symbols on such a map are a shorthand notation to represent data gathered from various observation stations. These symbols are called *weather codes.*

This shorthand notation compiles 18 items of data into a very small area called a station model. The circle at the center describes the overall appearance of the sky. Jutting from the circle is a wind arrow, its tail in the direction from which the wind comes and its feathers indicating wind speed. The other 15 weather elements are in standard position around the circle.

A weather map is covered with lines—*isobars*—that connect points of equal pressure. As air moves from a high-pressure region to a low-pressure region, it rises and cools and the moisture in it condenses into clouds. In the vicinity of the low (L on map), we see an extensive cloud cover. In the vicinity of the high (H on map), we see clear skies. In a high-pressure region, air sinks and warms adiabatically. Because sinking air does not produce clouds, we find clear skies and fair weather. The heavy lines on a weather map represent fronts. Because fronts generally mean a change in the weather, they are of great importance on weather maps.

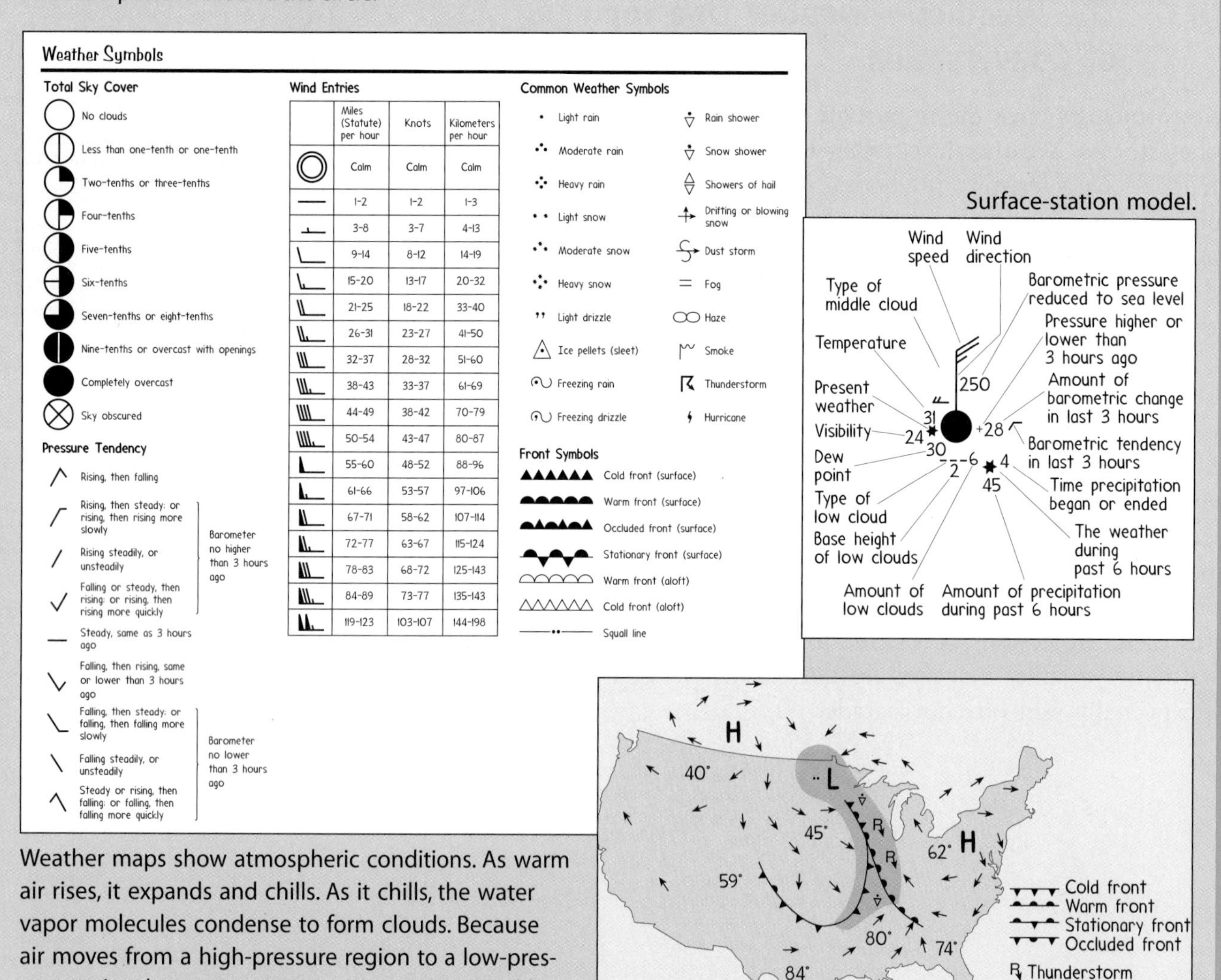

	Miles (Statute) per hour	Knots	Kilometers per hour
	Calm	Calm	Calm
	1-2	1-2	1-3
	3-8	3-7	4-13
	9-14	8-12	14-19
	15-20	13-17	20-32
	21-25	18-22	33-40
	26-31	23-27	41-50
	32-37	28-32	51-60
	38-43	33-37	61-69
	44-49	38-42	70-79
	50-54	43-47	80-87
	55-60	48-52	88-96
	61-66	53-57	97-106
	67-71	58-62	107-114
	72-77	63-67	115-124
	78-83	68-72	125-143
	84-89	73-77	135-143
	119-123	103-107	144-198

Weather maps show atmospheric conditions. As warm air rises, it expands and chills. As it chills, the water vapor molecules condense to form clouds. Because air moves from a high-pressure region to a low-pressure region, low-pressure zones are accompanied by cloud cover. In a high-pressure zone, air generally sinks. Because sinking air does not usually produce clouds, we find clear skies and fair weather.

Chapter Review

Key Terms and Matching Definitions

_____ clouds
_____ convectional lifting
_____ front
_____ frontal lifting
_____ humidity
_____ orographic lifting
_____ relative humidity
_____ temperature inversion

1. A measure of the amount of water vapor in the air.
2. The amount of water vapor in the air at a given temperature expressed as a percentage of the maximum amount of water vapor the air can hold at that temperature.
3. A condition in which the upper regions of the atmosphere are warmer than the lower regions.
4. The condensation of water droplets above the Earth's surface.
5. An air-circulation pattern in which air warmed by the ground rises while cooler air aloft sinks.
6. The lifting of air over a topographic barrier such as a mountain.
7. The contact zone between two different air masses.
8. The lifting that occurs as two air masses converge.

Review Questions

Water In the Atmosphere

1. Distinguish between humidity and relative humidity.
2. Why does relative humidity increase at night?
3. As air temperature decreases, does relative humidity increase, decrease, or stay the same?
4. What does saturation point have to do with dew point?
5. What happens to the water vapor in saturated air as the air cools?

Air Masses—Movement and Temperature Changes

6. Explain why warm air rises and cools as it expands.
7. Does a rising parcel of air get warmer, get cooler, or stay the same temperature?
8. What is an adiabatic process?
9. What is a temperature inversion? Give examples of where these inversions may occur.
10. What happens to the air pressure of an air parcel as it flows up the side of a mountain?

There Are Many Different Clouds

11. Explain how clouds form.
12. Rain or snow is most likely to be produced by which of the following cloud forms? a) cirrostratus, b) nimbostratus, c) altocumulus, d) stratocumulus
13. Are clouds having vertical development characteristic of stable air, stationary air, unstable air, or dry air?
14. Which type of clouds can become thunderheads?

Air Masses, Fronts, and Storms

15. Explain how convectional lifting plays a role in the formation of cumulus clouds.
16. Does a rain shadow occur on the windward side of a mountain range or on the leeward side? Explain.
17. Differentiate between a cold front and a warm front.

Weather Can Be Violent

18. What cloud form is associated with thunderstorms?

19. How do downdrafts form in thunderstorms?

20. Briefly describe how thunder and lightning develop.

The Weather—Number One Topic of Conversation

21. What information must be known to predict the weather?

22. The accuracy of weather forecasts depends on great quantities of data and thousands of calculations. If the number of data points were decreased, would accuracy also decrease?

Exercises

1. Why do clouds tend to form above mountain peaks?

2. Why does warm, moist air blowing over cold water result in fog?

3. Why does dew form on the ground during clear, calm summer nights?

4. Why does a July day in the Gulf of Mexico generally feel appreciably hotter than a July day in Arizona?

5. Would you expect a glass of water to evaporate more quickly on a windy, warm, dry summer day or a calm, cold, dry, winter day? Defend your answer.

6. Why does surface temperature increase on a clear, calm night as a low cloud cover moves overhead?

7. During a summer visit to Cancun, Mexico, you stay in an air-conditioned room. Getting ready to leave your room for the beach, you put on your sunglasses. The minute you step outside your sunglasses fog up. Why?

8. After a day of skiing in the Rocky Mountains, you decide to go indoors to get a warm cup of cocoa. As you enter the ski lodge, your eyeglasses fog up. Why?

9. Is it possible for the temperature of an air mass to change if no heat is added or subtracted? Explain.

10. Why is it necessary for an air mass to rise if it is to produce precipitation?

11. As an air mass moves first upslope and then downslope over a mountain, what happens to the air's temperature and moisture content?

12. The sky is overcast, and it is raining. What type of cloud is above you, nimbostratus or cumulonimbus? Defend your answer.

13. What accounts for the large spaces of blue sky between cumulus clouds?

14. Why don't cumulus clouds form over cool water?

15. What is the difference between rainfall that accompanies the passage of a warm front and rainfall that accompanies the passage of a cold front?

16. How do fronts cause clouds and precipitation?

17. Explain why freezing rain is more commonly associated with warm fronts than with cold fronts.

18. How does a rain-shadow desert form?

19. Why are clouds that form over water more efficient in producing precipitation than clouds that form over land?

20. What is the source of the enormous amount of energy released by a hurricane?

Suggested Web Sites

http://www.nws.noaa.gov/

http://www1.accuweather.com/adcbin/index.asp?partner=accuweather

http://lwf.ncdc.noaa.gov/oa/ncdc.html

http://wildwildweather.com/

Part 8 Astronomy

We began this book with the *physics* of the everyday world, then progressed to the microscopic world of *chemistry*. Then we explored *Earth science*, big by comparison. And now we conclude our explorations by studying the arena of the very big — the solar system, stars, and galaxies. We'll learn, for example, how the moon, seen here in stages of an eclipse, becomes red due to sunsets all over the world. To *astronomy*!

Chapter 39: The Solar System

Why does the moon go through phases from full to thin crescent while the sun remains round? What causes eclipses of the sun and of the moon? What makes the moon reddish during a lunar eclipse? Why do we see only one side of the moon? Do other planets have moons? Why do Saturn and other planets have rings? And what's the difference between a comet and a meteor? This chapter answers these questions and takes us on a brief tour of our solar system.

39.1 The Moon—Our Closest Celestial Neighbor

On July 20, 1969, Neil Armstrong was the first human to set foot on the moon. To date, 12 people have stood on the moon. We know more about the moon than any other celestial body. From nearly 400 kilograms of rock and soil samples brought back from lunar landings, we know the moon's age, its composition, and a lot about its history. Yet we still speculate about how it formed. Did it split off from the Earth while the Earth was forming? Did it materialize somewhere else and then fall into the Earth's gravitational grip? Or perhaps the Earth and moon are the result of a collision and merger of two very large planets in the making. How the moon formed has been a much-debated subject.

The moon is small, with a diameter about the distance from San Francisco to New York City. It once had a molten surface which cooled too rapidly for plate motion like the Earth's. In its early history it was intensely bombarded by meteoroids. A little more than 3 billion years ago, bombardment and volcanic activity filled basins with lava to produce its present surface. It has undergone very little change since. Its igneous crust is thicker than the Earth's. The moon is too small to have an atmosphere, and so without weather, the only eroding agents have been meteoroid impacts.

Figure 39.1
Edwin E. Aldrin, Jr., one of the three Apollo 11 astronauts, stands on the dusty lunar surface. Old Glory is rigged to appear flapping in the wind, for the moon is too small to have an atmosphere.

Figure 39.2
The Earth and moon as photographed in 1977 from the Voyager 1 spacecraft on its way to Jupiter and Saturn.

Figure 39.3
The moon in its various phases.

39.2 Phases of the Moon—Why Appearance Changes Nightly

Sunshine illuminates one-half of the moon's surface. The moon shows different amounts of its sunlit half as it circles around the Earth each month. These changes are the moon's **phases** (Figure 39.3). The moon cycle begins with the new moon. In this phase, its dark side faces us and we see darkness. This occurs when the sun, moon, and Earth are lined up, with the moon in between (position 1 in Figure 39.4).

During the next seven days, we see more and more of the moon's sunlit side (position 2 in Figure 39.4). The moon is going though its *waxing crescent* phase ("waxing" means *increasing*). At the *first quarter*, the angle between the sun, moon, and Earth is 90°. At this time we see half the sunlit part of the moon (position 3 in Figure 39.4).

During the next week, we see more and more of the sunlit part. The moon is going through its *waxing gibbous* phase (position 4 in Figure 39.4). ("Gibbous" means *more than half*.) We see a **full moon** when the sunlit side of the moon faces us squarely (position 5). We see the moon completely illuminated. The sun, Earth, and moon are lined up, with the Earth in between.

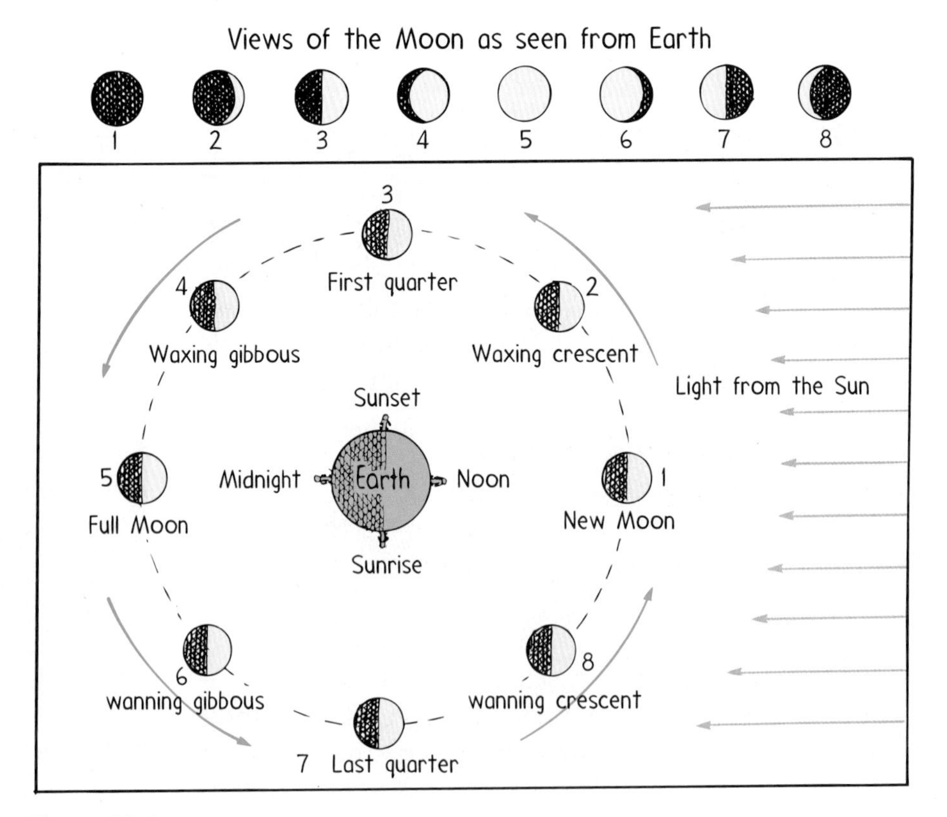

Figure 39.4
Sunlight illuminates only one-half of the moon. As the moon orbits the Earth, we see varying amounts of its sunlit side. One lunar phase cycle takes 29 1/2 days.

The cycle reverses during the following two weeks as we see less and less of the sunlit side while the moon continues in its orbit. This movement produces the *waning gibbous*, *last quarter*, and *waning crescent* phases. ("Waning" means *shrinking*.) The time for one complete cycle is about 29.5 days.*

Concept Check ✓

1. Can a full moon be seen at noon? Can a new moon be seen at midnight?
2. Astronomers prefer to view the stars when the moon is absent from the night sky. When, and how often, is the night sky moonless?

Check Your Answers

1. Inspection of Figure 39.4 shows that at noontime you would be on the wrong side of the Earth to see the full moon. Likewise, at midnight the new moon would be absent. The new moon is overhead in the daytime, not at nighttime.
2. At the time of the new moon and during the week on either side of the new moon, the night sky does not show the moon. Unless an astronomer wishes to study the moon, these dark nights are the best time for viewing other objects. Astronomers usually view the night skies during two-week periods every two weeks.

39.3 Eclipses—The Shadows of the Earth and the Moon

During any eclipse, the Earth, moon, and sun are exactly lined up.

The sun is 400 times larger in diameter than the moon but 400 times farther away from us. So from the Earth, the sun and moon appear the same size in the sky. It is this coincidence that lets us see solar eclipses. Both the Earth and the moon cast a shadow when sunlight shines on them. When either body crosses into the shadow cast by the other, an eclipse occurs.

A **solar eclipse** occurs during the time of a new moon when the moon is directly in front of the sun. The moon's shadow falls on part of the Earth. Since the sun is so large, rays of sunlight are not quite parallel, but taper (Figure 39.5). That's why the shadow on the Earth is so much smaller than the moon. If you're in the total shadow you experience darkness during the day—a total eclipse. It begins when the sun disappears behind the moon and ends when the

Figure 39.5
A solar eclipse occurs when the moon is directly between the sun and the Earth and the moon's shadow is cast on the Earth. Because of the small size of the moon and tapering of the solar rays, a solar eclipse occurs only on a small area of the Earth.

* The moon actually orbits the Earth once each 27.3 days relative to the stars. The 29.5-day cycle is relative to the sun and is due to the motion of the Earth-moon system as it revolves about the sun.

Figure 39.6
Eclipsed view of the sun, showing the corona, a pearly white halo of solar gases that extends several million kilometers beyond the sun's surface.

sun reappears on the other edge of the moon. The average time for a total eclipse is 2 or 3 minutes, with a maximum of 7.5 minutes. This time is brief because the moon is moving. If your location is off to the side of a total eclipse, you can see a partial eclipse. Then you still see part of the sun the entire time of the eclipse.

Interestingly, the darkness of a total eclipse is not complete because of the bright *corona* that surrounds the sun (Figure 39.6). A corona can only be seen when the sun is blocked, during an eclipse. You can't normally see it because it is overwhelmed by the sun's brightness (similar to the way you don't see stars in the daytime).

A **lunar eclipse** occurs when the moon passes into the Earth's shadow (Figure 39.9). Usually a lunar eclipse follows or lags a solar eclipse by two weeks. A lunar eclipse occurs only when the moon is full. All observers on the dark side of the Earth see a lunar eclipse at the same time, which may be partial or total. Fascinatingly, when the moon is fully eclipsed, it is still visible and is reddish in color.

Appearance of the Moon During a Lunar Eclipse

A fully eclipsed moon is visible. This is because the Earth's atmosphere acts as a lens and refracts light into the shadow region—enough to faintly illuminate the moon. Also, an eclipsed moon is reddish. To understand why, recall the reason for red sunsets in Chapter 15: The atmosphere scatters high-frequency light from sunlight and the low frequencies that aren't scattered produce a reddish color (the redness of sunsets). Beams of sunlight through the air travel the longest distance to our eyes when the sun is on the horizon—at sunset or sunrise. The longer "filtering path" at that time makes the light red.

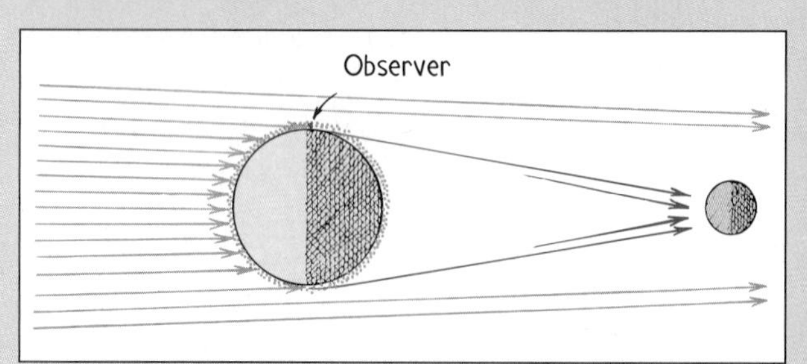

Figure 39.7
When the sun is low in the sky, the long path through the atmosphere to the observer filters high frequencies to make the sunlight reddish. Hence the red sunset. Light that continues past the observer travels through twice as much atmosphere and is even redder when it shines on the eclipsed moon.

The next time you view a sunset (or sunrise), quickly move your head to one side. When you do this, the light that would have met your eye instead misses and continues to the horizon behind you (Figure 39.7). If nothing is in the way, the light continues through the atmosphere and refracts into space. The light was reddish when it passed you, and gets even redder as it travels farther through the atmosphere before continuing into space. This red light shines on the eclipsed moon (Figure 39.8).

Figure 39.8
The fully eclipsed moon is often reddish because the Earth's atmosphere acts like a lens and refracts light from sunsets and sunrises all around the world onto the otherwise dark moon.

An eclipse requires exact alignment of the sun, Earth, and moon. This doesn't occur every month because the orbit of the moon about the Earth is slightly tilted to the Earth's orbit around the sun. So exact alignments only occur about twice per year—sometimes more frequent.

Figure 39.9
A lunar eclipse occurs when the Earth is directly between the moon and the sun and the Earth's shadow is cast on the moon.

Concept Check ✓

1. Does a solar eclipse occur at the time of a full moon or a new moon?
2. Does a lunar eclipse occur at the time of a full moon or a new moon?
3. Why are solar eclipses not seen monthly?
4. Which are more commonly seen—solar or lunar eclipses?

Check Your Answers

1. A solar eclipse occurs at the time of a new moon, when the moon is directly in front of the sun. Then the shadow of the moon falls on part of the Earth.
2. A lunar eclipse occurs at the time of a full moon, when the moon and sun are on opposite sides of the Earth. Then the shadow of the Earth falls on the full moon.
3. Solar eclipses don't occur each month because most of the time the shadow cast by the moon misses the Earth (due to the slight tilt of the moon's orbit around the Earth). When a solar eclipse does occur, the shadow is very small due to the tapering of sunlight. So relatively few people are in the shadow to witness the eclipse.
4. Lunar eclipses are more common. That's because the tapered shadow of the Earth completely covers the moon. This allows everybody on the night side of the Earth to see it at the same time. That's why nearly all your friends have seen a lunar eclipse, while relatively few of them have ever witnessed a solar eclipse.

Figure 39.10
The moon spins about its own polar axis just as often as it circles the Earth: once every 28 days. So as the moon circles the Earth, it spins so that the same side (shown yellow) always face the Earth. In each of the four successive positions shown here, the moon has spun 1/4 of a turn.

39.4 Why One Side of the Moon Always Faces Us

From the Earth we always see the same side of the moon. (The first humans to see its back side were Russian cosmonauts who orbited the moon in 1968.) The familiar facial features of the "man in the moon" are always turned toward us on Earth. Does this mean that the moon doesn't spin about its axis like the Earth does daily? No, for relative to the stars, the moon in fact does spin—although quite slowly, about once each 27 days. This monthly rate of spin matches the rate at which the moon revolves about the Earth. This explains why the same side of the moon always faces the Earth (Figure 39.10). This matching of monthly spin rate and orbital revolution rate is not a coincidence. Let's see why.

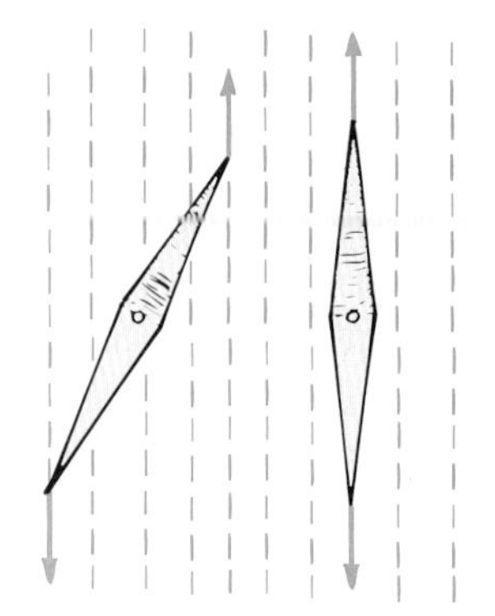

Figure 39.11
(a) When the compass needle is not aligned with the magnetic field (dashed lines), the forces represented by the blue arrows at either end produce a pair of torques that rotate the needle. (b) When the needle is aligned with the magnetic field, the forces no longer produce torques.

Think of a compass needle that lines up with a magnetic field. This line-up is caused by a *torque*—a "turning force with leverage" (like that produced by a child at the end of a see-saw). The compass needle on the left in Figure 39.11 rotates because of a pair of torques. The needle rotates counterclockwise until it lines up with the magnetic field. In a similar manner, the moon aligns with the Earth's gravitational field.

The side of the moon nearer to the Earth is gravitationally pulled more than the farther side. This stretches the moon out to a football shape (just as the moon does the same to Earth and gives us tides). If its long axis doesn't line up with the Earth's gravitational field, a torque acts (Figure 39.12). Like a compass in a magnetic field, it turns into alignment. So the moon lines up with the Earth in its monthly orbit. One hemisphere always faces us.

It's interesting to note that for most moons orbiting other planets, a single hemisphere faces its planet. We say the moons are "tidally locked." The long-range fate of the Earth is to be tidally locked to the sun.

Concept Check ✓

A friend says the moon does not spin about its axis, and evidence for a non-spinning moon is the fact that its same side always faces the Earth. What do you say?

Check Your Answer Help your friend tell the difference between *apparent* spin and *actual* spin. The moon has no apparent spin relative to the Earth. But from a broader point of view, from the stars, the moon clearly spins. It spins as slowly as it revolves about the Earth. So one side always faces the Earth as it spins. The fact that only one side of the moon faces the Earth is evidence that the moon *does* spin, not that it doesn't. The key concept is that the moon takes the same amount of time to complete one spin as it does to orbit the Earth.

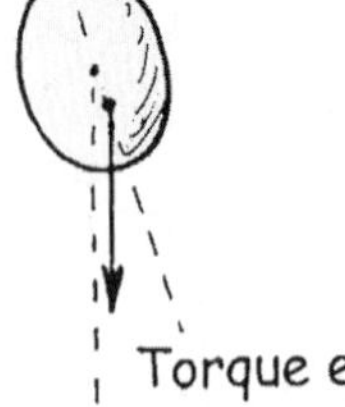

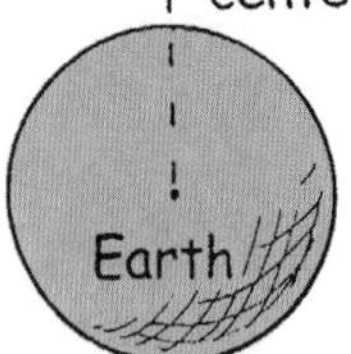

Figure 39.12
When the long axis of the moon is not aligned with the Earth's gravitational field, the Earth exerts a torque that tends to rotate the moon into alignment.

39.5 The Sun—Our Nearest and Most-Loved Star

The sun is our nearest star. Ancients who worshipped the sun seem to have realized that it is the source of all life on Earth. We are able to see, hear, touch, feel, and love only because every second 4.5 million tons of mass in the sun converts to radiant energy. A tiny fraction of this reaches the Earth.

The regions of the sun visible to us are its surface and its atmosphere. The sun's surface is a glowing 5800-kelvin plasma, perhaps about 500 kilometers thick. This transparent solar surface is the *photosphere* ("sphere of light"). On this surface there are relatively cool regions created by strong magnetic fields. These regions appear as **sunspots** when viewed from the Earth. These can be seen by the unaided eye through protective filters or when the sun is low enough on the horizon not to damage the eyes. Sunspots are typically twice the size of the Earth, they appear to move around because of the sun's rotation, and they last about a week or so. Often they cluster in groups (Figure 39.14).

Figure 39.13
In every second, 4.5 million tons of solar mass is converted to radiant energy in the sun. The sun is so massive, however, that in a million years only one ten-millionth of its mass is consumed.

The layer of the sun's atmosphere just above the photosphere is a 10,000-kilometer-thick transparent shell of plasma called the *chromosphere* ("sphere of color"). It is seen during an eclipse as a pinkish glow surrounding the sun. Beyond the chromosphere, there are streamers and filaments of outward-moving, high-temperature plasma which are curved by the sun's magnetic field. This outermost region of the sun's atmosphere is the *corona*, which mysteriously is hotter than the sun's surface. The corona extends several million kilometers from the sun until it merges into a hurricane of high-speed protons and electrons—the *solar wind*. It is the solar wind that powers the aurora borealis on Earth and produces the tails of comets. The solar wind also influences "space weather," and has a serious impact on the electronics and communications of Earth satellites.

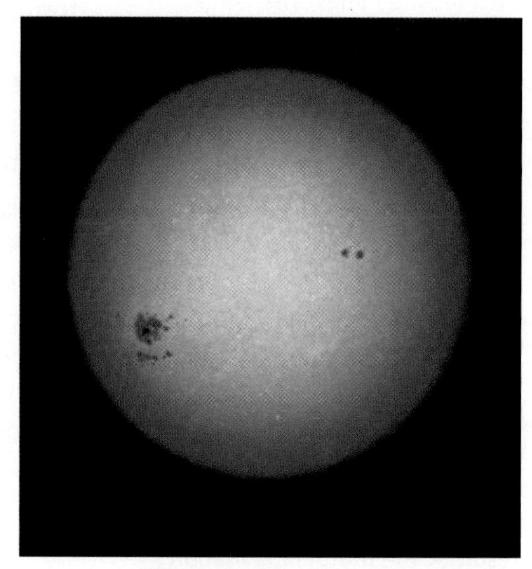

Figure 39.14
Sunspots on the solar surface are relatively cool regions. We say relatively cool because they are hotter than 4000 K. They look dark only by contrast with their 5800-K surroundings.

The sun spins slowly on its axis. Since the sun is a fluid rather than a solid, different latitudes of the sun spin at different rates. Equatorial regions spin once in 25 days, but higher latitudes take up to 36 days per complete spin. This differential spin means the surface near the equator pulls ahead of the surface farther north or south. The sun's differential spin wraps and distorts the solar magnetic field, which bursts out, forming the sunspots mentioned earlier. The number of sunspots reaches a maximum every 11 years (currently) and then a minimum after another 11 years. Hence there is a 22-year cycle of solar activity.

39.6 How Did the Solar System Form?

How the sun formed is not known for certain. It is generally believed to have originated from the gravitational contraction of a huge amount of interstellar matter nearly 5 billion years ago. The early universe was made primarily of hydrogen and helium. By the time the solar system formed some 10 billion or so years later, however, it contained small amounts of all the elements known today. All elements beyond hydrogen and helium were formed in the cores of stars that existed before the sun formed. When these stars underwent their death throes, their heavier elements were spewed into the interstellar mix, providing material for new stars.

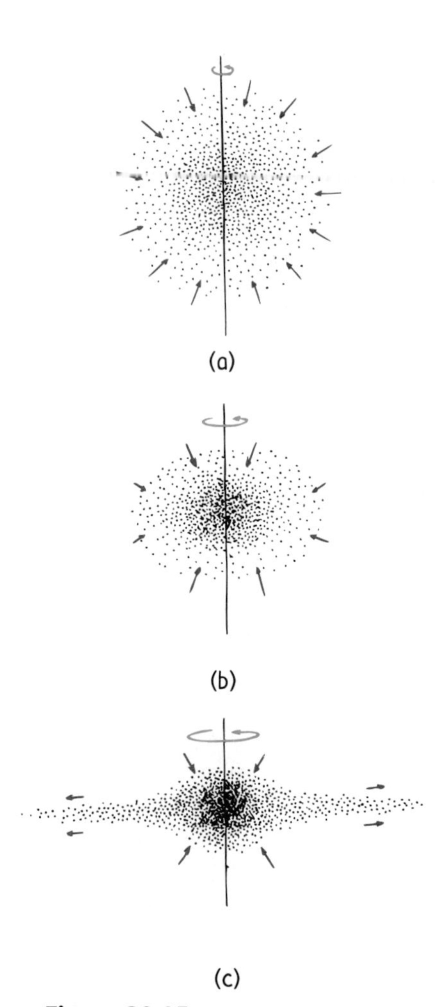

Figure 39.15
(a) A slowly rotating pocket of interstellar gas contracts as a result of the mutual gravitation between all the particles in it.
(b) The law of conservation of angular momentum accounts for the pocket speeding up.
(c) The increased momentum of individual particles and clusters of particles causes them to move in wider paths about the rotational axis, producing an overall disk shape.

As interstellar matter moves randomly in space, pockets of gas are believed to condense and dissolve repeatedly. They do this like wisps of fog that form and disperse in air. Exploding stars produce pressure waves and compress the gas pockets. Sometimes temporary clumps of condensed gas become permanent because the material within is held together by mutual gravity. A clump adds mass to itself by attracting neighboring material, and then, due to gravity, it falls in upon itself. Gravitational potential energy becomes thermal energy, and the clump's center becomes hotter. Continued gravitational contraction moves concentrated hot matter toward the center. Like an ice skater drawing arms inward when going into a spin, any rotational motion of the contraction speeds up. A faster spin flattens the matter into a disk shape (Figure 39.15). The increased surface area of the flattened disk radiates more of its energy into space. So the disk cools. In the formation of our solar system, this cooling likely led to the condensation of matter in swirling eddies—the birthplaces of the planets.

The center of the disk would have been too hot for matter to solidify. But farther out, heavier material probably solidified to become the four inner planets: Mercury, Venus, Earth, and Mars. Farther from the hot center, condensations of larger amounts of lighter matter, mostly hydrogen, formed the giant outer planets: Jupiter, Saturn, Uranus, and Neptune. Small and distant Pluto, as we shall see, is an exception to this creation hypothesis.

So we have a solar system, our home in the universe. Of the countless generations who have wondered about our place in the universe, only those of our lifetime have begun to understand it.

Concept Check ✓

What is the evidence that our sun is a second-or third-generation star?

Check Your Answer The first stars were made only of hydrogen and helium. Heavy elements arrived after these stars fused lighter elements into heavier elements in their cores. When stars explode, they spew heavy elements into space. These elements mix with others and make up the material for the formation of new stars. The abundance of heavy elements in the solar system is evidence of previous exploding stars. So the sun is young. There are many stars twice as old as the sun in our galaxy. Astronomers are not surprised to learn that these older stars have lower amounts of heavy elements.

39.7 Planets of the Solar System

The ancients could tell the difference between planets and stars because of their different movements in the sky. The stars remain relatively fixed in their patterns in the sky, but the planets wander.

Table 39.1

	Mean Distance from Sun (Earth-distances, Au)	Orbital Period (years)	Diameter (km)	Diameter (Earth = 1)	Mass (g)	Mass (Earth = 1)	Average Density (g/cm^3)
Sun			1,392,000	109.1	1.99×10^{33}	3.3×10^{5}	1.41
Mercury	0.39	0.24	4,880	0.38	3.3×10^{26}	0.06	5.4
Venus	0.72	0.62	12,100	0.95	4.9×10^{27}	0.81	5.2
Earth	1.00	1.00	12,760	1.00	6.0×10^{27}	1.00	5.5
Mars	1.52	1.88	6,800	0.53	6.4×10^{26}	0.11	3.9
Jupiter	5.20	11.86	142,800	11.19	1.9×10^{30}	317.73	1.3
Saturn	9.54	29.46	120,700	9.44	5.7×10^{29}	95.15	0.7
Uranus	19.18	84.0	50,800	3.98	8.7×10^{28}	14.65	1.3
Neptune	30.06	164.79	49,600	3.81	1.0×10^{29}	17.23	1.7
Pluto	39.44	247.70	2,300	0.18	10^{25}	0.002	1.9

The planets were called the *wanderers.* Today we know that planets are relatively cool bodies orbiting the sun. Planets emit no visible light of their own and, like the moon, simply reflect sunlight.

Very little was known about the planets a century or more ago. Detailed knowledge of planets today is enormous. How human knowledge advanced from only knowing planets as star-like spots that crossed the nighttime skies to what we understand today is a fascinating detective story. The journey of discovery was made possible first by careful observation with the naked eye, then with telescopes, and more recently with satellite probes launched from Earth. To keep this book brief, we leave this fascinating story to the suggested reading. We'll only look at a brief description of the planets. We divide the planets into two groups: the inner planets and the outer planets.

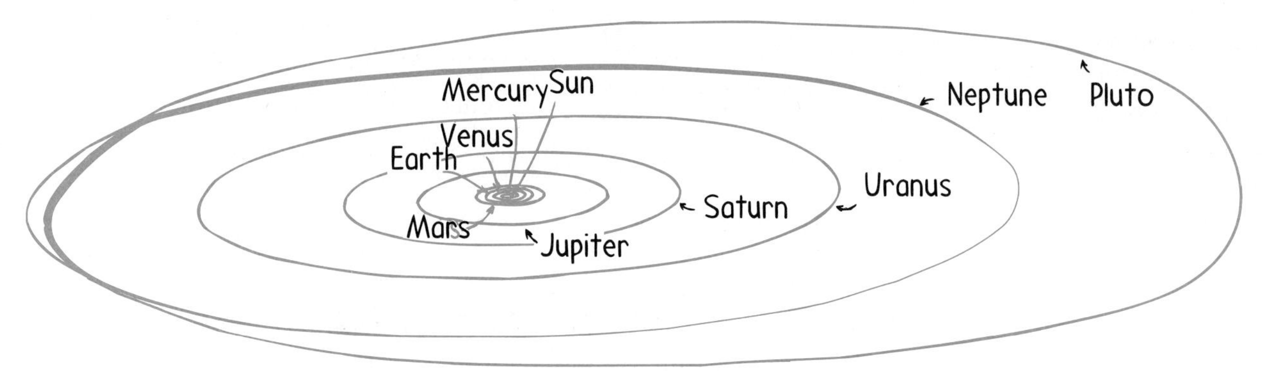

Figure 39.16
Scale drawing of the solar system, showing the four inner planets crowded around the sun and the five outer planets orbiting at greater distances.

39.8 The Inner Planets—Mercury, Venus, Earth, and Mars

Compared with the outer planets, the four planets nearest the sun are close together. These are Mercury, Venus, Earth, and Mars. These small and dense inner planets all have atmospheres. Each has a solid, mineral-containing crust and Earth-like composition. This is why they are called the *terrestrial planets.*

Mercury Mercury is the planet nearest the sun. It is somewhat larger than the moon and similar in appearance. Because of its closeness to the sun, it is the fastest planet. It completely circles the sun in only 88 Earth days. Thus one "year" on Mercury lasts for only 88 Earth days. Mercury spins about its axis only three times for each two revolutions about the sun. This makes its "daytime" very long and very hot, as high as 430°C.

Because of its smallness and weak gravitational field, Mercury holds very little atmosphere. Its atmosphere is about a trillionth as dense as the Earth's atmosphere. Without an insulating blanket of atmosphere and no winds to carry heat from one region to another, nighttime is very cold, about −170°C. Mercury isn't a very nice vacation spot!

Venus Venus is the second planet from the sun. Venus is often the first star-like object to appear after the sun goes down, so it is often called the evening "star." Nearby Mercury, although less bright, is also seen as an evening and morning "star." Both are seen near the sun at either sunup or sunset.

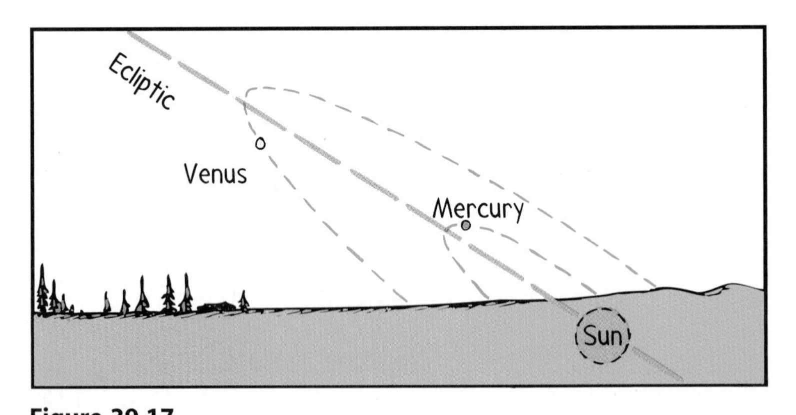

Figure 39.17
Because the orbits of Mercury and Venus lie inside the orbit of the Earth, they are always near the sun. Near sunset (or sunrise) they are visible in the sky as evening "stars" (or morning "stars").

Compared with the other planets, Venus most closely resembles the Earth. It is similar in size, density, and distance from the sun. A major difference is its very dense atmosphere, opaque cloud cover, and high average temperature (460°C)—too hot for oceans. Another difference between Venus and the Earth is in how the two planets spin about their axes. Venus takes 243 Earth days to make one full spin, and only 225 Earth days to make one revolution around the sun. This means that a day on Venus lasts longer than a year! Venus spins in a direction opposite to the direction of the Earth's spin. A space-traveling observer hovering about the solar system who sees the Earth spinning clockwise sees Venus spinning counterclockwise.

Because Venus has been thought of as almost an Earth twin, early speculations were that its surface was a steamy swamp inhabited by unfamiliar creatures. Speculations of life there ended after visits by space probes. In recent years, 17 probes have landed on the surface of Venus. There have been 18 flyby spacecraft (notably Pioneer Venus in 1978 and Magellan in 1993). From spacecraft data, we know that Venus has been very active volcanically and is an extremely harsh place. However, gathered evidence suggests that the surface temperature and atmosphere of Venus were initially very much like those of the Earth. Whereas most of the Earth's carbon dioxide is locked up in limestone formations and oceans, with very little in the atmosphere, the greater amount of sunlight on Venus has produced the opposite result on that planet. It is now the hottest place in the solar system. Much carbon dioxide escaped into its atmosphere, which increased the greenhouse effect, which released even more carbon dioxide, which now makes up about 96% of the atmosphere. Venus is a model for the Earth, for we wonder if a small temperature rise here on the Earth could trigger a similar irreversible chain reaction.

Figure 39.18
The Earth, the blue planet.

Earth The Earth is well described in Chapters 30-38, and so our treatment here is very brief. In answer to the question of what it must be like to live in outer space, we must not forget that we in fact *are* in outer space—gathered together on a hospitable planet in an inhospitable universe. Planet Earth is our home and deserves our greatest respect.

Ours is the blue planet, with more water surface than land. We're not too close to the sun and not too far away, with an average surface temperature delicately between that of freezing and boiling water. And the atmosphere is just dense enough to keep the oceans liquid. The insulating properties of our atmosphere and our relatively high daily spin rate allow only a brief and small lowering of temperature on the nighttime side of the Earth. So temperature extremes of day and night are favorable for life on Earth. Considering the harshness of most of the universe, the Earth is a very nice place to live. Our activities ought to be consistent with keeping it that way.

Mars Mars captures our fancy as another world, perhaps one with life. This is because of the similarities between the Earth and Mars: Mars is a little more than half the Earth's size, its mass is about 1/9 that of the Earth, and it has a core, mantle, crust, and a thin, nearly cloudless atmosphere. It has polar ice caps and seasons that are nearly twice as long as Earth's, because Mars takes nearly 2 Earth years to orbit the sun. When Mars is closest to the Earth, which occurs once every 15 to 17 years, its bright ruddy color outshines the brightest stars.

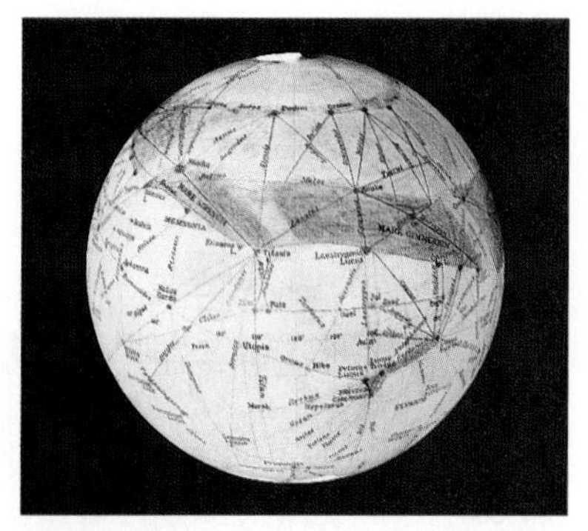

Figure 39.19
Martian canals mapped by astronomer Percival Lowell in the late 1800s. The canals proved to be optical illusions produced by the brain's ability to assemble vague markings into a coherent image (the same ability that enables us to see TV images rather than swarms of incoherent dots).

The Martian atmosphere is about 95% carbon dioxide, with only about 0.15% oxygen. So bring your own air supply if you plan to visit there. Also bring warm clothing, for its surface temperature at the equator goes from a comfortable 30°C in the day to a chilly −130°C at night. Night is only slightly longer than on the Earth, for the Martian day lasts 24 hours and 37.4 minutes. Never mind your raincoat, for there is far too little water vapor in the atmosphere

Figure 39.20
The rover Sojourner on the surface of Mars before setting out to explore the red planet.

for rain. Even the ice at the poles is primarily carbon dioxide. And don't give a second thought to waterproof footwear, for the low atmospheric pressure won't permit any puddles or lakes.

Dry sea beds indicate that water was abundant in the Martian past. Channels that appear to have been carved by water are also seen on the Martian surface. The largest of these channels fascinated early investigators, who perceived them through their telescopes as canals. This reinforced speculation of a Martian civilization. Except for questionable traces of life in Martian meteorites found in Antarctica, landings on Mars show that it has no life at the surface and no canals. The 1997 Pathfinder Mission shows it to be a very dry and windy place. Since the Martian atmosphere has a very low density, unequal heating produces Martian winds about ten times stronger than winds on the Earth.

Mars has two small moons—Phobos, the inner, and Deimos, the outer. Both are potato-shaped and have cratered surfaces. Phobos orbits in the same easterly direction that Mars spins (like our moon), at a distance of almost 6000 kilometers in a period of 7.5 hours. From Mars it appears about half the size of our moon. Deimos is about half the size of Phobos and orbits Mars in 30.3 hours at a distance of 20,000 kilometers. From the Martian surface it appears quite small.

39.9 The Outer Planets—Jupiter, Saturn, Uranus, and Neptune

The more widely spaced outer planets beyond Mars are much different from the inner planets. They're different in size, in composition, and in the way they were formed. Jupiter, Saturn, Uranus, and Neptune are gigantic, gaseous, low-density worlds. They are similar to Jupiter and are called *Jovian* planets. All have ring systems, Saturn's being the most prominent. Beyond these giants is outermost Pluto, much more different than any planet. Pluto is neither terrestrial nor Jovian. We consider the outer planets in turn.

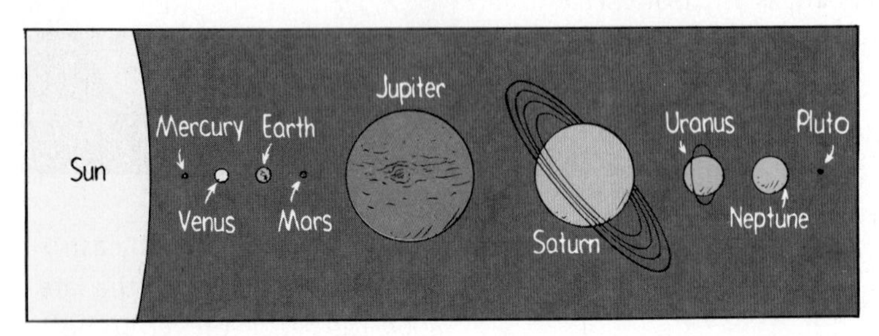

Figure 39.21
Relative sizes of the sun and planets. The sun's diameter is more than 10 times that of Jupiter.

Jupiter The largest of all the planets is Jupiter, whose yellow light in the night sky outshines the stars. In pre-spacecraft years, Jupiter was thought of as a failed star, for its composition is closer to that of the sun than to that of the terrestrial planets. Jupiter is more liquid than gaseous or solid. It spins rapidly about its axis in about ten hours, a speed that flattens it so that its equatorial diameter about 6% greater than its polar diameter. As with the sun, all parts do not rotate in unison. Equatorial regions complete a revolution several minutes before nearby higher and lower latitude regions. Jupiter doesn't have a hard surface crust that an astronaut could walk on. And if there were a place to stand, atmospheric pressure would be a crushing experience. The atmospheric pressure at Jupiter's surface is more than a million times more than the atmospheric pressure of the Earth. Jupiter's atmosphere is about 82% hydrogen, 17% helium, and 1% methane, ammonia, and other molecules.

Figure 39.22
Jupiter, with its moons Io and Europa, as seen from the Voyager I spacecraft in February 1979. The great red spot, lower left, is a cyclonic weather pattern of high winds and turbulence larger than the Earth. (NASA)

The average diameter of Jupiter is about eleven times greater than the Earth's. This means Jupiter's volume is more than a thousand times the Earth's. Jupiter's mass is greater than the combined masses of all the other planets. Due to its low density, however—about one-fourth the Earth's—Jupiter's mass is barely more than three hundred times the Earth's. Jupiter's core is a solid sphere about 20 times as massive as the entire Earth and is composed of iron, nickel, and other minerals.

More than half of Jupiter's volume is an ocean of liquid hydrogen. Beneath the hydrogen ocean lies an inner layer of hydrogen compressed into a sort of liquid metallic state. In it are abundant conduction electrons that flow to produce an enormous magnetic field. The strong magnetic field about Jupiter captures high-energy particles and produces radiation belts 400 million times as energetic as the Earth's Van Allen radiation belts. Radiation levels surrounding Jupiter are the highest ever recorded in space.

Surface temperatures are about the same day and night. Jupiter radiates about twice as much heat as it receives from the sun. The excess heat likely comes from internal heat generated long ago by gravitational contraction at the time the planet formed. When forming planets contract, gravitational potential energy is converted to thermal energy.

If you're planning to visit Jupiter, choose one of its moons instead. At least 28 of them, in addition to a faint ring, orbit Jupiter. Among the four largest moons, discovered by Galileo in 1610, Io and Europa are about the size of our moon, and Ganymede and Callisto are about as large as Mercury. The most interesting of Jupiter's moons seems to be Io, which has more volcanic activity than any other body in the solar system.

Saturn Saturn is one of the most remarkable objects in the sky. This is mainly because its rings are clearly visible with binoculars. Saturn is brighter than all but two stars and is second among the planets in mass and size. Saturn is twice as far from Earth as Jupiter.

Figure 39.23
Saturn surrounded by its famous rings believed to be composed of rocks and chunks of ice.

Its diameter, not counting its ring system, is nearly ten times that of the Earth and its mass nearly 100 times greater. It is composed primarily of hydrogen and helium and has the lowest density of any planet, only 0.7 times the density of water. That means Saturn would easily float in a bathtub if the bathtub were big enough. The low density and 10.2-hour rapid spin produce more polar flattening than any other planet. Like Jupiter, Saturn radiates about twice as much heat energy as it receives from the sun.

Saturn's rings are likely only a few kilometers thick. Four major rings have been known for many years, and spacecraft missions have detected others made up of hundreds of ringlets. The rings are made of chunks of frozen water and rocks, believed to be the remains of a moon that never formed or one ripped apart by tidal forces. All the rocks and ice making up the rings travel about Saturn in independent orbits. Inner parts of the ring travel faster than outer parts, just as any satellite near a planet travels faster than a more distant satellite.

Saturn has some twenty-three moons beyond its rings. The largest is Titan, 1.6 times larger than our moon and even larger than Mercury. It spins once each 16 days and has a methane atmosphere with atmospheric pressure likely greater than the Earth's. Its surface temperature is a cold −170°C. So bring a heavy coat and breathing gear if you plan to visit Titan. If that doesn't work out, try another of Saturn's large moons, Iapetus. One side of Iapetus is very bright and the other very dark. Try the region between these two extremes.

Uranus Uranus is twice as far from the Earth as Saturn is, and can barely be seen with the naked eye. Uranus was unknown to ancient astronomers. It has a diameter four times larger than the Earth's and a density slightly greater than that of water. So if you place Uranus in a giant bathtub, it would sink. The most unusual feature of Uranus is its tilt. Its axis is tilted 98° to the perpendicular of its orbital plane. So it lies on its side. Unlike Jupiter and Saturn, it appears to have no internal source of heat. Uranus is a cold place.

Uranus has at least 21 moons, in addition to a complicated faint ring system.

Neptune All planets are held in the solar system by their gravitational attraction to the sun. But the planets attract each other and everything else as well. When one planet is near another, the pull between them slightly disturbs their orbits. This disturbance is called a *perturbation.* Early in the nineteenth century, unexplained perturbations were observed for Uranus. Either the law of gravitation was failing at this great distance from the sun or an unknown eighth planet was perturbing Uranus. An Englishman and a Frenchman, J. C. Adams and Urbain Leverrier, independently calculated where an eighth planet should be. At about the same time, both sent letters to observatories in their countries with instructions to search a certain area of the sky. The request by Adams was delayed by misunderstandings at Greenwich. But Leverrier's request to the director of the Berlin observatory was heeded immediately. The planet Neptune was discovered one-half hour later.

Neptune's diameter is about 3.9 times the Earth's, its mass is 17 times greater, and its mean density is about a third that of the Earth. Its atmosphere is mainly hydrogen and helium, with some methane and ammonia. Like Jupiter and Saturn, it emits about 2.5 times as much heat energy as it receives from the sun.

Neptune has at least eight moons in addition to a ring system. The largest moon is Triton, which orbits Neptune in 5.9 days in a direction opposite the planet's eastward spin. Triton's diameter is three-quarters the size of our moon's diameter, and Titon has twice as much mass as our moon. It has bright polar caps and geysers of liquid nitrogen. A smaller moon, Nereid, takes nearly a year to orbit Neptune in a highly elongated elliptical path.

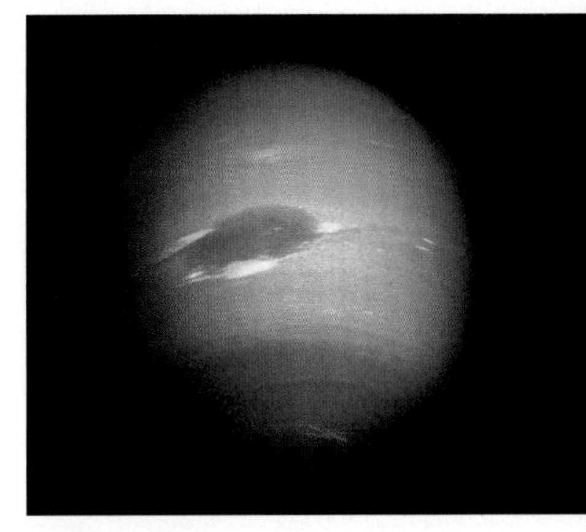

Figure 39.24
Cyclonic disturbances on Neptune in 1989 produced a great dark spot, which was larger than the Earth and similar to Jupiter's great red spot. It has now disappeared. (NASA)

Pluto The relative positions of stars on photographs do not change. That is, star images taken on any particular night will be in the same relative position as images taken nights later. But the images of planets will be in different places. Careful examination of such photographic plates resulted in the 1930 discovery of Pluto at the Lowell Observatory in Arizona. Because Pluto takes 248 years to make a single revolution about the sun, no one will see it in its discovered position again until 2178!

Whereas most of the planetary orbits in the solar system are nearly circular, Pluto's is the most elliptical and most steeply inclined to the planetary plane. Pluto's unique orbit allows it to sometimes be closer to the sun than Neptune. This occurred for twenty years, from 1979 to 1999. Since 1999, it has moved farther from the sun than Neptune.

Pluto is smaller than our moon. It has a diameter about one-fifth the Earth's diameter and a mass of about 0.002 that of Earth. Pluto has a moon named Charon that is about half the size of Pluto. Charon has a period of 6.4 days, which matches the rotational rate of Pluto, resulting in Charon appearing motionless in Pluto's sky.

A Pluto-like body discovered in 2002, named Quaoar, also has a moon. Pluto and Quaoar are among thousands of bodies beyond Neptune that orbit the sun in the disk-shaped region known as the Kuiper belt. Pluto is the largest of the Kuiper-belt objects. Whatever you learned in grade school, Pluto doesn't fit in the planetary family and is no longer considered a planet. To remain one, many other similar bodies in the Kuiper belt would share planetary status. Instead of being the pipsqueak of planets, Pluto now has the status of being the king of the Kuiper belt.

Planets are now being found beyond the solar system. More than 100 stars with planets have been identified. These distant planets are found by wobbles in their parent stars, and more recently by measuring miniscule reductions in light when they pass between their star and Earth. We search for planets that may be home to extraterrestrial life.

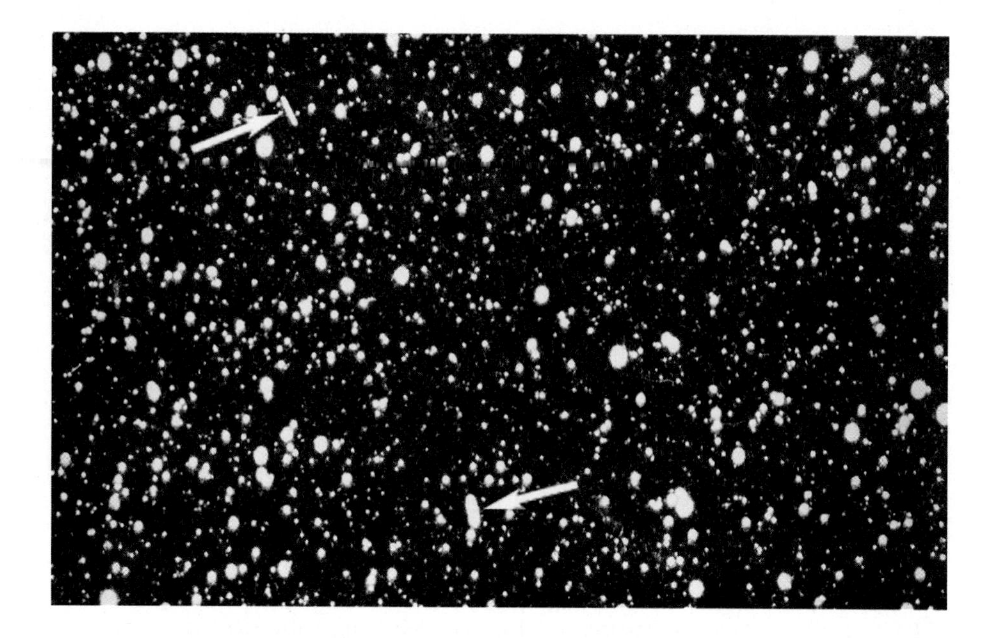

Figure 39.25
Asteroids leave blurred trails on time-exposure photographs of the stars. The images of two asteroids are indicated by the two white arrows.

A most interesting candidate is a star (47 Ursae Majoris) visible with the naked eye in the Great Bear constellation. This star is similar in size and chemical composition to our sun. It has at least two planets sweeping around it in nearly perfect circles, just as Earth moves around the sun. We wait in fascination for more findings.

Why does the moon show more craters than the Earth? Answer: Due to no weathering, the moon wears no "makeup."

39.10 Asteroids, Meteoroids, and Comets

There is an unusually large space between Mars and Jupiter. This is the *asteroid belt,* populated by tens of thousands of small rocky bodies called **asteroids** that orbit the sun. The smallest asteroids are irregular in shape, like boulders, and the larger ones are roughly spherical. Asteroids vary from grains of sand to rocky chunks hundreds of kilometers in diameter. The largest is Ceres, which is often called a *minor planet* and has a diameter of 750 kilometers. Asteroids are thought to be material that was unsuccessful in becoming a planet during the formation of the solar system. If the planet had formed, it would have been small, for the combined mass of all the asteroids is considerably less than the mass of the Earth's moon.

Although many asteroids neatly circle the sun, others do not. Collisions among asteroids are common, sending some of them helter-skelter. Some stray toward planet Earth. Asteroids smaller than a few hundred kilometers in diameter are called **meteoroids.** A **meteor** is a meteoroid that strikes the Earth's atmosphere, usually at an altitude of about 80 kilometers. A meteor is heated white-hot by friction with the atmosphere and is seen from the Earth as a flash of light—a "falling star." Most meteors we see are very small meteoroids, about the size of a grain of sand. Any meteor that survives its fiery decent through the atmosphere and reaches the ground is called a **meteorite.**

Figure 39.26
A meteor is produced when a meteoroid enters the Earth's atmosphere, usually about 80 km high. Most are grains of sand, seen as "falling stars."

Most meteorites are small and strike the Earth with no more energy than a falling hailstone. Some are big, however, and evidence of their impact is seen as craters. If the Earth were without weather and erosion, its surface would be as cratered as the moon's. Most impact craters on the Earth were eroded or covered by geologic processes long ago. More recent impacts, however, leave telltale marks (Figure 39.27). The most dramatic impact on record was near the Yucatan Peninsula in Mexico 65 million years ago, discussed in Chapter 36. The effects of that impact likely led to the extinction of dinosaurs and half the living species in the Cretaceous period.

Figure 39.27
The Barringer Crater in Arizona, made 25,000 years ago by an iron meteorite having a diameter of about 50 meters. The crater is 1.2 km across and 200 m deep.

Concept Check ✓

Using information from the Mesozoic Era in Chapter 36, how can meteor impact on Earth cause mass extinctions of life?

Check Your Answer As mentioned back in Chapter 36, a huge impact can produce a gigantic light-blocking cloud of dust for months or years. Photosynthesis cannot occur without sunlight, so food supplies diminish, and organisms perish. Other killing mechanisms include acid rain, tsunamis, wildfires, and a delayed greenhouse effect. No doubt about it—meteor impacts are not friendly.

Meteoroids come not only from the asteroid belt but from comets as well. A **comet** is a chunk of dust and ice that orbits the sun and becomes partly vaporized as it passes near the sun. Interestingly, most of the meteors we see are small particles of comet debris (Figure 39.28). Unlike meteors that shoot briefly across the sky, a comet moves slowly and gracefully to display one of nature's most beautiful astronomical spectacles (Figure 39.29).

Figure 39.28
When the Earth crosses the orbit of a comet, we see a meteor shower.

Whereas asteroids travel between the planets in roughly circular orbits, the orbits of comets are highly elliptical, extending far beyond Pluto's orbit. As a comet approaches the sun, solar heat vaporizes the ices. Escaping vapors glow to produce a fuzzy, luminous ball called a *coma.* A typical coma is a million kilometers in diameter. Within the bright coma is the solid part of the comet, the *nucleus.* A typical nucleus is a chunk of ice, dust, and other materials that measure a few kilometers across.

The solar wind and radiation pressure blow luminous vapors from the coma outward, away from the sun. That's why a comet has a long flowing *tail.* This tail can extend over 100 million kilometers. Most often the sun produces two tails on a comet: an *ion tail* and a *dust tail* (Figure 39.29). The ions are largely the remnants of water

Figure 39.29
The two tails of Comet West. A comet is always named after the person who first sees it. Guess who was the first person to see this comet?

Figure 39.30
Comet Hale-Bopp in 1997.

vapor, too massive to be affected by the pressure of sunlight. They flow with the high-speed solar wind, directly away from the sun. The dust tail is composed of micron-sized dust particles large enough to be affected by radiation pressure. The lower-speed dust tail curves, much as a water stream curves from the nozzle of a moving hose.

The density of the material in a comet's tail is quite low—less than the density of industrial vacuums in Earth laboratories. So compared with the Earth's atmosphere, the tails of a comet are "nothing at all." When a tail crosses the Earth directly, except for meteor showers high in the atmosphere, nothing at all changes at the Earth's surface. The incidence of a comet nucleus, however, is a different story. "Meteor" craters are formed by the impact of comets as well as meteors. Only the impact debris indicates the difference.

Comets are plentiful. There is almost always a comet in the sky, but most are too faint to be seen without the aid of a good telescope. About half a dozen new comets are discovered each year, many by amateur astronomers. Most comets have no visible tails, for their supply of ice is eventually exhausted. After about 100–1000 passes around the sun, a comet is pretty much burned out.

Chapter Review

Key Terms and Matching Definitions

_____ asteroid
_____ comet
_____ lunar eclipse
_____ meteor
_____ meteorite
_____ meteoroid
_____ moon phases
_____ solar eclipse
_____ sunspots

1. The cycles of change of the face of the moon as seen from the Earth: New (dark) to waxing to full to waning and back to new.
2. The phenomenon whereby the shadow of the moon falls on the Earth, producing a region of darkness in the daytime.
3. The phenomenon whereby the shadow of the Earth falls on the moon, producing relative darkness of the full Moon.
4. Temporary relatively cool, dark regions on the sun's surface.
5. A small rocky fragment that orbits the sun.
6. A very small asteroid.
7. A meteoroid once it enters the Earth's atmosphere; a "shooting star."
8. A meteoroid that survives passage through the Earth's atmosphere and hits the ground.
9. A body composed of ice and dust that orbits the sun, usually in a very eccentric orbit, and which has one or more luminous tails when it is close to the sun.

Review Questions

The Moon—Our Closest Celestial Neighbor

1. How does the Moon's rate of rotation about its own axis compare with its rate of revolution around the Earth?

Phases of the Moon—Why Appearance Changes Nightly

2. Where is the sun when you view a full moon?

Eclipses—The Shadows of the Earth and the Moon

3. In what alignment of sun, moon, and Earth does a solar eclipse occur?
4. In what alignment of sun, moon, and Earth does a lunar eclipse occur?
5. Why is totality during a lunar eclipse not altogether dark?

Why One Side of the Moon Always Faces Us

6. What does the spin rate of the moon have in common with its orbital rate about the Earth?

The Sun—Our Nearest and Most-Loved Star

7. What happens to the amount of the sun's mass as it "burns"?
8. What is the solar wind?
9. How does the rotation of the sun differ from the rotation of a solid body?

How Did the Solar System Form?

10. How old is the sun?
11. What happens to the speed of a spinning mass of gas when it contracts?

Planets of the Solar System

12. Why did the ancients call planets "wanderers"?
13. Why are days on Mercury very hot and nights very cold?
14. Why is Earth called the *blue planet*?
15. What gas makes up most of the Martian atmosphere?

The Outer Planets—Jupiter, Saturn, Uranus, and Neptune

16. What surface feature do Jupiter and the sun have in common?
17. Why does Jupiter bulge at the equator?
18. Which move faster, Saturn's inner rings or the outer ones?

Asteroids, Meteoroids, and Comets

19. What is a falling star?
20. Why do the tails of comets point away from the sun?

Exercises

1. Which is larger, the radius of the sun or the distance between the moon and the Earth? (See the inside front cover.)
2. Why does the moon lack an atmosphere? Defend your answer.
3. In what ways would a telescope mounted on the moon produce better views into space than telescopes mounted on the Earth?
4. Is the fact that we see only one side of the moon evidence that the moon spins or that it doesn't rotate? Defend your answer.
5. Photograph (a) shows the moon partially lit by the sun. Photograph (b) shows a ping-pong ball in sunlight. Compare the positions of the sun in the sky when each photograph was taken. Do the photos support or refute the claim that they were taken on the same day? Defend your answer.

a

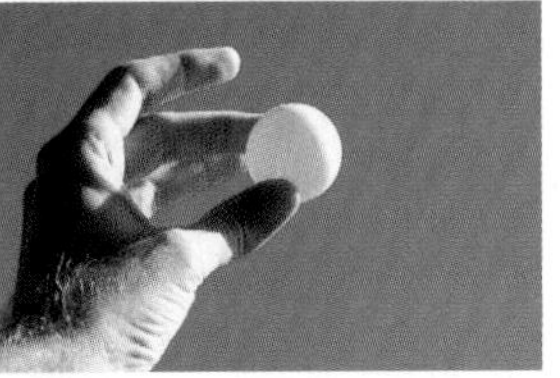
b

6. Why are there a lot more craters on the surface of the moon than on the surface of the Earth?

7. Why is it not totally dark in the location where a total solar eclipse occurs?

8. Because of the Earth's shadow, the partially eclipsed moon looks like a cookie with a bite taken out of it. Explain with a sketch how the curvature of the bite indicates the size of the Earth relative to the size of the moon. How does the tapering of the sun's rays affect the curvature of the bite?

9. What energy processes make the sun shine? In what sense can it be said that gravity is the prime source of solar energy?

10. Where are elements heavier than hydrogen and helium formed?

11. What is the cause of winds on Mars (and on almost every other planet, too)?

11. What is the major difference between the terrestrial and Jovian planets?

12. Why was Jupiter once thought to be a failed star?

13. Why are the seasons on Uranus different from the seasons on any other planet?

14. What were the similar historical circumstances that link the names of the planets Neptune and Pluto with the elements Neptunium and Plutonium?

15. By what investigative method was Pluto discovered?

16. Why are meteorites so much more easily found on Antarctica than on other continents?

17. A meteor is visible only once, but a comet may be visible at regular intervals throughout its lifetime. Why?

18. What would be the consequence of a comet's tail sweeping across the Earth?

19. Chances are about 50-50 that in any night sky there is at least one visible comet that has not been discovered. This keeps amateur astronomers busy looking night after night, for the discoverer of a comet gets the honor of having it named for him or her. With this high probability of comets in the sky, why aren't more of them found?

20. In terms of the conservation of energy, describe why comets eventually burn out.

Suggested Readings and Web Sites

Bronowski, Jacob. *The Ascent of Man.* Boston: Little Brown, 1973.

Sagan, Carl. *Cosmos.* New York: Random House, 1980.

Weiskopf, Victor. *Knowledge and Wonder: The Natural World as Man Knows It.* Cambridge, Mass: MIT Press, 1979.

http://sse.jpl.nasa.gov/features/planets/planetsfeat.html

http://planetscapes.com

http://lpl.arizona.edu/billa/tnp/

http://pds.jpl.nasa.gov/planets/

http://ethel.as.arizona.edu/~collins/astro/subjects/srchplanet1.html

http://origins.jpl.nasa.gov/library/exnps/ch01_0.html

http://exoplanets.org/

Chapter 40: The Stars

Why do we call the sun a *star*? Do stars experience a life cycle from birth to death? What's a black hole? And if black holes are invisible, how do scientists know they exist? Is it true there are more galaxies than there are stars in our Milky Way galaxy? What are the types of galaxies, and how did they come to be? And finally, what's the difference between astronomy and astrology? We conclude this book with the truly far-out subject of stars and galaxies. We will examine the life of stars—how they are born, how they live, and how they die.

40.1 The Constellations

When we look at the sky on a moonless night, we might guess we see many thousands of stars. But the unaided eye sees at most about 3000 stars, horizon to horizon. With the aid of a telescope, we can see more stars. But disappointingly, stars appear as point sources of light, with or without magnification. Telescopes can show the details of the moon or planets but no details of stars. This is because stars are very, very far away. Many of the brightest are 10 to 1000 light-years distant.* Because of their great distance, they all appear equally remote, as on the celestial sphere imagined by the ancients.

Figure 40.1
The constellations and Taurus represent figures from Greek mythology.

* Recall from Chapter 14 that one light-year is the distance light travels in one year, about 9.5×10^{12} km. Another unit of distance popular with astronomers is the *parsec*, which is the same as 3.26 light-years.

The roots of astronomy date back to prehistoric times when humans became familiar with star patterns in the night sky. These early astronomers divided the night sky into groups of stars, called *constellations.* The names of the constellations are mainly a carryover from the names assigned by early Greek, Babylonian, and Egyptian astronomers.

The grouping of stars and the significance given to them varied from culture to culture. In some cultures, the constellations stimulated storytelling and the creation of great myths. In others, the constellations honored heroes like Hercules and Orion or served as navigational aids for travelers and sailors. In some other cultures, they provided a guide for planting and harvesting crops—for the constellations were seen to move periodically in the sky, in step with the seasons. Some of the first calendars originated from the charts of this periodic movement.

Stars were thought to be points of light on a great revolving celestial sphere with the Earth at its center. Positions of the sphere were believed to affect earthly events, and so were carefully measured. Keen observations and logical reasoning gave birth to both astrology and later, astronomy. So we find that astronomy and astrology—science and pseudoscience, share the same roots. Today we know that the Earth orbits the sun, with its night side always away from the sun. We see in Figure 40.2 why the background of stars varies in the night sky throughout the year.

To better understand Figure 40.2, imagine placing a lamp on a table in the middle of your room. Then move counterclockwise around the table with your back toward the lamp at all times. You'll see different parts of the room as you walk. Likewise, an observer on the night side of the Earth sees different parts of the sky as the Earth orbits around the sun.

Figure 40.2
The night side of the Earth always faces away from the sun. As the Earth circles the sun, different parts of the universe are seen in the nighttime sky. Here the circle, representing one year, is divided into twelve parts—the monthly constellations. To Earth observers, the stars seen in the night sky appear to move in a yearly cycle.

Concept Check ✓

Looking at Figure 40.2, why are the background stars during a midday solar eclipse of constellations normally seen six months earlier or later?

Check Your Answer The sunlit side of the Earth faces constellations opposite to those seen at nighttime. But the only time these can be seen is when the sun's light is blocked. That's the time of a solar eclipse.

The Big Dipper and the North Star

Perhaps the most easily recognized star group in the northern hemisphere is the Big Dipper (Figure 40.3). Because of its great distance, it appears as a plane, but the seven stars actually lie at quite different distances from us. The Big Dipper and the larger groups that make up constellations, of course, would have entirely different patterns if viewed from other locations in the universe. Because of the variety of speed and directions of stars, the familiar patterns of all groups are temporary. We see in Figure 40.4 how the Big Dipper appeared from the Earth in the past and how it is projected to look in the future.

The Big Dipper is useful for locating the North Star (Polaris). The North Star is easily located by drawing a line through the two stars in the end of the bowl of the Big Dipper and extending the line upward about five times the distance between these two stars (Figure 40.5). Because it lies very close to the extended Earth's axis, the North Star appears to remain stationary as the Earth undergoes its daily spin. All the surrounding stars appear to move in circles around the North Star, as evidenced in long time exposure photographs (Figure 40.6).

For centuries, a test of good eyesight has been to locate which star in the Big Dipper is actually two closely spaced stars. The answer is the next-to-last star in the handle. Although this star and its companion star appear to be very closely spaced, they are actually quite far apart in space. They look close because they happen to lie along the same line of sight from Earth. Interestingly, the brighter star, Mizar, *is* actually a pair of stars—a **binary**—the first optical binary to be observed by a telescope. Even with excellent vision, you'd be unable to see the double star Mizar without the aid of a good-sized telescope.

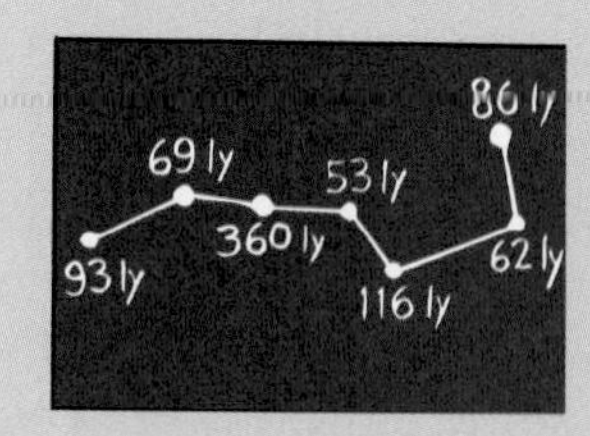

Figure 40.3
The familiar Big Dipper. The size of the dots represents the apparent brightness of the stars, which are not all the same distance from the Earth. Their distances are noted in light-years.

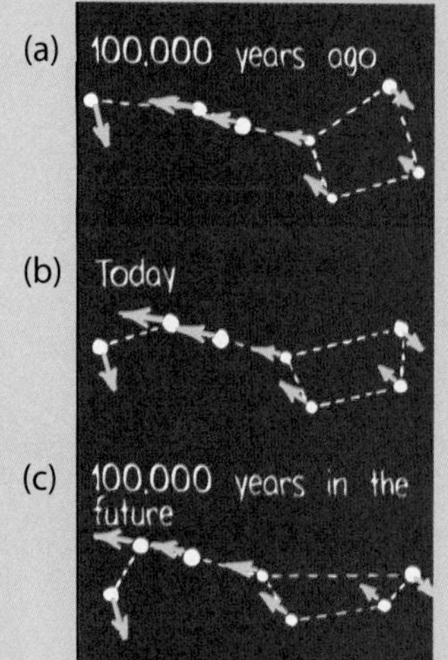

Figure 40.4
The pattern of the Big Dipper is temporary:
(a) we see its pattern 100,000 years ago;
(b) as it appears at present;
(c) in the future about 100,000 years from now.

Figure 40.5
The pair of stars in the end of the Dipper's bowl point to the North Star. The Earth rotates about its axis and therefore about the North Star, so over a 24-hour period the Big Dipper and other surrounding star groups make a complete revolution.

Figure 40.6
Time exposure of the northern night sky.

40.2 Birth of Stars

Interstellar space, the space between the stars, is not empty. It contains faint amounts of elements, primarily hydrogen and helium. To a lesser degree, interstellar space contains a wide variety of molecules ranging from ammonia to ethyl alcohol. Among these atoms and molecules are specks of interstellar dust. This dust is composed of carbon and silicates, which are sometimes coated with frozen ices of water, carbon dioxide, methane, and ammonia. These interstellar dust particles may play the same condensing role in star formation that similar dust particles play in cloud formation. The density of all this interstellar material is a million times lower than the densities of the highest vacuums achieved in earthbound laboratories.

To form a star, begin with a huge cloud of low-temperature interstellar material. The density of such a cloud will not be perfectly uniform. Regions of slightly greater gas density will have slightly more mass and a slightly greater gravitational field. Therefore they will more strongly attract neighboring particles. This attraction of neighbors increases the mass and gravitational field of the region, which then pulls more particles. In time we have a collection of matter many times the mass of the sun spreading out over a volume many times larger than the solar system. This is a forming star—a **protostar.**

Mutual gravitation between the gaseous particles in a protostar results in an overall contraction of this huge ball of gas. The density at the center greatly increases as matter is scrunched together. Pressure and temperature rise. When the central temperature reaches about 10 million kelvins, some of the hydrogen nuclei fuse to helium nuclei. As we learned in Chapter 20, this is a *thermonuclear reaction.* Hydrogen is converted to helium and enormous amounts of radiant and thermal energy are released. This ignition of nuclear fuel marks the change from protostar to star. Outward-moving radiant energy and gas exert an outward pressure on the contracting matter, finally becoming strong enough to stop the contraction. When outward radiation and gas pressures balance inward gravitational pressure, we have a stable star.

The material that makes up a star depends on how old the universe was when the star formed. The very first protostars had only primordial hydrogen and some helium. Stars run their life cycles and then, like living things, return their materials to the overall environment. Elements heavier than hydrogen and helium are manufactured in star cores, and when the stars' life cycles end, they spew their heavier elements into the interstellar mix. So the protostars that followed the earliest ones were enriched with heavier elements. The sun and its planets evolved from heavy elements, which are testimony that many stars lived and died before the solar system came to be. All atoms on the Earth heavier than helium were once part of another star. So we are quite literally made of star dust.

Concept Check ✓

How does thermonuclear fusion and gravitational contraction affect the physical size of a star?

Check Your Answer Thermonuclear fusion tends to blow the star outward and gravitation tends to pull it inward. The outward thermonuclear expansion and inward gravitational contraction produce an equilibrium that accounts for the star's size.

40.3 Life and Evolution of Stars

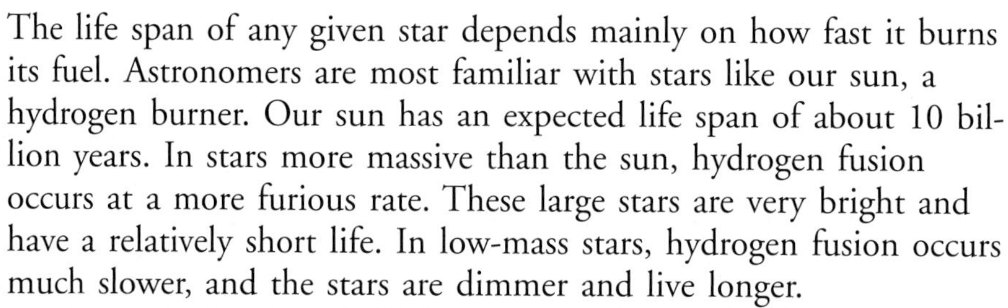

The life span of any given star depends mainly on how fast it burns its fuel. Astronomers are most familiar with stars like our sun, a hydrogen burner. Our sun has an expected life span of about 10 billion years. In stars more massive than the sun, hydrogen fusion occurs at a more furious rate. These large stars are very bright and have a relatively short life. In low-mass stars, hydrogen fusion occurs much slower, and the stars are dimmer and live longer.

Figure 40.7
The sun is about 5 billion years old, having spent about half its expected life span of 10 billion years.

Surprisingly, about half the stars seen in the sky do not live alone. They are actually two stars revolving around a common center (just as the Earth and the moon revolve around each other). These double stars are *binary stars.* By observing how the two stars in a binary revolve around their common center, we can calculate their masses. In fact, the only way to determine the mass of a star is to locate it in a binary system (the sun excluded, for its planets provide this information). Mass can be calculated since the size and orbital period of binary stars depend on their masses. So binaries help astronomers a great deal. They give astronomers an easy way to determine how much matter a particular star contains.

There has been speculation that our sun does not live a solitary life but is part of a binary system. If it is, its partner is very small and very distant. This partner star is thought to travel in a very large elliptical orbit, as much as three light-years away from the sun. At its closest approach, which would occur every 26 to 30 million years, it would pass near the fringes of the outermost planets. Its gravity would perturb billions of comets into the inner solar system, all within a few million years. This star has been named *Nemesis,* after the goddess of divine retribution, because some speculators credit it with triggering the meteorite impact that may have led to the demise of the dinosaurs 65 million years ago. This deadly companion has not been found and may not exist.

Recent findings indicate that the sun is not the only star with a planetary system. We can only wonder if some of these, like Earth, are at a distance from their star that is not too hot and not too cold—at a location that would support life similar to ours. This idea is appealing. That's why TPF (Terrestrial Planet Finder) is presently in progress. Our own civilization is so young that there has hardly been enough time for it to have come to the attention of others. The most conspicuous evidence of life on the Earth—radio, TV, and radar broadcast—has by now reached some 70 light-years into space, a distance encompassing only a few hundred stars.

In a similar way, most of the starlight from far-away stars has not reached the Earth yet. That's how far away most of the universe is. And light from most distant stars that does reach us is Doppler-shifted below the visible part of the spectrum and is invisible to us. Hence the night sky is black instead of ablaze with starlight!

40.4 Death of Stars

All luminous stars "burn" nuclear fuel. A star's life begins when it ignites its nuclear fuel, and it ends when its nuclear fires burn out. The first ignition in a star core is the fusion of hydrogen to helium. This may last from a few million to a few hundred billion years, depending on the star's mass. In the old age of an average-mass star like our sun, the burned-out hydrogen core that converts to helium contracts because of gravity. This contraction raises its temperature. This gain in temperature ignites both the helium in the core and the unfused hydrogen outside the core, and the star expands to become a **red giant.** Our sun will reach this stage about 5 billion years from now. On the route to attaining this stage, the sun will swell, becoming more luminous, and causing increasing temperatures on the Earth. First, it will strip the Earth of its atmosphere, and then boil the oceans dry. Ouch!

The cores of both solar-mass stars and those having mass lower than the sun are not hot enough to fuse carbon. So, lacking a source of nuclear energy, they shrink. As they shrink, their outer layers are sometimes cast off to form expanding shells that resemble smoke rings. They eventually disperse and mix with the interstellar material. This expanding shell is a **planetary nebula** (Figure 40.8). The shrunken core that remains blazes white-hot and is a **white dwarf.** Here matter is so compressed that a teaspoonful of it weighs tons.

Because the nuclear fires of a white dwarf have burned out, it is not actually a star anymore. It's more accurate to call it a *stellar remnant.* It may continue to radiate energy and change from white to yellow and then to red, until it slowly but ultimately fades to a cold, black lump of matter—a **black dwarf.** The density of a black dwarf is enormous. Into a volume no more than that of an average size planet is concentrated a mass hundreds of thousands times greater than that of the Earth. The black dwarf has a density comparable to that of an aircraft carrier squeezed into a quart jar!

Figure 40.8
The planetary ring nebula in the constellation Lyra, which can be seen through a modest telescope.

Astrology—A Famous Pseudoscience

There is more than one way to view the cosmos and its processes—astronomy is one way and astrology is another. Astrology is a belief system that began more than 2000 years ago in Babylonia. It has survived nearly unchanged since the second century AD, when some revisions were made by Egyptians and Greeks who believed that their gods moved heavenly bodies to influence the lives of people on the Earth.

Astrology today says that the position of the Earth in its orbit around the sun at the time of a person's birth, as well as the relative positions of the planets at that moment, has some influence over the person's life. The stars and planets are said to affect such personal things as character, marriage, friendships, wealth, and death.

The question is often raised as to whether the force of gravity exerted by these celestial bodies is a legitimate factor in human affairs. After all, the ocean tides are the result of the moon's and sun's gravitational pull, and the gravitational pull of one planet on another perturbs both their orbits. Because slight variations in gravity produce these effects, might not slight variations in the planetary positions at the time of birth affect a newborn? If the influence of stars and planets is gravitational, then credit must also be given to the effect of the gravitational pull between the newborn and the Earth. This pull is enormously greater than the combined pull of all the planets, even when they are all lined up in a row (as occasionally happens). And the gravitational influence of the hospital building on the newborn would surely exceed that of the distant planets! So planetary gravitation cannot be an underlying agent for astrology.

Astrology must therefore look to another realm for its basis, for all attempts to find physical explanations to support it have failed. Astrology is not a science, for it doesn't advance with new information as science does, nor are its predictions borne out by fact. So the realm of astrology may be spiritual, a religion of sorts. Or it may be a primitive psychology where the stars serve as a point of departure for musings about personality and personal decisions. Or astrology may be in the realm of numerology or phrenology—rigid and empty superstitions that prevail because of their focus on what is very important to each of us: ourselves.

A common position is that astrology is a harmless belief—a little fun at minimum cost. But is it harmless when believers are led to think their personalities are fixed by the stars at birth, that weak people will remain weak, that sad people will remain sad, that one's fate is dictated by the stars? We must also question the harm dealt people whose astrological signs are seen as incompatible with the signs of others. Astrology teaches, in a nutshell, that people are hostages to the stars. To say that this is a harmless belief is questionable. What do you think?

There is another possible fate for a white dwarf, if it is part of a binary and if its partner is close enough. A white dwarf, with its great mass, may gravitationally pull hydrogen from its companion star and deposit this material on its own surface as a very dense hydrogen layer. Continued compacting increases the temperature of this layer, which ignites to embroil the white dwarf's surface in a thermonuclear holocaust that we see as a **nova.** A nova is an event, not a stellar object. After a while, a nova subsides until enough matter again accumulates to repeat the event. A given nova flares up at irregular intervals that may range from decades to hundreds of thousands of years.

40.5 The Bigger They Are, the Harder They Fall—Supernova

How a star evolves depends on its mass. Low and medium-mass stars become white dwarfs, but more massive stars don't. When a star with a mass much greater than the sun's mass contracts, more heat is generated than for smaller stars. This heat keeps the star from shrinking to a white dwarf because carbon nuclei in the core fuses and releases energy while synthesizing heavier elements such as neon and magnesium. Gas pressure and radiation pressure halt further gravitational contraction until all the carbon is fused. Then the core contracts again to produce even higher temperatures and a new fusion series that produces even heavier elements. The fusion cycles repeat until the element iron is formed. The fusion of elements beyond iron requires an input of energy rather than a release of energy. (Recall from Chapter 20 that the fusion of elements beyond iron gains mass and absorbs energy.) With no release of energy from the iron core, the center of the star collapses without rekindling. The star begins its final collapse.

The collapse is catastrophic. When the core density is so great that all the nuclei are compressed against one another, the collapse momentarily comes to a halt. Then it explodes violently, hurling into space the elements previously manufactured over billions of years. The entire episode can last a few minutes. It is during this brief time that the heavy elements beyond iron are synthesized, as protons and neutrons mash with other nuclei to produce elements such as silver, gold, and uranium. Because the time available for synthesizing these heavy elements is so brief, they are not as abundant as iron and the lighter elements.

Such a stellar explosion is a **supernova,** one of nature's most spectacular events. Supernovae are fiery furnaces that generate the elements essential to life, for all the elements beyond iron that make up our bodies originated in far-off, long-ago supernovae. A supernova flares up to millions of times its former brightness. In 1054 AD, Chinese astronomers recorded their observation of a star so bright it could be seen by day as well as by night. This was a supernova, its glowing plasma remnants now making up the spectacular Crab nebula (Figure 40.9). A less spectacular but more recent supernova was witnessed in 1987. This gave astronomers an exciting firsthand look at one of these seldom seen events.

Figure 40.9
The Crab nebula, the remnant of a supernova explosion that was seen from the Earth in 1054 AD.

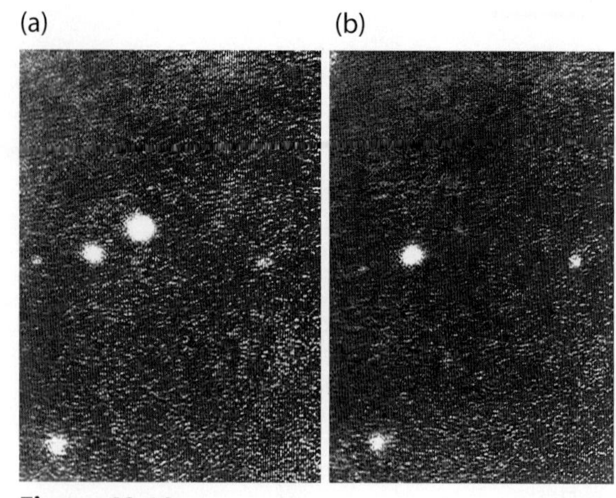

Figure 40.10
The pulsar in the Crab nebula rotates like a searchlight, beaming light and X-rays toward the Earth about 30 times a second, blinking on and off; (a) pulsar on, (b) pulsar off.

The inner part of a supernova star implodes to form a core compressed to *neutron density*. Incredibly, protons and electrons compress together to form a core of neutrons just a few kilometers wide. This superdense, central remnant of a supernova survives as a **neutron star.** In accord with the law of conservation of angular momentum, these tiny bodies, with densities hundreds of millions times that of white dwarfs, spin at fantastic speeds. Neutron stars are the explanation of **pulsars,** discovered in 1967 as rapidly varying sources of low-frequency radio emission. As a pulsar spins, the beam of radiation it emits sweeps around the sky. If the beam sweeps over the Earth, we detect pulses. Of the approximately 300 known pulsars, only a few have been found emitting X-ray or visible light. One is in the center of the Crab nebula (Figure 40.10). It has one of the highest rotational speeds of any pulsar studied, rotating more than 30 times per second. This is a relatively young pulsar, and it is theorized that X-ray and optical radiation is emitted only during a pulsar's early history.

Dying stars with cores greater than twenty to twenty-five solar masses collapse so violently that no physical forces are strong enough to halt continued contraction. The bigger they are, the harder they fall! The enormous gravitational field about the imploding concentration of mass makes explosion impossible. Collapse continues and the star disappears from the observable universe. What is left is a *black hole.*

40.6 Black Holes—The Fate of the Supergiants

A **black hole** is the remains of a supergiant star that has collapsed into itself. It is so dense and of such an intense gravitational field that light cannot escape from it.

We can see why gravity is so great near a black hole by noting the gravitational field change at the surface of any star that collapses. In accord with Newton's law of gravity, any object at the surface of a star, whether it is a large object or simply a particle, has weight that depends both on its own mass and on the mass of the star. But more importantly, the object's weight also depends on the distance between it and the center of the star. When the star collapses, the distance decreases. The object's weight increases without any change in its mass. How much? That depends on the amount of collapse. If a star collapses to half its original size, then the inverse-square law tells us that weight at its surface quadruples (Figure 40.11). If it collapses to a tenth its original size, weight at the surface is 100 times as much. Gravitation at the surface continues to increase as the collapse continues.

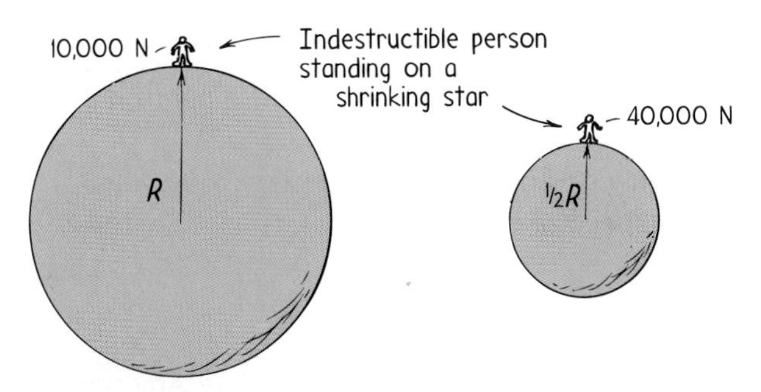

Figure 40.11
If a star collapses to half its radius and there is no mass change, gravitation at its surface increases by four (inverse-square law). If the star collapses to one-tenth its radius, gravitation at its surface increases a hundredfold.

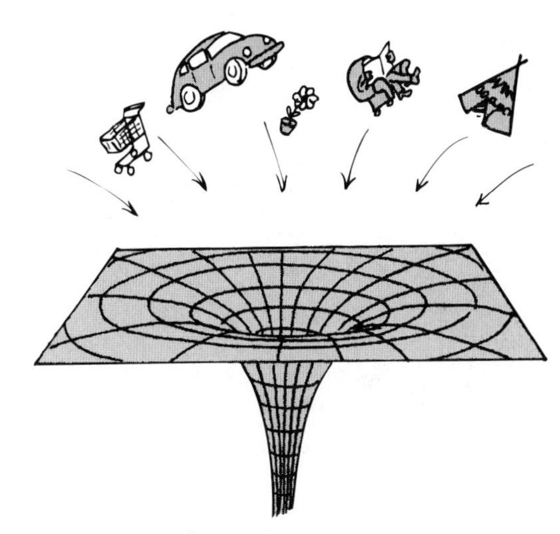

Figure 40.12
A black hole is the funneling and collapse of space itself, as indicated in this representation. Anything that falls into it is crushed out of existence.

Along with the increase in gravitational field, the escape speed also increases from the surface of the collapsing star. If our sun were to collapse to a radius of 3 kilometers, the escape speed from its surface would exceed the speed of light, and so nothing—not even light—could escape! The sun would be invisible. It would be a black hole. Fortunately for us, the sun has too little mass to develop into a black hole.

A black hole is no more massive than the star from which it is formed. For this reason, the gravitational field at a distance from the center that is equal to the star's original radius is no different after collapse than before. If you're traveling near a black hole, don't get closer than its original radius. It will be a one-way trip, for once beyond a certain point there's no turning back. You'd disappear from the observable universe.

Locating black holes is very difficult. One method is to look for a binary in which a luminous star appears to orbit about an invisible companion. If the two members of the binary are near each other, matter ejected by the luminous star accelerates into the neighboring invisible black hole and emits X-rays (Figure 40.13). Several black holes have been detected in the past 30 years. Massive black holes of 100 to 1000 solar masses are thought to exist at the centers of certain globular clusters.

Figure 40.13
A rendering of a black hole stealing matter from a companion star.

Concept Check ✓

1. What determines whether a star becomes a white dwarf, a neutron star, or a black hole?
2. If the sun somehow suddenly collapsed to a black hole, what change would occur in the orbital speed of planet Earth?

Check Your Answers

1. The mass of a star is the main factor that determines its fate. Any star having a mass up to six to eight solar masses becomes a white dwarf. Any star having a mass of eight to twenty solar masses becomes a neutron star. Any star having a mass of twenty to twenty-five solar masses sooner or later becomes a black hole.
2. None. This is best understood by Newton's law of gravitation, $F = G\frac{mM}{d^2}$.

   Note that nothing in the equation changes. Compression doesn't change the sun's initial mass M nor its distance from the Earth d. Because the Earth's mass m and G don't change either, the force F holding the Earth in its orbit remains unchanged. How about that!

40.7 Galaxies

A **galaxy** is a large group of stars, planetary nebulae, and interstellar gas and dust. Galaxies are the breeding grounds of stars. Our own star, the sun, is an ordinary star among 200 billion others in an ordinary galaxy known as the *Milky Way.* With unaided eyes we see the Milky Way as a faint band of light across the sky. The early Greeks called it the "milky circle," and the Romans called it the "milky road" or "milky way." The latter name has remained to this day.

Most astronomers believe that 10 to 15 billion years ago galaxies formed from huge clouds of primordial gas pulled together by gravity, similar to our description of the solar system's formation in the previous chapter. Formation begins with gravitational attraction between distant particles. Then contraction is accompanied by an increased rotational rate (like a skater who spins faster when arms are drawn in). In most cases rotation causes the galaxies to flatten to a disk. This is what occurred to our Milky Way galaxy. A most striking feature of our galaxy is the spiral arms that wind outward through the disk. These

Figure 40.14
A wide-angle photograph of the Milky Way, from the constellation Cassiopeia on the left to the constellation Sagittarius on the right. The dark lanes and blotches are interstellar gas and dust obscuring the light of background stars. (Steward Observatory)

arms are swarms of hot, blue stars and clusters of young stars, amidst clouds of dust and gas.

Masses of galaxies range from about a millionth the mass of our galaxy to some 50 times more. Galaxies are calculated to have much more mass than has been detected. This undetected mass is known as *dark matter.* The nature of this dark matter is still in question.

The millions of galaxies visible on long-exposure photographs can be separated into three main classes—*elliptical, spiral,* and *irregular.*

Elliptical galaxies are the most common galaxies in the universe. Because most of them are relatively dim, they are the most difficult to see. An exception is the gigantic elliptical galaxy M87 (Figure 40.15). Elliptical galaxies contain little gas and dust and are unable to create new stars.

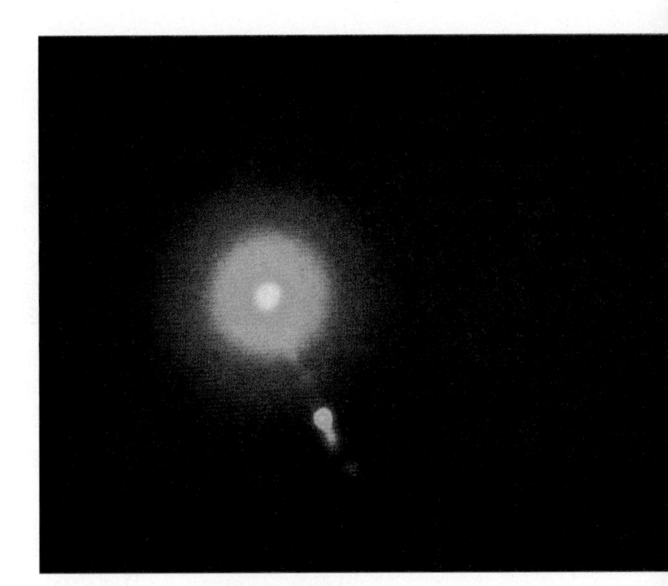

Figure 40.15
The giant elliptical galaxy M87, one of the most luminous galaxies in the sky, is located near the center of the Virgo cluster, some 50 million light-years from the Earth. It is about 25 times more massive than our Milky Way galaxy.

Irregular galaxies are normally small and faint. They are difficult to detect. They don't have spiral arms or dense centers. They contain large clouds of gas and dust mixed with both young and old stars. The irregular galaxy first described by the navigator on Magellan's voyage around the world in 1521 is our nearest neighboring galaxy—the *Magellanic Clouds.* This galaxy consists of two clouds, called the LMC and the SMC. The Large M Cloud is dotted with hot young stars having a combined mass of some 20 billion solar masses, and the Small M Cloud contains stars having a combined mass of about 2 billion solar masses (Figure 40.16). The combined mass is small for a galaxy. Irregular galaxies are probably as common as spiral galaxies.

Spiral galaxies appear as the most beautiful arrangements of stars in the heavens. They are bright with the light of newly formed stars, making them easy to see at great distances. How their spiral arms form is still being investigated. About two-thirds of all known galaxies are spirals, although they make up probably only 15 to 20% of all galaxies. We do not see the greater number of fainter elliptical galaxies thought to exist.

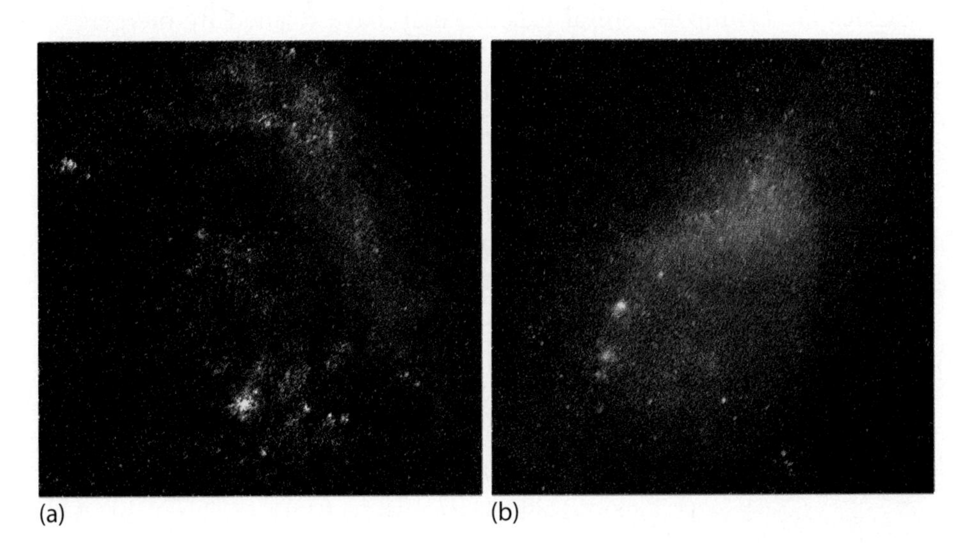

Figure 40.16
(a) The Large Magellanic Cloud and (b) neighboring small Magellanic Cloud are a pair of irregular galaxies. The Magellanic Clouds are our closest galactic neighbors, about 150 thousand light-years distant. They likely orbit the Milky Way galaxy.

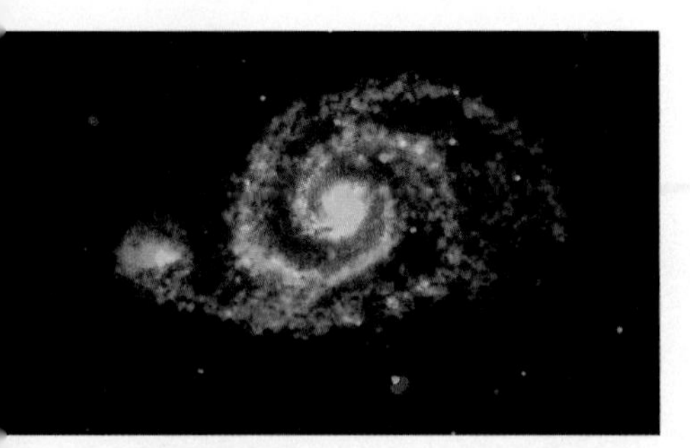

Figure 40.17
Spiral galaxy M83 in the southern constellation Centaurus, about 12 million light-years from the Earth.

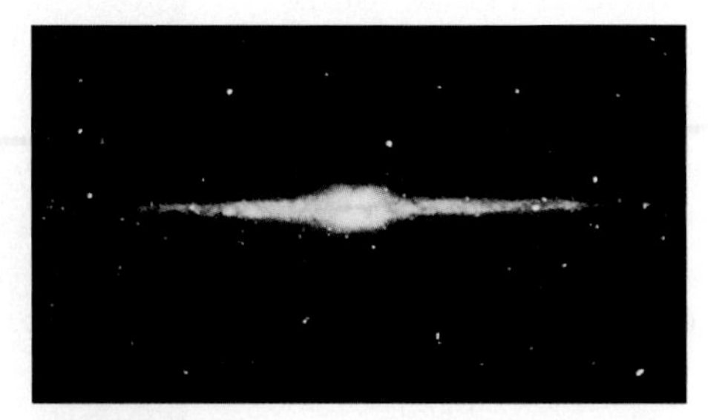

Figure 40.18
An edge-on view of our Milky Way galaxy, which makes 4 rotations every billion years. Our sun is about 5/8 the distance from the center to the outer visible rim.

We know what it's like to live in a spiral galaxy, for our Milky Way galaxy is a typical one. When we look at the Milky Way that crosses the night sky, we are looking through the disk of the galaxy. Interstellar dust obscures our view of most of the visible light there. Most information about our galaxy is acquired by infrared and radio telescopes. The nucleus seems to be crowded with stars and hot dust, and at the very center is thought to be a massive black hole that has a mass of a million suns. Don't go too near the center of the Milky Way.

Galaxies collide. The stars in a galaxy are normally so far apart, however, that physical collisions of individual stars are highly unlikely. But interstellar gases and dust collide violently, with matter stripped from one galaxy and deposited in another. These collisions also trigger the formation of new stars. Low-speed collisions can result in the merger of two galaxies. There is evidence that the Milky Way galaxy may be presently consuming the Magellanic Clouds. At high velocities, colliding galaxies can distort each other through tidal forces and create tails and bridges. The collisions of spiral galaxies are thought to form huge elliptical galaxies. Many large elliptical galaxies are believed to contain the merged remains of several spiral galaxies. Galaxies are cannibals. Spiral galaxies may have formed by mergers of many small dwarf galaxies.

Figure 40.19
The great spiral nebula in Andromeda, a spiral galaxy about 2.3 million light-years from Earth.

Galaxies are not the largest things in the universe. Galaxies come in **clusters.** And clusters of galaxies appear to be part of even larger clusters, the **superclusters.** It doesn't stop there; superclusters in turn seem to be part of a network of filaments surrounding empty voids. Comprehension of the universe at this scale becomes mind boggling.

40.8 The Big Bang

Most astronomers think that the universe began 10 to 15 billion years ago, when a primordial explosion called the **Big Bang** occurred. This is the standard model of the universe. The Big Bang marks the beginning of both space and time for our universe.

The space formed by the Big Bang was filled by extremely energetic high frequency radiation called the **primeval fireball.** Today this radiation survives as microwaves, which continually stretch out more and more as the universe expands.

When we look at the stars and far-away galaxies, we are looking backward in time because it takes a considerable length of time for the light from those galaxies to reach all the way to the Earth. The galaxies farthest away are the ones we are seeing as they existed longest ago.

The present expansion of the universe is evident in the Doppler red shift in the light we receive from galaxies. Recall from Chapter 13 that sound and light waves are stretched out when a source recedes and compressed when a source approaches. The visible light we see from distant galaxies is stretched out, which shows an increasing distance between us and the other galaxies in the universe. This does not, however, place our Milky Way galaxy in a central location. To see why not, consider ants on a balloon that expands. As the balloon is inflated, every ant sees every other ant moving farther away. This doesn't imply that each ant is in a central position. In an expanding universe, any observer sees all other galaxies receding.

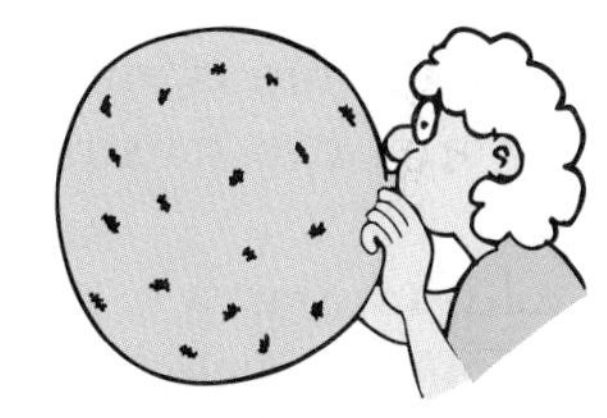

Figure 40.20
Every ant on the expanding balloon sees every other ant moving farther away. Each ant may therefore think it is at the center of the expansion. Not so!

Recent evidence indicates that the expansion rate of the universe is increasing. Why this is so is not known as this book goes to press. There is much we have to learn. There are indications of a dark unseen matter, and a dark unknown energy that we know very little about.

We don't view the world the same way the ancient Egyptians, Greeks, and Chinese did. It is not likely that people in the future will see the universe as we do. Our view of the universe may be incorrect, but most likely less wrong than the views of others before us. Our present view of the universe began with the findings of Copernicus, Galileo, and Newton. What they found was very much opposed by others at the time, mainly because established order was based on Aristotle's teachings. These "new" ideas were thought to diminish the role of humans in the universe, to undermine our importance. It was believed that people are important because we are higher than nature—apart from nature. We have expanded our vision since then by enormous effort, painstaking observation, and an ongoing desire to comprehend our surroundings. Seen from today's understanding of the universe, we find our importance not in being apart from nature, but in being very much a part of it. We are the part of nature that is becoming more and more conscious of itself.

Chapter Review

Key Terms and Definitions

_____ Big Bang
_____ binary
_____ black dwarf
_____ black hole
_____ cluster
_____ elliptical galaxy
_____ galaxy
_____ irregular galaxy
_____ neutron star
_____ nova
_____ planetary nebula
_____ primeval fireball
_____ protostar
_____ pulsar
_____ red giant
_____ spiral galaxy
_____ supercluster
_____ supernova
_____ white dwarf

1. The aggregation of matter that goes into and precedes the formation of a star.
2. A system of two stars that orbit around each other.
3. A cool giant star.
4. An expanding shell of gas ejected from a low-mass star during the latter stages of its evolution.
5. A dying star that has collapsed to the size of the Earth and is slowly cooling off.
6. The presumed final stage of a white dwarf.
7. A star that suddenly brightens, appearing as a "new" star; believed to be associated with eruptions from the surface of white dwarfs that are members of a binary system.
8. An exploding star, caused either by transfer of matter to a white dwarf or by gravitational collapse of a massive star, where enormous quantities of matter are emitted.
9. A small, dense star composed of tightly packed neutrons formed by the welding together of protons and electrons.
10. A neutron star that rapidly spins, sending short, precisely timed bursts of electromagnetic radiation.
11. The remains of a giant star that has collapsed upon itself, so dense and of such intense gravitational field that light cannot escape.
12. A large assembly of stars, numbering in the hundreds of millions to hundreds of billions, together with gas, dust, and other materials, that is held together by the forces of mutual gravitation.
13. A galaxy that is round or elliptical in outline. It contains little gas and dust, no disk or spiral arms, and few hot, bright stars.
14. A galaxy that has a chaotic appearance, large clouds of gas and dust, and no spiral arms.
15. A disk-shaped galaxy containing hot, bright stars and spiral arms. Our Milky Way galaxy is a spiral galaxy.
16. A grouping of more than one galaxy.
17. A grouping of an enormous number of galaxies.
18. The primordial explosion that marked the beginning of space and time.
19. Extremely high frequency radiation that filled space immediately following the Big Bang.

Review Questions

The Constellations

1. What are constellations?
2. Why does an observer at a given location see one set of constellations in the winter and a different set of constellations in the summer?

Birth of Stars

3. What process changes a protostar to a full-fledged star?
4. What are the outward forces that act on a star?
5. What are the inward forces that act on a star?
6. What do the outward and inward forces acting on a star have to do with its size?

Life and Evolution of Stars

7. Compare the lifetimes of high-mass and low-mass stars.
8. How common are binaries in the universe?
9. What is the goal of SETI programs?

Death of Stars

10. What event marks the birth of a star, and what event marks its death?

11. When will our sun reach the red-giant stage?
12. What is the relationship between a planetary nebula and a white dwarf?
13. What is the relationship between a white dwarf and a black dwarf?

The Bigger They Are, the Harder They Fall—Supernova

14. What is the relationship between a white dwarf and a nova?
15. What is the relationship between the heavy elements we find on the Earth today and supernovae?

Black Holes—The Fate of the Supergiants

16. What is the relationship between an ordinary star and a black hole?
17. How does the mass of a star before collapse compare with the mass of the black hole it becomes?
18. Being that black holes are invisible, what is the evidence for their existence?

Galaxies

19. What type of galaxy is the Milky Way galaxy?
20. What are the consequences of galaxies colliding?

Exercises

1. Why do we not see stars in the daytime?
2. Which figure in the chapter best shows that a constellation seen in the background of a solar eclipse is one that will be seen six months later in the night sky?
3. We see the constellations as distinct groups of stars. Discuss why they would look entirely different from some other location in the universe, far distant from the Earth.
4. The Big Dipper is sometimes right side up (can hold water), and at other times upside down (cannot hold water). What length of time is required for the Dipper to change from one position to the other?
5. In what sense are we all made of star dust?
6. How is the gold in a wedding ring evidence of ancient stars that ran their life cycles long before the solar system came into being?
7. Would you expect metals to be more abundant in old stars or new stars? Defend your answer.
8. Why is there a lower limit on the mass of a star? (What can't happen in a low-mass accumulation of hydrogen atoms and other interstellar material?)
9. What ordinarily keeps a star from collapsing?
10. How does a protostar differ from a star?
11. How does the energy of a protostar differ from the energy that powers a star?
12. Why do nuclear fusion reactions not occur on the outer layers of stars?
13. What is meant by the statement, "The bigger they are, the harder they fall," with respect to stellar evolution?
14. Why will the sun not be able to fuse carbon nuclei in its core?
15. Some stars contain fewer heavy elements than our sun contains. What does this indicate about the age of such stars relative to the age of our sun?
16. Which has the highest surface temperature: red star, white star, or blue star?
17. In what way is a black hole blacker than black ink?
18. What does it mean to say that galaxies are cannibals?
19. What is meant by saying that the universe does not exist in space? Change two words around to make the statement agree with the standard model of the universe.
20. In your own opinion, do you have to be at the center of your class to be special? Does the Earth have to be at the center of the universe to be special?

Appendix A: Systems of Measurement

Two major systems of measurement prevail in the world today: the *United States Customary System* (USCS, formerly called the British system of units), used in the United States of America and in Burma, and the *Système International* (SI) (known also as the international system and as the metric system), used everywhere else. Each system has its own standards of length, mass, and time. The units of length, mass, and time are sometimes called the *fundamental units* because, once they are selected, other quantities can be measured in terms of them.

United States Customary System

Based on the British Imperial System, the USCS is familiar to everyone in the United States. It uses the foot as the unit of length, the pound as the unit of weight or force, and the second as the unit of time. The USCS is presently being replaced by the international system—rapidly in science and technology (all Department of Defense contracts since 1988) and some sports (track and swimming), but so slowly in other areas and in some specialties it seems the change may never come. For example, we will continue to buy seats on the 50-yard line. Camera film is in millimeters, but computer disks are in inches.

For measuring time, there is no difference between the two systems except that in pure SI the only unit is the second (s, not sec) with prefixes; but in general, minute, hour, day, year, and so on, with two or more lettered abbreviations (h, not hr), are accepted in the USCS.

Systéme International

During the 1960 International Conference on Weights and Measures held in Paris, the SI units were defined and given status. Table A.1 shows SI units and their symbols. SI is based on the *metric system,* originated by French scientists after the French revolution in 1791. The orderliness of this system makes it useful for scientific work, and it is used by scientists all over the world. The metric system branches into two systems of units. In one of these the unit of length is the meter, the unit of mass is the kilogram, and the unit of time is the second. This is called the *meter-kilogram-second* (mks) system and is preferred in physics. The other branch is the *centimeter-gram-second* (cgs) system, which because of its smaller values is favored in chemistry. The cgs and mks units are related to each other as follows: 100 centimeters equal 1 meter; 1000 grams equal 1 kilogram. Table A.2 shows several units of length related to each other.

One major advantage of the metric system is that it uses the decimal system, where all units are related to smaller or larger units by dividing or multiplying by 10. The prefixes shown in Table A.3 are commonly used to show the relationship among units.

Table A.1

SI units

Quantity	Unit	Symbol
Length	meter	m
Mass	kilogram	kg
Time	second	s
Force	newton	N
Energy	joule	J
Current	ampere	A
Temperature	kelvin	K

Meter

The standard of length of the metric system orginally was defined in terms of the distance from the north pole to the equator. This distance was thought at the time to be close to 10,000 kilometers. One ten-millionth of this, the meter, was carefully determined and marked off by means of scratches on a bar of platinum-iridium alloy. This bar is kept at the International Bureau of Weights and Measures in France.

Table A.2

Table Conversions Between Different Units of Length

Unit of Length	Kilometer	Meter	Centimeter	Inch	Foot	Mile
1 kilometer	= 1	1000	100,000	39,370	3280.84	0.62140
1 meter	= 0.00100	1	100	39.370	3.28084	6.21×10^{-4}
1 centimeter	$= 1.0 \times 10^{-5}$	0.0100	1	0.39370	0.032808	6.21×10^{-6}
1 inch	$= 2.54 \times 10^{-5}$	0.02540	2.5400	1	0.08333	1.58×10^{-5}
1 foot	$= 3.05 \times 10^{-4}$	0.30480	30.480	12	1	1.89×10^{-4}
1 mile	= 1.60934	1609.34	160,934	63,360	5280	1

Table A.3

Some Prefixes

Prefix	Definition
micro-	One-millionth: a microsecond is one-millionth of a second
milli-	One-thousandth: a milligram is one-thousandth of a gram
centi-	One-hundredth: a centimeter is one-hundredth of a meter
kilo-	One thousand: a kilogram is 1000 grams
mega-	One million: a megahertz is 1 million hertz

The standard meter in France has since been calibrated in terms of the wavelength of light—it is 1,650,763.73 times the wavelength of orange light emitted by the atoms of the gas krypton-86. The meter is now defined as being the length of the path traveled by light in a vacuum during a time interval of 1/299,792,458 of a second.

Figure A.1
The standard kilogram.

Kilogram

The standard unit of mass, the kilogram, is a block of platinum-iridium alloy, also preserved at the International Bureau of Weights and Measures located in France (Figure A.1). The kilogram equals 1000 grams. A gram is the mass of 1 cubic centimeter (cc) of water at a temperature of 4°C. (The standard pound is defined in terms of the standard kilogram; the mass of an object that weighs 1 pound at the Earth's surface is equal to 0.4536 kilogram.)

Second

The official unit of time for both the USCS and the SI is the second. Until 1956, it was defined in terms of the mean solar day, which was divided into 24 hours. Each hour was divided into 60 minutes and each minute into 60 seconds. Thus there were 86,400 seconds per day, and the second was defined as 1/86,400 of the mean solar day. This proved unsatisfactory because the rate of rotation of the Earth is gradually becoming slower. In 1956, the mean solar day of the year 1900 was chosen as the standard on which to base the second. In 1964, the second was officially defined as the time taken by a cesium-133 atom to make 9,192,631,770 vibrations.

Newton

One newton is the force required to accelerate 1 kilogram at 1 meter per second per second. This unit is named after Sir Isaac Newton.

Joule

One joule is equal to the amount of work done by a force of 1 newton acting over a distance of 1 meter. In 1948, the joule was adopted as the unit of energy by the International Conference on Weights and Measures. Therefore, the specific heat of water at 15°C is now given as 4185.5 joules per kilogram Celsius degree. This figure is always associated with the mechanical equivalent of heat—4.1855 joules per calorie.

Ampere

The ampere is defined as the intensity of the constant electric current that, when maintained in two parallel conductors of infinite length and negligible cross section and placed 1 meter apart in a vacuum, would produce between them a force equal to 2×10^{-7} newton per meter length. In our treatment of electric current in this text, we have used the not-so-official but easier-to-comprehend definition of the ampere as being the rate of flow of 1 coulomb of charge per second, where 1 coulomb is the charge of 6.25×10^{18} electrons.

Kelvin

The fundamental unit of temperature is named after the scientist William Thomson, Lord Kelvin. The kelvin is defined to be 1/273.15 the thermodynamic temperature of the triple point of water (the fixed point at which ice, liquid water, and water vapor coexist in equilibrium). This definition was adopted in 1968 when it was decided to change the name *degree Kelvin* (°K) to *kelvin* (K). The temperature of melting ice at atmospheric pressure is 273.15 K. The temperature at which the vapor pressure of pure water is equal to standard atmospheric pressure is 373.15 K (the temperature of boiling water at standard atmospheric pressure).

Area

The unit of area is a square that has a standard unit of length as a side. In the USCS, it is a square with sides that are each 1 foot in length, called 1 square foot and written 1 ft^2. In the international system, it is a square with sides that are 1 meter in length, which makes a unit of area of 1 m^2. In the cgs system it is 1 cm^2. The area of a given surface is specified by the number of square feet, square meters, or square centimeters that would fit into it. The area of a rectangle equals the base times the height. The area of a circle is equal to πr^2, where $\pi = 3.14$ and r is the radius of the circle. Formulas for the surface areas of other objects can be found in geometry textbooks.

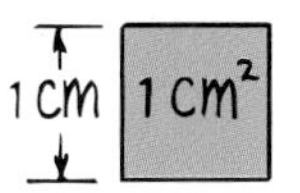

Figure A.2
Unit square.

Volume

The volume of an object refers to the space it occupies. The unit of volume is the space taken up by a cube that has a standard unit of length for its edge. In the USCS, one unit of volume is the space occupied by a cube 1 foot on an edge and is called 1 cubic foot, written 1 ft^3. In the metric system, it is the space occupied by a cube with sides of 1 meter (SI) or 1 centimeter (cgs). It is written 1 m^3 or 1 cm^3 (or cc). The volume of a given space is specified by the number of cubic feet, cubic meters, or cubic centimeters that will fill it.

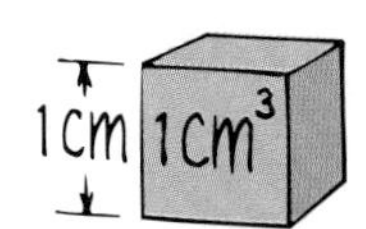

Figure A.3
Unit volume.

In the USCS, volumes can also be measured in quarts, gallons, and cubic inches as well as in cubic feet. There are 1728 ($12 \times 12 \times 12$) cubic inches in 1 ft^3. A U.S. gallon is a volume of 231 in^3. Four quarts equal 1 gallon. In the SI, volumes are also measured in liters. A liter is equal to 1000 cm^3.

Appendix B: Linear Motion

Speed

Speed is a measure of how fast something moves. It can be measured in a variety of units—for example, miles per hour, or meters per second. All speed units feature a unit of distance divided by a unit of time. Speaking precisely, speed is defined as the distance covered per unit of time. It has the following mathematical definition:

$$\text{Speed} = \frac{\text{distance}}{\text{time}}$$

The speed at any instant is called the *instantaneous speed.* When you look at a speedometer in a moving car, the speed reading shows instantaneous speed.

In planning a trip by car, the driver often wants to know the time of travel. In this case the driver is not concerned with instantaneous speed, but with the *average speed* for the trip. Average speed is defined:

$$\text{Average speed} = \frac{\text{total distance covered}}{\text{time interval}}$$

If a car goes 80 kilometers in a time of 1 hour, we say the average speed is 80 kilometers per hour. Likewise, if we travel 320 kilometers in 4 hours,

$$\text{Average speed} = \frac{\text{total distance covered}}{\text{time interval}} = \frac{320\text{ km}}{4\text{ h}} = 80\text{ km/h}$$

Consider a ball rolling down an incline (Figure B.1). It starts at rest and 1 second later it is rolling at 2 m/s. We see its average speed is not 2 m/s, but 1 m/s. Why? Because it rolled a total distance of 1 meter in 1 second.

If we know average speed and time of travel, distance traveled is easy to find. A simple rearrangement of the definition above gives

$$\text{Total distance covered} = \text{average speed} \times \text{time}$$

If your average speed is 80 kilometers per hour on a 4-hour trip, for example, you cover a total distance of 320 kilometers.

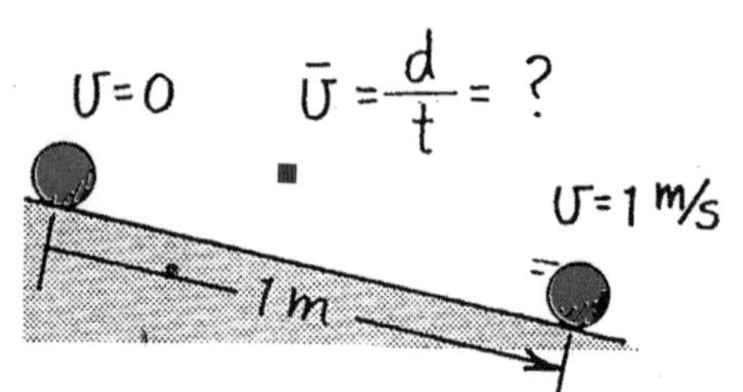

Figure B.1
The ball rolls 1 m down the incline in 1 s and reaches an instantaneous speed of 2 m/s. Its average speed, however, is 1 m/s. Do you see why?

Examples

1. What is the average speed of a cheetah that sprints 100 m in 4 seconds? How about if it sprints 50 m in 2 s?
Solution: In both cases the answer is 25 m/s:

$$\text{Average speed} = \frac{\text{distance covered}}{\text{time interval}} = \frac{100 \text{ meters}}{4 \text{ seconds}} = \frac{50 \text{ meters}}{2 \text{ seconds}} = 25 \text{ m/s}$$

2. If a car moves with an average speed of 60 km/h for an hour, it will travel a distance of 60 km.
   a. How far would it travel if it moved at this rate for 4 h?
   b. For 10 h?
Solution: The distance traveled is the average speed × time of travel, so
   a. Distance = 60 km/h × 4 h = 240 km
   b. Distance = 60 km/h × 10 hr = 600 km

Velocity

When we describe the speed and *direction* of motion, we are specifying **velocity.** Velocity is a vector quantity.

An object's velocity (or speed) can change. For example, a hummingbird speeds up, slows down, and darts back and forth in its search for food. On the other hand a car on "cruise control" moves at constant velocity as it rolls down a straight highway. When something moves at *constant* velocity (or constant speed), then it covers *equal distances* in equal intervals of time.

Constant velocity and constant speed, however, can be very different. Constant velocity means constant speed with no change in direction. A car that rounds a curve at a constant speed does not have a constant velocity—its velocity changes as its direction changes.

Example

During a certain period of time, the speedometer of a car reads a constant 60 km/h. Does this indicate a constant speed? A constant velocity?
Answer: The constant speedometer reading indicates a constant speed but not a constant velocity. The car may not be moving along a straight-line path, in which case it is *accelerating*.

Acceleration

Velocity changes if speed changes, direction changes, or both speed and direction change. How quickly velocity changes is **acceleration**:

$$\text{Acceleration} = \frac{\text{change of velocity}}{\text{time interval}}$$

Consider a car that increases its velocity at a steady rate. Suppose it's moving at 30 kilometers per hour. Then 1 second later it picks up to 35 kilometers per hour, and then to 40 kilometers per hour in the next second, and so on. Velocity changes by 5 kilometers per hour each second. This change of velocity is acceleration.

$$\text{Acceleration} = \frac{\text{change of velocity}}{\text{time interval}} = \frac{5 \text{ km/h}}{1 \text{ s}} = 5 \text{ km/h·s}$$

In this case the acceleration is 5 kilometers per hour second (abbreviated as 5 km/h·s).

Example

In 2.5 s a car increases its speed from 60 km/h to 65 km/h while a bicycle goes from rest to 5 km/h. Which undergoes the greater acceleration? What is the acceleration of each vehicle?

Solution: The accelerations of both the car and the bicycle are the same: 2 km/h·s.

$$\text{Acceleration}_{\text{car}} = \frac{\text{change of velocity}}{\text{time interval}} = \frac{65\text{ km/h} - 60\text{ km/h}}{2.5\text{ s}} = \frac{5\text{ km/h}}{2.5\text{ s}} = 2\text{ km/h}\cdot\text{s}$$

$$\text{Acceleration}_{\text{bike}} = \frac{\text{change of velocity}}{\text{time interval}} = \frac{5\text{ km/h} - 0\text{ km/h}}{2.5\text{ s}} = \frac{5\text{ km/h}}{2.5\text{ s}} = 2\text{ km/h}\cdot\text{s}$$

Actually, we have given only the magnitude of the acceleration here. To be complete we would also need to specify direction. Like velocity, acceleration consists of both speed *and* direction. The car's acceleration in the previous example might be 5 km/h·s in the forward direction, for instance. To keep things simple, however, we will focus here on amount of acceleration—its magnitude.

Acceleration of Free Fall

Falling objects pick up speed because of the force of gravity. When a falling object is free of all restraints—no friction, air or otherwise, and it falls under the influence of gravity alone, the object is in a state of **free fall.** The acceleration of free fall on the Earth is 10 m/s each second. We can shorten this to 10 m/s^2. This means that during each second of fall, the object gains a speed of 10 m/s.

The letter *g* is customarily used for free-fall acceleration (because the acceleration is due to *gravity*). The value of *g* is very different on the surface of the moon, and on other planets. Here on Earth, *g* varies slightly in different locations, with an average value of 9.8 meters per second each second, or, in shorter notation, 9.8 m/s^2. We round this off to 10 m/s^2 in this book. (Multiples of 10 are more obvious than multiples of 9.8.) Where accuracy is important, the value of 9.8 m/s^2 should be used.

How Fast

The formula for acceleration can be arranged for change in velocity. That is:

$$\text{Velocity acquired} = \text{acceleration} \times \text{time}$$

The instantaneous velocity v of an object falling from rest after a time t can be expressed in shorthand notation as

$$v = gt$$

If the object is initially moving downward at speed v_o the speed v after any elapsed time t is $v = v_o + gt$. That is, your speed is what you started with plus the speed you've gained in fall.

How Far

How *far* an object falls is different than how *fast* it falls. With his inclined planes Galileo found that the distance a uniformly accelerating object travels is proportional to the *square of the time.*

Study the inclined plane shown in Figure B.2. Note the distance covered by the ball in the first time interval is 1 m. Note in the second time interval the ball covers 3 m. This is because the average speed in this interval is 3 m/s. In the next 1-second interval the average speed is 5 m/s, so the distance covered is 5 m. Can you see that successive increments of distance increase as the sequence of odd numbers (1, 3, 5, 7, etc)? Nature clearly follows mathematical rules.

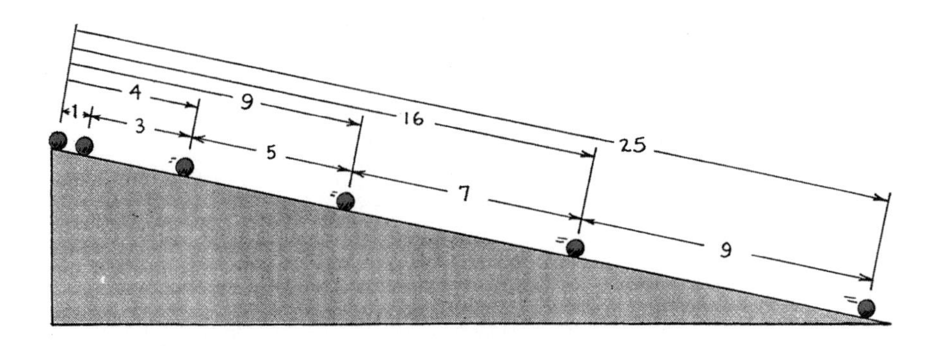

Figure B.2
If the ball covers 1 m during its first second, then in each successive seconds it will cover the odd-numbered sequence of 3, 5, 7, 9 m, and so on. Note that the total distance covered increases as the square of the total time.

More important, notice that the total distance the ball rolls from its starting point increases as the square of the time. At 2 s, it goes 4 m; at 3 s it goes 9 m; at 4 s it goes 16 m. At t seconds, it goes t^2.

If the incline is tipped to 90°, that is, vertical, then the ball will be in free fall. Then,

$$\begin{aligned}\text{Distance fallen} &= \text{average speed} \times \text{time interval}\\ &= \frac{\text{beginning speed} + \text{final speed}}{2} \times \text{time}\\ &= \frac{0 + gt}{2} \times t\\ &= \frac{1}{2}gt^2\end{aligned}$$

So for an object that freely falls from rest, distance of fall is

$$d = \frac{1}{2} gt^2.$$

If it has a head start, then add to this $d = v_0 t$, where v_o is the initial speed. Then

$$d = v_o t + \frac{1}{2} gt^2.$$

Table B.1

Velocity Acquired and Distance Fallen in Free Fall

Time of fall (seconds)	Velocity acquired (meters/second)	Distance fallen (meters)
0	0	0
1	10	5
2	20	20
3	30	45
4	40	80
5	50	125
.	.	
.	.	
.	.	
t	$10t$	$1/2\ 10\ t^2$

Examples

A cat steps off a ledge and drops to the ground in 1/2 second.

a. What is its speed on striking the ground?
b. What is its average speed during the 1/2 second?
c. How high is the ledge from the ground?

Solutions:

a. Speed: $v = gt = 10\text{ m/s}^2 \times 1/2\text{ s} = 5\text{ m/s}$.

b. Average speed: $\bar{v} = \frac{\text{initial } v + \text{final } v}{2} = \frac{0\text{ m/s} + 5\text{ m/s}}{2} = 2.5\text{ m/s}$.

We put a bar over the symbol to denote average speed: $\bar{v}$.

c. Distance: $d = \bar{v}t = 2.5\text{ m/s} \times 1/2\text{ s} = 1.25\text{ m}$.

Or equivalently,

$$d = 1/2\, gt^2 = 1/2 \times 10\text{ m/s}^2 \times (1/2\text{ s})^2 = 1/2 \times 10\text{ m/s}^2 \times 1/4\text{ s}^2 = 1.25\text{ m}.$$

Notice that we can find the distance by either of these equivalent relationships.

Appendix C: Vectors

Vectors and Scalars

A *vector* quantity is a directed quantity—one that must be specified not only by magnitude (size) but by direction as well. Recall from Chapter 2 that velocity is a vector quantity. Other examples are force, acceleration, and momentum. In contrast, a *scalar* quantity can be specified by magnitude alone. Some examples of scalar quantities are speed, time, temperature, and energy.

Vector quantities may be represented by arrows. The length of the arrow tells you the magnitude of the vector quantity, and the arrowhead tells you the direction of the vector quantity. Such an arrow drawn to scale and pointing appropriately is called a *vector.*

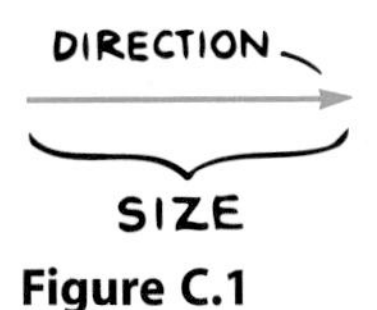

Figure C.1

Adding Vectors

Vectors that add together are called *component vectors.* The sum of component vectors is called a *resultant.*

To add two vectors, make a parallelogram with two component vectors acting as two of the adjacent sides (Figure C.2). (Here our parallelogram is a rectangle.) Then draw a diagonal from the origin of the vector pair; this is the resultant (Figure C.3).

Figure C.2

Caution: Do not try to mix vectors! We cannot add apples and oranges, so velocity vector combines only with velocity vector, force vector combines only with force vector, and acceleration vector combines only with acceleration vector—each on its own vector diagram. If you ever show different kinds of vectors on the same diagram, use different colors or some other method of distinguishing the different kinds of vectors.

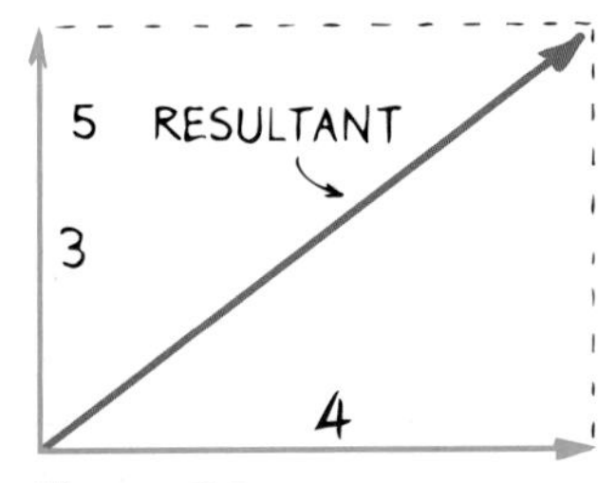

Figure C.3

Finding Components of Vectors

Recall from Chapter 8 that to find a pair of perpendicular components for a vector, first draw a dotted line through the tail end of the vector in the direction of one of the desired components. Second, draw another dotted line through the tail end of the vector at right angles to the first dotted line. Third, make a rectangle whose diagonal is the given vector. Draw in the two components. Here we let **F** stand for "total force," **U** stand for "upward force," and **S** stand for "sideways force."

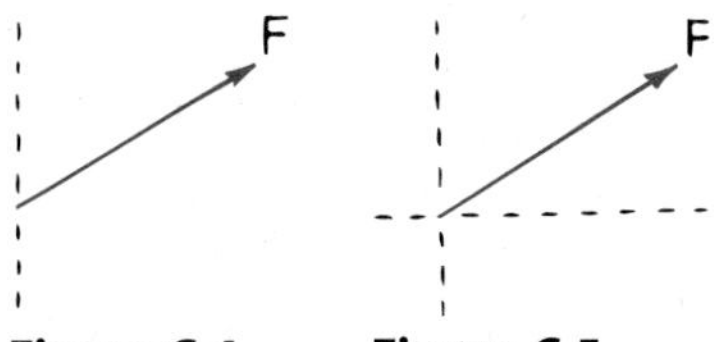

Figure C.4 **Figure C.5**

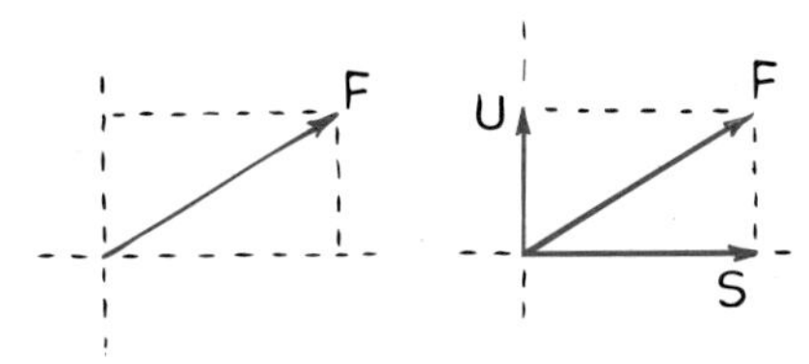

Figure C.6

Figure C.7

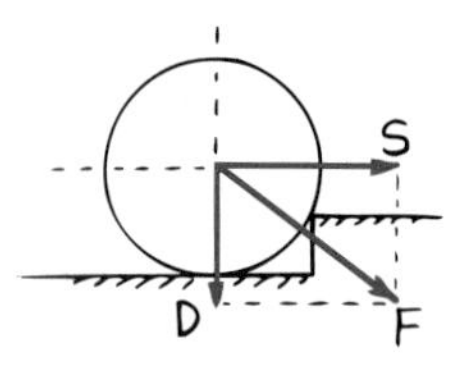

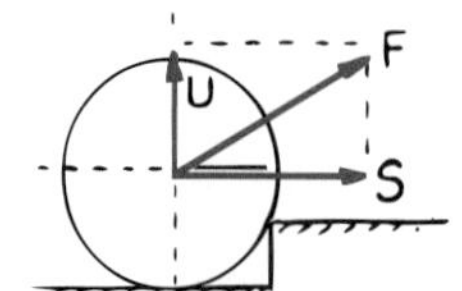

Figure C.8

Examples

1. Ernie Brown pushing a lawnmower applies a force that pushes the machine forward and also against the ground. In Figure C.7, **F** represents the force applied by Ernie. We can separate this force into two components. The vector **D** represents the downward component, and **S** is the sideways component, the force that moves the lawnmower forward. If we know the magnitude and direction of the vector **F**, we can estimate the magnitude of the components from the vector diagram.

2. Would it be easier to push or pull a wheelbarrow over a step? Figure C.8 shows a vector diagram for each case. When you push a wheelbarrow, part of the force is directed downward, which makes it harder to get over the step. When you pull, however, part of the pulling force is directed upward, which helps to lift the wheel over the step. Note that the vector diagram suggests that pushing the wheelbarrow may not get it over the step at all. Do you see that the height of the step, the radius of the wheel, and the angle of the applied force determine whether the wheelbarrow can be pushed over the step? We see how vectors help us analyze a situation so that we can see just what the problem is!

3. If we consider the components of the weight of an object rolling down an incline, we can see why its speed depends on the angle (Figure C.9). Note that the steeper the incline, the greater the component **S** becomes and the faster the object rolls. When the incline is vertical, **S** becomes equal to the weight, and the object attains maximum acceleration, 9.8 meters per second squared.

   There are two more force vectors that are not shown: the normal force **N**, which is equal and oppositely directed to **D**, and the friction force **f**, acting at the barrel-plane contact.

Figure C.9

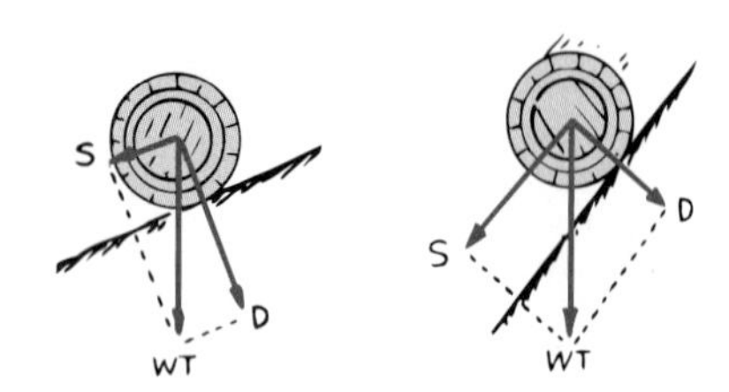

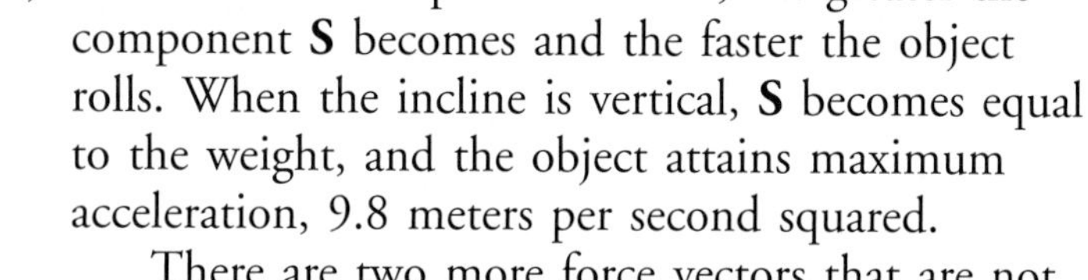

4. When moving air strikes the underside of an airplane wing, the force of air impact against the wing may be represented by a single vector perpendicular to the plane of the wing (Figure C.10). We represent the force vector as acting midway along the lower wing surface, where the dot is, and pointing above the wing to show the direction of the resulting wind impact force. This force can be broken up into two components, one sideways and the other up. The upward component, **U,** is called *lift.* The sideways component, **S,** is called *drag.* If the aircraft is to fly at constant velocity at constant altitude, then lift must equal the weight of the aircraft and the thrust of the plane's engines must equal drag. The magnitude of lift (and drag) can be altered by changing the speed of the airplane or by changing the angle (called *angle of attack*) between the wing and the horizontal.

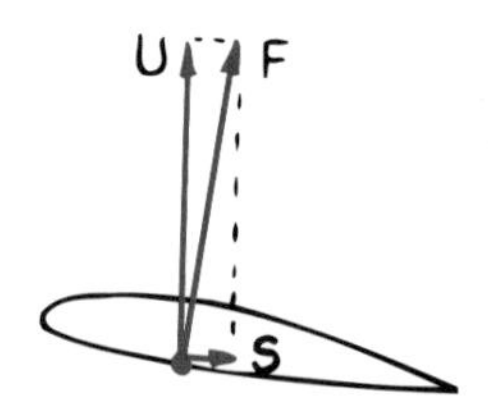

Figure C.10

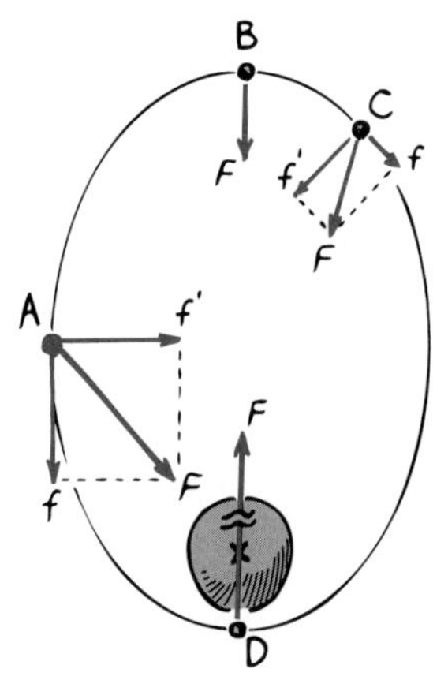

Figure C.11

5. Consider the satellite moving clockwise in Figure C.11. Everywhere in its orbital path, gravitational force $\boldsymbol{F}$ pulls it toward the center of the host planet. At position A we see $\boldsymbol{F}$ separated into two components: $\boldsymbol{f}$, which is tangent to the path of the projectile, and $\boldsymbol{f}'$, which is perpendicular to the path. The relative magnitudes of these components in comparison to the magnitude of $\boldsymbol{F}$ can be seen in the imaginary rectangle they compose; $\boldsymbol{f}$ and $\boldsymbol{f}'$ are the sides, and $\boldsymbol{F}$ is the diagonal. We see that component $\boldsymbol{f}$ is along the orbital path but against the direction of motion of the satellite. This force component reduces the speed of the satellite. The other component $\boldsymbol{f}'$, changes the direction of the satellite's motion and pulls it away from its tendency to go in a straight line. So the path of the satellite curves. The satellite loses speed until it reaches position B. At this farthest point from the planet (apogee), the gravitational force is somewhat weaker but perpendicular to the satellite's motion, and component $\boldsymbol{f}$ has reduced to zero. Component $\boldsymbol{f}'$, on the other hand, has increased and is now fully merged to become $\boldsymbol{F}$. Speed at this point is not enough for circular orbit, and the satellite begins to fall toward the planet. It picks up speed because the component $\boldsymbol{f}$ reappears and is in the direction of motion as shown in position C. The satellite picks up speed until it whips around to position D (perigee), where once again the direction of motion is perpendicular to the gravitational force, $\boldsymbol{f}'$ blends to full $\boldsymbol{F}$, and $\boldsymbol{f}$ is nonexistent. The speed is in excess of that needed for circular orbit at this distance, and it overshoots to repeat the cycle. Its loss in speed in going from D to B equals its gain in speed from B to D. Kepler discovered that planetary paths are elliptical, but never knew why. Do you?

6. Refer to the Polaroids held by Ludmila back in Chapter 16, in Figure 16.20. In the first picture (a), we see that light is transmitted through the pair of Polaroids because their axes are aligned. The emerging light can be represented as a vector aligned with the polarization axes of the Polaroids. When the Polaroids are crossed (b), no light emerges because light passing through the first Polaroid is perpendicular to the polarization axes of the second Polaroid, with no components along its axis. In the third picture (c), we see that light is transmitted when a third Polaroid is sandwiched at an angle between the crossed Polaroids. The explanation for this is shown in Figure C.12.

Figure C.12

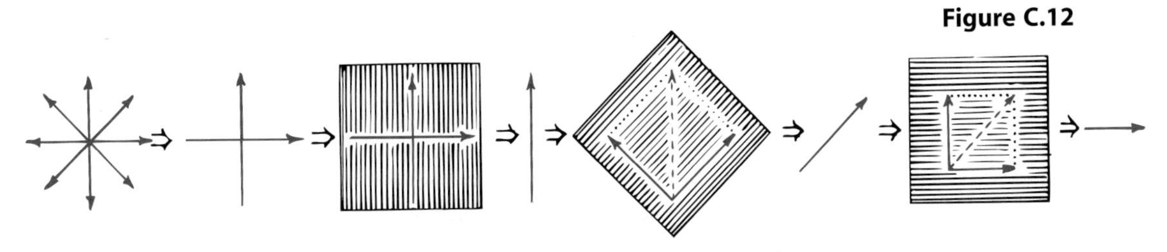

Sailboats

Sailors have always known that a sailboat can sail downwind, in the direction of the wind. Sailors have not always known, however, that a sailboat can sail upwind, against the wind. One reason for this has to do with a feature that is common only to recent sailboats—a finlike keel that extends deep beneath the bottom of the boat to ensure that the boat will knife through the water only in a forward (or backward) direction. Without a keel, a sailboat could be blown sideways.

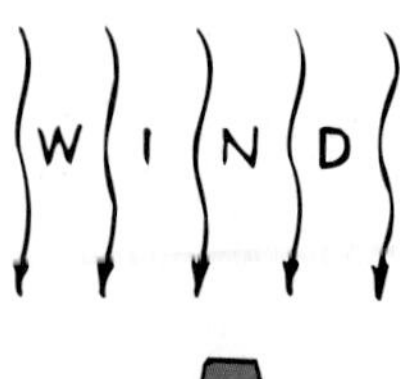

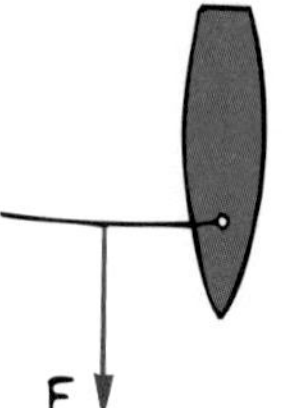

Figure C.13

Figure C.13 shows a sailboat sailing directly downwind. The force of wind impact against the sail accelerates the boat. Even if the drag of the water and all other resistance forces are negligible, the maximum speed of the boat is the wind speed. This is because the wind will not make impact against the sail if the boat is moving as fast as the wind. The wind would have no speed relative to the boat and the sail would simply sag. With no force, there is no acceleration. The force vector in Figure C.13 *decreases* as the boat travels faster. The force vector is maximum when the boat is at rest and the full impact of the wind fills the sail, and is minimum when the boat travels as fast as the wind. If the boat is somehow propelled to a speed faster than the wind (by way of a motor, for example), then air resistance against the front side of the sail will produce an oppositely directed force vector. This will slow the boat down. Hence, the boat when driven only by the wind cannot exceed wind speed.

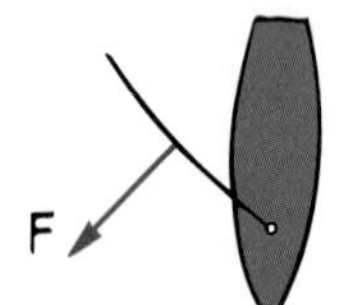

Figure C.14

If the sail is oriented at an angle, as shown in Figure C.14, the boat will move forward, but with less acceleration. There are two reasons for this:

1. The force on the sail is less because the sail does not intercept as much wind in this angular position.
2. The direction of the wind impact force on the sail is not in the direction of the boat's motion, but is perpendicular to the surface of the sail. Generally speaking, whenever any fluid (liquid or gas) interacts with a smooth surface, the force of interaction is perpendicular to the smooth surface.* The boat does not move in the same direction as the perpendicular force on the sail, but is constrained to move in a forward (or backward) direction by its keel.

We can better understand the motion of the boat by resolving the force of wind impact, ***F***, into perpendicular components. The important component is that which is parallel to the keel, which we

* You can do a simple exercise to see that this is so. Try bouncing one coin off another on a smooth surface, as shown. Note that the struck coin moves at right angles (perpendicular) to the contact edge. Note also that it makes no difference whether the projected coin moves along path A or path B.

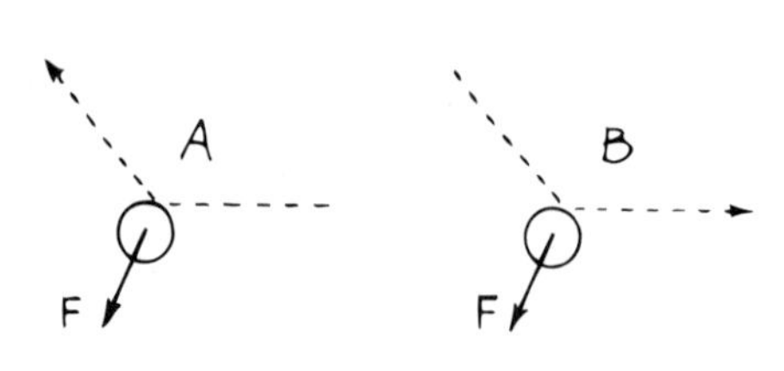

label ***K*,** and the other component is perpendicular to the keel, which we label ***T*.** It is the component ***K*,** as shown in Figure C.15 that is responsible for the forward motion of the boat. Component ***T*** is a useless force that tends to tip the boat over and move it sideways. This component force is offset by the deep keel. Again, the maximum speed of the boat can be no greater than the wind speed.

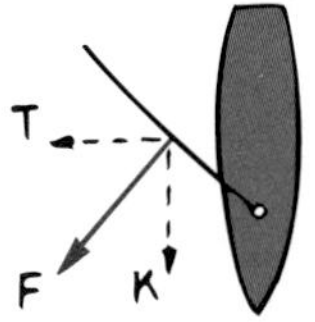

Figure C.15

Many sailboats sailing in directions other than exactly downwind (Figure C.16) with their sails properly oriented can exceed wind speed. In the case of a sailboat cutting across the wind, the wind may continue to make impact with the sail even after the boat exceeds wind speed. A surfer, in a similar way, exceeds the velocity of the propelling wave by angling his surfboard across the wave. Greater angles to the propelling medium (wind for the boat, water wave for the surfboard) result in greater speeds. A sailcraft can sail faster cutting across the wind than it can sailing downwind.

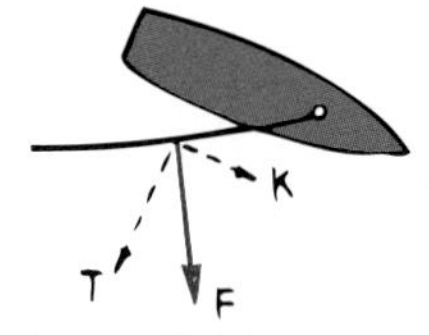

Figure C.16

As strange as it may seem, maximum speed for most sailcraft is attained by cutting into (against) the wind, that is, by angling the sailcraft in a direction upwind! Although a sailboat cannot sail directly upwind, it can reach a destination upwind by angling back and forth in a zigzag fashion. This is called *tacking*. Suppose the boat and sail are as shown in Figure C.17. Component ***K*** will push the boat along in a forward direction, angling into the wind. In the position shown, the boat can sail faster than the speed of the wind. This is because as the boat travels faster, the impact of wind is increased. This is similar to running in a rain that comes down at an angle. When you run into the direction of the downpour, the drops strike you harder and more frequently; but when you run away from the direction of the downpour, the drops don't strike you as hard or as frequently. In the same way, a boat sailing upwind experiences greater wind impact force, while a boat sailing downwind experiences a decreased wind impact force. In any case the boat reaches its terminal speed when opposing forces cancel the force of wind impact. The opposing forces consist mainly of water resistance against the hull of the boat. The hulls of racing boats are shaped to minimize this resistive force, which is the principal deterrent to high speeds.

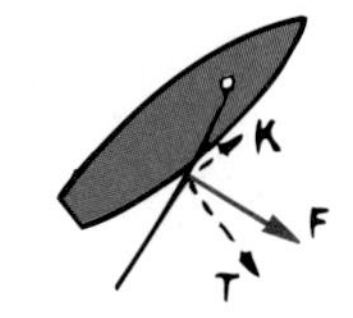

Figure C.17

Iceboats (sailcraft equipped with runners for traveling on ice) encounter no water resistance and can travel at several times the speed of the wind when they tack upwind. Although ice friction is nearly absent, an iceboat does not accelerate without limits. The terminal velocity of a sailcraft is determined not only by opposing friction forces but also by the change in relative wind direction. When the boat's orientation and speed are such that the wind seems to shift in direction, so the wind moves parallel to the sail rather than into it, forward acceleration ceases—at least in the case of a flat sail. In practice, sails are curved and produce an airfoil that is as important to sailcraft as it is to aircraft.

Appendix D: Physics of Fluids

Liquids and gases have the ability to flow; hence they are called *fluids.* In order to discuss the physics of fluids properly, we first need to understand two concepts—*density* and *pressure.*

Density

A basic property of materials—whether in the solid, liquid, or gaseous phases—is the measure of compactness: **density.**

$$\text{Density} = \frac{\text{mass}}{\text{volume}}$$

A loaf of bread has a certain mass, volume, and density. When squeezed, the volume decreases and its density increases—but its mass remains the same. Mass is measured in either grams or kilograms and volume in either cubic centimeters (cm^3) or cubic meters (m^3). Another unit of volume is the liter, which is 1000 cm^3. One gram per cubic centimeter = 1 kg per liter. The densities of a few materials are given in Table D.1.

Table D.1

Densities of Some Materials

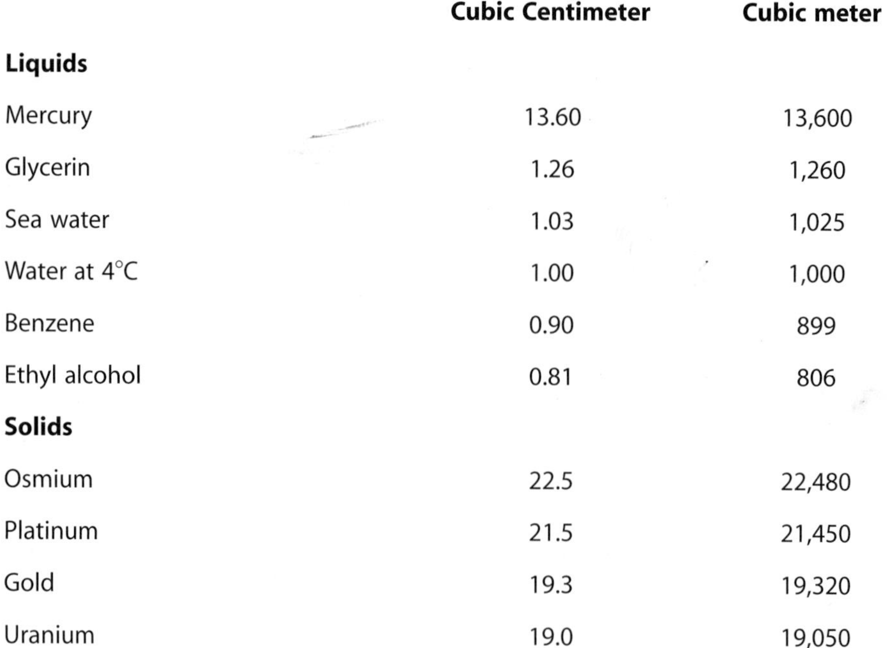

Material	Grams per Cubic Centimeter	Kilograms per Cubic meter
Liquids		
Mercury	13.60	13,600
Glycerin	1.26	1,260
Sea water	1.03	1,025
Water at 4°C	1.00	1,000
Benzene	0.90	899
Ethyl alcohol	0.81	806
Solids		
Osmium	22.5	22,480
Platinum	21.5	21,450
Gold	19.3	19,320
Uranium	19.0	19,050

continued

Figure D.1
A liter of water occupies a volume of 1000 cm^3, has a mass of 1 kg, and weighs 9.8 N. Its density may therefore be expressed as 1 kg/L and its weight density as 9.8 N/L. (Sea water is slightly denser, about 10 N/L).

Material	Grams per Cubic Centimeter	Kilograms per Cubic meter
Lead	11.3	11,344
Silver	10.5	10,500
Copper	8.9	8,920
Brass	8.6	8,560
Iron	7.8	7,800
Tin	7.3	7,280
Aluminum	2.7	2,702
Ice	0.92	917
Gases (atm. pressure at sea level)		
Dry air		
0°C	0.00129	1.29
10°C	0.00125	1.25
20°C	0.00121	1.21
30°C	0.00116	1.16
Hydrogen at 0°C	0.00090	0.090
Helium at 0°C	0.00178	0.178

A quantity known as *weight density* is commonly used when discussing liquid pressure.

$$\text{Weight density} = \frac{\text{weight}}{\text{volume}}$$

Weight density is common to British units, in which one cubic foot of fresh water (almost 7.5 gallons) weighs 62.4 pounds. So fresh water has a weight density of 62.4 lb/ft^3. Salt water is a bit denser, 64 lb/ft^3.

Pressure

Pressure is defined as the force exerted over a unit of area, such as a square meter or square foot:

$$\text{Pressure} = \frac{\text{force}}{\text{area}}$$

We see in Figure D.2 that force and pressure are different from each other. In the figure we see pressure due to the weight of a solid. Pressure occurs in fluids as well.

$$\text{Liquid pressure} = \text{weight density} \times \text{depth}$$

Figure D.2
Although the weight of both books is the same, the upright book exerts greater pressure against the table.

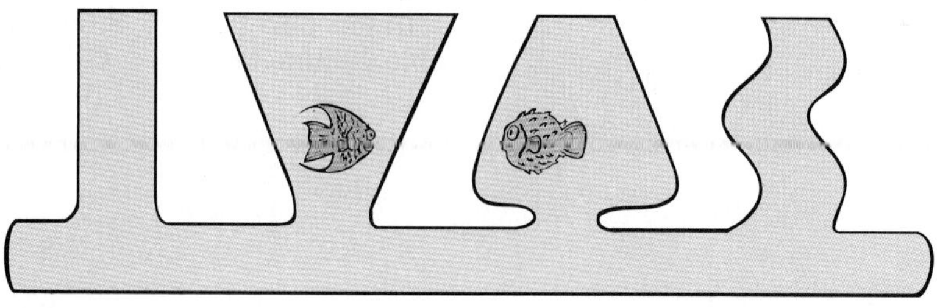

Figure D.3
The pressure exerted by a liquid is the same at any given depth below the surface, regardless of the shape of the containing vessel.

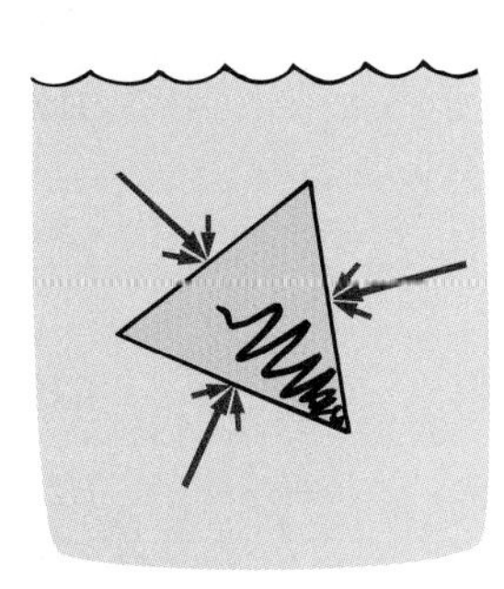

Figure D.4
The forces that produce pressure against a surface add up to a net force that is perpendicular to the surface.

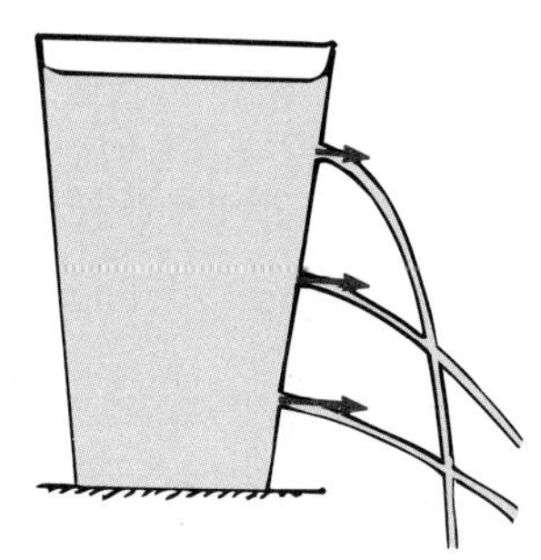

Figure D.5
Water pressure pushes perpendicularly against the sides of a container and increases with increasing depth.

It is important to note that pressure does not depend on the amount of liquid—but on its depth. You feel the same pressure a meter deep in a small pool as you do a meter deep in the middle of the ocean. The pressure is the same at the bottom of each of the connected vases in Figure D.4, for example. Depth, not volume, is the key to liquid pressure.

Pressure in a liquid at any point is exerted in equal amounts in all directions. For example, if you are submerged in water, no matter which way you tilt your head, you feel the same amount of water pressure on your ears.

When liquid presses against a surface, there is a net force directed perpendicular to the surface (Figure D.4). If there is a hole in the surface, the liquid spurts at right angles to the surface before curving downward due to gravity (Figure D.5). At greater depths the pressure is greater and the speed of the escaping liquid is greater.

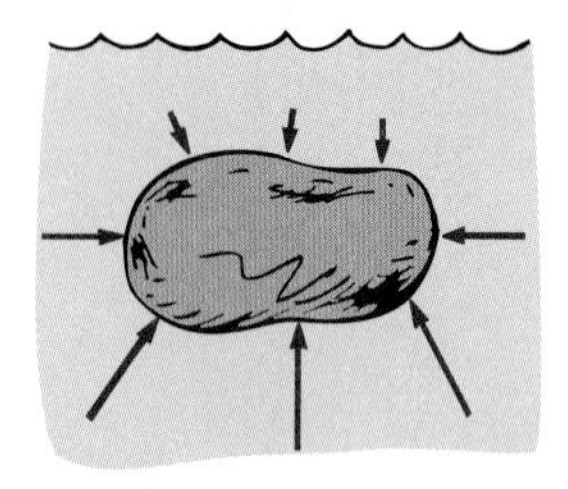

Figure D.6
The greater pressure against the bottom of a submerged object produces an upward buoyant force.

Buoyancy in a Liquid

When an object is submerged in water, the greater pressure on the bottom of the object results in an upward force called the **buoyant force.** We see why in Figure D.6. The arrows represent the forces at

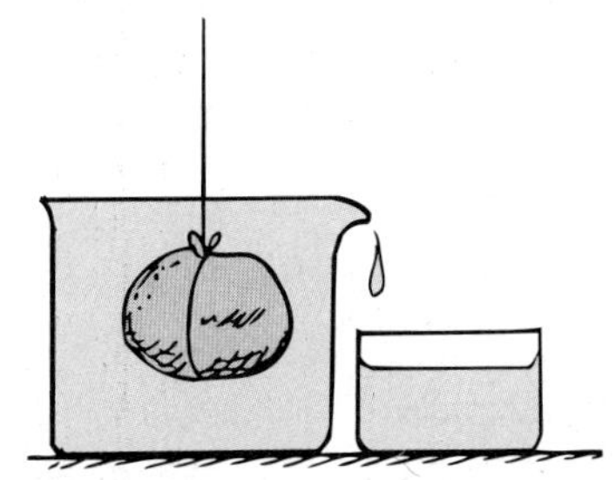

Figure D.7
When a stone is submerged, it displaces a volume of water equal to the volume of the stone.

Figure D.8
The increase in water level is the same as that which would occur if, instead of placing the stone in the container, we had poured in a volume of water equal to the stone's volume.

different places due to water pressure. Forces that produce pressures against opposite sides cancel one another because they are at the same depth. Pressure is greatest against the bottom of the object because the bottom is deeper (more pressure). Because the upward forces against the bottom are greater than the downward forces against the top, the forces do not cancel, and there is a net force upward. This net force is the buoyant force.

If the weight of the submerged object is greater than the buoyant force, the object sinks. If the weight is equal to the buoyant force, the object remains at any level, like a fish. If the buoyant force is greater than the weight of the completely submerged object, the object rises to the surface and floats.

Understanding buoyancy requires understanding the meaning of the expression "volume of water displaced." If a stone is placed in a container that is brimful of water, some water will overflow (Figure D.7). Water is *displaced* by the stone. A little thought tells you that the *volume of the stone*—that is, the amount of space (cubic centimeters) it takes up—is equal to the *volume of water displaced.* Place any object in a container partially filled with water, and the level of the surface rises (Figure D.8). By how much? By exactly the same amount as if we had added a volume of water equal to the volume of the immersed object. This is a good method for determining the volume of irregularly shaped objects: *A completely submerged object always displaces a volume of liquid equal to its own volume.*

Archimedes' Principle

The relationship between buoyancy and displaced liquid was first discovered in the third century BC by the Greek scientist Archimedes. It is stated as follows:

> **An immersed body is buoyed up by a force equal to the weight of the fluid it displaces.**

Figure D.9
Objects weigh more in air than in water. When submerged, this 3-N block appears to weigh only 1 N. The "missing" weight is equal to the weight of water displaced, 2 N, which equals the buoyant force.

This relationship is called **Archimedes' principle** and is true of all fluids, both liquids and gases. By *immersed,* we mean either *completely* or *partially submerged.* If we immerse a sealed 1-liter container halfway into a tub of water, it displaces a half-liter of water and is buoyed up by a force equal to the weight of a half-liter of water. If we immerse it completely (submerge it), a force equal to the weight of a full liter or 1 kilogram of water (which is 9.8 newtons) buoys it up. Unless the container is compressed, the buoyant force is 9.8 newtons at *any* depth, as long as the container is completely submerged.

A gas is a fluid and is also subject to Archimedes' principle. Any object less dense than air, a gas-filled balloon for example, rises in air.

Flotation

Iron is nearly eight times as dense as water and therefore sinks in water, but an iron ship floats. Why? Consider a solid 1-ton block of iron. When submerged it doesn't displace 1 ton of water because it's 8 times more compact than water. It displaces only 1/8 ton of water—certainly not enough to make it float. Suppose we reshape the same iron block into a bowl. It still weighs 1 ton, but when placed in water, it displaces a greater volume of water than before. Its larger volume displaces more water. The deeper it is immersed, the more water it displaces and the greater the buoyant force acting on it. When the buoyant force equals 1 ton, it sinks no farther. It floats.

When any object displaces a weight of water equal to its own weight, it floats. This is the **principle of flotation:**

> **A floating object displaces a weight of fluid equal to its own weight.**

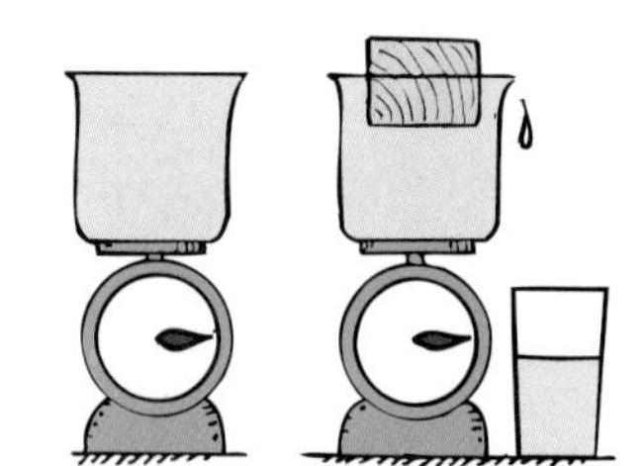

Figure D.10
A floating object displaces a weight of fluid equal to its own weight.

A 500-N friend floating in a swimming pool displaces 500 N of water. To accomplish this, your friend must be slightly less dense than water (which may or may not involve the use of a life preserver). Any floating object is less dense than the fluid it floats upon.

The density of ice, for example, is 0.9 that of water, so icebergs float in water. Interestingly, mountains are less dense than the semi-molten mantle beneath them. So they float. Just as most of an iceberg is below the water surface (90%), most of a mountain (about 85%) extends into the mantle. If you could shave off the top of an iceberg, the iceberg would be lighter and be buoyed up to nearly its original height before its top was shaved. Similarly, when mountains erode and wear away, they become lighter and are pushed up from below to float to nearly their original heights. So when a kilometer of mountain erodes away, about 0.85 kilometer of mountain pops up from below. That's why it takes so long for mountains to weather away.

Floating objects, whether mountains or ships, must displace a weight of fluid equal to their own weight. Thus, a 10,000-ton ship must be built wide enough to displace 10,000 tons of water before it sinks too deep in the water. The same holds true for vessels in the air. A dirigible or huge balloon that weighs 100 tons displaces at least 100 tons of air. If it displaces more, it rises; if it displaces less, it falls. If it displaces exactly its weight, it hovers at constant altitude.

Pressure in a Gas

There are similarities and there are differences between gases and liquids. The primary difference between a gas and a liquid is the distance between molecules. In a gas, the molecules are far apart and free from the cohesive forces that dominate their motions when in the liquid and solid phases. The motions of gas molecules are less restricted. A gas expands, fills all the space available to it, and exerts a pressure against its container. Only when the quantity of gas is very large, such as the Earth's atmosphere or a star, do gravitational forces limit the size or determine the shape of the mass of gas.

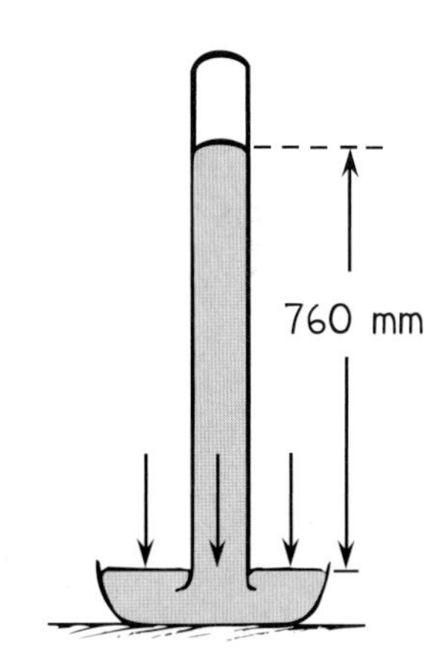

Figure D.11
A simple mercury barometer.

Atmospheric pressure is demonstrated with a **barometer.** A common barometer consists of a glass tube filled with mercury. When tipped upside down in a dish of mercury, atmospheric pressure holds up a vertical column of 76 centimeters. That's because the mercury in the column weighs as much as a column of air that extends from sea level to the "top" of the atmosphere. Like a seesaw, the mercury in the tube is balanced by equal pressure due to the air outside.

If water were used instead of mercury in a barometer, the column would be 13.6 times taller than a column of mercury. That's because water is 1/13.6 as dense as mercury. A water barometer therefore would have to be 10.3 meters (34 ft) tall—quite unpractical. Nevertheless, a practical application of the barometer effect is used in old fashioned farm-type pumps (Figure D.12). In these pumps, a pipe extends to well water below. When air pressure in the pipe is reduced, the atmospheric pressure on the water surface pushes water up into the pipe. Water is easily lifted from below by the pumping action.

Figure D.12
The atmosphere pushes water from below up a pipe that is evacuated of air by the pumping action. The ideal height to which water can be raised this way is 10.3 m, the height of a water barometer. Can you see why?

Boyle's Law

Pressure and volume in a confined gas are nicely related. "Pressure × volume" for a quantity of gas at any specified time is equal to any "different pressure × different volume" at any other time. In shorthand notation,

$$P_1V_1 = P_2V_2$$

where P_1 and V_1 represent the original pressure and volume, respectively, and P_2 and V_2 the second pressure and volume. This relationship is called **Boyle's Law,** after Robert Boyle, the seventeenth-century physicist who is credited with its discovery.

Boyle's law applies to ideal gases. An *ideal gas* is one in which the disturbing effects of the forces between molecules and the finite size of the individual molecules can be neglected. Air and other gases under normal pressure approach ideal-gas conditions.

A general law that takes temperature changes into account is

$$\frac{P_1V_1}{T_1} = \frac{P_2V_2}{T_2}$$

where T_1 and T_2 represent the initial and final *absolute* temperatures, measured in kelvins (Chapter 9).

Thus far we have treated pressure only as it applies to stationary fluids. Motion produces an additional influence.

Bernoulli's Principle

When a fluid flows through a narrow constriction, its speed increases. This is easily noticed by the increased speed of water that spurts from a garden hose when you narrow the opening of the nozzle. The fluid must speed up in the constricted region if the flow is to be continuous.

The Swiss scientist Daniel Bernoulli experimented with fluids in the eighteenth century. He wondered how the fluid gained this extra speed, and reasoned that it is acquired at the expense of a lowered internal pressure. His discovery, now called **Bernoulli's principle,** states:

When the speed of a fluid increases, pressure in the fluid decreases.

Bernoulli's principle is a consequence of the conservation of energy. In a steady flow of fluid, there are three kinds of energy: kinetic energy due to motion, gravitational potential energy due to elevation, and work done by pressure forces. In a steady fluid flow where no energy is added or removed, the sum of these forms of energy remains constant. If the elevation of the flowing fluid doesn't change, then an increase in speed means a decrease in pressure, and vice versa.

Bernoulli's principle is accurate only for steady flow. If the speed is too great, the flow may become turbulent and follow a changing, curling path known as an *eddy.* This type of flow exerts friction on the fluid and causes some of its energy to be transformed to heat. Then Bernoulli's principle does not hold.

Hold a sheet of paper in front of your mouth, as shown in Figure D.13. When you blow across the top surface, the paper rises. This is because the pressure of the moving air against the top of the paper is less than the pressure of the air at rest against the lower surface.

Figure D.13
The paper rises when Tim blows air across its top surface.

If we imagine the rising paper as an airplane wing, we can better understand the lifting force that supports a heavy airliner. In both cases a greater pressure below pushes the paper and wing into a region of lesser pressure above. The net upward pressure on a wing multiplied by the surface area of the wing gives the net lifting force. A blend of Bernoulli's principle and Newton's laws account for the air flight we see today. Quite awesome!

Appendix E: Exponential Growth and Doubling Time*

Try folding a piece of paper in half, then folding it again upon itself over and over for 9 times. You'll soon see that it gets too thick to keep folding. And if you could fold a fine piece of tissue paper upon itself 50 times, it would be more than 20 million kilometers thick! The continual doubling of a quantity builds up astronomically. Give a child a penny on his first birthday, two pennies on his second birthday, four pennies on his third birthday, and so on, doubling the number of pennies every birthday. When this child reaches age 30, he will have accumulated $10,737,418.23! One of the most important things we have trouble understanding is the process of exponential growth, and why it grows out of control.

When a quantity such as money in the bank, population, or the rate of consumption of a resource steadily grows at a fixed percent per year, the growth is said to be *exponential.* Money in the bank may grow at 5% or 6% per year; world population is presently growing at about 2% per year. The important thing about exponential growth is that the time required for the growing quantity to double in size (increase by 100%) is constant. For example, if the population of a growing city takes 10 years to double from 10,000 to 20,000 people and it continues with exponential growth, in the next 10 years the population will double to 40,000 and in the next 10 years to 80,000 and so on.

There is an important relationship between the percent growth rate and its *doubling time,* the time it takes to double a quantity:†

$$\text{doubling time} = \frac{69.2\%}{\text{percent growth per unit time}} \approx \frac{70\%}{\text{percent growth rate}}$$

So to estimate the doubling time for a steadily growing quantity, you simply divide 70% by the percentage growth rate. For example, if world population grows steadily at 2% per year, the world population will double in 35 years (since [70%]/[2%/year] = 35 years). A city planning commission that accepts what seems like a modest 3.5%-per-year growth rate may not realize that this means that

* This appendix is adapted from material written by University of Colorado physics professor Albert A. Bartlett, who asserts, "The greatest shortcoming of the human race is our inability to understand the exponential function."

† For exponential decay we speak about *half-life,* the time for a quantity to reduce to half its value. An example of this case is radioactive decay, treated in Chapter 19.

doubling will occur in 20 years (since [70%]/[3.5%/year] = 20 years). That means double capacity for such things as water supply, sewage-treatment plants, and other municipal services every 20 years.

Continued growth and continued doubling lead to enormous numbers. In two doubling times, a quantity will double twice ($2^2 = 4$), or quadruple in size; in three doubling times, its size will increase eightfold ($2^3 = 8$); in four doubling times, it will increase sixteenfold ($2^4 = 16$); and so on.

This is best illustrated by the story of the court mathematician in India who years ago invented the game of chess for his king. The king was so pleased with the game that he offered to repay the mathematician, whose request seemed modest enough. The mathematician requested a single grain of wheat on the first square of a chessboard, two grains on the second square, four on the third square, and so on, doubling the number of grains on each succeeding square until all squares had been used. At this rate there would be 2^{63} grains of wheat on the sixty-fourth square alone. The king soon saw that he could not fill this "modest" request, which amounted to more wheat than had been harvested in the entire history of the Earth!

As Table E.1 shows, the number of grains on any square is one grain more than the total of all grains on the preceding squares. This is true anywhere on the board. For example, when four grains are placed on the third square, that number of grains is one more than the total of three grains already on the board. The number of grains (eight) on the fourth square is one more than the total of seven grains already on the board. The same pattern occurs everywhere on the board. In any case of exponential growth, a greater quantity is represented in one doubling time than in all the preceding growth. This is important enough to be repeated in different words: Whenever steady growth occurs, the numerical count of a quantity that exists

Table E.1

Filling the Squares on the Chessboard

Square number	Grains on a square	Total grains thus far
1	1	1
2	2	3
3	$4 = 2^2$	7
4	$8 = 2^3$	15
5	$16 = 2^4$	31
6	$32 = 2^5$	63
7	$64 = 2^6$	127
•	•	•
•	•	•
•	•	•
10	2^9	about 1,000
•	•	•
•	•	•
•	•	•
20	2^{19}	about 1,000,000
•	•	•
•	•	•
•	•	•
30	2^{29}	about 1,000,000,000
•	•	•
•	•	•
•	•	•
40	2^{39}	about 1,000,000,000,000
•	•	•
•	•	•
•	•	•
64	2^{63}	$2^{64} - 1$ (more than ten billion billion!)

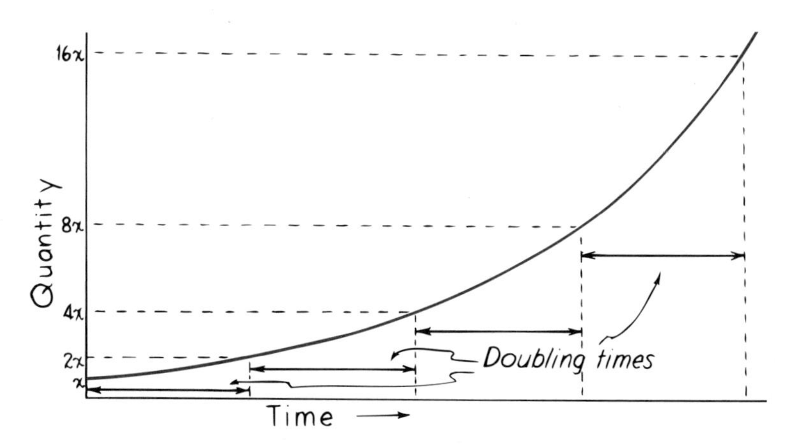

Figure E.1
Graph of a quantity that grows at an exponential rate. Notice that the quantity doubles during each of the successive equal time intervals marked on the horizontal scale. Each of these time intervals represents the doubling time.

after a single doubling time is one greater than the total count of that quantity in the entire history of growth.

Steady growth in a steadily expanding environment is one thing, but what happens when steady growth occurs in a finite environment? Consider the growth of bacteria that grow by division, so that one bacterium becomes two, the two divide to become four, the four divide to become eight, and so on. Suppose the division time for a certain kind of bacteria is one minute. This is then steady percentage growth—the number of bacteria grows exponentially with a doubling time of one minute. Further, suppose that one bacterium is put in a bottle at 11:00 a.m. and that growth continues steadily until the bottle becomes full of bacteria at 12 noon.

Concept Check ✓

When was the bottle half full?

Check Your Answer At 11:59 a.m., since the bacteria will double in number every minute!

It is startling to note that at 2 minutes before noon the bottle was only 1/4 full, and at 3 minutes before noon only 1/8 full. Table E.2 summarizes the amount of space left in the bottle in the last few minutes before noon. If bacteria could think, and if they were concerned about their future, at what time do you suppose they would sense they were running out of space? Would a serious problem have been evident at, say, 11:55 a.m., when the bottle was only 3% full (1/32) and had 97% open space (just yearning for development)? The point here is that there isn't much time between the moment the effects of growth become noticeable and the time when they become overwhelming.

Suppose that at 11:58 a.m. some farsighted bacteria see that they are running out of space and launch a full-scale search for new bottles. And further suppose they consider themselves lucky, for they find three new empty bottles. This is three times as much space as they have ever known. It may seem to the bacteria that their problems are solved—and just in time.

Table E.2

The Last Minutes in the Bottle

Time	Portion full	Portion empty
11:54 a.m.	1/64 (1.5%)	63/64 (98.5%)
11:55 a.m.	1/32 (3%)	31/32 (97%)
11:56 a.m.	1/16 (6%)	15/16 (94%)
11:57 a.m.	1/8 (12%)	7/8 (88%)
11:58 a.m.	1/4 (25%)	3/4 (75%)
11:59 a.m.	1/2 (50%)	1/2 (50%)
12:00 noon	Full (100%)	None (0%)

Figure E.2

Concept Check ✓

If the bacteria are able to migrate to the new bottles and their growth continues at the same rate, what time will it be when the three new bottles are filled to capacity?

Check Your Answer All four bottles will be filled to capacity at 12:02 p.m.!

Table E.3 illustrates that the discovery of the new bottles extends the resource by only two doubling times. In this example the resource is space—such as land area for a growing population. But it could be coal, oil, uranium, or any nonrenewable resource.

Table E.3

Effects of the Discovery of Three New Bottles

Time	Effect
11:58 a.m.	Bottle 1 is 1/4 full; bacteria divide into four bottles, each 1/16 full
11:59 a.m.	Bottles 1, 2, 3, and 4 are each 1/8 full
12:00 noon	Bottles 1, 2, 3, and 4 are each 1/4 full
12:01 p.m.	Bottles 1, 2, 3, and 4 are each 1/2 full
12:02 p.m.	Bottles 1, 2, 3, and 4 are each all full

Concept Check ✓

1. According to a French riddle, a lily pond starts with a single leaf. Each day the number of leaves doubles, until the pond is completely full on the thirtieth day. On what day was the pond half covered? One-quarter covered?
2. In 2000 the population grew to 6 billion (likely 7 billion in 2013, and 8 billion in 2027). At the 2000 world growth rate of 1.2% per year, how long will it take for the world population to reach 12 billion?
3. What annual percentage increase in world population would be required to double the world population in 100 years?

Check Your Answers

1. The pond was half covered on the 29th day, and was one-quarter covered on the 28th day!
2. 2058, for the doubling time (70%)/(1.2%/year) $\sim$ 58 years.
3. 0.7%, since (70%)/(0.7%/year) = 100 years. You can rearrange the equation so it reads percent growth rate = (70%)/(doubling time). Using the rearranged equation gives (70%)/(100 years) = 0.7%/year.

Growth in the empty bottles discovered by the bacteria can proceed unrestricted (until the bottles are full)—not typical of nature. Although bacteria and other organisms have the potential to multiply exponentially, limiting factors usually restrict the growth. The number of mice in a field, for example, depends not only on birthrate and food supply, but on the number of hawks and other predators in the vicinity. A "natural balance" of competing factors is struck. If the predators are removed, exponential growth of the mice population can proceed for a while. Remove certain plants from a region and others tend to exponential growth. All plants, animals, and creatures that inhabit the Earth are in states of balance—states that change with changing conditions. Hence the environmental adage, "You never change only one thing."

The consequences of unchecked exponential growth are staggering. It is very important to ask: Is growth really good? In answering this question, bear in mind that human growth is an early phase of life that continues normally through adolescence. Physical growth stops when physical maturity is reached. What do we say of growth that continues in the period of physical maturity? We say that such growth is obesity—or worse, cancer.

Figure E.3
A single grain of wheat placed on the first square of the chess board is doubled on the second square, and this number is doubled on the third square, and so on. Note that each square contains one more grain than all the preceding squares combined. Does enough wheat exist in the world to fill all 64 squares in this manner?

Questions to Ponder

1. In an economy that has a steady inflation rate of 7% per year, in how many years does a dollar lose half its value?
2. At a steady inflation rate of 7%, what will be the price every 10 years for the next 50 years for a theater ticket that now costs $20? For a coat that now costs $200? For a car that now costs $20,000? For a home that now costs $200,000?
3. If the population of a city with one overloaded sewage treatment plant grows steadily at 5% annually, how many overloaded sewage treatment plants will be necessary 42 years later?
4. If world population doubles in 40 years and world food production also doubles in 40 years, how many people then will be starving each year compared to now?
5. Suppose you get a prospective employer to agree to hire your services for a wage of a single penny for the first day, 2 pennies the second day, and doubling each day thereafter. If the employer keeps to the agreement for a month, what will be your total wages for the month?
6. In the previous question, how will your wages for only the 30th day compare to your total wages for the previous 29 days?

Appendix F: Safety

Hopefully, you will have the opportunity to perform many of the explorations and other activities presented in this textbook. In performing these, you should always keep the safety of yourself and others in mind. Most safety rules that must be followed involve common sense. For example, if you are ever unsure of a procedure or chemical, ask your teacher or parent who should always be there to help you.

To guide you to safe practices, in this textbook you will find several types of icons posted by selected activities. Here are the icons and what they indicate:

Wear approved safety goggles. Wear goggles when working with a chemical, solution, or when heating substances.

Wear gloves. Wear gloves when working with chemicals.

Flame/heat. Keep combustible items, such as paper towels, away from any open flame. Handle hot items with tongs, oven mitts, pot holders or the like. Do not put your hands or face over any boiling liquid. Use only heat-proof glass, and never point a heated test tube or other container at anyone. Turn off the heat source when you are finished with it.

Here are some specific safety rules that should be practiced at all times:

1. Do not eat, drink, or smoke in the laboratory.
2. Maintain a clean and orderly work space. Clean up spills at once or ask for assistance in doing so.

3. Do not perform unauthorized experiments. First obtain permission from your teacher or parent. It is most important that others know what you are doing and when you are doing it. Think of it this way: If you are doing an experiment alone and something goes terribly wrong, there will be no one there to help you. That's not using your common sense.
4. Do not taste any chemicals or directly breathe any chemical vapors.
5. Check all chemical labels for both name and concentration.
6. Do not grasp recently heated glassware, clamps, or other heated equipment because they remain hot for quite a while.
7. Discard all excess reagents or products in the proper waste containers.
8. If your skin comes in contact with a chemical, rinse under cold water for at least 15 minutes.
9. Do not work with flammable solvents near an open flame.
10. Assume any chemical is hazardous if you are unsure.

Glossary

Ablation The amount of snow mass a glacier loses in a year.

Absolute zero The lowest possible temperature; the temperature at which all particles have their minimum kinetic energy.

Acceleration The rate at which velocity changes with time; the change in velocity may be in magnitude or direction or both.

Accumulation The amount of snow added to a glacier in a year.

Acid A substance that produces or donates hydrogen ions in solution.

Acid-base reaction A reaction involving the transfer of a hydrogen ion, H+, from one reactant to another.

Acidic solution A solution in which the hydronium ion concentration is greater than the hydroxide ion concentration.

Actinides The inner transition metals of the seventh period.

Addition polymer A polymer formed simply by the joining together of monomer units.

Additive primary colors The three colors—red, blue, and green—that when added in certain proportions produce any other color in the visible-light part of the electromagnetic spectrum.

Adhesive forces Molecular interactions that arise between two different substances.

Adiabatic process A process, usually of expansion or compression, wherein no heat enters or leaves a system.

Alcohols A class of organic molecules that contain a hydroxyl group bonded to a saturated carbon.

Aldehydes A class of organic molecules containing a carbonyl group the carbon of which is bonded to one carbon atom and one hydrogen atom.

Alkali-earth metal A group 2 element.

Alkali metal A group 1 element.

Alloy A mixture of two or more metallic elements.

Alpha particle The nucleus of a helium atom, which consists of two neutrons and two protons, ejected by certain radioactive elements.

Alternating current (ac) Electrically charged particles that repeatedly reverse direction, vibrating about relatively fixed positions. In the United States the vibrational rate is 60 Hz.

Amides A class of organic molecules containing a carbonyl group, the carbon of which is bonded to one carbon atom and one nitrogen atom.

Amines A class of organic molecules containing the element nitrogen bonded to saturated carbon atoms.

Ampere The unit of electric current; 1 ampere = 1 coulomb per second (the flow of 6.25×10^{18} electrons per second); 1 A = 1 C/s.

Amphoteric substance A substance that can behave as either an acid or a base.

Amplitude For a wave or vibration, the maximum displacement on either side of the equilibrium (midpoint) position.

Angular unconformity An unconformity in which older tilted strata are overlain by horizontal younger strata.

Anticline A fold in strata that has relatively old rocks at its core, with rock age decreasing as you move horizontally away from the core fold.

Apparent weight The force you exert against a supporting floor or a weighing scale, wherein you are as heavy as you feel.

Archimedes' principle An immersed body is buoyed up by a force equal to the weight of the fluid it displaces.

Aromatic compound Any organic molecule containing a benzene ring.

Artesian system A system in which groundwater under pressure rises above the level of an aquifer.

Asteroid A small, rocky fragment that orbits the sun.

Asthenosphere A subdivision of the upper mantle situated below the lithosphere; a zone of plastic, easily deformed rock.

Atomic mass number The number associated with an atom that is the same as the number of protons plus neutrons in its nucleus.

Atomic mass unit A very small unit of mass used for atoms and molecules. One atomic mass unit (amu) is equal to $\frac{1}{12}$ the mass of the carbon-12 atom, or 1.661×10^{-24} grams.

Atomic nucleus The core of an atom, consisting of two basic subatomic particles—protons and neutrons.

Atomic number The number that designates the identity of an element, which is the number of protons in the nucleus of an atom; in a neutral atom, the atomic number is also the number of electrons in the atom.

Atomic orbital The region of space where electrons of a given energy are likely to be located.

Atmospheric pressure The pressure exerted against bodies immersed in the atmosphere resulting from the weight of air pressing down from above. At sea level, atmospheric pressure is about 101 kPa.

Atomic spectra The range of frequencies of light emitted by atoms.

Average speed Total distance traveled divided by time.

Atomic symbol An abbreviation for an element derived from the first one or two letters of the element's name.

Avogadro's number The number of atoms in the atomic mass of an element when expressed in grams. A very large number: 6.02×10^{23}.

Barometer Any device that measures atmospheric pressure.

Base A substance that produces hydroxide ions in solution or accepts hydrogen ions.

Basic solution A solution in which the hydroxide ion concentration is greater than the hydronium ion concentration.

Beats A series of alternate reinforcements and cancellations produced by the interference of two waves of slightly different frequency, heard as a throbbing effect in sound waves.

Bernoulli's principle When the speed of a fluid increases, pressure in the fluid decreases.

Beta particle An electron (or positron) emitted during the radioactive decay of certain nuclei.

Big Bang The primordial explosion of space at the beginning of time.

Binary A system of two stars that orbit about a common center of mass.

Black dwarf The presumed final stage of a white dwarf.

Black hole The remains of a giant star that has collapsed into itself, so dense and of such intense gravitational field that light cannot escape.

Body wave A seismic wave that travels through the Earth's interior.

Boiling Rapid evaporation that takes place within a liquid as well as at its surface.

Bond energy The amount of energy absorbed upon bond breaking and released upon bond formation. Each chemical bond has its own characteristic bond energy.

Bow wave The V-shaped wave made by an object moving across a liquid surface at a speed greater than the wave speed.

Boyle's law The product of pressure and volume is constant for a given mass of confined gas so long as temperature remains unchanged:

$$P_1V_1 = P_2V_2$$

Breeder reactor A fission reactor that is designed to breed more fissionable fuel than is put into it by converting non-fissionable isotopes to fissionable isotopes.

Buffer solution A solution of either a weak acid and one of its salts or a weak base and one of its salts that resists large change in pH.

Buoyant force The net upward force that a fluid exerts on an immersed object.

Calorie A unit of thermal energy, or heat. One calorie is the thermal energy required to raise the temperature of one gram of water 1 Celsius degree (1 cal = 4.184 J). One Calorie (with a capital *C*) is equal to one thousand calories, and is the unit used in describing the energy available from food.

Capillary action The rising of liquid into a small vertical space due to adhesion between the liquid and the sides of the container and to cohesive forces within the liquid.

Carbonyl group A carbon atom double-bonded to an oxygen atom, $C{=}O$. The carbonyl group is found in ketones, aldehydes, amides, carboxylic acids, and esters.

Carboxylic acids A class of organic molecules containing a carbonyl group the carbon of which is bonded to one carbon atom and one hydroxyl group.

Catalyst Any substance that serves to increase the rate of a chemical reaction.

Cenozoic The time of recent life. This period of time began 66 million years ago and is still current.

Chain reaction A self-sustaining reaction in which the products of one reaction event stimulate further reaction events.

Chemical change A change in which a substance changes its chemical identity. During a chemical change atoms are rearranged to give a new substance. Also referred to as a chemical reaction.

Chemical equation A representation of a chemical reaction showing the relative numbers of reactants and products.

Chemical equilibrium A dynamic state in which the rate of the forward chemical reaction is equal to the rate of the reverse chemical reaction. At chemical equilibrium the concentrations of reactants and products remain constant.

Chemical formula A notation used to denote the composition of a compound. In a chemical formula, the atomic symbols for the elements making up the compound are written along with numerical subscripts that indicate their proportions.

Chemical property The tendency of a substance to change chemical identity.

Chemical reaction Synonymous with *chemical change.*

Chemical weathering The breakdown of rocks on the Earth's surface by chemical means.

Chemistry The study of matter and of the transformations it can undergo.

Clouds The condensation of water droplets above the Earth's surface.

Cluster A grouping of more than one galaxy.

Cohesive forces The attractive forces within a substance.

Comet A body composed of ice and dust that orbits the sun, usually in a very eccentric orbit, and which has one or more luminous tails when it is close to the sun.

Complementary colors Any two colors that when added together produce white light.

Compound A material in which atoms of different elements are chemically held to one another.

Concentrated A solution containing a relatively large amount of solute.

Concentration A quantitative measure of the amount of solute in a solution.

Condensation The change of phase from gaseous to liquid.

Condensation polymer A polymer formed by the joining of monomers with the concomitant loss of a small molecule, such as water.

Conduction The transfer of heat energy by collisions between the particles in a substance (especially a solid).

Conformation The spatial orientation of a molecule, which changes as the single bonds in the molecule rotate.

Conservation of energy Energy cannot be created or destroyed; it may be transformed from one form into another, but the total amount of energy never changes. In an ideal machine, where no energy is transformed into heat, $\text{work}_{\text{input}} = \text{work}_{\text{output}}$ and $(Fd)_{\text{input}} = (Fd)_{\text{output}}$.

Conservation of momentum When no external net force acts on an object or a system of objects, no change of momentum takes place. Hence, the momentum before an event involving only internal forces is equal to the momentum after the event:

$$mv_{(\text{before event})} = mv_{(\text{after event})}$$

Continental margin The boundary between continental land and deep ocean basins, consisting of continental shelf, continental slope, and continental rise.

Convection The transfer of heat energy in a gas or liquid by means of currents in the heated fluid. The fluid moves, carrying energy with it.

Convectional lifting An air-circulation pattern in which air warmed by the ground rises, while cooler air aloft sinks.

Converging lens A lens that is thicker in the middle than at the edges, and refracts parallel rays passing through it to a focus.

Copolymer A polymer composed of at least two different types of monomers.

Core The central layer in the Earth's interior, divided into an outer liquid core and an inner solid core.

Coriolis effect The apparent deflection from a straight-line path observed in any body moving near the Earth's surface, caused by the Earth's rotation

Coulomb The SI unit of electrical charge. One coulomb (symbol C) is equal in magnitude to the total charge of 6.25×10^{18} electrons.

Coulomb's law For any two electrically charged bodies, the relationship among the electric force the bodies exert on each other, the charge on the two bodies, and the distance between them:

$$F = k\frac{q_1 q_2}{d^2}$$

If the charges are alike in sign, the force is repelling; if the charges are unlike, the force is attractive.

Covalent bond A chemical bond in which atoms are held together by their mutual attraction for two electrons they share.

Covalent compound An element or chemical compound in which atoms are held together by the covalent bond.

Covalent crystal A group of molecules arranged in an orderly fashion.

Critical mass The minimum mass of fissionable material in a reactor or nuclear bomb that will sustain a chain reaction.

Cross-cutting When an igneous intrusion or fault cuts through other rocks, the intrusion or fault is younger than the rock it cuts.

Crust The Earth's outermost layer.

Delta A flat-topped accumulation of sediments deposited where a stream flows into a standing body of water.

Density The amount of matter per unit volume:

$$\text{Density} = \frac{\text{mass}}{\text{volume}}$$

Weight density is weight per unit volume.

Diffraction The bending of light as it passes around an obstacle or through a narrow slit, causing the light to spread and to produce light and dark fringes.

Diffuse reflection Reflection in many directions from an irregular surface.

Dilute A solution containing a relatively small amount of solute.

Dipole A separation of charge.

Dipole-dipole interaction The molecular interaction involving dipoles.

Dipole-induced dipole interaction The molecular interaction involving a dipole and an induced dipole.

Direct current (dc) Electrically charged particles flowing in one direction only.

Dissolving The process of mixing a solute in a solvent.

Distillation The process of recollecting a vaporized substance.

Diverging lens A lens that is thinner in the middle than at the edges, causing parallel rays passing through it to diverge as if from a point.

Doppler effect The change in frequency of wave motion resulting from motion of the wave source or receiver.

Drift Generic term for all glacial deposits. Waterborne glacial deposits are known as *outwash.* Material deposited directly by melting ice is *till.*

Effective nuclear charge The nuclear charge experienced by outer-shell electrons, which is diminished by their distance from the nucleus and by the shielding effect of inner-shell electrons.

Efficiency The percent of the work put into a machine that is converted into useful work output.

Elastic collision A collision in which colliding objects rebound without lasting deformation or the generation of heat.

Electric current The flow of electric charge that transports energy from one place to another. Measured in amperes.

Electric field Defined as force per unit charge, it can be considered an "aura" surrounding charged objects and is a storehouse of electric energy. About a charged point, the strength of the electric field decreases with distance according to the inverse-square law.

Electric motor A device that uses a current-carrying coil forced to rotate in a magnetic field to convert electrical energy to mechanical energy.

Electric potential The electric potential energy per amount of charge (usually called *voltage*), measured in volts:

$$\text{Electric potential} = \frac{\text{energy}}{\text{charge}}$$

Electrical potential energy The energy a charge possesses by virtue of its location in an electric field.

Electric power The rate of energy transfer, or the rate of doing work; the amount of energy transferred per unit time, which electrically can be measured by

$$\text{Power} = \text{current} \times \text{voltage}$$

Measured in watts (or kilowatts), where 1 A × 1 V = 1 W.

Electrical resistance The property of a material that resists the flow of charged particles through it; measured in ohms (W).

Electrically polarized Term applied to an atom or molecule in which the charges are aligned so that one side has a slight excess of positive charge and the other side a slight excess of negative charge.

Electrochemistry A branch of chemistry concerned with the relationship between electrical energy and chemical change.

Electrode A conducting material used to establish electrical contact between metallic and nonmetallic parts of an electric circuit.

Electrolysis The use of electrical energy to produce chemical change.

Electromagnet A magnet whose field is produced by an electric current. Electromagnets are usually in the form of a wire coil with a piece of iron inside the coil.

Electromagnetic induction The induction of voltage when a magnetic field changes with time. If the magnetic field within a closed loop changes in any way, a voltage is induced in the loop:

$$\text{Voltage induced} \sim \frac{\text{number of loops} \times \text{mag. field change}}{\text{change in time}}$$

This is a statement of Faraday's law. The induction of voltage is the result of a more fundamental phenomenon: the induction of an electric *field,* as defined for the more general case below.

Electromagnetic wave A wave emitted by vibrating electrical charges (often electrons) and composed of vibrating electric and magnetic fields that regenerate one another.

Electromagnetic spectrum The range of electromagnetic waves extending in frequency from radio waves to gamma rays.

Electrons The negatively charged particles in an atom.

Electron affinity The ability of an atom to attract additional electrons.

Electronegativity The ability of an atom to attract a bonding pair of electrons to itself when bonded to another atom.

Element A fundamental material consisting of only one type of atom.

Elemental formula A notation that uses the atomic symbol and a numerical subscript to denote the composition of an element.

Ellipse An oval path. The sum of the distances from any point on the path to two points inside called foci is a constant.

Energy The property of a system that enables it to do work.

Equilibrium rule For any body in mechanical equilibrium, the vector sum of the forces on it are zero—that is, $\Sigma F = 0$.

Erosion The process by which rock particles are transported away by water, wind, or ice.

Escape speed The speed that a projectile, space probe, or similar object must reach in order to escape the gravitational influence of the Earth or celestial body to which it is attracted.

Esters A class of organic molecules containing a carbonyl group the carbon of which is bonded to one carbon atom and one oxygen atom that is also bonded to a carbon atom.

Ethers A class of organic molecules containing an oxygen atom bonded to two saturated carbon atoms.

Evaporation The change of phase from liquid to gaseous.

Exosphere The fifth atmospheric layer above the Earth's surface, extending from the thermosphere upward and out into interplanetary space.

Extrusive rocks Igneous rocks that form at the Earth's surface.

Faraday's law An electric field is induced in any region of space in which a magnetic field is changing with time. The magnitude of the induced electric field is proportional to the rate at which the magnetic field changes. The direction of the induced field is at right angles to the direction of the changing magnetic field.

Fault A fracture along which visible displacement can be detected on one side relative to the other.

Faunal succession Fossil organisms succeed one another in a definite, irreversible, determinable order.

First law of thermodynamics A restatement of the law of energy conservation, usually as it applies to systems involving changes in temperature: The heat added to a system equals the increase in the thermal energy of the system plus the external work the system does on its environment.

Fold A series of ripples that result from compressional deformation of the lithosphere.

Force Any influence that can cause an object to be accelerated, measured in newtons in the metric system and in pounds in the British system.

Forced vibration The setting up of vibrations in an object by a vibrating force.

Formula mass The mass of a chemical compound given in amu.

Fractional crystallization The process and sequence by which the different minerals in a magma crystallize at different temperatures as the magma cools.

Free fall A state of fall under the influence of only gravity—free from air resistance.

Freezing The change of phase from liquid to solid.

Frequency For a vibrating body or medium, the number of vibrations per unit time. For a wave, the number of crests that pass a particular point per unit time.

Friction The resistive force that opposes the motion or attempted motion of an object past another with which it is in contact, or through a fluid.

Front The contact zone between two air masses.

Frontal lifting The lifting that occurs as two air masses converge.

Functional group The essential heteroatom-containing structural feature found in all members of a class of compounds.

Fundamental frequency The lowest frequency of vibration of a musical note.

Galaxy A large assembly of stars, numbering in the hundreds of millions to hundreds of billions, together with gas, dust, and other materials, all held together by the forces of mutual gravitation.

Gamma ray High-frequency electromagnetic radiation emitted by the nuclei of radioactive atoms.

Gas Phase of matter in which molecules fill whatever space is available to them, taking neither definite shape nor definite volume.

Generator An electromagnetic induction device that produces electric current by rotating a coil within a stationary magnetic field; converts mechanical energy to electrical energy.

Glacier A large mass of ice formed by the compaction and recrystallization of snow, moving downslope under its own weight.

Gravitational red shift The lengthening of the waves of electromagnetic radiation escaping from a massive object.

Greenhouse effect Warming caused by short-wavelength radiant energy from the sun that easily enters the atmosphere and is absorbed by the Earth. This energy is then reradiated at longer wavelengths that cannot easily escape the Earth's atmosphere. Thus, the Earth's atmosphere and surface are warmed.

Groundwater Subsurface water in the zone of saturation.

Group A vertical column in the periodic table.

Gyre Circular or spiral whirl pattern, usually applied to very large current systems in the open ocean.

Half-life The time required for half the atoms in a sample of a radioactive isotope to decay.

Halogen A group 17 element.

Harmonic A frequency that is an integer multiple of the fundamental.

Heat The thermal energy that flows from an object at higher temperature to one at a lower temperature, commonly measured in calories or joules.

Heat engine A device that uses heat as input and supplies mechanical work as output or one that uses work as input and moves heat "uphill" from a cooler to a warmer place.

Heat of fusion The amount of thermal energy required to change a substance from solid to liquid or from liquid to solid.

Heat of vaporization The amount of thermal energy required to change a substance from liquid to gas or from gas to liquid.

Hertz The unit of frequency. One hertz (symbol Hz) equals one vibration per second.

Heteroatom Any atom other than carbon or hydrogen in an organic molecule.

Heterogeneous mixture A mixture in which the components can be seen as individual substances.

Homogeneous mixture A mixture composed of components so finely mixed that composition is the same throughout the mixture.

Humidity A measure of the amount of water vapor in the air.

Hydrocarbon A compound containing only carbon and hydrogen atoms.

Hydrogen bond A strong dipole-dipole interaction that involves a hydrogen atom chemically bonded to a strongly electronegative element, such as nitrogen, oxygen, or fluorine.

Hydrologic cycle The natural circulation of water from ocean to atmosphere to ground back to ocean.

Hydronium ion A water molecule after accepting a hydrogen ion, H_3O^+.

Igneous rocks Rocks formed by the cooling and crystallization of hot, molten rock material called magma.

Impulse The product of the force acting on an object and the time during which it acts.

Impulse-momentum relationship Impulse is equal to the change in the momentum of the object that the impulse acts on. In symbol notation:

$$Ft = \Delta mv$$

Impure The state of a material that consists of more than one element or compound. A chemical mixture is impure.

Inclusion Any inclusion (pieces of one rock type contained within another) is older than the rock containing it.

Induced dipole A dipole temporarily created in an otherwise nonpolar molecule. It is induced by a neighboring charge or dipole.

Inelastic collision A collision in which the colliding objects become distorted, generate heat, and possibly stick together.

Inertia The tendency of things to resist changes in motion.

Inner-shell shielding The tendency of inner-shell electrons to partially shield outer-shell electrons from the nuclear charge.

Inner transition metals Two subgroups – lanthanides and actinides – of the transition metals.

Insoluble Not capable of dissolving to any appreciable extent in a solvent.

Instantaneous speed Speed at any given instant.

Interference The result of superposing two or more waves of the same wavelength. Constructive interference results from crest-to-crest reinforcement; destructive interference results from crest-to-trough cancellation.

Intrusive rocks Rocks that crystallize below the Earth's surface.

Inverse-square law A law relating the intensity of an effect to the inverse square of the distance from the cause:

$$\text{Intensity} \sim 1/\text{distance}^2$$

Ion-dipole interaction The molecular interaction involving an ion and a dipole.

Ionic bond The electrical force of attraction that holds ions of opposite charge together.

Ionic compound Any chemical compound containing ions.

Ionic crystal A group of many ions held together in an orderly three-dimensional array.

Ionization energy The amount of energy required to pull an electron away from an atom.

Ionosphere An electrified region within the thermosphere and uppermost mesosphere where fairly large concentrations of ions and free electrons exist.

Isostasy The principle that dictates how high the crust stands above the mantle.

Isotopes Atoms whose nuclei have the same number of protons but different numbers of neutrons.

Ketones A class of organic molecules containing a carbonyl group the carbon of which is bonded to two carbon atoms.

Kilogram The fundamental SI unit of mass. One kilogram (symbol kg) is this amount of mass in 1 liter (l) of water at 4°C.

Kinetic energy Energy of motion, described by the relationship

$$\text{Kinetic energy} = \tfrac{1}{2}mv^2$$

Laminar flow Water flowing smoothly in a straight line with no mixing of sediment.

Lanthanides The inner transition metals of the sixth period.

Lava Magma once it reaches the Earth's surface.

Law of action and reaction (Newton's Third Law) Whenever one object exerts a force on a second object, the second object exerts an equal and opposite force on the first.

Law of inertia (Newton's First Law) Every material object continues in its state of rest, or of uniform motion in a straight line, unless it is compelled to change that state by forces impressed upon it.

Law of reflection The angle of an reflection equals the angle of incidence. The reflected and incident rays lie in a plane that is normal to the reflecting surface.

Law of universal gravitation Every body in the universe attracts every other body with a force that, for two bodies, is directly proportional to the product of their masses and inversely proportional to the square of the distance separating them:

$$F = \frac{Gm_1m_2}{d^2}$$

Leachate A solution formed by water that has percolated through soil containing water-soluble substances.

Liquid Phase of matter characterized by definite volume but no definite shape; a liquid takes on the shape of its container.

Lithosphere The entire crust plus the portion of the mantle above the asthenosphere.

Longitudinal wave A wave in which the medium vibrates in a direction parallel (longitudinal) to the direction in which the wave travels. Sound is an example of a longitudinal wave.

Lunar eclipse The phenomenon whereby the shadow of the Earth falls on the moon, producing relative darkness of the full moon.

Magma Molten rock from the Earth's interior.

Magnetic domains Clustered regions of aligned magnetic atoms. When these regions are aligned with one another, the substance containing them is a magnet.

Magnetic field The region of magnetic influence around either a magnetic pole or a moving charged particle.

Magnetic force (1) Between magnets, the attraction of unlike magnetic poles for each other and the repulsion between like magnetic poles. (2) Between a magnetic field and a moving charged particle, a deflecting force due to the motion of the particle.

Mantle The middle layer in the Earth's interior, between crust and core.

Mass The quantity of matter in an object. More specifically, it is the measurement of the inertia or sluggishness that an object exhibits in response to any effort made to start it, stop it, deflect it, or change in any way its state of motion.

Mechanical weathering The breakdown of rocks on the Earth's surface by physical means.

Mesozoic The time of middle life, from 245 million years ago to about 66 million years ago.

Metallic bond A chemical bond in which metal atoms are held together by their attraction to a common pool of electrons.

Maxwell's counterpart to Faraday's law A magnetic field is induced in any region of space in which an electric field is changing with time. The magnitude of the induced magnetic field is proportional to the rate at which the electric field changes. The direction of the induced magnetic field is at right angles to the changing electric field.

Mechanical deformation Metamorphism caused by stress, such as increased pressure.

Mechanical equilibrium The state of an object or system of objects for which any impressed forces cancel to zero and no acceleration occurs.

Melting The change of phase from solid to liquid.

Meniscus The curving of a liquid at the interface of its container.

Mesosphere The third atmospheric layer above the Earth's surface, extending from the top of the stratosphere to 80 km.

Metal An element that is generally shiny, opaque, malleable, ductile, and a good conductor of electricity and heat.

Metalloid One of six elements—boron, silicon, germanium, arsenic, antimony, tellurium—that exhibit both metallic and nonmetallic proerties.

Metamorphic rocks Rocks formed from pre-existing rocks that have been transformed by high temperature, high pressure or both.

Metamorphism The changing of one kind of rock into another kind as a result of high temperature, high pressure, or both.

Meteor A meteoroid once it enters the Earth's atmosphere; a "shooting star."

Meteorite A meteoroid that survives passage through the Earth's atmosphere and hits the ground.

Meteoroid A very small asteroid.

Mineral A naturally formed, inorganic solid composed of an ordered array of atoms chemically bonded to form a particular crystalline structure.

Mohorovicic discontinuity (Moho) The crust-mantle boundary, marking the depth at which the speed of P-waves traveling toward the Earth's center increases.

Mohs scale of hardness A ranking of the relative hardness of minerals.

Molarity A common unit of concentration measured by the number of moles in one liter of solution.

Mole A very large number equal to 6.02×10^{23}. This number is a unit commonly used when describing a number of molecules.

Molecule A group of atoms held tightly together by covalent bonds.

Momentum The product of the mass of an object and its velocity.

Monomer The small molecular unit from which a polymer is formed.

Moon phases The cycles of change of the face of the moon as seen from the Earth: new (dark) to waxing to full to waning and back to new.

Natural frequency A frequency at which an elastic object naturally tends to vibrate, so that minimum energy is required to produce a forced vibration or to continue vibration at that frequency.

Neutral solution A solution in which the hydronium ion concentration is equal to the hydroxide ion concentration.

Neutralization A reaction in which an acid and base combine to form a salt plus water.

Neutrons The electrically neutral particles in an atomic nucleus.

Neutron star A small, dense star composed of tightly packed neutrons formed by the welding together of protons and electrons.

Newton The SI unit of force. One newton (symbol N) is the force that will give an object of mass 1 kg an acceleration of $1 m/s^2$.

Newton's second law The acceleration of an object is directly proportional to the net force acting on the object, is in the direction of the net force, and is inversely proportional to the mass of the object.

Noble gas A group 18 element.

Noble gas shell A spherical region of space about the atomic nucleus where electrons of a similar energy level may be found.

Nonconformity Overlying sedimentary rocks on an eroded surface of intrusive igneous or metamorphic rocks.

Nonmetal An element that is nonmalleable, nonductile, and a poor conductor of electricity and heat.

Nonpolar The state of a chemical bond or molecule having no dipole.

Nova The sudden brightening of a white dwarf caused by the explosion of accumulated hydrogen gas on its surface.

Nuclear fission The splitting of the nucleus of a heavy atom, such as uranium-235, into two main parts, accompanied by the release of much energy.

Nuclear fusion The combination of the nuclei of light atoms to form heavier nuclei, with the release of much energy.

Nucleon A nuclear proton or neutron; the collective name for either or both.

Ohm's law The current in a circuit varies in direct proportion to the voltage across the circuit and inversely with the circuit's resistance:

$$\text{Current} = \frac{\text{voltage}}{\text{resistance}}$$

A potential difference of 1 V across a resistance of 1 Ω produces a current of 1 A.

Opaque The term applied to materials through which light cannot pass.

Organic chemistry The study of carbon compounds.

Organic compound Any compound featuring the element carbon covalently bonded to a variety of nonmetal atoms including itself.

Original horizontality Layers of sediment are deposited evenly, with each new layer laid down almost horizontally over the older sediment.

Orographic lifting The lifting of air over a topographic barrier such as a mountain.

Oxidation The process whereby a reactant loses one or more electrons.

Oxidation-reduction reaction A reaction involving the transfer of one or more electrons from one reactant to another.

Paleomagnetism The study of natural magnetization in a rock in order to determine the intensity and direction of the Earth's magnetic field at the time of the rock's formation.

Paleozoic The time of ancient life, from 544 million years ago to 245 million years ago.

Pangaea A single, large landmass that existed in the geologic past and was composed of all the present-day continents.

Parallel circuit An electric circuit in which electrical devices are connected in such a way that the same voltage acts across each one and any single one completes the circuit independently of all the others.

Partial melting The incomplete melting of rocks, resulting in magmas of different compositions.

Period The time required for a vibration or a wave to make a complete cycle; equal to 1/frequency.

Period A horizontal row in the periodic table.

Periodic table A highly organized chart listing all the known elements arranged in horizonatal rows called periods and vertical columns called groups.

Permeability The ability of a material to transmit fluid.

pH A measure of the acidity of a solution. The pH is equal to the negative logarithm of the hydronium ion concentration. At 25°C, neutral solutions have a pH of 7, acidic solutions have a pH less than 7, and basic solutions have a pH greater than 7.

Phenols A class of organic molecules that contain a hydroxyl group bonded to a benzene ring.

Photoelectric effect The emission of electrons from a metal surface when light shines on it.

Photon A particle of light, or the basic packet of electromagnetic radiation.

Physical change A change in which a substance changes its physical properties without changing its chemical identity.

Physical property Any physical characteristic of a substance, such as color, density, and hardness.

Planetary nebula An expanding shell of gas ejected from a low-mass star during the latter stages of its evolution.

Pluton A large intrusive body formed below the Earth's surface.

Polar The state of a chemical bond or molecule having a dipole.

Polarization The alignment of the transverse electric vectors that make up electromagnetic radiation. Such waves of aligned vibrations are said to be *polarized.*

Polymer A long molecule made of many repeating parts.

Polymorph Minerals that have the same chemical composition but different crystal structures.

Porosity The volume of open space in a rock or sediment relative to the total volume of solids plus open space.

Potential energy The stored energy that a body possesses because of its position.

Power The time rate of work:

$$\text{Power} = \frac{\text{work}}{\text{time}}$$

Precambrian The time of hidden life, which began about 4.6 billion years ago when the Earth formed, lasted until about 544 million years ago (beginning of Paleozoic), and makes up 85% of the Earth's history.

Precipitate A solute that has come out of solution.

Pressure The ratio of force to the area over which that force is distributed:

$$\text{Pressure} = \frac{\text{force}}{\text{area}}$$

$$\text{Liquid pressure} = \text{weight density} \times \text{depth}$$

Primary waves A longitudinal body wave; travels through solids, liquids, and gases and is the fastest seismic wave.

Primeval fireball The burst of energy during the Big Bang.

Principle of flotation A floating object displaces a weight of fluid equal to its own weight.

Principle of the conservation of mass A principle stating that matter is neither created nor destroyed in a chemical reaction, as far as we are able to detect.

Probability cloud The pattern of electron positions plotted over a period of time that shows the likelihood of an electron's position at a given time.

Products The new material formed by a chemical reaction. It appears after the arrow in the chemical equation.

Projectile Any object that moves through the air or space under the influence of gravity.

Protons The positively charged particles in an atomic nucleus.

Protostar The aggregation of matter that goes into and precedes the formation of a star.

Pulsar A neutron star that spins rapidly and in doing so sends out short, precisely timed bursts of electromagnetic radiation.

Pure The state of a material that consists of only a single elemental or compound.

Quantum mechanics The field of wave mechanics where atomic and subatomic particles are treated as waves.

Quarks The elementary constituent particles of building blocks of nuclear matter.

Radiometric dating A method of calculating the age of geologic materials based on the nuclear decay of naturally occurring radioactive isotopes.

Radiation The transfer of energy by means of electromagnetic waves.

Reactants The starting material for a chemical reaction. It appears before the arrow in the chemical equation.

Real image An image formed by light rays that converge at the location of the image.

Recrystallization Metamorphism caused by high temperatures.

Red giant A cool giant star above main sequence on the H-R diagram.

Reduction The process whereby a reactant gains one or more electrons.

Reflection The return of light rays from a surface in such a way that the angle at which a given ray is returned is equal to the angle at which it strikes the surface.

Refraction The bending of a wave as it passes either through a nonuniform medium or from one medium to another, caused by differences in wave speed.

Relative humidity The amount of water vapor in the air at a given temperature expressed as a percentage of the maximum amount of water vapor the air can hold at that temperature.

Resonance The response of a body when a forcing frequency matches its natural frequency.

Resultant The single vector that results when two or more vectors are combined.

Rift (rift valley) A long, narrow trough that forms as a result of divergence of two plates.

Rock cycle A sequence of events involving the formation, destruction, alteration, and reformation of rocks as a result of the generation and movement of magma, the weathering, erosion, transportation, and deposition of sediment, and the metamorphism of preexisting rocks.

Salinity The mass of salts dissolved in 1000 grams of seawater.

Salt An ionic compound formed from the reaction of an acid and a base.

Saturated A solution containing as much solute as will dissolve.

Saturated hydrocarbon A hydrocarbon containing no multiple covalent bonds; the carbon atoms are "saturated" with hydrogen atoms.

Sea floor spreading The divergence of two oceanic plates to form a rift in the sea floor.

Secondary wave A transverse body wave; cannot travel through liquids and so does not travel through the Earth's outer core.

Second law of thermodynamics Heat never spontaneously flows from a cold object to a hot one. Also, no machine can be completely efficient in converting energy to work; some input energy is dissipated as waste heat at lower temperature. And finally, all systems tend to become more and more disordered as time goes by.

Sedimentary rocks Rocks formed from the accumulation of weathered material (sediments) carried by water, wind, or ice.

Series circuit An electric circuit in which electrical devices are connected in such a way that the same electric current exists in all of them.

Shell A region of space about the atomic nucleus within which an electron may reside.

Shock wave The cone-shaped wave made by an object moving at supersonic speed through a fluid.

Solar constant The 1400 joules per square meter received from the sun each second at the top of the Earth's atmosphere. Expressed in terms of power, it is 1.4 kilowatts per square meter.

Solar eclipse The phenomenon whereby the shadow of the moon falls on the Earth, producing a region of darkness in the daytime.

Solid Phase of matter characterized by definite volume and shape.

Solubility The ability of a solute to dissolve, which depends not only on molecular interactions between the solute and the solvent, but upon the interactions among both solute molecules and among solvent molecules.

Soluble Capable of dissolving in a solvent.

Solute Any component in a solution that is not the solvent.

Solution A homogeneous mixture in which all components are of the same phase.

Solvent The component in a solution present in the largest amount.

Sonic boom The loud sound resulting from the incidence of a shock wave.

Sound wave A longitudinal vibratory disturbance that travels in a medium and can be heard by the human ear when in the approximate frequency range 20 - 20,000 hertz.

Specific heat capacity The quantity of heat per unit of mass required to raise the temperature of a substance by 1 Celsius degree.

Speed Distance traveled divided by time.

Standing wave A stationary wave pattern formed in a medium when two sets of identical waves pass through the medium in opposite directions.

Stoichiometry An aspect of chemistry involving the calculation of quantities of substances involved in chemical reactions.

Stratosphere The second atmospheric layer above the Earth's surface, extending from the top of the troposphere up to 50 kilometers.

Structural isomers Molecules that have the same molecular formula but different chemical structures.

Sublimation The change of phase from solid to gaseous, skipping the liquid phase.

Subtractive primary colors The three colors of absorbing pigments—magenta, yellow, and cyan—that when mixed in certain proportions reflect any other color in the visible-light part of the electromagnetic spectrum.

Sunspots Temporary relatively cool, dark regions on the Sun's surface.

Supercluster A grouping of an enormous number of galaxies.

Supernova A stellar explosion, caused either by transfer of matter to a white dwarf or by gravitational collapse of a massive star, where enormous quantities of matter are emitted.

Superposition In an undeformed sequence of sedimentary rocks, each bed or layer is older than the one above and younger than the one below.

Supersaturated A solution that contains more solute than it normally contains.

Surface tension The energy required to break through the surface of a liquid.

Surface wave A seismic wave that travels along the Earth's surface.

Suspension A homogeneous mixture in which different components are of different phases.

Syncline A fold in strata that has relatively young rocks at its core, with rock age increasing as you move horizontally away from the core fold.

Synthetic polymer A polymer not found in nature.

Temperature A measure of the hotness of an object, related to the average kinetic energy per molecule in the object, measured in degrees Celsius, degrees Fahrenheit, or kelvins.

Terminal speed The constant speed of a falling object where acceleration terminates because air resistance balances the weight. When direction is specified, we speak of terminal velocity.

Theory of plate tectonics The theory that the Earth's lithosphere is broken up into pieces (plates) that move over the asthenosphere; boundaries between plates are where most earthquakes and volcanoes occur and where lithosphere is created and recycled.

Thermal energy (*internal energy*) The total energy (kinetic plus potential) of the particles that make up a substance.

Thermodynamics The study of heat and its transformation to mechanical energy.

Thermonuclear fusion Nuclear fusion produced by high temperature.

Thermosphere The fourth atmospheric layer above the Earth's surface, extending from the top of the mesosphere to 500 km.

Transform fault A plate boundary formed by two plates that are sliding horizontally past each other.

Transition metals The elements of groups 3 through 12.

Transmutation The conversion of an atomic nucleus of one element into an atomic nucleus of another element through a loss or gain in the number of protons.

Transparent The term applied to materials through which light can pass in straight lines.

Transverse wave A wave in which the medium vibrates in a direction perpendicular (transverse) to the direction in which the wave travels. Light is an example of a transverse wave.

Troposphere The atmospheric layer closest to the Earth's surface, 16 kilometers high over the equator and 8 kilometers high over the poles, and containing 90 percent of the atmosphere's mass and essentially all its water vapor and clouds.

Turbulent flow Water flowing erratically in a jumbled manner, stirring up everything it touches.

Uncertainty principle It is not possible to measure exactly both the position and the momentum of a particle at the same time, and it is not possible to measure exactly both the energy of a particle and the time at which it has that energy.

Unconformity A break or gap in the geologic record, caused by an interruption in the sequence of deposition or by erosion of pre-existing rock.

Unsaturated A solution that will dissolve additional solute if added.

Unsaturated hydrocarbon A hydrocarbon containing at least one multiple covalent bond.

Valence electron Any electron in the outermost shell of an atom.

Vector An arrow drawn to scale that represents the magnitude and direction of a quantity having both magnitude and direction.

Velocity The speed of an object and specification of its direction of motion.

Virtual image An image formed by light rays that do not converge at the location of the image.

Volcano A central vent through which lava, gases, and ash erupt and flow.

Volt The unit of electric potential, a potential of 1 volt equals 1 joule of energy per 1 coulomb of charge; $1\ V = 1\ J / C$.

Water table The upper boundary of the zone of saturation, the area where every pore space is filled with water.

Wave A disturbance or vibration propagated from point to point in a medium or in space.

Wave speed The speed with which waves pass a particular point:

$$\text{Wave speed} = \text{frequency} \times \text{wavelength}$$

Wavelength The distance between successive crests, troughs, or identical parts of a wave.

Weight The gravitational force exerted on an object by the nearest most-massive body (locally, by the Earth).

White dwarf A dying star that has collapsed to the size of the Earth and is slowly cooling off; located at lower left of the H-R diagram.

Work The product of the force and the distance through which the force moves:

$$W = Fd$$

Work-energy theorem The work done on an object is equal to the energy gained by the object:

$$\text{Work} = \Delta E$$

Photo Credits

1: Charles A. Speigel
2 : Roger Ressmeyer/Starlight-Corbis
5, clockwise from top left: Lennard Lessin/Peter Arnold Inc; Ray Nelson/Phototake; David Parker/Photo Researchers; Ray Nelson/Phototake; Paul G. Hewitt.
15: Paul G. Hewitt
16: David Hewitt
17: The Granger Collection, NY
19: The Granger Collection, NY
20, top: Rick Lucas
20, center: Addison Wesley Longman
22: The Granger Collection, NY
26: Paul G. Hewitt
34: Larry Sergeant/Comstock
47: Fundamental Photographs
53: Bob Abrams
63: John Suchocki
64: John Suchocki
67: Paul G. Hewitt
69: R. Llewellyn/Superstock
70: Craig Aurness/Westlight
71: The Harold E. Edgerton 1992 Trust, Courtesy Palm Press Inc.
72: Glen Allison/Tony Stone Images
73: Paul G. Hewitt
74: Paul G. Hewitt
75: Paul G. Hewitt
78: Paul G. Hewitt
83: John Hancock
85: Courtesy NASA
87: Paul G. Hewitt
88: Paul G. Hewitt
100: John Hancock
101: Paul G. Hewitt
102: Courtesy NASA/Hubble Space Telescope
117: Courtesy NASA
126: Paul G. Hewitt
133: Olan Mills
134: G. Brad Lewis
141: Meidor Hu
144, top: Wide World Photos
144, center: Meidor Hu
144, bottom: Meidor Hu
145: Nuridsany et Perennou/Photo Researchers
151: Tracy Suchocki
152: Milt and Joan Mann/Cameramann International
153: Paul G. Hewitt
154, top: Nancy Rogers
154, bottom: Paul G. Hewitt
158: Robert D. Carey
159, center: Leonard Lessin/Peter Arnold, Inc.
159, bottom left: Paul G. Hewitt
159, bottom right: Pat Crowe/Animals, Animals
160: Charles Cook
168: Paul G. Hewitt
169: Bob Abrams
170: Steven Hunt/The Image Bank
175: Addison Wesley Longman
176: Zig Leszcbynkski/Animals, Animals
177: John Lightfoot
182: Addison Wesley Longman
183: Addison Wesley Longman
184: Addison Wesley Longman
191: Pekka Parvianinen, Polar Image, Finland
193: Richard Megna/Fundamental Photographs
194: Richard Megna/Fundamental Photographs
195: Meidor Hu
196: Richard Megna/Fundamental Photographs
197, left: John Suchokci
197, right: Courtesy Magplane Technology
199: Addison Wesley Longman
209: Paul G. Hewitt
210: Herman Schlenker/Photo Researchers
212: Gabe Palmer/Mug Shots/The Stock Market
215: John Suchocki
217: Terence McCarthy/San Francisco Symphony
218, top: Leslie A. Hewitt
218, bottom: Eric Meala/The Image Bank
221, top: AP/Worldwide
221, bottom: Education Development Center
222: Paul G. Hewitt
227: U.S. Navy
228: Palm Press/Courtesy of Estate of Dr. Harold Edgerton
233: Paul G. Hewitt
234: Bob Abrams
239: Paul G. Hewitt
241: Meidor Hu
244: Dave Vasquez
245: John Suchocki
246, left: Paul G. Hewitt
246, right: John Suchocki
247: Meidor Hu
249, top: Camarique/H. Armstrong Roberts
249, bottom: Don King/The Image Bank
251: Suzanne Lyons
253: Nina Leen/Life, Time Warner
255: Paul G. Hewitt
256, center: Roger Ressmeyer/Starlight
256, bottom: Courtesy Institute of Paper
259: Ted Mahieu
260: Robert Greenler
262: Paul G. Hewitt
265: Robert G. Fitzgerald
267: Paul G. Hewitt
273: Paul G. Hewitt
274: Education Development Center
275: Ken Kay/Fundamental Photographs
279: David Hewitt
282: Diane Schiumo/Fundamental Photographs
283: Paul G. Hewitt
285: Courtesy Albert Rose
289: John Suchocki
290: Alamaden Research Center, courtesy IBM
292, left: Rachel Epstein/SKA
292, center: Rachel Epstein/SKA
292, right: Tony Freeman/PhotoEdit
293: Courtesy IBM Research
294: John Suchocki
298, clockwise from top left: PhotoDisc; PhotoDisc, Peter Arnold, Inc.; PhotoDisc; Fundamental Photographs; PhotoDisc; Photo Researchers; Fundamental Photographs
299: Mark Martin/Photo Researchers
301: Rachel Epstein/SKA
306: Paul G. Hewitt
307: John Suchocki
308, left: Volker Steger/Peter Arnold
308, center: IBM/Peter Arnold
308, right: Alamaden Research Center, IBM Research Divisions
309, left: Rachel Epstein/SKA
309, right: Tom Pantages
311, top right: Phototake
311, top center: John Suchocki
311, bottom: John Suchocki
312, left: Tom Pantages
312, center: Alan J. Jircitano
313: Tom Pantages
316: Rachel Epstein/SKA
322: Breck P. Kent/Earth Scenes
324: International Atomic Energy Agency
325: Larry Mulvehill/Photo Researchers
326: Richard Megna/Fundamental Photographs
332: Richard Frishman/Stone
338: Comstock
342: Comstock
347: Plasma Physics Laboratory, Princeton University
348: Lawrence Livermore National Laboratory
351: John Suchocki
352: Mauritius/Phototake
353, top: NASA
353, bottom: PhotoDisc
355: John Beatty/Stone
356, top: Fundamental Photographs
356, center: Definitive Stock
357, top: PhotoDisc
357, bottom: Tom Pantages
358: John Suchocki
359: Stephen R. Swinburne/Stock, Boston
360: Tom Pantages
361: John Suchocki
363: Tom Pantages
372: John Suchocki
373, top: Stone
373, bottom: John Suchocki
375, top: Dave Bartruff/Stock, Boston
375, bottom: Tom Pantages
376, top: Georg Gerster/Photo Researchers
376, bottom: Tom Pantages
378 top, clockwise from top left: Kevin Adams/Nature Photography, Stone, Photodisc, Science Source/Photo Research, PhotoDisc, PhotoDisc
378, bottom: Brian Yarvin/Photo Researchers
380, clockwise from top left: Fred Ward/Black Star, Image Works, Rachel Epstein/SKA
384: Tom Pantages
388: Charles M. Falco/Photo Researchers
390: Science Photo Library/Photo Researchers
395, top: John Suchocki
395, center: M. Claye/Photo Researchers
395, bottom: E. R. Degginger/Photo Researchers
396: Dee Breger/Photo Researchers
399: Rachel Epstein/SKA
400: Vaughn Fleming/Photo Researchers
405, top: David Taylor/Photo Researchers
405, bottom: Natalie Fobes/Stone
410: Dr. & T.S. Scrichte/Photo Resource Hawaii
413: Rachel Epstein/SKA
414: Tom Pantages
416, top: Leonard Lessin/Peter Arnold, Inc.
416, bottom: John Suchocki
422, top: Rachel Epstein/SKA

422, center: Tom Pantages
422, bottom: Tom Pantages
425: David Harry
426, clockwise from top left: M.P. Gadomski/Photo Researchers; M.P. Gadomski/Photo Researchers; Rachel Epstein/SKA; Rachel Epstein/SKA
427, left to right: Rachel Epstein/SKA; S. Grant/PhotoEdit; Rachel Epstein/SKA; Rachel Epstein/SKA
430: Rachel Epstein/SKA
433: Tom Pantages
436: Tom Pantages
438, left: M. Bleier/Peter Arnold, Inc.
438, right: Will McIntyre/Photo Researchers
439: Tom Pantages
444: John-Peter Lahall/Photo Researchers
445: Tom Pantages
447: Tom Pantages
451: Lennard Lessin/Peter Arnold, Inc.
455: Xcellis Fuel Cell Engine, Inc.
456: C. Liu et al, "Single-Walled Carbon Nanotubes at Room Temperature," Science, November 5, 1999.
456: John Suchocki
458: Tom Pantages
459: Frank Siteman/Stock, Boston
460: Rachel Epstein/SKA
465: John Suchocki
470: Rachel Epstein/SKA
472: Rachel Epstein/SKA
473: Tom Pantages
488: Dr. J. Burgess/Photo Researchers
492: Kurt Hostettmann
493, left: David Nunuk/Photo Researchers
494, right: Tom & Pat Leeson/Photo Researchers
495: Sim/Visuals Unlimited
501: Alan D. Carey/Photo Researchers
503, clockwise from top left: Inga Spence/Visuals Unlimited; Adam Jones/Photo Researchers; Tim Hazael/Stone; Matt Meadows/Peter Arnold, Inc.; Cristina Pedrazzini/Photo Researchers
504: Labat/Jerrican/Photo Researchers
513: John Suchocki
518: John Suchocki
521: Rachel Epstein/SKA
523: Paolo Koch/Photo Researchers
525, top: John Suchocki
525, center: Tom Suchocki
525, bottom: Bettman/Corbis
526, top: Greenwood Communications AB
526, bottom: Leonard Lessin/Peter Arnold, Inc.
528: AtoFina Chemicals, Inc.
529, top: Post Street Archives
529, bottom: "Re-enactment of the Invention of DuPont Teflon fluoropolymer in April, 1938." Roy Plunkett is shown at right. Photo courtesy of DuPont.
531: Cambridge Display Technologies
535: John Suchocki
536: Susan Middleton, 1986/California Academy of Sciences
537, center: Lee Boltin
537, bottom left: E.R. Degginger
537, bottom right: Breck P. Kent/Earth Scenes
540, top: Paul Silverman/Fundamental Photographs
540, bottom: Mark A. Schneider/Visuals Unlimited
541: E.R. Degginger
542: Leslie A. Hewitt
544: Runk/Schoenberger/Grant Heilman
548, top to bottom: Visuals Unlimited; Paul Silverman/Fundamental Photographs; E.R. Degginer/Earth Scenes; E.R. Degginger
549: E.R. Degginger
552: Leslie A. Hewitt
553, clockwise from top left: Beth Davidson/Visuals Unlimited; Susan Middleton, 1986/California Academy of Science; Ken Lucas/Visuals Unlimited; Susan Middleton, 1986/California Academy of Science; A.J. Copley/Visuals Unlimited; Susan Middleton, 1986/California Academy of Science
555, top: Paul Dix/Impact Visuals/PNI
555, center: USGS
555, bottom: W.H. Hodge/Peter Arnold, Inc.
556: USGS
557: Breck P. Kent/Earth Scenes
558: Jim Wark/Peter Arnold, Inc.
559: Dr. Jeremy Burgess/Science Photo Library/Photo Researchers
560: Grant Heilman/Grant Heilman
561, top left: Runk/Schoenberger/Grant Heilman
561, top right: Paul Silverman/Fundamental Photographs
562, bottom: Barry L. Runk/Grant Heilman
562, clockwise from top left: A.J. Cunningham/Visuals Unlimited; Paul Silverman/Fundamental Photographs; Cabisco/Visuals Unlimited; Alex Kertish/Visuals Unlimited
565: NASA Photo Research/Grant Heilman
566, top left: Biologica/Phototake
566, top center: Gerald & Buff Corsi/Visuals Unlimited
566, top right: A.J. Copley/Visuals Unlimited
566, center: Jeffrey Scovil
566, bottom: Joyce Photo/Photo Researchers
571: NASA
579: John S. Shelton
584: John Guzman/Life Magazine/Time Warner
588: Marie Tharp
600: Visuals Unlimited
608: Leslie A. Hewitt
609, left: USGS
609, right: E.R. Degginger
620: Grant Heilman/Grant Heilman
621: Steve Lissau/Rainbow
622: Joseph Burke/Rainbow
623: David Hiser/Photographers/PNI
624, left: Paul G. Hewitt
624, right: Visuals Unlimited
625: John Lemker/Earth Scenes
627: E.R. Degginger
629: Courtesy NASA/E.R. Degginger
630, top: Adrienne T. Gibson/Earth Scenes
630, bottom left: Breck P. Kent/Earth Scenes
630, bottom right: W.H. Hodge/Peter Arnold, Inc.
631, left: Leslie A. Hewitt
631, center: E.R. Degginger
631, right: E.R. Degginger
635: NASA Photo Research/Grant Heilman
638: J. Silver/SuperStock
640, top: C. BruceFoster/AllStock/PNI
640, bottom: Bob Abrams
645: Roland Seitre/Peter Arnold, Inc.
646, center: Alex Kertisch/Visuals Unlimited
646, bottom: Tom McHugh/Photo Researchers
647: Courtesy Department of Library Services/American Museum of Natural History
648: Philip James/Corbis
659: NASA/Photo Researchers
665: E.R. Degginger
680: Michael Townsend/Tony Stone Worldwide
682: Mike J. Howell/Stock, Boston
687, clockwise from top left: Mark A. Schneider/Visuals Unlimited; McCutcheon/Visuals Unlimited; E. Webber/Visuals Unlimited; Mark A. Schneider/Visuals Unlimited
690: H. Armstrong Roberts
693, left: Thomas Nes/The Stock Market
693, right: William Bickel/University of Arizona
694: E.R. Degginger
695: Tony Arruza/Bruce Coleman/PNI
699: Reprinted with permission from Physics Today, 54(8). Copyright 2001, American Institute of Physics.
700: NASA
701: NASA
702: Lick Observatory
704, top: Roger Ressmeyer/Starlight
704, bottom: Dennis di Cicco
707: Mark Martin/NASA/Photo Researchers
711, top: NASA
711, bottom: Lowell Observatory
712: NASA/JPL/Phototake
713: JPL/NASA
714: NASA
715: JPL/NASA
716, top: Yerkes Observatory
716, bottom: Dennis Milon/SPL/Photo Researchers
717, top: John Sanford/SPL/Photo Researchers
717, center: Reverand Ronald Royer/SPL/Photo Researchers
717, bottom: Reverand Ronald Royer/SPL/Photo Researchers
718: A. Behrend/Phototake
719: John Suchocki
721: Paul G. Hewitt
722: Paolo Ragazzinni/Corbis
724: Roger Ressmeyer/Starlight
727: Hale Observatory/Photo Researchers
729: Ronald Royer/SPL/Photo Researchers
730: Lick Observatory
731: Julian Baum/New Scientist/JPL/Photo Researchers
732: Ronald Royer/SPL/Photo Researchers
733, top: Dr. Steve Gull/SPL/Photo Researchers
733, bottom left: National Optical Astronomy Observatories
733, bottom right: Royal Observatory, Edinburgh/SPL/Photo Researchers
734, top left: NRAO/SPL/Photo Researchers
734, top right: NASA/SS/Photo Researchers
734, bottom: Hale Observatory/Photo Researchers
758: David Hewitt
759: Paul G. Hewitt

Index

A

Aberrations, 268–69
Ablation, 610
Absolute zero, 136–38
Absorption, 157
Acceleration, 35–37
  air drag and, 46–48
  free-fall and equal, 44–46
  mass resists, 40–41
  Newton's second law of, 41–42
Accumulation, 609–10
Acid rain, 436–41
Acids, 426–38
  behavior with ions, 426–31
  pH scale, 435–36
  strength of, 431–34
Acoustics. *See also* terms under Sound
  dolphins and, 219
  Doppler effect, 225–26
  shock waves and sonic boom, 228–29
Adams, J. C., 714
Addition polymers, 514–19
Adiabatic processes, 683–84
Air. *See also* Atmosphere, Wind motion
  atoms in, 293
  convection currents, 154, 669
  lifting of, 690–91
  masses, 682, 689–92
  motion of, 669–71, 682
  moving dust and sand, 621
Air drag, 46–48, 122–23
Alcohol, 504–6
Alkaline
  bases, 426–37
  elements, 301
Alpha and beta particles, 323–24
Alternating current, 181, 200–201
Aluminum processing, 456–58
Alvarez, Luis and Walter, 650
American Chemistry Council, 354
Ampere, 174–75
Amphetamines, 499–500
Amphoteric behavior, 434–35
Amplitude, 211
Analgesics, 507–9
Anesthetics, 506–7
Antibiotics, 495
Applied research, 353–54
Aquifiers and springs, 604
Aristotle, 17
  Galileo and, 6–7, 17–19
  motion classification, 17
Armstrong, Neil, 701
Aromatic compounds, 472
Art and science, 8–9
Asbestos, 543
Asteroids and meteorites, 716–17
  dinosaur extinction, 650
Astrology, 8, 728
Atmosphere. *See also* Air, Wind motion
  adiabatic processes, 683–84
  circulation patterns, 672–74
  clouds, 686–89, 690–92
  composition and layers, 663–65
  coriolis effect, 670–71
  evolution of, 662
  lifting of air, 690–91
  stability, 684–85
  Sun, 707
  upper atmosphere circulation, 673
  water in, 681–82
  weather conditions, 692–95
Atmospheric refraction, 259–60
Atomic bomb, 347
Atomic isotopes, 296–97, 331–32
Atomic mass, 296–97
Atomic spectrum, 311–16
Atomic symbols, 292, 361
Atoms, 354–56
  atomic mass and number, 296–97
  atomic spectrum, 311–16
  bonding, 389–90
  electrical repulsion and, 294
  electronegativity, 401–2
  electrons and ions, 390–93
  elements and, 361
  identified by light, 310–13
  isotopes, 296–97, 331–32
  models of, 308–10, 315, 317–18, 389–90
  molecules and, 354–56
  nucleus and nucleons, 294–96
  periodic table, 291–94 (*See also* Periodic table)
  quantum number, 315–16
  scanning tunneling microscope (STM), 308
  valence electrons and shell, 317–18, 389–93

B

Baekeland, Leo, 525
Bakelite, 525
Barriers, 226–27
Basaltic lava, 554
Bases
  behavior with ions, 426–31
  strength of, 431–34
Basic research, 353–54
Batteries
  fuel cells, 454–56
  oxidation-reduction reactions in, 450–53
  types of, 450–52
  voltaic cells, 198
Beats, 223
Benzodiazepines, 504–6
Bevan, Edward, 526
Big Bang theory, 112, 735
Big dipper and the North star, 724
Black dwarf (star), 727
Black holes, 730–32
Blamer, Johann, 312–13
Bohr, Niels, 313–16
Bow waves, 226–27
Brandenberger, Jacques, 526
Breeder nuclear reactors, 342–43
Brown, Robert, 293

C

Caffeine, 501–2
Cameras, 262
Carbonate group, 546
Carbon compounds
  amide, 481
  carboxylic acid, 481–82
  conformations, 468
  ester, 482
  ketone, 480
  organic chemistry, 466
Carbon-14 dating, 333–34
Carbonyl group, 480–82
Cellophane, 526
Celluloid, 524–25
Celsius, Anders, 135
Celsius scale, 135–36
Cenozoic era, 651–55
  plate tectonics, 653–55
Chain reaction (nuclear), 340–41
Chalcogen elements, 301
Chemical bonds, 358, 389–90
  covalent, 397–400
Chemical equations, 364–67
Chemical formulas, 362–63
Chemical properties
  drug structure and model, 491–93
  of matter, 356–61
Chemical reactions, 358–59
  acid-base reactions, 430–31
  electrolysis, 456–58
  equations and, 364–67
  half reactions, 445–46
  neutralization reaction, 430–31
  oxidation-reduction reactions, 445–47, 447–50
Chemistry, 353–54
  law of mass conservation, 364–65
Chemotherapy, 494–95
Cleavage (minerals), 540–41
Climate
  Cenozoic era, 652–53
Clouds, 686–89
  color of, 249
  high clouds, 686
  low clouds, 688
  middle clouds, 686
  vertical development in, 688–89
Coal industry, 326–27
Coastal erosional landforms, 634–35
Cocaine, 500–501
Collisions, 77–78
Collodian, 524–25
Colors
  blue sky, 246–47
  clouds, 249
  complementary, 244–45
  mixing pigments, 245–46
  ocean water, 249
  primary, 243
  red sunsets, 248
  reflection and, 240–41
  reflection and interference, 278–80
  selective transmission, 241–42
  subtractive primaries, 246
Combustion, 460–61
Comets, 717–18
Complementary colors, 244–45
Compounds
  aromatic, 472
  carbon, 466
  covalent, 398
  formation from elements, 362–63
  hydrocarbons, 467–71
  ionic bonds and, 394–96
Conceptual models, 309
Condensation, 160–61
Condensation polymers, 519–23
  examples of, 517
  nylon, 519–20
Conduction, electrical, 174–78
Conduction, thermal, 152–53
Conservation of energy, 91–92, 141
Conservation of momentum, 74–79
Constellations, 722–24

Constructive interference, 221, 276–78
Continental drift, 585–89
Convection, 153–54
Converging lenses, 263–65
Cooling processes, 159
Copernicus, Nicolas, 3. *See also* Sun-centered universe
Coriolis effect, 670–71
Corrosion
 acids and, 433–34
 oxygen creates, 459–60
Coulomb, Charles, 172–73
Coulomb's law of electric force, 172–73
Covalent bonds
 compounds and, 397–400
 electronegativity in polar, 401–2
 nonpolar, 402–3
 polar, 401–3
Craters, 556–57
Cross, Charles, 526
Crystals
 form, 537–39
 polymorphs, 538–39
 recrystallization, 563

D

Density (minerals), 542–43
Deposition of sediments, 625–26
Depressants, 504–6
Desert surface water, 629–30
Destructive interference, 221, 276–78
Detergents, 421
Diffraction, 274–75
Diffuse reflection, 256–57
Dinosaurs, 650–51
Direct current, 181
Diverging lenses, 263–65
Doppler effect, 225–26
Drainage networks, 607–8
Drugs
 chemotherapy, 494
 classification and definition, 489–90
 derived from plants, 492–93
 lock and key model, 491–93
 synergistic effects, 490
Drug types
 analgesics, 507–9
 anesthetics, 506–7
 antibiotics, 495
 depressants, 504–6
 psychoactive drugs, 499–506
 synthetic drugs, 494–95
DuPont Corp., 526–27
Dynamic equilibrium, 29

E

Earth
 climate, 652
 continental drift, 585–89
 core, 575
 crust, 545, 576–77
 geologic time, 639
 humans change geology of, 654
 interior layers, 573–77
 mantle, 575–76
 mass measurement, 108–9
 moon, 110, 701–6
 orbital escape speed, 127–29
 Pangea supercontinent, 586–87, 649
 rock formation and plate tectonics, 547–48, 553–54, 567, 589–97, 596–97, 641–42
 satellites orbiting, 123–26
 seasons, 666–67
 terrestrial radiation, 156, 667–68
Earth-centered universe, 3, 29–30
Earth environment
 acid rain, 436–41
 extreme conditions for life, 165
 responsible care of, 354
Earth history
 Cenozoic era, 651–55
 climate, 652
 dinosaurs, 650–51
 life begins, 646–47
 mammals, 651–55
 Mesozoic era, 650–51
 Paleozoic era, 646–49
 Precambrian era, 644–45
 reptiles, 648–49
Earthquakes
 faults and folds, 577–81
 Richter scale, 580–81
 seismic waves, 572–74
Eastman, George, 525
Eastman Kodak Corp., 525
Einstein, Albert, 284, 343–45
Electrical repulsion, 294
Electrical resistance, 176–77
Electric charge
 polarization, 174, 401–3
 rules of like and unlike charges, 171–72
Electric circuits, 182–86
Electric current
 ampere and electric flow, 174–75
 chemical change and, 447–49
 direct and alternating current, 181
 galvanometer, 196–97
 magnetic fields and magnetism, 193–95, 198–200
 Ohm's law, 177–78
 resistance and ohms, 176–77
 superconductors, 178
 voltage, 175–76
Electric light filaments, 181
Electric motors, 197–98
Electric power
 electromagnetic induction and, 203–4
 energy and work, 181–82
 nuclear energy conversion to, 341–43
 technology and, 186
 watts, 85–86
Electric shock, 178–80
Electrochemistry, 447–49
Electrode, 451
Electrolysis, 456–58
Electromagnetic induction, 200–202
 electric power production and, 203–4
Electromagnetic spectrum, 235–36
Electromagnets, 194–95
Electronegativity, 401–2
Electronics technology, 172
Electrons
 becoming ions, 390–93
 oxidation and reduction, 445–47, 447–49
 valence, 317–18, 389–90
Electrostatic charge, 172
Elemental formula, 361. *See also* Atomic symbols
Elements, 291
 alkaline, 301
 atoms and, 361
 carbon-14 dating, 333–35
 classification as pure/impure, 377–79
 compound formation, 362–63
 metalloids, 299
 metals and nonmetals, 298
 periodic table, 291–94, 297–99 (*See also* Periodic table)
Ellipse, 126
Elliptical galaxies, 733
Elliptical orbits, 126–27
Energy
 conservation of, 91–92
 heat and, 139–40
 mass and, 343–45
 for phase change, 164–65
 thermal, 135
Energy efficiency, 95–96
Energy sources, 96–97
Energy transport, 212–13
Entomology, 165
Equilibrium, 26–27
 moving objects, 28–29
Equilibrium rule, 26–27
Erosion
 forming sediments, 559–60
 glaciers, 630–32
 of sedimentary rocks by surface water, 625–26
Erosional landforms, 630–32
Escape speed/velocity, 127–29
Ether group, 477–78
Evaporation
 boiling liquid, 162–63
 phase change as, 158–59
Exosphere, 665
Eyeglasses, 269
 rainbow diffraction glasses, 313

F

Fahrenheit, Gabriel D., 135
Fahrenheit scale, 135–36
Faraday, Michael, 200
Faraday's law of electromagnetic induction, 201–2
First law of inertia, 23–24
First law of thermodynamics, 140
Fissure eruptions, 554
Fluid flow, 153–54
Focal point and length, 264–65
Foliation in rocks, 565–66
Forces
 increasing/decreasing momentum, 71–73
 interaction and, 54–55
 lifting, 63
 net force, 24, 37
 sound and vibrations, 219
 support force, 27–28
Fossil fuels, 563
Fractional distillation, 469–70
Fracture (minerals), 540–41
Free-fall and acceleration, 44–46
Freezing, 163–64
Frequency, 211–12
 natural, 219
Friction, 18, 43–44
Fuel cells, 454–56

G

Galactic clusters and superclusters, 734
Galaxies, 732–34
Galilei, Galileo
 Aristotle's hypothesis, 6–7, 17
 inertia, 17–18
 speed and velocity, 19–20
Galvanometer, 198–200
Gamma aminobutyric acid (GABA), 498, 504–5
Gamma radiation, 323–24
Generators, 202–203
Geology. *See also* Rocks, Minerals
 faults and folds, 577–81
 Grand Canyon rock layering, 639–40
 humans change Earth's, 654
 hydrologic cycle, 601–2
 karst topography, 624
 metamorphic rocks, 563–66
 Pangea supercontinent, 586–87, 649
 radiometric dating, 642–44
 rock cycle, 566–67
 rock formation and plate tectonics, 547–48, 553–54, 567, 589–95, 596–97, 641–42

seafloor spreading, 588–89
sedimentary rock layering, 639–42
time scales, 639, 643–44
weathering, 559–60
Glaciers, 608–11
ablation, 610
erosion, 630–32
ice ages and glaciation, 652–53
mass balance and accumulation, 609–10
sedimentation, 632–33
types of, 631–32
Global warming, 668–69
Goodyear, Charles, 523
Grand Canyon, 639–40
Gravitational potential energy, 87
Gravity
defines shape of universe, 112
inverse-square law, 105–6
Isaac Newton and, 103
ocean tide effects of, 109–12
and orbital satellites, 123–25
specific, 542–43
universal gravitational constant, 107–8
Greenhouse effect, 668–69
Grey, Burl, 25–26
Groundwater, 602–5
formation of caverns and caves, 623–24
land subsidence and pumping, 622
sinkholes, 624

H

Halogen elements, 301
Hardness scale, 539–40
Heat and heating
condensation, 160–62
as energy, 139–40
specific heat capacity, 141–43
thermal energy and, 138–39
Heat transfer
conduction, 151–53
convection, 153–54
phase change, 158–61 (*See also* Phase change)
radiation, 155–58
Henry, Joseph, 200
Hertz, Heinrich, 212
Hertz (Hz), 212
Hess, H.H., 588
Heterogeneous mixtures, 377–78
Hewitt, Paul, 25
High clouds, 686
Homogeneous mixtures, 377–78
solutions, 379–84
suspensions, 378–79
Horse-cart problem, 60–62
Human body
electric shock effects on, 178–80
energy and, 97–98
Human eye, 265
Humidity, 681–82
Hyatt, John, 524–25
Hydrocarbons, 467–71
fractional distillation, 469–70
multiple bonds in unsaturated, 471–73
Hydrogen bomb, 347
Hydrologic cycle, 601–2
Hydronium ion, 428
Hydroxide ion, 428
Hydroxyl group, 475–77
Hypothesis, 4, 6–7

I

Igneous rocks, 547–48
formation from magma, 553–54
intrusive plutons, 557–58
Image formation, 266–69
Impulse, 70–73
Inclined planes and acceleration, 44–46
Inertia, 17–18
mass and, 38–39
momentum, impulse and, 70–71
Newton's first law, 23–24
Insulation and insulators, 151–53
Interaction force, 54–55
Interference
beats and, 223
constructive and destructive, 221, 276–78
reflected colors, 278–80
of waves, 221–24
Inverse-square law
Coulomb's law of electrical force, 172–73
Newton's law of gravity, 105–6
Ionic bonds and electron transfer, 393–96
Ionosphere, 665
Ions
behavior with acids and bases, 426–29
electrons becoming, 390–93
hydronium and hydroxide, 427
Irregular galaxies, 733
Isotopes
atomic, 296–97
carbon-14 dating, 333–34
radioactive, 331–32

J

Joule, 84–85
Jupiter (planet), 712–13

K

Karst topography, 624
Kelvin scale, 137–38
Kilogram, 40
Kinetic energy, 88–89
work-energy theorem, 89–90

L

Lasers, 348
Lava, 554
Lenses
aberrations in, 268–69
converging and diverging, 263–65
image formation, 266–69
Leverrier, Urbain, 714–15
Levers, 92–93
Life sciences
definition of, 11
energy and, 97–98
Lifting force, 63
Light
diffraction of, 274–75
electromagnetic waves, 202
identifies atoms, 310–13
mixing colors, 242–44
opaque materials, 239
photoelectric effect, 284
photon and quanta, 313–14
polarization and transverse waves, 280–83
reflection of, 254–57
refraction (bending) of, 257–59
speed of, 238
transparent materials, 236–38
wave-particle duality, 283–86
Light dispersion and rainbows, 260–63
Lights and mixing colors, 242–45
Lock and key molecular model, 491–93
Longitudinal waves, 214–15
Loudspeakers, 216
Low clouds, 688
Lunar eclipses, 703–5
Luster (minerals), 541

M

Machines multiplying forces, 92–94
Magellanic clouds, 733
Magma rock formation, 547–48, 553–54
Magnetic fields, 191–92
domains, 192–93
electric current and, 193–95
electromagnetic induction, 202
paleomagnetism, 587
as plasma containers, 347–48
Magnetic forces, 190–91
moving charges, 195–98
Magnetic poles, 190–91
Magnetic therapy, 205
Magnetism and electric current, 200–2
Mammals era (Cenozoic), 651–55
Mars (planet), 711–12
Pathfinder spacecraft, 712
Mass
acceleration, 40–41
action and reaction on, 58–59
energy and, 343–45
inertia and, 38–39
kilogram, newtons and, 40
law of motion, 41–42
measurement of Earth's, 108–9
Mass balance, 609–10
Mass conservation, 364–65
Mass number, 296–97
Mathematics and science, 4
Matter
chemical properties of, 356–61
classification as pure/impure, 377–79
Maxwell, James Clerk, 202, 283
Measurement metrics
hertz (Hz), 212
joule, 84–85
molarity, 383
mole, 382–83
Ohms, 176–77
Mechanical energy, 86–88
Mechanical equilibrium, 26–27
Medical drugs, 495
Melting, 163–64
Mercalli earthquake scale, 581
Mercury (planet), 710
Mesosphere, 665
Mesozoic era, 650–51
Metalloids, 299
Metals, 298
Metamorphic rocks, 563–66
foliation in, 565–66
Metamorphism, 563–65
Meteorites and asteroids, 716–17
dinosaur extinction, 650
Microscopes, 308
Middle clouds, 686
Milky Way galaxy, 732–34
Mineral identification
carbonate group, 546
chemical tests, 544
oxide group, 546
silicate group, 545–46
streak, 541–42
sulfide and sulfate groups, 546–47
Minerals, 537–44. *See also* Geology, Rocks
chemical sediments, 549
cleavage and fracture, 540–41
crystallization, 547–48
hardness scale, 539–40
magma formation of, 547–48
Mirages and atmospheric refraction, 259–60
Mirrors and law of reflection, 254–55
Mixtures
elements combining into, 373–75
heterogeneous and homogeneous, 377–78
physical separation of, 375–76
solutions in homogeneous mixtures, 379–84
Models, 309

Mohorovičić, Andrija, 573
Mohorovičić discontinuity, 573–74
Mohs scale of hardness, 539–40
Molarity (measurement), 383
Molecular polarity, 403–5
Molecules
atoms and, 354–56
covalent bonds and, 398
dipoles in, 401, 12–13
polarity, 403–5
structural isomers, 468
Mole (measurement), 382–83
Momentum
bouncing changes, 74–76
collisions and, 77–78
conserving, 74–77
force to increase/decrease, 71–73
impulse and inertia, 70–73
Monomers. *See also* Plastics and Polymers
polymers, 514
Moon
eclipses, 703–5
gravitational effects on Earth's ocean, 109–12
near-side locked to Earth, 705–6
phases, 702–3
size, 701–2
Motion
first law of inertia, 23–24
projectile, 118–21
relative, 21
second law of acceleration, 41–42
third law of action and reaction, 55–58
waves, 212–15
Moving objects
equilibrium, 28–29

N

Nanotechnology, 308
Natural frequency, 219
Natural philosophy, 10
Natural radioactivity, 325–27
Neptune (planet), 714–15
Nervous system, 496–99
Net force, 24
acceleration and, 37
law of motion, 41–42
Neurons
nervous system, 496–99
synaptic cleft, 497
Neurotransmitters
gamma aminobutyric acid (GABA), 498, 504–5
neurons, 497–99
Neutralization reaction, 430
alkaline solutions and, 440–41
in ocean water, 440–41
Neutrons, 294–96
Neutron star, 730
Newton, Isaac, 22
first law of inertia, 23–24
gravity, 103
law of universal gravitation, 104–8
second law of acceleration, 41–42, 45–46
third law of action and reaction, 55–58
Newtons, 40
Nicotine, 502–3
Nitrocellulose, 524–25
Noble gases, 301
Nonmetals, 298
Nuclear energy
fission, 339–41
fusion, 346–49
lasers and, 348
reactor conversion to electrical energy, 341–43
Nuclear fission, 339–41
Nuclear force
imbalance/radioactivity, 327–30
Nuclear fusion, 346–49
controlled fusion, 347–49
Nuclear radiation therapy, 325
Nuclear reactors
breeder, 342–43
conversion to electrical energy, 341–43
Nucleus, 294–5
nuclear fission, 339–41
nucleons, 295–96, 328
protons and neutrons, 294–96, 328–29
strong nuclear force, 327–30
Nylon
condensation polymer, 519–20
parachutes in World War II, 527–28

O

Ocean
circulation patterns, 674–77
coastal erosional landforms, 634–35
continental margin, 611
currents, 675–77
evolution of, 662
gravitational effects on tides, 109–12
land temperatures, 660–61
seafloor spreading and mapping, 588–89
specific heat capacity, 142–43
water acid neutralization, 440–41
waves, 612–14
Ohm, George Simon, 176–77
Ohms, 176–77
Ohm's law of electric current, 177–78
Opaque materials, 239
Optical illusions and mirages, 259–60
Optometry, 269
spiked stars illusion, 278
Orbital motion, 123–26
elliptical orbits, 126–27
escape speed/velocity, 127–29
Organic chemistry
carbon compounds, 466
functional groups, 474–79
Organic compounds, 466
amine group, 477–78
carbonyl group, 480–82
ether group, 477–78
heteroatom, 474
hydrocarbons, 467–71
hydroxyl group, 475–77
Overloading circuits, 185–86
Oxidation and electron loss, 445–47
Oxidation-reduction reactions
in batteries, 450–53
chemical reactions, 445–47
electrochemistry, 447–49
Oxide group (minerals), 546
Oxygen and combustion, 458–61

P

Paleomagnetism, 587
Paleozoic era, 646–49
plate tectonics, 649
Parabola, 119–21
Parallel circuits, 184–86
Parkes, Alexander, 524
Particles
light as, 283–86
mole (measurement), 382–83
Pelton, Lester A., 74
Period, 212
Periodic table
elements, 291–94
ion types, 392–93
organization of elements, 297–99
periods and groups, 299–303
Permeability, 602–3
Petroleum fractional distillation, 469–70
Phase change. *See also* Heat transfer, phase change
condensation, 160–62
energy for, 164–65
evaporation, 158–60, 162–63
melting and freezing, 163–64
Photoelectric effect, 284
Photography
celluloid polymer based film, 525
light photons in, 284–86
Photons (light), 284
pH scale, 435–36
Physical models, 309
Physical sciences
definition of, 11
mathematics and, 4
physical separation of mixtures, 375–76
Physiology
atoms and air, 293
human eye, 265
Pinhole Camera, 262
Pitch and sound waves, 215
Planetary atomic model, 315
Planetary escape speed / velocity, 128
Plants and natural drugs, 492–93
Plasma, 136
magnetic field containers, 347–48
Plastics, 514. *See also* Polymers
attitudes and environment, 530–31
bakelite, 525
cellophane, 526
development of, 523–26
rubber as predecessor, 523–24
World War II and polymers, 527–30
Plate boundary types
convergent, 592–95
divergent, 591–92
transform-fault, 595
Plate tectonics, 589–95
Cenozoic era, 653–55
Mesozoic era, 651
Paleozoic era, 649
plate boundary types, 590–95
plate motion measurement, 590
Precambrian era, 644–45
rock formation, 547–48, 553–54, 596–97
Plunkett, Roy, 529
Pluto (planet), 715–16
Polar covalent bonds, 401–3
Polarization, electric charge, 403–6
dipoles, 401, 411–12
Polarization, light, 280–83
Polar molecules
dipoles, 401, 411–12
electric charge, 174
induced dipole in nonpolar molecule, 413–14
soap and, 419–21
transverse light waves, 280–83
Polaroid filters, 281–83
Polyethylene, 515–16, 530
Polymers. *See also* Plastics
addition, 514–18
condensation, 517, 519–23
examples of, 516–17
polyvinyl chloride, 528–29
rayon, 526
synthetic rubber, 527–30
teflon, 529–30
Polymorphs and crystals, 538–39
Polyvinyl chloride (PVC), 528–29
Porosity, 602
Potential energy, 87–88, 175–76
Precambrian era, 644–45
Pressure and solubility, 415–17

Projectile
air drag effect on, 122–23
altitude and range, 122
motion of, 118–21
satellites, 123–26
Protons, 294–96
Pseudoscience, 8
astrology, 728
crystal power, 538
magnetic therapy, 205
Psychoactive drugs, 499–506
depressants, 504–6
stimulants, 499–503
Pulleys, 93–94
Pulsars, 730

Q

Quantum hypothesis, 313–16
Quantum mechanics, 286
Quantum number and atoms, 315–16

R

Radiant energy, 155–58. *See also* radiation
absorption, 157
reflection, 158
Radiation, 155–56. *See also* Radiant energy
exposure to, 326–27
gamma, 323–24
nuclear, 325
Radiation therapy, 325
Radioactivity
alpha and beta particles, 323–24
atomic isotopes, 331–32
decay and half-life, 328, 331–32
element transmutation, 330–31
gamma radiation, 323–24
natural phenomenon, 325–27
nuclear force imbalance, 327–30
Radiometric dating, 642–44
Rainbows and light dispersion, 260–63
Rainwater and acidity, 436–41
Rational thinking, 3
Rayon, 526
Recrystallization in rocks, 563
Red giant (stars), 727
Reduction and electron gain, 445–47
Reflection
diffuse, 256–57
interference and thin films, 278–80
law of, 254–55
radiant energy, 158
Refraction, 257–59
atmospheric, 259–60
lenses and, 263–65
Relative humidity, 681–82
Relative motion, 21
Religion and science, 8–10
construction of temples, 12
Resonance, 219–21
Richter earthquake scale, 580–81
Risk, technology and society, 11
Rock cycle, 566–67
Rockets
acceleration of, 59
satellite payloads, 125
Rock formation and layering
magma, 547–48, 553–54
plate tectonics, 596–97
unconformities, 641–42
Rocks. *See also* Minerals, Rocks
age determination, 640, 642–44
chemical sediments, 549
faults, 578–81
folds, 577–78
formation and plate tectonics, 596–97
groundwater dissolving, 623
igneous and sedimentary, 547–48
magma formation of, 547–48
metamorphic rocks, 563–66
rock cycle, 566–67
rock-forming mineral groups, 544–47
sedimentary layering, 639–42
weathering, 559–60
Rubber
synthetic, 527–30
vulcanization of, 523–24
Rydberg, Johannes, 312–13

S

Salts, 430–31
Satellite orbital motion, 123–26
Saturn (planet), 713–14
Scanning tunneling microscope (STM), 308
Schobein, Christian, 524
Science
art and religion, 8–10
history of, 3–7
limitations of, 7
technology and, 10
Scientific attitude, 7
Scientific law, 5
Scientific method, 4–7
Scientific theory, 5
Seafloor spreading, 588–89
Second law of acceleration, 41–42
Sedimentary rocks, 559–63
erosion by surface water of, 625–26
fossils, 562–63
layering and dating, 639–42
types of, 560–62
Sediments and sedimentation, 559–60
deposition of, 625–26
glaciers, 632–33
Seismic waves, 572–74
Selective reflection and colors, 240–42
Selective scattering, 246–47
Selective transmission and colors, 241–42
Semon, Waldo, 528–29
Series circuits, 182–84
Shell, atomic model, 317–18
Shock waves, 228–29
Silicate group (minerals), 545–46
Silver tarnish, 446–47
Sinkholes, 624
Sky color, 246–47
Soap, 419–21
Society
science, technology and, 11
Soil porosity, 602
Solar energy, 96–97, 666–69
Solar system
formation of, 707–8
planets, 708–16
Solubility, 414–16
Solubility and pressure/temperature changes, 417–19
Solutions
acids and bases in, 434–35
concentration of, 381–82
in homogeneous mixtures, 379–84
molarity measure of concentration, 382–83
saturated and unsaturated, 381–82
solubility, 414–16
solute precipitates, 417–19
solvents and solutes, 381, 414–16
Solvents
soaps and detergents, 419–21
in solutions, 381
Sonic boom, 228–29
Sound. *See also* Acoustics
forced vibrations and, 219
reflection of, 217
refraction of, 217–18
shock waves and sonic boom, 228–29
speed of, 216
Sound waves
longitudinal, 214–15
pitch, 215
Specific gravity, 542–43
Specific heat capacity and thermal inertia, 141–43
Spectral patterns and diffraction glasses, 313
Spectroscope, 311–12
Speed
of light, 238
of sound, 216
velocity, 19–20
of waves, 213
Spiral galaxies, 733–34
Springs, 604
Standing waves, 223–24
Stars
birth of, 725–26
constellations, 722–24
death of, 727–28
evolution of, 726–27
neutron star, 730
pulsars, 730
spiked illusion, 278
supernova, 729–30
Static equilibrium, 29
Stimulants
amphetamines, 499–500
caffeine, 501–2
cocaine, 500–501
nicotine, 502–3
Stratosphere, 664–65
Streak, 541–42
Streams and rivers, 605–8
deltas, 629
drainage networks, 607–8
valleys and floodplains, 626–28
Structural isomers, 468
Sublimation, 160
Subtractive primaries, 246
Sulfa drugs and synthetic drugs, 494–95
Sulfide and sulfate groups (minerals), 546–47
Sun
atmosphere, 707
color of sunsets, 248
sunspots, 706–7
Sun-centered universe, 3, 29–30
Superconducting electromagnets, 194–95
Superconductors, 178
Supernova, 729–30
Support force, 27–28
Surface water, 624–30
desert, 629–30
drainage networks, 607–8
erosion of sedimentary rocks, 625–26
valleys and floodplains, 626–28
Sympathetic vibrations, 219–21
Synaptic cleft and neurons, 497
Synergistic effects of drugs, 490
Synthetic drugs and sulfa drugs, 494–95
Synthetic rubber, 527–29

T

Tangential velocity, 103
Technology
electric power and, 186
science and, 10
society and risk assessment, 11
Technology, society and risk assessment, 11
Teflon, 529–30

Temperature, 135–36
absolute zero, 136–38
Celsius and Fahrenheit scales, 135–36
Kelvin scale, 137–38
solubility changes with, 417–19
Tension, 25–26
action and reaction, 62–64
Terminal speed and velocity, 47
Terrestrial radiation, 156, 667–68
Thermal energy, 135
heat and, 138–39
specific heat capacity, 141–43
Thermal expansion, 144–47
Thermal inertia and specific heat capacity, 141–43
Thermodynamics laws, 140–41
Thermometers, 135–36
Thermonuclear fusion, 246–47
Thermosphere, 665
Thin films and reflection/interference, 278–80
Third law of action and reaction, 55–58
horse-cart problem, 60–62
Transition metals, 301–2
Transparent materials, 236–38
Transverse waves, 214
Troposphere, 664
Tupper, Earl, 530
Tupperware, 530

U

Ultrasound, 218
Universal gravitation, 104–8
Universal gravitational constant, 107–8
Uranus (planet), 714

V

Valence electrons, 317–18
covalent bonds sharing, 397–400
ions and electric charge, 390–93
transfer of, 393–96
valence shell, 389–90
Valence shell, 389–93
Vector quantity, 20, 24–26
Velocity
acceleration and, 35–37
speed and, 19–20
tangential, 103
Venus (planet), 710–11
Vibrations
natural frequency and forced, 219
resonance and sympathetic, 219–21
Vibrations and waves, 211–12
Virtual image, 267
Volcanoes, 555–57
Voltage, 175–76
Voltaic cell batteries, 198
Volume, 38
Vulcanization and rubber, 523–24

W

Water
acid rain, 436–31
amphoteric behavior, 434–35
aquifiers and springs, 604
in atmosphere, 681–82
expansion of, 146–47
glaciers, 608–11
groundwater, 602–5
hydrologic cycle, 601–2
ocean, 440–41, 611–14 (*See also* Ocean)
quality, 614–17
specific heat capacity, 142–43
streams, 605–8, 626–30
surface, 624–30
Water supply
conservation, 616–17
contamination, 615–16
Water table, 603
Water wheels, 74
Watt, 85
Wavelength, 211
Wave motion, 212–16
Waves
barriers and bow waves, 226–27
Doppler effect, 225–26
interference of, 221–24
light as, 283–86
ocean, 612–14
plane polarized, 280–83
quantum mechanics, 286
refraction of, 613–14
seismic, 572–74
standing, 223–24
transverse and longitudinal, 214–16
Waves and vibrations, 211–12
Weather conditions
hurricanes, 694–95
thunderstorms, 692–93
tornados, 693–94
Weather forecasting, 695–96
Weathering, 559–60
Wegener, Alfred, 586–87
Weight, 38
White dwarf (star), 727–28
Wind motion, 621. See also Air, Atmosphere
Work and energy
electric power, 181–82
energy efficiency, 95–96
joule, 84–85
Work-energy theorem, 89–90
World War II, plastics and polymers, 527–30

Y

Young, Thomas, 276, 283